Energetics of gaseous ions

Journal of
Physical and Chemical Reference Data

David R. Lide, Jr., Editor

The *Journal of Physical and Chemical Reference Data* is published quarterly by the American Chemical Society and the American Institute of Physics for the National Bureau of Standards. The objective of the Journal is to provide critically evaluated physical and chemical property data, fully documented as to the original sources and the criteria used for evaluation. Critical reviews of measurement techniques, whose aim is to assess the accuracy of available data in a given technical area, are also included. One of the principal sources for the Journal is the National Standard Reference Data System (NSRDS), which is described more fully below. The Journal is not intended as a publication outlet for original experimental measurements such as are normally reported in the primary research literature, nor for review articles of a descriptive or primarily theoretical nature.

Supplements to the Journal are published at irregular intervals and are not included in subscriptions to the Journal. They contain compilations which are too lengthy for a journal format.

The Editor welcomes appropriate manuscripts for consideration by the Editorial Board. Potential contributors who are interested in preparing a compilation are invited to submit an outline of the nature and scope of the proposed compilation, with criteria for evaluation of the data and other pertinent factors, to:

> David R. Lide, Jr., Editor
> J. Phys. Chem. Ref. Data
> National Bureau of Standards
> Washington, D.C. 20234

The National Standard Reference Data System was established in 1963 as a means of coordinating on a national scale the production and dissemination of critically evaluated reference data in the physical sciences. Under the Standard Reference Data Act (Public Law 90-396) the National Bureau of Standards of the U.S. Department of Commerce has the primary responsibility in the Federal Government for providing reliable scientific and technical reference data. The Office of Standard Reference Data of NBS coordinates a complex of data evaluation centers, located in university, industrial, and other Government laboratories as well as within the National Bureau of Standards, which are engaged in the compilation and critical evaluation of numerical data on physical and chemical properties retrieved from the world scientific literature. The participants in this NBS-sponsored program, together with similar groups under private or other Government support which are pursuing the same ends, comprise the National Standard Reference Data System.

The primary focus of the NSRDS is on well-defined physical and chemical properties of well-characterized materials or systems. An effort is made to assess the accuracy of data reported in the primary research literature and to prepare compilations of critically evaluated data which will serve as reliable and convenient reference sources for the scientific and technical community.

Information for Contributors

All mathematical expressions and formulas, insofar as possible, should be typed, with unavailable Greek letters or symbols inserted neatly in ink. Distinguish between capitals and small letters. Avoid complicated superscripts and subscripts. Avoid repetition of a complicated expression by representing it by a symbol. Identify in the margin handwritten Greek letters and unusual symbols. Use fractional exponents instead of root signs, and use the solidus (/) for fractions wherever its use will save vertical space.

Literature references are to be numbered consecutively in brackets. They are to be placed at the end of the paper. A footnote appears after the first reference stating: Figures in brackets indicate literature references at the end of this paper. Literature citations of technical journals should be in the form indicated in the *Style Manual* of the American Institute of Physics (price $2.50 prepaid from American Institute of Physics, Dept. BN, 335 East 45 St., New York, NY 10017).

Tables may be typewritten on sheets separate from the running text. Each table must have a caption that will make the data in the table intelligible without reference to the text. Avoid complicated column headings.

Figures should, whenever practicable, be planned for reproduction in one-column width (8.7 cm). Line drawings should be India ink on white paper or tracing cloth. The original drawing or a high-quality glossy print should be submitted. To facilitate mailing and handling, original drawings should not exceed 22 by 28 cm in size. Lettering, preferably drawn with a mechanical set, should be of a size so that when reduced the smallest lower-case letters will not be less than about 1.5 mm. Avoid gross disparities in lettering size on a drawing. Consult the *Style Manual* of the American Institute of Physics for further discussion of figures. Type figure captions on a separate sheet of paper. Number figures consecutively. Identify each original drawing or photograph with the figure number and authors names, written where they will not be reproduced.

Subscriptions, renewals, and address changes should be addressed to Subscription Service Dept., American Chemical Society, Ohio State University, Columbus, OH 43210. **Single-copy, reprint orders, and orders for supplements** should be addressed to American Chemical Society, 1155 Sixteenth Street NW, Washington, D.C. 20036. Members of AIP member and affiliated societies requesting Member subscription rates should renew subscriptions and direct orders to American Institute of Physics, Dept. S/F, 335 East 45 Street, New York, NY 10017. Allow at least six weeks advance notice. For address changes please send both old and new addresses, and, if possible, include an address-stencil imprint from the mailing wrapper of a recent issue. Prices for reprints and supplements are listed on the last page.

Subscription Prices, 1977 (not incl. supplements)

	U.S.A., Canada, and Mexico	Elsewhere
Members (affiliated and member societies)	$25	$29
All others	$90	$94

The *Journal of Physical and Chemical Reference Data* appears in Sec. 1 of *Current Physics Microform* (CPM) along with 18 other primary journals published by the American Institute of Physics. CPM Sec. 1 is available on an annual subscription for $2100.

Second-class postage paid at New York, NY and additional mailing offices.

Journal of
**Physical and
Chemical
Reference Data**

Volume 6, 1977
Supplement No. 1

Energetics of gaseous ions

H. M. Rosenstock

K. Draxl

B. W. Steiner

J. T. Herron

National Bureau of Standards
Washington
D.C. 20234

Published by the **American Chemical Society**
and the **American Institute of Physics** for
the **National Bureau of Standards**

Library of Congress Catalog Card Number 77-71668

International Standard Book Numbers

0-88318-230-0 (Hard cover)

American Institute of Physics, Inc.

335 East 45th Street

New York, New York 10017

Printed in the United States of America

Foreword

The *Journal of Physical and Chemical Reference Data* is published jointly by the American Institute of Physics and the American Chemical Society for the National Bureau of Standards. Its objective is to provide critically evaluated physical and chemical property data, fully documented as to the original sources and the criteria used for evaluation. One of the principal sources of material for the journal is the National Standard Reference Data System (NSRDS), a program coordinated by NBS for the purpose of promoting the compilation and critical evaluation of property data.

The regular issues of the *Journal of Physical and Chemical Reference Data* are published quarterly and contain compilations and critical data reviews of moderate length. Longer monographs, volumes of collected tables, and other material unsuited to a periodical format are published separately as *Supplements* to the *Journal*. This monograph, "Energetics of Gaseous Ions," by H. M. Rosenstock, K. Draxl, B. W. Steiner, and J. T. Herron, is presented as Supplement No. 1 to Volume 6 of the *Journal of Physical and Chemical Reference Data*.

David R. Lide, Jr., Editor
Journal of Physical and Chemical Reference Data

J. Phys. Chem. Ref. Data, Vol. 6, Suppl. 1, 1977

III

Energetics of gaseous ions

H. M. Rosenstock, K. Draxl, B. W. Steiner, and J. T. Herron

National Bureau of Standards, Washington, DC 20234

Critically evaluated data are compiled and presented on ionization potentials, appearance potentials, electron affinities, and heats of formation for gaseous positive and negative ions. The positive ion literature is covered for the period 1955–1971 inclusive, and earlier literature on molecular Rydberg series is covered as well. The negative ion literature is covered through the end of 1973. The techniques employed in determining these data are critically discussed.

Key words: Appearance potential; compilation; critically evaluated data; electron affinity; heat of formation; ionization potential; negative ions; photodetachment; photoelectron spectroscopy; photoionization; positive ions; Rydberg series.

Energetics of gaseous ions

H. M. Rosenstock, K. Draxl, B. W. Steiner and J. T. Herron

National Bureau of Standards, Washington DC 20234

Critically evaluated data on energetics are presented on ionization potentials and heats of formation, electron affinities and heats of formation for positive and negative ions. The updates to literature is covered for the period 1971 to 1977. Items 2 and earlier are covered by the earlier compilation as well. The negative ion literature is covered through 1976, and for 1977, and the negative ion data are critically discussed.

Contents

J. Phys. Chem. Ref. Data, Vol. 6, Suppl. 1, 1977

I–3

I–4

J. Phys. Chem. Ref. Data, Vol. 6, Suppl. 1, 1977

1. Introduction

The present volume constitutes a complete revision and updating of "Ionization Potentials, Appearance Potentials, and Heats of Formation of Gaseous Positive Ions", NSRDS–NBS 26, which appeared in 1969. That volume covered the positive ion literature from 1955 to mid-1966, in effect covering the papers that appeared since the publication of "Electron Impact Phenomena and the Properties of Gaseous Ions" by F. H. Field and J. L. Franklin. Since 1966, more than one thousand new publications have appeared and some major advances in instrumentation, techniques and interpretation have occurred which have brought about a marked improvement in the quality and reliability of the information. In consequence it has been possible to approach some of the material in a more critical spirit, although the interpretations and comments (sometimes on the work of others) in the literature form the major basis for the evaluation.

In addition to an updated tabulation of positive ion data, we have added a tabulation of selected data on negative ions. The basis of the latter table is discussed in section 5. The closing date for literature coverage is the end of 1971 for the positive ions and the end of 1973 for the negative ions.

In comparison with NSRDS–NBS 26, the following changes of scope and emphasis in the positive ion table are to be noted: All citations of quantum mechanical calculations have been eliminated. This area of research is undergoing rapid development and a thorough job of compilation or evaluation was simply beyond the capabilities of the present evaluators. However, much theoretical material of this type is presented or cited in the papers on photoelectron spectroscopy. Whereas the preceding compilation covered the literature from 1955 on, in the present revision the literature coverage for molecular ionization potentials obtained from Rydberg series has been extended back essentially to its beginnings, using the recent book of Duncan [1] as a reference source. Much of the early literature on Rydberg series gave quite reliable information and for some systems it is still the only reliable information. In connection with ionic fragmentation processes, it is well known that the appearance potentials of fragment ions cannot be used to deduce heats of formation unless the state of excitation of the fragments and their kinetic energies of separation at threshold are known. Knowledge in this area is quite scant. Where available, this information is given for the processes, and referenced. With the increase in precision and accuracy of some of the data the calculation of ion heats of formation must be more carefully thought out. Where possible, heats of formation have been computed at 0 K. Also, the use of estimated thermochemical information has been discontinued. In view of the abundance of ionization potentials of good-to-excellent accuracy, it no longer

seems appropriate to calculate heats of formation of fragment ions unless the fragmentation threshold determination is likely to be of at least roughly comparable accuracy. This has essentially ruled out those data obtained by non-monoenergetic electron impact techniques which are of very doubtful and, worse, unpredictable reliability. An attempt has been made to select for full citation the best measurements for each process. Earlier, obsolete and inferior measurements are, however, given in reference to each process. These earlier references often give interesting and useful information of a qualitative or mechanistic nature. Thus, the present volume continues to serve as a guide to the experimental literature on positive ions.

Reference for Introduction

[1] Duncan, A. B. F., Rydberg Series in Atoms and Molecules (Academic Press, New York, 1971).

2. Acknowledgments

The authors gratefully acknowledge the support and encouragement given them by Dr. S. Rossmassler and Dr. D. R. Lide of the Office of Standard Reference Data of the National Bureau of Standards. Considerable assistance on the computer aspects of assembling and editing this volume was received from Dr. B. C. Duncan, Dr. D. Garvin, Mr. J. Hilsenrath, Mr. J. Koch, Mr. R. McClenon, Mrs. C. Messina, and Mr. R. Thompson. Dr. V. H. Dibeler and Dr. W. H. Evans made contributions to the literature search, and the late Dr. R. B. Parlin assisted in the abstracting. Many questions of data interpretation and presentation received helpful criticism from Dr. V. H. Dibeler, Dr. R. E. Huie, Mr. P. Krupenie, Dr. K. E. McCulloh and Dr. R. L. Stockbauer. Helpful criticism and a number of useful corrections were received from Dr. F. P. Lossing of the National Research Council, Ottawa, Canada. Considerable secretarial help was provided during the course of this work by Mrs. P. Davis, Mrs. K. Fadely, Miss R. McCoy, and Mrs. C. Schmidt. Last, and not least, all phases of the preparation of this book were crucially dependent on the careful, patient, and dedicated work of Miss Margaret Moore.

3. Energetics of Gaseous Positive Ions

Quantitative data leading to energetics of positive ions have been obtained by numerous techniques. Correspondingly, there are numerous methods of interpreting the data to yield ultimately the quantities known as ionization potentials, appearance potentials and heats of formation. In the following sections we summarize the various experimental techniques and the methods of data interpretation. Special attention is paid to highlighting the accuracies, problems, and limitations of the techniques and interpretations. An effort has been made to select references which discuss these matters in more detail.

This approach leads inevitably to a degree of optimism or pessimism on the part of the writers and the readers which depends strongly on the degree of accuracy desired or expected of the data. That, in turn, depends on the intended application of the data, and the purpose of the application. The body of data tabulated here varies in demonstrated or likely accuracy from 1 meV or better to data which may be in error by as much as 0.5 to 1 eV for singly charged species and 10 eV or more for some multiply charged species. It covers an enormous variety of chemical species. The types of uses and systems considered are also very large. We have focused the discussion to follow on the problems and barriers to attaining high accuracies, of the order of several hundredths of an electron volt (or several kilojoules per mole) or better. For the reader who has less stringent accuracy requirements we state that nearly all of the data are accurate to roughly 0.5 eV or better, with the probable exception of the electron impact experiments directed to determining ionization energies for producing highly charged species. We note that two classes of systems have been studied extensively by conventional electron impact: large organic molecules and high temperature species effusing from Knudsen cells. The measurement errors estimated by the original authors tend to be larger (more conservative, more pessimistic?) for the Knudsen cell experiments than for the organic experiments. Limited comparisons with demonstrably more reliable information indicate, in fact, no higher accuracy for one group than the other. In some instances, very nearly correct answers will be found in both areas. The difficulty is that the specific instances are hard to predict. The only generalization is that parent ionization potentials are more accurate and fragment ion appearance potentials less accurate.

3.1. Experimental Techniques

In this section we define and discuss the measurement techniques which provide the data presented in the Positive Ion Table (section 4). The nineteen distinct techniques can be grouped into several major categories, each of which represents a particular type of experimental approach and theory or rationale for interpretation of the data. In table 1 we give the categories, the techniques and the abbreviations used for them in the Positive Ion Table.

a. Optical Spectroscopy

In general, the most accurate data are obtained by spectroscopic techniques. The instrumentation and calibration have been refined to the point that in the visible and near uv, atomic absorption or emission lines can be measured with an accuracy of 0.001 to 0.0001 nm or better. A typical modern 21 or 35 ft. optical spectrograph will have a resolution of the order

TABLE 1. Experimental techniques for positive ion energetics

Category	Technique	Abbreviation
Optical spectroscopy	—Spectroscopic	S
Threshold experiment	—Photoionization, with or without mass analysis	PI
	—Electron monochromator	EM
	—Retarding potential difference	RPD
	—Energy distribution difference	EDD
	—Square root plot	SRP
	—nth root extrapolation	NRE
	—First derivative electron impact	FD
	—Second derivative electron impact	SD
	—Sequential mass spectrometry	SEQ
	—Electron impact other than above	EI
Electron spectroscopy	—Photoelectron spectroscopy	PE
	—Auger electron spectroscopy	AUG
	—Resonant photoionization	RPI
	—Penning ionization	PEN
Other	—Surface or thermal ionization	SI
	—Born-Haber cycle calculations	BH
	—Charge transfer spectrum	CTS
	—A derived value	D

of several hundred thousand or more depending on the grating order used. On the other hand, most molecular spectroscopic studies cited have been carried out with instrumentation of somewhat lower resolution. Wavelengths are generally reported to 0.01 nm and occasionally to 0.001 nm. Details of typical instrumentation and techniques are given in references 1, 2, and 3.

b. Threshold Experiments

The second class of experimental techniques may be termed threshold techniques, in which the objective is to directly determine the minimum energy necessary to form an ion (parent or fragment) from a neutral species, or sometimes from an ion of lesser charge. These techniques include photoionization and the various electron impact techniques.

b.1. Photoionization

In photoionization, experiments are generally carried out with a Seya-Namioka 70° 15' or near-normal incidence monochromator. Wavelength resolution of the monochromator used for ionization ranges from 0.04 to 0.2 to 0.3 nm in first order and the wavelength scale is known typically to 0.01 to 0.02 nm from calibration with known emission lines. This represents a best energy resolution of about 5 meV at 12 eV. The commonly used light sources are a hydrogen discharge and a helium discharge. The hydrogen discharge produces a many line spectrum in the 160 to 90 nm wavelength range. This limits the minimum distance between points at which data may be taken. The helium Hopfield continuum provides useful light intensity in the 110 to 58 nm wavelength region, and some workers have used the argon continuum for the 155 to 105 nm range. The experimental techniques are described in detail in references 4-8.

b.2. Electron Impact Techniques

This class of techniques is the one most widely used over the years in experiments directed to measurements of ionization and fragmentation energetics. Electron impact methods have been used for this purpose for nearly fifty years. There are two main problems with these techniques. There is a large energy spread in the electron beam, and the actual energy maximum of the beam can differ from the nominal energy expected from the applied electrode potentials. As a result, it is necessary to calibrate the nominal energy scale by carrying out experiments with a reference gas or gases and to minimize the difficulties presented by the electron energy spread. Many approaches have been developed to deal with these problems, with varying success. These techniques can be divided into three categories of progressively decreasing refinement and quality: a) monoenergetic, b) quasi-monoenergetic, and c) conventional or non-monoenergetic.

Monoenergetic

The major problem with electron impact techniques is the energy spread of the electrons in a conventional beam. This spread is several tenths of an electron volt or more, due to filament potential drop, filament temperature, field penetration and surface effects. A monoenergetic electron beam technique currently in use in one laboratory limits the electron energy spread by passage through a double hemispherical electron monochromator of known electron optical behavior [9–11]. Subsequent analysis of the beam indicates an energy spread of 70 meV at half maximum. The electron energy scale is calibrated by determining the ion yield curve of a rare gas, generally argon, in the ionization threshold region over a range of several volts. This calibration will be valid for determining threshold values for other species if their threshold behavior is similar enough. The rationale for this procedure is not really well established. However, where comparisons can be made, the several dozen results obtained by this technique agree with experimental photoionization thresholds to within about 30 meV. Some work has also been done using 127° cylindrical electrostatic analyzers to obtain electron beams of comparably narrow and defined energy spread [12–14]. Electron energy spreads of smaller than 50 meV at half maximum were obtained some years ago [13], but no results pertinent to the present compilation have been published.

Quasi-Monoenergetic

A radically different approach to overcoming the energy spread problem is that known as the RPD (Retarding Potential Difference) technique [15, 16]. In this method the electron gun is designed with a potential distribution such that the low energy portion of the electron energy distribution is not transmitted into the ion source. This cut-off potential is rapidly varied between two values differing by a small amount (of the order of 0.1 eV). The corresponding difference component of the ion signal is detected and amplified. This component corresponds in principle to the ion current produced by electrons having an energy distribution of that portion of the original distribution lying between the two cut-off voltages. In fact, the distribution is somewhat broader and not sharply defined [17]. In the years since its development this method has been widely adopted. However, not all the care and precautions of the originators have been followed.

A second quasi-monoenergetic method is the Energy Distribution Difference method (EDD) [18]. In this method the ion current measured at a nominal electron energy E is adjusted by subtracting from it a constant fraction, b, of the ion current measured at a nominal energy $E + \Delta E$. The originators of the method show that this difference current represents the ion current due to a considerably narrower electron energy distribution. The optimum values of the parameters b and ΔE depend on the form of the electron energy distribution.

Both of these methods have been successful in considerably reducing the lack of sharpness of experimental first ionization thresholds due to the large electron energy spread. However, a calibration with a gas of known ionization potential is required, and it is necessary to assume similar threshold behavior.

Non-Monoenergetic

Most of the remaining electron impact techniques are simply grouped under the category EI. They include the conventional techniques employing electron beams of broad energy and differing principally in the method of extrapolating to "threshold". The simplest and most unreliable method is the linear extrapolation method, which is frequently used in studies of high temperature species where the experiment is directed to the study of the thermodynamics of vaporization of the species rather than to the ionization energetics. This is unfortunate since, for these species, threshold data of even moderate accuracy would be very useful. Generally, a variety of more accurate extrapolation methods are used to correct for the broad electron energy distribution by careful comparison of the unknown and calibrant ion current in the threshold region. These include semi-log plot, extrapolated voltage difference, energy compensation and critical slope methods. These are described in references 19 and 20. Their accuracy ranges from less than 0.1 eV to more than 0.5 eV in an unpredictable manner.

Still another group of extrapolation techniques has been developed and applied to determination of threshold energies for multiple ionization. They are to be distinguished from the various other non-monoenergetic extrapolation methods and are cited separately in the Positive Ion Table. The Square Root Plot (SRP) and Nth Root Extrapolation (NRE) techniques are extrapolation procedures often employed to obtain threshold values for N-fold charged species. Their rationale is the assumption that near threshold the cross section for direct multiple ionization varies with the Nth power of the electron energy in excess of the threshold value, e.g., a quadratic law for double ionization, a cubic law for triple ionization [21], etc. (see section 3.2.d.).

Another experimental technique which has been devised to determine threshold energies for multiple ionization is Sequential Mass Spectrometry (SEQ) [22]. In this technique, the ion source is operated with a potential configuration which brings about space charge trapping of the positive ions formed by electron impact. Some of the ions undergo a second ionizing collision with an electron before drifting out of the trapping region. Because of this step-wise ionization one can determine the ionization potentials of species which have been previously ionized. Trapping times vary from 10^{-4} s [20] to almost one second [23]. While this technique offers considerable advantages in obtaining information on highly charged species, there are considerable difficulties in interpretation because of the large energy spread of the electrons [24].

c. Electron Spectroscopy

The third class of techniques has as a common principle the energy analysis of electrons ejected from molecules which have been excited by uv photons (Photoelectron Spectroscopy, PE), X-ray Photons (Auger Electron Spectroscopy, AUG) or electronically metastable atoms and molecules (Penning Ionization, PEN) of known energy. For a given excitation energy, the energy distribution of the ejected electrons will reflect the distribution of accessible energy levels of the target neutral atom or molecule according to the relation:

$$E_{ion} = h\nu - E_{electron}.$$

Photoelectron spectroscopy in particular has undergone a rapid evolution since the first experiments fifteen years ago [25, 26]. It has been reviewed in a number of articles [27–35].

c.1. Photoelectron Spectroscopy

The most widely used photon source in photoelectron spectroscopy is the helium resonance line of wavelength 58.4331 nm corresponding to an energy of 21.218 eV. More recently, experiments have been carried out with a helium discharge operated so that there is appreciable intensity of the He II line at 30.3781 nm (40.813 eV) as well as the 58.4 nm line. Also, some work is being done with the neon resonance doublet, at wavelengths of 73.589 and 74.370 nm (16.848 and 16.671 eV) and the argon resonance doublet at 104.8218 and 106.6660 nm (11.829 and 11.623 eV). When the light source consists of a doublet, the photoelectron spectrum will consist of a superposition of two spectra shifted with respect to one another by the energy difference of the two emission lines, and with relative intensities depending on the relative line intensities and on the relative ionization cross sections at the two wavelengths. If the light source contains even traces of impurities, there is the possibility that impurity emission lines can contribute spurious structure in the photoelectron spectrum out of all proportion to the trace concentration [36–38].

The energy analysis of the ejected photoelectrons has been carried out with a great variety of analyzers including:

    Cylindrical retarding grid
    Spherical retarding grid
    180° magnetic sector
    127° electrostatic sector
    Plane parallel electrostatic analyzer
    Double focussing electrostatic prism
    Double focussing hemispherical condenser

The relative merits of these devices have been discussed in a number of publications [31, 39–42]. The early devices had an energy resolution of one to several tenths of an electron volt. At present, the highest resolution attained is of the order of 10–15 meV [43, 44], with a 127° electrostatic analyzer. The factors limiting resolution have been discussed and estimated by Turner [45]. They include factors depending on the target, the light source, and the analyzer. The principal contribution of the target is the ejected electron velocity spread brought about by the thermal velocity distribution of the target and is typically of the order of 2 meV. For hydrogen and helium, however, it is much larger, roughly 20 meV. Compared to this, the Doppler broadening of the photon source emission line is negligible. However, self-reversal of the emission line may be significant; it is very dependent on experimental parameters of the discharge source [4, 46, 47]. Experience to date indicates that as yet no rotational structure can be clearly resolved but that spectrum line broadening attributable to rotational envelopes is sometimes observable [45]. Because of contact potentials the energy scale must be calibrated with a gas or gases of known ionization potential(s). Rare gases are frequently used. Details are given in reference 48.

c.2. Auger Electron Spectroscopy

The technique of Auger Electron Spectroscopy (AUG) is similar in principle to photoelectron spectroscopy. It, too, is based on energy analysis of ejected electrons. However, in this case the electron is ejected via an Auger cascade following prior inner shell ionization. The inner shell ionization is brought about by a high energy electron beam [49, 50] or a discrete X-ray source [50, 51]. The energy analysis is carried out by means of either a cylindrical mirror analyzer operated at an energy resolution of 0.12 to 0.16% [49, 51] or a double focussing electrostatic prism with 0.06 to 0.09% energy resolution [50, 52].

Since the Auger electrons have energies of several hundred volts or more, this percentage implies an energy resolution of several tenths of an electron volt. Calibration of the energy scale is carried out with known Auger electron energies of neon and argon [50, 53].

c.3. Resonant Photoionization

The resonant photoionization technique, also called threshold photoelectron spectroscopy, differs from photoelectron spectroscopy in that the photon energy is varied and only those photoelectrons are detected which lie in a narrow energy band corresponding to essentially zero energy of ejection [54]. Thus, in principle, a photon wavelength scan will directly indicate those ion states which are directly accessible, without interference by autoionization processes involving transitions to lower levels. In practice, due to finite energy resolution a signal is also obtained from auto-ionization processes which populate states close to the direct ionization threshold. The zero-energy electrons are selected by two different methods. In one, all photoelectrons are accelerated through a known potential drop and those whose energy corresponds to the drop itself are selected by a 127° electrostatic analyzer [54–56]. In the second method the electrons are accelerated through a uniform electrostatic field and non-zero energy photoelectrons intercepted by passage through channels of small angular aperture oriented parallel to the field (steradiancy analysis) [57, 58]. The accuracy of ionization threshold values is reported to be 10 to 20 meV [54, 55] and 2 to 5 meV [56–58], respectively.

c.4. Penning Ionization

The Penning ionization technique is also based on ejected electron energy analysis. In this technique, the ionizing agent is a beam of metastable neutral rare gas atoms with known excitation energy or energies. The excited species are produced by electron impact. The metastable atoms employed include mixtures of $He(2^3S)$ (19.818 eV) and $He(2^1S)$ (20.614 eV), $Ne(^3P_2)$ (16.619 eV) and $Ne(^3P_0)$ (16.715 eV), or $Ar(^3P_2)$ (11.548 eV) and $Ar(^3P_0)$ (11.723 eV). Because of the presence of two metastable states in a given atom beam the Penning electron spectrum consists of a shifted superposition of two spectra, each formed by one of the species. The relative intensity depends on the relative proportion of the two species and the relative cross sections for the process [62]. Electron energy analysis was carried out a first with a cylindrical retarding grid arrangement (Lozier tube) of 0.1 to 0.2 eV resolution [59–61].

More recently, a plane parallel retarding grid of 10 to 20 meV resolution [63] and a 127° electrostatic sector of better than 60 meV resolution [64] have been employed.

Energy calibration is carried out with a gas of known ionization potential or with photoelectrons produced by He 58.4 nm radiation [64, 65]. In the case of helium metastables, the $He(2^1S)$ component of the beam can be quenched with radiation from a helium lamp [62]. In interpreting the results, careful consideration has to be given to possible kinetic energy transfer from the projectile to the target species [62].

d. Surface Ionization

The surface ionization method has been applied to the determination of first ionization potentials of some metal atoms, especially of the lanthanide and actinide elements. The principle of the method is based on the assumption that the atoms in a beam, after impinging on a hot metal surface, will come to thermodynamic equilibrium, producing a surface concentration of atoms and ions whose composition can be described by the Saha-Langmuir equation:

$$N_+/N_0 = g_+/g_0 \exp\left[e(\phi - I)\right]/kT,$$

where N_+/N_0 is the fraction of the atoms which are ionized, g_+ and g_0 are the statistical weights of the ions and atoms, e the electronic charge, ϕ the work function of the metal, I the ionization potential, k the Boltzmann constant, and T the absolute temperature. The temperature dependence of the positive ion current will give the ionization potential if the work function is known. Complications inherent in the method include the effect of surface coverage or impurities on the work function, definition of the work function for a polycrystalline surface exhibiting a variety of crystal planes with different work functions, and an occasional lack of reproducibility of experimental results which is simply not understood. Also, the fundamental assumption is open to question. These factors are discussed in detail in references 66–68. These difficulties are somewhat reduced by the determination of ionization potential differences by bombarding the metal surface with a composite beam of atoms of known ionization potential and atoms of unknown ionization potential [69], or comparing the temperature dependence of positive and negative ion emission of the same atomic species [70]. Where comparisons can be made with more reliable methods, the relative determinations give ionization potentials within several tenths of an electron volt of the correct value. Recently, the aniline molecule has been surface-ionized and the temperature variation of the parent ion current gave very nearly the correct ionization potential. It is noteworthy that not all the parent molecules pyrolyzed or fragmented [71].

e. Born-Haber Cycle Calculations

Another method occasionally used for determination of ionization potentials is based on the Born-Haber cycle. For an ionic crystal the lattice energy of the crystal can be defined as the energy liberated when the crystal is formed from ions at infinite separation. This quantity may be in turn related to the standard heat of formation of the crystal, and the difference between the lattice energy and the standard heat of formation is the energy required to convert the elements from their standard states to ions at infinite separation. If all but one quantity is known, such as an ionization potential or electron affinity, this unknown can be determined from the cycle. The lattice energy may be calculated or estimated by interpolation from related solids [72]. In the present context, this method has been used in determining the third ionization potentials of atoms of the lanthanide series [73–75]. The accuracy of the method is stated to be several tenths of an electron volt [75].

f. Charge Transfer Spectra

The charge transfer spectrum method is a semi-empirical method often used in estimating ionization potentials of large molecules such as polyphenyls, fused ring systems, amines, and certain biochemical compounds. It is based on a semi-empirical theory developed by Mulliken to explain the absorption bands of electron donor-electron acceptor complexes in solution [76]. These bands arise from a transition from the ground state of the molecular complex to an excited state in which an electron is largely transferred from the donor to the acceptor molecule. The bands are not characteristic of the isolated donor or acceptor molecules. Hastings et al. [77] derived from the Mulliken theory a simple algebraic relation between the frequency of the maximum of the charge transfer band and the ionization potential of the donor, and correlated it with experimental information. The subject has been extensively reviewed by Briegleb [78, 79]. A limited comparison of ionization potentials derived by this method and more accurate methods indicates that the estimates are frequently correct to within several tenths of an electron volt.

3.2. Interpretation of the Data

a. Atomic Spectra and Rydberg Series

The interpretation of atomic spectra to give ionization potentials is a highly developed field with an extensive theoretical foundation. Here we will only summarize some aspects of data handling and discuss two topics of general importance to the interpretation of the results of other techniques as well. These topics are Rydberg series and autoionization.

The analysis of atomic Rydberg series is discussed in detail in the review article of Edlén [80].

An atom with one excited electron in an orbital of high principal quantum number can, in first approximation, be considered as hydrogen-like. The excited (or optical) electron sees the central force field of the nucleus screened by the remaining core electrons.

Accordingly, the energy levels of this system can be described by the Rydberg formula:

$$T_n = I - [2\pi^2 Z^2 e^4/ch^3 n^2] \cdot [mM/(m+M)] = I - RZ^2/n^2,$$

where I is the ionization potential, Z is the net charge of the nucleus and core electrons, n the principal quantum number, e the electronic charge, m the electron mass, M the mass of the nucleus plus core electrons, c the velocity of light, h Planck's constant and R the Rydberg constant (for mass M). This model has been refined by consideration of two other factors. First, while the electron is outside the core it will polarize the core, and second, the electron can penetrate the core to some extent and see more of the nuclear charge. Both of these factors increase the binding energy of the electron. The theoretical treatment of these factors [81] leads to the energy expression:

$$T_n = I - RZ^2/(n-\delta)^2 = I - RZ^2/n^{*2},$$

where δ is the quantum defect, and n^* is the effective quantum number. The quantum defect varies with principal quantum number, and is generally written as a power series in $t = 1/n^{*2}$:

$$\delta = a + bt + ct^2 + dt^3 + \ldots$$

It is seen from this expression that the quantum defect varies linearly with energy in the limit of high quantum numbers. Of the two factors affecting the quantum defect, the penetration effect is important for the penetrating s and p orbitals and the smaller polarization effect is important for the non-penetrating d, f, and g orbitals. Quantum defects are largest for s orbitals, smaller for p orbitals and smallest for d, f, and g orbitals. For examples see Edlén, [80] Kuhn [82] and the references cited in Moore's compilation [83].

The procedure which is generally employed to determine the ionization limit is to choose an estimate of the ionization limit, calculate the quantum defects from the term values of the series and adjust the ionization limit until one obtains a linear dependence of the quantum defect on energy at large quantum numbers [80]. Frequently, several sets of terms are obtained from experiments corresponding to $s, p, d \ldots$ series. Other things being equal, it is generally the practice to determine the ionization potential from the terms of highest orbital angular momentum, for which both penetration and polarization effects are smallest and, consequently, the linear dependence of quantum defect on term energy is expected for lower terms than, say, for an s series. The procedure and some detailed examples are given in a study of the spectrum of Ca II [84]. For example, in figure 1 is shown a plot of quantum defect vs term value for five members of the Rydberg series ($n=5$ to 9) of singly ionized calcium, taken from a study of Edlén and Risberg [84]. Since it is a g series, the Rydberg electron has very high angular momentum, the orbits are essentially non-penetrating

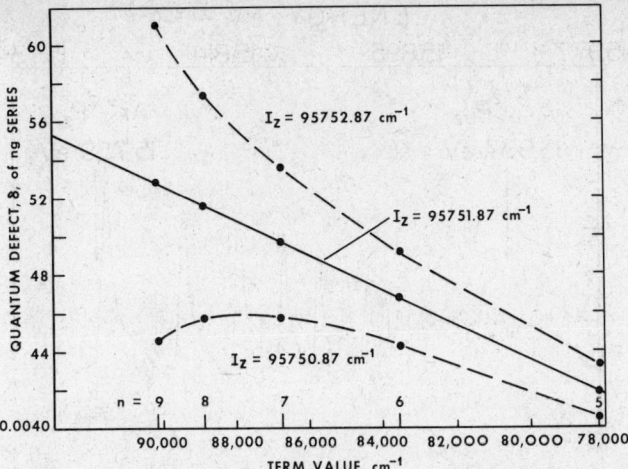

FIGURE 1. Dependence of the quantum defect, δ, on term values of the ng Rydberg series of Ca II, for three different assumed ionization limits. Adapted from data of Edlén and Risberg [84].

and the quantum defects are small, 0.004 to 0.006. The defects are plotted for the best value ionization limit of 95751.87 cm^{-1} and for ionization potentials assumed to be 1 cm^{-1} higher and lower, respectively.[1] It is seen that the plot is indeed linear for the correct value of the ionization potential and deviates significantly for values only slightly different from this. With this ionization potential the experimental term values can be computed with a maximum deviation of 0.005 cm^{-1} from the Rydberg formula with a linearly varying quantum defect. The authors [84] give an estimated error of several units in the last decimal place for this ionization limit. The basis for this error estimate and others in the literature is not clear. It seems to involve mainly the accuracy with which the energies of observed Rydberg series members agree with the values calculated from the formula. The most accurately known atomic series limits are those of the alkali metals and helium, all of which are stated to have errors of ± 0.01 cm^{-1} or less [83]. Of these, the most accurate is the lithium limit, with an error of ± 0.005 cm^{-1} [85].

There are several problems with the above approach to treatment of the data. First, assuming that the wavelength accuracy is constant in a given series of terms, the individual terms do not act equally as statistical estimators of the quantum defect. Second, if one has more experimentally determined series members than parameters in the quantum defect polynomial, the above procedure does not utilize all the information to the fullest. Third, there is no statistical basis for the error estimate. Seaton [86], has pointed out these problems and devised a least squares procedure for treating the data. He obtained for the ionization limit of helium 198310.76 ± 0.01 cm^{-1}, based on treatment of

the term values of seven Rydberg series tabulated by Martin [87], compared to Martin's value of 198310.81 cm^{-1} based on the mean of three series limits, and Herzberg's [88] value of 198310.82 ± 0.15 cm^{-1}. Seaton obtains different limits for the three series considered by Martin, and concludes there is a systematic error in the term values for series members with $n > 10$. This has not yet been verified.

The discussion and examples above are for essentially unperturbed Rydberg series. The situation is far more complex for terms and series which are perturbed by configuration interaction, and this is not uncommon. If an isolated group of terms is perturbed, this will show up as an irregularity in graphs of the energy dependence of the quantum defect. Examples are given by Edlén [80]. If many terms are perturbed, much more sophisticated analyses must be carried out. Also, the spin-orbit splitting of terms complicates the analysis. This problem is encountered in the analysis of rare earth spectra [89]. In some instances the analyses are not yet far enough along to yield deperturbed term values from which limits can be derived, especially for atoms of high atomic number. Frequently, the analysis and comparison of isoelectronic terms is helpful [80]. References to individual species are to be found in the compilation of Moore [83].

b. Autoionization

States corresponding to the excitation of a more tightly bound electron or the excitation of two electrons may lie at energies above the lowest ionization energy of the atom. These can spontaneously eject an electron, undergoing autoionization. The selection rules for this type of process are well known [90], and represent essentially conservation of angular momentum and of parity. The lifetimes of autoionizing levels can vary a great deal, ranging from $\sim 10^{-14}$ s all the way to $\sim 10^{-6}$ s. Autoionization processes have the following experimental consequences [91, 92]:

a. Since autoionization lifetimes are frequently shorter than radiative lifetimes, in such cases these states are not observed in emission spectra.

b. Because of the selection rules, certain terms of an atomic multiplet may autoionize and others live long enough to radiate.

c. In an arc emission source, some autoionizing states will be populated at high arc currents by ion-electron recombination and hence observed.

d. In absorption, the lines will not be sharp when the autoionizing level has a lifetime considerably shorter than the radiative lifetime.

Theoretical study of the autoionization process leads to the following important conclusions [93, 94]. First, a great variety of asymmetrical line shapes can occur, including "windows" in which an autoionizing state will be observed as a local *decrease* in the absorption coefficient. It is to be remembered that since autoionizing states lie above an ionization limit, they are always observed against a continuous background of absorption

___
[1] Although SI practice recommends joules and electron volts for energy units, it is useful to designate term values in cm^{-1} in spectroscopic discussions. The conversion factors are given elsewhere in the text.

due to direct ionization. Second, the higher members of Rydberg series with higher ionization potentials as their limit will always lie above the first ionization limit and often can autoionize. If one now considers the variation of the average absorption coefficient as one moves from below the limit to above the limit it is found theoretically that the variation is smooth and, in particular that there should be no "jump" or discontinuity at the limit. In an experiment measuring an absorption coefficient or a photoionization experiment one should not be able to detect a higher ionization "threshold". One can, of course, determine the wavelength of the autoionizing Rydberg series members insofar as they are resolved and carry out a conventional series analysis.

Various workers have in fact observed discontinuities at higher atomic ionization limits. These are experimental artifacts which arise as follows. In an autoionizing Rydberg series the level spacing converges with increasing energy, the individual autoionizing lines become less and less broadened and the absorption of a given line becomes concentrated in a progressively smaller wavelength range. As a result, unless the absorbing gas pressure is low enough the experimental absorption, which is of course averaged over the apparatus slit width, is no longer simply proportional to the average absorption coefficient. At and above the series limit, the absorption coefficient is no longer rapidly varying, and an abruptly increased absorption is observed. This type of artifact is observed in optical absorption experiments, which intrinsically require that a substantial fraction of the photons be absorbed. In contrast, in photoionization experiments generally a much smaller fractional absorption occurs, and no discontinuity is observed at the limit. An interesting example of this phenomenon is the argon absorption coefficient study of Hudson and Carter [95] as compared to the photoionization curves of Spohr, et al. [58] and of McCulloh [96] (see figure 2).

c. Molecular Spectra

The situation for molecules is still more complex due to added factors of vibrational and rotational structure. Approximately one hundred Rydberg series have been observed and analyzed [97]. In almost all instances, the series limits so obtained agree with photoionization threshold or later photoelectron spectroscopy measurements to within about 0.01 eV or better.

The most accurately known molecular ionization potential is that of hydrogen, which has been completely analyzed by Herzberg and Jungen [98] who obtained a value of 124417.2 ± 0.4 cm^{-1} (15.42541 ± 0.00005 eV) compared with the most accurate theoretical value of 124417.3 cm^{-1} calculated by Jeziorski and Kolos [99] and an upper bound of 124418 cm^{-1} based on observation and analysis of autoionizing rotational levels of Rydberg series in photoionization by Chupka and Berkowitz [100].

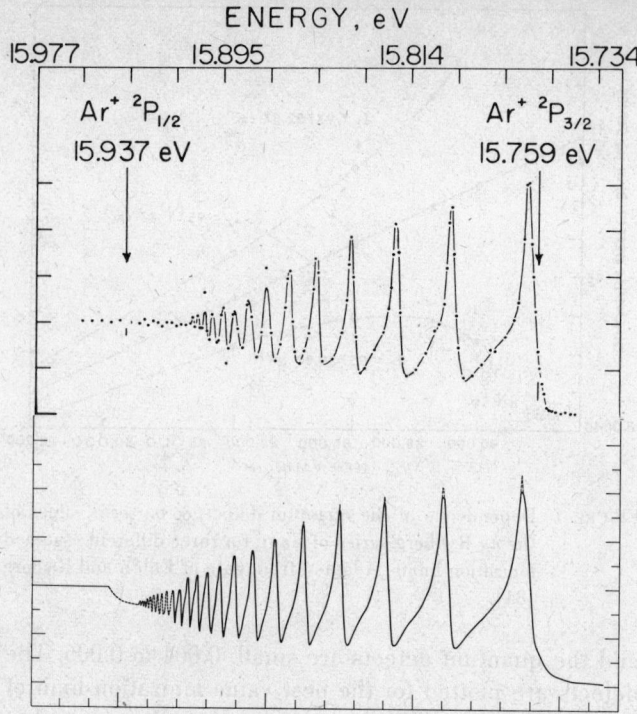

FIGURE 2. Comparison of photoionization yield (upper curve, from ref. [96]) and absorption cross section (lower curve, from ref. [95]) of argon in the ionization threshold region. Note the difference in appearance near the $^2P_{1/2}$ ionization limit.

The spectroscopic ionization potentials of other molecules are mostly based on analyses of Rydberg series. Interesting examples are described in a series of papers by Lindholm [101–103]. In these cases the rotational structure has not been resolved, leading to an uncertainty of roughly 5 meV due to difficulty in establishing the band origin. Since the geometry of the higher Rydberg states is often somewhat different from that of the ground state, the various rotational transitions to a particular vibrational level of a Rydberg state will produce a rotational envelope with a characteristic intensity distribution [104]. The appearance of this vibration-rotation band, e.g. shaded to the red or to the violet, depends on the sign and magnitude of the geometry change, as does the location of band origin corresponding to the (missing) · transition between the rotationless ground state and Rydberg state. The one case where this effect has been considered is the series converging to the $b\,^4\Sigma_g^-$ state of O_2^+ [105]. The correction amounts to ~ 10 cm^{-1}.

The identification of molecular Rydberg series is often not completely straightforward. Criteria which are used include the requirement that the transitions be strong and gradually decrease in intensity with increasing energy, and, evidently, that the series fit the Rydberg formula. Frequently a deviation of 50 to 100 cm^{-1} between the calculated and measured term values is considered acceptable. However, even then there are occasional erroneous series assignments. Examples of this are found in spectroscopic studies of methylacetylene [106], ethylene oxide [107], furan [108], and nitric oxide [109].

Sometimes emission spectra are observed which are due to ions undergoing radiative transition from an electronically excited state to the ion ground state. When properly analyzed, these give information on the separation between the ion ground state and one or another electronically excited state. If Rydberg series are observed for transitions with the electronically excited ion as a limiting state, then an independent check can be obtained to verify the analysis. A particularly good example is that of molecular nitrogen [110]. Three sets of Rydberg series have been observed and analyzed. The Worley-Jenkins series converges to the ion ground state, $X^2\Sigma_g^+$, with a series limit of 125666.8 cm^{-1}; the Worley series converges to the first excited $A^2\Pi_u$ state with an average limit of 134685 cm^{-1}; and the Hopfield series converges to the $B^2\Sigma_u^+$ state at 151233 cm^{-1}. In emission, the Meinel bands have been observed and assigned to the $A^2\Pi_u - X^2\Sigma_g^+$ transition. Analysis of the two emission band systems leads to a term difference of 9016 cm^{-1} between the A and X states, and 25566.0 cm^{-1} between the B and X states. This may be compared to 9018 cm^{-1} and 25566 cm^{-1} obtained from the Rydberg series limits [110].

The determination of the various ionization potentials of molecular oxygen from spectroscopic data represents a rather different combination of information and reasoning. The first five electronic states of the O_2^+ ion are in order of increasing energy, $X^2\Pi_g$, $a^4\Pi_u$, $A^2\Pi_u$, $b^4\Sigma_g^-$ and $B^2\Sigma_g^-$. No Rydberg series have been observed to date which converge to any of the first three states. The $X^2\Pi_g$ limit is based on a photoionization threshold measurement for the production of a vibrationally excited O_2^+ ion, presumably in the $v=1$ vibrational state [111]. The result was corrected to the $v=0$ vibrational state using the vibrational frequency 1843.34 cm^{-1} of O_2^+ ($X^2\Pi_g$) determined from analysis of the second negative band emission system [104], leading to an ionization potential of 12.063 ± 0.001 eV. The vibrational numbering of the second negative band system has since been revised [112–114] leading to a corrected vibrational frequency of 1876.40 cm^{-1} and a resulting ionization potential of 12.059 ± 0.001 eV. One possible check on this value is the oxygen ionization threshold determined from the long-lived $O_2(a^1\Delta_g)$ metastable molecule. This has been determined as 11.090 ± 0.001 eV by photoionization [115] and 11.09 ± 0.005 eV by photoelectron spectroscopy [116]. Combining this with the O_2 $a^1\Delta_g - X^3\Sigma_g^-$ term difference of 7918.1 cm^{-1} (0.9817 eV) [104], leads to an $O_2(X^3\Sigma_g^-)$ first ionization potential of 12.072 ± 0.005 eV. The differences among the various results are unexplained, and possibly reflect different effects of rotations on the observations and on their interpretation. A further possible check on the ground state ionization potential could be based on the dissociation energy of the O_2^+ ($X^2\Pi_g$) state combined with the accurately known molecular oxygen dissociation

energy and the atomic oxygen ionization potential. However, the O_2^+ dissociation limit is not accurately established and the results are inconclusive [113].

The $a^4\Pi_u$ series limit is based on the $b^4\Sigma_g^-$ series limit [105] and the term difference established from the first negative band $b^4\Sigma_g^- - a^4\Pi_u$ emission series [104]. The $A^2\Pi_u$ series limit is based on the $X^2\Pi_g$ corrected ionization threshold [112] and the term difference established from the second negative band $A^2\Pi_u - X^2\Pi_g$ emission series [113]. Further, the energies of other high lying states of O_2^+, the $b^4\Sigma_g^-, B^2\Sigma_g^-, c^4\Sigma_u^-$ and a possible $^2\Pi$ state have been determined from Rydberg series. The energies of the $b^4\Sigma_g^-$, $B^2\Sigma_g^-$, and $c^4\Sigma_u^-$ states have been confirmed by photoelectron spectroscopy; that of the $^2\Pi$ state has not (see the Positive Ion Table).

d. Ionization Threshold Laws

The determination of ionization thresholds is based upon assumptions concerning the energy dependence of the electron impact or photoionization cross section in the neighborhood of the threshold. Two cases are to be considered:

a. Determination of first ionization and multiple ionization thresholds.

b. Determination of higher ionization potentials.

For the first case, Wigner [117] and then Geltman [118] showed theoretically that, under certain restrictive assumptions, the cross section behavior in the threshold region is given by a power law of the form:

$$\sigma(E) = c(E - E_0)^{n-1},$$

where E is the ionizing electron or photon energy, E_0 the threshold energy, c a constant, and n the total number of outgoing electrons for the ionization process. This then leads to a linear threshold law for single ionization by electron impact, a quadratic threshold law for double ionization, and so on. For photoionization there would be a step function threshold for single ionization, a linear threshold for double ionization, etc. Morrison [119] showed how various derivatives of electron impact and photoionization curves could be used to obtain information on excited states of ions. His discussion was based implicitly on threshold laws of the above form.

In view of the widespread treatment and interpretation of ionization data tacitly based on use of these threshold laws, it is appropriate to inject a note of caution. The treatments of Wigner and Geltman do not give any information at all about how far above threshold the laws are expected to be valid. For practical purposes a range of validity of an electron volt or two would be desirable. However, there is no guarantee that the range of validity is this large; it may only be several millivolts, or even less, and it may vary from system to system. Thus, there is no theoretical reason to expect that there exist threshold laws of exactly the form, range of validity and universality desired by experimentalists.

The experimental evidence on photoionization of atoms is as follows. Among the rare gases, the photoionization cross sections of argon, krypton, and xenon are very nearly step functions above the $^2P_{1/2}$ threshold (the $^2P_{3/2}$ threshold region is overlaid with autoionizing terms of the Rydberg series converging to the $^2P_{1/2}$ limit) [120]. The photoionization cross section curve of neon exhibits a discontinuous onset but above that it has a pronounced positive slope [121]. Finally, the helium curve has a discontinuous onset followed by a negative slope, [121, 122] i.e. it resembles a saw-tooth. A large number of experimental results for other atoms, showing a wide variety of cross section behavior, are summarized and discussed by Marr [122]. A large number of these results have been more or less quantitatively confirmed by theoretical calculations. The simplest case, the photoionization cross section curve of the hydrogen atoms was accurately calculated many years ago and the energy dependence above the discontinuous threshold is proportional to $E^{-8/3}$ [123].

For electron impact ionization of atoms the situation is also complex. Following Wigner's general threshold law derivation, Bates and co-workers [124] carried out Born approximation calculations and also derived a linear threshold law for electron impact single ionization. However, Wannier [125] arrived by a classical phase space argument at a 1.127 power law. This argument has recently been put on a quantum mechanical basis by Rau [126]. It is now clear that there is no general linear threshold law for single ionization. The available experimental results indicate that near threshold the cross section energy dependence is not exactly linear, but close to the 1.127 power law [127–130]. In addition, for krypton and xenon which have large separations between the $^2P_{3/2}$ and $^2P_{1/2}$ ion states, there is evidence of more complicated behavior near threshold due to autoionization and, possibly, transient negative ion states. However, different experimenters obtain different results [131]. Studies on Na, K, and Mg on the other hand suggest a linear threshold law for single ionization [132, 133].

Experimental determinations of the threshold laws for multiple ionization have been carried out by a number of workers. However, none of these experiments employed truly monoenergetic electron beams. For helium, sodium, and potassium where there are no low lying excited doubly charged ion states, double ionization follows a quadratic threshold law [132, 133]. The experiments were carried out with both RPD and conventional electron beam sources. Studies of the other rare gases have also been carried out with these techniques [134–137]. The results are in conflict and the question of the form of the threshold law remains unresolved. These studies are complicated by the existence of excited multiply charged ion states near threshold, which may affect the experimental threshold behavior in a manner depending on the electron energy distribution. Studies on multiple ionization of other metal atoms

show that the ionization efficiency curves have structure which can be correlated with autoionization and Auger effects [138, 139]. However, these features in the ionization efficiency curves have not been employed in a predictive fashion, and the prospect for doing so is remote.

e. Photoionization Thresholds of Diatomic Molecules

In contrast to atoms, the photoionization thresholds of molecules are not sharp because an ionizing transition in the threshold region is in fact a sum of many individual transitions of closely similar energy from one or another rotational state of the molecule to one or another rotational state of the ion. The molecule will generally be in the vibrational ground state. However, a number of vibrational states of the ion will be accessible. In addition, the ion electronic state is often a doublet, with a spin-orbit splitting that may be small, comparable to, or larger than the spacing of the vibrational levels of the ion. Thus, the problem of deducing the energy difference between the ground state of the molecule and the rotationless, vibrationless, electronic ground state of the ion (i.e. the adiabatic ionization potential) requires some analysis. Aspects of this problem have been discussed in detail by Guyon and Berkowitz [140].

Since the energy resolution of almost all photoionization experiments is not enough to resolve rotational structure, it is useful to discuss vibrational effects first.

The direct photoionizing transition of a molecule obeys the Franck-Condon principle [141–143]. According to this principle, the position of the nuclei and their momenta are unchanged by the transition. As a result the vibrational levels of the ion that are accessible in the transition depend upon the position and shape of the ion potential curve relative to the neutral potential curve (see figure 3). If the ion curve is very similar in shape and has an equilibrium internuclear distance identical to that of the molecular curve, essentially only the ion vibrational ground state will be accessible from the molecule ground state. If the ion internuclear distance is a little greater or smaller than that of the molecule (several thousandths of a nanometer) additional ion vibrational levels become accessible and a vibrational progression will be observed, with an intense $(0 \leftarrow 0)$ transition and progressively less intense transitions to the higher vibrational levels. For larger differences, roughly 0.005 nm and more, the vibrational ground state $(0 \leftarrow 0)$ transition is no longer the most intense and the maximum transition probability shifts to a higher vibrational level. For still larger distance changes, the $(0 \leftarrow 0)$ transition may be so weak as to be unobservable, and a significant portion of the accessible transition region may lie above the dissociation limit. To a good approximation the relative transition probabilities to the various ion vibrational levels may be represented by squares of normalized vibrational overlap integrals, also called Franck-Condon factors. A graph of the distribution of relative transition probabilities is known as a

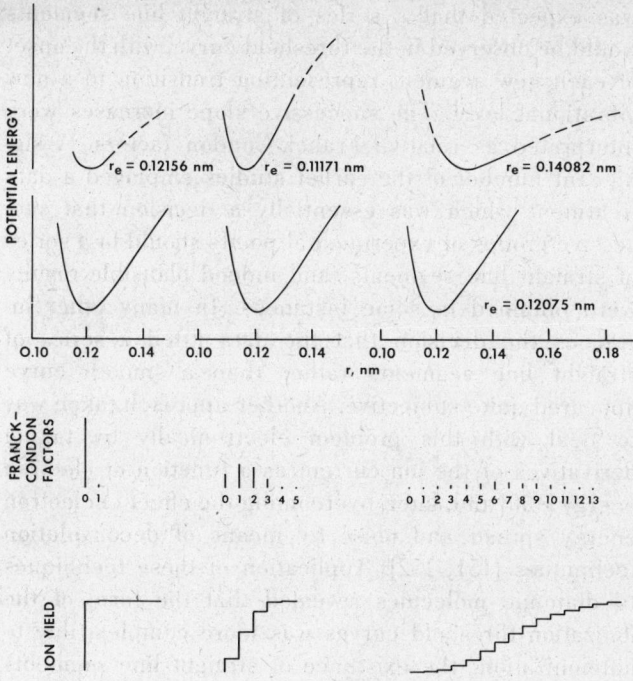

FIGURE 3. Franck-Condon factors and hypothetical direct photo-ionization yield curves expected for small, moderate and large bond length changes in a vertical ionization. Adapted from calculations by Krupenie [270].

Franck-Condon envelope. Thus, as the difference between the ion and neutral equilibrium internuclear distance changes from negligible to very large, the Franck-Condon envelope changes from a single vertical line to a monotone decreasing curve to a distorted bell shaped curve (see figure 3).

In direct photoionization, the various vibrational transitions manifest themselves as a staircase structure in the threshold region, with relative step heights corresponding to relative values of the Franck-Condon factors. This structure will occur if the photoionization threshold law is a step function. This appears to be the case for molecules studied thus far, in contrast to atoms. Experimentally, the sharp staircase structure would be smeared out into a series of sigmoidal onsets due to the triangular slit function of the monochromator. The point of inflection of the sigmoid occurs at the onset of the sharp step, and represents the point to be taken as the onset energy.

The effect of rotation on the direct ionization cross section curve shape is to produce some additional tailing of the onset and rounding of the step as a result of the spread of threshold energies of the individual rotational transitions. The exact shape of the threshold curve depends on the thermal population of rotational levels, the difference in moment of inertia of the molecule and ion, and the selection rules. The only case in which this effect has been analyzed carefully is in the photo-ionization of HF [140]. The analysis was only partially successful and the uncertainty in the estimated adiabatic ionization potential is somewhat less than 10 meV. For heavier molecules the uncertainty would be even less.

The next factor to be considered is vibrational hot bands. At room temperature most diatomic molecules have only small populations of vibrationally excited molecules. These will produce photoions below the adiabatic threshold through $(0 \leftarrow 1)$ transitions if, as is nearly always true, the ion and molecule bond lengths are not identical. In those cases the photoionization threshold region shows a small step followed by a large step, readily identified as the $(0 \leftarrow 0)$ transition, and a series of progressively smaller steps. This interpretation is readily confirmed since the energy difference between successive steps corresponds to the energy difference between the molecule ground and first vibrationally excited state. Further, changing the vibrational population by changing the sample temperature can be used to study and correct for hot band effects. In some cases, notably Br_2 and I_2, the vibrationally excited populations are so high and the vibrational progressions so long that this approach is essential to determining the adiabatic ionization potential [144].

Hot band problems are also widely encountered in the photoionization of high temperature species such as alkali and other metal halides [145]. The threshold curves have a more or less exponential tail, due principally to the Boltzmann distribution of vibrational populations [146]. The vibrationally excited species can have a lower ionization threshold for two reasons. First, $(0 \leftarrow 1)$, $(0 \leftarrow 2)$, $(0 \leftarrow 3)$, etc. transitions may occur as a result of differences between the ion and neutral bond distances. Second, only $(0 \leftarrow 0)$, $(1 \leftarrow 1)$, $(2 \leftarrow 2)$, etc. transitions may occur when there are only very small bond length changes. In this case, vibrational frequency differences between the ion and neutral will also produce progressively lower thresholds for the higher transitions. Consideration of these factors and plausible sets of Franck-Condon factors indicates that the threshold curve cannot be exactly exponential but will be nearly so. The adiabatic threshold is determined by plotting the data on a semilog scale and locating the point of departure from linearity. Numerous examples and detailed discussion are given in a review by Berkowitz [145].

The phenomenology of direct ionization sketched above has superimposed on it the effects of autoionization. In addition to the autoionization of high Rydberg states converging to the higher doublet limit of the ion ground state, and various terms converging to electronically excited ion states there is now a new possibility, namely, autoionization of Rydberg states belonging to series converging to vibrationally excited ions in their electronic ground state. The mechanisms of these processes have been discussed by Berry [147]. The effect of the autoionization is to change the expected staircase structure of direct ionization to a more complex one showing numerous peaks indicating autoionizing Rydberg levels converging to various vibrationally excited levels of the ion ground state. In NO for example, this additional structure is not very pronounced [148] whereas in the halogens [144] it is the dominant feature of the threshold region. Early photoionization work [146] on NO did not show this autoionization structure clearly, due probably to a poorer signal-to-noise ratio and fewer data points. A comparison is shown in figure 4.

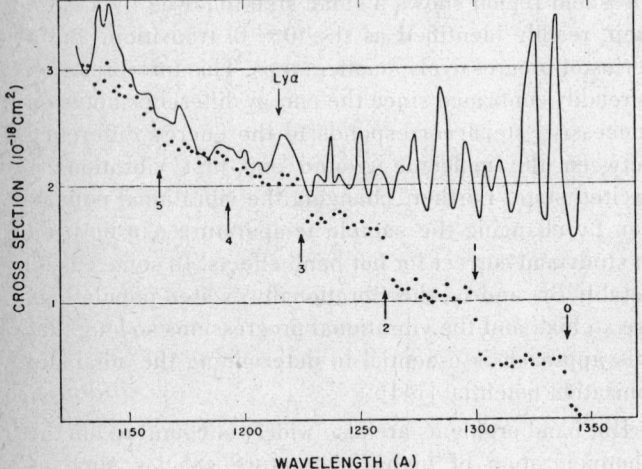

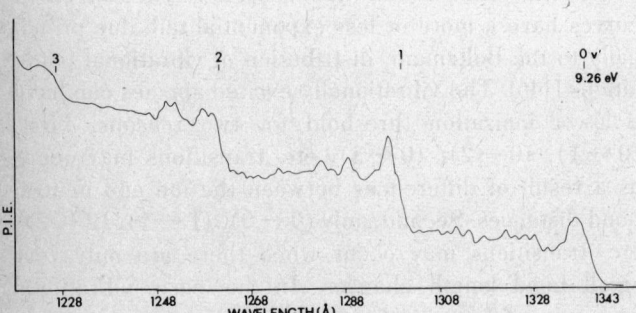

FIGURE 4. The ionization threshold region of NO. The upper curve is the total absorption cross section, and the experimental points the ionization cross section, determined in early experiments by Watanabe [146]. The lower curve is the photoionization yield curve recently determined by Killgoar et al. [148], showing pronounced autoionization peaks superposed on the staircase structure.

Lastly, Chupka and Berkowitz have drawn attention to the possibility that spurious lowering of ionization thresholds could result from collisional or electric field ionization of highly excited states produced by photon absorption slightly below the true ionization limit [149]. This can present a serious problem in very accurate threshold measurements, but no examples have yet been published.

f. Electron Impact Thresholds of Diatomic Molecules

Prior to the general recognition of the importance of autoionization it was expected that careful study of the electron impact ionization threshold region with mono-energetic or quasi-monoenergetic (RPD and EDD) techniques would give information on the Franck-Condon factors for vertical transition to the ion ground state. Assuming a linear threshold law for single ionization it

was expected that a series of straight line segments would be observed in the threshold curve, with the onset of each new segment representing transition to a new vibrational level. The successive slope increases were interpreted as relative Franck-Condon factors. A significant number of the earlier studies employed a data treatment which was essentially a decision that successive groups of experimental points should fit a series of straight line segments, and indeed plausible results were obtained in some instances. In many other instances the decision that the data fitted a series of straight line segments rather than a smooth curve appeared quite subjective. Another approach taken was to deal with this problem electronically by taking derivatives of the ion current as a function of electron energy [150] and, later, overcoming the effect of electron energy spread and noise by means of deconvolution techniques [151, 152]. Application of these techniques to diatomic molecules revealed that the form of the ionization threshold curves was more complex, due to autoionization; the existence of straight line segments was not confirmed. The manner in which a combination of direct ionization and autoionization determines the form of an electron impact threshold curve has not yet been analyzed in detail.

In practice, with nearly monoenergetic electrons one generally obtains smooth threshold curves of more or less similar shape no matter what molecule is studied, and with a rare gas threshold curve for calibration of the energy scale and as an indicator of the electron energy spread, one can obtain results in essential agreement ($\leqslant 50$ meV difference) with spectroscopic values [153, 154]. To date no procedure has been developed for correcting for hot bands or rotational effects.

There is no sound foundation for the still widely used conventional methods employing non-mono-energetic electron beams, a calibrating gas (generally a rare gas), and a variety of empirical extrapolation laws. These methods give results which are generally several tenths of an electron volt higher than the correct value, sometimes more and sometimes less. The problem is especially acute for species observed in high temperature Knudsen cell studies. Here hot band effects can be important, as discussed above, yet the electron impact results are still significantly higher than the true value. This is most likely a result of the conventional methods employed, especially the linear extrapolation and vanishing current techniques. Very careful studies have been carried out by both electron impact and photo-ionization on diatomic sulfur. These afford a basis for comparison. In spite of careful calibration procedures the electron impact result was more than 0.5 eV higher than the subsequent photoionization and spectroscopic results, which are in essential agreement [155, 156]. Additional comparisons and instances of closer agreement have been discussed by Hildenbrand [157]. Some other Knudsen cell studies undoubtedly give better answers. However, this will not become apparent until independent confirmation is obtained by more reliable techniques. At present, the only basis for assessing the likely inaccuracy of a particular measurement is the apparent care with which the measurement and calibration is carried out. The use of some diatomic and polyatomic gases of known ionization potential at high temperatures for calibration purposes might improve the degree of reliability of experimental results.

g. Photoelectron Ionization Thresholds

The basis for interpretation of photoelectron spectra has been extensively discussed in the books of Turner [32] and Eland [158]. In a photoelectron spectroscopy experiment, the ejected electron energy analyzer samples electrons arising from allowed electronic transitions to all vibronic levels of the ion which are Franck-Condon accessible and lie at or below the energy of the photon source line. With the assumption that the photoionization threshold law is a step function which is flat from the ionization threshold energy to the energy of the photon source line, the energy of the electrons ejected from any given electronic state will have a distribution determined by the Franck-Condon factors. With sufficient analyzer resolution one obtains for each accessible and allowed electronic transition a literal "bar graph" of experimental Franck-Condon factors. This enormously simplifies the determination of ionization potentials insofar as identification of the $(0 \leftarrow 0)$ transitions is concerned, not only for the first ionization potential but for higher ones as well. The spacing and Franck-Condon accessibility of higher electronic states is often of such a character that the accessible energy range of different states may overlap to some extent, i.e. there may be overlapping vibrational progressions. These aspects are well illustrated in figure 5 taken from the work of Edqvist et al. [44] on oxygen, carried out with an apparatus of 12 meV resolution (FWHM). An

important application of this aspect of photoelectron spectra is to problems where hot bands may be important. Recently the photoelectron spectrum of the SO radical has been observed [159]. The SO was produced by means of a gas discharge in an SO_2 rare gas mixture. This method of producing transient unstable species raises questions as to whether the species so generated are vibrationally excited or not. The photoelectron Franck-Condon factors clearly indicate that this is not a problem and that a good adiabatic ionization potential can be obtained. A similar case is the hot band problem in some halogen molecules which was discussed above in connection with photoionization. The photoelectron spectra at room temperature are of course also complicated by this factor. Here, too, experiments at different temperatures are very useful in correcting for hot band effects [160].

The above discussion on deriving experimental Franck-Condon factors is somewhat oversimplified. There is no reason to suppose that the molecular photoionization threshold law is strictly a step function. Small deviations from this step function form will bring about some differences in the electronic part of the transition probability to various vibrational levels of an electronic state, so that the experimental relative intensities of the vibrational transitions no longer accurately represent the Franck-Condon factors. As evidence of this it is to be noted that the relative integrated intensities (integrated over all the vibrational structure) of transitions to various electronic states do depend on the photon source line energy [161]. Further, the angular distribution of photoelectrons depends both on the type of electronic transition and on the energy with which the photoelectron is ejected in producing a given ion state, i.e. the incident photon energy and the specific vibrational level [162–164]. Unless the apparatus is set up to collect photoelectrons at a fixed special angle, the resulting information on electron energy distribution may be distorted [165]. Part of the data interpretation problem, of course, is concerned with proper assignment of the electronic transitions observed. This is an extensive and highly specialized subject on which the reader is referred to the books of Turner [32] and Eland [158] and to the various papers cited in the Positive Ion Table.

In comparison with photoionization, the photoelectron spectroscopy experiments suffer far less interference from autoionization phenomena. In photoionization, the wavelength scan sweeps over all accessible neutral states which can autoionize (and many do). The photoelectron experiment yields information on many states with one photon wavelength. If an accessible autoionizing neutral state happens to lie at the ionizing photon energy, the process will of course occur and will add its complications to the measurement and interpretation [166].

The best present-day photoelectron experiments are capable of producing data with a reported accuracy

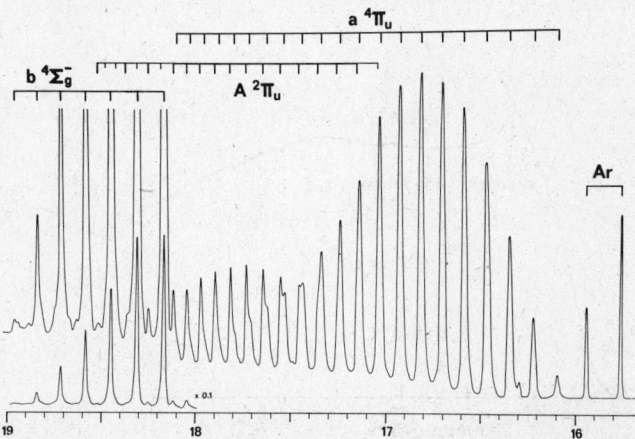

FIGURE 5. High resolution photoelectron spectrum of O_2, showing overlapping vibrational progressions from transitions to different electronic states of the ion. From work by Edqvist et al. [44].

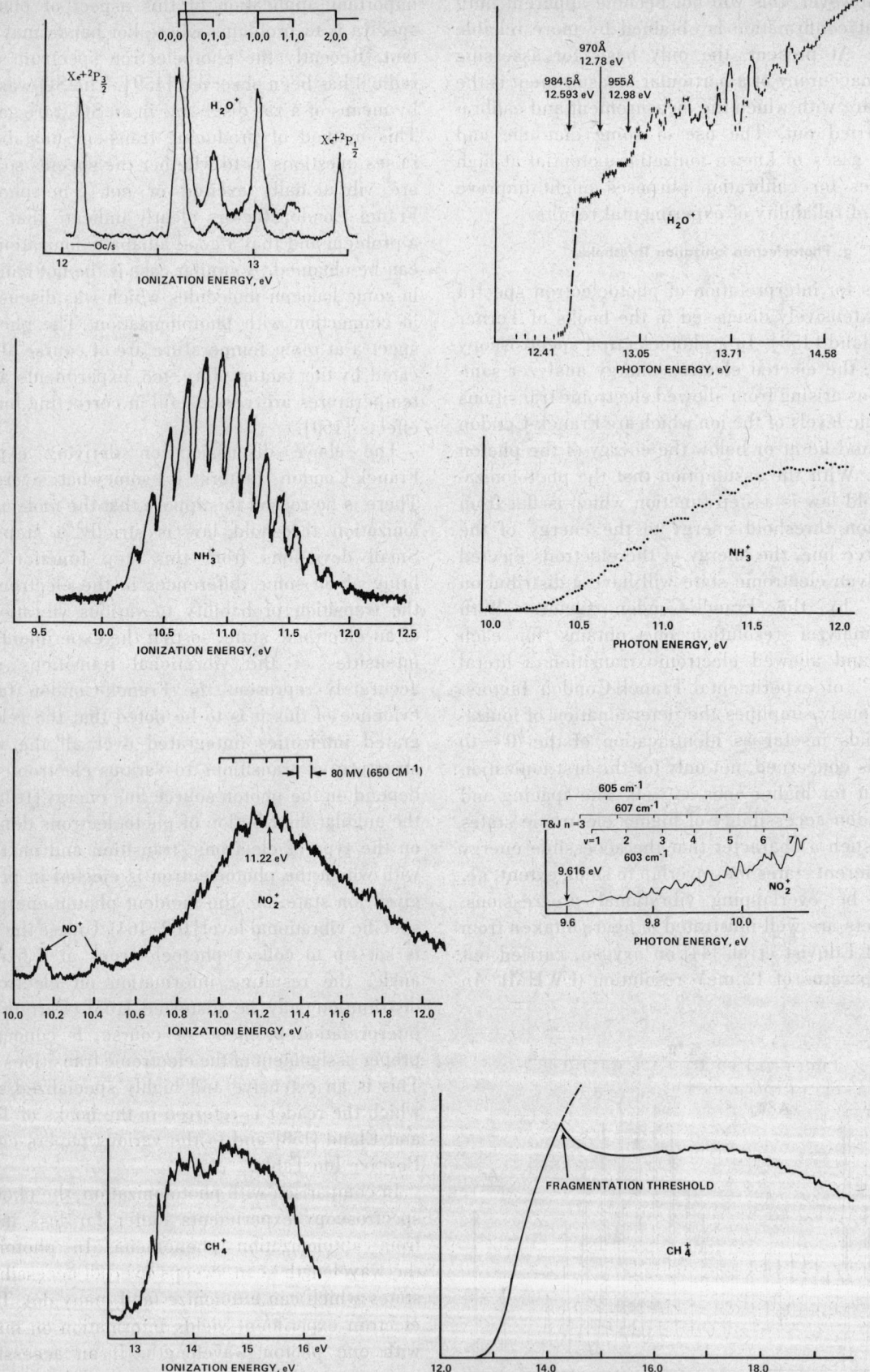

FIGURE 6. Photoelectron spectra and photoionization yield curves in the ionization threshold region for H_2O, refs. [271] and [272], NH_3, refs. [273] and [274], NO_2, refs. [275] and [174], and CH_4, refs. [180] and [168].

approaching 2 to 3 meV ($\sim$ 15 to 25 cm^{-1}). This accuracy is comparable to that of the best photoionization experiments (excepting the work on hydrogen [100], which is even better). Both in turn are beginning to approach the accuracy of many spectroscopic experiments. All have in common the problem of analyzing and correcting the experimental data for the contribution of various rotational transitions. With but three exceptions, H$_2$, N$_2$, and CO, spectroscopic data are measured on band heads, while photoionization data are measured on thresholds, and photoelectron data are measured on peak maxima of resolved vibrational levels. These three approaches are not simply related and remain to be analyzed in order to push accuracy and comparability beyond the 5 meV (40 cm^{-1}) limit.

h. Photoionization Thresholds of Polyatomic Molecules

The interpretation of photoionization thresholds of polyatomic molecules involves problems analogous to those for diatomic molecules. In addition to the form of the threshold law two aspects are especially significant, the Franck-Condon principle and hot band effects.

The experimental evidence for validity of the step function threshold law for direct ionization is not easy to assess because of the additional complexities introduced by Franck-Condon factors, hot bands, autoionization, and fragmentation. Where these complexities do not obscure things it appears that the step function threshold law is more or less closely followed. Exceptions include a number of ketones which have a variety of non-zero slopes above threshold [167], and methane, which appears to have an "overshoot" form [168]. It is not known how much of this overshoot is due to autoionization.

Whereas in the diatomic case one has only to consider the effect of a single internuclear distance change on the vibrational structure in direct photoionization, in the polyatomic case there can be changes in bond distances, bond angles and even in symmetry. The possibilities for vibrational structure are correspondingly more complex. In many instances the photoionization threshold curve still has a reasonably sharp onset which most probably corresponds to the adiabatic threshold. This is true for processes involving removal of a non-bonding electron, where little change in molecular geometry is expected. However, there are many cases in which significant geometry changes occur. These include ionization of tetrahedral and octahedral molecules, for which the equilibrium geometry of the ion ground state is considerably different from that of the neutral molecule due to the Jahn-Teller effect [169, 170]. In these instances, the photoionization threshold curve has a very gradual onset, and it is not immediately apparent whether the accessible (i.e. observed) Franck-Condon region includes the (0 ← 0) transition. A variety of ion yield curves are shown in figure 6. It is interesting to note that the values of the CH$_4$ ionization potential reported over the years have become progressively

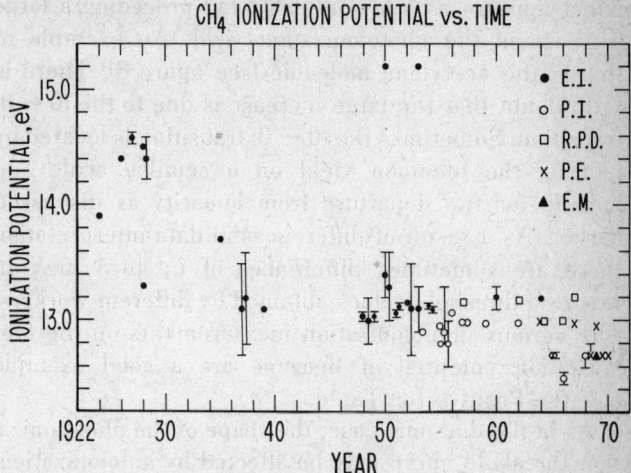

FIGURE 7. Variation over the years in the experimental values of the ionization potential of methane. Where no error bars are shown the limits of error are small.

lower and lower, due presumably to the progressive increase in apparatus sensitivity (see figure 7). The adiabatic ionization potential is still not firmly established. In other cases, ionization of tetrahedral molecules produces only fragment ions, indicating that the parent ion equilibrium geometry is so different that the Franck-Condon transition does not even reach a stable region of the ion potential surface.

In the case of NH$_3$, the removal of a "non-bonding" electron changes the molecular geometry from pyramidal to planar and the photoionization threshold exhibits only a weak (0 ← 0) transition [171]. Indeed, the identification of the (0 ← 0) transition is based on spectroscopic studies of the isotopic shift of vibrational Rydberg series terms [172]. A similar problem is found in the photoionization of the CF$_3$ radical [173], where the location of the (0 ← 0) transition is not yet firmly established. Also, it is not clear whether or not the experimentally generated radicals are vibrationally excited and show hot band effects.

Still another type of Franck-Condon problem arises in transitions between bent and linear (or vice versa) configurations, which produce long vibrational progressions. This is the case for NO$_2$, where the molecule is bent, with a bond angle of 134°, and the ion is linear. It is not yet established whether the photoionization threshold represents the (0 ← 0) transition. Only an upper limit has been obtained for the adiabatic ionization potential [174].

Large molecules often have low frequency normal modes which are appreciably populated at room temperature and whose effects on thresholds should be taken into account. However, the problem is very difficult. The photoionization curves of higher alkanes [175] and alkyl radicals [176] show a gradual onset which may be partially due to hot band effects. At present all that can be stated is that the adiabatic ionization potentials of these compounds are uncertain. Fortunately, there are many cases where the hot band effect appears as a low intensity tail preceding a large increase in the photoionization yield. An example of this is the acetylene molecule (see figure 8). There is little doubt that the large increase is due to the $(0 \leftarrow 0)$ transition. Sometimes the $(0 \leftarrow 0)$ transition is located by plotting the photoion yield on a semilog scale and looking for the departure from linearity as discussed earlier. As a result of differences in data interpretation there are sometimes differences of up to 5 meV in reported threshold values obtained by different workers. The various photoionization measurements on the first ionization potential of benzene are a good example (see the Positive Ion Table).

As in the diatomic case, the shape of the photoionization threshold curve may be affected by autoionization, which can somewhat complicate or even completely obscure the staircase structure expected from the Franck-Condon factors for direct ionization. Also, the observed structure is dependent on the photon energy resolution. Early work on photoionization of acetylene showed a staircase structure that could be readily related to Franck-Condon factors [177]. However, a recent study employing higher resolution (0.05 nm compared to 0.2 nm) demonstrated that the threshold ion yield curve shows a combination of direct ionization and a significant amount of superposed autoionization [178]. This is illustrated in figure 8. The experimental Franck-Condon factors for the estimated direct ionization were in good agreement with those determined by photoelectron spectroscopy and those theoretically calculated [171]. In another case, early work on photoionization of CO_2 near threshold suggested a significant deviation from the step function threshold law. Recent work at higher resolution demonstrated that in the threshold region there was considerable autoionization from an excited valence state of CO_2 [96] which strongly influenced the shape of the photoionization yield curve. Lastly, in the photoionization of methane there is considerable structure in the threshold curve. To what extent this is due to a complex series of vibrational progressions or to autoionization is unclear [179]. The photoelectron spectrum of methane also shows vibrational structure [180].

In summary, the determination of adiabatic ionization potentials from photoionization curves of polyatomic molecules is not always straightforward. In some cases, where there are large geometry differences between the ion and the molecule, the adiabatic value cannot be determined. These difficulties also manifest themselves in the uv spectra and in photoelectron spectra. However, in many other cases agreement among all three methods is better than 5 to 10 meV.

i. Electron Impact Thresholds of Polyatomic Molecules

It is evident from the above discussion that a wide variety of electron impact threshold behavior is to be expected for polyatomic molecules. This, when folded into a large electron energy spread (whose shape may

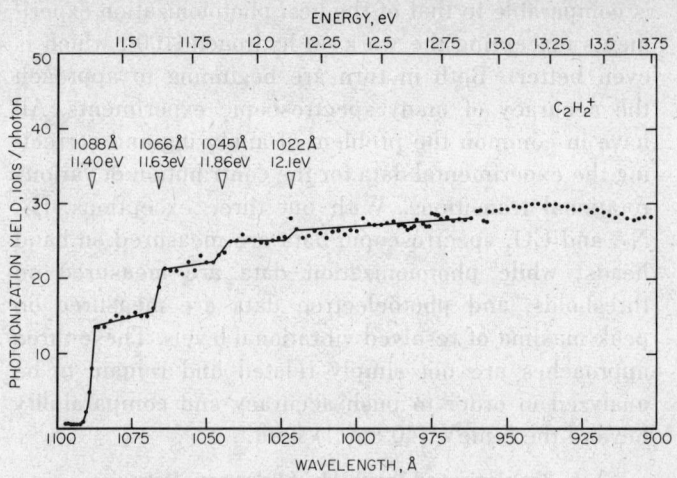

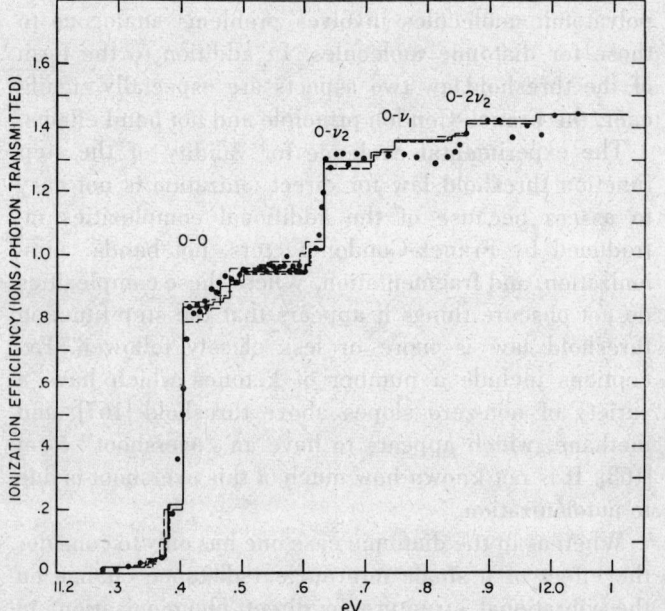

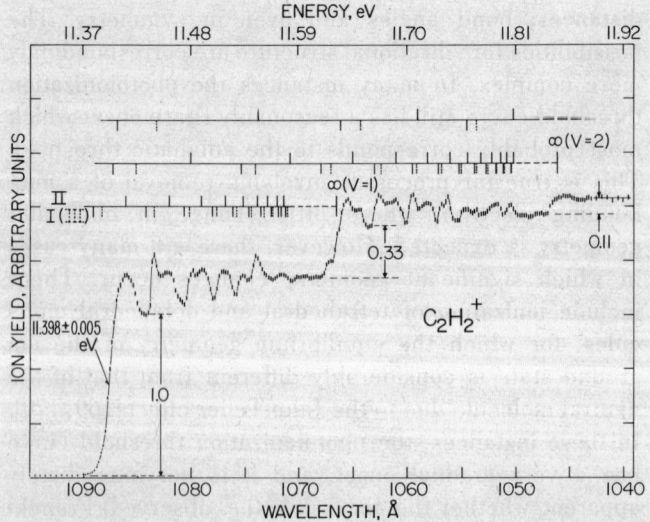

FIGURE 8. Photoionization yield curve in the threshold region of acetylene in three successive experiments of increasing refinement, refs. [276], [177], and [178].

vary with electron energy and the chemical nature of the gas sample) leads to a very ill-defined situation. The basis for the choice of calibrating gas and the meaning of the calibration process is not at all apparent. Reduction of the energy spread by means of an electron monochromator [154] appears to have overcome these difficulties in spite of the arbitrary nature of the calibration process. Essentially all of the ionization potentials determined by this technique are in satisfactory agreement (10 meV or better) with spectroscopic, photoelectron, and photoionization measurements with the exception of the results on methane [154] and benzyl radical [181, 182]. In methane the electron impact threshold curve does not have a sharp onset and the authors express the result as an upper limit ($I_z \leqslant 12.70$ eV) which is about 0.1 eV higher than the most recent photoionization ($\leqslant 12.615 \pm 0.010$ eV) [179] and photoelectron (< 12.616 eV) [170] upper limits. This is probably a matter of apparatus sensitivity. The ionization potential of the benzyl radical determined with an electron monochromator [181] ($\leqslant 7.27 \pm 0.03$ eV) is 0.4 eV lower than that obtained by photoionization [182] (7.63 eV). The reason for this is unknown. Other quasi-monoenergetic electron impact techniques (RPD and EDD) show a wide variety of agreement or disagreement which is, unfortunately, unpredictable.

j. Higher Ionization Thresholds

Numerous higher ionization thresholds have been determined by photoelectron spectroscopy and a small number by uv spectroscopy. The associated problems of Franck-Condon effects and assignment of the electronic transition have already been touched on. Some further aspects of photoionization and electron impact experiments should be discussed in this connection.

As mentioned above, molecular photoionization experiments reveal the presence of many autoionizing states which are members of Rydberg series converging to vibrationally and/or electronically excited ion states. In the discussion on atomic ionization potentials it was pointed out that in photoionization there would be no observable higher ionization threshold discontinuity. In the molecular case, however, the higher Rydberg states that autoionize also can undergo the competing process of predissociation into excited neutral species. This will lower the net ion yield per unit of absorption below a higher ionization threshold in comparison to the ion yield above the threshold where it is unity. Thus, in principle, higher ionization thresholds may be "directly" observable. However, the few observations of this type reported in the literature are not completely convincing [183]. Again, in principle, one could make the thresholds observable by raising the gas pressure to the point where true averaging of the absorption coefficient no longer occurs.

As for identification and interpretation of interesting features in the RPD and EDD electron impact curves above threshold no definitive information has yet been gained, the rationale is unclear and even the reality of the features is in dispute. One particular case that merits mention is the threshold curve for acetylene ion. Early photoionization work and Franck-Condon calculations indicated that the threshold photoion yield curve had a staircase structure due to a strong $(0 \leftarrow 0)$ transition and lesser excitation of the first and second carbon-carbon stretching overtones, with some minor additional transitions [177]. Later, the existence of these stretching transitions was confirmed by photoelectron spectroscopy [184]. An EDD study, however, concluded that the $(0 \leftarrow 0)$ transition was very weak and suggested that the Franck-Condon principle was violated [185]. Subsequently an electron monochromator study corroborated the results of the EDD study [186]. And more recently a different electron monochromator study disagreed with this work and essentially confirmed the Franck-Condon factors found in the photoionization and photoelectron spectroscopic work [187]. Finally, a recent photoionization study showed that autoionization was an important contributing process in the threshold region [178]. In view of the autoionization problem, the significance of the most recent electron monochromator results is not clear. One can only conclude that further careful work is required and that each molecule will present its own characteristic but interesting difficulties.

k. Multiple Ionization Thresholds

A number of molecules have doubly and triply charged ion states which are stable enough to permit ionization threshold measurements. Available experiments, employing conventional non-monoenergetic electron beams, indicate that double ionization follows a quadratic threshold law, whereas the results on triple ionization are inconclusive [188]. Also, high kinetic energy fragment ions, presumably arising from decomposition of doubly charged parent ions, have been shown to follow a quadratic electron impact threshold law [189]. It has also been shown that the appearance potential of these ions increases with increasing kinetic energy [190], as required from simple energy considerations.

l. Diatomic Fragmentation Thresholds

The idea that the fragmentation of diatomic ions could be described by the Franck-Condon principle was first advanced in 1928 by Condon and Smyth [191, 192]. The idea was that ionization produced vertical transitions from the molecule ground state to the various ion electronic states and that depending on the electron energy and the shape and location of each of the ion potential curves one would observe characteristic yields (large, small, none) of fragment ions with characteristic kinetic energy distributions. In the case of hydrogen, enough was known about the H_2^+ attractive and repulsive potential curves to lead to the prediction that little if any fragmentation would result from a vertical ionizing transition to the attractive $H_2^+(^2\Sigma_g^+)$ ground state, and that fragment ions would be produced exclusively by transitions to the repulsive $^2\Sigma_u^+$ state with a resultant kinetic energy of several electron volts. This was verified shortly thereafter by Bleakney [193]. The idea that fragmentation results from direct vertical transition to a repulsive curve or to an accessible part of an attractive curve which happens to lie above the dissociation limit is only a first approximation. Other factors must be considered. Fragmentation can occur via ion-pair production or via predissociation from an attractive ion state by curve crossing. Also, ion states may be populated by autoionization as well as by direct ionization.

In the autoionization case the Franck-Condon principle applies also, but in two steps. First, a particular region of an autoionizing electronic state becomes accessible by vertical transition from the molecule ground state. Second, for each populated autoionizing vibronic state a region of the ion curve becomes accessible since molecular autoionization is also governed by the Franck-Condon principle [166, 194]. The net result is, in effect, to broaden the accessible ion curve region beyond that expected from a direct ionization process. In order to derive ion thermochemical data from experimental fragmentation threshold data it is necessary to establish the nature of the dissociation process, the electronic states of the fragments, and the kinetic energy distribution of the fragments. Examples of the problems encountered are given below.

I.1. H_2

There has been very little recent work directed to measuring the dissociative ionization of hydrogen by electron impact and only one study by photoionization. This is probably due to the fact that the threshold energy for the dissociative ionization process is well known from the dissociation energy of the hydrogen molecule and the ionization potential of the hydrogen atom. In electron impact the threshold for formation of H^+ is in fact lower by about 0.8 eV due to the ion-pair process:

$$H_2 + e \rightarrow H^+ + H^- + e,$$

and there is a fairly sharp increase in proton current starting at the energy corresponding to dissociative ionization [195]. Recently the ion-pair process has been carefully studied in detail with photoionization [196]. Several thresholds were observed corresponding to predissociation of excited hydrogen molecules in different rotational states into ion pairs. With the assumption that the fragments had zero kinetic energy, and using accepted values for the dissociation energy of H_2 and the ionization potential of the hydrogen atom, a value was derived for the electron affinity of the hydrogen atom in essentially exact agreement with the best theoretical value (experiment, $\geq 0.754 \pm 0.002$ eV, theory, 0.75421 eV).

The kinetic energy distribution of protons from the dissociative ionization process has been carefully studied several times and is qualitatively in accord with expectations from the Franck-Condon principle [197, 198],

however, no quantitative agreement has yet been obtained [199, 200]. Important aspects of this problem have been discussed recently by Crowe and McConkey [201]. In contrast, the experimentally measured kinetic energy distribution of proton pairs formed by double ionization of molecular hydrogen with 1 keV electrons is in very close agreement with accurate Franck-Condon calculations [202].

I.2. N_2, O_2, CO, and NO

It is well known that if one has an accurately determined ion fragmentation threshold, has established the electronic states of the fragments at threshold, has measured the kinetic energy distribution of the fragments and established a meaningful threshold energy correction procedure, and knows the energy required to form the ion in its product state from the corresponding atom in the ground state, one can determine a bond dissociation energy. This line of reasoning is widely applied in more or less detail in mass spectrometric experiments for determining bond energies. A somewhat similar line of reasoning applies to spectroscopic determinations of bond dissociation energies. Ideally, one should obtain the same answer on the same system. In practice, very often the requisite information is not all at hand in sufficient detail and accuracy to provide a definitive answer. There is no better illustration of this than the years-long controversies concerning the dissociation energy of the nitrogen molecule and of the carbon monoxide molecule (the latter case is equivalent to the heat of sublimation of graphite controversy). The history of these controversies can be followed in the three successive editions of Gaydon's book [203].

Briefly, a number of spectroscopic observations led to accurate values of energy thresholds at which these molecules predissociated. However, the lack of definite information on the electronic states of the product atoms led to a set of highly accurate but widely differing bond energy values for each molecule, only one of which could be correct. Other spectroscopic studies directed to determining the bond dissociation energy by extrapolation of the energies of the converging vibrational levels were at first inconclusive. During the same period a number of very careful electron impact mass spectrometric studies were directed at this problem area as well [12, 197, 198, 204–209]. Much effort was devoted to measurement of fragment ion kinetic energies, and examination of the complications arising from ion-pair formation processes. The efforts were handicapped for some time by lack of accurate knowledge of certain electron affinities. All in all it took years of effort to arrive at a set of answers that was consistent with all spectroscopic, mass spectroscopic, thermochemical, kinetic and high temperature information. The literature of this problem is replete with reasonable, plausible, but ultimately incorrect information.

This story must be kept in mind when reporting or evaluating experiments on mass spectrometric bond dissociation energies. A very widespread assumption, almost never stated explicitly, is that dissociation products are formed in their ground state. Too few experimenters report the use of available capability to establish whether or not ion-pair processes are significant. And in the case of electron impact experiments the capability of observing negative ions is subject to possible interference from dissociative electron capture processes. In our opinion, the large body of bond energy data based on conventional non-monoenergetic electron impact measurements is subject to errors of up to 50 kJ mol^{-1}. The small body of results now being obtained by means of monoenergetic electron beam techniques is an order of magnitude more reliable. And this can be demonstrated by the self-consistency of the results and the agreement of experimental thresholds with photoionization measurements, which do not suffer from energy spread or energy calibration problems at this level of accuracy.

With these facts in mind we turn to a photoionization controversy. All that has changed is that the need for energy scale calibration and the broad energy distribution of the ionizing beam have been eliminated.

l.3. F_2, HF, and ClF

The photoionization controversy has centered about the dissociation energy of fluorine. In this case one set of photoionization studies by Dibeler et al. on fragmentation processes of the above molecules arrived at a seemingly self-consistent set of threshold values all supporting a value of 1.34 ± 0.03 eV for the dissociation energy of fluorine, lower than the generally accepted value of 1.594 ± 0.026 eV [210, 211]. The same observations were consistent with the accepted heats of formation of HF and ClF but suggested that the spectroscopically determined dissociation energy of HF, 5.86 ± 0.01 eV, was about 0.1 eV too high. These observations and conflicting results stimulated further photoionization studies on F_2 and HF by Berkowitz et al. which ultimately led to the conclusion that the HF and F_2 dissociation energies proposed in the earlier photoionization studies were incorrect and that the fragmentation energetics were in fact consistent with the accepted F_2 and HF dissociation energies [212]. The differences in the conclusions arrived at by the two groups of workers were due to the following deficiencies in the earlier study. Although Dibeler et al. observed an ion-pair process in the F_2 fragmentation threshold region, the onset of the F^+ ion yield curve was not corrected for the contribution of this process to the F^+ yield from dissociative ionization. Indeed, the intensity of the ion-pair process observed in that study was significantly lower than in the later study by Berkowitz et al., presumably due to the different ion source configurations employed. In addition, Dibeler et al. chose a location in the curved region at the foot of the threshold curve for a threshold value, whereas Berkowitz et al. corrected the ion-pair-corrected threshold curve for the effect of photon bandwidth and the

possible effect of rotational energy in lowering the fragmentation threshold [140]. The threshold values so deduced were 18.76 ± 0.03 eV [210] and 19.01 eV [212]. As for the dissociative ionization of HF, the two groups deduced thresholds for formation of $H^+ + F$ of 19.34 ± 0.03 eV [210] and 19.445 eV [212], respectively, differing principally because the latter value included the monochromator bandwidth and rotational population corrections. In addition, careful measurements were made of the kinetic energy of the protons from dissociative ionization of HF, and F^+ from the ion-pair process in F_2 [213]. There is no question that the results of Berkowitz et al. are to be preferred. The rotational correction model employed is one that tacitly assumes that the dissociation process is essentially a vertical transition from a population of rotational states of the molecule to the dissociation limit of the ion. However, the dissociation of HF^+ does not occur by this direct mechanism. Rather, it involves a predissociation of some sort, and the effect of rotations on this process is not clear.

Evidence for ion fragmentation has also been observed in HF by Brundle, using photoelectron spectroscopy [214]. In the first excited $^2\Sigma^+$ state of HF^+ the Franck-Condon envelope has vibrational structure only up to 19.400 ± 0.01 eV. Above this the envelope is smooth, indicating a very short lifetime. This threshold is consistent with the observations of Berkowitz et al. [212]. Lastly, an electron impact study on F_2 was recently carried out, leading to a value of 1.63 ± 0.1 eV for the F_2 bond dissociation energy [215], and confirming some of the kinetic energy measurements, thus supporting the accepted value. However, the experimental threshold data are not as precise. This controversy has been recently reviewed in detail by Berkowitz and Wahl [216].

m. Fragmentation Thresholds of Small Polyatomic Molecules

Over the past ten years numerous studies have been carried out on the fragmentation of triatomic and larger ions. These studies have established that the fragmentation processes are rather complex. In 1963, Dibeler and Rosenstock noted that the electron impact mass spectrum of H_2S showed evidence that at the fragmentation threshold some unimolecular ion decomposition processes took place which had lifetimes in the microsecond range (i.e. metastable transitions) [217]. They suggested that the decomposition process was a predissociation involving crossing of two potential surfaces. In 1964, Sharp and Rosenstock concluded on the basis of Franck-Condon calculations that fragmentation of the CO_2^+ ion could not possibly occur by direct vertical transition to the region above the dissociation limit of the ground state or first excited state of the ion [218]. In 1966, Fiquet-Fayard and Guyon showed, on the basis of adiabatic correlation rules, that the fragmentation of H_2O^+ and H_2S^+ occurred by predissociation of the accessible ion doublet states via the ion quartet states [219]. At the present time all available experimental evidence confirms the idea that the decomposition mechanism of small polyatomic ions is a predissociation, rather than a vertical transition to a region above the dissociation limit of the ion ground state. This of course raises questions as to whether the fragmentation products are formed in their ground states and whether the fragmentation occurs without kinetic energy at threshold. Two ions which have been studied in detail are CO_2^+ and N_2O^+.

The thermochemical threshold for the process $CO_2 \to O^+(^4S) + CO(X^1\Sigma^+, v=0)$ lies at 19.071 ± 0.002 eV, for $CO_2 \to O^+(^4S) + CO(X^1\Sigma^+, v=1)$ at 19.337 eV and for $CO_2 \to O(^3P_2) + CO^+(X^2\Sigma^+)$ at 19.466 eV, compared to the process $CO_2 \to CO_2^+ C^2\Sigma_g^+)$ which has a threshold of 19.389 eV [96]. It has been shown by photoelectron-photoion coincidence studies that the ground vibrational level of the C state of CO_2^+ is completely predissociated [220]. Most of the predissociations produce vibrationally excited CO and O^+ ions and some others ground state CO and O^+ with kinetic energy. Emission spectra have not been observed from the C state of CO_2^+, confirming the postulated predissociation mechanism [221]. A detailed photoionization study [96] showed that at the threshold of the C state the photoionization yield curve of the parent ion exhibited a sudden decrease, while the yield of the O^+ fragment ion increased at this threshold. The O^+ ion is observed at the expected thermochemical threshold and the fragmentation process is attributed to predissociation of Rydberg series members converging to the ground vibrational level of the ion C state, combined with or followed by autoionization of the O atom fragment. The ion yield curve for the CO^+ fragment shows an onset at its expected thermochemical threshold and the process is inferred to be predissociation of excited vibrational levels of the CO_2^+ C state.

The fragmentation of N_2O^+ has been the subject of considerable study, and illustrates once more the complexities encountered in interpreting fragmentation processes of small molecules. Dibeler et al. [222, 223] have measured the photoionization parent and fragment ion yield curves. Above the NO^+ threshold they observe additional sudden increases in ion yield suggesting the onset of fragmentation processes leading to excited nitrogen atoms (see figure 9). The experimental threshold values and the threshold values calculated from independent thermochemical and spectroscopic data are given in table 2.

TABLE 2. Comparison of experimental and thermochemical threshold values for fragmentation processes of N_2O^+

Fragmentation products	Experimental threshold, eV	Thermochemical threshold, eV
$NO^+(X^1\Sigma^+) + N(^4S)$	15.01	14.19
$NO^+(X^1\Sigma^+) + N(^2D)$	16.53	16.57
$NO^+(X^1\Sigma^+) + N(^2P)$	17.74	17.76
$N_2^+(X^2\Sigma_g^+) + O(^3P)$	17.27	17.25
$N_2(X^1\Sigma_g^+) + O^+(^4S)$	15.29	15.29
$NO(X^2\Pi) + N^+(^3P)$	20.06	19.46

The data suggest that the formation processes for only NO^+ and N^+ are accompanied by excess energy at threshold. Dibeler et al. suggested that the excess energy in the NO^+-forming process was vibrational excitation of the NO^+, the fragment ion being formed in the $v=3$ state for which the thermochemical threshold lies at 15.04 eV.

The fragmentation process $N_2O^+ \to NO^+ + N$ has been observed as a unimolecular delayed dissociation (metastable transition) in the mass spectrometer [224–226]. Coleman et al. [225] concluded that the dissociation process was a single decay process with a half-life of about 540 ns. Newton and Sciamanna [226], on the other hand, concluded that the delayed dissociation was a sum of two distinct decay processes with half-lives of ∼90 ns and ≥300 ns, respectively. The difference in conclusions is unexplained, but certainly involves difficulties in the numerical analysis of the data [227]. This type of problem has been discussed in detail in a recent review [241]. Both groups of workers have estimated the kinetic energy release accompanying the delayed dissociation by analyzing the effect of kinetic energy release on the peak width and shape of the metastable transition in a manner first proposed by Beynon et al. [228]. They showed that the kinetic energy release was roughly 0.6–1.0 eV. Thus at least some and possibly all of the

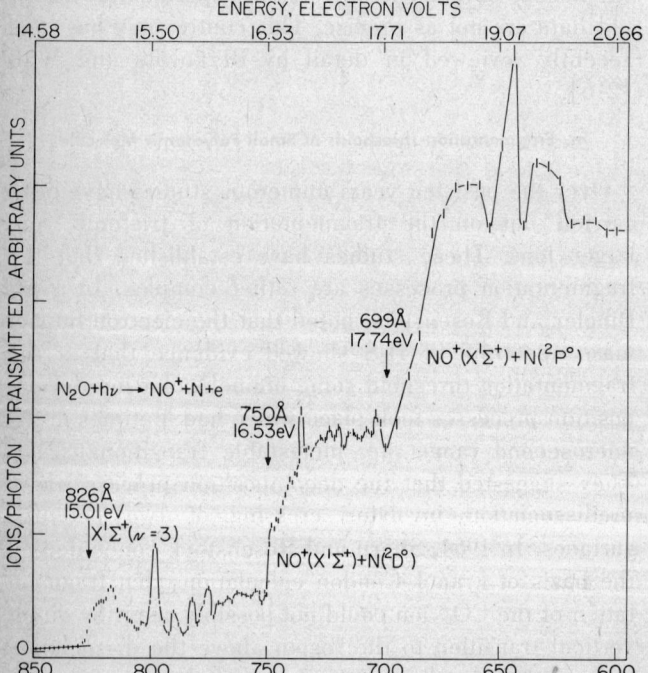

ENERGY, ELECTRON VOLTS

IONS/PHOTON TRANSMITTED, ARBITRARY UNITS

$N_2O + h\nu \to NO^+ + N + e$

699Å
17.74eV
$NO^+(X^1\Sigma^+) + N(^2P^\circ)$

750Å
16.53eV

826Å
15.01 eV
$X^1\Sigma^+(v=3)$

$NO^+(X^1\Sigma^+) + N(^2D^\circ)$

FIGURE 9. Photoion yield curve for the fragmentation process $N_2O + h\nu \to NO^+ + N + e$ showing discontinuities in the yield curve at energies corresponding to the formation of $N(^2D)$ and $N(^2P)$ neutral fragments, ref. [222].

excess energy of the fragmentation process appears as translational energy, and not as vibrational energy, as had been suggested by Dibeler et al.

The symmetries and energies (above the N_2O molecule ground state) of the first four doublet states of N_2O^+ are $X\,^2\Pi$, 12.89 eV; $A\,^2\Sigma^+$, 16.39 eV; $B\,^2\Pi$, 17.65 eV; and $C\,^2\Sigma^+$, 20.11 eV. These states are accessible by photoionization. From the extent of the Franck-Condon envelopes observed in photoelectron spectra it is clear that the NO^+ and O^+ thresholds lie above the vertically accessible region of the ion ground state and more than one electron volt below the A state. In fact, the NO^+ threshold coincides with an autoionizing Rydberg term which is the $n=3$, $v=1$ member of a Rydberg series converging to the ion A state. The autoionization peak observed in the parent ion photoionization yield curve indicates that this state can autoionize to produce stable N_2O^+ ions in the electronic ground state as well. Lorquet and Cadet [229] have given the adiabatic correlations for various fragments and show that both the $NO^+(X\,^1\Sigma^+) + N(^4S)$ and the $N_2(X\,^1\Sigma_g^+) + O^+(^4S)$ fragment pairs correlate with a $^4\Sigma^-$ state of N_2O^+. Thus the fragmentations involve a predissociation-autoionization process via a quartet state, and do not involve the ion ground state. In addition it should be mentioned that studies with isotopically labeled N_2O indicate that a significant fraction of the processes leading to formation of NO^+ involve loss of the central nitrogen atom [224].

As mentioned above, Dibeler et al. observed a sudden increase in the NO^+ yield at 16.53 eV, nearly coinciding with the thermochemical threshold for the formation of $NO^+(X\,^1\Sigma^+) + N(^2D)$ which lies at 16.572 eV. This energy also nearly coincides with that of the (100) vibrational level of the $N_2O^+\,A\,^2\Sigma^+$ state which is located at 16.56 eV [230]. Eland has studied the fragmentation of this state by photoion-photoelectron coincidence techniques [231]. He has demonstrated that the ground vibrational level of the N_2O^+ A state is completely stable (i.e. no fragmentation within a microsecond) whereas the (100) and (001) levels are partially predissociated. In addition, it is known that the (000) and (100) levels fluoresce, and the fluorescence lifetimes have been measured [232, 233].

From the branching ratio between fragmentation and fluorescence (i.e. observation of stable N_2O^+ ions) from the (100) vibrational level Eland deduced a half-life of 410 ± 90 ns for the fragmentation process. Adiabatic correlations indicate that formation of NO^+ $(X\,^1\Sigma^+) + N(^2D)$ also involves a predissociation via another state [229]. The half-life of the predissociation is in reasonable agreement with the mass spectrometric half-life measurements of the metastable transition mentioned above. However, it is not clear whether the two measurements refer to the same fragmentation process. In summary, fragmentation to produce NO^+ $(X\,^1\Sigma^+) + N(^2D)$ occurs at least in part through direct vertical transition from the molecule to the (100) level of the ion A state, followed by a competition between predissociation and fluorescence to various vibrational levels of the ion ground state. The agreement of the threshold value for this process with the thermochemical threshold is somewhat coincidental. Further details on the fragmentation processes are found in the recent literature [234, 235].

The fragmentation of the methane and acetylene ions illustrates still another aspect of interpreting fragmentation thresholds. In a careful study of the photoionization yield curve of the methyl ion fragment from methane Chupka showed that there was no sharp onset at threshold [168]. He pointed out that one had to take into account the contribution of rotational energy to the energy available for decomposition. He showed that for a quasi-diatomic system separating into an ion and a polarizable neutral fragment, the rotational energy would reduce the dissociation energy because of conservation of angular momentum. As a result of the original thermal distribution of rotational energy of the molecules being ionized, a sharp fragmentation threshold for non-rotating molecules would be replaced by a gradual smooth onset curve reflecting the rotational energy distribution (and the slit function). Chupka applied this simplified model to the room temperature photoionization yield curve of methyl ions from methane and deduced a threshold value for the fragmentation of non-rotating (0 K) molecules. Guyon and Berkowitz [140] later gave a more detailed discussion of this effect, again for diatomic systems. They showed that above threshold, the effect of the rotational energy contribution was to shift the photoionization yield curve to lower energy by an amount equal to kT. The model assumes that the dissociation occurs by direct transition to the dissociation limit of the ion rather than by predissociation. This is indeed the case for loss of H from the methane ion.

More recently Dibeler, Walker, and McCulloh [236] have studied the photoionization fragmentation of acetylene ion at various temperatures (298 K and 130 K). They showed that this thermal shift and rotational tailing were real and temperature dependent, and also were able to observe and correct for the effect of vibrational hot bands. As a result they were able to deduce an accurate threshold value for non-rotating molecules. The problem with acetylene, however, is that the mechanism of fragmentation is not yet established. The lowest fragmentation threshold, involving loss of a hydrogen atom, lies in the Franck-Condon region of the first excited state of the ion. Fiquet-Fayard has analyzed the adiabatic correlations for this fragmentation process, assuming only linear configurations, and concluded that fragmentation occurs by predissociation via a quartet state [237]. It is not clear how the effects of rotation will manifest themselves in this case.

It is clear from the above examples that photoionization studies of small polyatomic molecules, together with the results of other types of experiments, are revealing a diversity of fragmentation mechanisms and threshold behavior. The recent studies of metastable transitions in small molecules are very useful here [238]. There are no ready generalizations which will greatly simplify the task of interpreting fragmentation thresholds to yield accurate thermochemical data. As for electron impact studies of fragmentation thresholds, they suffer even more from unpredictable accuracy than is the case for photoionization thresholds. This is due to the above mentioned diversity of fragmentation mechanisms. And because of the much more limited information displayed in electron impact curves, the interpretations and the thermochemistry deduced from them are much less reliable than those obtained from photoionization. One question of great interest is whether significantly different threshold behavior should be found in electron impact processes because of possible direct transition to the quartet states which are not accessible in photoionization. This may be the case in the formation of NO^+ from N_2O by electron impact [239], where the threshold value is significantly lower than the photoionization fragmentation threshold. The discrepancy remains unexplained.

n. Fragmentation Thresholds of Large Molecules

Quantitative description of the fragmentation of large molecules is based on quasi-equilibrium theory [240–242]. According to this theory the fragmentation processes of the molecule-ion can be considered separately from the act producing the ionization and internal excitation energy distribution. Further, the fragmentation processes can be described as a series of competing, consecutive unimolecular reactions. Finally, the rate constants for the unimolecular reactions can be quantitatively calculated by means of activated complex theory. It is tacitly assumed that between the ionization event and the decomposition there is sufficient time for the excitation energy to randomize in spite of the fact that the ion does not undergo collisions (hence the term quasi-equilibrium). For simplicity it is generally assumed that prior to decomposition the ions produced in excited electronic states undergo internal conversion forming vibrationally excited ions in their electronic ground state. Indeed, photoelectron spectra show that in the direct ionization process vertical transition to the ion ground state is generally accompanied by very little vibrational excitation, i.e. the Franck-Condon envelope is narrow. The excitation energy effective in producing fragmentation then comes predominantly from transitions to excited electronic states and to some degree from autoionization processes. The Rydberg states converging to higher ionization limits can also predissociate into excited neutral species, a process which competes with autoionization. Thus one can expect considerable variation in the extent to which autoionization may contribute to ion excitation in threshold experiments. Experimental studies indicate that there is significant predissociation as well as autoionization, and there are associated isotope effects [243].

In almost all discussions and calculations of fragmentation of large molecule ions it is assumed that this process or set of competing processes occurs from the ion ground state. One possible exception is the fragmentation of benzene ion, which appears to occur via two non-competing sets of reactions [244].

The quasi-equilibrium theory of ion fragmentation introduces three distinct factors which have to be considered in deducing thermochemical data from experimental fragmentation thresholds:

a. The relation between activation energy for decomposition and the heat of reaction for the process.

b. The relation between activation energy and the minimum energy required to produce observable fragmentation in the mass spectrometer ion source, i.e. the kinetic shift.

c. The effect of competing and consecutive reaction paths on the shape and interpretation of fragmentation threshold curves.

n.1. Activation Energy and Heat of Reaction

It was pointed out many years ago that some ionic decomposition processes might occur via pathways involving the surmounting of a barrier on the potential surface [245, 246]. Surmounting such a barrier would require an activation energy greater than the heat of reaction for the process, i.e. there would be a non-zero activation energy for the reverse process. Examination of the (inaccurate) electron impact literature suggested that for simple bond rupture processes the activation energy and the heat of reaction were equal, whereas more complex fragmentation processes such as loss of H_2 or CH_4 occurred with activation energies up to an electron volt higher than the heat of reaction. There are no general rules. For example, accurate photoionization studies of the threshold for

$$C_2H_4 + h\nu \rightarrow C_2H_2^+ + H_2 + e,$$

indicate that there is no potential barrier for this four-center reaction, to an accuracy of about 10 meV [247]. However, loss of H_2 in ethane and propane requires excess energies of 0.22 eV and 0.74 eV at threshold, respectively [8]. There are other examples in the literature. Thus, the determination of thermochemical information from fragmentation threshold energetics is always subject to the uncertainty of equating the activation energy to the heat of reaction. In some instances, the existence of a potential barrier for a fragmentation process may be deduced from the observation of fragment ion kinetic energies.

Two approaches have been used in recent years. Beynon, Cooks, and co-workers have related the peak shapes of so-called metastable transitions or delayed unimolecular dissociation processes observed in the mass spectrometer to the kinetic energy released in the dissociation process [228, 248, 249].

A somewhat related method based on electrostatic

rather than magnetic analysis has been developed by Ottinger [250]. These metastable transitions represent unimolecular decompositions of ions which have an energy content such that the ions decompose after they have been accelerated out of the ion source and prior to deflection. Ions with slightly shorter lifetimes (slightly greater internal energy) will decompose in the ion source and hence will be detected as fragment ions. Thus the translational energy release observed for the metastable transition is, for all practical purposes, the translational energy release at the fragmentation threshold observed in the mass spectrometer. A very different method is based on the study of peak shapes observed in a time-of-flight mass spectrometer [251, 252]. Here the travel time distribution of a fragment ion is determined by its initial translational energy distribution, and this is studied as a function of electron energy and extrapolated to the fragmentation threshold. In some instances the translational energy is dependent on electron energy, in others it is not [253]. In principle the various methods should give the same results, but this is not always the case [254]. The reasons are not fully understood.

Work carried out along these lines indicates that fragmentation processes may be accompanied by kinetic energy release ranging from essentially zero to nearly one electron volt. The relation of this energy release to the height of the potential barrier is not straightforward. Instances are known where a large excess activation energy (determined from a threshold measurement and auxiliary thermochemical data) does not result in a fragmentation process with a correspondingly large translational energy release [255]. Thus, a correction for translational energy release may represent an underestimate of the excess activation energy. Further, studies have shown that the translational energy released in a fragmentation process is frequently a distribution rather than a single unique value [256]. The detailed procedures for correcting a threshold value for presence of a translational energy distribution remain to be worked out.

n.2. Kinetic Shift

In order to observe a fragmentation threshold in the mass spectrometer it is necessary that the fragmentation process occur prior to departure of the parent ion from the ion source. The residence time of an ion in the source is roughly several microseconds [257]. A necessary condition for fragmentation is that the ion contain enough excitation energy to equal or exceed the activation energy for the fragmentation process. However, if the activation energy is large and the number of degrees of freedom large the minimum rate of ion decomposition may be too slow to lead to observable fragmentation, i.e. fragmentation while the ion is still in the ion source. Additional excitation energy must be supplied to increase the decomposition rate. Thus, under some conditions the measured fragmentation threshold energy will overestimate the activation energy of the process. This difficulty was first pointed out by Friedman et al. [258] and discussed in detail by Chupka [259] and by Vestal [260]. Chupka pointed out that for these larger molecules, the kinetic shift effect is offset to some extent by the distribution of internal thermal energy of the molecule, all of which is assumed available for decomposition, in addition to the excitation energy imparted in the ionization process. As a result, the fragmentation threshold curves will vary with temperature [175, 261]. Vestal has given calculations which indicate the magnitude of the kinetic shift for a number of typical fragmentation processes.

For example the loss of H from propylene ion

$$C_3H_6^+ \rightarrow C_3H_5^+ + H,$$

for which an activation energy of 2.07 eV was assumed requires 0.19 eV additional energy to produce observable fragmentation in one microsecond. The estimates are, of course, dependent on the assumed value of the residence time and the assumed sensitivity of the mass spectrometer for detecting fragmentation "thresholds". Rosenstock et al. [244] have emphasized that for large molecule fragmentation processes the kinetics of decomposition lead to a gradual increase in fragment ion current so that there is no well defined threshold. They calculated the photoionization yield curve shapes for fragment ions of benzene in the threshold region and accounted quantitatively for kinetic shifts of various fragmentation processes in benzene.

Studies of the kinetic shift effect have also been carried out by electron impact techniques [262–264]. These are, however, subject to greater error resulting from the broad electron energy distribution. Although the kinetic shift phenomenon is widely recognized, it is seldom considered explicitly in experiments directed to determining thermochemical information from ion fragmentation processes.

n.3. Reaction Path and Reaction Competition

According to quasi-equilibrium theory, the fragmentation processes are a set of competing unimolecular reactions. Frequently the parent ion will decompose via a number of different competing reaction paths producing different first generation fragment ions. The energy dependence of the rate constant may be somewhat different for the various processes so that some fragment ions will be produced in experimentally detectable amounts only at energies somewhat in excess of the activation energy for the process. One indicator of this phenomenon is the observation that not all parent ion fragmentation reactions are observable as delayed dissociations (metastable transitions) [265]. At energies near the threshold for the energetically most favorable process the branching ratios for the other processes may be too small for detection or zero because they are below their threshold. One example of this is the fragmentation of n-butane, where the $C_2H_5^+$ ion is formed by a simple bond rupture process with a C_2H_5 radical as the accompanying neutral fragment. The process is not observed as a metastable transition because near the $C_2H_5^+$ threshold energy, the energetically more favored processes producing $C_3H_7^+$ and $C_3H_6^+$ ions have much higher rate constants [265]. Other examples are the pronounced effects of deuterium substitution on intensities of metastable transitions in propane [266]. These effects are due to rather small activation energy and frequency factor differences brought about by deuteration. The effect of reaction competition is then to shift the minimum observable decomposition rate for the less favored process to higher energies, increasing the kinetic shift.

Another factor must be considered in connection with thresholds for second generation fragment ions. First generation ions produced by fragmentation of parent ions containing a specific amount of energy will themselves contain a distribution of excitation energy. This is due to the large number of ways in which the excess energy of the parent can be partitioned between the daughter ion and the neutral fragment [267, 268]. As a result, there is no sharp threshold but instead a very gradual increase in second generation ion intensity with increasing energy. For all practical purposes this prevents the determination of meaningful threshold values for these processes.

3.3. Thermochemical Considerations

It should be apparent from the discussion above that the various types of ionization and fragmentation energy threshold measurements do not directly lead to room temperature heats of formation of the ions. The most nearly correct procedure is to attempt to make corrections for the effect of room temperature or, in the case of Knudsen cell experiments, elevated temperature, on the observations and arrive at an estimated 0 K energy requirement for the process [269]. Further, ions and their neutral counterparts have slightly different heat capacities and, in principle, this should be taken into account in computing room temperature heats of formation. In the tabulations to follow, we have given ionic heats of formation at absolute zero whenever possible. The tabulation of auxiliary thermochemistry, given in section 7, on which these values are partly based provides room temperature heats of formation as well, to facilitate estimates of room temperature ionic heats of formation should the need arise. In the ion tables room temperature values are given only when no 0 K neutral heats of formation are available.

The recent development of gas phase ionic equilibrium and steady state concentration measurements by high pressure mass spectrometry represents an important advance. These ion cyclotron resonance techniques are beginning to provide very precise relative heat and free energy values for gas phase ion species at and near room temperature. Detailed discussion of the relation of this class of measurements to those measurements described above will require considerable work in the future.

3.4. References for Section 3

[1] Davis, S. P., Diffraction Grating Spectrographs (Holt, Rinehart and Winston, New York, 1970), and references cited therein.

[2] Sawyer, R. A., Experimental Spectroscopy, 3rd Ed. (Dover Publications, New York, 1963).

[3] Harrison, G. R., Lord, R. C., and Loofbourow, J. R., Practical Spectroscopy (Prentice–Hall, New York, 1948).

[4] Samson, J. A. R., Techniques of Vacuum Ultraviolet Spectroscopy (John Wiley and Sons, New York, 1967).

[5] Hurzeler, H., Inghram, M. G., and Morrison, J. D., Photon impact studies of molecules using a mass spectrometer, J. Chem. Phys. **28**, 76 (1958).

[6] Dibeler, V. H., and Reese, R. M., Mass spectrometric study of photoionization. I. Apparatus and initial observations on acetylene, acetylene-d_2, benzene, and benzene-d_6, J. Res. NBS **68A**, 409 (1964).

[7] Berkowitz, J., and Chupka, W. A., Photoionization of high–temperature vapors. I. The iodides of sodium, magnesium, and thallium, J. Chem. Phys. **45**, 1287 (1966).

[8] Chupka, W. A., and Berkowitz, J., Photoionization of ethane, propane, and n-butane with mass analysis, J. Chem. Phys. **47**, 2921 (1967).

[9] Simpson, J. A., and Kuyatt, C. E., Design of low voltage electron guns, Rev. Sci. Instr. **34**, 265 (1963).

[10] Simpson, J. A., High resolution, low energy electron spectrometer, Rev. Sci. Instr. **35**, 1698 (1964).

[11] Maeda, K., Semeluk, G. P., and Lossing, F. P., A two–stage double hemispherical electron energy selector, Intern. J. Mass Spectrom. Ion Phys. **1**, 395 (1968).

[12] Clarke, E. M., Ionization probability curves using an electron selector, Can. J. Phys. **32**, 764 (1954).

[13] Marmet, P., and Kerwin, L., An improved electrostatic electron selector, Can. J. Phys. **38**, 787 (1960).

[14] Marmet, P., and Morrison, J. D., Secondary reactions in the ion chamber of a mass spectrometer, J. Chem. Phys. **36**, 1238 (1962).

[15] Fox, R. E., Hickam, W. M., Kjeldaas, T., Jr., and Grove, D. J., Ionization potentials and probabilities using a mass spectrometer, Phys. Rev. **84**, 859 (1951).

[16] Fox, R. E., Hickam, W. M., Grove, D. J., and Kjeldaas, T., Jr., Ionization in a mass spectrometer by monoenergetic electrons, Rev. Sci. Instr. **26**, 1101 (1955).

[17] Marmet, P., Effet des charges d'espace électroniques sur les courbes de probabilité d'ionisation des gaz, Can. J. Phys. **42**, 2102 (1964).

[18] Winters, R. E., Collins, J. H., and Courchene, W. L., Resolution of fine structure in ionization–efficiency curves, J. Chem. Phys. **45**, 1931 (1966).

[19] McDowell, C. A., The ionization and dissociation of molecules, in: Mass Spectrometry, ed. C. A. McDowell (McGraw–Hill Book Co. New York, 1963) Chap. 12.

[20] Kiser, R. W., Introduction to Mass Spectrometry and its Applications (Prentice–Hall, Englewood Cliffs, N. J., 1965) Chap. 8.

[21] Dorman, F. H., Morrison, J. D., and Nicholson, A. J. C., Probability of multiple ionization by electron impact, J. Chem. Phys. **31**, 1335 (1959).

[22] Baker, F. A., and Hasted, J. B., Electron collision studies with trapped positive ions, Phil. Trans. Roy. Soc. (London) **A261**, 33 (1966).

[23] Redhead, P. A., Multiple ionization of the rare gases by successive electron impacts (0–250 eV). I. Appearance potentials and metastable ion formation, Can. J. Phys. **45**, 1791 (1967).

[24] Cuthbert, J., Farren, J., and Prahallada Rao, B. S., Sequential mass spectrometry: nitrogen and the determination of appearance potentials, Proc. Phys. Soc. (London) **91**, 63 (1967).

[25] Vilesov, F. I., Kurbatov, B. L., and Terenin, A. N., Energy distribution of electrons in photoionization of aromatic amines in the gaseous phase, Dokl. Akad. Nauk SSSR **138**, 1329 (1961) [Engl. transl.: Soviet Phys. – Dokl. **6**, 490 (1961)].

[26] Turner, D. W., and Al–Joboury, M. I., Determination of ionization potentials by photoelectron energy measurement, J. Chem. Phys. **37**, 3007 (1962).

[27] Turner, D. W., Molecular photo–electron spectroscopy, in: Molecular Spectroscopy, ed. P. Hepple (Institute of Petroleum, London, 1968).

[28] Price, W. C., Developments in photo–electron spectroscopy, in: Molecular Spectroscopy, ed. P. Hepple (Institute of Petroleum, London, 1968).

[29] Turner, D. W., Molecular photoelectron spectroscopy, in: Physical Methods in Advanced Inorganic Chemistry, ed. H. A. O. Hill and P. Day (Interscience, London, 1968).

[30] Turner, D. W., Photoelectron spectroscopy, Ann. Rev. Phys. Chem. **21**, 107 (1970).

[31] Turner, D. W., Molecular photoelectron spectroscopy, Phil. Trans. Roy. Soc. (London) **A268**, 7 (1970).

[32] Turner, D. W., Baker, C., Baker, A. D., and Brundle, C. R., Molecular Photoelectron Spectroscopy (Wiley–Interscience, New York, 1970).

[33] Worley, S. D., Photoelectron spectroscopy in chemistry, Chem. Rev. **71**, 295 (1971).

[34] Brundle, C. R., Some recent advances in photoelectron spectroscopy, Appl. Spectry. **25**, 8 (1971).

[35] Betteridge, D., Electron spectroscopy. I. Ultraviolet photoexcitation, Anal. Chem. **44**, 100R (1972).

[36] Cairns, R. B., Harrison, H., and Schoen, R. I., Photoelectron spectroscopy with undispersed radiation, Appl. Opt. **9**, 605 (1970).

[37] Fuchs, V., and Hotop, H., Comment on the interpretation of the photo–electron spectrum of mercury, Chem. Phys. Letters **4**, 71 (1969).

[38] Kinsinger, J. A., Stebbings, W. L., Valenzi, R. A., and Taylor, J. W., Spectral evaluation of a sealed helium discharge lamp for studies in photoelectron spectroscopy, Anal. Chem. **44**, 773 (1972).

[39] Kuyatt, C. E., Measurement of electron scattering from a static gas target, in: Methods of Experimental Physics: Vol. 7. Atomic and Electron Physics—Atomic Interactions. Part A, ed. B. Bederson and W. L. Fite (Academic Press, New York, 1968).

[40] Hafner, H., Simpson, J. A., and Kuyatt, C. E., Comparison of the spherical deflector and the cylindrical mirror analyzer, Rev. Sci. Instr. **39**, 33 (1968).

[41] Heddle, D. W. O., A comparison of the étendue of electron spectrometers, J. Phys. E **4**, 589 (1971).

[42] Sar–El, H. Z., Criterion for comparing analyzers, Rev. Sci. Instr. **41**, 561 (1970).

[43] Turner, D. W., High resolution molecular photoelectron spectroscopy. I. Fine structure in the spectra of hydrogen and oxygen, Proc. Roy. Soc. (London) **A307**, 15 (1968).

[44] Edqvist, O., Lindholm, E., Selin, L. E., and Åsbrink, L., On the photoelectron spectrum of O_2, Physica Scripta **1**, 25 (1970).

[45] Turner, D. W., Limits to resolving power in photoelectron spectroscopy, Nature **213**, 795 (1967).

[46] Braun, W., and Carrington, T., Line emission sources for concentration measurements and photochemistry, J. Quant. Spectry. Radiative Transfer **9**, 1133 (1969).

[47] Braun, W., Bass, A. M., and Davis, D. D., Experimental test of a two–layer model characterizing emission–line profiles, J. Opt. Soc. Am. **60**, 166 (1970).

[48] Lloyd, D. R., Calibration of a He(I) photoelectron spectrometer, J. Phys. E **3**, 629 (1970).

[49] Körber, H., and Mehlhorn, W., Das K–Auger–Spektrum von Neon, Z. Physik **191**, 217 (1966).

[50] Moddeman, W. E., Carlson, T. A., Krause, M. O., Pullen, B. P., Bull, W. E., and Schweitzer, G. K., Determination of the $K–LL$ Auger spectra of N_2, O_2, CO, NO, H_2O, and CO_2, J. Chem. Phys. **55**, 2317 (1971).

[51] Mehlhorn, W., Die Feinstruktur des $L–MM$–Auger–Elektronenspektrums von Argon und der $K–LL$–Spektren von Stickstoff, Sauerstoff und Methan, Z. Physik **160**, 247 (1960).

[52] Pullen, B. P., Carlson, T. A., Moddeman, W. E., Schweitzer, G. K., Bull, W. E., and Grimm, F. A., Photoelectron spectra of methane, silane, germane, methyl fluoride, difluoromethane, and trifluoromethane, J. Chem. Phys. **53**, 768 (1970).

[53] Mehlhorn, W., and Stalherm, D., Die Auger–Spektren der L_2–und L_3–Schale von Argon, Z. Physik **217**, 294 (1968).

[54] Villarejo, D., Herm, R. R., and Inghram, M. G., Measurement of threshold electrons in the photoionization of Ar, Kr, and Xe, J. Chem. Phys. **46**, 4995 (1967).

[55] Villarejo, D., Measurement of threshold electrons in the photoionization of H_2 and D_2, J. Chem. Phys. **48**, 4014 (1968).

[56] Peatman, W. B., Borne, T. B., and Schlag, E. W., Photoionization resonance spectra. I. Nitric oxide and benzene, Chem. Phys. Letters **3**, 492 (1969).

[57] Baer, T., Peatman, W. B., and Schlag, E. W., Photoionization resonance studies with a steradiancy analyzer. II. The photoionization of CH_3I, Chem. Phys. Letters **4**, 243 (1969).

[58] Spohr, R., Guyon, P. M., Chupka, W. A., and Berkowitz, J., Threshold photoelectron detector for use in the vacuum ultraviolet, Rev. Sci. Instr. **42**, 1872 (1971).

[59] Čermák, V., Retarding–potential measurement of the kinetic energy of electrons released in Penning ionization, J. Chem. Phys. **44**, 3781 (1966).

[60] Čermák, V., Individual efficiency curves for the excitation of 2^3S and 2^1S states of helium by electron impact, J. Chem. Phys. **44**, 3774 (1966).

[61] Čermák, V., Penning ionization electron spectroscopy. I. Determination of ionization potentials of polyatomic molecules, Collection Czech. Chem. Commun. **33**, 2739 (1968).

[62] Hotop, H., Niehaus, A., and Schmeltekopf, A. L., Reactions of excited atoms and molecules with atoms and molecules. III. Relative cross sections for Penning– and associative ionization by He(2^1S)– and He(2^3S)–metastables, Z. Physik **229**, 1 (1969).

[63] Hotop, H., and Niehaus, A., Reactions of excited atoms and molecules with atoms and molecules. II. Energy analysis of Penning electrons, Z. Physik **228**, 68 (1969).

[64] Brion, C. E., McDowell, C. A., and Stewart, W. B., A comparison of vibrational transition probabilities for Penning ionization and photoionization of NO to the $X^1\Sigma^+$ state of NO^+, Chem. Phys. Letters **13**, 79 (1972).

[65] Hotop, H., and Niehaus, A., Vibrational excitation of $H_2^+(X^2\Sigma_g^+)$–ions formed in collisions with He(2^3S)–metastables, Chem. Phys. Letters **3**, 687 (1969).

[66] Zandberg, E. Ya., and Ionov, N. I., Surface Ionization (Poverkhnostnaya Ionizatsiya, translated by E. Harnik for the Israel Program for Scientific Translations, Jerusalem, 1971. Order from the National Technical Information Service, Springfield, Va. 22151, as TT70–50148).

[67] Kaminsky, M., Atomic and Ionic Impact Phenomena on Metal Surfaces (Academic Press, New York, 1965).

[68] Dresser, M. J., and Hudson, D. E., Surface ionization of some rare earths on tungsten, Phys. Rev. **137**, A673 (1965).

[69] Hertel, G. R., Surface ionization. III. The first ionization potentials of the lanthanides, J. Chem. Phys. **48**, 2053 (1968).

[70] Scheer, M. D., and Fine, J., Electron affinity of lithium, J. Chem. Phys. **50**, 4343 (1969).

[71] Zandberg, E. Ya., and Rasulev, U. Kh., Surface ionization of aniline molecules, Zh. Tekhn. Fiz. **38**, 1798 (1968) [Engl. transl.: Soviet Phys. – Tech. Phys. **13**, 1450 (1969)].

[72] Waddington, T. C., Lattice energies and their significance in inorganic chemistry, Advan. Inorg. Chem. Radiochem. **1**, 157 (1959).

[73] Johnson, D. A., Third ionization potentials and sublimation energies of the lanthanides, J. Chem. Soc. (A), 1525 (1969).

[74] Faktor, M. M., and Hanks, R., Calculation of the third ionisation potentials of the lanthanons, J. Inorg. Nucl. Chem. **31**, 1649 (1969).

[75] Morss, L. R., Thermochemistry of some chlorocomplex compounds of the rare earths. Third ionization potentials and hydration enthalpies of the trivalent ions, J. Phys. Chem. **75**, 392 (1971).

[76] Mulliken, R. S., Molecular compounds and their spectra. II, J. Am. Chem. Soc. **74**, 811 (1952).

[77] Hastings, S. H., Franklin, J. L., Schiller, J. C., and Matsen, F. A., Molecular complexes involving iodine, J. Am. Chem. Soc. **75**, 2900 (1953).

[78] Briegleb, G., Elektronen–Donator–Acceptor–Komplexe (Springer–Verlag, Berlin, 1961).

[79] Briegleb, G., and Czekalla, J., Die Bestimmung von Ionisierungsenergien aus den Spektren von Elektronenübergangskomplexen, Z. Elektrochem. **63**, 6 (1959).

[80] Edlén, B., Atomic spectra, in: Handbuch der Physik, Vol. XXVII, ed. S. Flügge (Springer–Verlag, Berlin, 1964).

[81] Hartree, D. R., The wave mechanics of an atom with a non–Coulomb central field. Part III. Term values and intensities in series in optical spectra, Proc. Cambridge Phil. Soc. **24**, 426 (1928).

[82] Kuhn, H. G., Atomic Spectra, 2nd Ed. (Academic Press, New York, 1969).

[83] Moore, C. E., Ionization potentials and ionization limits derived from the analyses of optical spectra, Natl. Stand. Ref. Data Ser., Natl. Bur. Stand. NSRDS–NBS 34 (1970).

[84] Edlén, B., and Risberg, P., The spectrum of singly–ionized calcium, Ca II, Arkiv Fysik **10**, 553 (1956).

[85] Johansson, I., The infrared spectrum of Li I, Arkiv Fysik **15**, 169 (1959).

[86] Seaton, M. J., Quantum defect theory. II. Illustrative one–channel and two–channel problems, Proc. Phys. Soc. (London) **88**, 815 (1966).

[87] Martin, W. C., Energy levels and spectrum of neutral helium (⁴He I), J. Res. NBS **64A**, 19 (1960).

[88] Herzberg, G., Ionization potentials and Lamb shifts of the ground states of ⁴He and ³He, Proc. Roy. Soc. (London) **A248**, 309 (1958).

[89] Reader, J., and Sugar, J., Ionization energies of the neutral rare earths, J. Opt. Soc. Am. **56**, 1189 (1966).

[90] Shenstone, A. G., Ultra–ionization potentials in mercury vapor, Phys. Rev. **38**, 873 (1931).

[91] Ref. [82], Chap. V.

[92] Marr, G. V., Photoionization Processes in Gases (Academic Press, New York, 1967) Chap. 7.

[93] Fano, U., Effects of configuration interaction on intensities and phase shifts, Phys. Rev. **124**, 1866 (1961).

[94] Fano, U., and Cooper, J. W., Line profiles in the far–uv absorption spectra of the rare gases, Phys. Rev. **137**, A1364 (1965).

[95] Hudson, R. D., and Carter, V. L., Bandwidth dependence of measured uv absorption cross sections of argon, J. Opt. Soc. Am. **58**, 227 (1968).

[96] McCulloh, K. E., Photoionization of carbon dioxide, J. Chem. Phys. **59**, 4250 (1973).

[97] Duncan, A. B. F., Rydberg Series in Atoms and Molecules (Academic Press, New York, 1971).

[98] Herzberg, G., and Jungen, C., Rydberg series and ionization potential of the H_2 molecule, J. Mol. Spectry. **41**, 425 (1972).

[99] Jeziorski, B., and Kolos, W., On the ionization potential of H_2, Chem. Phys. Letters **3**, 677 (1969).

[100] Chupka, W. A., and Berkowitz, J., High–resolution photoionization study of the H_2 molecule near threshold, J. Chem. Phys. **51**, 4244 (1969).

[101] Lindholm, E., Rydberg series in small molecules. I. Quantum defects in Rydberg series, Arkiv Fysik **40**, 97 (1969).

[102] Lindholm, E., Rydberg series in small molecules. II. Rydberg series in CO, Arkiv Fysik **40**, 103 (1969).

[103] Lindholm, E., Rydberg series in small molecules. III. Rydberg series in N_2, Arkiv Fysik **40**, 111 (1969).

[104] Herzberg, G., Molecular Spectra and Molecular Structure. I. Spectra of Diatomic Molecules, 2nd Ed. (D. Van Nostrand Co., New York, 1950).

[105] Yoshino, K., and Tanaka, Y., Rydberg absorption series and ionization energies of the oxygen molecule. I, J. Chem. Phys. **48**, 4859 (1968).

[106] Price, W. C., and Walsh, A. D., The absorption spectra of triple bond molecules in the vacuum ultra violet, Trans. Faraday Soc. **41**, 381 (1945).

[107] Lowrey, A., III, and Watanabe, K., Absorption and ionization coefficients of ethylene oxide, J. Chem. Phys. **28**, 208 (1958).

[108] Watanabe, K., and Nakayama, T., Absorption and photoionization coefficients of furan vapor, J. Chem. Phys. **29**, 48 (1958).

[109] Huber, K. P., The excited electronic states of the NO^+ ion, Can. J. Phys. **46**, 1691 (1968).

[110] Ogawa, M., and Tanaka, Y., Rydberg absorption series of N_2, Can. J. Phys. **40**, 1593 (1962).

[111] Samson, J. A. R., and Cairns, R. B., Ionization potential of O_2, J. Opt. Soc. Am. **56**, 769 (1966).

[112] Asundi, R. K., The first ionization potential of oxygen molecule, Current Sci. **37**, 160 (1968).

[113] Bhale, G. L., and Rao, P. R., Isotope shifts in the second negative bands of O_2^+, Proc. Indian Acad. Sci. **A67**, 350 (1968).

[114] Albritton, D. L., Schmeltekopf, A. L., and Zare, R. N., Evidence in support of the vibrational renumbering of the $O_2^+ {}^2\Pi_g$ ground state, J. Chem. Phys. **51**, 1667 (1969).

[115] McNeal, R. J., and Cook, G. R., Photoionization of O_2 in the metastable $a^1\Delta_g$ state, J. Chem. Phys. **45**, 3469 (1966).

[116] Jonathan, N., Smith, D. J., and Ross, K. J., High–resolution vacuum ultraviolet photoelectron spectroscopy of transient species: $O_2(^1\Delta_g)$, J. Chem. Phys. **53**, 3758 (1970).

[117] Wigner, E. P., On the behavior of cross sections near thresholds, Phys. Rev. **73**, 1002 (1948).

[118] Geltman, S., Theory of ionization probability near threshold, Phys. Rev. **102**, 171 (1956).

[119] Morrison, J. D., Studies of ionization efficiency. Part III. The detection and interpretation of fine structure, J. Chem. Phys. **21**, 1767 (1953).

[120] Dibeler, V. H., Reese, R. M., and Krauss, M., Mass spectrometric study of the photoionization of small molecules, Advan. Mass Spectrom. **3**, 471 (1966).

[121] Comes, F. J., and Elzer, A., Das Ionisationskontinuum von Helium, Neon und Argon, Z. Naturforsch. **19a**, 721 (1964).

[122] Ref. [92], Chap. 6.

[123] Bethe, H. A., and Salpeter, E. E., Quantum mechanics of one–and two–electron systems, in: Handbuch der Physik, Vol. XXXV, ed. S. Flügge (Springer–Verlag, Berlin, 1957) Sect. 71.

[124] Bates, D. R., Fundaminsky, A., Massey, H. S. W., and Leech, J. W., Excitation and ionization of atoms by electron impact—the Born and Oppenheimer approximations, Phil. Trans. Roy. Soc. (London) **A243**, 93 (1950).

[125] Wannier, G. H., The threshold law for single ionization of atoms or ions by electrons, Phys. Rev. **90**, 817 (1953).

[126] Rau, A. R. P., Two electrons in a Coulomb potential. Double–continuum wave functions and threshold law for electron–atom ionization, Phys. Rev. A **4**, 207 (1971).

[127] McGowan, J. W., and Clarke, E. M., Ionization of H(1s) near threshold, Phys. Rev. **167**, 43 (1968).

[128] Marchand, P., Paquet, C., and Marmet, P., Threshold behavior of the cross section for ionization of He and Ar by mono–energetic electrons, Phys. Rev. **180**, 123 (1969).

[129] McGowan, J. W., Some three–body atomic systems, Science **167**, 1083 (1970).

[130] Brion, C. E., and Thomas, G. E., The cross section for the ionization of helium by "mono–energetic" electrons, Intern. Conf. Phys. Electron. At. Collisions, 5th, Leningrad, 1967, Abstr. Pap., p. 53.

[131] Marmet, P., Bolduc, E., and Quemener, J. J., Autoionizing and negative ion states of Xe and Kr below the $^2P_{1/2}$ limits, J. Chem. Phys. **56**, 3463 (1972), and references cited therein.

[132] Dibeler, V. H., and Reese, R. M., Multiple ionization of sodium vapor by electron impact, J. Chem. Phys. **31**, 282 (1959).

[133] Kaneko, Y., Single and double ionization of Na, K and Mg by electron impact, J. Phys. Soc. Japan **16**, 2288 (1961).

[134] Fox, R. E., Study of multiple ionization in helium and xenon by electron impact, Advan. Mass Spectrom. **1**, 397 (1959).

[135] Krauss, M., Reese, R. M., and Dibeler, V. H., Multiple ionization of rare gases by electron impact, J. Res. NBS **63A**, 201 (1959).

[136] Dorman, F. H., and Morrison, J. D., Ionization potentials of multiply charged krypton, xenon, and mercury, J. Chem. Phys. **34**, 1407 (1961).

[137] Kiser, R. W., Studies of the shapes of ionization–efficiency curves of multiply charged monatomic ions. I. Instrumentation and relative electronic–transition probabilities for krypton and xenon ions, J. Chem. Phys. **36**, 2964 (1962).

[138] Kaneko, Y., and Kanomata, I., Single and double ionization of Ca by electron impact, J. Phys. Soc. Japan **18**, 1822 (1963).

[139] Abouaf, R., Mécanismes d'ionisation simple et multiple dans quelques vapeurs métalliques, par impact électronique, J. Phys. (Paris) Suppl. **32**, C4–128 (1971).

[140] Guyon, P. M., and Berkowitz, J., Interpretation of photoionization threshold behavior, J. Chem. Phys. **54**, 1814 (1971).

[141] Franck, J., Elementary processes of photochemical reactions, Trans. Faraday Soc. **21**, 536 (1926).

[142] Condon, E. U., Nuclear motions associated with electron transitions in diatomic molecules, Phys. Rev. **32**, 858 (1928).

[143] Condon, E. U., The Franck–Condon principle and related topics, Am. J. Phys. **15**, 365 (1947).

[144] Dibeler, V. H., Walker, J. A., McCulloh, K. E., and Rosenstock, H. M., Effect of hot bands on the ionization threshold of some diatomic halogen molecules, Intern. J. Mass Spectrom. Ion Phys. **7**, 209 (1971).

[145] Berkowitz, J., Photoionization mass spectrometry and photoelectron spectroscopy of high temperature vapors, in: Advances in High Temperature Chemistry, Vol. 3, ed. L. Eyring (Academic Press, New York, 1971).

[146] Watanabe, K., Photoionization and total absorption cross section of gases. I. Ionization potentials of several molecules. Cross sections of NH_3 and NO, J. Chem. Phys. **22**, 1564 (1954).

[147] Berry, R. S., Ionization of molecules at low energies, J. Chem. Phys. **45**, 1228 (1966).

[148] Killgoar, P. C., Jr., Leroi, G. E., Berkowitz, J., and Chupka, W. A., Photoionization mass spectrometric study of NO. A closer look at the threshold region, J. Chem. Phys. **58**, 803 (1973).

[149] Chupka, W. A., and Berkowitz, J., Photoionization of the H_2 molecule near threshold, J. Chem. Phys. **48**, 5726 (1968).

[150] Dorman, F. H., Morrison, J. D., and Nicholson, A. J. C., Threshold law for the probability of excitation by electron impact, J. Chem. Phys. **32**, 378 (1960).

[151] Morrison, J. D., On the optimum use of ionization–efficiency data, J. Chem. Phys. **39**, 200 (1963).

[152] Giessner, B. G., and Meisels, G. G., On numerical methods for the improvement of appearance potential measurements, J. Chem. Phys. **55**, 2269 (1971).

[153] Kerwin, L., Marmet, P., and Clarke, E. M., Recent work with the electrostatic electron selector, Advan. Mass Spectrom. **2**, 522 (1963).

[154] Lossing, F. P., and Semeluk, G. P., Threshold ionization efficiency curves for monoenergetic electron impact on H_2, D_2, CH_4 and CD_4, Intern. J. Mass Spectrom. Ion Phys. **2**, 408 (1969).

[155] Berkowitz, J., and Lifshitz, C., Photoionization of high–temperature vapors. II. Sulfur molecular species, J. Chem. Phys. **48**, 4346 (1968).

[156] Berkowitz, J., and Chupka, W. A., Vaporization processes involving sulfur, J. Chem. Phys. **40**, 287 (1964).

[157] Hildenbrand, D. L., First ionization potentials of the molecules BF, SiO and GeO, Intern. J. Mass Spectrom. Ion Phys. **7**, 255 (1971).

[158] Eland, J. H. D., Photoelectron Spectroscopy (Butterworths, London, 1974).

[159] Jonathan, N., Smith, D. J., and Ross, K. J., The high resolution photoelectron spectra of transient species: sulphur monoxide, Chem. Phys. Letters **9**, 217 (1971).

[160] Higginson, B. R., Lloyd, D. R., and Roberts, P. J., Variable temperature photoelectron spectroscopy. The adiabatic ionization potential of the iodine molecule, Chem. Phys. Letters **19**, 480 (1973).

[161] Price, W. C., Potts, A. W., and Streets, D. G., The dependence of photoionization cross section on the photoelectron energy, in: Electron Spectroscopy, ed. D. A. Shirley (North–Holland Pub. Co., Amsterdam, 1972).

[162] Berkowitz, J., and Ehrhardt, H., Angular distribution of valence–shell photoelectrons, Phys. Letters **21**, 531 (1966).

[163] Grimm, F. A., Photoelectron angular distributions, in: Electron Spectroscopy, ed. D. A. Shirley (North–Holland Pub. Co., Amsterdam, 1972).

[164] Carlson, T. A., McGuire, G. E., Jonas, A. E., Cheng, K. L., Anderson, C. P., Lu, C. C., and Pullen, B. P., Comprehensive examination of the angular distribution of photoelectron spectra from atoms and molecules, in: Electron Spectroscopy, ed. D. A. Shirley (North–Holland Pub. Co., Amsterdam, 1972).

[165] Samson, J. A. R., Angular distribution of photoelectrons, J. Opt. Soc. Am. **59**, 356 (1969).

[166] Smith, A. L., The role of autoionization in molecular photoelectron spectra, Phil. Trans. Roy. Soc. (London) **A268**, 169 (1970).

[167] Murad, E., and Inghram, M. G., Photoionization of aliphatic ketones, J. Chem. Phys. **40**, 3263 (1964).

[168] Chupka, W. A., Mass–spectrometric study of the photoionization of methane, J. Chem. Phys. **48**, 2337 (1968).

[169] Dixon, R. N., On the Jahn–Teller distortion of CH_4^+, Mol. Phys. **20**, 113 (1971).

[170] Rabalais, J. W., Bergmark, T., Werme, L. O., Karlsson, L., and Siegbahn, K., The Jahn–Teller effect in the electron spectrum of methane, Physica Scripta **3**, 13 (1971).

[171] Rosenstock, H. M., and Botter, R., Franck–Condon principle for the ionization of polyatomic molecules, in: Recent Developments in Mass Spectroscopy, ed. K. Ogata and T. Hayakawa (University Park Press, Baltimore, 1970).

[172] Walsh, A. D., and Warsop, P. A., The ultra–violet absorption spectrum of ammonia, Trans. Faraday Soc. **57**, 345 (1961).

[173] Lifshitz, C., and Chupka, W. A., Photoionization of the CF_3 free radical, J. Chem. Phys. **47**, 3439 (1967).

[174] Killgoar, P. C., Jr., Leroi, G. E., Chupka, W. A., and Berkowitz, J., Photoionization study of NO_2. I. The ionization potential, J. Chem. Phys. **59**, 1370 (1973).

[175] Steiner, B., Giese, C. F., and Inghram, M. G., Photoionization of alkanes. Dissociation of excited molecular ions, J. Chem. Phys. **34**, 189 (1961).

[176] Elder, F. A., Giese, C., Steiner, B., and Inghram, M., Photo–ionization of alkyl free radicals, J. Chem. Phys. **36**, 3292 (1962).

[177] Botter, R., Dibeler, V. H., Walker, J. A., and Rosenstock, H. M., Experimental and theoretical studies of photoionization–efficiency curves for C_2H_2 and C_2D_2, J. Chem. Phys. **44**, 1271 (1966).

[178] Dibeler, V. H., and Walker, J. A., Photoionization of acetylene near threshold, Intern. J. Mass Spectrom. Ion Phys. **11**, 49 (1973).

[179] Chupka, W. A., and Berkowitz, J., Photoionization of methane: ionization potential and proton affinity of CH_4, J. Chem. Phys. **54**, 4256 (1971).

[180] Baker, A. D., Baker, C., Brundle, C. R., and Turner, D. W., The electronic structures of methane, ethane, ethylene and formaldehyde studied by high–resolution molecular photoelectron spectroscopy, Intern. J. Mass Spectrom. Ion Phys. **1**, 285 (1968).

[181] Lossing, F. P., Free radicals by mass spectrometry. XLIII. Ionization potentials and ionic heats of formation for vinyl, allyl, and benzyl radicals, Can. J. Chem. **49**, 357 (1971).

[182] Elder, F. A., and Parr, A. C., Photoionization of the cycloheptatrienyl radical, J. Chem. Phys. **50**, 1027 (1969).

[183] Dibeler, V. H., and Walker, J. A., Mass–spectrometric study of photoionization. VI. O_2, CO_2, COS, and CS_2, J. Opt. Soc. Am. **57**, 1007 (1967).

[184] Baker, C., and Turner, D. W., High resolution molecular photoelectron spectroscopy. III. Acetylenes and aza–acetylenes, Proc. Roy. Soc. (London) **A308**, 19 (1968).

[185] Collins, J. H., Winters, R. E., and Engerholm, G. G., Fine structure in energy–distribution–difference ionization–efficiency curves, J. Chem. Phys. **49**, 2469 (1968).

[186] Brion, C. E., The Franck–Condon principle and the ionization of acetylene by electron impact, Chem. Phys. Letters **3**, 9 (1969).

[187] Lossing, F. P., Threshold ionization of acetylene by monoenergetic electron impact, Intern. J. Mass Spectrom. Ion Phys. **5**, 190 (1970).

[188] Dorman, F. H., and Morrison, J. D., Double and triple ionization in molecules induced by electron impact, J. Chem. Phys. **35**, 575 (1961).

[189] Olmsted, J., III, Street, K., Jr., and Newton, A. S., Excess–kinetic–energy ions in organic mass spectra, J. Chem. Phys. **40**, 2114 (1964).

[190] Ehrhardt, H., and Tekaat, T., Auftrittspotentialmessungen an ionisierten Molekülbruchstücken mit kinetischer Anfangsenergie, Z. Naturforsch. **19a**, 1382 (1964).

[191] Condon, E. U., and Smyth, H. D., The critical potentials of molecular hydrogen, Proc. Natl. Acad. Sci. **14**, 871 (1928).

[192] Condon, E. U., Complete dissociation of H_2, Phys. Rev. **35**, 658 (1930).

[193] Bleakney, W., The ionization of hydrogen by single electron impact, Phys. Rev. **35**, 1180 (1930).

[194] Berkowitz, J., and Chupka, W. A., Photoelectron spectroscopy of autoionization peaks, J. Chem. Phys. **51**, 2341 (1969).

[195] Curran, R. K., Negative ion formation in various gases at pressures up to .5 mm of Hg, Scientific Paper 62–908–113–P7, Westinghouse Research Laboratories, Pittsburgh (1962).

[196] McCulloh, K. E., and Walker, J. A., Photodissociative formation of ion pairs from molecular hydrogen and the electron affinity of the hydrogen atom, Chem. Phys. Letters **25**, 439 (1974).

[197] Hagstrum, H. D., and Tate, J. T., Ionization and dissociation of diatomic molecules by electron impact, Phys. Rev. **59**, 354 (1941).

[198] Hagstrum, H. D., Ionization by electron impact in CO, N_2, NO, and O_2, Rev. Mod. Phys. **23**, 185 (1951).

[199] Dunn, G. H., and Kieffer, L. J., Dissociative ionization of H_2: a study of angular distributions and energy distributions of resultant fast protons, Phys. Rev. **132**, 2109 (1963).

[200] Kieffer, L. J., and Dunn, G. H., Dissociative ionization of H_2 and D_2, Phys. Rev. **158**, 61 (1967).

[201] Crowe, A., and McConkey, J. W., Dissociative ionization by electron impact. I. Protons from H_2, J. Phys. B **6**, 2088 (1973).

[202] McCulloh, K. E., and Rosenstock, H. M., Experimental test of the Franck–Condon principle: double ionization of molecular hydrogen, J. Chem. Phys. **48**, 2084 (1968).

[203] Gaydon, A. G., Dissociation Energies and Spectra of Diatomic Molecules, 3rd Ed. (Chapman and Hall, London, 1968).

[204] Tate, J. T., and Lozier, W. W., The dissociation of nitrogen and carbon monoxide by electron impact, Phys. Rev. **39**, 254 (1932).

[205] Hagstrum, H. D., Reinterpretation of electron impact experiments in CO, N_2, NO, and O_2, J. Chem. Phys. **23**, 1178 (1955).

[206] Fineman, M. A., and Petrocelli, A. W., Electron impact study of CO using a Lozier apparatus, J. Chem. Phys. **36**, 25 (1962).

[207] Burns, J. F., On the heat of dissociation of N_2, J. Chem. Phys. **23**, 1347 (1955).

[208] Frost, D. C., and McDowell, C. A., The dissociation energy of the nitrogen molecule, Proc. Roy. Soc. (London) **A236**, 278 (1956).

[209] Cloutier, G. G., and Schiff, H. I., Electron impact study of nitric oxide using a modified retarding potential difference method, J. Chem. Phys. **31**, 793 (1959).

[210] Dibeler, V. H., Walker, J. A., and McCulloh, K. E., Photoionization study of the dissociation energy of fluorine and the heat of formation of hydrogen fluoride, J. Chem. Phys. **51**, 4230 (1969).

[211] Dibeler, V. H., Walker, J. A., and McCulloh, K. E., Photoionization study of chlorine monofluoride and the dissociation energy of fluorine, J. Chem. Phys. **53**, 4414 (1970).

[212] Berkowitz, J., Chupka, W. A., Guyon, P. M., Holloway, J. H., and Spohr, R., Photoionization mass spectrometric study of F_2, HF, and DF, J. Chem. Phys. **54**, 5165 (1971).

[213] Chupka, W. A., and Berkowitz, J., Kinetic energy of ions produced by photoionization of HF and F_2, J. Chem. Phys. **54**, 5126 (1971).

[214] Brundle, C. R., Ionization and dissociation energies of HF and DF and their bearing on $D_0^0(F_2)$, Chem. Phys. Letters **7**, 317 (1970).

[215] DeCorpo, J. J., Steiger, R. P., Franklin, J. L., and Margrave, J. L., Dissociation energy of F_2, J. Chem. Phys. **53**, 936 (1970).

[216] Berkowitz, J., and Wahl, A. C., The dissociation energy of fluorine, Advan. Fluorine Chem. **7**, 147 (1973).

[217] Dibeler, V. H., and Rosenstock, H. M., Mass spectra and metastable transitions of H_2S, HDS, and D_2S, J. Chem. Phys. **39**, 3106 (1963).

[218] Sharp, T. E., and Rosenstock, H. M., Franck–Condon factors for polyatomic molecules, J. Chem. Phys. **41**, 3453 (1964).

[219] Fiquet–Fayard, F., and Guyon, P. M., Préionisation et prédissociation dans la dissociation des molécules triatomiques par impact électronique, Mol. Phys. **11**, 17 (1966).

[220] Eland, J. H. D., Predissociation of triatomic ions studied by photoelectron–photoion coincidence spectroscopy and photoion kinetic energy analysis. I. CO_2^+, Intern. J. Mass Spectrom. Ion Phys. **9**, 397 (1972).

[221] Lee, L. C., and Judge, D. L., Cross sections for the production of $CO_2^+[A^2\Pi_u, B^2\Sigma_u^+ \to X^2\Pi_g]$ fluorescence by vacuum ultraviolet radiation, J. Chem. Phys. **57**, 4443 (1972).

[222] Dibeler, V. H., Walker, J. A., and Liston, S. K., Mass spectrometric study of photoionization. VII. Nitrogen dioxide and nitrous oxide, J. Res. NBS **71A**, 371 (1967).

[223] Dibeler, V. H., N_2O bond dissociation energy by photon impact, J. Chem. Phys. **47**, 2191 (1967).

[224] Begun, G. M., and Landau, L., Mass spectra and metastable transitions in isotopic nitrous oxides, J. Chem. Phys. **35**, 547 (1961).

[225] Coleman, R. J., Delderfield, J. S., and Reuben, B. G., The gas–phase decomposition of the nitrous oxide ion, Intern. J. Mass Spectrom. Ion Phys. **2**, 25 (1969).

[226] Newton, A. S., and Sciamanna, A. F., Metastable peaks in the mass spectra of N_2O and NO_2. II, J. Chem. Phys. **52**, 327 (1970).

[227] Lanczos, C., Applied Analysis (Prentice Hall, Englewood Cliffs, N. J., 1956) Chap. VI.

[228] Beynon, J. H., Saunders, R. A., and Williams, A. E., Dissociation of meta–stable ions in mass spectrometers with release of internal energy, Z. Naturforsch. **20a**, 180 (1965).

[229] Lorquet, J. C., and Cadet, C., Excited states of gaseous ions. I. Selection rules in photoelectron spectroscopy and photoionization. The case of N_2O^+, Intern. J. Mass Spectrom. Ion Phys. **7**, 245 (1971).

[230] Brundle, C. R., and Turner, D. W., Studies on the photoionisation of the linear triatomic molecules: N_2O, COS, CS_2 and CO_2 using high–resolution photoelectron spectroscopy, Intern. J. Mass Spectrom. Ion Phys. **2**, 195 (1969).

[231] Eland, J. H. D., Predissociation of N_2O^+ and COS^+ ions studied by photoelectron–photoion coincidence spectroscopy, Intern. J. Mass Spectrom. Ion Phys. **12**, 389 (1973).

[232] Fink, E. H., and Welge, K. H., Lebensdauern und Löschquerschnitte elektronisch angeregter Zustände von N_2O^+, NO, O_2^+, CO^+ und CO, Z. Naturforsch. **23a**, 358 (1968).

[233] Smith, W. H., Radiative lifetimes and total transition probabilities for some polyatomic species, J. Chem. Phys. **51**, 3410 (1969).

[234] Pipano, A., and Kaufman, J. J., *Ab initio* calculation of potential energy curves for the ion molecule reaction O +

[235] $N_2 \to NO^+ + N$, J. Chem. Phys. **56**, 5258 (1972).

[235] Coppens, P., Smets, J., Fishel, M. G., and Drowart, J., Mass spectrometric study of the photoionization of nitrous oxide in the wavelength interval 1000–600 Å, Intern. J. Mass Spectrom. Ion Phys. **14**, 57 (1974).

[236] Dibeler, V. H., Walker, J. A., and McCulloh, K. E., Observations on hot bands in the molecular and dissociative photoionization of acetylene and the heat of formation of the ethynyl ion, J. Chem. Phys. **59**, 2264 (1973).

[237] Fiquet–Fayard, F., Importance des prédissociations dans la fragmentation de l'ion acétylène, J. Chim. Phys. **64**, 320 (1967).

[238] Jones, E. G., Beynon, J. H., and Cooks, R. G., Kinetic energy release in the dissociation of some simple molecular ions. Water and hydrogen sulfide, J. Chem. Phys. **57**, 3207 (1972).

[239] Curran, R. K., and Fox, R. E., Mass spectrometer investigation of ionization of N_2O by electron impact, J. Chem. Phys. **34**, 1590 (1961).

[240] Rosenstock, H. M., Wallenstein, M. B., Wahrhaftig, A. L., and Eyring, H., Absolute rate theory for isolated systems and the mass spectra of polyatomic molecules, Proc. Natl. Acad. Sci. **38**, 667 (1952).

[241] Rosenstock, H. M., Theory of mass spectra—a general review, Advan. Mass Spectrom. **4**, 523 (1968).

[242] Wahrhaftig, A. L., Theory of mass spectra, in: MTP International Review of Science. Vol. 5. Mass Spectrometry, ed. A. Maccoll (University Park Press, Baltimore, 1972) Chap. 1.

[243] Person, J. C., and Nicole, P. P., Isotope effects in the photoionization yields and in the absorption cross sections for methanol, ethanol, methyl bromide, and ethyl bromide, J. Chem. Phys. **55**, 3390 (1971).

[244] Rosenstock, H. M., Larkins, J. T., and Walker, J. A., Interpretation of photoionization thresholds: quasiequilibrium theory and the fragmentation of benzene, Intern. J. Mass Spectrom. Ion Phys. **11**, 309 (1973).

[245] Wallenstein, M. B., Ph.D. Thesis (University of Utah, 1951).

[246] Rosenstock, H. M., Ph.D. Thesis (University of Utah, 1952).

[247] Botter, R., Dibeler, V. H., Walker, J. A., and Rosenstock, H. M., Mass–spectrometric study of photoionization. IV. Ethylene and 1,2–dideuteroethylene, J. Chem. Phys. **45**, 1298 (1966).

[248] Beynon, J. H., and Fontaine, A. E., Mass spectrometry: the shapes of "meta–stable peaks", Z. Naturforsch. **22a**, 334 (1967).

[249] Cooks, R. G., Beynon, J. H., Caprioli, R. M., and Lester, G. R., Metastable Ions (Elsevier Scientific Pub. Co., Amsterdam, 1973).

[250] Ottinger, C., Fragmentation energies of metastable organic ions, Phys. Letters **17**, 269 (1965); Erratum, *ibid.*, **19**, 256 (1965).

[251] Ferguson, R. E., McCulloh, K. E., and Rosenstock, H. M., Observation of the products of ionic collision processes and ion decomposition in a linear, pulsed time–of–flight mass spectrometer, J. Chem. Phys. **42**, 100 (1965).

[252] Franklin, J. L., Hierl, P. M., and Whan, D. A., Measurement of the translational energy of ions with a time–of–flight mass spectrometer, J. Chem. Phys. **47**, 3148 (1967).

[253] Haney, M. A., and Franklin, J. L., Correlation of excess energies of electron–impact dissociations with the translational energies of the products, J. Chem. Phys. **48**, 4093 (1968).

[254] Jones, E. G., Beynon, J. H., and Cooks, R. G., Kinetic energy release in unimolecular ionic reactions. Thermochemical aspects, J. Chem. Phys. **57**, 2652 (1972).

[255] Khodadadi, G., Botter, R., and Rosenstock, H. M., Effect of varying initial preparation on kinetic energy release in metastable transitions, Intern. J. Mass Spectrom. Ion Phys. **3**, 397 (1969).

[256] Terwilliger, D. T., Beynon, J. H., and Cooks, R. G., Kinetic energy distributions from the shapes of metastable peaks, Proc. Roy. Soc. (London) **A341**, 135 (1974).

[257] Meier, K., and Seibl, J., Measurement of ion residence times in a commercial electron impact ion source, Intern. J. Mass Spectrom. Ion Phys. **14**, 99 (1974).

[258] Friedman, L., Long, F. A., and Wolfsberg, M., Ionization efficiency curves and the statistical theory of mass spectra, J. Chem. Phys. **26**, 714 (1957).

[259] Chupka, W. A., Effect of unimolecular decay kinetics on the interpretation of appearance potentials, J. Chem. Phys. **30**, 191 (1959).

[260] Vestal, M., Ionic fragmentation processes, in: Fundamental Processes in Radiation Chemistry, ed. P. Ausloos (Interscience Publishers, New York, 1968).

[261] Chupka, W. A., Effect of thermal energy on ionization efficiency curves of fragment ions, J. Chem. Phys. **54**, 1936 (1971).

[262] Gross, M. L., Ion cyclotron resonance spectrometry. A means of evaluating "kinetic shifts", Org. Mass Spectrom. **6**, 827 (1972).

[263] Johnstone, R. A. W., and Mellon, F. A., Electron–impact ionization and appearance potentials, J. Chem. Soc. Faraday Trans. II **68**, 1209 (1972).

[264] Potzinger, P., and Bünau, G. v., Empirische Berücksichtigung von Überschussenergien bei der Auftrittspotentialbestimmung, Ber. Bunsenges. **73**, 466 (1969).

[265] Rosenstock, H. M., and Melton, C. E., Metastable transitions and collision–induced dissociations in mass spectra, J. Chem. Phys. **26**, 314 (1957).

[266] Ottinger, C., Metastable ions in the mass spectra of propane and deuterated propanes, J. Chem. Phys. **47**, 1452 (1967).

[267] Wallenstein, M. B., and Krauss, M., Interpretation of the appearance potentials of secondary ions, J. Chem. Phys. **34**, 929 (1961).

[268] Rosenstock, H. M., and Krauss, M., Current status of the statistical theory of mass spectra, Advan. Mass Spectrom. **2**, 251 (1963).

[269] Refaey, K. M. A., and Chupka, W. A., Photoionization of the lower aliphatic alcohols with mass analysis, J. Chem. Phys. **48**, 5205 (1968).

[270] Krupenie, P. H., The spectrum of molecular oxygen, J. Phys. Chem. Ref. Data **1**, 423 (1972).

[271] Brundle, C. R., and Turner, D. W., High resolution molecular photoelectron spectroscopy. II. Water and deuterium oxide, Proc. Roy. Soc. (London) **A307**, 27 (1968).

[272] Dibeler, V. H., Walker, J. A., and Rosenstock, H. M., Mass spectrometric study of photoionization. V. Water and ammonia, J. Res. NBS **70A**, 459 (1966).

[273] Weiss, M. J., and Lawrence, G. M., Photoelectron spectroscopy of NH_3 and ND_3 using molecular beams, J. Chem. Phys. **53**, 214 (1970).

[274] Chupka, W. A., and Russell, M. E., Ion–molecule reactions of NH_3^+ by photoionization, J. Chem. Phys. **48**, 1527 (1968).

[275] Brundle, C. R., Neumann, D., Price, W. C., Evans, D., Potts, A. W., and Streets, D. G., Electronic structure of NO_2 studied by photoelectron and vacuum–uv spectroscopy and Gaussian orbital calculations, J. Chem. Phys. **53**, 705 (1970).

[276] Dibeler, V. H., and Reese, R. M., Mass spectrometric study of the photoionization of acetylene and acetylene–d_2, J. Chem. Phys. **40**, 2034 (1964).

4. Tabulation of Positive Ion Data

4.1. Description of the Positive Ion Table

The Positive Ion Table gives the following information: empirical formula of the ion, electronic state where appropriate, the parent molecule studied, the nature of the neutral or other charged fragmentation product where known, the experimental threshold or other energy measured together with the likely error *given by the author*, the experimental technique used, and the reference or references. Also tabulated is the ionic heat of formation in kJ mol^{-1} computed in those instances where the measured threshold is, in our judgment, a reliable one. Our recommended value, if any, for the ionic heat of formation, in both kJ mol^{-1} and kcal mol^{-1}, is presented in the heading which begins each ion sub table.

The order of arrangement is based on the principle of latest atomic number in order of increasing charge and increasing elemental complexity. For a given ionic process the measurements are cited generally in order of decreasing reliability. When data are available for formation of an ion from different molecules, the processes are listed according to the principle of latest position for the molecule studied.

For the molecule studied one-line semi-structural formulas are given for all straight and branched chain compounds. For all ring compounds one-line condensed formulas are given along with a name. In naming these compounds we have attempted to follow the rules given in the Handbook of Chemistry and Physics [1].

In computing ionic heats of formation from fragmentation process threshold energies it is of course necessary to specify the identity of the other fragments. These are tabulated in the column headed "Other Products" for all fragmentation processes from which an ionic heat of formation is computed. In addition, for those other fragmentation processes not so used, the identity of the other fragments is given if in our opinion they are obvious, the evidence is convincing, or if they are unusual. Obvious other fragments include atoms such as H. However, when the fragment atom is a halogen, there is a possibility that this other fragment may be a negative ion. Examination of the literature reveals two kinds of information. First, those experiments where the charge of the halogen atom fragment was established experimentally, i.e. negative ion searched for and detected or not detected. Second, experiments in which the charge of the halogen is inferred from the threshold energies and auxiliary thermochemical information. We regard the latter as unconvincing, especially when the accuracy of the threshold measurement technique is low.

For polyatomic fragments, experience indicates that at threshold one generally finds the fragment in the most stable chemical form, e.g. CH_3 rather than $CH_2 + H$

or $CH + H_2$. However, there are exceptions. Recent unpublished photoionization work indicates that the CHO fragment is not formed in fragmentation of selected organic oxygen compounds such as glyoxal, CHOCHO, but rather appears as $CO + H$. Further there is rather convincing evidence for the processes

$$cyclo-C_2H_4O \rightarrow C_2H_2O^+ + 2H \qquad \text{(ref. 2)}$$

$$ONF_3 \rightarrow NO^+ + 3F \qquad \text{(ref. 3)}$$

$$NF_3 \rightarrow NF^+ + 2F \qquad \text{(ref. 3)}$$

For more elaborate fragmentation processes, even thermochemical information cannot always resolve ambiguities, for example, a choice between the $CH_3 + H_2$ and $CH_4 + H$, or $CH_3 + H_2O$ and $CH_4 + OH$ pairs. Here, the study of metastable transitions is of great help. Unfortunately these studies are often found in papers other than the ones containing the energy information. We have tended to be conservative in tabulating fragments. In some instances we have indicated larger neutral fragments by empirical formula only, since the structure is not known but it has been established that the other product consists of only one fragment (e.g. C_3H_7, C_4H_9, etc.). In any event, the information as presented by the authors is available in the references cited.

The ionization potential or appearance potential is given in electron volts. For this purpose all spectroscopic values have been computed using the conversion factor $1 \text{ eV} \doteq 8065.73 \text{ cm}^{-1}$. Heats of formation are calculated using the conversion factors $1 \text{ eV molecule}^{-1} = 96.4870 \text{ kJ mol}^{-1}$ and $1 \text{ kcal mol}^{-1} = 4.18400 \text{ kJ mol}^{-1}$.

The methods are described and evaluated in the preceding section. Wherever possible, heats of formation have been computed at absolute zero rather than at room temperature, for reasons discussed in the preceding section. Room temperature values are given only where insufficient auxiliary thermochemical data are available for the necessary computations. In these cases wherever possible we have used threshold values approximately corrected to 0 K rather than room temperature "thresholds". *In this compilation we assume the heat of formation of the electron is zero at all temperatures, in contrast to usage in other thermochemical compilations.* In the case of fragmentation processes, information on fragment kinetic energies has been included where available and sometimes used to correct the computed heats of formation. This is only approximate, for reasons discussed in the preceding section.

The heat of formation is calculated from those data which are in our judgment the most reliable. For this purpose we have used the sources explicitly cited in section 7 on auxiliary thermochemistry. We have arbitrarily taken information from the compilation of Stull, Sinke, and Westrum [4] for most of the heats of formation

of larger organic compounds. We are aware of occasional differences between these values and those recommended by Cox and Pilcher [5] in their compilation, but have made no attempt to reconcile these differences or to arrive at better values. We have arbitrarily limited the computed and recommended heats of formation to a maximum accuracy of 0.1 kJ mol^{-1} or 0.1 kcal mol^{-1} even in those few instances where the ionization potentials are of greater accuracy, such as for H_2, N_2 and a number of atoms. The temperature associated with the heat of formation is the same as that given in the heading. In a few cases heats of formation are listed in parentheses in column six. These are given for illustrative purposes only and are not used in arriving at recommended values given in the heading. In all instances the recommended values are simple average values, suitably rounded, of the column values. No compelling reason exists for proceeding otherwise. In a very few instances no ionic heats of formation are given because no accurate neutral heats of formation are available, even though the ionization potential is reasonably reliable (± 0.05 eV or better, generally from photoelectron spectroscopy). These instances are almost all to be found among the complex organic molecules.

Since the main thrust of this compilation is ion thermochemistry, the adiabatic ionization potential is the quantity of interest, in preference to the vertical ionization potential. In many instances the two quantities are one and the same, in other instances the vertical value is higher. Where available we have given the adiabatic value. A number of publications in photoelectron spectroscopy have emphasized determination of the vertical value and while the published spectra show clearly that the adiabatic value is lower, a number is given only for the former. In these instances the value tabulated here carries the designation (V) to indicate that it is a vertical value and that the adiabatic value is lower.

The development of photoelectron spectroscopy has led to many values of higher ionization potentials, corresponding to removal of more tightly bound electrons. We have tabulated these values for selected molecules, and only in those instances where an assignment is given for the excited states formed in the process. On the other hand there are numerous data on higher ionization potentials for many larger molecules. In those instances only the lowest ionization potential is tabulated here. Higher ionization potentials will be found in most of the photoelectron spectra and Penning ionization papers cited.

For some ionization processes, additional references are given in the form "See also". These refer to earlier work of lesser accuracy or, in the case of some electron impact work of low accuracy, to additional measurements of similar poor accuracy and similar nominal values. In a few instances some measurements of lower accuracy have been tabulated along with measurements of better quality to provide concrete examples of experimental disagreement.

References for Section 4.1

[1] Handbook of Chemistry and Physics, 50th Ed., ed. R. C. Weast (The Chemical Rubber Co., Cleveland, 1969).

[2] Gallegos, E. J., and Kiser, R. W., Electron impact spectroscopy of ethylene oxide and propylene oxide, J. Am. Chem. Soc. **83**, 773 (1961).

[3] Dibeler, V. H., and Walker, J. A., Mass spectrometric study of photoionization. XIV. Nitrogen trifluoride and trifluoramine oxide, Inorg. Chem. **8,** 1728 (1969).

[4] Stull, D. R., Westrum, E. F., Jr., and Sinke, G. C., The Chemical Thermodynamics of Organic Compounds (John Wiley and Sons, New York, 1969).

[5] Cox, J. D., and Pilcher, G., Thermochemistry of Organic and Organometallic Compounds (Academic Press, New York, 1970).

4.2. Index

Mn

Fe

Co

Kr

4.3. The Positive Ion Table

Ion	Reactant	Other products	Ionization or appearance potential (eV)	Method	Heat of formation (kJ mol^{-1})	Ref.
H$^+$	ΔH_{f0}° = 1528.0 kJ mol^{-1} (365.2 kcal mol^{-1})					
H$^+$	H		13.598	S	1528.0	2113, 3432
H$^+$	H		13.61	PE		3076

See also − EI: 87, 3294

Ion	Reactant	Other products	Ionization or appearance potential (eV)	Method	Heat of formation	Ref.
H$^+$	H$_2$	H$^-$	17.3	RPD		200
H$^+$	C$_3$H$_6$		19.5±1	EI		2542
H$^+$	C$_3$H$_8$		20.0±0.3	EI		1408

(Threshold value corrected for high fragment kinetic energy)

H$^+$	HCN	CN$^-$	15.18±0.02	PI		2602

(Threshold value approximately corrected to 0 K)

H$^+$	HCN	CN	19.00±0.01	PI		2602

(Threshold value approximately corrected to 0 K)

H$^+$	H$_2$O	OH$^-$	16.0±0.3	EI		3144
H$^+$	H$_2$O	OH	19.6±0.25	EI		3144

See also − EI: 2484

H$^+$	HF	F$^-$	16.061±0.005	PI		3274

(Position of peak maximum)

The fragments are formed with no kinetic energy at threshold, see W. A. Chupka and J. Berkowitz, J. Chem. Phys. **54**, 5126 (1971).

See also − PI: 2744
          EI: 286

H$^+$	HF	F	19.445±0.01	PI	(1528.2)	3217, 3274

(Threshold value approximately corrected to 0 K)

H$^+$	HF	F	≥19.44±0.02	PE		3086

See also − PI: 2744

H$^+$	(CH$_3$)$_2$SO		23±0.5	EI		3294
H$^+$	HCl	Cl$^-$	14.5	RPD		199

Ion	Reactant	Other products	Ionization or appearance potential (eV)	Method	Heat of formation	Ref.
D$^+$	ΔH_{f0}° = 1532.2 kJ mol^{-1} (366.2 kcal mol^{-1})					
D$^+$	D		13.602	S	1532.2	3432
D$^+$	D$_2$	D	25.3±0.2	SRP		1264

(High kinetic energy ion)

D$^+$	DF	F$^-$	16.134±0.005	PI		3274

(Position of peak maximum)

D$^+$	DF	F	19.513±0.005	PI		3274

(Threshold value approximately corrected to 0 K)

Ion	Reactant	Other products	Ionization or appearance potential (eV)	Method	Heat of formation	Ref.
T$^+$						
T$^+$	T		13.603	S		3432

4.3. The Positive Ion Table—Continued

Ion	Reactant	Other products	Ionization or appearance potential (eV)	Method	Heat of formation (kJ mol^{-1})	Ref.	

H_2^+ $\Delta H_{f0}^\circ = 1488.4$ kJ mol^{-1} (355.7 kcal mol^{-1})

Ion	Reactant	Other products	IP/AP (eV)	Method	Heat of formation	Ref.	
H_2^+	H_2		15.4256	S	1488.4	2762,	3077
H_2^+	H_2		15.4256±0.0001	S	1488.4		3018
H_2^+	H_2		15.4269±0.0016	S			2095
H_2^+	H_2		15.4225–15.4255	PI		2616,	2811
H_2^+	H_2		15.431±0.022	RPI			2831
H_2^+	H_2		15.439±0.015	PE		2810,	2875
H_2^+	H_2		15.43	PEN			2430
H_2^+	H_2		15.37±0.05	EM			3106
H_2^+	H_2		15.44±0.01	EM			2798

See also – S: 2655, 2891
           PI: 230, 1118, 1143, 2774
           PE: 248, 1050, 2830
           PEN: 3171
           EI: 119, 383, 1012, 1121, 1251, 2535, 3144

Ion	Reactant	Other products	IP/AP (eV)	Method	Heat of formation	Ref.	
H_2^+	C_2H_6	$C_2H_4^+$?	31.1±1	EI			3175
(High kinetic energy ion)							
H_2^+	C_3H_6		16±1	EI			2542
H_2^+	C_3H_8		16.4±0.5	EI			1408
H_2^+	H_2O	O	20.7±0.4	EI			2484

HD^+

Ion	Reactant	Other products	IP/AP (eV)	Method	Heat of formation	Ref.	
HD^+	HD		15.46	PI			1143

D_2^+ $\Delta H_{f0}^\circ = 1492.2$ kJ mol^{-1} (356.6 kcal mol^{-1})

Ion	Reactant	Other products	IP/AP (eV)	Method	Heat of formation	Ref.	
D_2^+	D_2		15.46±0.01	PI		1118,	1143
D_2^+	D_2		15.468±0.022	RPI			2831
D_2^+	D_2		15.47	PE			2855
D_2^+	D_2		15.43±0.01	EM			2798
D_2^+	D_2		15.5	RPD			1121

The zero–point energy difference should give IP(D_2) – IP(H_2) ≈ 0.039 eV, see ref. 2831, leading to
IP(D_2) = 15.465 eV when combined with Herzberg's (ref. 2762) and Takezawa's (ref. 3018) value for IP(H_2).
Surprisingly, no difference was found in the electron monochromator study of ref. 2798. The recommended
value for $\Delta H_{f0}^\circ(D_2^+)$ is based on IP(D_2) = 15.465 eV.

$He^+(^2S_{1/2})$ $\Delta H_{f0}^\circ = 2372.3$ kJ mol^{-1} (567.0 kcal mol^{-1})

Ion	Reactant	Other products	IP/AP (eV)	Method	Heat of formation	Ref.	
$He^+(^2S_{1/2})$	He		24.587	S	2372.3		2113
$He^+(^2S_{1/2})$	He		24.6±0.1	PI			163
$He^+(^2S_{1/2})$	He		24.59±0.03	RPD			1012

See also – EI: 1051, 1172, 2032

4.3. The Positive Ion Table—Continued

Ion	Reactant	Other products	Ionization or appearance potential (eV)	Method	Heat of forma- tion (kJ mol^{-1})	Ref.
He^{+2}		$\Delta H^{\circ}_{f0} = 7622.8$ kJ mol^{-1} (1821.9 kcal mol^{-1})				
He^{+2}	He		79.003	S	7622.8	2113
He^{+2}	He$^+$		54.416	S		2113, 3432
He^{+2}	He$^+$		54.5±0.2	SEQ		2473
He^{+2}	He$^+$		53±2	SEQ		2551
Li$^+$		$\Delta H^{\circ}_{f0} = 679.4$ kJ mol^{-1} (162.4 kcal mol^{-1})				
Li$^+$	Li		5.392	S	679.4	2113
See also – EI:	1112, 2487, 2565, 3257					
Li$^+$	(C$_2$H$_5$Li)$_4$		14±2	EI		1
Li$^+$	LiF	F	11.5	EI		2179
Li$^+$	LiCl	Cl	10.6	EI		2179
Li$^+$	LiBr	Br	9.9	EI		2179
Li$^+$	Li$_2$MoO$_4$		11.2±0.5	EI		3257
Li$^+$	LiI	I	9.5±0.3	EI		2001
Li$^+$	Li$_2$I$_2$?		11.6±0.3	EI		2001
Li$^+$	Li$_2$WO$_4$		11.7±1.0	EI		3257
Li$_2^+$						
Li$_2^+$	Li$_2$		5.15±0.1	PI		2633
Li$_2^+$	Li$_2$		4.94±0.1	EI		3416
Be$^+$		$\Delta H^{\circ}_{f0} = 1219.5$ kJ mol^{-1} (291.5 kcal mol^{-1})				
Be$^+$	Be		9.322	S	1219.5	2113
Be$^+$	Be		9.4±0.2	EI		2715
See also – EI:	1106, 2141					
Be$^+$	BeF$_2$		28.3±1.0	EI		2142
Be$^+$	BeCl$_2$		20.4±0.4	EI		2715
See also – EI:	2195					

4.3. The Positive Ion Table—Continued

Ion	Reactant	Other products	Ionization or appearance potential (eV)	Method	Heat of formation (kJ mol^{-1})	Ref.	
			B^+ $\quad\Delta H^\circ_{f0} = 1358.3$ kJ mol^{-1} (324.6 kcal mol^{-1})				
B^+	B		8.298	S	1358.3	2113,	3066
See also – EI:	1116, 2409, 3206						
B^+	B_2H_6		18.39±0.02	EI		2559	
See also – EI:	102, 209						
B^+	B_2D_6		18.6±0.1	EI		209	
B^+	$(CH_3)_3B$		23.1±0.3	EI		364	
B^+	$(C_2H_5)_3B$		30.1±0.5	EI		364	
B^+	BH_3CO		19.31±0.2	EI		2705	
B^+	B_4H_8CO		21.8±1	EI		3226	
B^+	$(CH_3O)_2BH$		23.4±0.5	EI		364	
B^+	$(CH_3O)_3B$		31.6±1.0	EI		364	
B^+	BF_3		30.1±0.5	EI		364	
B^+	BF_3		30.6±1	EI		440	
B^+	BF_3		31.3±0.4	EI		2040	
B^+	B_2S_3		17.0±0.2	EI		3428	
B^+	BCl?		13.6±0.2	EI		440	
(Probably formed by thermal decomposition of BCl_3)							
B^+	BCl_3		18.4±0.2	EI		440	
B^+	BCl_3		19.5±0.2	EI		206	
B^+	BCl_3		19.5±1.0	EI		364	
B^+	BBr_3		19.6±0.2	EI		206	
B^+	BI_3		16.6±0.5	EI		206	
See also – EI:	3199						
			B^{+2} $\quad\Delta H^\circ_{f0} = 3785.3$ kJ mol^{-1} (904.7 kcal mol^{-1})				
B^{+2}	B		33.452	S	3785.3	2113	
B^{+2}	B^+		25.154	S		2113,	3181
			B^{+3} $\quad\Delta H^\circ_{f0} = 7445.1$ kJ mol^{-1} (1779.4 kcal mol^{-1})				
B^{+3}	B		71.382	S	7445.1	2113	
B^{+3}	B^{+2}		37.930	S		2113,	2889
			B_2^+				
B_2^+	B_2H_6	$3H_2$	21.1±0.2	EI		209	
B_2^+	B_2H_6		26.3±0.5	EI		102	
B_2^+	B_2D_6	$3D_2$	21.8±0.2	EI		209	
			B_4^+				
B_4^+	B_4H_8CO		15.9±1	EI		3226	

4.3. The Positive Ion Table—Continued

Ion	Reactant	Other products	Ionization or appearance potential (eV)	Method	Heat of formation (kJ mol^{-1})	Ref.
BH$^+$		**$\Delta H_{f0}^\circ \sim 1389$ kJ mol^{-1} (332 kcal mol^{-1})**				
BH$^+$	BH		9.77±0.05	S	1389	1109
BH$^+$	BH		9.8±0.5	EI		2735
BH$^+$	BH$_3$	H$_2$?	13.66±0.02	EI		2559
BH$^+$	BH$_3$	H$_2$?	13.7±1.0	EI		2735
BH$^+$	B$_2$H$_6$		16.39±0.3	EI		2559

See also – EI: 102, 209

Ion	Reactant	Other products	Ionization or appearance potential (eV)	Method	Heat of formation	Ref.
BH$^+$	BH$_3$CO		15.16±0.2	EI		2705
BH$^+$	(CH$_3$O)$_2$BH		28.3±2.0	EI		364

BD$^+$						
BD$^+$	B$_2$D$_6$		14.8±0.1	EI		209

BH$_2^+$						
BH$_2^+$	BH$_2$		9.8±0.2	EI		2030
BH$_2^+$	BH$_3$	H	12.3±0.5	EI		2735
BH$_2^+$	BH$_3$	H	12.95±0.05	EI		2559
BH$_2^+$	B$_2$H$_6$		13.4±0.1	EI		209
BH$_2^+$	B$_2$H$_6$		13.5±0.5	EI		102
BH$_2^+$	B$_2$H$_6$		15.5±0.05	EI		2559
BH$_2^+$	BH$_3$CO		14.36±0.2	EI		2705
BH$_2^+$	B$_4$H$_8$CO		15.3±0.5	EI		3226

BD$_2^+$						
BD$_2^+$	B$_2$D$_6$		13.6±0.1	EI		209

BH$_3^+$						
BH$_3^+$	BH$_3$		12.32±0.1	RPD		2559
BH$_3^+$	BH$_3$		12.24±0.1	EI		2705
BH$_3^+$	BH$_3$		11.4±0.2	EI		2030
BH$_3^+$	BH$_3$		11.5±0.5	EI		2735
BH$_3^+$	BH$_3$		14.0±2	EI		2869
BH$_3^+$	B$_2$H$_6$	BH$_3$?	12.1±0.2	EI		102
BH$_3^+$	B$_2$H$_6$	BH$_3$?	13.1±0.2	EI		209
BH$_3^+$	B$_2$H$_6$	BH$_3$?	14.88±0.05	EI		2559
BH$_3^+$	BH$_3$CO	CO	13.70±0.2	EI		2705

BD$_3^+$						
BD$_3^+$	B$_2$D$_6$	BD$_3$?	12.7±0.2	EI		209

4.3. The Positive Ion Table—Continued

Ion	Reactant	Other products	Ionization or appearance potential (eV)	Method	Heat of formation (kJ mol^{-1})	Ref.
		B$_2$H$^+$				
B$_2$H$^+$	B$_2$H$_6$	2H$_2$+H	20.1±0.1	EI		209
B$_2$H$^+$	B$_2$H$_6$	2H$_2$+H	21.4±0.5	EI		102
		B$_2$D$^+$				
B$_2$D$^+$	B$_2$D$_6$	2D$_2$+D	18.3±0.1	EI		209
		B$_2$H$_2^+$				
B$_2$H$_2^+$	B$_2$H$_6$	2H$_2$	13.8±0.1	EI		209
B$_2$H$_2^+$	B$_2$H$_6$	2H$_2$	13.8±0.2	EI		102
B$_2$H$_2^+$	B$_2$H$_6$	2H$_2$	14.1±0.2	EI		1024
B$_2$H$_2^+$	B$_4$H$_8$CO		14.8±0.8	EI		3226
		B$_2$HD$^+$				
B$_2$HD$^+$	B$_2$HD$_5$?		14.0±0.1	EI		209
		B$_2$D$_2^+$				
B$_2$D$_2^+$	B$_2$D$_6$	2D$_2$	14.0±0.1	EI		209
		B$_2$H$_3^+$				
B$_2$H$_3^+$	B$_2$H$_6$	H$_2$+H	14.2±0.1	EI		209
B$_2$H$_3^+$	B$_2$H$_6$	H$_2$+H	14.3±0.2	EI		1024
B$_2$H$_3^+$	B$_2$H$_6$	H$_2$+H	14.8±0.5	EI		102
		B$_2$HD$_2^+$				
B$_2$HD$_2^+$	B$_2$HD$_5$?		14.2±0.1	EI		209
		B$_2$D$_3^+$				
B$_2$D$_3^+$	B$_2$D$_6$	D$_2$+D	14.3±0.1	EI		209
B$_2$D$_3^+$	B$_2$D$_6$	D$_2$+D	14.5±0.2	EI		1024
		B$_2$H$_4^+$				
B$_2$H$_4^+$	B$_2$H$_6$	H$_2$	12.3±0.1	EI		209
B$_2$H$_4^+$	B$_2$H$_6$	H$_2$	12.3±0.2	EI		1024
B$_2$H$_4^+$	B$_2$H$_6$	H$_2$	12.4±0.3	EI		102

4.3. The Positive Ion Table—Continued

Ion	Reactant	Other products	Ionization or appearance potential (eV)	Method	Heat of formation (kJ mol^{-1})	Ref.
			$B_2HD_3^+$			
$B_2HD_3^+$	B_2HD_5?		12.3±0.1	EI		209
			$B_2D_4^+$			
$B_2D_4^+$	B_2D_6	D_2	12.3±0.1	EI		209
$B_2D_4^+$	B_2D_6	D_2	12.3±0.2	EI		1024
			$B_2H_5^+$			
$B_2H_5^+$	B_2H_6	H	11.84±0.1	EI		2559
$B_2H_5^+$	B_2H_6	H	11.9±0.1	EI		209
$B_2H_5^+$	B_2H_6	H	12.0±0.1	EI		1024
$B_2H_5^+$	B_2H_6	H	12.0±0.3	EI		102
$B_2H_5^+$	B_4H_8CO		12.4±0.8	EI		3226
			$B_2HD_4^+$			
$B_2HD_4^+$	B_2HD_5?		12.0±0.1	EI		209
			$B_2D_5^+$			
$B_2D_5^+$	B_2D_6	D	12.0±0.1	EI		1024
$B_2D_5^+$	B_2D_6	D	12.1±0.1	EI		209
$B_2H_6^+(^2B_{2g})$	$\Delta H_{f0}^\circ \sim 1150$ kJ mol^{-1} (275 kcal mol^{-1})					
$B_2H_6^+(^2B_{2g})$	B_2H_6		11.38±0.01	PE	1149	3147
$B_2H_6^+(^2B_{2g})$	B_2H_6		11.41±0.02	PE	1152	3105
$B_2H_6^+(^2B_{2g})$	B_2H_6		11.37±0.01	PE	1148 2960,	3050
$B_2H_6^+(^2A_g)$	B_2H_6		<12.7	PE		3147
$B_2H_6^+(^2A_g)$	B_2H_6		12.83±0.04	PE		3105
$B_2H_6^+(^2A_g)$	B_2H_6		<12.88	PE		2960
$B_2H_6^+(^2B_{3u})$	B_2H_6		<13.6	PE		3147
$B_2H_6^+(^2B_{3u})$	B_2H_6		13.81±0.06	PE		3105
$B_2H_6^+(^2B_{3u})$	B_2H_6		~13.5	PE		2960
$B_2H_6^+(^2B_{2u})$	B_2H_6		<14.5	PE		3147
$B_2H_6^+(^2B_{2u})$	B_2H_6		14.42±0.04	PE		3105
$B_2H_6^+(^2B_{2u})$	B_2H_6		~14.5	PE		2960
$B_2H_6^+(^2B_{1u})$	B_2H_6		16.06±0.01	PE		3147
$B_2H_6^+(^2B_{1u})$	B_2H_6		16.08±0.01	PE		3105
$B_2H_6^+(^2B_{1u})$	B_2H_6		16.05	PE		2960

4.3. The Positive Ion Table—Continued

Ion	Reactant	Other products	Ionization or appearance potential (eV)	Method	Heat of formation (kJ mol^{-1})	Ref.
$B_2H_6^+(^2A_g)$	B_2H_6		20.2±0.1	PE		3147

The notation and assignments differ in all three studies. We have adopted the assignments of ref. 3147. There is also disagreement about the vibrational structure.

See also – EI: 102, 209

		$B_2D_6^+$				
$B_2D_6^+$	B_2D_6		12.0±0.1	EI		209
		B_3H^+				
B_3H^+	B_4H_{10}		16.5±0.8	EI		1119
		$B_3H_2^+$				
$B_3H_2^+$	B_4H_{10}		17.8±0.8	EI		1119
$B_3H_2^+$	B_4H_8CO		16.7±1	EI		3226
		$B_3H_3^+$				
$B_3H_3^+$	B_4H_{10}		14.2±0.3	EI		1119
$B_3H_3^+$	B_4H_8CO		14.8±0.5	EI		3226
		$B_3D_3^+$				
$B_3D_3^+$	B_4D_{10}		13.8±0.3	EI		1119
		$B_3H_4^+$				
$B_3H_4^+$	B_4H_8CO		14.3±0.2	EI		3226
		$B_3H_5^+$				
$B_3H_5^+$	B_4H_{10}		12.1±0.2	EI		1119
$B_3H_5^+$	B_4H_8CO		12.2±0.2	EI		3226
		$B_3D_5^+$				
$B_3D_5^+$	B_4D_{10}		12.8±0.3	EI		1119

4.3. The Positive Ion Table—Continued

Ion	Reactant	Other products	Ionization or appearance potential (eV)	Method	Heat of formation (kJ mol^{-1})	Ref.
			$B_3H_6^+$			
$B_3H_6^+$	B_4H_8CO		10.7±0.5	EI		3226
			$B_3H_7^+$			
$B_3H_7^+$	B_4H_8CO		11.5±0.5	EI		3226
			B_4H^+			
B_4H^+	B_5H_9		19.97±1.0	EI		205
B_4H^+	B_4H_8CO		13.0±0.8	EI		3226
			$B_4H_2^+$			
$B_4H_2^+$	B_5H_9		17.99±0.2	EI		205
$B_4H_2^+$	B_4H_8CO		13.1±0.8	EI		3226
			$B_4H_3^+$			
$B_4H_3^+$	B_4H_{10}	$3H_2+H$	14.0±0.4	EI		1119
$B_4H_3^+$	B_5H_9		15.96±0.5	EI		205
$B_4H_3^+$	B_4H_8CO		12.6±1	EI		3226
			$B_4D_3^+$			
$B_4D_3^+$	B_4D_{10}	$3D_2+D$	14.8±0.4	EI		1119
			$B_4H_4^+$			
$B_4H_4^+$	B_4H_{10}	$3H_2$	12.4±0.2	EI		1119
$B_4H_4^+$	B_5H_9		14.06±0.1	EI		205
$B_4H_4^+$	B_4H_8CO		12.6±0.3	EI		3226
			$B_4D_4^+$			
$B_4D_4^+$	B_4D_{10}	$3D_2$	12.5±0.2	EI		1119
$B_4D_4^+$	B_5D_9		13.7±0.2	EI		1024
$B_4D_4^+$	B_5D_{11}		12.4±0.5	EI		1024
			$B_4H_5^+$			
$B_4H_5^+$	B_4H_{10}	$2H_2+H$	12.5±0.3	EI		1119
$B_4H_5^+$	B_5H_9		15.07±0.3	EI		205
$B_4H_5^+$	B_4H_8CO		13.4±0.4	EI		3226

4.3. The Positive Ion Table—Continued

Ion	Reactant	Other products	Ionization or appearance potential (eV)	Method	Heat of forma-tion (kJ mol^{-1})	Ref.
		$B_4D_5^+$				
$B_4D_5^+$	B_4D_{10}	$2D_2+D$	12.5 ± 0.2	EI		1119
		$B_4H_6^+$				
$B_4H_6^+$	B_4H_{10}	$2H_2$	11.2 ± 0.1	EI		1119
$B_4H_6^+$	B_5H_9	BH_3?	12.25 ± 0.2	EI		205
$B_4H_6^+$	B_4H_8CO		11.2 ± 0.2	EI		3226
		$B_4D_6^+$				
$B_4D_6^+$	B_4D_{10}	$2D_2$	11.1 ± 0.1	EI		1119
$B_4D_6^+$	B_5D_9	BD_3?	12.4 ± 0.2	EI		1024
$B_4D_6^+$	B_5D_{11}		11.4 ± 0.5	EI		1024
		$B_4H_7^+$				
$B_4H_7^+$	B_4H_{10}	H_2+H	12.5 ± 0.2	EI		1119
$B_4H_7^+$	B_4H_8CO		11.5 ± 0.5	EI		3226
		$B_4D_7^+$				
$B_4D_7^+$	B_4D_{10}	D_2+D	12.2 ± 0.1	EI		1119
		$B_4H_8^+$				
$B_4H_8^+$	B_4H_8		10.19 ± 0.5	EI		3226
$B_4H_8^+$	B_4H_{10}	H_2	10.4 ± 0.1	EI		1119
$B_4H_8^+$	B_4H_8CO	CO	10.6 ± 0.5	EI		3226
		$B_4D_8^+$				
$B_4D_8^+$	B_4D_{10}	D_2	9.9 ± 0.1	EI		1119
		$B_4H_9^+$				
$B_4H_9^+$	B_4H_{10}	H	12.2 ± 0.2	EI		1119
$B_4H_9^+$	B_8H_{18}		11.2 ± 0.1	EI		2736
		$B_4D_9^+$				
$B_4D_9^+$	B_4D_{10}	D	11.9 ± 0.2	EI		1119

4.3. The Positive Ion Table—Continued

Ion	Reactant	Other products	Ionization or appearance potential (eV)	Method	Heat of formation (kJ mol^{-1})	Ref.
		$B_5H_4^+$				
$B_5H_4^+$	B_5H_9	$2H_2+H$	13.01 ± 0.3	EI		205
$B_5H_4^+$	B_5H_{11}	$3H_2+H$	14.2 ± 0.4	EI		1024
		$B_5H_5^+$				
$B_5H_5^+$	B_5H_9	$2H_2$	12.67 ± 0.03	EI		205
$B_5H_5^+$	B_5H_9	$2H_2$	12.8 ± 0.2	EI		1024
$B_5H_5^+$	B_5H_9	$2H_2$	12.5 ± 0.5	EI		3226
$B_5H_5^+$	B_5H_{11}	$3H_2$	12.7 ± 0.2	EI		1024
$B_5H_5^+$	B_6H_{10}		13.6 ± 0.2	EI		1024
		$B_5D_5^+$				
$B_5D_5^+$	B_5D_9	$2D_2$	12.4 ± 0.4	EI		1024
$B_5D_5^+$	B_5D_9	$2D_2$	12.92 ± 0.03	EI		205
$B_5D_5^+$	B_5D_{11}	$3D_2$	12.3 ± 0.2	EI		1024
		$B_5H_6^+$				
$B_5H_6^+$	B_5H_9	H_2+H	12.13 ± 0.3	EI		205
$B_5H_6^+$	B_5H_{11}	$2H_2+H$	12.6 ± 0.3	EI		1024
		$B_5D_6^+$				
$B_5D_6^+$	B_5D_9	D_2+D	12.4 ± 0.2	EI		1024
$B_5D_6^+$	B_5D_{11}	$2D_2+D$	12.2 ± 0.3	EI		1024
		$B_5H_7^+$				
$B_5H_7^+$	B_5H_9	H_2	11.43 ± 0.1	EI		205
$B_5H_7^+$	B_5H_9	H_2	11.6 ± 0.2	EI		1024
$B_5H_7^+$	B_5H_9	H_2	11.4 ± 0.5	EI		3226
$B_5H_7^+$	B_5H_{11}	$2H_2$	11.5 ± 0.2	EI		1024
$B_5H_7^+$	B_6H_{10}	$BH_3?$	12.0 ± 0.2	EI		1024
		$B_5D_7^+$				
$B_5D_7^+$	B_5D_9	D_2	10.93 ± 0.2	EI		205
$B_5D_7^+$	B_5D_9	D_2	11.2 ± 0.2	EI		1024
$B_5D_7^+$	B_5D_{11}	$2D_2$	11.1 ± 0.2	EI		1024

4.3. The Positive Ion Table—Continued

Ion	Reactant	Other products	Ionization or appearance potential (eV)	Method	Heat of formation (kJ mol^{-1})	Ref.
		$B_5H_8^+$				
$B_5H_8^+$	B_5H_{11}	H_2+H	12.0±0.3	EI		1024
$B_5H_8^+$	$B_{10}H_{16}$		11.6±0.2	EI		1102
$B_5H_8^+$	B_5H_8Br	Br	12.0±0.2	EI		1102
$B_5H_8^+$	B_5H_8I	I	11.1±0.1	EI		1102
		$B_5D_8^+$				
$B_5D_8^+$	B_5D_{11}	D_2+D	11.4±0.3	EI		1024
		$B_5H_9^+$				
$B_5H_9^+$	B_5H_9		10.54±0.01	RPD		3228
$B_5H_9^+$	B_5H_9		10.38±0.05	EI		205
$B_5H_9^+$	B_5H_9		10.5±0.1	EI		1024
$B_5H_9^+$	B_5H_9		10.5±0.5	EI		3226
$B_5H_9^+$	B_5H_9		10.8±0.5	EI		103
$B_5H_9^+$	B_5H_{11}	H_2	10.3±0.2	EI		1024
		$B_5D_9^+$				
$B_5D_9^+$	B_5D_9		9.77±0.1	EI		205
$B_5D_9^+$	B_5D_9		10.0±0.1	EI		1024
$B_5D_9^+$	B_5D_{11}	D_2	10.4±0.2	EI		1024
		$B_5H_{10}^+$				
$B_5H_{10}^+$	B_5H_{11}	H	11.8±0.4	EI		1024
		$B_5D_{10}^+$				
$B_5D_{10}^+$	B_5D_{11}	D	11.3±0.4	EI		1024
		$B_6H_4^+$				
$B_6H_4^+$	B_6H_{10}	$3H_2$	13.4±0.3	EI		1024
		$B_6H_5^+$				
$B_6H_5^+$	B_6H_{10}	$2H_2+H$	12.0±0.3	EI		1024
		$B_6H_6^+$				
$B_6H_6^+$	B_6H_{10}	$2H_2$	11.9±0.1	EI		1024

4.3. The Positive Ion Table—Continued

Ion	Reactant	Other products	Ionization or appearance potential (eV)	Method	Heat of formation (kJ mol^{-1})	Ref.
		$B_6H_7^+$				
$B_6H_7^+$	B_6H_{10}	H_2+H	11.5 ± 0.3	EI		1024
		$B_6H_8^+$				
$B_6H_8^+$	B_6H_{10}	H_2	11.2 ± 0.1	EI		1024
		$B_6H_9^+$				
$B_6H_9^+$	B_6H_{10}	H	11.1 ± 0.4	EI		1024
		$B_6H_{10}^+$				
$B_6H_{10}^+$	B_6H_{10}		9.3 ± 0.1	EI		1024
		$B_6D_{10}^+$				
$B_6D_{10}^+$	B_6D_{10}		9.7 ± 0.2	EI		1024
		$B_6H_{12}^+$				
$B_6H_{12}^+$	B_6H_{12}		9.75 ± 0.2	EI		3225
		$B_8H_{12}^+$				
$B_8H_{12}^+$	B_8H_{12}		9.52 ± 0.1	EI		3225
		$B_{10}H_6^+$				
$B_{10}H_6^+$	$B_{10}H_{14}$	$4H_2$	13.14 ± 0.3	EI		189
		$B_{10}H_7^+$				
$B_{10}H_7^+$	$B_{10}H_{14}$	$3H_2+H$	12.51 ± 0.5	EI		189
		$B_{10}H_8^+$				
$B_{10}H_8^+$	$B_{10}H_{14}$	$3H_2$	12.67 ± 0.3	EI		189
		$B_{10}H_{10}^+$				
$B_{10}H_{10}^+$	$B_{10}H_{14}$	$2H_2$	11.62 ± 0.2	EI		189

4.3. The Positive Ion Table—Continued

Ion	Reactant	Other products	Ionization or appearance potential (eV)	Method	Heat of formation (kJ mol^{-1})	Ref.
$B_{10}H_{11}^+$						
$B_{10}H_{11}^+$	$B_{10}H_{14}$	H_2+H	10.81±0.5	EI		189
$B_{10}H_{12}^+$						
$B_{10}H_{12}^+$	$B_{10}H_{14}$	H_2	10.87±0.2	EI		189
$B_{10}H_{14}^+$						
$B_{10}H_{14}^+$	$B_{10}H_{14}$		10.26±0.5	EI		189
$B_{10}H_{14}^+$	$B_{10}H_{14}$		11.0±0.5	EI		103
$B_{10}H_{16}^+$						
$B_{10}H_{16}^+$	$B_{10}H_{16}$		10.1±0.2	EI		1102
C^+ $\quad \Delta H_{f0}^\circ = 1797.6$ kJ mol^{-1} (429.6 kcal mol^{-1})						
C^+	C		11.260	S	1797.6	2113
C^+	C		11.3±0.2	EI		1155
C^+	C_2H_2		23.6	EI		3156
C^+	C_3H_6		24.5±0.5	EI		2542
C^+	CNC≡CCN		24±1.0	EI		154
C^+	CNC≡CC≡CCN		23.0±0.5	EI		154
C^+	CH_3CN		27.0±0.3	EI		131
C^+	CO	O^-	20.82±0.05	RPD	(1790)	2431
C^+	CO	O^-	20.89±0.09	RPD	(1796)	2180, 2191

See also – PI: 163
        EI: 200, 2014, 2016
        D: 6

Ion	Reactant	Other products	Ionization or appearance potential (eV)	Method	Heat of formation (kJ mol^{-1})	Ref.
C^+	CO	O	22.45±0.10	RPD	(1806)	2431
C^+	CO	O	22.57±0.20	RPD	(1817)	2180, 2191

See also – PI: 163
        EI: 200, 2014, 2016

Ion	Reactant	Other products	Ionization or appearance potential (eV)	Method	Heat of formation (kJ mol^{-1})	Ref.
C^+	CO_2	O_2	23.2±0.5	EI		2472
C^+	CO_2	2O	28.4±0.6	EI		2472
C^+	CO_2^+	2O	14.15±0.5	SEQ		2472
C^+	C_3O_2	2CO	18.7±0.3	EI		2687
C^+	CH_3NO_2		22.83±0.05	EI		90
C^+	CF_4		31.5±0.5	EI		24
C^+	CF_2O	2F+O?	36±1	EI		3236
C^+	CF_3OF	4F+O?	41±1	EI		3236
C^+	CS_2	S_2	20.0±0.1	EI		3402
C^+	CS_2	S_2	19.9±0.6	EI		2472
C^+	CS_2	2S	24.0±0.5	EI		2472

4.3. The Positive Ion Table—Continued

Ion	Reactant	Other products	Ionization or appearance potential (eV)	Method	Heat of formation (kJ mol^{-1})	Ref.
C^+	CS_2^+	2S	12.3±0.9	SEQ		2472
C^+	$(CH_3)_2SO$		22.9±0.5	EI		3294
C^+	CH_3Cl		26.25±0.1	EI		131
C^+	CH_2Cl_2		25.45±0.1	EI		131
C^+	$CHCl_3$		24.62±0.05	EI		131
C^+	CF_3Cl		31±1	EI		24
C^+	CBr_4		23.1±0.4	EI		1246
C^+	CH_2Br_2		24.52±0.05	EI		131
C^+	$CHBr_3$		23.55±0.05	EI		131
C^+	CF_3Br		33±1	EI		24
C^+	CF_3I		32.6±1	EI		24

C^{+2} $\Delta H_{f0}^{\circ} = 4150.3$ kJ mol^{-1} (991.9 kcal mol^{-1})

Ion	Reactant	Other products	Ionization or appearance potential (eV)	Method	Heat of formation (kJ mol^{-1})	Ref.
C^{+2}	C		35.643	S	4150.3	2113
C^{+2}	C^+		24.383	S		2113
C^{+2}	CO	O	54.2±0.2	EI		2431

(8.2±0.9 eV average translational energy of decomposition at threshold)

C_2^+ $\Delta H_{f0}^{\circ} = 1992$ kJ mol^{-1} (476 kcal mol^{-1})

Ion	Reactant	Other products	Ionization or appearance potential (eV)	Method	Heat of formation (kJ mol^{-1})	Ref.
C_2^+	C_2		12.0±0.6	EI	(~1987)	1155
C_2^+	C_2		~13	EI		2102
C_2^+	C_2H_2	H_2	18.2	EI	(~1983)	2102
C_2^+	C_2H_2	H_2	19.5	EI		3156
C_2^+	C_2H_2	2H?	23.6	FD		3131
C_2^+	C_2H_2	2H?	22.7	EI		2102
C_2^+	C_2H_2	2H?	23.3±0.5	EI		13
C_2^+	C_2H_2	2H?	23.8	EI		3156
C_2^+	C_2D_2	2D?	23.6	FD		3131
C_2^+	C_3H_6		28±1	EI		2542
C_2^+	C_2N_2	N_2	17.46±0.02	PI	1992	2621
C_2^+	C_2N_2	N_2	18.4±0.3	EI		154
C_2^+	CNC≡CCN		18.5±0.3	EI		154
C_2^+	CNC≡CC≡CCN		18.4±1.0	EI		154
C_2^+	CH≡CCN		18.6±0.2	EI		154
C_2^+	C_3O_2		24.5±0.7	EI		2687
C_2^+	$Ni(CO)_4$		39.9±2	EI		2579

C_3^+

Ion	Reactant	Other products	Ionization or appearance potential (eV)	Method	Heat of formation (kJ mol^{-1})	Ref.
C_3^+	C_3		12.1±0.3	EI	(~1979)	2996, 3208
C_3^+	C_3		12.6±0.6	EI		1155

See also – EI: 154

Ion	Reactant	Other products	Ionization or appearance potential (eV)	Method	Heat of formation (kJ mol^{-1})	Ref.
C_3^+	C_3H?		17.1±0.5	EI		154
C_3^+	$CH_3C≡CH$		26.0±1	EI		13
C_3^+	C_3H_6		27±1	EI		2542
C_3^+	CNC≡CCN		24.6±0.5	EI		154
C_3^+	CNC≡CC≡CCN		23.0±2.0	EI		154

4.3. The Positive Ion Table—Continued

Ion	Reactant	Other products	Ionization or appearance potential (eV)	Method	Heat of formation (kJ mol⁻¹)	Ref.
C_3^+	CH≡CCN		24.5±0.5	EI		154
C_3^+	C_3O_2		29.8±0.4	EI		2687

C_4^+

Ion	Reactant	Other products	Ionization or appearance potential (eV)	Method	Heat of formation (kJ mol⁻¹)	Ref.
C_4^+	C_4		12.6	EI		1155
C_4^+	CNC≡CCN	N_2	17.2±0.2	EI		154
C_4^+	CNC≡CC≡CCN		17.8±0.4	EI		154

C_5^+

Ion	Reactant	Other products	Ionization or appearance potential (eV)	Method	Heat of formation (kJ mol⁻¹)	Ref.
C_5^+	C_5		12.7±0.5	EI		154
C_5^+	C_5		12.5±1	EI		1155
C_5^+	CNC≡CC≡CCN		24.0±0.5	EI		154

C_6^+

Ion	Reactant	Other products	Ionization or appearance potential (eV)	Method	Heat of formation (kJ mol⁻¹)	Ref.
C_6^+	C_6		12.5±0.3	EI		154

CH^+ $\qquad \Delta H_{f0}^\circ = 1619$ kJ mol⁻¹ (387 kcal mol⁻¹)

Ion	Reactant	Other products	Ionization or appearance potential (eV)	Method	Heat of formation (kJ mol⁻¹)	Ref.
CH^+	CH		10.64±0.01	S	1619	3042
CH^+	CH_3	H_2	15.58±0.30	EI	(1649)	414
CH^+	C_2H_2	CH	21.9	RPD		3345

(0.28 eV average translational energy of decomposition at threshold)

Ion	Reactant	Other products	Ionization or appearance potential (eV)	Method	Heat of formation (kJ mol⁻¹)	Ref.
CH^+	C_2H_2	CH	21.5±0.2	EDD		3177

See also – EI: 2102, 2450, 3131, 3156

Ion	Reactant	Other products	Ionization or appearance potential (eV)	Method	Heat of formation (kJ mol⁻¹)	Ref.
CH^+	C_3H_6		22.5±0.5	EI		2542
CH^+	C_3H_8	$C_2H_5^+ + H_2$	26.0±1	EI		1408

(Threshold value corrected for high fragment kinetic energy)

Ion	Reactant	Other products	Ionization or appearance potential (eV)	Method	Heat of formation (kJ mol⁻¹)	Ref.
CH^+	CH_3CN		22.4±0.2	EI		131
CH^+	CH≡CCN		21.9±0.3	EI		154
CH^+	CH_3OH		22.31±0.09	EI		131
CH^+	$(CH_2)_2O$		22.8±0.4	EI		50
	(1,2–Epoxyethane)					
CH^+	CH≡CF	CF	19.2±0.2	EDD		3177
CH^+	$(CH_3)_2SO$		19.4±0.5	EI		3294
CH^+	CH_3Cl		22.5±0.06	EI		131
CH^+	CH≡CCl	CCl	18.7±0.2	EDD		3177
CH^+	CH_2Cl_2		21.72±0.04	EI		131
CH^+	$CHCl_3$		23.9±0.3	EI		43
CH^+	CH_3Br		21.41	EI		131
CH^+	CH≡CBr	CBr	20.3±0.2	EDD		3177
CH^+	CH_2Br_2		21.55±0.05	EI		131
CH^+	$CHBr_3$		21.70±0.05	EI		131
CH^+	CH_3I		21.2±0.2	EI		131

4.3. The Positive Ion Table—Continued

Ion	Reactant	Other products	Ionization or appearance potential (eV)	Method	Heat of formation (kJ mol^{-1})	Ref.

$$CH_2^+ \qquad \Delta H_{f0}^\circ \sim 1398 \text{ kJ mol}^{-1} \text{ (334 kcal mol}^{-1})$$

CH_2^+	CH_2		10.396±0.003	S		1078

Dibeler *et al.*, ref. 1128, combine this ionization potential with their appearance potential for CH_2^+ from CH_4 to yield $\Delta H_{f0}^\circ(CH_2) = 392.9$ kJ mol^{-1} (93.9 kcal mol^{-1}).

CH_2^+	CH_2		10.50±0.2	EI		2535

See also − EI: 87, 327

CH_2^+	CH_3	H	15.09±0.03	PI	1386	2618

  (Threshold value approximately corrected to 0 K)

See also − EI: 14

CH_2^+	CH_4	H_2	15.19±0.02	PI	1399	2605

  (Threshold value approximately corrected to 0 K)

CH_2^+	CH_4	H_2	15.16±0.04	PI	1396	1128

See also − EI: 160

Ion	Reactant	Other products	IP/AP (eV)	Method	ΔHf	Ref.
CH_2^+	C_2H_2	C	21	EI		2136
CH_2^+	C_2H_4	CH_2	18.05	PI	1409	2617

  (Threshold value approximately corrected to 0 K but may be subject to kinetic shift and reaction competition)

Ion	Reactant	Other products	IP/AP (eV)	Method	ΔHf	Ref.
CH_2^+	C_3H_6		17.0±0.5	EI		2542
CH_2^+	C_3H_6	C_2H_4	19.3	RPD		3345

  (Cyclopropane)
  (0.52 eV average translational energy of decomposition at threshold)

Ion	Reactant	Other products	IP/AP (eV)	Method	ΔHf	Ref.
CH_2^+	CH_3CN	HCN	14.94±0.02	PI	1400	2623

  (Threshold may be subject to kinetic shift and reaction competition)

Ion	Reactant	Other products	IP/AP (eV)	Method	ΔHf	Ref.
CH_2^+	CH_3CN	HCN	15.7	RPD		3345

  (0.12 eV average translational energy of decomposition at threshold)

Ion	Reactant	Other products	IP/AP (eV)	Method	ΔHf	Ref.
CH_2^+	CH_2N_2	N_2	12.3±0.1	EI		314

  (Diazomethane)

Ion	Reactant	Other products	IP/AP (eV)	Method	ΔHf	Ref.
CH_2^+	CH_2N_2	N_2	11.0±0.1	EI		314

  (Diazirine)

Ion	Reactant	Other products	IP/AP (eV)	Method	ΔHf	Ref.
CH_2^+	CH_3OH	H_2O	15.3	RPD		3345

  (0.13 eV average translational energy of decomposition at threshold)

Ion	Reactant	Other products	IP/AP (eV)	Method	ΔHf	Ref.
CH_2^+	$CH_2{=}CO$	CO	13.8±0.2	EI		2800
CH_2^+	$(CH_2)_2O$		16.5±0.4	EI		50

  (1,2−Epoxyethane)

Ion	Reactant	Other products	IP/AP (eV)	Method	ΔHf	Ref.
CH_2^+	C_3H_6O		18.8±0.5	EI		50

  (1,2−Epoxypropane)

Ion	Reactant	Other products	IP/AP (eV)	Method	ΔHf	Ref.
CH_2^+	$HCOOCH_3$		19.8	EI		3224
CH_2^+	CH_3COOCH_3		20.8	EI		3224
CH_2^+	$C_4H_8O_2$		21.3±0.5	EI		153

  (1,2−Epoxy−3−methoxypropane)

Ion	Reactant	Other products	IP/AP (eV)	Method	ΔHf	Ref.
CH_2^+	$CH_2{=}CF_2$	CF_2	17.8	EI		419
CH_2^+	CH_3CF_3		16.2±0.3	EI		1075
CH_2^+	CH_3BF_2		16.9±0.1	EI		1076
CH_2^+	$(CH_2)_2S$		20.4±0.5	EI		51

  (Ethylene sulfide)

4.3. The Positive Ion Table—Continued

Ion	Reactant	Other products	Ionization or appearance potential (eV)	Method	Heat of formation (kJ mol^{-1})	Ref.
CH_2^+	$(CH_3)_2SO$		15.5±0.3	EI		3294
CH_2^+	CH_3Cl	HCl	14.6±0.2	RPD		160
CH_2^+	CH_2Cl_2	Cl_2	17.0	RPD		3345
(0.36 eV average translational energy of decomposition at threshold)						
CH_2^+	C_3H_5OCl		21.6±0.5	EI		153
(1−Chloro−2,3−epoxypropane)						
CH_2^+	CH_3Br	HBr	14.9±0.2	RPD		160
CH_2^+	C_3H_5OBr		21.4±0.5	EI		153
(1−Bromo−2,3−epoxypropane)						
CH_2^+	CH_3I	HI	14.6±0.2	RPD		160

CD_2^+

Ion	Reactant	Other products	Ionization or appearance potential (eV)	Method	Heat of formation (kJ mol^{-1})	Ref.
CD_2^+	CD_4	D_2	15.25±0.04	PI		1128

CH_3^+ $\Delta H_{f0}^\circ = 1095$ kJ mol^{-1} (262 kcal mol^{-1})

Ion	Reactant	Other products	Ionization or appearance potential (eV)	Method	Heat of formation (kJ mol^{-1})	Ref.
CH_3^+	CH_3		9.842±0.002	S	1095	349
CH_3^+	CH_3		9.825±0.01	PI		2618
CH_3^+	CH_3		9.82±0.04	PI		1068
CH_3^+	CH_3		9.84±0.03	EM	3104,	3379
CH_3^+	CH_3		9.87±0.05	RPD		2776

The discrepancy between the spectroscopic and photoionization values is unexplained, see ref. 2618.

See also − EI: 87, 327, 414, 1129, 2158, 2464, 2535, 2904, 2961, 2986, 3202

Ion	Reactant	Other products	Ionization or appearance potential (eV)	Method	Heat of formation (kJ mol^{-1})	Ref.
CH_3^+	CH_4	H^-	13.50±0.05	PI	(1093)	2605
CH_3^+	CH_4	H	14.320±0.004	PI	(1099)	2605
(Threshold value corrected to 0 K)						
CH_3^+	CH_4	H	14.25±0.02	PI		1128
CH_3^+	CH_4	H	14.23±0.05	PI		2013
CH_3^+	CH_4	H	14.30	EM		3104
CH_3^+	CH_4	H	14.24±0.05	RPD		2776

See also − EI: 224, 1072, 1451, 2154, 3017

Ion	Reactant	Other products	Ionization or appearance potential (eV)	Method	Heat of formation (kJ mol^{-1})	Ref.
CH_3^+	C_2H_6	CH_3	13.46±0.05	RPD		2776

See also − EI: 160, 2421

Ion	Reactant	Other products	Ionization or appearance potential (eV)	Method	Heat of formation (kJ mol^{-1})	Ref.
CH_3^+	C_2H_6		30.3±0.2	SRP		1264
(High kinetic energy ion)						
CH_3^+	C_2H_5D		14.94?	EI		2421
CH_3^+	CH_3CD_3		15.15?	EI		2421
CH_3^+	$CH_3C\equiv CH$		15.4±0.5	EI		13
CH_3^+	C_3H_6	C_2H_3	14.9	RPD		3345
(0.10 eV average translational energy of decomposition at threshold)						

See also − EI: 2542

4.3. The Positive Ion Table—Continued

Ion	Reactant	Other products	Ionization or appearance potential (eV)	Method	Heat of formation (kJ mol^{-1})	Ref.
CH_3^+	C_3H_8	C_2H_5	14.0±0.5	EI		2021

See also – EI: 1408

| CH_3^+ | C_3H_8 | $C_2H_5^+$ | 22.0±0.5 | EI | | 1408 |

(Threshold value corrected for high fragment kinetic energy)

See also – EI: 2021, 2447

| CH_3^+ | C_3H_8 | $C_2H_3^+ + H_2$ | 25.0±0.5 | EI | | 1408 |

(Threshold value corrected for high fragment kinetic energy)

See also – EI: 1264

Ion	Reactant	Other products	IP/AP (eV)	Method	ΔHf	Ref.
CH_3^+	$CH_3CH=C=CH_2$		14.4±0.2	EI		462
CH_3^+	$CH_3C\equiv CCH_3$		17.6±0.5	EI		13
CH_3^+	$n-C_4H_{10}$		29.7±0.2	SRP		1264

(High kinetic energy ion)

| CH_3^+ | $iso-C_4H_{10}$ | | 29.4±0.2 | SRP | | 1264 |

(High kinetic energy ion)

| CH_3^+ | $neo-C_5H_{12}$ | | 13.14 | EI | | 2101 |
| CH_3^+ | $neo-C_5H_{12}$ | | 29.5±0.2 | SRP | | 1264 |

(High kinetic energy ion)

CH_3^+	$C_2H_5C\equiv CC\equiv CH$		18.50	EI		1197
CH_3^+	$CH_3C\equiv CC\equiv CCH_3$		25.70	EI		1197
CH_3^+	C_6H_6		28.2±0.2	SRP		1264

(Benzene)

(High kinetic energy ion)

| CH_3^+ | $n-C_7H_{16}$ | | 27.9±0.2 | SRP | | 1264 |

(High kinetic energy ion)

| CH_3^+ | $(CH_3)_3B$ | | 15.1±0.3 | EI | | 364 |
| CH_3^+ | CH_3NH_2 | NH_2 | 14.7 | RPD | | 3345 |

(0.25 eV average translational energy of decomposition at threshold)

| CH_3^+ | CH_3NC | CN | 14.76 | EDD | | 3214 |
| CH_3^+ | $(CH_2)_2NH$ | | 15.5±0.3 | EI | | 51 |

(Ethylenimine)

| CH_3^+ | $(CH_2)_3NH$ | | 14.4±1.0 | EI | | 52 |

(Trimethylenimine)

CH_3^+	$(CH_3)_3N$		14.0±0.1	EI		303
CH_3^+	CH_3NHNH_2		14.1±0.3	EI	424,	3216
CH_3^+	$CH_3N=NCH_3$		11.5±0.1	EI		304

See also – EI: 2549

CH_3^+	$(CH_3)_2NNH_2$		14.5±0.3	EI	424,	3216
CH_3^+	$CH_3NHNHCH_3$		13.9±0.3	EI	424,	3216
CH_3^+	$(CH_3)_2NNHCH_3$		14.0±0.5	EI	424,	3216
CH_3^+	$(CH_3)_2NN(CH_3)_2$		14±1	EI	424,	3216
CH_3^+	CH_3N_3		14.1±0.1	EI		340
CH_3^+	CH_3OH	OH	13.5	EI		46

See also – EI: 2018, 3176

4.3. The Positive Ion Table—Continued

Ion	Reactant	Other products	Ionization or appearance potential (eV)	Method	Heat of formation (kJ mol^{-1})	Ref.
CH_3^+	CH_3CHO	CHO?	14.53	RPD		3347

(≤ 0.04 eV average translational energy of decomposition at threshold)

See also – EI: 127, 298, 2883

Ion	Reactant	Other products	Ionization or appearance potential (eV)	Method	Heat of formation (kJ mol^{-1})	Ref.
CH_3^+	$(CH_2)_2O$ (1,2–Epoxyethane)		14.3±0.2	EI		50
CH_3^+	C_2H_5OH		14.70±0.10	RPD		3347

(0.11±0.02 eV average translational energy of decomposition at threshold)

Ion	Reactant	Other products	Ionization or appearance potential (eV)	Method	Heat of formation (kJ mol^{-1})	Ref.
CH_3^+	$(CH_3)_2O$		14.93±0.13	RPD		3347

(0.12±0.02 eV average translational energy of decomposition at threshold)

See also – EI: 2018

Ion	Reactant	Other products	Ionization or appearance potential (eV)	Method	Heat of formation (kJ mol^{-1})	Ref.
CH_3^+	$(CH_3)_2CO$		15.36	RPD		3347

(0.17±0.01 eV average translational energy of decomposition at threshold)

Ion	Reactant	Other products	Ionization or appearance potential (eV)	Method	Heat of formation (kJ mol^{-1})	Ref.
CH_3^+	$(CH_3)_2CO$		14.93	RPD		2977

See also – EI: 298, 2174, 2883

Ion	Reactant	Other products	Ionization or appearance potential (eV)	Method	Heat of formation (kJ mol^{-1})	Ref.
CH_3^+	C_3H_6O (1,2–Epoxypropane)		13.9±0.2	EI		50
CH_3^+	$iso-C_3H_7OH$		30.2±0.2	SRP		1264

(High kinetic energy ion)

Ion	Reactant	Other products	Ionization or appearance potential (eV)	Method	Heat of formation (kJ mol^{-1})	Ref.
CH_3^+	$C_2H_5OCH_3$		15.02	RPD		3347

(0.11±0.01 eV average translational energy of decomposition at threshold)

Ion	Reactant	Other products	Ionization or appearance potential (eV)	Method	Heat of formation (kJ mol^{-1})	Ref.
CH_3^+	$C_2H_5COCH_3$		15.49	RPD		2977

See also – EI: 298, 2883

Ion	Reactant	Other products	Ionization or appearance potential (eV)	Method	Heat of formation (kJ mol^{-1})	Ref.
CH_3^+	$n-C_3H_7COCH_3$		15.13	RPD		2977
CH_3^+	CH_3COOH		14.0±0.15	RPD		3347

(0.09±0.01 eV average translational energy of decomposition at threshold)

See also – EI: 171

Ion	Reactant	Other products	Ionization or appearance potential (eV)	Method	Heat of formation (kJ mol^{-1})	Ref.
CH_3^+	$HCOOCH_3$		13.71	RPD		3347

(≤ 0.10 eV average translational energy of decomposition at threshold)

See also – EI: 3224

Ion	Reactant	Other products	Ionization or appearance potential (eV)	Method	Heat of formation (kJ mol^{-1})	Ref.
CH_3^+	$HCOOC_2H_5$		11.07±0.04	EI		305
CH_3^+	CH_3COOCH_3		13.07±0.10	EI		305

See also – EI: 3176, 3224

Ion	Reactant	Other products	Ionization or appearance potential (eV)	Method	Heat of formation (kJ mol^{-1})	Ref.
CH_3^+	$HCOOCH_2CH_2CH_3$		11.94±0.02	EI		305
CH_3^+	$CH_3COOC_2H_5$		13.94±0.08	EI		305
CH_3^+	$C_4H_8O_2$ (1,2–Epoxy–3–methoxypropane)		16.0±0.3	EI		153
CH_3^+	$(CH_3O)_2BH$		13.4±0.5	EI		364
CH_3^+	$(CH_3O)_3B$		13.6±0.5	EI		364

See also – EI: 115

4.3. The Positive Ion Table—Continued

Ion	Reactant	Other products	Ionization or appearance potential (eV)	Method	Heat of formation (kJ mol^{-1})	Ref.
CH_3^+	CH_3NCO	NCO	14.6	RPD		3345
(0.33 eV average translational energy of decomposition at threshold)						
CH_3^+	CH_3COCHN_2		13.2±0.06	EI		2174
CH_3^+	CH_3NO_2	NO_2	12.6	RPD		2018
CH_3^+	CH_3NO_2	NO_2	13.6	RPD		3345
(0.18 eV average translational energy of decomposition at threshold)						
CH_3^+	CH_3ONO_2		15.5	EI		2456
CH_3^+	$C_2H_5ONO_2$		13.75±0.50	EI		1013

See also – EI: 2456

Ion	Reactant	Other products	IP	Method		Ref.
CH_3^+	$iso-C_3H_7ONO_2$		14.9	EI		2456
CH_3^+	CH_3F	F^-	12.56	PI		2637
CH_3^+	CH_3F	F	16.25	PI		2637

This ion is formed with kinetic energy, see V. H. Dibeler and R. M. Reese, J. Res. NBS **54**, 127 (1955).

See also – EI: 1136, 2154

Ion	Reactant	Other products	IP	Method		Ref.
CH_3^+	CH_3CHF_2		18.6	EI		1288
CH_3^+	CH_3CF_3		15.0±0.1	EI		1075
CH_3^+	CH_3BF_2		14.8±0.1	EI		1076
CH_3^+	CH_3COCF_3		14.60	EI		298
CH_3^+	CH_3SiH_3		15.1±0.3	EI		2182
CH_3^+	$(CH_3)_3SiH$		14.8±0.5	EI		83
CH_3^+	CH_3PH_2		14.8±0.2	EI		2045
CH_3^+	$(CH_3)_3P$		21.7±0.5	EI		1036
CH_3^+	$(CH_3O)_2CH_3PO$		16.8±0.1	EI		3211
CH_3^+	$(CH_3O)_3PO$		18.6±0.3	EI		3211
CH_3^+	$(CH_3)_2S$		13.0	EI		307

See also – EI: 84, 3202

Ion	Reactant	Other products	IP	Method		Ref.
CH_3^+	CH_3SCD_3		13.1	EI		307
CH_3^+	C_3H_6S		18.1±0.4	EI		188
	(Propylene sulfide)					
CH_3^+	$C_2H_5SCH_3$		17.6±0.5	EI		176
CH_3^+	$CH_3SCH_2CH=CH_2$		17.7±0.5	EI		186
CH_3^+	$n-C_3H_7SCH_3$		16.6±0.5	EI		176
CH_3^+	$iso-C_3H_7SCH_3$		19.4±0.5	EI		186
CH_3^+	CH_3SSCH_3		~12.9	EI		3202

See also – EI: 176, 3286

Ion	Reactant	Other products	IP	Method		Ref.
CH_3^+	CH_3NCS		15.3±0.3	EI		315
CH_3^+	C_2H_5NCS		19.6±0.5	EI		315
CH_3^+	$(CH_3)_2SO$		16.3±0.1	EI		3294
CH_3^+	CH_3SCF_3		14.24	EI		3202
CH_3^+	CH_3SSCF_3		~14	EI		3202
CH_3^+	CH_3Cl	Cl^-	10.07	PI	1399,	2637

See also – EI: 2776

4.3. The Positive Ion Table—Continued

Ion	Reactant	Other products	Ionization or appearance potential (eV)	Method	Heat of formation (kJ mol^{-1})	Ref.
CH_3^+	CH_3Cl	Cl	13.87	PI		2637

The thermochemical threshold for this process is 13.35 eV.

See also – EI: 364, 1136, 2154

Ion	Reactant	Other products	Ionization or appearance potential (eV)	Method	Heat of formation	Ref.
CH_3^+	C_2H_5Cl		15.9±0.3	EI		356
CH_3^+	$iso-C_3H_7Cl$		29.7±0.2	SRP		1264
(High kinetic energy ion)						
CH_3^+	CH_3COCH_2Cl		13.9±0.20	EI		2174
CH_3^+	C_3H_5OCl		14.6±0.5	EI		153
(1–Chloro–2,3–epoxypropane)						
CH_3^+	CH_3SiCl_3		15.0±0.2	EI		2182
CH_3^+	$(CH_3)_2Zn$		15.1±0.5	EI		2556
CH_3^+	$(CH_3)_4Ge$		20.1±0.5	EI		83
CH_3^+	CH_3Br	Br^-	9.60±0.05	RPD		2776
CH_3^+	CH_3Br	Br	12.77	PI		2637

The thermochemical threshold for this process is 12.78 eV compared to an estimated 0 K photoionization threshold of 12.80 eV.

See also – EI: 1136, 2154, 2973

Ion	Reactant	Other products	Ionization or appearance potential (eV)	Method	Heat of formation	Ref.
CH_3^+	C_2H_5Br		16.9±0.3	EI		356
CH_3^+	C_3H_5OBr		15.6±0.5	EI		153
(1–Bromo–2,3–epoxypropane)						
CH_3^+	CH_3I	I	12.22	EM		3104

The thermochemical threshold for this process is 12.23 eV.

See also – EI: 1136, 2154

Ion	Reactant	Other products	Ionization or appearance potential (eV)	Method	Heat of formation	Ref.
CH_3^+	C_2H_5I		16.3±0.3	EI		356
CH_3^+	$(CH_3)_2Hg$	CH_3+Hg?	12.55	PI		2983

See also – EI: 306

Ion	Reactant	Other products	Ionization or appearance potential (eV)	Method	Heat of formation	Ref.
CH_3^+	CH_3HgCl		14.8±0.2	EI		306

CH_2D^+

Ion	Reactant	Other products	Ionization or appearance potential (eV)	Method	Heat of formation	Ref.
CH_2D^+	C_2H_5D	CH_3	14.71±0.30	EI		2421
CH_2D^+	CH_3CD_3	CHD_2	14.94±0.14	EI		2421
CH_2D^+	CH_2DCl	Cl^-	10.10	PI		2637

CHD_2^+

Ion	Reactant	Other products	Ionization or appearance potential (eV)	Method	Heat of formation	Ref.
CHD_2^+	CH_3CD_3	CH_2D	14.97±0.10	EI		2421
CHD_2^+	CHD_2Cl	Cl^-	10.09	PI		2637

4.3. The Positive Ion Table—Continued

Ion	Reactant	Other products	Ionization or appearance potential (eV)	Method	Heat of formation (kJ mol^{-1})	Ref.
			CD$_3^+$			
CD$_3^+$	CD$_3$		9.832±0.002	S		349
CD$_3^+$	CD$_4$	D	14.38±0.03	PI		1128
CD$_3^+$	CH$_3$CD$_3$	CH$_3$	15.10±0.10	EI		2421
CD$_3^+$	C$_2$D$_6$	CD$_3$	15.54±0.10	EI		2421
CD$_3^+$	CD$_3$COOH		15.56	EI		171
CD$_3^+$	CD$_3$Cl	Cl	13.8	PI		2637
			CH$_4^+$(^{2}B$_2$) $\Delta H_{f0}^\circ \leqslant 1150$ kJ mol^{-1} (275 kcal mol^{-1})			
			CH$_4^+$(^{2}A$_1$) $\Delta H_{f0}^\circ = 2094$ kJ mol^{-1} (500 kcal mol^{-1})			
CH$_4^+$(^{2}B$_2$)	CH$_4$		≤12.615±0.010	PI	≤1150	3415
CH$_4^+$(^{2}B$_2$)	CH$_4$		12.704±0.008	PI		1253
CH$_4^+$(^{2}B$_2$)	CH$_4$		12.71±0.02	PI		1128
CH$_4^+$(^{2}B$_2$)	CH$_4$		12.55±0.05	PI		2013
CH$_4^+$(^{2}B$_2$)	CH$_4$		12.75±0.05	RPI		2857, 2858, 3293
CH$_4^+$(^{2}B$_2$)	CH$_4$		12.70	PE		2803
CH$_4^+$(^{2}B$_2$)	CH$_4$		12.75	PE		3092
CH$_4^+$(^{2}B$_2$)	CH$_4$		12.78	PE		3116
CH$_4^+$(^{2}B$_2$)	CH$_4$		12.9	PEN		2430
CH$_4^+$(^{2}B$_2$)	CH$_4$		≤12.70	EM		2798
CH$_4^+$(^{2}B$_2$)	CH$_4$		13.00±0.02	RPD		224
CH$_4^+$(^{2}B$_2$)	CH$_4$		12.99±0.05	RPD		2776

The ion ground state has a large Jahn–Teller distortion, see for example R. N. Dixon, Mol. Phys. **20**, 113 (1971), F. A. Grimm and J. Godoy, Chem. Phys. Letters **6**, 336 (1970) and refs. 3092, 3116, 3119. Consequently the onset is not sharp and the adiabatic value may be lower. Several PE studies (refs. 2803, 3092, 3119) have resolved vibrational structure near onset with a separation of ~1200 cm^{-1} (0.15 eV). This is just the difference between the PI threshold value given by Brehm, ref. 2013, and those determined by Nicholson, ref. 1253, and Dibeler *et al.*, ref. 1128.

See also – S: 138
            PI: 182, 230, 331, 416, 2605, 3115, 3132
            PE: 1130, 2801, 2829, 2843, 3072, 3119, 3132
            PEN: 2467
            EI: 289, 1072, 1129, 2136, 2154, 2414, 2535, 2575, 3435

Ion	Reactant	Other products	Ionization or appearance potential (eV)	Method	Heat of formation (kJ mol^{-1})	Ref.
CH$_4^+$(^{2}A$_1$)	CH$_4$		22.39	PE	2094	3092
CH$_4^+$(^{2}A$_1$)	CH$_4$		22.4	PE		3119
CH$_4^+$(^{2}A$_1$)	CH$_4$		23.1 (V)	PE		3072
CH$_4^+$(^{2}A$_1$)	CH$_4$		24	RPD		2414
CH$_4^+$(^{2}A$_1$)	CH$_4$		23.5–24	D		2846

Earlier electron impact work (refs. 289, 1072) gave values around 19.4 eV. These are due to autoionization or collision processes, see refs. 2414, 2575, 2846.

Ion	Reactant	Other products	Ionization or appearance potential (eV)	Method	Heat of formation (kJ mol^{-1})	Ref.
CH$_4^+$	C$_3$H$_6$		14.7±0.5	EI		2542
CH$_4^+$	(CH$_2$)$_2$O (1,2–Epoxyethane)	CO	12.3±0.2	EI		50

4.3. The Positive Ion Table—Continued

Ion	Reactant	Other products	Ionization or appearance potential (eV)	Method	Heat of formation (kJ mol^{-1})	Ref.
			CD$_4^+$			
CD$_4^+$(^{2}B$_2$)	CD$_4$		12.882±0.008	PI		1253
CD$_4^+$(^{2}B$_2$)	CD$_4$		12.87±0.02	PI		1128
CD$_4^+$(^{2}B$_2$)	CD$_4$		12.83	PE		3092
CD$_4^+$(^{2}B$_2$)	CD$_4$		13.1	EI		2136
CD$_4^+$(^{2}A$_1$)	CD$_4$		22.48	PE		3092
			CH$_4^{+2}$			
CH$_4^{+2}$	CH$_4$		≤40.7±0.8 (V)	AUG		3424
			C$_2$H$^+$			
C$_2$H$^+$	C$_2$H$_2$	H	17.22	PI		1400
C$_2$H$^+$	C$_2$H$_2$	H	17.3	EI		2102, 2136

This process is probably accompanied by considerable kinetic energy of decomposition, see R. Botter, R. Hagemann, G. Khodadadi and H. M. Rosenstock in "Recent Developments in Mass Spectroscopy," Proc. Intern. Conf. Mass Spectry., Kyoto, 1079 (1970).

See also – EI: 13, 1129, 1451, 2450, 3131, 3156

Ion	Reactant	Other products	Ionization or appearance potential (eV)	Method	Heat of formation (kJ mol^{-1})	Ref.
C$_2$H$^+$	CH$_3$C≡CH	CH$_3$	17.2±0.5	EI		13
C$_2$H$^+$	C$_3$H$_6$		20.5±1	EI		2542
C$_2$H$^+$	C$_3$H$_8$	CH$_3^+$+2H$_2$	30.4±0.5	EI		1408
(Threshold value corrected for high fragment kinetic energy)						
C$_2$H$^+$	CH≡CC≡CH	C$_2$H	20.1±0.5	EI		13
C$_2$H$^+$	CH≡CCN	CN	19.0±0.2	EI		154
C$_2$H$^+$	(CH$_2$)$_2$O (1,2−Epoxyethane)		24.0±0.3	EI		50
C$_2$H$^+$	CH$_3$COC≡CH	CH$_3$+CO?	17.95	EI		298
C$_2$H$^+$	C$_2$H$_5$SSC$_2$H$_5$		11.35	EI		307
			C$_2$D$^+$			
C$_2$D$^+$	C$_2$D$_2$	D	17.34	PI		1400

See also – EI: 3131

Ion	Reactant	Other products	Ionization or appearance potential (eV)	Method	Heat of formation (kJ mol^{-1})	Ref.
C$_2$H$_2^+$($^2\Pi_u$)	$\Delta H_{f0}^\circ = 1328$ kJ mol^{-1} (317 kcal mol^{-1})					
C$_2$H$_2^+$($^2\Sigma_g$)	$\Delta H_{f0}^\circ = 1806$ kJ mol^{-1} (432 kcal mol^{-1})					
C$_2$H$_2^+$($^2\Sigma_u$)	$\Delta H_{f0}^\circ = 2001$ kJ mol^{-1} (478 kcal mol^{-1})					
C$_2$H$_2^+$($^2\Pi_u$)	C$_2$H$_2$		11.41	S	1328	3145
(Average of two Rydberg series limits)						
C$_2$H$_2^+$($^2\Pi_u$)	C$_2$H$_2$		11.396±0.003	PI		1253
C$_2$H$_2^+$($^2\Pi_u$)	C$_2$H$_2$		11.400±0.005	PI		2013
C$_2$H$_2^+$($^2\Pi_u$)	C$_2$H$_2$		11.406±0.006	PI	54, 1118	1019, 1400

4.3. The Positive Ion Table—Continued

Ion	Reactant	Other products	Ionization or appearance potential (eV)	Method	Heat of formation (kJ mol^{-1})	Ref.
$C_2H_2^+(^2\Pi_u)$	C_2H_2		11.41±0.01	PI		162, 182, 416, 1022
$C_2H_2^+(^2\Pi_u)$	C_2H_2		11.395±0.015	PI		2726
$C_2H_2^+(^2\Pi_u)$	C_2H_2		11.40±0.01	PE		2805
$C_2H_2^+(^2\Pi_u)$	C_2H_2		11.39±0.02	EM		3075
$C_2H_2^+(^2\Pi_u)$	C_2H_2		11.39±0.05	RPD		2776
$C_2H_2^+(^2\Pi_u)$	C_2H_2		11.40±0.02	RPD		224
$C_2H_2^+(^2\Pi_u)$	C_2H_2		11.41±0.01	EDD		2695
$C_2H_2^+(^2\Sigma_g)$	C_2H_2		16.36±0.01	PE	1806	2805
$C_2H_2^+(^2\Sigma_u)$	C_2H_2		18.38±0.01	PE	2001	2805
$C_2H_2^+(^2\Sigma_u)$	C_2H_2		18.5 (V)	PE		3096
$C_2H_2^+(^2\Sigma_g)$	C_2H_2		23.5 (V)	PE		3096

See also – PI: 156, 2759
           PE: 1108, 1130, 2759, 2804
           PEN: 2430, 2466, 2467
           EI: 13, 166, 305, 2450, 2535, 2752, 2759, 3131, 3156, 3177

Ion	Reactant	Other products	Ionization or appearance potential (eV)	Method	Heat of formation (kJ mol^{-1})	Ref.
$C_2H_2^+$	C_2H_4	H_2	13.13±0.02	PI	(1328)	2617

(Threshold value approximately corrected to 0 K)

Ion	Reactant	Other products	AP (eV)	Method		Ref.
$C_2H_2^+$	C_2H_4	H_2	13.12±0.03	PI		2607
$C_2H_2^+$	C_2H_4	H_2	12.96±0.02	PI		2013

The three photoionization studies differ in their interpretation of the threshold. That of ref. 2617
is most convincing. The agreement of the $C_2H_2^+$ heat of formation with that derived from IP(C_2H_2) implies
that the metastable transition $C_2H_4^+ \rightarrow C_2H_2^+ + H_2$ should have no excess energy.

See also – EI: 166, 419

Ion	Reactant	Other products	AP (eV)	Method		Ref.
$C_2H_2^+$	C_2H_6	$2H_2$	15.35±0.50	EI		2421
$C_2H_2^+$	C_2H_5D	H_2+HD	15.49±0.25	EI		2421
$C_2H_2^+$	C_3H_6	CH_4	14.1	RPD		3345

(0.13 eV average translational energy of decomposition at threshold)

Ion	Reactant	Other products	AP (eV)	Method		Ref.
$C_2H_2^+$	C_3H_6	CH_4	13.6±0.5	EI		2542
$C_2H_2^+$	C_3H_6	CH_4	13.1	RPD		3345

               (Cyclopropane)
(0.07 eV average translational energy of decomposition at threshold)

See also – EI: 3191

Ion	Reactant	Other products	AP (eV)	Method		Ref.
$C_2H_2^+$	C_3H_8	CH_4+H_2	14.1±0.15	EI		1408
$C_2H_2^+$	C_3H_8	$CH_3^++H_2+H$	28.5±1	EI		1408

(Threshold value corrected for high fragment kinetic energy)

Ion	Reactant	Other products	AP (eV)	Method		Ref.
$C_2H_2^+$	$CH_2=CHCH=CH_2$		16.46±0.1	EI		2455
$C_2H_2^+$	C_4H_8		16.8	EI		2742

               (Cyclobutane)

Ion	Reactant	Other products	AP (eV)	Method		Ref.
$C_2H_2^+$	C_6H_6		19±0.4	RPD		2520

               (Benzene)

Ion	Reactant	Other products	AP (eV)	Method		Ref.
$C_2H_2^+$	C_6H_6		18.6±0.3	EI		1238

               (Benzene)

4.3. The Positive Ion Table—Continued

Ion	Reactant	Other products	Ionization or appearance potential (eV)	Method	Heat of formation (kJ mol^{-1})	Ref.
$C_2H_2^+$	C_6H_6 (Benzene)		32.6±0.2	SRP		1264
(High kinetic energy ion)						
$C_2H_2^+$	$(CH_2)_2NH$ (Ethylenimine)		16.4±0.4	EI		51
$C_2H_2^+$	C_2H_3CN	HCN	13.13±0.10	EI		1406
$C_2H_2^+$	C_2H_5CN		14.70	EI		2966
$C_2H_2^+$	$(CH_2)_3NH$ (Trimethylenimine)		16.6±0.5	EI		52
$C_2H_2^+$	$n-C_3H_7CN$		15.50	EI		2966
$C_2H_2^+$	$(CH_2)_4NH$ (Pyrrolidine)		17.3±1.0	EI		52
$C_2H_2^+$	$C_4H_4N_2$ (1,2–Diazine)		14.94±0.10	EI		1406
$C_2H_2^+$	$C_4H_4N_2$ (1,3–Diazine)		15.79±0.05	EI		1406
$C_2H_2^+$	$C_4H_4N_2$ (1,4–Diazine)		15.23±0.10	EI		1406
$C_2H_2^+$	$(CH_2)_2O$ (1,2–Epoxyethane)	H_2O?	15.7±0.3	EI		50
$C_2H_2^+$	C_3H_6O (1,2–Epoxypropane)		13.9±0.2	EI		50
$C_2H_2^+$	$(CH_2)_3O$ (1,3–Epoxypropane)		15.2±0.2	EI		52
$C_2H_2^+$	C_4H_6O (3,4–Epoxy–1–butene)		13.8±0.3	EI		153
$C_2H_2^+$	$(CH_2)_4O$ (1,4–Epoxybutane)		17.3±0.3	EI		52
$C_2H_2^+$	$HCOOC_2H_5$		14.9	EI		3224
$C_2H_2^+$	$C_4H_8O_2$ (1,2–Epoxy–3–methoxypropane)		16.2±0.3	EI		153
$C_2H_2^+$	C_2H_3F	HF	13.73±0.1	EI		419
$C_2H_2^+$	$CH_2=CF_2$		19.78±0.1	EI		419
$C_2H_2^+$	$C_2H_3CF_3$		13.3±0.15	EI		1075
$C_2H_2^+$	$C_2H_3BF_2$		13.75±0.1	EI		1076
$C_2H_2^+$	$(C_2H_5)_2PP(C_2H_5)_2$		12.0±0.3	EI		2948
$C_2H_2^+$	$(CH_2)_2S$ (Ethylene sulfide)		17.9±0.5	EI		51
$C_2H_2^+$	C_2H_5SH		14.7±0.3	EI		3286
$C_2H_2^+$	C_3H_6S (Propylene sulfide)		17.7±0.4	EI		188
$C_2H_2^+$	$(CH_2)_3S$ (Trimethylene sulfide)		17.1±0.4	EI		52
$C_2H_2^+$	$C_2H_5SCH_3$		17.8±0.5	EI		176
$C_2H_2^+$	$(C_2H_5)_2S$		21.7±0.3	EI		3286
$C_2H_2^+$	$C_2H_5SSC_2H_5$		19.5±0.5	EI		186
$C_2H_2^+$	C_2H_5NCS		18.1±0.2	EI		315
$C_2H_2^+$	$(CH_3)_2SO$		12±0.3	EI		3294
$C_2H_2^+$	C_2H_3Cl	HCl	13.8±0.3	EI		2793
$C_2H_2^+$	C_3H_5OCl (1–Chloro–2,3–epoxypropane)		16.6±0.1	EI		153
$C_2H_2^+$	C_3H_5OBr (1–Bromo–2,3–epoxypropane)		16.7±0.6	EI		153

4.3. The Positive Ion Table—Continued

Ion	Reactant	Other products	Ionization or appearance potential (eV)	Method	Heat of formation (kJ mol^{-1})	Ref.
			C$_2$HD$^+$			
C$_2$HD$^+$	CH$_3$CD$_3$		15.50±0.10	EI		2421
			C$_2$D$_2^+$			
C$_2$D$_2^+$($^2\Pi_u$)	C$_2$D$_2$		11.416±0.006	PI	54, 1118,	1019, 1400
C$_2$D$_2^+$($^2\Pi_u$)	C$_2$D$_2$		11.40±0.01	PE		2805
C$_2$D$_2^+$($^2\Sigma_g$)	C$_2$D$_2$		16.53±0.01	PE		2805
C$_2$D$_2^+$($^2\Sigma_u$)	C$_2$D$_2$		18.44±0.01	PE		2805

See also – PE: 2804

| C$_2$D$_2^+$ | C$_2$D$_6$ | 2D$_2$ | 15.70±0.20 | EI | | 2421 |

C$_2$H$_3^+$ $\Delta H_{f0}^\circ \sim 1125$ kJ mol^{-1} (269 kcal mol^{-1})

| C$_2$H$_3^+$ | C$_2$H$_3$ | | 8.95 | EM | | 3350 |

This ionization potential combined with the ion heat of formation implies an ethylene C–H bond energy of ~417 kJ mol^{-1} (100 kcal mol^{-1}) compared to the kinetic value of ≥452±8 kJ mol^{-1} (108±2 kcal mol^{-1}) determined by D. M. Golden and S. W. Benson, Chem. Rev. **69**, 125 (1969).

See also – EI: 70, 87, 1129, 2535

C$_2$H$_3^+$	C$_2$H$_4$	H	13.25±0.05	PI	1123	2617
(Threshold value approximately corrected to 0 K)						
C$_2$H$_3^+$	C$_2$H$_4$	H	13.37±0.03	PI		2013

See also – PI: 2607
          EI: 70, 166, 419

C$_2$H$_3^+$	C$_2$H$_6$	H$_2$+H	15.22±0.10	EI		2421
C$_2$H$_3^+$	C$_2$H$_5$D		15.27±0.15	EI		2421
C$_2$H$_3^+$	C$_3$H$_6$	CH$_3$	13.7±0.5	EI		2542

See also – EI: 194

C$_2$H$_3^+$	C$_3$H$_6$ (Cyclopropane)	CH$_3$	13.4	RPD		3345
(0.12 eV average translational energy of decomposition at threshold)						
C$_2$H$_3^+$	C$_3$H$_6$ (Cyclopropane)	CH$_3$	13.3±0.2	EI		3191
C$_2$H$_3^+$	C$_3$H$_8$		14.5±0.15	EI		1408
C$_2$H$_3^+$	C$_3$H$_8$	CH$_3^+$+H$_2$	25.0±0.5	EI		1408
(Threshold value corrected for high fragment kinetic energy)						

See also – EI: 1264

4.3. The Positive Ion Table—Continued

Ion	Reactant	Other products	Ionization or appearance potential (eV)	Method	Heat of formation (kJ mol⁻¹)	Ref.
$C_2H_3^+$	$CH_2=CHCH=CH_2$		15.68±0.2	EI		2455

See also – EI: 462

Ion	Reactant	Other products	IP	Method	Heat	Ref.
$C_2H_3^+$	$CH_3C{\equiv}CCH_3$		14.7±0.2	EI		13
$C_2H_3^+$	$1-C_4H_8$		13.6	EI		194
$C_2H_3^+$	C_4H_8 (Cyclobutane)		17.7	EI		2742
$C_2H_3^+$	$neo-C_5H_{12}$		17.95	EI		2101
$C_2H_3^+$	C_6H_6 (Benzene)		19±0.4	RPD		2520
$C_2H_3^+$	C_6H_6 (Benzene)		31.1±0.2	SRP		1264

(High kinetic energy ion)

Ion	Reactant	Other products	IP	Method	Heat	Ref.
$C_2H_3^+$	$(CH_2)_2NH$ (Ethylenimine)		16.9±0.3	EI		51
$C_2H_3^+$	$C_2H_5NH_2$		16.14	EI		2470
$C_2H_3^+$	C_2H_5CN		15.40	EI		2966
$C_2H_3^+$	$(CH_2)_3NH$ (Trimethylenimine)		16.6±0.5	EI		52
$C_2H_3^+$	$n-C_3H_7CN$		15.10	EI		2966
$C_2H_3^+$	$(CH_2)_4NH$ (Pyrrolidine)		16.7±0.3	EI		52
$C_2H_3^+$	$(C_2H_5)_2NH$		15.35	EI		2428
$C_2H_3^+$	$(CH_2)_2O$ (1,2-Epoxyethane)	OH	14.3±0.2	EI		50
$C_2H_3^+$	C_2H_5OH		14.7	EI		46

See also – EI: 3176

Ion	Reactant	Other products	IP	Method	Heat	Ref.
$C_2H_3^+$	$CH_2=CHCHO$	CHO?	13.64	RPD		3347

(0.09±0.02 eV average translational energy of decomposition at threshold)

See also – EI: 130

Ion	Reactant	Other products	IP	Method	Heat	Ref.
$C_2H_3^+$	$(CH_3)_2CO$		16.9	RPD		2883
$C_2H_3^+$	C_3H_6O (1,2-Epoxypropane)		14.3±0.1	EI		50
$C_2H_3^+$	$(CH_2)_3O$ (1,3-Epoxypropane)		14.9±0.3	EI		52
$C_2H_3^+$	$n-C_3H_7OH$		14.7	EI		46
$C_2H_3^+$	$iso-C_3H_7OH$		14.6	EI		46
$C_2H_3^+$	C_4H_6O (3,4-Epoxy-1-butene)	$CH_3CO?$	12.6±0.3	EI		153
$C_2H_3^+$	$(CH_2)_4O$ (1,4-Epoxybutane)		16.1±0.3	EI		52
$C_2H_3^+$	$HCOOC_2H_5$		15.0	EI		3224
$C_2H_3^+$	$CH_3COOC_2H_5$		15.32±0.20	EI		3176
$C_2H_3^+$	$C_4H_8O_2$ (1,2-Epoxy-3-methoxypropane)		16.3±0.2	EI		153
$C_2H_3^+$	C_2H_5NO		12.8±0.2	EDD		3180
$C_2H_3^+$	C_2H_3F	F	14.38±0.1	EI		419
$C_2H_3^+$	$C_2H_3CF_3$		14.20±0.05	EI		1075

4.3. The Positive Ion Table—Continued

Ion	Reactant	Other products	Ionization or appearance potential (eV)	Method	Heat of formation (kJ mol^{-1})	Ref.
$C_2H_3^+$	$C_2H_5CF_3$		15.3±0.1	EI		1075
$C_2H_3^+$	$C_2H_3BF_2$		14.25±0.05	EI		1076
$C_2H_3^+$	$iso-C_3H_7BF_2$		14.8±0.1	EI		1076
$C_2H_3^+$	$(CH_3)_3SiH$		15.3±0.5	EI		83
$C_2H_3^+$	$C_2H_5PH_2$		15.4±0.3	EI		2948
$C_2H_3^+$	$(C_2H_5)_2PH$		20.4±0.3	EI		2948
$C_2H_3^+$	$(C_2H_5)_3P$		22.1+0.3	EI		2948
$C_2H_3^+$	$(CH_3)_2PP(CH_3)_2$		20.7±0.3	EI		2948
$C_2H_3^+$	$(C_2H_5)_2PP(C_2H_5)_2$		13.2±0.3	EI		2948
$C_2H_3^+$	$(CH_2)_2S$ (Ethylene sulfide)	SH	14.0	RPD		3345

(0.10 eV average translational energy of decomposition at threshold)

See also – EI:　51

$C_2H_3^+$	C_2H_5SH		15.8±0.3	EI		3286
$C_2H_3^+$	$(CH_3)_2S$		14.7	EI		307

See also – EI:　84, 3202

$C_2H_3^+$	CH_3SCD_3		16.7	EI		307
$C_2H_3^+$	C_3H_6S (Propylene sulfide)		17.2±0.3	EI		188
$C_2H_3^+$	$(CH_2)_3S$ (Trimethylene sulfide)		16.7±0.2	EI		52
$C_2H_3^+$	$C_2H_5SCH_3$		16.0±0.4	EI		176
$C_2H_3^+$	$CH_3SCH_2CH=CH_2$		16.5±0.4	EI		186
$C_2H_3^+$	$(CH_2)_4S$ (Tetramethylene sulfide)		18.0±0.4	EI		52
$C_2H_3^+$	$n-C_3H_7SCH_3$		15.8±0.4	EI		176
$C_2H_3^+$	$iso-C_3H_7SCH_3$		16.5±0.5	EI		186
$C_2H_3^+$	$(C_2H_5)_2S$		16.7±0.5	EI		84

See also – EI:　3286

$C_2H_3^+$	CH_3SSCH_3		14.6±0.3	EI		3286
$C_2H_3^+$	$C_2H_5SSC_2H_5$		17.2±0.4	EI		186
$C_2H_3^+$	C_2H_5NCS		15.6±0.3	EI		315
$C_2H_3^+$	$(CH_3)_2SO$		15.9±0.1	EI		3294
$C_2H_3^+$	C_2H_3Cl	Cl	12.5	EM	~1129	3350

See also – EI:　2793

$C_2H_3^+$	C_3H_5OCl (1–Chloro–2,3–epoxypropane)		14.0±0.4	EI		153
$C_2H_3^+$	$(C_2H_5)_2Se$		19.0±0.3	EI		3285
$C_2H_3^+$	C_2H_3Br	Br	11.9	EM	~1123	3350
$C_2H_3^+$	C_3H_5OBr (1–Bromo–2,3–epoxypropane)		14.4±0.2	EI		153

4.3. The Positive Ion Table—Continued

Ion	Reactant	Other products	Ionization or appearance potential (eV)	Method	Heat of formation (kJ mol^{-1})	Ref.	
			C$_2$H$_2$D$^+$				
C$_2$H$_2$D$^+$	CHD=CHD	D	14.1	PI		2607	
C$_2$H$_2$D$^+$	CH$_3$CD$_3$		15.45±0.1	EI		2421	
C$_2$H$_2$D$^+$	CH$_3$SCD$_3$		15.05	EI		307	
			C$_2$HD$_2^+$				
C$_2$HD$_2^+$	CHD=CHD	H	14.1	PI		2607	
C$_2$HD$_2^+$	CH$_3$SCD$_3$		13.5	EI		307	
			C$_2$D$_3^+$				
C$_2$D$_3^+$	C$_2$D$_6$	D$_2$+D	15.60±0.1	EI		2421	
C$_2$D$_3^+$	CH$_3$SCD$_3$		12.1	EI		307	

Ion	Reactant	Other products	Ionization or appearance potential (eV)	Method	Heat of formation (kJ mol^{-1})	Ref.	
C$_2$H$_4^+$(^{2}B$_{3u}$)		$\Delta H_{f0}^\circ = 1075$ kJ mol^{-1} (257 kcal mol^{-1})					
C$_2$H$_4^+$(^{2}B$_{3g}$)		$\Delta H_{f0}^\circ \sim 1258$ kJ mol^{-1} (301 kcal mol^{-1})					
C$_2$H$_4^+$(^{2}A$_g$)		$\Delta H_{f0}^\circ \sim 1453$ kJ mol^{-1} (347 kcal mol^{-1})					
C$_2$H$_4^+$(^{2}B$_{2u}$)		$\Delta H_{f0}^\circ \sim 1579$ kJ mol^{-1} (377 kcal mol^{-1})					
C$_2$H$_4^+$(^{2}B$_{3u}$)	C$_2$H$_4$		10.51±0.03	S	1075		3353
C$_2$H$_4^+$(^{2}B$_{3u}$)	C$_2$H$_4$		10.50±0.01	PI	1074		2607
C$_2$H$_4^+$(^{2}B$_{3u}$)	C$_2$H$_4$		10.50±0.02	PI	1074		268
C$_2$H$_4^+$(^{2}B$_{3u}$)	C$_2$H$_4$		10.507±0.004	PI	1075		1253
C$_2$H$_4^+$(^{2}B$_{3u}$)	C$_2$H$_4$		10.511±0.005	PI	1075		2013
C$_2$H$_4^+$(^{2}B$_{3u}$)	C$_2$H$_4$		10.515±0.01	PI	1075	158,	182, 416
C$_2$H$_4^+$(^{2}B$_{3u}$)	C$_2$H$_4$		10.51	PE	1075		2803
C$_2$H$_4^+$(^{2}B$_{3u}$)	C$_2$H$_4$		10.51±0.05	PE	1075		2796
C$_2$H$_4^+$(^{2}B$_{3u}$)	C$_2$H$_4$		10.51±0.02	PE	1075	3028,	3061
C$_2$H$_4^+$(^{2}B$_{3u}$)	C$_2$H$_4$		10.50±0.05	RPD			2776
C$_2$H$_4^+$(^{2}B$_{3g}$)	C$_2$H$_4$		12.38	PE	1255		2803
C$_2$H$_4^+$(^{2}B$_{3g}$)	C$_2$H$_4$		12.40±0.05	PE	1257		2796
C$_2$H$_4^+$(^{2}B$_{3g}$)	C$_2$H$_4$		12.46±0.02	PE	1263	3028,	3061
C$_2$H$_4^+$(^{2}A$_g$)	C$_2$H$_4$		14.47	PE	1457		2803
C$_2$H$_4^+$(^{2}A$_g$)	C$_2$H$_4$		14.35?	PE	1445		2796
C$_2$H$_4^+$(^{2}A$_g$)	C$_2$H$_4$		14.46±0.02	PE	1456	3028,	3061
C$_2$H$_4^+$(^{2}B$_{2u}$)	C$_2$H$_4$		15.68	PE	1574		2803
C$_2$H$_4^+$(^{2}B$_{2u}$)	C$_2$H$_4$		15.76±0.05	PE	1581		2796
C$_2$H$_4^+$(^{2}B$_{2u}$)	C$_2$H$_4$		15.78±0.02	PE	1583	3028,	3061
C$_2$H$_4^+$(^{2}B$_{1u}$)	C$_2$H$_4$		18.87	PE			2803
C$_2$H$_4^+$(^{2}B$_{1u}$)	C$_2$H$_4$		18.46±0.05	PE			2796
C$_2$H$_4^+$(^{2}B$_{1u}$)	C$_2$H$_4$		18.87±0.02	PE		3028,	3061
C$_2$H$_4^+$(^{2}B$_{1u}$)	C$_2$H$_4$		19.5±1 (V)	PE			3096

4.3. The Positive Ion Table—Continued

Ion	Reactant	Other products	Ionization or appearance potential (eV)	Method	Heat of forma- tion (kJ mol⁻¹)	Ref.
$C_2H_4^+(^2A_g)$	C_2H_4		24.5±1 (V)	PE		3096

We use the assignments given in ref. 3028.

See also – S: 2059, 3145
    PI: 156, 297, 2059, 2617
    PE: 1130, 2843, 3132, 3382
    PEN: 2430, 2466, 2873
    EI: 166, 268, 419, 1129, 2187, 2535, 3435

Ion	Reactant	Other products	Ionization or appearance potential (eV)	Method	Heat of formation (kJ mol⁻¹)	Ref.
$C_2H_4^+$	C_2H_6	H_2	12.08±0.03	PI		2606

(Threshold value approximately corrected to 0 K)

$C_2H_4^+$	C_2H_6	H_2	12.24±0.10	EI		2421

The thermochemical threshold for this process is 11.85 eV. The corresponding metastable transition has a mean kinetic energy of 0.15–0.20 eV, see G. Khodadadi, R. Botter and H. M. Rosenstock, Intern. J. Mass Spectrom. Ion Phys. **3**, 397 (1969).

See also – PEN: 2873

$C_2H_4^+$	C_2H_5D	HD	12.52±0.10	EI		2421
$C_2H_4^+$	C_3H_6		12.4±0.5	EI		2542

See also – EI: 194

$C_2H_4^+$	C_3H_8	CH_4	11.72±0.02	PI		2606

(Threshold value approximately corrected to 0 K)

$C_2H_4^+$	C_3H_8	CH_4	11.70	RPD		2521
$C_2H_4^+$	C_3H_8	CH_4	11.5±0.1	EI		2521

(Appearance potential of the corresponding metastable transition)

The thermochemical threshold for this process is 11.35 eV. Very little of the excess energy appears as fragment kinetic energy, see J. Bracher, H. Ehrhardt, R. Fuchs, O. Osberghaus and R. Taubert, Advan. Mass Spectrom. **2**, 285 (1963) and G. Khodadadi, R. Botter and H. M. Rosenstock, Intern. J. Mass Spectrom. Ion Phys. **3**, 397 (1969).

See also – PEN: 2873
    EI: 1408

$C_2H_4^+$	C_3H_8	$CH_2^++H_2$?	27.2±0.5	EI		1408

(Threshold value corrected for high fragment kinetic energy)

$C_2H_4^+$	$CH_2=CHCH=CH_2$	C_2H_2	12.45±0.1	PI	(1085*)	2013

This ionic heat of formation may be compared to $\Delta H_{f298}^\circ(C_2H_4^+) = 1066$ kJ mol⁻¹ obtained from the spectroscopic ionization potential tabulated above.

See also – EI: 2455

$C_2H_4^+$	1–C_4H_8		11.7±0.2	SD		2941

See also – EI: 194

*ΔH_{f298}°

4.3. The Positive Ion Table—Continued

Ion	Reactant	Other products	Ionization or appearance potential (eV)	Method	Heat of formation (kJ mol^{-1})	Ref.
$C_2H_4^+$	cis-2-C_4H_8		11.7±0.25	SD		2941

See also – EI: 194

Ion	Reactant	Other products	Ionization or appearance potential (eV)	Method	Heat of formation (kJ mol^{-1})	Ref.
$C_2H_4^+$	trans-2-C_4H_8		11.8±0.25	SD		2941
$C_2H_4^+$	iso-C_4H_8		12.0±0.25	SD		2941
$C_2H_4^+$	$C_3H_5CH_3$ (Methylcyclopropane)		12.5±0.2	SD		2941
$C_2H_4^+$	C_4H_8 (Cyclobutane)		11.0±0.15	SD		2941

See also – EI: 2742

Ion	Reactant	Other products	Ionization or appearance potential (eV)	Method	Heat of formation (kJ mol^{-1})	Ref.
$C_2H_4^+$	n-C_4H_{10}	C_2H_6	~11.65	PI		2606

(Threshold value approximately corrected to 0 K)

The thermochemical threshold for this process is 11.48 eV.

Ion	Reactant	Other products	Ionization or appearance potential (eV)	Method	Heat of formation (kJ mol^{-1})	Ref.
$C_2H_4^+$	$(CH_2)_2NH$ (Ethylenimine)	NH	13.3±0.2	EI		51
$C_2H_4^+$	C_2H_5CN	HCN	12.40±0.05	EI		2704

See also – EI: 2966

Ion	Reactant	Other products	Ionization or appearance potential (eV)	Method	Heat of formation (kJ mol^{-1})	Ref.
$C_2H_4^+$	C_2H_5NC		12.83	EDD		3214
$C_2H_4^+$	$CH_3N=NCH_3$		12.9±0.3	EI		2549
$C_2H_4^+$	C_2H_5OH	H_2O	12.0	PI		2647

(Threshold value approximately corrected to 0 K)

The thermochemical threshold for this process is 10.91 eV.

Ion	Reactant	Other products	Ionization or appearance potential (eV)	Method	Heat of formation (kJ mol^{-1})	Ref.
$C_2H_4^+$	C_3H_6O (1,2-Epoxypropane)		11.6±0.2	EI		50
$C_2H_4^+$	$(CH_2)_3O$ (1,3-Epoxypropane)		12.4±0.3	EI		52
$C_2H_4^+$	n-C_3H_7OH		~11.9	PI		2647
$C_2H_4^+$	$(CH_2)_5O$ (1,5-Epoxypentane)		~13.8	EI		2694
$C_2H_4^+$	$HCOOC_2H_5$		11.2	EI		3224
$C_2H_4^+$	$C_4H_8O_2$ (1,3-Dioxane)		13.16	EI		2422
$C_2H_4^+$	$CH_3CH=NOH$		12.9±0.2	EDD		3180
$C_2H_4^+$	$C_2H_5CF_3$		13.0±0.2	EI		1075
$C_2H_4^+$	$C_2H_5BF_2$		12.08±0.01	EI		1076
$C_2H_4^+$	$(C_2H_5)_2PP(C_2H_5)_2$		10.0±0.3	EI		2948
$C_2H_4^+$	C_2H_5SH		13.0±0.3	EI		3286
$C_2H_4^+$	$(CH_2)_3S$ (Trimethylene sulfide)		13.6±0.2	EI		52
$C_2H_4^+$	$(C_2H_5)_2S$		14.5±0.3	EI		3286
$C_2H_4^+$	CH_3SSCH_3		15.6±0.3	EI		3286

4.3. The Positive Ion Table—Continued

Ion	Reactant	Other products	Ionization or appearance potential (eV)	Method	Heat of formation (kJ mol⁻¹)	Ref.
$C_2H_4^+$	$C_2H_5SSC_2H_5$		13.2±0.3	EI		3286
$C_2H_4^+$	$(CH_3)_2SO$		13.7±0.3	EI		3294
$C_2H_4^+$	C_2H_5Cl	HCl	11.33	EI		3201
$C_2H_4^+$	C_3H_5OCl (1–Chloro–2,3–epoxypropane)		13.6±0.4	EI		153
$C_2H_4^+$	$C_2H_5SiCl_3$		12.48±0.05	EI		2182
$C_2H_4^+$	$(C_2H_5)_2Se$		15.1±0.3	EI		3285

$C_2H_3D^+$

Ion	Reactant	Other products	IP/AP (eV)	Method	ΔH	Ref.
$C_2H_3D^+$	C_2H_5D	H_2	12.36±0.05	EI		2421
$C_2H_3D^+$	CH_3CD_3	D_2	12.75±0.10	EI		2421
$C_2H_3D^+$	$CH_3CD_2CH_3$	CH_3D	11.99	RPD		2907

$C_2H_2D_2^+$

Ion	Reactant	Other products	IP/AP (eV)	Method	ΔH	Ref.
$C_2H_2D_2^+$	CHD=CHD		10.50±0.01	PI		2607
$C_2H_2D_2^+$	CH_3CD_3	HD	12.35±0.06	EI		2421
$C_2H_2D_2^+$	$CH_3CD_2CH_3$	CH_4	11.78	RPD		2907

$C_2HD_3^+$

Ion	Reactant	Other products	IP/AP (eV)	Method	ΔH	Ref.
$C_2HD_3^+$	CH_3CD_3	H_2	12.81±0.01	EI		2421
$C_2HD_3^+$	$CD_3CH_2CD_3$	CHD_3	12.13	RPD		2907

$C_2D_4^+$

Ion	Reactant	Other products	IP/AP (eV)	Method	ΔH	Ref.
$C_2D_4^+(^2B_{3u})$	C_2D_4		10.52±0.03	S		3353
$C_2D_4^+(^2B_{3u})$	C_2D_4		10.52±0.02	PE		3028, 3061
$C_2D_4^+(^2B_{3g})$	C_2D_4		12.48±0.02	PE		3028, 3061
$C_2D_4^+(^2A_g)$	C_2D_4		14.45±0.02	PE		3028, 3061
$C_2D_4^+(^2B_{2u})$	C_2D_4		15.83±0.02	PE		3028, 3061
$C_2D_4^+(^2B_{1u})$	C_2D_4		18.90±0.02	PE		3028, 3061
$C_2D_4^+$	C_2D_6	D_2	12.58±0.08	EI		2421
$C_2D_4^+$	C_3D_8	CD_4	11.87	RPD		2521
$C_2D_4^+$	C_3D_8	CD_4	11.65±0.1	EI		2521

(Appearance potential of the corresponding metastable transition)

$C_2H_5^+$ $\Delta H^\circ_{f298} \sim 917$ kJ mol⁻¹ (219 kcal mol⁻¹)

Ion	Reactant	Other products	IP/AP (eV)	Method	ΔH	Ref.
$C_2H_5^+$	C_2H_5		≤8.4	PI	≤918	1068
$C_2H_5^+$	C_2H_5		8.38±0.05	EM	916	3104, 3379
$C_2H_5^+$	C_2H_5		8.34±0.05	RPD		2776

See also – EI: 59, 87, 1129, 2158, 2535, 2719, 2904, 2986

Ion	Reactant	Other products	IP/AP (eV)	Method	ΔH	Ref.
$C_2H_5^+$	C_2H_6	H⁻	12.00±0.05	PI		2606

(Threshold value approximately corrected to 0 K)

Ion	Reactant	Other products	Ionization or appearance potential (eV)	Method	Heat of formation (kJ mol⁻¹)	Ref.
$C_2H_5^+$	C_2H_6	H	12.65±0.08	PI	(918)	2606

(Threshold value approximately corrected to 0 K)

| $C_2H_5^+$ | C_2H_6 | H | 12.66±0.05 | RPD | | 2776 |

See also – EI: 160, 195, 1451, 2421

| $C_2H_5^+$ | C_3H_6 | | 12.6±0.5 | EI | | 2542 |
| $C_2H_5^+$ | C_3H_8 | CH_3 | 11.90±0.08 | PI | (902) | 2606 |

(Threshold value approximately corrected to 0 K)

| $C_2H_5^+$ | C_3H_8 | CH_3 | 12.02±0.05 | RPD | | 2776 |

See also – EI: 195, 1408, 1451, 2521

| $C_2H_5^+$ | C_3H_8 | CH_3^+ | 21±2 | EI | | 1408 |

(Threshold value corrected for high fragment kinetic energy)

| $C_2H_5^+$ | C_3H_8 | CH^++H_2 | 26.9±0.5 | EI | | 1408 |

(Threshold value corrected for high fragment kinetic energy)

$C_2H_5^+$	$C_2H_5C≡CH$	C_2H	12.9±0.1	EI		13
$C_2H_5^+$	$cis-2-C_4H_8$	C_2H_3	12.25	EI		194, 195
$C_2H_5^+$	C_4H_8		13.8	EI		2742
	(Cyclobutane)					
$C_2H_5^+$	$n-C_4H_{10}$	C_2H_5	12.55	EI		195

See also – PI: 2606

$C_2H_5^+$	$iso-C_4H_{10}$	C_2H_5	13.80	EI		195
$C_2H_5^+$	$n-C_5H_{12}$		28.1±0.2	SRP		1264
	(High kinetic energy ion)					
$C_2H_5^+$	$neo-C_5H_{12}$		13.81	EI		2101
$C_2H_5^+$	$n-C_7H_{16}$		12.89	RPD		2977
$C_2H_5^+$	$n-C_7H_{16}$		24.3±0.2	SRP		1264
	(High kinetic energy ion)					
$C_2H_5^+$	$n-C_8H_{18}$		13.44	RPD		2977
$C_2H_5^+$	$n-C_9H_{20}$		13.20	RPD		2977
$C_2H_5^+$	$C_2H_5NH_2$		13.5±0.2	EI		2470
$C_2H_5^+$	C_2H_5NC	CN	12.94	EDD		3214
$C_2H_5^+$	$n-C_3H_7CN$		12.97	EI		2966
$C_2H_5^+$	$(C_2H_5)_2NH$		14.85	EI		2428
$C_2H_5^+$	$C_2H_5N=NC_2H_5$		10.45±0.2	EI		304
$C_2H_5^+$	C_2H_5OH	OH	12.7	PI		2647

(Threshold value approximately corrected to 0 K)

The thermochemical threshold for this process is about 12.3 eV.

See also – EI: 2018, 3176

$C_2H_5^+$	$(CH_2)_3O$		12.6±0.2	EI		52
	(1,3–Epoxypropane)					
$C_2H_5^+$	$n-C_3H_7OH$	CH_2OH	12.3	PI		2647

(Threshold value approximately corrected to 0 K)

The thermochemical threshold for this process is about 11.8 eV, see ref. 2647.

4.3. The Positive Ion Table—Continued

Ion	Reactant	Other products	Ionization or appearance potential (eV)	Method	Heat of formation (kJ mol⁻¹)	Ref.
$C_2H_5^+$	$C_2H_5COCH_3$		12.88	RPD		2977
See also – EI:	298, 2883					
$C_2H_5^+$	$(CH_2)_4O$ (1,4–Epoxybutane)		15.8±0.2	EI		52
$C_2H_5^+$	$(C_2H_5)_2O$		11.98±0.1	RPD		2776
$C_2H_5^+$	$(C_2H_5)_2CO$		13.04	RPD		2977
$C_2H_5^+$	C_2H_5COOH		12.90	EI		3435
$C_2H_5^+$	$HCOOC_2H_5$		12.0	EI		3224
$C_2H_5^+$	$CH_3COOC_2H_5$		12.1	RPD		2018
See also – EI:	305, 3176					
$C_2H_5^+$	C_2H_5NO	NO	13.4±0.2	EDD		3180
$C_2H_5^+$	$C_2H_5NO_2$	NO_2	11.0	RPD		2018
$C_2H_5^+$	$C_2H_5ONO_2$		11.86±0.25	EI		1013
$C_2H_5^+$	$n-C_5H_{11}F$		14.67	EI		2029
$C_2H_5^+$	$C_2H_5CF_3$		12.82±0.02	EI		1075
$C_2H_5^+$	$C_2H_5BF_2$		13.1±0.2	EI		1076
$C_2H_5^+$	$C_2H_5SiH_3$		12.6±0.2	EI		2182
$C_2H_5^+$	$C_2H_5PH_2$		12.5±0.3	EI		2948
$C_2H_5^+$	$(C_2H_5)_2PH$		14.5±0.3	EI		2948
$C_2H_5^+$	$(C_2H_5)_3P$		18.5±0.3	EI		2948
$C_2H_5^+$	$(CH_3)_2PP(CH_3)_2$		14.3±0.3	EI		2948
$C_2H_5^+$	$(C_2H_5)_2PP(C_2H_5)_2$		11.5±0.3	EI		2948
$C_2H_5^+$	C_2H_5SH		12.1±0.3	EI		3286
$C_2H_5^+$	$C_2H_5SCH_3$		14.1±0.2	EI		176
$C_2H_5^+$	$n-C_3H_7SCH_3$		15.3±0.5	EI		176
$C_2H_5^+$	$(C_2H_5)_2S$		14.5±0.3	EI		84
See also – EI:	3286					
$C_2H_5^+$	$C_2H_5SCH=CHC\equiv CH$		13.9±0.3	EI		2949
$C_2H_5^+$	$C_2H_5SC\equiv CCH=CH_2$		13.8±0.3	EI		2949
$C_2H_5^+$	$C_6H_5SC_2H_5$ (Ethylthiobenzene)		13.7	EI		307
$C_2H_5^+$	$C_2H_5SSC_2H_5$		14.2±0.2	EI		186
See also – EI:	3286					
$C_2H_5^+$	C_2H_5NCS		12.9±0.2	EI		315
$C_2H_5^+$	C_2H_5Cl	Cl	11.83±0.06	EI		3201
See also – EI:	160, 356					
$C_2H_5^+$	$n-C_3H_7Cl$		12.48±0.1	EI		72
$C_2H_5^+$	$C_2H_5SiCl_3$		12.77±0.05	EI		2182
$C_2H_5^+$	$(C_2H_5)_2Se$		13.2±0.3	EI		3285
$C_2H_5^+$	C_2H_5Br	Br	11.15	EI		2973
See also – EI:	160, 356					

4.3. The Positive Ion Table—Continued

Ion	Reactant	Other products	Ionization or appearance potential (eV)	Method	Heat of formation (kJ mol^{-1})	Ref.
$C_2H_5^+$	C_2H_5I	I	11.0±0.3	EI		356
See also – EI: 160						
$C_2H_5^+$	$(C_2H_5)_2Hg$		10.25±0.1	EI		306

$C_2H_4D^+$

$C_2H_4D^+$	C_2H_5D	H	12.86±0.1	EI		2421

$C_2H_3D_2^+$

$C_2H_3D_2^+$	$CH_3CD_2CH_3$	CH_3	12.26	RPD		2907

$C_2H_2D_3^+$

$C_2H_2D_3^+$	CH_3CD_3	H	12.52±0.08	EI		2421
$C_2H_2D_3^+$	$CD_3CH_2CD_3$	CD_3	12.28	RPD		2907

$C_2D_5^+$

$C_2D_5^+$	C_2D_6	D	13.52±0.03	EI		2421
$C_2D_5^+$	C_3D_8	CD_3	12.23	RPD		2521
$C_2D_5^+$	$C_2D_5NH_2$		13.4	EI		2470

$C_2H_6^+$ $\Delta H_{f0}^\circ \sim 1041$ kJ mol^{-1} (249 kcal mol^{-1})

$C_2H_6^+$	C_2H_6		11.521±0.007	PI	1042	1253
$C_2H_6^+$	C_2H_6		11.45±0.05	RPI	1036	3293
$C_2H_6^+$	C_2H_6		11.51	PE	1041	2843
$C_2H_6^+$	C_2H_6		11.56	PE	1046	2803
$C_2H_6^+$	C_2H_6		11.55	PEN		2466
$C_2H_6^+$	C_2H_6		11.66±0.05	RPD		2776

For vibrational structure see refs. 2606, 2803.

See also – PI: 182, 416, 1120, 2606
PE: 1130, 2801, 2829, 3072, 3096
PEN: 2430, 2467, 2873
EI: 160, 2421, 2535, 3338

$C_2H_6^+$	$(CH_2)_3O$ (1,3–Epoxypropane)	CO	10.8±0.3	EI		52
$C_2H_6^+$	$(C_2H_5)_2PP(C_2H_5)_2$		10.5±0.3	EI		2948

4.3. The Positive Ion Table—Continued

Ion	Reactant	Other products	Ionization or appearance potential (eV)	Method	Heat of formation (kJ mol⁻¹)	Ref.
			$C_2H_5D^+$			
$C_2H_5D^+$	C_2H_5D		11.58±0.04	EI		2421
			$C_2H_3D_3^+$			
$C_2H_3D_3^+$	CH_3CD_3		11.78±0.03	EI		2421
			$C_2D_6^+$			
$C_2D_6^+$	C_2D_6		11.73±0.06	EI		2421
			$C_2H_6^{+2}$			
$C_2H_6^{+2}$	C_2H_6		≤34.8±0.8 (V)	AUG		3424
			C_3H^+			
C_3H^+	$CH_2=C=CH_2$	H_2+H	18.56±0.05	EI		2455
C_3H^+	$CH_3C\equiv CH$	H_2+H	15.4±0.3	EI		13
C_3H^+	$CH_3C\equiv CH$	H_2+H	17.5±0.5	EI		3191
C_3H^+	C_3H_6	$2H_2+H$	20.2±0.5	EI		3191
C_3H^+	C_3H_6	$2H_2+H$	20.5±0.5	EI		2542
C_3H^+	C_3H_6 (Cyclopropane)	$2H_2+H$	19.7±0.5	EI		3191
C_3H^+	$CH_2=CHC\equiv CH$		18.71	EI		2102
C_3H^+	$CH_2=CHCH=CH_2$		12.44	EI		2102
C_3H^+	$C_2H_5C\equiv CH$		12.59	EI		2102
C_3H^+	$CH\equiv CCN$	N	18.0±0.2	EI		154
C_3H^+	C_3H_6S (Propylene sulfide)		22.2±0.5	EI		188
C_3H^+	$CH_3SCH_2CH=CH_2$		16.6±0.5	EI		186
			C_3H^{+2}			
C_3H^{+2}	C_3H_6		47±1	EI		2542
			$C_3H_2^+$			
$C_3H_2^+$	$CH_2=C=CH_2$	H_2	14.34±0.08	EI		2455
$C_3H_2^+$	$CH_3C\equiv CH$	H_2	14.0±0.1	EI		13
$C_3H_2^+$	C_3H_6	$2H_2$	16.5±1	EI		2542
$C_3H_2^+$	$trans\text{-}CH_2=CHCH=CHCH_3$		20.86±0.1	EI		2455
$C_3H_2^+$	C_6H_8 (1,3–Cyclohexadiene)		23.51±0.50	EI		2751
$C_3H_2^+$	C_6H_8 (1,4–Cyclohexadiene)		19.21±0.15	EI		2751

4.3. The Positive Ion Table—Continued

Ion	Reactant	Other products	Ionization or appearance potential (eV)	Method	Heat of formation (kJ mol^{-1})	Ref.
$C_3H_2^+$	C_8H_8 (syn-Tricyclo[4.2.0.0^{2,5}]octa-3,7-diene)		16.91±0.10	EI		2914
$C_3H_2^+$	C_8H_8 (anti-Tricyclo[4.2.0.0^{2,5}]octa-3,7-diene)		17.00±0.35	EI		2914
$C_3H_2^+$	C_4H_6O (3,4-Epoxy-1-butene)		15.8±0.5	EI		153
$C_3H_2^+$	C_3H_6S (Propylene sulfide)		19.2±0.4	EI		188
$C_3H_2^+$	$CH_3SCH_2CH=CH_2$		20.3±0.5	EI		186

$$C_3H_2^{+2}$$

Ion	Reactant	Other products	Ionization or appearance potential (eV)	Method	Heat of formation (kJ mol^{-1})	Ref.
$C_3H_2^{+2}$	$CH_2=C=CH_2$		32.52±0.2	EI		2455
$C_3H_2^{+2}$	C_3H_6		33.3±0.5	EI		2542

$$CH_2C\equiv CH^+ \qquad \Delta H^\circ_{f298} \sim 1175 \text{ kJ mol}^{-1} \text{ (281 kcal mol}^{-1}\text{)}$$
$$cyclo-C_3H_3^+ \qquad \Delta H^\circ_{f298} \sim 1075 \text{ kJ mol}^{-1} \text{ (257 kcal mol}^{-1}\text{)}$$

Ion	Reactant	Other products	Ionization or appearance potential (eV)	Method	Heat of formation (kJ mol^{-1})	Ref.
$C_3H_3^+$	$CH_2C\equiv CH$		8.68	EM	1175	3380
$C_3H_3^+$	C_3H_3		8.20±0.5	EI		2535
$C_3H_3^+$	$CH_2=C=CH_2$	H	11.48±0.02	PI	1082	2644
$C_3H_3^+$	$CH_2=C=CH_2$	H	11.48±0.03	PI	1082	2724
$C_3H_3^+$	$CH_2=C=CH_2$	H	11.47	EM	1081	3380

See also – EI: 165, 2455

$C_3H_3^+$	$CH_3C\equiv CH$	H	11.55±0.02	PI	1082	2644
$C_3H_3^+$	$CH_3C\equiv CH$	H	11.56±0.03	PI	1083	2724
$C_3H_3^+$	$CH_3C\equiv CH$	H	11.60	EM	1087	3380

See also – EI: 13, 17, 462

$C_3H_3^+$	C_3H_4 (cis-Cyclopropene)	H	10.54	EM	1075	3380

See also – EI: 165

$C_3H_3^+$	C_3H_6	H_2+H	14.21	EI		194
$C_3H_3^+$	C_3H_6	H_2+H	14.3±0.5	EI		2542
$C_3H_3^+$	C_3H_6 (Cyclopropane)	H_2+H	13.73±0.1	EI		3191
$C_3H_3^+$	$CH_3CH=C=CH_2$	CH_3	10.86±0.04	PI	1068	2644

See also – EI: 462

$C_3H_3^+$	$CH_2=CHCH=CH_2$	CH_3	11.40±0.02	PI	1068	2644
$C_3H_3^+$	$CH_2=CHCH=CH_2$	CH_3	11.39±0.03	PI	1067	2724
$C_3H_3^+$	$CH_2=CHCH=CH_2$	CH_3	11.35±0.05	PI	1063	2013

See also – EI: 462, 2455

4.3. The Positive Ion Table—Continued

Ion	Reactant	Other products	Ionization or appearance potential (eV)	Method	Heat of formation (kJ mol^{-1})	Ref.
$C_3H_3^+$	$C_2H_5C{\equiv}CH$	CH_3	10.84	EM	1069	3380

See also — EI: 13, 462

$C_3H_3^+$	$CH_3C{\equiv}CCH_3$	CH_3	11.04	EM	1069	3380

See also — EI: 13

$C_3H_3^+$	$1-C_4H_8$		13.82	EI		195
$C_3H_3^+$	$cis-2-C_4H_8$		13.75	EI		195
$C_3H_3^+$	C_4H_8 (Cyclobutane)		14.8	EI		2742
$C_3H_3^+$	$trans-CH_2{=}CHCH{=}CHCH_3$		15.16±0.02	EI		2455
$C_3H_3^+$	$neo-C_5H_{12}$		17.08	EI		2101
$C_3H_3^+$	$CH{\equiv}CCH{=}CHCH{=}CH_2$		14.57	EI		1197
$C_3H_3^+$	$C_2H_5C{\equiv}CC{\equiv}CH$		12.20	EI		1197
$C_3H_3^+$	$CH_3C{\equiv}CCH_2C{\equiv}CH$		12.05	EI		1197
$C_3H_3^+$	$CH_3C{\equiv}CC{\equiv}CCH_3$		11.99	EI		1197
$C_3H_3^+$	$CH{\equiv}CCH_2CH_2C{\equiv}CH$		12.17	EI		1197
$C_3H_3^+$	C_6H_6 (Benzene)		14.7±0.1	RPD		2520

See also — EI: 1197, 1238, 2103

$C_3H_3^+$	$CH_2{=}CHCH{=}CHCH{=}CH_2$	$C_2H_2+CH_3$?	14.65±0.10	EI		2751

(0.30 eV average translational energy of decomposition at threshold)

$C_3H_3^+$	C_6H_8 (1,3–Cyclohexadiene)		14.87±0.10	EI		2751
$C_3H_3^+$	C_6H_8 (1,4–Cyclohexadiene)		15.20±0.10	EI		2751
$C_3H_3^+$	$CH_3CH{=}CHCH{=}CHCH_3$		14.92±0.03	EI		2455
$C_3H_3^+$	$C_5H_8{=}CH_2$ (Methylenecyclopentane)		13.44±0.20	EDD		2738
$C_3H_3^+$	$C_5H_7CH_3$ (1–Methylcyclopentene)		13.23±0.21	EDD		2738
$C_3H_3^+$	$C_5H_7CH_3$ (3–Methylcyclopentene)		13.24±0.32	EDD		2738
$C_3H_3^+$	C_6H_{10} (Cyclohexene)		13.45±0.18	EDD		2738
$C_3H_3^+$	$(C_3H_5)_2$ (Bicyclopropyl)		11.99±0.18	EDD		2738
$C_3H_3^+$	C_6H_{10} (Bicyclo[3.1.0]hexane)		12.98±0.19	EDD		2738
$C_3H_3^+$	$C_6H_{10}{=}CH_2$ (Methylenecyclohexane)		13.90±0.13	EDD		2558
$C_3H_3^+$	$C_6H_9CH_3$ (1–Methylcyclohexene)		14.06±0.13	EDD		2558
$C_3H_3^+$	$C_6H_9CH_3$ (3–Methylcyclohexene)		13.85±0.08	EDD		2558
$C_3H_3^+$	$C_6H_9CH_3$ (4–Methylcyclohexene)		14.02±0.15	EDD		2558
$C_3H_3^+$	C_7H_{12} (Bicyclo[2.2.1]heptane)		14.03±0.09	EDD		2558

4.3. The Positive Ion Table—Continued

Ion	Reactant	Other products	Ionization or appearance potential (eV)	Method	Heat of formation (kJ mol^{-1})	Ref.
$C_3H_3^+$	C_7H_{12} (Bicyclo[4.1.0]heptane)		12.53±0.12	EDD		2558
$C_3H_3^+$	$C_6H_5CH=CH_2$ (Ethenylbenzene)		14.90±0.10	EI		2914
(0.07 eV average translational energy of decomposition at threshold)						
$C_3H_3^+$	C_8H_8 (Cyclooctatetraene)		13.40±0.10	EI		2914
$C_3H_3^+$	C_8H_8 (Cyclobutenobenzene)		14.16±0.15	EI		2914
(0.11 eV average translational energy of decomposition at threshold)						
$C_3H_3^+$	C_8H_8 (Bicyclo[2.2.2]octatriene)		13.64±0.25	EI		2914
$C_3H_3^+$	C_8H_8 (syn–Tricyclo[4.2.0.0^{2,5}]octa–3,7–diene)		11.87±0.15	EI		2914
$C_3H_3^+$	C_8H_8 (anti–Tricyclo[4.2.0.0^{2,5}]octa–3,7–diene)		11.95±0.10	EI		2914
$C_3H_3^+$	C_8H_8 (Cubane)		10.01±0.10	EI		2914

See also – EI: 2105

Ion	Reactant	Other products	Ionization or appearance potential (eV)	Method	Heat of formation (kJ mol^{-1})	Ref.
$C_3H_3^+$	$(CH_3)_2C=CHCH=C(CH_3)_2$		26.52±0.1	EI		2455
$C_3H_3^+$	$(CH_2)_4NH$ (Pyrrolidine)		18.9±0.4	EI		52
$C_3H_3^+$	C_5H_5N (Pyridine)		14.00±0.10	EI		1406
$C_3H_3^+$	$(CH_2)_3O$ (1,3–Epoxypropane)		14.5±0.2	EI		52
$C_3H_3^+$	$n–C_3H_7OH$		15.6±0.3	EI		46
$C_3H_3^+$	C_4H_6O (3,4–Epoxy–1–butene)		13.5±0.3	EI		153
$C_3H_3^+$	$(CH_2)_4O$ (1,4–Epoxybutane)		18.7±0.6	EI		52
$C_3H_3^+$	$C_4H_8O_2$ (1,2–Epoxy–3–methoxypropane)		15.9±0.4	EI		153
$C_3H_3^+$	C_6H_5F (Fluorobenzene)		14.27±0.1	EI		2103
$C_3H_3^+$	C_3H_6S (Propylene sulfide)		15.9±0.2	EI		188
$C_3H_3^+$	$(CH_2)_3S$ (Trimethylene sulfide)		15.3±0.4	EI		52
$C_3H_3^+$	C_4H_4S (Thiophene)		12.8±0.2	EI		2166
$C_3H_3^+$	$CH_3SCH_2CH=CH_2$		16.5±0.4	EI		186
$C_3H_3^+$	$(CH_2)_4S$ (Tetramethylene sulfide)		17.2±0.2	EI		52
$C_3H_3^+$	$n–C_3H_7SCH_3$		18.4±0.5	EI		176
$C_3H_3^+$	$iso–C_3H_7SCH_3$		21.0±0.5	EI		186
$C_3H_3^+$	$CH_3SCH=CHC≡CH$		13.4±0.3	EI		2949
$C_3H_3^+$	$CH_3SC≡CCH=CH_2$		13.4±0.3	EI		2949
$C_3H_3^+$	C_6H_5SH (Mercaptobenzene)		20.0±0.3	EI		3286
$C_3H_3^+$	$C_2H_5SCH=CHC≡CH$		14.8±0.3	EI		2949
$C_3H_3^+$	$C_2H_5SC≡CCH=CH_2$		15.2±0.3	EI		2949

4.3. The Positive Ion Table—Continued

Ion	Reactant	Other products	Ionization or appearance potential (eV)	Method	Heat of formation (kJ mol⁻¹)	Ref.
$C_3H_3^+$	$CH_3C{\equiv}CCl$	Cl	11.0±0.2	EI		13
$C_3H_3^+$	$(C_5H_5)_2TiCl_2$		18.0±0.5	EI		2479
	(Bis(cyclopentadienyl)titanium dichloride)					
$C_3H_3^+$	$C_5H_5V(CO)_4$		13.5±0.3	EI		1381
	(Cyclopentadienylvanadium tetracarbonyl)					
$C_3H_3^+$	$C_5H_5Mn(CO)_3$		20.3±0.4	EI		1381
	(Cyclopentadienylmanganese tricarbonyl)					
$C_3H_3^+$	$CH_3C{\equiv}CBr$	Br	11.1±0.2	EI		13
$C_3H_3^+$	$(C_5H_5)_2ZrCl_2$		19.5±0.4	EI		2479
	(Bis(cyclopentadienyl)zirconium dichloride)					

The heat of formation of the propargyl ion is ~1175 kJ mol⁻¹. For all other $C_3H_3^+$ ions for which accurate appearance potentials have been measured the computed heats of formation are on the average about 1075 kJ mol⁻¹. This value is assigned to the cyclic $C_3H_3^+$ structure on the basis of its greater stability. For details see ref. 3380.

$C_3HD_2^+$

Ion	Reactant	Other products	Ionization or appearance potential (eV)	Method	Heat of formation	Ref.
$C_3HD_2^+$	$CD_3C{\equiv}CH$	D	12.22±0.05	EI		17

$C_3D_3^+$

Ion	Reactant	Other products	Ionization or appearance potential (eV)	Method	Heat of formation	Ref.
$C_3D_3^+$	$CD_3C{\equiv}CH$	H	12.16±0.06	EI		17

$C_3H_3^{+2}$

Ion	Reactant	Other products	Ionization or appearance potential (eV)	Method	Heat of formation	Ref.
$C_3H_3^{+2}$	$CH_2{=}C{=}CH_2$	H	34.57±0.1	EI		2455
$C_3H_3^{+2}$	C_3H_6		34.1±0.5	EI		2542

$CH_2{=}C{=}CH_2^+$ $\Delta H^\circ_{f298} \sim 1124$ kJ mol⁻¹ (269 kcal mol⁻¹)

$CH_3C{\equiv}CH^+$ $\Delta H^\circ_{f298} = 1185$ kJ mol⁻¹ (283 kcal mol⁻¹)

$C_3H_4^+$ (cis–Cyclopropene) $\Delta H^\circ_{f298} \sim 1209$ kJ mol⁻¹ (289 kcal mol⁻¹)

Ion	Reactant	Other products	Ionization or appearance potential (eV)	Method	Heat of formation	Ref.
$C_3H_4^+$	$CH_2{=}C{=}CH_2$		9.53±0.03	PI	1112	2724
$C_3H_4^+$	$CH_2{=}C{=}CH_2$		9.62±0.04	PI	1120	2644
$C_3H_4^+$	$CH_2{=}C{=}CH_2$		9.69	PE	1127	3058
$C_3H_4^+$	$CH_2{=}C{=}CH_2$		9.83	PE	1141	2843
$C_3H_4^+$	$CH_2{=}C{=}CH_2$		9.62	EM	1120	3380

The ionization potential is uncertain due to Jahn–Teller distortion and hot band effects.

See also – S: 3124
       EI: 462, 2455

Ion	Reactant	Other products	Ionization or appearance potential (eV)	Method	Heat of formation	Ref.
$C_3H_4^+$	$CH_3C{\equiv}CH$		10.36	S	1185	162, 1022
(Average of three Rydberg series limits)						
$C_3H_4^+$	$CH_3C{\equiv}CH$		10.349±0.015	PI	1184	2726, 2965
$C_3H_4^+$	$CH_3C{\equiv}CH$		10.365±0.015	PI	1186	3098
$C_3H_4^+$	$CH_3C{\equiv}CH$		10.36±0.01	PI	1185	162, 182, 416, 1022

4.3. The Positive Ion Table—Continued

Ion	Reactant	Other products	Ionization or appearance potential (eV)	Method	Heat of formation (kJ mol^{-1})	Ref.
$C_3H_4^+$	$CH_3C{\equiv}CH$		10.36±0.02	PI	1185	2724
$C_3H_4^+$	$CH_3C{\equiv}CH$		10.38±0.02	PI	1187	2644
$C_3H_4^+$	$CH_3C{\equiv}CH$		10.37±0.01	PE	1186	2805
$C_3H_4^+$	$CH_3C{\equiv}CH$		10.37	PE	1186	2851
$C_3H_4^+$	$CH_3C{\equiv}CH$		10.36	EM	1185	3380

See also – S: 3152
      EI: 13, 17

Ion	Reactant	Other products	Ionization or appearance potential (eV)	Method	Heat of formation (kJ mol^{-1})	Ref.
$C_3H_4^+$	C_3H_4 (cis–Cyclopropene)		9.70	PE	1212	3330
$C_3H_4^+$	C_3H_4 (cis–Cyclopropene)		9.64	EM	1206	3380

See also – EI: 62, 87, 1129, 2535

Ion	Reactant	Other products	Ionization or appearance potential (eV)	Method	Heat of formation (kJ mol^{-1})	Ref.
$C_3H_4^+$	C_3H_6	H_2	12.3±0.5	EI		2542
$C_3H_4^+$	C_3H_6	H_2	12.52	EI		194, 195
$C_3H_4^+$	C_3H_6 (Cyclopropane)	H_2	11.57±0.1	EI		3191
$C_3H_4^+$	trans–$CH_2{=}CHCH{=}CHCH_3$	C_2H_4	12.63±0.02	EI		2455
$C_3H_4^+$	C_6H_8 (1,3–Cyclohexadiene)		14.52±0.10	EI		2751
$C_3H_4^+$	C_6H_8 (1,4–Cyclohexadiene)		13.95±0.10	EI		2751
$C_3H_4^+$	C_4H_6O (3,4–Epoxy–1–butene)		11.3±0.3	EI		153
$C_3H_4^+$	$(CH_2)_4O$ (1,4–Epoxybutane)		15.2±0.3	EI		52
$C_3H_4^+$	C_3H_6S (Propylene sulfide)		14.4±0.3	EI		188

$C_3HD_3^+$

Ion	Reactant	Other products	Ionization or appearance potential (eV)	Method	Heat of formation (kJ mol^{-1})	Ref.
$C_3HD_3^+$	$CD_3C{\equiv}CH$		10.62±0.05	EI		17

$C_3D_4^+$

Ion	Reactant	Other products	Ionization or appearance potential (eV)	Method	Heat of formation (kJ mol^{-1})	Ref.
$C_3D_4^+$	$CD_3C{\equiv}CD$		10.375±0.015	PI		3098

$C_3H_4^{+2}$

Ion	Reactant	Other products	Ionization or appearance potential (eV)	Method	Heat of formation (kJ mol^{-1})	Ref.
$C_3H_4^{+2}$	$CH_2{=}C{=}CH_2$		30.24±0.2	EI		2455
$C_3H_4^{+2}$	C_3H_6		30.1±0.5	EI		2542

4.3. The Positive Ion Table—Continued

Ion	Reactant	Other products	Ionization or appearance potential (eV)	Method	Heat of formation (kJ mol^{-1})	Ref.

$$CH_2\!=\!CHCH_2^+ \qquad \Delta H^\circ_{f298} \sim 946 \text{ kJ mol}^{-1} \ (226 \text{ kcal mol}^{-1})$$

$C_3H_5^+$	$CH_2\!=\!CHCH_2$		8.07 ± 0.03	EM	949	3350
$C_3H_5^+$	$CH_2\!=\!CHCH_2$		8.05 ± 0.1	EI		123

This radical, identified in ref. 123 as cyclopropyl, is now considered to have isomerized to allyl, see ref. 3380.

See also – EI: 87, 1129, 3368

$C_3H_5^+$	C_3H_6	H	11.88	EM	949	3350, 3380

See also – PEN: 3348
           EI: 194, 195, 1451, 2542, 3368

$C_3H_5^+$	C_3H_6 (Cyclopropane)	H	11.49	EM	944	3380

See also – PEN: 3348
           EI: 123, 3191

$C_3H_5^+$	C_3H_8	$H_2\!+\!H$	14.76	EI		195
$C_3H_5^+$	$1\!-\!C_4H_8$	CH_3	11.28	EM	946	3350, 3380

See also – EI: 194, 195, 1451, 2941

$C_3H_5^+$	$cis\!-\!2\!-\!C_4H_8$	CH_3	11.33	EM	944	3350, 3380

See also – EI: 194, 195, 2941

$C_3H_5^+$	$iso\!-\!C_4H_8$	CH_3	11.45	EM	946	3380

See also – EI: 2941

$C_3H_5^+$	$C_3H_5CH_3$ (Methylcyclopropane)	CH_3	11.02	EM		3380

See also – EI: 2941

$C_3H_5^+$	C_4H_8 (Cyclobutane)	CH_3	11.00	EM	946	3380

See also – EI: 2742, 2941

$C_3H_5^+$	$n\!-\!C_4H_{10}$		13.40	EI		195
$C_3H_5^+$	$iso\!-\!C_4H_{10}$		14.55	EI		195
$C_3H_5^+$	$trans\!-\!CH_2\!=\!CHCH\!=\!CHCH_3$		14.20 ± 0.02	EI		2455
$C_3H_5^+$	$neo\!-\!C_5H_{12}$		13.13	EI		2101
$C_3H_5^+$	$CH_3CH\!=\!CHCH\!=\!CHCH_3$		13.06 ± 0.02	EI		2455
$C_3H_5^+$	$CH_2\!=\!CHCH_2CH_2CH\!=\!CH_2$		$\geqslant10.9$	EI		3368
$C_3H_5^+$	$C_5H_8\!=\!CH_2$ (Methylenecyclopentane)		12.03 ± 0.13	EDD		2738
$C_3H_5^+$	$C_5H_7CH_3$ (1–Methylcyclopentene)		12.45 ± 0.13	EDD		2738

4.3. The Positive Ion Table—Continued

Ion	Reactant	Other products	Ionization or appearance potential (eV)	Method	Heat of formation (kJ mol^{-1})	Ref.
$C_3H_5^+$	$C_5H_7CH_3$ (3−Methylcyclopentene)		12.28±0.17	EDD		2738
$C_3H_5^+$	C_6H_{10} (Cyclohexene)		12.12±0.12	EDD		2738
$C_3H_5^+$	$(C_3H_5)_2$ (Bicyclopropyl)		10.64±0.14	EDD		2738
$C_3H_5^+$	C_6H_{10} (Bicyclo[3.1.0]hexane)		11.65±0.14	EDD		2738
$C_3H_5^+$	$C_6H_{10}=CH_2$ (Methylenecyclohexane)		13.25±0.11	EDD		2558
$C_3H_5^+$	$C_6H_9CH_3$ (1−Methylcyclohexene)		13.46±0.09	EDD		2558
$C_3H_5^+$	$C_6H_9CH_3$ (3−Methylcyclohexene)		13.29±0.11	EDD		2558
$C_3H_5^+$	$C_6H_9CH_3$ (4−Methylcyclohexene)		13.38±0.11	EDD		2558
$C_3H_5^+$	C_7H_{12} (Bicyclo[2.2.1]heptane)		13.22±0.12	EDD		2558
$C_3H_5^+$	C_7H_{12} (Bicyclo[4.1.0]heptane)		11.90±0.10	EDD		2558
$C_3H_5^+$	$n-C_7H_{16}$		12.7±0.1	PI		2013
$C_3H_5^+$	$(CH_3)_2C=CHCH=C(CH_3)_2$		14.27±0.03	EI		2455
$C_3H_5^+$	$C_2H_5CH=C=CHCH_2C(CH_3)_3$		15.0	EI		3008
$C_3H_5^+$	C_3H_5CN (Cyclopropanecarboxylic acid nitrile)	CN	12.70±0.15	EI		202
$C_3H_5^+$	$n-C_3H_7NC$		13.18	EDD		3214
$C_3H_5^+$	$n-C_4H_9CN$		13.73	EI		2966
$C_3H_5^+$	$tert-C_4H_9CN$		13.50	EDD		3214
$C_3H_5^+$	$n-C_4H_9NC$		13.35	EDD		3214
$C_3H_5^+$	$(CH_2)_3O$ (1,3−Epoxypropane)	OH	11.8±0.2	EI		52
$C_3H_5^+$	$n-C_3H_7OH$		12.6	PI		2647
(Threshold value approximately corrected to 0 K)						
$C_3H_5^+$	C_4H_6O (3,4−Epoxy−1−butene)	CO+H?	11.1±0.2	EI		153
$C_3H_5^+$	$(CH_2)_4O$ (1,4−Epoxybutane)		13.72	EI		2694

See also − EI: 52

Ion	Reactant	Other products	Ionization or appearance potential (eV)	Method	Heat of formation (kJ mol^{-1})	Ref.
$C_3H_5^+$	$(C_2H_5)_2O$		11.6	EI		2971
$C_3H_5^+$	$(CH_2)_5O$ (1,5−Epoxypentane)		~12.8	EI		2694
$C_3H_5^+$	$n-C_5H_{11}F$		14.12	EI		2029
$C_3H_5^+$	$(C_2H_5)_3P$		16.4±0.3	EI		2948
$C_3H_5^+$	C_3H_6S (Propylene sulfide)	SH?	11.5±0.2	EI		188
$C_3H_5^+$	$(CH_2)_3S$ (Trimethylene sulfide)	SH?	12.2±0.2	EI		52
$C_3H_5^+$	$CH_3SCH_2CH=CH_2$		12.7±0.3	EI		186
$C_3H_5^+$	$(CH_2)_4S$ (Tetramethylene sulfide)		15.5±0.2	EI		52
$C_3H_5^+$	$n-C_3H_7SCH_3$		14.8±0.2	EI		176
$C_3H_5^+$	$iso-C_3H_7SCH_3$		15.2±0.2	EI		186

4.3. The Positive Ion Table—Continued

Ion	Reactant	Other products	Ionization or appearance potential (eV)	Method	Heat of formation (kJ mol^{-1})	Ref.
			$C_3H_5^{+2}$			
$C_3H_5^{+2}$	C_3H_6	H	31.1 ± 0.5	EI		2542
	$C_3H_6^+$		$\Delta H^\circ_{f298} = 960$ kJ mol^{-1} (229 kcal mol^{-1})			
	$C_3H_6^+$ (Cyclopropane)		$\Delta H^\circ_{f298} \leq 999$ kJ mol^{-1} (239 kcal mol^{-1})			
$C_3H_6^+$	C_3H_6		9.74	S	960	133
$C_3H_6^+$	C_3H_6		9.727 ± 0.010	PI	959	1253
$C_3H_6^+$	C_3H_6		9.73 ± 0.01	PI	959	133, 182, 416
$C_3H_6^+$	C_3H_6		9.73 ± 0.02	PI	959	1120
$C_3H_6^+$	C_3H_6		9.73	PI	959	168
$C_3H_6^+$	C_3H_6		9.74 ± 0.01	PI	960	3098
$C_3H_6^+$	C_3H_6		9.69	PE		2843
$C_3H_6^+$	C_3H_6		9.76	PEN		2430, 2466
$C_3H_6^+$	C_3H_6		9.72	EM		3380

See also − S: 3353
         PE: 3359
         EI: 194, 195, 411, 1129, 2535, 3201

Ion	Reactant	Other products	Ionization or appearance potential (eV)	Method	Heat of formation (kJ mol^{-1})	Ref.
$C_3H_6^+$	C_3H_6 (Cyclopropane)		10.06 ± 0.03	PI		182
$C_3H_6^+$	C_3H_6 (Cyclopropane)		≤9.8	PE	≤999	2808

(Value estimated from fig. 7 of this reference)

Ion	Reactant	Other products	Ionization or appearance potential (eV)	Method	Heat of formation (kJ mol^{-1})	Ref.
$C_3H_6^+$	C_3H_6 (Cyclopropane)		10.1	PEN		2430
$C_3H_6^+$	C_3H_6 (Cyclopropane)		≤9.93	EM		3380

See also − PI: 416
         EI: 123, 3191

Ion	Reactant	Other products	Ionization or appearance potential (eV)	Method	Heat of formation (kJ mol^{-1})	Ref.
$C_3H_6^+$	C_3H_8	H_2	11.75 ± 0.05	PI		2606

(Threshold value approximately corrected to 0 K)

The thermochemical threshold for this process is 11.02 eV. The excess is due probably to reaction competition and kinetic energy of fragmentation.

See also − EI: 195

Ion	Reactant	Other products	Ionization or appearance potential (eV)	Method	Heat of formation (kJ mol^{-1})	Ref.
$C_3H_6^+$	$n-C_4H_{10}$	CH_4	11.16 ± 0.03	PI		1120

(Threshold value approximately corrected to 0 K)

Ion	Reactant	Other products	Ionization or appearance potential (eV)	Method	Heat of formation (kJ mol^{-1})	Ref.
$C_3H_6^+$	$n-C_4H_{10}$	CH_4	11.18	PI		2606

(Threshold value approximately corrected to 0 K)

The thermochemical threshold for this process is 10.48 eV.

See also − EI: 195

4.3. The Positive Ion Table—Continued

Ion	Reactant	Other products	Ionization or appearance potential (eV)	Method	Heat of formation (kJ mol⁻¹)	Ref.
$C_3H_6^+$	$iso-C_4H_{10}$	CH_4	10.93 ± 0.03	PI		1120

(Threshold value approximately corrected to 0 K)

The thermochemical threshold for this process is 10.56 eV.

See also – EI: 195

Ion	Reactant	Other products	Ionization or appearance potential (eV)	Method	Heat of formation (kJ mol⁻¹)	Ref.
$C_3H_6^+$	$trans-CH_2=CHCH=CHCH_3$	C_2H_2	12.73 ± 0.03	EI		2455
$C_3H_6^+$	$C_3H_4(CH_3)_2$ (1,1–Dimethylcyclopropane)	C_2H_4	11.23 ± 0.04	EI		1146
$C_3H_6^+$	$C_3H_4(CH_3)_2$ (cis–1,2–Dimethylcyclopropane)	C_2H_4	11.26 ± 0.02	EI		1146
$C_3H_6^+$	$C_3H_4(CH_3)_2$ (trans–1,2–Dimethylcyclopropane)	C_2H_4	11.29 ± 0.05	EI		1146
$C_3H_6^+$	$n-C_5H_{12}$	C_2H_6	10.99 ± 0.02	PI		1120

(Threshold value approximately corrected to 0 K)

Ion	Reactant	Other products	Ionization or appearance potential (eV)	Method	Heat of formation (kJ mol⁻¹)	Ref.
$C_3H_6^+$	$iso-C_5H_{12}$	C_2H_6	10.84 ± 0.025	PI		1120

(Threshold value approximately corrected to 0 K)

Ion	Reactant	Other products	Ionization or appearance potential (eV)	Method	Heat of formation (kJ mol⁻¹)	Ref.
$C_3H_6^+$	$n-C_6H_{14}$	C_3H_8	11.00 ± 0.035	PI		1120

(Threshold value approximately corrected to 0 K)

Ion	Reactant	Other products	Ionization or appearance potential (eV)	Method	Heat of formation (kJ mol⁻¹)	Ref.
$C_3H_6^+$	$iso-C_6H_{14}$	C_3H_8	10.91 ± 0.05	PI		1120

(Threshold value approximately corrected to 0 K)

Ion	Reactant	Other products	Ionization or appearance potential (eV)	Method	Heat of formation (kJ mol⁻¹)	Ref.
$C_3H_6^+$	$(CH_3)_2CHCH(CH_3)_2$	C_3H_8	10.695 ± 0.02	PI		1120

(Threshold value approximately corrected to 0 K)

Ion	Reactant	Other products	Ionization or appearance potential (eV)	Method	Heat of formation (kJ mol⁻¹)	Ref.
$C_3H_6^+$	$n-C_7H_{16}$	C_4H_{10}	10.97 ± 0.08	PI		1120

(Threshold value approximately corrected to 0 K)

Ion	Reactant	Other products	Ionization or appearance potential (eV)	Method	Heat of formation (kJ mol⁻¹)	Ref.
$C_3H_6^+$	$n-C_7H_{16}$	C_4H_{10}	10.65 ± 0.1	PI		2013
$C_3H_6^+$	$iso-C_3H_7CN$		12.25	EDD		3214
$C_3H_6^+$	$n-C_3H_7NC$		12.13	EDD		3214
$C_3H_6^+$	$n-C_3H_7OH$	H_2O	10.50	PI		11

(Threshold value approximately corrected for thermal energy and kinetic shift)

Ion	Reactant	Other products	Ionization or appearance potential (eV)	Method	Heat of formation (kJ mol⁻¹)	Ref.
$C_3H_6^+$	$n-C_3H_7OH$	H_2O	10.65 ± 0.03	PI		2647

(Threshold value approximately corrected to 0 K)

Ion	Reactant	Other products	Ionization or appearance potential (eV)	Method	Heat of formation (kJ mol⁻¹)	Ref.
$C_3H_6^+$	$iso-C_3H_7OH$	H_2O	~12.0	PI		2647
$C_3H_6^+$	$(CH_2)_4O$ (1,4–Epoxybutane)		11.54	EI		2694

See also – EI: 52

Ion	Reactant	Other products	Ionization or appearance potential (eV)	Method	Heat of formation (kJ mol⁻¹)	Ref.
$C_3H_6^+$	$C_5H_{10}O_2$ (1,3–Dioxepane)		12.28	EI		2694
$C_3H_6^+$	$n-C_5H_{11}F$		11.47	EI		2029
$C_3H_6^+$	$iso-C_3H_7BF_2$		11.48 ± 0.02	EI		1076
$C_3H_6^+$	$iso-C_3H_7SiH_3$		10.81 ± 0.04	EI		2182
$C_3H_6^+$	$(C_2H_5)_2PP(C_2H_5)_2$		11.3 ± 0.3	EI		2948
$C_3H_6^+$	$n-C_3H_7SCH_3$		12.5 ± 0.4	EI		176
$C_3H_6^+$	$iso-C_3H_7SCH_3$		13.5 ± 0.2	EI		186
$C_3H_6^+$	$n-C_3H_7Cl$	HCl	11.03	EI		3201
$C_3H_6^+$	$iso-C_3H_7Cl$	HCl	$10.3?$	EI		3201
$C_3H_6^+$	$iso-C_3H_7SiCl_3$		10.92 ± 0.1	EI		2182

4.3. The Positive Ion Table—Continued

Ion	Reactant	Other products	Ionization or appearance potential (eV)	Method	Heat of formation (kJ mol^{-1})	Ref.
			C$_3$D$_6^+$			
C$_3$D$_6^+$	C$_3$D$_6$		9.755±0.01	PI		3098

	n-C$_3$H$_7^+$	$\Delta H_{f298}^\circ \sim 868$ kJ mol^{-1} (207 kcal mol^{-1})				
	iso-C$_3$H$_7^+$	$\Delta H_{f298}^\circ \sim 800$ kJ mol^{-1} (191 kcal mol^{-1})				
C$_3$H$_7^+$	n-C$_3$H$_7$		≤8.1	PI	≤868	1068
C$_3$H$_7^+$	n-C$_3$H$_7$		8.10±0.05	EM	868	3104, 3379
C$_3$H$_7^+$	n-C$_3$H$_7$		8.15±0.1	RPD		2158
C$_3$H$_7^+$	n-C$_3$H$_7$		8.13±0.05	RPD		2776

See also − EI: 141, 145, 2719

C$_3$H$_7^+$	iso-C$_3$H$_7$		≤7.5	PI	≤797	1068
C$_3$H$_7^+$	iso-C$_3$H$_7$		7.55±0.05	EM	802	3104, 3379
C$_3$H$_7^+$	iso-C$_3$H$_7$		7.52±0.1	RPD		2158
C$_3$H$_7^+$	iso-C$_3$H$_7$		7.57±0.05	RPD		2776

See also − EI: 2719

C$_3$H$_7^+$	C$_3$H$_8$	H$^-$	~11.0	PI		2606
(Threshold value approximately corrected to 0 K)						
C$_3$H$_7^+$	C$_3$H$_8$	H	11.59±0.01	PI	(796)	2606
(Threshold value approximately corrected to 0 K)						
C$_3$H$_7^+$	C$_3$H$_8$	H	11.585±0.03	PI	(796)	1120
(Threshold value approximately corrected to 0 K)						
C$_3$H$_7^+$	C$_3$H$_8$	H	11.52	RPD		2521
C$_3$H$_7^+$	C$_3$H$_8$	H	11.57±0.05	RPD		2776

See also − PEN: 2873
         EI: 195

C$_3$H$_7^+$	n-C$_4$H$_{10}$	CH$_3$	11.19±0.02	PI	(811)	1120
(Threshold value approximately corrected to 0 K)						
C$_3$H$_7^+$	n-C$_4$H$_{10}$	CH$_3$	11.18	PI	(810)	2606
(Threshold value approximately corrected to 0 K)						
C$_3$H$_7^+$	n-C$_4$H$_{10}$	CH$_3$	11.10±0.05	RPD		2776

See also − EI: 195

C$_3$H$_7^+$	iso-C$_4$H$_{10}$	CH$_3$	11.23±0.03	PI	(807)	1120
(Threshold value approximately corrected to 0 K)						
C$_3$H$_7^+$	iso-C$_4$H$_{10}$	CH$_3$	11.16±0.05	RPD		2776

See also − EI: 195

C$_3$H$_7^+$	n-C$_5$H$_{12}$	C$_2$H$_5$	11.105±0.05	PI	(818)	1120
(Threshold value approximately corrected to 0 K)						
C$_3$H$_7^+$	iso-C$_5$H$_{12}$	C$_2$H$_5$	11.145±0.05	PI	(813)	1120
(Threshold value approximately corrected to 0 K)						
C$_3$H$_7^+$	n-C$_6$H$_{14}$	C$_3$H$_7$	11.33±0.055	PI		1120
(Threshold value approximately corrected to 0 K)						

4.3. The Positive Ion Table—Continued

Ion	Reactant	Other products	Ionization or appearance potential (eV)	Method	Heat of formation (kJ mol^{-1})	Ref.
$C_3H_7^+$	$n-C_6H_{14}$	C_3H_7	11.42	RPD		2977
$C_3H_7^+$	$iso-C_6H_{14}$	C_3H_7	~11.35±0.10	PI		1120
(Threshold value approximately corrected to 0 K)						
$C_3H_7^+$	$(CH_3)_2CHCH(CH_3)_2$	C_3H_7	11.395±0.07	PI		1120
(Threshold value approximately corrected to 0 K)						
$C_3H_7^+$	$n-C_7H_{16}$		11.05±0.05	PI		2013
$C_3H_7^+$	$n-C_7H_{16}$		11.58	RPD		2977
$C_3H_7^+$	$(CH_3)_2C=CHCH=C(CH_3)_2$		12.23±0.06	EI		2455
$C_3H_7^+$	$n-C_8H_{18}$		11.89	RPD		2977
$C_3H_7^+$	$n-C_9H_{20}$		12.17	RPD		2977
$C_3H_7^+$	$n-C_3H_7NC$	CN	12.43	EDD		3214
$C_3H_7^+$	$n-C_4H_9CN$		12.50	EI		2966
$C_3H_7^+$	$n-C_4H_9NC$		12.80	EDD		3214
$C_3H_7^+$	$(CH_3)_2NNHCH_3$		10.7±0.3	EI	424,	3216
$C_3H_7^+$	$(CH_3)_2NN(CH_3)_2$		10.9±0.2	EI	424,	3216
$C_3H_7^+$	$(CH_3)_2CHN=NCH(CH_3)_2$		9.35±0.1	EI		304
$C_3H_7^+$	$n-C_3H_7OH$	OH	11.6±0.1	PI	(~823)	2647

See also – EI: 2018

Ion	Reactant	Other products	Ionization or appearance potential (eV)	Method	Heat of formation (kJ mol^{-1})	Ref.
$C_3H_7^+$	$iso-C_3H_7OH$	OH	11.6	PI	(~808)	2647
(Threshold value approximately corrected to 0 K)						
$C_3H_7^+$	$(n-C_3H_7)_2O$		11.97±0.1	RPD		2776
$C_3H_7^+$	$(iso-C_3H_7)_2O$		11.33±0.1	RPD		2776
$C_3H_7^+$	$(n-C_3H_7)_2CO$		11.80	RPD		2977
$C_3H_7^+$	$HCOOCH_2CH_2CH_3$		11.63±0.05	EI		305
$C_3H_7^+$	$HCOO(CH_2)_3CH_3$		12.24±0.12	EI		305
$C_3H_7^+$	$CH_3COOCH_2CH_2CH_3$		11.41±0.04	EI		305
$C_3H_7^+$	$CH_3COOCH(CH_3)_2$		11.12±0.08	EI		305
$C_3H_7^+$	$C_5H_{10}O_2$ (1,3–Dioxepane)		12.36	EI		2694
$C_3H_7^+$	$CH_3COO(CH_2)_3CH_3$		11.56±0.10	EI		305
$C_3H_7^+$	$n-C_3H_7NO_2$		10.6	RPD		2018
$C_3H_7^+$	$n-C_3H_7ONO_2$		11.8	EI		2456
$C_3H_7^+$	$n-C_4H_9ONO_2$		11.5	EI		2456
$C_3H_7^+$	$n-C_5H_{11}F$		12.02	EI		2029
$C_3H_7^+$	$iso-C_3H_7BF_2$		12.05±0.05	EI		1076
$C_3H_7^+$	$iso-C_3H_7SiH_3$		11.33±0.03	EI		2182
$C_3H_7^+$	$n-C_3H_7SCH_3$		12.3±0.4	EI		176
$C_3H_7^+$	$iso-C_3H_7SCH_3$		12.7±0.2	EI		186
$C_3H_7^+$	$(n-C_3H_7)_2S$		12.0	EI		307
$C_3H_7^+$	$n-C_3H_7Cl$	Cl?	11.13±0.03	EI		3201
$C_3H_7^+$	$iso-C_3H_7Cl$	Cl?	10.99±0.05	EI		3201

See also – EI: 160, 2776

Ion	Reactant	Other products	Ionization or appearance potential (eV)	Method	Heat of formation (kJ mol^{-1})	Ref.
$C_3H_7^+$	$n-C_4H_9Cl$		11.92±0.1	EI		72
$C_3H_7^+$	$iso-C_4H_9Cl$		11.26±0.1	EI		72
$C_3H_7^+$	$iso-C_3H_7SiCl_3$		11.36±0.1	EI		2182
$C_3H_7^+$	$n-C_3H_7Br$	Br?	11.3±0.2	RPD		160
$C_3H_7^+$	$n-C_3H_7I$	I?	10.4±0.2	RPD		160
$C_3H_7^+$	$(iso-C_3H_7)_2Hg$		9.65±0.1	EI		306

4.3. The Positive Ion Table—Continued

Ion	Reactant	Other products	Ionization or appearance potential (eV)	Method	Heat of formation (kJ mol^{-1})	Ref.
			C$_3$H$_6$D$^+$			
C$_3$H$_6$D$^+$	CH$_3$CD$_2$CH$_3$	D	11.58±0.05	RPD		2776
C$_3$H$_6$D$^+$	CH$_3$CD$_2$CH$_3$	D	11.63	RPD		2907
			C$_3$H$_5$D$_2^+$			
C$_3$H$_5$D$_2^+$	CH$_3$CD$_2$CH$_3$	H	11.63	RPD		2907
			C$_3$H$_2$D$_5^+$			
C$_3$H$_2$D$_5^+$	CD$_3$CH$_2$CD$_3$	D	11.57?	RPD		2907
			C$_3$HD$_6^+$			
C$_3$HD$_6^+$	CD$_3$CH$_2$CD$_3$	H	11.56	RPD		2907
			C$_3$D$_7^+$			
C$_3$D$_7^+$	C$_3$D$_8$	D	11.74	RPD		2521
C$_3$D$_7^+$	C$_3$D$_8$	D	11.6±0.1	EI		2521

(Appearance potential of the corresponding metastable transition)

C$_3$H$_8^+$ $\Delta H_{f298}^\circ \leqslant 953$ kJ mol^{-1} (228 kcal mol^{-1})

Ion	Reactant	Other products	Ionization or appearance potential (eV)	Method	Heat of formation (kJ mol^{-1})	Ref.
C$_3$H$_8^+$	C$_3$H$_8$		10.95±0.05	PI	953	2606
	(Value estimated from fig. 4 of this reference)					
C$_3$H$_8^+$	C$_3$H$_8$		10.97	PI	955	3115
C$_3$H$_8^+$	C$_3$H$_8$		10.94±0.05	RPI	952	3293
C$_3$H$_8^+$	C$_3$H$_8$		11.06	PE		2843
C$_3$H$_8^+$	C$_3$H$_8$		11.07	PE		1130
C$_3$H$_8^+$	C$_3$H$_8$		11.07	PE		2829
C$_3$H$_8^+$	C$_3$H$_8$		11.12	PEN	2430,	2466
C$_3$H$_8^+$	C$_3$H$_8$		11.09±0.05	RPD		2776
C$_3$H$_8^+$	C$_3$H$_8$		11.22	RPD		2521

The ionization potential may not be adiabatic.

See also – PI: 182, 416, 1253
           PE: 3060
           PEN: 2467
           EI: 195, 2521

Ion	Reactant	Other products	Ionization or appearance potential (eV)	Method	Heat of formation (kJ mol^{-1})	Ref.
			C$_3$H$_6$D$_2^+$			
C$_3$H$_6$D$_2^+$	CH$_3$CD$_2$CH$_3$		11.29	RPD		2907

4.3. The Positive Ion Table—Continued

Ion	Reactant	Other products	Ionization or appearance potential (eV)	Method	Heat of formation (kJ mol^{-1})	Ref.
			C$_3$H$_2$D$_6^+$			
C$_3$H$_2$D$_6^+$	CD$_3$CH$_2$CD$_3$		11.27	RPD		2907
			C$_3$D$_8^+$			
C$_3$D$_8^+$	C$_3$D$_8$		11.40	RPD		2521
			C$_4$H$^+$			
C$_4$H$^+$	CH≡CC≡CH	H	12.1±0.3	EI		13
C$_4$H$^+$	CH$_2$=CHC≡CH	H$_2$+H	12.13	EI		2102
C$_4$H$^+$	CH$_2$=CHCH=CH$_2$	2H$_2$+H	15.75	EI		2102
C$_4$H$^+$	C$_2$H$_5$C≡CH	2H$_2$+H	13.20	EI		2102

CH≡CC≡CH$^+$($^2\Pi_g$) $\Delta H^\circ_{f298} = 1455$ kJ mol^{-1} (348 kcal mol^{-1})
CH≡CC≡CH$^+$($^2\Pi_u$) $\Delta H^\circ_{f298} = 1690$ kJ mol^{-1} (404 kcal mol^{-1})
CH≡CC≡CH$^+$($^2\Sigma_u$) $\Delta H^\circ_{f298} = 2075$ kJ mol^{-1} (496 kcal mol^{-1})

Ion	Reactant	Other products	Ionization or appearance potential (eV)	Method	Heat of formation (kJ mol^{-1})	Ref.
C$_4$H$_2^+$($^2\Pi_g$)	CH≡CC≡CH		10.180±0.003	S	1455	2669
C$_4$H$_2^+$($^2\Pi_g$)	CH≡CC≡CH		10.17±0.01	PE	1454	2804, 2805
C$_4$H$_2^+$($^2\Pi_u$)	CH≡CC≡CH		12.62±0.04	S	1690	2669
C$_4$H$_2^+$($^2\Pi_u$)	CH≡CC≡CH		12.62±0.01	PE	1690	2804, 2805
C$_4$H$_2^+$($^2\Sigma_u$)	CH≡CC≡CH		16.61±0.01	PE	2075	2804, 2805
C$_4$H$_2^+$($^2\Sigma_g$)	CH≡CC≡CH		19.8 (V)	PE		2804, 2805

See also – S: 3152
EI: 13, 2535

Ion	Reactant	Other products	Ionization or appearance potential (eV)	Method	Heat of formation (kJ mol^{-1})	Ref.
C$_4$H$_2^+$	CH$_2$=CHC≡CH	H$_2$	12.84	EI		2102
C$_4$H$_2^+$	CH$_2$=CHCH=CH$_2$		16.87±0.05	EI		2455
C$_4$H$_2^+$	CH$_3$C≡CCH$_3$		16.7±0.3	EI		13
C$_4$H$_2^+$	CH≡CCH=CHCH=CH$_2$		17.55	EI		1197
C$_4$H$_2^+$	C$_2$H$_5$C≡CC≡CH		14.15	EI		1197
C$_4$H$_2^+$	CH$_3$C≡CCH$_2$C≡CH		15.10	EI		1197
C$_4$H$_2^+$	CH≡CCH$_2$CH$_2$C≡CH		15.02	EI		1197
C$_4$H$_2^+$	C$_6$H$_6$ (Benzene)		17.5±0.3	RPD		2520
C$_4$H$_2^+$	CH$_2$=CHCH=CHCH=CH$_2$	C$_2$H$_2$+2H$_2$	16.46±0.15	RPD		2751
(0.31 eV average translational energy of decomposition at threshold)						
C$_4$H$_2^+$	C$_6$H$_8$ (1,3–Cyclohexadiene)	C$_2$H$_2$+2H$_2$	19.81±0.10	EI		2751
(0.10 eV average translational energy of decomposition at threshold)						
C$_4$H$_2^+$	C$_6$H$_8$ (1,4–Cyclohexadiene)	C$_2$H$_2$+2H$_2$	17.82±0.25	EI		2751
(0.20 eV average translational energy of decomposition at threshold)						
C$_4$H$_2^+$	CH$_3$CH=CHCH=CHCH$_3$		23.64±0.2	EI		2455

4.3. The Positive Ion Table—Continued

Ion	Reactant	Other products	Ionization or appearance potential (eV)	Method	Heat of formation (kJ mol^{-1})	Ref.
$C_4H_2^+$	$C_6H_5CH=CH_2$ (Ethenylbenzene)	$2C_2H_2+H_2$	20.22 ± 0.10	RPD		2914
(0.07 eV average translational energy of decomposition at threshold)						
$C_4H_2^+$	$C_6H_5CH=CH_2$ (Ethenylbenzene)	$2C_2H_2+H_2$	19.85 ± 0.25	EI		2914
$C_4H_2^+$	C_8H_8 (Cyclooctatetraene)	$2C_2H_2+H_2$	17.11 ± 0.10	RPD		2914
(0.12 eV average translational energy of decomposition at threshold)						
$C_4H_2^+$	C_8H_8 (Cyclobutenobenzene)	$2C_2H_2+H_2$	17.69 ± 0.30	EI		2914
(0.05 eV average translational energy of decomposition at threshold)						
$C_4H_2^+$	C_8H_8 (Bicyclo[2.2.2]octatriene)	$2C_2H_2+H_2$	17.20 ± 0.30	EI		2914
$C_4H_2^+$	C_8H_8 (syn–Tricyclo[4.2.0.0^{2,5}]octa–3,7–diene)	$2C_2H_2+H_2$	15.80 ± 0.10	EI		2914
(0.14 eV average translational energy of decomposition at threshold)						
$C_4H_2^+$	C_8H_8 ($anti$–Tricyclo[4.2.0.0^{2,5}]octa–3,7–diene)	$2C_2H_2+H_2$	16.78 ± 0.10	EI		2914
(0.14 eV average translational energy of decomposition at threshold)						
$C_4H_2^+$	C_8H_8 (Cubane)	$2C_2H_2+H_2$	14.33 ± 0.20	EI		2914

See also – EI: 2105

Ion	Reactant	Other products	Ionization or appearance potential (eV)	Method	Heat of formation (kJ mol^{-1})	Ref.
$C_4H_2^+$	C_5H_5N (Pyridine)	$HCN+H_2$	16.17 ± 0.10	EI		1406
$C_4H_2^+$	$C_4H_4N_2$ (1,2–Diazine)	N_2+H_2	13.67 ± 0.10	EI		1406
$C_4H_2^+$	$CH_3SC\equiv CCH=CH_2$		19.3 ± 0.3	EI		2949
$C_4H_2^+$	C_6H_5SH (Mercaptobenzene)		21.0 ± 0.3	EI		3286
$C_4H_2^+$	$C_2H_5SCH=CHC\equiv CH$		16.4 ± 0.3	EI		2949
$C_4H_2^+$	$C_6H_5SSC_6H_5$ (Diphenyl disulfide)		23.6 ± 0.3	EI		3286

$C_4D_2^+$

Ion	Reactant	Other products	Ionization or appearance potential (eV)	Method	Heat of formation (kJ mol^{-1})	Ref.
$C_4D_2^+(^2\Pi_g)$	$CD\equiv CC\equiv CD$		10.180 ± 0.003	S		2669
$C_4D_2^+(^2\Pi_g)$	$CD\equiv CC\equiv CD$		10.18 ± 0.01	PE		2804, 2805
$C_4D_2^+(^2\Pi_u)$	$CD\equiv CC\equiv CD$		12.62 ± 0.04	S		2669
$C_4D_2^+(^2\Pi_u)$	$CD\equiv CC\equiv CD$		12.62 ± 0.01	PE		2804, 2805
$C_4D_2^+(^2\Sigma_u)$	$CD\equiv CC\equiv CD$		16.74 ± 0.01	PE		2804, 2805
$C_4D_2^+(^2\Sigma_g)$	$CD\equiv CC\equiv CD$		19.8 (V)	PE		2804, 2805

4.3. The Positive Ion Table—Continued

Ion	Reactant	Other products	Ionization or appearance potential (eV)	Method	Heat of formation (kJ mol^{-1})	Ref.
			C$_4$H$_3^+$			
C$_4$H$_3^+$	CH$_2$=CHC≡CH	H	12.59	EI		2102
C$_4$H$_3^+$	CH$_2$=CHCH=CH$_2$	H$_2$+H	16.25±0.05	EI		2455
C$_4$H$_3^+$	C$_2$H$_5$C≡CH	H$_2$+H	14.6±0.2	EI		13
C$_4$H$_3^+$	CH$_3$C≡CCH$_3$	H$_2$+H	15.1±0.2	EI		13
C$_4$H$_3^+$	trans-CH$_2$=CHCH=CHCH$_3$	CH$_3$+H$_2$	16.36±0.08	EI		2455
C$_4$H$_3^+$	CH≡CCH=CHCH=CH$_2$		18.27	EI		1197
C$_4$H$_3^+$	C$_2$H$_5$C≡CC≡CH		15.50	EI		1197
C$_4$H$_3^+$	CH$_3$C≡CCH$_2$C≡CH		15.70	EI		1197
C$_4$H$_3^+$	CH$_3$C≡CC≡CCH$_3$		15.04	EI		1197
C$_4$H$_3^+$	CH≡CCH$_2$CH$_2$C≡CH		15.45	EI		1197
C$_4$H$_3^+$	C$_6$H$_6$ (Benzene)		17.6±0.1	RPD		2520

See also – EI: 1197

Ion	Reactant	Other products	Ionization or appearance potential (eV)	Method	Heat of formation (kJ mol^{-1})	Ref.
C$_4$H$_3^+$	CH$_2$=CHCH=CHCH=CH$_2$	C$_2$H$_2$+H$_2$+H?	16.54±0.15	RPD		2751
	(0.21 eV average translational energy of decomposition at threshold)					
C$_4$H$_3^+$	C$_6$H$_8$ (1,3–Cyclohexadiene)	C$_2$H$_2$+H$_2$+H?	17.62±0.10	EI		2751
	(0.13 eV average translational energy of decomposition at threshold)					
C$_4$H$_3^+$	C$_6$H$_8$ (1,4–Cyclohexadiene)	C$_2$H$_2$+H$_2$+H?	17.08±0.10	EI		2751
	(0.09 eV average translational energy of decomposition at threshold)					
C$_4$H$_3^+$	CH$_3$CH=CHCH=CHCH$_3$		18.32±0.04	EI		2455
C$_4$H$_3^+$	C$_6$H$_5$CH=CH$_2$ (Ethenylbenzene)	2C$_2$H$_2$+H	19.61±0.10	EI		2914
	(0.06 eV average translational energy of decomposition at threshold)					
C$_4$H$_3^+$	C$_8$H$_8$ (Cyclooctatetraene)	2C$_2$H$_2$+H	18.16±0.25	EI		2914
	(0.08 eV average translational energy of decomposition at threshold)					
C$_4$H$_3^+$	C$_8$H$_8$ (Cyclobutenobenzene)	2C$_2$H$_2$+H	18.68±0.10	EI		2914
	(0.18 eV average translational energy of decomposition at threshold)					
C$_4$H$_3^+$	C$_8$H$_8$ (Bicyclo[2.2.2]octatriene)	2C$_2$H$_2$+H	18.04±0.10	EI		2914
C$_4$H$_3^+$	C$_8$H$_8$ (syn–Tricyclo[4.2.0.0^{2,5}]octa-3,7–diene)	2C$_2$H$_2$+H	16.38±0.10	EI		2914
	(0.10 eV average translational energy of decomposition at threshold)					
C$_4$H$_3^+$	C$_8$H$_8$ (anti–Tricyclo[4.2.0.0^{2,5}]octa-3,7–diene)	2C$_2$H$_2$+H	16.41±0.10	EI		2914
	(0.10 eV average translational energy of decomposition at threshold)					
C$_4$H$_3^+$	C$_8$H$_8$ (Cubane)	2C$_2$H$_2$+H	14.96±0.10	EI		2914

See also – EI: 2105

4.3. The Positive Ion Table—Continued

Ion	Reactant	Other products	Ionization or appearance potential (eV)	Method	Heat of formation (kJ mol^{-1})	Ref.
$C_4H_3^+$	C_5H_5N (Pyridine)	HCN+H	16.61±0.10	EI		1406
$C_4H_3^+$	$C_4H_4N_2$ (1,2-Diazine)	N$_2$+H	13.84±0.10	EI		1406
$C_4H_3^+$	$CH_3SCH=CHC\equiv CH$		15.3±0.3	EI		2949
$C_4H_3^+$	$CH_3SC\equiv CCH=CH_2$		15.6±0.3	EI		2949
$C_4H_3^+$	C_6H_5SH (Mercaptobenzene)		18.0±0.3	EI		3286
$C_4H_3^+$	$C_2H_5SCH=CHC\equiv CH$		16.4±0.3	EI		2949
$C_4H_3^+$	$C_2H_5SC\equiv CCH=CH_2$		16.5±0.3	EI		2949
$C_4H_3^+$	$C_6H_5SSC_6H_5$ (Diphenyl disulfide)		20.5±0.3	EI		3286

$C_4H_3^{+2}$

Ion	Reactant	Other products	Ionization or appearance potential (eV)	Method	Heat of formation (kJ mol^{-1})	Ref.
$C_4H_3^{+2}$	$CH_2=CHCH=CH_2$		35.90±0.2	EI		2455

$C_4H_4^+$

Ion	Reactant	Other products	Ionization or appearance potential (eV)	Method	Heat of formation (kJ mol^{-1})	Ref.
$C_4H_4^+$	$CH_2=C=C=CH_2$		9.25	EI		2723
$C_4H_4^+$	$CH_2=C=C=CH_2$		9.4	EI		2712
$C_4H_4^+$	$CH_2=CHC\equiv CH$		9.87	EI		411
$C_4H_4^+$	$CH_2=CHC\equiv CH$		9.9	EI		2723
$C_4H_4^+$	$CH_2=CHC\equiv CH$		9.9	EI	2712,	2752
$C_4H_4^+$	$CH_2=CHC\equiv CH$		9.9	EI		3009

See also – EI: 2535

Ion	Reactant	Other products	Ionization or appearance potential (eV)	Method	Heat of formation (kJ mol^{-1})	Ref.
$C_4H_4^+$	C_4H_4 (Cyclobutadiene?)		8.2	EI		2712
$C_4H_4^+$	C_4H_4 (Cyclobutadiene?)		9.55	EI		2723
$C_4H_4^+$	C_4H_4 (Cyclobutadiene)		8.5	D	2843,	3056

There is disagreement on the identity of the C_4H_4 isomer, see refs. 2712, 2723, 2752, 3056.

4.3. The Positive Ion Table—Continued

Ion	Reactant	Other products	Ionization or appearance potential (eV)	Method	Heat of formation (kJ mol^{-1})	Ref.
$C_4H_4^+$	$CH_2=CHCH=CH_2$	H_2	13.84±0.07	EI		2455
$C_4H_4^+$	$C_2H_5C\equiv CH$	H_2	10.9±0.2	EI		13
$C_4H_4^+$	$CH_3C\equiv CCH_3$	H_2	14.0±0.1	EI		13
$C_4H_4^+$	$trans-CH_2=CHCH=CHCH_3$		21.12±0.08	EI		2455
$C_4H_4^+$	$CH\equiv CCH=CHCH=CH_2$	C_2H_2	14.77	EI		1197
$C_4H_4^+$	$C_2H_5C\equiv CC\equiv CH$	C_2H_2	12	EI		1197
$C_4H_4^+$	$CH_3C\equiv CCH_2C\equiv CH$	C_2H_2	11.82	EI		1197
$C_4H_4^+$	$CH_3C\equiv CC\equiv CCH_3$	C_2H_2	12.12	EI		1197
$C_4H_4^+$	$CH\equiv CCH_2CH_2C\equiv CH$	C_2H_2	11.4	EI		1197
$C_4H_4^+$	C_6H_6 (Benzene)	C_2H_2	14.5±0.2	RPD		2520

See also – EI: 1197, 1238, 2103, 2833

Ion	Reactant	Other products	Ionization or appearance potential (eV)	Method	Heat of formation (kJ mol^{-1})	Ref.
$C_4H_4^+$	$CH_2=CHCH=CHCH=CH_2$	$C_2H_2+H_2$	12.82±0.10	RPD		2751

(0.21 eV average translational energy of decomposition at threshold)

Ion	Reactant	Other products	Potential	Method		Ref.
$C_4H_4^+$	C_6H_8 (1,3–Cyclohexadiene)	$C_2H_2+H_2$	13.91±0.20	EI		2751

(0.10 eV average translational energy of decomposition at threshold)

$C_4H_4^+$	C_6H_8 (1,4–Cyclohexadiene)	$C_2H_2+H_2$	13.55±0.10	EI		2751

(0.05 eV average translational energy of decomposition at threshold)

$C_4H_4^+$	$CH_3CH=CHCH=CHCH_3$		15.41±0.08	EI		2455
$C_4H_4^+$	$C_6H_5CH=CH_2$ (Ethenylbenzene)	$2C_2H_2$	17.25±0.15	EI		2914

(0.07 eV average translational energy of decomposition at threshold)

$C_4H_4^+$	C_8H_8 (Cyclooctatetraene)	$2C_2H_2$	15.10±0.10	RPD		2914

(0.11 eV average translational energy of decomposition at threshold)

$C_4H_4^+$	C_8H_8 (Cyclobutenobenzene)	$2C_2H_2$	16.01±0.15	EI		2914

(0.19 eV average translational energy of decomposition at threshold)

$C_4H_4^+$	C_8H_8 (Bicyclo[2.2.2]octatriene)	$2C_2H_2$	15.31±0.20	EI		2914
$C_4H_4^+$	C_8H_8 (syn–Tricyclo[4.2.0.0^{2,5}]octa–3,7–diene)	$2C_2H_2$	13.85±0.10	EI		2914

(0.13 eV average translational energy of decomposition at threshold)

$C_4H_4^+$	C_8H_8 ($anti$–Tricyclo[4.2.0.0^{2,5}]octa–3,7–diene)	$2C_2H_2$	13.90±0.10	EI		2914

(0.13 eV average translational energy of decomposition at threshold)

$C_4H_4^+$	C_8H_8 (Cubane)	$2C_2H_2$	12.88±0.15	EI		2914

See also – EI: 2105

$C_4H_4^+$	$C_{10}H_8$ (Naphthalene)		19.6±0.20	EI		2112
$C_4H_4^+$	$C_{10}H_8$ (Azulene)		17.8±0.10	EI		2112

4.3. The Positive Ion Table—Continued

Ion	Reactant	Other products	Ionization or appearance potential (eV)	Method	Heat of formation (kJ mol^{-1})	Ref.
$C_4H_4^+$	C_5H_5N (Pyridine)	HCN	13.28	EI		2833

See also – EI: 1406, 3013

Ion	Reactant	Other products	Ionization or appearance potential (eV)	Method	Heat of formation (kJ mol^{-1})	Ref.
$C_4H_4^+$	$C_4H_4N_2$ (1,2–Diazine)	N_2	11.64±0.05	EI		1406
$C_4H_4^+$	$C_5H_4O_2$ (2–Oxa–3–oxobicyclo[2.2.0]hex–5–ene)		10.2	EI		2712
$C_4H_4^+$	C_6H_5F (Fluorobenzene)		17.00±0.1	EI		2103
$C_4H_4^+$	C_6H_5SH (Mercaptobenzene)		17.5±0.3	EI		3286
$C_4H_4^+$	C_6H_5Cl (Chlorobenzene)		17.57±0.1	EI		2103
$C_4H_4^+$	C_6H_5Br (Bromobenzene)		16.77±0.1	EI		2103

$$C_4H_5^+ \qquad \Delta H^\circ_{f\,298} \sim 1003 \text{ kJ mol}^{-1} \ (240 \text{ kcal mol}^{-1})$$

Ion	Reactant	Other products	Ionization or appearance potential (eV)	Method	Heat of formation (kJ mol^{-1})	Ref.
$C_4H_5^+$	$CH_3CH=C=CH_2$	H	11.04±0.04	PI	1009	2644
$C_4H_5^+$	$CH_2=CHCH=CH_2$	H	11.39±0.05	PI	991	2013
$C_4H_5^+$	$CH_2=CHCH=CH_2$	H	11.56±0.04	PI	1008	2644

The two photoionization studies differ in their interpretation of the threshold.

See also – EI: 2455

Ion	Reactant	Other products	Ionization or appearance potential (eV)	Method	Heat of formation (kJ mol^{-1})	Ref.
$C_4H_5^+$	$C_2H_5C\equiv CH$	H	11.6±0.1	EI		13
$C_4H_5^+$	$CH_3C\equiv CCH_3$	H	12.1±0.1	EI		13
$C_4H_5^+$	$trans$–$CH_2=CHCH=CHCH_3$	CH_3	12.57±0.05	EI		2455
$C_4H_5^+$	$CH_2=CHCH=CHCH=CH_2$	C_2H_2+H	13.60±0.10	EI		2751
(0.23 eV average translational energy of decomposition at threshold)						
$C_4H_5^+$	C_6H_8 (1,3–Cyclohexadiene)	C_2H_2+H	14.69±0.10	EI		2751
(0.04 eV average translational energy of decomposition at threshold)						
$C_4H_5^+$	C_6H_8 (1,4–Cyclohexadiene)	C_2H_2+H	14.48±0.10	EI		2751
(0.05 eV average translational energy of decomposition at threshold)						
$C_4H_5^+$	$CH_3CH=CHCH=CHCH_3$		13.23±0.05	EI		2455
$C_4H_5^+$	$C_5H_8=CH_2$ (Methylenecyclopentane)		12.69±0.17	EDD		2738
$C_4H_5^+$	$C_5H_7CH_3$ (1–Methylcyclopentene)		13.14±0.09	EDD		2738
$C_4H_5^+$	$C_5H_7CH_3$ (3–Methylcyclopentene)		12.69±0.20	EDD		2738
$C_4H_5^+$	C_6H_{10} (Cyclohexene)		13.31±0.15	EDD		2738
$C_4H_5^+$	$(C_3H_5)_2$ (Bicyclopropyl)		11.22±0.14	EDD		2738
$C_4H_5^+$	C_6H_{10} (Bicyclo[3.1.0]hexane)		12.62±0.11	EDD		2738

4.3. The Positive Ion Table—Continued

Ion	Reactant	Other products	Ionization or appearance potential (eV)	Method	Heat of formation (kJ mol^{-1})	Ref.
$C_4H_5^+$	$C_6H_{10}=CH_2$ (Methylenecyclohexane)		13.10±0.10	EDD		2558
$C_4H_5^+$	$C_6H_9CH_3$ (1-Methylcyclohexene)		13.22±0.10	EDD		2558
$C_4H_5^+$	$C_6H_9CH_3$ (3-Methylcyclohexene)		13.42±0.07	EDD		2558
$C_4H_5^+$	$C_6H_9CH_3$ (4-Methylcyclohexene)		13.43±0.13	EDD		2558
$C_4H_5^+$	C_7H_{12} (Bicyclo[2.2.1]heptane)		13.44±0.11	EDD		2558
$C_4H_5^+$	C_7H_{12} (Bicyclo[4.1.0]heptane)		12.42±0.10	EDD		2558
$C_4H_5^+$	$(CH_3)_2C=CHCH=C(CH_3)_2$		14.60±0.04	EI		2455
$C_4H_5^+$	$CH_3SCH=CHC\equiv CH$		13.0±0.3	EI		2949
$C_4H_5^+$	$CH_3SC\equiv CCH=CH_2$		14.0±0.3	EI		2949
$C_4H_5^+$	$C_2H_5SCH=CHC\equiv CH$		13.1±0.3	EI		2949
$C_4H_5^+$	$C_2H_5SC\equiv CCH=CH_2$		13.7±0.3	EI		2949

$$CH_3CH=C=CH_2^+ \quad \Delta H_{f298}^\circ = 1053 \text{ kJ mol}^{-1} (252 \text{ kcal mol}^{-1})$$
$$trans-CH_2=CHCH=CH_2^+ \quad \Delta H_{f298}^\circ = 985 \text{ kJ mol}^{-1} (235 \text{ kcal mol}^{-1})$$
$$C_2H_5C\equiv CH^+ \quad \Delta H_{f298}^\circ = 1147 \text{ kJ mol}^{-1} (274 \text{ kcal mol}^{-1})$$
$$CH_3C\equiv CCH_3^+ \quad \Delta H_{f298}^\circ = 1069 \text{ kJ mol}^{-1} (255 \text{ kcal mol}^{-1})$$
$$C_4H_6^+ (cis-\text{Cyclobutene}) \quad \Delta H_{f298}^\circ = 1040 \text{ kJ mol}^{-1} (248 \text{ kcal mol}^{-1})$$

Ion	Reactant	Other products	Ionization or appearance potential (eV)	Method	Heat of formation (kJ mol^{-1})	Ref.	
$C_4H_6^+$	$CH_3CH=C=CH_2$		9.23±0.02	PI	1053	2644	

See also – EI: 462

Ion	Reactant	Other products	Ionization or appearance potential (eV)	Method	Heat of formation (kJ mol^{-1})	Ref.	
$C_4H_6^+$	$cis-CH_2=CHCH=CH_2$		~9.2	D		2842	
$C_4H_6^+$ (Average of two Rydberg series limits)	$trans-CH_2=CHCH=CH_2$		9.062±0.01	S	985	3352	
$C_4H_6^+$	$trans-CH_2=CHCH=CH_2$		9.070±0.01	PI	985	158,	182, 416
$C_4H_6^+$	$trans-CH_2=CHCH=CH_2$		9.075±0.005	PI	986	2013	
$C_4H_6^+$	$trans-CH_2=CHCH=CH_2$		9.06±0.02	PI	984	2724	
$C_4H_6^+$	$trans-CH_2=CHCH=CH_2$		9.07±0.02	PI	985	2644	
$C_4H_6^+$	$trans-CH_2=CHCH=CH_2$		9.06	PE	984	3120	
$C_4H_6^+$	$trans-CH_2=CHCH=CH_2$		9.07	PE	985	2842,	2843, 3056
$C_4H_6^+$	$trans-CH_2=CHCH=CH_2$		9.09±0.05	PE	987	2796	
$C_4H_6^+$	$trans-CH_2=CHCH=CH_2$		9.09±0.03	EI		2455	

See also – PE: 1130
PEN: 2430
EI: 224, 462, 2531, 2752

4.3. The Positive Ion Table—Continued

Ion	Reactant	Other products	Ionization or appearance potential (eV)	Method	Heat of formation (kJ mol^{-1})	Ref.
$C_4H_6^+$	$C_2H_5C{\equiv}CH$		10.18	S	1147	1022
(Average of three Rydberg series limits)						
$C_4H_6^+$	$C_2H_5C{\equiv}CH$		10.18±0.01	PI	1147	162, 182, 416, 1022
$C_4H_6^+$	$C_2H_5C{\equiv}CH$		10.18	EM	1147	3380
See also – EI: 13						
$C_4H_6^+$	$CH_3C{\equiv}CCH_3$		9.56	EM	1069	3380
See also – EI: 13						
$C_4H_6^+$	C_4H_6 (cis−Cyclobutene)		9.43	PE	1040	3330
$C_4H_6^+$	trans−$CH_2{=}CHCH{=}CHCH_3$		13.10±0.1	EI		2455
$C_4H_6^+$	$CH_2{=}CHCH{=}CHCH{=}CH_2$	C_2H_2	12.25±0.30	EI		2751
$C_4H_6^+$	C_6H_8 (1,3−Cyclohexadiene)	C_2H_2	12.60±0.10	EI		2751
$C_4H_6^+$	C_6H_8 (1,4−Cyclohexadiene)	C_2H_2	12.17±0.10	EI		2751
$C_4H_6^+$	$CH_3CH{=}CHCH{=}CHCH_3$		11.53±0.07	EI		2455
$C_4H_6^+$	$C_5H_8{=}CH_2$ (Methylenecyclopentane)		10.88±0.07	EDD		2738
$C_4H_6^+$	$C_5H_7CH_3$ (1−Methylcyclopentene)		11.02±0.12	EDD		2738
$C_4H_6^+$	$C_5H_7CH_3$ (3−Methylcyclopentene)		10.69±0.07	EDD		2738
$C_4H_6^+$	C_6H_{10} (Cyclohexene)		10.67±0.06	EDD		2738
$C_4H_6^+$	$(C_3H_5)_2$ (Bicyclopropyl)		9.34±0.06	EDD		2738
$C_4H_6^+$	C_6H_{10} (Bicyclo[3.1.0]hexane)		10.28±0.05	EDD		2738
$C_4H_6^+$	$C_6H_{10}{=}CH_2$ (Methylenecyclohexane)		10.94±0.05	EDD		2558
$C_4H_6^+$	$C_6H_9CH_3$ (1−Methylcyclohexene)		11.02±0.08	EDD		2558
$C_4H_6^+$	$C_6H_9CH_3$ (3−Methylcyclohexene)		10.94±0.04	EDD		2558
$C_4H_6^+$	$C_6H_9CH_3$ (4−Methylcyclohexene)		10.82±0.12	EDD		2558
$C_4H_6^+$	C_7H_{12} (Bicyclo[2.2.1]heptane)		11.12±0.03	EDD		2558
$C_4H_6^+$	C_7H_{12} (Bicyclo[4.1.0]heptane)		10.10±0.09	EDD		2558
$C_4H_6^+$	$(CH_2)_4S$ (Tetramethylene sulfide)		11.9±0.2	EI		52

4.3. The Positive Ion Table—Continued

Ion	Reactant	Other products	Ionization or appearance potential (eV)	Method	Heat of formation (kJ mol⁻¹)	Ref.

$$C_4H_7^+ \qquad \Delta H°_{f298} \sim 863 \text{ kJ mol}^{-1} \text{ (206 kcal mol}^{-1}\text{)}$$

Ion	Reactant	Other products	IP/AP (eV)	Method	ΔHf (kJ mol⁻¹)	Ref.
$C_4H_7^+$	$CH_3CHCH=CH_2$		7.54	EM	855	3380

See also – EI: 108

Ion	Reactant	Other products	IP/AP (eV)	Method	ΔHf	Ref.
$C_4H_7^+$	$CH_2=C(CH_3)CH_2$		7.89	EM		3380

See also – EI: 108

Ion	Reactant	Other products	IP/AP (eV)	Method	ΔHf	Ref.
$C_4H_7^+$	C_4H_7		7.88±0.05	EI		123

This radical, identified in ref. 123 as cyclobutyl, may have isomerized to 2–methylallyl. F. P. Lossing, private communication.

Ion	Reactant	Other products	IP/AP (eV)	Method	ΔHf	Ref.
$C_4H_7^+$	$1-C_4H_8$	H	11.26	EM	868	3380

See also – EI: 194, 195

Ion	Reactant	Other products	IP/AP (eV)	Method	ΔHf	Ref.
$C_4H_7^+$	$cis-2-C_4H_8$	H	11.32	EM	867	3380

See also – EI: 194, 195

Ion	Reactant	Other products	IP/AP (eV)	Method	ΔHf	Ref.
$C_4H_7^+$	$iso-C_4H_8$	H	11.41	EM	866	3380
$C_4H_7^+$	$C_3H_5CH_3$ (Methylcyclopropane)	H	11.02	EM		3380
$C_4H_7^+$	C_4H_8 (Cyclobutane)	H	10.91	EM	861	3380

See also – EI: 123, 2742

Ion	Reactant	Other products	IP/AP (eV)	Method	ΔHf	Ref.
$C_4H_7^+$	$CH_3CH_2CH_2CH=CH_2$	CH_3	10.64	EM	863	3380
$C_4H_7^+$	$trans-C_2H_5CH=CHCH_3$	CH_3	10.68	EM	856	3380
$C_4H_7^+$	$(CH_3)_2CHCH=CH_2$	CH_3	10.74	EM	865	3380
$C_4H_7^+$	$C_2H_5C(CH_3)=CH_2$	CH_3	10.85	EM	868	3380
$C_4H_7^+$	$(CH_3)_2C=CHCH_3$	CH_3	10.84	EM	861	3380
$C_4H_7^+$	$C_3H_5C_2H_5$ (Ethylcyclopropane)	CH_3	10.34	EM		3380
$C_4H_7^+$	$C_3H_4(CH_3)_2$ (1,1-Dimethylcyclopropane)	CH_3	10.47	EM		3380
$C_4H_7^+$	$C_3H_4(CH_3)_2$ (1,1-Dimethylcyclopropane)	CH_3	11.37±0.02	EI		1146
$C_4H_7^+$	$C_3H_4(CH_3)_2$ (cis-1,2-Dimethylcyclopropane)	CH_3	11.32±0.03	EI		1146
$C_4H_7^+$	$C_3H_4(CH_3)_2$ (trans-1,2-Dimethylcyclopropane)	CH_3	11.38±0.04	EI		1146

4.3. The Positive Ion Table—Continued

Ion	Reactant	Other products	Ionization or appearance potential (eV)	Method	Heat of formation (kJ mol^{-1})	Ref.
$C_4H_7^+$	C_5H_{10} (Cyclopentane)	CH_3	11.14	EM	855	3380
$C_4H_7^+$	$CH_3CH=CHCH=CHCH_3$		13.19±0.1	EI		2455
$C_4H_7^+$	$C_6H_{10}=CH_2$ (Methylenecyclohexane)		11.34±0.04	EDD		2558
$C_4H_7^+$	$C_6H_9CH_3$ (1–Methylcyclohexene)		11.48±0.05	EDD		2558
$C_4H_7^+$	$C_6H_9CH_3$ (3–Methylcyclohexene)		11.22±0.05	EDD		2558
$C_4H_7^+$	$C_6H_9CH_3$ (4–Methylcyclohexene)		11.24±0.04	EDD		2558
$C_4H_7^+$	C_7H_{12} (Bicyclo[2.2.1]heptane)		10.60±0.10	EDD		2558
$C_4H_7^+$	C_7H_{12} (Bicyclo[4.1.0]heptane)		10.27±0.08	EDD		2558
$C_4H_7^+$	$n-C_7H_{16}$		11.5±0.1	PI		2013
$C_4H_7^+$	$(CH_3)_2C=CHCH=C(CH_3)_2$		13.03±0.02	EI		2455
$C_4H_7^+$	$n-C_4H_9CN$	HCN+H	14.43	EI		2966
$C_4H_7^+$	$n-C_4H_9NC$		13.00	EDD		3214
$C_4H_7^+$	$(CH_2)_5O$ (1,5–Epoxypentane)	CH_2O+H	~12.2	EI		2694
$C_4H_7^+$	$n-C_5H_{11}F$		11.42	EI		2029
$C_4H_7^+$	$(CH_2)_4S$ (Tetramethylene sulfide)	SH	12.4±0.2	EI		52
$C_4H_7^+$	$CH_3CH=CHCH_2I$	I	9.15±0.05	EI		108
$C_4H_7^+$	$CH_2=C(CH_3)CH_2I$	I	9.40±0.05	EI		108

$$1-C_4H_8^+ \qquad \Delta H^\circ_{f298} \sim 925 \text{ kJ mol}^{-1} \text{ (221 kcal mol}^{-1})$$
$$2-C_4H_8^+ \qquad \Delta H^\circ_{f298} \sim 871 \text{ kJ mol}^{-1} \text{ (208 kcal mol}^{-1})$$
$$iso-C_4H_8^+ \qquad \Delta H^\circ_{f298} \sim 871 \text{ kJ mol}^{-1} \text{ (208 kcal mol}^{-1})$$
$$C_4H_8^+ \text{(Cyclobutane)} \qquad \Delta H^\circ_{f298} = 997 \text{ kJ mol}^{-1} \text{ (238 kcal mol}^{-1})$$

Ion	Reactant	Other products	Ionization or appearance potential (eV)	Method	Heat of formation (kJ mol^{-1})	Ref.
$C_4H_8^+$	$1-C_4H_8$		9.58±0.01	PI	924	133, 182, 416
$C_4H_8^+$	$1-C_4H_8$		9.61±0.02	PI	927	1120
$C_4H_8^+$	$1-C_4H_8$		9.59	PE	925	2843
$C_4H_8^+$	$1-C_4H_8$		9.62	PEN		2466
$C_4H_8^+$	$1-C_4H_8$		9.58	EM	924	3380

See also – PEN: 2430
EI: 62, 194, 195, 2941

Ion	Reactant	Other products	Ionization or appearance potential (eV)	Method	Heat of formation (kJ mol^{-1})	Ref.
$C_4H_8^+$	$cis-2-C_4H_8$		9.119	S	873	2663
$C_4H_8^+$	$cis-2-C_4H_8$		9.13±0.01	PI	874	182
$C_4H_8^+$	$cis-2-C_4H_8$		9.13	PI	874	168
$C_4H_8^+$	$cis-2-C_4H_8$		9.12	PE	873	2843
$C_4H_8^+$	$cis-2-C_4H_8$		9.10	EM	871	3380

See also – EI: 62, 194, 195, 2535, 2941

4.3. The Positive Ion Table—Continued

Ion	Reactant	Other products	Ionization or appearance potential (eV)	Method	Heat of formation (kJ mol^{-1})	Ref.
$C_4H_8^+$	trans-2-C_4H_8		9.137	S	870	2663
(Average of two Rydberg series limits)						
$C_4H_8^+$	trans-2-C_4H_8		9.13±0.01	PI	870	182
$C_4H_8^+$	trans-2-C_4H_8		9.13	PI	870	168
$C_4H_8^+$	trans-2-C_4H_8		9.11	PE	868	3073, 3087
$C_4H_8^+$	trans-2-C_4H_8		9.12	PE	869	2843

See also – S: 3353
       EI: 62,411,2941

Ion	Reactant	Other products	Ionization or appearance potential (eV)	Method	Heat of formation (kJ mol^{-1})	Ref.
$C_4H_8^+$	iso-C_4H_8		9.23±0.02	PI	874	182
$C_4H_8^+$	iso-C_4H_8		9.23	PI	874	168
$C_4H_8^+$	iso-C_4H_8		9.17	PE	868	2843
$C_4H_8^+$	iso-C_4H_8		9.19	EM	870	3380

See also – EI: 62, 2941, 3201

Ion	Reactant	Other products	Ionization or appearance potential (eV)	Method	Heat of formation (kJ mol^{-1})	Ref.
$C_4H_8^+$	$C_3H_5CH_3$ (Methylcyclopropane)		9.46	EM		3380

See also – EI: 2941

Ion	Reactant	Other products	Ionization or appearance potential (eV)	Method	Heat of formation (kJ mol^{-1})	Ref.
$C_4H_8^+$	C_4H_8 (Cyclobutane)		≤10.3	PI		3429
$C_4H_8^+$	C_4H_8 (Cyclobutane)		10.06	EM	997	3380

See also – EI: 123, 2742, 2941

Ion	Reactant	Other products	Ionization or appearance potential (eV)	Method	Heat of formation (kJ mol^{-1})	Ref.
$C_4H_8^+$	n-C_5H_{12}	CH_4	10.93±0.03	PI		1120
(Threshold value approximately corrected to 0 K)						
$C_4H_8^+$	iso-C_5H_{12}	CH_4	10.735±0.02	PI		1120
(Threshold value approximately corrected to 0 K)						
$C_4H_8^+$	neo-C_5H_{12}	CH_4	10.39±0.02	PI		1120
(Threshold value approximately corrected to 0 K)						
$C_4H_8^+$	n-C_6H_{14}	C_2H_6	11.00±0.015	PI		1120
(Threshold value approximately corrected to 0 K)						
$C_4H_8^+$	iso-C_6H_{14}	C_2H_6	10.65±0.015	PI		1120
(Threshold value approximately corrected to 0 K)						
$C_4H_8^+$	$(C_2H_5)_2CHCH_3$	C_2H_6	10.58±0.015	PI		1120
(Threshold value approximately corrected to 0 K)						
$C_4H_8^+$	$C_2H_5C(CH_3)_3$	C_2H_6	10.23±0.015	PI		1120
(Threshold value approximately corrected to 0 K)						
$C_4H_8^+$	n-C_7H_{16}	C_3H_8	10.97±0.03	PI		1120
(Threshold value approximately corrected to 0 K)						
$C_4H_8^+$	n-C_7H_{16}	C_3H_8	10.56±0.05	PI		2013
$C_4H_8^+$	n-C_8H_{18}	C_4H_{10}	11.19±0.07	PI		1120
(Threshold value approximately corrected to 0 K)						
$C_4H_8^+$	n-C_4H_9NC		12.00	EDD		3214
$C_4H_8^+$	$(CH_2)_5O$ (1,5-Epoxypentane)	CH_2O?	11.88	EI		2694
$C_4H_8^+$	tert-$C_4H_9SiH_3$		9.89±0.05	EI		2182
$C_4H_8^+$	n-C_4H_9Cl	HCl	10.95	EI		3201
$C_4H_8^+$	sec-C_4H_9Cl	HCl	10.71	EI		3201

4.3. The Positive Ion Table—Continued

Ion	Reactant	Other products	Ionization or appearance potential (eV)	Method	Heat of formation (kJ mol^{-1})	Ref.
$C_4H_8^+$	iso–C_4H_9Cl	HCl	10.77	EI		3201
$C_4H_8^+$	tert–C_4H_9Cl	HCl	9.56?	EI		3201
$C_4H_8^+$	tert–$C_4H_9SiCl_3$		10.26±0.2	EI		2182

At threshold the various $C_4H_8^+$ ions have different structures, see S. G. Lias and P. Ausloos, J. Res. NBS **75A**, 591 (1971) and ref. 2663.

$C_4D_8^+$

$C_4D_8^+$	trans–2–C_4D_8		9.168	S		2663
(Average of two Rydberg series limits)						

	n–$C_4H_9^+$	$\Delta H_{f298}^\circ \sim$ 839 kJ mol^{-1} (200 kcal mol^{-1})				
	sec–$C_4H_9^+$	$\Delta H_{f298}^\circ \sim$ 768 kJ mol^{-1} (183 kcal mol^{-1})				
	iso–$C_4H_9^+$	$\Delta H_{f298}^\circ \sim$ 830 kJ mol^{-1} (198 kcal mol^{-1})				
	$tert$–$C_4H_9^+$	$\Delta H_{f298}^\circ \sim$ 697 kJ mol^{-1} (167 kcal mol^{-1})				
$C_4H_9^+$	n–C_4H_9		8.01±0.05	EM	839	3104, 3379
$C_4H_9^+$	n–C_4H_9		8.01±0.05	RPD		2776

See also – EI: 141, 145, 2719

$C_4H_9^+$	sec–C_4H_9		7.41±0.05	EM	768	3104, 3379

See also – EI: 141, 145, 2719, 3178

$C_4H_9^+$	iso–C_4H_9		8.01±0.05	EM	830	3104, 3379

See also – EI: 141, 145, 2719, 3178, 3182

$C_4H_9^+$	$tert$–C_4H_9		6.93±0.05	EM	697	3104, 3379

See also – EI: 141, 145, 2719, 3178

$C_4H_9^+$	n–C_4H_{10}	H$^-$	10.9±0.1	PI		2606
(Threshold value approximately corrected to 0 K)						
$C_4H_9^+$	n–C_4H_{10}	H	11.65±0.1	PI	(780)	2606
(Threshold value approximately corrected to 0 K)						

See also – EI: 195

$C_4H_9^+$	iso–C_4H_{10}	H	11.6	EI		195
$C_4H_9^+$	n–C_5H_{12}	CH_3	11.055±0.07	PI	(778)	1120
(Threshold value approximately corrected to 0 K)						
$C_4H_9^+$	n–C_5H_{12}	CH_3	10.98±0.05	EM	(771)	3104
$C_4H_9^+$	iso–C_5H_{12}	CH_3	11.145±0.07	PI	(779)	1120
(Threshold value approximately corrected to 0 K)						
$C_4H_9^+$	neo–C_5H_{12}	CH_3	10.565±0.02	PI	(711)	1120
(Threshold value approximately corrected to 0 K)						
$C_4H_9^+$	neo–C_5H_{12}	CH_3	10.56	EM	(711)	3104

See also – EI: 2101, 2980

4.3. The Positive Ion Table—Continued

Ion	Reactant	Other products	Ionization or appearance potential (eV)	Method	Heat of formation (kJ mol⁻¹)	Ref.
$C_4H_9^+$	$n-C_6H_{14}$	C_2H_5	11.025±0.07	PI	(789)	1120
(Threshold value approximately corrected to 0 K)						
$C_4H_9^+$	$n-C_6H_{14}$	C_2H_5	11.05	RPD		2977
$C_4H_9^+$	$iso-C_6H_{14}$	C_2H_5	10.73±0.02	PI	(753)	1120
(Threshold value approximately corrected to 0 K)						
$C_4H_9^+$	$(C_2H_5)_2CHCH_3$	C_2H_5	10.95±0.07	PI	(777)	1120
(Threshold value approximately corrected to 0 K)						
$C_4H_9^+$	$C_2H_5C(CH_3)_3$	C_2H_5	10.60±0.025	PI	(730)	1120
(Threshold value approximately corrected to 0 K)						
$C_4H_9^+$	$n-C_7H_{16}$	$n-C_3H_7$?	11.185±0.07	PI	(805)	1120
(Threshold value approximately corrected to 0 K)						
$C_4H_9^+$	$n-C_7H_{16}$	$n-C_3H_7$?	10.56±0.05	PI	(745)	2013
$C_4H_9^+$	$n-C_7H_{16}$	$n-C_3H_7$?	10.72	RPD		2977
$C_4H_9^+$	$n-C_8H_{18}$	$n-C_4H_9$?	11.40±0.07	PI	(826)	1120
(Threshold value approximately corrected to 0 K)						
$C_4H_9^+$	$n-C_8H_{18}$	$n-C_4H_9$?	11.12	RPD		2977
$C_4H_9^+$	$n-C_9H_{20}$		11.15	RPD		2977

Although the structure of the larger neutral fragments is uncertain, the results above suggest that no primary butyl fragment ions are formed. See also W. A. Chupka, J. Chem. Phys. **54**, 1936 (1971) and refs. 2977, 3104.

Ion	Reactant	Other products	Ionization or appearance potential (eV)	Method	Heat of formation (kJ mol⁻¹)	Ref.
$C_4H_9^+$	$C_2H_5CH=C=CHCH_2C(CH_3)_3$		11.7±0.05	EI		3007, 3008
$C_4H_9^+$	$cis-(CH_3)_3CCH=CHC(CH_3)_3$		11.12±0.07	EI		2533
$C_4H_9^+$	$trans-(CH_3)_3CCH=CHC(CH_3)_3$		11.23±0.05	EI		2533
$C_4H_9^+$	$(CH_3)_3CCH=C=CH(CH_2)_4CH_3$		12.0±0.05	EI		3007
$C_4H_9^+$	$(CH_3)_3CCH=C=CHCH_2C(CH_3)_3$		11.6±0.05	EI		3007
$C_4H_9^+$	$tert-C_4H_9NC$	CN	12.15	EDD		3214
$C_4H_9^+$	$n-C_4H_9COCH_3$		11.92	RPD		2977
$C_4H_9^+$	$(n-C_4H_9)_2O$		11.75±0.1	RPD		2776
$C_4H_9^+$	$(n-C_4H_9)_2CO$		11.82	RPD		2977
$C_4H_9^+$	$CH_3COO(CH_2)_3CH_3$		11.31±0.10	EI		305
$C_4H_9^+$	$n-C_4H_9ONO$		10.6	RPD		2018
$C_4H_9^+$	$tert-C_4H_9SiH_3$		10.25±0.02	EI		2182
$C_4H_9^+$	$tert-C_4H_9Si(CH_3)_3$		11.88	RPD		3187
$C_4H_9^+$	$n-C_4H_9Cl$	Cl?	10.95±0.05	EI		3201
$C_4H_9^+$	$sec-C_4H_9Cl$	Cl?	10.99±0.05	EI		3201
$C_4H_9^+$	$iso-C_4H_9Cl$	Cl?	11.40?	EI		3201
$C_4H_9^+$	$tert-C_4H_9Cl$	Cl?	10.80±0.07	EI		3201
$C_4H_9^+$	$tert-C_4H_9SiCl_3$		10.72±0.1	EI		2182
$C_4H_9^+$	$tert-C_4H_9Sn(CH_3)_3$		10.03±0.23	EI		2720
$C_4H_9^+$	$(n-C_4H_9)_2Hg$		10.55±0.1	EI		306

4.3. The Positive Ion Table—Continued

Ion	Reactant	Other products	Ionization or appearance potential (eV)	Method	Heat of formation (kJ mol^{-1})	Ref.

$n-C_4H_{10}^+$ $\quad \Delta H_{f298}^\circ \leq 887$ kJ mol^{-1} (212 kcal mol^{-1})

$iso-C_4H_{10}^+$ $\quad \Delta H_{f298}^\circ \leq 879$ kJ mol^{-1} (210 kcal mol^{-1})

Ion	Reactant	Other products	Ionization or appearance potential (eV)	Method	Heat of formation (kJ mol^{-1})	Ref.
$C_4H_{10}^+$	$n-C_4H_{10}$		10.55±0.05	PI		1120
(Value estimated from fig. 7 of this reference)						
$C_4H_{10}^+$	$n-C_4H_{10}$		10.55±0.05	PI		2606
(Value estimated from fig. 6 of this reference)						
$C_4H_{10}^+$	$n-C_4H_{10}$		10.63±0.03	PI		182, 416
$C_4H_{10}^+$	$n-C_4H_{10}$		10.50	PE	887	1130
$C_4H_{10}^+$	$n-C_4H_{10}$		10.67	PE		2843

See also – PE: 　3060
　　　　　PEN: 2430, 2466, 2467
　　　　　EI: 　195, 2587, 3356

Ion	Reactant	Other products	Ionization or appearance potential (eV)	Method	Heat of formation (kJ mol^{-1})	Ref.
$C_4H_{10}^+$	$iso-C_4H_{10}$		10.57	PI		182
$C_4H_{10}^+$	$iso-C_4H_{10}$		~10.5	PI	~879	1120
(Value estimated from fig. 12 of this reference)						
$C_4H_{10}^+$	$iso-C_4H_{10}$		10.69	PE		2843
$C_4H_{10}^+$	$iso-C_4H_{10}$		10.78	PE		1130
$C_4H_{10}^+$	$iso-C_4H_{10}$		10.79	PE		2829

See also – PEN: 　2430
　　　　　EI: 　62, 195, 3356

The ionization potentials are probably not adiabatic, see ref. 1120.

$C_5H_2^{+2}$

Ion	Reactant	Other products	Ionization or appearance potential (eV)	Method	Heat of formation (kJ mol^{-1})	Ref.
$C_5H_2^{+2}$	$trans-CH_2=CHCH=CHCH_3$		32.64±0.1	EI		2455

$C_5H_3^+$

Ion	Reactant	Other products	Ionization or appearance potential (eV)	Method	Heat of formation (kJ mol^{-1})	Ref.
$C_5H_3^+$	$CH_3C\equiv CC\equiv CCH_3$	CH_3	13.55	EI		1197
$C_5H_3^+$	C_6H_6 (Benzene)	CH_3	15.7±0.1	RPD		2520

See also – EI: 　2498

Ion	Reactant	Other products	Ionization or appearance potential (eV)	Method	Heat of formation (kJ mol^{-1})	Ref.
$C_5H_3^+$	$CH_2=CHCH=CHCH=CH_2$	CH_3+H_2?	14.95±0.15	EI		2751
(0.18 eV average translational energy of decomposition at threshold)						
$C_5H_3^+$	C_6H_8 (1,3–Cyclohexadiene)	CH_3+H_2?	15.44±0.10	EI		2751
(0.07 eV average translational energy of decomposition at threshold)						
$C_5H_3^+$	C_6H_8 (1,4–Cyclohexadiene)	CH_3+H_2?	16.12±0.10	EI		2751
(0.20 eV average translational energy of decomposition at threshold)						
$C_5H_3^+$	$C_6H_5CH=CH_2$ (Ethenylbenzene)	$C_2H_2+CH_3$?	17.74±0.10	EI		2914
(0.09 eV average translational energy of decomposition at threshold)						
$C_5H_3^+$	C_8H_8 (Cyclooctatetraene)	$C_2H_2+CH_3$?	16.41±0.15	EI		2914
(0.15 eV average translational energy of decomposition at threshold)						

4.3. The Positive Ion Table—Continued

Ion	Reactant	Other products	Ionization or appearance potential (eV)	Method	Heat of formation (kJ mol^{-1})	Ref.
$C_5H_3^+$	C_8H_8 (Cyclobutenobenzene)	$C_2H_2+CH_3$?	17.34±0.10	EI		2914
(0.20 eV average translational energy of decomposition at threshold)						
$C_5H_3^+$	C_8H_8 (Bicyclo[2.2.2]octatriene)		16.11±0.10	EI		2914
$C_5H_3^+$	C_8H_8 ($syn-$Tricyclo[4.2.0.0^{2,5}]octa–3,7–diene)	$C_2H_2+CH_3$?	14.07±0.10	EI		2914
(0.24 eV average translational energy of decomposition at threshold)						
$C_5H_3^+$	C_8H_8 ($anti-$Tricyclo[4.2.0.0^{2,5}]octa–3,7–diene)	$C_2H_2+CH_3$?	14.36±0.10	EI		2914
(0.24 eV average translational energy of decomposition at threshold)						
$C_5H_3^+$	C_8H_8 (Cubane)		13.60±0.10	EI		2914

See also – EI: 2105

Ion	Reactant	Other products	Ionization or appearance potential (eV)	Method	Heat of formation	Ref.
$C_5H_3^+$	C_6H_5SH (Mercaptobenzene)		22.8±0.3	EI		3286
$C_5H_3^+$	$C_6H_5SSC_6H_5$ (Diphenyl disulfide)		22.2±0.3	EI		3286

$C_5H_5^+$

Ion	Reactant	Other products	Ionization or appearance potential (eV)	Method	Heat of formation	Ref.
$C_5H_5^+$	C_5H_5 (Cyclopentadienyl radical)		8.56	EI		2732, 2940
$C_5H_5^+$	C_5H_5 (Cyclopentadienyl radical)		8.69±0.1	EI		68
$C_5H_5^+$	$CH\equiv CCH=CHCH_3$	H	11.57	EI		2541
(Metastable transition indicates zero kinetic energy release)						
$C_5H_5^+$	$CH_2=C(CH_3)C\equiv CH$	H	11.6	EI		3335
$C_5H_5^+$	C_5H_6 (Cyclopentadiene)	H	11.9±0.5	SD		1451
$C_5H_5^+$	C_5H_6 (Cyclopentadiene)	H	12.62	EI		2541
(Metastable transition indicates zero kinetic energy release)						
$C_5H_5^+$	C_5H_6 (Cyclopentadiene)	H	12.6	EI		68
$C_5H_5^+$	C_5H_6 (Cyclopentadiene)	H	12.9	EI		3335
$C_5H_5^+$	$trans-CH_2=CHCH=CHCH_3$	H_2+H	13.90±0.05	EI		2455
$C_5H_5^+$	$CH_2=C(CH_3)CH=CH_2$	H_2+H	13.9	EI		3335
$C_5H_5^+$	$CH_2=CHCH=CHCH=CH_2$	CH_3	12.25±0.10	EI		2751
(0.06 eV average translational energy of decomposition at threshold)						
$C_5H_5^+$	$C_2H_5CH=CHC\equiv CH$	CH_3	12.2	EI		3335
$C_5H_5^+$	$CH_3CH=C(CH_3)C\equiv CH$	CH_3	12.4	EI		3335
$C_5H_5^+$	$C_5H_5CH_3$ (Methylcyclopentadiene)	CH_3	13.5	EI		3335

4.3. The Positive Ion Table—Continued

Ion	Reactant	Other products	Ionization or appearance potential (eV)	Method	Heat of formation (kJ mol^{-1})	Ref.
$C_5H_5^+$	C_6H_8 (1,3-Cyclohexadiene)	CH_3	13.02±0.10	EI		2751
(0.13 eV average translational energy of decomposition at threshold)						
$C_5H_5^+$	C_6H_8 (1,4-Cyclohexadiene)	CH_3	13.41±0.10	EI		2751
(0.07 eV average translational energy of decomposition at threshold)						
$C_5H_5^+$	C_6H_8 (1,4-Cyclohexadiene)	CH_3	13.6	EI		3335
$C_5H_5^+$	$CH_3CH=CHCH=CHCH_3$		13.79±0.04	EI		2455
$C_5H_5^+$	$C_5H_8=CH_2$ (Methylenecyclopentane)		13.45±0.19	EDD		2738
$C_5H_5^+$	$C_5H_7CH_3$ (1-Methylcyclopentene)		13.45±0.19	EDD		2738
$C_5H_5^+$	$C_5H_7CH_3$ (3-Methylcyclopentene)		13.35±0.05	EDD		2738
$C_5H_5^+$	C_6H_{10} (Cyclohexene)		13.57±0.11	EDD		2738
$C_5H_5^+$	$(C_3H_5)_2$ (Bicyclopropyl)		12.20±0.13	EDD		2738
$C_5H_5^+$	C_6H_{10} (Bicyclo[3.1.0]hexane)		13.20±0.12	EDD		2738
$C_5H_5^+$	$C_6H_5CH_3$ (Toluene)		16.7	EI		3335
$C_5H_5^+$	C_7H_8 (Cycloheptatriene)		16.0	EI		219
$C_5H_5^+$	C_7H_8 (Cycloheptatriene)		16.66	EI		3335
$C_5H_5^+$	C_7H_8 (Bicyclo[3.2.0]hepta-2,6-diene)		14.89	EI		219
$C_5H_5^+$	C_7H_8 (Bicyclo[3.2.0]hepta-2,6-diene)		14.9	EI		3335
$C_5H_5^+$	$C_6H_5NH_2$ (Aniline)	HCN+H	15.24	EI		2541
(Metastable transitions indicate zero kinetic energy release)						
$C_5H_5^+$	C_6H_5OH (Phenol)	CO+H	14.25	EI		2541
(Metastable transitions indicate ~0.2 eV kinetic energy release)						
$C_5H_5^+$	$C_6H_5OCH_3$ (Methoxybenzene)		13.5	EI		3335
$C_5H_5^+$	$CH_3SCH=CHC\equiv CH$		11.5±0.3	EI		2949
$C_5H_5^+$	C_6H_5SH (Mercaptobenzene)		17.3±0.3	EI		3286
$C_5H_5^+$	$(C_5H_5)_2TiCl_2$ (Bis(cyclopentadienyl)titanium dichloride)		13.0±0.4	EI		2479
$C_5H_5^+$	$(C_5H_5)_2ZrCl_2$ (Bis(cyclopentadienyl)zirconium dichloride)		13.6±0.2	EI		2479

$C_5H_5^{+2}$

Ion	Reactant	Other products	Ionization or appearance potential (eV)	Method	Heat of formation (kJ mol^{-1})	Ref.
$C_5H_5^{+2}$	$trans$-$CH_2=CHCH=CHCH_3$		30.75±0.2	EI		2455

4.3. The Positive Ion Table—Continued

Ion	Reactant	Other products	Ionization or appearance potential (eV)	Method	Heat of formation (kJ mol^{-1})	Ref.

$C_5H_6^+$ (Cyclopentadiene) $\Delta H^\circ_{f\,298} = 960$ kJ mol^{-1} (229 kcal mol^{-1})

$C_5H_6^+$	$CH_2=CHC\equiv CCH_3$		8.1	EI		3009
$C_5H_6^+$	$CH\equiv CCH=CHCH_3$		9.14	EI		2541
$C_5H_6^+$	C_5H_6 (Cyclopentadiene)		8.574±0.01	PI	961	2877
$C_5H_6^+$	C_5H_6 (Cyclopentadiene)		8.566±0.01	PE	960	3411
$C_5H_6^+$	C_5H_6 (Cyclopentadiene)		8.55	PE	959	2842, 2843

Price and Walsh, ref. 3351, found a Rydberg series giving an ionization potential of 8.62 eV. Their absorption spectrum has been reinterpreted by Derrick *et al.*, ref. 3411, to give limits of 8.566 and 10.620 eV, in agreement with the photoelectron spectra.

See also – PE: 3246, 3330
        EI: 68, 2163, 2541

$C_5H_6^+$	*trans*–$CH_2=CHCH=CHCH_3$	H_2	12.34±0.06	EI		2455
$C_5H_6^+$	C_7H_8 (Bicyclo[3.2.0]hepta–2,6–diene)		10.02	EI		219
$C_5H_6^+$	C_7H_{10} (1,3–Cycloheptadiene)		10.70	EI		219
$C_5H_6^+$	C_7H_{10} (Bicyclo[2.2.1]hept–2–ene)		9.58±0.15	EI		2155
$C_5H_6^+$	C_7H_{10} (Bicyclo[3.2.0]hept–6–ene)		10.15	EI		219
$C_5H_6^+$	$C_6H_5NH_2$ (Aniline)	HCN	12.3±0.1	PI		1160

See also – EI: 2541, 2972, 3238

$C_5H_6^+$	C_6H_5OH (Phenol)	CO	11.67	EI		3238

See also – EI: 2541

$C_5H_6^+$	C_6H_5SH (Mercaptobenzene)		13.4±0.3	EI		3286
$C_5H_6^+$	C_7H_9Cl (*endo*–5–Chlorobicyclo[2.2.1]hept–2–ene)		9.75±0.15	EI		2155
$C_5H_6^+$	C_7H_9Cl (*exo*–5–Chlorobicyclo[2.2.1]hept–2–ene)		9.77±0.15	EI		2155
$C_5H_6^+$	C_7H_9Cl (3–Chloronortricyclene)		10.15±0.15	EI		2155

$C_5H_5D^+$

$C_5H_5D^+$	C_6H_5SD (Mercapto–d_1–benzene)		11.9±0.2	EI		1039

$C_5H_6^{+2}$

$C_5H_6^{+2}$	*trans*–$CH_2=CHCH=CHCH_3$		25.41±0.2	EI		2455

4.3. The Positive Ion Table—Continued

Ion	Reactant	Other products	Ionization or appearance potential (eV)	Method	Heat of formation (kJ mol^{-1})	Ref.
		$C_5H_7^+$				
$C_5H_7^+$	$CH_2=CHCHCH=CH_2$		7.76	EI		2543
$C_5H_7^+$	C_5H_7 (3–Cyclopentenyl radical)		7.54	EI		2543
$C_5H_7^+$	$trans-CH_2=CHCH=CHCH_3$	H	10.93±0.02	EI		2455
$C_5H_7^+$	C_5H_8 (Cyclopentene)	H	11.19	EI		2543
$C_5H_7^+$	C_5H_8 (Bicyclo[1.1.1]pentane)	H	10.60	EI		2560
$C_5H_7^+$	$CH_3CH=CHCH=CHCH_3$	CH_3	10.72±0.05	EI		2455
$C_5H_7^+$	$CH_2=CHCH(CH_3)CH=CH_2$	CH_3	9.77	EI		2543
$C_5H_7^+$	$C_5H_8=CH_2$ (Methylenecyclopentane)	CH_3	9.65±0.03	EDD		2738
$C_5H_7^+$	$C_5H_7CH_3$ (1–Methylcyclopentene)	CH_3	9.99±0.09	EDD		2738
$C_5H_7^+$	$C_5H_7CH_3$ (3–Methylcyclopentene)	CH_3	9.67±0.11	EDD		2738
$C_5H_7^+$	$C_5H_7CH_3$ (3–Methylcyclopentene)	CH_3	10.52	EI		2543
$C_5H_7^+$	C_6H_{10} (Cyclohexene)	CH_3	10.18±0.12	EDD		2738
$C_5H_7^+$	$(C_3H_5)_2$ (Bicyclopropyl)	CH_3	9.12±0.08	EDD		2738
$C_5H_7^+$	C_6H_{10} (Bicyclo[3.1.0]hexane)	CH_3	9.68±0.07	EDD		2738
$C_5H_7^+$	$C_6H_{10}=CH_2$ (Methylenecyclohexane)	C_2H_5	10.45±0.11	EDD		2558
$C_5H_7^+$	$C_6H_9CH_3$ (1–Methylcyclohexene)	C_2H_5	10.47±0.07	EDD		2558
$C_5H_7^+$	$C_6H_9CH_3$ (3–Methylcyclohexene)	C_2H_5	10.43±0.05	EDD		2558
$C_5H_7^+$	$C_6H_9CH_3$ (4–Methylcyclohexene)	C_2H_5	10.43±0.10	EDD		2558
$C_5H_7^+$	C_7H_{12} (Bicyclo[2.2.1]heptane)	C_2H_5	10.60±0.10	EDD		2558
$C_5H_7^+$	C_7H_{12} (Bicyclo[4.1.0]heptane)	C_2H_5	9.53±0.07	EDD		2558
$C_5H_7^+$	$CH_3CH=C=CH(CH_2)_3CH_3$		13.8	EI		3008
$C_5H_7^+$	$C_2H_5CH=C=CHCH_2CH_2CH_3$		11.6	EI		3008
$C_5H_7^+$	$(CH_3)_2C=CHCH=C(CH_3)_2$	$C_2H_4+CH_3$	12.88±0.03	EI		2455
$C_5H_7^+$	$CH_3CH=C=CHCH_2CH_2CH(CH_3)_2$		12.7	EI		3008
$C_5H_7^+$	$C_2H_5CH=C=CHCH_2CH(CH_3)_2$		11.7	EI		3008
$C_5H_7^+$	C_9H_{16} (cis–Hexahydroindan)	$C_2H_5+C_2H_4$	12.21±0.05	EI	1184,	2028
$C_5H_7^+$	C_9H_{16} (trans–Hexahydroindan)	$C_2H_5+C_2H_4$	12.19±0.03	EI	1184,	2028
$C_5H_7^+$	$(C_5H_7)_2$ (3,3'–Bicyclopentenyl)	$cyclo-C_5H_7$	9.75	EI		2543
$C_5H_7^+$	$C_2H_5CH=C=CH(CH_2)_4CH_3$		11.9	EI		3008
$C_5H_7^+$	$C_2H_5CH=C=CHCH_2C(CH_3)_3$		13.5	EI		3008
$C_5H_7^+$	$C_6H_{11}OH$ (Cyclohexanol)	CH_3+H_2O	10.9	EI		3022

4.3. The Positive Ion Table—Continued

Ion	Reactant	Other products	Ionization or appearance potential (eV)	Method	Heat of formation (kJ mol^{-1})	Ref.
	$CH_2=CHCH=CHCH_3^+$	$\Delta H^\circ_{f298} \sim$	905 kJ mol^{-1} (216 kcal mol^{-1})			
	$CH_2=C(CH_3)CH=CH_2^+$	$\Delta H^\circ_{f298} =$	929 kJ mol^{-1} (222 kcal mol^{-1})			
	$C_5H_8^+$ (Cyclopentene)	$\Delta H^\circ_{f298} =$	902 kJ mol^{-1} (216 kcal mol^{-1})			
	$C_5H_8^+$ (Spiropentane)	$\Delta H^\circ_{f298} =$	1097 kJ mol^{-1} (262 kcal mol^{-1})			
$C_5H_8^+$	$C_2H_5CH=C=CH_2$		9.42	EI		62
$C_5H_8^+$	$cis-CH_2=CHCH=CHCH_3$		8.59	PE	907	2842, 2843

See also – EI: 62

| $C_5H_8^+$ | $trans-CH_2=CHCH=CHCH_3$ | | 8.56 | PE | 904 | 2842, 2843 |

See also – EI: 62, 2455

$C_5H_8^+$	$CH_3CH=C=CHCH_3$		9.26	EI		62
$C_5H_8^+$	$CH_2=CHCH_2CH=CH_2$		9.58	EI		62
$C_5H_8^+$	$CH_2=C(CH_3)CH=CH_2$		8.844±0.01	S	929	3352
(Average of two Rydberg series limits)						
$C_5H_8^+$	$CH_2=C(CH_3)CH=CH_2$		8.845±0.005	PI	929	182

See also – EI: 2411

$C_5H_8^+$	$C_4H_6=CH_2$ (Methylenecyclobutane)		9.16±0.02	PI		2877
$C_5H_8^+$	$C_4H_6=CH_2$ (Methylenecyclobutane)		9.12	PE		3343
$C_5H_8^+$	C_5H_8 (Cyclopentene)		9.01±0.01	PI	902	182
$C_5H_8^+$	C_5H_8 (Cyclopentene)		9.02±0.01	PI	903	2877
$C_5H_8^+$	C_5H_8 (Cyclopentene)		9.00	PE	901	2843
$C_5H_8^+$	C_5H_8 (Cyclopentene)		9.01±0.01	PE	902	3158
$C_5H_8^+$	C_5H_8 (cis–Cyclopentene)		9.01	PE	902	3330

See also – EI: 62, 411, 2531, 3342

$C_5H_8^+$	C_5H_8 (Bicyclo[1.1.1]pentane)		9.65	EI		2560
$C_5H_8^+$	C_5H_8 (Spiropentane)		9.45	PE	1097	2843, 2951
$C_5H_8^+$	$C_6H_{10}=CH_2$ (Methylenecyclohexane)	C_2H_4	10.46±0.08	EDD		2558
$C_5H_8^+$	$C_6H_9CH_3$ (1–Methylcyclohexene)	C_2H_4	10.56±0.15	EDD		2558
$C_5H_8^+$	$C_6H_9CH_3$ (3–Methylcyclohexene)	C_2H_4	10.50±0.04	EDD		2558
$C_5H_8^+$	$C_6H_9CH_3$ (4–Methylcyclohexene)	C_2H_4	10.39±0.05	EDD		2558
$C_5H_8^+$	C_7H_{12} (Bicyclo[2.2.1]heptane)	C_2H_4	10.30±0.07	EDD		2558

4.3. The Positive Ion Table—Continued

Ion	Reactant	Other products	Ionization or appearance potential (eV)	Method	Heat of formation (kJ mol^{-1})	Ref.
$C_5H_8^+$	C_7H_{12} (Bicyclo[4.1.0]heptane)	C_2H_4	9.37±0.07	EDD		2558

For a discussion of these fragment ion structures see ref. 2558.

Ion	Reactant	Other products	Ionization or appearance potential (eV)	Method	Heat of formation (kJ mol^{-1})	Ref.
$C_5H_8^+$	$CH_3CH=C=CH(CH_2)_3CH_3$		9.2	EI		3008
$C_5H_8^+$	$CH_3CH=C=CHCH_2CH_2CH(CH_3)_2$		9.9	EI		3008
$C_5H_8^+$	$C_4H_4(CH_3)_2(CH=CH_2)_2$ (trans-1,2-Diethenyl-1,2-dimethylcyclobutane)		9.4±0.1	EI		2411
$C_5H_8^+$	$C_2H_5CH=C=CHCH_2C(CH_3)_3$		12.6	EI		3008
$C_5H_8^+$	C_5H_9OH (Cyclopentanol)	H_2O	9.49	RPD		2999
$C_5H_8^+$	C_5H_9F (Fluorocyclopentane)	HF	10.56	EI		2029
$C_5H_8^+$	C_5H_9Cl (Chlorocyclopentane)	HCl	10.53	EI		2029

$C_5H_8^{+2}$

Ion	Reactant	Other products	Ionization or appearance potential (eV)	Method	Heat of formation (kJ mol^{-1})	Ref.
$C_5H_8^{+2}$	trans-$CH_2=CHCH=CHCH_3$		24.51±0.1	EI		2455

$C_5H_9^+$

Ion	Reactant	Other products	Ionization or appearance potential (eV)	Method	Heat of formation (kJ mol^{-1})	Ref.
$C_5H_9^+$	$C_2H_5CH=CHCH_2$		7.65	EI		2543
$C_5H_9^+$	C_5H_9 (Cyclopentyl radical)		7.79±0.03	EI		123
$C_5H_9^+$	$C_5H_9CH_3$ (Methylcyclopentane)	CH_3	10.95	EI		123
$C_5H_9^+$	$C_6H_{10}(CH_3)_2$ (cis-1,2-Dimethylcyclohexane)		12.12±0.05	EI		1145
$C_5H_9^+$	$C_6H_{10}(CH_3)_2$ (trans-1,2-Dimethylcyclohexane)		12.00±0.05	EI		1145
$C_5H_9^+$	cis-$(CH_3)_3CCH=CHC(CH_3)_3$		11.70±0.03	EI		2533
$C_5H_9^+$	trans-$(CH_3)_3CCH=CHC(CH_3)_3$		12.06±0.05	EI		2533

$$CH_3CH_2CH_2CH=CH_2^+ \qquad \Delta H^\circ_{f298} = 895 \text{ kJ mol}^{-1} \text{ (214 kcal mol}^{-1}\text{)}$$
$$C_2H_5CH=CHCH_3^+ \qquad \Delta H^\circ_{f298} \sim 829 \text{ kJ mol}^{-1} \text{ (198 kcal mol}^{-1}\text{)}$$
$$(CH_3)_2CHCH=CH_2^+ \qquad \Delta H^\circ_{f298} = 889 \text{ kJ mol}^{-1} \text{ (213 kcal mol}^{-1}\text{)}$$
$$C_2H_5C(CH_3)=CH_2^+ \qquad \Delta H^\circ_{f298} = 844 \text{ kJ mol}^{-1} \text{ (202 kcal mol}^{-1}\text{)}$$
$$(CH_3)_2C=CHCH_3^+ \qquad \Delta H^\circ_{f298} = 795 \text{ kJ mol}^{-1} \text{ (190 kcal mol}^{-1}\text{)}$$
$$C_5H_{10}^+ \text{ (Cyclopentane)} \qquad \Delta H^\circ_{f298} \sim 936 \text{ kJ mol}^{-1} \text{ (224 kcal mol}^{-1}\text{)}$$

Ion	Reactant	Other products	Ionization or appearance potential (eV)	Method	Heat of formation (kJ mol^{-1})	Ref.
$C_5H_{10}^+$	$CH_3CH_2CH_2CH=CH_2$		9.50±0.02	PI	896	182
$C_5H_{10}^+$	$CH_3CH_2CH_2CH=CH_2$		9.50±0.02	PI	896	1120
$C_5H_{10}^+$	$CH_3CH_2CH_2CH=CH_2$		9.48	EM	894	3380

See also — EI: 62

Ion	Reactant	Other products	Ionization or appearance potential (eV)	Method	Heat of formation (kJ mol^{-1})	Ref.
$C_5H_{10}^+$	cis-$C_2H_5CH=CHCH_3$		9.11	EI		62
$C_5H_{10}^+$	trans-$C_2H_5CH=CHCH_3$		8.92	EM	829	3380
$C_5H_{10}^+$	trans-$C_2H_5CH=CHCH_3$		9.06	EI		62

4.3. The Positive Ion Table—Continued

Ion	Reactant	Other products	Ionization or appearance potential (eV)	Method	Heat of formation (kJ mol^{-1})	Ref.
C$_5$H$_{10}^+$	(CH$_3$)$_2$CHCH=CH$_2$		9.51±0.03	PI	889	182
C$_5$H$_{10}^+$	(CH$_3$)$_2$CHCH=CH$_2$		9.52	PE	890	2843
C$_5$H$_{10}^+$	(CH$_3$)$_2$CHCH=CH$_2$		9.52	EM	890	3380

See also – EI: 62

Ion	Reactant	Other products	Ionization or appearance potential (eV)	Method	Heat of formation (kJ mol^{-1})	Ref.
C$_5$H$_{10}^+$	C$_2$H$_5$C(CH$_3$)=CH$_2$		9.12±0.02	PI	844	182
C$_5$H$_{10}^+$	C$_2$H$_5$C(CH$_3$)=CH$_2$		9.12±0.02	PI	844	2877
C$_5$H$_{10}^+$	C$_2$H$_5$C(CH$_3$)=CH$_2$		9.12	EM	844	3380

See also – EI: 62

Ion	Reactant	Other products	Ionization or appearance potential (eV)	Method	Heat of formation (kJ mol^{-1})	Ref.
C$_5$H$_{10}^+$	(CH$_3$)$_2$C=CHCH$_3$		8.67±0.02	PI	794	182
C$_5$H$_{10}^+$	(CH$_3$)$_2$C=CHCH$_3$		8.68	PI	795	168
C$_5$H$_{10}^+$	(CH$_3$)$_2$C=CHCH$_3$		8.70	EM	797	3380
C$_5$H$_{10}^+$	(CH$_3$)$_2$C=CHCH$_3$		8.85±0.04	RPD		2410

See also – S: 3353
        EI: 62

Ion	Reactant	Other products	Ionization or appearance potential (eV)	Method	Heat of formation (kJ mol^{-1})	Ref.
C$_5$H$_{10}^+$	C$_3$H$_5$C$_2$H$_5$ (Ethylcyclopropane)		9.50	EM		3380
C$_5$H$_{10}^+$	C$_3$H$_4$(CH$_3$)$_2$ (1,1–Dimethylcyclopropane)		9.08	EM		3380
C$_5$H$_{10}^+$	C$_3$H$_4$(CH$_3$)$_2$ (1,1–Dimethylcyclopropane)		9.76±0.02	EI		1146
C$_5$H$_{10}^+$	C$_3$H$_4$(CH$_3$)$_2$ (cis–1,2–Dimethylcyclopropane)		9.76±0.02	EI		1146
C$_5$H$_{10}^+$	C$_3$H$_4$(CH$_3$)$_2$ (trans–1,2–Dimethylcyclopropane)		9.73±0.02	EI		1146
C$_5$H$_{10}^+$	C$_5$H$_{10}$ (Cyclopentane)		10.53±0.05	PI	939	182
C$_5$H$_{10}^+$	C$_5$H$_{10}$ (Cyclopentane)		10.49	PE	935	2843
C$_5$H$_{10}^+$	C$_5$H$_{10}$ (Cyclopentane)		10.50±0.01	PE	936	3158
C$_5$H$_{10}^+$	C$_5$H$_{10}$ (Cyclopentane)		10.49	EM	935	3380

See also – EI: 123

4.3. The Positive Ion Table—Continued

Ion	Reactant	Other products	Ionization or appearance potential (eV)	Method	Heat of formation (kJ mol⁻¹)	Ref.
$C_5H_{10}^+$	$n-C_6H_{14}$	CH_4	11.005±0.055	PI		1120
(Threshold value approximately corrected to 0 K)						
$C_5H_{10}^+$	$iso-C_6H_{14}$	CH_4	10.835±0.025	PI		1120
(Threshold value approximately corrected to 0 K)						
$C_5H_{10}^+$	$(C_2H_5)_2CHCH_3$	CH_4	10.70±0.055	PI		1120
(Threshold value approximately corrected to 0 K)						
$C_5H_{10}^+$	$C_2H_5C(CH_3)_3$	CH_4	10.28±0.02	PI		1120
(Threshold value approximately corrected to 0 K)						
$C_5H_{10}^+$	$(CH_3)_2CHCH(CH_3)_2$	CH_4	10.54±0.03	PI		1120
(Threshold value approximately corrected to 0 K)						
$C_5H_{10}^+$	$n-C_7H_{16}$	C_2H_6	11.035±0.025	PI		1120
(Threshold value approximately corrected to 0 K)						
$C_5H_{10}^+$	$n-C_7H_{16}$	C_2H_6	10.40±0.05	PI		2013
$C_5H_{10}^+$	$n-C_7H_{16}$	C_2H_6	10.33	RPD		2999
$C_5H_{10}^+$	$C_6H_{10}(CH_3)_2$		11.62±0.08	EI		1145
	($cis-1,2-$Dimethylcyclohexane)					
$C_5H_{10}^+$	$C_6H_{10}(CH_3)_2$		11.60±0.07	EI		1145
	($trans-1,2-$Dimethylcyclohexane)					
$C_5H_{10}^+$	$n-C_8H_{18}$	C_3H_8	11.08±0.03	PI		1120
(Threshold value approximately corrected to 0 K)						
$C_5H_{10}^+$	$cis-(CH_3)_3CCH=CHC(CH_3)_3$		9.15±0.06	EI		2533
$C_5H_{10}^+$	$trans-(CH_3)_3CCH=CHC(CH_3)_3$		9.60±0.05	EI		2533
$C_5H_{10}^+$	$n-C_5H_{11}F$	HF	10.07	EI		2029
$C_5H_{10}^+$	$n-C_5H_{11}Cl$	HCl	10.63	EI		2029

$C_5H_{11}^+$

Ion	Reactant	Other products	Ionization or appearance potential (eV)	Method	Heat of formation (kJ mol⁻¹)	Ref.
$C_5H_{11}^+$	$n-C_3H_7CHCH_3$		7.73±0.1	EI		151
$C_5H_{11}^+$	$(C_2H_5)_2CH$		7.86±0.05	EI		151
$C_5H_{11}^+$	$C_2H_5C(CH_3)_2$		7.12±0.1	EI		151
$C_5H_{11}^+$	$neo-C_5H_{11}$		8.33±0.1	EI		151
$C_5H_{11}^+$	$n-C_6H_{14}$	CH_3	11.045±0.085	PI		1120
(Threshold value approximately corrected to 0 K)						
$C_5H_{11}^+$	$iso-C_6H_{14}$	CH_3	10.865±0.085	PI		1120
(Threshold value approximately corrected to 0 K)						
$C_5H_{11}^+$	$(C_2H_5)_2CHCH_3$	CH_3	10.86±0.085	PI		1120
(Threshold value approximately corrected to 0 K)						
$C_5H_{11}^+$	$C_2H_5C(CH_3)_3$	CH_3	10.555±0.045	PI		1120
(Threshold value approximately corrected to 0 K)						
$C_5H_{11}^+$	$(CH_3)_2CHCH(CH_3)_2$	CH_3	10.72±0.085	PI		1120
(Threshold value approximately corrected to 0 K)						
$C_5H_{11}^+$	$n-C_7H_{16}$	C_2H_5	10.96±0.085	PI		1120
(Threshold value approximately corrected to 0 K)						
$C_5H_{11}^+$	$n-C_7H_{16}$	C_2H_5	10.43±0.05	PI		2013
$C_5H_{11}^+$	$n-C_7H_{16}$	C_2H_5	10.66	RPD		2977

4.3. The Positive Ion Table—Continued

Ion	Reactant	Other products	Ionization or appearance potential (eV)	Method	Heat of formation (kJ mol^{-1})	Ref.
$C_5H_{11}^+$	$n-C_8H_{18}$	C_3H_7	11.22±0.085	PI		1120
(Threshold value approximately corrected to 0 K)						
$C_5H_{11}^+$	$n-C_8H_{18}$	C_3H_7	11.03	RPD		2977
$C_5H_{11}^+$	$n-C_9H_{20}$	C_4H_9	11.10	RPD		2977
$C_5H_{11}^+$	$n-C_5H_{11}COCH_3$		11.16	EI		2977
$C_5H_{11}^+$	$n-C_5H_{11}COC_2H_5$		11.5±0.3	EI		2740
$C_5H_{11}^+$	$n-C_5H_{11}OOH$		11.3±0.1	EI		2464
$C_5H_{11}^+$	$n-C_3H_7CH(OOH)CH_3$		10.7±0.1	EI		2464
$C_5H_{11}^+$	$(C_2H_5)_2CHOOH$		10.9±0.1	EI		2464

$n-C_5H_{12}^+$ $\Delta H_{f298}^\circ = 853$ kJ mol^{-1} (204 kcal mol^{-1})
$iso-C_5H_{12}^+$ $\Delta H_{f298}^\circ = 841$ kJ mol^{-1} (201 kcal mol^{-1})
$neo-C_5H_{12}^+$ $\Delta H_{f298}^\circ \leqslant 835$ kJ mol^{-1} (200 kcal mol^{-1})

Ion	Reactant	Other products	Ionization or appearance potential (eV)	Method	Heat of formation (kJ mol^{-1})	Ref.
$C_5H_{12}^+$	$n-C_5H_{12}$		10.35	PI	852	182
$C_5H_{12}^+$	$n-C_5H_{12}$		10.37	PE	854	2843

See also – PE: 3060

$C_5H_{12}^+$	$iso-C_5H_{12}$		10.32	PI	841	182
$C_5H_{12}^+$	$iso-C_5H_{12}$		10.32	PE	841	2843

See also – EI: 62

$C_5H_{12}^+$	$neo-C_5H_{12}$		10.35	PI	833	182
$C_5H_{12}^+$	$neo-C_5H_{12}$		10.40	PE	837	2843

The ionization potential is probably not adiabatic. The parent ion is not observed in electron impact mass spectra.

C_6H^+

C_6H^+	C_6H_6		29±2	RPD		2520
	(Benzene)					
C_6H^+	$C_6H_5SCH_3$		14.4	EI		307
	(Methylthiobenzene)					

$C_6H_2^+$

$C_6H_2^+$	C_6H_2		9.8±0.1	EI		87

(Fragment from electron impact induced decomposition of benzene)

See also – EI: 2635

4.3. The Positive Ion Table—Continued

Ion	Reactant	Other products	Ionization or appearance potential (eV)	Method	Heat of formation (kJ mol^{-1})	Ref.

$C_6H_3^+$

Ion	Reactant	Other products	Ionization or appearance potential (eV)	Method	Heat of formation (kJ mol^{-1})	Ref.
$C_6H_3^+$	$C_2H_5C{\equiv}CC{\equiv}CH$	H_2+H	17.92	EI		1197
$C_6H_3^+$	$CH_3C{\equiv}CC{\equiv}CCH_3$	H_2+H	17.99	EI		1197
$C_6H_3^+$	$C_{10}H_8$ (Naphthalene)		20.77±0.01	EI		2112
$C_6H_3^+$	$C_{10}H_8$ (Azulene)		19.2±0.15	EI		2112
$C_6H_3^+$	C_6H_5SH (Mercaptobenzene)		20.6±0.3	EI		3286
$C_6H_3^+$	C_6H_4FCl (1–Chloro–2–fluorobenzene)		16.67	EI		1185
$C_6H_3^+$	C_6H_4FCl (1–Chloro–3–fluorobenzene)		16.78	EI		1185
$C_6H_3^+$	C_6H_4FCl (1–Chloro–4–fluorobenzene)		16.81	EI		1185

$C_6H_4^+$

Ion	Reactant	Other products	Ionization or appearance potential (eV)	Method	Heat of formation (kJ mol^{-1})	Ref.
$C_6H_4^+$	$CH{\equiv}CCH{=}CHC{\equiv}CH$		9.60±0.2	RPD		2492
$C_6H_4^+$	C_6H_4 (Benzyne)		9.45±0.2	RPD		2492

See also – EI: 29, 87, 2635

Ion	Reactant	Other products	Ionization or appearance potential (eV)	Method	Heat of formation (kJ mol^{-1})	Ref.
$C_6H_4^+$	$CH{\equiv}CCH{=}CHCH{=}CH_2$	H_2	13.72	EI		1197
$C_6H_4^+$	$C_2H_5C{\equiv}CC{\equiv}CH$	H_2	11.07	EI		1197
$C_6H_4^+$	$CH_3C{\equiv}CCH_2C{\equiv}CH$	H_2	11.02	EI		1197
$C_6H_4^+$	$CH_3C{\equiv}CC{\equiv}CCH_3$	H_2	11.35	EI		1197
$C_6H_4^+$	$CH{\equiv}CCH_2CH_2C{\equiv}CH$	H_2	11.17	EI		1197
$C_6H_4^+$	C_6H_6 (Benzene)	H_2	14.2±0.2	RPD		2520
$C_6H_4^+$	C_6H_6 (Benzene)	H_2	14.09±0.07	EI		1238

See also – EI: 1197, 2103

4.3. The Positive Ion Table—Continued

Ion	Reactant	Other products	Ionization or appearance potential (eV)	Method	Heat of formation (kJ mol^{-1})	Ref.
$C_6H_4^+$	$C_{10}H_8$ (Naphthalene)		18.2±0.15	EI		2112
$C_6H_4^+$	$C_{10}H_8$ (Azulene)		16.7±0.15	EI		2112
$C_6H_4^+$	$(C_6H_5)_2$ (Biphenyl)		18.05±0.3	EI		1238
$C_6H_4^+$	$C_{14}H_{10}$ (Phenanthrene)		29±1	RPD		2540
$C_6H_4^+$	C_6H_5CN (Benzoic acid nitrile)	HCN	14.60	EI		3238

See also – EI: 2420, 2972

Ion	Reactant	Other products	Ionization or appearance potential (eV)	Method	Heat of formation (kJ mol^{-1})	Ref.
$C_6H_4^+$	$C_{10}H_6O_2$ (1,4–Naphthoquinone)		15.70±0.2	RPD		2492
$C_6H_4^+$	$C_8H_4O_3$ (1,2–Benzenedicarboxylic acid anhydride)		13.24±0.2	RPD		2492
$C_6H_4^+$	$C_6H_4(COOCH_3)_2$ (1,2–Benzenedicarboxylic acid dimethyl ester)		14.73±0.2	RPD		2492
$C_6H_4^+$	C_6H_5F (Fluorobenzene)	HF	15.37±0.1	EI		2103
$C_6H_4^+$	C_6H_5SH (Mercaptobenzene)		17.2±0.3	EI		3286
$C_6H_4^+$	C_6H_5Cl (Chlorobenzene)	HCl	14.87±0.2	EI		2103
$C_6H_4^+$	C_6H_5Br (Bromobenzene)	HBr	14.20±0.2	EI		2103
$C_6H_4^+$	$C_6H_4I_2$ (1,2–Diiodobenzene)		13.61±0.2	RPD		2492
$C_6H_4^+$	$C_6H_4I_2$ (1,3–Diiodobenzene)		13.40±0.2	RPD		2492
$C_6H_4^+$	$C_6H_4I_2$ (1,4–Diiodobenzene)		13.83±0.2	RPD		2492

$C_6H_2D_2^+$

Ion	Reactant	Other products	Ionization or appearance potential (eV)	Method	Heat of formation (kJ mol^{-1})	Ref.
$C_6H_2D_2^+$	$C_6H_2D_3NC$ (Isocyanobenzene-2,4,6-d_3)		13.6±0.6	EI		2919

$C_6HD_3^+$

Ion	Reactant	Other products	Ionization or appearance potential (eV)	Method	Heat of formation (kJ mol^{-1})	Ref.
$C_6HD_3^+$	$C_6H_2D_3NC$ (Isocyanobenzene-2,4,6-d_3)		13.6±0.6	EI		2919

4.3. The Positive Ion Table—Continued

Ion	Reactant	Other products	Ionization or appearance potential (eV)	Method	Heat of formation (kJ mol^{-1})	Ref.
			C$_6$H$_5^+$			
C$_6$H$_5^+$	C$_6$H$_5$ (Phenyl radical)		9.20	EI	(1188)	1079

See also — EI: 87, 2635

Ion	Reactant	Other products	Ionization or appearance potential (eV)	Method	Heat of formation (kJ mol^{-1})	Ref.
C$_6$H$_5^+$	CH≡CCH=CHCH=CH$_2$	H	13.52	EI		1197
C$_6$H$_5^+$	C$_2$H$_5$C≡CC≡CH	H	11.50	EI		1197
C$_6$H$_5^+$	CH$_3$C≡CCH$_2$C≡CH	H	11.50	EI		1197
C$_6$H$_5^+$	CH$_3$C≡CC≡CCH$_3$	H	11.57	EI		1197
C$_6$H$_5^+$	CH≡CCH$_2$CH$_2$C≡CH	H	11.47	EI		1197
C$_6$H$_5^+$	C$_6$H$_6$ (Benzene)	H	13.80±0.03	PI	(1196)	3212
C$_6$H$_5^+$	C$_6$H$_6$ (Benzene)	H	13.8±0.1	PI	(~1196)	2013
C$_6$H$_5^+$	C$_6$H$_6$ (Benzene)	H	14.1±0.1	RPD		2520

See also — PI: 3025
          EI: 301, 1197, 1238, 2103, 2538, 3344

Ion	Reactant	Other products	Ionization or appearance potential (eV)	Method	Heat of formation (kJ mol^{-1})	Ref.
C$_6$H$_5^+$	CH$_2$=CHCH=CHCH=CH$_2$	H$_2$+H	13.11±0.10	EI		2751
C$_6$H$_5^+$	C$_2$H$_5$CH=CHC≡CH	H$_2$+H	12.7	EI		3335
C$_6$H$_5^+$	CH$_3$CH=C(CH$_3$)C≡CH	H$_2$+H	12.7	EI		3335
C$_6$H$_5^+$	C$_5$H$_5$CH$_3$ (Methylcyclopentadiene)	H$_2$+H	14.0	EI		3335
C$_6$H$_5^+$	C$_6$H$_8$ (1,3−Cyclohexadiene)	H$_2$+H	13.92±0.10	EI		2751
C$_6$H$_5^+$	C$_6$H$_8$ (1,4−Cyclohexadiene)	H$_2$+H	13.92±0.10	EI		2751
C$_6$H$_5^+$	C$_6$H$_5$CH$_3$ (Toluene)	CH$_3$	13.7±0.1	EI		301, 3344
C$_6$H$_5^+$	C$_7$H$_{10}$ (Bicyclo[2.2.1]hept-2-ene)		13.8±0.3	EI		2155
C$_6$H$_5^+$	C$_6$H$_5$CH=CH$_2$ (Ethenylbenzene)	C$_2$H$_2$+H	16.02±0.10	EI		2914
C$_6$H$_5^+$	C$_8$H$_8$ (Cyclooctatetraene)	C$_2$H$_2$+H	14.58±0.10	EI		2914
C$_6$H$_5^+$	C$_8$H$_8$ (Cyclobutenobenzene)	C$_2$H$_2$+H	15.58±0.10	EI		2914
C$_6$H$_5^+$	C$_8$H$_8$ (Bicyclo[2.2.2]octatriene)	C$_2$H$_2$+H	14.49±0.10	EI		2914
C$_6$H$_5^+$	C$_8$H$_8$ (syn−Tricyclo[4.2.0.0^{2,5}]octa−3,7−diene)	C$_2$H$_2$+H	12.41±0.10	EI		2914
C$_6$H$_5^+$	C$_8$H$_8$ (anti−Tricyclo[4.2.0.0^{2,5}]octa−3,7−diene)	C$_2$H$_2$+H	12.59±0.10	EI		2914
C$_6$H$_5^+$	C$_8$H$_8$ (Cubane)	C$_2$H$_2$+H	10.93±0.10	EI		2914

See also — EI: 2105

4.3. The Positive Ion Table—Continued

Ion	Reactant	Other products	Ionization or appearance potential (eV)	Method	Heat of formation (kJ mol⁻¹)	Ref.
$C_6H_5^+$	$(CH_3)_2C{=}CHCH{=}C(CH_3)_2$		16.05±0.1	EI		2455
$C_6H_5^+$	$C_{10}H_8$ (Naphthalene)		18.45±0.05	EI		2112
$C_6H_5^+$	$C_{10}H_8$ (Azulene)		16.9±0.10	EI		2112
$C_6H_5^+$	$(C_6H_5)_2$ (Biphenyl)		18.2±0.5	EI		1238
$C_6H_5^+$	$C_6H_5C{\equiv}CC_6H_5$ (Diphenylacetylene)		20.7±0.1	EI		1238
$C_6H_5^+$	$C_6H_5N(CH_3)_2$ (N,N–Dimethylaniline)		15.7±0.1	EI	303,	3344
$C_6H_5^+$	C_6H_5CHO (Benzenecarbonal)	CHO?	13.51±0.12	EI		130

See also – EI: 308, 1237

Ion	Reactant	Other products	Ionization or appearance potential (eV)	Method	Heat of formation (kJ mol⁻¹)	Ref.
$C_6H_5^+$	$C_6H_5OC{\equiv}CH$ (Phenoxyacetylene)		12.2±0.1	EI		13
$C_6H_5^+$	$C_6H_5COCH_3$ (Acetophenone)		13.42±0.07	EI		2174

See also – EI: 308, 1237, 3344

Ion	Reactant	Other products	Ionization or appearance potential (eV)	Method	Heat of formation (kJ mol⁻¹)	Ref.
$C_6H_5^+$	$C_6H_5CH_2COCH_3$ (Benzyl methyl ketone)		13.66±0.02	EI		2174
$C_6H_5^+$	$(C_6H_5)_2O$ (Diphenyl ether)		14.85±0.05	EI		1237
$C_6H_5^+$	$(C_6H_5)_2CO$ (Benzophenone)		16.22±0.07	EI		1237
$C_6H_5^+$	$C_6H_5COOC_6H_5$ (Benzoic acid phenyl ester)		15.46±0.05	EI		1237
$C_6H_5^+$	$C_6H_5COCOC_6H_5$ (Diphenylglyoxal)		15.12±0.2	EI		1237
$C_6H_5^+$	$(C_6H_5O)_2CO$ (Carbonic acid diphenyl ester)		12.1±0.1	EI		1237
$C_6H_5^+$	$C_6H_5CONH_2$ (Benzoic acid amide)		13.5±0.1	EI		1168
$C_6H_5^+$	$C_6H_5COCHN_2$ (α–Diazoacetophenone)		14.07±0.14	EI		2174
$C_6H_5^+$	$C_6H_5NO_2$ (Nitrobenzene)		12.16	EI		3238
$C_6H_5^+$	C_6H_5F (Fluorobenzene)	F?	14.5±0.1	EI	301,	3344

See also – EI: 2103

Ion	Reactant	Other products	Ionization or appearance potential (eV)	Method	Heat of formation (kJ mol⁻¹)	Ref.
$C_6H_5^+$	$C_6H_5CF_3$ (α,α,α–Trifluorotoluene)		15.2±0.1	EI		301

See also – EI: 3344

Ion	Reactant	Other products	Ionization or appearance potential (eV)	Method	Heat of formation (kJ mol⁻¹)	Ref.
$C_6H_5^+$	$C_6H_5COCF_3$ (α,α,α–Trifluoroacetophenone)		12.0	EI		308

4.3. The Positive Ion Table—Continued

Ion	Reactant	Other products	Ionization or appearance potential (eV)	Method	Heat of formation (kJ mol^{-1})	Ref.
$C_6H_5^+$	C_6H_5SH (Mercaptobenzene)	SH	14.7±0.3	EI		3286
$C_6H_5^+$	C_6H_5SD (Mercapto-d_1-benzene)	SD	13.3±0.2	EI		1039
$C_6H_5^+$	$C_2H_5SCH=CHC\equiv CH$		12.6±0.3	EI		2949
$C_6H_5^+$	C_6H_5Cl (Chlorobenzene)	Cl?	12.55±0.07	PI	(1141)	3212
$C_6H_5^+$	C_6H_5Cl (Chlorobenzene)	Cl?	13.2±0.1	EI		301, 3344

See also – EI: 2103, 2972, 3230, 3238

Ion	Reactant	Other products	Ionization or appearance potential (eV)	Method	Heat of formation (kJ mol^{-1})	Ref.
$C_6H_5^+$	C_7H_9Cl (endo-5-Chlorobicyclo[2.2.1]hept-2-ene)		13.0±0.3	EI		2155
$C_6H_5^+$	C_7H_9Cl (exo-5-Chlorobicyclo[2.2.1]hept-2-ene)		13.0±0.3	EI		2155
$C_6H_5^+$	C_7H_9Cl (3-Chloronortricyclene)		12.7±0.3	EI		2155
$C_6H_5^+$	C_6H_5Br (Bromobenzene)	Br?	11.75±0.05	PI	(1127)	3212
$C_6H_5^+$	C_6H_5Br (Bromobenzene)	Br?	12.02	EI		3238

See also – EI: 301, 2103, 3230, 3344

Ion	Reactant	Other products	Ionization or appearance potential (eV)	Method	Heat of formation (kJ mol^{-1})	Ref.
$C_6H_5^+$	C_6H_5I (Iodobenzene)	I?	11.06±0.04	PI	(1123)	3212
$C_6H_5^+$	C_6H_5I (Iodobenzene)	I?	11.46	EI		3238

See also – EI: 2103, 3230, 3238

The heat of formation of phenyl ion is very uncertain. Taking $D(C_6H_5-H) = 435$ kJ mol^{-1} (J. A. Kerr, Chem. Rev. **66**, 465 (1966)), the direct electron impact ionization of phenyl radical leads to a heat of formation of about 1188 kJ mol^{-1}, close to that from the photoionization of benzene. However, none of the photoionization thresholds have been corrected for kinetic shifts which may amount to roughly 50 kJ mol^{-1}. Further, the low thresholds of the halobenzenes may be due in part to ion–pair processes. Taking into consideration that the radical ionization potential determined by electron impact is probably too high by as much as 0.5 eV, we suggest a value for $\Delta H_{f298}^{\circ}(C_6H_5^+)$ of about 1140 kJ mol^{-1} (272 kcal mol^{-1}).

4.3. The Positive Ion Table—Continued

Ion	Reactant	Other products	Ionization or appearance potential (eV)	Method	Heat of formation (kJ mol^{-1})	Ref.
	C$_6$H$_6^+$ (Benzene)		**$\Delta H^\circ_{f298} = 975$ kJ mol^{-1} (233 kcal mol^{-1})**			
C$_6$H$_6^+$	CH$_2$=CHC≡CCH=CH$_2$		~10.5	S		3010
C$_6$H$_6^+$	CH≡CCH=CHCH=CH$_2$		9.50	EI		1197
C$_6$H$_6^+$	C$_2$H$_5$C≡CC≡CH		9.25	EI		1197
C$_6$H$_6^+$	CH$_3$C≡CCH$_2$C≡CH		9.75	EI		1197
C$_6$H$_6^+$	CH$_3$C≡CC≡CCH$_3$		9.20	EI		1197
C$_6$H$_6^{+*}$	CH$_3$C≡CC≡CCH$_3$		11.51±0.02	S		3152
C$_6$H$_6^+$	CH≡CCH$_2$CH$_2$C≡CH		10.35	EI		1197
C$_6$H$_6^+$	C$_4$H$_2$(=CH$_2$)$_2$ (3,4–Dimethylenecyclobutene)		8.80	PE		3292
C$_6$H$_6^+$	C$_5$H$_4$=CH$_2$ (5–Methylenecyclopentadiene)		8.36	PE		3292
C$_6$H$_6^+$(^{2}E$_{1g}$)	C$_6$H$_6$ (Benzene) (Average of four Rydberg series limits)		9.247±0.002	S	975	422, 423, 3376
C$_6$H$_6^+$(^{2}E$_{1g}$)	C$_6$H$_6$ (Benzene) (Average of two Rydberg series limits)		9.242±0.005	S		344, 1114
C$_6$H$_6^+$(^{2}E$_{1g}$)	C$_6$H$_6$ (Benzene)		9.248	S		1115
C$_6$H$_6^+$(^{2}E$_{2g}$)	C$_6$H$_6$ (Benzene)		11.489	S		1115
See also – S: 344, 1114						
C$_6$H$_6^+$(^{2}A$_{1g}$)	C$_6$H$_6$ (Benzene)		16.84	S		1115
C$_6$H$_6^+$(^{2}E$_{1g}$)	C$_6$H$_6$ (Benzene)		9.241±0.001	PE		3080
C$_6$H$_6^+$(^{2}E$_{1g}$)	C$_6$H$_6$ (Benzene)		9.24	PE		2806
C$_6$H$_6^+$(^{2}E$_{1g}$)	C$_6$H$_6$ (Benzene)		9.25±0.02	PE		2838
C$_6$H$_6^+$(^{2}E$_{1g}$)	C$_6$H$_6$ (Benzene)		9.24±0.01	PE		2843, 2844, 2942
C$_6$H$_6^+$(^{2}E$_{1g}$)	C$_6$H$_6$ (Benzene)		9.25	PE		1130
C$_6$H$_6^+$(^{2}E$_{2g}$)	C$_6$H$_6$ (Benzene)		11.490	PE		3080
C$_6$H$_6^+$(^{2}E$_{2g}$)	C$_6$H$_6$ (Benzene)		11.48	PE		2806
C$_6$H$_6^+$(^{2}E$_{2g}$)	C$_6$H$_6$ (Benzene)		11.51±0.04	PE		2838
C$_6$H$_6^+$(^{2}E$_{2g}$)	C$_6$H$_6$ (Benzene)		11.50	PE		2843, 2844, 2942
C$_6$H$_6^+$(^{2}E$_{2g}$)	C$_6$H$_6$ (Benzene)		11.49	PE		1130

4.3. The Positive Ion Table—Continued

Ion	Reactant	Other products	Ionization or appearance potential (eV)	Method	Heat of formation (kJ mol^{-1})	Ref.
$C_6H_6^+(^2A_{2u})$	C_6H_6 (Benzene)		12.24 (V)	PE		2806
$C_6H_6^+(^2A_{2u})$	C_6H_6 (Benzene)		12.19?	PE		1130
$C_6H_6^+(^2E_{1u})$	C_6H_6 (Benzene)		13.88±0.04	PE		2838
$C_6H_6^+(^2E_{1u})$	C_6H_6 (Benzene)		13.87	PE	2843,	2844, 2942
$C_6H_6^+(^2E_{1u})$	C_6H_6 (Benzene)		13.67	PE		1130
$C_6H_6^+(^2B_{2u})$	C_6H_6 (Benzene)		14.87±0.06?	PE		2838
$C_6H_6^+(^2B_{2u})$	C_6H_6 (Benzene)		14.44?	PE		1130
$C_6H_6^+(^2B_{1u})$	C_6H_6 (Benzene)		15.4	PE		3080
$C_6H_6^+(^2B_{1u})$	C_6H_6 (Benzene)		15.54±0.06?	PE		2838
$C_6H_6^+(^2B_{1u})$	C_6H_6 (Benzene)		15.46?	PE	2843,	2844, 2942
$C_6H_6^+(^2A_{1g})$	C_6H_6 (Benzene)		16.9	PE		3080
$C_6H_6^+(^2A_{1g})$	C_6H_6 (Benzene)		16.83	PE		2806
$C_6H_6^+(^2A_{1g})$	C_6H_6 (Benzene)		16.84±0.05	PE		2838
$C_6H_6^+(^2A_{1g})$	C_6H_6 (Benzene)		16.84	PE	2843,	2844, 2942
$C_6H_6^+(^2A_{1g})$	C_6H_6 (Benzene)		16.73	PE		1130
$C_6H_6^+(^2E_{2g})$	C_6H_6 (Benzene)		18.22±0.08	PE		2838
$C_6H_6^+(^2E_{2g})$	C_6H_6 (Benzene)		18.43?	PE	2843,	2844, 2942
$C_6H_6^+(^2E_{2g})$	C_6H_6 (Benzene)		18.75	PE		1130
$C_6H_6^+(^2E_{2g})$	C_6H_6 (Benzene)		19.0 (V)	PE		3081

4.3. The Positive Ion Table—Continued

Ion	Reactant	Other products	Ionization or appearance potential (eV)	Method	Heat of forma-tion (kJ mol^{-1})	Ref.
$C_6H_6^+(^2E_{1u})$	C_6H_6 (Benzene)		22.5 (V)	PE		3081
$C_6H_6^+(^2A_{1g})$	C_6H_6 (Benzene)		28.8 (V)	PE		3081

The photoelectron spectrum of benzene has been the subject of much debate concerning both the form of the spectrum and the assignment of orbitals, see the discussion and references in S. D. Worley, Chem. Rev. **71**, 294 (1971). We have based our assignments on the independent work of B. Jonsson and E. Lindholm, Arkiv Fysik **39**, 65 (1969), R. M. Stevens, E. Switkes, E. A. Laws and W. N. Lipscomb, J. Am. Chem. Soc. **93**, 2603 (1971) and refs. 3080, 3081. The best published photoelectron spectra are to be found in refs. 2806, 3080, 3081.

Ion	Reactant	Other products	Ionization or appearance potential (eV)	Method	Heat of forma-tion (kJ mol^{-1})	Ref.
$C_6H_6^+(^2E_{1g})$	C_6H_6 (Benzene)		9.246±0.005	PI		2013
$C_6H_6^+(^2E_{1g})$	C_6H_6 (Benzene)		9.241±0.006	PI		1253
$C_6H_6^+(^2E_{1g})$	C_6H_6 (Benzene)		9.242±0.01	PI	54,	1118
$C_6H_6^+(^2E_{1g})$	C_6H_6 (Benzene)		9.245±0.01	PI	158,	182, 416
$C_6H_6^+(^2E_{1g})$	C_6H_6 (Benzene)		9.24±0.01	PI		3212
$C_6H_6^+(^2E_{1g})$	C_6H_6 (Benzene)		9.25±0.01	PI		2682
$C_6H_6^+(^2E_{1g})$	C_6H_6 (Benzene)		9.25±0.01	PI		2877
$C_6H_6^+(^2E_{1g})$	C_6H_6 (Benzene)		9.241	RPI		2773
$C_6H_6^+(^2E_{1g})$	C_6H_6 (Benzene)		9.26±0.02	RPD		2538
$C_6H_6^+(^2E_{1g})$	C_6H_6 (Benzene)		9.20±0.04	RPD		2407
$C_6H_6^+(^2E_{1g})$	C_6H_6 (Benzene)		9.36±0.05	RPD		3223

The RPI study of ref. 2773 also shows evidence of a state at 10.388±0.003 eV. This has been observed in some photoelectron spectra and not in others. It is probably due to autoionization, see ref. 2890.

See also – S: 2666, 3153
PI: 54, 190, 1142, 1159, 1160, 1166, 2612, 3025
PE: 1159, 2015, 2822, 2829, 2890, 3109, 3331
PEN: 2430, 2466
EI: 218, 301, 381, 383, 413, 1066, 1079, 1132, 1197, 1238, 2158, 2163, 2458, 2463, 2498, 2530, 2718, 2752, 2865, 2868, 2981, 2989, 3157, 3174
CTS: 1064, 2562, 2947

4.3. The Positive Ion Table—Continued

Ion	Reactant	Other products	Ionization or appearance potential (eV)	Method	Heat of formation (kJ mol^{-1})	Ref.
$C_6H_6^+$	$CH_2{=}CHCH{=}CHCH{=}CH_2$	H_2	10.77±0.10	RPD		2751
$C_6H_6^+$	C_6H_8 (1,3–Cyclohexadiene)	H_2	10.12±0.10	RPD		2751
$C_6H_6^+$	C_6H_8 (1,3–Cyclohexadiene)	H_2	9.88±0.10	EI		2751
$C_6H_6^+$	C_6H_8 (1,4–Cyclohexadiene)	H_2	9.86±0.05	RPD		2751
$C_6H_6^+$	C_6H_8 (1,4–Cyclohexadiene)	H_2	9.61±0.10	EI		2751
$C_6H_6^+$	$C_6H_5CH{=}CH_2$ (Ethenylbenzene)	C_2H_2	12.38±0.05	RPD		2914
(0.28 eV average translational energy of decomposition at threshold)						
$C_6H_6^+$	$C_6H_5CH{=}CH_2$ (Ethenylbenzene)	C_2H_2	12.30±0.10	EI		2914
$C_6H_6^+$	C_8H_8 (Cyclooctatetraene)	C_2H_2	9.70±0.12	RPD		2914
(0.07 eV average translational energy of decomposition at threshold)						
$C_6H_6^+$	C_8H_8 (Cyclobutenobenzene)	C_2H_2	11.55±0.10	EI		2914
(0.30 eV average translational energy of decomposition at threshold)						
$C_6H_6^+$	C_8H_8 (Bicyclo[2.2.2]octatriene)	C_2H_2	10.50±0.10	EI		2914
$C_6H_6^+$	C_8H_8 (syn–Tricyclo[4.2.0.0^{2,5}]octa–3,7–diene)	C_2H_2	9.09±0.10	EI		2914
(0.29 eV average translational energy of decomposition at threshold)						
$C_6H_6^+$	C_8H_8 ($anti$–Tricyclo[4.2.0.0^{2,5}]octa–3,7–diene)	C_2H_2	9.01±0.10	EI		2914
(0.29 eV average translational energy of decomposition at threshold)						
$C_6H_6^+$	C_8H_8 (Cubane)	C_2H_2	9.00±0.10	EI		2914

See also – EI: 2105

Ion	Reactant	Other products	Ionization or appearance potential (eV)	Method	Heat of formation (kJ mol^{-1})	Ref.
$C_6H_6^+$	$C_6H_5C_2H_5$ (Ethylbenzene)	C_2H_4?	11.0±0.1	PI		2612
$C_6H_6^+$	$C_{10}H_8$ (Naphthalene)		15.20±0.05	EI		2112
$C_6H_6^+$	$C_{10}H_8$ (Azulene)		13.86±0.05	EI		2112
$C_6H_6^+$	$C_6H_5OCH_3$ (Methoxybenzene)		11.30	EI		3238
$C_6H_6^+$	C_6H_5SH (Mercaptobenzene)		14.6±0.3	EI		3286
$C_6H_6^+$	$C_6H_5SCH_3$ (Methylthiobenzene)		12.0	EI		307
$C_6H_6^+$	$(C_6H_6)_2Cr$ (Bis(benzene)chromium)		10.1	EI		2981

$C_6H_5D^+$

Ion	Reactant	Other products	Ionization or appearance potential (eV)	Method	Heat of formation (kJ mol^{-1})	Ref.
$C_6H_5D^+$	C_6H_5D (Benzene–d_1)		9.44	EI		413

4.3. The Positive Ion Table—Continued

Ion	Reactant	Other products	Ionization or appearance potential (eV)	Method	Heat of formation (kJ mol^{-1})	Ref.
			$C_6D_6^+$			
$C_6D_6^+(^2E_{1g})$	C_6D_6 (Benzene-d_6)		9.251±0.002	S	422,	423, 3376
(Average of four Rydberg series limits)						
$C_6D_6^+(^2E_{1g})$	C_6D_6 (Benzene-d_6)		9.246±0.005	S		1114
(Based upon the Rydberg series limit for benzene and an average isotopic shift of 33 cm^{-1})						
$C_6D_6^+(^2E_{1g})$	C_6D_6 (Benzene-d_6)		9.251	S		1115
$C_6D_6^+(^2E_{1g})$	C_6D_6 (Benzene-d_6)		9.245±0.01	PI	54,	1118
$C_6D_6^+(^2E_{2g})$	C_6D_6 (Benzene-d_6)		11.520	S		1115
$C_6D_6^+(^2A_{1g})$	C_6D_6 (Benzene-d_6)		16.87	S		1115

See also – PE: 3080

Ion	Reactant	Other products	Ionization or appearance potential (eV)	Method	Heat of formation	Ref.
			$C_6H_6^{+2}$			
$C_6H_6^{+2}$	C_6H_6 (Benzene)		≤26.1±0.8 (V)	AUG		3424
			$C_6H_5D^{+2}$			
$C_6H_5D^{+2}$	C_6H_5D (Benzene-d_1)		26.0±0.2	FD		212
			$C_6H_5D^{+3}$			
$C_6H_5D^{+3}$	C_6H_5D (Benzene-d_1)		44±5	NRE		212
			$C_6H_7^+$			
$C_6H_7^+$	$C_5H_4CH_3$ (Methylcyclopentadienyl radical)		8.54	EI		126
$C_6H_7^+$	$CH_2=CHCH=CHCH=CH_2$	H	9.96±0.15	EI		2751
$C_6H_7^+$	$C_2H_5CH=CHC\equiv CH$	H	10.7	EI		3335
$C_6H_7^+$	$CH_3CH=C(CH_3)C\equiv CH$	H	10.7	EI		3335
$C_6H_7^+$	$C_5H_5CH_3$ (Methylcyclopentadiene)	H	11.3	EI		3335
$C_6H_7^+$	C_6H_8 (1,3–Cyclohexadiene)	H	10.82±0.10	EI		2751

See also – EI: 3192, 3335, 3339

4.3. The Positive Ion Table—Continued

Ion	Reactant	Other products	Ionization or appearance potential (eV)	Method	Heat of formation (kJ mol^{-1})	Ref.
$C_6H_7^+$	C_6H_8 (1,4–Cyclohexadiene)	H	10.94±0.10	EI		2751

See also – EI: 3339

Ion	Reactant	Other products	Ionization or appearance potential (eV)	Method	Heat of formation (kJ mol^{-1})	Ref.
$C_6H_7^+$	$CH_3CH=CHCH=CHCH_3$	H_2+H	13.42±0.02	EI		2455
$C_6H_7^+$	$C_5H_8=CH_2$ (Methylenecyclopentane)	H_2+H	11.66±0.16	EDD		2738
$C_6H_7^+$	$C_5H_7CH_3$ (1–Methylcyclopentene)	H_2+H	12.40±0.17	EDD		2738
$C_6H_7^+$	$C_5H_7CH_3$ (3–Methylcyclopentene)	H_2+H	11.75±0.04	EDD		2738
$C_6H_7^+$	C_6H_{10} (Cyclohexene)	H_2+H	12.13±0.10	EDD		2738
$C_6H_7^+$	$(C_3H_5)_2$ (Bicyclopropyl)	H_2+H	10.97±0.19	EDD		2738
$C_6H_7^+$	C_6H_{10} (Bicyclo[3.1.0]hexane)	H_2+H	11.60±0.15	EDD		2738
$C_6H_7^+$	$C_5H_4(CH_3)_2$ (1,2–Dimethylcyclopentadiene)	CH_3	10.7	EI		3335
$C_6H_7^+$	$C_5H_4(CH_3)_2$ (5,5–Dimethylcyclopentadiene)	CH_3	10.75	EI		3335
$C_6H_7^+$	$C_6H_7CH_3$ (1–Methyl–1,3–cyclohexadiene)	CH_3	10.67	EI		3339
$C_6H_7^+$	$C_6H_7CH_3$ (5–Methyl–1,3–cyclohexadiene)	CH_3	10.6	EI		3335

See also – EI: 3192

Ion	Reactant	Other products	Ionization or appearance potential (eV)	Method	Heat of formation (kJ mol^{-1})	Ref.
$C_6H_7^+$	$C_6H_7CH_3$ (1–Methyl–1,4–cyclohexadiene)	CH_3	10.61	EI		3335
$C_6H_7^+$	C_7H_{10} (1,3–Cycloheptadiene)	CH_3	10.1	EI		3335
$C_6H_7^+$	C_7H_{10} (Bicyclo[2.2.1]hept–2–ene)	CH_3	11.2±0.15	EI		2155
$C_6H_7^+$	C_7H_{10} (Bicyclo[3.2.0]hept–6–ene)	CH_3	9.8	EI		3335
$C_6H_7^+$	$(CH_3)_2C=CHCH=C(CH_3)_2$		14.42±0.05	EI		2455
$C_6H_7^+$	$C_6H_4(CH_3)C_2H_5$ (1–Ethyl–4–methylbenzene)		15.3	EI		3335
$C_6H_7^+$	$C_6H_5CH_2OH$ (α–Hydroxytoluene)	CHO?	10.9	EI		3335
$C_6H_7^+$	$C_6H_5OCH_3$ (Methoxybenzene)	CO+H?	12.1	EI		3335
$C_6H_7^+$	$C_6H_4(OCH_3)CH_3$ (3–Methoxytoluene)		12.9	EI		3335
$C_6H_7^+$	$C_6H_4(OCH_3)CH_3$ (4–Methoxytoluene)		12.9	EI		3335
$C_6H_7^+$	$C_7H_7OCH_3$ (7–Methoxycycloheptatriene)		11.1	EI		3335
$C_6H_7^+$	C_7H_9Cl (endo–5–Chlorobicyclo[2.2.1]hept–2–ene)		10.9±0.15	EI		2155

Ion	Reactant	Other products	Ionization or appearance potential (eV)	Method	Heat of formation (kJ mol^{-1})	Ref.
$C_6H_7^+$	C_7H_9Cl (*exo*−5−Chlorobicyclo[2.2.1]hept−2−ene)		10.8±0.15	EI		2155
$C_6H_7^+$	C_7H_9Cl (3−Chloronortricyclene)		10.25±0.15	EI		2155

$C_6H_8^+$ (1,3−Cyclohexadiene) $\Delta H^\circ_{f298} = 904$ kJ mol^{-1} (216 kcal mol^{-1})

Ion	Reactant	Other products	Ionization or appearance potential (eV)	Method	Heat of formation (kJ mol^{-1})	Ref.
$C_6H_8^+$	$CH_2=CHCH=CHCH=CH_2$		8.27±0.05	S		3010
$C_6H_8^+$	$CH_2=CHCH=CHCH=CH_2$		8.42±0.05	RPD		2751
$C_6H_8^+$	$C_5H_5CH_3$ (1−Methylcyclopentadiene)		8.43±0.05	EI		2163
$C_6H_8^+$	$C_5H_5CH_3$ (2−Methylcyclopentadiene)		8.46±0.05	EI		2163
$C_6H_8^+$	C_6H_8 (1,3−Cyclohexadiene)		8.25±0.03	PI	904	2877
$C_6H_8^+$	C_6H_8 (1,3−Cyclohexadiene)		8.25	PE	904	3330
$C_6H_8^+$	C_6H_8 (1,3−Cyclohexadiene)		8.28±0.05	RPD		2751

See also − EI: 2543

Ion	Reactant	Other products	Ionization or appearance potential (eV)	Method	Heat of formation (kJ mol^{-1})	Ref.
$C_6H_8^+$	C_6H_8 (1,4−Cyclohexadiene)		8.82±0.02	PI		2877
$C_6H_8^+$	C_6H_8 (1,4−Cyclohexadiene)		8.80	PE		3327
$C_6H_8^+$	C_6H_8 (1,4−Cyclohexadiene)		8.65±0.05	RPD		2751

See also − EI: 2543

$C_6H_9^+$

Ion	Reactant	Other products	Ionization or appearance potential (eV)	Method	Heat of formation (kJ mol^{-1})	Ref.
$C_6H_9^+$	C_6H_9 (3−Cyclohexenyl radical)		7.54	EI		2543
$C_6H_9^+$	C_6H_9 (4−Cyclohexenyl radical)		7.54	EI		2543
$C_6H_9^+$	$CH_3CH=CHCH=CHCH_3$	H	10.95±0.03	EI		2455
$C_6H_9^+$	$C_5H_8=CH_2$ (Methylenecyclopentane)	H	10.20±0.03	EDD		2738
$C_6H_9^+$	$C_5H_7CH_3$ (1−Methylcyclopentene)	H	10.59±0.13	EDD		2738
$C_6H_9^+$	$C_5H_7CH_3$ (3−Methylcyclopentene)	H	10.35±0.17	EDD		2738
$C_6H_9^+$	C_6H_{10} (Cyclohexene)	H	10.62±0.07	EDD		2738
$C_6H_9^+$	$(C_3H_5)_2$ (Bicyclopropyl)	H	9.33±0.15	EDD		2738
$C_6H_9^+$	C_6H_{10} (Bicyclo[3.1.0]hexane)	H	10.20±0.10	EDD		2738

4.3. The Positive Ion Table—Continued

Ion	Reactant	Other products	Ionization or appearance potential (eV)	Method	Heat of formation (kJ mol^{-1})	Ref.
$C_6H_9^+$	$C_6H_{10}=CH_2$ (Methylenecyclohexane)	CH_3	10.27±0.08	EDD		2558
$C_6H_9^+$	$C_6H_9CH_3$ (1−Methylcyclohexene)	CH_3	10.27±0.09	EDD		2558
$C_6H_9^+$	$C_6H_9CH_3$ (3−Methylcyclohexene)	CH_3	9.98±0.07	EDD		2558
$C_6H_9^+$	$C_6H_9CH_3$ (4−Methylcyclohexene)	CH_3	10.12±0.10	EDD		2558
$C_6H_9^+$	C_7H_{12} (Bicyclo[2.2.1]heptane)	CH_3	10.17±0.06	EDD		2558
$C_6H_9^+$	C_7H_{12} (Bicyclo[4.1.0]heptane)	CH_3	9.30±0.09	EDD		2558
$C_6H_9^+$	$(CH_3)_2C=CHCH=C(CH_3)_2$		11.68±0.1	EI		2455
$C_6H_9^+$	C_9H_{16} (cis−Hexahydroindan)		11.99±0.05	EI		1184, 2028
$C_6H_9^+$	C_9H_{16} (trans−Hexahydroindan)		12.07±0.04	EI		1184, 2028

$$CH_2=C(CH_3)C(CH_3)=CH_2^+ \qquad \Delta H^\circ_{f298} = 888 \text{ kJ mol}^{-1} \text{ (212 kcal mol}^{-1}\text{)}$$
$$C_6H_{10}^+ \text{ (1−Methylcyclopentene)} \qquad \Delta H^\circ_{f298} = 819 \text{ kJ mol}^{-1} \text{ (196 kcal mol}^{-1}\text{)}$$
$$C_6H_{10}^+ \text{ (Cyclohexene)} \qquad \Delta H^\circ_{f298} = 858 \text{ kJ mol}^{-1} \text{ (205 kcal mol}^{-1}\text{)}$$

Ion	Reactant	Other products	Ionization or appearance potential (eV)	Method	Heat of formation (kJ mol^{-1})	Ref.
$C_6H_{10}^+$	$CH_3CH=CHCH=CHCH_3$		8.48±0.05	EI		2455
$C_6H_{10}^+$	$CH_2=C(CH_3)C(CH_3)=CH_2$		8.709	S	888	3352
(Average of two Rydberg series limits)						
$C_6H_{10}^+$	$CH≡CC(CH_3)_3$		10.31±0.04	RPD		2408
$C_6H_{10}^+$	$C_5H_8=CH_2$ (Methylenecyclopentane)		8.94±0.01	PI		2877
$C_6H_{10}^+$	$C_5H_8=CH_2$ (Methylenecyclopentane)		8.51±0.01	PE		3158
$C_6H_{10}^+$	$C_5H_8=CH_2$ (Methylenecyclopentane)		9.05±0.02	EDD		2738
$C_6H_{10}^+$	$C_5H_7CH_3$ (1−Methylcyclopentene)		8.54±0.01	PE	819	3158
$C_6H_{10}^+$	$C_5H_7CH_3$ (1−Methylcyclopentene)		8.62±0.02	EDD		2738
$C_6H_{10}^+$	$C_5H_7CH_3$ (3−Methylcyclopentene)		8.99±0.04	EDD		2738

4.3. The Positive Ion Table—Continued

Ion	Reactant	Other products	Ionization or appearance potential (eV)	Method	Heat of formation (kJ mol⁻¹)	Ref.	
$C_6H_{10}^+$	C_6H_{10} (Cyclohexene)		8.945±0.01	PI	858	182,	416
$C_6H_{10}^+$	C_6H_{10} (Cyclohexene)		8.945±0.01	PI	858		2877
$C_6H_{10}^+$	C_6H_{10} (cis–Cyclohexene)		8.94	PE			3330
$C_6H_{10}^+$	C_6H_{10} (Cyclohexene)		8.92	PE			3343
$C_6H_{10}^+$	C_6H_{10} (Cyclohexene)		8.99	RPD			2999
$C_6H_{10}^+$	C_6H_{10} (Cyclohexene)		8.92±0.02	EDD			2738

See also – S: 3353
        PE: 1130, 3327
        EI: 411, 3342

Ion	Reactant	Other products	Ionization or appearance potential (eV)	Method	Heat of formation (kJ mol⁻¹)	Ref.	
$C_6H_{10}^+$	$(C_3H_5)_2$ (Bicyclopropyl)		9.04	PE			2951
$C_6H_{10}^+$	$(C_3H_5)_2$ (Bicyclopropyl)		8.80±0.02	EDD			2738
$C_6H_{10}^+$	C_6H_{10} (Bicyclo[3.1.0]hexane)		9.16±0.02	EDD			2738
$C_6H_{10}^+$	$C_6H_{10}(CH_3)_2$ (cis–1,2–Dimethylcyclohexane)	C_2H_6	10.62±0.04	EI			1145
$C_6H_{10}^+$	$C_6H_{10}(CH_3)_2$ (trans–1,2–Dimethylcyclohexane)	C_2H_6	10.87±0.08	EI			1145
$C_6H_{10}^+$	C_9H_{16} (cis–Hexahydroindan)	C_3H_6	10.79±0.03	EI		1184,	2028
$C_6H_{10}^+$	C_9H_{16} (trans–Hexahydroindan)	C_3H_6	11.00±0.02	EI		1184,	2028
$C_6H_{10}^+$	$C_{10}H_{18}$ (cis–Decalin)		10.89±0.02	EI		1182,	1183, 3337
$C_6H_{10}^+$	$C_{10}H_{18}$ (trans–Decalin)		11.29±0.02	EI		1182,	1183, 3337
$C_6H_{10}^+$	cis–$(CH_3)_3CCH=CHC(CH_3)_3$	C_4H_{10}	8.52±0.03	EI			2533
$C_6H_{10}^+$	trans–$(CH_3)_3CCH=CHC(CH_3)_3$	C_4H_{10}	8.76±0.08	EI			2533
$C_6H_{10}^+$	$C_6H_{11}OH$ (Cyclohexanol)	H_2O	9.47	RPD			2999

See also – EI: 3022

4.3. The Positive Ion Table—Continued

Ion	Reactant	Other products	Ionization or appearance potential (eV)	Method	Heat of formation (kJ mol^{-1})	Ref.

$C_6H_{11}^+$

Ion	Reactant	Other products	Ionization or appearance potential (eV)	Method	Heat of formation (kJ mol^{-1})	Ref.
$C_6H_{11}^+$	C_6H_{11} (Cyclohexyl radical)		7.66±0.05	EI		123
$C_6H_{11}^+$	C_6H_{12} (Cyclohexane)	H	11.66	EI		123
$C_6H_{11}^+$	$C_6H_{11}CH_3$ (Methylcyclohexane)	CH_3	10.95	EI		123
$C_6H_{11}^+$	$C_6H_{10}(CH_3)_2$ (cis–1,2–Dimethylcyclohexane)	C_2H_5	11.06±0.02	EI		1145
$C_6H_{11}^+$	$C_6H_{10}(CH_3)_2$ (trans–1,2–Dimethylcyclohexane)	C_2H_5	11.27±0.02	EI		1145
$C_6H_{11}^+$	cis–$(CH_3)_3CCH=CHC(CH_3)_3$		10.28±0.04	EI		2533
$C_6H_{11}^+$	trans–$(CH_3)_3CCH=CHC(CH_3)_3$		10.68±0.04	EI		2533

$1–C_6H_{12}^+$ $\Delta H_{f298}^\circ = 871$ kJ mol^{-1} (208 kcal mol^{-1})
trans–$3–C_6H_{12}^+$ $\Delta H_{f298}^\circ = 809$ kJ mol^{-1} (193 kcal mol^{-1})
$(CH_3)_2C=C(CH_3)_2^+$ $\Delta H_{f298}^\circ = 742$ kJ mol^{-1} (177 kcal mol^{-1})
$C_6H_{12}^+$ (Cyclohexane) $\Delta H_{f298}^\circ \sim 827$ kJ mol^{-1} (198 kcal mol^{-1})

Ion	Reactant	Other products	Ionization or appearance potential (eV)	Method	Heat of formation (kJ mol^{-1})	Ref.
$C_6H_{12}^+$	$1–C_6H_{12}$		9.45±0.02	PI	870	1120
$C_6H_{12}^+$	$1–C_6H_{12}$		9.46±0.02	PI	871	182
$C_6H_{12}^+$	trans–$3–C_6H_{12}$		8.945±0.01	PI	809	2877
$C_6H_{12}^+$	$(CH_3)_3CCH=CH_2$		9.62±0.04	RPD		2410
$C_6H_{12}^+$	$(CH_3)_2C=C(CH_3)_2$		8.30	PI	742	168

See also – S: 3353

Ion	Reactant	Other products	Ionization or appearance potential (eV)	Method	Heat of formation (kJ mol^{-1})	Ref.
$C_6H_{12}^+$	$C_5H_9CH_3$ (Methylcyclopentane)		10.45	EI		123
$C_6H_{12}^+$	C_6H_{12} (Cyclohexane)		9.88±0.02	PI	830	182, 416
$C_6H_{12}^+$	C_6H_{12} (Cyclohexane)		9.79	PE	821	1130
$C_6H_{12}^+$	C_6H_{12} (Cyclohexane)		9.81	PE	823	2843
$C_6H_{12}^+$	C_6H_{12} (Cyclohexane)		9.89	PE	831	3343

The higher ionization potentials given in refs. 1130, 2843, 3327, 3343 are in disaccord.

See also – PE: 3327
         EI: 123

4.3. The Positive Ion Table—Continued

Ion	Reactant	Other products	Ionization or appearance potential (eV)	Method	Heat of formation (kJ mol^{-1})	Ref.
$C_6H_{12}^+$	$n-C_7H_{16}$	CH_4	11.145±0.035	PI		1120
(Threshold value approximately corrected to 0 K)						
$C_6H_{12}^+$	$C_6H_{10}(CH_3)_2$ (cis-1,2-Dimethylcyclohexane)	C_2H_4	11.17±0.02	EI		1145
$C_6H_{12}^+$	$C_6H_{10}(CH_3)_2$ (trans-1,2-Dimethylcyclohexane)	C_2H_4	11.25±0.04	EI		1145
$C_6H_{12}^+$	$n-C_8H_{18}$	C_2H_6	10.81±0.03	PI		1120
(Threshold value approximately corrected to 0 K)						
$C_6H_{12}^+$	$n-C_8H_{18}$	C_2H_6	10.29	RPD		2999
$C_6H_{12}^+$	$cis-(CH_3)_3CCH=CHC(CH_3)_3$	C_4H_8	8.88±0.02	EI		2533
$C_6H_{12}^+$	$trans-(CH_3)_3CCH=CHC(CH_3)_3$	C_4H_8	9.15±0.06	EI		2533
$C_6H_{12}^+$	$n-C_{12}H_{26}$		10.40	RPD		2999

$C_6H_{13}^+$

Ion	Reactant	Other products	Ionization or appearance potential (eV)	Method	Heat of formation (kJ mol^{-1})	Ref.
$C_6H_{13}^+$	$n-C_7H_{16}$	CH_3	10.925±0.105	PI		1120
(Threshold value approximately corrected to 0 K)						
$C_6H_{13}^+$	$n-C_7H_{16}$	CH_3	10.7±0.1	PI		2013
$C_6H_{13}^+$	$n-C_8H_{18}$	C_2H_5	10.91±0.035	PI		1120
(Threshold value approximately corrected to 0 K)						
$C_6H_{13}^+$	$n-C_9H_{20}$		10.63	RPD		2977
$C_6H_{13}^+$	$n-C_6H_{13}COCH_3$		10.21	RPD		2977
$C_6H_{13}^+$	$n-C_6H_{13}OOH$		10.4±0.1	EI		2464
$C_6H_{13}^+$	$n-C_4H_9CH(OOH)CH_3$		10.0±0.1	EI		2464
$C_6H_{13}^+$	$n-C_3H_7CH(OOH)C_2H_5$		9.9±0.1	EI		2464

	$n-C_6H_{14}^+$	$\Delta H_{f298}^\circ \leqslant 815$ kJ mol^{-1} (195 kcal mol^{-1})				
	$iso-C_6H_{14}^+$	$\Delta H_{f298}^\circ \leqslant 802$ kJ mol^{-1} (192 kcal mol^{-1})				
	$(C_2H_5)_2CHCH_3^+$	$\Delta H_{f298}^\circ \leqslant 801$ kJ mol^{-1} (191 kcal mol^{-1})				
	$C_2H_5C(CH_3)_3^+$	$\Delta H_{f298}^\circ \leqslant 785$ kJ mol^{-1} (188 kcal mol^{-1})				
	$(CH_3)_2CHCH(CH_3)_2^+$	$\Delta H_{f298}^\circ \leqslant 789$ kJ mol^{-1} (189 kcal mol^{-1})				
$C_6H_{14}^+$	$n-C_6H_{14}$		10.18	PI	815	182
$C_6H_{14}^+$	$n-C_6H_{14}$		10.27	PE		2843

See also – PE: 3060
EI: 2977

Ion	Reactant	Other products	Ionization or appearance potential (eV)	Method	Heat of formation (kJ mol^{-1})	Ref.
$C_6H_{14}^+$	$iso-C_6H_{14}$		10.12	PI	802	182
$C_6H_{14}^+$	$(C_2H_5)_2CHCH_3$		10.08	PI	801	182
$C_6H_{14}^+$	$C_2H_5C(CH_3)_3$		10.06	PI	785	182
$C_6H_{14}^+$	$(CH_3)_2CHCH(CH_3)_2$		10.02	PI	789	182

The ionization potentials may not be adiabatic.

4.3. The Positive Ion Table—Continued

Ion	Reactant	Other products	Ionization or appearance potential (eV)	Method	Heat of formation (kJ mol^{-1})	Ref.
			$C_7H_5^+$			
$C_7H_5^+$	$(C_6H_5)_2$ (Biphenyl)		20.85±0.2	EI		1238
			$C_7H_6^+$			
$C_7H_6^+$	$C_5H_4=C=CH_2$ (5−Vinylidenecyclopentadiene)		8.88	EI		3296
$C_7H_6^+$	$C_6H_4(OH)C\equiv CH$ (2−Ethynylphenol)	CO	11.25	EI		2541
(Metastable transition indicates zero kinetic energy release)						
$C_7H_6^+$	C_8H_6O (Benzofuran)	CO	12.52	EI		2541
(Metastable transition indicates zero kinetic energy release)						
$C_7H_6^+$	$C_9H_6O_2$ (Coumarin)	2CO	13.68	EI		2541
(Metastable transitions indicate ~0.3 eV kinetic energy release)						
			$C_7H_7^+$			
$C_7H_7^+$	$C_5H_4CH=CH_2$ (Ethenylcyclopentadienyl radical)		8.44	EI		126
$C_7H_7^+$	$C_6H_5CH_2$ (Benzyl radical)		7.63	PI		2632
$C_7H_7^+$	$C_6H_5CH_2$ (Benzyl radical)		≤7.27±0.03	EM	(≤890)	3350
$C_7H_7^+$	$C_6H_5CH_2$ (Benzyl radical)		7.76±0.08	EI		69
$C_7H_7^+$	$C_6H_5CH_2$ (Benzyl radical)		7.45±0.1	D		2612
The disagreement between the PI and EM results is unexplained.						
$C_7H_7^+$	C_7H_7 (Cycloheptatrienyl radical)		6.240±0.01	S	(900)	2189
$C_7H_7^+$	C_7H_7 (Cycloheptatrienyl radical)		6.236±0.005	PI		2632
$C_7H_7^+$	C_7H_7 (Cycloheptatrienyl radical)		6.60±0.1	EI		68
$C_7H_7^+$	$C_6H_5CH_3$ (Toluene)	H	11.55±0.05	PI	(946)	3025
$C_7H_7^+$	$C_6H_5CH_3$ (Toluene)	H	11.7±0.1	RPD		2538

See also − PI: 2612
         EI: 2109, 3017, 3230, 3238, 3287

Ion	Reactant	Other products	Ionization or appearance potential (eV)	Method	Heat of formation (kJ mol^{-1})	Ref.
$C_7H_7^+$	C_7H_8 (Cycloheptatriene)	H	≤10.0	PI	(≤929)	2612
$C_7H_7^+$	C_7H_8 (Cycloheptatriene)	H	10.1±0.2	EI		68

See also − EI: 219, 2108, 2109

4.3. The Positive Ion Table—Continued

Ion	Reactant	Other products	Ionization or appearance potential (eV)	Method	Heat of formation (kJ mol^{-1})	Ref.
$C_7H_7^+$	C_7H_8 (Bicyclo[2.2.1]hepta–2,5–diene)	H	9.6	EI		2109
$C_7H_7^+$	C_7H_8 (Bicyclo[2.2.1]hepta–2,5–diene)	H	9.75	EI		2185
$C_7H_7^+$	C_7H_8 (Bicyclo[3.2.0]hepta–2,6–diene)	H	9.66	EI		219
$C_7H_7^+$	C_7H_8 (Spiro[2.4]hepta–4,6–diene)	H	10.45±0.1	EI		1122
$C_7H_7^+$	C_7H_8 (Quadricyclene)	H	9.56	EI		2185
$C_7H_7^+$	C_7H_{10} (1,3–Cycloheptadiene)	H_2+H	13.37	EI		219
$C_7H_7^+$	C_7H_{10} (Bicyclo[2.2.1]hept–2–ene)	H_2+H	13.6±0.3	EI		2155
$C_7H_7^+$	$CH_2=C(CH_3)C\equiv CC(CH_3)=CH_2$	CH_3	10.5±0.1	EI		1122
$C_7H_7^+$	$C_6H_5C_2H_5$ (Ethylbenzene)	CH_3	10.9±0.1	PI	(~939)	2612

See also – EI: 135, 1122, 3230, 3238

Ion	Reactant	Other products	Ionization or appearance potential (eV)	Method	Heat of formation (kJ mol^{-1})	Ref.
$C_7H_7^+$	$C_6H_4(CH_3)_2$ (1,2–Dimethylbenzene)	CH_3	11.10±0.05	PI	(948)	3025
$C_7H_7^+$	$C_6H_4(CH_3)_2$ (1,2–Dimethylbenzene)	CH_3	11.2±0.1	RPD		2538
$C_7H_7^+$	$C_6H_4(CH_3)_2$ (1,3–Dimethylbenzene)	CH_3	11.25±0.1	PI	(960)	2612
$C_7H_7^+$	$C_6H_4(CH_3)_2$ (1,3–Dimethylbenzene)	CH_3	11.4±0.1	RPD		2538

See also – EI: 1122

Ion	Reactant	Other products	Ionization or appearance potential (eV)	Method	Heat of formation (kJ mol^{-1})	Ref.
$C_7H_7^+$	$C_6H_4(CH_3)_2$ (1,4–Dimethylbenzene)	CH_3	11.05±0.05	PI	(942)	3025
$C_7H_7^+$	$C_6H_4(CH_3)_2$ (1,4–Dimethylbenzene)	CH_3	11.3±0.1	RPD		2538

See also – PI: 2612
            EI: 3230

Ion	Reactant	Other products	Ionization or appearance potential (eV)	Method	Heat of formation (kJ mol^{-1})	Ref.
$C_7H_7^+$	$C_7H_7CH_3$ (7–Methylcycloheptatriene)	CH_3	9.5±0.2	EI		68
$C_7H_7^+$	$C_7H_7CH_3$ (7–Methylcycloheptatriene)	CH_3	10.0±0.1	EI		1122
$C_7H_7^+$	$C_7H_7CH_3$ (1–Methylspiro[2.4]hepta–4,6–diene)	CH_3	9.7±0.1	EI		1122
$C_7H_7^+$	$C_7H_7CH_3$ (4–Methylspiro[2.4]hepta–4,6–diene)	CH_3	9.8±0.1	EI		1122
$C_7H_7^+$	$C_7H_7CH_3$ (5–Methylspiro[2.4]hepta–4,6–diene)	CH_3	9.8±0.1	EI		1122
$C_7H_7^+$	$C_6H_5C_3H_7$ (Propylbenzene)		11.64	EI		3338
$C_7H_7^+$	$C_6H_5CH_2CH_2C_6H_5$ (1,2–Diphenylethane)	C_7H_7	10.6±0.2	EI		3288
$C_7H_7^+$	$(C_7H_7)_2$ (7,7'–Bicycloheptatrienyl)	C_7H_7	8.09±0.05	PI		2632

4.3. The Positive Ion Table—Continued

Ion	Reactant	Other products	Ionization or appearance potential (eV)	Method	Heat of formation (kJ mol^{-1})	Ref.
$C_7H_7^+$	$C_{18}H_{30}$ (1–Phenyldodecane)		11.82±0.1	EI		2153
$C_7H_7^+$	$C_{18}H_{30}$ (3–Phenyldodecane)		12.05±0.1	EI		2153
$C_7H_7^+$	$C_{19}H_{32}$ (7–Phenyltridecane)		12.21±0.1	EI		2153
$C_7H_7^+$	$C_{26}H_{46}$ (1–Phenyleicosane)		11.83±0.1	EI		2153
$C_7H_7^+$	$C_{26}H_{46}$ (2–Phenyleicosane)		12.28±0.1	EI		2153
$C_7H_7^+$	$C_{26}H_{46}$ (3–Phenyleicosane)		12.24±0.1	EI		2153
$C_7H_7^+$	$C_{26}H_{46}$ (4–Phenyleicosane)		12.33±0.1	EI		2153
$C_7H_7^+$	$C_{26}H_{46}$ (5–Phenyleicosane)		12.34±0.1	EI		2153
$C_7H_7^+$	$C_{26}H_{46}$ (7–Phenyleicosane)		12.60±0.1	EI		2153
$C_7H_7^+$	$C_{26}H_{46}$ (9–Phenyleicosane)		12.09±0.1	EI		2153
$C_7H_7^+$	$C_6H_5CH_2ND_2$ (α–Amino-d_2–toluene)	ND_2	11.7±0.1	PI		1160
$C_7H_7^+$	$C_6H_5CH_2NHCH_3$ (Benzyl methyl amine)		12.61	EI		3338
$C_7H_7^+$	$C_6H_5CH_2CH_2C_6H_4NH_2$ (1–(3–Aminophenyl)–2–phenylethane)		~13	EI		3288
$C_7H_7^+$	$C_6H_5CH_2CH_2C_6H_4NH_2$ (1–(4–Aminophenyl)–2–phenylethane)		~15	EI		3288
$C_7H_7^+$	$C_6H_5CH_2CH_2C_6H_4CN$ (1–(3–Cyanophenyl)–2–phenylethane)		10.3±0.2	EI		3288
$C_7H_7^+$	$C_6H_5CH_2OH$ (α–Hydroxytoluene)	OH	11.7	EI		3237
$C_7H_7^+$	$C_6H_5CH_2OCH_3$ (α–Methoxytoluene)		11.65	EI		3338
$C_7H_7^+$	$C_6H_5CH_2OCH_3$ (α–Methoxytoluene)		11.78	EI		3287
$C_7H_7^+$	$C_6H_4(OCH_3)CH_3$ (4–Methoxytoluene)		12.59	EI		3287
$C_7H_7^+$	$C_7H_7OCH_3$ (7–Methoxycycloheptatriene)		11.23	EI		3287
$C_7H_7^+$	$C_6H_5CH_2OC_6H_5$ (Benzyl phenyl ether)		9.7	EI		2737
$C_7H_7^+$	$C_6H_5CH_2CH_2C_6H_4OH$ (1–(3–Hydroxyphenyl)–2–phenylethane)		10.9±0.2	EI		3288
$C_7H_7^+$	$C_6H_5CH_2CH_2C_6H_4OH$ (1–(4–Hydroxyphenyl)–2–phenylethane)		~15	EI		3288
$C_7H_7^+$	$C_6H_5CH_2OC_6H_4CH_3$ (Benzyl 3–tolyl ether)		9.9	EI		2737
$C_7H_7^+$	$C_6H_5CH_2OC_6H_4CH_3$ (Benzyl 4–tolyl ether)		9.8	EI		2737

4.3. The Positive Ion Table—Continued

Ion	Reactant	Other products	Ionization or appearance potential (eV)	Method	Heat of formation (kJ mol^{-1})	Ref.
$C_7H_7^+$	$C_6H_5CH_2CH_2C_6H_4OCH_3$ (1–(4–Methoxyphenyl)–2–phenylethane)		~15	EI		3288
$C_7H_7^+$	$C_6H_5(CH_2)_3COOCH_3$ (4–Phenylbutanoic acid methyl ester)		13.41±0.2	EI		2497
$C_7H_7^+$	$C_6H_5CH_2OC_6H_4OH$ (Benzyl 3–hydroxyphenyl ether)		9.9	EI		2737
$C_7H_7^+$	$C_6H_5CH_2OC_6H_4OH$ (Benzyl 4–hydroxyphenyl ether)		9.9	EI		2737
$C_7H_7^+$	$C_6H_5CH_2OC_6H_4CHO$ (Benzyl 3–formylphenyl ether)		9.7	EI		2737
$C_7H_7^+$	$C_6H_5CH_2OC_6H_4CHO$ (Benzyl 4–formylphenyl ether)		9.7	EI		2737
$C_7H_7^+$	$C_6H_5CH_2OC_6H_4OCH_3$ (Benzyl 3–methoxyphenyl ether)		10.1	EI		2737
$C_7H_7^+$	$C_6H_5CH_2OC_6H_4OCH_3$ (Benzyl 4–methoxyphenyl ether)		10.0	EI		2737
$C_7H_7^+$	$C_6H_5CH_2OC_6H_4NH_2$ (3–Aminophenyl benzyl ether)		9.9	EI		2737
$C_7H_7^+$	$C_6H_5CH_2OC_6H_4NH_2$ (4–Aminophenyl benzyl ether)		9.9	EI		2737
$C_7H_7^+$	$C_6H_5CH_2CH_2C_6H_4NO_2$ (1–(4–Nitrophenyl)–2–phenylethane)		10.4±0.2	EI		3288
$C_7H_7^+$	$C_6H_5CH_2OC_6H_4NO_2$ (Benzyl 3–nitrophenyl ether)		9.9	EI		2737
$C_7H_7^+$	$C_6H_5CH_2OC_6H_4NO_2$ (Benzyl 4–nitrophenyl ether)		10.0	EI		2737
$C_7H_7^+$	$C_6H_5CH_2CH_2C_6H_4F$ (1–(4–Fluorophenyl)–2–phenylethane)		11.0±0.2	EI		3288
$C_7H_7^+$	$C_6H_5SCH_3$ (Methylthiobenzene)	SH	12.0	EI		307
$C_7H_7^+$	$C_6H_5CH_2SCH_3$ (α–Methylthiotoluene)		10.58	EI		3338
$C_7H_7^+$	$C_6H_5CH_2Cl$ (α–Chlorotoluene)	Cl?	10.40±0.05	PI	(901)	2612

See also – EI: 3230

Ion	Reactant	Other products	Ionization or appearance potential (eV)	Method	Heat of formation (kJ mol^{-1})	Ref.
$C_7H_7^+$	$C_6H_4ClCH_3$ (2–Chlorotoluene)	Cl?	11.8	EI		3230
$C_7H_7^+$	$C_6H_4ClCH_3$ (3–Chlorotoluene)	Cl?	11.32±0.1	EI		2972
$C_7H_7^+$	$C_6H_4ClCH_3$ (3–Chlorotoluene)	Cl?	11.9	EI		3230
$C_7H_7^+$	$C_6H_4ClCH_3$ (4–Chlorotoluene)	Cl?	11.14±0.1	EI		2972
$C_7H_7^+$	$C_6H_4ClCH_3$ (4–Chlorotoluene)	Cl?	11.7	EI		3230
$C_7H_7^+$	C_7H_9Cl (endo–5–Chlorobicyclo[2.2.1]hept–2–ene)		12.5±0.3	EI		2155
$C_7H_7^+$	C_7H_9Cl (exo–5–Chlorobicyclo[2.2.1]hept–2–ene)		12.6±0.3	EI		2155
$C_7H_7^+$	C_7H_9Cl (3–Chloronortricyclene)		12.2±0.3	EI		2155

4.3. The Positive Ion Table—Continued

Ion	Reactant	Other products	Ionization or appearance potential (eV)	Method	Heat of formation (kJ mol⁻¹)	Ref.
$C_7H_7^+$	$C_6H_5CH_2CH_2C_6H_4Cl$ (1–(3–Chlorophenyl)–2–phenylethane)		10.4±0.2	EI		3288
$C_7H_7^+$	$C_6H_5CH_2OC_6H_4Cl$ (Benzyl 3–chlorophenyl ether)		9.6	EI		2737
$C_7H_7^+$	$C_6H_5CH_2OC_6H_4Cl$ (Benzyl 4–chlorophenyl ether)		9.7	EI		2737
$C_7H_7^+$	$C_6H_5CH_2Br$ (α–Bromotoluene)	Br?	9.1	EI		3230

See also – EI: 2973

Ion	Reactant	Other products	Ionization or appearance potential (eV)	Method	Heat of formation (kJ mol⁻¹)	Ref.
$C_7H_7^+$	$C_6H_4BrCH_3$ (2–Bromotoluene)	Br?	11.2	EI		3230
$C_7H_7^+$	$C_6H_4BrCH_3$ (3–Bromotoluene)	Br?	11.3	EI		3230
$C_7H_7^+$	$C_6H_4BrCH_3$ (4–Bromotoluene)	Br?	11.30	EI		3238

See also – EI: 3230

Ion	Reactant	Other products	Ionization or appearance potential (eV)	Method	Heat of formation (kJ mol⁻¹)	Ref.
$C_7H_7^+$	$C_6H_5CH_2CH_2C_6H_4Br$ (1–(4–Bromophenyl)–2–phenylethane)		10.5±0.2	EI		3288
$C_7H_7^+$	$C_6H_5CH_2I$ (α–Iodotoluene)	I?	9.2	EI		3230
$C_7H_7^+$	$C_6H_4ICH_3$ (2–Iodotoluene)	I?	11.3	EI		3230
$C_7H_7^+$	$C_6H_4ICH_3$ (3–Iodotoluene)	I?	11.3	EI		3230
$C_7H_7^+$	$C_6H_4ICH_3$ (4–Iodotoluene)	I?	11.3	EI		3230
$C_7H_7^+$	$C_6H_5CH_2CH_2C_6H_4I$ (1–(4–Iodophenyl)–2–phenylethane)		10.6±0.2	EI		3288

The $C_7H_7^+$ ion from most of these compounds is believed to have the seven–membered tropylium ion structure, see H. M. Grubb and S. Meyerson in "Mass Spectrometry of Organic Ions," ed. F. W. McLafferty (Academic Press, New York, 1963) Chap. 10. From the heat of formation of cycloheptatriene (182 kJ mol⁻¹), a C–H bond energy equal to that of the secondary bond in pentadiene which is 334 kJ mol⁻¹ (D. M. Golden and S. W. Benson, Chem. Rev. **69**, 125 (1969)), and the spectroscopic ionization potential of cycloheptatrienyl radical given above (6.24 eV), one obtains $\Delta H^\circ_{f298}(C_7H_7^+) = 900$ kJ mol⁻¹. From the heat of formation of benzyl radical, 188 kJ mol⁻¹ (S. W. Benson and H. E. O'Neal, "Kinetic Data on Gas Phase Unimolecular Reactions," NSRDS–NBS 21 (U.S. Government Printing Office, Washington, D.C., 1970)), and the ionization potential determined by electron monochromator techniques (7.27 eV), we obtain $\Delta H^\circ_{f298}(C_7H_7^+) = 890$ kJ mol⁻¹, which is equal to the cycloheptatrienyl value within the estimated errors of all the individual data. Except for the case of α–chlorotoluene, whose low threshold may be due in part to an ion–pair process, these values are significantly lower than values based on fragmentation processes, even the photoionization of toluene which gives 946 kJ mol⁻¹ and of cycloheptatriene which gives 929 kJ mol⁻¹. The threshold values for these processes have not been corrected for kinetic shift which is expected to be significant here and will lead to too high a value. We suggest, therefore, a heat of formation of about 895 kJ mol⁻¹ (214 kcal mol⁻¹) as the best value.

4.3. The Positive Ion Table—Continued

Ion	Reactant	Other products	Ionization or appearance potential (eV)	Method	Heat of formation (kJ mol^{-1})	Ref.
			$C_7H_5D_2^+$			
$C_7H_5D_2^+$	$C_6H_5CD_2$ (Benzyl–α,α–d_2 radical)		7.71	EI		124

	$C_7H_8^+$ (Toluene)		$\Delta H^\circ_{f298} = 901$ kJ mol^{-1} (215 kcal mol^{-1})			
	$C_7H_8^+$ (Cycloheptatriene)		$\Delta H^\circ_{f298} \sim 982$ kJ mol^{-1} (235 kcal mol^{-1})			
$C_7H_8^+$	$C_6H_5CH_3$ (Toluene)		8.82±0.05	S	901	3153
(Average of two Rydberg series limits)						
$C_7H_8^+$	$C_6H_5CH_3$ (Toluene)		8.821±0.01	PI	901 158,	182, 416
$C_7H_8^+$	$C_6H_5CH_3$ (Toluene)		8.82±0.02	PI	901	3025
$C_7H_8^+$	$C_6H_5CH_3$ (Toluene)		8.82	PI	901	168
$C_7H_8^+$	$C_6H_5CH_3$ (Toluene)		8.82	PE	901	2843
$C_7H_8^+$	$C_6H_5CH_3$ (Toluene)		8.80±0.04	RPD		2407

See also – S: 344
         PI: 1142, 1159, 1166, 2612
         PE: 1159, 2806, 3331
         EI: 381, 383, 1066, 2025, 2109, 2158, 2163, 2463, 2538, 2865, 2989, 3157, 3223, 3230, 3238, 3287

$C_7H_8^+$	C_7H_8 (Cycloheptatriene)		8.28	S	981	3099
$C_7H_8^+$	C_7H_8 (Cycloheptatriene)		8.20±0.05	PI	973	2612
$C_7H_8^+$	C_7H_8 (Cycloheptatriene)		8.40	PE	992	2951

See also – EI: 68, 219, 2109

| $C_7H_8^+$ | C_7H_8 (Bicyclo[2.2.1]hepta–2,5–diene) | | 8.42±0.02 | PI | | 2877 |
| $C_7H_8^+$ | C_7H_8 (Bicyclo[2.2.1]hepta–2,5–diene) | | 8.42 | PE | | 3343 |

See also – PE: 2951, 3327
         EI: 219, 2109, 2185, 2531

$C_7H_8^+$	C_7H_8 (Bicyclo[3.2.0]hepta–2,6–diene)		8.92	EI		219
$C_7H_8^+$	C_7H_8 (Spiro[3.3]hepta–2,5–diene)		9.02	PE		2951
$C_7H_8^+$	C_7H_8 (Quadricyclene)		8.70	EI		2185
$C_7H_8^+$	C_7H_{10} (1,3–Cycloheptadiene)	H_2	10.8	EI		219

4.3. The Positive Ion Table—Continued

Ion	Reactant	Other products	Ionization or appearance potential (eV)	Method	Heat of formation (kJ mol^{-1})	Ref.
$C_7H_8^+$	C_7H_{10} (Bicyclo[3.2.0]hept–6–ene)	H_2	9.4	EI		219
$C_7H_8^+$	$C_{18}H_{30}$ (1–Phenyldodecane)		10.75±0.1	EI		2153
$C_7H_8^+$	$C_{26}H_{46}$ (1–Phenyleicosane)		11.35±0.1	EI		2153

$C_7H_8^{+2}$

Ion	Reactant	Other products	Ionization or appearance potential (eV)	Method	Heat of formation (kJ mol^{-1})	Ref.
$C_7H_8^{+2}$	$C_6H_5CH_3$ (Toluene)		23.5±0.2	FD		212

$C_7H_8^{+3}$

Ion	Reactant	Other products	Ionization or appearance potential (eV)	Method	Heat of formation (kJ mol^{-1})	Ref.
$C_7H_8^{+3}$	$C_6H_5CH_3$ (Toluene)		42±5	NRE		212

$C_7H_9^+$

Ion	Reactant	Other products	Ionization or appearance potential (eV)	Method	Heat of formation (kJ mol^{-1})	Ref.
$C_7H_9^+$	$C_5H_4(CH_3)_2$ (1,2–Dimethylcyclopentadiene)	H	11.05	EI		3335
$C_7H_9^+$	$C_6H_7CH_3$ (1–Methyl–1,4–cyclohexadiene)	H	11.03	EI		3335
$C_7H_9^+$	C_7H_{10} (1,3–Cycloheptadiene)	H	11.30	EI		219
$C_7H_9^+$	C_7H_{10} (1,3–Cycloheptadiene)	H	11.30	EI		3335
$C_7H_9^+$	C_7H_{10} (Bicyclo[2.2.1]hept–2–ene)	H	11.5±0.15	EI		2155
$C_7H_9^+$	C_7H_{10} (Bicyclo[3.2.0]hept–6–ene)	H	10.67	EI		219
$C_7H_9^+$	C_7H_{10} (Bicyclo[3.2.0]hept–6–ene)	H	10.7	EI		3335
$C_7H_9^+$	$C_5H_3(CH_3)_3$ (1,2,3–Trimethylcyclopentadiene)	CH_3	10.8	EI		3335
$C_7H_9^+$	$C_6H_4(OCH_3)CH_3$ (3–Methoxytoluene)	CO+H?	11.7	EI		3335
$C_7H_9^+$	C_7H_9Cl (endo–5–Chlorobicyclo[2.2.1]hept–2–ene)	Cl	11.0±0.15	EI		2155
$C_7H_9^+$	C_7H_9Cl (exo–5–Chlorobicyclo[2.2.1]hept–2–ene)	Cl	11.1±0.15	EI		2155
$C_7H_9^+$	C_7H_9Cl (3–Chloronortricyclene)	Cl	10.7±0.15	EI		2155

4.3. The Positive Ion Table—Continued

Ion	Reactant	Other products	Ionization or appearance potential (eV)	Method	Heat of formation (kJ mol^{-1})	Ref.

$$C_7H_{10}^+$$

Ion	Reactant	Other products	Ionization or appearance potential (eV)	Method	Heat of formation (kJ mol^{-1})	Ref.
$C_7H_{10}^+$	$C_5H_4(CH_3)_2$ (1,2–Dimethylcyclopentadiene)		8.1±0.1	EI		2163
$C_7H_{10}^+$	$C_5H_4(CH_3)_2$ (5,5–Dimethylcyclopentadiene)		8.22±0.05	EI		2163
$C_7H_{10}^+$	C_7H_{10} (cis,cis–1,3–Cycloheptadiene)		8.31 (V)	PE		3330
$C_7H_{10}^+$	C_7H_{10} (1,3–Cycloheptadiene)		8.40	EI		219

See also – PE: 3328

Ion	Reactant	Other products	Ionization or appearance potential (eV)	Method	Heat of formation (kJ mol^{-1})	Ref.
$C_7H_{10}^+$	C_7H_{10} (1,4–Cycloheptadiene)		8.85±0.03	PI		2877
$C_7H_{10}^+$	C_7H_{10} (Bicyclo[2.2.1]hept–2–ene)		8.81±0.02	PI		2877
$C_7H_{10}^+$	C_7H_{10} (Bicyclo[2.2.1]hept–2–ene)		8.82	PE		3343
$C_7H_{10}^+$	C_7H_{10} (Bicyclo[2.2.1]hept–2–ene)		8.83	PE		2951
$C_7H_{10}^+$	C_7H_{10} (Bicyclo[2.2.1]hept–2–ene)		8.95±0.15	EI		2155
$C_7H_{10}^+$	C_7H_{10} (Bicyclo[2.2.1]hept–2–ene)		8.98±0.1	EI		2531
$C_7H_{10}^+$	C_7H_{10} (Bicyclo[2.2.1]hept–2–ene)		9.05	EI		3341

See also – PE: 3327, 3328

Ion	Reactant	Other products	Ionization or appearance potential (eV)	Method	Heat of formation (kJ mol^{-1})	Ref.
$C_7H_{10}^+$	C_7H_{10} (Bicyclo[3.2.0]hept–6–ene)		9.37	EI		219
$C_7H_{10}^+$	C_7H_{10} (Nortricyclene)		9.02	PE		2951

$$C_7H_{11}^+$$

Ion	Reactant	Other products	Ionization or appearance potential (eV)	Method	Heat of formation (kJ mol^{-1})	Ref.
$C_7H_{11}^+$	$(CH_3)_2C=CHCH=C(CH_3)_2$	CH_3	10.80±0.03	EI		2455
$C_7H_{11}^+$	C_9H_{16} (cis–Hexahydroindan)	C_2H_5	11.71±0.05	EI	1184,	2028
$C_7H_{11}^+$	C_9H_{16} (trans–Hexahydroindan)	C_2H_5	11.80±0.05	EI	1184,	2028
$C_7H_{11}^+$	$C_7H_{11}Br$ (endo–2–Bromobicyclo[2.2.1]heptane)	Br	10.26±0.05	EI		2749
$C_7H_{11}^+$	$C_7H_{11}Br$ (exo–2–Bromobicyclo[2.2.1]heptane)	Br	10.27±0.05	EI		2749

4.3. The Positive Ion Table—Continued

Ion	Reactant	Other products	Ionization or appearance potential (eV)	Method	Heat of formation (kJ mol^{-1})	Ref.
	$C_7H_{12}^+$ (*cis*-Cycloheptene)		$\Delta H_{f298}^\circ = 847$ kJ mol^{-1} (202 kcal mol^{-1})			
$C_7H_{12}^+$	C_5H_8=CHCH$_3$ (Ethylidenecyclopentane)		8.49±0.02	PI		2877
$C_7H_{12}^+$	C_5H_8=CHCH$_3$ (Ethylidenecyclopentane)		8.51	PE		3343
$C_7H_{12}^+$	C_6H_{10}=CH$_2$ (Methylenecyclohexane)		8.97±0.01	PI		2877
$C_7H_{12}^+$	C_6H_{10}=CH$_2$ (Methylenecyclohexane)		8.94	PE		3343
$C_7H_{12}^+$	C_6H_{10}=CH$_2$ (Methylenecyclohexane)		9.04±0.03	EDD		2558
$C_7H_{12}^+$	$C_6H_9CH_3$ (1-Methylcyclohexene)		8.67±0.02	EDD		2558
$C_7H_{12}^+$	$C_6H_9CH_3$ (3-Methylcyclohexene)		8.94±0.03	EDD		2558

See also – EI: 2543

Ion	Reactant	Other products	Ionization or appearance potential (eV)	Method	Heat of formation (kJ mol^{-1})	Ref.
$C_7H_{12}^+$	$C_6H_9CH_3$ (4-Methylcyclohexene)		8.91±0.01	PI		182
$C_7H_{12}^+$	$C_6H_9CH_3$ (4-Methylcyclohexene)		8.92±0.02	EDD		2558

See also – EI: 2543

Ion	Reactant	Other products	Ionization or appearance potential (eV)	Method	Heat of formation (kJ mol^{-1})	Ref.
$C_7H_{12}^+$	C_7H_{12} (*cis*-Cycloheptene)		8.87	PE	847	3330

See also – EI: 2531, 3342

Ion	Reactant	Other products	Ionization or appearance potential (eV)	Method	Heat of formation (kJ mol^{-1})	Ref.
$C_7H_{12}^+$	C_7H_{12} (Bicyclo[2.2.1]heptane)		9.74	PE		3343
$C_7H_{12}^+$	C_7H_{12} (Bicyclo[2.2.1]heptane)		9.80	PE		2951
$C_7H_{12}^+$	C_7H_{12} (Bicyclo[2.2.1]heptane)		9.93±0.02	EDD		2558

See also – PE: 3327

Ion	Reactant	Other products	Ionization or appearance potential (eV)	Method	Heat of formation (kJ mol^{-1})	Ref.
$C_7H_{12}^+$	C_7H_{12} (Bicyclo[4.1.0]heptane)		9.03±0.02	EDD		2558
$C_7H_{12}^+$	C_9H_{16} (*cis*-Hexahydroindan)	C_2H_4	10.96±0.03	EI	1184,	2028
$C_7H_{12}^+$	C_9H_{16} (*trans*-Hexahydroindan)	C_2H_4	11.16±0.03	EI	1184,	2028
$C_7H_{12}^+$	$C_{10}H_{18}$ (*cis*-Decalin)	C_3H_6	10.72±0.02	EI	1182,	1183, 3337
$C_7H_{12}^+$	$C_{10}H_{18}$ (*trans*-Decalin)	C_3H_6	11.04±0.02	EI	1182,	1183, 3337

4.3. The Positive Ion Table—Continued

Ion	Reactant	Other products	Ionization or appearance potential (eV)	Method	Heat of formation (kJ mol⁻¹)	Ref.

C$_7$H$_{13}^+$

Ion	Reactant	Other products	Ionization or appearance potential (eV)	Method	Heat of formation	Ref.
C$_7$H$_{13}^+$	C$_6$H$_{10}$(CH$_3$)$_2$ (*cis*–1,2–Dimethylcyclohexane)	CH$_3$	10.78±0.02	EI		1145
C$_7$H$_{13}^+$	C$_6$H$_{10}$(CH$_3$)$_2$ (*trans*–1,2–Dimethylcyclohexane)	CH$_3$	10.84±0.02	EI		1145
C$_7$H$_{13}^+$	*cis*–(CH$_3$)$_3$CCH=CHC(CH$_3$)$_3$	C$_3$H$_7$	9.59±0.03	EI		2533
C$_7$H$_{13}^+$	*trans*–(CH$_3$)$_3$CCH=CHC(CH$_3$)$_3$	C$_3$H$_7$	10.04±0.05	EI		2533

C$_7$H$_{14}^+$ (Methylcyclohexane) $\Delta H_{f298}^\circ = 796$ kJ mol⁻¹ (190 kcal mol⁻¹)

Ion	Reactant	Other products	Ionization or appearance potential (eV)	Method	Heat of formation	Ref.
C$_7$H$_{14}^+$	C$_6$H$_{11}$CH$_3$ (Methylcyclohexane)		9.85±0.03	PI	796	182

See also – EI: 123

C$_7$H$_{15}^+$

Ion	Reactant	Other products	Ionization or appearance potential (eV)	Method	Heat of formation	Ref.
C$_7$H$_{15}^+$	*n*–C$_8$H$_{18}$	CH$_3$	~10.90±0.10	PI		1120
(Threshold value approximately corrected to 0 K)						
C$_7$H$_{15}^+$	*n*–C$_7$H$_{15}$COCH$_3$		10.00±0.1	RPD		2977
C$_7$H$_{15}^+$	*n*–C$_7$H$_{15}$OOH		10.6±0.1	EI		2464
C$_7$H$_{15}^+$	*n*–C$_5$H$_{11}$CH(OOH)CH$_3$		9.7±0.1	EI		2464

n–C$_7$H$_{16}^+$ $\Delta H_{f298}^\circ \sim 776$ kJ mol⁻¹ (185 kcal mol⁻¹)

Ion	Reactant	Other products	Ionization or appearance potential (eV)	Method	Heat of formation	Ref.
C$_7$H$_{16}^+$	*n*–C$_7$H$_{16}$		9.90±0.05	PI	767	2013
C$_7$H$_{16}^+$	*n*–C$_7$H$_{16}$		10.08	PI	785	182
C$_7$H$_{16}^+$	*n*–C$_7$H$_{16}$		10.16	PE		2829

See also – PE: 1130

C$_8$H$_5^+$

Ion	Reactant	Other products	Ionization or appearance potential (eV)	Method	Heat of formation	Ref.
C$_8$H$_5^+$	C$_{10}$H$_8$ (Naphthalene)		18.07±0.05	EI		2112
C$_8$H$_5^+$	C$_{10}$H$_8$ (Azulene)		16.3±0.15	EI		2112

4.3. The Positive Ion Table—Continued

Ion	Reactant	Other products	Ionization or appearance potential (eV)	Method	Heat of formation (kJ mol⁻¹)	Ref.
	$C_8H_6^+$ (Ethynylbenzene)		$\Delta H^\circ_{f298} = 1178$ kJ mol⁻¹ (282 kcal mol⁻¹)			
$C_8H_6^+$	$CH_3C\equiv CC\equiv CCH_3$		8.60±0.01	PE		2805
$C_8H_6^+$	$C_6H_5C\equiv CH$ (Ethynylbenzene)		8.815±0.005	PI	1178	182
$C_8H_6^+$	$C_6H_5CH=CH_2$ (Ethenylbenzene)	H_2	12.72±0.10	EI		2914
$C_8H_6^+$	C_8H_8 (Cyclooctatetraene)	H_2	11.70±0.10	EI		2914
$C_8H_6^+$	C_8H_8 (Cyclobutenobenzene)	H_2	11.84±0.10	RPD		2914
$C_8H_6^+$	C_8H_8 (Bicyclo[2.2.2]octatriene)	H_2	11.97±0.20	EI		2914
$C_8H_6^+$	C_8H_8 (syn–Tricyclo[4.2.0.0²,⁵]octa–3,7–diene)	H_2	10.03±0.10	EI		2914
$C_8H_6^+$	C_8H_8 (anti–Tricyclo[4.2.0.0²,⁵]octa–3,7–diene)	H_2	9.86±0.10	EI		2914
$C_8H_6^+$	C_8H_8 (Cubane)	H_2	8.92±0.10	EI		2914

See also – EI: 2105

Ion	Reactant	Other products	Ionization or appearance potential (eV)	Method	Heat of formation (kJ mol⁻¹)	Ref.
$C_8H_6^+$	$C_{10}H_8$ (Naphthalene)	C_2H_2	15.4±0.10	EI		2112
$C_8H_6^+$	$C_{10}H_8$ (Azulene)	C_2H_2	13.6±0.10	EI		2112
$C_8H_6^+$	$(C_6H_5)_2$ (Biphenyl)		18.10±0.05	EI		1238
$C_8H_6^+$	$C_6H_5C\equiv CC_6H_5$ (Diphenylacetylene)		17.8±0.1	EI		1238

$C_8H_7^+$

Ion	Reactant	Other products	Ionization or appearance potential (eV)	Method	Heat of formation (kJ mol⁻¹)	Ref.
$C_8H_7^+$	$C_6H_5CH=CH_2$ (Ethenylbenzene)	H	12.41±0.10	EI		2914
$C_8H_7^+$	C_8H_8 (Cyclooctatetraene)	H	10.90±0.10	EI		2914
$C_8H_7^+$	C_8H_8 (Cyclobutenobenzene)	H	11.94±0.10	EI		2914
$C_8H_7^+$	C_8H_8 (Bicyclo[2.2.2]octatriene)	H	10.63±0.10	EI		2914
$C_8H_7^+$	C_8H_8 (syn–Tricyclo[4.2.0.0²,⁵]octa–3,7–diene)	H	9.12±0.10	EI		2914
$C_8H_7^+$	C_8H_8 (anti–Tricyclo[4.2.0.0²,⁵]octa–3,7–diene)	H	9.13±0.10	EI		2914
$C_8H_7^+$	C_8H_8 (Cubane)	H	8.96±0.10	EI		2914

See also – EI: 2105

4.3. The Positive Ion Table—Continued

Ion	Reactant	Other products	Ionization or appearance potential (eV)	Method	Heat of formation (kJ mol^{-1})	Ref.
	$C_8H_8^+$ (Ethenylbenzene)		$\Delta H^\circ_{f298} = 962$ kJ mol^{-1} (230 kcal mol^{-1})			
	$C_8H_8^+$ (Cyclooctatetraene)		$\Delta H^\circ_{f298} = 1071$ kJ mol^{-1} (256 kcal mol^{-1})			
$C_8H_8^+$	$C_6H_5CH=CH_2$ (Ethenylbenzene)		8.47±0.02	PI	965	182
$C_8H_8^+$	$C_6H_5CH=CH_2$ (Ethenylbenzene)		8.43±0.01	PE	961	2843
$C_8H_8^+$	$C_6H_5CH=CH_2$ (Ethenylbenzene)		8.42	PE	960	3321
$C_8H_8^+$	$C_6H_5CH=CH_2$ (Ethenylbenzene)		8.53	CTS		2798

(Average of two values)

See also – EI: 1066, 2914

Ion	Reactant	Other products	Ionization or appearance potential (eV)	Method	Heat of formation (kJ mol^{-1})	Ref.
$C_8H_8^+$	C_8H_8 (Cyclooctatetraene)		7.99±0.02	PI	1069	182
$C_8H_8^+$	C_8H_8 (Cyclooctatetraene)		8.04	PE	1074	1130
$C_8H_8^+$	C_8H_8 (Cyclooctatetraene)		8.21	PE		2841, 2951

See also – PE: 2796
       EI: 381, 2752, 2865, 2914

Ion	Reactant	Other products	Ionization or appearance potential (eV)	Method	Heat of formation (kJ mol^{-1})	Ref.
$C_8H_8^+$	C_8H_8 (Cyclobutenobenzene)		8.74±0.05	RPD		2914
$C_8H_8^+$	C_8H_8 (Bicyclo[2.2.2]octatriene)		8.23	PE		3291
$C_8H_8^+$	C_8H_8 (Bicyclo[2.2.2]octatriene)		8.24	PE		2951
$C_8H_8^+$	C_8H_8 (Bicyclo[2.2.2]octatriene)		7.95±0.10	EI		2914
$C_8H_8^+$	C_8H_8 (syn–Tricyclo[4.2.0.0^{2,5}]octa–3,7–diene)		8.20±0.10	EI		2914

See also – EI: 2723, 2752

Ion	Reactant	Other products	Ionization or appearance potential (eV)	Method	Heat of formation (kJ mol^{-1})	Ref.
$C_8H_8^+$	C_8H_8 ($anti$–Tricyclo[4.2.0.0^{2,5}]octa–3,7–diene)		8.27±0.10	EI		2914
$C_8H_8^+$	C_8H_8 (Cubane)		8.74	PE		2843, 2951
$C_8H_8^+$	C_8H_8 (Cubane)		8.64±0.10	EI		2914
$C_8H_8^+$	C_8H_8 (Cubane)		8.74±0.15	EI		2105

4.3. The Positive Ion Table—Continued

Ion	Reactant	Other products	Ionization or appearance potential (eV)	Method	Heat of formation (kJ mol^{-1})	Ref.
$C_8H_8^+$	$C_{10}H_{12}$ (1,2,3,4–Tetrahydronaphthalene)		11.31	EI		3338
$C_8H_8^+$	$C_{26}H_{46}$ (1–Phenyleicosane)		12.2±0.1	EI		2153
$C_8H_8^+$	$C_9H_{11}N$ (1,2,3,4–Tetrahydroisoquinoline)		11.10	EI		3338
$C_8H_8^+$	$C_9H_{10}O$ (Isochroman)		9.95	EI		3338
$C_8H_8^+$	$C_6H_5(CH_2)_3CHO$ (4–Phenylbutanal)	CH_2CHOH	9.60±0.2	EI		2522
$C_8H_8^+$	$C_6H_5(CH_2)_3COCH_3$ (5–Phenyl–2–pentanone)	$CH_2C(CH_3)OH$	9.58±0.2	EI		2522
$C_8H_8^+$	$C_6H_5CH_2CH_2COOH$ (3–Phenylpropanoic acid)	$HCOOH$	9.77±0.2	EI		2522
$C_8H_8^+$	$C_6H_5(CH_2)_3COOH$ (4–Phenylbutanoic acid)	$CH_2=CO+H_2O$	10.55±0.2	EI		2522
$C_8H_8^+$	$C_6H_5CH_2CH_2COOCH_3$ (3–Phenylpropanoic acid methyl ester)	$HCOOCH_3$	9.87±0.2	EI		2522
$C_8H_8^+$	$C_6H_5(CH_2)_3COOCH_3$ (4–Phenylbutanoic acid methyl ester)	$CH_2=CO+CH_3OH$	10.69±0.2	EI		2522

See also – EI: 2497

Ion	Reactant	Other products	Ionization or appearance potential (eV)	Method	Heat of formation (kJ mol^{-1})	Ref.
$C_8H_8^+$	$C_9H_{10}S$ (Isothiochroman)		10.87	EI		3338

$C_8H_9^+$

Ion	Reactant	Other products	Ionization or appearance potential (eV)	Method	Heat of formation (kJ mol^{-1})	Ref.
$C_8H_9^+$	$C_6H_4(CH_3)CH_2$ (3–Methylbenzyl radical)		7.65±0.03	EI		69
$C_8H_9^+$	$C_6H_4(CH_3)CH_2$ (4–Methylbenzyl radical)		7.46±0.03	EI		69
$C_8H_9^+$	$CH_2=C(CH_3)C≡CC(CH_3)=CH_2$	H	10.4±0.1	EI		1122
$C_8H_9^+$	$C_6H_5C_2H_5$ (Ethylbenzene)	H	12.1±0.1	PI		2612
$C_8H_9^+$	$C_6H_5C_2H_5$ (Ethylbenzene)	H	11.4±0.1	EI		1122
$C_8H_9^+$	$C_6H_4(CH_3)_2$ (1,2–Dimethylbenzene)	H	11.30±0.05	PI	(891)	3025

See also – EI: 2538

Ion	Reactant	Other products	Ionization or appearance potential (eV)	Method	Heat of formation (kJ mol^{-1})	Ref.
$C_8H_9^+$	$C_6H_4(CH_3)_2$ (1,3–Dimethylbenzene)	H	11.7±0.1	PI		2612
$C_8H_9^+$	$C_6H_4(CH_3)_2$ (1,3–Dimethylbenzene)	H	11.84±0.1	EI		2970

See also – EI: 2538

4.3. The Positive Ion Table—Continued

Ion	Reactant	Other products	Ionization or appearance potential (eV)	Method	Heat of formation (kJ mol⁻¹)	Ref.
$C_8H_9^+$	$C_6H_4(CH_3)_2$ (1,4–Dimethylbenzene)	H	11.35±0.05	PI	(895)	3025
$C_8H_9^+$	$C_6H_4(CH_3)_2$ (1,4–Dimethylbenzene)	H	11.86±0.1	EI		2970

See also – PI: 2612
         EI: 2538

Ion	Reactant	Other products	Ionization or appearance potential (eV)	Method	Heat of formation (kJ mol⁻¹)	Ref.
$C_8H_9^+$	$C_7H_7CH_3$ (7–Methylcycloheptatriene)	H	11.0±0.1	EI		1122
$C_8H_9^+$	$C_7H_7CH_3$ (1–Methylspiro[2.4]hepta–4,6–diene)	H	10.65±0.1	EI		1122
$C_8H_9^+$	$C_7H_7CH_3$ (4–Methylspiro[2.4]hepta–4,6–diene)	H	9.9±0.1	EI		1122
$C_8H_9^+$	$C_7H_7CH_3$ (5–Methylspiro[2.4]hepta–4,6–diene)	H	9.9±0.1	EI		1122
$C_8H_9^+$	C_8H_{10} (1,3,5–Cyclooctatriene)	H	10.8	EI		2973
$C_8H_9^+$	$C_6H_5C_3H_7$ (Isopropylbenzene)	CH_3	10.65	EI		3238
$C_8H_9^+$	$C_6H_4(CH_3)C_2H_5$ (1–Ethyl–4–methylbenzene)	CH_3	11.2±0.1	EI		1122
$C_8H_9^+$	$C_6H_4(CD_3)C_2H_5$ (1–Ethyl–4–methyl–d_3–benzene)	CD_3	11.2±0.1	EI		2144
$C_8H_9^+$	$C_6H_4(CH_3)CH_2CD_3$ (1–Ethyl–2,2,2–d_3–4–methylbenzene)	CD_3	11.2±0.1	EI		2144
$C_8H_9^+$	$C_{26}H_{46}$ (2–Phenyleicosane)		10.83±0.1	EI		2153
$C_8H_9^+$	$C_{26}H_{46}$ (3–Phenyleicosane)		11.28±0.1	EI		2153
$C_8H_9^+$	$C_6H_5CH_2CH_2COCH_3$ (4–Phenyl–2–butanone)	CH_3CO	10.37±0.2	EI		2522
$C_8H_9^+$	$C_6H_5CH(CH_3)CH_2COOCH_3$ (3–Phenylbutanoic acid methyl ester)		11.43±0.2	EI		2497
$C_8H_9^+$	$C_6H_4(CH_3)(CH_2)_3COOCH_3$ (4–(4–Tolyl)butanoic acid methyl ester)		12.50±0.2	EI		2497
$C_8H_9^+$	$C_6H_5CH_2CH_2Br$ (1–Bromo–2–phenylethane)	Br	10.1	EI		2973
$C_8H_9^+$	$C_6H_4BrC_2H_5$ (1–Bromo–4–ethylbenzene)	Br	10.80	EI		3238
$C_8H_9^+$	$C_6H_4(CH_3)CH_2Br$ (1–Bromomethyl–3–methylbenzene)	Br	9.44±0.1	EI		2970
$C_8H_9^+$	$C_6H_4(CH_3)CH_2Br$ (1–Bromomethyl–4–methylbenzene)	Br	9.44±0.1	EI		2970
$C_8H_9^+$	C_8H_9Br (7–Bromo–1,3,5–cyclooctatriene)	Br	9.6	EI		2973

This body of data is completely discordant. Tait, Shannon and Harrison, ref. 2970, suggest that the ion has a substituted tropylium structure, Akopyan and Vilesov, ref. 2612, suggest there are structural differences. Kinetic shift effects have not been considered. We suggest a tentative value of $\Delta H^\circ_{f298}(C_8H_9^+) \approx 893$ kJ mol⁻¹ (213 kcal mol⁻¹).

4.3. The Positive Ion Table—Continued

Ion	Reactant	Other products	Ionization or appearance potential (eV)	Method	Heat of formation (kJ mol^{-1})	Ref.
	C$_8$H$_{10}^+$ (Ethylbenzene)		$\Delta H_{f298}^\circ = 875$ kJ mol^{-1} (209 kcal mol^{-1})			
	C$_8$H$_{10}^+$ (1,2–Dimethylbenzene)		$\Delta H_{f298}^\circ = 845$ kJ mol^{-1} (202 kcal mol^{-1})			
	C$_8$H$_{10}^+$ (1,3–Dimethylbenzene)		$\Delta H_{f298}^\circ = 843$ kJ mol^{-1} (201 kcal mol^{-1})			
	C$_8$H$_{10}^+$ (1,4–Dimethylbenzene)		$\Delta H_{f298}^\circ = 834$ kJ mol^{-1} (199 kcal mol^{-1})			
C$_8$H$_{10}^+$	CH$_2$=CHCH=CHCH=CHCH=CH$_2$		~7.8	D		3010
C$_8$H$_{10}^+$	CH$_2$=C(CH$_3$)C≡CC(CH$_3$)=CH$_2$		8.95±0.1	EI		1122
C$_8$H$_{10}^+$	C$_6$H$_5$C$_2$H$_5$ (Ethylbenzene)		8.77±0.01	S	876	344
C$_8$H$_{10}^+$	C$_6$H$_5$C$_2$H$_5$ (Ethylbenzene)		8.76±0.01	PI	875	182
C$_8$H$_{10}^+$	C$_6$H$_5$C$_2$H$_5$ (Ethylbenzene)		8.75±0.05	PI	874	2612

See also – EI: 2522, 3230, 3238
　　　　　CTS: 2562

Ion	Reactant	Other products	Ionization or appearance potential (eV)	Method	Heat of formation (kJ mol^{-1})	Ref.
C$_8$H$_{10}^+$	C$_6$H$_4$(CH$_3$)$_2$ (1,2–Dimethylbenzene)		8.58±0.01	S	847	344
C$_8$H$_{10}^+$	C$_6$H$_4$(CH$_3$)$_2$ (1,2–Dimethylbenzene)		8.555	PI	844	168
C$_8$H$_{10}^+$	C$_6$H$_4$(CH$_3$)$_2$ (1,2–Dimethylbenzene)		8.56±0.01	PI	845	182, 416
C$_8$H$_{10}^+$	C$_6$H$_4$(CH$_3$)$_2$ (1,2–Dimethylbenzene)		8.56±0.02	PI	845	1142, 1159, 1166, 3025

See also – S: 3153
　　　　　PE: 1159
　　　　　EI: 1066, 2163, 2538, 2865, 2989

Ion	Reactant	Other products	Ionization or appearance potential (eV)	Method	Heat of formation (kJ mol^{-1})	Ref.
C$_8$H$_{10}^+$	C$_6$H$_4$(CH$_3$)$_2$ (1,3–Dimethylbenzene)		8.56	PI	843	168
C$_8$H$_{10}^+$	C$_6$H$_4$(CH$_3$)$_2$ (1,3–Dimethylbenzene)		8.56±0.01	PI	843	182, 416
C$_8$H$_{10}^+$	C$_6$H$_4$(CH$_3$)$_2$ (1,3–Dimethylbenzene)		8.55±0.05	PI	842	2612

See also – S: 344, 3153
　　　　　PI: 1142, 1159, 1166
　　　　　PE: 1159
　　　　　EI: 1066, 1122, 2538, 2865, 2989

Ion	Reactant	Other products	Ionization or appearance potential (eV)	Method	Heat of formation (kJ mol^{-1})	Ref.
C$_8$H$_{10}^+$	C$_6$H$_4$(CH$_3$)$_2$ (1,4–Dimethylbenzene)		8.48	S	836	344
C$_8$H$_{10}^+$	C$_6$H$_4$(CH$_3$)$_2$ (1,4–Dimethylbenzene)		8.445	PI	833	168
C$_8$H$_{10}^+$	C$_6$H$_4$(CH$_3$)$_2$ (1,4–Dimethylbenzene)		8.445±0.015	PI	833	158, 182, 416
C$_8$H$_{10}^+$	C$_6$H$_4$(CH$_3$)$_2$ (1,4–Dimethylbenzene)		8.44±0.02	PI	832	1142, 1159, 1166, 3025

4.3. The Positive Ion Table—Continued

Ion	Reactant	Other products	Ionization or appearance potential (eV)	Method	Heat of formation (kJ mol⁻¹)	Ref.
$C_8H_{10}^+$	$C_6H_4(CH_3)_2$ (1,4–Dimethylbenzene)		8.52	CTS		2909

See also – S: 3153
PI: 2612
PE: 1159, 2806
EI: 1066, 2163, 2538, 2865, 2989, 3223, 3230

Ion	Reactant	Other products	Ionization or appearance potential (eV)	Method	Heat of formation (kJ mol⁻¹)	Ref.
$C_8H_{10}^+$	$C_7H_7CH_3$ (7–Methylcycloheptatriene)		8.39±0.1	EI		1122
$C_8H_{10}^+$	$C_7H_7CH_3$ (1–Methylspiro[2.4]hepta–4,6–diene)		8.40±0.1	EI		1122
$C_8H_{10}^+$	$C_7H_7CH_3$ (4–Methylspiro[2.4]hepta–4,6–diene)		8.02±0.1	EI		1122
$C_8H_{10}^+$	$C_7H_7CH_3$ (5–Methylspiro[2.4]hepta–4,6–diene)		8.07±0.1	EI		1122
$C_8H_{10}^+$	C_8H_{10} (Bicyclo[2.2.2]octa–2,5–diene)		8.87 (V)	PE		3327
$C_8H_{10}^+$	C_8H_{10} (Bicyclo[3.2.1]octa–2,6–diene)		8.44±0.01	PI		2877
$C_8H_{10}^+$	C_8H_{10} (Bicyclo[5.1.0]octa–2,5–diene)		8.43	PE		3329
$C_8H_{10}^+$	$C_6H_5CH_2CH_2CHO$ (3–Phenylpropanal)	CO	9.68±0.2	EI		2522

$C_8H_{12}^+$

Ion	Reactant	Other products	Ionization or appearance potential (eV)	Method	Heat of formation (kJ mol⁻¹)	Ref.
$C_8H_{12}^+$	$C_5H_3(CH_3)_3$ (1,2,3–Trimethylcyclopentadiene)		7.96±0.05	EI		2163
$C_8H_{12}^+$	$C_5H_3(CH_3)_3$ (1,5,5–Trimethylcyclopentadiene)		8.00±0.1	EI		2163
$C_8H_{12}^+$	$C_6H_9CH=CH_2$ (4–Ethenylcyclohexene)		8.93±0.02	PI		182
$C_8H_{12}^+$	C_8H_{12} (cis,cis–1,3–Cyclooctadiene)		8.68 (V)	PE		3330
$C_8H_{12}^+$	C_8H_{12} (1,4–Cyclooctadiene)		8.64±0.03	PI		2877
$C_8H_{12}^+$	C_8H_{12} (1,5–Cyclooctadiene)		9.1±0.1	EI		2698
$C_8H_{12}^+$	C_8H_{12} (Bicyclo[2.2.2]oct–2–ene)		8.92	PE		2951

See also – PE: 3327

Ion	Reactant	Other products	Ionization or appearance potential (eV)	Method	Heat of formation (kJ mol⁻¹)	Ref.
$C_8H_{12}^+$	C_8H_{12} (Bicyclo[3.2.1]oct–2–ene)		8.76±0.02	PI		2877
$C_8H_{12}^+$	C_8H_{12} (Tricyclo[3.2.1.0^{3,6}]octane)		8.75	PE		2951

4.3. The Positive Ion Table—Continued

Ion	Reactant	Other products	Ionization or appearance potential (eV)	Method	Heat of formation (kJ mol^{-1})	Ref.

$C_8H_{13}^+$

$C_8H_{13}^+$	$C_6H_8C_2H_5$? (6–Ethyl–2–cyclohexen–1–yl radical?)		7.5	EI		3002
$C_8H_{13}^+$	$(CH_3)_3CCH=C=CH(CH_2)_4CH_3$		10.1±0.05	EI		3007
$C_8H_{13}^+$	$(CH_3)_3CCH=C=CHCH_2C(CH_3)_3$		10.0±0.05	EI		3007

$C_8H_{14}^+$

$C_8H_{14}^+$	$(CH_3)_2C=CHCH=C(CH_3)_2$		7.91±0.04	EI		2455
$C_8H_{14}^+$	$C_6H_{10}=CHCH_3$ (Ethylidenecyclohexane)		8.47±0.02	PI		2877
$C_8H_{14}^+$	$C_6H_{10}=CHCH_3$ (Ethylidenecyclohexane)		8.41	PE		3343
$C_8H_{14}^+$	C_8H_{14} (cis–Cyclooctene)		8.82	PE		3330

See also – EI: 2531, 3342

$C_8H_{14}^+$	C_8H_{14} (Bicyclo[2.2.2]octane)		9.53	PE		2951

See also – PE: 2962, 3327

$1-C_8H_{16}^+$ $\Delta H^\circ_{f298} = 827$ kJ mol^{-1} (198 kcal mol^{-1})

$C_8H_{16}^+$	$1–C_8H_{16}$		9.43±0.01	PI	827	2877
$C_8H_{16}^+$	$C_6H_{10}(CH_3)_2$ (cis–1,2–Dimethylcyclohexane)		10.08±0.02	EI		1145
$C_8H_{16}^+$	$C_6H_{10}(CH_3)_2$ (trans–1,2–Dimethylcyclohexane)		10.08±0.03	EI		1145

$(CH_3)_2CHCH_2C(CH_3)_3^+$ $\Delta H^\circ_{f298} \sim 730$ kJ mol^{-1} (174 kcal mol^{-1})

$C_8H_{18}^+$	$n–C_8H_{18}$		10.25	RPD		2977
$C_8H_{18}^+$	$(CH_3)_2CHCH_2C(CH_3)_3$		9.86	PI	727	182
$C_8H_{18}^+$	$(CH_3)_2CHCH_2C(CH_3)_3$		9.91	PE	732	1130

$C_9H_5^+$

$C_9H_5^+$	$C_6H_5C\equiv CC_6H_5$ (Diphenylacetylene)		21.3±0.2	EI		1238

4.3. The Positive Ion Table—Continued

Ion	Reactant	Other products	Ionization or appearance potential (eV)	Method	Heat of formation (kJ mol^{-1})	Ref.
			$C_9H_7^+$			
$C_9H_7^+$	C_9H_7 (Indenyl radical)		8.35	EI		126
$C_9H_7^+$	$C_6H_4(CH_3)C \equiv CH$ (2−Ethynyltoluene)	H	11.36	EI		2541
(Metastable transition indicates zero kinetic energy release)						
$C_9H_7^+$	C_9H_8 (Indene)	H	12.53	EI		2541
(Metastable transition indicates zero kinetic energy release)						
$C_9H_7^+$	$(C_6H_5)_2$ (Biphenyl)		16.08±0.05	EI		1238
$C_9H_7^+$	$C_6H_5C \equiv CC_6H_5$ (Diphenylacetylene)		17.5±0.1	EI		1238
$C_9H_7^+$	$C_{14}H_{10}$ (Phenanthrene)		23.9±0.2	RPD		2540
$C_9H_7^+$	$C_{10}H_7NH_2$ (1−Aminonaphthalene)	HCN+H	14.15	EI		2541
(Metastable transitions indicate zero kinetic energy release)						
$C_9H_7^+$	$C_{10}H_7NH_2$ (2−Aminonaphthalene)	HCN+H	14.28	EI		2541
(Metastable transitions indicate zero kinetic energy release)						
$C_9H_7^+$	$C_{10}H_7OH$ (1−Hydroxynaphthalene)	CO+H	13.68	EI		2541
(Metastable transitions indicate ~0.1 eV kinetic energy release)						
$C_9H_7^+$	$C_{10}H_7OH$ (2−Hydroxynaphthalene)	CO+H	13.91	EI		2541
(Metastable transitions indicate ~0.2 eV kinetic energy release)						

Occolowitz and White, ref. 2541, conclude that the $C_9H_7^+$ ions from indene, 2−ethynyltoluene and the substituted naphthalenes have an ethynyltropylium structure rather than indenyl.

Ion	Reactant	Other products	Ionization or appearance potential (eV)	Method	Heat of formation (kJ mol^{-1})	Ref.
	$C_9H_8^+$ (Indene)	$\Delta H_{f298}^\circ = 948$ kJ mol^{-1} (227 kcal mol^{-1})				
$C_9H_8^+$	$C_6H_4(CH_3)C \equiv CH$ (2−Ethynyltoluene)		8.90	EI		2541
$C_9H_8^+$	C_9H_8 (Indene)		8.13±0.05	PE	948	2847
$C_9H_8^+$	C_9H_8 (Indene)		8.14±0.01	PE	949	2942
$C_9H_8^+$	C_9H_8 (Indene)		8.62	EI		2541
$C_9H_8^+$	$C_{10}H_7NH_2$ (1−Aminonaphthalene)	HCN	12.56	EI		2541
(Metastable transition indicates zero kinetic energy release)						
$C_9H_8^+$	$C_{10}H_7NH_2$ (2−Aminonaphthalene)	HCN	12.59	EI		2541
(Metastable transition indicates zero kinetic energy release)						
$C_9H_8^+$	$C_{10}H_7OH$ (1−Hydroxynaphthalene)	CO	11.73	EI		2541
(Metastable transition indicates ~0.1 eV kinetic energy release)						

4.3. The Positive Ion Table—Continued

Ion	Reactant	Other products	Ionization or appearance potential (eV)	Method	Heat of formation (kJ mol^{-1})	Ref.
C$_9$H$_8^+$	C$_{10}$H$_7$OH (2-Hydroxynaphthalene)	CO	11.27	EI		2541

(Metastable transition indicates ~0.2 eV kinetic energy release)

Occolowitz and White, ref. 2541, conclude that the C$_9$H$_8^+$ fragment ions above have a 2-ethynyltoluene structure rather than indene.

	C$_9$H$_{10}^+$ (Isopropenylbenzene)	$\Delta H_{f298}^\circ = 919$ kJ mol^{-1} (220 kcal mol^{-1})				
C$_9$H$_{10}^+$	C$_6$H$_5$C(CH$_3$)=CH$_2$ (Isopropenylbenzene)		8.35±0.01	PI	919	182
C$_9$H$_{10}^+$	C$_9$H$_{10}$ (Indan)		9.05±0.05	EI		3342
C$_9$H$_{10}^+$	C$_9$H$_{10}$ (Bicyclo[4.2.1]nona-2,4,7-triene)		8.36 (V)	PE		3328
C$_9$H$_{10}^+$	C$_6$H$_5$CH(CH$_3$)CH$_2$COOCH$_3$ (3-Phenylbutanoic acid methyl ester)		9.72±0.2	EI		2497
C$_9$H$_{10}^+$	C$_6$H$_4$(CH$_3$)(CH$_2$)$_3$COOCH$_3$ (4-(4-Tolyl)butanoic acid methyl ester)		10.65±0.2	EI		2497

	C$_9$H$_{11}^+$					
C$_9$H$_{11}^+$	C$_6$H$_5$C$_4$H$_9$ (tert-Butylbenzene)	CH$_3$	10.26	EI		3238
C$_9$H$_{11}^+$	C$_6$H$_3$(CH$_3$)$_2$CH$_2$CD$_3$ (1,2-Dimethyl-4-ethyl-2,2,2-d$_3$-benzene)	CD$_3$	10.5	EI		2144
C$_9$H$_{11}^+$	C$_6$H$_3$(CH$_3$)$_2$CH$_2$CD$_3$ (1,4-Dimethyl-3-ethyl-2,2,2-d$_3$-benzene)	CD$_3$	10.5	EI		2144
C$_9$H$_{11}^+$	C$_6$H$_3$(CH$_3$)$_2$CH$_2$CD$_3$ (1,3-Dimethyl-5-ethyl-2,2,2-d$_3$-benzene)	CD$_3$	10.5	EI		2144
C$_9$H$_{11}^+$	C$_{18}$H$_{30}$ (3-Phenyldodecane)		10.42±0.1	EI		2153
C$_9$H$_{11}^+$	C$_{26}$H$_{46}$ (3-Phenyleicosane)		10.69±0.1	EI		2153
C$_9$H$_{11}^+$	C$_{26}$H$_{46}$ (4-Phenyleicosane)		11.36±0.1	EI		2153
C$_9$H$_{11}^+$	C$_6$H$_4$(CH$_3$)CH(CH$_3$)CH$_2$COOCH$_3$ (3-(4-Tolyl)butanoic acid methyl ester)		10.95±0.2	EI		2497
C$_9$H$_{11}^+$	C$_6$H$_5$(CH$_2$)$_3$Br (1-Bromo-3-phenylpropane)	Br?	10.1	EI		2973

	C$_9$H$_8$D$_3^+$					
C$_9$H$_8$D$_3^+$	C$_6$H$_3$(CH$_3$)$_2$CH$_2$CD$_3$ (1,2-Dimethyl-4-ethyl-2,2,2-d$_3$-benzene)	CH$_3$	10.5	EI		2144
C$_9$H$_8$D$_3^+$	C$_6$H$_3$(CH$_3$)$_2$CH$_2$CD$_3$ (1,4-Dimethyl-3-ethyl-2,2,2-d$_3$-benzene)	CH$_3$	10.5	EI		2144
C$_9$H$_8$D$_3^+$	C$_6$H$_3$(CH$_3$)$_2$CH$_2$CD$_3$ (1,3-Dimethyl-5-ethyl-2,2,2-d$_3$-benzene)	CH$_3$	10.5	EI		2144

4.3. The Positive Ion Table—Continued

Ion	Reactant	Other products	Ionization or appearance potential (eV)	Method	Heat of formation (kJ mol^{-1})	Ref.	
	$C_9H_{12}^+$ (Propylbenzene)		$\Delta H^\circ_{f\,298} = 849$ kJ mol^{-1} (203 kcal mol^{-1})				
	$C_9H_{12}^+$ (Isopropylbenzene)		$\Delta H^\circ_{f\,298} = 842$ kJ mol^{-1} (201 kcal mol^{-1})				
	$C_9H_{12}^+$ (1,2,3–Trimethylbenzene)		$\Delta H^\circ_{f\,298} = 809$ kJ mol^{-1} (193 kcal mol^{-1})				
	$C_9H_{12}^+$ (1,2,4–Trimethylbenzene)		$\Delta H^\circ_{f\,298} = 784$ kJ mol^{-1} (187 kcal mol^{-1})				
	$C_9H_{12}^+$ (1,3,5–Trimethylbenzene)		$\Delta H^\circ_{f\,298} = 794$ kJ mol^{-1} (190 kcal mol^{-1})				
$C_9H_{12}^+$	$C_6H_5C_3H_7$ (Propylbenzene)		8.72 ± 0.01	PI	849	182,	416

See also – EI: 2522, 3338

Ion	Reactant	Other products	Ionization or appearance potential (eV)	Method	Heat of formation (kJ mol^{-1})	Ref.	
$C_9H_{12}^+$	$C_6H_5C_3H_7$ (Isopropylbenzene)		~8.76	S			344
$C_9H_{12}^+$	$C_6H_5C_3H_7$ (Isopropylbenzene)		8.69 ± 0.01	PI	842	182,	416

See also – EI: 3238

Ion	Reactant	Other products	Ionization or appearance potential (eV)	Method	Heat of formation (kJ mol^{-1})	Ref.	
$C_9H_{12}^+$	$C_6H_3(CH_3)_3$ (1,2,3–Trimethylbenzene)		8.48 ± 0.01	PI	809	168,	3154

See also – EI: 2163

Ion	Reactant	Other products	Ionization or appearance potential (eV)	Method	Heat of formation (kJ mol^{-1})	Ref.	
$C_9H_{12}^+$	$C_6H_3(CH_3)_3$ (1,2,4–Trimethylbenzene)		8.27 ± 0.01	PI	784	168,	3154

See also – EI: 2163

Ion	Reactant	Other products	Ionization or appearance potential (eV)	Method	Heat of formation (kJ mol^{-1})	Ref.	
$C_9H_{12}^+$	$C_6H_3(CH_3)_3$ (1,3,5–Trimethylbenzene)		8.39 ± 0.01	PI	793	168,	3154
$C_9H_{12}^+$	$C_6H_3(CH_3)_3$ (1,3,5–Trimethylbenzene)		8.40 ± 0.01	PI	794		182
$C_9H_{12}^+$	$C_6H_3(CH_3)_3$ (1,3,5–Trimethylbenzene)		8.41 ± 0.02	PI	795	1142,	1159, 1166
$C_9H_{12}^+$	$C_6H_3(CH_3)_3$ (1,3,5–Trimethylbenzene)		8.47	CTS			3403
	(Average of two values)						
$C_9H_{12}^+$	$C_6H_3(CH_3)_3$ (1,3,5–Trimethylbenzene)		8.55	CTS			2909

See also – PI: 190, 416
       PE: 1159
       EI: 1066, 2163, 2511, 2865

Ion	Reactant	Other products	Ionization or appearance potential (eV)	Method	Heat of formation (kJ mol^{-1})	Ref.	
$C_9H_{12}^+$	C_9H_{12} (cis,cis,cis–1,4,7–Cyclononatriene)		8.42 ± 0.03	PI			2877
$C_9H_{12}^+$	C_9H_{12} (cis,cis,cis–1,4,7–Cyclononatriene)		8.45	PE			3328

See also – EI: 3341

Ion	Reactant	Other products	Ionization or appearance potential (eV)	Method	Heat of formation (kJ mol^{-1})	Ref.	
$C_9H_{12}^+$	C_9H_{12} (Bicyclo[4.3.0]nona–3,7–diene)		8.78	PE			2951

4.3. The Positive Ion Table—Continued

Ion	Reactant	Other products	Ionization or appearance potential (eV)	Method	Heat of formation (kJ mol^{-1})	Ref.
$C_9H_{14}^+$						
$C_9H_{14}^+$	$C_5H_2(CH_3)_4$ (1,2,3,4–Tetramethylcyclopentadiene)		7.8±0.1	EI		2163
$C_9H_{14}^+$	$C_5H_2(CH_3)_4$ (1,4,5,5–Tetramethylcyclopentadiene)		7.84±0.05	EI		2163
$C_9H_{14}^+$	C_9H_{14} (1,4–Cyclononadiene)		8.60±0.03	PI		2877
$C_9H_{16}^+$						
$C_9H_{16}^+$	C_9H_{16} (cis–Hexahydroindan)		10.13±0.03	EI		1184, 2028
$C_9H_{16}^+$	C_9H_{16} (trans–Hexahydroindan)		10.18±0.03	EI		1184, 2028
$C_9H_{17}^+$						
$C_9H_{17}^+$	cis–$(CH_3)_3CCH=CHC(CH_3)_3$	CH_3	10.53±0.02	EI		2533
$C_9H_{17}^+$	trans–$(CH_3)_3CCH=CHC(CH_3)_3$	CH_3	10.70±0.06	EI		2533
$C_9H_{20}^+$						
$C_9H_{20}^+$	n–C_9H_{20}		10.19	RPD		2977
$C_{10}H_6^+$						
$C_{10}H_6^+$	$C_{10}H_8$ (Naphthalene)	H_2	16.2±0.15	EI		2112
$C_{10}H_6^+$	$C_{10}H_8$ (Azulene)	H_2	14.7±0.10	EI		2112
$C_{10}H_6^+$	$C_6H_5C{\equiv}CC_6H_5$ (Diphenylacetylene)	$2C_2H_2$	18.23±0.1	EI		1238
$C_{10}H_6^+$	$C_{14}H_{10}$ (Phenanthrene)	$2C_2H_2$	20.80±0.3	RPD		2540
$C_{10}H_6^+$	$C_{14}H_{10}$ (Phenanthrene)	$2C_2H_2$	20.9±0.3	EI		1238
$C_{10}H_7^+$						
$C_{10}H_7^+$	$C_{10}H_8$ (Naphthalene)	H	15.4±0.10	EI		2112
$C_{10}H_7^+$	$C_{10}H_8$ (Azulene)	H	14.0±0.10	EI		2112

4.3. The Positive Ion Table—Continued

Ion	Reactant	Other products	Ionization or appearance potential (eV)	Method	Heat of forma-tion (kJ mol⁻¹)	Ref.
	$C_{10}H_8^+$ (Naphthalene)	$\Delta H^\circ_{f298} = 936$ kJ mol⁻¹ (224 kcal mol⁻¹)				
	$C_{10}H_8^+$ (Azulene)	$\Delta H^\circ_{f298} = 996$ kJ mol⁻¹ (238 kcal mol⁻¹)				
$C_{10}H_8^+$	$C_{10}H_8$ (Naphthalene)		8.136±0.005	S	936	2686
(Average of five Rydberg series limits)						
$C_{10}H_8^+$	$C_{10}H_8$ (Naphthalene)		8.133	S	936	2661
(Average of three Rydberg series limits)						
$C_{10}H_8^+$	$C_{10}H_8$ (Naphthalene)		8.12±0.01	PI	934	182, 416
$C_{10}H_8^+$	$C_{10}H_8$ (Naphthalene)		8.14	PI	936	2661
$C_{10}H_8^+$	$C_{10}H_8$ (Naphthalene)		8.14±0.02	PI	936	1166
$C_{10}H_8^+$	$C_{10}H_8$ (Naphthalene)		8.15±0.01	PI	937	2651
$C_{10}H_8^+$	$C_{10}H_8$ (Naphthalene)		8.11±0.01	PE	933	2840, 2843, 2844, 2942
$C_{10}H_8^+$	$C_{10}H_8$ (Naphthalene)		8.12±0.05	PE	934	2847
$C_{10}H_8^+$	$C_{10}H_8$ (Naphthalene)		8.16±0.03	EDD		3174
$C_{10}H_8^+$	$C_{10}H_8$ (Naphthalene)		8.08	CTS		2909
$C_{10}H_8^+$	$C_{10}H_8$ (Naphthalene)		8.10	CTS		2947
$C_{10}H_8^+$	$C_{10}H_8$ (Naphthalene)		8.15	CTS		1064
$C_{10}H_8^+$	$C_{10}H_8$ (Naphthalene)		8.16	CTS		2910

See also – S: 344
        EI: 413, 2112, 2458, 2538, 2847, 3011, 3284
        CTS: 2562

Ion	Reactant	Other products	Ionization or appearance potential (eV)	Method	Heat of forma-tion (kJ mol⁻¹)	Ref.
$C_{10}H_8^+$	$C_{10}H_8$ (Azulene)		7.408	S	995	1459, 2661
(Average of two Rydberg series limits)						
$C_{10}H_8^+$	$C_{10}H_8$ (Azulene)		7.431±0.006	S	997	1420
(Average of two Rydberg series limits)						

4.3. The Positive Ion Table—Continued

Ion	Reactant	Other products	Ionization or appearance potential (eV)	Method	Heat of formation (kJ mol^{-1})	Ref.	
$C_{10}H_8^+$	$C_{10}H_8$ (Azulene)		7.41	PI	995	2636,	2661
$C_{10}H_8^+$	$C_{10}H_8$ (Azulene)		7.42±0.05	PE	996		2847
$C_{10}H_8^+$	$C_{10}H_8$ (Azulene)		7.43±0.01	PE	997	2843,	2942
$C_{10}H_8^+$	$C_{10}H_8$ (Azulene)		7.4–7.5	CTS			2037
See also – EI: 2112, 2458, 2847							
$C_{10}H_8^+$	$(C_6H_5)_2$ (Biphenyl)		14.81±0.05	EI			1238

$C_{10}D_8^+$

Ion	Reactant	Other products	Ionization or appearance potential (eV)	Method	Heat of formation (kJ mol^{-1})	Ref.
$C_{10}D_8^+$	$C_{10}D_8$ (Naphthalene–d_8) (Average of five Rydberg series limits)		8.138±0.008	S		2686

$C_{10}H_8^{+2}$

Ion	Reactant	Other products	Ionization or appearance potential (eV)	Method	Heat of formation (kJ mol^{-1})	Ref.
$C_{10}H_8^{+2}$	$C_{10}H_8$ (Naphthalene)		22.8±0.2	FD		212
$C_{10}H_8^{+2}$	$C_{10}H_8$ (Naphthalene)		22.7	EI		413

$C_{10}H_8^{+3}$

Ion	Reactant	Other products	Ionization or appearance potential (eV)	Method	Heat of formation (kJ mol^{-1})	Ref.
$C_{10}H_8^{+3}$	$C_{10}H_8$ (Naphthalene)		40±5	NRE		212

$C_{10}H_{10}^+$

Ion	Reactant	Other products	Ionization or appearance potential (eV)	Method	Heat of formation (kJ mol^{-1})	Ref.	
$C_{10}H_{10}^+$	$(C_5H_5)_2$ (5,5′–Bicyclopentadienyl)		7.75	EI	2732,	2940	
$C_{10}H_{10}^+$	$C_{10}H_{10}$ (Bullvalene)		8.05	PE		3329	
$C_{10}H_{10}^+$	$C_{10}H_{10}$ (Bullvalene)		8.13	PE		2951	

4.3. The Positive Ion Table—Continued

Ion	Reactant	Other products	Ionization or appearance potential (eV)	Method	Heat of formation (kJ mol⁻¹)	Ref.	

$$C_{10}H_{12}^+$$

Ion	Reactant	Other products	Ionization or appearance potential (eV)	Method	Heat of formation (kJ mol⁻¹)	Ref.	
$C_{10}H_{12}^+$	$C_{10}H_{12}$ (1,2,3,4–Tetrahydronaphthalene)		8.73	EI			3338
$C_{10}H_{12}^+$	$C_{10}H_{12}$ (1,2,3,4–Tetrahydronaphthalene)		9.14±0.05	EI			3342
$C_{10}H_{12}^+$	$C_{10}H_{12}$ (1,4,5,8–Tetrahydronaphthalene)		8.2	PE			3326
$C_{10}H_{12}^+$	$C_{10}H_{12}$ (Dihydrobullvalene)		8.02	PE			3329
$C_{10}H_{12}^+$	$C_{10}H_{12}$ (Tricyclo[5.2.1.0^{2,6}]deca–3,8–diene)		8.79±0.05	PE			3246
$C_{10}H_{12}^+$	$C_6H_4(CH_3)CH(CH_3)CH_2COOCH_3$ (3–(4–Tolyl)butanoic acid methyl ester)		9.74±0.2	EI			2497

$$C_{10}H_{13}^+$$

Ion	Reactant	Other products	Ionization or appearance potential (eV)	Method	Heat of formation (kJ mol⁻¹)	Ref.	
$C_{10}H_{13}^+$	$C_6H_4(C_3H_7)CH_2$ (4–Isopropylbenzyl radical)		7.42±0.1	EI			69
$C_{10}H_{13}^+$	$C_{18}H_{30}$ (1–Phenyldodecane)		11.07±0.1	EI			2153
$C_{10}H_{13}^+$	$C_{26}H_{46}$ (1–Phenyleicosane)		10.86±0.1	EI			2153
$C_{10}H_{13}^+$	$C_{26}H_{46}$ (4–Phenyleicosane)		10.58±0.1	EI			2153

$C_{10}H_{14}^+$ (Butylbenzene)		$\Delta H^\circ_{f298} = 825$ kJ mol⁻¹ (197 kcal mol⁻¹)
$C_{10}H_{14}^+$ (sec-Butylbenzene)		$\Delta H^\circ_{f298} = 820$ kJ mol⁻¹ (196 kcal mol⁻¹)
$C_{10}H_{14}^+$ (Isobutylbenzene)		$\Delta H^\circ_{f298} = 817$ kJ mol⁻¹ (195 kcal mol⁻¹)
$C_{10}H_{14}^+$ (tert-Butylbenzene)		$\Delta H^\circ_{f298} = 815$ kJ mol⁻¹ (195 kcal mol⁻¹)
$C_{10}H_{14}^+$ (1,2,4,5-Tetramethylbenzene)		$\Delta H^\circ_{f298} = 730$ kJ mol⁻¹ (174 kcal mol⁻¹)

Ion	Reactant	Other products	Ionization or appearance potential (eV)	Method	Heat of formation (kJ mol⁻¹)	Ref.	
$C_{10}H_{14}^+$	$C_6H_5C_4H_9$ (Butylbenzene)		8.69±0.01	PI	825	182,	416
$C_{10}H_{14}^+$	$C_6H_5C_4H_9$ (Butylbenzene)		8.69±0.01	PI	825		3154
$C_{10}H_{14}^+$	$C_6H_5C_4H_9$ (sec–Butylbenzene)		8.68±0.01	PI	820		182
$C_{10}H_{14}^+$	$C_6H_5C_4H_9$ (sec–Butylbenzene)		8.68±0.01	PI	820		3154
$C_{10}H_{14}^+$	$C_6H_5C_4H_9$ (Isobutylbenzene)		8.69±0.01	PI	817		3154

4.3. The Positive Ion Table—Continued

Ion	Reactant	Other products	Ionization or appearance potential (eV)	Method	Heat of formation (kJ mol^{-1})	Ref.
$C_{10}H_{14}^+$	$C_6H_5C_4H_9$ (*tert*–Butylbenzene)		8.68±0.01	PI	815	182
$C_{10}H_{14}^+$	$C_6H_5C_4H_9$ (*tert*–Butylbenzene)		8.68±0.01	PI	815	3154

See also – EI: 2407, 2463, 3238

Ion	Reactant	Other products	Ionization or appearance potential (eV)	Method	Heat of formation (kJ mol^{-1})	Ref.
$C_{10}H_{14}^+$	$C_6H_2(CH_3)_4$ (1,2,3,5–Tetramethylbenzene)		8.47±0.05	EI		2163
$C_{10}H_{14}^+$	$C_6H_2(CH_3)_4$ (1,2,4,5–Tetramethylbenzene)		8.025±0.005	PI	729	182
$C_{10}H_{14}^+$	$C_6H_2(CH_3)_4$ (1,2,4,5–Tetramethylbenzene)		8.03	PI	730	168
$C_{10}H_{14}^+$	$C_6H_2(CH_3)_4$ (1,2,4,5–Tetramethylbenzene)		8.05±0.02	PI	731	1142, 1159, 1166
$C_{10}H_{14}^+$	$C_6H_2(CH_3)_4$ (1,2,4,5–Tetramethylbenzene)		8.50±0.05	EI		2163
$C_{10}H_{14}^+$	$C_6H_2(CH_3)_4$ (1,2,4,5–Tetramethylbenzene)		8.37	CTS		2978

(Average of three values)

See also – PE: 1159

Ion	Reactant	Other products	Ionization or appearance potential (eV)	Method	Heat of formation (kJ mol^{-1})	Ref.
$C_{10}H_{14}^+$	$C_{10}H_{14}$ (Tetrahydrobullvalene)		7.95	PE		3329

$C_{10}H_{16}^+$

Ion	Reactant	Other products	Ionization or appearance potential (eV)	Method	Heat of formation (kJ mol^{-1})	Ref.
$C_{10}H_{16}^+$	$C_5H(CH_3)_5$ (1,2,3,5,5– and 1,2,4,5,5–Pentamethylcyclopentadiene mixture)		7.77±0.05	EI		2163
$C_{10}H_{16}^+$	$C_{10}H_{16}$ (*cis,trans*–1,5–Cyclodecadiene)		8.90 (V)	PE		3330
$C_{10}H_{16}^+$	$C_{10}H_{16}$ (*cis,cis*–1,6–Cyclodecadiene)		8.68 (V)	PE		3330
$C_{10}H_{16}^+$	$C_{10}H_{16}$ (*trans,trans*–1,6–Cyclodecadiene)		8.05 (V)	PE		3330
$C_{10}H_{16}^+$	$C_{10}H_{16}$ (α–Pinene)		8.07	PE		1130
$C_{10}H_{16}^+$	$C_{10}H_{16}$ (Adamantane)		9.25	PE		2843, 2951
$C_{10}H_{16}^+$	$C_{10}H_{16}$ (Hexahydrobullvalene)		8.71	PE		3329

4.3. The Positive Ion Table—Continued

Ion	Reactant	Other products	Ionization or appearance potential (eV)	Method	Heat of formation (kJ mol^{-1})	Ref.
	$C_{10}H_{18}^+$ (*cis*-Decalin)		$\Delta H_{f298}^\circ = 738$ kJ mol^{-1} (176 kcal mol^{-1})			
	$C_{10}H_{18}^+$ (*trans*-Decalin)		$\Delta H_{f298}^\circ = 720$ kJ mol^{-1} (172 kcal mol^{-1})			
$C_{10}H_{18}^+$	$(CH_3)_3CC\equiv CC(CH_3)_3$		9.19±0.04	RPD		2408
$C_{10}H_{18}^+$	$C_{10}H_{18}$ (*cis*-Cyclodecene)		8.80	PE		3330
$C_{10}H_{18}^+$	$C_{10}H_{18}$ (*trans*-Cyclodecene)		8.80	PE		3330
$C_{10}H_{18}^+$	$C_{10}H_{18}$ (*cis*-Decalin)		9.40	PE	738	2840, 2843

See also – EI: 1182, 1183, 3337

Ion	Reactant	Other products	Ionization or appearance potential (eV)	Method	Heat of formation (kJ mol^{-1})	Ref.
$C_{10}H_{18}^+$	$C_{10}H_{18}$ (*trans*-Decalin)		9.35	PE	720	2840, 2843

See also – EI: 1182, 1183, 3337

$$C_{10}H_{20}^+$$

Ion	Reactant	Other products	Ionization or appearance potential (eV)	Method	Heat of formation (kJ mol^{-1})	Ref.
$C_{10}H_{20}^+$	*cis*-$(CH_3)_3CCH=CHC(CH_3)_3$		8.68±0.02	PI		2877

See also – EI: 2533

Ion	Reactant	Other products	Ionization or appearance potential (eV)	Method	Heat of formation (kJ mol^{-1})	Ref.
$C_{10}H_{20}^+$	*trans*-$(CH_3)_3CCH=CHC(CH_3)_3$		8.734±0.01	PI		2877

See also – EI: 2410, 2533

$$C_{11}H_7^+$$

Ion	Reactant	Other products	Ionization or appearance potential (eV)	Method	Heat of formation (kJ mol^{-1})	Ref.
$C_{11}H_7^+$	$C_6H_5C\equiv CC_6H_5$ (Diphenylacetylene)		17.52±0.1	EI		1238
$C_{11}H_7^+$	$C_{14}H_{10}$ (Phenanthrene)		21.10±0.2	RPD		2540
$C_{11}H_7^+$	$C_{14}H_{10}$ (Phenanthrene)		21.1±0.3	EI		1238

$$C_{11}H_9^+$$

Ion	Reactant	Other products	Ionization or appearance potential (eV)	Method	Heat of formation (kJ mol^{-1})	Ref.
$C_{11}H_9^+$	$C_{10}H_7CH_2$ (1-Naphthylmethyl radical)		7.35±0.1	EI		71
$C_{11}H_9^+$	$C_{10}H_7CH_2$ (2-Naphthylmethyl radical)		7.56±0.05	EI		69, 71
$C_{11}H_9^+$	$C_{10}H_7CH_3$ (1-Methylnaphthalene)	H	12.4±0.1	RPD		2538
$C_{11}H_9^+$	$C_{10}H_7CH_3$ (2-Methylnaphthalene)	H	13.2±0.2	RPD		2538
$C_{11}H_9^+$	$(C_6H_5O)_2CO$ (Carbonic acid diphenyl ester)		13.90±0.05	EI		1237

4.3. The Positive Ion Table—Continued

Ion	Reactant	Other products	Ionization or appearance potential (eV)	Method	Heat of formation (kJ mol^{-1})	Ref.	
	$C_{11}H_{10}^+$ (1–Methylnaphthalene)		$\Delta H_{f298}^\circ = 885$ kJ mol^{-1} (211 kcal mol^{-1})				
	$C_{11}H_{10}^+$ (2–Methylnaphthalene)		$\Delta H_{f298}^\circ = 884$ kJ mol^{-1} (211 kcal mol^{-1})				
$C_{11}H_{10}^+$	$C_{10}H_7CH_3$ (1–Methylnaphthalene)		7.96±0.01	PI	885	182,	416
$C_{11}H_{10}^+$	$C_{10}H_7CH_3$ (1–Methylnaphthalene)		7.98	CTS			2909

See also – EI: 2538

Ion	Reactant	Other products	Ionization or appearance potential (eV)	Method	Heat of formation (kJ mol^{-1})	Ref.
$C_{11}H_{10}^+$	$C_{10}H_7CH_3$ (2–Methylnaphthalene)		7.955±0.01	PI	884	182

See also – S: 344
      EI: 2538
     CTS: 2562

Ion	Reactant	Other products	Ionization or appearance potential (eV)	Method	Heat of formation (kJ mol^{-1})	Ref.
$C_{11}H_{10}^+$	$(C_6H_5O)_2CO$ (Carbonic acid diphenyl ester)		12.41±0.1	EI		1237

$C_{11}H_{14}^+$

Ion	Reactant	Other products	Ionization or appearance potential (eV)	Method	Heat of formation (kJ mol^{-1})	Ref.
$C_{11}H_{14}^+$	$C_{11}H_{14}$ (Cycloheptenobenzene)		9.10±0.05	EI		3342

$C_{11}H_{15}^+$

Ion	Reactant	Other products	Ionization or appearance potential (eV)	Method	Heat of formation (kJ mol^{-1})	Ref.
$C_{11}H_{15}^+$	$C_{26}H_{46}$ (5–Phenyleicosane)		10.43±0.1	EI		2153

$C_{11}H_{16}^+$ (Pentamethylbenzene) $\Delta H_{f298}^\circ = 690$ kJ mol^{-1} (165 kcal mol^{-1})

Ion	Reactant	Other products	Ionization or appearance potential (eV)	Method	Heat of formation (kJ mol^{-1})	Ref.	
$C_{11}H_{16}^+$	$C_6H_5C_5H_{11}$ (Neopentylbenzene)		8.41±0.04	RPD			2407
$C_{11}H_{16}^+$	$C_6H(CH_3)_5$ (Pentamethylbenzene)		7.92±0.02	PI	690	1142,	1166
$C_{11}H_{16}^+$	$C_6H(CH_3)_5$ (Pentamethylbenzene)		7.92	PI	690		168

See also – EI: 2163

$C_{11}H_{18}^+$

Ion	Reactant	Other products	Ionization or appearance potential (eV)	Method	Heat of formation (kJ mol^{-1})	Ref.
$C_{11}H_{18}^+$	$C_5(CH_3)_6$ (Hexamethylcyclopentadiene)		7.74±0.05	EI		2163
$C_{11}H_{18}^+$	$C_{11}H_{18}$ (2–Methylene–1,7,7–trimethylbicyclo[2.2.1]heptane)		8.62±0.01	PI		2877
$C_{11}H_{18}^+$	$C_{10}H_{15}CH_3$ (1–Methyladamantane)		9.24	PE		2951
$C_{11}H_{18}^+$	$C_{10}H_{15}CH_3$ (2–Methyladamantane)		9.24	PE		2951

4.3. The Positive Ion Table—Continued

Ion	Reactant	Other products	Ionization or appearance potential (eV)	Method	Heat of formation (kJ mol⁻¹)	Ref.
			$C_{12}H_6^+$			
$C_{12}H_6^+$	$(C_6H_4CN)_2$ (4,4'–Dicyanobiphenyl)	2HCN	17.35	EI		3295
$C_{12}H_6^+$	$C_{16}H_6O_6$ (3,3',4,4'–Biphenyltetracarboxylic acid dianhydride)	$2CO_2+2CO$	18.47	EI		3295
			$C_{12}H_7^+$			
$C_{12}H_7^+$	$C_6H_5C\equiv CC_6H_5$ (Diphenylacetylene)		17.46±0.06	EI		1238
$C_{12}H_7^+$	$C_{14}H_{10}$ (Phenanthrene)		18.80±0.1	RPD		2540
$C_{12}H_7^+$	$C_{14}H_{10}$ (Phenanthrene)		19.63±0.05	EI		1238

See also – EI: 2538

Ion	Reactant	Other products	Ionization or appearance potential (eV)	Method	Heat of formation (kJ mol⁻¹)	Ref.
$C_{12}H_8^+$ (Biphenylene)		$\Delta H^{\circ}_{f298} = 1211$ kJ mol⁻¹ (290 kcal mol⁻¹)				
$C_{12}H_8^+$ (Acenaphthylene)		$\Delta H^{\circ}_{f298} = 1032$ kJ mol⁻¹ (247 kcal mol⁻¹)				
$C_{12}H_8^+$	$C_{12}H_8$ (Biphenylene)		7.56±0.05	PE	1211	2796

See also – EI: 2458, 2847

Ion	Reactant	Other products	Ionization or appearance potential (eV)	Method	Heat of formation (kJ mol⁻¹)	Ref.
$C_{12}H_8^+$	$C_{12}H_8$ (Acenaphthylene)		8.02±0.01	PE	1032	2942
$C_{12}H_8^+$	$(C_6H_5)_2$ (Biphenyl)		16.89±0.08	EI		1238
$C_{12}H_8^+$	$C_6H_5C\equiv CC_6H_5$ (Diphenylacetylene)		15.58±0.05	EI		1238
$C_{12}H_8^+$	$C_{14}H_{10}$ (Phenanthrene)		15.65±0.2	RPD		2540
$C_{12}H_8^+$	$C_{14}H_{10}$ (Phenanthrene)		16.63±0.05	EI		1238

See also – EI: 2538

Ion	Reactant	Other products	Ionization or appearance potential (eV)	Method	Heat of formation (kJ mol⁻¹)	Ref.
$C_{12}H_8^+$	$(C_6H_5)_2CO$ (Benzophenone)		17.48±0.12	EI		1237
			$C_{12}H_8^{+2}$			
$C_{12}H_8^{+2}$	$(C_6H_5)_2$ (Biphenyl)		22.0±1.0	EI		1238
$C_{12}H_8^{+2}$	$C_6H_5C\equiv CC_6H_5$ (Diphenylacetylene)		20.5±0.1	EI		1238

4.3. The Positive Ion Table—Continued

Ion	Reactant	Other products	Ionization or appearance potential (eV)	Method	Heat of formation (kJ mol^{-1})	Ref.
			C$_{12}$H$_9^+$			
C$_{12}$H$_9^+$	(C$_6$H$_5$)$_2$ (Biphenyl)	H	14.36±0.05	EI		1238
C$_{12}$H$_9^+$	(C$_6$H$_5$)$_2$CO (Benzophenone)	CO+H?	15.28±0.05	EI		1237

C$_{12}$H$_{10}^+$ (Biphenyl) $\Delta H_{f298}^{\circ} \sim 976$ kJ mol^{-1} (233 kcal mol^{-1})
C$_{12}$H$_{10}^+$ (Acenaphthene) $\Delta H_{f298}^{\circ} = 902$ kJ mol^{-1} (216 kcal mol^{-1})

Ion	Reactant	Other products	Ionization or appearance potential (eV)	Method	Heat of formation (kJ mol^{-1})	Ref.
C$_{12}$H$_{10}^+$	(C$_6$H$_5$)$_2$ (Biphenyl)		8.27±0.01	PI	980	182
C$_{12}$H$_{10}^+$	(C$_6$H$_5$)$_2$ (Biphenyl)		8.20±0.05	PE	973	2847
C$_{12}$H$_{10}^+$	(C$_6$H$_5$)$_2$ (Biphenyl)		8.23±0.01	PE	976	2942
C$_{12}$H$_{10}^+$	(C$_6$H$_5$)$_2$ (Biphenyl)		8.22±0.15	EI		2458
C$_{12}$H$_{10}^+$	(C$_6$H$_5$)$_2$ (Biphenyl)		8.4	CTS		2978
(Average of three values)						
C$_{12}$H$_{10}^+$	(C$_6$H$_5$)$_2$ (Biphenyl)		8.46	CTS		2562
C$_{12}$H$_{10}^+$	(C$_6$H$_5$)$_2$ (Biphenyl)		8.64	CTS		2909

See also – PE: 3290
EI: 1238, 2438, 2634, 2847, 3286

Ion	Reactant	Other products	Ionization or appearance potential (eV)	Method	Heat of formation (kJ mol^{-1})	Ref.
C$_{12}$H$_{10}^+$	C$_{12}$H$_{10}$ (Acenaphthene)		7.73±0.01	PE	902	2942
C$_{12}$H$_{10}^+$	C$_{12}$H$_{10}$ (Acenaphthene)		7.66	CTS		2909
C$_{12}$H$_{10}^+$	(C$_6$H$_5$)$_2$CO (Benzophenone)	CO	12.24±0.13	EI		1237

Ion	Reactant	Other products	Ionization or appearance potential (eV)	Method	Heat of formation (kJ mol^{-1})	Ref.
			C$_{12}$H$_{10}^{+2}$			
C$_{12}$H$_{10}^{+2}$	(C$_6$H$_5$)$_2$ (Biphenyl)		21.9±0.3	EI		1238

Ion	Reactant	Other products	Ionization or appearance potential (eV)	Method	Heat of formation (kJ mol^{-1})	Ref.
			C$_{12}$H$_{11}^+$			
C$_{12}$H$_{11}^+$	C$_{10}$H$_6$(CH$_3$)$_2$ (1,4–Dimethylnaphthalene)	H	12.9±0.1	RPD		2538
C$_{12}$H$_{11}^+$	C$_{10}$H$_7$CH$_2$CH$_2$OH (2–(4–Azulyl)ethanol)	OH	10.3	EI		3333

4.3. The Positive Ion Table—Continued

Ion	Reactant	Other products	Ionization or appearance potential (eV)	Method	Heat of formation (kJ mol^{-1})	Ref.
		$C_{12}H_{12}^+$				
$C_{12}H_{12}^+$	$C_{10}H_6(CH_3)_2$ (1,4–Dimethylnaphthalene)		7.78±0.03	RPD		2538
$C_{12}H_{12}^+$	$C_{10}H_6(CH_3)_2$ (2,3–Dimethylnaphthalene)		8.20±0.05	RPD		2538
$C_{12}H_{12}^+$	$C_{10}H_6(CH_3)_2$ (2,3–Dimethylnaphthalene)		8.11	CTS		2562
		$C_{12}H_{14}^+$				
$C_{12}H_{14}^+$	$C_{12}H_{14}$ (1,2:3,4–Dicyclopentenobenzene)		8.66±0.02	EI		3342
		$C_{12}H_{16}^+$				
$C_{12}H_{16}^+$	$C_{12}H_{16}$ (Cyclooctenobenzene)		8.97±0.03	EI		3342

$C_{12}H_{18}^+$ (Hexamethylbenzene) $\Delta H_{f298}^\circ = 652$ kJ mol^{-1} (156 kcal mol^{-1})

Ion	Reactant	Other products	Ionization or appearance potential (eV)	Method	Heat of formation (kJ mol^{-1})	Ref.
$C_{12}H_{18}^+$	$(CH_3)_3CC{\equiv}CC{\equiv}CC(CH_3)_3$		8.82±0.04	RPD		2408
$C_{12}H_{18}^+$	$C_6(CH_3)_6$ (Hexamethylbenzene)		7.85±0.02	PI	652	1142
$C_{12}H_{18}^+$	$C_6(CH_3)_6$ (Hexamethylbenzene)		7.85	PI	652	168
$C_{12}H_{18}^+$	$C_6(CH_3)_6$ (Hexamethylbenzene)		7.87	EI		2511
$C_{12}H_{18}^+$	$C_6(CH_3)_6$ (Hexamethylbenzene)		7.95	CTS		2978

(Average of three values)

See also – CTS: 2562

Ion	Reactant	Other products	Ionization or appearance potential (eV)	Method	Heat of formation (kJ mol^{-1})	Ref.
		$C_{12}H_{20}^+$				
$C_{12}H_{20}^+$	$C_{12}H_{20}$ (2–Ethylidene–1,7,7–trimethylbicyclo[2.2.1]heptane)		8.22±0.01	PI		2877
		$C_{12}H_{22}^+$				
$C_{12}H_{22}^+$	$(C_6H_{11})_2$ (Bicyclohexyl)		9.41	PE		2951
		$C_{13}H_7^+$				
$C_{13}H_7^+$	$C_{14}H_{10}$ (Phenanthrene)		20.0±0.3	RPD		2540

4.3. The Positive Ion Table—Continued

Ion	Reactant	Other products	Ionization or appearance potential (eV)	Method	Heat of formation (kJ mol^{-1})	Ref.
			$C_{13}H_8^+$			
$C_{13}H_8^+$	$C_{14}H_8N_2O$ (9−Diazo−10−oxophenanthrene)	$CO+N_2$	14.7±0.2	EI		2995
			$C_{13}H_9^+$			
$C_{13}H_9^+$	$C_{13}H_9$ (Fluorenyl radical)		7.07	EI		126
$C_{13}H_9^+$	$C_{13}H_{10}$ (Fluorene)	H	12.5±0.1	EDD		2974
$C_{13}H_9^+$	$(C_6H_5)_2CH_2$ (Diphenylmethane)	H_2+H	14.9±0.1	EDD		2974
$C_{13}H_9^+$	$C_6H_5CH=CHC_6H_5$ (1,2−Diphenylethene)	CH_3	11.3±0.1	EDD		2974
$C_{13}H_9^+$	$C_{14}H_{12}$ (9,10−Dihydrophenanthrene)	CH_3	11.8±0.1	EDD		2974
$C_{13}H_9^+$	$C_{14}H_9CH_3$ (1−Methylphenanthrene)		15.8±0.2	RPD		2538
$C_{13}H_9^+$	$C_{14}H_9CH_3$ (2−Methylphenanthrene)		13.8±0.2	RPD		2538
$C_{13}H_9^+$	$C_{14}H_9CH_3$ (3−Methylphenanthrene)		13.4±0.2	RPD		2538
$C_{13}H_9^+$	$C_6H_5CH_2CH=CHC_6H_5$ (1,3−Diphenylpropene)	C_2H_5	13.75	EI		2498
$C_{13}H_9^+$	$C_{13}H_{10}N_4$ (1−Benzylideneamino−1,2,3−benzotriazole)		9.6±0.1	EDD		2974

A value for the heat of formation of $C_{13}H_9^+$ computed in ref. 2974 is based on an appearance potential of 9.0 eV for this process. Which appearance potential is the correct one is not known.

Ion	Reactant	Other products	Ionization or appearance potential (eV)	Method	Heat of formation (kJ mol^{-1})	Ref.
			$C_{13}H_{10}^+$			
$C_{13}H_{10}^+$	$C_{13}H_{10}$ (Fluorene)		7.93±0.01	PE		2942
$C_{13}H_{10}^+$	$C_{13}H_{10}$ (Fluorene)		7.78	CTS		2562
$C_{13}H_{10}^+$	$C_{13}H_{10}$ (Fluorene)		8.42	CTS		2911

See also − EI: 126

Ion	Reactant	Other products	Ionization or appearance potential (eV)	Method	Heat of formation (kJ mol^{-1})	Ref.
			$C_{13}H_{11}^+$			
$C_{13}H_{11}^+$	$(C_6H_5)_2CH$ (Diphenylmethyl radical)		7.32±0.1	EI		71
$C_{13}H_{11}^+$	$C_6H_5CH_2CH=CHC_6H_5$ (1,3−Diphenylpropene)	C_2H_3	12.40	EI		2498
$C_{13}H_{11}^+$	$C_6H_5C_6H_4(CH_2)_3COOCH_3$ (4−(4−Biphenylyl)butanoic acid methyl ester)		12.15±0.2	EI		2497

4.3. The Positive Ion Table—Continued

Ion	Reactant	Other products	Ionization or appearance potential (eV)	Method	Heat of formation (kJ mol^{-1})	Ref.
			$C_{13}H_{12}^+$			
$C_{13}H_{12}^+$	$(C_6H_5)_2CH_2$ (Diphenylmethane)		9.1 (V)	PE		3290
			$C_{13}H_{19}^+$			
$C_{13}H_{19}^+$	$C_6H_4(C_4H_9)_2$ (1,2–Di–$tert$–butylbenzene)	CH_3	10.64±0.07	EI		2416
$C_{13}H_{19}^+$	$C_6H_4(C_4H_9)_2$ (1,3–Di–$tert$–butylbenzene)	CH_3	11.70±0.07	EI		2416
$C_{13}H_{19}^+$	$C_6H_4(C_4H_9)_2$ (1,4–Di–$tert$–butylbenzene)	CH_3	11.69±0.07	EI		2416
$C_{13}H_{19}^+$	$C_{19}H_{32}$ (7–Phenyltridecane)		10.15±0.1	EI		2153
$C_{13}H_{19}^+$	$C_{26}H_{46}$ (7–Phenyleicosane)		10.28±0.1	EI		2153
			$C_{14}H_7^+$			
$C_{14}H_7^+$	$C_{14}H_{10}$ (Phenanthrene)	H_2+H	18.15±0.2	RPD		2540
			$C_{14}H_8^+$			
$C_{14}H_8^+$	$C_6H_5C{\equiv}CC_6H_5$ (Diphenylacetylene)	H_2	16.66±0.05	EI		1238
$C_{14}H_8^+$	$C_{14}H_{10}$ (Phenanthrene)	H_2	16.20±0.2	RPD		2540
$C_{14}H_8^+$	$C_{14}H_{10}$ (Phenanthrene)	H_2	18.58±0.1	EI		1238

See also – EI: 2538

Ion	Reactant	Other products	Ionization or appearance potential (eV)	Method	Heat of formation (kJ mol^{-1})	Ref.
$C_{14}H_8^+$	$C_{15}H_8N_2O$ (9–Diazo–4,5–methylene–10–oxophenanthrene)	$CO+N_2$	13.8±0.2	EI		2995
			$C_{14}H_9^+$			
$C_{14}H_9^+$	$C_6H_5C{\equiv}CC_6H_5$ (Diphenylacetylene)	H	15.13±0.1	EI		1238
$C_{14}H_9^+$	$C_{14}H_{10}$ (Phenanthrene)	H	15.45±0.1	RPD		2540
$C_{14}H_9^+$	$C_{14}H_{10}$ (Phenanthrene)	H	16.25±0.1	EI		1238

See also – EI: 2538, 2539

4.3. The Positive Ion Table—Continued

Ion	Reactant	Other products	Ionization or appearance potential (eV)	Method	Heat of formation (kJ mol^{-1})	Ref.
			$C_{14}H_{10}^+$			
$C_{14}H_{10}^+$	$C_6H_5C\equiv CC_6H_5$ (Diphenylacetylene)		7.91	PE		3412
$C_{14}H_{10}^+$	$C_6H_5C\equiv CC_6H_5$ (Diphenylacetylene)		8.85±0.05	EI		1238
$C_{14}H_{10}^+$	$C_{14}H_{10}$ (Anthracene)		7.15	S		2652
(Average of five Rydberg series limits)						
$C_{14}H_{10}^+$	$C_{14}H_{10}$ (Anthracene)		7.414	S		2661
(Average of three Rydberg series limits)						
$C_{14}H_{10}^+$	$C_{14}H_{10}$ (Anthracene)		7.5	PI		2661
$C_{14}H_{10}^+$	$C_{14}H_{10}$ (Anthracene)		7.41	PE		3412
$C_{14}H_{10}^+$	$C_{14}H_{10}$ (Anthracene)		7.55	EI		413
$C_{14}H_{10}^+$	$C_{14}H_{10}$ (Anthracene)		7.2	CTS		2037
$C_{14}H_{10}^+$	$C_{14}H_{10}$ (Anthracene)		7.23	CTS		3000
$C_{14}H_{10}^+$	$C_{14}H_{10}$ (Anthracene)		7.37	CTS	2562,	2947
$C_{14}H_{10}^+$	$C_{14}H_{10}$ (Anthracene)		7.40	CTS		2909
$C_{14}H_{10}^+$	$C_{14}H_{10}$ (Anthracene)		7.4	CTS		2978
$C_{14}H_{10}^+$	$C_{14}H_{10}$ (Anthracene)		7.42	CTS		1064
$C_{14}H_{10}^+$	$C_{14}H_{10}$ (Anthracene)		7.43	CTS		2910

The series assignments of refs. 2652 and 2661 are uncertain. We recommend a tentative value for the ionization potential of about 7.5 eV and an ion heat of formation at 298 K of approximately 954 kJ mol^{-1} (228 kcal mol^{-1}).

See also – EI: 3011

4.3. The Positive Ion Table—Continued

Ion	Reactant	Other products	Ionization or appearance potential (eV)	Method	Heat of formation (kJ mol^{-1})	Ref.
$C_{14}H_{10}^+$	$C_{14}H_{10}$ (Phenanthrene)		7.69	S		2661
$C_{14}H_{10}^+$	$C_{14}H_{10}$ (Phenanthrene)		7.75	PI		2661
$C_{14}H_{10}^+$	$C_{14}H_{10}$ (Phenanthrene)		7.91±0.01	PE		2942
$C_{14}H_{10}^+$	$C_{14}H_{10}$ (Phenanthrene)		7.92	PE		3412
$C_{14}H_{10}^+$	$C_{14}H_{10}$ (Phenanthrene)		8.03±0.01	RPD	2538,	2539, 2540
$C_{14}H_{10}^+$	$C_{14}H_{10}$ (Phenanthrene)		8.03	EI		413
$C_{14}H_{10}^+$	$C_{14}H_{10}$ (Phenanthrene)		8.10±0.04	EI		1238
$C_{14}H_{10}^+$	$C_{14}H_{10}$ (Phenanthrene)		7.6	CTS		2910
$C_{14}H_{10}^+$	$C_{14}H_{10}$ (Phenanthrene)		8.02	CTS		3000
$C_{14}H_{10}^+$	$C_{14}H_{10}$ (Phenanthrene) (Average of three values)		8.03	CTS		2978
$C_{14}H_{10}^+$	$C_{14}H_{10}$ (Phenanthrene)		8.07	CTS		1064
$C_{14}H_{10}^+$	$C_{14}H_{10}$ (Phenanthrene)		8.08	CTS		2911
$C_{14}H_{10}^+$	$C_{14}H_{10}$ (Phenanthrene)		8.09	CTS		2947
$C_{14}H_{10}^+$	$C_{14}H_{10}$ (Phenanthrene)		8.22	CTS		2909

We recommend a tentative value for the ionization potential of about 7.8 eV and an ion heat of formation at 298 K of approximately 960 kJ mol^{-1} (229 kcal mol^{-1}).

See also – EI: 3011

$C_{14}H_{10}^+$	$C_6H_5CH_2CH=CHC_6H_5$ (1,3–Diphenylpropene)		15.05	EI		2498

$C_{14}H_8D_2^+$

$C_{14}H_8D_2^+$	$C_{14}H_8D_2$ (Phenanthrene–9,10–d_2)		8.03	RPD		2539

$C_{14}D_{10}^+$

$C_{14}D_{10}^+$	$C_{14}D_{10}$ (Phenanthrene–d_{10})		8.05	RPD	2539,	2540

4.3. The Positive Ion Table—Continued

Ion	Reactant	Other products	Ionization or appearance potential (eV)	Method	Heat of formation (kJ mol⁻¹)	Ref.

$$C_{14}H_{10}^{+2}$$

Ion	Reactant	Other products	Ionization or appearance potential (eV)	Method	Heat of formation	Ref.
$C_{14}H_{10}^{+2}$	$C_6H_5C\equiv CC_6H_5$ (Diphenylacetylene)		23.35±0.1	EI		1238
$C_{14}H_{10}^{+2}$	$C_{14}H_{10}$ (Anthracene)		21.1	EI		413
$C_{14}H_{10}^{+2}$	$C_{14}H_{10}$ (Phenanthrene)		23.1	EI		413

$$C_{14}H_{11}^{+}$$

Ion	Reactant	Other products	Ionization or appearance potential (eV)	Method	Heat of formation	Ref.
$C_{14}H_{11}^{+}$	$C_6H_5CH_2CH=CHC_6H_5$ (1,3–Diphenylpropene)	CH_3	11.95	EI		2498

$$C_{14}H_{12}^{+}$$

Ion	Reactant	Other products	Ionization or appearance potential (eV)	Method	Heat of formation	Ref.
$C_{14}H_{12}^{+}$	$C_6H_5CH=CHC_6H_5$ (1,2–Diphenylethene)		7.60	CTS		2562
$C_{14}H_{12}^{+}$	$C_6H_5CH=CHC_6H_5$ (1,2–Diphenylethene) (Average of three values)		7.95	CTS		2978
$C_{14}H_{12}^{+}$	$C_{14}H_{12}$ (9,10–Dihydrophenanthrene)		8.08±0.06	EI		3340
$C_{14}H_{12}^{+}$	$C_6H_5C_6H_4(CH_2)_3COOCH_3$ (4–(4–Biphenylyl)butanoic acid methyl ester)		10.90±0.2	EI		2497

$$C_{14}H_{13}^{+}$$

Ion	Reactant	Other products	Ionization or appearance potential (eV)	Method	Heat of formation	Ref.
$C_{14}H_{13}^{+}$	$C_6H_5C_6H_4CH(CH_3)CH_2COOCH_3$ (3–(4–Biphenylyl)butanoic acid methyl ester)		10.87±0.2	EI		2497

$$C_{14}H_{14}^{+}$$

Ion	Reactant	Other products	Ionization or appearance potential (eV)	Method	Heat of formation	Ref.
$C_{14}H_{14}^{+}$	$C_6H_5CH_2CH_2C_6H_5$ (1,2–Diphenylethane)		9.1 (V)	PE		3290
$C_{14}H_{14}^{+}$	$C_6H_5CH_2CH_2C_6H_5$ (1,2–Diphenylethane)		8.7±0.1	EI		3288

$$C_{14}H_{18}^{+}$$

Ion	Reactant	Other products	Ionization or appearance potential (eV)	Method	Heat of formation	Ref.
$C_{14}H_{18}^{+}$	$C_{14}H_{18}$ (1,2,3,4,5,6,7,8–Octahydrophenanthrene)		8.79±0.02	EI		3342

$$C_{14}H_{20}^{+}$$

Ion	Reactant	Other products	Ionization or appearance potential (eV)	Method	Heat of formation	Ref.
$C_{14}H_{20}^{+}$	$C_{14}H_{20}$ (Congressane)		8.93	PE		2843, 2951

4.3. The Positive Ion Table—Continued

Ion	Reactant	Other products	Ionization or appearance potential (eV)	Method	Heat of formation (kJ mol^{-1})	Ref.
			$C_{14}H_{22}^+$			
$C_{14}H_{22}^+$	$C_6H_4(C_4H_9)_2$ (1,2-Di-*tert*-butylbenzene)		8.60±0.07	EI		2416
$C_{14}H_{22}^+$	$C_6H_4(C_4H_9)_2$ (1,3-Di-*tert*-butylbenzene)		8.71±0.07	EI		2416
$C_{14}H_{22}^+$	$C_6H_4(C_4H_9)_2$ (1,4-Di-*tert*-butylbenzene)		8.74±0.07	EI		2416
			$C_{14}H_{24}^+$			
$C_{14}H_{24}^+$	$C_{10}H_{12}(CH_3)_4$ (1,3,5,7-Tetramethyladamantane)		9.23	PE		2951
			$C_{15}H_9^+$			
$C_{15}H_9^+$	$C_{14}H_9CH_3$ (1-Methylphenanthrene)	H_2+H	18.0±0.2	RPD		2538
$C_{15}H_9^+$	$C_{14}H_9CH_3$ (2-Methylphenanthrene)	H_2+H	16.6±0.2	RPD		2538
$C_{15}H_9^+$	$C_{14}H_9CH_3$ (3-Methylphenanthrene)	H_2+H	16.1±0.2	RPD		2538
$C_{15}H_9^+$	$C_{14}H_9CH_3$ (9-Methylphenanthrene)	H_2+H	16.5±0.2	RPD		2538
$C_{15}H_9^+$	$C_{14}H_8(CH_3)_2$ (2,3-Dimethylphenanthrene)		15.6±0.1	RPD		2538
$C_{15}H_9^+$	$C_{14}H_8(CH_3)_2$ (3,6-Dimethylphenanthrene)		18.2±0.1	RPD		2538
			$C_{15}H_{10}^+$			
$C_{15}H_{10}^+$	$C_{14}H_9CH_3$ (1-Methylphenanthrene)	H_2	14.7±0.2	RPD		2538
$C_{15}H_{10}^+$	$C_{14}H_9CH_3$ (2-Methylphenanthrene)	H_2	13.1±0.2	RPD		2538
$C_{15}H_{10}^+$	$C_{14}H_9CH_3$ (3-Methylphenanthrene)	H_2	13.6±0.2	RPD		2538
$C_{15}H_{10}^+$	$C_{14}H_9CH_3$ (9-Methylphenanthrene)	H_2	13.3±0.2	RPD		2538
$C_{15}H_{10}^+$	$C_{14}H_8(CH_3)_2$ (2,3-Dimethylphenanthrene)		15.1±0.1	RPD		2538
$C_{15}H_{10}^+$	$C_{14}H_8(CH_3)_2$ (2,5-Dimethylphenanthrene)		16.4±0.2	RPD		2538
$C_{15}H_{10}^+$	$C_{14}H_8(CH_3)_2$ (3,6-Dimethylphenanthrene)		16.2±0.1	RPD		2538
$C_{15}H_{10}^+$	$C_{14}H_8(CH_3)_2$ (4,5-Dimethylphenanthrene)		15.3±0.2	RPD		2538
$C_{15}H_{10}^+$	$C_{14}H_8(CH_3)_2$ (9,10-Dimethylphenanthrene)		18.0±0.1	RPD		2538

4.3. The Positive Ion Table—Continued

Ion	Reactant	Other products	Ionization or appearance potential (eV)	Method	Heat of formation (kJ mol^{-1})	Ref.
			$C_{15}H_{11}^+$			
$C_{15}H_{11}^+$	$C_{14}H_9CH_3$ (1-Methylphenanthrene)	H	13.1±0.1	RPD		2538, 2539, 2540
$C_{15}H_{11}^+$	$C_{14}H_9CH_3$ (2-Methylphenanthrene)	H	12.9±0.1	RPD		2538, 2539, 2540
$C_{15}H_{11}^+$	$C_{14}H_9CH_3$ (3-Methylphenanthrene)	H	12.8±0.1	RPD		2538, 2539, 2540
$C_{15}H_{11}^+$	$C_{14}H_9CH_3$ (4-Methylphenanthrene)	H	12.2±0.2	RPD		2538
$C_{15}H_{11}^+$	$C_{14}H_9CH_3$ (9-Methylphenanthrene)	H	12.4±0.1	RPD		2538, 2539
$C_{15}H_{11}^+$	$C_{14}H_8(CH_3)_2$ (2,3-Dimethylphenanthrene)	CH_3	12.3±0.2	RPD		2538, 2539
$C_{15}H_{11}^+$	$C_{14}H_8(CH_3)_2$ (2,7-Dimethylphenanthrene)	CH_3	13.10	RPD		2539

See also – EI: 2538

Ion	Reactant	Other products	Ionization or appearance potential (eV)	Method	Heat of formation (kJ mol^{-1})	Ref.
$C_{15}H_{11}^+$	$C_{14}H_8(CH_3)_2$ (3,6-Dimethylphenanthrene)	CH_3	13.0±0.05	RPD		2538, 2539
$C_{15}H_{11}^+$	$C_{14}H_8(CH_3)_2$ (4,5-Dimethylphenanthrene)	CH_3	10.8±0.1	RPD		2538
$C_{15}H_{11}^+$	$C_{14}H_8(CH_3)_2$ (9,10-Dimethylphenanthrene)	CH_3	12.0±0.1	RPD		2538, 2539

			$C_{15}H_{12}^+$			
$C_{15}H_{12}^+$	$C_{14}H_9CH_3$ (1-Methylphenanthrene)		7.70±0.03	RPD		2538, 2539, 2540
$C_{15}H_{12}^+$	$C_{14}H_9CH_3$ (2-Methylphenanthrene)		7.70	RPD		2539, 2540

See also – EI: 2538

$C_{15}H_{12}^+$	$C_{14}H_9CH_3$ (3-Methylphenanthrene)		7.90	RPD		2539, 2540

See also – EI: 2538

$C_{15}H_{12}^+$	$C_{14}H_9CH_3$ (4-Methylphenanthrene)		7.70±0.02	RPD		2538
$C_{15}H_{12}^+$	$C_{14}H_9CH_3$ (9-Methylphenanthrene)		7.46±0.03	RPD		2538, 2539

			$C_{15}H_9D_3^+$			
$C_{15}H_9D_3^+$	$C_{14}H_9CD_3$ (3-Methyl-d_3-phenanthrene)		7.85	RPD		2539

4.3. The Positive Ion Table—Continued

Ion	Reactant	Other products	Ionization or appearance potential (eV)	Method	Heat of formation (kJ mol^{-1})	Ref.
			$C_{15}H_{13}^+$			
$C_{15}H_{13}^+$	$C_6H_5CH_2CH=CHC_6H_5$ (1,3–Diphenylpropene)	H	11.65	EI		2498
			$C_{15}H_{14}^+$			
$C_{15}H_{14}^+$	$C_6H_5CH_2CH=CHC_6H_5$ (1,3–Diphenylpropene)		8.30	EI		2498
$C_{15}H_{14}^+$	$C_6H_5C_6H_4CH(CH_3)CH_2COOCH_3$ (3–(4–Biphenylyl)butanoic acid methyl ester)		9.91±0.2	EI		2497
			$C_{15}H_{18}^+$			
$C_{15}H_{18}^+$	$C_{15}H_{18}$ (Tricyclopentenobenzene)		8.36±0.02	EI		3342
			$C_{15}H_{23}^+$			
$C_{15}H_{23}^+$	$C_{26}H_{46}$ (9–Phenyleicosane)		10.30±0.1	EI		2153
			$C_{16}H_9^+$			
$C_{16}H_9^+$	$C_{14}H_8(CH_3)_2$ (4,5–Dimethylphenanthrene)	$2H_2+H$	12.5±0.2	RPD		2538

$C_{16}H_{10}^+$ (Fluoranthene) $\Delta H_{f298}^\circ = 1047$ kJ mol^{-1} (250 kcal mol^{-1})

Ion	Reactant	Other products	Ionization or appearance potential (eV)	Method	Heat of formation (kJ mol^{-1})	Ref.
$C_{16}H_{10}^+$	$C_{16}H_{10}$ (Fluoranthene)		7.80±0.01	PE	1047	2942
$C_{16}H_{10}^+$	$C_{16}H_{10}$ (Fluoranthene)		7.72	CTS		2562
$C_{16}H_{10}^+$	$C_{16}H_{10}$ (Pyrene)		7.72±0.3	EI		1069
$C_{16}H_{10}^+$	$C_{16}H_{10}$ (Pyrene)		7.31	CTS		2037
$C_{16}H_{10}^+$	$C_{16}H_{10}$ (Pyrene)		7.48	CTS		2909
$C_{16}H_{10}^+$	$C_{16}H_{10}$ (Pyrene) (Average of two values)		7.53	CTS		2978
$C_{16}H_{10}^+$	$C_{16}H_{10}$ (Pyrene)		7.55	CTS	2562,	2947
$C_{16}H_{10}^+$	$C_{16}H_{10}$ (Pyrene)		7.58	CTS		3000
$C_{16}H_{10}^+$	$C_{16}H_{10}$ (Pyrene)		7.70	CTS		1064

4.3. The Positive Ion Table—Continued

Ion	Reactant	Other products	Ionization or appearance potential (eV)	Method	Heat of formation (kJ mol^{-1})	Ref.
$C_{16}H_{10}^{+}$	$C_{16}H_{10}$ (Pyrene)		7.72	CTS		2910
$C_{16}H_{10}^{+}$	$C_{14}H_8(CH_3)_2$ (4,5–Dimethylphenanthrene)	$2H_2$	13.0±0.2	RPD		2538

$$C_{16}H_{10}^{+2}$$

Ion	Reactant	Other products	Ionization or appearance potential (eV)	Method	Heat of formation (kJ mol^{-1})	Ref.
$C_{16}H_{10}^{+2}$	$C_{16}H_{10}$ (Pyrene)		24.00±0.5	EI		1069

$$C_{16}H_{12}^{+}$$

Ion	Reactant	Other products	Ionization or appearance potential (eV)	Method	Heat of formation (kJ mol^{-1})	Ref.
$C_{16}H_{12}^{+}$	$C_{14}H_8(CH_3)_2$ (2,5–Dimethylphenanthrene)	H_2	12.1±0.2	RPD		2538

$$C_{16}H_{13}^{+}$$

Ion	Reactant	Other products	Ionization or appearance potential (eV)	Method	Heat of formation (kJ mol^{-1})	Ref.
$C_{16}H_{13}^{+}$	$C_{14}H_8(CH_3)_2$ (2,3–Dimethylphenanthrene)	H	13.1±0.1	RPD		2538
$C_{16}H_{13}^{+}$	$C_{14}H_8(CH_3)_2$ (2,5–Dimethylphenanthrene)	H	13.0±0.2	RPD		2538
$C_{16}H_{13}^{+}$	$C_{14}H_8(CH_3)_2$ (2,7–Dimethylphenanthrene)	H	14.2±0.1	RPD		2538
$C_{16}H_{13}^{+}$	$C_{14}H_8(CH_3)_2$ (3,6–Dimethylphenanthrene)	H	13.8±0.2	RPD		2538

$$C_{16}H_{14}^{+}$$

Ion	Reactant	Other products	Ionization or appearance potential (eV)	Method	Heat of formation (kJ mol^{-1})	Ref.
$C_{16}H_{14}^{+}$	$C_6H_5(CH{=}CH)_2C_6H_5$ (1,4–Diphenyl–1,3–butadiene)		7.75	CTS		2978
$C_{16}H_{14}^{+}$	$C_{14}H_8(CH_3)_2$ (2,3–Dimethylphenanthrene)		7.80±0.03	RPD		2538, 2539
$C_{16}H_{14}^{+}$	$C_{14}H_8(CH_3)_2$ (2,5–Dimethylphenanthrene)		7.83±0.04	RPD		2538
$C_{16}H_{14}^{+}$	$C_{14}H_8(CH_3)_2$ (2,7–Dimethylphenanthrene)		7.98±0.05	RPD		2538, 2539
$C_{16}H_{14}^{+}$	$C_{14}H_8(CH_3)_2$ (3,6–Dimethylphenanthrene)		7.60±0.03	RPD		2538, 2539
$C_{16}H_{14}^{+}$	$C_{14}H_8(CH_3)_2$ (4,5–Dimethylphenanthrene)		7.53±0.04	RPD		2538
$C_{16}H_{14}^{+}$	$C_{14}H_8(CH_3)_2$ (9,10–Dimethylphenanthrene)		8.01±0.05	RPD		2538, 2539

$$C_{16}H_{16}^{+}$$

Ion	Reactant	Other products	Ionization or appearance potential (eV)	Method	Heat of formation (kJ mol^{-1})	Ref.
$C_{16}H_{16}^{+}$	$C_{16}H_{16}$ ([2.2]Paracyclophane)		~7.8	PE		3290

4.3. The Positive Ion Table—Continued

Ion	Reactant	Other products	Ionization or appearance potential (eV)	Method	Heat of formation (kJ mol^{-1})	Ref.
		$C_{16}H_{22}^+$				
$C_{16}H_{22}^+$	$C_{16}H_{22}$ (1,2:3,4–Dicycloheptenobenzene)		8.39±0.02	EI		3342
		$C_{16}H_{25}^+$				
$C_{16}H_{25}^+$	$C_{18}H_{30}$ (3–Phenyldodecane)		9.26±0.1	EI		2153
		$C_{17}H_{11}^+$				
$C_{17}H_{11}^+$	$C_6H_4(C_6H_5)_2$ (o–Terphenyl)		12.0±0.1	EI		2438
$C_{17}H_{11}^+$	$C_6H_5(C_6H_4)_2C_6H_5$ (o–Quaterphenyl)		13.0±0.2	EI		2438
		$C_{17}H_{27}^+$				
$C_{17}H_{27}^+$	$C_6H_3(C_4H_9)_3$ (1,2,4–Tri–tert–butylbenzene)	CH_3	11.14±0.07	EI		2416
$C_{17}H_{27}^+$	$C_6H_3(C_4H_9)_3$ (1,3,5–Tri–tert–butylbenzene)	CH_3	11.91±0.07	EI		2416
		$C_{18}H_{10}^+$				
$C_{18}H_{10}^+$	$C_6H_5(C_6H_4)_2C_6H_5$ (o–Quaterphenyl)		19.0±0.4	EI		2438
$C_{18}H_{10}^+$	$C_6H_5(C_6H_4)_2C_6H_5$ (m–Quaterphenyl)		18.3±0.3	EI		2438
		$C_{18}H_{12}^+$				
$C_{18}H_{12}^+$	$C_{18}H_{12}$ (Chrysene)		8.01±0.3	EI		1069
$C_{18}H_{12}^+$	$C_{18}H_{12}$ (Chrysene)		7.68	CTS		2909
$C_{18}H_{12}^+$	$C_{18}H_{12}$ (Chrysene)		7.72	CTS		3000
$C_{18}H_{12}^+$	$C_{18}H_{12}$ (Chrysene) (Average of two values)		7.78	CTS		2978
$C_{18}H_{12}^+$	$C_{18}H_{12}$ (Chrysene)		7.80	CTS	2562,	2947
$C_{18}H_{12}^+$	$C_{18}H_{12}$ (Chrysene)		7.82	CTS		1064
$C_{18}H_{12}^+$	$C_{18}H_{12}$ (Chrysene)		7.83	CTS		2910

4.3. The Positive Ion Table—Continued

Ion	Reactant	Other products	Ionization or appearance potential (eV)	Method	Heat of formation (kJ mol⁻¹)	Ref.
$C_{18}H_{12}^+$	$C_{18}H_{12}$ (3,4–Benzophenanthrene)		7.76	CTS		3000
$C_{18}H_{12}^+$	$C_{18}H_{12}$ (3,4–Benzophenanthrene)		7.86	CTS		2910
$C_{18}H_{12}^+$	$C_{18}H_{12}$ (9,10–Benzophenanthrene)		8.19±0.3	EI		1069
$C_{18}H_{12}^+$	$C_{18}H_{12}$ (9,10–Benzophenanthrene)		7.95	CTS		2978
(Average of two values)						
$C_{18}H_{12}^+$	$C_{18}H_{12}$ (9,10–Benzophenanthrene)		8.08	CTS		2909
$C_{18}H_{12}^+$	$C_{18}H_{12}$ (9,10–Benzophenanthrene)		8.09	CTS	2562,	2947
$C_{18}H_{12}^+$	$C_{18}H_{12}$ (9,10–Benzophenanthrene)		8.13	CTS		3000
$C_{18}H_{12}^+$	$C_{18}H_{12}$ (9,10–Benzophenanthrene)		8.17	CTS		2910
$C_{18}H_{12}^+$	$C_{18}H_{12}$ (1,2–Benzanthracene)		7.53±0.3	EI		1069
$C_{18}H_{12}^+$	$C_{18}H_{12}$ (1,2–Benzanthracene)		7.35	CTS		3000
$C_{18}H_{12}^+$	$C_{18}H_{12}$ (1,2–Benzanthracene)		7.45	CTS		2947
$C_{18}H_{12}^+$	$C_{18}H_{12}$ (1,2–Benzanthracene)		7.52	CTS		1064
$C_{18}H_{12}^+$	$C_{18}H_{12}$ (1,2–Benzanthracene)		7.53	CTS		2910
$C_{18}H_{12}^+$	$C_{18}H_{12}$ (1,2–Benzanthracene)		7.56	CTS		2909
$C_{18}H_{12}^+$	$C_{18}H_{12}$ (1,2–Benzanthracene)		7.6	CTS		2978

See also – EI: 3011

Ion	Reactant	Other products	Ionization or appearance potential (eV)	Method	Heat of formation (kJ mol⁻¹)	Ref.
$C_{18}H_{12}^+$	$C_{18}H_{12}$ (Naphthacene)		6.95±0.3	EI		1069
$C_{18}H_{12}^+$	$C_{18}H_{12}$ (Naphthacene)		6.64	CTS		3000
$C_{18}H_{12}^+$	$C_{18}H_{12}$ (Naphthacene)		6.94	CTS		1064
$C_{18}H_{12}^+$	$C_{18}H_{12}$ (Naphthacene)		6.95	CTS		2910
$C_{18}H_{12}^+$	$C_{18}H_{12}$ (Naphthacene)		7.00	CTS		2947
$C_{18}H_{12}^+$	$C_{18}H_{12}$ (Naphthacene)		7.0	CTS		2978

See also – S: 3046

Ion	Reactant	Other products	Ionization or appearance potential (eV)	Method	Heat of formation (kJ mol⁻¹)	Ref.
$C_{18}H_{12}^+$	$C_6H_4(C_6H_5)_2$ (o–Terphenyl)	H_2	11.7±0.1	EI		2438
$C_{18}H_{12}^+$	$C_6H_5(C_6H_4)_2C_6H_5$ (o–Quaterphenyl)		13.0±0.1	EI		2438

4.3. The Positive Ion Table—Continued

Ion	Reactant	Other products	Ionization or appearance potential (eV)	Method	Heat of formation (kJ mol^{-1})	Ref.
$C_{18}H_{12}^{+}$	$C_6H_5(C_6H_4)_2C_6H_5$ (m–Quaterphenyl)		16.7±0.3	EI		2438
$C_{18}H_{12}^{+}$	$C_6H_5(C_6H_4)_4C_6H_5$ (p–Hexaphenyl)		18.5±0.3	EI		2438

$$C_{18}H_{12}^{+2}$$

Ion	Reactant	Other products	Ionization or appearance potential (eV)	Method	Heat of formation (kJ mol^{-1})	Ref.
$C_{18}H_{12}^{+2}$	$C_{18}H_{12}$ (Chrysene)		23.35±0.5	EI		1069
$C_{18}H_{12}^{+2}$	$C_{18}H_{12}$ (9,10–Benzophenanthrene)		24.10±1.0	EI		1069
$C_{18}H_{12}^{+2}$	$C_{18}H_{12}$ (1,2–Benzanthracene)		22.03±0.5	EI		1069
$C_{18}H_{12}^{+2}$	$C_{18}H_{12}$ (Naphthacene)		22.14±0.5	EI		1069

$$C_{18}H_{13}^{+}$$

Ion	Reactant	Other products	Ionization or appearance potential (eV)	Method	Heat of formation (kJ mol^{-1})	Ref.
$C_{18}H_{13}^{+}$	$C_6H_4(C_6H_5)_2$ (o–Terphenyl)	H	11.7±0.1	EI		2438
$C_{18}H_{13}^{+}$	$C_6H_5(C_6H_4)_2C_6H_5$ (o–Quaterphenyl)		12.8±0.1	EI		2438
$C_{18}H_{13}^{+}$	$C_6H_5(C_6H_4)_4C_6H_5$ (p–Hexaphenyl)		19.5±0.3	EI		2438

$$C_{18}H_{14}^{+}$$

Ion	Reactant	Other products	Ionization or appearance potential (eV)	Method	Heat of formation (kJ mol^{-1})	Ref.
$C_{18}H_{14}^{+}$	$C_6H_4(C_6H_5)_2$ (o–Terphenyl)		8.64±0.05	EI		2438
$C_{18}H_{14}^{+}$	$C_6H_4(C_6H_5)_2$ (o–Terphenyl)		8.43	CTS		2562
$C_{18}H_{14}^{+}$	$C_6H_4(C_6H_5)_2$ (m–Terphenyl)		8.80±0.05	EI		2438
$C_{18}H_{14}^{+}$	$C_6H_4(C_6H_5)_2$ (p–Terphenyl)		8.78±0.05	EI		2438
$C_{18}H_{14}^{+}$	$C_6H_4(C_6H_5)_2$ (p–Terphenyl)		8.29	CTS		2562

See also – S: 2661

$$C_{18}H_{14}^{+2}$$

Ion	Reactant	Other products	Ionization or appearance potential (eV)	Method	Heat of formation (kJ mol^{-1})	Ref.
$C_{18}H_{14}^{+2}$	$C_6H_4(C_6H_5)_2$ (o–Terphenyl)		21.5±0.4	EI		2438
$C_{18}H_{14}^{+2}$	$C_6H_4(C_6H_5)_2$ (m–Terphenyl)		21.5±0.4	EI		2438
$C_{18}H_{14}^{+2}$	$C_6H_4(C_6H_5)_2$ (p–Terphenyl)		21.5±0.4	EI		2438

4.3. The Positive Ion Table—Continued

Ion	Reactant	Other products	Ionization or appearance potential (eV)	Method	Heat of formation (kJ mol⁻¹)	Ref.

$$C_{18}H_{16}^+$$

Ion	Reactant	Other products	Ionization or appearance potential (eV)	Method	Heat of formation (kJ mol⁻¹)	Ref.
$C_{18}H_{16}^+$	$C_6H_5(CH=CH)_3C_6H_5$ (1,6–Diphenyl–1,3,5–hexatriene)		7.6	CTS		2978

$$C_{18}H_{18}^+$$

Ion	Reactant	Other products	Ionization or appearance potential (eV)	Method	Heat of formation (kJ mol⁻¹)	Ref.
$C_{18}H_{18}^+$	$C_{18}H_{18}$ (1,6–Diphenylhexadiene)		8.2	CTS		2978

$$C_{18}H_{24}^+$$

Ion	Reactant	Other products	Ionization or appearance potential (eV)	Method	Heat of formation (kJ mol⁻¹)	Ref.
$C_{18}H_{24}^+$	$C_{18}H_{24}$ (Perhydro–9,10–benzophenanthrene)		8.60±0.03	EI		3342

$$C_{18}H_{26}^+$$

Ion	Reactant	Other products	Ionization or appearance potential (eV)	Method	Heat of formation (kJ mol⁻¹)	Ref.
$C_{18}H_{26}^+$	$C_{18}H_{26}$ (1,2:3,4–Dicyclooctenobenzene)		7.93±0.03	EI		3342

$$C_{18}H_{29}^+$$

Ion	Reactant	Other products	Ionization or appearance potential (eV)	Method	Heat of formation (kJ mol⁻¹)	Ref.
$C_{18}H_{29}^+$	$C_{26}H_{46}$ (9–Phenyleicosane)		10.12±0.1	EI		2153

$$C_{18}H_{30}^+$$

Ion	Reactant	Other products	Ionization or appearance potential (eV)	Method	Heat of formation (kJ mol⁻¹)	Ref.
$C_{18}H_{30}^+$	$C_{18}H_{30}$ (1–Phenyldodecane)		9.05±0.1	EI		2153
$C_{18}H_{30}^+$	$C_{18}H_{30}$ (3–Phenyldodecane)		8.95±0.1	EI		2153
$C_{18}H_{30}^+$	$C_6H_3(C_4H_9)_3$ (1,2,4–Tri–*tert*–butylbenzene)		8.60±0.07	EI		2416
$C_{18}H_{30}^+$	$C_6H_3(C_4H_9)_3$ (1,3,5–Tri–*tert*–butylbenzene)		8.56±0.07	EI		2416

$$C_{19}H_{14}^+$$

Ion	Reactant	Other products	Ionization or appearance potential (eV)	Method	Heat of formation (kJ mol⁻¹)	Ref.
$C_{19}H_{14}^+$	$C_{18}H_{11}CH_3$ (3–Methyl–1,2–benzanthracene)		7.43	CTS		2947
$C_{19}H_{14}^+$	$C_{18}H_{11}CH_3$ (4–Methyl–1,2–benzanthracene)		7.41	CTS		2947
$C_{19}H_{14}^+$	$C_{18}H_{11}CH_3$ (5–Methyl–1,2–benzanthracene)		7.39	CTS		2947
$C_{19}H_{14}^+$	$C_{18}H_{11}CH_3$ (6–Methyl–1,2–benzanthracene)		7.37	CTS		2947
$C_{19}H_{14}^+$	$C_{18}H_{11}CH_3$ (7–Methyl–1,2–benzanthracene)		7.37	CTS		2947

4.3. The Positive Ion Table—Continued

Ion	Reactant	Other products	Ionization or appearance potential (eV)	Method	Heat of formation (kJ mol^{-1})	Ref.
$C_{19}H_{14}^{+}$	$C_{18}H_{11}CH_3$ (10–Methyl–1,2–benzanthracene)		7.29	CTS		2947
$C_{19}H_{14}^{+}$	$C_{18}H_{11}CH_3$ (2'–Methyl–1,2–benzanthracene)		7.39	CTS		2947
$C_{19}H_{14}^{+}$	$C_{18}H_{11}CH_3$ (3'–Methyl–1,2–benzanthracene)		7.43	CTS		2947
$C_{19}H_{14}^{+}$	$C_{18}H_{11}CH_3$ (4'–Methyl–1,2–benzanthracene)		7.43	CTS		2947

$C_{19}H_{32}^{+}$

Ion	Reactant	Other products	Ionization or appearance potential (eV)	Method	Heat of formation (kJ mol^{-1})	Ref.
$C_{19}H_{32}^{+}$	$C_{19}H_{32}$ (7–Phenyltridecane)		8.91±0.1	EI		2153

$C_{20}H_{12}^{+}$

Ion	Reactant	Other products	Ionization or appearance potential (eV)	Method	Heat of formation (kJ mol^{-1})	Ref.
$C_{20}H_{12}^{+}$	$C_{20}H_{12}$ (Perylene)		7.10±0.1	EI		2489
$C_{20}H_{12}^{+}$	$C_{20}H_{12}$ (Perylene)		6.83	CTS		3000
$C_{20}H_{12}^{+}$	$C_{20}H_{12}$ (Perylene)		6.85	CTS		2037
$C_{20}H_{12}^{+}$	$C_{20}H_{12}$ (Perylene)		7.03	CTS		2947
$C_{20}H_{12}^{+}$	$C_{20}H_{12}$ (Perylene)		7.06	CTS		2909
$C_{20}H_{12}^{+}$	$C_{20}H_{12}$ (Perylene)		7.10	CTS		1064
$C_{20}H_{12}^{+}$	$C_{20}H_{12}$ (Perylene)		7.11	CTS		2910
$C_{20}H_{12}^{+}$	$C_{20}H_{12}$ (Perylene)		7.15	CTS		2978
$C_{20}H_{12}^{+}$	$C_{20}H_{12}$ (1,2–Benzopyrene)		7.15	CTS		3000
$C_{20}H_{12}^{+}$	$C_{20}H_{12}$ (1,2–Benzopyrene)		7.19	CTS	2562,	2947
$C_{20}H_{12}^{+}$	$C_{20}H_{12}$ (1,2–Benzopyrene)		7.37	CTS		2910
$C_{20}H_{12}^{+}$	$C_{20}H_{12}$ (4,5–Benzopyrene)		7.56	CTS		2947
$C_{20}H_{12}^{+}$	$C_{20}H_{12}$ (4,5–Benzopyrene)		7.60	CTS		3000
$C_{20}H_{12}^{+}$	$C_{20}H_{12}$ (4,5–Benzopyrene)		7.73	CTS		2910

$C_{20}H_{12}^{+2}$

Ion	Reactant	Other products	Ionization or appearance potential (eV)	Method	Heat of formation (kJ mol^{-1})	Ref.
$C_{20}H_{12}^{+2}$	$C_{20}H_{12}$ (Perylene)		20.0	EI		2489

4.3. The Positive Ion Table—Continued

Ion	Reactant	Other products	Ionization or appearance potential (eV)	Method	Heat of formation (kJ mol^{-1})	Ref.
		$C_{20}H_{16}^{+}$				
$C_{20}H_{16}^{+}$	$C_{18}H_{10}(CH_3)_2$ (9,10–Dimethyl–1,2–benzanthracene)		7.43	CTS		2947
		$C_{20}H_{18}^{+}$				
$C_{20}H_{18}^{+}$	$C_6H_5(CH=CH)_4C_6H_5$ (1,8–Diphenyl–1,3,5,7–octatetraene)		7.5	D		2978
		$C_{20}H_{33}^{+}$				
$C_{20}H_{33}^{+}$	$C_{26}H_{46}$ (7–Phenyleicosane)		9.94±0.1	EI		2153
		$C_{21}H_{16}^{+}$				
$C_{21}H_{16}^{+}$	$C_{20}H_{13}CH_3$ (20–Methylcholanthrene)		7.66	CTS		2562
		$C_{21}H_{30}^{+}$				
$C_{21}H_{30}^{+}$	$C_{21}H_{30}$ (Tricycloheptenobenzene)		8.75±0.05	EI		3342
		$C_{22}H_{12}^{+}$				
$C_{22}H_{12}^{+}$	$C_{22}H_{12}$ (4,5:10,11–Dibenzochrysene)		6.84	CTS		3000
$C_{22}H_{12}^{+}$	$C_{22}H_{12}$ (4,5:10,11–Dibenzochrysene)		7.01	CTS		2909
$C_{22}H_{12}^{+}$	$C_{22}H_{12}$ (4,5:10,11–Dibenzochrysene)		7.10	CTS		1064
$C_{22}H_{12}^{+}$	$C_{22}H_{12}$ (4,5:10,11–Dibenzochrysene)		7.11	CTS		2910
$C_{22}H_{12}^{+}$	$C_{22}H_{12}$ (1,12–Benzoperylene)		7.13	CTS		3000
$C_{22}H_{12}^{+}$	$C_{22}H_{12}$ (1,12–Benzoperylene)		7.35	CTS		2910
		$C_{22}H_{14}^{+}$				
$C_{22}H_{14}^{+}$	$C_{22}H_{14}$ (1,2:3,4–Dibenzanthracene)		7.43	CTS		3000
$C_{22}H_{14}^{+}$	$C_{22}H_{14}$ (1,2:3,4–Dibenzanthracene)		7.6	CTS		2910

4.3. The Positive Ion Table—Continued

Ion	Reactant	Other products	Ionization or appearance potential (eV)	Method	Heat of formation (kJ mol⁻¹)	Ref.
$C_{22}H_{14}^+$	$C_{22}H_{14}$ (1,2:3,4–Dibenzanthracene)		7.61	CTS	2562,	2947
$C_{22}H_{14}^+$	$C_{22}H_{14}$ (1,2:5,6–Dibenzanthracene)		7.59±0.1	EI		2489
$C_{22}H_{14}^+$	$C_{22}H_{14}$ (1,2:5,6–Dibenzanthracene)		7.42	CTS		3000
$C_{22}H_{14}^+$	$C_{22}H_{14}$ (1,2:5,6–Dibenzanthracene)		7.57	CTS		1064
$C_{22}H_{14}^+$	$C_{22}H_{14}$ (1,2:5,6–Dibenzanthracene)		7.58	CTS		2910
$C_{22}H_{14}^+$	$C_{22}H_{14}$ (1,2:5,6–Dibenzanthracene)		7.80	CTS	2562,	2947
$C_{22}H_{14}^+$	$C_{22}H_{14}$ (1,2:6,7–Dibenzanthracene)		6.74	CTS		3000
$C_{22}H_{14}^+$	$C_{22}H_{14}$ (1,2:6,7–Dibenzanthracene)		7.03	CTS		2910
$C_{22}H_{14}^+$	$C_{22}H_{14}$ (1,2:7,8–Dibenzanthracene)		7.42	CTS		3000
$C_{22}H_{14}^+$	$C_{22}H_{14}$ (1,2:7,8–Dibenzanthracene)		7.58	CTS		2910
$C_{22}H_{14}^+$	$C_{22}H_{14}$ (1,2:7,8–Dibenzanthracene)		7.68	CTS	2562,	2947
$C_{22}H_{14}^+$	$C_{22}H_{14}$ (Pentacene)		6.55±0.1	EI		2489
$C_{22}H_{14}^+$	$C_{22}H_{14}$ (Pentacene)		6.23	CTS		3000
$C_{22}H_{14}^+$	$C_{22}H_{14}$ (Pentacene)		6.61	CTS		1064
$C_{22}H_{14}^+$	$C_{22}H_{14}$ (Pentacene)		6.62	CTS		2910
$C_{22}H_{14}^+$	$C_{22}H_{14}$ (Picene)		7.80±0.1	EI		2489
$C_{22}H_{14}^+$	$C_{22}H_{14}$ (Picene)		7.62	CTS		3000
$C_{22}H_{14}^+$	$C_{22}H_{14}$ (Picene)		7.75	CTS		2910
$C_{22}H_{14}^+$	$C_{22}H_{14}$ (Picene)		7.80	CTS		2947
$C_{22}H_{14}^+$	$C_{22}H_{14}$ (1,2:3,4–Dibenzophenanthrene)		7.57	CTS		3000
$C_{22}H_{14}^+$	$C_{22}H_{14}$ (1,2:5,6–Dibenzophenanthrene)		7.71	CTS		3000
$C_{22}H_{14}^+$	$C_{22}H_{14}$ (1,2:5,6–Dibenzophenanthrene)		7.82	CTS		2910
$C_{22}H_{14}^+$	$C_{22}H_{14}$ (1,2:6,7–Dibenzophenanthrene)		7.29	CTS		3000

4.3. The Positive Ion Table—Continued

Ion	Reactant	Other products	Ionization or appearance potential (eV)	Method	Heat of formation (kJ mol^{-1})	Ref.
$C_{22}H_{14}^{+}$	$C_{22}H_{14}$ (2,3:5,6–Dibenzophenanthrene)		7.11	CTS		3000
$C_{22}H_{14}^{+}$	$C_{22}H_{14}$ (2,3:5,6–Dibenzophenanthrene)		7.33	CTS		2910
$C_{22}H_{14}^{+}$	$C_{22}H_{14}$ (2,3:6,7–Dibenzophenanthrene)		7.35	CTS		3000
$C_{22}H_{14}^{+}$	$C_{22}H_{14}$ (2,3:6,7–Dibenzophenanthrene)		7.53	CTS		2910
		$C_{22}H_{14}^{+2}$				
$C_{22}H_{14}^{+2}$	$C_{22}H_{14}$ (1,2:5,6–Dibenzanthracene)		20.8	EI		2489
$C_{22}H_{14}^{+2}$	$C_{22}H_{14}$ (Pentacene)		19.6	EI		2489
$C_{22}H_{14}^{+2}$	$C_{22}H_{14}$ (Picene)		21.5	EI		2489
		$C_{22}H_{20}^{+}$				
$C_{22}H_{20}^{+}$	$C_6H_5(CH=CH)_5C_6H_5$ (1,10–Diphenyl–1,3,5,7,9–decapentaene)		7.4	D		2978
		$C_{22}H_{37}^{+}$				
$C_{22}H_{37}^{+}$	$C_{26}H_{46}$ (5–Phenyleicosane)		9.99±0.1	EI		2153
		$C_{23}H_{13}^{+}$				
$C_{23}H_{13}^{+}$	$C_6H_5(C_6H_4)_2C_6H_5$ (o–Quaterphenyl)		15.5±0.2	EI		2438
		$C_{23}H_{15}^{+}$				
$C_{23}H_{15}^{+}$	$C_6H_5(C_6H_4)_2C_6H_5$ (o–Quaterphenyl)		12.7±0.1	EI		2438
		$C_{23}H_{39}^{+}$				
$C_{23}H_{39}^{+}$	$C_{26}H_{46}$ (4–Phenyleicosane)		9.76±0.1	EI		2153

4.3. The Positive Ion Table—Continued

Ion	Reactant	Other products	Ionization or appearance potential (eV)	Method	Heat of formation (kJ mol⁻¹)	Ref.
			$C_{24}H_{12}^+$			
$C_{24}H_{12}^+$	$C_{24}H_{12}$ (Coronene)		7.68±0.05	EDD		2634
$C_{24}H_{12}^+$	$C_{24}H_{12}$ (Coronene)		7.65±0.1	EI		2489
$C_{24}H_{12}^+$	$C_{24}H_{12}$ (Coronene)		7.44	CTS		2947
$C_{24}H_{12}^+$	$C_{24}H_{12}$ (Coronene)		7.50	CTS		3000
$C_{24}H_{12}^+$	$C_{24}H_{12}$ (Coronene)		7.6	CTS		2978
$C_{24}H_{12}^+$	$C_{24}H_{12}$ (Coronene)		7.64	CTS		1064
$C_{24}H_{12}^+$	$C_{24}H_{12}$ (Coronene)		7.65	CTS		2910
			$C_{24}H_{12}^{+2}$			
$C_{24}H_{12}^{+2}$	$C_{24}H_{12}$ (Coronene)		21.0	EI		2489
			$C_{24}H_{14}^+$			
$C_{24}H_{14}^+$	$C_{24}H_{14}$ (1,2:4,5–Dibenzopyrene)		7.20	CTS		3000
$C_{24}H_{14}^+$	$C_{24}H_{14}$ (1,2:4,5–Dibenzopyrene)		7.27	CTS	2562,	2947
$C_{24}H_{14}^+$	$C_{24}H_{14}$ (1,2:4,5–Dibenzopyrene)		7.41	CTS		2910
$C_{24}H_{14}^+$	$C_{24}H_{14}$ (1,2:6,7–Dibenzopyrene)		6.75	CTS		3000
$C_{24}H_{14}^+$	$C_{24}H_{14}$ (1,2:6,7–Dibenzopyrene)		7.04	CTS		2910
$C_{24}H_{14}^+$	$C_{24}H_{14}$ (1,2:7,8–Dibenzopyrene)		7.06	CTS		3000
$C_{24}H_{14}^+$	$C_{24}H_{14}$ (1,2:7,8–Dibenzopyrene)		7.30	CTS		2910
$C_{24}H_{14}^+$	$C_{24}H_{14}$ (1,2:9,10–Dibenzopyrene)		7.03	CTS		3000
$C_{24}H_{14}^+$	$C_{24}H_{14}$ (1,2:9,10–Dibenzopyrene)		7.27	CTS		2910
$C_{24}H_{14}^+$	$C_{24}H_{14}$ (4,5:9,10–Dibenzopyrene)		7.62	CTS		3000
$C_{24}H_{14}^+$	$C_{24}H_{14}$ (4,5:9,10–Dibenzopyrene)		7.75	CTS		2910

4.3. The Positive Ion Table—Continued

Ion	Reactant	Other products	Ionization or appearance potential (eV)	Method	Heat of formation (kJ mol^{-1})	Ref.
$C_{24}H_{14}^+$	$C_{24}H_{14}$ (1,12:4,5–Dibenzonaphthacene)		7.47	CTS		3000
$C_{24}H_{14}^+$	$C_{24}H_{14}$ (1,12:4,5–Dibenzonaphthacene)		7.60	CTS		2910
$C_{24}H_{14}^+$	$C_{24}H_{14}$ (1,2–Benzoperylene)		6.51	CTS		3000
$C_{24}H_{14}^+$	$C_{24}H_{14}$ (1,2–Benzoperylene)		6.84	CTS		2910
$C_{24}H_{14}^+$	$C_{24}H_{14}$ (Naphtho[2,3–a]pyrene)		6.70	CTS		3000
$C_{24}H_{14}^+$	$C_{24}H_{14}$ (Naphtho[2,3–a]pyrene)		7.00	CTS		2910

$C_{24}H_{17}^+$

Ion	Reactant	Other products	Ionization or appearance potential (eV)	Method	Heat of formation (kJ mol^{-1})	Ref.
$C_{24}H_{17}^+$	$C_6H_5(C_6H_4)_2C_6H_5$ (o–Quaterphenyl)	H	11.7±0.1	EI		2438

$C_{24}H_{18}^+$

Ion	Reactant	Other products	Ionization or appearance potential (eV)	Method	Heat of formation (kJ mol^{-1})	Ref.
$C_{24}H_{18}^+$	$C_6H_5(C_6H_4)_2C_6H_5$ (o–Quaterphenyl)		8.52±0.05	EI		2438
$C_{24}H_{18}^+$	$C_6H_5(C_6H_4)_2C_6H_5$ (m–Quaterphenyl)		8.51±0.05	EI		2438
$C_{24}H_{18}^+$	$C_6H_5(C_6H_4)_2C_6H_5$ (p–Quaterphenyl)		8.08±0.05	EI		2438

$C_{24}H_{18}^{+2}$

Ion	Reactant	Other products	Ionization or appearance potential (eV)	Method	Heat of formation (kJ mol^{-1})	Ref.
$C_{24}H_{18}^{+2}$	$C_6H_5(C_6H_4)_2C_6H_5$ (o–Quaterphenyl)		20.5±0.4	EI		2438
$C_{24}H_{18}^{+2}$	$C_6H_5(C_6H_4)_2C_6H_5$ (m–Quaterphenyl)		20.5±0.4	EI		2438
$C_{24}H_{18}^{+2}$	$C_6H_5(C_6H_4)_2C_6H_5$ (p–Quaterphenyl)		20.2±0.4	EI		2438

$C_{24}H_{36}^+$

Ion	Reactant	Other products	Ionization or appearance potential (eV)	Method	Heat of formation (kJ mol^{-1})	Ref.
$C_{24}H_{36}^+$	$C_{24}H_{36}$ (Tricyclooctenobenzene)		8.52±0.10	EI		3342

$C_{24}H_{41}^+$

Ion	Reactant	Other products	Ionization or appearance potential (eV)	Method	Heat of formation (kJ mol^{-1})	Ref.
$C_{24}H_{41}^+$	$C_{26}H_{46}$ (3–Phenylicosane)		9.68±0.1	EI		2153

4.3. The Positive Ion Table—Continued

Ion	Reactant	Other products	Ionization or appearance potential (eV)	Method	Heat of formation (kJ mol^{-1})	Ref.
			$C_{25}H_{43}^+$			
$C_{25}H_{43}^+$	$C_{26}H_{46}$ (2–Phenyleicosane)		10.14±0.1	EI		2153
			$C_{26}H_{14}^+$			
$C_{26}H_{14}^+$	$C_{26}H_{14}$ (1,12:2,3–Dibenzoperylene)		7.20	CTS		3000
$C_{26}H_{14}^+$	$C_{26}H_{14}$ (1,12:2,3–Dibenzoperylene)		7.41	CTS		2910
$C_{26}H_{14}^+$	$C_{26}H_{14}$ (3,4:9,10–Dibenzoperylene)		6.82	CTS		3000
$C_{26}H_{14}^+$	$C_{26}H_{14}$ (3,4:9,10–Dibenzoperylene)		7.10	CTS		2910
			$C_{26}H_{16}^+$			
$C_{26}H_{16}^+$	$C_{26}H_{16}$ (Bifluorenylidene)		8.5±0.2	EI		3336
$C_{26}H_{16}^+$	$C_{26}H_{16}$ (1,2–Benzopentacene)		6.32	CTS		3000
$C_{26}H_{16}^+$	$C_{26}H_{16}$ (1,2–Benzopentacene)		6.69	CTS		2910
$C_{26}H_{16}^+$	$C_{26}H_{16}$ (3,4–Benzopentaphene)		7.28	CTS		3000
$C_{26}H_{16}^+$	$C_{26}H_{16}$ (3,4–Benzopentaphene)		7.47	CTS		2910
$C_{26}H_{16}^+$	$C_{26}H_{16}$ (2,3:8,9–Dibenzochrysene)		6.92	CTS		3000
$C_{26}H_{16}^+$	$C_{26}H_{16}$ (5,6:11,12–Dibenzochrysene)		7.42	CTS		3000
$C_{26}H_{16}^+$	$C_{26}H_{16}$ (5,6:11,12–Dibenzochrysene)		7.58	CTS		2910
$C_{26}H_{16}^+$	$C_{26}H_{16}$ (1,2:3,4–Dibenzonaphthacene)		6.80	CTS		3000
$C_{26}H_{16}^+$	$C_{26}H_{16}$ (1,2:3,4–Dibenzonaphthacene)		7.08	CTS		2910
$C_{26}H_{16}^+$	$C_{26}H_{16}$ (1,2:7,8–Dibenzonaphthacene)		6.82	CTS		3000
$C_{26}H_{16}^+$	$C_{26}H_{16}$ (1,2:7,8–Dibenzonaphthacene)		7.09	CTS		2910
$C_{26}H_{16}^+$	$C_{26}H_{16}$ (Hexaphene)		6.81	CTS		3000

4.3. The Positive Ion Table—Continued

Ion	Reactant	Other products	Ionization or appearance potential (eV)	Method	Heat of formation (kJ mol^{-1})	Ref.
$C_{26}H_{16}^+$	$C_{26}H_{16}$ (Hexaphene)		7.09	CTS		2910
$C_{26}H_{16}^+$	$C_{26}H_{16}$ (1,2:3,4:5,6–Tribenzanthracene)		7.47	CTS		3000

$C_{26}H_{46}^+$

Ion	Reactant	Other products	Ionization or appearance potential (eV)	Method	Heat of formation (kJ mol^{-1})	Ref.
$C_{26}H_{46}^+$	$C_{26}H_{46}$ (1–Phenyleicosane)		9.34±0.1	EI		2153
$C_{26}H_{46}^+$	$C_{26}H_{46}$ (2–Phenyleicosane)		9.22±0.1	EI		2153
$C_{26}H_{46}^+$	$C_{26}H_{46}$ (3–Phenyleicosane)		8.95±0.1	EI		2153
$C_{26}H_{46}^+$	$C_{26}H_{46}$ (4–Phenyleicosane)		9.01±0.1	EI		2153
$C_{26}H_{46}^+$	$C_{26}H_{46}$ (5–Phenyleicosane)		9.04±0.1	EI		2153
$C_{26}H_{46}^+$	$C_{26}H_{46}$ (7–Phenyleicosane)		8.97±0.1	EI		2153
$C_{26}H_{46}^+$	$C_{26}H_{46}$ (9–Phenyleicosane)		9.06±0.1	EI		2153

$C_{28}H_{14}^+$

Ion	Reactant	Other products	Ionization or appearance potential (eV)	Method	Heat of formation (kJ mol^{-1})	Ref.
$C_{28}H_{14}^+$	$C_{28}H_{14}$ (Phenanthro[1,10,9,8–$opqra$]perylene)		6.42	CTS		1064

$C_{28}H_{16}^+$

Ion	Reactant	Other products	Ionization or appearance potential (eV)	Method	Heat of formation (kJ mol^{-1})	Ref.
$C_{28}H_{16}^+$	$C_{28}H_{16}$ (6,7:13,14–Dibenzopentaphene)		7.17	CTS		3000
$C_{28}H_{16}^+$	$C_{28}H_{16}$ (6,7:13,14–Dibenzopentaphene)		7.38	CTS		2910
$C_{28}H_{16}^+$	$C_{28}H_{16}$ (1,2:7,8–Dibenzoperylene)		6.34	CTS		3000
$C_{28}H_{16}^+$	$C_{28}H_{16}$ (1,2:7,8–Dibenzoperylene)		6.70	CTS		2910
$C_{28}H_{16}^+$	$C_{28}H_{16}$ (1,2:10,11–Dibenzoperylene)		6.51	CTS		3000
$C_{28}H_{16}^+$	$C_{28}H_{16}$ (1,2:10,11–Dibenzoperylene)		6.84	CTS		2910
$C_{28}H_{16}^+$	$C_{28}H_{16}$ (2,3:8,9–Dibenzoperylene)		6.84	CTS		3000
$C_{28}H_{16}^+$	$C_{28}H_{16}$ (2,3:8,9–Dibenzoperylene)		7.11	CTS		2910

4.3. The Positive Ion Table—Continued

Ion	Reactant	Other products	Ionization or appearance potential (eV)	Method	Heat of formation (kJ mol⁻¹)	Ref.
$C_{30}H_{14}^+$						
$C_{30}H_{14}^+$	$C_{30}H_{14}$ (2,3:4,5–Dibenzocoronene)		6.37	CTS		3000
$C_{30}H_{14}^+$	$C_{30}H_{14}$ (2,3:4,5–Dibenzocoronene)		6.73	CTS		2910
$C_{30}H_{16}^+$						
$C_{30}H_{16}^+$	$C_{30}H_{16}$ (Naphthaceno[2,1,12,11–$opqra$]naphthacene)		6.19	CTS		3000
$C_{30}H_{16}^+$	$C_{30}H_{16}$ (Naphthaceno[2,1,12,11–$opqra$]naphthacene)		6.58	CTS		2910
$C_{30}H_{16}^+$	$C_{30}H_{16}$ (Pyranthrene)		6.69	CTS		3000
$C_{30}H_{16}^+$	$C_{30}H_{16}$ (Pyranthrene)		6.98	CTS		1064
$C_{30}H_{16}^+$	$C_{30}H_{16}$ (Pyranthrene)		6.99	CTS		2910
$C_{30}H_{16}^+$	$C_{30}H_{16}$ (1,12:2,3:8,9–Tribenzoperylene)		7.03	CTS		3000
$C_{30}H_{16}^+$	$C_{30}H_{16}$ (1,12:2,3:8,9–Tribenzoperylene)		7.27	CTS		2910
$C_{30}H_{18}^+$						
$C_{30}H_{18}^+$	$C_{30}H_{18}$ (1,2:3,4–Dibenzopentacene)		6.36	CTS		3000
$C_{30}H_{18}^+$	$C_{30}H_{18}$ (1,2:3,4–Dibenzopentacene)		6.72	CTS		2910
$C_{30}H_{18}^+$	$C_{30}H_{18}$ (1,2:8,9–Dibenzopentacene)		6.95±0.1	EI		2489
$C_{30}H_{18}^+$	$C_{30}H_{18}$ (1,2:8,9–Dibenzopentacene)		6.42	CTS		3000
$C_{30}H_{18}^+$	$C_{30}H_{18}$ (1,2:8,9–Dibenzopentacene)		6.77	CTS		2910
$C_{30}H_{18}^+$	$C_{30}H_{18}$ (3,4:9,10–Dibenzopentaphene)		7.28	CTS		3000
$C_{30}H_{18}^+$	$C_{30}H_{18}$ (3,4:9,10–Dibenzopentaphene)		7.47	CTS		2910
$C_{30}H_{18}^+$	$C_{30}H_{18}$ (Naphtho[2,3–c]pentaphene)		7.13	CTS		3000
$C_{30}H_{18}^+$	$C_{30}H_{18}$ (Naphtho[2,3–c]pentaphene)		7.35	CTS		2910

4.3. The Positive Ion Table—Continued

Ion	Reactant	Other products	Ionization or appearance potential (eV)	Method	Heat of formation (kJ mol^{-1})	Ref.
		$C_{30}H_{18}^{+2}$				
$C_{30}H_{18}^{+2}$	$C_{30}H_{18}$ (1,2:8,9–Dibenzopentacene)		19.0	EI		2489
		$C_{30}H_{22}^{+}$				
$C_{30}H_{22}^{+}$	$C_6H_5(C_6H_4)_3C_6H_5$ (m–Quinquephenyl)		8.45±0.05	EI		2438
$C_{30}H_{22}^{+}$	$C_6H_5(C_6H_4)_3C_6H_5$ (p–Quinquephenyl)		8.18±0.05	EI		2438
		$C_{30}H_{22}^{+2}$				
$C_{30}H_{22}^{+2}$	$C_6H_5(C_6H_4)_3C_6H_5$ (m–Quinquephenyl)		20.0±0.4	EI		2438
$C_{30}H_{22}^{+2}$	$C_6H_5(C_6H_4)_3C_6H_5$ (p–Quinquephenyl)		19.6±0.4	EI		2438
		$C_{32}H_{14}^{+}$				
$C_{32}H_{14}^{+}$	$C_{32}H_{14}$ (Ovalene)		7.24±0.1	EI		2489
$C_{32}H_{14}^{+}$	$C_{32}H_{14}$ (Ovalene)		7.01	CTS		1064
		$C_{32}H_{14}^{+2}$				
$C_{32}H_{14}^{+2}$	$C_{32}H_{14}$ (Ovalene)		19.6	EI		2489
		$C_{32}H_{18}^{+}$				
$C_{32}H_{18}^{+}$	$C_{32}H_{18}$ (16,17–Benzoheptaphene)		6.74	CTS		3000
$C_{32}H_{18}^{+}$	$C_{32}H_{18}$ (16,17–Benzoheptaphene)		7.03	CTS		2910
$C_{32}H_{18}^{+}$	$C_{32}H_{18}$ (Naphthaceno[2,1,12–qra]naphthacene)		6.23	CTS		3000
$C_{32}H_{18}^{+}$	$C_{32}H_{18}$ (Naphthaceno[2,1,12–qra]naphthacene)		6.62	CTS		2910
		$C_{34}H_{18}^{+}$				
$C_{34}H_{18}^{+}$	$C_{34}H_{18}$ (1,14:5,6:7,8:12,13–Tetrabenzopentacene)		6.06	CTS		3000

4.3. The Positive Ion Table—Continued

Ion	Reactant	Other products	Ionization or appearance potential (eV)	Method	Heat of formation (kJ mol^{-1})	Ref.
$C_{34}H_{18}^+$	$C_{34}H_{18}$ (4,5:6,7:8,9:13,14–Tetrabenzopentaphene)		6.04	CTS		3000
$C_{34}H_{18}^+$	$C_{34}H_{18}$ (4,5:6,7:8,9:13,14–Tetrabenzopentaphene)		6.46	CTS		2910
$C_{34}H_{18}^+$	$C_{34}H_{18}$ (1,2:3,4:7,8:9,10–Tetrabenzoperylene)		6.77	CTS		3000
$C_{34}H_{18}^+$	$C_{34}H_{18}$ (1,2:3,4:7,8:9,10–Tetrabenzoperylene)		7.06	CTS		2910

$$C_{34}H_{20}^+$$

Ion	Reactant	Other products	Ionization or appearance potential (eV)	Method	Heat of formation (kJ mol^{-1})	Ref.
$C_{34}H_{20}^+$	$C_{34}H_{20}$ (Violanthrene)		6.86	CTS		1064
$C_{34}H_{20}^+$	$C_{34}H_{20}$ (Isoviolanthrene)		6.76	CTS		1064

$$C_{36}H_{18}^+$$

| $C_{36}H_{18}^+$ | $C_{36}H_{18}$ (Decacyclene) | | 7.27±0.1 | EI | | 2489 |

$$C_{36}H_{18}^{+2}$$

| $C_{36}H_{18}^{+2}$ | $C_{36}H_{18}$ (Decacyclene) | | 20.1 | EI | | 2489 |

$$C_{36}H_{22}^+$$

| $C_{36}H_{22}^+$ | $C_6H_5(C_6H_4)_4C_6H_5$ (p–Hexaphenyl) | $2H_2$ | 15.6±0.2 | EI | | 2438 |

$$C_{36}H_{26}^+$$

| $C_{36}H_{26}^+$ | $C_6H_5(C_6H_4)_4C_6H_5$ (p–Hexaphenyl) | | 7.67±0.05 | EI | | 2438 |

$$C_{36}H_{26}^{+2}$$

| $C_{36}H_{26}^{+2}$ | $C_6H_5(C_6H_4)_4C_6H_5$ (p–Hexaphenyl) | | 19.5±0.4 | EI | | 2438 |

$$C_{42}H_{18}^+$$

| $C_{42}H_{18}^+$ | $C_{42}H_{18}$ (1,12:2,3:4,5:6,7:8,9:10,11–Hexabenzocoronene) | | 7.05±0.1 | EI | | 2489 |

4.3. The Positive Ion Table—Continued

Ion	Reactant	Other products	Ionization or appearance potential (eV)	Method	Heat of formation (kJ mol^{-1})	Ref.
		$C_{42}H_{18}^{+2}$				
$C_{42}H_{18}^{+2}$	$C_{42}H_{18}$ (1,12:2,3:4,5:6,7:8,9:10,11−Hexabenzocoronene)		19.6	EI		2489
		$C_{48}H_{34}^{+}$				
$C_{48}H_{34}^{+}$	$C_6H_5(C_6H_4)_6C_6H_5$ (m−Octaphenyl)		8.28±0.05	EI		2438
		$C_{48}H_{34}^{+2}$				
$C_{48}H_{34}^{+2}$	$C_6H_5(C_6H_4)_6C_6H_5$ (m−Octaphenyl)		20.3±0.4	EI		2438
		CB^{+}				
CB^{+}	CB		10.0±0.6	EI		2176
CB^{+}	CB		10.5	EI		1116
		CB_2^{+}				
CB_2^{+}	CB_2		10.2±0.6	EI		2176
CB_2^{+}	CB_2		10.7	EI		1116
		C_2B^{+}				
C_2B^{+}	C_2B		10.4±0.6	EI		2176
C_2B^{+}	C_2B		10.7	EI		1116
		$C_2H_5Li_2^{+}$				
$C_2H_5Li_2^{+}$	$(C_2H_5Li)_4$		11.7±0.5	EI		1
		$C_4H_{10}Li_3^{+}$				
$C_4H_{10}Li_3^{+}$	$(C_2H_5Li)_4$		11.7±0.5	EI		1
		$C_6H_{15}Li_4^{+}$				
$C_6H_{15}Li_4^{+}$	$(C_2H_5Li)_4$	C_2H_5	8.0±0.5	EI		1
		$C_8H_{20}Li_5^{+}$				
$C_8H_{20}Li_5^{+}$	$(C_2H_5Li)_6$		12.5±0.5	EI		1

4.3. The Positive Ion Table—Continued

Ion	Reactant	Other products	Ionization or appearance potential (eV)	Method	Heat of formation (kJ mol^{-1})	Ref.
			$C_{10}H_{25}Li_6^+$			
$C_{10}H_{25}Li_6^+$	$(C_2H_5Li)_6$	C_2H_5	7.7±0.5	EI		1
			CH_2Be^+			
CH_2Be^+	$(CH_3)_2Be$	CH_4	11.92±0.05	EI		2874, 2913
			CH_3Be^+			
CH_3Be^+	$(CH_3)_2Be$	CH_3	12.67±0.02	EI		2874, 2913
			$C_2H_4Be^+$			
$C_2H_4Be^+$	$(C_2H_5)_2Be$	C_2H_6	10.35±0.03	EI		2874, 2913
			$C_2H_5Be^+$			
$C_2H_5Be^+$	$(C_2H_5)_2Be$	C_2H_5	11.51±0.05	EI		2874, 2913
			$C_2H_6Be^+$			
$C_2H_6Be^+$	$(CH_3)_2Be$		10.67±0.07	EI		2874, 2913
			$C_3H_6Be^+$			
$C_3H_6Be^+$	$(n-C_3H_7)_2Be$	C_3H_8	9.86±0.05	EI		2874, 2913
$C_3H_6Be^+$	$(iso-C_3H_7)_2Be$	C_3H_8	9.60±0.01	EI		2874, 2913
			$C_3H_7Be^+$			
$C_3H_7Be^+$	$(n-C_3H_7)_2Be$	C_3H_7	10.81±0.05	EI		2874, 2913
$C_3H_7Be^+$	$(iso-C_3H_7)_2Be$	C_3H_7	10.65±0.01	EI		2874, 2913
			$C_4H_8Be^+$			
$C_4H_8Be^+$	$(iso-C_4H_9)_2Be$	C_4H_{10}	9.14±0.03	EI		2874, 2913
			$C_4H_9Be^+$			
$C_4H_9Be^+$	$(iso-C_4H_9)_2Be$	C_4H_9	10.00±0.05	EI		2874, 2913

4.3. The Positive Ion Table—Continued

Ion	Reactant	Other products	Ionization or appearance potential (eV)	Method	Heat of formation (kJ mol^{-1})	Ref.
			$C_4H_{10}Be^+$			
$C_4H_{10}Be^+$	$(C_2H_5)_2Be$		9.46 ± 0.05	EI		2874, 2913
			$C_6H_{14}Be^+$			
$C_6H_{14}Be^+$	$(n-C_3H_7)_2Be$		8.71 ± 0.06	EI		2874, 2913
$C_6H_{14}Be^+$	$(iso-C_3H_7)_2Be$		8.80 ± 0.02	EI		2874, 2913
			$C_8H_{18}Be^+$			
$C_8H_{18}Be^+$	$(iso-C_4H_9)_2Be$		8.74 ± 0.05	EI		2874, 2913
			CH_3B^+			
CH_3B^+	$(CH_3)_3B$		17.0 ± 0.5	EI		364
			$CH_{11}B_5^+$			
$CH_{11}B_5^+$	$1-B_5H_8CH_3$		9.80 ± 0.02	RPD		3228
$CH_{11}B_5^+$	$2-B_5H_8CH_3$		10.25 ± 0.02	RPD		3228
			$C_2H_6B^+$			
$C_2H_6B^+$	$(CH_3)_3B$	CH_3	10.3 ± 0.2	EI		364
			$C_2H_{13}B_5^+$			
$C_2H_{13}B_5^+$	$1-B_5H_8C_2H_5$		9.67 ± 0.05	RPD		3228
$C_2H_{13}B_5^+$	$2-B_5H_8C_2H_5$		9.87 ± 0.01	RPD		3228
			$C_2H_{18}B_{10}^+$			
$C_2H_{18}B_{10}^+$	$B_{10}H_{13}C_2H_5$		9.0 ± 0.5	EI		103
			$C_3H_9B^+$			
$C_3H_9B^+$	$(CH_3)_3B$		10.4	PE		3359
$C_3H_9B^+$	$(CH_3)_3B$		8.8 ± 0.2	EI		364
			$C_4H_{10}B^+$			
$C_4H_{10}B^+$	$(C_2H_5)_3B$	C_2H_5	9.6 ± 0.2	EI		364

4.3. The Positive Ion Table—Continued

Ion	Reactant	Other products	Ionization or appearance potential (eV)	Method	Heat of formation (kJ mol^{-1})	Ref.
			$C_6H_9B^+$			
$C_6H_9B^+$	$(C_2H_3)_3B$		9.7	PE		3359
			$C_6H_{15}B^+$			
$C_6H_{15}B^+$	$(C_2H_5)_3B$		9.6	PE		3359
$C_6H_{15}B^+$	$(C_2H_5)_3B$		9.0±0.2	EI		364
$C_6H_{15}B^+$	$(C_2H_5)_3B$		9.66±0.1	EI		2513, 3227
			N^+ $\Delta H^\circ_{f0} = 1873.2$ kJ mol^{-1} (447.7 kcal mol^{-1})			
N^+	N		14.534	S	1873.2	2113, 2681
N^+	N		14.55	PE		3076
N^+	N		14.56	EI		2780

See also – PI: 2760
        EI: 2, 78, 154, 1133

N^+	N_2	N	24.32±0.02	RPD	(1876)	49
N^+	N_2	N	24.32±0.03	RPD	(1876)	2431

The fragments are formed with no kinetic energy at threshold according to refs. 2021, 2431. See, however, ref. 2823.

See also – PI: 163
        EI: 78, 364, 2021, 2465, 2471, 2484, 2642, 2765, 2895

N^+	N_2	N^+	48±2	RPD		2431

(~9 eV average translational energy of decomposition at threshold)

N^+	NH_3	H_2+H	22.6±0.1	EI		132
N^+	HN_3		19.7±0.3	EI		340
N^+	CNC≡CCN		26.0±1.0	EI		154
N^+	CNC≡CC≡CCN		<19	EI		154
N^+	NO	O^-	19.55±0.04	RPD	(1871)	328
N^+	NO	O^-	19.94±0.14	RPD		2431

(0.6±0.2 eV average translational energy of decomposition at threshold)

The discrepancy between refs. 328 and 2431 is unexplained.

See also – EI: 200, 1378
        D: 6

N^+	NO	O	21.11±0.04	RPD	(1880)	328
N^+	NO	O	21.78±0.11	RPD		2431

(0.9±0.2 eV average translational energy of decomposition at threshold)

The discrepancy between refs. 328 and 2431 is unexplained.

See also – PI: 163
        EI: 200, 2021

4.3. The Positive Ion Table—Continued

Ion	Reactant	Other products	Ionization or appearance potential (eV)	Method	Heat of formation (kJ mol^{-1})	Ref.
N$^+$	NO	O$^+$	34.1±0.7	EI		2021
N$^+$	N$_2$O	NO	20.06	PI		2629

The thermochemical threshold for this process is 19.46 eV.

See also – PI: 163
         EI: 2018

Ion	Reactant	Other products	Ionization or appearance potential (eV)	Method	Heat of formation (kJ mol^{-1})	Ref.
N$^+$	NF$_3$		22.2±0.2	EI		401
N$^+$	PN	P	20.0±0.5	EI		2465

N^{+2} $\Delta H_{f0}^\circ = 4729.3$ kJ mol^{-1} (1130.3 kcal mol^{-1})

Ion	Reactant	Other products	Ionization or appearance potential (eV)	Method	Heat of formation (kJ mol^{-1})	Ref.
N^{+2}	N		44.135	S	4729.3	2113
N^{+2}	N$^+$		29.601	S		2113
N^{+2}	N$_2$	N	63.63±0.20	EI		2431
(10.0±0.8 eV average translational energy of decomposition at threshold)						
N^{+2}	N$_2$	N	54.2±0.5	EI		2471
N^{+2}	N$_2$	N	58.6$^{+0.3}_{-2.0}$	NRE		2823
N^{+2}	N$_2$	N	63.2±0.5	EI		2471
(High kinetic energy ion)						
N^{+2}	N$_2$	N	65	NRE		2823
(High kinetic energy ion)						

See also – EI: 2765

Ion	Reactant	Other products	Ionization or appearance potential (eV)	Method	Heat of formation (kJ mol^{-1})	Ref.
N^{+2}	N$_2^+$	N	38.5±0.7	SEQ		2471
N^{+2}	NO	O	56.0±0.2	EI		2431
(3.8±1 eV average translational energy of decomposition at threshold)						
N^{+2}	NO	O	57.6±1.0	NRE		2823
N^{+2}	N$_2$O		57.1±1.0	NRE		3138
N^{+2}	NO$_2$		51.5±1	NRE		3138

N$_2^+(X^2\Sigma_g^+)$ $\Delta H_{f0}^\circ = 1503.3$ kJ mol^{-1} (359.3 kcal mol^{-1})
N$_2^+(A^2\Pi_{3/2u})$ $\Delta H_{f0}^\circ = 1610.7$ kJ mol^{-1} (385.0 kcal mol^{-1})
N$_2^+(A^2\Pi_{1/2u})$ $\Delta H_{f0}^\circ = 1611.7$ kJ mol^{-1} (385.2 kcal mol^{-1})
N$_2^+(B^2\Sigma_u^+)$ $\Delta H_{f0}^\circ = 1809.1$ kJ mol^{-1} (432.4 kcal mol^{-1})
N$_2^+(C^2\Sigma_u^+)$ $\Delta H_{f0}^\circ = 2275.5$ kJ mol^{-1} (543.8 kcal mol^{-1})

Ion	Reactant	Other products	Ionization or appearance potential (eV)	Method	Heat of formation (kJ mol^{-1})	Ref.
N$_2^+(X^2\Sigma_g^+)$	N$_2$		15.5802	S	1503.3	3301
N$_2^+(X^2\Sigma_g^+)$	N$_2$		15.5803	S	1503.3	3143
N$_2^+(A^2\Pi_{3/2u})$	N$_2$		16.6933	S	1610.7	3143
N$_2^+(A^2\Pi_{1/2u})$	N$_2$		16.7035	S	1611.7	3143
N$_2^+(B^2\Sigma_u^+)$	N$_2$		18.7501	S	1809.1	3143
N$_2^+(C^2\Sigma_u^+)$	N$_2$		23.583	S	2275.5	—
(Based on X$^2\Sigma_g^+$ limit above and second negative bands from ref. 3309)						
N$_2^+(D^2\Pi_g)$	N$_2$		22.31	S	2153	—
(Based on average of A$^2\Pi_{3/2u}$ and A$^2\Pi_{1/2u}$ limits above and Janin–d'Incan bands from ref. 3302)						
N$_2^+(X^2\Sigma_g^+)$	N$_2$		15.58±0.01	PE		3171
N$_2^+(A^2\Pi_u)$	N$_2$		16.695±0.01	PE		3171
N$_2^+(B^2\Sigma_u^+)$	N$_2$		18.75±0.015	PE		3171

4.3. The Positive Ion Table—Continued

Ion	Reactant	Other products	Ionization or appearance potential (eV)	Method	Heat of formation (kJ mol^{-1})	Ref.
$N_2^+(X^2\Sigma_g^+)$	N_2		15.58	PI		3031
$N_2^+(X^2\Sigma_g^+)$	N_2		15.56	PEN		2430
$N_2^+(A^2\Pi_u)$	N_2		16.7	PEN		2430
$N_2^+(B^2\Sigma_u^+)$	N_2		18.8	PEN		2430

See also – S: 2100, 2654, 3303
PI: 163, 2033
PE: 248, 2792, 2812, 2813, 2816, 2829, 2830, 2843, 2855, 3116
PEN: 2467, 3171
EI: 218, 383, 1012, 2431, 2465, 2471, 2557, 2895, 3131, 3133, 3435

Ion	Reactant	Other products	Ionization or appearance potential (eV)	Method	Heat of formation (kJ mol^{-1})	Ref.
N_2^+	N_2H_4	$2H_2$	16.2±0.1	EI	424,	3216
N_2^+	HN_3	NH	16.0±0.1	EI		340
N_2^+?	CH_3NHNH_2		13.2±0.3	EI	424,	3216
N_2^+	$CH_3N{=}NCH_3$		16.3±0.3	EI		2549
N_2^+?	$(CH_3)_2NNH_2$		13.2±0.1	EI	424,	3216
N_2^+?	$CH_3NHNHCH_3$		12.5±0.2	EI	424,	3216
N_2^+?	$(CH_3)_2NNHCH_3$		13.2±0.2	EI	424,	3216
N_2^+?	$(CH_3)_2NN(CH_3)_2$		13.1±0.2	EI	424,	3216
N_2^+	N_2O	O	17.29	PI	2619,	2624, 2629

See also – PI: 163
EI: 2697

$$N_2^{+2} \qquad \Delta H_{f0}^\circ \sim 4126 \text{ kJ mol}^{-1} \text{ (986 kcal mol}^{-1})$$

Ion	Reactant	Other products	Ionization or appearance potential (eV)	Method	Heat of formation (kJ mol^{-1})	Ref.
N_2^{+2}	N_2		42.9	AUG	~4139	3304
N_2^{+2}	N_2		44.2±0.5	RPD		2431
N_2^{+2}	N_2		42.7±0.2	NRE	~4120	2823
N_2^{+2}	N_2		42.7±0.1	FD	~4120	2785, 3130
N_2^{+2}	N_2		43.5±0.3	FD		212
N_2^{+2}	N_2		43.2±0.5	EI		2471
N_2^{+2}	N_2^+		27.2±0.5	SEQ		2471

$$N_2^{+3}$$

Ion	Reactant	Other products	Ionization or appearance potential (eV)	Method	Heat of formation (kJ mol^{-1})	Ref.
N_2^{+3}	N_2		84	AUG		3304

$$N_3^+$$

Ion	Reactant	Other products	Ionization or appearance potential (eV)	Method	Heat of formation (kJ mol^{-1})	Ref.
N_3^+	HN_3	H	16.0±0.2	EI		340
N_3^+	CH_3N_3	CH_3	17.6±0.5	EI		340

$$NH^+$$

Ion	Reactant	Other products	Ionization or appearance potential (eV)	Method	Heat of formation (kJ mol^{-1})	Ref.
NH^+	NH		13.10±0.05	EI		132
NH^+	NH		13.1±0.2	EI		2454

4.3. The Positive Ion Table—Continued

Ion	Reactant	Other products	Ionization or appearance potential (eV)	Method	Heat of formation (kJ mol^{-1})	Ref.	
NH$^+$	NH		12.8	EI		2768,	2771, 2772
NH$^+$	NH$_3$	H$_2$	17.1±0.1	EI			132
NH$^+$	HN$_3$	N$_2$	14.4±0.2	EI			340
NH$^+$	HNCO	CO	17.26±0.15	EI			3365
NH$^+$	HNCO	CO	<17.7	EI		2797,	3012

$$\text{NH}_2^+ \qquad \Delta H_{f0}^{\circ} \sim 1263 \text{ kJ mol}^{-1} \text{ (302 kcal mol}^{-1})$$

Ion	Reactant	Other products	Ionization or appearance potential (eV)	Method	Heat of formation (kJ mol^{-1})	Ref.	
NH$_2^+$	NH$_2$		11.4±0.1	EI			34
NH$_2^+$	NH$_2$		11.7	EI		2768,	2771, 2772
NH$_2^+$	NH$_2$		11.22	D			2631
NH$_2^+$	NH$_3$	H	15.73±0.02	PI	1263		2631
NH$_2^+$	NH$_3$	H	16.0±0.1	EI			34
NH$_2^+$	N$_2$H$_4$	NH$_2$	13.9±0.4	EI			34
NH$_2^+$	CH$_3$NH$_2$	CH$_3$	15.7	RPD			3345
(0.12 eV average translational energy of decomposition at threshold)							
NH$_2^+$	(CH$_3$)$_2$NNH$_2$		13.3±0.1	EI			303

$$\text{NH}_3^+(^2A_1) \qquad \Delta H_{f0}^{\circ} = 941 \text{ kJ mol}^{-1} \text{ (225 kcal mol}^{-1})$$
$$\text{NH}_3^+(^2E_1) \qquad \Delta H_{f0}^{\circ} \sim 1401 \text{ kJ mol}^{-1} \text{ (335 kcal mol}^{-1})$$

Ion	Reactant	Other products	Ionization or appearance potential (eV)	Method	Heat of formation (kJ mol^{-1})	Ref.	
NH$_3^+(^2A_1)$	NH$_3$		10.166	S	942		3053
NH$_3^+(^2A_1)$	NH$_3$		10.162±0.008	PI	941		2631
NH$_3^+(^2A_1)$	NH$_3$		10.154±0.01	PI	941		159
NH$_3^+(^2A_1)$	NH$_3$		10.16±0.02	PI	941		2727
NH$_3^+(^2A_1)$	NH$_3$		10.17	PI	942		3030
NH$_3^+(^2A_1)$	NH$_3$		10.175±0.01	PE	943		3054
NH$_3^+(^2A_1)$	NH$_3$		10.14	PE	939	3027,	3061
NH$_3^+(^2A_1)$	NH$_3$		10.16	PE	941		1130
NH$_3^+(^2E_1)$	NH$_3$		14.94±0.03	PE	1402		3054
NH$_3^+(^2E_1)$	NH$_3$		14.92	PE	1401	3027,	3061
NH$_3^+(^2E_1)$	NH$_3$		15.02	PE			1130

See also – S: 138, 1070
      PI: 155, 158, 182, 297, 416
      PE: 2853, 2854
      PEN: 2430
      EI: 14, 218, 411, 463, 2486

Ion	Reactant	Other products	Ionization or appearance potential (eV)	Method	Heat of formation (kJ mol^{-1})	Ref.
NH$_3^+$	C$_2$H$_5$NH$_2$		12.99	EI		2470

$$\text{NH}_2\text{D}^+$$

Ion	Reactant	Other products	Ionization or appearance potential (eV)	Method	Heat of formation (kJ mol^{-1})	Ref.
NH$_2$D$^+$	CD$_3$CH$_2$NH$_2$		13.3	EI		2470

4.3. The Positive Ion Table—Continued

Ion	Reactant	Other products	Ionization or appearance potential (eV)	Method	Heat of formation (kJ mol^{-1})	Ref.
			NHD$_2^+$			
NHD$_2^+$	CH$_3$ND$_2$	CH$_2$	13.40±0.4	EI		2429
			ND$_3^+$			
ND$_3^+(^2A_1)$	ND$_3$		10.180±0.01	PE		3054
ND$_3^+(^2A_1)$	ND$_3$		10.17	PE	3027,	3061
ND$_3^+(^2E_1)$	ND$_3$		14.94±0.03	PE		3054
ND$_3^+(^2E_1)$	ND$_3$		15.15	PE	3027,	3061
			NH$_3^{+2}$			
NH$_3^{+2}$	NH$_3$		33.7±0.2	FD		212
			N$_2$H$^+$			
N$_2$H$^+$	N$_2$H$_4$	H$_2$+H	14.8±0.3	EI	424,	3216
N$_2$H$^+$	HN$_3$	N	13.8±0.2	EI		340
N$_2$H$^+$?	CH$_3$NHNH$_2$		13.3±0.3	EI	424,	3216
			N$_2$H$_2^+$			
N$_2$H$_2^+$	N$_2$H$_2$		9.85±0.1	EI	33,	34
N$_2$H$_2^+$	N$_2$H$_2$		9.9	EI	2768,	2771, 2772
N$_2$H$_2^+$	N$_2$H$_4$	H$_2$	10.98±0.2	EI	33,	34
N$_2$H$_2^+$	N$_2$H$_4$	H$_2$	11.9±0.2	EI	424,	3216
N$_2$H$_2^+$?	CH$_3$NHNH$_2$		11.2±0.2	EI	424,	3216
N$_2$H$_2^+$	(CH$_3$)$_2$NNH$_2$	C$_2$H$_6$	8.6±0.1	PI		1141
N$_2$H$_2^+$?	(CH$_3$)$_2$NNH$_2$		12.9±0.1	EI	424,	3216
N$_2$H$_2^+$?	CH$_3$NHNHCH$_3$		11.0±0.2	EI	424,	3216
N$_2$H$_2^+$?	(CH$_3$)$_2$NNHCH$_3$		11.9±1.0	EI	424,	3216
N$_2$H$_2^+$?	(CH$_3$)$_2$NN(CH$_3$)$_2$		11.9±0.2	EI	424,	3216
N$_2$H$_2^+$	n-C$_4$H$_9$N(CH$_3$)NH$_2$	C$_5$H$_{12}$	9.5±0.2	PI		1141

4.3. The Positive Ion Table—Continued

Ion	Reactant	Other products	Ionization or appearance potential (eV)	Method	Heat of formation (kJ mol^{-1})	Ref.
			N$_2$H$_3^+$			
N$_2$H$_3^+$	N$_2$H$_3$		7.88±0.2	EI		34
N$_2$H$_3^+$	N$_2$H$_3$		7.6	EI	2768,	2771, 2772
N$_2$H$_3^+$	N$_2$H$_4$	H	10.6±0.1	PI		1141
N$_2$H$_3^+$	N$_2$H$_4$	H	11.18±0.1	EI		34
N$_2$H$_3^+$	N$_2$H$_4$	H	11.3±0.1	EI	424,	3216
N$_2$H$_3^+$	N$_2$H$_4$	H	11.3	EI		1455
N$_2$H$_3^+$	CH$_3$NHNH$_2$	CH$_3$	9.5±0.1	PI		1141
N$_2$H$_3^+$	CH$_3$NHNH$_2$	CH$_3$	10.7±0.3	EI	424,	3216
N$_2$H$_3^+$?	CH$_3$NHNHCH$_3$		10.2±0.2	EI	424,	3216
N$_2$H$_3^+$?	(CH$_3$)$_2$NNHCH$_3$		11.7±1.0	EI	424,	3216
			N$_2$H$_4^+$			
N$_2$H$_4^+$	N$_2$H$_4$		8.36±0.03	PI		2173
(Value obtained without mass analysis)						
N$_2$H$_4^+$	N$_2$H$_4$		8.74±0.06	PI	1141,	2173
(Value obtained with mass analysis)						
N$_2$H$_4^+$	N$_2$H$_4$		9.56±0.02	PI		1166
N$_2$H$_4^+$	N$_2$H$_4$		8.93	PE		3235
N$_2$H$_4^+$	N$_2$H$_4$		9.00±0.1	EI	424,	3216
N$_2$H$_4^+$	N$_2$H$_4$		8.8	EI	2768,	2771, 2772
N$_2$H$_4^+$	(CH$_3$)$_2$NNHCH$_3$		11.9±0.2	EI	424,	3216
N$_2$H$_4^+$	(CH$_3$)$_2$NN(CH$_3$)$_2$		12.3±0.1	EI	424,	3216
	HN$_3^+$ $\Delta H_{f0}^\circ = 1337$ kJ mol^{-1} (319 kcal mol^{-1})					
N$_3$H$^+$	HN$_3$		10.740±0.005	PE	1337	3067

See also − S: 3103
        EI: 340

			N$_3$H$_3^+$			
N$_3$H$_3^+$	N$_3$H$_3$		9.6±0.1	EI		34

4.3. The Positive Ion Table—Continued

Ion	Reactant	Other products	Ionization or appearance potential (eV)	Method	Heat of formation (kJ mol^{-1})	Ref.

$$CN^+ \qquad \Delta H^\circ_{f0} = 1794 \text{ kJ mol}^{-1} \text{ (429 kcal mol}^{-1}\text{)}$$

Ion	Reactant	Other products	AP (eV)	Method	ΔHf	Ref.
CN^+	CN		14.5 ± 0.2	EI		154
CN^+	CN		14.2 ± 0.3	EI		2145
CN^+	CN		14.03 ± 0.02	D		2602
CN^+	CN		$\leqslant14.20\pm0.02$	D		2621
CN^+	C_2N_2	CN	20.42 ± 0.02	PI		2621
CN^+	C_2N_2	CN	20.4	RPD		3345

(0.21 eV average translational energy of decomposition at threshold)

See also – EI: 154, 2145

Ion	Reactant	Other products	AP (eV)	Method	ΔHf	Ref.
CN^+	$CNC{\equiv}CCN$		19.2 ± 0.3	EI		154
CN^+	$CNC{\equiv}CC{\equiv}CCN$		20.0 ± 1.0	EI		154
CN^+	HCN	H	19.43 ± 0.01	PI	1794	2602

(Threshold value approximately corrected to 0 K)

Ion	Reactant	Other products	AP (eV)	Method	ΔHf	Ref.
CN^+	HCN	H	19.40 ± 0.02	PI		2623
CN^+	$CH{\equiv}CCN$	C_2H	19.8 ± 0.2	EI		154
CN^+	C_3H_5CN		19.5 ± 0.4	EI		202

(Cyclopropanecarboxylic acid nitrile)

Ion	Reactant	Other products	AP (eV)	Method	ΔHf	Ref.
CN^+	FCN	F	19.21 ± 0.02	PI		2621
CN^+	$ClCN$	Cl^-	15.5 ± 0.2	EI		73
CN^+	$ClCN$	Cl	18.50 ± 0.02	PI	(1802)	2621
CN^+	$ClCN$	Cl	18.3 ± 0.2	EI		73
CN^+	$BrCN$	Br^-	14.6 ± 0.1	EI		73
CN^+	$BrCN$	Br	18.3 ± 0.1	EI		73
CN^+	ICN	I	18.1 ± 0.1	EI		73

$$C_2N^+$$

Ion	Reactant	Other products	AP (eV)	Method	ΔHf	Ref.
C_2N^+	C_2N		~13	EI		154
C_2N^+	C_2N_2	N	19.5 ± 0.1	EI		154
C_2N^+	$CNC{\equiv}CCN$		18.1 ± 0.4	EI		154
C_2N^+	$CNC{\equiv}CC{\equiv}CCN$		17.0 ± 0.1	EI		154
C_2N^+	CH_3NC		18.28	EDD		3214
C_2N^+	$(CH_2)_2NH$	$H_2+3H?$	23.0 ± 0.4	EI		51

(Ethylenimine)

Ion	Reactant	Other products	AP (eV)	Method	ΔHf	Ref.
C_2N^+	$CH{\equiv}CCN$	$CH?$	18.0 ± 0.5	EI		154

$$C_3N^+$$

Ion	Reactant	Other products	AP (eV)	Method	ΔHf	Ref.
C_3N^+	C_3N		~14.4	EI		154
C_3N^+	$CNC{\equiv}CCN$	CN	18.4 ± 0.2	EI		154
C_3N^+	$CNC{\equiv}CC{\equiv}CCN$		22.0 ± 0.5	EI		154
C_3N^+	$CH{\equiv}CCN$	H	18.2 ± 0.3	EI		154
C_3N^+	$CH_2{=}CHCN$		21.6 ± 0.1	EI		2954

$$C_4N^+$$

Ion	Reactant	Other products	AP (eV)	Method	ΔHf	Ref.
C_4N^+	C_4N		11.9 ± 0.5	EI		154
C_4N^+	$CNC{\equiv}CCN$	N	18.8 ± 0.5	EI		154
C_4N^+	$CNC{\equiv}CC{\equiv}CCN$		19.0 ± 1.0	EI		154

4.3. The Positive Ion Table—Continued

Ion	Reactant	Other products	Ionization or appearance potential (eV)	Method	Heat of formation (kJ mol^{-1})	Ref.
			C_5N^+			
C_5N^+	CNC≡CC≡CCN	CN	17.3±0.2	EI		154
			C_6N^+			
C_6N^+	C_6N		12.2±0.1	EI		154
C_6N^+	CNC≡CC≡CCN	N	19.2±0.3	EI		154

		$C_2N_2^+$ $\qquad \Delta H_{f0}^\circ = 1597$ kJ mol^{-1} (382 kcal mol^{-1})				
$C_2N_2^+$	C_2N_2		13.374±0.008	PI	1597	2621
$C_2N_2^+$	C_2N_2		13.36±0.01	PE	1596	2805

See also − S: 3152
          EI: 154

			$C_4N_2^+$			
$C_4N_2^+$	CNC≡CCN		11.81±0.01	PE		2805
$C_4N_2^+$	CNC≡CCN		11.4±0.2	EI		154

These results are unusual since electron impact almost always gives a higher value than photoelectron spectroscopy.

			$C_6N_2^+$			
$C_6N_2^+$	CNC≡CC≡CCN		11.4±0.2	EI		154

			BH_4N^+			
BH_4N^+	BH_2NH_2		11.0±0.1	EI		3184

			BH_6N^+			
BH_6N^+	BH_3NH_3		10.33±0.04 (V)	PE		3044

			$B_3H_6N_3^+$			
$B_3H_6N_3^+$	$B_3H_6N_3$ (Borazine)		9.88±0.02	PE		3105
$B_3H_6N_3^+$	$B_3H_6N_3$ (Borazine)		10.01±0.01	PE		3078
$B_3H_6N_3^+$	$B_3H_6N_3$ (Borazine)		9.77	EI		2511

4.3. The Positive Ion Table—Continued

Ion	Reactant	Other products	Ionization or appearance potential (eV)	Method	Heat of formation (kJ mol^{-1})	Ref.
HCN^+ $\Delta H_{f0}^\circ = 1447$ kJ mol^{-1} (346 kcal mol^{-1})						
CHN^+	HCN		13.59±0.01	PI	1447	2623
CHN^+	HCN		13.60±0.01	PE	1448	2805
CHN^+	HCN		13.73±0.09	EI		411
See also – EI: 282						
CHN^+	CH_3N_3	N_2+H_2?	13.6±0.5	EI		340
CDN^+						
CDN^+	DCN		13.60±0.01	PE		2805
CH_2N^+						
CH_2N^+	CH_3NH_2	H_2+H	15.21±0.3	EI		2429
CH_2N^+	CH_3ND_2		15.14±0.08	EI		2429
CH_2N^+	$C_2H_5NH_2$		15.45	EI		2470
CH_2N^+	$CD_3CH_2NH_2$		14.5±0.5	EI		2470
CH_2N^+	$C_2D_5NH_2$		15.1	EI		2470
CH_2N^+	C_2H_5CN		14.88	EI		2966
CH_2N^+	$(CH_2)_3NH$ (Trimethylenimine)		13.1±0.2	EI		52
CH_2N^+	$(CH_2)_4NH$ (Pyrrolidine)		13.9±0.2	EI		52
CH_2N^+	$(C_2H_5)_2NH$		14.64	EI		2428
CH_2N^+	$CH_3N=NCH_3$		14.7±0.3	EI		2549
CH_2N^+	CH_3N_3	N_2+H	10.5±0.1	EI		340
$CHDN^+$						
$CHDN^+$	CH_3ND_2		14.92±0.20	EI		2429
$CHDN^+$	$CH_3CD_2NH_2$		13.7±0.2	EI		2470
$CHDN^+$	$C_2D_5NH_2$		14.9	EI		2470
CH_3N^+						
CH_3N^+	CH_3NH_2	H_2	13.30±0.2	EI		2429
CH_3N^+	$C_2H_5NH_2$	CH_4	13.5±0.2	EI		2470
CH_3N^+	$CD_3CH_2NH_2$	CHD_3	13.3±0.5	EI		2470
CH_2DN^+						
CH_2DN^+	CH_3ND_2	HD	12.50±0.10	EI		2429
CH_2DN^+	$C_2D_5NH_2$	CD_4	12.87	EI		2470

4.3. The Positive Ion Table—Continued

Ion	Reactant	Other products	Ionization or appearance potential (eV)	Method	Heat of formation (kJ mol⁻¹)	Ref.

$$\text{CH}_2\text{NH}_2^+ \qquad \Delta H^\circ_{f298} \sim 745 \text{ kJ mol}^{-1} \text{ (178 kcal mol}^{-1}\text{)}$$

Ion	Reactant	Other products	IP/AP (eV)	Method	ΔH_f	Ref.
CH_4N^+	CH_3NH_2	H	10.3±0.1	RPD	(753)	3017
CH_4N^+	CH_3NH_2	H	10.82±0.15	EI		2429
CH_4N^+	$\text{C}_2\text{H}_5\text{NH}_2$	CH_3	9.71	PI	747	11
(Threshold value approximately corrected for thermal energy and kinetic shift)						
CH_4N^+	$\text{C}_2\text{H}_5\text{NH}_2$	CH_3	10.2±0.1	EI		2470
CH_4N^+	$\text{CH}_3\text{CD}_2\text{NH}_2$	CHD_2	10.0	EI		2470
CH_4N^+	$\text{CD}_3\text{CH}_2\text{NH}_2$	CD_3	9.8±0.1	EI		2470
CH_4N^+	$(\text{CH}_2)_3\text{NH}$		12.3±0.2	EI		52
(Trimethylenimine)						
CH_4N^+	$n\text{–C}_3\text{H}_7\text{NH}_2$	C_2H_5	9.54	PI	743	11
(Threshold value approximately corrected for thermal energy and kinetic shift)						
CH_4N^+	$(\text{CH}_2)_4\text{NH}$		12.7±0.2	EI		52
(Pyrrolidine)						
CH_4N^+	$(\text{C}_2\text{H}_5)_2\text{NH}$	$\text{C}_2\text{H}_4+\text{CH}_3$	13.10±0.10	EI		2428
CH_4N^+	$\text{CH}_2(\text{NH}_2)\text{COOH}$		10.23±0.09	EI		2587
CH_4N^+	$\text{CH}_2(\text{NH}_2)\text{COOH}$		10.1±0.2	EI		88

The ion structure is probably CH_2NH_2^+, see ref. 2429.

$$\text{CH}_3\text{DN}^+$$

Ion	Reactant	Other products	IP/AP (eV)	Method	ΔH_f	Ref.
CH_3DN^+	CH_3ND_2	D	11.52±0.60	EI		2429
CH_3DN^+	$\text{CD}_3\text{CH}_2\text{NH}_2$	CHD_2	9.82±0.1	EI		2470
CH_3DN^+	$(\text{C}_2\text{H}_5)_2\text{ND}$	$\text{C}_2\text{H}_4+\text{CH}_3$	13.15±0.1	EI		2428

$$\text{CH}_2\text{D}_2\text{N}^+$$

Ion	Reactant	Other products	IP/AP (eV)	Method	ΔH_f	Ref.
$\text{CH}_2\text{D}_2\text{N}^+$	CH_3ND_2	H	10.10±0.30	EI		2429
$\text{CH}_2\text{D}_2\text{N}^+$	$\text{CH}_3\text{CD}_2\text{NH}_2$	CH_3	10.3±0.1	EI		2470
$\text{CH}_2\text{D}_2\text{N}^+$	$\text{CD}_3\text{CH}_2\text{NH}_2$	$\text{CH}_2\text{D}?$	11.6±0.3	EI		2470
$\text{CH}_2\text{D}_2\text{N}^+$	$\text{C}_2\text{D}_5\text{NH}_2$	CD_3	10.13	EI		2470
$\text{CH}_2\text{D}_2\text{N}^+$	$\text{C}_6\text{H}_5\text{CH}_2\text{ND}_2$		11.0±0.1	PI		1160
(α–Amino–d_2–toluene)						

$$\text{CH}_3\text{NH}_2^+ \qquad \Delta H^\circ_{f298} = 843 \text{ kJ mol}^{-1} \text{ (202 kcal mol}^{-1}\text{)}$$

Ion	Reactant	Other products	IP/AP (eV)	Method	ΔH_f	Ref.
CH_5N^+	CH_3NH_2		8.97±0.02	PI	843	159, 182, 416
CH_5N^+	CH_3NH_2		8.99	PE	844	3320
CH_5N^+	CH_3NH_2		9.36±0.02	EI		2429
CH_5N^+	CH_3NH_2		9.29	CTS		2562

See also – PE: 1130
        PEN: 2430
        EI: 14, 383, 384, 1072

Ion	Reactant	Other products	IP/AP (eV)	Method	ΔH_f	Ref.
CH_5N^+	CH_3NHNH_2	NH	11.3±0.1	PI		1141

4.3. The Positive Ion Table—Continued

Ion	Reactant	Other products	Ionization or appearance potential (eV)	Method	Heat of formation (kJ mol^{-1})	Ref.
			CH$_3$D$_2$N$^+$			
CH$_3$D$_2$N$^+$	CH$_3$ND$_2$		9.27±0.05	EI		2429

C$_2$HN$^+$ $\Delta H_{f0}^\circ \sim 1551$ kJ mol^{-1} (371 kcal mol^{-1})

Ion	Reactant	Other products	Ionization or appearance potential (eV)	Method	Heat of formation	Ref.
C$_2$HN$^+$	CH$_3$CN	H$_2$	15.1±0.1	PI	~1551	2623
C$_2$HN$^+$	CH$_3$NC	H$_2$	14.46	EDD	1551	3214
C$_2$HN$^+$	(CH$_2$)$_2$NH (Ethylenimine)		18.1±0.6	EI		51

C$_2$H$_2$N$^+$

Ion	Reactant	Other products	Ionization or appearance potential (eV)	Method	Heat of formation	Ref.
C$_2$H$_2$N$^+$	CH$_2$CN		10.87±0.1	EI	(1263*)	125
C$_2$H$_2$N$^+$	CH$_3$CN	H	14.01±0.02	PI	(1221*)	2623
C$_2$H$_2$N$^+$	CH$_3$CN	H	14.28±0.05	EI	(1247*)	125
C$_2$H$_2$N$^+$	CH$_3$CN	H	13.54±0.08	EI	(1176*)	2704

See also – EI: 2966, 3017

Ion	Reactant	Other products	Ionization or appearance potential (eV)	Method	Heat of formation	Ref.
C$_2$H$_2$N$^+$	CH$_3$NC	H	13.21	EDD	(1206*)	3214
C$_2$H$_2$N$^+$	(CH$_2$)$_2$NH (Ethylenimine)		17.0±0.2	EI		51
C$_2$H$_2$N$^+$	C$_2$H$_5$NC	CH$_3$	14.17	EDD		3214

The results of refs. 2623 and 2704 are unusual since electron impact almost always gives a higher value than photoionization.

*ΔH_{f298}°

CH$_3$CN$^+$(^{2}E) $\Delta H_{f0}^\circ \sim 1270$ kJ mol^{-1} (304 kcal mol^{-1})
CH$_3$NC$^+$ $\Delta H_{f0}^\circ = 1240$ kJ mol^{-1} (296 kcal mol^{-1})

Ion	Reactant	Other products	Ionization or appearance potential (eV)	Method	Heat of formation	Ref.
C$_2$H$_3$N$^+$(^{2}E)	CH$_3$CN		12.19±0.01	PI	1271	2623
C$_2$H$_3$N$^+$(^{2}E)	CH$_3$CN		12.205±0.004	PI	1272	1253
C$_2$H$_3$N$^+$(^{2}E)	CH$_3$CN		12.22±0.01	PI	1274	182
C$_2$H$_3$N$^+$(^{2}E)	CH$_3$CN		12.12	PE	1264	2851
C$_2$H$_3$N$^+$(^{2}E)	CH$_3$CN		12.20 (V)	PE	1272	3045
C$_2$H$_3$N$^+$(^{2}E)	CH$_3$CN		12.23±0.05	EI		2704
C$_2$H$_3$N$^+$(^{2}A$_1$)	CH$_3$CN		13.11	PE	1359	2851
C$_2$H$_3$N$^+$(^{2}A$_1$)	CH$_3$CN		13.14	PE	1362	3045
C$_2$H$_3$N$^+$(^{2}A$_1$)	CH$_3$CN		15.12	PE	1553	2851
C$_2$H$_3$N$^+$(^{2}A$_1$)	CH$_3$CN		15.11	PE	1552	3045
C$_2$H$_3$N$^+$(^{2}E)	CH$_3$CN		16.98	PE	1733	2851

The assignments are those of ref. 2851.

See also – EI: 286, 411, 2171, 2966

4.3. The Positive Ion Table—Continued

Ion	Reactant	Other products	Ionization or appearance potential (eV)	Method	Heat of formation (kJ mol^{-1})	Ref.
$C_2H_3N^+$	CH_3NC		11.24	PE	1240	3420
$C_2H_3N^+$	CH_3NC		11.83	EDD		3214
See also – EI: 2171						
$C_2H_3N^+$	$(CH_2)_2NH$ (Ethylenimine)		15.2±0.3	EI		51
$C_2H_3N^+$	$n-C_3H_7CN$		12.46	EI		2966
$C_2H_3N^+$	$n-C_3H_7NC$		12.05	EDD		3214
$C_2H_3N^+$	$n-C_4H_9CN$		12.26	EI		2966
$C_2H_3N^+$	$n-C_4H_9NC$		12.75	EDD		3214
$C_2H_3N^+$	$CH_3CH=NOH$	H_2O	12.9±0.2	EDD		3180

$C_2D_3N^+$

Ion	Reactant	Other products	Ionization or appearance potential (eV)	Method	Heat of formation (kJ mol^{-1})	Ref.
$C_2D_3N^+(^2E)$	CD_3CN		12.23 (V)	PE		3045
$C_2D_3N^+(^2E)$	CD_3CN		12.29±0.05	EI		2704
$C_2D_3N^+(^2A_1)$	CD_3CN		13.14	PE		3045
$C_2D_3N^+(^2A_1)$	CD_3CN		15.17	PE		3045
$C_2D_3N^+$	CD_3NC		11.25	PE		3420

$C_2H_4N^+$

Ion	Reactant	Other products	Ionization or appearance potential (eV)	Method	Heat of formation (kJ mol^{-1})	Ref.
$C_2H_4N^+$	$(CH_2)_2NH$ (Ethylenimine)	H	12.2±0.1	EI		51
$C_2H_4N^+$	$(CH_2)_3NH$ (Trimethylenimine)	CH_3	11.9±0.2	EI		52
$C_2H_4N^+$	$iso-C_3H_7CN$		12.70	EDD		3214
$C_2H_4N^+$	$(CH_2)_4NH$ (Pyrrolidine)	C_2H_5	13.0±0.2	EI		52
$C_2H_4N^+$	$tert-C_4H_9CN$		12.20	EDD		3214
$C_2H_4N^+$?	$(CH_3)_2NNH_2$		12.8±0.2	EI	424,	3216
$C_2H_4N^+$?	$CH_3NHNHCH_3$		12.1±0.5	EI	424,	3216
$C_2H_4N^+$?	$(CH_3)_2NNHCH_3$		11.7±0.2	EI	424,	3216
$C_2H_4N^+$?	$(CH_3)_2NN(CH_3)_2$		12.2±0.2	EI	424,	3216
$C_2H_4N^+$	$CH_3CH=NOH$	OH	11.3±0.2	EDD		3180

4.3. The Positive Ion Table—Continued

Ion	Reactant	Other products	Ionization or appearance potential (eV)	Method	Heat of formation (kJ mol^{-1})	Ref.
	$C_2H_5N^+$ (Ethylenimine)	$\Delta H^\circ_{f298} \sim 1045$ kJ mol^{-1} (250 kcal mol^{-1})				
$C_2H_5N^+$	$CH_3N=CH_2$		9.8±0.1	EI		2452
$C_2H_5N^+$	$(CH_2)_2NH$ (Ethylenimine)		9.52	PE	1045	3235
$C_2H_5N^+$	$(CH_2)_2NH$ (Ethylenimine)		9.94±0.15	EI		51
$C_2H_5N^+$	$(CH_2)_2NH$ (Ethylenimine)		9.8	EI		218
See also – PE: 2808						
$C_2H_5N^+$	$(CH_2)_4NH$ (Pyrrolidine)		12.3±0.2	EI		52
$C_2H_5N^+$?	$(CH_3)_2NNH_2$		8.8±0.1	PI		1141
$C_2H_5N^+$?	$(CH_3)_2NNH_2$		12.5±0.2	EI		424, 3216
	$C_2H_6N^+$					
$C_2H_6N^+$	$(CH_3)_2N$		9.42±0.1	EI		2452
$C_2H_6N^+$	$C_2H_5NH_2$	H	11.96±0.1	EI		2470
$C_2H_6N^+$	$(CH_3)_2NH$	H	10.1±0.1	RPD		3017
$C_2H_6N^+$	$(CH_3)_3N$	CH_3	12.3±0.1	EDD		2452
$C_2H_6N^+$	$(CH_3)_3N$	CH_3	12.3±0.1	EI		303
$C_2H_6N^+$	$(C_2H_5)_2NH$		13.65±0.08	EI		2428
$C_2H_6N^+$	$n-C_3H_7CH(NH_2)CH_3$		10.43±0.13	EI		2587
$C_2H_6N^+$?	$(CH_3)_2NNH_2$		9.0±0.2	PI		1141
$C_2H_6N^+$?	$(CH_3)_2NNH_2$		10.45±0.05	EDD		2452
$C_2H_6N^+$?	$(CH_3)_2NNH_2$		10.7±0.1	EI		303
$C_2H_6N^+$?	$(CH_3)_2NNH_2$		10.9±0.2	EI		424, 3216
$C_2H_6N^+$?	$(CH_3)_2NNHCH_3$		11.1±0.2	EI		424, 3216
$C_2H_6N^+$?	$(CH_3)_2NN(CH_3)_2$		11.2±0.1	EI		303
$C_2H_6N^+$?	$(CH_3)_2NN(CH_3)_2$		11.2±0.2	EI		424, 3216
$C_2H_6N^+$	$(CH_3)_2NN=NN(CH_3)_2$		9.85±0.1	EI		303
$C_2H_6N^+$	$HCON(CH_3)_2$		11.6±0.1	EI		303
$C_2H_6N^+$	$CH_3CON(CH_3)_2$		12.4±0.1	EI		303
$C_2H_6N^+$	$(CH_3)_2NNO$	NO?	10.3±0.05	EDD		2452
$C_2H_6N^+$	$(CH_3)_2NNO$	NO?	10.9±0.1	EI		303
$C_2H_6N^+$	$(CH_3)_2NNO_2$	NO_2?	10.68±0.05	EDD		2452
$C_2H_6N^+$	$(CH_3)_2NNO_2$	NO_2?	11.1±0.1	EI		303
	$C_2H_4D_2N^+$					
$C_2H_4D_2N^+$	$CH_3CD_2NH_2$	H	11.50±0.10	EI		2470
$C_2H_4D_2N^+$	$CD_3CH_2NH_2$	D	11.64±0.15	EI		2470
	$C_2H_3D_3N^+$					
$C_2H_3D_3N^+$	$CD_3CH_2NH_2$	H	11.80±0.13	EI		2470

4.3. The Positive Ion Table—Continued

Ion	Reactant	Other products	Ionization or appearance potential (eV)	Method	Heat of formation (kJ mol^{-1})	Ref.
		$C_2H_2D_4N^+$				
$C_2H_2D_4N^+$	$C_2D_5NH_2$	D	11.90±0.20	EI		2470
		$C_2HD_5N^+$				
$C_2HD_5N^+$	$C_2D_5NH_2$	H	17.30	EI		2470
	$C_2H_5NH_2^+$	$\Delta H^{\circ}_{f298} = 808$ kJ mol^{-1} (193 kcal mol^{-1})				
	$(CH_3)_2NH^+$	$\Delta H^{\circ}_{f298} = 777$ kJ mol^{-1} (186 kcal mol^{-1})				
$C_2H_7N^+$	$C_2H_5NH_2$		8.86±0.02	PI	808	159, 182
$C_2H_7N^+$	$C_2H_5NH_2$		9.19	PE		1130
$C_2H_7N^+$	$C_2H_5NH_2$		8.74	CTS		2562
See also – PI: 11, 86						
EI: 14, 2470						
$C_2H_7N^+$	$(CH_3)_2NH$		8.24±0.02	PI	777	159, 182
$C_2H_7N^+$	$(CH_3)_2NH$		8.25	PE	778	3320
$C_2H_7N^+$	$(CH_3)_2NH$		8.36	PE		1130
See also – PE: 3410						
EI: 14, 384, 2452, 3338						
$C_2H_7N^+$?	$(CH_3)_2NNH_2$		11.2±0.2	PI		1141
		$C_2H_5D_2N^+$				
$C_2H_5D_2N^+$	$CH_3CD_2NH_2$		9.37±0.10	EI		2470
		$C_2H_4D_3N^+$				
$C_2H_4D_3N^+$	$CD_3CH_2NH_2$		9.13±0.05	EI		2470
		$C_2H_2D_5N^+$				
$C_2H_2D_5N^+$	$C_2D_5NH_2$		9.37	EI		2470
		C_3HN^+				
C_3HN^+	$CH{\equiv}CCN$		11.60±0.01	PE		2805
C_3HN^+	$CH{\equiv}CCN$		11.6±0.2	EI		154
C_3HN^+	$CH_2{=}CHCN$		16.4±0.1	EI		2954
C_3HN^+	$CH_2{=}CFCN$		16.3±0.1	EI		2954
		C_3HN^{+2}				
C_3HN^{+2}	$CH{\equiv}CCN$		32.3±0.2	EI		154

4.3. The Positive Ion Table—Continued

Ion	Reactant	Other products	Ionization or appearance potential (eV)	Method	Heat of formation (kJ mol^{-1})	Ref.
			$C_3H_2N^+$			
$C_3H_2N^+$	CH_2=CHCN	H	13.82±0.08	EI		1406
$C_3H_2N^+$	$C_4H_4N_2$ (1,3–Diazine)		15.01±0.10	EI		1406
$C_3H_2N^+$	$C_4H_4N_2$ (1,4–Diazine)		15.25±0.10	EI		1406

Ion	Reactant	Other products	Ionization or appearance potential (eV)	Method	Heat of formation (kJ mol^{-1})	Ref.
	CH_2=CHCN$^+$	$\Delta H^\circ_{f298} = 1238$ kJ mol^{-1} (296 kcal mol^{-1})				
$C_3H_3N^+$	CH_2=CHCN		10.91±0.01	PI	1238	182
$C_3H_3N^+$	CH_2=CHCN		10.91	PE	1238	3045

See also – EI: 1406, 2954

Ion	Reactant	Other products	Ionization or appearance potential (eV)	Method	Heat of formation (kJ mol^{-1})	Ref.
$C_3H_3N^+$	C_5H_5N (Pyridine)	C_2H_2	13.84±0.10	EI		1406
$C_3H_3N^+$	$C_4H_4N_2$ (1,3–Diazine)	HCN	12.87±0.10	EI		1406
$C_3H_3N^+$	$C_4H_4N_2$ (1,4–Diazine)	HCN	12.81±0.10	EI		1406

Ion	Reactant	Other products	Ionization or appearance potential (eV)	Method	Heat of formation (kJ mol^{-1})	Ref.
			$C_3H_4N^+$			
$C_3H_4N^+$	CH_3CHCN		9.76±0.1	EI		125
$C_3H_4N^+$	CH_2CH_2CN		9.85±0.1	EI		125
$C_3H_4N^+$	C_2H_5CN	H	12.55±0.05	EI		2704
$C_3H_4N^+$	C_2H_5CN	H	13.00	EI		2966
$C_3H_4N^+$	C_2H_5NC	H	12.51	EDD		3214
$C_3H_4N^+$	$iso-C_3H_7CN$		13.25	EDD		3214
$C_3H_4N^+$	$n-C_3H_7NC$		12.09	EDD		3214
$C_3H_4N^+$	$n-C_4H_9CN$		12.12	EI		2966
$C_3H_4N^+$	$n-C_4H_9NC$		12.80	EDD		3214

Ion	Reactant	Other products	Ionization or appearance potential (eV)	Method	Heat of formation (kJ mol^{-1})	Ref.
	$C_2H_5CN^+$	$\Delta H^\circ_{f298} = 1194$ kJ mol^{-1} (285 kcal mol^{-1})				
$C_3H_5N^+$	C_2H_5CN		11.84±0.02	PI	1193	182
$C_3H_5N^+$	C_2H_5CN		11.85	PE	1194	3045

See also – EI: 2171, 2704, 2966

Ion	Reactant	Other products	Ionization or appearance potential (eV)	Method	Heat of formation (kJ mol^{-1})	Ref.
$C_3H_5N^+$	C_2H_5NC		11.38	EDD		3214
$C_3H_5N^+$	C_2H_5NC		11.2±0.1	EI		2171
$C_3H_5N^+$	$n-C_4H_9CN$		12.49	EI		2966
$C_3H_5N^+$	$n-C_4H_9NC$		11.80	EDD		3214

Ion	Reactant	Other products	Ionization or appearance potential (eV)	Method	Heat of formation (kJ mol^{-1})	Ref.
			$C_3D_5N^+$			
$C_3D_5N^+$	C_2D_5CN		11.75±0.05	EI		2704

4.3. The Positive Ion Table—Continued

Ion	Reactant	Other products	Ionization or appearance potential (eV)	Method	Heat of formation (kJ mol^{-1})	Ref.
			$C_3H_6N^+$			
$C_3H_6N^+$	$(CH_2)_3NH$ (Trimethylenimine)	H	11.4±0.2	EI		52
			$C_3H_7N^+$			
$C_3H_7N^+$	$trans-CH_3CH=NCH_3$		9.1±0.1	PE		3087
See also – PE: 3073						
$C_3H_7N^+$	$(CH_2)_3NH$ (Trimethylenimine)		9.1±0.15	EI		52
$C_3H_7N^+$	$(CH_2)_3NH$ (Trimethylenimine)		8.9	EI		218
			$C_3H_8N^+$			
$C_3H_8N^+$	$(CH_3)_3N$	H	9.8±0.1	RPD		3017
$C_3H_8N^+$	$(C_2H_5)_2NH$	CH_3	9.55±0.10	EI		2428
			$C_3H_7DN^+$			
$C_3H_7DN^+$	$(C_2H_5)_2ND$		9.53±0.01	EI		2428
	$n-C_3H_7NH_2^+$	$\Delta H^{\circ}_{f298} = 777$ kJ mol^{-1} (186 kcal mol^{-1})				
	$iso-C_3H_7NH_2^+$	$\Delta H^{\circ}_{f298} = 758$ kJ mol^{-1} (181 kcal mol^{-1})				
	$(CH_3)_3N^+$	$\Delta H^{\circ}_{f298} = 730$ kJ mol^{-1} (174 kcal mol^{-1})				
$C_3H_9N^+$	$n-C_3H_7NH_2$		8.78±0.02	PI	777	159, 182
See also – PI: 11, 86						
$C_3H_9N^+$	$iso-C_3H_7NH_2$		8.72±0.03	PI	758	159, 182
$C_3H_9N^+$	$iso-C_3H_7NH_2$		8.86	PE		1130
$C_3H_9N^+$	$(CH_3)_3N$		7.82±0.02	PI	731	159, 182, 416
$C_3H_9N^+$	$(CH_3)_3N$		7.80	PE	729	3320
See also – PE: 1130, 3410 EI: 14, 384 CTS: 2562						
			$C_4H_4N^+$			
$C_4H_4N^+$	$C_6H_5NH_2$ (Aniline)		12.3±0.1	PI		1160

4.3. The Positive Ion Table—Continued

Ion	Reactant	Other products	Ionization or appearance potential (eV)	Method	Heat of formation (kJ mol^{-1})	Ref.

C$_4$H$_5$N$^+$ (Pyrrole) $\Delta H^\circ_{f298} = 900$ kJ mol^{-1} (215 kcal mol^{-1})

Ion	Reactant	Other products	IP/AP (eV)	Method	ΔH_f	Ref.
C$_4$H$_5$N$^+$	CH$_2$=CHCH$_2$CN		10.39±0.01	PI		182
C$_4$H$_5$N$^+$	CH$_2$=CHCH$_2$CN		10.18	PE		3045
C$_4$H$_5$N$^+$	C$_3$H$_5$CN (Cyclopropanecarboxylic acid nitrile)		11.2±0.2	EI		202
C$_4$H$_5$N$^+$(^{2}A$_2$)	C$_4$H$_5$N (Pyrrole)		8.209	S	900	3423
C$_4$H$_5$N$^+$(^{2}A$_2$)	C$_4$H$_5$N (Pyrrole)		8.20±0.01?	PI		182
C$_4$H$_5$N$^+$(^{2}A$_2$)	C$_4$H$_5$N (Pyrrole)		8.20±0.01	PI		3354
C$_4$H$_5$N$^+$(^{2}A$_2$)	C$_4$H$_5$N (Pyrrole)		8.209	PE	900	3423
C$_4$H$_5$N$^+$(^{2}B$_1$)	C$_4$H$_5$N (Pyrrole)		9.200	S		3423
C$_4$H$_5$N$^+$(^{2}B$_1$)	C$_4$H$_5$N (Pyrrole)		9.20	PE		3423

Additional higher ionization potentials are given in ref. 3423.

See also – PE: 2796, 2975, 3109, 3246, 3349
      EI: 381, 2865, 3174, 3233

C$_4$H$_6$N$^+$

Ion	Reactant	Other products	IP/AP (eV)	Method	ΔH_f	Ref.
C$_4$H$_6$N$^+$	(CH$_3$)$_2$CCN		9.15±0.1	EI		125
C$_4$H$_6$N$^+$	n–C$_3$H$_7$CN	H	13.00	EI		2966
C$_4$H$_6$N$^+$	iso-C$_3$H$_7$CN	H	12.55	EDD		3214
C$_4$H$_6$N$^+$	n–C$_3$H$_7$NC	H	11.96	EDD		3214
C$_4$H$_6$N$^+$	tert–C$_4$H$_9$CN	CH$_3$	12.60	EDD		3214

n–C$_3$H$_7$CN$^+$ $\Delta H^\circ_{f298} \sim 1160$ kJ mol^{-1} (277 kcal mol^{-1})

Ion	Reactant	Other products	IP/AP (eV)	Method	ΔH_f	Ref.
C$_4$H$_7$N$^+$	n–C$_3$H$_7$CN		11.67±0.05	PI	1160	182
C$_4$H$_7$N$^+$	n–C$_3$H$_7$NC		11.33	EDD		3214
C$_4$H$_7$N$^+$	n–C$_3$H$_7$NC		11.1±0.1	EI		2171

C$_4$H$_8$N$^+$

Ion	Reactant	Other products	IP/AP (eV)	Method	ΔH_f	Ref.
C$_4$H$_8$N$^+$	(CH$_2$)$_4$NH (Pyrrolidine)	H	11.0±0.2	EI		52

4.3. The Positive Ion Table—Continued

Ion	Reactant	Other products	Ionization or appearance potential (eV)	Method	Heat of formation (kJ mol⁻¹)	Ref.

$$\mathbf{C_4H_9N^+}$$

Ion	Reactant	Other products	Ionization or appearance potential (eV)	Method	Heat of formation (kJ mol⁻¹)	Ref.
$C_4H_9N^+$	$CH_3CH=NC_2H_5$		9.29	PE		1130
$C_4H_9N^+$	C_4H_9N (2,2–Dimethylethylenimine)		8.94	PE		3235
$C_4H_9N^+$	$(CH_2)_4NH$ (Pyrrolidine)		8.41	PE		1130

See also – EI: 52, 218

$$\mathbf{C_4H_{10}N^+}$$

Ion	Reactant	Other products	Ionization or appearance potential (eV)	Method	Heat of formation (kJ mol⁻¹)	Ref.
$C_4H_{10}N^+$	$C_4H_{10}N$		5.9–6.7	SI		3229

This radical ion results from the surface ionization of $(C_2H_5)_2NH$, see ref. 3229.

Ion	Reactant	Other products	Ionization or appearance potential (eV)	Method	Heat of formation (kJ mol⁻¹)	Ref.
$C_4H_{10}N^+$	$n-C_3H_7CH(NH_2)CH_3$	CH_3	9.92±0.16	EI		2587

$n-C_4H_9NH_2^+$	$\Delta H^\circ_{f298} \sim 748 \text{ kJ mol}^{-1} \text{ (179 kcal mol}^{-1})$
$sec-C_4H_9NH_2^+$	$\Delta H^\circ_{f298} \sim 735 \text{ kJ mol}^{-1} \text{ (176 kcal mol}^{-1})$
$tert-C_4H_9NH_2^+$	$\Delta H^\circ_{f298} \sim 714 \text{ kJ mol}^{-1} \text{ (171 kcal mol}^{-1})$
$(C_2H_5)_2NH^+$	$\Delta H^\circ_{f298} = 700 \text{ kJ mol}^{-1} \text{ (167 kcal mol}^{-1})$

Ion	Reactant	Other products	Ionization or appearance potential (eV)	Method	Heat of formation (kJ mol⁻¹)	Ref.
$C_4H_{11}N^+$	$n-C_4H_9NH_2$		8.71±0.03	PI	748	159, 182
$C_4H_{11}N^+$	$n-C_4H_9NH_2$		8.79	PE		1130

See also – EI: 2587
            CTS: 2562

Ion	Reactant	Other products	Ionization or appearance potential (eV)	Method	Heat of formation (kJ mol⁻¹)	Ref.
$C_4H_{11}N^+$	$sec-C_4H_9NH_2$		8.70	PI	735	182
$C_4H_{11}N^+$	$iso-C_4H_9NH_2$		8.70	PI		182
$C_4H_{11}N^+$	$tert-C_4H_9NH_2$		8.64	PI	714	182
$C_4H_{11}N^+$	$tert-C_4H_9NH_2$		8.83	PE		1130

See also – CTS: 2562

Ion	Reactant	Other products	Ionization or appearance potential (eV)	Method	Heat of formation (kJ mol⁻¹)	Ref.
$C_4H_{11}N^+$	$(C_2H_5)_2NH$		8.01±0.01	PI	700	159, 182
$C_4H_{11}N^+$	$(C_2H_5)_2NH$		8.51	PE		1130

See also – EI: 14, 2428
            CTS: 2562

Ion	Reactant	Other products	Ionization or appearance potential (eV)	Method	Heat of formation (kJ mol⁻¹)	Ref.
$C_4H_{11}N^+$	$(C_2H_5)_2NNH_2$	NH	11.2±0.1	PI		1141

$$\mathbf{C_4H_{10}DN^+}$$

Ion	Reactant	Other products	Ionization or appearance potential (eV)	Method	Heat of formation (kJ mol⁻¹)	Ref.
$C_4H_{10}DN^+$	$(C_2H_5)_2ND$		8.31±0.01	EI		2428

4.3. The Positive Ion Table—Continued

Ion	Reactant	Other products	Ionization or appearance potential (eV)	Method	Heat of formation (kJ mol^{-1})	Ref.

$C_5H_3N^+$

| $C_5H_3N^+$ | C_5H_5N (Pyridine) | H_2 | 12.42±0.10 | EI | | 1406 |

$C_5H_4N^+$

| $C_5H_4N^+$ | C_5H_5N (Pyridine) | H | 14.00±0.10 | EI | | 1406 |

$C_5H_2D_2N^+$

| $C_5H_2D_2N^+$ | $C_6H_5CH_2ND_2$ (α−Amino−d_2−toluene) | | 11.0±0.2 | PI | | 1160 |

$C_5H_5N^+$ (Pyridine)	$\Delta H^{\circ}_{f298} = 1034$ kJ mol^{-1} (247 kcal mol^{-1})					
$C_5H_5N^+$	C_5H_5N (Pyridine)		9.266	S	1034	1115
$C_5H_5N^+$	C_5H_5N (Pyridine)		9.23±0.03	PI		416
$C_5H_5N^+$	C_5H_5N (Pyridine)		9.20±0.05	PI		1160
$C_5H_5N^+$	C_5H_5N (Pyridine)		9.30±0.01	PI		3325
$C_5H_5N^+$	C_5H_5N (Pyridine)		9.10±0.01	PI		2789
$C_5H_5N^+$	C_5H_5N (Pyridine)		9.28	PE		1130
$C_5H_5N^+$	C_5H_5N (Pyridine)		9.31	PE		2844
$C_5H_5N^+$	C_5H_5N (Pyridine)		9.10	PE		2789

Higher ionization potentials are given in refs. 1115, 1130, 2789, 2829, 2844, 2968, 3323 but they are not well established. Various PE studies are in disagreement among themselves and with the spectroscopic values of ref. 1115.

See also – S: 3153
PI: 182, 3323
PE: 2829, 2968, 3323
EI: 217, 383, 1406, 2481, 2865, 2989, 3013
CTS: 2562

$C_5H_6N^+$

$C_5H_6N^+$	$C_5H_4NH_2$ (Aminocyclopentadienyl radical)		7.55	EI		126
$C_5H_6N^+$	$C_4H_4NCH_3$ (N−Methylpyrrole)	H	11.2±0.1	PI		3354
$C_5H_6N^+$	$C_4H_4NCH_3$ (2−Methylpyrrole)	H	10.25±0.1	PI		3354

4.3. The Positive Ion Table—Continued

Ion	Reactant	Other products	Ionization or appearance potential (eV)	Method	Heat of formation (kJ mol⁻¹)	Ref.
C₅H₇N⁺						
$C_5H_7N^+$	$C_4H_4NCH_3$ (N–Methylpyrrole)		8.09±0.01	PI		3354
$C_5H_7N^+$	$C_4H_4NCH_3$ (N–Methylpyrrole)		7.95±0.05	PE		3246
$C_5H_7N^+$	$C_4H_4NCH_3$ (2–Methylpyrrole)		7.78±0.01	PI		3354
$C_5H_7N^+$	$C_4H_4NCH_3$ (3–Methylpyrrole)		7.90±0.02	PI		3354
C₅H₈N⁺						
$C_5H_8N^+$	$n-C_4H_9CN$	H	12.47	EI		2966
$C_5H_8N^+$	$tert-C_4H_9CN$	H	12.05	EDD		3214
C₅H₉N⁺						
$C_5H_9N^+$	$n-C_4H_9NC$		11.71±0.05	EI		2481
$C_5H_9N^+$	$tert-C_4H_9NC$		10.50	EDD		3214
C₅H₁₁N⁺						
$C_5H_{11}N^+$	$(CH_3)_2C=NC_2H_5$		8.83	PE		1130
$C_5H_{11}N^+$	$(CH_2)_5NH$ (Piperidine)		8.7	EI		218
$C_5H_{11}N^+$	$(CH_2)_5NH$ (Piperidine)		9.15	CTS		2031
C₅H₁₂N⁺						
$C_5H_{12}N^+$	$n-C_4H_9N(CH_3)NH_2$	NH_2	9.0±0.1	PI		1141
$C_5H_{12}N^+$	$n-C_4H_9CH(NH_2)COOH$		9.76±0.05	EI		2587
$C_5H_{12}N^+$	$sec-C_4H_9CH(NH_2)COOH$		9.9±0.2	EI		88
C₅H₁₃N⁺						
$C_5H_{13}N^+$	$n-C_3H_7CH(NH_2)CH_3$		9.31±0.15	EI		2587
$C_5H_{13}N^+$	$n-C_4H_9N(CH_3)NH_2$	NH	10.5±0.1	PI		1141
C₆H₄N⁺						
$C_6H_4N^+$	C_5H_4CN (Cyanocyclopentadienyl radical)		9.44	EI		126

4.3. The Positive Ion Table—Continued

Ion	Reactant	Other products	Ionization or appearance potential (eV)	Method	Heat of formation (kJ mol⁻¹)	Ref.

$C_6H_6N^+$

Ion	Reactant	Other products	Ionization or appearance potential (eV)	Method	Heat of formation (kJ mol⁻¹)	Ref.
$C_6H_6N^+$	C_6H_5NH (Anilino radical)		8.26 ± 0.1	EI		1011
$C_6H_6N^+$	$C_5H_4NCH_2$ (2–Pyridylmethyl radical)		8.17 ± 0.1	EI		1011
$C_6H_6N^+$	$C_5H_4NCH_2$ (3–Pyridylmethyl radical)		7.92 ± 0.1	EI		1011
$C_6H_6N^+$	$C_5H_4NCH_2$ (4–Pyridylmethyl radical)		8.40 ± 0.15	EI		1011
$C_6H_6N^+$	$C_5H_4NCH_3$ (2–Methylpyridine)	H	12.38 ± 0.1	EI		1011
$C_6H_6N^+$	$C_5H_4NCH_3$ (3–Methylpyridine)	H	12.31 ± 0.1	EI		1011
$C_6H_6N^+$	$C_5H_4NCH_3$ (4–Methylpyridine)	H	12.22 ± 0.1	EI		1011
$C_6H_6N^+$	$C_6H_4ClNH_2$ (3–Chloroaniline)	Cl	12.25 ± 0.1	EI		2972
$C_6H_6N^+$	$C_6H_4ClNH_2$ (4–Chloroaniline)	Cl	12.37	EI		3238
$C_6H_6N^+$	$C_6H_4ClNH_2$ (4–Chloroaniline)	Cl	12.50 ± 0.1	EI		2972
$C_6H_6N^+$	$C_6H_4BrNH_2$ (4–Bromoaniline)	Br	11.95	EI		3238

$C_6H_7N^+$ (Aniline)　　　$\Delta H^\circ_{f298} \sim 829$ kJ mol^{-1} (198 kcal mol^{-1})
$C_6H_7N^+$ (2–Methylpyridine)　$\Delta H^\circ_{f298} \sim 969$ kJ mol^{-1} (232 kcal mol^{-1})
$C_6H_7N^+$ (3–Methylpyridine)　$\Delta H^\circ_{f298} \sim 978$ kJ mol^{-1} (234 kcal mol^{-1})
$C_6H_7N^+$ (4–Methylpyridine)　$\Delta H^\circ_{f298} \sim 974$ kJ mol^{-1} (233 kcal mol^{-1})

Ion	Reactant	Other products	Ionization or appearance potential (eV)	Method	Heat of formation (kJ mol⁻¹)	Ref.
$C_6H_7N^+$	$C_6H_5NH_2$ (Aniline)		7.70 ± 0.02	PI	830	159, 182, 416
$C_6H_7N^+$	$C_6H_5NH_2$ (Aniline)		7.67 ± 0.03	PI	827	1160
$C_6H_7N^+$	$C_6H_5NH_2$ (Aniline)		7.69 ± 0.02	PI	829	1159, 1166
$C_6H_7N^+$	$C_6H_5NH_2$ (Aniline)		7.68?	PE		2796
$C_6H_7N^+$	$C_6H_5NH_2$ (Aniline)		7.61 ± 0.05	SI		2741

See also – PE:　1159, 1160, 2806
　　　　　　EI:　1066, 2458, 2865, 2972, 3223
　　　　　　CTS:　1281, 2562, 2909, 2978

Ion	Reactant	Other products	Ionization or appearance potential (eV)	Method	Heat of formation (kJ mol⁻¹)	Ref.
$C_6H_7N^+$	$C_5H_4NCH_3$ (2–Methylpyridine)		9.02 ± 0.03	PI	969	182

See also – EI:　217, 2865, 2989

Ion	Reactant	Other products	Ionization or appearance potential (eV)	Method	Heat of formation (kJ mol⁻¹)	Ref.
$C_6H_7N^+$	$C_5H_4NCH_3$ (3–Methylpyridine)		9.04 ± 0.03	PI	978	182

See also – EI:　2865, 2989

4.3. The Positive Ion Table—Continued

Ion	Reactant	Other products	Ionization or appearance potential (eV)	Method	Heat of formation (kJ mol^{-1})	Ref.
$C_6H_7N^+$	$C_5H_4NCH_3$ (4–Methylpyridine)		9.04±0.03	PI	974	182

See also – EI: 217, 2865, 2989

$C_6H_7N^+$	$C_6H_5NHCOCH_3$ (N–Phenylacetic acid amide)		8.88±0.15	EI		1126

See also – EI: 3406

$C_6H_8N^+$

$C_6H_8N^+$	$C_4H_3N(CH_3)_2$ (2,4–Dimethylpyrrole)	H	10.15±0.1	PI		3354

$C_6H_9N^+$

$C_6H_9N^+$	$C_4H_3N(CH_3)_2$ (2,4–Dimethylpyrrole)		7.54±0.02	PI		3354

$C_6H_{10}N^+$

$C_6H_{10}N^+$	$n-C_5H_{11}CN$	H	12.62	EI		2966

$C_6H_{13}N^+$

$C_6H_{13}N^+$	$n-C_3H_7CH=NC_2H_5$		9.00	PE		1130
$C_6H_{13}N^+$	$iso-C_3H_7CH=NC_2H_5$		8.94	PE		1130
$C_6H_{13}N^+$	$C_6H_{11}NH_2$ (Aminocyclohexane)		8.86	PE		1130
$C_6H_{13}N^+$	$(CH_2)_6NH$ (Azacycloheptane)		8.5	EI		218

$(C_2H_5)_3N^+$ $\Delta H^°_{f298} \sim 624$ kJ mol^{-1} (149 kcal mol^{-1})

$C_6H_{15}N^+$	$(n-C_3H_7)_2NH$		7.84±0.02	PI		159, 182
$C_6H_{15}N^+$	$(iso-C_3H_7)_2NH$		7.73±0.03	PI		159, 182
$C_6H_{15}N^+$	$(C_2H_5)_3N$		7.50±0.02	PI	624	159, 182
$C_6H_{15}N^+$	$(C_2H_5)_3N$		7.84	PE		1130

See also – EI: 14, 2158
        CTS: 2031, 2562

$C_7H_4N^+$

$C_7H_4N^+$	C_6H_4ClCN (3–Chlorobenzoic acid nitrile)	Cl	13.83±0.1	EI		2972

4.3. The Positive Ion Table—Continued

Ion	Reactant	Other products	Ionization or appearance potential (eV)	Method	Heat of formation (kJ mol^{-1})	Ref.
$C_7H_4N^+$	C_6H_4ClCN (4–Chlorobenzoic acid nitrile)	Cl	13.97±0.1	EI		2972
$C_7H_4N^+$	C_6H_4BrCN (4–Bromobenzoic acid nitrile)	Br	13.21	EI		3238

$C_7H_5N^+$ (Benzoic acid nitrile) $\qquad \Delta H^\circ_{f298} = 1155$ kJ mol^{-1} (276 kcal mol^{-1})

Ion	Reactant	Other products	Ionization or appearance potential (eV)	Method	Heat of formation (kJ mol^{-1})	Ref.
$C_7H_5N^+$	C_6H_5CN (Benzoic acid nitrile)		9.705±0.01	PI	1155	182
$C_7H_5N^+$	C_6H_5CN (Benzoic acid nitrile)		10.02 (V)	PE		2806

See also – EI: 1066, 2420, 2972, 3223, 3238

Ion	Reactant	Other products	Ionization or appearance potential (eV)	Method	Heat of formation (kJ mol^{-1})	Ref.
$C_7H_5N^+$	C_6H_5NC (Isocyanobenzene)		9.70±0.05	EI		2481
$C_7H_5N^+$	$C_6H_5CONH_2$ (Benzoic acid amide)	H_2O	10.19±0.10	EI		1126
$C_7H_5N^+$	$(C_6H_5)_2C_2N_2O$ (3,5–Diphenyl–1,2,4–oxadiazole)		10.2±0.1	EI		1125

$C_7H_2D_3N^+$

Ion	Reactant	Other products	Ionization or appearance potential (eV)	Method	Heat of formation (kJ mol^{-1})	Ref.
$C_7H_2D_3N^+$	$C_6H_2D_3NC$ (Isocyanobenzene–2,4,6–d_3)		9.9±0.2	EI		2919

$C_7H_7N^+$

Ion	Reactant	Other products	Ionization or appearance potential (eV)	Method	Heat of formation (kJ mol^{-1})	Ref.
$C_7H_7N^+$	$C_6H_5CH_2NH_2$ (α–Aminotoluene)	H_2	9.35±0.07	PI		1147

$C_7H_8N^+$

Ion	Reactant	Other products	Ionization or appearance potential (eV)	Method	Heat of formation (kJ mol^{-1})	Ref.
$C_7H_8N^+$	$C_6H_5CH_2NH_2$ (α–Aminotoluene)	H	9.21±0.07	PI		1147
$C_7H_8N^+$	$C_6H_5CH_2NH_2$ (α–Aminotoluene)	H	9.3±0.1	PI		1160
$C_7H_8N^+$	$C_6H_5NHCH_3$ (N–Methylaniline)	H	11.0±0.1	PI		1160
$C_7H_8N^+$	$C_6H_4(NH_2)CH_3$ (4–Aminotoluene)	H	10.80	EI		3238
$C_7H_8N^+$	$C_6H_5CH_2CH_2C_6H_4NH_2$ (1–(3–Aminophenyl)–2–phenylethane)		11.0±0.2	EI		3288
$C_7H_8N^+$	$C_6H_5CH_2CH_2C_6H_4NH_2$ (1–(4–Aminophenyl)–2–phenylethane)		9.1±0.2	EI		3288
$C_7H_8N^+$	$C_6H_4(NH_2)CH_2CH_2C_6H_4NH_2$ (1,2–Bis(4–aminophenyl)ethane)		9.3±0.2	EI		3288
$C_7H_8N^+$	$C_6H_4(NH_2)(CH_2)_3COOCH_3$ (4–(4–Aminophenyl)butanoic acid methyl ester)		10.92±0.2	EI		2497

4.3. The Positive Ion Table—Continued

Ion	Reactant	Other products	Ionization or appearance potential (eV)	Method	Heat of formation (kJ mol⁻¹)	Ref.
$C_7H_8N^+$	$C_6H_4(NH_2)CH_2CH_2C_6H_4NO_2$ (1–(4–Aminophenyl)–2–(4–nitrophenyl)ethane)		9.2±0.2	EI		3288

$C_7H_9N^+$

Ion	Reactant	Other products	Ionization or appearance potential (eV)	Method	Heat of formation (kJ mol⁻¹)	Ref.
$C_7H_9N^+$	$C_6H_5CH_2NH_2$ (α–Aminotoluene)		8.64±0.05	PI	1147,	1160, 1166
$C_7H_9N^+$	$C_6H_5CH_2NH_2$ (α–Aminotoluene)		8.73	PE		3235

See also – PI: 159
EI: 2025, 2458

$C_7H_9N^+$	$C_6H_5NHCH_3$ (N–Methylaniline)		7.30±0.05	PI		1160
$C_7H_9N^+$	$C_6H_5NHCH_3$ (N–Methylaniline)		7.34±0.02	PI	1159,	1166
$C_7H_9N^+$	$C_6H_5NHCH_3$ (N–Methylaniline)		7.73 (V)	PE		2806

See also – PE: 1159, 1160
EI: 2458
CTS: 1281, 2978

$C_7H_9N^+$	$C_6H_4(NH_2)CH_3$ (2–Aminotoluene)		7.68±0.1	EI		2458
$C_7H_9N^+$	$C_6H_4(NH_2)CH_3$ (2–Aminotoluene)		7.59±0.1	CTS		2485
$C_7H_9N^+$	$C_6H_4(NH_2)CH_3$ (2–Aminotoluene)		7.69	CTS		2909
$C_7H_9N^+$	$C_6H_4(NH_2)CH_3$ (2–Aminotoluene)		7.75	CTS		2978

See also – EI: 1066

$C_7H_9N^+$	$C_6H_4(NH_2)CH_3$ (3–Aminotoluene)		7.50±0.02	PI		1166
$C_7H_9N^+$	$C_6H_4(NH_2)CH_3$ (3–Aminotoluene)		7.57±0.1	EI		2458
$C_7H_9N^+$	$C_6H_4(NH_2)CH_3$ (3–Aminotoluene)		7.68±0.1	CTS		2485
$C_7H_9N^+$	$C_6H_4(NH_2)CH_3$ (3–Aminotoluene)		7.75	CTS		2978

See also – EI: 1066, 2025

4.3. The Positive Ion Table—Continued

Ion	Reactant	Other products	Ionization or appearance potential (eV)	Method	Heat of formation (kJ mol^{-1})	Ref.
$C_7H_9N^+$	$C_6H_4(NH_2)CH_3$ (4−Aminotoluene)		7.78 (V)	PE		2806
$C_7H_9N^+$	$C_6H_4(NH_2)CH_3$ (4−Aminotoluene)		7.60±0.1	EI		2458
$C_7H_9N^+$	$C_6H_4(NH_2)CH_3$ (4−Aminotoluene)		7.57±0.1	CTS		2485
$C_7H_9N^+$	$C_6H_4(NH_2)CH_3$ (4−Aminotoluene)		7.58	CTS		2909
$C_7H_9N^+$	$C_6H_4(NH_2)CH_3$ (4−Aminotoluene)		7.65	CTS		2978

See also – EI: 1066, 3223

Ion	Reactant	Other products	Ionization or appearance potential (eV)	Method	Heat of formation (kJ mol^{-1})	Ref.
$C_7H_9N^+$	$C_5H_3N(CH_3)_2$ (2,3−Dimethylpyridine)		8.85±0.02	PI		182

See also – EI: 3232

Ion	Reactant	Other products	Ionization or appearance potential (eV)	Method	Heat of formation (kJ mol^{-1})	Ref.
$C_7H_9N^+$	$C_5H_3N(CH_3)_2$ (2,4−Dimethylpyridine)		8.85±0.03	PI		182

See also – EI: 3232

Ion	Reactant	Other products	Ionization or appearance potential (eV)	Method	Heat of formation (kJ mol^{-1})	Ref.
$C_7H_9N^+$	$C_5H_3N(CH_3)_2$ (2,6−Dimethylpyridine)		8.85±0.02	PI		182

See also – EI: 3232

Ion	Reactant	Other products	Ionization or appearance potential (eV)	Method	Heat of formation (kJ mol^{-1})	Ref.
$C_7H_9N^+$	$C_6H_4(CH_3)NHCOCH_3$ (N−(3−Tolyl)acetic acid amide)		10.85±0.2	EI		3406
$C_7H_9N^+$	$C_6H_4(CH_3)NHCOCH_3$ (N−(4−Tolyl)acetic acid amide)		10.62±0.2	EI		3406

$C_7H_{11}N^+$

Ion	Reactant	Other products	Ionization or appearance potential (eV)	Method	Heat of formation (kJ mol^{-1})	Ref.
$C_7H_{11}N^+$	$C_6H_{11}NC$ (Isocyanocyclohexane)		10.72±0.05	EI		2481

$C_7H_{13}N^+$

Ion	Reactant	Other products	Ionization or appearance potential (eV)	Method	Heat of formation (kJ mol^{-1})	Ref.
$C_7H_{13}N^+$	$C_7H_{13}N$ (1−Azabicyclo[2.2.2]octane)		8.02 (V)	PE		2962

$C_8H_6N^+$

Ion	Reactant	Other products	Ionization or appearance potential (eV)	Method	Heat of formation (kJ mol^{-1})	Ref.
$C_8H_6N^+$	$C_6H_4(CN)CH_2$ (3−Cyanobenzyl radical)		8.58±0.1	EI		69
$C_8H_6N^+$	$C_6H_4(CN)CH_2$ (4−Cyanobenzyl radical)		8.36±0.1	EI		69
$C_8H_6N^+$	$C_6H_5CH_2CH_2C_6H_4CN$ (1−(3−Cyanophenyl)−2−phenylethane)		~15	EI		3288

4.3. The Positive Ion Table—Continued

Ion	Reactant	Other products	Ionization or appearance potential (eV)	Method	Heat of formation (kJ mol^{-1})	Ref.
$C_8H_6N^+$	$C_6H_4(CN)(CH_2)_3COOCH_3$ (4–(4–Cyanophenyl)butanoic acid methyl ester)		13.85±0.2	EI		2497

$C_8H_7N^+$ (Indole) $\quad \Delta H^\circ_{f298} = 936$ kJ mol^{-1} (224 kcal mol^{-1})

Ion	Reactant	Other products	Ionization or appearance potential (eV)	Method	Heat of formation (kJ mol^{-1})	Ref.
$C_8H_7N^+$	$C_6H_5CH_2CN$ (Phenylacetic acid nitrile)		9.50±0.04	RPD		3223
$C_8H_7N^+$	$C_6H_5CH_2CN$ (Phenylacetic acid nitrile)		9.40±0.05	EI		2025
$C_8H_7N^+$	$C_6H_4(CH_3)CN$ (3–Methylbenzoic acid nitrile)		9.58±0.05	RPD		3223
$C_8H_7N^+$	$C_6H_4(CH_3)CN$ (3–Methylbenzoic acid nitrile)		9.66±0.05	EI		2025
$C_8H_7N^+$	$C_6H_4(CH_3)CN$ (4–Methylbenzoic acid nitrile)		9.56±0.05	RPD		3223
$C_8H_7N^+$	$C_6H_4(CH_3)CN$ (4–Methylbenzoic acid nitrile)		9.76	EI		1066
$C_8H_7N^+$	$C_6H_5CH_2NC$ (α–Isocyanotoluene)		9.61±0.05	EI		2481
$C_8H_7N^+$	$C_6H_4(CH_3)NC$ (Isocyanotoluene)		9.63±0.05	EI		2481
$C_8H_7N^+$	C_8H_7N (Indole)		7.75±0.05	PE	934	2796
$C_8H_7N^+$	C_8H_7N (Indole)		7.78	PE	937	3349

See also – CTS: 2562

$C_8H_9N^+$

Ion	Reactant	Other products	Ionization or appearance potential (eV)	Method	Heat of formation (kJ mol^{-1})	Ref.
$C_8H_9N^+$	$C_6H_4(NH_2)(CH_2)_3COOCH_3$ (4–(4–Aminophenyl)butanoic acid methyl ester)		10.52±0.2	EI		2497

$C_8H_{10}N^+$

Ion	Reactant	Other products	Ionization or appearance potential (eV)	Method	Heat of formation (kJ mol^{-1})	Ref.
$C_8H_{10}N^+$	$C_6H_5N(CH_3)_2$ (N,N–Dimethylaniline)	H	10.75±0.05	PI		1160

See also – EI: 3238

Ion	Reactant	Other products	Ionization or appearance potential (eV)	Method	Heat of formation (kJ mol^{-1})	Ref.
$C_8H_{10}N^+$	$C_6H_4(NH_2)CH_2CH_2Br$ (1–(4–Aminophenyl)–2–bromoethane)	Br	9.9	EI		2973

4.3. The Positive Ion Table—Continued

Ion	Reactant	Other products	Ionization or appearance potential (eV)	Method	Heat of formation (kJ mol^{-1})	Ref.
			C$_8$H$_{11}$N$^+$			
C$_8$H$_{11}$N$^+$	C$_6$H$_5$CH$_2$NHCH$_3$ (Benzyl methyl amine)		8.65	EI		3338
C$_8$H$_{11}$N$^+$	C$_6$H$_5$NHC$_2$H$_5$ (N–Ethylaniline)		7.56	CTS		1281
C$_8$H$_{11}$N$^+$	C$_6$H$_5$NHC$_2$H$_5$ (N–Ethylaniline)		7.5	CTS		2978
C$_8$H$_{11}$N$^+$	C$_6$H$_4$(NH$_2$)C$_2$H$_5$ (2–Ethylaniline)		7.57±0.1	CTS		2485
C$_8$H$_{11}$N$^+$	C$_6$H$_4$(NH$_2$)C$_2$H$_5$ (4–Ethylaniline)		7.62±0.1	CTS		2485
C$_8$H$_{11}$N$^+$	C$_6$H$_5$N(CH$_3$)$_2$ (N,N–Dimethylaniline)		7.14±0.03	PI	1159,	1166
C$_8$H$_{11}$N$^+$	C$_6$H$_5$N(CH$_3$)$_2$ (N,N–Dimethylaniline)		7.10±0.05	PI		1160
C$_8$H$_{11}$N$^+$	C$_6$H$_5$N(CH$_3$)$_2$ (N,N–Dimethylaniline)		7.51 (V)	PE		2806
C$_8$H$_{11}$N$^+$	C$_6$H$_5$N(CH$_3$)$_2$ (N,N–Dimethylaniline)		7.31	CTS		2909
C$_8$H$_{11}$N$^+$	C$_6$H$_5$N(CH$_3$)$_2$ (N,N–Dimethylaniline)		7.2	CTS		2037

See also – PE: 1159, 1160
    EI: 2458, 2950, 3238
    CTS: 1281, 2562, 2978

Ion	Reactant	Other products	Ionization or appearance potential (eV)	Method	Heat of formation (kJ mol^{-1})	Ref.
C$_8$H$_{11}$N$^+$	C$_6$H$_4$(CH$_3$)NHCH$_3$ (N,2–Dimethylaniline)		7.58±0.1	EI		2458
C$_8$H$_{11}$N$^+$	C$_6$H$_4$(CH$_3$)NHCH$_3$ (N,3–Dimethylaniline)		7.45±0.1	EI		2458
C$_8$H$_{11}$N$^+$	C$_6$H$_4$(CH$_3$)NHCH$_3$ (N,4–Dimethylaniline)		7.58±0.1	EI		2458
C$_8$H$_{11}$N$^+$	C$_6$H$_3$(CH$_3$)$_2$NH$_2$ (2,3–Dimethylaniline)		7.51±0.1	CTS		2485
C$_8$H$_{11}$N$^+$	C$_6$H$_3$(CH$_3$)$_2$NH$_2$ (2,4–Dimethylaniline)		7.40±0.1	CTS		2485
C$_8$H$_{11}$N$^+$	C$_6$H$_3$(CH$_3$)$_2$NH$_2$ (2,5–Dimethylaniline)		7.50±0.1	CTS		2485
C$_8$H$_{11}$N$^+$	C$_6$H$_3$(CH$_3$)$_2$NH$_2$ (2,6–Dimethylaniline)		7.46±0.1	CTS		2485
C$_8$H$_{11}$N$^+$	C$_6$H$_3$(CH$_3$)$_2$NH$_2$ (3,5–Dimethylaniline)		7.61±0.1	CTS		2485

4.3. The Positive Ion Table—Continued

Ion	Reactant	Other products	Ionization or appearance potential (eV)	Method	Heat of formation (kJ mol^{-1})	Ref.
		C$_8$H$_{13}$N$^+$				
C$_8$H$_{13}$N$^+$	C$_4$H$_4$NC$_4$H$_9$ (N−Butylpyrrole)		7.87±0.02	PI		3354
		C$_8$H$_{19}$N$^+$				
C$_8$H$_{19}$N$^+$	(n−C$_4$H$_9$)$_2$NH		7.69±0.03	PI	159,	182
		C$_9$H$_7$N$^+$				
C$_9$H$_7$N$^+$	C$_9$H$_7$N (Quinoline)		8.62±0.01	PI		2651
C$_9$H$_7$N$^+$	C$_9$H$_7$N (Quinoline)		8.62	PE		2844
C$_9$H$_7$N$^+$	C$_9$H$_7$N (Quinoline)		8.67±0.05	PE		2847

See also − EI: 3174
        CTS: 2485, 2562

Ion	Reactant	Other products	Ionization or appearance potential (eV)	Method	Heat of formation	Ref.
C$_9$H$_7$N$^+$	C$_9$H$_7$N (Isoquinoline)		8.55±0.02	PI		2651
C$_9$H$_7$N$^+$	C$_9$H$_7$N (Isoquinoline)		8.54	PE		2844
C$_9$H$_7$N$^+$	C$_9$H$_7$N (Isoquinoline)		8.53±0.05	PE		2847
C$_9$H$_7$N$^+$	C$_6$H$_4$(CN)(CH$_2$)$_3$COOCH$_3$ (4−(4−Cyanophenyl)butanoic acid methyl ester)		11.20±0.2	EI		2497
		C$_9$H$_8$N$^+$				
C$_9$H$_8$N$^+$	C$_6$H$_4$(CN)CH(CH$_3$)CH$_2$COOCH$_3$ (3−(4−Cyanophenyl)butanoic acid methyl ester)		11.53±0.2	EI		2497
		C$_9$H$_9$N$^+$				
C$_9$H$_9$N$^+$	C$_6$H$_4$(CH$_3$)CH$_2$CN (3−Tolylacetic acid nitrile)		9.18±0.04	RPD		3223
C$_9$H$_9$N$^+$	C$_6$H$_4$(CH$_3$)CH$_2$CN (4−Tolylacetic acid nitrile)		9.16±0.06	RPD		3223
C$_9$H$_9$N$^+$	C$_6$H$_3$(CH$_3$)$_2$CN (3,4−Dimethylbenzoic acid nitrile)		9.16±0.04	RPD		3223

4.3. The Positive Ion Table—Continued

Ion	Reactant	Other products	Ionization or appearance potential (eV)	Method	Heat of formation (kJ mol^{-1})	Ref.
			$C_9H_{11}N^+$			
$C_9H_{11}N^+$	$C_9H_{11}N$ (1,2,3,4–Tetrahydroquinoline)		7.61	EI		3338
$C_9H_{11}N^+$	$C_9H_{11}N$ (1,2,3,4–Tetrahydroisoquinoline)		8.63	EI		3338
			$C_9H_{13}N^+$			
$C_9H_{13}N^+$	$C_6H_5NHC_3H_7$ (N–Propylaniline)		7.54	CTS		1281
$C_9H_{13}N^+$	$C_6H_5NHC_3H_7$ (N–Propylaniline)		7.5	CTS		2978
$C_9H_{13}N^+$	$C_6H_5NHC_3H_7$ (N–Isopropylaniline)		7.5	CTS		2978
$C_9H_{13}N^+$	$C_6H_4(NH_2)C_3H_7$ (4–Isopropylaniline)		7.68±0.1	CTS		2485
$C_9H_{13}N^+$	$C_6H_5N(CH_3)C_2H_5$ (N–Ethyl–N–methylaniline)		7.37	CTS		1281
$C_9H_{13}N^+$	$C_6H_4(CH_3)N(CH_3)_2$ (N,N,2–Trimethylaniline)		7.37	CTS		1281
$C_9H_{13}N^+$	$C_6H_4(CH_3)N(CH_3)_2$ (N,N,3–Trimethylaniline)		7.35	CTS		1281
$C_9H_{13}N^+$	$C_6H_4(CH_3)N(CH_3)_2$ (N,N,4–Trimethylaniline)		7.48 (V)	PE		2806
$C_9H_{13}N^+$	$C_6H_4(CH_3)N(CH_3)_2$ (N,N,4–Trimethylaniline)		7.33	CTS		1281
			$C_9H_{21}N^+$			
$C_9H_{21}N^+$	$(n-C_3H_7)_3N$		7.23?	PI	159,	182
			$C_{10}H_9N^+$			
$C_{10}H_9N^+$	$C_{10}H_7NH_2$ (1–Aminonaphthalene)		7.26±0.1	CTS		2485
$C_{10}H_9N^+$	$C_{10}H_7NH_2$ (1–Aminonaphthalene)		7.39	CTS		2909
$C_{10}H_9N^+$	$C_{10}H_7NH_2$ (1–Aminonaphthalene)		7.4	CTS		2978
$C_{10}H_9N^+$	$C_{10}H_7NH_2$ (2–Aminonaphthalene)		7.37±0.1	CTS		2485
$C_{10}H_9N^+$	$C_{10}H_7NH_2$ (2–Aminonaphthalene)		7.5	CTS		2978
$C_{10}H_9N^+$	$C_6H_4(CN)CH(CH_3)CH_2COOCH_3$ (3–(4–Cyanophenyl)butanoic acid methyl ester)		9.51±0.2	EI		2497

4.3. The Positive Ion Table—Continued

Ion	Reactant	Other products	Ionization or appearance potential (eV)	Method	Heat of formation (kJ mol^{-1})	Ref.
			$C_{10}H_{15}N^+$			
$C_{10}H_{15}N^+$	$C_6H_5NHC_4H_9$ (N−Butylaniline)		7.53	CTS		1281
$C_{10}H_{15}N^+$	$C_6H_5NHC_4H_9$ (N−Butylaniline)		7.5	CTS		2978
$C_{10}H_{15}N^+$	$C_6H_4(NH_2)C_4H_9$ (4−tert−Butylaniline)		7.72±0.1	CTS		2485
$C_{10}H_{15}N^+$	$C_6H_5N(C_2H_5)_2$ (N,N−Diethylaniline)		7.51 (V)	PE		2806
$C_{10}H_{15}N^+$	$C_6H_5N(C_2H_5)_2$ (N,N−Diethylaniline)		6.99	CTS		1281
$C_{10}H_{15}N^+$	$C_6H_5N(C_2H_5)_2$ (N,N−Diethylaniline)		7.15	CTS		2978
$C_{10}H_{15}N^+$	$C_6H_4(C_2H_5)N(CH_3)_2$ (N,N−Dimethyl−4−ethylaniline)		7.38	CTS		1281
$C_{10}H_{15}N^+$	$C_6H_3(CH_3)_2N(CH_3)_2$ (N,N,2,4−Tetramethylaniline)		7.17	CTS		1281
$C_{10}H_{15}N^+$	$C_6H_3(CH_3)_2N(CH_3)_2$ (N,N,2,6−Tetramethylaniline)		7.22	CTS		1281
$C_{10}H_{15}N^+$	$C_6H_3(CH_3)_2N(CH_3)_2$ (N,N,3,5−Tetramethylaniline)		7.25	CTS		1281
			$C_{11}H_{15}N^+$			
$C_{11}H_{15}N^+$	$C_6H_5NHC_5H_9$ (N−Cyclopentylaniline)		7.45	CTS		2978
			$C_{11}H_{17}N^+$			
$C_{11}H_{17}N^+$	$C_6H_5NHC_5H_{11}$ (N−Pentylaniline)		7.5	CTS		2978
$C_{11}H_{17}N^+$	$C_6H_4(CH_3)N(C_2H_5)_2$ (N,N−Diethyl−4−methylaniline)		6.93	CTS		1281
$C_{11}H_{17}N^+$	$C_6H_4(C_3H_7)N(CH_3)_2$ (N,N−Dimethyl−4−isopropylaniline)		7.41	CTS		1281
			$C_{12}H_9N^+$			
$C_{12}H_9N^+$	$C_{12}H_9N$ (Carbazole)		7.6±0.1	EDD		2974
$C_{12}H_9N^+$	$C_{12}H_9N$ (Carbazole)		7.2±0.1	EI		3011

See also − CTS: 2911

Ion	Reactant	Other products	Ionization or appearance potential (eV)	Method	Heat of formation	Ref.
$C_{12}H_9N^+$	$C_{12}H_9N_3$ (1−Phenyl−1,2,3−benzotriazole)	N_2	9.65±0.1	EDD		2974
$C_{12}H_9N^+$	$C_{13}H_{10}N_4$ (2−Benzylideneamino−2,1,3−benzotriazole)		10.4±0.1	EDD		2974

4.3. The Positive Ion Table—Continued

Ion	Reactant	Other products	Ionization or appearance potential (eV)	Method	Heat of formation (kJ mol⁻¹)	Ref.
	$C_{12}H_{11}N^+$ (Diphenylamine)	**$\Delta H^\circ_{f298} \sim 901$ kJ mol⁻¹ (215 kcal mol⁻¹)**				
$C_{12}H_{11}N^+$	$(C_6H_5)_2NH$ (Diphenylamine)		7.25±0.03	PI	901	1140
See also – CTS: 2909, 2978						
$C_{12}H_{11}N^+$	$C_6H_5C_6H_4NH_2$ (2–Aminobiphenyl)		7.55±0.1	CTS		2485
$C_{12}H_{11}N^+$	$C_6H_5C_6H_4NH_2$ (4–Aminobiphenyl)		7.49±0.1	CTS		2485
	$C_{12}H_{17}N^+$					
$C_{12}H_{17}N^+$	$C_6H_5NHC_6H_{11}$ (N–Cyclohexylaniline)		7.45	CTS		2978
	$C_{12}H_{19}N^+$					
$C_{12}H_{19}N^+$	$C_6H_5NHC_6H_{13}$ (N–Hexylaniline)		7.5	CTS		2978
$C_{12}H_{19}N^+$	$C_6H_5N(C_3H_7)_2$ (N,N–Dipropylaniline)		6.96	CTS		1281
$C_{12}H_{19}N^+$	$C_6H_5N(C_3H_7)_2$ (N,N–Dipropylaniline)		7.15	CTS		2978
$C_{12}H_{19}N^+$	$C_6H_4(C_4H_9)N(CH_3)_2$ (N,N–Dimethyl–4–tert–butylaniline)		7.43	CTS		1281
	$C_{13}H_7N^+$					
$C_{13}H_7N^+$	$(C_6H_4CN)_2$ (4,4'–Dicyanobiphenyl)	HCN	15.65	EI		3295
	$C_{13}H_9N^+$					
$C_{13}H_9N^+$	$C_{13}H_9N$ (Acridine)		7.39	CTS		2562
$C_{13}H_9N^+$	$C_{13}H_9N$ (Acridine)		7.98±0.1	CTS		2485
$C_{13}H_9N^+$	$C_{13}H_9N$ (Acridine)		8.04	CTS		2909
$C_{13}H_9N^+$	$C_{13}H_9N$ (3,4–Benzoquinoline)		8.38±0.1	CTS		2485
$C_{13}H_9N^+$	$C_{13}H_9N$ (5,6–Benzoquinoline)		8.38±0.1	CTS		2485
$C_{13}H_9N^+$	$C_{13}H_9N$ (6,7–Benzoquinoline)		7.58±0.1	CTS		2485
$C_{13}H_9N^+$	$C_{13}H_9N$ (7,8–Benzoquinoline)		8.34±0.1	CTS		2485

4.3. The Positive Ion Table—Continued

Ion	Reactant	Other products	Ionization or appearance potential (eV)	Method	Heat of formation (kJ mol⁻¹)	Ref.
$C_{13}H_9N^+$	$C_{13}H_9N$ (5,6–Benzoisoquinoline)		8.32±0.1	CTS		2485
$C_{13}H_9N^+$	$C_{14}H_9NO_2$ (N–Phenylphthalic acid imide)	CO_2	11.0	EI		2412

$C_{13}H_{11}N^+$

Ion	Reactant	Other products	Ionization or appearance potential (eV)	Method	Heat of formation (kJ mol⁻¹)	Ref.
$C_{13}H_{11}N^+$	$C_{12}H_8NCH_3$ (N–Methylcarbazole)		7.5±0.1	EI		3011
$C_{13}H_{11}N^+$	$C_{13}H_{11}N$ (9,10–Dihydroacridine)		7.24±0.03	PI		2728

$C_{13}H_{19}N^+$

Ion	Reactant	Other products	Ionization or appearance potential (eV)	Method	Heat of formation (kJ mol⁻¹)	Ref.
$C_{13}H_{19}N^+$	$C_6H_5NHC_7H_{13}$ (N–Cycloheptylaniline)		7.45	CTS		2978

$C_{13}H_{21}N^+$

Ion	Reactant	Other products	Ionization or appearance potential (eV)	Method	Heat of formation (kJ mol⁻¹)	Ref.
$C_{13}H_{21}N^+$	$C_6H_5NHC_7H_{15}$ (N–Heptylaniline)		7.5	CTS		2978

$C_{14}H_9N^+$

Ion	Reactant	Other products	Ionization or appearance potential (eV)	Method	Heat of formation (kJ mol⁻¹)	Ref.
$C_{14}H_9N^+$	$C_{14}H_9N$ (4,5–Iminophenanthrene)		7.6±0.1	EI		3011

$C_{14}H_{11}N^+$

Ion	Reactant	Other products	Ionization or appearance potential (eV)	Method	Heat of formation (kJ mol⁻¹)	Ref.
$C_{14}H_{11}N^+$	$C_{14}H_9NH_2$ (2–Aminophenanthrene)		7.55±0.1	CTS		2485
$C_{14}H_{11}N^+$	$C_{14}H_9NH_2$ (9–Aminophenanthrene)		7.19±0.1	CTS		2485

$C_{14}H_{13}N^+$

Ion	Reactant	Other products	Ionization or appearance potential (eV)	Method	Heat of formation (kJ mol⁻¹)	Ref.
$C_{14}H_{13}N^+$	$C_6H_5CH=CHC_6H_4NH_2$ (1–(2–Aminophenyl)–2–phenylethene)		7.39±0.1	CTS		2485

$C_{14}H_{15}N^+$

Ion	Reactant	Other products	Ionization or appearance potential (eV)	Method	Heat of formation (kJ mol⁻¹)	Ref.
$C_{14}H_{15}N^+$	$C_6H_5CH_2CH_2C_6H_4NH_2$ (1–(3–Aminophenyl)–2–phenylethane)		7.9±0.1 (V)	PE		3290
$C_{14}H_{15}N^+$	$C_6H_5CH_2CH_2C_6H_4NH_2$ (1–(3–Aminophenyl)–2–phenylethane)		7.6±0.1	EI		3288
$C_{14}H_{15}N^+$	$C_6H_5CH_2CH_2C_6H_4NH_2$ (1–(4–Aminophenyl)–2–phenylethane)		7.7±0.1	EI		3288

4.3. The Positive Ion Table—Continued

Ion	Reactant	Other products	Ionization or appearance potential (eV)	Method	Heat of formation (kJ mol^{-1})	Ref.
$C_{14}H_{15}N^+$	$(C_6H_5CH_2)_2NH$ (Dibenzylamine)		8.22	PE		3235
$C_{14}H_{15}N^+$	$C_6H_5N(CH_3)CH_2C_6H_5$ (N–Benzyl–N–methylaniline)		7.44±0.1	EI		2458

$C_{14}H_{23}N^+$

Ion	Reactant	Other products	Ionization or appearance potential (eV)	Method	Heat of formation (kJ mol^{-1})	Ref.
$C_{14}H_{23}N^+$	$C_6H_5NHC_8H_{17}$ (N–Octylaniline)		7.5	CTS		2978
$C_{14}H_{23}N^+$	$C_6H_5N(C_4H_9)_2$ (N,N–Dibutylaniline)		6.95	CTS		1281
$C_{14}H_{23}N^+$	$C_6H_5N(C_4H_9)_2$ (N,N–Dibutylaniline)		7.15	CTS		2978

$C_{15}H_{13}N^+$

Ion	Reactant	Other products	Ionization or appearance potential (eV)	Method	Heat of formation (kJ mol^{-1})	Ref.
$C_{15}H_{13}N^+$	$C_6H_5CH_2CH_2C_6H_4CN$ (1–(3–Cyanophenyl)–2–phenylethane)		8.9±0.1	EI		3288

$C_{15}H_{15}N^+$

Ion	Reactant	Other products	Ionization or appearance potential (eV)	Method	Heat of formation (kJ mol^{-1})	Ref.
$C_{15}H_{15}N^+$	$C_{12}H_8NC_3H_7$ (N–Isopropylcarbazole)		7.80	CTS		3299
(Average of two values)						
$C_{15}H_{15}N^+$	$C_{12}H_8NC_3H_7$ (N–Isopropylcarbazole)		7.4	CTS		3355
(Average of three values)						

$C_{15}H_{17}N^+$

Ion	Reactant	Other products	Ionization or appearance potential (eV)	Method	Heat of formation (kJ mol^{-1})	Ref.
$C_{15}H_{17}N^+$	$(C_6H_5CH_2)_2NCH_3$ (Dibenzyl methyl amine)		7.85	PE		3235

$C_{15}H_{25}N^+$

Ion	Reactant	Other products	Ionization or appearance potential (eV)	Method	Heat of formation (kJ mol^{-1})	Ref.
$C_{15}H_{25}N^+$	$C_6H_5NHC_9H_{19}$ (N–Nonylaniline)		7.5	CTS		2978

$C_{16}H_{11}N^+$

Ion	Reactant	Other products	Ionization or appearance potential (eV)	Method	Heat of formation (kJ mol^{-1})	Ref.
$C_{16}H_{11}N^+$	$C_{16}H_{11}N$ (1,2–Benzocarbazole)		7.1±0.1	EI		3011
$C_{16}H_{11}N^+$	$C_{16}H_{11}N$ (2,3–Benzocarbazole)		7.05±0.1	EI		3011
$C_{16}H_{11}N^+$	$C_{16}H_{11}N$ (3,4–Benzocarbazole)		7.3±0.1	EI		3011
$C_{16}H_{11}N^+$	$C_{16}H_9NH_2$ (1–Aminopyrene)		6.82±0.1	CTS		2485
$C_{16}H_{11}N^+$	$C_{16}H_9NH_2$ (2–Aminopyrene)		7.43±0.1	CTS		2485

4.3. The Positive Ion Table—Continued

Ion	Reactant	Other products	Ionization or appearance potential (eV)	Method	Heat of formation (kJ mol^{-1})	Ref.
		C$_{16}$H$_{13}$N$^+$				
C$_{16}$H$_{13}$N$^+$	C$_{10}$H$_7$NHC$_6$H$_5$ (1–Naphthyl phenyl amine)		7.12	CTS		2909
C$_{16}$H$_{13}$N$^+$	C$_{10}$H$_7$NHC$_6$H$_5$ (2–Naphthyl phenyl amine)		7.15	CTS		2909
		C$_{16}$H$_{27}$N$^+$				
C$_{16}$H$_{27}$N$^+$	C$_6$H$_5$NHC$_{10}$H$_{21}$ (N–Decylaniline)		7.5	CTS		2978
C$_{16}$H$_{27}$N$^+$	C$_6$H$_5$N(C$_5$H$_{11}$)$_2$ (N,N–Dipentylaniline)		7.1	CTS		2978
		C$_{17}$H$_{11}$N$^+$				
C$_{17}$H$_{11}$N$^+$	C$_{17}$H$_{11}$N (1,2–Benzacridine)		8.07±0.1	CTS		2485
C$_{17}$H$_{11}$N$^+$	C$_{17}$H$_{11}$N (3,4–Benzacridine)		8.07±0.1	CTS		2485
		C$_{17}$H$_{29}$N$^+$				
C$_{17}$H$_{29}$N$^+$	C$_6$H$_5$NHC$_{11}$H$_{23}$ (N–Undecylaniline)		7.5	CTS		2978
		C$_{18}$H$_{13}$N$^+$				
C$_{18}$H$_{13}$N$^+$	C$_{18}$H$_{11}$NH$_2$ (6–Aminochrysene)		6.99±0.1	CTS		2485
C$_{18}$H$_{13}$N$^+$	C$_{12}$H$_8$NC$_6$H$_5$ (N–Phenylcarbazole)		7.7±0.1	EI		3011
		C$_{18}$H$_{15}$N$^+$				
C$_{18}$H$_{15}$N$^+$	(C$_6$H$_5$)$_3$N (Triphenylamine)		6.86±0.03	PI		1140
		C$_{18}$H$_{31}$N$^+$				
C$_{18}$H$_{31}$N$^+$	C$_6$H$_5$NHC$_{12}$H$_{25}$ (N–Dodecylaniline)		7.5	CTS		2978
C$_{18}$H$_{31}$N$^+$	C$_6$H$_5$N(C$_6$H$_{13}$)$_2$ (N,N–Dihexylaniline)		7.1	CTS		2978

4.3. The Positive Ion Table—Continued

Ion	Reactant	Other products	Ionization or appearance potential (eV)	Method	Heat of formation (kJ mol^{-1})	Ref.
		C$_{19}$H$_{12}$N$^+$				
C$_{19}$H$_{12}$N$^+$	C$_{20}$H$_{13}$NO$_2$ (N−(2−Biphenylyl)phthalic acid imide)		9.8	EI		2412
		C$_{19}$H$_{21}$N$^+$				
C$_{19}$H$_{21}$N$^+$	C$_{19}$H$_{21}$N (Nortriptylene)		8.39±0.11	CTS		2987
		C$_{20}$H$_{13}$N$^+$				
C$_{20}$H$_{13}$N$^+$	C$_{20}$H$_{13}$N (1,2:5,6−Dibenzocarbazole)		6.9±0.1	EI		3011
C$_{20}$H$_{13}$N$^+$	C$_{20}$H$_{13}$N (1,2:7,8−Dibenzocarbazole)		7.1±0.1	EI		3011
C$_{20}$H$_{13}$N$^+$	C$_{20}$H$_{13}$N (3,4:5,6−Dibenzocarbazole)		7.1±0.1	EI		3011
C$_{20}$H$_{13}$N$^+$	C$_{20}$H$_{13}$N (Naphtho[2,3−b]carbazole)		6.95±0.1	EI		3011
C$_{20}$H$_{13}$N$^+$	C$_{20}$H$_{13}$N (Naphtho[2,3−c]carbazole)		7.0±0.1	EI		3011
		C$_{20}$H$_{23}$N$^+$				
C$_{20}$H$_{23}$N$^+$	C$_{20}$H$_{23}$N (Amitriptylene)		8.32±0.08	CTS		2987
		C$_{21}$H$_{15}$N$^+$				
C$_{21}$H$_{15}$N$^+$	C$_{20}$H$_{12}$NCH$_3$ (N−Methyl−2,3:6,7−dibenzocarbazole)		6.95±0.1	EI		3011
C$_{21}$H$_{15}$N$^+$	C$_{20}$H$_{12}$NCH$_3$ (N−Methylnaphtho[2,3−b]carbazole)		7.2±0.1	EI		3011
		C$_{21}$H$_{21}$N$^+$				
C$_{21}$H$_{21}$N$^+$	C$_{21}$H$_{21}$N (Cyproheptadine)		7.94±0.06	CTS		2987
		C$_{22}$H$_{39}$N$^+$				
C$_{22}$H$_{39}$N$^+$	C$_6$H$_5$N(C$_8$H$_{17}$)$_2$ (N,N−Dioctylaniline)		7.1	CTS		2978

4.3. The Positive Ion Table—Continued

Ion	Reactant	Other products	Ionization or appearance potential (eV)	Method	Heat of formation (kJ mol^{-1})	Ref.
			C$_{24}$H$_{15}$N$^+$			
C$_{24}$H$_{15}$N$^+$	C$_{24}$H$_{15}$N (Benzo[b]naphtho[2,3–h]carbazole)		7.1±0.1	EI		3011
C$_{24}$H$_{15}$N$^+$	C$_{24}$H$_{15}$N (Anthra[2,3–b]carbazole)		6.5±0.1	EI		3011
C$_{24}$H$_{15}$N$^+$	C$_{24}$H$_{15}$N (Anthra[2,3–c]carbazole)		7.3±0.1	EI		3011
			C$_{26}$H$_{47}$N$^+$			
C$_{26}$H$_{47}$N$^+$	C$_6$H$_5$N(C$_{10}$H$_{21}$)$_2$ (N,N–Didecylaniline)		7.1	CTS		2978
			CHN$_2^+$			
CHN$_2^+$	CH$_2$N$_2$ (Diazomethane)	H	14.8±0.1	EI		314
CHN$_2^+$	CH$_2$N$_2$ (Diazirine)	H	14.2±0.1	EI		314
	CH$_2$N$_2^+$ (Diazomethane)	$\Delta H_{f298}^\circ = 1061$ kJ mol^{-1} (254 kcal mol^{-1})				
CH$_2$N$_2^+$	CH$_2$N$_2$ (Diazomethane)		8.999±0.001	S	1061	1169
CH$_2$N$_2^+$	CH$_2$N$_2$ (Diazomethane)		9.03±0.05	EI		314
CH$_2$N$_2^+$	CH$_2$N$_2$ (Diazomethane)		9.2±0.3	EI		464
CH$_2$N$_2^+$	CH$_2$N$_2$ (Diazirine)		10.18±0.05	EI		314
CH$_2$N$_2^+$	CH$_3$NHNH$_2$		15.2±0.2	EI		424, 3216
			CH$_3$N$_2^+$			
CH$_3$N$_2^+$	CH$_3$NHNH$_2$	H$_2$+H	9.2±0.2	PI		1141
CH$_3$N$_2^+$	CH$_3$NHNH$_2$	H$_2$+H	11.9±0.3	EI		424, 3216
CH$_3$N$_2^+$	CH$_3$N=NCH$_3$	CH$_3$	9.0±0.1	EI		304
CH$_3$N$_2^+$	CH$_3$N=NCH$_3$	CH$_3$	9.5±0.3	EI		2549
CH$_3$N$_2^+$	CH$_3$NHNHCH$_3$		9.7±0.5	EI		424, 3216
			CH$_4$N$_2^+$			
CH$_4$N$_2^+$	CH$_3$NNH		9.28±0.1	EI		3178

See also – EI: 67

CH$_4$N$_2^+$	CH$_3$NHNH$_2$	H$_2$	9.4±0.1	PI		1141
CH$_4$N$_2^+$	CH$_3$NHNH$_2$	H$_2$	9.92±0.1	EI		3178

4.3. The Positive Ion Table—Continued

Ion	Reactant	Other products	Ionization or appearance potential (eV)	Method	Heat of formation (kJ mol^{-1})	Ref.
$CH_4N_2^+$	CH_3NHNH_2	H_2	10.4±0.2	EI	424,	3216
$CH_4N_2^+$	$CH_3NHNHCH_3$	CH_4	9.7±0.3	EI	424,	3216
$CH_4N_2^+$	$n-C_4H_9N(CH_3)NH_2$		9.6±0.1	PI		1141

$$CH_5N_2^+$$

Ion	Reactant	Other products	Ionization or appearance potential (eV)	Method	Heat of formation (kJ mol^{-1})	Ref.
$CH_5N_2^+$	CH_3NHNH		7.3±0.2	EI		3178
$CH_5N_2^+$	CH_3NHNH_2	H	9.2±0.1	PI		1141
$CH_5N_2^+$	CH_3NHNH_2	H	10.18±0.1	EI		3178
$CH_5N_2^+$	CH_3NHNH_2	H	10.2±0.1	EI	424,	3216
$CH_5N_2^+$	$(CH_3)_2NNH_2$	CH_3	8.4±0.1	PI		1141
$CH_5N_2^+$	$(CH_3)_2NNH_2$	CH_3	9.7±0.2	EI	424,	3216
$CH_5N_2^+$	$CH_3NHNHCH_3$	CH_3	9.1±0.2	EI	424,	3216
$CH_5N_2^+$	$n-C_4H_9N(CH_3)NH_2$		9.0±0.1	PI		1141

$$CH_6N_2^+$$

Ion	Reactant	Other products	Ionization or appearance potential (eV)	Method	Heat of formation (kJ mol^{-1})	Ref.
$CH_6N_2^+$	CH_3NHNH_2		7.67±0.02	PI		2173
(Value obtained without mass analysis)						
$CH_6N_2^+$	CH_3NHNH_2		8.00±0.06	PI	1141,	2173
(Value obtained with mass analysis)						
$CH_6N_2^+$	CH_3NHNH_2		8.67	PE		3235
$CH_6N_2^+$	CH_3NHNH_2		8.63±0.1	EI	424,	3216

$$C_2H_5N_2^+$$

Ion	Reactant	Other products	Ionization or appearance potential (eV)	Method	Heat of formation (kJ mol^{-1})	Ref.
$C_2H_5N_2^+$	$(CH_3)_2NNHCH_3$		11.1±0.4	EI	424,	3216
$C_2H_5N_2^+$	$(CH_3)_2NN(CH_3)_2$		12.4±0.2	EI	424,	3216

$$C_2H_6N_2^+$$

Ion	Reactant	Other products	Ionization or appearance potential (eV)	Method	Heat of formation (kJ mol^{-1})	Ref.
$C_2H_6N_2^+$	$trans-CH_3N=NCH_3$		8.7±0.1	PE		3087
$C_2H_6N_2^+$	$CH_3N=NCH_3$		8.65±0.1	EI		304
$C_2H_6N_2^+$	$CH_3N=NCH_3$		8.65±0.2	EI		2549

See also – PE: 3073

Ion	Reactant	Other products	Ionization or appearance potential (eV)	Method	Heat of formation (kJ mol^{-1})	Ref.
$C_2H_6N_2^+$	$(CH_3)_2NNH_2$	H_2	9.5±0.1	PI		1141
$C_2H_6N_2^+$	$(CH_3)_2NN(CH_3)_2$	C_2H_6	10.5±0.1	EI	424,	3216
$C_2H_6N_2^+$	$n-C_4H_9N(CH_3)NH_2$	C_3H_8	9.5±0.2	PI		1141

$$C_2H_7N_2^+$$

Ion	Reactant	Other products	Ionization or appearance potential (eV)	Method	Heat of formation (kJ mol^{-1})	Ref.
$C_2H_7N_2^+$	$(CH_3)_2NNH$		6.4±0.2	EI		3178
$C_2H_7N_2^+$	$(CH_3)_2NNH$		6.6±0.3	EI		67
$C_2H_7N_2^+$	$(CH_3)_2NNH_2$	H	8.7±0.2	PI		1141
$C_2H_7N_2^+$	$(CH_3)_2NNH_2$	H	10.08±0.1	EI		3178
$C_2H_7N_2^+$	$(CH_3)_2NNH_2$	H	10.0±0.3	EI		67
$C_2H_7N_2^+$	$(CH_3)_2NNH_2$	H	10.2±0.2	EI	424,	3216

4.3. The Positive Ion Table—Continued

Ion	Reactant	Other products	Ionization or appearance potential (eV)	Method	Heat of formation (kJ mol^{-1})	Ref.	
$C_2H_7N_2^+$	$CH_3NHNHCH_3$	H	9.3±0.2	EI		424,	3216
$C_2H_7N_2^+$	$(CH_3)_2NNHCH_3$	CH_3	9.4±0.1	EI		424,	3216
$C_2H_7N_2^+$	$n-C_4H_9N(CH_3)NH_2$		9.1±0.1	PI			1141

$(CH_3)_2NNH_2^+$ $\Delta H^\circ_{f298} \sim 883$ kJ mol^{-1} (211 kcal mol^{-1})

$CH_3NHNHCH_3^+$ $\Delta H^\circ_{f298} \sim 888$ kJ mol^{-1} (212 kcal mol^{-1})

Ion	Reactant	Other products	Ionization or appearance potential (eV)	Method	Heat of formation (kJ mol^{-1})	Ref.	
$C_2H_8N_2^+$	$(CH_3)_2NNH_2$		7.46±0.02	PI			2173
(Value obtained without mass analysis)							
$C_2H_8N_2^+$	$(CH_3)_2NNH_2$		7.67±0.05	PI		1141,	2173
(Value obtained with mass analysis)							
$C_2H_8N_2^+$	$(CH_3)_2NNH_2$		8.28	PE	883		3235
$C_2H_8N_2^+$	$(CH_3)_2NNH_2$		8.12±0.1	EI		424,	3216
$C_2H_8N_2^+$	$CH_3NHNHCH_3$		8.22	PE	888		3235
$C_2H_8N_2^+$	$CH_3NHNHCH_3$		7.75±0.1	EI		424,	3216

$C_3H_4N_2^+$ (Pyrazole) $\Delta H^\circ_{f298} \sim 1076$ kJ mol^{-1} (257 kcal mol^{-1})

Ion	Reactant	Other products	Ionization or appearance potential (eV)	Method	Heat of formation (kJ mol^{-1})	Ref.
$C_3H_4N_2^+$	$C_3H_4N_2$ (Pyrazole)		9.27±0.05	PE	1076	3246

See also – PE: 2975

$C_3H_7N_2^+$

Ion	Reactant	Other products	Ionization or appearance potential (eV)	Method	Heat of formation (kJ mol^{-1})	Ref.	
$C_3H_7N_2^+$	$(CH_3)_2NNHCH_3$	H_2+H	10.7±0.1	EI		424,	3216
$C_3H_7N_2^+$	$(CH_3)_2NN(CH_3)_2$		10.7±0.1	EI		424,	3216

$C_3H_8N_2^+$

Ion	Reactant	Other products	Ionization or appearance potential (eV)	Method	Heat of formation (kJ mol^{-1})	Ref.	
$C_3H_8N_2^+$	$(CH_3)_2NNHCH_3$	H_2	8.2±0.1	EI		424,	3216
$C_3H_8N_2^+$	$(CH_3)_2NN(CH_3)_2$	CH_4	8.9±0.1	EI		424,	3216

$C_3H_9N_2^+$

Ion	Reactant	Other products	Ionization or appearance potential (eV)	Method	Heat of formation (kJ mol^{-1})	Ref.	
$C_3H_9N_2^+$	$(CH_3)_2NNHCH_3$	H	8.9±0.1	EI		424,	3216
$C_3H_9N_2^+$	$(C_2H_5)_2NNH_2$	CH_3	8.0±0.1	PI			1141
$C_3H_9N_2^+$	$(CH_3)_2NN(CH_3)_2$	CH_3	9.1±0.1	EI		424,	3216

$C_3H_{10}N_2^+$

Ion	Reactant	Other products	Ionization or appearance potential (eV)	Method	Heat of formation (kJ mol^{-1})	Ref.	
$C_3H_{10}N_2^+$	$(CH_3)_2NNHCH_3$		7.93±0.1	EI		424,	3216

$C_4H_3N_2^+$

Ion	Reactant	Other products	Ionization or appearance potential (eV)	Method	Heat of formation (kJ mol^{-1})	Ref.
$C_4H_3N_2^+$	$C_4H_4N_2$ (1,3–Diazine)	H	13.01±0.10	EI		1406

4.3. The Positive Ion Table—Continued

Ion	Reactant	Other products	Ionization or appearance potential (eV)	Method	Heat of formation (kJ mol^{-1})	Ref.
$C_4H_3N_2^+$	$C_4H_4N_2$ (1,4–Diazine)	H	13.68±0.10	EI		1406

	$C_4H_4N_2^+$(1,2–Diazine)	$\Delta H_{f298}^\circ = 1119$ kJ mol^{-1} (267 kcal mol^{-1})				
	$C_4H_4N_2^+$(1,3–Diazine)	$\Delta H_{f298}^\circ = 1099$ kJ mol^{-1} (263 kcal mol^{-1})				
	$C_4H_4N_2^+$(1,4–Diazine)	$\Delta H_{f298}^\circ \sim 1089$ kJ mol^{-1} (260 kcal mol^{-1})				
$C_4H_4N_2^+$	$C_4H_4N_2$ (1,2–Diazine)		8.71±0.01	PI	1119	2651
$C_4H_4N_2^+$	$C_4H_4N_2$ (1,2–Diazine)		8.90	PE		2844

See also – EI: 1406, 3232

| $C_4H_4N_2^+$ | $C_4H_4N_2$ (1,3–Diazine) | | 9.35±0.01 | PI | 1099 | 2651 |
| $C_4H_4N_2^+$ | $C_4H_4N_2$ (1,3–Diazine) | | 9.42 | PE | | 2844 |

See also – EI: 1406, 3232

$C_4H_4N_2^+$	$C_4H_4N_2$ (1,4–Diazine)		9.29±0.03	S		3332
$C_4H_4N_2^+$	$C_4H_4N_2$ (1,4–Diazine)		9.29±0.01	PI	1092	2651
$C_4H_4N_2^+$	$C_4H_4N_2$ (1,4–Diazine)		9.28±0.05	PE		2796
$C_4H_4N_2^+$	$C_4H_4N_2$ (1,4–Diazine)		9.216	PE	1085	3080
$C_4H_4N_2^+$	$C_4H_4N_2$ (1,4–Diazine)		9.36	PE		2844

The discrepancy between the PI result, ref. 2651, and the PE result, ref. 3080, is not understood.

See also – EI: 1406, 3232

$C_4H_{10}N_2^+$

| $C_4H_{10}N_2^+$ | $(C_2H_5)_2NNH_2$ | H_2 | 8.3±0.2 | PI | | 1141 |

$C_4H_{11}N_2^+$

| $C_4H_{11}N_2^+$ | $(C_2H_5)_2NNH_2$ | H | 8.9±0.1 | PI | | 1141 |
| $C_4H_{11}N_2^+$ | $n-C_4H_9N(CH_3)NH_2$ | CH_3 | 8.0±0.1 | PI | | 1141 |

$C_4H_{12}N_2^+$

$C_4H_{12}N_2^+$	$(C_2H_5)_2NNH_2$		7.59±0.05	PI		1141
$C_4H_{12}N_2^+$	$(CH_3)_2NN(CH_3)_2$		7.93	PE		3235
$C_4H_{12}N_2^+$	$(CH_3)_2NN(CH_3)_2$		7.76±0.05	EI	424,	3216

4.3. The Positive Ion Table—Continued

Ion	Reactant	Other products	Ionization or appearance potential (eV)	Method	Heat of formation (kJ mol^{-1})	Ref.

$$C_5H_6N_2^+$$

Ion	Reactant	Other products	Ionization or appearance potential (eV)	Method	Heat of formation (kJ mol^{-1})	Ref.
$C_5H_6N_2^+$	$C_5H_4NNH_2$ (4–Aminopyridine)		8.97±0.05	EI		217

$$C_5H_{12}N_2^+$$

| $C_5H_{12}N_2^+$ | $n-C_4H_9N(CH_3)NH_2$ | H_2 | 8.0±0.2 | PI | | 1141 |

$$C_5H_{13}N_2^+$$

| $C_5H_{13}N_2^+$ | $n-C_4H_9N(CH_3)NH_2$ | H | 8.0±0.3 | PI | | 1141 |

$$C_5H_{14}N_2^+$$

| $C_5H_{14}N_2^+$ (Value obtained without mass analysis) | $n-C_4H_9N(CH_3)NH_2$ | | 7.51±0.02 | PI | | 2173 |
| $C_5H_{14}N_2^+$ (Value obtained with mass analysis) | $n-C_4H_9N(CH_3)NH_2$ | | 7.62±0.05 | PI | | 1141, 2173 |

$C_6H_8N_2^+$ (Phenylhydrazine)　　　$\Delta H^\circ_{f298} \sim 946$ kJ mol^{-1} (226 kcal mol^{-1})

Ion	Reactant	Other products	Ionization or appearance potential (eV)	Method	Heat of formation (kJ mol^{-1})	Ref.
$C_6H_8N_2^+$	$C_6H_4(NH_2)_2$ (1,2–Diaminobenzene)		8.00	EI		1066
$C_6H_8N_2^+$	$C_6H_4(NH_2)_2$ (1,2–Diaminobenzene)		7.35±0.1	CTS		2485
$C_6H_8N_2^+$	$C_6H_4(NH_2)_2$ (1,2–Diaminobenzene)		7.36	CTS		2909
$C_6H_8N_2^+$	$C_6H_4(NH_2)_2$ (1,2–Diaminobenzene)		7.45	CTS		2978
$C_6H_8N_2^+$	$C_6H_4(NH_2)_2$ (1,3–Diaminobenzene)		7.96	EI		1066
$C_6H_8N_2^+$	$C_6H_4(NH_2)_2$ (1,3–Diaminobenzene)		7.48±0.1	CTS		2485
$C_6H_8N_2^+$	$C_6H_4(NH_2)_2$ (1,4–Diaminobenzene)		7.58	EI		1066
$C_6H_8N_2^+$	$C_6H_4(NH_2)_2$ (1,4–Diaminobenzene)		7.04±0.1	CTS		2485
$C_6H_8N_2^+$	$C_6H_4(NH_2)_2$ (1,4–Diaminobenzene)		7.15	CTS		2978
$C_6H_8N_2^+$	$C_6H_5NHNH_2$ (Phenylhydrazine)		7.64±0.02	PI	941	1166
$C_6H_8N_2^+$	$C_6H_5NHNH_2$ (Phenylhydrazine)		7.74	PE	951	3235

4.3. The Positive Ion Table—Continued

Ion	Reactant	Other products	Ionization or appearance potential (eV)	Method	Heat of formation (kJ mol^{-1})	Ref.
			C$_6$H$_{12}$N$_2^+$			
C$_6$H$_{12}$N$_2^+$	C$_6$H$_{12}$N$_2$ (1,4–Diazabicyclo[2.2.2]octane)		7.52 (V)	PE		2962
			C$_7$H$_6$N$_2^+$			
C$_7$H$_6$N$_2^+$	C$_6$H$_4$(NH$_2$)CN (3–Aminobenzoic acid nitrile)		8.61±0.05	RPD		3223
C$_7$H$_6$N$_2^+$	C$_6$H$_4$(NH$_2$)CN (4–Aminobenzoic acid nitrile)		8.64±0.04	RPD		3223
			C$_7$H$_{10}$N$_2^+$			
C$_7$H$_{10}$N$_2^+$	C$_6$H$_5$CH$_2$NHNH$_2$ (Benzylhydrazine)		8.64	PE		3235
C$_7$H$_{10}$N$_2^+$	C$_6$H$_5$N(CH$_3$)NH$_2$ (1–Methyl–1–phenylhydrazine)		7.43	PE		3235
			C$_8$H$_6$N$_2^+$			
C$_8$H$_6$N$_2^+$	C$_8$H$_6$N$_2$ (Cinnoline)		8.95±0.01	PI		2651
C$_8$H$_6$N$_2^+$	C$_8$H$_6$N$_2$ (Cinnoline)		8.51	PE		2844
C$_8$H$_6$N$_2^+$	C$_8$H$_6$N$_2$ (Quinazoline)		9.02	PE		2844
C$_8$H$_6$N$_2^+$	C$_8$H$_6$N$_2$ (Phthalazine)		9.22±0.01	PI		2651
C$_8$H$_6$N$_2^+$	C$_8$H$_6$N$_2$ (Phthalazine)		8.68	PE		2844
C$_8$H$_6$N$_2^+$	C$_8$H$_6$N$_2$ (Quinoxaline)		9.02±0.01	PI		2651
C$_8$H$_6$N$_2^+$	C$_8$H$_6$N$_2$ (Quinoxaline)		8.99	PE		2844
			C$_8$H$_8$N$_2^+$			
C$_8$H$_8$N$_2^+$	C$_6$H$_4$(NH$_2$)CH$_2$CN (3–Aminophenylacetic acid nitrile)		8.31±0.05	RPD		3223
C$_8$H$_8$N$_2^+$	C$_6$H$_4$(NH$_2$)CH$_2$CN (4–Aminophenylacetic acid nitrile)		8.26±0.04	RPD		3223
			C$_8$H$_{12}$N$_2^+$			
C$_8$H$_{12}$N$_2^+$	C$_6$H$_4$(NH$_2$)N(CH$_3$)$_2$ (1–Amino–4–dimethylaminobenzene)		≤6.46±0.02	PI		3372

4.3. The Positive Ion Table—Continued

Ion	Reactant	Other products	Ionization or appearance potential (eV)	Method	Heat of formation (kJ mol^{-1})	Ref.
		$\mathbf{C_8H_{20}N_2^+}$				
$C_8H_{20}N_2^+$	$(n-C_4H_9)_2NNH_2$ (Value obtained without mass analysis)		7.47±0.05	PI		2173
		$\mathbf{C_9H_{10}N_2^+}$				
$C_9H_{10}N_2^+$	$C_6H_4(CN)N(CH_3)_2$ (4−Dimethylaminobenzoic acid nitrile)		7.99±0.04	RPD		3223
		$\mathbf{C_{10}H_{14}N_2^+}$				
$C_{10}H_{14}N_2^+$	$C_6H_5N=NC_4H_9$ (Phenyl tert−butyl diimine)		7.75±0.02	RPD		3254
		$\mathbf{C_{10}H_{16}N_2^+}$				
$C_{10}H_{16}N_2^+$	$C_6H_4(N(CH_3)_2)_2$ (1,4−Bis(dimethylamino)benzene)		≤6.20±0.02	PI		3372
See also − CTS: 2037, 2978						
		$\mathbf{C_{14}H_8N_2^+}$				
$C_{14}H_8N_2^+$	$(C_6H_4CN)_2$ (4,4′−Dicyanobiphenyl)		10.40	EI		3295
		$\mathbf{C_{14}H_{16}N_2^+}$				
$C_{14}H_{16}N_2^+$	$C_6H_4(NH_2)CH_2CH_2C_6H_4NH_2$ (1,2−Bis(4−aminophenyl)ethane)		7.6±0.1	EI		3288
		$\mathbf{C_{18}H_{10}N_2^+}$				
$C_{18}H_{10}N_2^+$	$C_{18}H_{10}N_2$ (Acenaphtho[1,2−b]quinoxaline)		8.63±0.1	EI		2489
		$\mathbf{C_{18}H_{10}N_2^{+2}}$				
$C_{18}H_{10}N_2^{+2}$	$C_{18}H_{10}N_2$ (Acenaphtho[1,2−b]quinoxaline)		21.0	EI		2489
		$\mathbf{C_{18}H_{22}N_2^+}$				
$C_{18}H_{22}N_2^+$	$C_{18}H_{22}N_2$ (Pertofran)		7.39±0.06	CTS		2987

4.3. The Positive Ion Table—Continued

Ion	Reactant	Other products	Ionization or appearance potential (eV)	Method	Heat of formation (kJ mol⁻¹)	Ref.
$C_{19}H_{24}N_2^+$						
$C_{19}H_{24}N_2^+$	$C_{19}H_{24}N_2$ (Tofranil)		7.35±0.05	CTS		2987
$C_{19}H_{24}N_2^+$	$C_{19}H_{24}N_2$ (Tofranil)		8.03	CTS		2562
$C_{22}H_{12}N_2^+$						
$C_{22}H_{12}N_2^+$	$C_{22}H_{12}N_2$ (8,9−Benzacenaphtho[1,2−b]quinoxaline)		8.07±0.1	EI		2489
$C_{22}H_{12}N_2^{+2}$						
$C_{22}H_{12}N_2^{+2}$	$C_{22}H_{12}N_2$ (8,9−Benzacenaphtho[1,2−b]quinoxaline)		20.5	EI		2489
$C_{24}H_{14}N_2^+$						
$C_{24}H_{14}N_2^+$	$C_{24}H_{14}N_2$ (1,2:3,4:6,7−Tribenzophenazine)		7.79±0.1	EI		2489
$C_{24}H_{14}N_2^{+2}$						
$C_{24}H_{14}N_2^{+2}$	$C_{24}H_{14}N_2$ (1,2:3,4:6,7−Tribenzophenazine)		20.0	EI		2489
CHN_3^{+2}						
CHN_3^{+2}	CH_3N_3		~34	EI		340
$CH_3N_3^+$						
$CH_3N_3^+$	CH_3N_3		9.5±0.1	EI		340
$CH_5N_3^+$						
$CH_5N_3^+$	$NH=C(NH_2)_2$		9.10±0.05	EI		2515
$C_2H_6N_3^+$						
$C_2H_6N_3^+$	$(CH_3)_2NN=NN(CH_3)_2$		9.65±0.1	EI		303

4.3. The Positive Ion Table—Continued

Ion	Reactant	Other products	Ionization or appearance potential (eV)	Method	Heat of formation (kJ mol^{-1})	Ref.
		$C_2H_7N_3^+$				
$C_2H_7N_3^+$	$NH=C(NH_2)NHCH_3$		8.60±0.05	EI		2515
		$C_3H_3N_3^+$				
$C_3H_3N_3^+$	$C_3H_3N_3$ (1,3,5–Triazine)		10.07±0.05	EI		3232
		$C_3H_9N_3^+$				
$C_3H_9N_3^+$	$NH=C(NH_2)N(CH_3)_2$		8.18±0.05	EI		2515
$C_3H_9N_3^+$	$CH_3N=C(NH_2)NHCH_3$		8.39±0.05	EI		2515
		$C_4H_{11}N_3^+$				
$C_4H_{11}N_3^+$	$CH_3N=C(NH_2)N(CH_3)_2$		8.06±0.05	EI		2515
$C_4H_{11}N_3^+$	$CH_3N=C(NHCH_3)_2$		8.15±0.05	EI		2515
		$C_5H_{13}N_3^+$				
$C_5H_{13}N_3^+$	$NH=C(N(CH_3)_2)_2$		8.12±0.05	EI		2515
$C_5H_{13}N_3^+$	$CH_3N=C(NHCH_3)N(CH_3)_2$		7.97±0.05	EI		2515
		$C_6H_{15}N_3^+$				
$C_6H_{15}N_3^+$	$CH_3N=C(N(CH_3)_2)_2$		7.84±0.05	EI		2515
		$C_5H_4N_4^+$				
$C_5H_4N_4^+$	$C_5H_4N_4$ (Purine)		9.68±0.1	EI		2514, 3336
		$C_6H_{12}N_4^+$				
$C_6H_{12}N_4^+$	$C_6H_{12}N_4$ (Hexamethylenetetramine)		8.26	PE		2843
		$C_{10}H_{20}N_4^+$				
$C_{10}H_{20}N_4^+$	$C_{10}H_{20}N_4$ (1,1',3,3'–Tetramethyl–$\Delta^{2,2'}$–bi(imidazolidine))		≤5.41±0.02	PI		3372

4.3. The Positive Ion Table—Continued

Ion	Reactant	Other products	Ionization or appearance potential (eV)	Method	Heat of formation (kJ mol^{-1})	Ref.
			$C_{10}H_{24}N_4^+$			
$C_{10}H_{24}N_4^+$	$((CH_3)_2N)_2C=C(N(CH_3)_2)_2$		$\leq 5.36\pm0.02$	PI		3372
			$C_5H_5N_5^+$			
$C_5H_5N_5^+$	$C_5H_3N_4NH_2$ (6–Aminopurine)		8.91 ± 0.1	EI		2514, 3336
			$C_8H_8BN^+$			
$C_8H_8BN^+$	C_8H_8BN (10,9–Borazaronaphthalene)		8.24	PE		2840, 2843
			$C_8H_{16}BN^+$			
$C_8H_{16}BN^+$	$C_8H_{16}BN$ (9–Aza–10–boradecalin)		8.47	PE		2840, 2843
			$C_{10}H_{15}BN_2^{+2}$			
$C_{10}H_{15}BN_2^{+2}$	$C_{10}H_{15}BN_2$ (1,3–Dimethyl–2–phenyl–1,3,2–diazaborolidine)		23.5 ± 1	EI		1418
			$C_3H_{12}BN_3^+$			
$C_3H_{12}BN_3^+$	$(CH_3NH)_3B$		8.04 ± 0.08	EI		2863
			$C_3H_{12}B_3N_3^+$			
$C_3H_{12}B_3N_3^+$	$C_3H_{12}B_3N_3$ (B–Trimethylborazine)		9.30	EI		2511
$C_3H_{12}B_3N_3^+$	$C_3H_{12}B_3N_3$ (N–Trimethylborazine)		9.07	EI		2511
			$C_6H_{18}BN_3^+$			
$C_6H_{18}BN_3^+$	$((CH_3)_2N)_3B$		7.57 ± 0.05	EI		2513
$C_6H_{18}BN_3^+$	$((CH_3)_2N)_3B$		7.75	EI		3227
$C_6H_{18}BN_3^+$	$((CH_3)_2N)_3B$		7.9 ± 0.1	EI		2863
			$C_6H_{18}BN_3^{+2}$			
$C_6H_{18}BN_3^{+2}$	$((CH_3)_2N)_3B$		21 ± 1	EI		1418

4.3. The Positive Ion Table—Continued

Ion	Reactant	Other products	Ionization or appearance potential (eV)	Method	Heat of formation (kJ mol^{-1})	Ref.

$C_6H_{18}B_3N_3^+$

Ion	Reactant	Other products	Ionization or appearance potential (eV)	Method	Heat of formation (kJ mol^{-1})	Ref.
$C_6H_{18}B_3N_3^+$	$C_6H_{18}B_3N_3$ (Hexamethylborazine)		8.77	EI		2511

$O^+(^4S_{3/2})$ $\Delta H_{f0}^\circ = 1560.7$ kJ mol^{-1} (373.0 kcal mol^{-1})

Ion	Reactant	Other products	Ionization or appearance potential (eV)	Method	Heat of formation (kJ mol^{-1})	Ref.
$O^+(^4S_{3/2})$	O		13.618	S	1560.7	2113
$O^+(^4S)$	O		13.62	PE		3076

See also – EI: 79, 2021, 2128, 2130, 2472, 2518

Ion	Reactant	Other products	Ionization or appearance potential (eV)	Method	Heat of formation (kJ mol^{-1})	Ref.
$O^+(^2D_{5/2})$	O		16.942	S	1881.5	2113
$O^+(^2D)$	O		16.96	PE		3076
$O^+(^2P_{3/2})$	O		18.635	S	2044.8	2113
O^+	O_2	O^-	17.272±0.024	PI	(1561)	2614
O^+	O_2	O^-	17.25±0.01	PI	(1559)	2624, 2627

See also – PI: 163
           EI: 79, 200, 288, 2014
           D: 6

Ion	Reactant	Other products	Ionization or appearance potential (eV)	Method	Heat of formation (kJ mol^{-1})	Ref.
O^+	O_2	O	18.8±0.4	PI		163
O^+	O_2	O	18.99±0.05	RPD		288

See also – EI: 79, 200, 2014

Ion	Reactant	Other products	Ionization or appearance potential (eV)	Method	Heat of formation (kJ mol^{-1})	Ref.
O^+	H_2O	H_2	19.0±0.2	EI		2484
O^+	H_2O	2H?	26.5±0.3	EI		2484
O^+	H_2O	2H?	29.15±0.25	EI		3144
O^+	D_2O	2D?	29.25±0.3	EI		3144
O^+	H_2O_2	H_2O	17.0±1.0	EI		37
O^+	CO	C^-	23.20±0.05	RPD		2431
O^+	CO	C^-	23.41±0.17	RPD		2180, 2191

The thermochemical threshold for this process is 23.46 eV.

See also – EI: 200, 1378

Ion	Reactant	Other products	Ionization or appearance potential (eV)	Method	Heat of formation (kJ mol^{-1})	Ref.
O^+	CO	C	24.65±0.05	RPD		2431
O^+	CO	C	24.78±0.23	RPD		2180, 2191

The thermochemical threshold for this process is 24.73 eV.

Ion	Reactant	Other products	Ionization or appearance potential (eV)	Method	Heat of formation (kJ mol^{-1})	Ref.
O^+	CO_2	CO	19.10±0.01	PI	(1564)	2624, 2627

See also – PI: 163
           EI: 2021, 2472

Ion	Reactant	Other products	Ionization or appearance potential (eV)	Method	Heat of formation (kJ mol^{-1})	Ref.
O^+	C_3O_2		25.8±0.5	EI		2687
O^+	NO	N	20.11±0.03	RPD	(1559)	328
O^+	NO	N	20.46±0.10	RPD		2431

The fragments are formed with 0.5±0.3 eV kinetic energy at threshold according to ref. 2431.
See, however, ref. 2823.

See also – PI: 163
           EI: 2021

4.3. The Positive Ion Table—Continued

Ion	Reactant	Other products	Ionization or appearance potential (eV)	Method	Heat of formation (kJ mol^{-1})	Ref.
O$^+$	N$_2$O	N$_2$	15.31	PI	(1563)	2629

See also – PI: 163
        EI: 58, 1451, 2697

Ion	Reactant	Other products	Ionization or appearance potential (eV)	Method	Heat of formation (kJ mol^{-1})	Ref.
O$^+$	NO$_2$	NO	16.82	PI	(1569)	2629

See also – PI: 163
        EI: 2018, 2021

Ion	Reactant	Other products	Ionization or appearance potential (eV)	Method	Heat of formation (kJ mol^{-1})	Ref.
O$^+$	CH$_3$NO$_2$		14.50±0.16	EI		90
O$^+$	CF$_2$O		35±1	EI		3236
O$^+$	CF$_3$OF		36±1	EI		3236
O$^+$	SO$_2$	SO	20.6	EI		418
O$^+$	(CH$_3$)$_2$SO		15.8±0.5	EI		3294
O$^+$	ClO$_3$F		22±1	EI		53
O$^+$	POCl$_3$		13±2	EI		1101

O^{+2} $\Delta H_{f0}^{\circ} = 4949.0$ kJ mol^{-1} (1182.8 kcal mol^{-1})

Ion	Reactant	Other products	Ionization or appearance potential (eV)	Method	Heat of formation (kJ mol^{-1})	Ref.
O^{+2}	O		48.734	S	4949.0	2113
O^{+2}	O$^+$(^{4}S$_{3/2}$)		35.116	S		2113
O^{+2}	O$^+$(^{4}S)		35.7	SEQ		2474
O^{+2}	O$^+$(^{2}D$_{5/2}$)		31.792	S		2113
O^{+2}	O$^+$(^{2}D)		32.7	SEQ		2474
O^{+2}	O$^+$(^{2}P$_{3/2}$)		30.099	S		2113
O^{+2}	O$^+$(^{2}P)		30.6	SEQ		2474
O^{+2}	O$_2$	O	52.7±0.5	NRE		2474
O^{+2}	CO	C	61.3±0.3	EI		2431

(1.6±0.7 eV average translational energy of decomposition at threshold)

Ion	Reactant	Other products	Ionization or appearance potential (eV)	Method	Heat of formation (kJ mol^{-1})	Ref.
O^{+2}	NO	N	61.62±0.15	EI		2431

(5.4±1 eV average translational energy of decomposition at threshold)

Ion	Reactant	Other products	Ionization or appearance potential (eV)	Method	Heat of formation (kJ mol^{-1})	Ref.
O^{+2}	NO	N	62.2±1.0	NRE		2823
O^{+2}	N$_2$O	2N?	60.5±1.0	NRE		3138
O^{+2}	NO$_2$	NO	52.1±1	NRE		3138

O^{+3} $\Delta H_{f0}^{\circ} = 10249.4$ kJ mol^{-1} (2449.7 kcal mol^{-1})

Ion	Reactant	Other products	Ionization or appearance potential (eV)	Method	Heat of formation (kJ mol^{-1})	Ref.
O^{+3}	O		103.668	S	10249.4	2113
O^{+3}	O^{+2}		54.934	S		2113
O^{+3}	O$_2$	O?	130±3	EI		2474

O^{+4} $\Delta H_{f0}^{\circ} = 17718.7$ kJ mol^{-1} (4234.9 kcal mol^{-1})

Ion	Reactant	Other products	Ionization or appearance potential (eV)	Method	Heat of formation (kJ mol^{-1})	Ref.
O^{+4}	O		181.080	S	17718.7	2113
O^{+4}	O^{+3}		77.412	S		2113, 3166

O^{+5} $\Delta H_{f0}^{\circ} = 28708.1$ kJ mol^{-1} (6861.4 kcal mol^{-1})

Ion	Reactant	Other products	Ionization or appearance potential (eV)	Method	Heat of formation (kJ mol^{-1})	Ref.
O^{+5}	O		294.976	S	28708.1	2113
O^{+5}	O^{+4}		113.896	S		2113, 3166

4.3. The Positive Ion Table—Continued

Ion	Reactant	Other products	Ionization or appearance potential (eV)	Method	Heat of formation (kJ mol^{-1})	Ref.	
$O_2^+(X^2\Pi_{3/2g})$		$\Delta H_{f0}^\circ = 1165$ kJ mol^{-1} (278 kcal mol^{-1})					
$O_2^+(X^2\Pi_{1/2g})$		$\Delta H_{f0}^\circ = 1167$ kJ mol^{-1} (279 kcal mol^{-1})					
$O_2^+(a^4\Pi_u)$		$\Delta H_{f0}^\circ = 1553.8$ kJ mol^{-1} (371.4 kcal mol^{-1})					
$O_2^+(A^2\Pi_u)$		$\Delta H_{f0}^\circ = 1646$ kJ mol^{-1} (394 kcal mol^{-1})					
$O_2^+(b^4\Sigma_g^-)$		$\Delta H_{f0}^\circ = 1753.2$ kJ mol^{-1} (419.0 kcal mol^{-1})					
$O_2^+(B^2\Sigma_g^-)$		$\Delta H_{f0}^\circ = 1958.3$ kJ mol^{-1} (468.0 kcal mol^{-1})					
$O_2^+(c^4\Sigma_u^-)$		$\Delta H_{f0}^\circ = 2370.1$ kJ mol^{-1} (566.5 kcal mol^{-1})					
$O_2^+(X^2\Pi_g)$	O_2		12.059±0.001	S		2048, 3305,	2758, 3306
$O_2^+(a^4\Pi_u)$	O_2		16.104	S	1553.8	—	
(Based on $b^4\Sigma_g^-$ limit below and first negative bands from ref. 3309)							
$O_2^+(A^2\Pi_u)$	O_2		17.064	S	1646	—	
(Based on average of $X^2\Pi_{3/2g}$ and $X^2\Pi_{1/2g}$ limits below and second negative bands from ref. 3306)							
$O_2^+(b^4\Sigma_g^-)$	O_2		18.1702±0.0002	S	1753.2	2678	
$O_2^+(B^2\Sigma_g^-)$	O_2		20.2960±0.0009	S	1958.3	2678	
$O_2^+*?$	O_2		20.996±0.005	S		2678	
$O_2^+(c^4\Sigma_u^-)$	O_2		24.564±0.004	S	2370.1	3307	
(Average of two Rydberg series limits)							
$O_2^+(X^2\Pi_g)$	O_2		12.065±0.003	PI		1032	
$O_2^+(X^2\Pi_g)$	O_2		12.078±0.005	PI		2013	
$O_2^+(X^2\Pi_g)$	O_2		12.072±0.008	PI		2624,	2627
$O_2^+(X^2\Pi_g)$	O_2		12.075±0.01	PI		182,	416
$O_2^+(X^2\Pi_{3/2g})$	O_2		12.071±0.005	PE	1165	3068	
$O_2^+(X^2\Pi_{1/2g})$	O_2		12.095±0.005	PE	1167	3068	
$O_2^+(X^2\Pi_g)$	O_2		12.070±0.005	PE		2827	
$O_2^+(X^2\Pi_g)$	O_2		12.075	PE		2792	
$O_2^+(X^2\Pi_g)$	O_2		12.08	PE		2830	
$O_2^+(a^4\Pi_u)$	O_2		16.101±0.002	PE		3068	
$O_2^+(a^4\Pi_u)$	O_2		16.11	PE		2792	
$O_2^+(a^4\Pi_u)$	O_2		16.12	PE		2830	
$O_2^+(A^2\Pi_u)$	O_2		17.045±0.004	PE		3068,	3069
$O_2^+(b^4\Sigma_g^-)$	O_2		18.171±0.003	PE		3068	
$O_2^+(b^4\Sigma_g^-)$	O_2		18.19	PE		2792	
$O_2^+(b^4\Sigma_g^-)$	O_2		18.17	PE		2830	
$O_2^+(B^2\Sigma_g^-)$	O_2		20.296±0.004	PE		3068	
$O_2^+(B^2\Sigma_g^-)$	O_2		20.33	PE		2792	
$O_2^+(B^2\Sigma_g^-)$	O_2		20.29	PE		2827,	2830
$O_2^+(^2\Pi_u)$	O_2		23.5±0.3	PE		3095	
$O_2^+(^2\Pi_u)$	O_2		24.0 (V)	PE		3068	
$O_2^+(^2\Pi_u)$	O_2		~23.7 (V)	PE		3064	
$O_2^+(c^4\Sigma_u^-)$	O_2		24.577±0.012	PE		3068	
$O_2^+(c^4\Sigma_u^-)$	O_2		24.6±0.3	PE		3095	

4.3. The Positive Ion Table—Continued

Ion	Reactant	Other products	Ionization or appearance potential (eV)	Method	Heat of formation (kJ mol^{-1})	Ref.
O_2^{+*}	O_2		27.3±0.3	PE		3095
O_2^{+*}	O_2		27.4 (V)	PE		3068

See general theoretical discussion of PE results in ref. 3064.

See also – S: 2099, 3006, 3303
        PI: 163, 230, 297, 3093
        PE: 1108, 1130, 2812, 2817
        PEN: 2430
        EM: 116, 1094
        EI: 3, 31, 79, 119, 218, 287, 288, 364, 383, 1029, 2136, 2188, 2444, 2518, 2557, 3131, 3244, 3414

Ion	Reactant	Other products	Ionization or appearance potential (eV)	Method	Heat of formation (kJ mol^{-1})	Ref.
$O_2^+(X^2\Pi_g)$	$O_2(a^1\Delta_g)$		11.090±0.001	PI		2767
$O_2^+(X^2\Pi_g)$	$O_2(a^1\Delta_g)$		11.09±0.005	PE		3107
O_2^+	H_2O_2	H_2	15.8±0.5	EI		37
O_2^+	SO_2	S	17.5±0.3	EI		418
O_2^+	ClO_3F		15±1	EI		54

$O_2^{+2}(X^1\Sigma_g^+)$ $\Delta H_{f0}^\circ \sim 3595$ kJ mol^{-1} (859 kcal mol^{-1})

Ion	Reactant	Other products	Ionization or appearance potential (eV)	Method	Heat of formation (kJ mol^{-1})	Ref.
$O_2^{+2}(X^1\Sigma_g^+)$	O_2		37.4	AUG	~3609	3304
$O_2^{+2}(X^1\Sigma_g^+)$	O_2		36.3±0.5	NRE	~3502	2785
$O_2^{+2}(X^1\Sigma_g^+)$	O_2		37.2±0.5	NRE	~3589	2474
$O_2^{+2}(X^1\Sigma_g^+)$	O_2		37.4±0.2	EI	~3609	2474
$O_2^{+2}(X^1\Sigma_g^+)$	O_2^+		25.9±0.2	SEQ	~3665	2474
$O_2^{+2}(A^3\Sigma_u^+)$	O_2^+		29.9±0.2	SEQ		2474

O_3^+ $\Delta H_{f0}^\circ \sim 1368$ kJ mol^{-1} (327 kcal mol^{-1})

Ion	Reactant	Other products	Ionization or appearance potential (eV)	Method	Heat of formation (kJ mol^{-1})	Ref.
O_3^+	O_3		12.67	PI	1368	3146, 3308
O_3^+	O_3		12.89±0.10	RPD		20
O_3^+	O_3		12.80±0.05	EI		77

See also – PI: 416
        PE: 1441
        EI: 2516

OH^+ $\Delta H_{f0}^\circ = 1287$ kJ mol^{-1} (308 kcal mol^{-1})

Ion	Reactant	Other products	Ionization or appearance potential (eV)	Method	Heat of formation (kJ mol^{-1})	Ref.
OH^+	OH		13.18±0.1	EI		67, 2786
OH^+	OH		12.94	D		2631
OH^+	H_2O	H	18.05	PI	1287	2631
OH^+	H_2O	H	18.19±0.1	EI		67, 2786

See also – PI: 427
        EI: 2066, 2484, 3144

4.3. The Positive Ion Table—Continued

Ion	Reactant	Other products	Ionization or appearance potential (eV)	Method	Heat of formation (kJ mol⁻¹)	Ref.
OH⁺	H₂O₂	OH	15.35±0.10	EI		37
OH⁺	H₂O₂	OH	15.60±0.08	EI		2066
OH⁺	CH₃COOH		15.1	EI		298

$$H_2O^+(^2B_1) \quad \Delta H_{f0}^\circ = \;\;978 \text{ kJ mol}^{-1} \text{ (234 kcal mol}^{-1}\text{)}$$
$$H_2O^+(^2A_1) \quad \Delta H_{f0}^\circ \sim 1083 \text{ kJ mol}^{-1} \text{ (259 kcal mol}^{-1}\text{)}$$
$$H_2O^+(^2B_2) \quad \Delta H_{f0}^\circ = 1423 \text{ kJ mol}^{-1} \text{ (340 kcal mol}^{-1}\text{)}$$

Ion	Reactant	Other products	Ionization or appearance potential (eV)	Method	Heat of formation (kJ mol⁻¹)	Ref.
$H_2O^+(^2B_1)$	H₂O		12.62±0.02	S	979	3140
$H_2O^+(^2B_1)$	H₂O		12.614±0.005	PI	978	2013
$H_2O^+(^2B_1)$	H₂O		12.597±0.010	PI	977	1253
$H_2O^+(^2B_1)$	H₂O		12.593±0.01	PI	976	2631
$H_2O^+(^2B_1)$	H₂O		12.616±0.01	PE	978	2836
$H_2O^+(^2B_1)$	H₂O		12.65	PEN		2430
$H_2O^+(^2B_1)$	H₂O		12.60±0.01	RPD		463
$H_2O^+(^2A_1)$	H₂O		13.7	PE	~1083	2836
$H_2O^+(^2A_1)$	H₂O		14.3	PEN		2430
$H_2O^+(^2B_2)$	H₂O		17.22	PE	1423	2836
$H_2O^+(^2B_2)$	H₂O		18.0	PEN		2430

See also – PI: 182, 230, 416, 427, 1103
         PE: 1130, 2801, 2813, 2855
         EI: 97, 1372, 2060, 2066, 2136, 2786, 2991, 3128, 3129, 3144, 3190, 3414

Ion	Reactant	Other products	Ionization or appearance potential (eV)	Method	Heat of formation (kJ mol⁻¹)	Ref.
H₂O⁺	H₂O₂	O	14.09±0.10	EI		37
H₂O⁺	C₂H₅OH	C₂H₄	13.06	RPD		2999
H₂O⁺	C₅H₉OH (Cyclopentanol)	C₅H₈	13.23	RPD		2999

$$D_2O^+(^2B_1) \quad \Delta H_{f0}^\circ = \;\;972 \text{ kJ mol}^{-1} \text{ (232 kcal mol}^{-1}\text{)}$$
$$D_2O^+(^2A_1) \quad \Delta H_{f0}^\circ \sim 1076 \text{ kJ mol}^{-1} \text{ (257 kcal mol}^{-1}\text{)}$$
$$D_2O^+(^2B_2) \quad \Delta H_{f0}^\circ = 1419 \text{ kJ mol}^{-1} \text{ (339 kcal mol}^{-1}\text{)}$$

Ion	Reactant	Other products	Ionization or appearance potential (eV)	Method	Heat of formation (kJ mol⁻¹)	Ref.
$D_2O^+(^2B_1)$	D₂O		12.637±0.005	PI	973	2013
$D_2O^+(^2B_1)$	D₂O		12.624±0.01	PE	972	2836
$D_2O^+(^2A_1)$	D₂O		13.7	PE	~1076	2836
$D_2O^+(^2B_2)$	D₂O		17.26	PE	1419	2836

See also – PE: 1130

H_2O^{+2}

Ion	Reactant	Other products	Ionization or appearance potential (eV)	Method	Heat of formation (kJ mol⁻¹)	Ref.
(H_2O^{+2})	H₂O		39.2	AUG		3304

No H_2O^{+2} ions have been directly observed by mass spectrometric techniques.

4.3. The Positive Ion Table—Continued

Ion	Reactant	Other products	Ionization or appearance potential (eV)	Method	Heat of formation (kJ mol⁻¹)	Ref.	

HO_2^+

Ion	Reactant	Other products	Ionization or appearance potential (eV)	Method	Heat of formation	Ref.	
HO_2^+	HO_2		11.53±0.02	EI		31,	36
HO_2^+	H_2O_2	H	15.36±0.05	EI		36,	37

See also – EI: 31

$H_2O_2^+$

| $H_2O_2^+$ | H_2O_2 | | 10.92±0.05 | EI | | | 37 |
| $H_2O_2^+$ | H_2O_2 | | 11.26±0.05 | EI | | | 2066 |

LiO^+

| LiO^+ | LiO | | 8.6±1.0 | EI | | | 3257 |
| LiO^+ | LiO | | 9.0±0.5 | EI | | | 2565 |

Li_2O^+

Li_2O^+	Li_2O		6.3±1.0	EI			3257
Li_2O^+	Li_2O		6.8±0.2	EI			318
Li_2O^+	Li_2O		6.9±0.3	EI			1112
Li_2O^+	Li_2O		7.0±0.5	EI			2565
Li_2O^+	Li_2MoO_4	MoO_3?	10.8±1.0	EI			3257
Li_2O^+	Li_2WO_4	WO_3?	11.1±1.0	EI			3257

BeO^+

| BeO^+ | BeO | | 10.1±0.4 | EI | | | 1106 |
| BeO^+ | BeO | | 10.4 | EI | | | 3197 |

Be_2O^+

| Be_2O^+ | Be_2O | | 10.5±0.5 | EI | | | 1106 |

$Be_2O_2^+$

| $Be_2O_2^+$ | Be_2O_2 | | 10.5 | EI | | | 3197 |
| $Be_2O_2^+$ | Be_2O_2 | | 11.1±0.4 | EI | | | 1106 |

$Be_3O_2^+$

| $Be_3O_2^+$ | Be_3O_2 | | 12.5±1.0 | EI | | | 1106 |

4.3. The Positive Ion Table—Continued

Ion	Reactant	Other products	Ionization or appearance potential (eV)	Method	Heat of formation (kJ mol^{-1})	Ref.
			Be$_3$O$_3^+$			
Be$_3$O$_3^+$	Be$_3$O$_3$		10.7±0.4	EI		1106
Be$_3$O$_3^+$	Be$_3$O$_3$		11.0	EI		3197
			Be$_4$O$_4^+$			
Be$_4$O$_4^+$	Be$_4$O$_4$		11.0	EI		3197
			Be$_5$O$_5^+$			
Be$_5$O$_5^+$	Be$_5$O$_5$		~11	EI		3197
			Be$_6$O$_6^+$			
Be$_6$O$_6^+$	Be$_6$O$_6$		~11	EI		3197
			BO$^+$			
BO$^+$	BO		12.8±1	EI		3206
BO$^+$	BO		13.5±0.5	EI		3199
			BO$_2^+$			
BO$_2^+$	(CH$_3$O)$_3$B		17.5±0.3	EI		115
			B$_2$O$_2^+$			
B$_2$O$_2^+$	B$_2$O$_2$		14±0.5	EI		3199
			BO$_3^+$			
BO$_3^+$	(CH$_3$O)$_3$B		12.7±1.0	EI		364
BO$_3^+$	(CH$_3$O)$_3$B		13.2±0.3	EI		115
			B$_2$O$_3^+$			
B$_2$O$_3^+$	B$_2$O$_3$		13.7–14	EI		3128
B$_2$O$_3^+$	B$_2$O$_3$		14±0.5	EI		3199

4.3. The Positive Ion Table—Continued

Ion	Reactant	Other products	Ionization or appearance potential (eV)	Method	Heat of formation (kJ mol^{-1})	Ref.

	$CO^+(X^2\Sigma^+)$	$\Delta H^\circ_{f0} = 1238$ kJ mol^{-1} (296 kcal mol^{-1})				
	$CO^+(A^2\Pi_i)$	$\Delta H^\circ_{f0} = 1482$ kJ mol^{-1} (354 kcal mol^{-1})				
	$CO^+(B^2\Sigma^+)$	$\Delta H^\circ_{f0} = 1785$ kJ mol^{-1} (427 kcal mol^{-1})				
$CO^+(X^2\Sigma^+)$	CO		14.013±0.004	S	1238	2098
$CO^+(A^2\Pi_i)$	CO		16.537	S	1482	2098
$CO^+(B^2\Sigma^+)$	CO		19.675±0.004	S	1785	2098
$CO^+(X^2\Sigma^+)$	CO		13.985	PI		1382
$CO^+(X^2\Sigma^+)$	CO		14.01	PE		3171
$CO^+(X^2\Sigma^+)$	CO		14.00	PE		2830
$CO^+(X^2\Sigma^+)$	CO		14.01	PE		2792
$CO^+(X^2\Sigma^+)$	CO		14.0 (V)	PE		3094
$CO^+(A^2\Pi_i)$	CO		16.54±0.01	PE		3171
$CO^+(A^2\Pi_i)$	CO		16.54	PE		2830
$CO^+(A^2\Pi_i)$	CO		16.55	PE		2792
$CO^+(A^2\Pi_i)$	CO		17.2 (V)	PE		3094
$CO^+(B^2\Sigma^+)$	CO		19.67±0.01	PE		3171
$CO^+(B^2\Sigma^+)$	CO		19.65	PE		2830
$CO^+(B^2\Sigma^+)$	CO		19.69	PE		2792
$CO^+(B^2\Sigma^+)$	CO		19.8 (V)	PE		3094
$CO^+(C^2\Sigma^+)$	CO		38.9 (V)	PE		3094

See also – PI: 163, 182, 416, 2956
        PE: 1108, 1130, 2812
        PEN: 2430, 2467, 3171
        EI: 3, 127, 1012, 1029, 1051, 1172, 2023, 2431, 2453, 2557, 2687, 2883

Ion	Reactant	Other products	Ionization or appearance potential (eV)	Method	Heat of formation (kJ mol^{-1})	Ref.
CO^+	CO_2	O^-?	19.5±0.2	PI		163
CO^+	C_3O_2		20.1±0.3	EI		2687
CO^+	HCHO		18.7±0.2	EI		204
CO^+	DCDO		18.8±0.3	EI		204
CO^+	CH_3OH	$2H_2$	14.31±0.05	EI		3176
CO^+	CH_3OH	$2H_2$	13.7	EI		46
CO^+	$CH_2=CO$		13.57±0.3	EI		2800
CO^+	CH_3CHO	CH_4	14.0±0.1	RPD		2576
CO^+	CH_3CHO	CH_4	13.9±0.1	SD		1404

See also – EI: 2883

Ion	Reactant	Other products	Ionization or appearance potential (eV)	Method	Heat of formation (kJ mol^{-1})	Ref.
CO^+	$(CH_2)_2O$ (1,2–Epoxyethane)	CH_4	12.6±0.4	EI		50
CO^+	CH_3COOH	CH_3OH	15.3±0.1	RPD		2576

CO^{+2} $\Delta H^\circ_{f0} \sim 3751$ kJ mol^{-1} (896 kcal mol^{-1})

Ion	Reactant	Other products	Ionization or appearance potential (eV)	Method	Heat of formation (kJ mol^{-1})	Ref.
CO^{+2}	CO		39.9	AUG	~3736	3304
CO^{+2}	CO		40.2	AUG	~3765	3304
CO^{+2}	CO		41.5±0.4	NRE		3139

4.3. The Positive Ion Table—Continued

Ion	Reactant	Other products	Ionization or appearance potential (eV)	Method	Heat of formation (kJ mol^{-1})	Ref.
CO^{+2}	CO		41.8±0.3	FD		212
CO^{+2}	$CO^+(A^2\Pi_i)$		25.0±0.4	SEQ		2016

C_2O^+

C_2O^+	C_3O_2	CO	15.1±0.1	EI		2687
C_2O^+	$Ni(CO)_4$		31.7±1	EI		2579

C_3O^+

C_3O^+	C_3O_2		15.9±0.3	EI		2687

$CO_2^+(X^2\Pi_{3/2g})$	$\Delta H_{f0}^{\circ} =$	935 kJ mol^{-1} (224 kcal mol^{-1})			
$CO_2^+(X^2\Pi_{1/2g})$	$\Delta H_{f0}^{\circ} =$	938 kJ mol^{-1} (224 kcal mol^{-1})			
$CO_2^+(A^2\Pi_{3/2u})$	$\Delta H_{f0}^{\circ} =$	1277 kJ mol^{-1} (305 kcal mol^{-1})			
$CO_2^+(A^2\Pi_{1/2u})$	$\Delta H_{f0}^{\circ} =$	1278 kJ mol^{-1} (306 kcal mol^{-1})			
$CO_2^+(B^2\Sigma_u^+)$	$\Delta H_{f0}^{\circ} =$	1351 kJ mol^{-1} (323 kcal mol^{-1})			
$CO_2^+(C^2\Sigma_g^+)$	$\Delta H_{f0}^{\circ} =$	1478 kJ mol^{-1} (353 kcal mol^{-1})			

Ion	Reactant	Other products	Ionization or appearance potential (eV)	Method	Heat of formation	Ref.
$CO_2^+(X^2\Pi_{3/2g})$	CO_2		13.769±0.03	S	935	148
$CO_2^+(X^2\Pi_{1/2g})$	CO_2		13.792±0.03	S	938	148
$CO_2^+(A^2\Pi_{3/2u})$	CO_2		17.312	S	1277	1179
$CO_2^+(A^2\Pi_{1/2u})$	CO_2		17.323	S	1278	1179
$CO_2^+(B^2\Sigma_u^+)$	CO_2		18.076	S	1351	1179

(Average of two Rydberg series limits)

See also – S: 148

$CO_2^+(C^2\Sigma_g^+)$	CO_2		19.389	S	1478	148

(Average of three Rydberg series limits)

$CO_2^+(X^2\Pi_g)$	CO_2		13.788±0.005	PE		2901
$CO_2^+(X^2\Pi_g)$	CO_2		13.78±0.01	PE		2848
$CO_2^+(X^2\Pi_g)$	CO_2		13.77	PE		2855
$CO_2^+(A^2\Pi_u)$	CO_2		17.323±0.005	PE		2901
$CO_2^+(A^2\Pi_u)$	CO_2		17.32±0.02	PE		2848
$CO_2^+(A^2\Pi_u)$	CO_2		17.32	PE		2855
$CO_2^+(B^2\Sigma_u^+)$	CO_2		18.082±0.005	PE		2901
$CO_2^+(B^2\Sigma_u^+)$	CO_2		18.05±0.02	PE		2848
$CO_2^+(B^2\Sigma_u^+)$	CO_2		18.05	PE		2855
$CO_2^+(C^2\Sigma_g^+)$	CO_2		19.400±0.005	PE		2901
$CO_2^+(C^2\Sigma_g^+)$	CO_2		19.36±0.03	PE		2848
$CO_2^+(X^2\Pi_{3/2g})$	CO_2		13.767±0.003	PI		3048
$CO_2^+(X^2\Pi_{1/2g})$	CO_2		13.786±0.003	PI		3048

4.3. The Positive Ion Table—Continued

Ion	Reactant	Other products	Ionization or appearance potential (eV)	Method	Heat of formation (kJ mol^{-1})	Ref.
$CO_2^+(X^2\Pi_g)$	CO_2		13.75 ± 0.05	EM		2427

See also – S: 409, 410
PI: 163, 182, 230, 416, 2624, 2627, 3032
PE: 92, 1130, 2839, 2856, 2875, 3019, 3026
PEN: 2430, 2467
EI: 3, 164, 169, 1029, 3131, 3435

Ion	Reactant	Other products	Ionization or appearance potential (eV)	Method	Heat of formation (kJ mol^{-1})	Ref.
CO_2^{+2}		$\Delta H_{f0}^\circ \sim 3188$ kJ mol^{-1} (762 kcal mol^{-1})				
CO_2^{+2}	CO_2		37.4	AUG	~3215	3304
CO_2^{+2}	CO_2		37.8	AUG	~3254	3304
CO_2^{+2}	CO_2		38.0 ± 0.2	NRE	~3273	2476
CO_2^{+2}	CO_2		36.4 ± 0.3	FD	~3119	212
CO_2^{+2}	CO_2		36.3 ± 0.5	EI	~3109	2472
CO_2^{+2}	CO_2^+		23.0 ± 0.6	SEQ	~3156	2472

Ion	Reactant	Other products	Ionization or appearance potential (eV)	Method	Heat of formation (kJ mol^{-1})	Ref.
$C_2O_2^+$						
$C_2O_2^+$	C_3O_2		15.8	EI		2687
$C_2O_2^+$	$Ni(CO)_4$		21.6 ± 0.5	EI		2579

Ion	Reactant	Other products	Ionization or appearance potential (eV)	Method	Heat of formation (kJ mol^{-1})	Ref.
$C_3O_2^+$		$\Delta H_{f0}^\circ = 926$ kJ mol^{-1} (221 kcal mol^{-1})				
$C_3O_2^+$	C_3O_2		10.60	S	926	2668, 2717
$C_3O_2^+$	C_3O_2		10.60 ± 0.03	PI	926	2717
$C_3O_2^+$	C_3O_2		10.60	PE	926	2807

See also – EI: 2687

Ion	Reactant	Other products	Ionization or appearance potential (eV)	Method	Heat of formation (kJ mol^{-1})	Ref.
$C_3O_2^{+2}$						
$C_3O_2^{+2}$	C_3O_2		33.0 ± 0.2	EI		2687

Ion	Reactant	Other products	Ionization or appearance potential (eV)	Method	Heat of formation (kJ mol^{-1})	Ref.
	$NO^+(X^1\Sigma^+)$	$\Delta H_{f0}^\circ = 983.6$ kJ mol^{-1} (235.1 kcal mol^{-1})				
	$NO^+(a^3\Sigma^+)$	$\Delta H_{f0}^\circ = 1600$ kJ mol^{-1} (382 kcal mol^{-1})				
	$NO^+(b^3\Pi)$	$\Delta H_{f0}^\circ = 1687.4$ kJ mol^{-1} (403.3 kcal mol^{-1})				
	$NO^+(w^3\Delta)$	$\Delta H_{f0}^\circ = 1717$ kJ mol^{-1} (410 kcal mol^{-1})				
	$NO^+(b'^3\Sigma^-)$	$\Delta H_{f0}^\circ = 1786$ kJ mol^{-1} (427 kcal mol^{-1})				
	$NO^+(A'^1\Sigma^-)$	$\Delta H_{f0}^\circ = 1809$ kJ mol^{-1} (432 kcal mol^{-1})				
	$NO^+(W^1\Delta)$	$\Delta H_{f0}^\circ \sim 1833$ kJ mol^{-1} (438 kcal mol^{-1})				
	$NO^+(A^1\Pi)$	$\Delta H_{f0}^\circ = 1858$ kJ mol^{-1} (444 kcal mol^{-1})				
$NO^+(X^1\Sigma^+)$	NO		9.2639 ± 0.0006	S	983.6	2764
$NO^+(X^1\Sigma^+)$	NO		9.267 ± 0.005	S		1217
$NO^+(X^1\Sigma^+)$	NO		9.266 ± 0.008	S		1148
$NO^+(X^1\Sigma^+)$	NO		9.267	RPI		2773

4.3. The Positive Ion Table—Continued

Ion	Reactant	Other products	Ionization or appearance potential (eV)	Method	Heat of formation (kJ mol⁻¹)	Ref.	
NO⁺(X¹Σ⁺)	NO		9.250±0.005	PI		1032,	1253
NO⁺(X¹Σ⁺)	NO		9.25±0.02	PI		158,	182, 416
NO⁺(X¹Σ⁺)	NO		9.27	PE			3418
NO⁺(X¹Σ⁺)	NO		9.28±0.03	RPD			2431
NO⁺(X¹Σ⁺)	NO		9.25±0.15	EDD			3172
NO⁺(X¹Σ⁺)	NO		9.25±0.02	EI			2991
NO⁺(a³Σ⁺)	NO		15.649	PE	1600		3123
NO⁺(a³Σ⁺)	NO		15.67	PE			3418
NO⁺(b³Π)	NO		16.558±0.001	S	1687.4		1217
NO⁺(b³Π)	NO		16.558	PE	1687.4		3123
NO⁺(b³Π)	NO		16.585	PE			3418
NO⁺(w³Δ)	NO		16.860	PE	1717		3123
NO⁺(w³Δ)	NO		16.89	PE			3418
NO⁺(b'³Σ⁻)	NO		17.585	PE	1786		3123
NO⁺(b'³Σ⁻)	NO		~17.64	PE			3418
NO⁺(A'¹Σ⁻)	NO		17.820	PE	1809		3123
NO⁺(A'¹Σ⁻)	NO		~17.86	PE			3418
NO⁺(W¹Δ)	NO		~18.07	PE	~1833		3123
NO⁺(W¹Δ)	NO		~18.09	PE			3418
NO⁺(A¹Π)	NO		18.328±0.005	S	1858		1217
NO⁺(A¹Π)	NO		18.322	PE	1858		3123
NO⁺(A¹Π)	NO		18.335	PE	1859		3418
NO⁺(c³Π)	NO		20.41	PE			3123
NO⁺(B¹Π)	NO		21.72	PE			3123
NO⁺(B'¹Σ⁺)	NO		<22.5	PE			3123

The term assignments are those of Edqvist *et al.*, ref. 3123. The assignments of the states above 20 eV are questionable, see H. Lefebvre–Brion, Chem. Phys. Letters **9**, 463 (1971). See also K. P. Huber, Can. J. Phys. **46**, 1691 (1968) and discussion in refs. 3123, 3418.

See also – S: 3303, 3319
        PI: 86, 157, 163, 227, 228, 297
        PE: 1108, 1130, 1415, 2809, 2810, 2812, 2817, 2824, 2825, 2830, 2875, 3109
        PEN: 2430, 2467
        EI: 58, 328

4.3. The Positive Ion Table—Continued

Ion	Reactant	Other products	Ionization or appearance potential (eV)	Method	Heat of formation (kJ mol^{-1})	Ref.
NO$^+$	N$_2$O	N(^{4}S$°$)	15.01	PI		2624, 2629
NO$^+$	N$_2$O	N(^{4}S$°$)	13.75±0.10	RPD		2697
NO$^+$	N$_2$O	N(^{4}S$°$)	14.3±0.3	RPD		2693

The thermochemical threshold for this process is 14.19 eV. Kinetic energy measurements of the fragments are conflicting. Curran and Fox, ref. 2697, report zero kinetic energy for NO$^+$ up to 3 eV above threshold. However, Newton and Sciamanna, ref. 3138, and Coleman, Delderfield and Reuben, ref. 2693, report a metastable transition for N$_2$O$^+$ → NO$^+$ + N with an appearance potential of 15.1±0.2 or 15.7±0.5 eV, and a total kinetic energy of 1.05±0.05 eV. The discrepancies are unexplained.

Ion	Reactant	Other products	Ionization or appearance potential (eV)	Method	Heat of formation (kJ mol^{-1})	Ref.
NO$^+$	N$_2$O	N(^{2}D$°$)?	16.53	PI		2624, 2629
NO$^+$	N$_2$O	N(^{2}P$°$)?	17.74	PI		2624, 2629

See also – PI: 163
　　　　　 EI: 58, 3135

Ion	Reactant	Other products	Ionization or appearance potential (eV)	Method	Heat of formation (kJ mol^{-1})	Ref.
NO$^+$	NO$_2$	O	12.34	PI	(980)	2624, 2629

Newton and Sciamanna, ref. 3138, have observed metastable transitions corresponding to this process, with total kinetic energies of 0.5 and 1.1 eV. Whether these occur at threshold is problematic.

See also – PI: 163
　　　　　 EI: 139

Ion	Reactant	Other products	Ionization or appearance potential (eV)	Method	Heat of formation (kJ mol^{-1})	Ref.
NO$^+$	HNCO		15.76±0.15	EI		3365
NO$^+$	C$_2$H$_5$NO	C$_2$H$_5$	12.3±0.2	EDD		3180
NO$^+$	(CH$_3$)$_2$NNO		11.1±0.1	EI		303
NO$^+$	CH$_3$ONO	CH$_3$O	11.07±0.06	RPD		1139
NO$^+$	CH$_3$ONO	CH$_3$O	11.15	RPD		3347

(0.07±0.02 eV average translational energy of decomposition at threshold)

Ion	Reactant	Other products	Ionization or appearance potential (eV)	Method	Heat of formation (kJ mol^{-1})	Ref.
NO$^+$	C$_2$H$_5$ONO	C$_2$H$_5$O	11.28±0.05	RPD		2776
NO$^+$	C$_2$H$_5$ONO	C$_2$H$_5$O	11.69±0.05	RPD		3347

(0.07±0.01 eV average translational energy of decomposition at threshold)

Ion	Reactant	Other products	Ionization or appearance potential (eV)	Method	Heat of formation (kJ mol^{-1})	Ref.
NO$^+$	n–C$_4$H$_9$ONO		11.48±0.05	RPD		2776
NO$^+$	C$_6$H$_4$(NO$_2$)OH (3–Nitrophenol)		10.5±0.2	EI		2833
NO$^+$	FNO	F	11.6±0.2	EDD		3172
NO$^+$	ONF$_3$	3F?	15.21±0.02	PI		3038

NO^{+2}　　　$\Delta H°_{f0} \sim 3486$ kJ mol^{-1} (833 kcal mol^{-1})

Ion	Reactant	Other products	Ionization or appearance potential (eV)	Method	Heat of formation	Ref.
NO^{+2}	NO		34.7	AUG	~3438	3304
NO^{+2}	NO		35.7	AUG	~3534	3304
NO^{+2}	NO		38.3±0.5	NRE		2823
NO^{+2}	NO		39.8±0.3	FD		212

4.3. The Positive Ion Table—Continued

Ion	Reactant	Other products	Ionization or appearance potential (eV)	Method	Heat of formation (kJ mol^{-1})	Ref.
	$N_2O^+(X^2\Pi_{3/2})$	$\Delta H^\circ_{f0} = 1330$ kJ mol^{-1} (318 kcal mol^{-1})				
	$N_2O^+(X^2\Pi_{1/2})$	$\Delta H^\circ_{f0} = 1333$ kJ mol^{-1} (319 kcal mol^{-1})				
	$N_2O^+(A^2\Sigma^+)$	$\Delta H^\circ_{f0} = 1667$ kJ mol^{-1} (398 kcal mol^{-1})				
	$N_2O^+(B^2\Pi)$	$\Delta H^\circ_{f0} \sim 1793$ kJ mol^{-1} (429 kcal mol^{-1})				
	$N_2O^+(C^2\Sigma^+)$	$\Delta H^\circ_{f0} = 2025$ kJ mol^{-1} (484 kcal mol^{-1})				
$N_2O^+(X^2\Pi_{3/2})$	N_2O		12.894	S	1330	149
$N_2O^+(X^2\Pi_{1/2})$	N_2O		12.931	S	1333	149
$N_2O^+(A^2\Sigma^+)$	N_2O		16.392	S	1667	149
$N_2O^+(C^2\Sigma^+)$	N_2O		20.105	S	2025	149
(Average of two Rydberg series limits)						
$N_2O^+(X^2\Pi)$	N_2O		12.893±0.005	PE		2901
$N_2O^+(X^2\Pi)$	N_2O		12.891±0.008	PE		2875, 2902
$N_2O^+(A^2\Sigma^+)$	N_2O		16.389±0.005	PE		2901
$N_2O^+(A^2\Sigma^+)$	N_2O		16.410	PE		2875, 2902
$N_2O^+(B^2\Pi)$	N_2O		17.650±0.005	PE	1788	2901
$N_2O^+(B^2\Pi)$	N_2O		17.753	PE	1798	2875, 2902
$N_2O^+(C^2\Sigma^+)$	N_2O		20.113±0.005	PE		2901
$N_2O^+(C^2\Sigma^+)$	N_2O		20.147	PE		2875, 2902
$N_2O^+(X^2\Pi_{3/2})$	N_2O		12.888±0.007	PI		2624, 2629
$N_2O^+(X^2\Pi_{3/2})$	N_2O		12.882±0.008	PI		1253
$N_2O^+(X^2\Pi_{3/2})$	N_2O		12.89	PI		3033
$N_2O^+(X^2\Pi)$	N_2O		12.8±0.05	EM		2427

Other ionization potentials reported in refs. 2875, 2902 appear doubtful.

See also – S: 409, 410
PI: 157, 163, 182, 416
PE: 92, 1130, 2855, 2856, 3148
PEN: 2430
EI: 58, 1451, 2018, 2693, 2697, 3135

N_2O^{+2}

Ion	Reactant	Other products	Ionization or appearance potential (eV)	Method	Heat of formation (kJ mol^{-1})	Ref.
N_2O^{+2}	N_2O		36.4±0.5	NRE		3138

4.3. The Positive Ion Table—Continued

Ion	Reactant	Other products	Ionization or appearance potential (eV)	Method	Heat of formation (kJ mol^{-1})	Ref.
$NO_2^+(X^1A_1)$		$\Delta H^\circ_{f0} \leq 977$ kJ mol^{-1} (233 kcal mol^{-1})				
$NO_2^+(^3B_2)$		$\Delta H^\circ_{f0} = 1276$ kJ mol^{-1} (305 kcal mol^{-1})				
$NO_2^+(^3A_2)$		$\Delta H^\circ_{f0} = 1348$ kJ mol^{-1} (322 kcal mol^{-1})				
$NO_2^+(^1A_2)$		$\Delta H^\circ_{f0} = 1394$ kJ mol^{-1} (333 kcal mol^{-1})				
$NO_2^+(^1B_2)$		$\Delta H^\circ_{f0} \sim 1426$ kJ mol^{-1} (341 kcal mol^{-1})				
$NO_2^+(^3B_2)$		$\Delta H^\circ_{f0} = 1856$ kJ mol^{-1} (444 kcal mol^{-1})				
$NO_2^+(^3A_1)$		$\Delta H^\circ_{f0} = 2087$ kJ mol^{-1} (499 kcal mol^{-1})				
$NO_2^+(X^1A_1)$	NO_2		9.75 ± 0.01	PI	977	2624, 2629

Ionization involves a bent–linear transition with a broad Franck–Condon envelope and weak onset. Thus the ionization potential may not be adiabatic. No reliable value is obtained from photoelectron spectroscopy, see refs. 2745, 2821, 3090, 3122, 3215, 3312.

Ion	Reactant	Other products	Ionization or appearance potential (eV)	Method	Heat of formation (kJ mol^{-1})	Ref.
$NO_2^+(^3B_2)$	NO_2		12.863	PE	1277	3215
$NO_2^+(^3B_2)$	NO_2		12.85	PE	1276	3090
$NO_2^+(^3A_2)$	NO_2		13.60	PE	1348	3215
$NO_2^+(^3A_2)$	NO_2		13.60	PE	1348	3090
$NO_2^+(^1A_2)$	NO_2		14.070	PE	1394	3215
$NO_2^+(^1A_2)$	NO_2		14.07	PE	1394	3090
$NO_2^+(^1B_2)$	NO_2		14.446	PE	1430	3215
$NO_2^+(^1B_2)$	NO_2		14.37	PE	1423	3090
$NO_2^+(^3B_1?)$	NO_2		17.069	PE		3215
$NO_2^+(^3A_1,^3B_1?)$	NO_2		16.99	PE		3090
$NO_2^+(^1B_1?)$	NO_2		17.13	PE		3215
$NO_2^+(^3A_1,^3B_1?)$	NO_2		17.06	PE		3090
$NO_2^+(^3A_1?)$	NO_2		~17.5	PE		3215
$NO_2^+(^3B_2)$	NO_2		18.86 ± 0.05	S	1856	1097
$NO_2^+(^3B_2)$	NO_2		18.864	PE	1856	3215
$NO_2^+(^3B_2)$	NO_2		18.86	PE	1856	3090
$NO_2^+(^1A_1?)$	NO_2		~20.7	PE		3215
$NO_2^+(^1B_2?)$	NO_2		~20.8	PE		3090
$NO_2^+(^3A_1)$	NO_2		21.26	PE	2087	3215
$NO_2^+(^3A_1)$	NO_2		21.26	PE	2087	3090
$NO_2^+{}^*$	NO_2		~23.2	PE		3215

See also – S: 117, 3142
       PI: 61, 117, 163, 182, 416
       PE: 1130, 2821, 3122, 3312
       EI: 58, 139, 218, 2018
       D: 2745

4.3. The Positive Ion Table—Continued

Ion	Reactant	Other products	Ionization or appearance potential (eV)	Method	Heat of forma- tion (kJ mol^{-1})	Ref.
NO$_2^+$	CH$_3$NO$_2$	CH$_3$	12.60±0.10	RPD		1241
NO$_2^+$	(CH$_3$)$_2$NNO$_2$		14.6±0.1	EI		303
NO$_2^+$	CH$_3$ONO$_2$	CH$_3$O	12.3	EI		2456
NO$_2^+$	C$_2$H$_5$ONO$_2$	C$_2$H$_5$O	11.40±0.12	EI	1004,	1013
NO$_2^+$	C$_2$H$_5$ONO$_2$	C$_2$H$_5$	12.3	EI		2456
NO$_2^+$	iso-C$_3$H$_7$ONO$_2$	C$_3$H$_7$O	11.8	EI		2456
NO$_2^+$	iso-C$_4$H$_9$ONO$_2$	C$_4$H$_9$O	12.7	EI		2456

BHO$_2^+$

Ion	Reactant	Other products	Ionization or appearance potential (eV)	Method	Heat of formation (kJ mol^{-1})	Ref.
BHO$_2^+$	BHO$_2$		12.6±0.2	EI		3209

B$_2$HO$_2^+$

Ion	Reactant	Other products	Ionization or appearance potential (eV)	Method	Heat of formation (kJ mol^{-1})	Ref.
B$_2$HO$_2^+$	B$_3$H$_3$O$_3$ (Boroxine)		>20	EI		2175

B$_3$H$_2$O$_3^+$

Ion	Reactant	Other products	Ionization or appearance potential (eV)	Method	Heat of formation (kJ mol^{-1})	Ref.
B$_3$H$_2$O$_3^+$	B$_3$H$_3$O$_3$ (Boroxine)	H	14.5±0.5	EI		2175

B$_3$H$_3$O$_3^+$

Ion	Reactant	Other products	Ionization or appearance potential (eV)	Method	Heat of formation (kJ mol^{-1})	Ref.
B$_3$H$_3$O$_3^+$	B$_3$H$_3$O$_3$ (Boroxine)		13.5±0.5	EI		2175

CHO$^+$ $\Delta H_{f298}^\circ \sim 815$ kJ mol^{-1} (195 kcal mol^{-1})

Ion	Reactant	Other products	Ionization or appearance potential (eV)	Method	Heat of formation (kJ mol^{-1})	Ref.
CHO$^+$	CHO		9.83±0.18	EI		128
CHO$^+$	CHO		10.03±0.17	EI		128

See also – EI: 127

CHO$^+$	HCHO	H	11.95±0.06	PI	~818	2724

See also – EI: 127, 204, 2550, 2883, 3224

CHO$^+$	CH$_3$OH	H$_2$+H	14.0±0.2	RPD		2905

See also – EI: 46, 3176

CHO$^+$	CH$_2$=CO	CH	16.07±0.4	EI		2800
CHO$^+$	CH$_3$CHO	CH$_3$	11.79±0.03	PI	~814	2724

Haney and Franklin, ref. 3347, have determined that this process has about 0.15 eV average translational energy of decomposition at threshold. The heat of formation given is corrected for this energy.
Matthews and Warneck, ref. 2724, did not correct for this but suggested that ΔH_{f298}°(CH$_3$CHO) might be in error.

See also – EI: 130, 298, 2550, 2576, 2883, 3347

J. Phys. Chem. Ref. Data, Vol. 6, Suppl. 1, 1977

4.3. The Positive Ion Table—Continued

Ion	Reactant	Other products	Ionization or appearance potential (eV)	Method	Heat of formation (kJ mol^{-1})	Ref.
CHO$^+$	(CH$_2$)$_2$O (1,2–Epoxyethane)	CH$_3$	12.2±0.1	EI		50
CHO$^+$	CH$_2$=CHCHO	C$_2$H$_3$	13.30±0.10	RPD		3347

(0.13 eV average translational energy of decomposition at threshold)

See also – EI: 130

Ion	Reactant	Other products	Ionization or appearance potential (eV)	Method	Heat of formation (kJ mol^{-1})	Ref.
CHO$^+$	C$_3$H$_6$O (1,2–Epoxypropane)		11.8±0.2	EI		50
CHO$^+$	(CH$_2$)$_3$O (1,3–Epoxypropane)		12.6±0.2	EI		52
CHO$^+$	C$_4$H$_6$O (3,4–Epoxy–1–butene)		12.9±0.6	EI		153
CHO$^+$	C$_6$H$_5$CHO (Benzenecarbonal)		13.67±0.13	EI		130

See also – EI: 127

Ion	Reactant	Other products	Ionization or appearance potential (eV)	Method	Heat of formation (kJ mol^{-1})	Ref.
CHO$^+$	HCOOH	OH	12.79±0.03	PI	~812	2724

Haney and Franklin, ref. 3347, have determined that this process has about 0.05 eV average translational energy of decomposition at threshold. The heat of formation given is corrected for this energy.

See also – EI: 127, 3347

Ion	Reactant	Other products	Ionization or appearance potential (eV)	Method	Heat of formation (kJ mol^{-1})	Ref.
CHO$^+$	CHOCHO	CHO?	12.72±0.12	EI		128
CHO$^+$	HCOOCH$_3$	CH$_3$O	13.47±0.05	RPD		3347

(0.17 eV average translational energy of decomposition at threshold)

See also – EI: 305, 3224

Ion	Reactant	Other products	Ionization or appearance potential (eV)	Method	Heat of formation (kJ mol^{-1})	Ref.
CHO$^+$	CH$_3$COCHO	CH$_3$CO	12.48±0.05	EI		128
CHO$^+$	HCOOC$_2$H$_5$		11.39±0.01	EI		305
CHO$^+$	CH$_3$COOCH$_3$		13.95±0.08	EI		305

See also – EI: 3224

Ion	Reactant	Other products	Ionization or appearance potential (eV)	Method	Heat of formation (kJ mol^{-1})	Ref.
CHO$^+$	HCOOCH$_2$CH$_2$CH$_3$		11.56±0.06	EI		305
CHO$^+$	C$_4$H$_8$O$_2$ (1,2–Epoxy–3–methoxypropane)		14.4±0.2	EI		153

4.3. The Positive Ion Table—Continued

Ion	Reactant	Other products	Ionization or appearance potential (eV)	Method	Heat of formation (kJ mol^{-1})	Ref.
CHO$^+$	HCOO(CH$_2$)$_3$CH$_3$		11.47±0.11	EI		305
CHO$^+$	C$_3$H$_6$O$_3$ (1,3,5–Trioxane)		13.59±0.05	RPD		3324
CHO$^+$	(CH$_3$O)$_3$B		19.1±0.3	EI		115
CHO$^+$	HNCO	N	15.76±0.15	EI		3365

(0.5 eV average translational energy of decomposition at threshold)

See also – EI: 2797, 3012

CHO$^+$	HCON(CH$_3$)$_2$		14.3±0.1	EI		303
CHO$^+$	(CH$_3$O)$_2$CH$_3$PO		18.9±0.8	EI		3211
CHO$^+$	(CH$_3$O)$_3$PO		17.3±0.4	EI		3211
CHO$^+$	(CH$_3$)$_2$SO		15±0.1	EI		3294
CHO$^+$	C$_3$H$_5$OCl (1–Chloro–2,3–epoxypropane)		12.0±0.5	EI		153
CHO$^+$	C$_3$H$_5$OBr (1–Bromo–2,3–epoxypropane)		11.8±0.2	EI		153

CDO$^+$

CDO$^+$	DCDO	D	13.10±0.12	EI		127, 204
CDO$^+$	CD$_3$OH		14.2±0.3	RPD		2905
CDO$^+$	DCOOH	OH	13.19±0.03	EI		2550

HCHO$^+$(^{2}B$_2$) $\Delta H^\circ_{f0} = $ 936 kJ mol^{-1} (224 kcal mol^{-1})
HCHO$^+$(^{2}B$_1$) $\Delta H^\circ_{f0} = $ 1247 kJ mol^{-1} (298 kcal mol^{-1})

CH$_2$O$^+$(^{2}B$_2$)	HCHO		10.88±0.01	S	936	3141
(Average of two Rydberg series limits)						
CH$_2$O$^+$(^{2}B$_2$)	HCHO		10.87±0.01	PI	935	182, 416
CH$_2$O$^+$(^{2}B$_2$)	HCHO		10.87±0.01	PI	935	3425
CH$_2$O$^+$(^{2}B$_2$)	HCHO		10.88±0.02	PI	936	2724
CH$_2$O$^+$(^{2}B$_2$)	HCHO		10.884	PE	937	2803
CH$_2$O$^+$(^{2}B$_2$)	HCHO		10.86±0.02	RPD		2883
CH$_2$O$^+$(^{2}B$_1$)	HCHO		14.095	PE	1247	2803
CH$_2$O$^+$(^{2}B$_2$?)	HCHO		15.854	PE		2803
CH$_2$O$^+$(^{2}A$_1$?)	HCHO		16.254	PE		2803

The assignment of the third and fourth ionization potentials is uncertain, see refs. 2803, 3425.

See also – PI: 1166, 2729
　　　　　PE: 2835, 3101
　　　　　EI: 127, 204, 286, 2649, 3224

4.3. The Positive Ion Table—Continued

Ion	Reactant	Other products	Ionization or appearance potential (eV)	Method	Heat of formation (kJ mol^{-1})	Ref.
CH_2O^+	CH_3OH	H_2	12.45	PI		2647

(Threshold value approximately corrected to 0 K)

The thermochemical threshold for this process is 11.67 eV.

See also – EI: 46

CH_2O^+	C_2H_5OH	CH_4	11.70	PI		2647

(Threshold value approximately corrected to 0 K)

The thermochemical threshold for this process is 11.26 eV.

See also – EI: 46

CH_2O^+	C_3H_6O (1,2–Epoxypropane)	C_2H_4	11.6±0.3	EI		50
CH_2O^+	$HCOOCH_3$		13.6	EI		3224
CH_2O^+	$C_4H_8O_2$ (1,2–Epoxy–3–methoxypropane)		10.9±0.2	EI		153
CH_2O^+	$C_4H_8O_2$ (1,3–Dioxane)		11.82	EI		2422
CH_2O^+	$C_2H_5ONO_2$		11.76±0.65	EI		1013
CH_2O^+	$(CH_3)_2SO$		10.9±0.5	EI		3294

CD_2O^+

$CD_2O^+(^2B_2)$	DCDO		10.904	PE		2803
$CD_2O^+(^2B_1)$	DCDO		14.095	PE		2803
$CD_2O^+(^2B_2?)$	DCDO		15.846	PE		2803

See also – EI: 127, 204

CH_2OH^+ $\Delta H^\circ_{f0} = 720$ kJ mol^{-1} (172 kcal mol^{-1})

CH_3O^+	CH_2OH		8.14±0.15	EI		2452
CH_3O^+	CH_3OH	H	11.67±0.03	PI	720	2647

(Threshold value approximately corrected to 0 K)

CH_3O^+	CH_3OH	H	11.66±0.04	PI	719	2915
CH_3O^+	CH_3OH	H	11.67	RPD		2905

Studies on CH_3OD indicate that at threshold the ion structure is CH_2OD^+, see ref. 2915.

See also – EI: 46, 97, 1100, 2709, 3017, 3176

CH_3O^+	C_2H_5OH	CH_3	11.25	PI	(722)	2647

(Threshold value approximately corrected to 0 K)

Haney and Franklin, ref. 3347, have determined that this process has about 0.04 eV average translational energy of decomposition at threshold. It is not known what fraction of this represents rotational effects. No correction has been made for this in the derived heat of formation.

See also – EI: 46, 97, 1100, 2709, 3176, 3347

4.3. The Positive Ion Table—Continued

Ion	Reactant	Other products	Ionization or appearance potential (eV)	Method	Heat of formation (kJ mol⁻¹)	Ref.
CH_3O^+	$(CH_3)_2O$	CH_3	11.95±0.05	RPD		3347

(0.20 eV average translational energy of decomposition at threshold)

See also – EI: 1100, 2018, 2709, 3224, 3435

Ion	Reactant	Other products	Ionization or appearance potential (eV)	Method	Heat of formation (kJ mol⁻¹)	Ref.
CH_3O^+	C_3H_6O (1,2–Epoxypropane)		13.4±0.2	EI		50
CH_3O^+	$(CH_2)_3O$ (1,3–Epoxypropane)		13.3±0.2	EI		52
CH_3O^+	$n–C_3H_7OH$	C_2H_5	~11.11	PI		11

(Threshold value approximately corrected for thermal energy and kinetic shift)

CH_3O^+	$n–C_3H_7OH$	C_2H_5	~11.3	PI		2647

(Value estimated from fig. 6 of this reference, that given in table I is evidently in error)

See also – EI: 46, 97, 1100, 2709

CH_3O^+	$iso–C_3H_7OH$		12.5	EI		46

See also – EI: 2709

CH_3O^+	$C_2H_5OCH_3$	C_2H_5	12.50	RPD		3347

(0.07 eV average translational energy of decomposition at threshold)

See also – EI: 2709

CH_3O^+	C_4H_6O (3,4–Epoxy–1–butene)		13.3±0.5	EI		153
CH_3O^+	$n–C_4H_9OH$		11.46	RPD		97

See also – EI: 1100

CH_3O^+	$sec–C_4H_9OH$		12.5	EI		2709
CH_3O^+	$(C_2H_5)_2O$		12.1	EI		2709

See also – EI: 2196

CH_3O^+	$HCOOCH_3$	CHO?	12.23	RPD		3347

(0.12 eV average translational energy of decomposition at threshold)

See also – EI: 210, 305, 1100, 3224

CH_3O^+	$HCOOC_2H_5$		12.02±0.1	Ei		210

See also – EI: 1059, 3224

4.3. The Positive Ion Table—Continued

Ion	Reactant	Other products	Ionization or appearance potential (eV)	Method	Heat of formation (kJ mol⁻¹)	Ref.
CH_3O^+	CH_3COOCH_3		12.52±0.10	EI		305

See also – EI: 1100, 3176, 3224

Ion	Reactant	Other products	Ionization or appearance potential (eV)	Method	Heat of formation (kJ mol⁻¹)	Ref.
CH_3O^+	$HCOOCH(CH_3)_2$		13.45±0.1	EI		210
CH_3O^+	$C_4H_8O_2$ (1,2–Epoxy–3–methoxypropane)		13.9±0.4	EI		153
CH_3O^+	$C_3H_6O_3$ (1,3,5–Trioxane)		11.49±0.05	RPD		3324
CH_3O^+	$(CH_3O)_3B$		12.7±0.2	EI		115
CH_3O^+	CH_3ONO	NO	10.96	RPD		3347

(0.26 eV average translational energy of decomposition at threshold)

See also – EI: 1100

Ion	Reactant	Other products	Ionization or appearance potential (eV)	Method	Heat of formation (kJ mol⁻¹)	Ref.
CH_3O^+	$(CH_3O)_2CH_3PO$		13.8±0.2	EI		3211
CH_3O^+	$(CH_3O)_3PO$		17.3±0.4	EI		3211
CH_3O^+	$(CH_3)_2SO$		12.2	EI		2685

See also – EI: 3294

Ion	Reactant	Other products	Ionization or appearance potential (eV)	Method	Heat of formation (kJ mol⁻¹)	Ref.
CH_3O^+	CH_2ClCH_2OH		11.51±0.1	EI		72
CH_3O^+	C_3H_5OCl (1–Chloro–2,3–epoxypropane)		13.4±0.2	EI		153
CH_3O^+	C_3H_5OBr (1–Bromo–2,3–epoxypropane)		12.5±0.2	EI		153

CH_2DO^+

Ion	Reactant	Other products	Ionization or appearance potential (eV)	Method	Heat of formation (kJ mol⁻¹)	Ref.
CH_2DO^+	CH_3OD	H	11.676±0.04	PI		2915
CH_2DO^+	C_2H_5OD	CH_3	11.2	PI		3325
CH_2DO^+	$DCOOCH(CH_3)_2$		13.53±0.1	EI		210

CHD_2O^+

Ion	Reactant	Other products	Ionization or appearance potential (eV)	Method	Heat of formation (kJ mol⁻¹)	Ref.
CHD_2O^+	CD_3OH	D	11.60	PI		3047
CHD_2O^+	CD_3OH	D	11.85	RPD		2905
CHD_2O^+	$C_2H_5OCD_3$		13.03±0.1	EI		2971
CHD_2O^+	$iso-C_3H_7OCD_3$		12.90±0.1	EI		2971
CHD_2O^+	$sec-C_4H_9OCD_3$		12.98±0.1	EI		2971

4.3. The Positive Ion Table—Continued

Ion	Reactant	Other products	Ionization or appearance potential (eV)	Method	Heat of formation (kJ mol^{-1})	Ref.
			CH$_3$OH$^+$　　$\Delta H^\circ_{f0} = 856$ kJ mol^{-1} (205 kcal mol^{-1})			
CH$_4$O$^+$	CH$_3$OH		10.85±0.02	PI	857	158, 182, 416
CH$_4$O$^+$	CH$_3$OH		10.829±0.015	PI	855	2915, 2965
CH$_4$O$^+$	CH$_3$OH		10.84±0.02	PI	856	2647
CH$_4$O$^+$	CH$_3$OH		10.83	PE	855	1130, 2801
CH$_4$O$^+$	CH$_3$OH		10.83	PE	855	2843
CH$_4$O$^+$	CH$_3$OH		10.85	PE	857	3374
CH$_4$O$^+$	CH$_3$OH		10.85±0.02	PE	857	3289
CH$_4$O$^+$	CH$_3$OH		10.85	PEN		2430
CH$_4$O$^+$	CH$_3$OH		10.85	RPD		2905

See also — PI:　　297
　　　　　　 PE:　　3132
　　　　　　 EI:　　28, 46, 164, 383, 384, 1072, 2018, 2060, 3176

Ion	Reactant	Other products	Ionization or appearance potential (eV)	Method	Heat of formation (kJ mol^{-1})	Ref.
CH$_4$O$^+$	HCOOCH$_3$	CO	11.53±0.1	EI		210

See also — EI:　　3224

Ion	Reactant	Other products	Ionization or appearance potential (eV)	Method	Heat of formation (kJ mol^{-1})	Ref.
			CH$_3$DO$^+$			
CH$_3$DO$^+$	CH$_3$OD		10.84±0.02	PI		2915
CH$_3$DO$^+$	CH$_3$OD		11.13	RPD		2905

Ion	Reactant	Other products	Ionization or appearance potential (eV)	Method	Heat of formation (kJ mol^{-1})	Ref.
			CHD$_3$O$^+$			
CHD$_3$O$^+$	CD$_3$OH		10.98	RPD		2905

Ion	Reactant	Other products	Ionization or appearance potential (eV)	Method	Heat of formation (kJ mol^{-1})	Ref.
			CD$_4$O$^+$			
CD$_4$O$^+$	CD$_3$OD		10.98	RPD		2905

Ion	Reactant	Other products	Ionization or appearance potential (eV)	Method	Heat of formation (kJ mol^{-1})	Ref.
			C$_2$HO$^+$			
C$_2$HO$^+$	CH$_2$=CO	H	14.91±0.3	EI		2800
C$_2$HO$^+$	C$_4$H$_8$O$_2$ (1,2–Epoxy–3–methoxypropane)		13.0±0.3	EI		153

4.3. The Positive Ion Table—Continued

Ion	Reactant	Other products	Ionization or appearance potential (eV)	Method	Heat of formation (kJ mol^{-1})	Ref.
	CH$_2$= CO$^+$	**$\Delta H^\circ_{f0} \sim 871$ kJ mol^{-1} (208 kcal mol^{-1})**				
C$_2$H$_2$O$^+$	CH$_2$=CO		9.607±0.02	S	869	3151
C$_2$H$_2$O$^+$	CH$_2$=CO		9.64	PE	872	3058
See also – EI: 2800						
C$_2$H$_2$O$^+$	CH$_3$CHO	H$_2$	10.7±0.1	RPD		2576
C$_2$H$_2$O$^+$	(CH$_2$)$_2$O (1,2–Epoxyethane)	2H?	14.0±0.3	EI	(~879)	50
C$_2$H$_2$O$^+$	(CH$_3$)$_2$CO	CH$_4$	10.7±0.1	RPD		2576
See also – EI: 1404						
C$_2$H$_2$O$^+$	C$_3$H$_6$O (1,2–Epoxypropane)		12.7±0.2	EI		50
C$_2$H$_2$O$^+$	C$_4$H$_6$O (3,4–Epoxy–1–butene)		9.8±0.4	EI		153
C$_2$H$_2$O$^+$	n–C$_4$H$_9$OH		11.23	RPD		97
C$_2$H$_2$O$^+$	CH$_3$COOCH$_3$		11.81±0.15	EI		3176
See also – EI: 3224						
C$_2$H$_2$O$^+$	C$_3$H$_6$O$_2$ (1,3–Dioxolane)		13.20	EI		2422
C$_2$H$_2$O$^+$	C$_4$H$_8$O$_2$ (1,2–Epoxy–3–methoxypropane)		12.3±0.3	EI		153
C$_2$H$_2$O$^+$	C$_3$H$_5$OCl (1–Chloro–2,3–epoxypropane)		12.1±0.1	EI		153
	CH$_3$CO$^+$	**$\Delta H^\circ_{f298} \sim 630$ kJ mol^{-1} (151 kcal mol^{-1})**				
C$_2$H$_3$O$^+$	CH$_3$CO		8.05±0.17	EI		128
C$_2$H$_3$O$^+$	CH$_2$CHO?		10.85	EI		2998
C$_2$H$_3$O$^+$	CH$_3$CHO	H	10.89	PI	(667)	2728
C$_2$H$_3$O$^+$	CH$_3$CHO	H	10.75±0.08	RPD	(653)	2576
C$_2$H$_3$O$^+$	CH$_3$CHO	H	10.5±0.2	SD		1404
See also – EI: 128, 130, 298, 2883, 2998						
C$_2$H$_3$O$^+$	(CH$_2$)$_2$O (1,2–Epoxyethane)	H	12.1±0.2	EI		50
C$_2$H$_3$O$^+$	C$_2$H$_5$OH	H$_2$+H	14.5	EI		46

4.3. The Positive Ion Table—Continued

Ion	Reactant	Other products	Ionization or appearance potential (eV)	Method	Heat of formation (kJ mol^{-1})	Ref.
$C_2H_3O^+$	$(CH_3)_2CO$	CH_3	10.37	PI	~630	1099
$C_2H_3O^+$	$(CH_3)_2CO$	CH_3	10.42	PI		2728
$C_2H_3O^+$	$(CH_3)_2CO$	CH_3	10.28	EDD		3174
$C_2H_3O^+$	$(CH_3)_2CO$	CH_3	10.2±0.1	SD		1404

Haney and Franklin, ref. 3347, have determined that this process has about 0.11 eV average translational energy of decomposition at threshold. The heat of formation given is corrected for this energy.

See also – PI: 95
       EI: 128, 298, 2174, 2548, 2576, 2883, 2977, 3224, 3347

Ion	Reactant	Other products	Ionization or appearance potential (eV)	Method	Heat of formation (kJ mol^{-1})	Ref.
$C_2H_3O^+$	$CH_2=CHOCH_3$	CH_3	11.44	EI		3435
$C_2H_3O^+$	C_3H_6O (1,2–Epoxypropane)	CH_3	10.9±0.2	EI		50
$C_2H_3O^+$	$CH_3COC\equiv CH$		11.85	EI		298
$C_2H_3O^+$	$CH_3COCH=CH_2$		12.40	EI		298
$C_2H_3O^+$	C_4H_6O (3,4–Epoxy–1–butene)		10.5±0.2	EI		153
$C_2H_3O^+$	$C_2H_5COCH_3$	C_2H_5	10.3	PI	(648)	95
$C_2H_3O^+$	$C_2H_5COCH_3$	C_2H_5	10.97	RPD		2977

See also – EI: 298, 2883

Ion	Reactant	Other products	Ionization or appearance potential (eV)	Method	Heat of formation (kJ mol^{-1})	Ref.
$C_2H_3O^+$	$(CH_2)_4O$ (1,4–Epoxybutane)		12.8±0.2	EI		52
$C_2H_3O^+$	$n–C_3H_7COCH_3$		11.54	RPD		2977

See also – EI: 298

Ion	Reactant	Other products	Ionization or appearance potential (eV)	Method	Heat of formation (kJ mol^{-1})	Ref.
$C_2H_3O^+$	$iso–C_3H_7COCH_3$	$iso–C_3H_7$	10.4	PI	(~667)	95
$C_2H_3O^+$	$n–C_4H_9COCH_3$		10.8	PI		95
$C_2H_3O^+$	$n–C_4H_9COCH_3$		11.65	RPD		2977
$C_2H_3O^+$	$n–C_5H_{11}COCH_3$		11.83	RPD		2977
$C_2H_3O^+$	$C_6H_5COCH_3$ (Acetophenone)		11.40±0.28	EI		2174

See also – EI: 298

Ion	Reactant	Other products	Ionization or appearance potential (eV)	Method	Heat of formation (kJ mol^{-1})	Ref.
$C_2H_3O^+$	$n–C_6H_{13}COCH_3$		12.04	RPD		2977
$C_2H_3O^+$	$n–C_7H_{15}COCH_3$		12.24	RPD		2977
$C_2H_3O^+$	CH_3COOH	OH	11.4±0.15	RPD		2576
$C_2H_3O^+$	CH_3COOH	OH	11.75	RPD		3347

    (0.10 eV average translational energy of decomposition at threshold)

See also – EI: 171, 298

4.3. The Positive Ion Table—Continued

Ion	Reactant	Other products	Ionization or appearance potential (eV)	Method	Heat of formation (kJ mol⁻¹)	Ref.
$C_2H_3O^+$	CH_3COCHO	CHO?	10.65±0.12	EI		128
$C_2H_3O^+$	$HCOOC_2H_5$		12.2	EI		3224
$C_2H_3O^+$	CH_3COOCH_3	CH_3O	11.37±0.05	RPD		3347

(0.07 eV average translational energy of decomposition at threshold)

See also – EI: 305, 3176, 3224

Ion	Reactant	Other products	Ionization or appearance potential (eV)	Method	Heat of formation (kJ mol⁻¹)	Ref.
$C_2H_3O^+$	$C_3H_6O_2$ (1,3–Dioxolane)		14.14	EI		2422
$C_2H_3O^+$	$CH_3COCOCH_3$	CH_3CO	9.88	PI	(649)	1099

See also – EI: 128, 298

Ion	Reactant	Other products	Ionization or appearance potential (eV)	Method	Heat of formation (kJ mol⁻¹)	Ref.
$C_2H_3O^+$	$CH_3COOC_2H_5$		11.75±0.07	EI		305

See also – EI: 298, 1059, 3176

Ion	Reactant	Other products	Ionization or appearance potential (eV)	Method	Heat of formation (kJ mol⁻¹)	Ref.
$C_2H_3O^+$	$C_4H_8O_2$ (1,2–Epoxy–3–methoxypropane)		13.1±0.2	EI		153
$C_2H_3O^+$	$C_4H_8O_2$ (1,4–Dioxane)		12.92	EI		2422
$C_2H_3O^+$	$CH_3COOCOCH_3$		10.19±0.02	PI		3015
$C_2H_3O^+$	$C_5H_{10}O_3$ (1,3,6–Trioxocane)		13.16	EI		2422
$C_2H_3O^+$	$CH_3CON(CH_3)_2$		11.35±0.1	EI		303
$C_2H_3O^+$	CH_3COCHN_2		10.46±0.05	EI		2174
$C_2H_3O^+$	CH_3COF	F	12.3	EI		298
$C_2H_3O^+$	CH_3COCF_3	CF_3	11.45	EI		298
$C_2H_3O^+$	CH_3COCl	Cl	11.25	RPD		3347

(0.16 eV average translational energy of decomposition at threshold)

See also – EI: 298

Ion	Reactant	Other products	Ionization or appearance potential (eV)	Method	Heat of formation (kJ mol⁻¹)	Ref.
$C_2H_3O^+$	CH_3COCH_2Cl		10.29±0.04	EI		2174
$C_2H_3O^+$	CH_3COBr	Br	10.60	EI		298

$C_2D_3O^+$

Ion	Reactant	Other products	Ionization or appearance potential (eV)	Method	Heat of formation (kJ mol⁻¹)	Ref.
$C_2D_3O^+$	$n-C_3H_7CD_2COCD_3$		12.2±0.2	EI		2766
$C_2D_3O^+$	CD_3COOH	OH	12.97	EI		171

4.3. The Positive Ion Table—Continued

Ion	Reactant	Other products	Ionization or appearance potential (eV)	Method	Heat of formation (kJ mol^{-1})	Ref.
	CH$_3$CHO$^+$		$\Delta H_{f0}^\circ = 831.6$ kJ mol^{-1} (198.8 kcal mol^{-1})			
	C$_2$H$_4$O$^+$ (1,2–Epoxyethane)		$\Delta H_{f0}^\circ = 979.4$ kJ mol^{-1} (234.1 kcal mol^{-1})			
C$_2$H$_4$O$^+$	CH$_3$CHO		10.2291±0.0007	S	831.6	3020
C$_2$H$_4$O$^+$	CH$_3$CHO		10.21±0.01	PI	182,	416
C$_2$H$_4$O$^+$	CH$_3$CHO		10.20±0.03	PI		1166
C$_2$H$_4$O$^+$	CH$_3$CHO		10.25±0.03	PI		86
C$_2$H$_4$O$^+$	CH$_3$CHO		10.20±0.02	PI		2724
C$_2$H$_4$O$^+$	CH$_3$CHO		10.22±0.01	PI		2728
C$_2$H$_4$O$^+$	CH$_3$CHO		10.22±0.01	PE		3289
C$_2$H$_4$O$^+$	CH$_3$CHO		10.20	PE		2843

See also – EI: 127, 130, 286, 1404, 2026, 2576, 2649, 2883, 2998

Ion	Reactant	Other products	Ionization or appearance potential (eV)	Method	Heat of formation (kJ mol^{-1})	Ref.
C$_2$H$_4$O$^+$(^{2}B$_2$)	(CH$_2$)$_2$O (1,2–Epoxyethane)		10.566	S	979.4	101
C$_2$H$_4$O$^+$(^{2}B$_2$)	(CH$_2$)$_2$O (1,2–Epoxyethane)		10.565±0.01	PI	101,	182, 416
C$_2$H$_4$O$^+$(^{2}B$_2$)	(CH$_2$)$_2$O (1,2–Epoxyethane)		10.57	PE		2808
C$_2$H$_4$O$^+$(^{2}A$_1$)	(CH$_2$)$_2$O (1,2–Epoxyethane)		11.7 (V)	PE		2808
C$_2$H$_4$O$^+$(^{2}B$_1$)	(CH$_2$)$_2$O (1,2–Epoxyethane)		13.7 (V)	PE		2808
C$_2$H$_4$O$^+$(^{2}A$_2$)	(CH$_2$)$_2$O (1,2–Epoxyethane)		~14.2 (V)	PE		2808
C$_2$H$_4$O$^+$(^{2}A$_1$)	(CH$_2$)$_2$O (1,2–Epoxyethane)		16.6 (V)	PE		2808
C$_2$H$_4$O$^+$(^{2}B$_2$)	(CH$_2$)$_2$O (1,2–Epoxyethane)		17.4 (V)	PE		2808

See also – PE: 1130
         PEN: 2430
          EI: 50

Ion	Reactant	Other products	Ionization or appearance potential (eV)	Method	Heat of formation (kJ mol^{-1})	Ref.
C$_2$H$_4$O$^+$	iso–C$_3$H$_7$OH	CH$_4$	10.27±0.03	PI		2647

(Threshold value approximately corrected to 0 K)

The thermochemical threshold for this process is 10.50 eV, assuming the ion to have the acetaldehyde structure. This leads Refaey and Chupka, ref. 2647, to postulate instead formation of the oxonium ion, CH$_2$CHOH$^+$.

Ion	Reactant	Other products	Ionization or appearance potential (eV)	Method	Heat of formation (kJ mol^{-1})	Ref.
C$_2$H$_4$O$^+$	(CH$_2$)$_4$O (1,4–Epoxybutane)	C$_2$H$_4$	12.27	EI		2694
C$_2$H$_4$O$^+$	C$_3$H$_6$O$_2$ (1,3–Dioxolane)		11.56	EI		2422
C$_2$H$_4$O$^+$	C$_4$H$_8$O$_2$ (1,4–Dioxane)		10.90	EI		2422
C$_2$H$_4$O$^+$	C$_5$H$_{10}$O$_3$ (1,3,6–Trioxocane)		10.03	EI		2422

4.3. The Positive Ion Table—Continued

Ion	Reactant	Other products	Ionization or appearance potential (eV)	Method	Heat of formation (kJ mol⁻¹)	Ref.

CH$_3$CHOH$^+$ $\quad$ ΔH$^\circ_{f0}$ ~ 608 kJ mol⁻¹ (145 kcal mol⁻¹)

Ion	Reactant	Other products	Ionization or appearance potential (eV)	Method	Heat of formation (kJ mol⁻¹)	Ref.
C$_2$H$_5$O$^+$	C$_2$H$_5$O		9.11±0.05	RPD		2776
C$_2$H$_5$O$^+$	C$_2$H$_5$OH	H	10.78±0.02	PI	607	2647

(Threshold value approximately corrected to 0 K)

See also – EI: 46, 97, 1100, 2709, 2776, 3176

Ion	Reactant	Other products	AP (eV)	Method	ΔHf	Ref.
C$_2$H$_5$O$^+$	(CH$_3$)$_2$O	H	11.42±0.01	RPD		1139

See also – EI: 1100, 2709, 3017, 3224, 3435

Ion	Reactant	Other products	AP (eV)	Method	ΔHf	Ref.
C$_2$H$_5$O$^+$	n–C$_3$H$_7$OH	CH$_3$	11.1±0.1	PI		2647
C$_2$H$_5$O$^+$	n–C$_3$H$_7$OH	CH$_3$	11.1	EI		46

See also – EI: 97

Ion	Reactant	Other products	AP (eV)	Method	ΔHf	Ref.
C$_2$H$_5$O$^+$	iso–C$_3$H$_7$OH	CH$_3$	10.40	PI	610	2647

(Threshold value approximately corrected to 0 K)

Ion	Reactant	Other products	AP (eV)	Method	ΔHf	Ref.
C$_2$H$_5$O$^+$	iso–C$_3$H$_7$OH	CH$_3$	10.70	RPD		3347

(0.03 eV average translational energy of decomposition at threshold)

See also – EI: 46, 97, 2709

Ion	Reactant	Other products	AP (eV)	Method	ΔHf	Ref.
C$_2$H$_5$O$^+$	C$_2$H$_5$OCH$_3$	CH$_3$	11.30	RPD		3347

(0.10 eV average translational energy of decomposition at threshold)

Ion	Reactant	Other products	AP (eV)	Method	ΔHf	Ref.
C$_2$H$_5$O$^+$	C$_2$H$_5$OCH$_3$	CH$_3$	10.96	EI		2709

See also – EI: 1139

Ion	Reactant	Other products	AP (eV)	Method	ΔHf	Ref.
C$_2$H$_5$O$^+$	sec–C$_4$H$_9$OH	C$_2$H$_5$	10.4	EI		2709
C$_2$H$_5$O$^+$	(C$_2$H$_5$)$_2$O	C$_2$H$_5$	11.8	EI		2709

See also – EI: 1100, 2196, 2776

Ion	Reactant	Other products	AP (eV)	Method	ΔHf	Ref.
C$_2$H$_5$O$^+$	(CH$_2$)$_5$O (1,5–Epoxypentane)		11.56	EI		2694
C$_2$H$_5$O$^+$	HCOOC$_2$H$_5$	CHO?	11.34	RPD		3347

(0.10 eV average translational energy of decomposition at threshold)

See also – EI: 210, 305, 1100, 3224

Ion	Reactant	Other products	AP (eV)	Method	ΔHf	Ref.
C$_2$H$_5$O$^+$	DCOOC$_2$H$_5$	CDO?	11.55±0.1	EI		210

4.3. The Positive Ion Table—Continued

Ion	Reactant	Other products	Ionization or appearance potential (eV)	Method	Heat of forma-tion (kJ mol^{-1})	Ref.
$C_2H_5O^+$	$C_3H_6O_2$ (1,3−Dioxolane)		12.45	EI		2422
$C_2H_5O^+$	$(CH_3O)_2CH_2$	CH_3O	11.1	EI		2709

See also − EI: 1139

$C_2H_5O^+$	$HCOOCH(CH_3)_2$		11.53±0.1	EI		210
$C_2H_5O^+$	$CH_3COOC_2H_5$		10.8±0.1	EI		1100

See also − EI: 305, 2709, 3176

$C_2H_5O^+$	$C_4H_8O_2$ (1,2−Epoxy−3−methoxypropane)		12.1±0.15	EI		153
$C_2H_5O^+$	C_2H_5ONO	NO	10.43±0.10	RPD		3347

(0.10 eV average translational energy of decomposition at threshold)

See also − EI: 2776

$C_2H_5O^+$	$(CH_3O)_2CH_3PO$		16.3±0.1	EI		3211
$C_2H_5O^+$	$(CH_3O)_3PO$		14.0±0.2	EI		3211
$C_2H_5O^+$	$(C_2H_5O)_3PO$		16.8±0.2	EI		3211
$C_2H_5O^+$	$(CH_3)_2SO$		14±0.3	EI		3294
$C_2H_5O^+$	CH_3OCH_2Cl	Cl	10.79	RPD		3347

(0.14 eV average translational energy of decomposition at threshold)

$C_2H_4DO^+$

$C_2H_4DO^+$	C_2H_5OD	H	10.8	PI		3325
$C_2H_4DO^+$	CH_3CD_2OH	D	10.8	PI		3047
$C_2H_4DO^+$	$DCOOC_2H_5$		12.33±0.1	EI		210

	$C_2H_5OH^+$	$\Delta H^\circ_{f0} = 793$ kJ mol^{-1} (189 kcal mol^{-1})				
	$(CH_3)_2O^+$	$\Delta H^\circ_{f0} \sim 795$ kJ mol^{-1} (190 kcal mol^{-1})				
$C_2H_6O^+$	C_2H_5OH		10.48±0.05	PI	794	182
$C_2H_6O^+$	C_2H_5OH		10.47±0.02	PI	793	2647
$C_2H_6O^+$	C_2H_5OH		10.46±0.02	PE	792	3289
$C_2H_6O^+$	C_2H_5OH		10.46	PE	792	2843

See also − PI: 158, 416
         PE: 1130, 3374
         PEN: 2430, 2873
         EI: 28, 46, 97, 383, 384, 2018, 2060, 3176

$C_2H_6O^+$	$(CH_3)_2O$		9.96±0.05	S	795	2170
$C_2H_6O^+$	$(CH_3)_2O$		10.00±0.02	PI	799	182, 416
$C_2H_6O^+$	$(CH_3)_2O$		9.94±0.01	PE	793	3289
$C_2H_6O^+$	$(CH_3)_2O$		9.94	PE	793	2843

See also − PEN: 2430
         EI: 2018, 3338, 3435

4.3. The Positive Ion Table—Continued

Ion	Reactant	Other products	Ionization or appearance potential (eV)	Method	Heat of formation (kJ mol^{-1})	Ref.
			C$_2$H$_5$DO$^+$			
C$_2$H$_5$DO$^+$	C$_2$H$_5$OD		10.49	PI		3325
			C$_3$HO$^+$			
C$_3$HO$^+$	CH$_3$COC≡CH	CH$_3$	11.0	EI		298
			C$_3$H$_3$O$^+$			
C$_3$H$_3$O$^+$	CH$_3$COCH=CH$_2$	CH$_3$	10.85	EI		298
			C$_3$H$_4$O$^+$			
C$_3$H$_4$O$^+$	CH$_2$=CHCHO		10.103±0.006	S		3383
(Average of three Rydberg series limits)						
C$_3$H$_4$O$^+$	CH$_2$=CHCHO		10.10±0.01	PI	182,	416
C$_3$H$_4$O$^+$	CH$_2$=CHCHO		10.14±0.06	EI		130
See also – EI: 384						
C$_3$H$_4$O$^+$	(CH$_3$)$_2$CO	H$_2$	15.2±0.15	RPD		2576
C$_3$H$_4$O$^+$	C$_2$H$_5$COOH	H$_2$O	11.57	EI		3435
C$_3$H$_4$O$^+$	CH$_3$COCHN$_2$	N$_2$	9.86±0.03	EI		2174
C$_2$H$_5$CO$^+$ $\Delta H^\circ_{f298} = 602$ kJ mol^{-1} (144 kcal mol^{-1})						
C$_3$H$_5$O$^+$	(CH$_3$)$_2$CO	H	13.1±0.2	RPD	2548,	2576
C$_3$H$_5$O$^+$	C$_3$H$_6$O (1,2–Epoxypropane)	H	11.5±0.3	EI		50
C$_3$H$_5$O$^+$	C$_2$H$_5$COCH$_3$	CH$_3$	10.18	PI	602	1099
C$_3$H$_5$O$^+$	C$_2$H$_5$COCH$_3$	CH$_3$	10.60	RPD		2977
See also – PI: 95 EI: 298, 2883						
C$_3$H$_5$O$^+$	n–C$_3$H$_7$COCH$_3$	C$_2$H$_5$	10.58	RPD		2977
C$_3$H$_5$O$^+$	C$_6$H$_{11}$OH (Cyclohexanol)		11.5	EI		3022
C$_3$H$_5$O$^+$	n–C$_5$H$_{11}$COC$_2$H$_5$		12.0±0.3	EI		2740
C$_3$H$_5$O$^+$	C$_2$H$_5$COOH	OH	12.20	EI		3435
C$_3$H$_5$O$^+$	C$_4$H$_8$O$_2$ (1,2–Epoxy–3–methoxypropane)	CH$_3$O	11.2±0.2	EI		153
C$_3$H$_5$O$^+$	C$_4$H$_8$O$_2$ (1,3–Dioxane)		12.17	EI		2422
C$_3$H$_5$O$^+$	C$_2$H$_5$COCOCH$_3$	CH$_3$CO	9.67	PI		1099
C$_3$H$_5$O$^+$	C$_3$H$_5$OCl (1–Chloro–2,3–epoxypropane)	Cl	11.4±0.3	EI		153
C$_3$H$_5$O$^+$	C$_3$H$_5$OBr (1–Bromo–2,3–epoxypropane)	Br	10.8±0.1	EI		153

4.3. The Positive Ion Table—Continued

Ion	Reactant	Other products	Ionization or appearance potential (eV)	Method	Heat of formation (kJ mol⁻¹)	Ref.
$C_2H_5CHO^+$		$\Delta H_{f298}^\circ = 770$ kJ mol^{-1} (184 kcal mol^{-1})				
$(CH_3)_2CO^+$		$\Delta H_{f298}^\circ = 719$ kJ mol^{-1} (172 kcal mol^{-1})				
$C_3H_6O^+$ (1,2-Epoxypropane)		$\Delta H_{f298}^\circ = 893$ kJ mol^{-1} (214 kcal mol^{-1})				
$C_3H_6O^+$ (1,3-Epoxypropane)		$\Delta H_{f298}^\circ = 852$ kJ mol^{-1} (204 kcal mol^{-1})				
$C_3H_6O^+$	$CH_2=CHCH_2OH$		9.67±0.05?	PI		182
$C_3H_6O^+$	$CH_2=CHOCH_3$		8.93±0.02	PI		182

See also – EI: 3435

Ion	Reactant	Other products	Ionization or appearance potential (eV)	Method	Heat of formation (kJ mol⁻¹)	Ref.
$C_3H_6O^+$	C_2H_5CHO		9.98±0.01	PI	771	182
$C_3H_6O^+$	C_2H_5CHO		9.94	PE		2843
$C_3H_6O^+$	C_2H_5CHO		9.97±0.01	PE	770	3289

See also – EI: 130, 2522

Ion	Reactant	Other products	Ionization or appearance potential (eV)	Method	Heat of formation (kJ mol⁻¹)	Ref.
$C_3H_6O^+$	$(CH_3)_2CO$		9.705	S	719	158
$C_3H_6O^+$	$(CH_3)_2CO$		9.690±0.01	PI		158, 182, 416
$C_3H_6O^+$	$(CH_3)_2CO$		9.68±0.02	PI		95
$C_3H_6O^+$	$(CH_3)_2CO$		9.71±0.03	PI		1166
$C_3H_6O^+$	$(CH_3)_2CO$		9.71±0.01	PI		2728
$C_3H_6O^+$	$(CH_3)_2CO$		9.67	PE		1130
$C_3H_6O^+$	$(CH_3)_2CO$		9.68	PE		2843
$C_3H_6O^+$	$(CH_3)_2CO$		9.71±0.01	PE		3289
$C_3H_6O^+$	$(CH_3)_2CO$		9.74±0.03	EDD		3174
$C_3H_6O^+$	$(CH_3)_2CO$		9.7±0.1	SD		1404

See also –   S: 3065
           PI: 86
           PEN: 2430
           EI: 384, 1254, 1256, 2026, 2174, 2433, 2548, 2649, 2883, 2977
           CTS: 2562

Ion	Reactant	Other products	Ionization or appearance potential (eV)	Method	Heat of formation (kJ mol⁻¹)	Ref.
$C_3H_6O^+$	C_3H_6O (1,2-Epoxypropane)		10.22±0.02	PI	893	182

See also – EI: 50

Ion	Reactant	Other products	Ionization or appearance potential (eV)	Method	Heat of formation (kJ mol⁻¹)	Ref.
$C_3H_6O^+$	$(CH_2)_3O$ (1,3-Epoxypropane)		9.668±0.005	S	852	2169

See also – EI: 52, 218

4.3. The Positive Ion Table—Continued

Ion	Reactant	Other products	Ionization or appearance potential (eV)	Method	Heat of formation (kJ mol^{-1})	Ref.
$C_3H_6O^+$	$n-C_3H_7COCH_3$	C_2H_4	10.07	PI		95

The thermochemical threshold for the process leading to $(CH_3)_2CO^+$ is 10.67 eV. Murad and Inghram, ref. 95, suggest that the ion is formed in a more stable enol form by rearrangement.

$C_3H_6O^+$	$n-C_4H_9COCH_3$	C_3H_6	10.00	PI		95

The thermochemical threshold for the process leading to $(CH_3)_2CO^+$ is 10.56 eV. Murad and Inghram, ref. 95, suggest that the ion is formed in a more stable enol form by rearrangement.

$C_3H_6O^+$	$iso-C_4H_9COCH_3$	C_3H_6	10.1	PI		95
$C_3H_6O^+$	$C_4H_8O_2$ (1,2-Epoxy-3-methoxypropane)	CH_2O?	10.2±0.2	EI		153
$C_3H_6O^+$	$C_4H_8O_2$ (1,3-Dioxane)	CH_2O?	11.12	EI		2422
$C_3H_6O^+$	$C_4H_8O_2$ (1,4-Dioxane)	CH_2O	11	EI		2422
$C_3H_6O^+$	$(CH_3)_2C(OH)CH_2COCH_3$	$(CH_3)_2CO$?	9.60±0.03	PI		3015

Shigorin, et al., ref. 3015, suggest that the ion has an enol structure.

$C_3H_6O^+$	$C_5H_{10}O_3$ (1,3,6-Trioxocane)		11.4	EI		2422

$C_3HD_5O^+$

Ion	Reactant	Other products	Ionization or appearance potential (eV)	Method	Heat of formation (kJ mol^{-1})	Ref.
$C_3HD_5O^+$	$n-C_3H_7CD_2COCD_3$	C_3H_6	10.4±0.2	EI		2766

$C_2H_5CHOH^+$ $\Delta H^\circ_{f0} \sim 585$ kJ mol^{-1} (140 kcal mol^{-1})
$(CH_3)_2COH^+$ $\Delta H^\circ_{f0} \sim 525$ kJ mol^{-1} (125 kcal mol^{-1})

$C_3H_7O^+$	$n-C_3H_7O$		9.20±0.05	RPD		2776
$C_3H_7O^+$	$iso-C_3H_7O$		9.20±0.05	RPD		2776
$C_3H_7O^+$	$n-C_3H_7OH$	H	10.72	PI	585	2647
(Threshold value approximately corrected to 0 K)						
$C_3H_7O^+$	$n-C_3H_7OH$	H	10.69	RPD		97

See also – PI: 11
        EI: 46, 2776

$C_3H_7O^+$	$iso-C_3H_7OH$	H	10.6	PI	(~559)	2647
(Threshold value approximately corrected to 0 K)						
$C_3H_7O^+$	$iso-C_3H_7OH$	H	11.85	RPD		97

See also – EI: 46

$C_3H_7O^+$	$C_2H_5OCH_3$	H	10.3	EI		2709

See also – EI: 2971

4.3. The Positive Ion Table—Continued

Ion	Reactant	Other products	Ionization or appearance potential (eV)	Method	Heat of formation (kJ mol^{-1})	Ref.
$C_3H_7O^+$	sec-C_4H_9OH	CH_3	10.7	EI		2709
$C_3H_7O^+$	tert-C_4H_9OH	CH_3	9.87	PI	525	3325
$C_3H_7O^+$	tert-C_4H_9OH	CH_3	10.2	EI		2709
$C_3H_7O^+$	iso-$C_3H_7OCH_3$	CH_3	10.34±0.1	EI		2971
$C_3H_7O^+$	$(C_2H_5)_2O$	CH_3	10.3	EI		2709

See also – EI:　　2196, 2971

$C_3H_7O^+$	$(CH_3)_2C(OH)C_2H_5$	C_2H_5	10.0	EI		2709
$C_3H_7O^+$	sec-$C_4H_9OCH_3$	C_2H_5	10.10±0.1	EI		2971
$C_3H_7O^+$	iso-$C_4H_9OC_2H_5$		10.31±0.1	EI		2971
$C_3H_7O^+$	$(n-C_3H_7)_2O$		12.93±0.1	RPD		2776
$C_3H_7O^+$	$(iso-C_3H_7)_2O$		11.59±0.05	RPD		2776
$C_3H_7O^+$	$DCOOCH_2CH_2CH_3$	CDO?	11.22	EI		3435
$C_3H_7O^+$	$C_4H_8O_2$ (1,3–Dioxane)	CHO?	12.17	EI		2422
$C_3H_7O^+$	$(CH_3O)_2CHCH_3$	CH_3O	10.63±0.04	RPD		1139
$C_3H_7O^+$	$CH_3COOCH_2CH_2CH_3$	CH_3CO	11.64±0.03	EI		305
$C_3H_7O^+$	$CH_3COOCH(CH_3)_2$	CH_3CO	10.65	EI		2709

See also – EI:　　305

$C_3H_7O^+$	$(CH_3)_2C(OH)CH_2COCH_3$	CH_3COCH_2?	9.71±0.02	PI		3015

$n-C_3H_7OH^+$　　　$\Delta H^\circ_{f0} \sim 753$ kJ mol^{-1} (180 kcal mol^{-1})
$iso-C_3H_7OH^+$　　$\Delta H^\circ_{f0} \sim 731$ kJ mol^{-1} (175 kcal mol^{-1})

$C_3H_8O^+$	$n-C_3H_7OH$		10.20	PI	751	182
$C_3H_8O^+$	$n-C_3H_7OH$		10.22±0.04	PI	753	2647
$C_3H_8O^+$	$n-C_3H_7OH$		10.25	PE	756	2843
$C_3H_8O^+$	$n-C_3H_7OH$		10.32±0.02	PE		3289

The higher ionization potentials given in refs. 2843, 3374 are in serious disagreement.

See also – PI:　　11, 86
　　　　　　PE:　　3374
　　　　　　EI:　　46, 97, 384, 2018

$C_3H_8O^+$	$iso-C_3H_7OH$		10.12±0.03	PI	728	2647
$C_3H_8O^+$	$iso-C_3H_7OH$		10.15±0.05?	PI	731	416
$C_3H_8O^+$	$iso-C_3H_7OH$		10.18	PE	734	2843
$C_3H_8O^+$	$iso-C_3H_7OH$		10.29±0.02	PE		3289

The higher ionization potentials given in refs. 2843, 3374 are in serious disagreement.

See also – PI:　　182
　　　　　　PE:　　3374
　　　　　　EI:　　97, 384

4.3. The Positive Ion Table—Continued

Ion	Reactant	Other products	Ionization or appearance potential (eV)	Method	Heat of formation (kJ mol^{-1})	Ref.	

$C_4H_4O^+$ **(Furan)** $\Delta H^\circ_{f298} = 823$ kJ mol^{-1} (197 kcal mol^{-1})

Ion	Reactant		IP	Method	Hf	Ref.	
$C_4H_4O^+(^2A_2)$	C_4H_4O (Furan)		8.883	S	822	3421	
$C_4H_4O^+(^2A_2)$	C_4H_4O (Furan)		8.89±0.01	PI	823	161,	182, 416
$C_4H_4O^+(^2A_2)$	C_4H_4O (Furan)		8.883	PE	822	3421	
$C_4H_4O^+(^2A_2)$	C_4H_4O (Furan)		8.89±0.05	PE		3246	
$C_4H_4O^+(^2A_2)$	C_4H_4O (Furan)		8.87±0.03	EDD		3174	
$C_4H_4O^+(^2B_1)$	C_4H_4O (Furan)		10.308	S		3421	
$C_4H_4O^+(^2B_1)$	C_4H_4O (Furan)		10.308	PE		3421	
$C_4H_4O^+(^2B_1)$	C_4H_4O (Furan)		10.30±0.05	PE		3246	

Additional higher ionization potentials are given in refs. 3246, 3421.

See also – S: 161, 3351
PE: 2796, 2975, 3109
EI: 381, 383, 411, 2865, 3233

$CH_3CH=CHCHO^+$ $\Delta H^\circ_{f298} = 838$ kJ mol^{-1} (200 kcal mol^{-1})

Ion	Reactant	Other products	IP	Method	Hf	Ref.	
$C_4H_6O^+$	$CH_3CH=CHCHO$		9.73±0.01	PI	838	182,	416

See also – EI: 384

$C_4H_6O^+$	C_4H_6O (3,4–Epoxy–1–butene)		9.7±0.3	EI		153	
$C_4H_6O^+$	C_4H_6O (Cyclobutanone)		9.354	S		3361	
$C_4H_6O^+$	$CH_3COOC_2H_5$	H_2O	10.45	EI		3435	

4.3. The Positive Ion Table—Continued

Ion	Reactant	Other products	Ionization or appearance potential (eV)	Method	Heat of formation (kJ mol^{-1})	Ref.

$C_4H_7O^+$

Ion	Reactant	Other products	Ionization or appearance potential (eV)	Method	Heat of formation (kJ mol^{-1})	Ref.
$C_4H_7O^+$	$(CH_2)_4O$ (1,4-Epoxybutane)	H	10.44	EI		2694

See also – EI: 52

Ion	Reactant	Other products	Ionization or appearance potential (eV)	Method	Heat of formation (kJ mol^{-1})	Ref.
$C_4H_7O^+$	$n-C_3H_7COCH_3$	CH_3	10.03	PI	(567*)	95

See also – EI: 2977

Ion	Reactant	Other products	Ionization or appearance potential (eV)	Method	Heat of formation (kJ mol^{-1})	Ref.
$C_4H_7O^+$	$iso-C_3H_7COCH_3$	CH_3	9.94	PI	(554*)	95
$C_4H_7O^+$	$n-C_4H_9COCH_3$	C_2H_5	10.03	PI	(580*)	95
$C_4H_7O^+$	$n-C_3H_7COC_2H_5$	C_2H_5	10.6±0.3	EI		2740
$C_4H_7O^+$	$(n-C_3H_7)_2CO$		10.61	RPD		2977
$C_4H_7O^+$	$n-C_3H_7COOCH_3$	CH_3O	11.20±0.2	EI		2497

See also – EI: 2496

Ion	Reactant	Other products	Ionization or appearance potential (eV)	Method	Heat of formation (kJ mol^{-1})	Ref.
$C_4H_7O^+$	$C_5H_{10}O_2$ (1,3-Dioxepane)	CH_3O	10.55	EI		2694

*ΔH°_{f298}

$C_4H_2D_5O^+$

Ion	Reactant	Other products	Ionization or appearance potential (eV)	Method	Heat of formation (kJ mol^{-1})	Ref.
$C_4H_2D_5O^+$	$n-C_3H_7CD_2COCD_3$	C_2H_5	10.9±0.2	EI		2766

$n-C_3H_7CHO^+$ $\Delta H^\circ_{f298} \sim 740$ kJ mol^{-1} (177 kcal mol^{-1})
$iso-C_3H_7CHO^+$ $\Delta H^\circ_{f298} \sim 719$ kJ mol^{-1} (172 kcal mol^{-1})
$C_2H_5COCH_3^+$ $\Delta H^\circ_{f298} = 676$ kJ mol^{-1} (162 kcal mol^{-1})
$C_4H_8O^+$ (1,4-Epoxybutane) $\Delta H^\circ_{f298} = 725$ kJ mol^{-1} (173 kcal mol^{-1})

Ion	Reactant	Other products	Ionization or appearance potential (eV)	Method	Heat of formation (kJ mol^{-1})	Ref.
$C_4H_8O^+$	$n-C_3H_7CHO$		9.86±0.02	PI	746	182
$C_4H_8O^+$	$n-C_3H_7CHO$		9.73±0.03	PE	734	3289

See also – EI: 2522

Ion	Reactant	Other products	Ionization or appearance potential (eV)	Method	Heat of formation (kJ mol^{-1})	Ref.
$C_4H_8O^+$	$iso-C_3H_7CHO$		9.74±0.03	PI	721	182
$C_4H_8O^+$	$iso-C_3H_7CHO$		9.69±0.01	PE	716	3289
$C_4H_8O^+$	$C_2H_5COCH_3$		9.48±0.02	PI	676	95
$C_4H_8O^+$	$C_2H_5COCH_3$		9.53±0.01	PI		182
$C_4H_8O^+$	$C_2H_5COCH_3$		9.55±0.03	PI		1166
$C_4H_8O^+$	$C_2H_5COCH_3$		9.45±0.1	PI		86
$C_4H_8O^+$	$C_2H_5COCH_3$		9.51	PE		2843
$C_4H_8O^+$	$C_2H_5COCH_3$		9.54±0.01	PE		3289

See also – PI: 158, 416
EI: 411, 2522, 2649, 2883, 2977

4.3. The Positive Ion Table—Continued

Ion	Reactant	Other products	Ionization or appearance potential (eV)	Method	Heat of formation (kJ mol^{-1})	Ref.
$C_4H_8O^+$	$(CH_2)_4O$ (1,4–Epoxybutane)		9.42±0.01	S	725	2169
$C_4H_8O^+$	$(CH_2)_4O$ (1,4–Epoxybutane)		9.54?	PI		182

See also – EI: 52, 218, 2694
CTS: 2562

Ion	Reactant	Other products	Ionization or appearance potential (eV)	Method	Heat of formation (kJ mol^{-1})	Ref.
$C_4H_8O^+$	$n-C_3H_7COC_2H_5$	C_2H_4	10.2±0.3	EI		2740
$C_4H_8O^+$	$n-C_4H_9COC_2H_5$	C_3H_6	10.3±0.3	EI		2740
$C_4H_8O^+$	$n-C_5H_{11}COC_2H_5$		10.5±0.3	EI		2740
$C_4H_8O^+$	$C_5H_{10}O_2$ (1,3–Dioxepane)	CH_2O	10.21	EI		2694

$C_4H_9O^+$

Ion	Reactant	Other products	Ionization or appearance potential (eV)	Method	Heat of formation (kJ mol^{-1})	Ref.
$C_4H_9O^+$	$n-C_4H_9O$		9.22±0.05	RPD		2776
$C_4H_9O^+$	$n-C_5H_{11}COC_2H_5$		9.8±0.3	EI		2740
$C_4H_9O^+$	$(n-C_4H_9)_2O$		12.96±0.1	RPD		2776
$C_4H_9O^+$	$(CH_3O)_2C(CH_3)_2$	CH_3O	10.28±0.05	RPD		1139
$C_4H_9O^+$	$n-C_4H_9ONO$	NO	9.94±0.05	RPD		2776

$n-C_4H_9OH^+$	ΔH_{f0}° ~ 725 kJ mol^{-1} (173 kcal mol^{-1})		
$iso-C_4H_9OH^+$	ΔH_{f298}° ~ 690 kJ mol^{-1} (165 kcal mol^{-1})		
$tert-C_4H_9OH^+$	ΔH_{f0}° ~ 680 kJ mol^{-1} (162 kcal mol^{-1})		
$(C_2H_5)_2O^+$	ΔH_{f298}° = 666 kJ mol^{-1} (159 kcal mol^{-1})		

Ion	Reactant	Other products	Ionization or appearance potential (eV)	Method	Heat of formation (kJ mol^{-1})	Ref.
$C_4H_{10}O^+$	$n-C_4H_9OH$		10.04	PI	723	182
$C_4H_{10}O^+$	$n-C_4H_9OH$		10.09±0.02	PE	728	3289
$C_4H_{10}O^+$	$n-C_4H_9OH$		10.37	PE		3374

See also – EI: 97

Ion	Reactant	Other products	Ionization or appearance potential (eV)	Method	Heat of formation (kJ mol^{-1})	Ref.
$C_4H_{10}O^+$	$iso-C_4H_9OH$		10.09±0.02	PE	690	3289
$C_4H_{10}O^+$	$tert-C_4H_9OH$		9.97±0.02	PE	680	3289
$C_4H_{10}O^+$	$tert-C_4H_9OH$		10.23	PE		3374
$C_4H_{10}O^+$	$(C_2H_5)_2O$		9.53±0.02	PI	667	182, 416
$C_4H_{10}O^+$	$(C_2H_5)_2O$		9.51	PE	665	2843
$C_4H_{10}O^+$	$(C_2H_5)_2O$		9.50±0.01	PE	664	3289

See also – PI: 1166
PE: 1130
EI: 2196, 2776

4.3. The Positive Ion Table—Continued

Ion	Reactant	Other products	Ionization or appearance potential (eV)	Method	Heat of formation (kJ mol⁻¹)	Ref.

$$\text{C}_5\text{H}_6\text{O}^+$$

Ion	Reactant	Other products	Ionization or appearance potential (eV)	Method	Heat of formation	Ref.
$\text{C}_5\text{H}_6\text{O}^+$	$\text{C}_4\text{H}_3\text{OCH}_3$ (2–Methylfuran)		8.39±0.01	PI		182
$\text{C}_5\text{H}_6\text{O}^+$	$\text{C}_4\text{H}_3\text{OCH}_3$ (2–Methylfuran)		8.31±0.09	EI		411
$\text{C}_5\text{H}_6\text{O}^+$	$\text{C}_5\text{H}_6\text{O}$ (2–Cyclopenten–1–one)		9.34±0.02 (V)	PE		3407
$\text{C}_5\text{H}_6\text{O}^+$	$\text{C}_5\text{H}_6\text{O}$ (3–Cyclopenten–1–one)		9.44±0.02 (V)	PE		3407

$$\text{C}_5\text{H}_8\text{O}^+ \text{ (Cyclopentanone)} \quad \Delta\text{H}^\circ_{f298} = 701 \text{ kJ mol}^{-1} \text{ (168 kcal mol}^{-1})$$
$$\text{C}_5\text{H}_8\text{O}^+ \text{ (Dihydropyran)} \quad \Delta\text{H}^\circ_{f298} = 680 \text{ kJ mol}^{-1} \text{ (162 kcal mol}^{-1})$$

Ion	Reactant	Other products	Ionization or appearance potential (eV)	Method	Heat of formation	Ref.
$\text{C}_5\text{H}_8\text{O}^+$	$\text{C}_5\text{H}_8\text{O}$ (Cyclopentanone)		9.26±0.01	PI	701	182
$\text{C}_5\text{H}_8\text{O}^+$	$\text{C}_5\text{H}_8\text{O}$ (Cyclopentanone)		9.25±0.02	PE	700	3407
$\text{C}_5\text{H}_8\text{O}^+$	$\text{C}_5\text{H}_8\text{O}$ (Cyclopentanone)		9.28±0.01	PE	703	3289
$\text{C}_5\text{H}_8\text{O}^+$	$\text{C}_5\text{H}_8\text{O}$ (Dihydropyran)		8.34±0.01	PI	680	182

See also – CTS: 2031

$$\text{C}_5\text{H}_9\text{O}^+$$

Ion	Reactant	Other products	Ionization or appearance potential (eV)	Method	Heat of formation	Ref.
$\text{C}_5\text{H}_9\text{O}^+$	$(\text{CH}_2)_5\text{O}$ (1,5–Epoxypentane)	H	11.22	EI		2694
$\text{C}_5\text{H}_9\text{O}^+$	n–$\text{C}_4\text{H}_9\text{COCH}_3$	CH_3	9.66	PI	(510*)	95

See also – EI: 2977

Ion	Reactant	Other products	Ionization or appearance potential (eV)	Method	Heat of formation	Ref.
$\text{C}_5\text{H}_9\text{O}^+$	iso–$\text{C}_4\text{H}_9\text{COCH}_3$	CH_3	9.80	PI		95
$\text{C}_5\text{H}_9\text{O}^+$	n–$\text{C}_4\text{H}_9\text{COC}_2\text{H}_5$	C_2H_5	10.8±0.3	EI		2740
$\text{C}_5\text{H}_9\text{O}^+$	n–$\text{C}_5\text{H}_{11}\text{COC}_2\text{H}_5$		10.9±0.3	EI		2740
$\text{C}_5\text{H}_9\text{O}^+$	$(n$–$\text{C}_4\text{H}_9)_2\text{CO}$		10.86	RPD		2977
$\text{C}_5\text{H}_9\text{O}^+$	n–$\text{C}_4\text{H}_9\text{COOCH}_3$	CH_3O	11.18±0.2	EI		2497

*$\Delta\text{H}^\circ_{f298}$

$$\text{C}_5\text{H}_7\text{D}_2\text{O}^+$$

Ion	Reactant	Other products	Ionization or appearance potential (eV)	Method	Heat of formation	Ref.
$\text{C}_5\text{H}_7\text{D}_2\text{O}^+$	n–$\text{C}_3\text{H}_7\text{CD}_2\text{COCD}_3$	CD_3	10.2±0.2	EI		2766

$$\text{C}_5\text{H}_4\text{D}_5\text{O}^+$$

Ion	Reactant	Other products	Ionization or appearance potential (eV)	Method	Heat of formation	Ref.
$\text{C}_5\text{H}_4\text{D}_5\text{O}^+$	n–$\text{C}_3\text{H}_7\text{CD}_2\text{COCD}_3$	CH_3	9.5±0.2	EI		2766

4.3. The Positive Ion Table—Continued

Ion	Reactant	Other products	Ionization or appearance potential (eV)	Method	Heat of formation (kJ mol^{-1})	Ref.
n-C$_4$H$_9$CHO$^+$		$\Delta H^\circ_{f298} \sim 717$ kJ mol^{-1} (171 kcal mol^{-1})				
n-C$_3$H$_7$COCH$_3^+$		$\Delta H^\circ_{f298} = 647$ kJ mol^{-1} (155 kcal mol^{-1})				
iso-C$_3$H$_7$COCH$_3^+$		$\Delta H^\circ_{f298} = 635$ kJ mol^{-1} (152 kcal mol^{-1})				
(C$_2$H$_5$)$_2$CO$^+$		$\Delta H^\circ_{f298} = 640$ kJ mol^{-1} (153 kcal mol^{-1})				
C$_5$H$_{10}$O$^+$ (1,5–Epoxypentane)		$\Delta H^\circ_{f298} = 669$ kJ mol^{-1} (160 kcal mol^{-1})				
C$_5$H$_{10}$O$^+$	n-C$_4$H$_9$CHO		9.82±0.05	PI	720	182
C$_5$H$_{10}$O$^+$	n-C$_4$H$_9$CHO		9.77±0.01	PE	715	3289
C$_5$H$_{10}$O$^+$	iso-C$_4$H$_9$CHO		9.71±0.05	PI		182
C$_5$H$_{10}$O$^+$	iso-C$_4$H$_9$CHO		9.68±0.02	PE		3289
C$_5$H$_{10}$O$^+$	$tert$-C$_4$H$_9$CHO		9.51±0.01	PE		3289
C$_5$H$_{10}$O$^+$	n-C$_3$H$_7$COCH$_3$		9.37±0.02	PI	645	95
C$_5$H$_{10}$O$^+$	n-C$_3$H$_7$COCH$_3$		9.39±0.02	PI	647	182
C$_5$H$_{10}$O$^+$	n-C$_3$H$_7$COCH$_3$		9.40±0.01	PE	648	3289

See also – PI: 1166
        EI: 2433, 2522, 2977

Ion	Reactant	Other products	Ionization or appearance potential (eV)	Method	Heat of formation (kJ mol^{-1})	Ref.
C$_5$H$_{10}$O$^+$	iso-C$_3$H$_7$COCH$_3$		9.30±0.02	PI	635	95
C$_5$H$_{10}$O$^+$	iso-C$_3$H$_7$COCH$_3$		9.32±0.02	PI	637	182
C$_5$H$_{10}$O$^+$	iso-C$_3$H$_7$COCH$_3$		9.30±0.01	PE	635	3289

See also – EI: 2433

Ion	Reactant	Other products	Ionization or appearance potential (eV)	Method	Heat of formation (kJ mol^{-1})	Ref.
C$_5$H$_{10}$O$^+$	(C$_2$H$_5$)$_2$CO		9.32±0.01	PI	641	182
C$_5$H$_{10}$O$^+$	(C$_2$H$_5$)$_2$CO		9.31±0.02	PE	640	3289

See also – EI: 2977, 3231

Ion	Reactant	Other products	Ionization or appearance potential (eV)	Method	Heat of formation (kJ mol^{-1})	Ref.
C$_5$H$_{10}$O$^+$	(CH$_2$)$_5$O (1,5–Epoxypentane)		9.25±0.01	S	669	2169
C$_5$H$_{10}$O$^+$	(CH$_2$)$_5$O (1,5–Epoxypentane)		9.26±0.03	PI	670	182

See also – EI: 218, 2694

Ion	Reactant	Other products	Ionization or appearance potential (eV)	Method	Heat of formation (kJ mol^{-1})	Ref.
C$_5$H$_{10}$O$^+$	n-C$_5$H$_{11}$COC$_2$H$_5$	C$_3$H$_6$	10.4±0.3	EI		2740

C$_5$H$_{11}$O$^+$

Ion	Reactant	Other products	Ionization or appearance potential (eV)	Method	Heat of formation (kJ mol^{-1})	Ref.
C$_5$H$_{11}$O$^+$	n-C$_5$H$_{11}$OOH	OH	10.0±0.1	EI		2464
C$_5$H$_{11}$O$^+$	n-C$_3$H$_7$CH(OOH)CH$_3$	OH	10.4±0.1	EI		2464
C$_5$H$_{11}$O$^+$	(C$_2$H$_5$)$_2$CHOOH	OH	10.2±0.1	EI		2464

4.3. The Positive Ion Table—Continued

Ion	Reactant	Other products	Ionization or appearance potential (eV)	Method	Heat of formation (kJ mol^{-1})	Ref.

$$C_6H_5O^+$$

Ion	Reactant	Other products	Ionization or appearance potential (eV)	Method	Heat of formation (kJ mol^{-1})	Ref.
$C_6H_5O^+$	C_6H_5O (Phenoxy radical)		8.84	EI		1079
$C_6H_5O^+$	$C_6H_5OCH_3$ (Methoxybenzene)	CH_3	11.86±0.1	EI		2970
$C_6H_5O^+$	$C_6H_5OCH_3$ (Methoxybenzene)	CH_3	11.92±0.1	EI		1079
$C_6H_5O^+$	$C_6H_5OC\equiv CH$ (Phenoxyacetylene)	C_2H	9.5±0.1	EI		13
$C_6H_5O^+$	C_6H_4BrOH (4–Bromophenol)	Br	12.17	EI		3238

$$C_6H_6O^+ \text{ (Phenol)} \qquad \Delta H^\circ_{f298} = 724 \text{ kJ mol}^{-1} \text{ (173 kcal mol}^{-1}\text{)}$$

Ion	Reactant	Other products	Ionization or appearance potential (eV)	Method	Heat of formation (kJ mol^{-1})	Ref.
$C_6H_6O^+$	C_6H_5OH (Phenol)		8.50±0.01	PI	724	182, 416
$C_6H_6O^+$	C_6H_5OH (Phenol)		8.52±0.02	PI	726	1166
$C_6H_6O^+$	C_6H_5OH (Phenol)		8.52	PE	726	2843
$C_6H_6O^+$	C_6H_5OH (Phenol)		8.48±0.05	PE	722	2796

See also – PE: 2806
EI: 1066, 2865, 3174, 3223, 3238

Ion	Reactant	Other products	Ionization or appearance potential (eV)	Method	Heat of formation (kJ mol^{-1})	Ref.
$C_6H_6O^+$	$C_6H_5OC_2H_5$ (Ethoxybenzene)	C_2H_4	10.73	EI		2706

See also – EI: 2945

Ion	Reactant	Other products	Ionization or appearance potential (eV)	Method	Heat of formation (kJ mol^{-1})	Ref.
$C_6H_6O^+$	$C_6H_5OC_3H_7$ (Propoxybenzene)	C_3H_6	10.21	EI		2706
$C_6H_6O^+$	$C_6H_5OC_4H_9$ (Butoxybenzene)		10.05	EI		2706
$C_6H_6O^+$	$(C_6H_5)_2O$ (Diphenyl ether)		13.88±0.15	EI		1237

$$C_6H_8O^+$$

Ion	Reactant	Other products	Ionization or appearance potential (eV)	Method	Heat of formation (kJ mol^{-1})	Ref.
$C_6H_8O^+$	$C_4H_2O(CH_3)_2$ (2,3–Dimethylfuran)		8.01±0.09	EI		411

4.3. The Positive Ion Table—Continued

Ion	Reactant	Other products	Ionization or appearance potential (eV)	Method	Heat of formation (kJ mol^{-1})	Ref.
	$C_6H_{10}O^+$ (Cyclohexanone)		$\Delta H^\circ_{f298} = 653$ kJ mol^{-1} (156 kcal mol^{-1})			
$C_6H_{10}O^+$	$(CH_3)_2C=CHCOCH_3$		9.08±0.03	PI		182
$C_6H_{10}O^+$	$(CH_3)_2C=CHCOCH_3$		8.89±0.05	EI		384
$C_6H_{10}O^+$	$C_6H_{10}O$ (Cyclohexanone)		9.14±0.01	PI	652	182
$C_6H_{10}O^+$	$C_6H_{10}O$ (Cyclohexanone)		9.16±0.01	PE	654	3289

See also – EI: 431

	$C_6H_{11}O^+$					
$C_6H_{11}O^+$	$n-C_5H_{11}COCH_3$	CH_3	9.83	RPD		2977
$C_6H_{11}O^+$	$n-C_5H_{11}COC_2H_5$	C_2H_5	10.2±0.3	EI		2740

	$n-C_4H_9COCH_3^+$		$\Delta H^\circ_{f298} = 623$ kJ mol^{-1} (149 kcal mol^{-1})			
	$tert-C_4H_9COCH_3^+$		$\Delta H^\circ_{f298} = 593$ kJ mol^{-1} (142 kcal mol^{-1})			
$C_6H_{12}O^+$	$n-C_4H_9COCH_3$		9.37±0.02	PI	624	95
$C_6H_{12}O^+$	$n-C_4H_9COCH_3$		9.34±0.03	PI	621	182
$C_6H_{12}O^+$	$n-C_4H_9COCH_3$		9.36±0.02	PE	623	3289

See also – EI: 1256, 2433, 2977

$C_6H_{12}O^+$	$sec-C_4H_9COCH_3$		9.69	EI	1254,	2433
$C_6H_{12}O^+$	$iso-C_4H_9COCH_3$		9.30±0.02	PI		95
$C_6H_{12}O^+$	$iso-C_4H_9COCH_3$		9.30±0.03	PI		182
$C_6H_{12}O^+$	$iso-C_4H_9COCH_3$		9.34±0.01	PE		3289

See also – EI: 1256, 2433

$C_6H_{12}O^+$	$tert-C_4H_9COCH_3$		9.17±0.03	PI	595	182
$C_6H_{12}O^+$	$tert-C_4H_9COCH_3$		9.14±0.01	PE	592	3289

See also – EI: 1254, 2433

$C_6H_{12}O^+$	$C_6H_{11}OH$ (Cyclohexanol)		10.0	EI		3022

See also – D: 2908

	$C_6H_7D_5O^+$					
$C_6H_7D_5O^+$	$iso-C_3H_7CD_2COCD_3$		9.35	EI	1256,	2433

4.3. The Positive Ion Table—Continued

Ion	Reactant	Other products	Ionization or appearance potential (eV)	Method	Heat of formation (kJ mol^{-1})	Ref.
			$C_6H_{13}O^+$			
$C_6H_{13}O^+$	$n-C_6H_{13}OOH$	OH	10.2 ± 0.1	EI		2464
$C_6H_{13}O^+$	$n-C_4H_9CH(OOH)CH_3$	OH	9.7 ± 0.1	EI		2464
$C_6H_{13}O^+$	$n-C_3H_7CH(OOH)C_2H_5$	OH	9.3 ± 0.1	EI		2464

$(n-C_3H_7)_2O^+$ $\Delta H^\circ_{f298} \sim 604$ kJ mol^{-1} (144 kcal mol^{-1})
$(iso-C_3H_7)_2O^+$ $\Delta H^\circ_{f298} \sim 569$ kJ mol^{-1} (136 kcal mol^{-1})

Ion	Reactant	Other products	Ionization or appearance potential (eV)	Method	Heat of formation (kJ mol^{-1})	Ref.
$C_6H_{14}O^+$	$(n-C_3H_7)_2O$		9.27 ± 0.05	PI	602	182
$C_6H_{14}O^+$	$(n-C_3H_7)_2O$		9.32 ± 0.01	PE	606	3289
$C_6H_{14}O^+$	$(iso-C_3H_7)_2O$		9.20 ± 0.05	PI	569	182
$C_6H_{14}O^+$	$(iso-C_3H_7)_2O$		9.16 ± 0.05	RPD		2776

Ion	Reactant	Other products	Ionization or appearance potential (eV)	Method	Heat of formation (kJ mol^{-1})	Ref.
			$C_7H_5O^+$			
$C_7H_5O^+$	C_6H_5CHO (Benzenecarbonal)	H	10.99	EI		3238

See also – EI: 130, 308, 1237

Ion	Reactant	Other products	Ionization or appearance potential (eV)	Method	Heat of formation (kJ mol^{-1})	Ref.
$C_7H_5O^+$	$C_6H_5COCH_3$ (Acetophenone)	CH_3	9.91	EI		3334

See also – EI: 308, 1237, 2174, 2967, 3238

Ion	Reactant	Other products	Ionization or appearance potential (eV)	Method	Heat of formation (kJ mol^{-1})	Ref.
$C_7H_5O^+$	$C_6H_5COCD_3$ (Acetophenone-$\alpha,\alpha,\alpha-d_3$)	CD_3	10.45	EI		308
$C_7H_5O^+$	$(C_6H_5)_2CO$ (Benzophenone)		12.00 ± 0.05	EI		1237
$C_7H_5O^+$	$C_6H_5COOCH_3$ (Benzoic acid methyl ester)	CH_3O	10.80	EI		3238

See also – EI: 308

Ion	Reactant	Other products	Ionization or appearance potential (eV)	Method	Heat of formation (kJ mol^{-1})	Ref.
$C_7H_5O^+$	$C_6H_5COOC_6H_5$ (Benzoic acid phenyl ester)		10.01 ± 0.07	EI		1237
$C_7H_5O^+$	$C_6H_5COCOC_6H_5$ (Diphenylglyoxal)		9.70 ± 0.05	EI		1237
$C_7H_5O^+$	$C_6H_5CONH_2$ (Benzoic acid amide)	NH_2	9.9 ± 0.1	EI		1168
$C_7H_5O^+$	$C_6H_5CONHC_6H_5$ (N-Phenylbenzoic acid amide)		10.6 ± 0.1	EI		2918
$C_7H_5O^+$	$C_6H_5COCHN_2$ (α-Diazoacetophenone)		10.42 ± 0.18	EI		2174
$C_7H_5O^+$	$(C_6H_5)_2C_2N_2O$ (2,5-Diphenyl-1,3,4-oxadiazole)		12.1 ± 0.2	EI		1125
$C_7H_5O^+$	$C_6H_5CONHC_6H_4OCH_3$ (N-(3-Methoxyphenyl)benzoic acid amide)		10.8 ± 0.1	EI		2918
$C_7H_5O^+$	$C_6H_5CONHC_6H_4OCH_3$ (N-(4-Methoxyphenyl)benzoic acid amide)		11.2 ± 0.1	EI		2918

4.3. The Positive Ion Table—Continued

Ion	Reactant	Other products	Ionization or appearance potential (eV)	Method	Heat of formation (kJ mol^{-1})	Ref.
$C_7H_5O^+$	$C_6H_5CONHC_6H_4NO_2$ (N-(3-Nitrophenyl)benzoic acid amide)		10.2±0.1	EI		2918
$C_7H_5O^+$	$C_6H_5CONHC_6H_4NO_2$ (N-(4-Nitrophenyl)benzoic acid amide)		10.2±0.1	EI		2918
$C_7H_5O^+$	C_6H_5COF (Benzoic acid fluoride)	F	11.5	EI		308
$C_7H_5O^+$	$C_6H_5COCF_3$ (α,α,α-Trifluoroacetophenone)	CF_3	10.05	EI		308
$C_7H_5O^+$	C_6H_5COCl (Benzoic acid chloride)	Cl	10.5	EI		308

See also – EI: 130

Ion	Reactant	Other products	Ionization or appearance potential (eV)	Method	Heat of formation (kJ mol^{-1})	Ref.
$C_7H_5O^+$	C_6H_5COBr (Benzoic acid bromide)	Br	10.0	EI		308

$C_7H_6O^+$

Ion	Reactant	Other products	Ionization or appearance potential (eV)	Method	Heat of formation (kJ mol^{-1})	Ref.
$C_7H_6O^+$	C_6H_5CHO (Benzenecarbonal)		9.51±0.02?	PI		416
$C_7H_6O^+$	C_6H_5CHO (Benzenecarbonal)		9.53±0.03	PI		182
$C_7H_6O^+$	C_6H_5CHO (Benzenecarbonal)		9.60±0.02	PI		1166
$C_7H_6O^+$	C_6H_5CHO (Benzenecarbonal)		9.80 (V)	PE		2806
$C_7H_6O^+$	C_6H_5CHO (Benzenecarbonal)		9.53±0.03	RPD		2463

See also – EI: 127, 130, 308, 1237, 2026, 2718, 3238

Ion	Reactant	Other products	Ionization or appearance potential (eV)	Method	Heat of formation (kJ mol^{-1})	Ref.
$C_7H_6O^+$	C_7H_6O (2,4,6-Cycloheptatrien-1-one)		9.68±0.02	EI		431

$C_7H_7O^+$

Ion	Reactant	Other products	Ionization or appearance potential (eV)	Method	Heat of formation (kJ mol^{-1})	Ref.
$C_7H_7O^+$	$C_6H_4(OH)CH_3$ (3-Methylphenol)	H	12.33±0.1	EI		2970
$C_7H_7O^+$	$C_6H_4(OH)CH_3$ (4-Methylphenol)	H	12.41±0.1	EI		2970
$C_7H_7O^+$	$C_6H_4(OH)C_2H_5$ (3-Ethylphenol)	CH_3	10.80±0.1	EI		2970
$C_7H_7O^+$	$C_6H_4(OH)C_2H_5$ (4-Ethylphenol)	CH_3	10.83±0.1	EI		2970
$C_7H_7O^+$	$C_6H_5CH_2OCH_3$ (α-Methoxytoluene)	CH_3	11.47	EI		3287
$C_7H_7O^+$	$C_6H_4(OCH_3)CH_3$ (3-Methoxytoluene)	CH_3	11.33±0.1	EI		2970
$C_7H_7O^+$	$C_6H_4(OCH_3)CH_3$ (4-Methoxytoluene)	CH_3	10.83±0.1	EI		2970

See also – EI: 3287

4.3. The Positive Ion Table—Continued

Ion	Reactant	Other products	Ionization or appearance potential (eV)	Method	Heat of formation (kJ mol^{-1})	Ref.
$C_7H_7O^+$	$C_7H_7OCH_3$ (7–Methoxycycloheptatriene)	CH_3	10.26	EI		3287
$C_7H_7O^+$	$C_6H_5CH_2CH_2C_6H_4OH$ (1–(3–Hydroxyphenyl)–2–phenylethane)		10.8±0.2	EI		3288
$C_7H_7O^+$	$C_6H_5CH_2CH_2C_6H_4OH$ (1–(4–Hydroxyphenyl)–2–phenylethane)		9.8±0.2	EI		3288
$C_7H_7O^+$	$C_6H_5CH_2OC_6H_4CH_3$ (Benzyl 3–tolyl ether)		11.9	EI		2737
$C_7H_7O^+$	$C_6H_5CH_2OC_6H_4CH_3$ (Benzyl 4–tolyl ether)		11.8	EI		2737
$C_7H_7O^+$	$C_6H_4(OH)(CH_2)_3COOCH_3$ (4–(4–Hydroxyphenyl)butanoic acid methyl ester)		11.69±0.2	EI		2497

$C_7H_8O^+$ (Methoxybenzene) $\Delta H^\circ_{f298} = 720$ kJ mol^{-1} (172 kcal mol^{-1})

Ion	Reactant	Other products	Ionization or appearance potential (eV)	Method	Heat of formation (kJ mol^{-1})	Ref.
$C_7H_8O^+$	$C_6H_5OCH_3$ (Methoxybenzene)		8.20±0.02	PI	719	416
$C_7H_8O^+$	$C_6H_5OCH_3$ (Methoxybenzene)		8.22±0.02	PI	721	182
$C_7H_8O^+$	$C_6H_5OCH_3$ (Methoxybenzene)		8.21	PE	720	2843

See also – PE: 2806
         EI: 1066, 2865, 3223, 3238

Ion	Reactant	Other products	Ionization or appearance potential (eV)	Method	Heat of formation (kJ mol^{-1})	Ref.
$C_7H_8O^+$	$C_6H_5CH_2OH$ (α–Hydroxytoluene)		9.14±0.05	EI		2025

See also – D: 2908

Ion	Reactant	Other products	Ionization or appearance potential (eV)	Method	Heat of formation (kJ mol^{-1})	Ref.
$C_7H_8O^+$	$C_6H_4(OH)CH_3$ (2–Methylphenol)		8.93	EI		1066
$C_7H_8O^+$	$C_6H_4(OH)CH_3$ (3–Methylphenol)		8.98	EI		1066
$C_7H_8O^+$	$C_6H_4(OH)CH_3$ (3–Methylphenol)		8.52±0.05	EI		2025
$C_7H_8O^+$	$C_6H_4(OH)CH_3$ (4–Methylphenol)		8.97	EI		1066
$C_7H_8O^+$	C_7H_8O (Bicyclo[2.2.1]hept–2–en–5–one)		8.90±0.02 (V)	PE		3408
$C_7H_8O^+$	C_7H_8O (Bicyclo[2.2.1]hept–2–en–7–one)		9.19±0.02 (V)	PE		3408
$C_7H_8O^+$	C_7H_8O (Nortricyclone)		9.01	PE		2951

$C_7H_{10}O^+$

Ion	Reactant	Other products	Ionization or appearance potential (eV)	Method	Heat of formation (kJ mol^{-1})	Ref.
$C_7H_{10}O^+$	$C_7H_{10}O$ (Bicyclo[2.2.1]heptan–2–one)		8.94±0.02 (V)	PE		3408
$C_7H_{10}O^+$	$C_7H_{10}O$ (Bicyclo[2.2.1]heptan–7–one)		9.01±0.02 (V)	PE		3408

4.3. The Positive Ion Table—Continued

Ion	Reactant	Other products	Ionization or appearance potential (eV)	Method	Heat of formation (kJ mol^{-1})	Ref.
	$C_7H_{12}O^+$ (Cycloheptanone)		**$\Delta H^\circ_{f298} = 668$ kJ mol^{-1} (160 kcal mol^{-1})**			
$C_7H_{12}O^+$	$C_7H_{12}O$ (Cycloheptanone)		9.49±0.01	PE	668	3289
	$C_7H_{13}O^+$					
$C_7H_{13}O^+$	$n-C_6H_{13}COCH_3$	CH_3	9.85	RPD		2977
	$(iso-C_3H_7)_2CO^+$		**$\Delta H^\circ_{f298} = 553$ kJ mol^{-1} (132 kcal mol^{-1})**			
$C_7H_{14}O^+$	$n-C_5H_{11}COCH_3$		9.33±0.03	PI		182
$C_7H_{14}O^+$	$n-C_5H_{11}COCH_3$		9.79	RPD		2977
$C_7H_{14}O^+$	$n-C_4H_9COC_2H_5$		9.15±0.02	PE		3289
$C_7H_{14}O^+$	$(n-C_3H_7)_2CO$		9.15±0.02	PE		3289
$C_7H_{14}O^+$	$(n-C_3H_7)_2CO$		9.84	RPD		2977
$C_7H_{14}O^+$	$(iso-C_3H_7)_2CO$		8.96±0.01	PE	553	3289
	$C_7H_{15}O^+$					
$C_7H_{15}O^+$	$n-C_7H_{15}OOH$	OH	10.3±0.1	EI		2464
$C_7H_{15}O^+$	$n-C_5H_{11}CH(OOH)CH_3$	OH	9.7±0.1	EI		2464
	$C_8H_6O^+$					
$C_8H_6O^+$	C_8H_6O (Benzofuran)		8.29±0.05	PE		2796
$C_8H_6O^+$	C_8H_6O (Benzofuran)		8.42	PE		3349

See also – EI: 2541

Ion	Reactant	Other products	Ionization or appearance potential (eV)	Method	Heat of formation (kJ mol^{-1})	Ref.
$C_8H_6O^+$	C_8H_6O (Benzocyclobutenone)		8.99	EI		3296
$C_8H_6O^+$	$C_6H_4(OH)C\equiv CH$ (2-Ethynylphenol)		8.71	EI		2541
$C_8H_6O^+$	$C_9H_6O_2$ (Coumarin)	CO	11.1	EI		2946

See also – EI: 2541

Ion	Reactant	Other products	Ionization or appearance potential (eV)	Method	Heat of formation (kJ mol^{-1})	Ref.
$C_8H_6O^+$	$C_6H_5COCHN_2$ (α–Diazoacetophenone)	N_2	10.08±0.11	EI		2174

4.3. The Positive Ion Table—Continued

Ion	Reactant	Other products	Ionization or appearance potential (eV)	Method	Heat of formation (kJ mol⁻¹)	Ref.

$$\mathbf{C_8H_7O^+}$$

Ion	Reactant	Other products	Ionization or appearance potential (eV)	Method	Heat of formation	Ref.
$C_8H_7O^+$	$C_6H_5CH_2COCH_3$ (Benzyl methyl ketone)	CH_3	9.90±0.17	EI		2174
$C_8H_7O^+$	$C_6H_4(CH_3)COCH_3$ (3−Methylacetophenone)	CH_3	10.10±0.1	EI		2967
$C_8H_7O^+$	$C_6H_4(CH_3)COCH_3$ (4−Methylacetophenone)	CH_3	9.72	EI		3334

See also − EI: 2967, 3238

$$\mathbf{C_8H_8O^+} \text{ (Acetophenone)} \qquad \Delta H^\circ_{f298} \sim 808 \text{ kJ mol}^{-1} \text{ (193 kcal mol}^{-1}\text{)}$$

Ion	Reactant	Other products	Ionization or appearance potential (eV)	Method	Heat of formation	Ref.
$C_8H_8O^+$	$C_6H_5COCH_3$ (Acetophenone)		9.27±0.03?	PI	808	182
$C_8H_8O^+$	$C_6H_5COCH_3$ (Acetophenone)		9.15±0.03	RPD		2463

See also − EI: 308, 1237, 2025, 2026, 2174, 2967, 3238, 3334

Ion	Reactant	Other products	Ionization or appearance potential (eV)	Method	Heat of formation	Ref.
$C_8H_8O^+$	$C_6H_4(CH_3)CHO$ (4−Methylbenzenecarbonal)		9.33±0.05	EI		2026
$C_8H_8O^+$	$C_6H_4(OH)(CH_2)_3COOCH_3$ (4−(4−Hydroxyphenyl)butanoic acid methyl ester)		10.65±0.2	EI		2497

$$\mathbf{C_8H_9O^+}$$

Ion	Reactant	Other products	Ionization or appearance potential (eV)	Method	Heat of formation	Ref.
$C_8H_9O^+$	$C_6H_4(OCH_3)CH_2$ (4−Methoxybenzyl radical)		6.82±0.1	EI		69
$C_8H_9O^+$	$C_6H_5CH_2OCH_3$ (α−Methoxytoluene)	H	10.65±0.1	EI		122

See also − EI: 3287

Ion	Reactant	Other products	Ionization or appearance potential (eV)	Method	Heat of formation	Ref.
$C_8H_9O^+$	$C_6H_4(OCH_3)CH_3$ (3−Methoxytoluene)	H	12.13±0.1	EI		122
$C_8H_9O^+$	$C_6H_4(OCH_3)CH_3$ (4−Methoxytoluene)	H	11.91	EI		3287

See also − EI: 122

Ion	Reactant	Other products	Ionization or appearance potential (eV)	Method	Heat of formation	Ref.
$C_8H_9O^+$	$C_7H_7OCH_3$ (7−Methoxycycloheptatriene)	H	9.70±0.1	EI		122

See also − EI: 3287

Ion	Reactant	Other products	Ionization or appearance potential (eV)	Method	Heat of formation	Ref.
$C_8H_9O^+$	$C_6H_4(OCH_3)C_2H_5$ (1−Ethyl−4−methoxybenzene)	CH_3	10.80±0.1	EI		122
$C_8H_9O^+$	$C_6H_5CH_2CH_2C_6H_4OCH_3$ (1−(4−Methoxyphenyl)−2−phenylethane)		9.6±0.2	EI		3288

4.3. The Positive Ion Table—Continued

Ion	Reactant	Other products	Ionization or appearance potential (eV)	Method	Heat of formation (kJ mol^{-1})	Ref.
$C_8H_9O^+$	$C_6H_4(OCH_3)(CH_2)_3COOCH_3$ (4–(4–Methoxyphenyl)butanoic acid methyl ester)		11.11±0.2	EI		2497
$C_8H_9O^+$	$C_6H_4(OCH_3)CH_2Cl$ (α–Chloro–3–methoxytoluene)	Cl	9.85±0.1	EI		2970
$C_8H_9O^+$	$C_6H_4(OCH_3)CH_2Cl$ (α–Chloro–4–methoxytoluene)	Cl	8.69±0.1	EI		2970

$C_8H_7D_2O^+$

Ion	Reactant	Other products	Ionization or appearance potential (eV)	Method	Heat of formation (kJ mol^{-1})	Ref.
$C_8H_7D_2O^+$	$C_6H_4(OCH_3)CD_3$ (1–Methoxy–3–methyl–d_3–benzene)	D	12.10±0.1	EI		122
$C_8H_7D_2O^+$	$C_6H_4(OCH_3)CD_3$ (1–Methoxy–4–methyl–d_3–benzene)	D	12.10±0.1	EI		122

$C_8H_6D_3O^+$

Ion	Reactant	Other products	Ionization or appearance potential (eV)	Method	Heat of formation (kJ mol^{-1})	Ref.
$C_8H_6D_3O^+$	$C_6H_4(OCH_3)CD_3$ (1–Methoxy–3–methyl–d_3–benzene)	H	12.10±0.1	EI		122
$C_8H_6D_3O^+$	$C_6H_4(OCH_3)CD_3$ (1–Methoxy–4–methyl–d_3–benzene)	H	12.10±0.1	EI		122
$C_8H_6D_3O^+$	$C_6H_4(OCD_3)C_2H_5$ (1–Ethyl–3–methoxy–d_3–benzene)	CH_3	11.70±0.1	EI		122
$C_8H_6D_3O^+$	$C_6H_4(OCD_3)C_2H_5$ (1–Ethyl–4–methoxy–d_3–benzene)	CH_3	10.90±0.1	EI		122

$C_8H_{10}O^+$ (Ethoxybenzene) $\Delta H^\circ_{f298} = 674$ kJ mol^{-1} (161 kcal mol^{-1})

Ion	Reactant	Other products	Ionization or appearance potential (eV)	Method	Heat of formation (kJ mol^{-1})	Ref.
$C_8H_{10}O^+$	$C_6H_5CH_2OCH_3$ (α–Methoxytoluene)		8.85±0.03	PI		416
$C_8H_{10}O^+$	$C_6H_5CH_2OCH_3$ (α–Methoxytoluene)		8.83±0.05	EI		2025
$C_8H_{10}O^+$	$C_6H_5CH_2OCH_3$ (α–Methoxytoluene)		8.76	EI		3287

See also – EI: 122, 3338

Ion	Reactant	Other products	Ionization or appearance potential (eV)	Method	Heat of formation (kJ mol^{-1})	Ref.
$C_8H_{10}O^+$	$C_6H_4(OCH_3)CH_3$ (2–Methoxytoluene)		8.1±0.15	CTS		3373
$C_8H_{10}O^+$	$C_6H_4(OCH_3)CH_3$ (3–Methoxytoluene)		8.31±0.05	EI		2025
$C_8H_{10}O^+$	$C_6H_4(OCH_3)CH_3$ (3–Methoxytoluene)		8.1±0.15	CTS		3373

See also – EI: 122

Ion	Reactant	Other products	Ionization or appearance potential (eV)	Method	Heat of formation (kJ mol^{-1})	Ref.
$C_8H_{10}O^+$	$C_6H_4(OCH_3)CH_3$ (4–Methoxytoluene)		7.83	EI		3287
$C_8H_{10}O^+$	$C_6H_4(OCH_3)CH_3$ (4–Methoxytoluene)		8.0±0.15	CTS		3373

See also – EI: 122

4.3. The Positive Ion Table—Continued

Ion	Reactant	Other products	Ionization or appearance potential (eV)	Method	Heat of formation (kJ mol^{-1})	Ref.
$C_8H_{10}O^+$	$C_6H_5OC_2H_5$ (Ethoxybenzene)		8.13±0.02	PI	674	182

See also – EI: 2945

$C_8H_{10}O^+$	$C_7H_7OCH_3$ (7–Methoxycycloheptatriene)		7.23	EI		3287

See also – EI: 122

$$C_8H_{15}O^+$$

$C_8H_{15}O^+$	$n-C_7H_{15}COCH_3$	CH_3	10.03	RPD		2977

$$C_8H_{16}O^+$$

$C_8H_{16}O^+$	$n-C_6H_{13}COCH_3$		9.75	RPD		2977

$$(tert-C_4H_9)_2O^+ \qquad \Delta H^\circ_{f298} = 498 \text{ kJ mol}^{-1} \text{ (119 kcal mol}^{-1})$$

$C_8H_{18}O^+$	$(n-C_4H_9)_2O$		9.28±0.05	RPD		2776
$C_8H_{18}O^+$	$(tert-C_4H_9)_2O$		8.94±0.01	PE	498	3289

$$C_9H_{10}O^+$$

$C_9H_{10}O^+$	$C_6H_5CH_2CH_2CHO$ (3–Phenylpropanal)		9.13±0.1	EI		2522
$C_9H_{10}O^+$	$C_6H_5CH_2COCH_3$ (Benzyl methyl ketone)		9.14±0.09	EI		2174
$C_9H_{10}O^+$	$C_6H_5COC_2H_5$ (Ethyl phenyl ketone)		9.27±0.05	EI		2025
$C_9H_{10}O^+$	$C_6H_4(CH_3)COCH_3$ (3–Methylacetophenone)		9.15±0.05	EI		2035

See also – EI: 2967

$C_9H_{10}O^+$	$C_6H_4(CH_3)COCH_3$ (4–Methylacetophenone)		9.14	EI		3238

See also – EI: 2967, 3334

$C_9H_{10}O^+$	$C_9H_{10}O$ (Chroman)		8.43	EI		3338
$C_9H_{10}O^+$	$C_9H_{10}O$ (Isochroman)		8.90	EI		3338
$C_9H_{10}O^+$	$C_6H_4(OCH_3)(CH_2)_3COOCH_3$ (4–(4–Methoxyphenyl)butanoic acid methyl ester)		10.55±0.2	EI		2497

4.3. The Positive Ion Table—Continued

Ion	Reactant	Other products	Ionization or appearance potential (eV)	Method	Heat of formation (kJ mol^{-1})	Ref.
			C$_9$H$_{11}$O$^+$			
C$_9$H$_{11}$O$^+$	C$_{10}$H$_{14}$O (Perillaldehyde)	CH$_3$	10.4	RPD		2979
C$_9$H$_{11}$O$^+$	C$_{10}$H$_{14}$O (Isopiperitenone)	CH$_3$	10.0	RPD		2979
C$_9$H$_{11}$O$^+$	C$_{10}$H$_{14}$O (Carvone)	CH$_3$	10.1	RPD		2979
C$_9$H$_{11}$O$^+$	C$_{10}$H$_{14}$O (Eucarvone)	CH$_3$	11.1	RPD		2979
C$_9$H$_{11}$O$^+$	C$_{10}$H$_{14}$O (Myrtenal)	CH$_3$	10.7	RPD		2979
C$_9$H$_{11}$O$^+$	C$_{10}$H$_{14}$O (Verbenone)	CH$_3$	10.8	RPD		2979
C$_9$H$_{11}$O$^+$	C$_6$H$_4$(OCH$_3$)CH(CH$_3$)CH$_2$COOCH$_3$ (3–(4–Methoxyphenyl)butanoic acid methyl ester)		10.61±0.2	EI		2497
C$_9$H$_{11}$O$^+$	C$_6$H$_4$(OCH$_3$)CH$_2$CH$_2$Br (1–Bromo–2–(3–methoxyphenyl)ethane)	Br	10.1	EI		2973
C$_9$H$_{11}$O$^+$	C$_6$H$_4$(OCH$_3$)CH$_2$CH$_2$Br (1–Bromo–2–(4–methoxyphenyl)ethane)	Br	9.9	EI		2973
			C$_9$H$_{13}$O$^+$			
C$_9$H$_{13}$O$^+$	C$_{10}$H$_{16}$O (Caranone)	CH$_3$	10.5	RPD		2979
			C$_9$H$_{15}$O$^+$			
C$_9$H$_{15}$O$^+$	C$_{10}$H$_{18}$O (Isomenthone)	CH$_3$	11.4	RPD		2979
	(tert–C$_4$H$_9$)$_2$CO$^+$	$\Delta H^\circ_{f298} = 495$ kJ mol^{-1} (118 kcal mol^{-1})				
C$_9$H$_{18}$O$^+$	n–C$_7$H$_{15}$COCH$_3$		9.32±0.02	PE		3289
See also – EI: 2977						
C$_9$H$_{18}$O$^+$	(n–C$_4$H$_9$)$_2$CO		9.71	RPD		2977
C$_9$H$_{18}$O$^+$	(tert–C$_4$H$_9$)$_2$CO		8.71±0.02	PE	495	3289
			C$_{10}$H$_{10}$O$^+$			
C$_{10}$H$_{10}$O$^+$	C$_{10}$H$_{10}$O (Bullvalone)		8.72	PE		2951
C$_{10}$H$_{10}$O$^+$	C$_6$H$_5$(CH$_2$)$_3$COOH (4–Phenylbutanoic acid)	H$_2$O	9.23±0.2	EI		2522
C$_{10}$H$_{10}$O$^+$	C$_6$H$_5$(CH$_2$)$_3$COOCH$_3$ (4–Phenylbutanoic acid methyl ester)	CH$_3$OH	9.12±0.2	EI		2522
See also – EI: 2497						

4.3. The Positive Ion Table—Continued

Ion	Reactant	Other products	Ionization or appearance potential (eV)	Method	Heat of formation (kJ mol^{-1})	Ref.
			$C_{10}H_{11}O^+$			
$C_{10}H_{11}O^+$	$C_6H_5CH(CH_3)CH_2COOCH_3$ (3–Phenylbutanoic acid methyl ester)	CH_3O	11.02±0.2	EI		2497
$C_{10}H_{11}O^+$	$C_6H_5(CH_2)_3COOCH_3$ (4–Phenylbutanoic acid methyl ester)	CH_3O	11.11±0.3	EI		2496

See also – EI: 2497

Ion	Reactant	Other products	Ionization or appearance potential (eV)	Method	Heat of formation (kJ mol^{-1})	Ref.
			$C_{10}H_{12}O^+$			
$C_{10}H_{12}O^+$	$C_6H_5(CH_2)_3CHO$ (4–Phenylbutanal)		8.83±0.1	EI		2522
$C_{10}H_{12}O^+$	$C_6H_5COC_3H_7$ (1–Phenyl–1–butanone)		9.38±0.2	EI		2534, 2567
$C_{10}H_{12}O^+$	$C_6H_5CH_2CH_2COCH_3$ (4–Phenyl–2–butanone)		9.00±0.1	EI		2522
$C_{10}H_{12}O^+$	$C_6H_4(OCH_3)CH(CH_3)CH_2COOCH_3$ (3–(4–Methoxyphenyl)butanoic acid methyl ester)		10.24±0.2	EI		2497

Ion	Reactant	Other products	Ionization or appearance potential (eV)	Method	Heat of formation (kJ mol^{-1})	Ref.
			$C_{10}H_{14}O^+$			
$C_{10}H_{14}O^+$	$C_6H_5OC_4H_9$ (tert–Butoxybenzene)		8.75 (V)	PE		2806
$C_{10}H_{14}O^+$	$C_{10}H_{14}O$ (Perillaldehyde)		10.10	RPD		2979
$C_{10}H_{14}O^+$	$C_{10}H_{14}O$ (Isopiperitenone)		9.53	RPD		2979
$C_{10}H_{14}O^+$	$C_{10}H_{14}O$ (Carvone)		9.77	RPD		2979
$C_{10}H_{14}O^+$	$C_{10}H_{14}O$ (Eucarvone)		9.62	RPD		2979
$C_{10}H_{14}O^+$	$C_{10}H_{14}O$ (Myrtenal)		9.36	RPD		2979
$C_{10}H_{14}O^+$	$C_{10}H_{14}O$ (Verbenone)		9.83	RPD		2979
$C_{10}H_{14}O^+$	$C_{10}H_{14}O$ (Adamantanone)		8.76	PE		2951

Ion	Reactant	Other products	Ionization or appearance potential (eV)	Method	Heat of formation (kJ mol^{-1})	Ref.
			$C_{10}H_{16}O^+$			
$C_{10}H_{16}O^+$	$C_{10}H_{16}O$ (Caranone)		9.74	RPD		2979
$C_{10}H_{16}O^+$	$C_{10}H_{15}OH$ (1–Adamantanol)		9.23	PE		2951
$C_{10}H_{16}O^+$	$C_{10}H_{15}OH$ (2–Adamantanol)		9.25	PE		2951

4.3. The Positive Ion Table—Continued

Ion	Reactant	Other products	Ionization or appearance potential (eV)	Method	Heat of formation (kJ mol^{-1})	Ref.

$C_{10}H_{18}O^+$

Ion	Reactant	Other products	IP (eV)	Method	ΔHf	Ref.
$C_{10}H_{18}O^+$	$C_{10}H_{18}O$ (Isomenthone)		9.86	RPD		2979

$C_{11}H_{12}O^+$

Ion	Reactant	Other products	IP (eV)	Method	ΔHf	Ref.
$C_{11}H_{12}O^+$	$C_6H_4(CH_3)(CH_2)_3COOCH_3$ (4–(4–Tolyl)butanoic acid methyl ester)	CH_3OH	9.50±0.2	EI		2497

$C_{11}H_{13}O^+$

Ion	Reactant	Other products	IP (eV)	Method	ΔHf	Ref.
$C_{11}H_{13}O^+$	$C_6H_4(C_4H_9)COOCH_3$ (4–tert–Butylbenzoic acid methyl ester)	CH_3O	11.02	EI		3238
$C_{11}H_{13}O^+$	$C_6H_4(CH_3)CH(CH_3)CH_2COOCH_3$ (3–(4–Tolyl)butanoic acid methyl ester)	CH_3O	11.16±0.2	EI		2497
$C_{11}H_{13}O^+$	$C_6H_4(CH_3)(CH_2)_3COOCH_3$ (4–(4–Tolyl)butanoic acid methyl ester)	CH_3O	11.18±0.2	EI		2497

$C_{11}H_{14}O^+$

Ion	Reactant	Other products	IP (eV)	Method	ΔHf	Ref.
$C_{11}H_{14}O^+$	$C_6H_5(CH_2)_3COCH_3$ (5–Phenyl–2–pentanone)		8.95±0.1	EI		2522
$C_{11}H_{14}O^+$	$C_6H_5COC_4H_9$ (Phenyl tert–butyl ketone)		9.04±0.04	RPD		2407, 2463
$C_{11}H_{14}O^+$	$C_6H_4(CH_3)COC_3H_7$ (1–(4–Tolyl)–1–butanone)		8.75±0.2	EI		2534

$C_{11}H_{18}O^+$

Ion	Reactant	Other products	IP (eV)	Method	ΔHf	Ref.
$C_{11}H_{18}O^+$	$C_{10}H_{14}(CH_3)OH$ (2–Methyl–2–adamantanol)		9.22	PE		2951

$C_{11}H_{22}O^+$

Ion	Reactant	Other products	IP (eV)	Method	ΔHf	Ref.
$C_{11}H_{22}O^+$	$n–C_9H_{19}COCH_3$		9.29±0.01	PE		3289

$C_{12}H_8O^+$ (Dibenzofuran) $\Delta H_{f298}^{\circ} \sim 846$ kJ mol^{-1} (202 kcal mol^{-1})

Ion	Reactant	Other products	IP (eV)	Method	ΔHf	Ref.
$C_{12}H_8O^+$	$C_{12}H_8O$ (Dibenzofuran)		7.9±0.05	PE	~846	2796
$C_{12}H_8O^+$	$C_{12}H_8O$ (Dibenzofuran)		8.22	PE		3349

See also – CTS: 2911

4.3. The Positive Ion Table—Continued

Ion	Reactant	Other products	Ionization or appearance potential (eV)	Method	Heat of formation (kJ mol⁻¹)	Ref.
C₁₂H₉O⁺						
$C_{12}H_9O^+$	$(C_6H_5)_2O$ (Diphenyl ether)	H	12.90±0.05	EI		1237
$C_{12}H_9O^+$	$(C_6H_5O)_2CO$ (Carbonic acid diphenyl ester)		12.51±0.05	EI		1237
C₁₂H₁₀O⁺						
$C_{12}H_{10}O^+$	$(C_6H_5)_2O$ (Diphenyl ether)		8.10?	PE		2796
$C_{12}H_{10}O^+$	$(C_6H_5)_2O$ (Diphenyl ether)		8.82±0.05	EI		1237
$C_{12}H_{10}O^+$	$(C_6H_5O)_2CO$ (Carbonic acid diphenyl ester)	CO_2	10.78±0.05	EI		1237
C₁₂H₁₁O⁺						
$C_{12}H_{11}O^+$	$C_{10}H_7CH_2CH_2OH$ (2–(4–Azulyl)ethanol)	H	10.3	EI		3333
$C_{12}H_{11}O^+$	$C_6H_4(CH_3)SO_3CH_2CH_2C_{10}H_7$ (4–Toluenesulfonic acid 4–azulylethyl ester)		10.5	EI		3333
C₁₂H₁₂O⁺						
$C_{12}H_{12}O^+$	$C_{10}H_7CH_2CH_2OH$ (2–(4–Azulyl)ethanol)		7.1	EI		3333
C₁₃H₆O⁺						
$C_{13}H_6O^+$	$C_{16}H_6O_6$ (3,3′,4,4′–Biphenyltetracarboxylic acid dianhydride)	$2CO_2+CO$	15.45	EI		3295
C₁₃H₁₀O⁺						
$C_{13}H_{10}O^+$	$(C_6H_5)_2CO$ (Benzophenone)		9.46±0.05	EI		1237
$C_{13}H_{10}O^+$	$(C_6H_5)_2CO$ (Benzophenone)		9.35±0.04	EI		2026
C₁₃H₁₁O⁺						
$C_{13}H_{11}O^+$	$C_6H_5OC_6H_4(CH_2)_3COOCH_3$ (4–(4–Phenoxyphenyl)butanoic acid methyl ester)		11.27±0.2	EI		2497

4.3. The Positive Ion Table—Continued

Ion	Reactant	Other products	Ionization or appearance potential (eV)	Method	Heat of formation (kJ mol⁻¹)	Ref.
		$C_{13}H_{12}O^+$				
$C_{13}H_{12}O^+$	$C_6H_5CH_2OC_6H_5$ (Benzyl phenyl ether)		8.4	EI		2737
		$C_{14}H_8O^+$				
$C_{14}H_8O^+$	$C_{14}H_8N_2O$ (9−Diazo−10−oxophenanthrene)	N_2	8.6±0.2	EI		2995
		$C_{14}H_{10}O^+$				
$C_{14}H_{10}O^+$	$C_{14}H_9OH$ (1−Hydroxyanthracene)		7.70	EI		2706
$C_{14}H_{10}O^+$	$C_{14}H_9OH$ (2−Hydroxyanthracene)		7.73	EI		2706
$C_{14}H_{10}O^+$	$C_{14}H_{10}O$ (9,10−Dihydro−9−oxoanthracene)		9.43	EI		2706
$C_{14}H_{10}O^+$	$C_{14}H_9OC_2H_5$ (9−Ethoxyanthracene)	C_2H_4	9.86	EI		2706
		$C_{14}H_{12}O^+$				
$C_{14}H_{12}O^+$	$C_6H_5COC_6H_4CH_3$ (4−Methylbenzophenone)		9.13±0.05	EI		2026
$C_{14}H_{12}O^+$	$C_6H_5OC_6H_4(CH_2)_3COOCH_3$ (4−(4−Phenoxyphenyl)butanoic acid methyl ester)		10.49±0.2	EI		2497
		$C_{14}H_{14}O^+$				
$C_{14}H_{14}O^+$	$C_6H_5CH_2CH_2C_6H_4OH$ (1−(3−Hydroxyphenyl)−2−phenylethane)		8.3±0.1	EI		3288
$C_{14}H_{14}O^+$	$C_6H_5CH_2CH_2C_6H_4OH$ (1−(4−Hydroxyphenyl)−2−phenylethane)		8.3±0.1	EI		3288
$C_{14}H_{14}O^+$	$C_6H_5CH_2OC_6H_4CH_3$ (Benzyl 3−tolyl ether)		8.4	EI		2737
$C_{14}H_{14}O^+$	$C_6H_5CH_2OC_6H_4CH_3$ (Benzyl 4−tolyl ether)		8.2	EI		2737
		$C_{15}H_8O^+$				
$C_{15}H_8O^+$	$C_{15}H_8N_2O$ (9−Diazo−4,5−methylene−10−oxophenanthrene)	N_2	8.6±0.2	EI		2995
		$C_{15}H_{16}O^+$				
$C_{15}H_{16}O^+$	$C_6H_5CH_2CH_2C_6H_4OCH_3$ (1−(4−Methoxyphenyl)−2−phenylethane)		8.1±0.1	EI		3288

4.3. The Positive Ion Table—Continued

Ion	Reactant	Other products	Ionization or appearance potential (eV)	Method	Heat of formation (kJ mol^{-1})	Ref.
			C$_{16}$H$_{14}$O$^+$			
C$_{16}$H$_{14}$O$^+$	C$_{14}$H$_9$OC$_2$H$_5$ (1–Ethoxyanthracene)		7.52	EI		2706
C$_{16}$H$_{14}$O$^+$	C$_{14}$H$_9$OC$_2$H$_5$ (2–Ethoxyanthracene)		7.44	EI		2706
C$_{16}$H$_{14}$O$^+$	C$_{14}$H$_9$OC$_2$H$_5$ (9–Ethoxyanthracene)		7.44	EI		2706
			C$_{16}$H$_{15}$O$^+$			
C$_{16}$H$_{15}$O$^+$	C$_6$H$_5$C$_6$H$_4$(CH$_2$)$_3$COOCH$_3$ (4–(4–Biphenylyl)butanoic acid methyl ester)	CH$_3$O	11.23±0.2	EI		2497
			C$_{16}$H$_{16}$O$^+$			
C$_{16}$H$_{16}$O$^+$	C$_6$H$_5$C$_6$H$_4$COC$_3$H$_7$ (1–(4–Biphenylyl)–1–butanone)		8.44±0.2	EI		2534
			C$_{17}$H$_{18}$O$^+$			
C$_{17}$H$_{18}$O$^+$	C$_6$H$_5$CH$_2$C$_6$H$_4$COC$_3$H$_7$ (1–(4–Benzylphenyl)–1–butanone)		8.69±0.2	EI		2567
			CHO$_2^+$			
CHO$_2^+$	DCOOH	D	12.8±0.1	EI		2646
CHO$_2^+$	CH$_3$COOH	CH$_3$	12.27±0.05	RPD		3347
(Zero average translational energy of decomposition at threshold)						
CHO$_2^+$	CH$_3$COOH	CH$_3$	12.9±0.1	RPD		2576
See also – EI:　171						
CHO$_2^+$	CD$_3$COOH	CD$_3$	14.08	EI		171
CHO$_2^+$	HCOOCH$_3$	CH$_3$	15.9	EI		3224
CHO$_2^+$	C$_2$H$_5$COOH	C$_2$H$_5$	12.84	EI		3435
CHO$_2^+$	HCOOCH$_2$CH$_2$CH$_3$		12.34±0.04	EI		305
			CDO$_2^+$			
CDO$_2^+$	DCOOH	H	12.4±0.1	EI		2646
CDO$_2^+$	DCOOC$_2$H$_5$	C$_2$H$_5$	12.15	EI		3435
	HCOOH$^+$	$\Delta H^\circ_{f298} = 715$ kJ mol^{-1} (171 kcal mol^{-1})				
CH$_2$O$_2^+$	HCOOH		11.33	S	715	3434
CH$_2$O$_2^+$	HCOOH		11.05±0.01	PI	182,	416

4.3. The Positive Ion Table—Continued

Ion	Reactant	Other products	Ionization or appearance potential (eV)	Method	Heat of formation (kJ mol^{-1})	Ref.
$CH_2O_2^+$	HCOOH		11.16±0.03	PI		2724
$CH_2O_2^+$	HCOOH		11.33	PE	715	2837

See also – PE: 3132
EI: 127, 2649, 3435

$CHDO_2^+$

$CHDO_2^+$	DCOOH		11.57	EI		3435

$CH_3O_2^+$

$CH_3O_2^+$	$HCOOC_2H_5$		11.3±0.1	EI		1100
$CH_3O_2^+$	$HCOOC_2H_5$		11.60±0.1	EI	210,	2709

See also – EI: 1059, 2778, 3224

$CH_3O_2^+$	$HCOOCH_2CH_2CH_3$		11.0±0.1	EI		1100
$CH_3O_2^+$	$HCOOCH(CH_3)_2$		10.87±0.1	EI	210,	2709

$C_2H_2O_2^+$

$C_2H_2O_2^+$	CHOCHO		9.48±0.08	EI		128

$C_2H_3O_2^+$

$C_2H_3O_2^+$	$HCOOCH_3$	H	12.3	EI		3224
$C_2H_3O_2^+$	$HCOOC_2H_5$	CH_3	11.51±0.1	EI		1059
$C_2H_3O_2^+$	CH_3COOCH_3	CH_3	12.35±0.03	EI		3222

See also – EI: 305, 3176, 3224

$C_2H_3O_2^+$	$CH_3COO(CH_2)_3CH_3$		12.33±0.12	EI		305
$C_2H_3O_2^+$	$(CH_3O)_2CO$	CH_3O	12.15±0.08	EI		3222
$C_2H_3O_2^+$	$CH_2(CN)COOCH_3$		11.72±0.03	EI		3222
$C_2H_3O_2^+$	$ClCOOCH_3$	Cl	11.78±0.05	EI		3222
$C_2H_3O_2^+$	$CH_2ClCOOCH_3$		11.50±0.05	EI		3222

CH_3COOH^+ $\Delta H_{f0}^\circ = 581$ kJ mol^{-1} (139 kcal mol^{-1})
$HCOOCH_3^+$ $\Delta H_{f298}^\circ = 693$ kJ mol^{-1} (166 kcal mol^{-1})

$C_2H_4O_2^+$	$(HCHO)_2$		10.51±0.03	PI		1166
$C_2H_4O_2^+$	CH_3COOH		10.35±0.03	PI	580	416
$C_2H_4O_2^+$	CH_3COOH		10.37±0.03	PI	582	182

See also – EI: 171, 384, 2026, 2576, 2649

4.3. The Positive Ion Table—Continued

Ion	Reactant	Other products	Ionization or appearance potential (eV)	Method	Heat of formation (kJ mol^{-1})	Ref.
$C_2H_4O_2^+$	HCOOCH$_3$		10.815±0.005	PI	693	182

See also – PI: 190
          EI: 210, 305, 411, 1100, 3224

$C_2H_4O_2^+$	CH$_3$COOC$_2$H$_5$	C$_2$H$_4$	11.15±0.1	EI		1059
$C_2H_4O_2^+$	CH$_3$COOCOCH$_3$	CH$_2$=CO?	9.65±0.02	PI		3015

This threshold appears to be too low when compared to the $C_2H_4O_2^+$ parent ion thresholds.

$C_2HD_3O_2^+$

Ion	Reactant	Other products	Ionization or appearance potential (eV)	Method	Heat of formation (kJ mol^{-1})	Ref.
$C_2HD_3O_2^+$	CD$_3$COOH		10.71	EI		171

$C_2H_5O_2^+$

Ion	Reactant	Other products	Ionization or appearance potential (eV)	Method	Heat of formation (kJ mol^{-1})	Ref.
$C_2H_5O_2^+$	CH$_3$COOC$_2$H$_5$		10.80±0.1	EI		1413, 2709

See also – EI: 1059, 1100, 3176, 3435

$C_2H_5O_2^+$	CH$_3$COOCH$_2$CH$_2$CH$_3$		10.48±0.07	EI		305

See also – EI: 1100

$C_2H_5O_2^+$	CH$_3$COOCH(CH$_3$)$_2$		10.42±0.1	EI		1413, 2709

See also – EI: 305

$C_2H_5O_2^+$	C$_3$H$_6$O$_3$ (1,3,5–Trioxane)	CHO?	10.79±0.05	RPD		3324

$C_3H_4O_2^+$ (3–Hydroxypropanoic acid lactone) $\Delta H_{f298}^\circ = 653$ kJ mol^{-1} (156 kcal mol^{-1})

Ion	Reactant	Other products	Ionization or appearance potential (eV)	Method	Heat of formation (kJ mol^{-1})	Ref.
$C_3H_4O_2^+$	CH$_3$COCHO		9.60±0.06	EI		128
$C_3H_4O_2^+$	C$_3$H$_4$O$_2$ (3–Hydroxypropanoic acid lactone)		9.70±0.01	PI	653	182

$C_3H_5O_2^+$

Ion	Reactant	Other products	Ionization or appearance potential (eV)	Method	Heat of formation (kJ mol^{-1})	Ref.
$C_3H_5O_2^+$	C$_2$H$_5$COOH	H	11.70	EI		3435
$C_3H_5O_2^+$	HCOOC$_2$H$_5$	H	11.05±0.1	EI		1059
$C_3H_5O_2^+$	C$_3$H$_6$O$_2$ (1,3–Dioxolane)	H	10.38	EI		2422
$C_3H_5O_2^+$	CH$_3$COOC$_2$H$_5$	CH$_3$	10.95±0.1	EI		1059
$C_3H_5O_2^+$	CH$_3$COOCH$_2$CH$_2$CH$_3$	C$_2$H$_5$	11.29±0.04	EI		305
$C_3H_5O_2^+$	CH$_3$COO(CH$_2$)$_3$CH$_3$		11.70±0.05	EI		305
$C_3H_5O_2^+$	C$_4$H$_8$O$_3$ (1,3,5–Trioxepane)	CH$_3$O	11	EI		2422

4.3. The Positive Ion Table—Continued

Ion	Reactant	Other products	Ionization or appearance potential (eV)	Method	Heat of formation (kJ mol⁻¹)	Ref.
			$C_3H_4DO_2^+$			
$C_3H_4DO_2^+$	$HCOOCD_2CH_3$	D	11.05±0.1	EI		1059
			$C_3H_3D_2O_2^+$			
$C_3H_3D_2O_2^+$	$HCOOCD_2CH_3$	H	10.97±0.1	EI		1059
	$C_2H_5COOH^+$	$\Delta H^\circ_{f298} = 534$ kJ mol⁻¹ (128 kcal mol⁻¹)				
	$HCOOC_2H_5^+$	$\Delta H^\circ_{f298} = 652$ kJ mol⁻¹ (156 kcal mol⁻¹)				
	$CH_3COOCH_3^+$	$\Delta H^\circ_{f298} = 581$ kJ mol⁻¹ (139 kcal mol⁻¹)				
$C_3H_6O_2^+$	C_2H_5COOH		10.24±0.03	PI	534	182
See also – EI: 2522						
$C_3H_6O_2^+$	$HCOOC_2H_5$		10.61±0.01	PI	652	182
See also – PI: 190						
EI: 210, 305, 1059, 1100, 3224						
$C_3H_6O_2^+$	CH_3COOCH_3		10.27±0.02	PI	581	182
See also – EI: 305, 1100, 2025, 2026, 2649, 3176, 3224						
$C_3H_6O_2^+$	$C_3H_6O_2$ (1,3–Dioxolane)		10.02	EI		2422
$C_3H_6O_2^+$	$n-C_3H_7COOCH_3$	C_2H_4	10.97±0.2	EI		2497
$C_3H_6O_2^+$	$C_5H_{10}O_2$ (1,3–Dioxepane)	C_2H_4	10.0	EI		2694
$C_3H_6O_2^+$	$n-C_4H_9COOCH_3$	C_3H_6	10.95±0.2	EI		2497
$C_3H_6O_2^+$	$C_6H_5(CH_2)_3COOCH_3$ (4–Phenylbutanoic acid methyl ester)		11.48±0.2	EI		2497
$C_3H_6O_2^+$	$C_6H_4(CH_3)(CH_2)_3COOCH_3$ (4–(4–Tolyl)butanoic acid methyl ester)		11.50±0.2	EI		2497
$C_3H_6O_2^+$	$C_6H_5C_6H_4(CH_2)_3COOCH_3$ (4–(4–Biphenylyl)butanoic acid methyl ester)		11.49±0.2	EI		2497
$C_3H_6O_2^+$	$CH_3CO(CH_2)_3COOCH_3$		11.50±0.2	EI		2497
$C_3H_6O_2^+$	$C_6H_5OC_6H_4(CH_2)_3COOCH_3$ (4–(4–Phenoxyphenyl)butanoic acid methyl ester)		11.20±0.2	EI		2497
$C_3H_6O_2^+$	$CH_3OCO(CH_2)_3COOCH_3$		11.59±0.2	EI		2497
$C_3H_6O_2^+$	$C_6H_4(CN)(CH_2)_3COOCH_3$ (4–(4–Cyanophenyl)butanoic acid methyl ester)		11.08±0.2	EI		2497
$C_3H_6O_2^+$	$C_6H_4(NO_2)(CH_2)_3COOCH_3$ (4–(4–Nitrophenyl)butanoic acid methyl ester)		11.07±0.2	EI		2497
$C_3H_6O_2^+$	$C_6H_4F(CH_2)_3COOCH_3$ (4–(4–Fluorophenyl)butanoic acid methyl ester)		11.34±0.2	EI		2497
			$C_3H_4D_2O_2^+$			
$C_3H_4D_2O_2^+$	$HCOOCD_2CH_3$		10.75±0.1	EI		1059

4.3. The Positive Ion Table—Continued

Ion	Reactant	Other products	Ionization or appearance potential (eV)	Method	Heat of formation (kJ mol^{-1})	Ref.

$$C_3H_7O_2^+$$

Ion	Reactant	Other products	Ionization or appearance potential (eV)	Method	Heat of formation (kJ mol^{-1})	Ref.
$C_3H_7O_2^+$	$(CH_3O)_2CH_2$	H	10.38±0.03	RPD		1139
$C_3H_7O_2^+$	$(CH_3O)_2CHCH_3$	CH_3	10.34±0.07	RPD		1139
$C_3H_7O_2^+$	$C_2H_5COOC_2H_5$		10.4±0.1	EI		1100

See also – EI: 1059, 1413, 2709

Ion	Reactant	Other products	Ionization or appearance potential (eV)	Method	Heat of formation (kJ mol^{-1})	Ref.
$C_3H_7O_2^+$	$C_2H_5COOCH(CH_3)_2$		10.40±0.1	EI		1413, 2709
$C_3H_7O_2^+$	$C_4H_8O_3$ (1,3,5–Trioxepane)	CHO?	9.98	EI		2422
$C_3H_7O_2^+$	$(CH_3O)_3CH$	CH_3O	10.36±0.06	RPD		1139

$(CH_3O)_2CH_2^+$ $\Delta H^\circ_{f298} \sim 616$ kJ mol^{-1} (147 kcal mol^{-1})

Ion	Reactant	Other products	Ionization or appearance potential (eV)	Method	Heat of formation (kJ mol^{-1})	Ref.
$C_3H_8O_2^+$	$(CH_3O)_2CH_2$		10.00±0.05	PI	616	182

$C_4H_5O_2^+$ $\Delta H^\circ_{f298} \sim 467$ kJ mol^{-1} (112 kcal mol^{-1})

Ion	Reactant	Other products	Ionization or appearance potential (eV)	Method	Heat of formation (kJ mol^{-1})	Ref.
$C_4H_5O_2^+$	$CH_3COCH_2COCH_3$	CH_3	10.24	PI	467	3015
$C_4H_5O_2^+$	$CH_3COCH_2COCH_3$	CH_3	10.7±0.1	EI		2731, 2959
$C_4H_5O_2^+$	$CF_3COCH_2COCH_3$	CF_3	10.6±0.2	EI		2959

$CH_3COCOCH_3^+$ $\Delta H^\circ_{f298} = 564$ kJ mol^{-1} (135 kcal mol^{-1})
$CH_3COOCH=CH_2^+$ $\Delta H^\circ_{f298} \sim 571$ kJ mol^{-1} (136 kcal mol^{-1})

Ion	Reactant	Other products	Ionization or appearance potential (eV)	Method	Heat of formation (kJ mol^{-1})	Ref.
$C_4H_6O_2^+$	$CH_3COCOCH_3$		9.23±0.03	PI	563	182
$C_4H_6O_2^+$	$CH_3COCOCH_3$		9.25±0.03	PI	565	416

See also – EI: 128

Ion	Reactant	Other products	Ionization or appearance potential (eV)	Method	Heat of formation (kJ mol^{-1})	Ref.
$C_4H_6O_2^+$	$CH_3COOCH=CH_2$		9.19±0.05?	PI	571	182

$$C_4H_7O_2^+$$

Ion	Reactant	Other products	Ionization or appearance potential (eV)	Method	Heat of formation (kJ mol^{-1})	Ref.
$C_4H_7O_2^+$	$C_4H_8O_2$ (1,3–Dioxane)	H	10.42	EI		2422
$C_4H_7O_2^+$	$C_4H_8O_2$ (1,4–Dioxane)	H	10.95	EI		2422
$C_4H_7O_2^+$	$CH_3COOCH(CH_3)_2$	CH_3	11.34±0.07	EI		305

4.3. The Positive Ion Table—Continued

Ion	Reactant	Other products	Ionization or appearance potential (eV)	Method	Heat of formation (kJ mol^{-1})	Ref.

$n-C_3H_7COOH^+$ $\quad\Delta H^\circ_{f298} \sim 510$ kJ mol^{-1} (122 kcal mol^{-1})
$CH_3COOC_2H_5^+$ $\quad\Delta H^\circ_{f298} = 532$ kJ mol^{-1} (127 kcal mol^{-1})
$C_4H_8O_2^+$ (1,4–Dioxane) $\quad\Delta H^\circ_{f298} = 566$ kJ mol^{-1} (135 kcal mol^{-1})

Ion	Reactant	Other products	IP/AP (eV)	Method	ΔHf	Ref.
$C_4H_8O_2^+$	$n-C_3H_7COOH$		10.16±0.05	PI	510	182
See also – EI: 2522						
$C_4H_8O_2^+$	$iso-C_3H_7COOH$		10.02±0.05	PI		182
$C_4H_8O_2^+$	$HCOOCH_2CH_2CH_3$		10.54±0.01	PI		182
$C_4H_8O_2^+$	$CH_3COOC_2H_5$		10.09±0.02	PI	531	416
$C_4H_8O_2^+$	$CH_3COOC_2H_5$		10.11±0.02	PI	533	182
See also – EI: 305, 1059, 1100, 2018, 2025, 3176, 3435						
$C_4H_8O_2^+$	$C_2H_5COOCH_3$		10.15±0.03	PI		182
See also – EI: 2522						
$C_4H_8O_2^+$	$C_4H_8O_2$ (1,3–Dioxane)		10.33	EI		2422
$C_4H_8O_2^+$	$C_4H_8O_2$ (1,4–Dioxane)		9.13±0.03	PI	566	182
See also – EI: 218, 2422						
$C_4H_8O_2^+$	$C_5H_{10}O_3$ (1,3,6–Trioxocane)	CH_2O	9.92	EI		2422

$C_4H_9O_2^+$

| $C_4H_9O_2^+$ | $(CH_3O)_3CCH_3$ | CH_3O | 10.37±0.02 | RPD | | 1139 |

$(CH_3O)_2CHCH_3^+$ $\quad\Delta H^\circ_{f298} = 541$ kJ mol^{-1} (129 kcal mol^{-1})

| $C_4H_{10}O_2^+$ | $(CH_3O)_2CHCH_3$ | | 9.65±0.03 | PI | 541 | 182 |

$C_5H_4O_2^+$ (Furfural) $\quad\Delta H^\circ_{f298} = 738$ kJ mol^{-1} (176 kcal mol^{-1})

$C_5H_4O_2^+$	$C_5H_4O_2$ (Furfural)		9.21±0.01	PI	738	182
$C_5H_4O_2^+$	$C_5H_4O_2$ (Furfural)		9.22	PE	739	2843
See also – EI: 411						

$CH_3COCH_2COCH_3^+$ $\quad\Delta H^\circ_{f298} \sim 475$ kJ mol^{-1} (114 kcal mol^{-1})

$C_5H_8O_2^+$	$CH_3COCH_2COCH_3$		8.87±0.03	PI	477	182
$C_5H_8O_2^+$	$CH_3COCH_2COCH_3$		8.82±0.02	PI	472	3015
See also – EI: 2460, 2731, 2959						

4.3. The Positive Ion Table—Continued

Ion	Reactant	Other products	Ionization or appearance potential (eV)	Method	Heat of formation (kJ mol^{-1})	Ref.
			$C_5H_9O_2^+$			
$C_5H_9O_2^+$	$C_5H_{10}O_2$ (1,3–Dioxepane)	H	10.04	EI		2694
$C_5H_9O_2^+$	$(CH_3)_2C(OH)CH_2COCH_3$	CH_3	9.50±0.07	PI		3015
	$CH_3COOCH(CH_3)_2^+$	$\Delta H^\circ_{f298} = 482$ kJ mol^{-1} (115 kcal mol^{-1})				
	$C_2H_5COOC_2H_5^+$	$\Delta H^\circ_{f298} = 539$ kJ mol^{-1} (129 kcal mol^{-1})				
$C_5H_{10}O_2^+$	$HCOO(CH_2)_3CH_3$		10.50±0.02	PI		182
$C_5H_{10}O_2^+$	$HCOOCH_2CH(CH_3)_2$		10.46±0.02	PI		182
$C_5H_{10}O_2^+$	$CH_3COOCH_2CH_2CH_3$		10.04±0.03	PI		182
$C_5H_{10}O_2^+$	$CH_3COOCH(CH_3)_2$		9.99±0.03	PI	482	182
$C_5H_{10}O_2^+$	$C_2H_5COOC_2H_5$		10.00±0.02	PI	539	182
See also – EI: 1100						
$C_5H_{10}O_2^+$	$n–C_3H_7COOCH_3$		10.07±0.03	PI		182
See also – EI: 2496, 2497, 2522						
$C_5H_{10}O_2^+$	$iso–C_3H_7COOCH_3$		9.98±0.02	PI		182
$C_5H_{10}O_2^+$	$C_5H_{10}O_2$ (1,3–Dioxepane)		9.45	EI		2694
			$C_5H_{11}O_2^+$			
$C_5H_{11}O_2^+$	$C_5H_{11}O_2$ (Pentyl peroxy radical)		7.9±0.2	EI		2464
$C_5H_{11}O_2^+$	$n–C_5H_{11}OOH$	H	10.3±0.1	EI		2464
$C_5H_{11}O_2^+$	$n–C_3H_7CH(OOH)CH_3$	H	10.1±0.1	EI		2464
$C_5H_{11}O_2^+$	$(C_2H_5)_2CHOOH$	H	10.8±0.1	EI		2464
			$C_5H_{12}O_2^+$			
$C_5H_{12}O_2^+$	$(C_2H_5O)_2CH_2$		9.70±0.05	PI		182
	$C_6H_4O_2^+$ (1,4–Benzoquinone)	$\Delta H^\circ_{f298} = 815$ kJ mol^{-1} (195 kcal mol^{-1})				
$C_6H_4O_2^+$	$C_6H_4O_2$ (1,4–Benzoquinone)		9.67±0.02	PI	815	1166
$C_6H_4O_2^+$	$C_6H_4O_2$ (1,4–Benzoquinone)		9.95	PE		2843
			$C_6H_5O_2^+$			
$C_6H_5O_2^+$	$C_6H_4(OH)OCH_3$ (3–Methoxyphenol)	CH_3	11.92±0.1	EI		2970
$C_6H_5O_2^+$	$C_6H_4(OH)OCH_3$ (4–Methoxyphenol)	CH_3	11.01±0.1	EI		2970

4.3. The Positive Ion Table—Continued

Ion	Reactant	Other products	Ionization or appearance potential (eV)	Method	Heat of formation (kJ mol⁻¹)	Ref.

$$C_6H_6O_2^+$$

Ion	Reactant	Other products	Ionization or appearance potential (eV)	Method	Heat of formation (kJ mol⁻¹)	Ref.
$C_6H_6O_2^+$	$C_6H_4(OH)OC_2H_5$ (3–Ethoxyphenol)	C_2H_4	11.03±0.15	EI		2945
$C_6H_6O_2^+$	$C_6H_4(OH)OC_2H_5$ (4–Ethoxyphenol)	C_2H_4	10.84±0.15	EI		2945

$$C_6H_8O_2^+$$

$C_6H_8O_2^+$	$CH_3CO(CH_2)_3COOCH_3$	CH_3OH	9.75±0.2	EI		2497

$$C_6H_9O_2^+$$

$C_6H_9O_2^+$	$CH_3CO(CH_2)_3COOCH_3$	CH_3O	11.05±0.3	EI		2496

See also – EI: 2497

$$CH_3COO(CH_2)_3CH_3^+ \quad \Delta H^\circ_{f298} = 437 \text{ kJ mol}^{-1} \text{ (104 kcal mol}^{-1}\text{)}$$

Ion	Reactant	Other products	Ionization or appearance potential (eV)	Method	Heat of formation (kJ mol⁻¹)	Ref.
$C_6H_{12}O_2^+$	$CH_3COO(CH_2)_3CH_3$		9.56±0.03	PI	437	1166
$C_6H_{12}O_2^+$	$CH_3COO(CH_2)_3CH_3$		10.01?	PI		182
$C_6H_{12}O_2^+$	$CH_3COOCH_2CH(CH_3)_2$		9.97?	PI		182
$C_6H_{12}O_2^+$	$CH_3COOCH(CH_3)C_2H_5$		9.91±0.03	PI		182
$C_6H_{12}O_2^+$	$n–C_4H_9COOCH_3$		10.40±0.2	EI		2497

$$C_6H_{13}O_2^+$$

$C_6H_{13}O_2^+$	$n–C_6H_{13}OOH$	H	10.8±0.1	EI		2464
$C_6H_{13}O_2^+$	$n–C_4H_9CH(OOH)CH_3$	H	9.6±0.1	EI		2464
$C_6H_{13}O_2^+$	$n–C_3H_7CH(OOH)C_2H_5$	H	9.6±0.1	EI		2464

$$C_7H_5O_2^+$$

$C_7H_5O_2^+$	$C_6H_4(OH)COCH_3$ (4–Hydroxyacetophenone)	CH_3	9.84	EI		3334
$C_7H_5O_2^+$	$C_6H_4(OH)COCH_3$ (4–Hydroxyacetophenone)	CH_3	10.42	EI		3238
$C_7H_5O_2^+$	$C_6H_5CH_2OC_6H_4CHO$ (Benzyl 3–formylphenyl ether)		11.5	EI		2737
$C_7H_5O_2^+$	$C_6H_5CH_2OC_6H_4CHO$ (Benzyl 4–formylphenyl ether)		11.1	EI		2737

4.3. The Positive Ion Table—Continued

Ion	Reactant	Other products	Ionization or appearance potential (eV)	Method	Heat of formation (kJ mol^{-1})	Ref.
			$C_7H_6O_2^+$			
$C_7H_6O_2^+$	C_6H_5COOH (Benzoic acid)		9.73±0.09	EI		2026
$C_7H_6O_2^+$	$C_6H_4(OH)CHO$ (4–Hydroxybenzenecarbonal)		9.32±0.02	EI		2026
$C_7H_6O_2^+$	$C_7H_6O_2$ (2–Hydroxy–2,4,6–cycloheptatrien–1–one)		9.86±0.02	EI		431
			$C_7H_7O_2^+$			
$C_7H_7O_2^+$	$C_6H_4(OCH_3)_2$ (1,3–Dimethoxybenzene)	CH_3	11.57±0.1	EI		2970
$C_7H_7O_2^+$	$C_6H_4(OCH_3)_2$ (1,4–Dimethoxybenzene)	CH_3	10.37±0.1	EI		2970
$C_7H_7O_2^+$	$C_6H_5CH_2OC_6H_4OCH_3$ (Benzyl 3–methoxyphenyl ether)		10.4	EI		2737
$C_7H_7O_2^+$	$C_6H_5CH_2OC_6H_4OCH_3$ (Benzyl 4–methoxyphenyl ether)		10.6	EI		2737
			$C_7H_8O_2^+$			
$C_7H_8O_2^+$	$C_6H_4(OCH_3)OC_2H_5$ (1–Ethoxy–3–methoxybenzene)	C_2H_4	10.87±0.15	EI		2945
$C_7H_8O_2^+$	$C_6H_4(OCH_3)OC_2H_5$ (1–Ethoxy–4–methoxybenzene)	C_2H_4	10.52±0.15	EI		2945
			$C_7H_{15}O_2^+$			
$C_7H_{15}O_2^+$	$C_7H_{15}O_2$ (Heptyl peroxy radical)		8.3±0.2	EI		2464
$C_7H_{15}O_2^+$	$n–C_7H_{15}OOH$	H	11.1±0.1	EI		2464
$C_7H_{15}O_2^+$	$n–C_5H_{11}CH(OOH)CH_3$	H	10.7±0.1	EI		2464
			$C_8H_6O_2^+$			
$C_8H_6O_2^+$	$C_6H_4(CHO)_2$ (1,4–Benzenedicarboxaldehyde)		10.13±0.01	EI		2718
$C_8H_6O_2^+$	$C_9H_6O_3$ (7–Hydroxycoumarin)	CO	10.6	EI		2946

4.3. The Positive Ion Table—Continued

Ion	Reactant	Other products	Ionization or appearance potential (eV)	Method	Heat of formation (kJ mol^{-1})	Ref.
		$C_8H_7O_2^+$				
$C_8H_7O_2^+$	$C_6H_4(OCH_3)COCH_3$ (3–Methoxyacetophenone)	CH_3	10.67±0.1	EI		2967
$C_8H_7O_2^+$	$C_6H_4(OCH_3)COCH_3$ (4–Methoxyacetophenone)	CH_3	10.19±0.1	EI		2967

See also – EI: 3238

		$C_8H_8O_2^+$				
$C_8H_8O_2^+$	$C_6H_4(OCH_3)CHO$ (4–Methoxybenzenecarbonal)		8.87 (V)	PE		2806
$C_8H_8O_2^+$	$C_6H_4(OCH_3)CHO$ (4–Methoxybenzenecarbonal)		8.60±0.03	EI		2026
$C_8H_8O_2^+$	$C_6H_5COCH_2OH$ (α–Hydroxyacetophenone)		9.33±0.05	EI		2025
$C_8H_8O_2^+$	$C_6H_4(OH)COCH_3$ (3–Hydroxyacetophenone)		8.67±0.05	EI		2025
$C_8H_8O_2^+$	$C_6H_4(OH)COCH_3$ (4–Hydroxyacetophenone)		8.70±0.03	EI		2026

See also – EI: 3238, 3334

$C_8H_8O_2^+$	$C_6H_5COOCH_3$ (Benzoic acid methyl ester)		9.35±0.06	EI		2026

See also – EI: 308, 3238

		$C_8H_{10}O_2^+$				
$C_8H_{10}O_2^+$	$C_6H_4(OH)OC_2H_5$ (3–Ethoxyphenol)		8.49±0.15	EI		2945
$C_8H_{10}O_2^+$	$C_6H_4(OH)OC_2H_5$ (4–Ethoxyphenol)		8.25±0.15	EI		2945
$C_8H_{10}O_2^+$	$C_6H_4(OCH_3)_2$ (1,2–Dimethoxybenzene)		7.8±0.15	CTS		3373
$C_8H_{10}O_2^+$	$C_6H_4(OCH_3)_2$ (1,3–Dimethoxybenzene)		8.0±0.15	CTS		3373
$C_8H_{10}O_2^+$	$C_6H_4(OCH_3)_2$ (1,4–Dimethoxybenzene)		7.90 (V)	PE		2806
$C_8H_{10}O_2^+$	$C_6H_4(OCH_3)_2$ (1,4–Dimethoxybenzene)		7.7±0.15	CTS		3373

See also – CTS: 3367

4.3. The Positive Ion Table—Continued

Ion	Reactant	Other products	Ionization or appearance potential (eV)	Method	Heat of formation (kJ mol^{-1})	Ref.
		C$_9$H$_{10}$O$_2^+$				
C$_9$H$_{10}$O$_2^+$	C$_6$H$_5$COCH$_2$OCH$_3$ (α−Methoxyacetophenone)		8.60±0.05	EI		2025
C$_9$H$_{10}$O$_2^+$	C$_6$H$_4$(OCH$_3$)COCH$_3$ (3−Methoxyacetophenone)		8.53±0.05	EI		2025

See also − EI: 2967

Ion	Reactant	Other products	IP/AP	Method	Heat	Ref.
C$_9$H$_{10}$O$_2^+$	C$_6$H$_4$(OCH$_3$)COCH$_3$ (4−Methoxyacetophenone)		8.62±0.05	EI		2026

See also − EI: 2967, 3238, 3334

Ion	Reactant	Other products	IP/AP	Method	Heat	Ref.
C$_9$H$_{10}$O$_2^+$	C$_6$H$_5$CH$_2$CH$_2$COOH (3−Phenylpropanoic acid)		8.95±0.1	EI		2522
C$_9$H$_{10}$O$_2^+$	C$_6$H$_4$(CH$_3$)COOCH$_3$ (2−Methylbenzoic acid methyl ester)		8.5	EI		3231
C$_9$H$_{10}$O$_2^+$	C$_6$H$_4$(CH$_3$)COOCH$_3$ (4−Methylbenzoic acid methyl ester)		8.94±0.04	EI		2026

Ion	Reactant	Other products	IP/AP	Method	Heat	Ref.
		C$_9$H$_{12}$O$_2^+$				
C$_9$H$_{12}$O$_2^+$	C$_6$H$_4$(OCH$_3$)OC$_2$H$_5$ (1−Ethoxy−3−methoxybenzene)		8.24±0.15	EI		2945
C$_9$H$_{12}$O$_2^+$	C$_6$H$_4$(OCH$_3$)OC$_2$H$_5$ (1−Ethoxy−4−methoxybenzene)		8.03±0.15	EI		2945

Ion	Reactant	Other products	IP/AP	Method	Heat	Ref.
		C$_{10}$H$_{11}$O$_2^+$				
C$_{10}$H$_{11}$O$_2^+$	C$_6$H$_4$(OH)(CH$_2$)$_3$COOCH$_3$ (4−(4−Hydroxyphenyl)butanoic acid methyl ester)	CH$_3$O	11.16±0.2	EI		2497

Ion	Reactant	Other products	IP/AP	Method	Heat	Ref.
		C$_{10}$H$_{12}$O$_2^+$				
C$_{10}$H$_{12}$O$_2^+$	C$_6$H$_4$(OH)COC$_3$H$_7$ (1−(4−Hydroxyphenyl)−1−butanone)		8.65±0.2	EI		2534
C$_{10}$H$_{12}$O$_2^+$	C$_6$H$_5$(CH$_2$)$_3$COOH (4−Phenylbutanoic acid)		9.00±0.1	EI		2522
C$_{10}$H$_{12}$O$_2^+$	C$_6$H$_5$CH$_2$CH$_2$COOCH$_3$ (3−Phenylpropanoic acid methyl ester)		9.05±0.1	EI		2522

Ion	Reactant	Other products	IP/AP	Method	Heat	Ref.
		C$_{11}$H$_{13}$O$_2^+$				
C$_{11}$H$_{13}$O$_2^+$	C$_6$H$_4$(C$_4$H$_9$)COOCH$_3$ (4−tert−Butylbenzoic acid methyl ester)	CH$_3$	10.10	EI		3238
C$_{11}$H$_{13}$O$_2^+$	C$_6$H$_4$(OCH$_3$)(CH$_2$)$_3$COOCH$_3$ (4−(4−Methoxyphenyl)butanoic acid methyl ester)	CH$_3$O	11.10±0.5	EI		2496

See also − EI: 2497

4.3. The Positive Ion Table—Continued

Ion	Reactant	Other products	Ionization or appearance potential (eV)	Method	Heat of formation (kJ mol^{-1})	Ref.
		$C_{11}H_{14}O_2^+$				
$C_{11}H_{14}O_2^+$	$C_6H_4(OCH_3)COC_3H_7$ (1–(4–Methoxyphenyl)–1–butanone)		8.33±0.2	EI		2534
$C_{11}H_{14}O_2^+$	$C_6H_5CH(CH_3)CH_2COOCH_3$ (3–Phenylbutanoic acid methyl ester)		8.70±0.2	EI		2497
$C_{11}H_{14}O_2^+$	$C_6H_5(CH_2)_3COOCH_3$ (4–Phenylbutanoic acid methyl ester)		8.57±0.3	EI		2496

See also – EI: 2497, 2522

Ion	Reactant	Other products	Ionization or appearance potential (eV)	Method	Heat of formation (kJ mol^{-1})	Ref.
		$C_{12}H_8O_2^+$				
$C_{12}H_8O_2^+$	$C_{12}H_8O_2$ (Diphenylene dioxide)		8.10±0.03	RPD		2538
$C_{12}H_8O_2^+$	$C_{12}H_8O_2$ (Diphenylene dioxide)		7.7	CTS		3300
		$C_{12}H_{16}O_2^+$				
$C_{12}H_{16}O_2^+$	$C_6H_4(C_4H_9)COOCH_3$ (4–tert–Butylbenzoic acid methyl ester)		9.38	EI		3238
$C_{12}H_{16}O_2^+$	$C_6H_4(CH_3)CH(CH_3)CH_2COOCH_3$ (3–(4–Tolyl)butanoic acid methyl ester)		8.22±0.2	EI		2497
$C_{12}H_{16}O_2^+$	$C_6H_4(CH_3)(CH_2)_3COOCH_3$ (4–(4–Tolyl)butanoic acid methyl ester)		8.51±0.2	EI		2497
		$C_{13}H_{10}O_2^+$				
$C_{13}H_{10}O_2^+$	$C_6H_5COC_6H_4OH$ (4–Hydroxybenzophenone)		8.59±0.05	EI		2026
$C_{13}H_{10}O_2^+$	$C_6H_5COOC_6H_5$ (Benzoic acid phenyl ester)		8.98±0.05	EI		1237
		$C_{13}H_{12}O_2^+$				
$C_{13}H_{12}O_2^+$	$C_6H_5CH_2OC_6H_4OH$ (Benzyl 3–hydroxyphenyl ether)		8.4	EI		2737
$C_{13}H_{12}O_2^+$	$C_6H_5CH_2OC_6H_4OH$ (Benzyl 4–hydroxyphenyl ether)		8.2	EI		2737
		$C_{14}H_{10}O_2^+$				
$C_{14}H_{10}O_2^+$	$C_6H_5COCOC_6H_5$ (Diphenylglyoxal)		8.78±0.05	EI		1237

4.3. The Positive Ion Table—Continued

Ion	Reactant	Other products	Ionization or appearance potential (eV)	Method	Heat of formation (kJ mol^{-1})	Ref.
			$C_{14}H_{12}O_2^+$			
$C_{14}H_{12}O_2^+$	$C_6H_5CH_2OC_6H_4CHO$ (Benzyl 3-formylphenyl ether)		8.6	EI		2737
$C_{14}H_{12}O_2^+$	$C_6H_5CH_2OC_6H_4CHO$ (Benzyl 4-formylphenyl ether)		8.7	EI		2737
			$C_{14}H_{14}O_2^+$			
$C_{14}H_{14}O_2^+$	$C_6H_5CH_2OC_6H_4OCH_3$ (Benzyl 3-methoxyphenyl ether)		8.3	EI		2737
$C_{14}H_{14}O_2^+$	$C_6H_5CH_2OC_6H_4OCH_3$ (Benzyl 4-methoxyphenyl ether)		8.0	EI		2737
			$C_{16}H_{15}O_2^+$			
$C_{16}H_{15}O_2^+$	$C_6H_5OC_6H_4(CH_2)_3COOCH_3$ (4-(4-Phenoxyphenyl)butanoic acid methyl ester)	CH_3O	11.17±0.2	EI		2497
			$C_{16}H_{16}O_2^+$			
$C_{16}H_{16}O_2^+$	$C_6H_5OC_6H_4COC_3H_7$ (1-(4-Phenoxyphenyl)-1-butanone)		8.55±0.2	EI		2534
			$C_{17}H_{18}O_2^+$			
$C_{17}H_{18}O_2^+$	$C_6H_5C_6H_4CH(CH_3)CH_2COOCH_3$ (3-(4-Biphenylyl)butanoic acid methyl ester)		8.00±0.2	EI		2497
$C_{17}H_{18}O_2^+$	$C_6H_5C_6H_4(CH_2)_3COOCH_3$ (4-(4-Biphenylyl)butanoic acid methyl ester)		8.27±0.2	EI		2497
			$C_{22}H_{26}O_2^+$			
$C_{22}H_{26}O_2^+$	$C_6H_4(COC_3H_7)CH_2CH_2C_6H_4COC_3H_7$ (1,2-Bis(4-butyrylphenyl)ethane)		8.91±0.2	EI		2567
		$C_3H_3O_3^+$ $\Delta H_{f298}^{\circ} \sim 260$ kJ mol^{-1} (62 kcal mol^{-1})				
$C_3H_3O_3^+$	$CH_3COOCOCH_3$	CH_3	10.14±0.02	PI	260	3015
			$C_3H_5O_3^+$			
$C_3H_5O_3^+$	$C_3H_6O_3$ (1,3,5-Trioxane)	H	10.59±0.05	RPD		3324

4.3. The Positive Ion Table—Continued

Ion	Reactant	Other products	Ionization or appearance potential (eV)	Method	Heat of formation (kJ mol^{-1})	Ref.
			$C_3H_6O_3^+$			
$C_3H_6O_3^+$	$C_3H_6O_3$ (1,3,5–Trioxane)		10.59 ± 0.05	RPD		3324
			$C_4H_8O_3^+$			
$C_4H_8O_3^+$	$CH_3OCH_2COOCH_3$		9.56 ± 0.05	EI		2025
$C_4H_8O_3^+$	$C_4H_8O_3$ (1,3,5–Trioxepane)		9.62	EI		2422
			$C_4H_9O_3^+$			
$C_4H_9O_3^+$	$(CH_3O)_3CH$	H	10.39 ± 0.05	RPD		1139
$C_4H_9O_3^+$	$(CH_3O)_3CCH_3$	CH_3	10.39 ± 0.10	RPD		1139
$C_4H_9O_3^+$	$(CH_3O)_4C$	CH_3O	10.32 ± 0.10	RPD		1139
			$C_5H_{10}O_3^+$			
$C_5H_{10}O_3^+$	$C_5H_{10}O_3$ (1,3,6–Trioxocane)		9.23	EI		2422
			$C_6H_8O_3^+$			
$C_6H_8O_3^+$	$CH_3OCO(CH_2)_3COOCH_3$	CH_3OH	9.71 ± 0.2	EI		2497
			$C_6H_9O_3^+$			
$C_6H_9O_3^+$	$CH_3OCO(CH_2)_3COOCH_3$	CH_3O	11.05 ± 0.3	EI		2496

See also – EI: 2497

Ion	Reactant	Other products	Ionization or appearance potential (eV)	Method	Heat of formation (kJ mol^{-1})	Ref.
			$C_7H_{12}O_3^+$			
$C_7H_{12}O_3^+$	$CH_3CO(CH_2)_3COOCH_3$		9.63 ± 0.2	EI		2497

See also – EI: 2496

Ion	Reactant	Other products	Ionization or appearance potential (eV)	Method	Heat of formation (kJ mol^{-1})	Ref.
			$C_8H_8O_3^+$			
$C_8H_8O_3^+$	$C_6H_4(OH)COOCH_3$ (2–Hydroxybenzoic acid methyl ester)		7.65	EI		3231

4.3. The Positive Ion Table—Continued

Ion	Reactant	Other products	Ionization or appearance potential (eV)	Method	Heat of formation (kJ mol⁻¹)	Ref.
$C_9H_{10}O_3^+$						
$C_9H_{10}O_3^+$	$C_6H_4(OCH_3)COOCH_3$ (4–Methoxybenzoic acid methyl ester)		8.43±0.04	EI		2026
$C_9H_{12}O_3^+$						
$C_9H_{12}O_3^+$	$C_6H_3(OCH_3)_3$ (1,2,3–Trimethoxybenzene)		8.2±0.15	CTS		3373
$C_9H_{12}O_3^+$	$C_6H_3(OCH_3)_3$ (1,2,4–Trimethoxybenzene)		7.49	CTS		3367
$C_9H_{12}O_3^+$	$C_6H_3(OCH_3)_3$ (1,2,4–Trimethoxybenzene)		7.5±0.15	CTS		3373
$C_9H_{12}O_3^+$	$C_6H_3(OCH_3)_3$ (1,3,5–Trimethoxybenzene)		7.96	CTS		3367
$C_{11}H_{14}O_3^+$						
$C_{11}H_{14}O_3^+$	$C_6H_4(OH)(CH_2)_3COOCH_3$ (4–(4–Hydroxyphenyl)butanoic acid methyl ester)		8.45±0.2	EI		2497
$C_{12}H_{16}O_3^+$						
$C_{12}H_{16}O_3^+$	$C_6H_4(OCH_3)CH(CH_3)CH_2COOCH_3$ (3–(4–Methoxyphenyl)butanoic acid methyl ester)		7.69±0.2	EI		2497
$C_{12}H_{16}O_3^+$	$C_6H_4(OCH_3)(CH_2)_3COOCH_3$ (4–(4–Methoxyphenyl)butanoic acid methyl ester)		7.8±0.5	EI		2496

See also – EI: 2497

Ion	Reactant	Other products	Ionization or appearance potential (eV)	Method	Heat of formation (kJ mol⁻¹)	Ref.
$C_{13}H_{10}O_3^+$						
$C_{13}H_{10}O_3^+$	$(C_6H_5O)_2CO$ (Carbonic acid diphenyl ester)		9.01±0.05	EI		1237
$C_{14}H_6O_3^+$						
$C_{14}H_6O_3^+$	$C_{16}H_6O_6$ (3,3′,4,4′–Biphenyltetracarboxylic acid dianhydride)	CO_2+CO	14.36	EI		3295
$C_{17}H_{18}O_3^+$						
$C_{17}H_{18}O_3^+$	$C_6H_5OC_6H_4(CH_2)_3COOCH_3$ (4–(4–Phenoxyphenyl)butanoic acid methyl ester)		7.94±0.2	EI		2497

4.3. The Positive Ion Table—Continued

Ion	Reactant	Other products	Ionization or appearance potential (eV)	Method	Heat of formation (kJ mol⁻¹)	Ref.
$C_{10}H_{10}O_4^+$						
$C_{10}H_{10}O_4^+$	$C_6H_4(COOCH_3)_2$ (1,2–Benzenedicarboxylic acid dimethyl ester)		9.64±0.07	EI		2718
$C_{10}H_{10}O_4^+$	$C_6H_4(COOCH_3)_2$ (1,3–Benzenedicarboxylic acid dimethyl ester)		9.84±0.09	EI		2718
$C_{10}H_{10}O_4^+$	$C_6H_4(COOCH_3)_2$ (1,4 Benzenedicarboxylic acid dimethyl ester)		9.78±0.03	EI		2718
$C_{10}H_{14}O_4^+$						
$C_{10}H_{14}O_4^+$	$C_6H_2(OCH_3)_4$ (1,2,4,5–Tetramethoxybenzene)		7.25	CTS		3367
$C_{14}H_{10}O_4^+$						
$C_{14}H_{10}O_4^+$	$C_6H_5OCOCOOC_6H_5$ (Oxalic acid diphenyl ester)		7.94	CTS		2562
$C_{15}H_6O_4^+$						
$C_{15}H_6O_4^+$	$C_{16}H_6O_6$ (3,3',4,4'–Biphenyltetracarboxylic acid dianhydride)	CO_2	11.36	EI		3295
$C_6H_{11}O_5^+$						
$C_6H_{11}O_5^+$	$C_7H_{14}O_6$ (α–Methyl–(D)–glucoside)	CH_3O	12.9±0.16	EI		2036
$C_6H_{11}O_5^+$	$C_7H_{14}O_6$ (β–Methyl–(D)–glucoside)	CH_3O	13.5±0.16	EI		2036
$C_{16}H_6O_6^+$						
$C_{16}H_6O_6^+$	$C_{16}H_6O_6$ (3,3',4,4'–Biphenyltetracarboxylic acid dianhydride)		9.80	EI		3295
DNO^+						
DNO^+	DNO		10.29±0.14	EI		2510
HNO_3^+						
HNO_3^+	HNO_3		11.03±0.01?	PI		1253

4.3. The Positive Ion Table—Continued

Ion	Reactant	Other products	Ionization or appearance potential (eV)	Method	Heat of formation (kJ mol^{-1})	Ref.
		CNO$^+$				
CNO$^+$	HNCO	H	<16.1	EI	2797,	3012
CNO$^+$	HNCO	H	16.66±0.15	EI		3365
		CH$_3$BO$^+$				
CH$_3$BO$^+$	BH$_3$CO		11.92±0.02 (V)	PE		3044
		CH$_5$B$_3$O$^+$				
CH$_5$B$_3$O$^+$	B$_4$H$_8$CO	BH$_3$	11.5±0.5	EI		3226
		CH$_6$B$_4$O$^+$				
CH$_6$B$_4$O$^+$	B$_4$H$_8$CO	H$_2$	10.3±0.3	EI		3226
		CH$_7$B$_4$O$^+$				
CH$_7$B$_4$O$^+$	B$_4$H$_8$CO	H	10.5±0.5	EI		3226
		CH$_8$B$_4$O$^+$				
CH$_8$B$_4$O$^+$	B$_4$H$_8$CO		10.2±0.5	EI		3226
		C$_2$H$_6$BO$^+$				
C$_2$H$_6$BO$^+$	(CH$_3$O)$_3$B		16.6±0.3	EI		115
		C$_2$H$_5$BO$_2^+$				
C$_2$H$_5$BO$_2^+$	(CH$_3$O)$_3$B		13.2±0.2	EI		115
		C$_2$H$_6$BO$_2^+$				
C$_2$H$_6$BO$_2^+$	(CH$_3$O)$_2$BH	H	9.0±0.2	EI		364
C$_2$H$_6$BO$_2^+$	(CH$_3$O)$_3$B	CH$_3$O	9.6±0.2	EI		364
C$_2$H$_6$BO$_2^+$	(CH$_3$O)$_3$B	CH$_3$O	13.0±0.2	EI		115
		C$_2$H$_7$BO$_2^+$				
C$_2$H$_7$BO$_2^+$	(CH$_3$O)$_2$BH		9.7±1.0	EI		364

4.3. The Positive Ion Table—Continued

Ion	Reactant	Other products	Ionization or appearance potential (eV)	Method	Heat of formation (kJ mol^{-1})	Ref.
			$C_2H_6BO_3^+$			
$C_2H_6BO_3^+$	$(CH_3O)_3B$	CH_3	12.1±0.2	EI		115
			$C_3H_9BO_3^+$			
$C_3H_9BO_3^+$	$(CH_3O)_3B$		8.9±0.2	EI		364
$C_3H_9BO_3^+$	$(CH_3O)_3B$		10.62	EI		3227
$C_3H_9BO_3^+$	$(CH_3O)_3B$		10.8±0.3	EI		115
			$C_6H_{15}BO_3^+$			
$C_6H_{15}BO_3^+$	$(C_2H_5O)_3B$		10.13	EI		3227
$C_6H_{15}BO_3^+$	$(C_2H_5O)_3B$		10.47±0.12	EI		2702
			$C_9H_{21}BO_3^+$			
$C_9H_{21}BO_3^+$	$(n-C_3H_7O)_3B$		10.02	EI		3227
$C_9H_{21}BO_3^+$	$(n-C_3H_7O)_3B$		10.62±0.15	EI		2702
			$C_{12}H_{27}BO_3^+$			
$C_{12}H_{27}BO_3^+$	$(n-C_4H_9O)_3B$		10.72±0.74	EI		2702
			$CHNO^+$			
$CHNO^+$	$HNCO$		11.60±0.01	PE		2797, 3012, 3067

See also – EI: 3365

Ion	Reactant	Other products	Ionization or appearance potential (eV)	Method	Heat of formation (kJ mol^{-1})	Ref.
			CH_2NO^+			
CH_2NO^+	C_2H_5NO	CH_3	10.55±0.2	EDD		3180
CH_2NO^+	$CH_3CH=NOH$	CH_3	13.0±0.2	EDD		3180

$HCONH_2^+$ $\Delta H_{f\,298}^\circ \sim 802$ kJ mol^{-1} (192 kcal mol^{-1})

Ion	Reactant	Other products	Ionization or appearance potential (eV)	Method	Heat of formation (kJ mol^{-1})	Ref.
CH_3NO^+	CH_3NO		10.8±0.3	EI		2894
CH_3NO^+	$HCONH_2$		10.24	S	802	3108
CH_3NO^+	$HCONH_2$		10.25±0.02	PI	803	182
CH_3NO^+	$HCONH_2$		10.13	PE		2837

See also – CTS: 2562

4.3. The Positive Ion Table—Continued

Ion	Reactant	Other products	Ionization or appearance potential (eV)	Method	Heat of formation (kJ mol^{-1})	Ref.
			C$_2$H$_3$NO$^+$			
C$_2$H$_3$NO$^+$	C$_4$H$_7$NO (4–Aminobutanoic acid lactam)	C$_2$H$_4$	10.7	PI		2728
			C$_2$H$_4$NO$^+$			
C$_2$H$_4$NO$^+$	C$_2$H$_5$NO	H	11.2±0.2	EDD		3180
C$_2$H$_4$NO$^+$	CH$_3$CH=NOH	H	10.2±0.2	EDD		3180
			C$_2$H$_5$NO$^+$			
C$_2$H$_5$NO$^+$	C$_2$H$_5$NO		10.3±0.2	EDD		3180
C$_2$H$_5$NO$^+$	CH$_3$CH=NOH		10.65±0.2	EDD		3180
C$_2$H$_5$NO$^+$	HCONHCH$_3$		9.79	PE		2837
C$_2$H$_5$NO$^+$	CH$_3$CONH$_2$		9.77±0.02	PI		182

See also – EI: 384, 2649

Ion	Reactant	Other products	Ionization or appearance potential (eV)	Method	Heat of formation (kJ mol^{-1})	Ref.
			C$_3$H$_3$NO$^+$			
C$_3$H$_3$NO$^+$	C$_3$H$_3$NO (Isoxazole)		9.99±0.05	PE		3246

See also – PE: 2975

HCON(CH$_3$)$_2^+$ $\Delta H^\circ_{f298} = 689$ kJ mol^{-1} (165 kcal mol^{-1})

Ion	Reactant	Other products	Ionization or appearance potential (eV)	Method	Heat of formation (kJ mol^{-1})	Ref.
C$_3$H$_7$NO$^+$	HCON(CH$_3$)$_2$		9.12±0.02	PI	688	182
C$_3$H$_7$NO$^+$	HCON(CH$_3$)$_2$		9.14	PE	690	2837

See also – CTS: 2562

Ion	Reactant	Other products	Ionization or appearance potential (eV)	Method	Heat of formation (kJ mol^{-1})	Ref.
C$_3$H$_7$NO$^+$	CH$_3$CONHCH$_3$		8.90±0.02	PI		182

See also – EI: 2649

Ion	Reactant	Other products	Ionization or appearance potential (eV)	Method	Heat of formation (kJ mol^{-1})	Ref.
			C$_4$H$_6$NO$^+$			
C$_4$H$_6$NO$^+$	C$_4$H$_7$NO (4–Aminobutanoic acid lactam)	H	9.9	PI		2728
			C$_4$H$_7$NO$^+$			
C$_4$H$_7$NO$^+$	C$_4$H$_7$NO (4–Aminobutanoic acid lactam)		9.32±0.02	PI		2728

4.3. The Positive Ion Table—Continued

Ion	Reactant	Other products	Ionization or appearance potential (eV)	Method	Heat of formation (kJ mol⁻¹)	Ref.

$$\mathbf{C_4H_9NO^+}$$

Ion	Reactant	Other products	Ionization or appearance potential (eV)	Method	Heat of formation (kJ mol⁻¹)	Ref.
$C_4H_9NO^+$	$CH_3CON(CH_3)_2$		8.81±0.03	PI		182
$C_4H_9NO^+$	$CH_3CON(CH_3)_2$		8.81	CTS		2562

$$\mathbf{C_5H_5NO^+}$$

$C_5H_5NO^+$	C_5H_4NOH (4–Hydroxypyridine)		9.70±0.05	EI		217

$$\mathbf{C_5H_9NO^+}$$

$C_5H_9NO^+$	C_5H_9NO (5–Aminopentanoic acid lactam)		9.15±0.02	PI		2728

$$\mathbf{C_5H_{11}NO^+}$$

$C_5H_{11}NO^+$	$HCON(C_2H_5)_2$		8.89±0.02	PI		182

$$\mathbf{C_6H_5NO^+}$$

$C_6H_5NO^+$	C_6H_5NO (Nitrosobenzene)		8.87 (V)	PE		2806
$C_6H_5NO^+$	C_5H_4NCHO (2–Pyridinecarboxaldehyde)		9.75±0.05	EI		217
$C_6H_5NO^+$	C_5H_4NCHO (4–Pyridinecarboxaldehyde)		10.12±0.05	EI		217

$$\mathbf{C_6H_6NO^+}$$

$C_6H_6NO^+$	$C_6H_4(NH_2)OCH_3$ (4–Methoxyaniline)	CH_3	9.67±0.1	EI		2970

See also – EI: 3238

$C_6H_6NO^+$	$C_6H_5CH_2OC_6H_4NH_2$ (3–Aminophenyl benzyl ether)		10.1	EI		2737
$C_6H_6NO^+$	$C_6H_5CH_2OC_6H_4NH_2$ (4–Aminophenyl benzyl ether)		9.7	EI		2737

$$\mathbf{C_6H_7NO^+}$$

$C_6H_7NO^+$	$C_6H_4(NH_2)OC_2H_5$ (3–Ethoxyaniline)	C_2H_4	10.71±0.15	EI		2945
$C_6H_7NO^+$	$C_6H_4(NH_2)OC_2H_5$ (4–Ethoxyaniline)	C_2H_4	10.43±0.15	EI		2945

4.3. The Positive Ion Table—Continued

Ion	Reactant	Other products	Ionization or appearance potential (eV)	Method	Heat of formation (kJ mol⁻¹)	Ref.
	C₆H₁₁NO⁺ (6–Aminohexanoic acid lactam)		**ΔH°f₂₉₈ = 629 kJ mol⁻¹ (150 kcal mol⁻¹)**			
$C_6H_{11}NO^+$	$C_6H_{11}NO$ (6–Aminohexanoic acid lactam)		9.07±0.02	PI	629	2728
			C₇H₁₃NO⁺			
$C_6H_{13}NO^+$	$CH_3CON(C_2H_5)_2$		8.60±0.02	PI		182
			C₇H₄NO⁺			
$C_7H_4NO^+$	$C_6H_4(CN)OCH_3$ (4–Methoxybenzoic acid nitrile)	CH_3	12.68±0.1	EI		2970
			C₇H₅NO⁺			
$C_7H_5NO^+$	C_6H_5NCO (Isocyanatobenzene)		8.77±0.02	PI		182
$C_7H_5NO^+$	$C_6H_4(CN)OH$ (3–Hydroxybenzoic acid nitrile)		9.39±0.05	RPD		3223
$C_7H_5NO^+$	$C_6H_4(CN)OH$ (4–Hydroxybenzoic acid nitrile)		9.40±0.05	RPD		3223
$C_7H_5NO^+$	$C_6H_4(CN)OC_2H_5$ (3–Ethoxybenzoic acid nitrile)	C_2H_4	11.31±0.15	EI		2945
$C_7H_5NO^+$	$C_6H_4(CN)OC_2H_5$ (4–Ethoxybenzoic acid nitrile)	C_2H_4	11.19±0.15	EI		2945
$C_7H_5NO^+$	$(C_6H_5)_2C_2N_2O$ (3,5–Diphenyl–1,2,4–oxadiazole)		10.8±0.1	EI		1125
			C₇H₆NO⁺			
$C_7H_6NO^+$	$C_6H_4(NH_2)COCH_3$ (3–Aminoacetophenone)	CH_3	11.01±0.2	EI		2967
$C_7H_6NO^+$	$C_6H_4(NH_2)COCH_3$ (4–Aminoacetophenone)	CH_3	9.34	EI		3334
$C_7H_6NO^+$	$C_6H_4(NH_2)COCH_3$ (4–Aminoacetophenone)	CH_3	10.20±0.2	EI		2967

See also – EI: 3238

Ion	Reactant	Other products	Ionization or appearance potential (eV)	Method	Heat of formation (kJ mol⁻¹)	Ref.
$C_7H_6NO^+$	$C_6H_4(NH_2)COOCH_3$ (4–Aminobenzoic acid methyl ester)	CH_3O	10.91	EI		3238

4.3. The Positive Ion Table—Continued

Ion	Reactant	Other products	Ionization or appearance potential (eV)	Method	Heat of formation (kJ mol^{-1})	Ref.
			C$_7$H$_7$NO$^+$			
C$_7$H$_7$NO$^+$	C$_6$H$_5$CONH$_2$ (Benzoic acid amide)		9.4±0.2	EI		1168
See also – EI: 1126						
C$_7$H$_7$NO$^+$	C$_6$H$_4$(NH$_2$)CHO (4–Aminobenzenecarbonal)		8.25±0.02	EI		2026
C$_7$H$_7$NO$^+$	C$_7$H$_7$NO (2–Amino–2,4,6–cycloheptatrien–1–one)		9.43±0.02	EI		431
			C$_7$H$_9$NO$^+$			
C$_7$H$_9$NO$^+$	C$_6$H$_4$(NH$_2$)OCH$_3$ (2–Methoxyaniline)		7.46±0.1	CTS		2485
C$_7$H$_9$NO$^+$	C$_6$H$_4$(NH$_2$)OCH$_3$ (4–Methoxyaniline)		7.82	EI		1066
C$_7$H$_9$NO$^+$	C$_6$H$_4$(NH$_2$)OCH$_3$ (4–Methoxyaniline)		7.41±0.1	CTS		2485
C$_7$H$_9$NO$^+$	C$_6$H$_4$(OCH$_3$)NHCOCH$_3$ (N–(3–Methoxyphenyl)acetic acid amide)		10.68±0.2	EI		3406
C$_7$H$_9$NO$^+$	C$_6$H$_4$(OCH$_3$)NHCOCH$_3$ (N–(4–Methoxyphenyl)acetic acid amide)		10.68±0.2	EI		3406
			C$_8$H$_4$NO$^+$			
C$_8$H$_4$NO$^+$	C$_6$H$_4$(CN)COOCH$_3$ (4–Cyanobenzoic acid methyl ester)	CH$_3$O	11.60	EI		3238
			C$_8$H$_7$NO$^+$			
C$_8$H$_7$NO$^+$	C$_6$H$_4$(CN)OCH$_3$ (3–Methoxybenzoic acid nitrile)		9.05±0.04	RPD		3223
C$_8$H$_7$NO$^+$	C$_6$H$_4$(CN)OCH$_3$ (4–Methoxybenzoic acid nitrile)		9.08±0.05	RPD		3223
C$_8$H$_7$NO$^+$	C$_6$H$_4$(OH)CH$_2$CN (4–Hydroxyphenylacetic acid nitrile)		9.01±0.05	RPD		3223
			C$_8$H$_9$NO$^+$			
C$_8$H$_9$NO$^+$	C$_6$H$_5$NHCOCH$_3$ (N–Phenylacetic acid amide)		8.39±0.10	EI		1126
See also – EI: 3406						

4.3. The Positive Ion Table—Continued

Ion	Reactant	Other products	Ionization or appearance potential (eV)	Method	Heat of formation (kJ mol⁻¹)	Ref.
$C_8H_9NO^+$	$C_6H_4(NH_2)COCH_3$ (3−Aminoacetophenone)		8.09±0.05	EI		2025
$C_8H_9NO^+$	$C_6H_4(NH_2)COCH_3$ (3−Aminoacetophenone)		8.56±0.2	EI		2967
$C_8H_9NO^+$	$C_6H_4(NH_2)COCH_3$ (4−Aminoacetophenone)		8.17±0.02	EI		2026
$C_8H_9NO^+$	$C_6H_4(NH_2)COCH_3$ (4−Aminoacetophenone)		8.33±0.1	EI		2967

See also − EI: 3238, 3334

$C_8H_{11}NO^+$

Ion	Reactant	Other products	Ionization or appearance potential (eV)	Method	Heat of formation (kJ mol⁻¹)	Ref.
$C_8H_{11}NO^+$	$C_6H_4(NH_2)OC_2H_5$ (3−Ethoxyaniline)		7.91±0.15	EI		2945
$C_8H_{11}NO^+$	$C_6H_4(NH_2)OC_2H_5$ (4−Ethoxyaniline)		7.67±0.15	EI		2945
$C_8H_{11}NO^+$	$C_6H_4(NH_2)OC_2H_5$ (4−Ethoxyaniline)		7.38±0.1	CTS		2485

$C_9H_7NO^+$

Ion	Reactant	Other products	Ionization or appearance potential (eV)	Method	Heat of formation (kJ mol⁻¹)	Ref.
$C_9H_7NO^+$	$C_6H_5COCH_2CN$ (α−Cyanoacetophenone)		9.56±0.05	EI		2025

$C_9H_9NO^+$

Ion	Reactant	Other products	Ionization or appearance potential (eV)	Method	Heat of formation (kJ mol⁻¹)	Ref.
$C_9H_9NO^+$	$C_6H_4(CN)OC_2H_5$ (3−Ethoxybenzoic acid nitrile)		8.92±0.15	EI		2945
$C_9H_9NO^+$	$C_6H_4(CN)OC_2H_5$ (4−Ethoxybenzoic acid nitrile)		8.89±0.15	EI		2945
$C_9H_9NO^+$	$C_6H_4(OCH_3)CH_2CN$ (4−Methoxyphenylacetic acid nitrile)		8.77±0.05	RPD		3223

$C_9H_{11}NO^+$

Ion	Reactant	Other products	Ionization or appearance potential (eV)	Method	Heat of formation (kJ mol⁻¹)	Ref.
$C_9H_{11}NO^+$	$C_6H_4(CH_3)NHCOCH_3$ (N−(3−Tolyl)acetic acid amide)		8.29±0.2	EI		3406
$C_9H_{11}NO^+$	$C_6H_4(CH_3)NHCOCH_3$ (N−(4−Tolyl)acetic acid amide)		8.24±0.2	EI		3406

$C_9H_{13}NO^+$

Ion	Reactant	Other products	Ionization or appearance potential (eV)	Method	Heat of formation (kJ mol⁻¹)	Ref.
$C_9H_{13}NO^+$	$C_6H_4(OCH_3)N(CH_3)_2$ (N,N−Dimethyl−methoxyaniline)		≤6.75±0.02	PI		3372

4.3. The Positive Ion Table—Continued

Ion	Reactant	Other products	Ionization or appearance potential (eV)	Method	Heat of formation (kJ mol⁻¹)	Ref.
		$C_{10}H_{12}NO^+$				
$C_{10}H_{12}NO^+$	$C_6H_4(NH_2)(CH_2)_3COOCH_3$ (4–(4–Aminophenyl)butanoic acid methyl ester)	CH_3O	11.26±0.2	EI		2497
		$C_{10}H_{13}NO^+$				
$C_{10}H_{13}NO^+$	$C_6H_4(NH_2)COC_3H_7$ (1–(3–Aminophenyl)–1–butanone)		8.06±0.2	EI		2534
$C_{10}H_{13}NO^+$	$C_6H_4(NH_2)COC_3H_7$ (1–(4–Aminophenyl)–1–butanone)		8.01±0.2	EI		2534
		$C_{11}H_9NO^+$				
$C_{11}H_9NO^+$	$C_6H_4(CN)(CH_2)_3COOCH_3$ (4–(4–Cyanophenyl)butanoic acid methyl ester)	CH_3OH	9.56±0.2	EI		2497
		$C_{11}H_{10}NO^+$				
$C_{11}H_{10}NO^+$	$C_6H_4(CN)CH(CH_3)CH_2COOCH_3$ (3–(4–Cyanophenyl)butanoic acid methyl ester)	CH_3O	11.18±0.2	EI		2497
$C_{11}H_{10}NO^+$	$C_6H_4(CN)(CH_2)_3COOCH_3$ (4–(4–Cyanophenyl)butanoic acid methyl ester)	CH_3O	11.22±0.2	EI		2497
		$C_{13}H_9NO^+$				
$C_{13}H_9NO^+$	$C_{13}H_9NO$ (9,10–Dihydro–9–oxoacridine)		7.60±0.03	PI		2728
		$C_{13}H_{11}NO^+$				
$C_{13}H_{11}NO^+$	$C_6H_5CONHC_6H_5$ (N–Phenylbenzoic acid amide)		8.1±0.1	EI		2918
		$C_{13}H_{13}NO^+$				
$C_{13}H_{13}NO^+$	$C_6H_5CH_2OC_6H_4NH_2$ (3–Aminophenyl benzyl ether)		8.0	EI		2737
$C_{13}H_{13}NO^+$	$C_6H_5CH_2OC_6H_4NH_2$ (4–Aminophenyl benzyl ether)		7.6	EI		2737
		$C_{14}H_{11}NO^+$				
$C_{14}H_{11}NO^+$	$C_{14}H_{11}NO$ (9–Hydro–10–methyl–9–oxoacridine)		7.53±0.02	PI		2728

4.3. The Positive Ion Table—Continued

Ion	Reactant	Other products	Ionization or appearance potential (eV)	Method	Heat of formation (kJ mol^{-1})	Ref.
			C$_{15}$H$_{13}$NO$^+$			
C$_{15}$H$_{13}$NO$^+$	C$_{13}$H$_9$CH$_2$CONH$_2$ (2–Fluorenylacetic acid amide)		8.34	CTS		2562
C$_{15}$H$_{13}$NO$^+$	C$_{15}$H$_{13}$NO (10–Ethyl–9–hydro–9–oxoacridine)		7.49±0.03	PI		2728
			C$_{17}$H$_{11}$NO$^+$			
C$_{17}$H$_{11}$NO$^+$	C$_{17}$H$_{11}$NO (Naphtho[2,3–a]–4–oxo–3–phenylazacyclobutene)		7.5±0.1	EI		2964
C$_{17}$H$_{11}$NO$^+$	C$_{17}$H$_{11}$N$_3$O (Naphtho[2,3–e]–4–hydro–4–oxo–3–phenyl–1,2,3–triazine)	N$_2$	9.45±0.1	EI		2964
			C$_{18}$H$_{21}$NO$^+$			
C$_{18}$H$_{21}$NO$^+$	C$_6$H$_4$(NH$_2$)CH$_2$CH$_2$C$_6$H$_4$COC$_3$H$_7$ (1–(4–Aminophenyl)–2–(4–butyrylphenyl)ethane)		8.14±0.2	EI		2567
			C$_{19}$H$_{13}$NO$^+$			
C$_{19}$H$_{13}$NO$^+$	C$_{19}$H$_{13}$NO (9–Hydro–9–oxo–10–phenylacridine)		7.46±0.02	PI		2728
			CH$_4$N$_2$O$^+$			
CH$_4$N$_2$O$^+$	(NH$_2$)$_2$CO		10.27±0.05	EI		2515, 2867
			C$_2$HN$_2$O$^+$			
C$_2$HN$_2$O$^+$	CH$_3$COCHN$_2$	CH$_3$	11.42±0.08	EI		2174
C$_2$HN$_2$O$^+$	CH$_2$ClCOCHN$_2$	CH$_2$Cl	10.95±0.15	EI		2174
C$_2$HN$_2$O$^+$	CCl$_3$COCHN$_2$	CCl$_3$	10.12±0.08	EI		2174
			C$_2$H$_6$N$_2$O$^+$			
C$_2$H$_6$N$_2$O$^+$	CH$_3$NHCONH$_2$		9.73±0.05	EI		2515, 2867
			C$_3$H$_4$N$_2$O$^+$			
C$_3$H$_4$N$_2$O$^+$	CH$_3$COCHN$_2$		9.40±0.03	EI		2174
			C$_3$H$_8$N$_2$O$^+$			
C$_3$H$_8$N$_2$O$^+$	(CH$_3$)$_2$NCONH$_2$		9.10±0.05	EI		2515, 2867
C$_3$H$_8$N$_2$O$^+$	(CH$_3$NH)$_2$CO		9.42±0.05	EI		2515, 2867

4.3. The Positive Ion Table—Continued

Ion	Reactant	Other products	Ionization or appearance potential (eV)	Method	Heat of formation (kJ mol^{-1})	Ref.
		$C_4H_{10}N_2O^+$				
$C_4H_{10}N_2O^+$	$(CH_3)_2NCONHCH_3$		8.94±0.05	EI		2515, 2867
		$C_5H_{12}N_2O^+$				
$C_5H_{12}N_2O^+$	$((CH_3)_2N)_2CO$		8.74±0.05	EI		2515, 2867
		$C_8H_5N_2O^+$				
$C_8H_5N_2O^+$	$C_6H_5C_2N_2OC_3F_7$ (2–Perfluoropropyl–5–phenyl–1,3,4–oxadiazole)		11.8±0.2	EI		2156
$C_8H_5N_2O^+$	$C_6H_5C_2N_2OC_7F_{15}$ (2–Perfluoroheptyl–5–phenyl–1,3,4–oxadiazole)		12.1±0.1	EI		2156
		$C_8H_6N_2O^+$				
$C_8H_6N_2O^+$	$C_6H_5COCHN_2$ (α–Diazoacetophenone)		9.22±0.04	EI		2174
		$C_{14}H_{10}N_2O^+$				
$C_{14}H_{10}N_2O^+$	$(C_6H_5)_2C_2N_2O$ (3,5–Diphenyl–1,2,4–oxadiazole)		9.2±0.1	EI		1125
$C_{14}H_{10}N_2O^+$	$(C_6H_5)_2C_2N_2O$ (2,5–Diphenyl–1,3,4–oxadiazole)		8.9±0.3	EI		1125
		$C_{15}H_{12}N_2O^+$				
$C_{15}H_{12}N_2O^+$	$C_{15}H_{12}N_2O$ (Tegretol)		8.07±0.08	CTS		2987
		$C_4H_5N_3O^+$				
$C_4H_5N_3O^+$	$C_4H_5N_3O$ (Cytosine)		8.90±0.2	EI		2514, 3336
		$C_5H_4N_4O^+$				
$C_5H_4N_4O^+$	$C_5H_4N_4O$ (Hypoxanthine)		9.17±0.1	EI		2514, 3336
		$CH_2NO_2^+$				
$CH_2NO_2^+$	CH_3NO_2	H	11.97±0.02	EI		90

4.3. The Positive Ion Table—Continued

Ion	Reactant	Other products	Ionization or appearance potential (eV)	Method	Heat of formation (kJ mol^{-1})	Ref.
	CH$_3$NO$_2^+$	**$\Delta H_{f0}^{\circ} \sim 1011$ kJ mol^{-1} (242 kcal mol^{-1})**				
CH$_3$NO$_2^+$	CH$_3$NO$_2$		11.08±0.03	PI	1008	182
CH$_3$NO$_2^+$	CH$_3$NO$_2$		11.130±0.006	PI	1013	1253
CH$_3$NO$_2^+$	CH$_3$NO$_2$		11.23±0.01	PE		2701

See also – PEN: 2430
       EI: 90, 2018

Ion	Reactant	Other products	Ionization or appearance potential (eV)	Method	Heat of formation (kJ mol^{-1})	Ref.
	C$_2$H$_5$NO$_2^+$	**$\Delta H_{f298}^{\circ} \sim 953$ kJ mol^{-1} (228 kcal mol^{-1})**				
	C$_2$H$_5$ONO$^+$	**$\Delta H_{f298}^{\circ} = 912$ kJ mol^{-1} (218 kcal mol^{-1})**				
C$_2$H$_5$NO$_2^+$	C$_2$H$_5$NO$_2$		10.88±0.05	PI	951	182
C$_2$H$_5$NO$_2^+$	C$_2$H$_5$NO$_2$		10.92±0.01	PE	955	2701

See also – EI: 2018

Ion	Reactant	Other products	Ionization or appearance potential (eV)	Method	Heat of formation (kJ mol^{-1})	Ref.
C$_2$H$_5$NO$_2^+$	C$_2$H$_5$ONO		10.53±0.01	PE	912	2701
C$_2$H$_5$NO$_2^+$	CH$_2$(NH$_2$)COOH		9.25±0.10	EI		2587
C$_2$H$_5$NO$_2^+$	CH$_2$(NH$_2$)COOH		9.30	CTS		2562

See also – EI: 88

Ion	Reactant	Other products	Ionization or appearance potential (eV)	Method	Heat of formation (kJ mol^{-1})	Ref.
	n–C$_3$H$_7$NO$_2^+$	**$\Delta H_{f298}^{\circ} \sim 915$ kJ mol^{-1} (219 kcal mol^{-1})**				
	iso–C$_3$H$_7$NO$_2^+$	**$\Delta H_{f298}^{\circ} \sim 896$ kJ mol^{-1} (214 kcal mol^{-1})**				
C$_3$H$_7$NO$_2^+$	n–C$_3$H$_7$NO$_2$		10.81±0.03	PI	918	182
C$_3$H$_7$NO$_2^+$	n–C$_3$H$_7$NO$_2$		10.75±0.01	PE	913	2701

See also – EI: 2018

Ion	Reactant	Other products	Ionization or appearance potential (eV)	Method	Heat of formation (kJ mol^{-1})	Ref.
C$_3$H$_7$NO$_2^+$	iso–C$_3$H$_7$NO$_2$		10.71±0.05	PI	893	182
C$_3$H$_7$NO$_2^+$	iso–C$_3$H$_7$NO$_2$		10.77±0.01	PE	899	2701
C$_3$H$_7$NO$_2^+$	n–C$_3$H$_7$ONO		10.34±0.01	PE		2701
C$_3$H$_7$NO$_2^+$	iso–C$_3$H$_7$ONO		10.23±0.01	PE		2701
C$_3$H$_7$NO$_2^+$	CH$_3$CH(NH$_2$)COOH		9.63	CTS		2562

Ion	Reactant	Other products	Ionization or appearance potential (eV)	Method	Heat of formation (kJ mol^{-1})	Ref.
	C$_4$H$_5$NO$_2^+$					
C$_4$H$_5$NO$_2^+$	CH$_2$(CN)COOCH$_3$		10:87±0.05	EI		2025

Ion	Reactant	Other products	Ionization or appearance potential (eV)	Method	Heat of formation (kJ mol^{-1})	Ref.
	n–C$_4$H$_9$NO$_2^+$	**$\Delta H_{f298}^{\circ} = 889$ kJ mol^{-1} (213 kcal mol^{-1})**				
	sec–C$_4$H$_9$NO$_2^+$	**$\Delta H_{f298}^{\circ} = 870$ kJ mol^{-1} (208 kcal mol^{-1})**				
C$_4$H$_9$NO$_2^+$	n–C$_4$H$_9$NO$_2$		10.71±0.01	PE	889	2701
C$_4$H$_9$NO$_2^+$	sec–C$_4$H$_9$NO$_2$		10.71±0.01	PE	870	2701

4.3. The Positive Ion Table—Continued

Ion	Reactant	Other products	Ionization or appearance potential (eV)	Method	Heat of formation (kJ mol⁻¹)	Ref.

$C_5H_9NO_2^+$

Ion	Reactant	Other products	Ionization or appearance potential (eV)	Method	Heat of formation (kJ mol⁻¹)	Ref.
$C_5H_9NO_2^+$	$C_5H_9NO_2$ (Proline)		9.36	CTS		2562

$C_6H_5NO_2^+$

Ion	Reactant	Other products	Ionization or appearance potential (eV)	Method	Heat of formation (kJ mol⁻¹)	Ref.
$C_6H_5NO_2^+$	$C_6H_5NO_2$ (Nitrobenzene)		9.92?	PI		182
$C_6H_5NO_2^+$	$C_6H_5NO_2$ (Nitrobenzene)		9.86±0.05	PE		3174
$C_6H_5NO_2^+$	$C_6H_5NO_2$ (Nitrobenzene)		10.26 (V)	PE		2806
$C_6H_5NO_2^+$	$C_6H_5NO_2$ (Nitrobenzene)		9.90±0.03	EDD		3174
$C_6H_5NO_2^+$	$C_6H_5NO_2$ (Nitrobenzene)		10.16±0.04	RPD		3223

See also – EI: 1066, 3231, 3238

$C_6H_{13}NO_2^+$

Ion	Reactant	Other products	Ionization or appearance potential (eV)	Method	Heat of formation (kJ mol⁻¹)	Ref.
$C_6H_{13}NO_2^+$	$n-C_4H_9CH(NH_2)COOH$		9.09±0.11	EI		2587
$C_6H_{13}NO_2^+$	$sec-C_4H_9CH(NH_2)COOH$		9.5±0.2	EI		88

$C_7H_6NO_2^+$

Ion	Reactant	Other products	Ionization or appearance potential (eV)	Method	Heat of formation (kJ mol⁻¹)	Ref.
$C_7H_6NO_2^+$	$C_6H_4(NO_2)CH_2$ (3–Nitrobenzyl radical)		8.56±0.1	EI		69
$C_7H_6NO_2^+$	$C_6H_5CH_2CH_2C_6H_4NO_2$ (1–(4–Nitrophenyl)–2–phenylethane)		~16	EI		3288
$C_7H_6NO_2^+$	$C_6H_4(NH_2)CH_2CH_2C_6H_4NO_2$ (1–(4–Aminophenyl)–2–(4–nitrophenyl)ethane)		~15	EI		3288
$C_7H_6NO_2^+$	$C_6H_4(NO_2)(CH_2)_3COOCH_3$ (4–(4–Nitrophenyl)butanoic acid methyl ester)		14.01±0.2	EI		2497
$C_7H_6NO_2^+$	$C_6H_4(NO_2)CH_2CH_2C_6H_4NO_2$ (1,2–Bis(4–nitrophenyl)ethane)		11.3±0.2	EI		3288

$C_7H_7NO_2^+$

Ion	Reactant	Other products	Ionization or appearance potential (eV)	Method	Heat of formation (kJ mol⁻¹)	Ref.
$C_7H_7NO_2^+$	$C_6H_4(NO_2)CH_3$ (3–Nitrotoluene)		9.65±0.05	EI		2025
$C_7H_7NO_2^+$	$C_6H_4(NO_2)CH_3$ (4–Nitrotoluene)		9.76±0.05	RPD		3223

See also – EI: 1066

Ion	Reactant	Other products	Ionization or appearance potential (eV)	Method	Heat of formation (kJ mol⁻¹)	Ref.
$C_7H_7NO_2^+$	$C_6H_4(NH_2)COOH$ (2–Aminobenzoic acid)		8.29	CTS		2562
$C_7H_7NO_2^+$	$C_6H_4(COOH)NHCOCH_3$ (N–(4–Carboxyphenyl)acetic acid amide)		10.58±0.2	EI		3406

4.3. The Positive Ion Table—Continued

Ion	Reactant	Other products	Ionization or appearance potential (eV)	Method	Heat of formation (kJ mol^{-1})	Ref.
			$C_8H_7NO_2^+$			
$C_8H_7NO_2^+$	$C_6H_4(NO_2)(CH_2)_3COOCH_3$ (4−(4−Nitrophenyl)butanoic acid methyl ester)		11.22±0.2	EI		2497
			$C_8H_8NO_2^+$			
$C_8H_8NO_2^+$	$C_6H_4(NO_2)CH(CH_3)CH_2COOCH_3$ (3−(4−Nitrophenyl)butanoic acid methyl ester)		11.55±0.2	EI		2497
$C_8H_8NO_2^+$	$C_6H_4(NO_2)CH_2CH_2Br$ (1−Bromo−2−(4−nitrophenyl)ethane)	Br	10.3	EI		2973
			$C_8H_9NO_2^+$			
$C_8H_9NO_2^+$	$C_6H_4(NH_2)COOCH_3$ (4−Aminobenzoic acid methyl ester)		8.08±0.01	EI		2026

See also − EI: 3238

Ion	Reactant	Other products	Ionization or appearance potential (eV)	Method	Heat of formation (kJ mol^{-1})	Ref.
			$C_9H_7NO_2^+$			
$C_9H_7NO_2^+$	$C_6H_4(CN)COOCH_3$ (4−Cyanobenzoic acid methyl ester)		9.72	EI		3238
			$C_9H_9NO_2^+$			
$C_9H_9NO_2^+$	$C_6H_3(OCH_3)_2CN$ (3,4−Dimethoxybenzoic acid nitrile)		8.72±0.06	RPD		3223
$C_9H_9NO_2^+$	$C_6H_4(NO_2)CH(CH_3)CH_2COOCH_3$ (3−(4−Nitrophenyl)butanoic acid methyl ester)		9.47±0.2	EI		2497
			$C_9H_{11}NO_2^+$			
$C_9H_{11}NO_2^+$	$C_6H_4(OCH_3)NHCOCH_3$ (N−(3−Methoxyphenyl)acetic acid amide)		7.96±0.2	EI		3406
$C_9H_{11}NO_2^+$	$C_6H_4(OCH_3)NHCOCH_3$ (N−(4−Methoxyphenyl)acetic acid amide)		8.10±0.2	EI		3406
			$C_{11}H_{15}NO_2^+$			
$C_{11}H_{15}NO_2^+$	$C_6H_4(NH_2)(CH_2)_3COOCH_3$ (4−(4−Aminophenyl)butanoic acid methyl ester)		7.85±0.2	EI		2497
			$C_{12}H_{13}NO_2^+$			
$C_{12}H_{13}NO_2^+$	$C_6H_4(CN)CH(CH_3)CH_2COOCH_3$ (3−(4−Cyanophenyl)butanoic acid methyl ester)		9.20±0.2	EI		2497

4.3. The Positive Ion Table—Continued

Ion	Reactant	Other products	Ionization or appearance potential (eV)	Method	Heat of formation (kJ mol^{-1})	Ref.
$C_{12}H_{13}NO_2^+$	$C_6H_4(CN)(CH_2)_3COOCH_3$ (4–(4–Cyanophenyl)butanoic acid methyl ester)		9.25±0.2	EI		2497
$C_{14}H_9NO_2^+$						
$C_{14}H_9NO_2^+$	$C_{14}H_9NO_2$ (N–Phenylphthalic acid imide)		8.8	EI		2412
$C_{14}H_{13}NO_2^+$						
$C_{14}H_{13}NO_2^+$	$C_6H_5CH_2CH_2C_6H_4NO_2$ (1–(4–Nitrophenyl)–2–phenylethane)		8.9±0.1	EI		3288
$C_{14}H_{13}NO_2^+$	$C_6H_5CONHC_6H_4OCH_3$ (N–(3–Methoxyphenyl)benzoic acid amide)		7.8±0.1	EI		2918
$C_{14}H_{13}NO_2^+$	$C_6H_5CONHC_6H_4OCH_3$ (N–(4–Methoxyphenyl)benzoic acid amide)		7.6±0.1	EI		2918
$C_{20}H_{13}NO_2^+$						
$C_{20}H_{13}NO_2^+$	$C_{20}H_{13}NO_2$ (N–(2–Biphenylyl)phthalic acid imide)		8.5	EI		2412
$C_{20}H_{13}NO_2^+$	$C_{20}H_{13}NO_2$ (N–(4–Biphenylyl)phthalic acid imide)		8.4	EI		2412
$C_4H_4N_2O_2^+$						
$C_4H_4N_2O_2^+$	$C_4H_4N_2O_2$ (Uracil)		9.82±0.1	EI		2514, 3336
$C_5H_6N_2O_2^+$						
$C_5H_6N_2O_2^+$	$C_5H_6N_2O_2$ (5–Methyluracil)		9.43±0.1	EI		2514, 3336
$C_6H_6N_2O_2^+$						
$C_6H_6N_2O_2^+$	$C_6H_4(NO_2)NH_2$ (2–Nitroaniline)		8.66	EI		1066
$C_6H_6N_2O_2^+$	$C_6H_4(NO_2)NH_2$ (3–Nitroaniline)		8.80	EI		1066
$C_6H_6N_2O_2^+$	$C_6H_4(NO_2)NH_2$ (4–Nitroaniline)		8.85	EI		1066
$C_6H_6N_2O_2^+$	$C_6H_4(NO_2)NHCOCH_3$ (N–(3–Nitrophenyl)acetic acid amide)		10.49±0.2	EI		3406
$C_6H_6N_2O_2^+$	$C_6H_4(NO_2)NHCOCH_3$ (N–(4–Nitrophenyl)acetic acid amide)		10.44±0.2	EI		3406

4.3. The Positive Ion Table—Continued

Ion	Reactant	Other products	Ionization or appearance potential (eV)	Method	Heat of formation (kJ mol⁻¹)	Ref.
			$C_7H_4N_2O_2^+$			
$C_7H_4N_2O_2^+$	$C_6H_4(NO_2)CN$ (2–Nitrobenzoic acid nitrile)		10.52±0.05	RPD		3223
$C_7H_4N_2O_2^+$	$C_6H_4(NO_2)CN$ (3–Nitrobenzoic acid nitrile)		10.57±0.04	RPD		3223
$C_7H_4N_2O_2^+$	$C_6H_4(NO_2)CN$ (4–Nitrobenzoic acid nitrile)		10.59±0.05	RPD		3223
			$C_8H_6N_2O_2^+$			
$C_8H_6N_2O_2^+$	$C_6H_4(NO_2)CH_2CN$ (4–Nitrophenylacetic acid nitrile)		10.11±0.04	RPD		3223
			$C_{11}H_{12}N_2O_2^+$			
$C_{11}H_{12}N_2O_2^+$	$C_{11}H_{12}N_2O_2$ (Tryptophan)		8.43	CTS		2562
			$C_{14}H_{14}N_2O_2^+$			
$C_{14}H_{14}N_2O_2^+$	$C_6H_4(NH_2)CH_2CH_2C_6H_4NO_2$ (1–(4–Aminophenyl)–2–(4–nitrophenyl)ethane)		7.8±0.1	EI		3288
			$C_3H_3N_3O_2^+$			
$C_3H_3N_3O_2^+$	$C_3H_3N_3O_2$ (6–Azauracil)		10.18±0.1	EI		2514, 3336
			$C_5H_4N_4O_2^+$			
$C_5H_4N_4O_2^+$	$C_5H_4N_4O_2$ (Xanthine)		9.30±0.2	EI		2514, 3336
			$C_8H_{10}N_4O_2^+$			
$C_8H_{10}N_4O_2^+$	$C_8H_{10}N_4O_2$ (Caffeine)		8.50	CTS		2562
			$CH_2NO_3^+$			
$CH_2NO_3^+$	CH_3ONO_2	H	11.7	EI		2456
$CH_2NO_3^+$	$C_2H_5ONO_2$	CH_3	10.13±0.11	EI		1013
$CH_2NO_3^+$	$C_2H_5ONO_2$	CH_3	11.4	EI		2456
$CH_2NO_3^+$	iso–$C_4H_9ONO_2$		11.5	EI		2456

4.3. The Positive Ion Table—Continued

Ion	Reactant	Other products	Ionization or appearance potential (eV)	Method	Heat of formation (kJ mol⁻¹)	Ref.

$CH_3ONO_2^+$ $\Delta H^\circ_{f298} = 988$ kJ mol^{-1} (236 kcal mol^{-1})

Ion	Reactant	Other products	Ionization or appearance potential (eV)	Method	Heat of formation (kJ mol⁻¹)	Ref.
$CH_3NO_3^+$	CH_3ONO_2		11.53±0.01	PE	988	2701

$C_2H_5NO_3^+$

Ion	Reactant	Other products	Ionization or appearance potential (eV)	Method	Heat of formation (kJ mol⁻¹)	Ref.
$C_2H_5NO_3^+$	$C_2H_5ONO_2$		11.22?	PI		182

$n-C_3H_7ONO_2^+$ $\Delta H^\circ_{f298} = 894$ kJ mol^{-1} (214 kcal mol^{-1})

Ion	Reactant	Other products	Ionization or appearance potential (eV)	Method	Heat of formation (kJ mol⁻¹)	Ref.
$C_3H_7NO_3^+$	$n-C_3H_7ONO_2$		11.07±0.02	PI	894	182

$C_6H_5NO_3^+$

Ion	Reactant	Other products	Ionization or appearance potential (eV)	Method	Heat of formation (kJ mol⁻¹)	Ref.
$C_6H_5NO_3^+$	$C_6H_4(NO_2)OH$ (4–Nitrophenol)		9.52	EI		1066
$C_6H_5NO_3^+$	$C_6H_4(NO_2)OC_2H_5$ (1–Ethoxy–3–nitrobenzene)	C_2H_4	10.91±0.15	EI		2945
$C_6H_5NO_3^+$	$C_6H_4(NO_2)OC_2H_5$ (1–Ethoxy–4–nitrobenzene)	C_2H_4	10.97±0.15	EI		2945

$C_7H_4NO_3^+$

Ion	Reactant	Other products	Ionization or appearance potential (eV)	Method	Heat of formation (kJ mol⁻¹)	Ref.
$C_7H_4NO_3^+$	$C_6H_4(NO_2)COCH_3$ (3–Nitroacetophenone)	CH_3	10.67±0.2	EI		2967
$C_7H_4NO_3^+$	$C_6H_4(NO_2)COCH_3$ (4–Nitroacetophenone)	CH_3	10.32	EI		3334
$C_7H_4NO_3^+$	$C_6H_4(NO_2)COCH_3$ (4–Nitroacetophenone)	CH_3	10.73±0.1	EI		2967
$C_7H_4NO_3^+$	$C_6H_4(NO_2)COOCH_3$ (4–Nitrobenzoic acid methyl ester)	CH_3O	11.52	EI		3238

$C_7H_5NO_3^+$

Ion	Reactant	Other products	Ionization or appearance potential (eV)	Method	Heat of formation (kJ mol⁻¹)	Ref.
$C_7H_5NO_3^+$	$C_6H_4(NO_2)CHO$ (4–Nitrobenzenecarbonal)		10.27±0.01	EI		2026

$C_8H_7NO_3^+$

Ion	Reactant	Other products	Ionization or appearance potential (eV)	Method	Heat of formation (kJ mol⁻¹)	Ref.
$C_8H_7NO_3^+$	$C_6H_4(NO_2)COCH_3$ (3–Nitroacetophenone)		9.89±0.05	EI		2025

See also – EI: 2967

Ion	Reactant	Other products	Ionization or appearance potential (eV)	Method	Heat of formation (kJ mol⁻¹)	Ref.
$C_8H_7NO_3^+$	$C_6H_4(NO_2)COCH_3$ (4–Nitroacetophenone)		10.07±0.02	EI		2026

See also – EI: 2967, 3334

4.3. The Positive Ion Table—Continued

Ion	Reactant	Other products	Ionization or appearance potential (eV)	Method	Heat of formation (kJ mol^{-1})	Ref.
		$C_8H_9NO_3^+$				
$C_8H_9NO_3^+$	$C_6H_4(NO_2)OC_2H_5$ (1−Ethoxy−3−nitrobenzene)		9.11±0.15	EI		2945
$C_8H_9NO_3^+$	$C_6H_4(NO_2)OC_2H_5$ (1−Ethoxy−4−nitrobenzene)		9.22±0.15	EI		2945
		$C_9H_9NO_3^+$				
$C_9H_9NO_3^+$	$C_6H_4(COOH)NHCOCH_3$ (N−(4−Carboxyphenyl)acetic acid amide)		8.70±0.2	EI		3406
		$C_{10}H_{10}NO_3^+$				
$C_{10}H_{10}NO_3^+$	$C_6H_4(NO_2)CH(CH_3)CH_2COOCH_3$ (3−(4−Nitrophenyl)butanoic acid methyl ester)	CH_3O	11.11±0.2	EI		2497
$C_{10}H_{10}NO_3^+$	$C_6H_4(NO_2)(CH_2)_3COOCH_3$ (4−(4−Nitrophenyl)butanoic acid methyl ester)	CH_3O	11.18±0.2	EI		2497

See also − EI: 2496

Ion	Reactant	Other products	Ionization or appearance potential (eV)	Method	Heat of formation (kJ mol^{-1})	Ref.
		$C_{10}H_{11}NO_3^+$				
$C_{10}H_{11}NO_3^+$	$C_6H_4(NO_2)COC_3H_7$ (1−(3−Nitrophenyl)−1−butanone)		9.88±0.2	EI		2534
$C_{10}H_{11}NO_3^+$	$C_6H_4(NO_2)COC_3H_7$ (1−(4−Nitrophenyl)−1−butanone)		9.86±0.2	EI		2534
		$C_{13}H_{11}NO_3^+$				
$C_{13}H_{11}NO_3^+$	$C_6H_5CH_2OC_6H_4NO_2$ (Benzyl 3−nitrophenyl ether)		9.0	EI		2737
$C_{13}H_{11}NO_3^+$	$C_6H_5CH_2OC_6H_4NO_2$ (Benzyl 4−nitrophenyl ether)		9.1	EI		2737
		$C_{18}H_{19}NO_3^+$				
$C_{18}H_{19}NO_3^+$	$C_6H_4(NO_2)CH_2CH_2C_6H_4COC_3H_7$ (1−(4−Butyrylphenyl)−2−(4−nitrophenyl)ethane)		9.10±0.2	EI		2567
		$C_8H_8N_2O_3^+$				
$C_8H_8N_2O_3^+$	$C_6H_4(NO_2)NHCOCH_3$ (N−(3−Nitrophenyl)acetic acid amide)		8.84±0.2	EI		3406
$C_8H_8N_2O_3^+$	$C_6H_4(NO_2)NHCOCH_3$ (N−(4−Nitrophenyl)acetic acid amide)		9.05±0.2	EI		3406

4.3. The Positive Ion Table—Continued

Ion	Reactant	Other products	Ionization or appearance potential (eV)	Method	Heat of formation (kJ mol^{-1})	Ref.
C$_{13}$H$_{10}$N$_2$O$_3^+$						
C$_{13}$H$_{10}$N$_2$O$_3^+$	C$_6$H$_5$CONHC$_6$H$_4$NO$_2$ (N−(3−Nitrophenyl)benzoic acid amide)		8.5±0.1	EI		2918
C$_{13}$H$_{10}$N$_2$O$_3^+$	C$_6$H$_5$CONHC$_6$H$_4$NO$_2$ (N−(4−Nitrophenyl)benzoic acid amide)		8.6±0.1	EI		2918
C$_9$H$_{12}$N$_4$O$_3^+$						
C$_9$H$_{12}$N$_4$O$_3^+$	C$_9$H$_{12}$N$_4$O$_3$ (Tetramethyluric acid)		7.87±0.1	EI	2514,	3336
C$_9$H$_{12}$N$_4$O$_3^+$	C$_9$H$_{12}$N$_4$O$_3$ (Tetramethyluric acid)		7.91	CTS		2562
C$_8$H$_7$NO$_4^+$						
C$_8$H$_7$NO$_4^+$	C$_6$H$_4$(NO$_2$)COOCH$_3$ (4−Nitrobenzoic acid methyl ester)		9.70	EI		3238

See also – EI:　2026

Ion	Reactant	Other products	Ionization or appearance potential (eV)	Method	Heat of formation	Ref.
C$_{11}$H$_{13}$NO$_4^+$						
C$_{11}$H$_{13}$NO$_4^+$	C$_6$H$_4$(NO$_2$)CH(CH$_3$)CH$_2$COOCH$_3$ (3−(4−Nitrophenyl)butanoic acid methyl ester)		9.36±0.2	EI		2497
C$_{11}$H$_{13}$NO$_4^+$	C$_6$H$_4$(NO$_2$)(CH$_2$)$_3$COOCH$_3$ (4−(4−Nitrophenyl)butanoic acid methyl ester)		8.88±0.3	EI		2496
C$_{11}$H$_{13}$NO$_4^+$	C$_6$H$_4$(NO$_2$)(CH$_2$)$_3$COOCH$_3$ (4−(4−Nitrophenyl)butanoic acid methyl ester)		9.30±0.2	EI		2497
C$_{14}$H$_{12}$N$_2$O$_4^+$						
C$_{14}$H$_{12}$N$_2$O$_4^+$	C$_6$H$_4$(NO$_2$)CH$_2$CH$_2$C$_6$H$_4$NO$_2$ (1,2−Bis(4−nitrophenyl)ethane)		9.5±0.1	EI		3288
C$_{33}$H$_{40}$N$_2$O$_9^+$						
C$_{33}$H$_{40}$N$_2$O$_9^+$	C$_{33}$H$_{40}$N$_2$O$_9$ (Reserpine)		7.88	CTS		2562

4.3. The Positive Ion Table—Continued

Ion	Reactant	Other products	Ionization or appearance potential (eV)	Method	Heat of formation (kJ mol^{-1})	Ref.

F^+ $\Delta H_{f0}^\circ = 1757.9$ kJ mol^{-1} (420.1 kcal mol^{-1})

Ion	Reactant	Other products	Ionization or appearance potential (eV)	Method	Heat of formation (kJ mol^{-1})	Ref.
F^+	F		17.422	S	1757.9	2113

See also – EI: 2165

| F^+ | F_2 | F^- | 15.6 | PI | | 3274 |
| F^+ | F_2 | F^- | 15.48 | PI | | 2630, 2744 |

See also – EI: 3149

| F^+ | F_2 | F | 19.008 | PI | (1757.1) | 3217, 3274 |

(Threshold value approximately corrected to 0 K)

See also – PI: 2630, 2744
       EI: 3149

| F^+ | BF_3 | | 31.5±2 | EI | | 440 |
| F^+ | CF_4 | | 24.0±1.0 | EI | | 2157 |

See also – EI: 24

F^+	C_2F_4		29.5±1.0	EI		2571
F^+	C_2F_6		22.6	EI		1062
F^+	C_6F_6 (Hexafluorobenzene)		29.2±0.5	EI		1132
F^+	C_3F_8		23.5	EI		1062
F^+	C_4F_8 (Octafluorocyclobutane)		24.0	EI		1062
F^+	NF_3		25±1	EI		401
F^+	CF_2O		38±1	EI		3236
F^+	CF_3OF		36±1	EI		3236
F^+	CF_3Cl		31±1	EI		24
F^+	ClO_3F		27±3	EI		53
F^+	SF_5Cl		33.8±0.3	EI		2777
F^+	CF_3Br		29±1	EI		24
F^+	CF_3I		33±1	EI		24

F^{+2} $\Delta H_{f0}^\circ = 5132.0$ kJ mol^{-1} (1226.6 kcal mol^{-1})

| F^{+2} | F | | 52.392 | S | 5132.0 | 2113 |
| F^{+2} | F^+ | | 34.970 | S | | 2113, 3164 |

F^{+3} $\Delta H_{f0}^\circ = 11182.5$ kJ mol^{-1} (2672.7 kcal mol^{-1})

| F^{+3} | F | | 115.099 | S | 11182.5 | 2113 |
| F^{+3} | F^{+2} | | 62.707 | S | | 2113, 3161 |

4.3. The Positive Ion Table—Continued

Ion	Reactant	Other products	Ionization or appearance potential (eV)	Method	Heat of formation (kJ mol^{-1})	Ref.

$$F_2^+(^2\Pi_{3/2g}) \qquad \Delta H_{f0}^\circ = 1514 \text{ kJ mol}^{-1} \text{ (362 kcal mol}^{-1}\text{)}$$

Ion	Reactant		Potential	Method	Heat	Ref.
$F_2^+(^2\Pi_g)$	F_2		15.7	S		355
$F_2^+(^2\Pi_{3/2g})$	F_2		15.686±0.006	PI	1513	3274
(Threshold value corrected for hot bands)						
$F_2^+(^2\Pi_{3/2g})$	F_2		15.69±0.01	PI	1514	2630, 2744,
(Threshold value corrected for hot bands)						3413
$F_2^+(^2\Pi_{3/2g})$	F_2		15.70	PE		3275
$F_2^+(^2\Pi_{1/2g})$	F_2		15.74	PE		3275
$F_2^+(^2\Pi_u)$	F_2		18.4	PE		3275
$F_2^+(^2\Sigma?)$	F_2		21? (V)	PE		3275

See also – PE: 2815
            EI: 75

$$HF^+(^2\Pi_{3/2}) \qquad \Delta H_{f0}^\circ = 1273 \text{ kJ mol}^{-1} \text{ (304 kcal mol}^{-1}\text{)}$$
$$HF^+(^2\Sigma^+) \qquad \Delta H_{f0}^\circ = 1574 \text{ kJ mol}^{-1} \text{ (376 kcal mol}^{-1}\text{)}$$

Ion	Reactant		Potential	Method	Heat	Ref.
$HF^+(^2\Pi_{3/2})$	HF		16.007±0.010	PI	1273	3274
(Threshold value corrected for hot bands)						
$HF^+(^2\Pi)$	HF		15.92±0.01	PI		2744
$HF^+(^2\Pi)$	HF		16.045±0.01	PE		3086
$HF^+(^2\Pi)$	HF		16.05±0.01	PE		2819, 2820
$HF^+(^2\Pi)$	HF		16.06±0.01	PE		2815
$HF^+(^2\Sigma^+)$	HF		19.146	S	1576	3274
$HF^+(^2\Sigma^+)$	HF		19.118±0.005	RPI	1574	3274
$HF^+(^2\Sigma^+)$	HF		19.092±0.02	PE	1571	3086
$HF^+(^2\Sigma^+)$	HF		18.6	PE		2819, 2820

See also – EI: 463, 2436

$$DF^+$$

Ion	Reactant		Potential	Method	Heat	Ref.
$DF^+(^2\Pi_{3/2})$	DF		16.030±0.010	PI		3274
(Threshold value corrected for hot bands)						
$DF^+(^2\Pi)$	DF		16.053±0.01	PE		3086
$DF^+(^2\Pi)$	DF		16.06±0.01	PE		2820
$DF^+(^2\Sigma^+)$	DF		19.194	S		3274
$DF^+(^2\Sigma^+)$	DF		19.100±0.02	PE		3086
$DF^+(^2\Sigma^+)$	DF		18.90±0.05	PE		2819

$$LiF^+$$

Ion	Reactant		Potential	Method	Heat	Ref.
LiF^+	LiF		11.3	EI		2179

4.3. The Positive Ion Table—Continued

Ion	Reactant	Other products	Ionization or appearance potential (eV)	Method	Heat of formation (kJ mol⁻¹)	Ref.
			Li₂F⁺			
Li_2F^+	Li_2F_2	F	11.5	EI		2179
			BeF⁺			
BeF^+	BeF		9.1±0.5	EI		2141
BeF^+	BeF_2	F	15.4±0.4	EI		2142
BeF^+	BeF_2	F	15.5	EI		2141
			BeF₂⁺			
BeF_2^+	BeF_2		14.5±0.4	EI		2142
BeF_2^+	BeF_2		14.7±0.4	EI		2141
		BF⁺ $\Delta H_{f0}^\circ \sim 937$ kJ mol⁻¹ (224 kcal mol⁻¹)				
BF^+	BF		$11.115^{+0.004}_{-0.002}$	S	947	2872
BF^+	BF		~11.45	S		3414
BF^+	BF		11.06±0.10	EI		3414

See also – EI: 440, 1297, 2432

Ion	Reactant	Other products	Ionization or appearance potential (eV)	Method	Heat of formation (kJ mol⁻¹)	Ref.
BF^+	BF_3	2F	25.2±0.2	EI		440
BF^+	B_2F_4	BF_3	12.75±0.01	PI	927	2626
		BF₂⁺ $\Delta H_{f0}^\circ = 314$ kJ mol⁻¹ (75 kcal mol⁻¹)				
BF_2^+	BF_3	F	15.81±0.04	PI	314	2626

See also – EI: 362, 364, 440, 1297, 2040, 2432, 3200

Ion	Reactant	Other products	Ionization or appearance potential (eV)	Method	Heat of formation (kJ mol⁻¹)	Ref.
BF_2^+	B_2F_4	BF_2	12.94±0.01	PI		2626
BF_2^+	CH_3BF_2	CH_3	13.62±0.02	EI		1076
BF_2^+	$CH_2=CHBF_2$	C_2H_3	14.8±0.1	EI		1076
BF_2^+	$C_2H_5BF_2$	C_2H_5	14.3±0.2	EI		1076
BF_2^+	$iso-C_3H_7BF_2$	C_3H_7	14.6±0.2	EI		1076
BF_2^+	Si_2BF_7		14.67±0.1	RPD		2525
		BF₃⁺ $\Delta H_{f0}^\circ = 367$ kJ mol⁻¹ (88 kcal mol⁻¹)				
BF_3^+	BF_3		15.55±0.04	PI	366	2626
BF_3^+	BF_3		15.57±0.02	PE	368	3375

See also – PE: 2802, 2834, 3119
         EI: 362, 364, 440, 2040, 2512, 2513, 3200, 3227

4.3. The Positive Ion Table—Continued

Ion	Reactant	Other products	Ionization or appearance potential (eV)	Method	Heat of formation (kJ mol^{-1})	Ref.

$B_2F_3^+$ $\Delta H_{f0}^\circ = -28$ kJ mol^{-1} (-7 kcal mol^{-1})

$B_2F_3^+$	B_2F_4	F	15.40 ± 0.01	PI	-28	2626

$B_2F_4^+$ $\Delta H_{f0}^\circ = -273$ kJ mol^{-1} (-65 kcal mol^{-1})

$B_2F_4^+$	B_2F_4		12.07 ± 0.01	PI	-273	2626

CF^+ $\Delta H_{f0}^\circ \sim 1126$ kJ mol^{-1} (269 kcal mol^{-1})

CF^+	CF		$\sim8.91\pm0.12$	S	~1111	3110
CF^+	CF		~8.91	S	~1111	2149
CF^+	CF		9.23 ± 0.08	D	1142	2746

(Based upon AP(CF$^+$/C$_2$F$_4$) $-$ AP(CF$_3^+$/C$_2$F$_4$) $= 0.06\pm0.01$ eV and a "best value" of IP(CF$_3$) $= 9.17\pm0.08$ eV)

See also – EI: 129

CF^+	CF_4		22.6 ± 0.5	EI		24

See also – EI: 129, 2157

CF^+	C_2F_4	CF_3	13.76 ± 0.01	PI	1140	2746

(Threshold value approximately corrected to 0 K)

See also – EI: 419, 2586, 2601

CF^+	C_2F_6		16.75	PI		2643
CF^+	C_2F_6		16.1	EI		2572
CF^+	C_3F_6		18.1	EI		1290
CF^+	C_3F_6 (Hexafluorocyclopropane)		17.3	EI		1290
CF^+	C_6F_6 (Hexafluorobenzene)		17.3 ± 0.3	EI		1132
CF^+	C_3F_8		13.5?	EI		2572
CF^+	$C_6F_{11}CF_3$ (Perfluoromethylcyclohexane)		15.93 ± 0.2	EI		2192
CF^+	CH≡CF	CH	17.0 ± 0.1	EDD		3177
CF^+	C_2H_3F		15.43	EI		419
CF^+	CH_2F_2		18.8	EI		1288
CF^+	$CH_2=CF_2$		15.23	EI		419
CF^+	CHF_3		20.2 ± 0.4	EI		43

See also – EI: 1288

CF^+	C_2HF_3		15.2	EI		419
CF^+	FCN	N	16.35	PI	(1130)	2621

4.3. The Positive Ion Table—Continued

Ion	Reactant	Other products	Ionization or appearance potential (eV)	Method	Heat of formation (kJ mol⁻¹)	Ref.
CF^+	$CF_2=CFCN$		13.3 ± 0.1	EI		2954
CF^+	CF_2O		27.0 ± 0.3	EI		3236
CF^+	CF_3OF		31.1 ± 0.5	EI		3236
CF^+	$CH_2=CFCN$		14.2 ± 0.1	EI		2954
CF^+	$CF_2=CHCN$		13.7 ± 0.1	EI		2954
CF^+	CF_3SSCF_3		~16	EI		3202
CF^+	CF_3Cl		22.6 ± 0.5	EI		24
CF^+	$CF_2ClCF=CF_2$		17.8	EI		1290
CF^+	CHF_2Cl		17.30 ± 0.15	EI		43
CF^+	$CHFCl_2$		16.9 ± 0.2	EI		43
CF^+	CF_3Br		22.9 ± 0.5	EI		24
CF^+	CF_3I		19.9 ± 0.3	EI		439

See also – EI: 24

C_3F^+

Ion	Reactant	Other products	Ionization or appearance potential (eV)	Method	Heat of formation (kJ mol⁻¹)	Ref.
C_3F^+	C_6F_6 (Hexafluorobenzene)		22 ± 1?	EI		1132

C_5F^+

Ion	Reactant	Other products	Ionization or appearance potential (eV)	Method	Heat of formation (kJ mol⁻¹)	Ref.
C_5F^+	C_6F_6 (Hexafluorobenzene)		29.0 ± 0.5	EI		1132

CF_2^+ $\Delta H^\circ_{f0} \sim 939$ kJ mol⁻¹ (224 kcal mol⁻¹)

Ion	Reactant	Other products	Ionization or appearance potential (eV)	Method	Heat of formation (kJ mol⁻¹)	Ref.
CF_2^+	CF_2		11.7 ± 0.2	EDD	(~946)	2795
CF_2^+	CF_2		11.7 ± 0.1	EI		2164
CF_2^+	CF_2		11.8 ± 0.3	EI		2601
CF_2^+	CF_2		11.86 ± 0.1	EI		1252

See also – EI: 129

Ion	Reactant	Other products	Ionization or appearance potential (eV)	Method	Heat of formation (kJ mol⁻¹)	Ref.
CF_2^+	CF_4		20.3 ± 0.5	EI		24

See also – EI: 129, 1288, 2157

Ion	Reactant	Other products	Ionization or appearance potential (eV)	Method	Heat of formation (kJ mol⁻¹)	Ref.
CF_2^+	C_2F_4 (Threshold value approximately corrected to 0 K)	CF_2	14.63 ± 0.04	PI	939	2746
CF_2^+	C_2F_4	CF_2	15.0 ± 0.3	EI		2601
CF_2^+	C_2F_4	CF_2	15.26 ± 0.05	EI		1252

See also – EI: 214, 419, 2586

4.3. The Positive Ion Table—Continued

Ion	Reactant	Other products	Ionization or appearance potential (eV)	Method	Heat of formation (kJ mol⁻¹)	Ref.
CF_2^+	C_3F_6		19.8	EI		1290
CF_2^+	C_3F_6 (Hexafluorocyclopropane)		17.4	EI		1290
CF_2^+	$C_6F_{11}CF_3$ (Perfluoromethylcyclohexane)		13.93±0.2	EI		2192
CF_2^+	CH_2F_2	H_2	14.8±0.4	EI		2160

See also – EI: 1288

CF_2^+	CHF_3		14.7±0.4	EI		2160

See also – EI: 43, 1288

CF_2^+	C_2HF_3		19.28	EI		419
CF_2^+	CF_2O	O	25.6±0.3	EI		3236
CF_2^+	CF_3OF		28.0±0.5	EI		3236
CF_2^+	CF_3Cl		20±1	EI		24
CF_2^+	CHF_2Cl		16.1±0.3	EI		43
CF_2^+	CF_3Br		18.3±0.1	EI		439

See also – EI: 24

CF_2^+	CF_3I		17.95±0.1	EI		439

See also – EI: 24

CF_2^{+2}

CF_2^{+2}	CF_4		44.3±0.3	EI		2157

$C_2F_2^+$

$C_2F_2^+$	C_2F_2		11.4±0.5	EDD		2795
$C_2F_2^+$	C_2HF_3		14.83	EI		419
$C_2F_2^+$	$CF_2=CHCN$		15.9±0.1	EI		2954

$C_3F_2^+$

$C_3F_2^+$	C_6F_6 (Hexafluorobenzene)		18.9±0.5	EI		1132

$C_4F_2^+$

$C_4F_2^+$	C_6F_6 (Hexafluorobenzene)		18±1?	EI		1132

4.3. The Positive Ion Table—Continued

Ion	Reactant	Other products	Ionization or appearance potential (eV)	Method	Heat of formation (kJ mol^{-1})	Ref.
			$C_5F_2^+$			
$C_5F_2^+$	C_6F_6 (Hexafluorobenzene)		22 ± 1?	EI		1132

CF_3^+ $\Delta H_{f0}^\circ \sim 417$ kJ mol^{-1} (100 kcal mol^{-1})

Ion	Reactant	Other products	Ionization or appearance potential (eV)	Method	Heat of formation (kJ mol^{-1})	Ref.
CF_3^+	CF_3		9.25 ± 0.04	PI	2638,	2746
CF_3^+	CF_3		9.11	D		2746

Walter, *et al.*, ref. 2746, recommend as "best value" 9.17 ± 0.08 eV for the CF_3 adiabatic ionization potential. This leads to $\Delta H_{f0}^\circ(CF_3^+) = 417$ kJ mol^{-1} adopted above.

Ion	Reactant	Other products	Ionization or appearance potential (eV)	Method	Heat of formation (kJ mol^{-1})	Ref.
CF_3^+	CF_3		9.8 ± 0.2	EDD		2795

See also – EI: 129, 141, 441

Ion	Reactant	Other products	Ionization or appearance potential (eV)	Method	Heat of formation (kJ mol^{-1})	Ref.
CF_3^+	CF_4	F	$\leqslant15.35$	PI	($\leqslant477$)	2746
CF_3^+	CF_4	F	15.52 ± 0.02	PI		2643
CF_3^+	CF_4	F	15.56 ± 0.01	PI		1235

See also – EI: 24, 129, 1288, 2157

Ion	Reactant	Other products	Ionization or appearance potential (eV)	Method	Heat of formation (kJ mol^{-1})	Ref.
CF_3^+	C_2F_4 (Threshold value approximately corrected to 0 K)	CF	13.70 ± 0.02	PI	(415)	2746

See also – EI: 419, 1378, 2586, 2601

Ion	Reactant	Other products	Ionization or appearance potential (eV)	Method	Heat of formation (kJ mol^{-1})	Ref.
CF_3^+	C_2F_6	CF_3	13.62 ± 0.015	PI	(446)	2643

See also – EI: 1062, 1288, 1419, 2572

Ion	Reactant	Other products	Ionization or appearance potential (eV)	Method	Heat of formation (kJ mol^{-1})	Ref.
CF_3^+	C_3F_6		15.0 ± 0.1	EI		1067

See also – EI: 1290

Ion	Reactant	Other products	Ionization or appearance potential (eV)	Method	Heat of formation (kJ mol^{-1})	Ref.
CF_3^+	C_3F_6 (Hexafluorocyclopropane)		15.38	EI		1290
CF_3^+	C_6F_6 (Hexafluorobenzene)		17.1?	EI		1132
CF_3^+	C_3F_8		13.22	PI		2643
CF_3^+	C_3F_8		13.4 ± 0.1	RPD		2790

See also – EI: 1062, 1419, 2572

Ion	Reactant	Other products	Ionization or appearance potential (eV)	Method	Heat of formation (kJ mol^{-1})	Ref.
CF_3^+	C_4F_8 (Octafluorocyclobutane)		15.7	EI		1062
CF_3^+	$n-C_4F_{10}$		13.22	PI		2643

See also – EI: 1419

Ion	Reactant	Other products	Ionization or appearance potential (eV)	Method	Heat of formation (kJ mol^{-1})	Ref.
CF_3^+	$C_6F_{11}CF_3$ (Perfluoromethylcyclohexane)		14.4 ± 0.2	EI		2192

4.3. The Positive Ion Table—Continued

Ion	Reactant	Other products	Ionization or appearance potential (eV)	Method	Heat of formation (kJ mol^{-1})	Ref.
CF_3^+	CHF_3	H	14.14	PI	(467)	2643
CF_3^+	CHF_3	H	14.03±0.06	RPD		1139

See also – EI: 43, 441, 1288

Ion	Reactant	Other products	Ionization or appearance potential (eV)	Method	Heat of formation (kJ mol^{-1})	Ref.
CF_3^+	CH_3CF_3	CH_3	13.90±0.03	EI		1075
CF_3^+	$CH_2=CHCF_3$		15.0±0.2	EI		1075
CF_3^+	$C_2H_5CF_3$		14.8±0.1	EI		1075
CF_3^+	CF_3OF	OF?	14.8±0.2	EI		3236
CF_3^+	$(CF_3)_2CO$		14.26±0.10	EI		2864
CF_3^+	$(CF_3)_2S$		12.54±0.06	EI		3202
CF_3^+	CF_3SSCF_3		12.07±0.13	EI		3202
CF_3^+	CH_3SCF_3		13.07±0.09	EI		3202
CF_3^+	CH_3SSCF_3		12.50±0.08	EI		3202
CF_3^+	CF_3NSF_2		14.23±0.05	RPD		2443
CF_3^+	CF_3Cl	Cl	12.63	PI	(409)	2643

A discrepancy exists in ref. 2643 between the threshold wavelength and its equivalent energy; the wavelength has been assumed correct.

See also – EI: 24, 441

Ion	Reactant	Other products	Ionization or appearance potential (eV)	Method	Heat of formation (kJ mol^{-1})	Ref.
CF_3^+	$CF_2ClCF=CF_2$		15.63	EI		1290
CF_3^+	CF_3Br	Br	11.71	PI	(381)	2643

A discrepancy exists in ref. 2643 between the threshold wavelength and its equivalent energy; the wavelength has been assumed correct.

See also – EI: 24, 439, 441, 1131

Ion	Reactant	Other products	Ionization or appearance potential (eV)	Method	Heat of formation (kJ mol^{-1})	Ref.
CF_3^+	CF_3I	I	10.89±0.01	PI	(360)	2643

See also – EI: 24, 439, 1111

CF_3^{+2}

Ion	Reactant	Other products	Ionization or appearance potential (eV)	Method	Heat of formation (kJ mol^{-1})	Ref.
CF_3^{+2}	CF_4	F	42.7±0.3	EI		2157

$C_2F_3^+$ $\Delta H_{f0}^\circ \sim 796$ kJ mol^{-1} (190 kcal mol^{-1})

Ion	Reactant	Other products	Ionization or appearance potential (eV)	Method	Heat of formation (kJ mol^{-1})	Ref.
$C_2F_3^+$	C_2F_4	F	15.84±0.02	PI	796	2746

(Threshold value approximately corrected to 0 K)

See also – EI: 419, 2586

Ion	Reactant	Other products	Ionization or appearance potential (eV)	Method	Heat of formation (kJ mol^{-1})	Ref.
$C_2F_3^+$	C_3F_6	CF_3	16.1±0.2	EI		1067

See also – EI: 1290

Ion	Reactant	Other products	Ionization or appearance potential (eV)	Method	Heat of formation (kJ mol^{-1})	Ref.
$C_2F_3^+$	C_3F_6 (Hexafluorocyclopropane)		18.3	EI		1290

4.3. The Positive Ion Table—Continued

Ion	Reactant	Other products	Ionization or appearance potential (eV)	Method	Heat of forma-tion (kJ mol^{-1})	Ref.
$C_2F_3^+$	$n-C_4F_{10}$		15.65	PI		2643
$C_2F_3^+$	$C_6F_{11}CF_3$ (Perfluoromethylcyclohexane)		14.1±0.2	EI		2192

$C_3F_3^+$

$C_3F_3^+$	C_6F_6 (Hexafluorobenzene)		16.8±0.3	EI		1132
$C_3F_3^+$	$C_6F_{11}CF_3$ (Perfluoromethylcyclohexane)		16.6±0.2	EI		2192

$C_5F_3^+$

$C_5F_3^+$	C_6F_6 (Hexafluorobenzene)		15.8±0.2	EI		1132

CF_4^+

The stable region of the CF_4^+ ion ground state surface is not accessible by a vertical transition from the CF_4 molecule; no CF_4^+ ions have been experimentally observed. One upper bound on the adiabatic ionization potential is the $CF_4 \rightarrow CF_3^+ + F$ photoionization threshold, ≤15.35 eV. The photoelectron spectra generally give a higher value. However, for a stable CF_4^+ ion the heat of formation must be less than the heat of formation of $CF_3^+ + F$. This gives an upper bound of 14.7 eV for the ionization potential. Photoelectron spectra are given in refs. 2850, 3059, 3063, 3092, 3117, 3119, 3362.

$C_2F_4^+$ $\Delta H_{f0}^\circ = 321$ kJ mol^{-1} (77 kcal mol^{-1})

Ion	Reactant	Other products	IP/AP (eV)	Method	Heat of formation	Ref.
$C_2F_4^+$	C_2F_4		10.12	PI	321	168
$C_2F_4^+$	C_2F_4		10.12±0.01	PI	321	2746
$C_2F_4^+$	C_2F_4		10.11	PE	320	2885

See also − EI: 214, 419, 2586, 2601, 2795

$C_2F_4^+$	C_2F_6		20.7	EI		2572
$C_2F_4^+$	C_3F_6		12.5±0.1	EI		1067

See also − EI: 1288, 1290

$C_2F_4^+$	C_3F_6 (Hexafluorocyclopropane)		11.85	EI		1290
$C_2F_4^+$	C_3F_8		13.5±0.1	RPD		2790

See also − EI: 2572

$C_2F_4^+$	C_4F_8 (Octafluorocyclobutane)		12.25	EI		1062

See also − EI: 1290

4.3. The Positive Ion Table—Continued

Ion	Reactant	Other products	Ionization or appearance potential (eV)	Method	Heat of formation (kJ mol^{-1})	Ref.
$C_2F_4^+$	$C_6F_{11}CF_3$ (Perfluoromethylcyclohexane)		12.4±0.2	EI		2192
$C_3F_4^+$						
$C_3F_4^+$	$C_6F_{11}CF_3$ (Perfluoromethylcyclohexane)		11.9±0.2	EI		2192
$C_5F_4^+$						
$C_5F_4^+$	C_6F_6 (Hexafluorobenzene)		16.1±0.3	EI		1132
$C_2F_5^+$						
$C_2F_5^+$	C_2F_5		9.98±0.1	EI		2164
$C_2F_5^+$	C_2F_6	F	15.46±0.02	PI	(69*)	2643

See also – EI: 1062, 1419, 2572

Ion	Reactant	Other products	Ionization or appearance potential (eV)	Method	Heat of formation (kJ mol^{-1})	Ref.
$C_2F_5^+$	C_3F_8	CF_3	13.32	PI	(7*)	2643
$C_2F_5^+$	C_3F_8	CF_3	13.9±0.1	RPD		2790

See also – EI: 1062, 1419, 2572

Ion	Reactant	Other products	Ionization or appearance potential (eV)	Method	Heat of formation (kJ mol^{-1})	Ref.
$C_2F_5^+$	$n-C_4F_{10}$		13.05	PI		2643

See also – EI: 1419

Ion	Reactant	Other products	Ionization or appearance potential (eV)	Method	Heat of formation (kJ mol^{-1})	Ref.
$C_2F_5^+$	$C_6F_{11}CF_3$ (Perfluoromethylcyclohexane)		14.4±0.2	EI		2192

*$\Delta H^\circ_{f\,298}$

			$C_3F_5^+$			
$C_3F_5^+$	C_3F_6	F	14.8±0.3	EI		1067

See also – EI: 1290

Ion	Reactant	Other products	Ionization or appearance potential (eV)	Method	Heat of formation (kJ mol^{-1})	Ref.
$C_3F_5^+$	C_3F_6 (Hexafluorocyclopropane)	F	14.14	EI		1290
$C_3F_5^+$	C_4F_8 (Octafluorocyclobutane)		12.25	EI		1062

See also – EI: 1290

Ion	Reactant	Other products	Ionization or appearance potential (eV)	Method	Heat of formation (kJ mol^{-1})	Ref.
$C_3F_5^+$	$n-C_4F_{10}$		15.65	PI		2643

4.3. The Positive Ion Table—Continued

Ion	Reactant	Other products	Ionization or appearance potential (eV)	Method	Heat of formation (kJ mol^{-1})	Ref.
$C_3F_5^+$	$C_6F_{11}CF_3$ (Perfluoromethylcyclohexane)		13.9±0.2	EI		2192
$C_3F_5^+$	$CF_2ClCF=CF_2$	Cl	11.22	EI		1290

$C_4F_5^+$

$C_4F_5^+$	$C_6F_{11}CF_3$ (Perfluoromethylcyclohexane)		14.9±0.2	EI		2192

$C_5F_5^+$

$C_5F_5^+$	C_6F_6 (Hexafluorobenzene)	CF	17.2±0.2	EI		1132

$C_6F_5^+$

$C_6F_5^+$	C_6F_6 (Hexafluorobenzene)	F	16.9±0.1	EI		301

See also – EI: 1132

$C_6F_5^+$	$C_6F_5COCF_3$ (Perfluoroacetophenone)		16.0	EI		308
$C_6F_5^+$	$C_6F_5COCH_3$ (2,3,4,5,6–Pentafluoroacetophenone)		16.7	EI		308
$C_6F_5^+$	$C_6F_5CONH_2$ (2,3,4,5,6–Pentafluorobenzoic acid amide)		16.3±0.3	EI		1168
$C_6F_5^+$	C_6F_5Cl (Chloropentafluorobenzene)	Cl	15.9±0.1	EI		301
$C_6F_5^+$	C_6F_5Br (Bromopentafluorobenzene)	Br	15.2±0.1	EI		301

$C_3F_6^+$

$C_3F_6^+$	C_3F_6		10.3±0.2	EI		1067
$C_3F_6^+$	C_3F_6		11.11	EI		1290

See also – EI: 1123, 2795

$C_3F_6^+$	C_3F_6 (Hexafluorocyclopropane)		11.3	EI		1123, 1290

$C_4F_6^+$

$C_4F_6^+$	$CF_2=CFCF=CF_2$		~9.5	PE		3120
$C_4F_6^+$	$C_6F_{11}CF_3$ (Perfluoromethylcyclohexane)		13.4±0.2	EI		2192

4.3. The Positive Ion Table—Continued

Ion	Reactant	Other products	Ionization or appearance potential (eV)	Method	Heat of formation (kJ mol^{-1})	Ref.
	$C_6F_6^+$ (Hexafluorobenzene)		$\Delta H_{f298}^\circ \sim 1$ kJ mol^{-1} (0 kcal mol^{-1})			
$C_6F_6^+$	C_6F_6 (Hexafluorobenzene)		9.97	PI	5	168
$C_6F_6^+$	C_6F_6 (Hexafluorobenzene)		9.88±0.05	PE	−3	2838
See also − EI: 301, 1127, 1132						
		$C_3F_7^+$				
$C_3F_7^+$	$n-C_3F_7$		10.06±0.1	EI		2164
$C_3F_7^+$	$iso-C_3F_7$		10.5±0.1	EI		2164
$C_3F_7^+$	C_3F_8	F	15.44±0.02	PI	(−338*)	2643
$C_3F_7^+$	C_3F_8	F	15.7±0.1	RPD		2790
See also − EI: 1062, 1419, 2572						
$C_3F_7^+$	$n-C_4F_{10}$	CF_3	13.30	PI	(−405*)	2643
See also − EI: 1419						
*ΔH_{f298}°						
		$C_4F_7^+$				
$C_4F_7^+$	$C_6F_{11}CF_3$ (Perfluoromethylcyclohexane)		15.9±0.2	EI		2192
		$C_5F_7^+$				
$C_5F_7^+$	$C_6F_{11}CF_3$ (Perfluoromethylcyclohexane)		11.9±0.2	EI		2192
		$C_{11}F_7^+$				
$C_{11}F_7^+$	$(C_6F_5)_2$ (Perfluorobiphenyl)		17.2±0.1	EI		1127
	$C_3F_8^+$ $\Delta H_{f298}^\circ \sim -457$ kJ mol^{-1} (−109 kcal mol^{-1})					
$C_3F_8^+$	C_3F_8		13.38	PE	−457	2843
		$C_4F_8^+$				
$C_4F_8^+$	$2-C_4F_8$		11.25	PE		2843
$C_4F_8^+$	$C_6F_{11}CF_3$ (Perfluoromethylcyclohexane)		11.9±0.2	EI		2192

4.3. The Positive Ion Table—Continued

Ion	Reactant	Other products	Ionization or appearance potential (eV)	Method	Heat of formation (kJ mol^{-1})	Ref.
		C$_7$F$_8^+$				
C$_7$F$_8^+$	C$_6$F$_5$CF$_3$ (Perfluorotoluene)		10.4±0.1	EI		301
		C$_{12}$F$_8^+$				
C$_{12}$F$_8^+$	(C$_6$F$_5$)$_2$ (Perfluorobiphenyl)		18.4±0.3	EI		1127
		C$_4$F$_9^+$				
C$_4$F$_9^+$	n–C$_4$F$_{10}$	F	15.42	PI	(−750*)	2643
*ΔH$^\circ_{f298}$						
		C$_5$F$_9^+$				
C$_5$F$_9^+$	C$_6$F$_{11}$CF$_3$ (Perfluoromethylcyclohexane)		13.9±0.2	EI		2192
		C$_6$F$_9^+$				
C$_6$F$_9^+$	C$_6$F$_{11}$CF$_3$ (Perfluoromethylcyclohexane)		12.9±0.2	EI		2192
		C$_{12}$F$_9^+$				
C$_{12}$F$_9^+$	(C$_6$F$_5$)$_2$ (Perfluorobiphenyl)	F	16.7±0.1	EI		1127
		C$_5$F$_{10}^+$				
C$_5$F$_{10}^+$	C$_5$F$_{10}$ (Decafluorocyclopentane)		11.7	EI		299
C$_5$F$_{10}^+$	C$_6$F$_{11}$CF$_3$ (Perfluoromethylcyclohexane)		15.9±0.2	EI		2192
		C$_6$F$_{10}^+$				
C$_6$F$_{10}^+$	C$_6$F$_{11}$CF$_3$ (Perfluoromethylcyclohexane)		12.4±0.2	EI		2192

4.3. The Positive Ion Table—Continued

Ion	Reactant	Other products	Ionization or appearance potential (eV)	Method	Heat of formation (kJ mol⁻¹)	Ref.
			$C_{12}F_{10}^+$			
$C_{12}F_{10}^+$	$(C_6F_5)_2$ (Perfluorobiphenyl)		10.0 ± 0.1	EI		1127
			$C_6F_{11}^+$			
$C_6F_{11}^+$	$C_6F_{11}CF_3$ (Perfluoromethylcyclohexane)		13.9 ± 0.2	EI		2192
			$C_{17}F_{11}^+$			
$C_{17}F_{11}^+$	$C_6F_4(C_6F_5)_2$ (Perfluoro−p−terphenyl)		18.2 ± 0.1	EI		1127
			$C_6F_{12}^+$			
$C_6F_{12}^+$	C_6F_{12} (Dodecafluorocyclohexane)		13.2	EI		299
			$C_{18}F_{12}^+$			
$C_{18}F_{12}^+$	$C_6F_4(C_6F_5)_2$ (Perfluoro−p−terphenyl)		19.9	EI		1127
			$C_7F_{13}^+$			
$C_7F_{13}^+$	$C_6F_{11}CF_3$ (Perfluoromethylcyclohexane)	F	15.4 ± 0.2	EI		2192
			$C_{18}F_{13}^+$			
$C_{18}F_{13}^+$	$C_6F_4(C_6F_5)_2$ (Perfluoro−p−terphenyl)	F	17.5 ± 0.3	EI		1127
			$C_7F_{14}^+$			
$C_7F_{14}^+$	$n-C_5F_{11}CF=CF_2$		$10.48\pm0.02?$	PI		182
			$C_{18}F_{14}^+$			
$C_{18}F_{14}^+$	$C_6F_4(C_6F_5)_2$ (Perfluoro−p−terphenyl)		9.85 ± 0.3	EI		1127

4.3. The Positive Ion Table—Continued

Ion	Reactant	Other products	Ionization or appearance potential (eV)	Method	Heat of formation (kJ mol^{-1})	Ref.
		$C_{24}F_{18}^+$				
$C_{24}F_{18}^+$	$C_6F_5(C_6F_4)_2C_6F_5$ (Perfluoro−p−quaterphenyl)		9.9±0.1	EI		1127
		NF^+ $\Delta H_{f0}^\circ \sim 1420$ kJ mol^{-1} (339 kcal mol^{-1})				
NF^+	NF_2	F^-	11.8±0.2	EI		76
NF^+	NF_2	F	15.5±0.2	EI		76
NF^+	NF_2	F	15.0±0.2	EI		100
NF^+	$cis-N_2F_2$	NF	16.9±0.2	EI		76
NF^+	$trans-N_2F_2$	NF	17.0±0.2	EI		76
NF^+	NF_3	$2F$	17.54±0.02	PI	1420	3038
NF^+	NF_3	$2F$	17.9±0.3	EI		401
		N_2F^+				
N_2F^+	$cis-N_2F_2$	F	14.0±0.2	EI		76
N_2F^+	$trans-N_2F_2$	F	13.9±0.2	EI		76
		$NF_2^+(^1A_1)$ $\Delta H_{f0}^\circ = 1167$ kJ mol^{-1} (279 kcal mol^{-1})				
$NF_2^+(^1A_1)$	NF_2		11.62±0.02	PE	1167	3363
$NF_2^+(^1A_1)$	NF_2		11.79±0.12	EDD		3173
$NF_2^+(^3B_1)$	NF_2		14.05±0.02	PE		3363

Additional higher ionization potentials are given in ref. 3363.

See also − EI: 76, 100

$NF_2^+(^1A_1)$	NF_3	F	14.12±0.01	PI	1167	3038

See also − EI: 76, 100, 401

$NF_2^+(^1A_1)$	N_2F_4	NF_2	12.63±0.1	EDD	(1177)	3173
$NF_2^+(^1A_1)$	N_2F_4	NF_2	12.7±0.2	EI		74
NF_2^+	NF_2NO	NO	12.12±0.12	EDD		3173
NF_2^+	SF_5NF_2		16.3±0.2	EI		1144
NF_2^+	FSO_2NF_2		14.6±0.3	EI		1144
NF_2^+	FSO_2ONF_2		13.3±0.1	EI		1144
		$N_2F_2^+$				
$N_2F_2^+$	$trans-N_2F_2$		13.1±0.1	EI		76
		NF_3^+ $\Delta H_{f0}^\circ = 1135$ kJ mol^{-1} (271 kcal mol^{-1})				
NF_3^+	NF_3		13.00±0.02	PI	1135	3038
NF_3^+	NF_3		13.73±0.03 (V)	PE		3083

4.3. The Positive Ion Table—Continued

Ion	Reactant	Other products	Ionization or appearance potential (eV)	Method	Heat of formation (kJ mol⁻¹)	Ref.
NF_3^+	NF_3		13.73 (V)	PE		3119

See also − S: 2662
         EI: 76, 401

$N_2F_3^+$

Ion	Reactant	Other products	Ionization or appearance potential (eV)	Method	Heat of formation	Ref.
$N_2F_3^+$	N_2F_4	F^-	12.0	EI		76
$N_2F_3^+$	N_2F_4	F	15.6	EI		76

$N_2F_4^+$

Ion	Reactant	Other products	Ionization or appearance potential (eV)	Method	Heat of formation	Ref.
$N_2F_4^+$	N_2F_4		12.84 (V)	PE		3363
$N_2F_4^+$	N_2F_4		12.04±0.10	EI		74

OF^+

Ion	Reactant	Other products	Ionization or appearance potential (eV)	Method	Heat of formation	Ref.
OF^+	OF_2	F	15.8±0.2	RPD		42, 2516
OF^+	OF_2	F	15.8±0.2	EI		2047
OF^+	O_2F_2	OF	17.5±0.2	RPD		42
OF^+	CF_3OF		38±1	EI		3236

O_2F^+

Ion	Reactant	Other products	Ionization or appearance potential (eV)	Method	Heat of formation	Ref.
O_2F^+	O_2F		12.6±0.2	EI		2143
O_2F^+	O_2F_2	F	14.0±0.1	RPD		42
O_2F^+	O_2F_2	F	14.0±0.1	EI		2143

$OF_2^+(^2B_1)$ $\Delta H_{f0}^\circ = 1247$ kJ mol⁻¹ (298 kcal mol⁻¹)

Ion	Reactant	Other products	Ionization or appearance potential (eV)	Method	Heat of formation	Ref.
$OF_2^+(^2B_1)$	OF_2		13.13	PE	1247	3404

See also − EI: 286, 2047, 2516

CHF^+

Ion	Reactant	Other products	Ionization or appearance potential (eV)	Method	Heat of formation	Ref.
CHF^+	CH_2F_2		17.7	EI		1288
CHF^+	C_2HF_3	CF_2?	15.38	EI		419

CH_2F^+ $\Delta H_{f298}^\circ \sim 838$ kJ mol⁻¹ (200 kcal mol⁻¹)

Ion	Reactant	Other products	Ionization or appearance potential (eV)	Method	Heat of formation	Ref.
CH_2F^+	CH_2F		9.35	EI		141
CH_2F^+	CH_3F	H	13.37	PI	838	2637
CH_2F^+	CH_3F	H	13.25±0.06	RPD		1139

See also − EI: 160, 3017

4.3. The Positive Ion Table—Continued

Ion	Reactant	Other products	Ionization or appearance potential (eV)	Method	Heat of formation (kJ mol⁻¹)	Ref.
CH_2F^+	CH_2F_2	F	15.28	EI		1288
CH_2F^+	$CH_2=CF_2$	CF	15.08	EI		419
CH_2F^+	CH_3CF_3		15.6±0.2	EI		1075
CH_2F^+	$C_2H_5CF_3$		15.7±0.3	EI		1075

$$CH_3F^+(^2E) \qquad \Delta H^\circ_{f298} \sim 975 \text{ kJ mol}^{-1} \text{ (233 kcal mol}^{-1}\text{)}$$

Ion	Reactant	Other products	Ionization or appearance potential (eV)	Method	Heat of formation (kJ mol⁻¹)	Ref.
$CH_3F^+(^2E)$	CH_3F		12.50	PI	972	2637
$CH_3F^+(^2E)$	CH_3F		12.54	PE	976	3092
$CH_3F^+(^2E)$	CH_3F		12.54	PE	976	3116

See also − S: 3346
         PE: 3119
         EI: 289, 2154

$$CD_3F^+$$

Ion	Reactant	Other products	Ionization or appearance potential (eV)	Method	Heat of formation (kJ mol⁻¹)	Ref.
$CD_3F^+(^2E)$	CD_3F		12.67	PE		3092

$$C_2HF^+$$

Ion	Reactant	Other products	Ionization or appearance potential (eV)	Method	Heat of formation (kJ mol⁻¹)	Ref.
$C_2HF^+(^2\Pi)$	$CH\equiv CF$		11.26	PE		3071
$C_2HF^+(^2\Pi)$	$CH\equiv CF$		11.5±0.1	EDD		3177
C_2HF^+	C_2H_3F	H_2	14.04	EI		419
C_2HF^+	$CH_2=CF_2$	HF	14.44	EI		419
C_2HF^+	C_2HF_3		20.0	EI		419
C_2HF^+	$CH_2=CFCN$		13.6±0.1	EI		2954

$$C_2HF^{+2}$$

Ion	Reactant	Other products	Ionization or appearance potential (eV)	Method	Heat of formation (kJ mol⁻¹)	Ref.
C_2HF^{+2}	$CH\equiv CF$		31.5±0.1	EDD		3177

$$C_2H_2F^+$$

Ion	Reactant	Other products	Ionization or appearance potential (eV)	Method	Heat of formation (kJ mol⁻¹)	Ref.
$C_2H_2F^+$	C_2H_3F	H	14.02	EI		419
$C_2H_2F^+$	$CH_2=CF_2$	F	14.80	EI		419
$C_2H_2F^+$	CH_3CF_3		15.8±0.2	EI		1075

$$C_2H_3F^+$$

Ion	Reactant	Other products	Ionization or appearance potential (eV)	Method	Heat of formation (kJ mol⁻¹)	Ref.
$C_2H_3F^+$	C_2H_3F		10.37±0.02	PI		268
$C_2H_3F^+$	C_2H_3F		10.37	PI		168
$C_2H_3F^+$	C_2H_3F		10.37	PE		2885

See also − EI: 268, 419

Ion	Reactant	Other products	Ionization or appearance potential (eV)	Method	Heat of formation (kJ mol⁻¹)	Ref.
$C_2H_3F^+$	CH_3CHF_2		14.8	EI		1288
$C_2H_3F^+$	$CH_2=CHCF_3$		13.85±0.02	EI		1075

4.3. The Positive Ion Table—Continued

Ion	Reactant	Other products	Ionization or appearance potential (eV)	Method	Heat of formation (kJ mol^{-1})	Ref.
			$C_2H_4F^+$			
$C_2H_4F^+$	CH_3CHF_2	F	14.9	EI		1288
			$C_3H_2F^+$			
$C_3H_2F^+$	C_6H_5F (Fluorobenzene)		15.77±0.1	EI		2103
			$C_3H_4F^+$			
$C_3H_4F^+$	$C_2H_5CF_3$		15.8±0.1	EI		1075
			$C_4H_3F^+$			
$C_4H_3F^+$	C_6H_5F (Fluorobenzene)	C_2H_2	14.73	EI		3238
See also – EI: 2103						
			$C_5H_4F^+$			
$C_5H_4F^+$	C_5H_4F (Fluorocyclopentadienyl radical)		8.82	EI		126
			$C_6H_4F^+$			
$C_6H_4F^+$	C_6H_5F (Fluorobenzene)	H	14.1	EI		3230
See also – EI: 2103						
$C_6H_4F^+$	$C_6H_4F_2$ (1,4–Difluorobenzene)	F	15.5±0.1	EI		301
$C_6H_4F^+$	C_6H_4ClF (1–Chloro–2–fluorobenzene)	Cl	13.94	EI		1185
$C_6H_4F^+$	C_6H_4ClF (1–Chloro–3–fluorobenzene)	Cl	13.35±0.1	EI		2972
See also – EI: 1185						
$C_6H_4F^+$	C_6H_4ClF (1–Chloro–4–fluorobenzene)	Cl	13.26±0.1	EI		2972
See also – EI: 1185						

4.3. The Positive Ion Table—Continued

Ion	Reactant	Other products	Ionization or appearance potential (eV)	Method	Heat of formation (kJ mol^{-1})	Ref.
	$C_6H_5F^+$ (Fluorobenzene)		$\Delta H^\circ_{f298} = 771$ kJ mol^{-1} (184 kcal mol^{-1})			
$C_6H_5F^+$	C_6H_5F (Fluorobenzene)		9.200±0.005	S	771	344
$C_6H_5F^+$	C_6H_5F (Fluorobenzene)		9.20	S		3371
$C_6H_5F^+$	C_6H_5F (Fluorobenzene)		9.182	PI		2682
$C_6H_5F^+$	C_6H_5F (Fluorobenzene)		9.195±0.01	PI		182
$C_6H_5F^+$	C_6H_5F (Fluorobenzene)		9.20	PI		168
$C_6H_5F^+$	C_6H_5F (Fluorobenzene)		9.21±0.04	PE		2838
$C_6H_5F^+*$	C_6H_5F (Fluorobenzene)		9.86	S		3371
$C_6H_5F^+*$	C_6H_5F (Fluorobenzene)		9.87±0.07	PE		2838

Additional higher ionization potentials are given in ref. 2838.

See also – PI: 416
          PE: 2806, 2822, 3331
          EI: 301, 3223, 3230, 3238

			$C_7H_6F^+$			
$C_7H_6F^+$	$C_6H_4FCH_2$ (3–Fluorobenzyl radical)		8.18±0.06	EI		69
$C_7H_6F^+$	$C_6H_4FCH_2$ (4–Fluorobenzyl radical)		7.78±0.1	EI		69
$C_7H_6F^+$	$C_6H_5CH_2F$ (α–Fluorotoluene)	H	12.2	EI		3230
$C_7H_6F^+$	$C_6H_4FCH_3$ (2–Fluorotoluene)	H	12.3	EI		3230
$C_7H_6F^+$	$C_6H_4FCH_3$ (3–Fluorotoluene)	H	11.92±0.1	EI		2970

See also – EI: 3230

$C_7H_6F^+$	$C_6H_4FCH_3$ (4–Fluorotoluene)	H	11.89±0.1	EI		2970

See also – EI: 3230

$C_7H_6F^+$	$C_6H_5CH_2CH_2C_6H_4F$ (1–(4–Fluorophenyl)–2–phenylethane)		10.7±0.2	EI		3288
$C_7H_6F^+$	$C_6H_4F(CH_2)_3COOCH_3$ (4–(4–Fluorophenyl)butanoic acid methyl ester)		13.42±0.2	EI		2497
$C_7H_6F^+$	$C_6H_4FCH_2Br$ (α–Bromo–3–fluorotoluene)	Br	9.85±0.1	EI		2970

4.3. The Positive Ion Table—Continued

Ion	Reactant	Other products	Ionization or appearance potential (eV)	Method	Heat of formation (kJ mol⁻¹)	Ref.
$C_7H_6F^+$	$C_6H_4FCH_2Br$ (α–Bromo–4–fluorotoluene)	Br	9.93±0.1	EI		2970

$C_7H_7F^+$ (4-Fluorotoluene) $\Delta H^\circ_{f298} = 700$ kJ mol^{-1} (167 kcal mol^{-1})

Ion	Reactant	Other products	Ionization or appearance potential (eV)	Method	Heat of formation (kJ mol⁻¹)	Ref.
$C_7H_7F^+$	$C_6H_5CH_2F$ (α–Fluorotoluene)		9.4	EI		3230
$C_7H_7F^+$	$C_6H_4FCH_3$ (2–Fluorotoluene)		8.915±0.01	PI		182

See also – EI: 3230

Ion	Reactant	Other products	Ionization or appearance potential (eV)	Method	Heat of formation (kJ mol⁻¹)	Ref.
$C_7H_7F^+$	$C_6H_4FCH_3$ (3–Fluorotoluene)		8.915±0.01	PI		182

See also – EI: 2025, 3230

Ion	Reactant	Other products	Ionization or appearance potential (eV)	Method	Heat of formation (kJ mol⁻¹)	Ref.
$C_7H_7F^+$	$C_6H_4FCH_3$ (4–Fluorotoluene)		8.785±0.01	PI	700	182

See also – EI: 3223, 3230

$C_8H_7F^+$

Ion	Reactant	Other products	Ionization or appearance potential (eV)	Method	Heat of formation (kJ mol⁻¹)	Ref.
$C_8H_7F^+$	$C_6H_4F(CH_2)_3COOCH_3$ (4–(4–Fluorophenyl)butanoic acid methyl ester)		10.67±0.2	EI		2497

$C_8H_8F^+$

Ion	Reactant	Other products	Ionization or appearance potential (eV)	Method	Heat of formation (kJ mol⁻¹)	Ref.
$C_8H_8F^+$	$C_6H_4FCH(CH_3)CH_2COOCH_3$ (3–(4–Fluorophenyl)butanoic acid methyl ester)		10.95±0.2	EI		2497

$C_9H_9F^+$

Ion	Reactant	Other products	Ionization or appearance potential (eV)	Method	Heat of formation (kJ mol⁻¹)	Ref.
$C_9H_9F^+$	$C_6H_4FCH(CH_3)CH_2COOCH_3$ (3–(4–Fluorophenyl)butanoic acid methyl ester)		9.70±0.2	EI		2497

$C_{14}H_{13}F^+$

Ion	Reactant	Other products	Ionization or appearance potential (eV)	Method	Heat of formation (kJ mol⁻¹)	Ref.
$C_{14}H_{13}F^+$	$C_6H_5CH_2CH_2C_6H_4F$ (1–(4–Fluorophenyl)–2–phenylethane)		8.8±0.1	EI		3288

CHF_2^+

Ion	Reactant	Other products	Ionization or appearance potential (eV)	Method	Heat of formation (kJ mol⁻¹)	Ref.
CHF_2^+	CHF_2		9.45	EI		141
CHF_2^+	CH_2F_2	H	13.14±0.02	RPD		1139
CHF_2^+	CH_2F_2	H	13.1	EI		1288
CHF_2^+	CH_3CHF_2	CH_3	13.21	EI		1288

4.3. The Positive Ion Table—Continued

Ion	Reactant	Other products	Ionization or appearance potential (eV)	Method	Heat of formation (kJ mol^{-1})	Ref.
CHF_2^+	CHF_3	F	15.75	EI		1288
CHF_2^+	CHF_3	F	16.4±0.3	EI		43
CHF_2^+	C_2HF_3	CF	14.22	EI		419
CHF_2^+	$CH_2=CHCF_3$		14.9±0.1	EI		1075
CHF_2^+	$C_2H_5CF_3$		15.9±0.1	EI		1075
CHF_2^+	CHF_2Cl	Cl	12.59±0.15	EI		43

$$CH_2F_2^+(^2B_2) \qquad \Delta H_{f0}^\circ = 787 \text{ kJ mol}^{-1} \text{ (188 kcal mol}^{-1}\text{)}$$

$CH_2F_2^+(^2B_2)$	CH_2F_2		12.70	PE	786	3116
$CH_2F_2^+(^2B_2)$	CH_2F_2		12.72	PE	788	3092

See also − S: 3346
          PE: 3119
          EI: 1288

$$CD_2F_2^+$$

$CD_2F_2^+(^2B_2)$	CD_2F_2		12.79	PE		3092

$$C_2HF_2^+$$

$C_2HF_2^+$	$CH_2=CF_2$	H	16.67	EI		419
$C_2HF_2^+$	C_2HF_3	F	16.13	EI		419

$$CH_2=CF_2^+ \qquad \Delta H_{f0}^\circ = 673 \text{ kJ mol}^{-1} \text{ (161 kcal mol}^{-1}\text{)}$$

$C_2H_2F_2^+$	$CH_2=CF_2$		10.30	PI	672	168
$C_2H_2F_2^+$	$CH_2=CF_2$		10.31±0.02	PI	673	268
$C_2H_2F_2^+$	$CH_2=CF_2$		10.31	PE	673	2885

See also − EI: 268, 419

$C_2H_2F_2^+$	CH_3CHF_2	H_2	16.5	EI		1288
$C_2H_2F_2^+$	CH_3CF_3	HF	11.2±0.1	EI		1075

$$C_2H_3F_2^+$$

$C_2H_3F_2^+$	CH_3CHF_2	H	12.33	EI		1288
$C_2H_3F_2^+$	CH_3CF_3	F	14.9±0.2	EI		1075

$$C_2H_4F_2^+$$

$C_2H_4F_2^+$	CH_3CHF_2		12.68	EI		1288

4.3. The Positive Ion Table—Continued

Ion	Reactant	Other products	Ionization or appearance potential (eV)	Method	Heat of formation (kJ mol^{-1})	Ref.
			C$_3$HF$_2^+$			
C$_3$HF$_2^+$	CH$_2$=CHCF$_3$		14.8±0.2	EI		1075
			C$_3$H$_2$F$_2^+$			
C$_3$H$_2$F$_2^+$	CH$_2$=CHCF$_3$		13.8±0.1	EI		1075
			C$_3$H$_3$F$_2^+$			
C$_3$H$_3$F$_2^+$	CH$_2$=CHCF$_3$		13.3±0.15	EI		1075
C$_3$H$_3$F$_2^+$	C$_2$H$_5$CF$_3$		13.6±0.1	EI		1075
			C$_3$H$_4$F$_2^+$			
C$_3$H$_4$F$_2^+$	C$_2$H$_5$CF$_3$		12.53±0.04	EI		1075
			C$_3$H$_5$F$_2^+$			
C$_3$H$_5$F$_2^+$	C$_2$H$_5$CF$_3$	F	14.9±0.2	EI		1075
			C$_4$H$_2$F$_2^+$			
C$_4$H$_2$F$_2^+$	C$_6$H$_4$F$_2$ (1,2-Difluorobenzene)	C$_2$H$_2$	15.27	EI		1185
C$_4$H$_2$F$_2^+$	C$_6$H$_4$F$_2$ (1,3-Difluorobenzene)	C$_2$H$_2$	15.30	EI		1185
C$_4$H$_2$F$_2^+$	C$_6$H$_4$F$_2$ (1,4-Difluorobenzene)	C$_2$H$_2$	15.27	EI		1185
			C$_6$H$_3$F$_2^+$			
C$_6$H$_3$F$_2^+$	C$_6$H$_3$F$_3$ (1,2,4-Trifluorobenzene)	F	15.2±0.1	EI		301

C$_6$H$_4$F$_2^+$(**1,2-Difluorobenzene**) $\Delta H^\circ_{f298} = 604$ kJ mol^{-1} (144 kcal mol^{-1})
C$_6$H$_4$F$_2^+$(**1,4-Difluorobenzene**) $\Delta H^\circ_{f298} = 576$ kJ mol^{-1} (138 kcal mol^{-1})

Ion	Reactant	Other products	Ionization or appearance potential (eV)	Method	Heat of formation (kJ mol^{-1})	Ref.
C$_6$H$_4$F$_2^+$	C$_6$H$_4$F$_2$ (1,2-Difluorobenzene)		9.31	PI	604	168
C$_6$H$_4$F$_2^+$	C$_6$H$_4$F$_2$ (1,2-Difluorobenzene)		9.74±0.02	EI		1185
C$_6$H$_4$F$_2^+$	C$_6$H$_4$F$_2$ (1,3-Difluorobenzene)		9.78±0.02	EI		1185
C$_6$H$_4$F$_2^+$	C$_6$H$_4$F$_2$ (1,4-Difluorobenzene)		9.15	PI	576	168

4.3. The Positive Ion Table—Continued

Ion	Reactant	Other products	Ionization or appearance potential (eV)	Method	Heat of formation (kJ mol^{-1})	Ref.
$C_6H_4F_2^+$	$C_6H_4F_2$ (1,4–Difluorobenzene)		9.15±0.06	PE	576	2838

See also – PE: 2806
           EI: 301, 1185

	$CHF_3^+(^2A_1)$	**$\Delta H_{i0}^\circ = 656$ kJ mol^{-1} (157 kcal mol^{-1})**				
$CHF_3^+(^2A_1)$	CHF_3		~13.84	S		3346
$CHF_3^+(^2A_1)$	CHF_3		13.86	PE	656	3116
$CHF_3^+(^2A_1)$	CHF_3		≥13.8	PE		3092

See also – PE: 3119

	$C_2HF_3^+$	**$\Delta H_{i298}^\circ = 483$ kJ mol^{-1} (115 kcal mol^{-1})**				
$C_2HF_3^+$	C_2HF_3		10.14	PI	483	168

See also – EI: 419

		$C_2H_2F_3^+$				
$C_2H_2F_3^+$	CF_3CH_2		10.6±0.1	EI		2164

		$C_3H_2F_3^+$				
$C_3H_2F_3^+$	$CH_2=CHCF_3$	H	12.69±0.05	EI		1075

		$C_3H_3F_3^+$				
$C_3H_3F_3^+$	$CH_2=CHCF_3$		10.9?	PI		168
$C_3H_3F_3^+$	$CH_2=CHCF_3$		11.24±0.04	EI		1075

		$C_6H_2F_3^+$				
$C_6H_2F_3^+$	$C_6H_2F_4$ (1,2,4,5–Tetrafluorobenzene)	F	15.9±0.1	EI		301

		$C_6H_3F_3^+$				
$C_6H_3F_3^+$	$C_6H_3F_3$ (1,2,4–Trifluorobenzene)		9.37	PI		168
$C_6H_3F_3^+$	$C_6H_3F_3$ (1,2,4–Trifluorobenzene)		9.30±0.05	PE		2838
$C_6H_3F_3^+$	$C_6H_3F_3$ (1,3,5–Trifluorobenzene)		9.3?	PI		168

4.3. The Positive Ion Table—Continued

Ion	Reactant	Other products	Ionization or appearance potential (eV)	Method	Heat of formation (kJ mol⁻¹)	Ref.

$C_7H_4F_3^+$

Ion	Reactant	Other products	Ionization or appearance potential (eV)	Method	Heat of formation	Ref.
$C_7H_4F_3^+$	$C_6H_4ClCF_3$ (3−Chloro−α,α,α−trifluorotoluene)	Cl	12.99±0.1	EI		2972
$C_7H_4F_3^+$	$C_6H_4ClCF_3$ (4−Chloro−α,α,α−trifluorotoluene)	Cl	12.89±0.1	EI		2972

$C_7H_5F_3^+$ (α,α,α−Trifluorotoluene) $\Delta H^\circ_{f298} = 334$ kJ mol⁻¹ (80 kcal mol⁻¹)

Ion	Reactant	Other products	Ionization or appearance potential (eV)	Method	Heat of formation	Ref.
$C_7H_5F_3^+$	$C_6H_5CF_3$ (α,α,α−Trifluorotoluene)		9.685±0.005	S	334	344
$C_7H_5F_3^+$	$C_6H_5CF_3$ (α,α,α−Trifluorotoluene)		9.68±0.02	PI	334	182

See also – PE: 2806, 3331
          EI: 301

$C_7H_{11}F_3^+$

Ion	Reactant	Other products	Ionization or appearance potential (eV)	Method	Heat of formation	Ref.
$C_7H_{11}F_3^+$	$C_6H_{11}CF_3$ (Trifluoromethylcyclohexane)		10.46±0.02	PI		182

$C_4H_2F_4^+$

Ion	Reactant	Other products	Ionization or appearance potential (eV)	Method	Heat of formation	Ref.
$C_4H_2F_4^+$	trans−CF_2=CHCH=CF_2		8.98	PE		3120

$C_6HF_4^+$

Ion	Reactant	Other products	Ionization or appearance potential (eV)	Method	Heat of formation	Ref.
$C_6HF_4^+$	C_6HF_5 (Pentafluorobenzene)	F	16.5±0.1	EI		301

$C_6H_2F_4^+$

Ion	Reactant	Other products	Ionization or appearance potential (eV)	Method	Heat of formation	Ref.
$C_6H_2F_4^+$	$C_6H_2F_4$ (1,2,3,4−Tetrafluorobenzene)		9.61	PI		168
$C_6H_2F_4^+$	$C_6H_2F_4$ (1,2,3,5−Tetrafluorobenzene)		9.55	PI		168
$C_6H_2F_4^+$	$C_6H_2F_4$ (1,2,4,5−Tetrafluorobenzene)		9.39	PI		168

$C_6HF_5^+$ (Pentafluorobenzene) $\Delta H^\circ_{f298} = 143$ kJ mol⁻¹ (34 kcal mol⁻¹)

Ion	Reactant	Other products	Ionization or appearance potential (eV)	Method	Heat of formation	Ref.
$C_6HF_5^+$	C_6HF_5 (Pentafluorobenzene)		9.84	PI	143	168

See also – EI: 301

4.3. The Positive Ion Table—Continued

Ion	Reactant	Other products	Ionization or appearance potential (eV)	Method	Heat of formation (kJ mol⁻¹)	Ref.
			$C_7H_3F_5^+$			
$C_7H_3F_5^+$	$C_6F_5CH_3$ (2,3,4,5,6−Pentafluorotoluene)		9.6±0.1	EI		301
			$C_8H_4F_6^+$			
$C_8H_4F_6^+$	$C_6H_4(CF_3)_2$ (1,4−Bis(trifluoromethyl)benzene)		10.43±0.05	EDD		2634
See also − PE: 3331						
			CNF^+			
CNF^+	FCN		13.32±0.01	PI	(1309)	2621
			C_2NF^+			
C_2NF^+	$CF_2=CFCN$		15.1±0.1	EI		2954
			$C_2NF_2^+$			
$C_2NF_2^+$	$CF_2=CFCN$		14.9±0.1	EI		2954
			$C_3NF_2^+$			
$C_3NF_2^+$	$CF_2=CFCN$	F	14.3±0.1	EI		2954
$C_3NF_2^+$	$CF_2=CHCN$	H	15.9±0.1	EI		2954
			$C_3NF_3^+$			
$C_3NF_3^+$	$CF_2=CFCN$		10.6±0.1	EI		2954
			$C_6NF_{15}^+$			
$C_6NF_{15}^+$	$(C_2F_5)_3N$		11.7?	PI		182
			BOF^+			
BOF^+	BOF		13.4±0.5	EI		2040

4.3. The Positive Ion Table—Continued

Ion	Reactant	Other products	Ionization or appearance potential (eV)	Method	Heat of formation (kJ mol^{-1})	Ref.
$B_2O_2F_2^+$						
$B_2O_2F_2^+$	$B_3O_3F_3$		16.60±0.2	EI		3200
See also – EI:	2040					
$B_3O_3F_2^+$						
$B_3O_3F_2^+$	$B_3O_3F_3$	F	16.15±0.2	EI		3200
See also – EI:	2040					
$B_2OF_3^+$						
$B_2OF_3^+$	$B_3O_3F_3$	BO_2	15.35±0.2	EI		3200
See also – EI:	2040					
$B_3O_3F_3^+$						
$B_3O_3F_3^+$	$B_3O_3F_3$		13.91±0.1	EI		3200
See also – EI:	2040					
$B_2OF_4^+$						
$B_2OF_4^+$	B_2OF_4		14.12±0.1	EI		3200
COF^+						
COF^+	CF_2O	F	16.0±0.1	EI		3236
COF^+	CF_3OF		20.0±0.3	EI		3236
COF_2^+						
COF_2^+	CF_2O		13.17±0.1	EI		2893
COF_2^+	CF_2O		14.6±0.1	EI		3236
COF_2^+	CF_3OF		18.4±0.2	EI		3236
COF_3^+						
COF_3^+	CF_3OF	F	16.1±0.2	EI		3236
$C_2OF_3^+$						
$C_2OF_3^+$	$(CF_3)_2CO$		12.04±0.12	EI		2864

4.3. The Positive Ion Table—Continued

Ion	Reactant	Other products	Ionization or appearance potential (eV)	Method	Heat of formation (kJ mol^{-1})	Ref.
			$C_7OF_5^+$			
$C_7OF_5^+$	$C_6F_5COCF_3$ (Perfluoroacetophenone)		11.15	EI		308
$C_7OF_5^+$	C_6F_5CHO (2,3,4,5,6–Pentafluorobenzenecarbonal)	H	11.6	EI		308
$C_7OF_5^+$	$C_6F_5COCH_3$ (2,3,4,5,6–Pentafluoroacetophenone)		11.25	EI		308
$C_7OF_5^+$	$C_6F_5CONH_2$ (2,3,4,5,6–Pentafluorobenzoic acid amide)		11.3±0.2	EI		1168
			$C_3OF_6^+$			
$C_3OF_6^+$	$(CF_3)_2CO$		11.68	PE		2843
			$C_8OF_8^+$			
$C_8OF_8^+$	$C_6F_5COCF_3$ (Perfluoroacetophenone)		11.05	EI		308
			NOF_2^+			
NOF_2^+	ONF_3	F	13.59±0.01	PI		3038
			NOF_3^+			
NOF_3^+	ONF_3		13.26±0.01	PI		3038

See also – PE: 3083, 3358

Ion	Reactant	Other products	Ionization or appearance potential (eV)	Method	Heat of formation (kJ mol^{-1})	Ref.
			CH_2BF^+			
CH_2BF^+	CH_3BF_2		13.38±0.02	EI		1076
			CH_3BF^+			
CH_3BF^+	CH_3BF_2	F	15.05±0.1	EI		1076
			$C_2H_2BF^+$			
$C_2H_2BF^+$	$CH_2=CHBF_2$		12.69±0.05	EI		1076
$C_2H_2BF^+$	$C_2H_5BF_2$		12.0±0.2	EI		1076

4.3. The Positive Ion Table—Continued

Ion	Reactant	Other products	Ionization or appearance potential (eV)	Method	Heat of formation (kJ mol^{-1})	Ref.
			$C_2H_3BF^+$			
$C_2H_3BF^+$	$CH_2=CHBF_2$	F	11.9 ± 0.2	EI		1076
$C_2H_3BF^+$	$C_2H_5BF_2$		12.0 ± 0.2	EI		1076
$C_2H_3BF^+$	$iso-C_3H_7BF_2$		11.8 ± 0.1	EI		1076
			$C_2H_4BF^+$			
$C_2H_4BF^+$	$C_2H_5BF_2$		12.6 ± 0.2	EI		1076
			$C_2H_5BF^+$			
$C_2H_5BF^+$	$C_2H_5BF_2$	F	12.1 ± 0.2	EI		1076
			$CH_3BF_2^+$			
$CH_3BF_2^+$	CH_3BF_2		12.54 ± 0.03	EI		1076
			$C_2H_3BF_2^+$			
$C_2H_3BF_2^+$	$CH_2=CHBF_2$		11.06 ± 0.03	EI		1076
$C_2H_3BF_2^+$	$iso-C_3H_7BF_2$	CH_4	12.80 ± 0.05	EI		1076
			$C_2H_4BF_2^+$			
$C_2H_4BF_2^+$	$iso-C_3H_7BF_2$	CH_3	12.58 ± 0.05	EI		1076
			$C_2H_5BF_2^+$			
$C_2H_5BF_2^+$	$C_2H_5BF_2$		11.8 ± 0.05	EI		1076
			$B_3H_3N_3F_3^+$			
$B_3H_3N_3F_3^+$	$B_3H_3N_3F_3$ (B−Trifluoroborazine)		10.46 ± 0.01	PE		3105
			C_3HNF^+			
C_3HNF^+	$CF_2=CHCN$	F	13.1 ± 0.1	EI		2954
			$C_3H_2NF^+$			
$C_3H_2NF^+$	$CH_2=CFCN$		10.7 ± 0.1	EI		2954

4.3. The Positive Ion Table—Continued

Ion	Reactant	Other products	Ionization or appearance potential (eV)	Method	Heat of formation (kJ mol^{-1})	Ref.
		$C_6H_6NF^+$				
$C_6H_6NF^+$	$C_6H_4FNH_2$ (4−Fluoroaniline)		7.87±0.1	CTS		2485
		$C_7H_4NF^+$				
$C_7H_4NF^+$	C_6H_4FCN (3−Fluorobenzoic acid nitrile)		10.03±0.05	RPD		3223
$C_7H_4NF^+$	C_6H_4FCN (4−Fluorobenzoic acid nitrile)		9.99±0.05	RPD		3223
		$C_8H_6NF^+$				
$C_8H_6NF^+$	$C_6H_4FCH_2CN$ (4−Fluorophenylacetic acid nitrile)		9.65±0.04	RPD		3223
		$C_8H_{10}NF^+$				
$C_8H_{10}NF^+$	$C_6H_4FN(CH_3)_2$ (N,N−Dimethyl−4−fluoroaniline)		7.50	CTS		1281
		$C_3HNF_2^+$				
$C_3HNF_2^+$	$CF_2=CHCN$		10.6±0.1	EI		2954
		$C_8H_4NF_3^+$				
$C_8H_4NF_3^+$	$C_6H_4(CF_3)CN$ (4−Trifluoromethylbenzoic acid nitrile)		10.61±0.05	RPD		3223
		$C_8H_5NF_6^+$				
$C_8H_5NF_6^+$	$C_6H_5N(CF_3)_2$ (N,N−Bis(trifluoromethyl)aniline)		10.00 (V)	PE		2806
		$C_2H_5OF^+$				
$C_2H_5OF^+$	CH_2FCH_2OH		11.05	PE		3374
		$C_3H_5OF^+$				
$C_3H_5OF^+$	C_3H_5OF (1,2−Epoxy−3−fluoropropane)		10.74	PE		3374

4.3. The Positive Ion Table—Continued

Ion	Reactant	Other products	Ionization or appearance potential (eV)	Method	Heat of formation (kJ mol^{-1})	Ref.
		C$_6$H$_4$OF$^+$				
C$_6$H$_4$OF$^+$	C$_6$H$_4$FOCH$_3$ (1–Fluoro–4–methoxybenzene)	CH$_3$	11.53±0.1	EI		2970
		C$_6$H$_5$OF$^+$				
C$_6$H$_5$OF$^+$	C$_6$H$_4$FOH (2–Fluorophenol)		8.66±0.01	PI		182
		C$_7$H$_4$OF$^+$				
C$_7$H$_4$OF$^+$	C$_6$H$_4$FCOCH$_3$ (3–Fluoroacetophenone)	CH$_3$	10.37±0.1	EI		2967
C$_7$H$_4$OF$^+$	C$_6$H$_4$FCOCH$_3$ (4–Fluoroacetophenone)	CH$_3$	10.20±0.2	EI		2967

See also – EI: 3238

Ion	Reactant	Other products	Ionization or appearance potential (eV)	Method	Heat of formation (kJ mol^{-1})	Ref.
		C$_7$H$_5$OF$^+$				
C$_7$H$_5$OF$^+$	C$_6$H$_5$COF (Benzoic acid fluoride)		10.6	EI		308
		C$_8$H$_7$OF$^+$				
C$_8$H$_7$OF$^+$	C$_6$H$_4$FCOCH$_3$ (3–Fluoroacetophenone)		9.76±0.1	EI		2967
C$_8$H$_7$OF$^+$	C$_6$H$_4$FCOCH$_3$ (4–Fluoroacetophenone)		9.57±0.2	EI		2967

See also – EI: 3238

Ion	Reactant	Other products	Ionization or appearance potential (eV)	Method	Heat of formation (kJ mol^{-1})	Ref.
		C$_{10}$H$_9$OF$^+$				
C$_{10}$H$_9$OF$^+$	C$_6$H$_4$F(CH$_2$)$_3$COOCH$_3$ (4–(4–Fluorophenyl)butanoic acid methyl ester)	CH$_3$OH	9.41±0.2	EI		2497
		C$_{10}$H$_{10}$OF$^+$				
C$_{10}$H$_{10}$OF$^+$	C$_6$H$_4$FCH(CH$_3$)CH$_2$COOCH$_3$ (3–(4–Fluorophenyl)butanoic acid methyl ester)	CH$_3$O	11.15±0.2	EI		2497
C$_{10}$H$_{10}$OF$^+$	C$_6$H$_4$F(CH$_2$)$_3$COOCH$_3$ (4–(4–Fluorophenyl)butanoic acid methyl ester)	CH$_3$O	11.20±0.2	EI		2497

4.3. The Positive Ion Table—Continued

Ion	Reactant	Other products	Ionization or appearance potential (eV)	Method	Heat of formation (kJ mol⁻¹)	Ref.

$$C_{10}H_{11}OF^+$$

$C_{10}H_{11}OF^+$	$C_6H_4FCOC_3H_7$ (1–(4–Fluorophenyl)–1–butanone)		9.09±0.2	EI		2534

$$C_{11}H_{13}O_2F^+$$

$C_{11}H_{13}O_2F^+$	$C_6H_4FCH(CH_3)CH_2COOCH_3$ (3–(4–Fluorophenyl)butanoic acid methyl ester)		8.45±0.2	EI		2497
$C_{11}H_{13}O_2F^+$	$C_6H_4F(CH_2)_3COOCH_3$ (4–(4–Fluorophenyl)butanoic acid methyl ester)		8.76±0.2	EI		2497

$$C_2H_3OF_3^+$$

$C_2H_3OF_3^+$	CF_3CH_2OH		13.8	D		2908

$$C_3H_3OF_3^+$$

$C_3H_3OF_3^+$	CH_3COCF_3		10.67±0.01	PE		3289

$$C_7H_5OF_3^+$$

$C_7H_5OF_3^+$	$C_6H_5OCF_3$ (Trifluoromethoxybenzene)		10.00 (V)	PE		2806

$$C_8H_5OF_3^+$$

$C_8H_5OF_3^+$	$C_6H_5COCF_3$ (α,α,α–Trifluoroacetophenone)		10.25	EI		308

$$C_4H_2O_2F_3^+$$

$C_4H_2O_2F_3^+$	$CF_3COCH_2COCH_3$	CH_3	11.7±0.1	EI		2959
$C_4H_2O_2F_3^+$	$CF_3COCH_2COCF_3$	CF_3	11.2±0.1	EI		2959

$$C_5H_5O_2F_3^+$$

$C_5H_5O_2F_3^+$	$CF_3COCH_2COCH_3$		9.8±0.1	EI		2959

See also – EI: 2580

$$C_8H_3OF_5^+$$

$C_8H_3OF_5^+$	$C_6F_5COCH_3$ (2,3,4,5,6–Pentafluoroacetophenone)		11.25	EI		308

4.3. The Positive Ion Table—Continued

Ion	Reactant	Other products	Ionization or appearance potential (eV)	Method	Heat of formation (kJ mol⁻¹)	Ref.

$$C_8H_5OF_5^+$$

| $C_8H_5OF_5^+$ | $C_6H_5OC_2F_5$ (Pentafluoroethoxybenzene) | | 9.97 (V) | PE | | 2806 |

$$C_5H_2O_2F_6^+$$

| $C_5H_2O_2F_6^+$ | $CF_3COCH_2COCF_3$ | | 10.55±0.05 | EI | | 2959 |
| See also – EI: 2580 | | | | | | |

$$C_5H_3OF_7^+$$

| $C_5H_3OF_7^+$ | $n-C_3F_7COCH_3$ | | 10.58±0.03 | PI | | 182 |

$$C_2H_6BNF_2^+$$

| $C_2H_6BNF_2^+$ | $(CH_3)_2NBF_2$ | | 9.71 | EI | | 3227 |

$$C_2H_7BNF_3^+$$

| $C_2H_7BNF_3^+$ | $(CH_3)_2NHBF_3$ | | 12.18 (V) | PE | | 3410 |

$$C_3H_9BNF_3^+$$

| $C_3H_9BNF_3^+$ | $(CH_3)_3NBF_3$ | | 12.21 (V) | PE | | 3410 |

$$C_2H_6BO_2F^+$$

| $C_2H_6BO_2F^+$ | $(CH_3O)_2BF$ | | 10.92 | EI | | 3227 |

$$CH_3BOF_2^+$$

| $CH_3BOF_2^+$ | CH_3OBF_2 | | 11.97 | EI | | 3227 |

$$C_7H_2NOF_5^+$$

| $C_7H_2NOF_5^+$ | $C_6F_5CONH_2$ (2,3,4,5,6–Pentafluorobenzoic acid amide) | | 10.0±0.1 | EI | | 1168 |

$$C_{19}H_{10}N_4O_2F_6^+$$

| $C_{19}H_{10}N_4O_2F_6^+$ | $(C_6H_5C_2N_2O)_2C_3F_6$ (1,3–Bis(2–phenyl–1,3,4–oxadiazol–5–yl)perfluoropropane) | | 9.5±0.1 | EI | | 2156 |

4.3. The Positive Ion Table—Continued

Ion	Reactant	Other products	Ionization or appearance potential (eV)	Method	Heat of forma-tion (kJ mol⁻¹)	Ref.

$$\mathbf{C_{11}H_5N_2OF_7^+}$$

Ion	Reactant	Other products	Ionization or appearance potential (eV)	Method	Heat of formation (kJ mol⁻¹)	Ref.
$C_{11}H_5N_2OF_7^+$	$C_6H_5C_2N_2OC_3F_7$ (2–Perfluoropropyl–5–phenyl–1,3,4–oxadiazole)		9.8±0.2	EI		2156

$$\mathbf{C_{15}H_5N_2OF_{15}^+}$$

| $C_{15}H_5N_2OF_{15}^+$ | $C_6H_5C_2N_2OC_7F_{15}$ (2–Perfluoroheptyl–5–phenyl–1,3,4–oxadiazole) | | 9.9±0.1 | EI | | 2156 |

| $Ne^+(^2P_{3/2})$ | | $\Delta H_{f0}^\circ = 2080.6$ kJ mol⁻¹ (497.3 kcal mol⁻¹) | | | | |
| $Ne^+(^2P_{1/2})$ | | $\Delta H_{f0}^\circ = 2090.0$ kJ mol⁻¹ (499.5 kcal mol⁻¹) | | | | |

| $Ne^+(^2P_{3/2})$ | Ne | | 21.564 | S | 2080.6 | 2113 |
| $Ne^+(^2P_{1/2})$ | Ne | | 21.661 | S | 2090.0 | 2113 |

See also – PI: 163
EI: 52, 383, 1012, 1051, 1172, 2032, 3435

| Ne^{+2} | | $\Delta H_{f0}^\circ = 6032.9$ kJ mol⁻¹ (1441.9 kcal mol⁻¹) | | | | |

Ne^{+2}	Ne		62.526	S	6032.9	2113
Ne^{+2}	Ne		62.5	EI		1240
Ne^{+2}	Ne^+		40.962	S		2113
Ne^{+2}	Ne^+		41±2	SEQ		2551

| Ne^{+3} | | $\Delta H_{f0}^\circ = 12155$ kJ mol⁻¹ (2905 kcal mol⁻¹) | | | | |

Ne^{+3}	Ne		125.98	S	12155	2113
Ne^{+3}	Ne		129	EI		1240
Ne^{+3}	Ne^{+2}		63.45	S		2113
Ne^{+3}	Ne^{+2}		63±2	SEQ		2551

| Ne^{+4} | | $\Delta H_{f0}^\circ = 21525$ kJ mol⁻¹ (5145 kcal mol⁻¹) | | | | |

Ne^{+4}	Ne		223.09	S	21525	2113
Ne^{+4}	Ne		246	EI		1240
Ne^{+4}	Ne^{+3}		97.11	S		2113
Ne^{+4}	Ne^{+3}		94±2	SEQ		2551

| Ne^{+5} | | $\Delta H_{f0}^\circ = 33703$ kJ mol⁻¹ (8055 kcal mol⁻¹) | | | | |

Ne^{+5}	Ne		349.30	S	33703	2113
Ne^{+5}	Ne		900	EI		1240
Ne^{+5}	Ne^{+4}		126.21	S		2113
Ne^{+5}	Ne^{+4}		127±2	SEQ		2551

4.3. The Positive Ion Table—Continued

Ion	Reactant	Other products	Ionization or appearance potential (eV)	Method	Heat of formation (kJ mol^{-1})	Ref.

Na^+			$\Delta H^\circ_{f0} = 603.9$ kJ mol^{-1} (144.3 kcal mol^{-1})			
Na^+	Na		5.139	S	603.9	2113

See also – EI: 2487, 3257

Na^+	NaF	F	9.9±0.2	EI		2436
Na^+	NaCl	Cl	9.5±0.2	EI		2406
Na^+	NaI	I	8.138	PI		2610

(Threshold value corrected for hot bands)

| Na^+ | NaI | I | 8.7±0.3 | EI | | 2001 |

Na^{+2}			$\Delta H^\circ_{f0} = 5166.3$ kJ mol^{-1} (1234.8 kcal mol^{-1})			
Na^{+2}	Na		52.425	S	5166.3	2113
Na^{+2}	Na		52±1	NRE		99
Na^{+2}	Na^+		47.286	S		2113

Na^{+3}			$\Delta H^\circ_{f0} = 12079$ kJ mol^{-1} (2887 kcal mol^{-1})			
Na^{+3}	Na		124.07	S	12079	2113
Na^{+3}	Na		125±2	NRE		99
Na^{+3}	Na^{+2}		71.64	S		2113

Na_2^+			$\Delta H^\circ_{f0} = 613$ kJ mol^{-1} (146 kcal mol^{-1})			
Na_2^+	Na_2		~4.87	S		3179, 3183
Na_2^+	Na_2		4.90±0.01	PI	613	1189
Na_2^+	Na_2		4.9±0.1	PI		2585, 2633
Na_2^+	Na_2		4.79±0.1	EI		3416

Na_3^+						
Na_3^+	Na_3		3.9±0.1	PI		2585, 2633

Na_4^+						
Na_4^+	Na_4		4.2±0.1	PI		2585, 2633

Na_5^+						
Na_5^+	Na_5		3.9–4.3	PI		2585

Na_6^+						
Na_6^+	Na_6		4.05–4.35	PI		2585

4.3. The Positive Ion Table—Continued

Ion	Reactant	Other products	Ionization or appearance potential (eV)	Method	Heat of formation (kJ mol^{-1})	Ref.
			Na$_7^+$			
Na$_7^+$	Na$_7$		3.9–4.15	PI		2585
			Na$_8^+$			
Na$_8^+$	Na$_8$		4.0	PI		2585
			NaO$^+$			
NaO$^+$	NaO		6.5±0.7	EI		3186
			Na$_2$O$^+$			
Na$_2$O$^+$	Na$_2$O		5.5±0.5	EI		3186

Mg$^+$ $\Delta H_{f0}^\circ = 884.2$ kJ mol^{-1} (211.3 kcal mol^{-1})

Ion	Reactant	Other products	Ionization or appearance potential (eV)	Method	Heat of formation (kJ mol^{-1})	Ref.
Mg$^+$	Mg		7.646	S	884.2	2113, 3126
Mg$^+$	Mg		7.52±0.2	SI		3021

See also – EI: 1104, 2432, 2620, 2990

Ion	Reactant	Other products	Ionization or appearance potential (eV)	Method	Heat of formation (kJ mol^{-1})	Ref.
Mg$^+$	MgCl$_2$		16.31±0.10	EI		2991
Mg$^+$	MgCl$_2$		17.5±0.5	EI		178
Mg$^+$	MgBr$_2$		15.5±1	EI		178

			MgF$^+$			
MgF$^+$	MgF		~7.68	S		3082
MgF$^+$	MgF		7.5±0.5	EI		2432
MgF$^+$	MgF		7.8±0.3	EI	1104,	2148
MgF$^+$	MgF		8.0±0.5	EI		2620
MgF$^+$	MgF$_2$	F	13.5±0.4	EI		178
MgF$^+$	MgF$_2$	F	13.5±1.0	EI		2432
MgF$^+$	MgF$_2$	F	13.7±0.4	EI		1104

			MgF$_2^+$			
MgF$_2^+$	MgF$_2$		13.3±0.3	EI		2432
MgF$_2^+$	MgF$_2$		13.5±0.4	EI		178
MgF$_2^+$	MgF$_2$		14.0±0.5	EI		2620

			Mg$_2$F$_3^+$			
Mg$_2$F$_3^+$	Mg$_2$F$_4$	F	14.0±0.5	EI		178

4.3. The Positive Ion Table—Continued

Ion	Reactant	Other products	Ionization or appearance potential (eV)	Method	Heat of formation (kJ mol^{-1})	Ref.	
			Al^+ $\quad \Delta H_{f0}^{\circ} = 901.6$ kJ mol^{-1} (215.5 kcal mol^{-1})				
Al^+	Al		5.986	S	901.6	2113,	2199
See also − S:	3270						
EI:	1104, 2128, 2141, 2165, 2518, 2699, 2707, 2715, 2990, 3199						
Al^+	AlF	F^-?	9.2±0.3	EI			2148
Al^+	$(CH_3)_3Al$		14.6±0.2	EI			2556
Al^+	$AlCl_3$		22.1±0.5	EI			2897
Al^+	$AlBr_3$		18.3±0.5	EI			2897
			AlC^+				
AlC^+	AlCN	N	~15	EI			2439
			AlN^+				
AlN^+	AlCN	C	~15	EI			2439
			AlO^+				
AlO^+	AlO		9.5±0.5	EI		2128,	2518
			Al_2O^+				
Al_2O^+	Al_2O		7.7±0.5	EI		2128,	2518
Al_2O^+	Al_2O		7.9±0.3	EI			2699
Al_2O^+	Al_2O		8.5±0.5	EI			3199
			$Al_2O_2^+$				
$Al_2O_2^+$	Al_2O_2		9.9±0.5	EI			2128
			AlF^+				
AlF^+	AlF		8.9±0.6	EI			2142
AlF^+	AlF		9.2	EI			2141
AlF^+	AlF		9.4±0.5	EI			2432
See also − EI:	1104, 2148						
			AlF_2^+				
AlF_2^+	AlF_2		9±1	EI			2148
AlF_2^+	AlF_3	F	15.2±0.3	EI		1104,	2148
AlF_2^+	AlF_3	F	15.5±1.0	EI			2432

4.3. The Positive Ion Table—Continued

Ion	Reactant	Other products	Ionization or appearance potential (eV)	Method	Heat of formation (kJ mol^{-1})	Ref.
			CH$_3$Al$^+$			
CH$_3$Al$^+$	(CH$_3$)$_3$Al		13.9±0.3	EI		2556
			C$_2$H$_6$Al$^+$			
C$_2$H$_6$Al$^+$	(CH$_3$)$_3$Al	CH$_3$	10.1±0.3	EI		2556
			C$_3$H$_9$Al$^+$			
C$_3$H$_9$Al$^+$	(CH$_3$)$_3$Al		9.09±0.26	EI		2556
			CNAl$^+$			
CNAl$^+$	AlCN		7.4±0.3	EI		2439
			BOAl$^+$			
BOAl$^+$	AlBO		8.5±0.5	EI		3199
			BO$_2$Al$^+$			
BO$_2$Al$^+$	AlBO$_2$		9.5±0.5	EI		3199
			C$_{10}$H$_{14}$O$_4$Al$^+$			
C$_{10}$H$_{14}$O$_4$Al$^+$	(CH$_3$COCHCOCH$_3$)$_3$Al (Tris(2,4–pentanedionato)aluminum)		9.1±0.2	EI		2460, 2959
			C$_{22}$H$_{38}$O$_4$Al$^+$			
C$_{22}$H$_{38}$O$_4$Al$^+$	((CH$_3$)$_3$CCOCHCOC(CH$_3$)$_3$)$_3$Al (Tris(2,2,6,6–tetramethyl–3,5–heptanedionato)aluminum)		13.6±0.5	EI		2524
			C$_{15}$H$_{21}$O$_6$Al$^+$			
C$_{15}$H$_{21}$O$_6$Al$^+$	(CH$_3$COCHCOCH$_3$)$_3$Al (Tris(2,4–pentanedionato)aluminum)		7.95±0.05	EI		2460, 2959
C$_{15}$H$_{21}$O$_6$Al$^+$	(CH$_3$COCHCOCH$_3$)$_3$Al (Tris(2,4–pentanedionato)aluminum)		8.27±0.13	EI		2580
			C$_{25}$H$_{39}$O$_6$Al^{+2}			
C$_{25}$H$_{39}$O$_6$Al^{+2}	((CH$_3$)$_3$CCOCHCOC(CH$_3$)$_3$)$_3$Al (Tris(2,2,6,6–tetramethyl–3,5–heptanedionato)aluminum)		27.8±0.5	EI		2524

4.3. The Positive Ion Table—Continued

Ion	Reactant	Other products	Ionization or appearance potential (eV)	Method	Heat of formation (kJ mol^{-1})	Ref.
			C$_{29}$H$_{48}$O$_6$Al$^+$			
C$_{29}$H$_{48}$O$_6$Al$^+$	((CH$_3$)$_3$CCOCHCOC(CH$_3$)$_3$)$_3$Al (Tris(2,2,6,6–tetramethyl–3,5–heptanedionato)aluminum)		12.2±0.5	EI		2524
			C$_{33}$H$_{57}$O$_6$Al$^+$			
C$_{33}$H$_{57}$O$_6$Al$^+$	((CH$_3$)$_3$CCOCHCOC(CH$_3$)$_3$)$_3$Al (Tris(2,2,6,6–tetramethyl–3,5–heptanedionato)aluminum)		10.9±0.5	EI		2524
			C$_{10}$H$_8$O$_4$F$_6$Al$^+$			
C$_{10}$H$_8$O$_4$F$_6$Al$^+$	(CF$_3$COCHCOCH$_3$)$_3$Al (Tris(1,1,1–trifluoro–2,4–pentanedionato)aluminum)		10.2±0.1	EI		2959
			C$_{15}$H$_{12}$O$_6$F$_9$Al$^+$			
C$_{15}$H$_{12}$O$_6$F$_9$Al$^+$	(CF$_3$COCHCOCH$_3$)$_3$Al (Tris(1,1,1–trifluoro–2,4–pentanedionato)aluminum)		9.05±0.1	EI		2959
			C$_{10}$H$_2$O$_4$F$_{12}$Al$^+$			
C$_{10}$H$_2$O$_4$F$_{12}$Al$^+$	(CF$_3$COCHCOCF$_3$)$_3$Al (Tris(1,1,1,5,5,5–hexafluoro–2,4–pentanedionato)aluminum)		11.2±0.1	EI		2959
			C$_{14}$H$_3$O$_6$F$_{15}$Al$^+$			
C$_{14}$H$_3$O$_6$F$_{15}$Al$^+$	(CF$_3$COCHCOCF$_3$)$_3$Al (Tris(1,1,1,5,5,5–hexafluoro–2,4–pentanedionato)aluminum)	CF$_3$	10.7±0.1	EI		2959
			C$_{15}$H$_3$O$_6$F$_{18}$Al$^+$			
C$_{15}$H$_3$O$_6$F$_{18}$Al$^+$	(CF$_3$COCHCOCF$_3$)$_3$Al (Tris(1,1,1,5,5,5–hexafluoro–2,4–pentanedionato)aluminum)		10.21 (V)	PE		3169
C$_{15}$H$_3$O$_6$F$_{18}$Al$^+$	(CF$_3$COCHCOCF$_3$)$_3$Al (Tris(1,1,1,5,5,5–hexafluoro–2,4–pentanedionato)aluminum)		9.80±0.1	EI		2959
C$_{15}$H$_3$O$_6$F$_{18}$Al$^+$	(CF$_3$COCHCOCF$_3$)$_3$Al (Tris(1,1,1,5,5,5–hexafluoro–2,4–pentanedionato)aluminum)		10.30±0.11	EI		2580

4.3. The Positive Ion Table—Continued

Ion	Reactant	Other products	Ionization or appearance potential (eV)	Method	Heat of formation (kJ mol^{-1})	Ref.
			Si$^+$ $\quad\Delta H^\circ_{f0} = 1237.8$ kJ mol^{-1} (295.8 kcal mol^{-1})			
Si$^+$	Si		8.151	S	1237.8	2113
See also − EI: 1116, 2707, 2714						
Si$^+$	SiH$_4$		13.56±0.08	RPD		2899
Si$^+$	SiH$_4$		11.7±0.2	EI		2116
Si$^+$	Si$_2$H$_6$		15.2±0.3	EI		2133
Si$^+$	(CH$_3$)$_3$SiH		13.7±0.3	EI		83
Si$^+$	(CH$_3$)$_4$Si		17.9	EI		2980
Si$^+$	(C$_2$H$_5$)$_4$Si		28.0	EI		2980
			Si$_2^+$			
Si$_2^+$	Si$_2$		7.3	EI		1116
Si$_2^+$	Si$_2$		7.4±0.3	EI		333
Si$_2^+$	Si$_2$H$_6$		13.0	RPD		2899
Si$_2^+$	Si$_2$H$_6$		12.2±0.3	EI		2133
Si$_2^+$	Si$_2$C	C	~15	EI		2943
			Si$_3^+$			
Si$_3^+$	Si$_3$		8.0±0.3	EI		333
			SiH$^+$			
SiH$^+$	SiH		8.01±0.08	D		3219
SiH$^+$	SiH$_4$		15.3±0.3	RPD		2899
SiH$^+$	SiH$_4$		16.1±0.2	EI		2116
SiH$^+$	(CH$_3$)$_3$SiH		14.2±0.2	EI		83
SiH$^+$	(CH$_3$)$_4$Si		18.2	EI		2980
SiH$^+$	(C$_2$H$_5$)$_4$Si		26.8	EI		2980
			SiH$_2^+$			
SiH$_2^+$	SiH$_4$	H$_2$	11.90±0.02	RPD		2899
SiH$_2^+$	SiH$_4$	H$_2$	11.91±0.02	EI		2182
SiH$_2^+$	SiH$_4$	H$_2$	12.1±0.2	EI		2116
SiH$_2^+$	Si$_2$H$_6$	SiH$_4$	11.95±0.10	RPD		2899
SiH$_2^+$	Si$_2$H$_6$	SiH$_4$	11.94±0.04	EI		2183
SiH$_2^+$	CH$_3$SiH$_3$	CH$_4$	11.50±0.05	RPD	2757,	2898
SiH$_2^+$	CH$_3$SiH$_3$	CH$_4$	11.62±0.1	EI		2182
SiH$_2^+$	C$_2$H$_5$SiH$_3$	C$_2$H$_6$	12.0±0.1	EI		2182
SiH$_2^+$	(C$_2$H$_5$)$_4$Si		25.7	EI		2980

4.3. The Positive Ion Table—Continued

Ion	Reactant	Other products	Ionization or appearance potential (eV)	Method	Heat of formation (kJ mol^{-1})	Ref.
		SiH$_3^+$				
SiH$_3^+$	SiH$_4$	H	12.30±0.03	RPD		2899
SiH$_3^+$	SiH$_4$	H	11.8±0.2	EI	173,	2116
SiH$_3^+$	SiH$_4$	H	11.81±0.09	EI		2002
SiH$_3^+$	SiH$_4$	H	12.40±0.02	EI		2182
SiH$_3^+$	Si$_2$H$_6$	SiH$_3$	11.95±0.15	RPD		2899
SiH$_3^+$	Si$_2$H$_6$	SiH$_3$	11.3±0.2	EI		173
SiH$_3^+$	Si$_2$H$_6$	SiH$_3$	11.31±0.12	EI		2002
SiH$_3^+$	Si$_2$H$_6$	SiH$_3$	11.85±0.05	EI		2183
SiH$_3^+$	CH$_3$SiH$_3$	CH$_3$	12.80±0.1	EI		2182
SiH$_3^+$	C$_2$H$_5$SiH$_3$	C$_2$H$_5$	12.8±0.2	EI		2182
SiH$_3^+$	iso–C$_3$H$_7$SiH$_3$		13.1±0.2	EI		2182
SiH$_3^+$	(CH$_3$)$_3$SiH		14.3±0.5	EI		83
SiH$_3^+$	tert–C$_4$H$_9$SiH$_3$		13.7±0.2	EI		2182
SiH$_3^+$	(CH$_3$)$_4$Si		16.5	EI		2980
SiH$_3^+$	(C$_2$H$_5$)$_4$Si		20.6	EI		2980
SiH$_3^+$	H$_3$SiPH$_2$		11.5±0.2	EI		173
SiH$_3^+$	H$_3$GeSiH$_3$		12.01±0.09	EI		2002
SiH$_4^+$		$\Delta H_{f0}^\circ = 1168$ kJ mol^{-1} (279 kcal mol^{-1})				
SiH$_4^+$	SiH$_4$		11.66	PE	1168	3116
		Si$_2$H$^+$				
Si$_2$H$^+$	Si$_2$H$_6$	2H$_2$+H	12.90±0.2	RPD		2899
		Si$_2$H$_2^+$				
Si$_2$H$_2^+$	Si$_2$H$_6$	2H$_2$	11.80±0.10	RPD		2899
		Si$_2$H$_3^+$				
Si$_2$H$_3^+$	Si$_2$H$_6$	H$_2$+H	12.50±0.10	RPD		2899
		Si$_2$H$_4^+$				
Si$_2$H$_4^+$	Si$_2$H$_6$	H$_2$	10.85±0.10	RPD		2899
		Si$_2$H$_5^+$				
Si$_2$H$_5^+$	Si$_2$H$_6$	H	11.40±0.10	RPD		2899
		Si$_2$H$_6^+$				
Si$_2$H$_6^+$	Si$_2$H$_6$		10.15±0.10	RPD		2899
Si$_2$H$_6^+$	Si$_2$H$_6$		10.6±0.3	EI		2133

4.3. The Positive Ion Table—Continued

Ion	Reactant	Other products	Ionization or appearance potential (eV)	Method	Heat of formation (kJ mol⁻¹)	Ref.
		SiB⁺				
SiB⁺	SiB		7.8	EI		1116
		SiC⁺				
SiC⁺	SiC		9.0	EI		1116
SiC⁺	SiC		9.2±0.4	EI		333
		SiC₂⁺				
SiC₂⁺	SiC₂		10.2±0.3	EI		333, 1116
		Si₂C⁺				
Si₂C⁺	Si₂C		9.1	EI		1116
Si₂C⁺	Si₂C		9.2±0.3	EI		333
		Si₂C₂⁺				
Si₂C₂⁺	Si₂C₂		8.2±0.3	EI		333
		Si₂C₃⁺				
Si₂C₃⁺	Si₂C₃		9.2±0.3	EI		333
		Si₃C⁺				
Si₃C⁺	Si₃C		8.2±0.3	EI		333
		Si₂N⁺				
Si₂N⁺	Si₂N		9.4±0.3	EI		2599
		SiO⁺				
SiO⁺	SiO		~11.67	S		2714, 3414

(Based on a reinterpretation of the absorption spectrum of SiO which had previously led to a "provisional" value for the ionization potential of 10.51 eV, see ref. 2150)

Ion	Reactant	Other products	Ionization or appearance potential (eV)	Method	Heat of formation (kJ mol⁻¹)	Ref.
SiO⁺	SiO		11.58±0.10	EI		3414
SiO⁺	SiO		11.6±0.2	EI		2714

4.3. The Positive Ion Table—Continued

Ion	Reactant	Other products	Ionization or appearance potential (eV)	Method	Heat of formation (kJ mol^{-1})	Ref.
SiF^+			$\Delta H_{f0}^\circ \sim 705$ kJ mol^{-1} (168 kcal mol^{-1})			
SiF^+	SiF		7.26	S	705	2149
SiF^+	SiF_4		28.75±0.1	RPD		2525
SiF^+	Si_2F_6		14.16±0.1	RPD		2525
SiF^+	Si_2BF_7		11.56±0.1	RPD		2525
SiF_2^+						
SiF_2^+	SiF_2		11.29±0.1	RPD		2525
SiF_2^+	SiF_4		27.35±0.1	RPD		2525
SiF_2^+	Si_2F_6		13.02±0.1	RPD		2525
SiF_2^+	Si_3F_8		13.45±0.1	RPD		2525
SiF_2^+	Si_2BF_7		13.21±0.1	RPD		2525
SiF_3^+						
SiF_3^+	SiF_4	F?	16.20±0.1	RPD		2525
SiF_3^+	Si_2F_6		14.30±0.1	RPD		2525
SiF_3^+	Si_3F_8		15.51±0.1	RPD		2525
SiF_3^+	Si_2BF_7		15.42±0.1	RPD		2525
SiF_4^+			$\Delta H_{f0}^\circ \sim -79$ kJ mol^{-1} (-19 kcal mol^{-1})			
SiF_4^+	SiF_4		15.81±0.02	PE	-84	3362
SiF_4^+	SiF_4		15.92	PE	-73	3063
SiF_4^+	SiF_4		15.71±0.1	RPD		2525

See also – PE: 3059
          EI: 74

Ion	Reactant	Other products	Ionization or appearance potential (eV)	Method	Heat of formation (kJ mol^{-1})	Ref.
$Si_2F_4^+$						
$Si_2F_4^+$	Si_3F_8		11.43±0.1	RPD		2525
$Si_2F_4^+$	Si_2BF_7		11.32±0.1	RPD		2525
$Si_2F_5^+$						
$Si_2F_5^+$	Si_2F_6	F?	12.89±0.1	RPD		2525
$Si_2F_5^+$	Si_3F_8		11.77±0.1	RPD		2525
$Si_2F_5^+$	Si_2BF_7		12.97±0.1	RPD		2525
$Si_3F_7^+$						
$Si_3F_7^+$	Si_3F_8	F?	15.62±0.1	RPD		2525

4.3. The Positive Ion Table—Continued

Ion	Reactant	Other products	Ionization or appearance potential (eV)	Method	Heat of formation (kJ mol^{-1})	Ref.
			Si$_3$F$_8^+$			
Si$_3$F$_8^+$	Si$_3$F$_8$		10.84±0.1	RPD		2525
			CHSi$^+$			
CHSi$^+$	(CH$_3$)$_3$SiH		11.7±0.5	EI		83
			CH$_2$Si$^+$			
CH$_2$Si$^+$	(CH$_3$)$_3$SiH		10.6±0.3	EI		83
			CH$_3$Si$^+$			
CH$_3$Si$^+$	CH$_3$SiH$_3$	H$_2$+H	14.05±0.05	RPD	2757,	2898
CH$_3$Si$^+$	(CH$_3$)$_2$SiH$_2$		14.00±0.15	RPD		2898
CH$_3$Si$^+$	(CH$_3$)$_3$SiH		13.40±0.10	RPD		2898
CH$_3$Si$^+$	(CH$_3$)$_3$SiH		12.4±0.3	EI		83
CH$_3$Si$^+$	(CH$_3$)$_4$Si		17.1±0.4	EI		82
CH$_3$Si$^+$	(CH$_3$)$_4$Si		17.3	EI		2980
			CH$_4$Si$^+$			
CH$_4$Si$^+$	CH$_3$SiH$_3$	H$_2$	11.45±0.05	RPD	2757,	2898
CH$_4$Si$^+$	(CH$_3$)$_2$SiH$_2$		10.85±0.05	RPD		2898
CH$_4$Si$^+$	(CH$_3$)$_3$SiH		11.0±0.3	EI		83
CH$_4$Si$^+$	(CH$_3$)$_4$Si		16.3	EI		2980
			CH$_5$Si$^+$			
CH$_5$Si$^+$	CH$_3$SiH$_3$	H	11.80±0.05	RPD	2757,	2898
CH$_5$Si$^+$	(CH$_3$)$_2$SiH$_2$	CH$_3$	11.51±0.05	RPD		2898
CH$_5$Si$^+$	(CH$_3$)$_3$SiH		12.8±0.5	EI		83
CH$_5$Si$^+$	(CH$_3$)$_4$Si	C$_2$H$_4$+CH$_3$	13.81±0.02	PI		3039
(Threshold value approximately corrected for hot bands)						
CH$_5$Si$^+$	(CH$_3$)$_4$Si	C$_2$H$_4$+CH$_3$	14.4	EI		2980
			C$_2$H$_5$Si$^+$			
C$_2$H$_5$Si$^+$	(C$_2$H$_5$)$_4$Si		19.4	EI		2980
			C$_2$H$_6$Si$^+$			
C$_2$H$_6$Si$^+$	(CH$_3$)$_2$SiH$_2$	H$_2$	10.71±0.05	RPD		2898
C$_2$H$_6$Si$^+$	(CH$_3$)$_3$SiH		10.50±0.05	RPD		2898
C$_2$H$_6$Si$^+$	(CH$_3$)$_3$SiH		10.3±0.2	EI		83
C$_2$H$_6$Si$^+$	(CH$_3$)$_4$Si		13.4	EI		2980
C$_2$H$_6$Si$^+$	(CH$_3$)$_4$Si		13.9±0.3	EI		82

4.3. The Positive Ion Table—Continued

Ion	Reactant	Other products	Ionization or appearance potential (eV)	Method	Heat of formation (kJ mol⁻¹)	Ref.
			C₂H₇Si⁺			
$C_2H_7Si^+$	$(CH_3)_2SiH_2$	H	11.12±0.05	RPD		2898
$C_2H_7Si^+$	$(CH_3)_2SiH_2$	H	11.94±0.04	RPD		1421
$C_2H_7Si^+$	$(CH_3)_3SiH$	CH_3	10.91±0.05	RPD		2898
$C_2H_7Si^+$	$(CH_3)_3SiH$	CH_3	11.70±0.06	RPD		1421

See also – EI: 83, 2689

Ion	Reactant	Other products	Ionization or appearance potential (eV)	Method	Heat of formation	Ref.
$C_2H_7Si^+$	$(CH_3)_4Si$		13.3	EI		2980
$C_2H_7Si^+$	$(C_2H_5)_4Si$		11.8	EI		2980

			C₂H₈Si⁺			
$C_2H_8Si^+$	$C_2H_5SiH_3$		10.18±0.05	EI		2182

			C₃H₉Si⁺			
$C_3H_9Si^+$	$(CH_3)_3SiH$	H	10.52±0.05	RPD		2898
$C_3H_9Si^+$	$(CH_3)_3SiH$	H	10.78±0.07	RPD		1421, 3187
$C_3H_9Si^+$	$(CH_3)_3SiH$	H	10.6±0.1	EI		2401, 2689
$C_3H_9Si^+$	$(CH_3)_3SiH$	H	10.72±0.1	EI		2055

See also – EI: 83

$C_3H_9Si^+$	$(CH_3)_4Si$	CH_3	10.09±0.02	PI		3039
(Threshold value approximately corrected for hot bands)						
$C_3H_9Si^+$	$(CH_3)_4Si$	CH_3	10.25±0.05	RPD		2898
$C_3H_9Si^+$	$(CH_3)_4Si$	CH_3	10.63±0.13	RPD		1421, 3187
$C_3H_9Si^+$	$(CH_3)_4Si$	CH_3	10.4±0.1	EI		2401, 2689
$C_3H_9Si^+$	$(CH_3)_4Si$	CH_3	10.53±0.1	EI		2055
$C_3H_9Si^+$	$(CH_3)_4Si$	CH_3	10.53±0.20	EI		2720

See also – EI: 82, 2980

$C_3H_9Si^+$	$(CH_3)_3SiC_2H_5$		10.53±0.09	RPD		1421, 3187
$C_3H_9Si^+$	$(CH_3)_3SiC_2H_5$		10.34±0.11	EI		2055
$C_3H_9Si^+$	$iso-C_3H_7Si(CH_3)_3$		10.56±0.16	RPD		1421, 3187
$C_3H_9Si^+$	$tert-C_4H_9Si(CH_3)_3$		10.53±0.09	RPD		1421, 3187
$C_3H_9Si^+$	$C_6H_5CH_2Si(CH_3)_3$ (α–Trimethylsilyltoluene)		10.05±0.1	EI		2055
$C_3H_9Si^+$	$(CH_3)_3SiSi(CH_3)_3$		10.69±0.04	RPD		1421, 3187
$C_3H_9Si^+$	$(CH_3)_3SiSi(CH_3)_3$		10.0±0.1	EI		2401, 2689
$C_3H_9Si^+$	$(CH_3)_3SiSi(CH_3)_3$		10.03±0.1	EI		2055, 2413
$C_3H_9Si^+$	$(CH_3)_3SiSi(CH_3)_3$		10.22±0.18	EI		2720

See also – EI: 2475

$C_3H_9Si^+$	$(CH_3)_3SiSi(CH_3)_2C_2H_5$		10.1±0.1	EI		2418
$C_3H_9Si^+$	$(CH_3)_3SiN(C_2H_5)_2$		12.61±0.03	RPD		1421
$C_3H_9Si^+$	$(CH_3)_3SiOCH_3$		12.43±0.18	RPD		1421

4.3. The Positive Ion Table—Continued

Ion	Reactant	Other products	Ionization or appearance potential (eV)	Method	Heat of formation (kJ mol^{-1})	Ref.
$C_3H_9Si^+$	$(CH_3)_3SiOSi(CH_3)_3$		15.36±0.13	RPD		1421
$C_3H_9Si^+$	$(CH_3)_3SiF$	F^-?	11.7±0.5	RPD		1421
$C_3H_9Si^+$	$(CH_3)_3SiCl$	Cl?	12.40±0.06	RPD		1421
$C_3H_9Si^+$	$(CH_3)_3SiCl$	Cl?	12.42±0.03	RPD		3187
$C_3H_9Si^+$	$(CH_3)_3SiCl$	Cl?	10.9±0.1	EI		2401, 2689
$C_3H_9Si^+$	$(CH_3)_3SiCl$	Cl?	11.5±0.2	EI		2055
$C_3H_9Si^+$	$(CH_3)_3SiGe(CH_3)_3$		10.19±0.12	EI		2720
$C_3H_9Si^+$	$(CH_3)_3SiBr$	Br?	10.69±0.06	RPD		1421
$C_3H_9Si^+$	$(CH_3)_3SiBr$	Br?	10.5±0.1	EI		2401, 2689
$C_3H_9Si^+$	$(CH_3)_3SiSn(CH_3)_3$		10.18±0.26	EI		2720
$C_3H_9Si^+$	$(CH_3)_3SiI$	I?	10.1±0.1	EI		2401, 2689
$C_3H_9Si^+$	$(CH_3)_3SiHgSi(CH_3)_3$		8.56±0.1	EI		2055

$C_3H_{10}Si^+$

Ion	Reactant	Other products	Ionization or appearance potential (eV)	Method	Heat of formation (kJ mol^{-1})	Ref.
$C_3H_{10}Si^+$	iso-$C_3H_7SiH_3$		9.85±0.1	EI		2182
$C_3H_{10}Si^+$	$(CH_3)_3SiH$		10.11	PI		2413
$C_3H_{10}Si^+$	$(CH_3)_3SiH$		9.8±0.3	EI		83

See also – EI: 2689

$C_4H_{11}Si^+$

Ion	Reactant	Other products	Ionization or appearance potential (eV)	Method	Heat of formation (kJ mol^{-1})	Ref.
$C_4H_{11}Si^+$	$(CH_3)_3SiC_2H_5$	CH_3	11.41±0.06	RPD		1421
$C_4H_{11}Si^+$	$(C_2H_5)_4Si$		12.5	EI		2980
$C_4H_{11}Si^+$	$(CH_3)_3SiSi(CH_3)_2C_2H_5$		9.7±0.1	EI		2418

$(CH_3)_4Si^+$ $\Delta H^\circ_{f298} = 712$ kJ mol^{-1} (170 kcal mol^{-1})

Ion	Reactant	Other products	Ionization or appearance potential (eV)	Method	Heat of formation (kJ mol^{-1})	Ref.
$C_4H_{12}Si^+$	tert-$C_4H_9SiH_3$		9.5±0.2	EI		2182
$C_4H_{12}Si^+$	$(CH_3)_4Si$		9.86±0.02	PI	712	3039
(Threshold value approximately corrected for hot bands)						
$C_4H_{12}Si^+$	$(CH_3)_4Si$		9.74±0.05	RPD		2898
$C_4H_{12}Si^+$	$(CH_3)_4Si$		9.98±0.03	RPD		1421
$C_4H_{12}Si^+$	$(CH_3)_4Si$		9.85±0.16	EI		2720
$C_4H_{12}Si^+$	$(CH_3)_4Si$		9.9±0.1	EI		2689

See also – EI: 82, 218, 2900, 2980

$C_5H_{10}Si^+$

Ion	Reactant	Other products	Ionization or appearance potential (eV)	Method	Heat of formation (kJ mol^{-1})	Ref.
$C_5H_{10}Si^+$	$(CH_3)_3SiC \equiv CH$		10.14±0.04	RPD		2408

4.3. The Positive Ion Table—Continued

Ion	Reactant	Other products	Ionization or appearance potential (eV)	Method	Heat of formation (kJ mol^{-1})	Ref.
		$C_5H_{12}Si^+$				
$C_5H_{12}Si^+$	$(CH_3)_3SiCH=CH_2$		9.82±0.04	RPD		2410
$C_5H_{12}Si^+$	$C_5H_{12}Si$ (1,1–Dimethylsilacyclobutane)		8.83±0.07	EI		2547
		$C_5H_{14}Si^+$				
$C_5H_{14}Si^+$	$(CH_3)_3SiC_2H_5$		9.70±0.01	RPD		1421
		$C_6H_{12}Si^+$				
$C_6H_{12}Si^+$	$(CH_3)_3SiCH_2C\equiv CH$		9.04±0.04	RPD		2408
		$C_6H_{14}Si^+$				
$C_6H_{14}Si^+$	$(CH_3)_3SiCH_2CH=CH_2$		8.85±0.04	RPD		2410
		$C_6H_{15}Si^+$				
$C_6H_{15}Si^+$	$(C_2H_5)_4Si$		11.0	EI		2980
		$C_6H_{16}Si^+$				
$C_6H_{16}Si^+$	$iso-C_3H_7Si(CH_3)_3$		9.50±0.03	RPD		1421
		$C_7H_{18}Si^+$				
$C_7H_{18}Si^+$	$tert-C_4H_9Si(CH_3)_3$		9.34±0.06	RPD		1421
		$C_8H_{20}Si^+$				
$C_8H_{20}Si^+$	$(C_2H_5)_4Si$		10.5	EI		2980
		$C_9H_{14}Si^+$				
$C_9H_{14}Si^+$	$C_6H_5Si(CH_3)_3$ (Trimethylsilylbenzene)		8.72±0.04	RPD		2407, 2463
		$C_9H_{18}Si^+$				
$C_9H_{18}Si^+$	$(CH_3)_3CC\equiv CSi(CH_3)_3$		9.35±0.04	RPD		2408

4.3. The Positive Ion Table—Continued

Ion	Reactant	Other products	Ionization or appearance potential (eV)	Method	Heat of formation (kJ mol^{-1})	Ref.
		$C_9H_{20}Si^+$				
$C_9H_{20}Si^+$	trans$-(CH_3)_3CCH=CHSi(CH_3)_3$		9.08±0.04	RPD		2410
		$C_{10}H_{16}Si^+$				
$C_{10}H_{16}Si^+$	$C_6H_5CH_2Si(CH_3)_3$ ($\alpha-$Trimethylsilyltoluene)		7.96±0.04	RPD		2407, 2463
		$C_{11}H_{14}Si^+$				
$C_{11}H_{14}Si^+$	$C_6H_5C\equiv CSi(CH_3)_3$ ($\alpha-$Trimethylsilylethynylbenzene)		8.16	RPD		2950
		$C_5H_{15}Si_2^+$				
$C_5H_{15}Si_2^+$	$(CH_3)_3SiSi(CH_3)_3$	CH_3	10.74±0.08	RPD		1421
		$C_6H_{16}Si_2^+$				
$C_6H_{16}Si_2^+$	$C_6H_{16}Si_2$ (1,1,3,3-Tetramethyl-1,3-disilacyclobutane)		8.56±0.07	EI		2547
		$C_6H_{18}Si_2^+$				
$C_6H_{18}Si_2^+$	$(CH_3)_3SiSi(CH_3)_3$		8.79±0.08	RPD		1421
$C_6H_{18}Si_2^+$	$(CH_3)_3SiSi(CH_3)_3$		8.00±0.01	EI		2900
$C_6H_{18}Si_2^+$	$(CH_3)_3SiSi(CH_3)_3$		8.35±0.12	EI		2720
		$C_8H_{18}Si_2^+$				
$C_8H_{18}Si_2^+$	$(CH_3)_3SiC\equiv CSi(CH_3)_3$		9.63±0.04	RPD		2408
		$C_8H_{20}Si_2^+$				
$C_8H_{20}Si_2^+$	$((CH_3)_3Si)_2C=CH_2$		9.25±0.04	RPD		2410
$C_8H_{20}Si_2^+$	trans$-(CH_3)_3SiCH=CHSi(CH_3)_3$		9.32±0.04	RPD		2410
		$C_9H_{20}Si_2^+$				
$C_9H_{20}Si_2^+$	$(CH_3)_3SiC\equiv CCH_2Si(CH_3)_3$		8.95±0.04	RPD		2408
		$C_{10}H_{18}Si_2^+$				
$C_{10}H_{18}Si_2^+$	$(CH_3)_3SiC\equiv CC\equiv CSi(CH_3)_3$		9.23±0.04	RPD		2408

4.3. The Positive Ion Table—Continued

Ion	Reactant	Other products	Ionization or appearance potential (eV)	Method	Heat of formation (kJ mol⁻¹)	Ref.
		$C_{10}H_{22}Si_2^+$				
$C_{10}H_{22}Si_2^+$	$(CH_3)_3SiCH_2C{\equiv}CCH_2Si(CH_3)_3$		8.85±0.04	RPD		2408
		$C_{10}H_{24}Si_2^+$				
$C_{10}H_{24}Si_2^+$	$trans{-}(CH_3)_3SiCH_2CH{=}CHCH_2Si(CH_3)_3$		7.95±0.04	RPD		2410
		$C_{11}H_{20}Si_2^+$				
$C_{11}H_{20}Si_2^+$	$C_6H_5Si_2(CH_3)_5$ ($\beta,\beta,\gamma,\gamma{-}$Tetramethyl$-\beta,\gamma{-}$disilapropylbenzene)		7.82	RPD		2950
		$C_{12}H_{18}Si_2^+$				
$C_{12}H_{18}Si_2^+$	$(CH_3)_3SiC{\equiv}CC{\equiv}CC{\equiv}CSi(CH_3)_3$		8.90±0.04	RPD		2408
		$C_{12}H_{22}Si_2^+$				
$C_{12}H_{22}Si_2^+$	$(CH_3)_3SiCH_2C{\equiv}CC{\equiv}CCH_2Si(CH_3)_3$		8.38±0.04	RPD		2408
$C_{12}H_{22}Si_2^+$	$C_6H_4(Si(CH_3)_3)_2$ (1,4$-$Bis(trimethylsilyl)benzene)		8.25	RPD		2950
		$C_{12}H_{28}Si_2^+$				
$C_{12}H_{28}Si_2^+$	$((CH_3)_3Si)_2C{=}CHC(CH_3)_3$		8.72±0.04	RPD		2410
		$C_{13}H_{24}Si_2^+$				
$C_{13}H_{24}Si_2^+$	$C_6H_5CH(Si(CH_3)_3)_2$ ($\alpha,\alpha{-}$Bis(trimethylsilyl)toluene)		7.63±0.04	RPD		2407
		$C_{14}H_{26}Si_2^+$				
$C_{14}H_{26}Si_2^+$	$C_6H_4(CH_2Si(CH_3)_3)_2$ (1,2$-$Bis(2,2$-$dimethyl$-$2$-$silapropyl)benzene)		7.74	RPD		2950
$C_{14}H_{26}Si_2^+$	$C_6H_4(CH_2Si(CH_3)_3)_2$ (1,4$-$Bis(2,2$-$dimethyl$-$2$-$silapropyl)benzene)		7.25	RPD		2950
		$C_8H_{20}Si_3^+$				
$C_8H_{20}Si_3^+$	$C_8H_{20}Si_3$ (2,2,6,6$-$Tetramethyl$-$2,4,6$-$trisilaspiro[3.3]heptane)		7.78±0.06	EI		2547

4.3. The Positive Ion Table—Continued

Ion	Reactant	Other products	Ionization or appearance potential (eV)	Method	Heat of formation (kJ mol^{-1})	Ref.
		$C_8H_{24}Si_3^+$				
$C_8H_{24}Si_3^+$	$Si_3(CH_3)_8$		7.53±0.01	EI		2900
		$C_9H_{24}Si_3^+$				
$C_9H_{24}Si_3^+$	$C_9H_{24}Si_3$ (1,1,3,3,5,5–Hexamethyl–1,3,5–trisilacyclohexane)		9.39±0.03	EI		2547
		$C_{11}H_{28}Si_3^+$				
$C_{11}H_{28}Si_3^+$	$((CH_3)_3Si)_2C=CHSi(CH_3)_3$		8.85±0.04	RPD		2410
		$C_{16}H_{32}Si_3^+$				
$C_{16}H_{32}Si_3^+$	$C_6H_5C(Si(CH_3)_3)_3$ (α,α,α–Tris(trimethylsilyl)toluene)		7.52±0.04	RPD		2407
		$C_{17}H_{34}Si_3^+$				
$C_{17}H_{34}Si_3^+$	$C_6H_5CH_2C(Si(CH_3)_3)_3$ (α,α,α–Tris(trimethylsilyl)ethylbenzene)		8.22±0.04	RPD		2407
		$C_{10}H_{30}Si_4^+$				
$C_{10}H_{30}Si_4^+$	n–$Si_4(CH_3)_{10}$		7.29±0.01	EI		2900
$C_{10}H_{30}Si_4^+$	$CH_3Si(Si(CH_3)_3)_3$		7.41±0.01	EI		2900
		$C_{11}H_{32}Si_4^+$				
$C_{11}H_{32}Si_4^+$	n–$Si_3(CH_3)_7CH_2Si(CH_3)_3$		7.38±0.01	EI		2900
		$C_{12}H_{32}Si_4^+$				
$C_{12}H_{32}Si_4^+$	$C_{12}H_{32}Si_4$ (1,1,3,3,5,5,7,7–Octamethyl–1,3,5,7–tetrasilacyclooctane)		9.21±0.04	EI		2547
		$C_{10}H_{30}Si_5^+$				
$C_{10}H_{30}Si_5^+$	$Si_5(CH_3)_{10}$ (Decamethylcyclopentasilane)		7.18±0.01	EI		2900

4.3. The Positive Ion Table—Continued

Ion	Reactant	Other products	Ionization or appearance potential (eV)	Method	Heat of formation (kJ mol^{-1})	Ref.
		$C_{12}H_{36}Si_5^+$				
$C_{12}H_{36}Si_5^+$	$n-Si_5(CH_3)_{12}$		7.11±0.01	EI		2900
$C_{12}H_{36}Si_5^+$	$Si(Si(CH_3)_3)_4$		7.41±0.01	EI		2900
		$C_{12}H_{36}Si_6^+$				
$C_{12}H_{36}Si_6^+$	$Si_6(CH_3)_{12}$ (Dodecamethylcyclohexasilane)		7.29±0.01	EI		2900
		$C_{14}H_{42}Si_6^+$				
$C_{14}H_{42}Si_6^+$	$n-Si_6(CH_3)_{14}$		7.02±0.02	EI		2900
		$C_{14}H_{42}Si_7^+$				
$C_{14}H_{42}Si_7^+$	$Si_7(CH_3)_{14}$ (Tetradecamethylcycloheptasilane)		7.39±0.01	EI		2900
		$C_{18}H_{54}Si_8^+$				
$C_{18}H_{54}Si_8^+$	$n-Si_8(CH_3)_{18}$		6.82±0.02	EI		2900
		$BCSi^+$				
$BCSi^+$	BCSi		9.9	EI		1116
		$CNSi^+$				
$CNSi^+$	SiCN		8.7±0.5	EI		3004
		$SiHF_2^+$				
$SiHF_2^+$	$SiHF_3$	F?	13.4±0.3	EI		2896
		$Si_2HF_4^+$				
$Si_2HF_4^+$	Si_2HF_5	F?	11.1±0.3	EI		2896
		$SiBF_4^+$				
$SiBF_4^+$	Si_2BF_7		11.95±0.1	RPD		2525

4.3. The Positive Ion Table—Continued

Ion	Reactant	Other products	Ionization or appearance potential (eV)	Method	Heat of formation (kJ mol^{-1})	Ref.
			$Si_2BF_6^+$			
$Si_2BF_6^+$	Si_2BF_7	F?	15.36±0.1	RPD		2525
			$C_7H_{19}NSi^+$			
$C_7H_{19}NSi^+$	$(CH_3)_3SiN(C_2H_5)_2$		8.06±0.02	RPD		1421
			$C_{11}H_{19}NSi^+$			
$C_{11}H_{19}NSi^+$	$C_6H_4(N(CH_3)_2)Si(CH_3)_3$ (1–Dimethylamino–4–trimethylsilylbenzene)		6.73	RPD		2950
			$C_9H_{14}N_2Si^+$			
$C_9H_{14}N_2Si^+$	$C_6H_5N=NSi(CH_3)_3$ (Phenyl trimethylsilyl diimine)		7.05±0.02	RPD		3254
			$C_3H_9OSi^+$			
$C_3H_9OSi^+$	$(CH_3)_3SiOCH_3$	CH_3	10.25±0.05	RPD		1421
			$C_4H_{12}OSi^+$			
$C_4H_{12}OSi^+$	$(CH_3)_3SiOCH_3$		9.79±0.04	RPD		1421
			$C_{10}H_{14}OSi^+$			
$C_{10}H_{14}OSi^+$	$C_6H_5COSi(CH_3)_3$ (Phenyl trimethylsilyl ketone)		8.20±0.04	RPD		2407, 2463
			$C_5H_{15}OSi_2^+$			
$C_5H_{15}OSi_2^+$	$(CH_3)_3SiOSi(CH_3)_3$	CH_3	10.20±0.07	RPD		1421
			$C_6H_{18}OSi_2^+$			
$C_6H_{18}OSi_2^+$	$(CH_3)_3SiOSi(CH_3)_3$		9.59±0.04	RPD		1421

4.3. The Positive Ion Table—Continued

Ion	Reactant	Other products	Ionization or appearance potential (eV)	Method	Heat of formation (kJ mol^{-1})	Ref.
			CNOSi$^+$			
CNOSi$^+$	OSiCN		7.4±0.5	EI		3004
			C$_2$H$_6$FSi$^+$			
C$_2$H$_6$FSi$^+$	(CH$_3$)$_3$SiF	CH$_3$	11.11±0.05	RPD		1421
			C$_3$H$_9$FSi$^+$			
C$_3$H$_9$FSi$^+$	(CH$_3$)$_3$SiF		10.55±0.06	RPD		1421
			CH$_3$F$_2$Si$^+$			
CH$_3$F$_2$Si$^+$	CH$_3$SiF$_2$Co(CO)$_4$		17.0±0.4	EI		2581
P$^+$		**$\Delta H^\circ_{f0} \sim 1326$ kJ mol^{-1} (317 kcal mol^{-1})**				
P$^+$	P		10.486	S	~1326	2113
See also – EI: 8, 2699						
P$^+$	P$_2$	P	15.1±0.3	EI		2699
See also – EI: 8, 2465						
P$^+$	P$_4$		14	EI		8
See also – EI: 3370						
P$^+$	PH$_3$	H$_2$+H	15.9	EI		2486
See also – EI: 1033, 1036, 2116						
P$^+$	P$_2$H$_4$		16.7±0.3	EI		2133
P$^+$	P$_2$H$_4$		19.4±0.5	EI		2486
P$^+$	(CH$_3$)$_2$PH		11.9	EI		1036
P$^+$	(C$_2$H$_5$)$_2$PP(C$_2$H$_5$)$_2$		9.8±0.3	EI		2948
P$^+$	PCl$_3$		21.0±0.5	EI		1101
P$^+$	PCl$_3$		21.2±0.5	EI	192,	2506
P$^+$	POCl$_3$		21.4	EI		1101
P$^+$	POCl$_3$		28.1±0.5	EI		2506
P$^+$	PBr$_3$		20.1±0.5	EI		2506

4.3. The Positive Ion Table—Continued

Ion	Reactant	Other products	Ionization or appearance potential (eV)	Method	Heat of formation (kJ mol⁻¹)	Ref.
P⁺²		**ΔH°f₀ ~ 3229 kJ mol⁻¹ (772 kcal mol⁻¹)**				
P⁺²	P		30.211	S	~3229	2113
P⁺²	P⁺		19.725	S		2113
See also – EI: 8						
P⁺²	P₄		38	EI		8
P⁺²	PH₃		42±2	NRE		1036
P⁺²	PH₃		42±2	EI		1036
P₂⁺						
P₂⁺	P₂		9.6±0.3	EI		2465
P₂⁺	P₂		11.0±0.3	EI		2699
See also – EI: 8						
P₂⁺	P₄		14.3	EI		8
See also – EI: 3370						
P₂⁺	P₂H₂		11.9±0.4	EI		2486
P₂⁺	P₂H₄		12.2	EI		1033
P₂⁺	P₂H₄		13.2±0.2	EI		2486
P₂⁺	P₂H₄		13.7±0.3	EI		2133
P₂⁺	P₂Cl₄		19.7±0.4	EI		192
P₃⁺						
P₃⁺	P₃		11.5	EI		8
P₃⁺	P₄	P	14.3±0.3	EI		2465
P₃⁺	P₄	P	14.5	EI		8
See also – EI: 3370						
P₄⁺		**ΔH°f₀ = 942 kJ mol⁻¹ (225 kcal mol⁻¹)**				
P₄⁺	P₄		9.08±0.05	PI	942	2880
P₄⁺	P₄		9.2±0.3	EI		2465
See also – EI: 8, 3370						
PH⁺						
PH⁺	PH₃	H₂	12.4±0.2	EI		2116
PH⁺	PH₃	H₂	12.6±0.2	EI		2486
See also – EI: 1033, 1036, 2493						
PH⁺	P₂H₄		17.4±0.5	EI		2486
PH⁺	(C₂H₅)₂PP(C₂H₅)₂		11.0±0.3	EI		2948

4.3. The Positive Ion Table—Continued

Ion	Reactant	Other products	Ionization or appearance potential (eV)	Method	Heat of formation (kJ mol^{-1})	Ref.

PH^{+2}

Ion	Reactant	Other products	Ionization or appearance potential (eV)	Method	Heat of formation (kJ mol^{-1})	Ref.
PH^{+2}	PH_3		25.1	NRE		1036
PH^{+2}	PH_3		21.2	EI		1036

PH_2^+ $\Delta H_{f0}^\circ = 1097$ kJ mol^{-1} (262 kcal mol^{-1})

Ion	Reactant	Other products	Ionization or appearance potential (eV)	Method	Heat of formation (kJ mol^{-1})	Ref.
PH_2^+	PH_2		9.83±0.02	EM		2862

This ionization potential combined with $\Delta H_{f0}^\circ(PH_2^+)$ leads to $\Delta H_{f0}^\circ(PH_2) = 149$ kJ mol^{-1} (36 kcal mol^{-1}).

Ion	Reactant	Other products	Ionization or appearance potential (eV)	Method	Heat of formation (kJ mol^{-1})	Ref.
PH_2^+	PH_2		9.96	EI		2862
PH_2^+	PH_3	H	13.47±0.05	EM	1097	2862
PH_2^+	PH_3	H	13.2±0.2	EI		173
PH_2^+	PH_3	H	13.2±0.2	EI		2486
PH_2^+	PH_3	H	13.6	EI		2862

See also – EI: 1033, 1036, 2116, 2493

Ion	Reactant	Other products	Ionization or appearance potential (eV)	Method	Heat of formation (kJ mol^{-1})	Ref.
PH_2^+	P_2H_4		12.5±0.2	EI		173
PH_2^+	P_2H_4		15.3±0.5	EI		2486
PH_2^+	$(C_2H_5)_2PP(C_2H_5)_2$		9.3±0.3	EI		2948
PH_2^+	H_3SiPH_2		13.1±0.2	EI		173

PH_2^{+2}

Ion	Reactant	Other products	Ionization or appearance potential (eV)	Method	Heat of formation (kJ mol^{-1})	Ref.
PH_2^{+2}	PH_3	H	34.0	NRE		1036
PH_2^{+2}	PH_3	H	32.7	EI		1036

PH_3^+ $\Delta H_{f0}^\circ \sim 976$ kJ mol^{-1} (233 kcal mol^{-1})

Ion	Reactant	Other products	Ionization or appearance potential (eV)	Method	Heat of formation (kJ mol^{-1})	Ref.
PH_3^+	PH_3		9.98	PI	976	1091
PH_3^+	PH_3		10.13±0.02	PE		3085
PH_3^+	PH_3		10.28	PE		2853
PH_3^+	PH_3		9.97±0.02	EM	975	2862
PH_3^+	PH_3		10.05±0.05	RPD		2486
PH_3^+	PH_3		10.30±0.10	RPD		2493
PH_3^+	PH_3		10.10	EDD		2493

The ionization potential is somewhat uncertain. It is difficult to identify the 0–0 transition and to correct for hot bands, see ref. 3085.

See also – EI: 218, 1033, 1036, 1407, 2116, 2493, 2597

Ion	Reactant	Other products	Ionization or appearance potential (eV)	Method	Heat of formation (kJ mol^{-1})	Ref.
PH_3^+	$C_2H_5PH_2$	C_2H_4	11.2±0.2	EI		2045
PH_3^+	$C_2H_5PH_2$	C_2H_4	12.0±0.3	EI		2948
PH_3^+	$(C_2H_5)_2PH$		12.8±0.3	EI		2948
PH_3^+	$(C_2H_5)_3P$		14.2±0.3	EI		2948
PH_3^+	$(C_2H_5)_3P$		14.7±0.2	EI		2045
PH_3^+	$(CH_3)_2PP(CH_3)_2$		17.0±0.3	EI		2948
PH_3^+	$(C_2H_5)_2PP(C_2H_5)_2$		11.1±0.3	EI		2948

4.3. The Positive Ion Table—Continued

Ion	Reactant	Other products	Ionization or appearance potential (eV)	Method	Heat of formation (kJ mol^{-1})	Ref.
			PH$_2$D$^+$			
PH$_2$D$^+$	PH$_2$D		10.1±0.2	EI		1033
			PHD$_2^+$			
PHD$_2^+$	PHD$_2$		10.2±0.2	EI		1033
			PD$_3^+$			
PD$_3^+$	PD$_3$		10.1±0.2	EI		1033
			PH$_3^{+2}$			
PH$_3^{+2}$	PH$_3$		15.6	NRE		1036
PH$_3^{+2}$	PH$_3$		15.0	EI		1036
			PH$_4^+$			
PH$_4^+$	PH$_3$+H$^+$			BH	(728)	1414
(Based on a proton affinity for PH$_3$ of 813.8±21 kJ mol^{-1} (194.5±5 kcal mol^{-1}))						
PH$_4^+$	P$_3$H$_5$		10.0±0.2	EI		2753
PH$_4^+$	(CH$_3$)$_3$P		14.2±0.2	EI		2045
PH$_4^+$	(C$_2$H$_5$)$_3$P		14.7±0.3	EI		2045
PH$_4^+$	(C$_2$H$_5$)$_2$PP(C$_2$H$_5$)$_2$		9.7±0.3	EI		2948
			P$_2$H$^+$			
P$_2$H$^+$	P$_2$H$_2$	H	13.3±0.4	EI		2486
P$_2$H$^+$	P$_2$H$_4$		13.2	EI		1033
P$_2$H$^+$	P$_2$H$_4$		13.6±0.3	EI		2133
P$_2$H$^+$	P$_2$H$_4$		14.6±0.3	EI		2486
P$_2$H$^+$	(CH$_3$)$_2$PP(CH$_3$)$_2$		15.8±0.3	EI		2948
P$_2$H$^+$	(C$_2$H$_5$)$_2$PP(C$_2$H$_5$)$_2$		13.8±0.3	EI		2948
P$_2$H$^+$	(n-C$_4$H$_9$)$_2$PP(n-C$_4$H$_9$)$_2$		17.0±0.3	EI		2948
			P$_2$H$_2^+$			
P$_2$H$_2^+$	P$_2$H$_2$		10.2±0.2	EI		2486
See also – EI: 1407						
P$_2$H$_2^+$	P$_2$H$_4$	H$_2$	10.5	EI		1033
P$_2$H$_2^+$	P$_2$H$_4$	H$_2$	11.1±0.2	EI		2486
P$_2$H$_2^+$	P$_2$H$_4$	H$_2$	12.7±0.3	EI		2133
P$_2$H$_2^+$	(C$_2$H$_5$)$_2$PP(C$_2$H$_5$)$_2$		11.4±0.3	EI		2948
P$_2$H$_2^+$	(n-C$_4$H$_9$)$_2$PP(n-C$_4$H$_9$)$_2$		15.2±0.3	EI		2948

4.3. The Positive Ion Table—Continued

Ion	Reactant	Other products	Ionization or appearance potential (eV)	Method	Heat of formation (kJ mol^{-1})	Ref.
			$P_2H_3^+$			
$P_2H_3^+$	P_2H_3		9.1	EI		1033
$P_2H_3^+$	P_2H_4	H	11.3±0.3	EI		2133
$P_2H_3^+$	P_2H_4	H	12.2±0.2	EI	2486,	2753
See also – EI: 1407						
$P_2H_3^+$	P_3H_5		10.6±0.1	EI		2753
$P_2H_3^+$	$(C_2H_5)_2PP(C_2H_5)_2$		11.8±0.3	EI		2948
			$P_2H_4^+$			
$P_2H_4^+$	P_2H_4		9.17±0.05	RPD		2486
$P_2H_4^+$	P_2H_4		8.7±0.3	EI		1033
$P_2H_4^+$	P_2H_4		10.6±0.3	EI		2133
See also – EI: 3431						
$P_2H_4^+$	$(C_2H_5)_2PP(C_2H_5)_2$		11.3±0.3	EI		2948
			$P_2H_3D^+$			
$P_2H_3D^+$	P_2H_3D		8.4±0.3	EI		1033
			$P_2H_2D_2^+$			
$P_2H_2D_2^+$	$P_2H_2D_2$		8.2±0.5	EI		1033
			$P_3H_5^+$			
$P_3H_5^+$	P_3H_5		8.7±0.1	EI		2753
See also – EI: 3431						
			PC^+			
PC^+	CH_3PH_2		14.5±0.3	EI		2045
PC^+	$C_2H_5PH_2$		12.0±0.3	EI		2045
PC^+	$(CH_3)_3P$		13.2±0.3	EI		2045
PC^+	$(C_2H_5)_3P$		19.1±0.5	EI		2045
			PN^+			
PN^+	PN		11.8±0.1	RPD	2465,	2570
See also – EI: 2465, 2916						

4.3. The Positive Ion Table—Continued

Ion	Reactant	Other products	Ionization or appearance potential (eV)	Method	Heat of formation (kJ mol^{-1})	Ref.
		PN^{+2}				
PN^{+2}	PN		40.7 ± 0.5	EI		2465
See also − EI:	2570					
		PO^+				
PO^+	P_4O_6		18.0 ± 0.5	EI		3370
PO^+	P_4O_{10}		23.6 ± 0.5	EI		3370
PO^+	$(CH_3O)_2PHO$		16.3	EI		1036
PO^+	$(CH_3O)_2CH_3PO$		19.4 ± 0.3	EI		3211
PO^+	$(CH_3O)_3PO$		19.6 ± 0.3	EI		3211
PO^+	$POCl_3$		14.5 ± 0.5	EI		1101
PO^+	$POCl_3$		16.6 ± 0.4	EI		2506
		PO_2^+				
PO_2^+	P_4O_6		23.0 ± 0.5	EI		3370
PO_2^+	P_4O_{10}		22.5 ± 0.5	EI		3370
		PO_3^+				
PO_3^+	P_4O_{10}		33.0 ± 0.5	EI		3370
PO_3^+	$(CH_3O)_2PHO$		13.5	EI		1036
		P_2O^+				
P_2O^+	P_4O_6		23.6 ± 0.5	EI		3370
		$P_2O_2^+$				
$P_2O_2^+$	P_4O_6		18.8 ± 0.5	EI		3370
$P_2O_2^+$	P_4O_{10}		29.5 ± 0.5	EI		3370
		$P_2O_3^+$				
$P_2O_3^+$	P_4O_6		17.5 ± 0.5	EI		3370
$P_2O_3^+$	P_4O_{10}		28.0 ± 0.5	EI		3370
		$P_2O_4^+$				
$P_2O_4^+$	P_4O_{10}		24.3 ± 0.5	EI		3370
		$P_2O_5^+$				
$P_2O_5^+$	P_4O_{10}		22.9 ± 0.5	EI		3370

4.3. The Positive Ion Table—Continued

Ion	Reactant	Other products	Ionization or appearance potential (eV)	Method	Heat of formation (kJ mol^{-1})	Ref.
			$P_3O_3^+$			
$P_3O_3^+$	P_4O_6		18.8±0.5	EI		3370
			$P_3O_4^+$			
$P_3O_4^+$	P_4O_6		16.1±0.5	EI		3370
$P_3O_4^+$	P_4O_{10}		34.0±0.5	EI		3370
			$P_3O_5^+$			
$P_3O_5^+$	P_4O_6		15.0±0.5	EI		3370
$P_3O_5^+$	P_4O_{10}		32.9±0.5	EI		3370
			$P_3O_6^+$			
$P_3O_6^+$	P_4O_{10}		22.3±0.5	EI		3370
			$P_3O_7^+$			
$P_3O_7^+$	P_4O_{10}		17.8±0.5	EI		3370
			$P_4O_6^+$			
$P_4O_6^+$	P_4O_6		10.6±0.5	EI		3370
$P_4O_6^+$	P_4O_{10}		32.8±0.5	EI		3370
			$P_4O_7^+$			
$P_4O_7^+$	P_4O_7		12.1±0.5	EI		3370
			$P_4O_8^+$			
$P_4O_8^+$	P_4O_8		12.4±0.5	EI		3370
			$P_4O_9^+$			
$P_4O_9^+$	P_4O_9		13.1±0.5	EI		3370
$P_4O_9^+$	P_4O_{10}	O	20.0±0.5	EI		3370
			$P_4O_{10}^+$			
$P_4O_{10}^+$	P_4O_{10}		13.6±0.5	EI		3370

4.3. The Positive Ion Table—Continued

Ion	Reactant	Other products	Ionization or appearance potential (eV)	Method	Heat of formation (kJ mol⁻¹)	Ref.
			PF⁺			
PF^+	PSF_3		22.9 ± 0.4	EI		2506
			PF$_2^+$			
PF_2^+	PSF_3		17.2 ± 0.3	EI		2506
			PF$_3^+$			
PF_3^+	PF_3		9.71	PI		1091
PF_3^+	PF_3		12.28 (V)	PE		3119
PF_3^+	PF_3		12.3 (V)	PE	3043,	3084
PF_3^+	PF_3		12.31 (V)	PE		3070

The disagreement between the photoionization value and those obtained by photoelectron spectroscopy is not understood.

Ion	Reactant	Other products	Ionization or appearance potential (eV)	Method	Heat of formation (kJ mol⁻¹)	Ref.
PF_3^+	PSF_3	S?	14.3 ± 0.2	EI		2506
			AlP⁺			
AlP^+	AlP		8.4 ± 0.4	EI		2152
			CHP⁺			
CHP^+	CH_3PH_2		14.7 ± 0.3	EI		2045
CHP^+	$C_2H_5PH_2$		13.1 ± 0.5	EI		2045
CHP^+	$C_2H_5PH_2$		17.0 ± 0.3	EI		2948
CHP^+	$(CH_3)_2PH$		14.1	EI		1036
CHP^+	$(CH_3)_3P$		18.4 ± 0.2	EI		2045
			CH$_2$P⁺			
CH_2P^+	CH_3PH_2		14.7 ± 0.2	EI		2045
CH_2P^+	$C_2H_5PH_2$		12.7 ± 0.4	EI		2045
CH_2P^+	$C_2H_5PH_2$		15.1 ± 0.3	EI		2948
CH_2P^+	$(CH_3)_3P$		16.1 ± 0.4	EI		2045
CH_2P^+	$(CH_3)_3P$		17 ± 1	EI		1036
CH_2P^+	$(C_2H_5)_2PH$		18.1 ± 0.3	EI		2948
CH_2P^+	$(C_2H_5)_3P$		21.1 ± 0.3	EI		2948
CH_2P^+	$(CH_3)_2PP(CH_3)_2$		19.3 ± 0.3	EI		2948
CH_2P^+	$(C_2H_5)_2PP(C_2H_5)_2$		14.7 ± 0.3	EI		2948

4.3. The Positive Ion Table—Continued

Ion	Reactant	Other products	Ionization or appearance potential (eV)	Method	Heat of formation (kJ mol⁻¹)	Ref.
CH_3P^+						
CH_3P^+	CH_3PH_2	H_2	12.2±0.2	EI		2045
CH_3P^+	$C_2H_5PH_2$		12.0±0.2	EI		2045
CH_3P^+	$C_2H_5PH_2$		13.2±0.3	EI		2948
CH_3P^+	$(CH_3)_2PH$		11.9	EI		1036
CH_3P^+	$(CH_3)_3P$		14.0±0.3	EI		2045
CH_3P^+	$(C_2H_5)_2PH$		16.8±0.3	EI		2948
CH_3P^+	$(C_2H_5)_3P$		17.9±0.5	EI		2045
CH_3P^+	$(C_2H_5)_3P$		21.4±0.3	EI		2948
CH_3P^+	$(CH_3)_2PP(CH_3)_2$		13.5±0.3	EI		2948
CH_3P^+	$(C_2H_5)_2PP(C_2H_5)_2$		10.9±0.3	EI		2948
CH_4P^+						
CH_4P^+	CH_3PH_2	H	11.6±0.12	EI		2045
CH_4P^+	$C_2H_5PH_2$	CH_3	12.2±0.2	EI		2045
CH_4P^+	$C_2H_5PH_2$	CH_3	12.3±0.3	EI		2948
CH_4P^+	$(CH_3)_3P$		14.7±0.2	EI		2045
CH_4P^+	$(C_2H_5)_2PH$		14.4±0.3	EI		2948
CH_4P^+	$(C_2H_5)_3P$		15.8±0.2	EI		2045
CH_4P^+	$(C_2H_5)_3P$		16.2±0.3	EI		2948
CH_4P^+	$(CH_3)_2PP(CH_3)_2$		15.2±0.3	EI		2948
CH_4P^+	$(C_2H_5)_2PP(C_2H_5)_2$		13.1±0.3	EI		2948
CH_5P^+						
CH_5P^+	CH_3PH_2		9.72±0.15	EI		2045
CH_5P^+	$(CH_3)_2PP(CH_3)_2$		16.8±0.3	EI		2948
CH_5P^+	$(C_2H_5)_2PP(C_2H_5)_2$		10.7±0.3	EI		2948
CH_5P^+	$(n-C_4H_9)_2PP(n-C_4H_9)_2$		13.3±0.3	EI		2948
C_2HP^+						
C_2HP^+	$C_2H_5PH_2$		24.2±0.3	EI		2948
$C_2H_2P^+$						
$C_2H_2P^+$	$C_2H_5PH_2$		15.8±0.3	EI		2045
$C_2H_2P^+$	$C_2H_5PH_2$		17.4±0.3	EI		2948
$C_2H_2P^+$	$(CH_3)_3P$		16.7±0.2	EI		2045
$C_2H_2P^+$	$(C_2H_5)_3P$		16.5±0.3	EI		2045
$C_2H_3P^+$						
$C_2H_3P^+$	$C_2H_5PH_2$		12.9±0.4	EI		2045
$C_2H_3P^+$	$C_2H_5PH_2$		13.7±0.3	EI		2948
$C_2H_3P^+$	$(C_2H_5)_3P$		16.7±0.2	EI		2045

4.3. The Positive Ion Table—Continued

Ion	Reactant	Other products	Ionization or appearance potential (eV)	Method	Heat of formation (kJ mol^{-1})	Ref.
			C$_2$H$_4$P$^+$			
C$_2$H$_4$P$^+$	C$_2$H$_5$PH$_2$		12.9±0.3	EI		2045
C$_2$H$_4$P$^+$	C$_2$H$_5$PH$_2$		14.6±0.3	EI		2948
C$_2$H$_4$P$^+$	(CH$_3$)$_3$P		14.0±0.2	EI		2045
C$_2$H$_4$P$^+$	(CH$_3$)$_3$P		15±1	EI		1036
C$_2$H$_4$P$^+$	(C$_2$H$_5$)$_2$PH		15.6±0.3	EI		2948
C$_2$H$_4$P$^+$	(C$_2$H$_5$)$_3$P		16.0±0.2	EI		2045
C$_2$H$_4$P$^+$	(C$_2$H$_5$)$_3$P		17.5±0.3	EI		2948
C$_2$H$_4$P$^+$	(CH$_3$)$_2$PP(CH$_3$)$_2$		15.2±0.3	EI		2948
C$_2$H$_4$P$^+$	(C$_2$H$_5$)$_2$PP(C$_2$H$_5$)$_2$		12.5±0.3	EI		2948
			C$_2$H$_5$P$^+$			
C$_2$H$_5$P$^+$	C$_2$H$_5$PH$_2$	H$_2$	12.0±0.2	EI		2045
C$_2$H$_5$P$^+$	C$_2$H$_5$PH$_2$	H$_2$	12.4±0.3	EI		2948
C$_2$H$_5$P$^+$	(C$_2$H$_5$)$_2$PH		12.0±0.3	EI		2948
C$_2$H$_5$P$^+$	(C$_2$H$_5$)$_3$P		13.4±0.5	EI		2045
C$_2$H$_5$P$^+$	(C$_2$H$_5$)$_3$P		14.6±0.3	EI		2948
C$_2$H$_5$P$^+$	(CH$_3$)$_2$PP(CH$_3$)$_2$		13.6±0.3	EI		2948
C$_2$H$_5$P$^+$	(C$_2$H$_5$)$_2$PP(C$_2$H$_5$)$_2$		12.8±0.3	EI		2948
			C$_2$H$_6$P$^+$			
C$_2$H$_6$P$^+$	C$_2$H$_5$PH$_2$	H	12.0±0.3	EI		2045
C$_2$H$_6$P$^+$	C$_2$H$_5$PH$_2$	H	12.5±0.3	EI		2948
C$_2$H$_6$P$^+$	(CH$_3$)$_2$PH	H	12.2	EI		1036
C$_2$H$_6$P$^+$	(CH$_3$)$_3$P	CH$_3$	11.7±0.2	EI		2045
C$_2$H$_6$P$^+$	(CH$_3$)$_3$P	CH$_3$	11.8±0.2	EI		1036
C$_2$H$_6$P$^+$	(C$_2$H$_5$)$_2$PH		13.7±0.3	EI		2948
C$_2$H$_6$P$^+$	(C$_2$H$_5$)$_3$P		14.0±0.2	EI		2045
C$_2$H$_6$P$^+$	(C$_2$H$_5$)$_3$P		14.2±0.3	EI		2948
C$_2$H$_6$P$^+$	(CH$_3$)$_2$PP(CH$_3$)$_2$		13.3±0.3	EI		2948
C$_2$H$_6$P$^+$	(C$_2$H$_5$)$_2$PP(C$_2$H$_5$)$_2$		14.0±0.3	EI		2948
C$_2$H$_6$P$^+$	(n−C$_4$H$_9$)$_2$PP(n−C$_4$H$_9$)$_2$		17.8±0.3	EI		2948
			C$_2$H$_7$P$^+$			
C$_2$H$_7$P$^+$	C$_2$H$_5$PH$_2$		9.47±0.5	EI		2045
C$_2$H$_7$P$^+$	C$_2$H$_5$PH$_2$		9.5±0.3	EI		2948
C$_2$H$_7$P$^+$	(CH$_3$)$_2$PH		9.7	EI		1036
C$_2$H$_7$P$^+$	(C$_2$H$_5$)$_2$PH		10.9±0.3	EI		2948
C$_2$H$_7$P$^+$	(C$_2$H$_5$)$_3$P		12.3±0.3	EI		2948
C$_2$H$_7$P$^+$	(C$_2$H$_5$)$_3$P		12.7±0.2	EI		2045
			C$_3$H$_6$P$^+$			
C$_3$H$_6$P$^+$	(C$_2$H$_5$)$_2$PH		13.2±0.3	EI		2948
C$_3$H$_6$P$^+$	(C$_2$H$_5$)$_3$P		14.9±0.3	EI		2948
C$_3$H$_6$P$^+$	(CH$_3$)$_2$PP(CH$_3$)$_2$		11.3±0.3	EI		2948
C$_3$H$_6$P$^+$	(C$_2$H$_5$)$_2$PP(C$_2$H$_5$)$_2$		12.5±0.3	EI		2948

4.3. The Positive Ion Table—Continued

Ion	Reactant	Other products	Ionization or appearance potential (eV)	Method	Heat of formation (kJ mol^{-1})	Ref.
			$C_3H_8P^+$			
$C_3H_8P^+$	$(CH_3)_3P$	H	11.0±0.1	RPD		3017
$C_3H_8P^+$	$(CH_3)_3P$	H	10.2±0.5	EI		2045
$C_3H_8P^+$	$(CH_3)_3P$	H	11.8±0.2	EI		1036
$C_3H_8P^+$	$(C_2H_5)_2PH$	CH_3	12.1±0.3	EI		2948
$C_3H_8P^+$	$(C_2H_5)_3P$		13.8±0.5	EI		2045
			$C_3H_9P^+$			
$C_3H_9P^+$	$(CH_3)_3P$		8.60±0.2	EI		2045
$C_3H_9P^+$	$(CH_3)_3P$		9.2±0.5	EI		1036
			$C_4H_8P^+$			
$C_4H_8P^+$	$(C_2H_5)_3P$		14.5±0.3	EI		2948
			$C_4H_{10}P^+$			
$C_4H_{10}P^+$	$(C_2H_5)_2PH$	H	13.2±0.3	EI		2948
$C_4H_{10}P^+$	$(C_2H_5)_3P$	C_2H_5	11.4±0.3	EI		2045
$C_4H_{10}P^+$	$(C_2H_5)_3P$	C_2H_5	12.4±0.3	EI	2708,	2948
$C_4H_{10}P^+$	$(C_2H_5)_2PP(C_2H_5)_2$		13.3±0.3	EI	2708,	2948
			$C_4H_{11}P^+$			
$C_4H_{11}P^+$	$(C_2H_5)_2PH$		8.5±0.3	EI		2948
$C_4H_{11}P^+$	$(C_2H_5)_3P$		10.4±0.3	EI		2948
$C_4H_{11}P^+$	$(C_2H_5)_3P$		10.7±0.3	EI		2045
			$C_5H_{12}P^+$			
$C_5H_{12}P^+$	$(C_2H_5)_3P$	CH_3	11.8±0.3	EI		2948
$C_5H_{12}P^+$	$(C_2H_5)_3P$	CH_3	12.0±0.2	EI		2045
			$C_6H_{14}P^+$			
$C_6H_{14}P^+$	$(C_2H_5)_3P$	H	12.0±0.3	EI		2948
			$C_6H_{15}P^+$			
$C_6H_{15}P^+$	$(C_2H_5)_3P$		8.0±0.3	EI		2948
$C_6H_{15}P^+$	$(C_2H_5)_3P$		8.18±0.05	EI		2481
$C_6H_{15}P^+$	$(C_2H_5)_3P$		8.27±0.24	EI		2045
			$C_8H_{18}P^+$			
$C_8H_{18}P^+$	$(n-C_4H_9)_2PP(n-C_4H_9)_2$		11.5±0.3	EI		2948

4.3. The Positive Ion Table—Continued

Ion	Reactant	Other products	Ionization or appearance potential (eV)	Method	Heat of formation (kJ mol^{-1})	Ref.
			C$_9$H$_{21}$P$^+$			
C$_9$H$_{21}$P$^+$	(iso–C$_3$H$_7$)$_3$P		7.75±0.1	EI		2531, 2597
			C$_{12}$H$_{27}$P$^+$			
C$_{12}$H$_{27}$P$^+$	(n–C$_4$H$_9$)$_3$P		8.00±0.05	EI		2481
	C$_{18}$H$_{15}$P$^+$ (Triphenylphosphine)	$\Delta H^\circ_{f298} \sim$ 1039 kJ mol^{-1} (248 kcal mol^{-1})				
C$_{18}$H$_{15}$P$^+$	(C$_6$H$_5$)$_3$P (Triphenylphosphine)		7.36±0.05	PI	1039	1140
C$_{18}$H$_{15}$P$^+$	(C$_6$H$_5$)$_3$P (Triphenylphosphine)		7.83	EI		2597
C$_{18}$H$_{15}$P$^+$	(C$_6$H$_5$)$_3$P (Triphenylphosphine)		8.2±0.05	EI		2481
			CH$_3$P$_2^+$			
CH$_3$P$_2^+$	(CH$_3$)$_2$PP(CH$_3$)$_2$		13.5±0.3	EI		2948
CH$_3$P$_2^+$	(C$_2$H$_5$)$_2$PP(C$_2$H$_5$)$_2$		14.1±0.3	EI		2948
			CH$_4$P$_2^+$			
CH$_4$P$_2^+$	(n–C$_4$H$_9$)$_2$PP(n–C$_4$H$_9$)$_2$		13.8±0.3	EI		2948
			CH$_5$P$_2^+$			
CH$_5$P$_2^+$	(n–C$_4$H$_9$)$_2$PP(n–C$_4$H$_9$)$_2$		18.0±0.3	EI		2948
			C$_2$H$_5$P$_2^+$			
C$_2$H$_5$P$_2^+$	(CH$_3$)$_2$PP(CH$_3$)$_2$		13.5±0.3	EI		2948
C$_2$H$_5$P$_2^+$	(C$_2$H$_5$)$_2$PP(C$_2$H$_5$)$_2$		8.9±0.3	EI		2948
C$_2$H$_5$P$_2^+$	(n–C$_4$H$_9$)$_2$PP(n–C$_4$H$_9$)$_2$		12.6±0.3	EI		2948
			C$_2$H$_6$P$_2^+$			
C$_2$H$_6$P$_2^+$	(CH$_3$)$_2$PP(CH$_3$)$_2$		14.3±0.3	EI		2948
C$_2$H$_6$P$_2^+$	(C$_2$H$_5$)$_2$PP(C$_2$H$_5$)$_2$		16.7±0.3	EI		2948
C$_2$H$_6$P$_2^+$	(n–C$_4$H$_9$)$_2$PP(n–C$_4$H$_9$)$_2$		12.8±0.3	EI		2948
			C$_2$H$_7$P$_2^+$			
C$_2$H$_7$P$_2^+$	(CH$_3$)$_2$PP(CH$_3$)$_2$		13.2±0.3	EI		2948
C$_2$H$_7$P$_2^+$	(C$_2$H$_5$)$_2$PP(C$_2$H$_5$)$_2$		13.9±0.3	EI		2948

4.3. The Positive Ion Table—Continued

Ion	Reactant	Other products	Ionization or appearance potential (eV)	Method	Heat of formation (kJ mol⁻¹)	Ref.
			$C_2H_8P_2^+$			
$C_2H_8P_2^+$	$(CH_3)_2PP(CH_3)_2$		10.0 ± 0.3	EI		2948
$C_2H_8P_2^+$	$(C_2H_5)_2PP(C_2H_5)_2$		12.0 ± 0.3	EI		2948
			$C_3H_7P_2^+$			
$C_3H_7P_2^+$	$(C_2H_5)_2PP(C_2H_5)_2$		13.0 ± 0.3	EI		2948
$C_3H_7P_2^+$	$(n-C_4H_9)_2PP(n-C_4H_9)_2$		12.6 ± 0.3	EI		2948
			$C_3H_8P_2^+$			
$C_3H_8P_2^+$	$(C_2H_5)_2PP(C_2H_5)_2$		11.5 ± 0.3	EI		2948
$C_3H_8P_2^+$	$(n-C_4H_9)_2PP(n-C_4H_9)_2$		11.9 ± 0.3	EI		2948
			$C_3H_9P_2^+$			
$C_3H_9P_2^+$	$(CH_3)_2PP(CH_3)_2$	CH_3	10.8 ± 0.3	EI		2948
			$C_3H_{10}P_2^+$			
$C_3H_{10}P_2^+$	$(CH_3)_2PP(CH_3)_2$		10.6 ± 0.3	EI		2948
			$C_4H_{10}P_2^+$			
$C_4H_{10}P_2^+$	$(n-C_4H_9)_2PP(n-C_4H_9)_2$		12.1 ± 0.3	EI		2948
			$C_4H_{11}P_2^+$			
$C_4H_{11}P_2^+$	$(C_2H_5)_2PP(C_2H_5)_2$		12.3 ± 0.3	EI		2948
$C_4H_{11}P_2^+$	$(n-C_4H_9)_2PP(n-C_4H_9)_2$		13.5 ± 0.3	EI		2948
			$C_4H_{12}P_2^+$			
$C_4H_{12}P_2^+$	$(CH_3)_2PP(CH_3)_2$		8.6 ± 0.3	EI		2948
$C_4H_{12}P_2^+$	$(C_2H_5)_2PP(C_2H_5)_2$		11.6 ± 0.3	EI		2948
$C_4H_{12}P_2^+$	$(n-C_4H_9)_2PP(n-C_4H_9)_2$		12.8 ± 0.3	EI		2948
			$C_5H_{11}P_2^+$			
$C_5H_{11}P_2^+$	$(C_2H_5)_2PP(C_2H_5)_2$		11.9 ± 0.3	EI		2948

4.3. The Positive Ion Table—Continued

Ion	Reactant	Other products	Ionization or appearance potential (eV)	Method	Heat of formation (kJ mol^{-1})	Ref.
		$C_5H_{13}P_2^+$				
$C_5H_{13}P_2^+$	$(C_2H_5)_2PP(C_2H_5)_2$		12.1±0.3	EI		2948
$C_5H_{13}P_2^+$	$(n-C_4H_9)_2PP(n-C_4H_9)_2$		9.9±0.3	EI		2948
		$C_6H_{13}P_2^+$				
$C_6H_{13}P_2^+$	$(n-C_4H_9)_2PP(n-C_4H_9)_2$		11.3±0.3	EI		2948
		$C_6H_{14}P_2^+$				
$C_6H_{14}P_2^+$	$(C_2H_5)_2PP(C_2H_5)_2$		10.2±0.3	EI		2948
		$C_6H_{15}P_2^+$				
$C_6H_{15}P_2^+$	$(C_2H_5)_2PP(C_2H_5)_2$	C_2H_5	10.4±0.3	EI		2948
		$C_6H_{16}P_2^+$				
$C_6H_{16}P_2^+$	$(C_2H_5)_2PP(C_2H_5)_2$		9.8±0.3	EI		2948
		$C_6H_{17}P_2^+$				
$C_6H_{17}P_2^+$	$(C_2H_5)_2PP(C_2H_5)_2$		9.4±0.3	EI		2948
		$C_7H_{15}P_2^+$				
$C_7H_{15}P_2^+$	$(n-C_4H_9)_2PP(n-C_4H_9)_2$		14.7±0.3	EI		2948
		$C_7H_{16}P_2^+$				
$C_7H_{16}P_2^+$	$(n-C_4H_9)_2PP(n-C_4H_9)_2$		9.8±0.3	EI		2948
		$C_7H_{17}P_2^+$				
$C_7H_{17}P_2^+$	$(C_2H_5)_2PP(C_2H_5)_2$	CH_3	11.5±0.3	EI		2948
$C_7H_{17}P_2^+$	$(n-C_4H_9)_2PP(n-C_4H_9)_2$		12.0±0.3	EI		2948
		$C_8H_{18}P_2^+$				
$C_8H_{18}P_2^+$	$(n-C_4H_9)_2PP(n-C_4H_9)_2$		8.1±0.3	EI		2948

4.3. The Positive Ion Table—Continued

Ion	Reactant	Other products	Ionization or appearance potential (eV)	Method	Heat of formation (kJ mol^{-1})	Ref.
			$C_8H_{19}P_2^+$			
$C_8H_{19}P_2^+$	$(C_2H_5)_2PP(C_2H_5)_2$	H	8.2 ± 0.3	EI		2948
$C_8H_{19}P_2^+$	$(n-C_4H_9)_2PP(n-C_4H_9)_2$		12.2 ± 0.3	EI		2948
			$C_8H_{20}P_2^+$			
$C_8H_{20}P_2^+$	$(C_2H_5)_2PP(C_2H_5)_2$		7.8 ± 0.3	EI		2948
See also – EI: 2708						
$C_8H_{20}P_2^+$	$(n-C_4H_9)_2PP(n-C_4H_9)_2$		10.3 ± 0.3	EI		2948
			$C_9H_{19}P_2^+$			
$C_9H_{19}P_2^+$	$(n-C_4H_9)_2PP(n-C_4H_9)_2$		9.4 ± 0.3	EI		2948
			$C_{12}H_{27}P_2^+$			
$C_{12}H_{27}P_2^+$	$(n-C_4H_9)_2PP(n-C_4H_9)_2$		10.6 ± 0.3	EI		2948
			$C_{12}H_{28}P_2^+$			
$C_{12}H_{28}P_2^+$	$(n-C_4H_9)_2PP(n-C_4H_9)_2$		8.0 ± 0.3	EI		2948
			$C_{16}H_{36}P_2^+$			
$C_{16}H_{36}P_2^+$	$(n-C_4H_9)_2PP(n-C_4H_9)_2$		7.1 ± 0.3	EI		2948
			$PH_2O_2^+$			
$PH_2O_2^+$	$(CH_3O)_3PO$		15.1 ± 0.3	EI		3211
			PHO_3^+			
PHO_3^+	$(CH_3O)_2PHO$		11.2	EI		1036
			$PH_2O_3^+$			
$PH_2O_3^+$	$(C_2H_5O)_3PO$		17.8 ± 0.3	EI		3211
			$PH_3O_3^+$			
$PH_3O_3^+$	$(C_2H_5O)_3PO$		14.4 ± 0.2	EI		3211

4.3. The Positive Ion Table—Continued

Ion	Reactant	Other products	Ionization or appearance potential (eV)	Method	Heat of formation (kJ mol^{-1})	Ref.
		PH$_4$O$_3^+$				
PH$_4$O$_3^+$	(C$_2$H$_5$O)$_3$PO		14.5±0.3	EI		3211
		PH$_4$O$_4^+$				
PH$_4$O$_4^+$	(C$_2$H$_5$O)$_3$PO		14.3±0.3	EI		3211
		P$_4$H$_2$O$_7^+$				
P$_4$H$_2$O$_7^+$	P$_4$H$_2$O$_7$		11.9±0.5	EI		3370
		P$_4$H$_2$O$_9^+$				
P$_4$H$_2$O$_9^+$	P$_4$H$_2$O$_9$		13.5±0.5	EI		3370
		P$_4$H$_2$O$_{10}^+$				
P$_4$H$_2$O$_{10}^+$	P$_4$H$_2$O$_{10}$		14.0±0.5	EI		3370
		C$_3$F$_9$P$^+$				
C$_3$F$_9$P$^+$	(CF$_3$)$_3$P		11.3±0.1	EI		1007
		P$_3$N$_3$F$_6^+$				
P$_3$N$_3$F$_6^+$	(NPF$_2$)$_3$ (Cyclo–tris(difluorophosphonitrile))		11.4±0.1 (V)	PE		2952
P$_3$N$_3$F$_6^+$	(NPF$_2$)$_3$ (Cyclo–tris(difluorophosphonitrile))		11.64±0.05	EI		2425
		P$_4$N$_4$F$_8^+$				
P$_4$N$_4$F$_8^+$	(NPF$_2$)$_4$ (Cyclo–tetrakis(difluorophosphonitrile))		10.7±0.1 (V)	PE		2952
P$_4$N$_4$F$_8^+$	(NPF$_2$)$_4$ (Cyclo–tetrakis(difluorophosphonitrile))		10.86±0.05	EI		2425
		P$_5$N$_5$F$_{10}^+$				
P$_5$N$_5$F$_{10}^+$	(NPF$_2$)$_5$ (Cyclo–pentakis(difluorophosphonitrile))		11.4±0.1 (V)	PE		2952

See also – EI: 2425

4.3. The Positive Ion Table—Continued

Ion	Reactant	Other products	Ionization or appearance potential (eV)	Method	Heat of formation (kJ mol^{-1})	Ref.
		$P_6N_6F_{12}^+$				
$P_6N_6F_{12}^+$	$(NPF_2)_6$ (Cyclo−hexakis(difluorophosphonitrile))		10.9±0.1 (V)	PE		2952
See also − EI: 2425						
		$P_7N_7F_{14}^+$				
$P_7N_7F_{14}^+$	$(NPF_2)_7$ (Cyclo−heptakis(difluorophosphonitrile))		11.3±0.1 (V)	PE		2952
		$P_8N_8F_{16}^+$				
$P_8N_8F_{16}^+$	$(NPF_2)_8$ (Cyclo−octakis(difluorophosphonitrile))		10.9±0.1 (V)	PE		2952
		POF_3^+ $\quad \Delta H_{f0}^\circ = 28$ kJ mol^{-1} (7 kcal mol^{-1})				
POF_3^+	POF_3		12.75	PE	28	3358
See also − PE: 3084						
		$SiPH_5^+$				
$SiPH_5^+$	H_3SiPH_2		10.0±0.2	EI		173
		$C_6H_{18}N_3P_3^+$				
$C_6H_{18}N_3P_3^+$	$(NP(CH_3)_2)_3$ (Cyclo−tris(dimethylphosphonitrile))		8.35±0.05	EI		2952
		$C_{12}H_{36}N_9P_3^+$				
$C_{12}H_{36}N_9P_3^+$	$(NP(N(CH_3)_2)_2)_3$ (Cyclo−tris(bis(dimethylamino)phosphonitrile))		7.85±0.05	EI		2952
		$C_8H_{24}N_4P_4^+$				
$C_8H_{24}N_4P_4^+$	$(NP(CH_3)_2)_4$ (Cyclo−tetrakis(dimethylphosphonitrile))		7.99±0.05	EI		2952
		$C_{16}H_{48}N_{12}P_4^+$				
$C_{16}H_{48}N_{12}P_4^+$	$(NP(N(CH_3)_2)_2)_4$ (Cyclo−tetrakis(bis(dimethylamino)phosphonitrile))		7.45±0.05	EI		2952

4.3. The Positive Ion Table—Continued

Ion	Reactant	Other products	Ionization or appearance potential (eV)	Method	Heat of formation (kJ mol⁻¹)	Ref.
			CH_4OP^+			
CH_4OP^+	$(CH_3O)_2CH_3PO$		16.9±0.2	EI		3211
			$CH_4O_2P^+$			
$CH_4O_2P^+$	$(CH_3O)_2CH_3PO$		13.4±0.3	EI		3211
$CH_4O_2P^+$	$(CH_3O)_3PO$		15.1±0.2	EI		3211
			$CH_5O_2P^+$			
$CH_5O_2P^+$	$(CH_3O)_2CH_3PO$		11.9	EI		3211
$CH_5O_2P^+$	$(CH_3O)_3PO$		13.9±0.4	EI		3211
			$C_2H_6O_2P^+$			
$C_2H_6O_2P^+$	$(CH_3O)_2CH_3PO$		12.5±0.1	EI		3211
			$C_2H_7O_2P^+$			
$C_2H_7O_2P^+$	$(CH_3O)_2CH_3PO$	CH_2O	11.5±0.2	EI		3211
			$CH_4O_3P^+$			
$CH_4O_3P^+$	$(CH_3O)_2PHO$		11.9	EI		1036
$CH_4O_3P^+$	$(CH_3O)_3PO$		14.4±0.3	EI		3211
			$C_2H_6O_3P^+$			
$C_2H_6O_3P^+$	$(CH_3O)_2PHO$	H	12.7	EI		1036
$C_2H_6O_3P^+$	$(CH_3O)_2CH_3PO$	CH_3	13.3±0.3	EI		3211
$C_2H_6O_3P^+$	$(CH_3O)_3PO$		14.1	EI		3211
$C_2H_6O_3P^+$	$(C_2H_5O)_3PO$		15.5±0.4	EI		3211
			$C_2H_7O_3P^+$			
$C_2H_7O_3P^+$	$(CH_3O)_2PHO$		10.5	EI		1036
$C_2H_7O_3P^+$	$(CH_3O)_3PO$	CH_2O	11.9±0.2	EI		3211
$C_2H_7O_3P^+$	$(C_2H_5O)_3PO$		13.5±0.3	EI		3211
			$C_2H_8O_3P^+$			
$C_2H_8O_3P^+$	$(C_2H_5O)_3PO$		13.6±0.2	EI		3211

4.3. The Positive Ion Table—Continued

Ion	Reactant	Other products	Ionization or appearance potential (eV)	Method	Heat of formation (kJ mol⁻¹)	Ref.
		$C_3H_9O_3P^+$				
$C_3H_9O_3P^+$	$(CH_3O)_3P$		8.92	EI		3211
$C_3H_9O_3P^+$	$(CH_3O)_3P$		9.00±0.05	EI	2481,	2544
$C_3H_9O_3P^+$	$(CH_3O)_2CH_3PO$		10.48±0.20	EI		3211
		$C_4H_{10}O_3P^+$				
$C_4H_{10}O_3P^+$	$(C_2H_5O)_3PO$		12.7±0.2	EI		3211
		$C_4H_{11}O_3P^+$				
$C_4H_{11}O_3P^+$	$(C_2H_5O)_3PO$		11.6±0.4	EI		3211
		$C_4H_{12}O_3P^+$				
$C_4H_{12}O_3P^+$	$(C_2H_5O)_3PO$		12.3±0.2	EI		3211
		$C_6H_{15}O_3P^+$				
$C_6H_{15}O_3P^+$	$(C_2H_5O)_3P$		8.40±0.1	EI		2453
		$C_9H_{21}O_3P^+$				
$C_9H_{21}O_3P^+$	$(iso-C_3H_7O)_3P$		8.05	EI		2597
$C_9H_{21}O_3P^+$	$(iso-C_3H_7O)_3P$		8.46±0.05	EI		2481
		$C_{12}H_{27}O_3P^+$				
$C_{12}H_{27}O_3P^+$	$(n-C_4H_9O)_3P$		8.44±0.05	EI		2481
		$C_{18}H_{15}O_3P^+$				
$C_{18}H_{15}O_3P^+$	$(C_6H_5O)_3P$ (Triphenoxyphosphine)		8.60	EI		2597
		$C_2H_6O_4P^+$				
$C_2H_6O_4P^+$	$(C_2H_5O)_3PO$		13.5±0.3	EI		3211
		$C_2H_7O_4P^+$				
$C_2H_7O_4P^+$	$(C_2H_5O)_3PO$		12.8±0.2	EI		3211

4.3. The Positive Ion Table—Continued

Ion	Reactant	Other products	Ionization or appearance potential (eV)	Method	Heat of formation (kJ mol^{-1})	Ref.
		$C_2H_8O_4P^+$				
$C_2H_8O_4P^+$	$(C_2H_5O)_3PO$		12.7±0.2	EI		3211
		$C_3H_9O_4P^+$				
$C_3H_9O_4P^+$	$(CH_3O)_3PO$		10.77±0.30	EI		3211
		$C_4H_{12}O_4P^+$				
$C_4H_{12}O_4P^+$	$(C_2H_5O)_3PO$	C_2H_3	11.5±0.3	EI		3211
		$C_5H_{12}O_4P^+$				
$C_5H_{12}O_4P^+$	$(C_2H_5O)_3PO$	CH_3	11.9±0.3	EI		3211
		$C_6H_{15}O_4P^+$				
$C_6H_{15}O_4P^+$	$(C_2H_5O)_3PO$		10.06±0.27	EI		3211
		$BH_3F_3P^+$				
$BH_3F_3P^+$	PF_3BH_3		12.0 (V)	PE		3043
		$C_6H_{18}N_3O_6P_3^+$				
$C_6H_{18}N_3O_6P_3^+$	$(NP(OCH_3)_2)_3$ (Cyclo−tris(dimethoxyphosphonitrile))		9.29±0.05	EI		2952
		$C_{36}H_{30}N_3O_6P_3^+$				
$C_{36}H_{30}N_3O_6P_3^+$	$(NP(OC_6H_5)_2)_3$ (Cyclo−tris(diphenoxyphosphonitrile))		8.83±0.05	EI		2952
		$C_8H_{24}N_4O_8P_4^+$				
$C_8H_{24}N_4O_8P_4^+$	$(NP(OCH_3)_2)_4$ (Cyclo−tetrakis(dimethoxyphosphonitrile))		8.83±0.05	EI		2952
		$C_{48}H_{40}N_4O_8P_4^+$				
$C_{48}H_{40}N_4O_8P_4^+$	$(NP(OC_6H_5)_2)_4$ (Cyclo−tetrakis(diphenoxyphosphonitrile))		8.70±0.05	EI		2952

4.3. The Positive Ion Table—Continued

Ion	Reactant	Other products	Ionization or appearance potential (eV)	Method	Heat of formation (kJ mol^{-1})	Ref.
		$C_{12}H_{12}N_3O_6F_{18}P_3^+$				
$C_{12}H_{12}N_3O_6F_{18}P_3^+$	$(NP(OCH_2CF_3)_2)_3$ (Cyclo–tris(bis(2,2,2–trifluoroethoxy)phosphonitrile))		10.43±0.05	EI		2952
		$C_{16}H_{16}N_4O_8F_{24}P_4^+$				
$C_{16}H_{16}N_4O_8F_{24}P_4^+$	$(NP(OCH_2CF_3)_2)_4$ (Cyclo–tetrakis(bis(2,2,2–trifluoroethoxy)phosphonitrile))		10.01±0.05	EI		2952
	S^+	$\Delta H^\circ_{f0} = 1276$ kJ mol^{-1} (305 kcal mol^{-1})				
S^+	S		10.360	S	1276	2113

See also – EI: 2472

S^+	S_2	S	14.74±0.01	PI	(1274)	2609

   (Threshold value approximately corrected to 0 K)

See also – EI: 2569

S^+	H_2S	H_2	13.40±0.01	PI		2622
S^+	H_2S	H_2	13.36±0.01	PI	(1271)	2622

  (Appearance potential of the corresponding metastable transition)

See also – EI: 3134

S^+	CS_2	CS	14.81±0.03	PI		2624, 2627

See also – EI: 2472, 2528, 3402

S^+	SO_2	O_2	17.5±0.3	EI		418
S^+	$(CH_2)_2S$ (Ethylene sulfide)	C_2H_4	13.1±0.2	EI		51
S^+	C_2H_5SH		14.2±0.3	EI		3286
S^+	CH_3SSCH_3		15.0±0.3	EI		3286
S^+	COS	CO	13.65±0.03	PI	(1289)	2624, 2627
S^+	CF_3SSCF_3		~13	EI		3202
S^+	$(CH_3)_2SO$		10.8±0.3	EI		3294
S^+	SF_5Cl		33.2±0.5	EI		2777
S^+	SnS	Sn	16.5±2.0	EI		2139
S^+	PbS	Pb	16.0±2.0	EI		2139

	S_2^+	$\Delta H^\circ_{f0} \sim 1031$ kJ mol^{-1} (246 kcal mol^{-1})				
S_2^+	S_2		9.40±0.05	S	1035	2878
S_2^+	S_2		~9.32	S	~1027	3111
S_2^+	S_2		9.36±0.02	PI	1031	2603, 2609

  (Threshold value approximately corrected to 0 K)

See also – EI: 319, 2022, 2139, 2172, 2469, 2482, 2569, 2747, 3155, 3249, 3402, 3428

4.3. The Positive Ion Table—Continued

Ion	Reactant	Other products	Ionization or appearance potential (eV)	Method	Heat of formation (kJ mol⁻¹)	Ref.
S_2^+	S_6		12.8±1.0	EI		1035
S_2^+	CS_2	C	14.9±0.3	EI		3402
S_2^+	CS_2	C	18.2±0.9	EI		2472
S_2^+	CS_2^+	C	9.6±0.6	SEQ		2472
S_2^+	CH_3SSCH_3		15.01+0.13	EI		3202
S_2^+	CH_3SSCH_3		15.4±0.3	EI		176
S_2^+	CH_3SSCH_3		15.9±0.3	EI		3286
S_2^+	$C_4H_8S_2$ (1,4–Dithiane)	$2C_2H_4$	12.4±0.2	EI		2969
S_2^+	$C_2H_5SSC_2H_5$		14.0±0.3	EI		3286
S_2^+	$C_2H_5SSC_2H_5$		14.9±0.4	EI		186
S_2^+	CH_3SSSCH_3		14.4±0.3	EI		84
S_2^+	CF_3SSCF_3		13.22±0.17	EI		3202

$$S_3^+ \qquad \Delta H^\circ_{f298} = 1067 \text{ kJ mol}^{-1} \text{ (255 kcal mol}^{-1}\text{)}$$

Ion	Reactant	Other products	Ionization or appearance potential (eV)	Method	Heat of formation (kJ mol⁻¹)	Ref.
S_3^+	S_3		9.68±0.03	PI	1067	2603
(Threshold value approximately corrected to 0 K)						
S_3^+	S_3		9.9±0.4	EI		2482
S_3^+	S_6		13.3±0.5	EI		1035
S_3^+	S_8		12.6±0.5	EI		1035

$$S_4^+ \qquad \Delta H^\circ_{f298} \sim 1126 \text{ kJ mol}^{-1} \text{ (269 kcal mol}^{-1}\text{)}$$

Ion	Reactant	Other products	Ionization or appearance potential (eV)	Method	Heat of formation (kJ mol⁻¹)	Ref.
S_4^+	S_4		10.4±0.5	EI	(~1140)	2172

See also – EI: 2482

Ion	Reactant	Other products	Ionization or appearance potential (eV)	Method	Heat of formation (kJ mol⁻¹)	Ref.
S_4^+	S_6	S_2	11.94±0.05	PI	1126	2603
(Threshold value approximately corrected to 0 K)						

See also – EI: 1035

Ion	Reactant	Other products	Ionization or appearance potential (eV)	Method	Heat of formation (kJ mol⁻¹)	Ref.
S_4^+	S_8		12.5±0.3	EI		1035

$$S_5^+ \qquad \Delta H^\circ_{f298} = 954 \text{ kJ mol}^{-1} \text{ (228 kcal mol}^{-1}\text{)}$$

Ion	Reactant	Other products	Ionization or appearance potential (eV)	Method	Heat of formation (kJ mol⁻¹)	Ref.
S_5^+	S_5		8.60±0.05	PI	954	2603
(Threshold value approximately corrected to 0 K)						

See also – EI: 2482

Ion	Reactant	Other products	Ionization or appearance potential (eV)	Method	Heat of formation (kJ mol⁻¹)	Ref.
S_5^+	S_8	S_3	10.2	PI	(954)	2603
(Threshold value approximately corrected to 0 K)						

See also – EI: 1035

4.3. The Positive Ion Table—Continued

Ion	Reactant	Other products	Ionization or appearance potential (eV)	Method	Heat of formation (kJ mol^{-1})	Ref.
S_6^+		$\Delta H_{f298}^\circ = 971$ kJ mol^{-1} (232 kcal mol^{-1})				
S_6^+	S_6		9.00±0.03	PI	971	2603
(Threshold value approximately corrected to 0 K)						
S_6^+	S_6		8.5±0.3	EI		2172
S_6^+	S_6		9.7±0.3	EI		1035
See also – EI: 2482						
S_6^+	S_8	S_2	≤11.0	PI	(≤1035)	2603
S_6^+	S_8	S_2	10.1±0.3	EI		1035
S_7^+		$\Delta H_{f298}^\circ = 950$ kJ mol^{-1} (227 kcal mol^{-1})				
S_7^+	S_7		8.67±0.03	PI	950	2603
(Threshold value approximately corrected to 0 K)						
S_7^+	S_7		8.7±0.2	EI		2482
S_7^+	S_7		9.3±0.3	EI		1035
S_8^+		$\Delta H_{f0}^\circ = 978$ kJ mol^{-1} (234 kcal mol^{-1})				
S_8^+	S_8		9.04±0.03	PI	978	2603
(Threshold value approximately corrected to 0 K)						
S_8^+	S_8		9.6±0.2	EI		1035
S_8^+	S_8		7.3±0.3	EI		2172
See also – EI: 2482, 3155						
HS^+		$\Delta H_{f0}^\circ \sim 1145$ kJ mol^{-1} (274 kcal mol^{-1})				
HS^+	HS		10.41±0.03	S	~1147	2665
HS^+	HS		10.50±0.1	EI		120
HS^+	H_2S	H	14.27±0.02	PI	1143	2622
HS^+	H_2S	H	14.43±0.1	EI		120
See also – EI: 3134						
HS^+	$(CH_3)_2SO$		15.8±0.1	EI		3294
$H_2S^+(^2B_1)$		$\Delta H_{f0}^\circ = 993$ kJ mol^{-1} (237 kcal mol^{-1})				
$H_2S^+(^2B_1)$	H_2S		10.47±0.01	S	993	387, 3140, 3317
(Average of two Rydberg series limits)						
$H_2S^+(^2B_1)$	H_2S		10.43±0.01	PI		2622
$H_2S^+(^2B_1)$	H_2S		10.458±0.01	PI		158, 182, 416, 1103
$H_2S^+(^2B_1)$	H_2S		10.43	PE		3035, 3170
$H_2S^+(^2B_1)$	H_2S		10.42	PE		1130
$H_2S^+(^2B_1)$	H_2S		10.45±0.03	RPD		463

4.3. The Positive Ion Table—Continued

Ion	Reactant	Other products	Ionization or appearance potential (eV)	Method	Heat of formation (kJ mol⁻¹)	Ref.
$H_2S^+(^2A_1)$	H_2S		12.81	PE		3170
$H_2S^+(^2A_1)$	H_2S		12.76	PE		3035
$H_2S^+(^2A_1)$	H_2S		12.62	PE		1130
$H_2S^+(^2B_2)$	H_2S		14.79	PE		3170
$H_2S^+(^2B_2)$	H_2S		14.91	PE		3035
$H_2S^+(^2B_2)$	H_2S		14.82	PE		1130
H_2S^+*	H_2S		18.0?	PE		3035
H_2S^+*	H_2S		18.00?	PE		1130
H_2S^+*	H_2S		20.8	PE		3035
H_2S^+*	H_2S		20.12	PE		1130

See also — EI: 383, 2483, 3134, 3426

Ion	Reactant	Other products	Ionization or appearance potential (eV)	Method	Heat of formation (kJ mol⁻¹)	Ref.
H_2S^+	$(CH_2)_2S$ (Ethylene sulfide)	C_2H_2	13.4	RPD		3345
(0.07 eV average translational energy of decomposition at threshold)						
H_2S^+	$(CH_2)_2S$ (Ethylene sulfide)	C_2H_2	13.4±0.1	EI		51
H_2S^+	C_2H_5SH	C_2H_4	11.8±0.3	EI		3286
H_2S^+	$(CH_3)_2S$		14.29±0.04	EI		3202
H_2S^+	$(CH_3)_2SO$		11±0.1	EI		3294

HDS⁺

Ion	Reactant	Other products	Ionization or appearance potential (eV)	Method	Heat of formation (kJ mol⁻¹)	Ref.
HDS^+	HDS		10.3±0.2	EI		3134

D₂S⁺

Ion	Reactant	Other products	Ionization or appearance potential (eV)	Method	Heat of formation (kJ mol⁻¹)	Ref.
D_2S^+	D_2S		10.2±0.2	EI		3134

H₂S₂⁺

Ion	Reactant	Other products	Ionization or appearance potential (eV)	Method	Heat of formation (kJ mol⁻¹)	Ref.
$H_2S_2^+$	$C_2H_5SSC_2H_5$		11.9	EI		307
$H_2S_2^+$	$C_2H_5SSC_2H_5$		12.2±0.2	EI	186,	191
$H_2S_2^+$	$C_2H_5SSC_2H_5$		13.1±0.3	EI		3286

BS⁺

Ion	Reactant	Other products	Ionization or appearance potential (eV)	Method	Heat of formation (kJ mol⁻¹)	Ref.
BS^+	B_2S_3		17.2±0.2	EI		3428

BS₂⁺

Ion	Reactant	Other products	Ionization or appearance potential (eV)	Method	Heat of formation (kJ mol⁻¹)	Ref.
BS_2^+	B_2S_3	BS	13.6±0.2	EI		3428
BS_2^+	$(HBS_2)_3$?		16.4±0.3	EI	2483,	3426

4.3. The Positive Ion Table—Continued

Ion	Reactant	Other products	Ionization or appearance potential (eV)	Method	Heat of formation (kJ mol^{-1})	Ref.
		B$_2$S$_2^+$				
B$_2$S$_2^+$	B$_2$S$_3$	S	12.5±0.2	EI		3428
		B$_2$S$_3^+$				
B$_2$S$_3^+$	B$_2$S$_3$		9.4±0.2	EI		3428
B$_2$S$_3^+$	B$_2$S$_3$		10.4±0.2	EI		2147
B$_2$S$_3^+$	(HBS$_2$)$_3$		13.1±0.3	EI		2483, 3426
		B$_2$S$_3^{+2}$				
B$_2$S$_3^{+2}$	B$_2$S$_3$		24.4±0.2	EI		3428
		B$_3$S$_3^+$				
B$_3$S$_3^+$	(HBS$_2$)$_3$		13.0±0.3	EI		2483, 3426
		B$_2$S$_4^+$				
B$_2$S$_4^+$	(HBS$_2$)$_3$?		13.0±0.3	EI		2483, 3426
		B$_3$S$_4^+$				
B$_3$S$_4^+$	B$_4$S$_6$	BS$_2$	11.0±0.2	EI		3428
B$_3$S$_4^+$	(HBS$_2$)$_3$		15.0±0.3	EI		2483, 3426
		B$_3$S$_5^+$				
B$_3$S$_5^+$	(HBS$_2$)$_3$		15.2±0.3	EI		2483, 3426
		B$_4$S$_6^+$				
B$_4$S$_6^+$	B$_4$S$_6$		9.5±0.8	EI		2147
B$_4$S$_6^+$	B$_4$S$_6$		10.0±0.2	EI		3428
		B$_8$S$_{14}^+$				
B$_8$S$_{14}^+$	B$_8$S$_{16}$	S$_2$	11.0	EI		3427
		B$_8$S$_{16}^+$				
B$_8$S$_{16}^+$	B$_8$S$_{16}$		8.5	EI		3427

4.3. The Positive Ion Table—Continued

Ion	Reactant	Other products	Ionization or appearance potential (eV)	Method	Heat of formation (kJ mol^{-1})	Ref.

CS^+

Ion	Reactant	Other products	Ionization or appearance potential (eV)	Method	Heat of formation (kJ mol^{-1})	Ref.
CS^+	CS		~11.65	S	(~1394)	2878
CS^+	CS		11.8±0.2	EI		319
CS^+	CS		11.71±0.03	D		2624, 2627

See also – EI: 3402

Ion	Reactant	Other products	Ionization or appearance potential (eV)	Method	Heat of formation (kJ mol^{-1})	Ref.
CS^+	CS_2	S	16.16±0.01	PI	(1399)	2624, 2627
CS^+	CS_2	S	16.15±0.10	RPD		2528

See also – EI: 2472, 3402

Ion	Reactant	Other products	Ionization or appearance potential (eV)	Method	Heat of formation (kJ mol^{-1})	Ref.
CS^+	CS_2^+	S	9.6±0.6	SEQ		2472
CS^+	C_2H_5SH	CH_4+H_2	11.7±0.3	EI		3286
CS^+	CH_3SSCH_3	H_2S+CH_4	12.0±0.3	EI		3286
CS^+	$C_2H_5SSC_2H_5$	$C_2H_5SH+CH_4$	11.4±0.3	EI		3286
CS^+	CH_3NCS		15.6±0.4	EI		315
CS^+	C_2H_5NCS		16.1±0.5	EI		315
CS^+	$(CH_3)_2SO$		11.3±0.3	EI		3294

$CS_2^+(X^2\Pi_{3/2g})$	$\Delta H_{f0}^\circ = 1089$ kJ mol^{-1} (260 kcal mol^{-1})
$CS_2^+(X^2\Pi_{1/2g})$	$\Delta H_{f0}^\circ = 1094$ kJ mol^{-1} (262 kcal mol^{-1})
$CS_2^+(A^2\Pi_{3/2u})$	$\Delta H_{f0}^\circ = 1329$ kJ mol^{-1} (318 kcal mol^{-1})
$CS_2^+(A^2\Pi_{1/2u})$	$\Delta H_{f0}^\circ = 1331$ kJ mol^{-1} (318 kcal mol^{-1})
$CS_2^+(B^2\Sigma_u^+)$	$\Delta H_{f0}^\circ = 1513$ kJ mol^{-1} (362 kcal mol^{-1})
$CS_2^+(C^2\Sigma_g^+)$	$\Delta H_{f0}^\circ = 1679$ kJ mol^{-1} (401 kcal mol^{-1})

Ion	Reactant	Other products	Ionization or appearance potential (eV)	Method	Heat of formation (kJ mol^{-1})	Ref.
$CS_2^+(X^2\Pi_{3/2g})$	CS_2		10.080	S	1089	149
$CS_2^+(X^2\Pi_{3/2g})$	CS_2		10.079	S	1089	3313
$CS_2^+(X^2\Pi_{1/2g})$	CS_2		10.134	S	1094	149
$CS_2^+(X^2\Pi_{1/2g})$	CS_2		10.133	S	1094	3313
$CS_2^+(A^2\Pi_{3/2u})$	CS_2		12.563	S	1329	3049
$CS_2^+(A^2\Pi_{1/2u})$	CS_2		12.586	S	1331	3049
$CS_2^+(B^2\Sigma_u^+)$ (Average of three Rydberg series limits)	CS_2		14.475	S	1513	3049
$CS_2^+(B^2\Sigma_u^+)$ (Average of two Rydberg series limits)	CS_2		14.476	S	1513	149
$CS_2^+(C^2\Sigma_g^+)$ (Average of three Rydberg series limits)	CS_2		16.188	S	1678	3049
$CS_2^+(C^2\Sigma_g^+)$ (Average of three Rydberg series limits)	CS_2		16.190	S	1679	149
$CS_2^+*?$ (Average of two Rydberg series limits)	CS_2		19.381	S		3049
$CS_2^+*?$	CS_2		19.389	S		149

(Limit decreased by 1000 cm^{-1} as suggested in ref. 3049)

This state was searched for but not found with photoelectron spectroscopy, see ref. 2901.

4.3. The Positive Ion Table—Continued

Ion	Reactant	Other products	Ionization or appearance potential (eV)	Method	Heat of formation (kJ mol⁻¹)	Ref.
$CS_2^+(X^2\Pi_{3/2g})$	CS_2		10.068±0.005	PE		2901
$CS_2^+(X^2\Pi_{3/2g})$	CS_2		10.084±0.01	PE		2848
$CS_2^+(X^2\Pi_{1/2g})$	CS_2		10.122±0.005	PE		2901
$CS_2^+(X^2\Pi_{1/2g})$	CS_2		10.136±0.01	PE		2848
$CS_2^+(A^2\Pi_u)$	CS_2		12.694±0.005	PE		2901
$CS_2^+(A^2\Pi_u)$	CS_2		12.63±0.02	PE		2848
$CS_2^+(B^2\Sigma_u^+)$	CS_2		14.478±0.005	PE		2901
$CS_2^+(B^2\Sigma_u^+)$	CS_2		14.47±0.01	PE		2848
$CS_2^+(C^2\Sigma_g^+)$	CS_2		16.196±0.005	PE		2901
$CS_2^+(C^2\Sigma_g^+)$	CS_2		16.19±0.01	PE		2848
$CS_2^+(D^2\Sigma_u^+?)$	CS_2		~17.0 (V)	PE		2901
$CS_2^+(X^2\Pi_{3/2g})$	CS_2		10.059±0.008	PI	2624,	2627
$CS_2^+(X^2\Pi_{3/2g})$	CS_2		10.075	PI		2528
$CS_2^+(X^2\Pi_{1/2g})$	CS_2		10.112±0.008	PI	2624,	2627
$CS_2^+(X^2\Pi_{1/2g})$	CS_2		10.13	PI		2528

See also – S: 409, 410
PI: 158, 182, 190, 416
PE: 92, 1130, 2839, 2856, 2875
PEN: 2430
EI: 164, 169, 2528, 3174, 3201, 3402

Ion	Reactant	Other products	Ionization or appearance potential (eV)	Method	Heat of formation (kJ mol⁻¹)	Ref.
CS_2^+	$C_2H_5SSC_2H_5$		10.0±0.3	EI		3286

CS_2^{+2}

Ion	Reactant	Other products	Ionization or appearance potential (eV)	Method	Heat of formation (kJ mol⁻¹)	Ref.
CS_2^{+2}	CS_2		27.45±0.2	NRE		1040
CS_2^{+2}	CS_2		25.5±0.3	EI		2472
CS_2^{+2}	CS_2^+		16.4±0.3	SEQ		2472

CS_2^{+3}

Ion	Reactant	Other products	Ionization or appearance potential (eV)	Method	Heat of formation (kJ mol⁻¹)	Ref.
CS_2^{+3}	CS_2		53.6±0.5	NRE		1040

NS^+

Ion	Reactant	Other products	Ionization or appearance potential (eV)	Method	Heat of formation (kJ mol⁻¹)	Ref.
NS^+	NS		9.85±0.28	D		2888
NS^+	NSF	F	11.80±0.05	RPD		2443
NS^+?	CH_3NCS		12.5±0.2	EI		315

4.3. The Positive Ion Table—Continued

Ion	Reactant	Other products	Ionization or appearance potential (eV)	Method	Heat of formation (kJ mol⁻¹)	Ref.

$$SO^+ \qquad \Delta H^\circ_{f0} = 1004 \text{ kJ mol}^{-1} \text{ (240 kcal mol}^{-1}\text{)}$$

Ion	Reactant	Other products	IP/AP (eV)	Method	ΔHf (kJ mol⁻¹)	Ref.
SO^+	SO		10.0 ± 0.1	S		2761
SO^+	SO		10.34 ± 0.02	PE	1004	3241
$SO^+{}^*$	SO		14.96 ± 0.02	PE		3241
$SO^+{}^*$	SO		$\sim18?$	PE		3241
SO^+	SO_2	O	15.81 ± 0.02	PI	(984)	2622

See also – EI: 418, 2022, 2172

Ion	Reactant	Other products	IP/AP (eV)	Method	ΔHf (kJ mol⁻¹)	Ref.
SO^+	S_2O	S	14.5 ± 0.2	EI		2022, 2172
SO^+	SO_2F_2	2F+O	24.3 ± 0.3	EI		418
SO^+	$(CH_3)_2SO$		11 ± 0.1	EI		3294

$$SO_2^+({}^2A_1) \qquad \Delta H^\circ_{f0} = 895 \text{ kJ mol}^{-1} \text{ (214 kcal mol}^{-1}\text{)}$$

Ion	Reactant	Other products	IP/AP (eV)	Method	ΔHf (kJ mol⁻¹)	Ref.
$SO_2^+({}^2A_1)$	SO_2		12.34	S	896	3318
$SO_2^+({}^2A_1)$	SO_2		12.32 ± 0.01	PI	894	2622
$SO_2^+({}^2A_1)$	SO_2		12.34 ± 0.02	PI	896	182, 416, 3318
$SO_2^+({}^2A_1)$	SO_2		12.30 ± 0.01	PE	893	2848
$SO_2^+{}^*$	SO_2		13.01 ± 0.05	PE		2848
$SO_2^+{}^*$	SO_2		$13.24^{+0.05}_{-0.15}$	PE		2848
$SO_2^+{}^*$	SO_2		15.986 ± 0.005	PE		2848
$SO_2^+{}^*$	SO_2		16.326 ± 0.005	PE		2848
$SO_2^+{}^*$	SO_2		16.7?	PE		2848

The assignment of the excited states is somewhat uncertain, see ref. 2848.

See also – S: 3313
           PI: 297
           PE: 1130
           PEN: 2430
           EI: 418, 2022, 2172, 2487

Ion	Reactant	Other products	IP/AP (eV)	Method	ΔHf (kJ mol⁻¹)	Ref.
SO_2^+	SO_2F_2	2F?	19.9 ± 0.3	EI		418

$$SO_3^+$$

Ion	Reactant	Other products	IP/AP (eV)	Method	ΔHf (kJ mol⁻¹)	Ref.
SO_3^+	SO_3		11.0 ± 0.5	EI		2487

4.3. The Positive Ion Table—Continued

Ion	Reactant	Other products	Ionization or appearance potential (eV)	Method	Heat of formation (kJ mol^{-1})	Ref.
			S_2O^+			
S_2O^+	S_2O		10.3 ± 0.1	EI		2022, 2172
			SF^+			
SF^+	SF		~14	EI		3220
SF^+	SF_2?		~18	EI		3220
SF^+	NSF	N	15.45 ± 0.1	RPD		2443
			SF_2^+			
SF_2^+	NSF_3		15.70 ± 0.05	RPD		2443
SF_2^+	$NCNSF_2$		12.1 ± 0.1	RPD		2443
SF_2^+	CF_3NSF_2		12.95 ± 0.1	RPD		2443
			SF_3^+			
SF_3^+	SF_4	F	12.70 ± 0.03	RPD		2443
SF_3^+	SF_6	F_2+F?	19.80 ± 0.10	RPD		3040, 3041
SF_3^+	NSF_3	N	13.46 ± 0.03	RPD		2443
SF_3^+	SF_5NF_2	NF_2+F_2?	16.0 ± 0.2	EI		1144
			SF_4^+			
SF_4^+	SF_4		12.28 ± 0.03	RPD		2443
SF_4^+	SF_6	2F?	18.50 ± 0.10	RPD		3040, 3041
SF_4^+	SF_5NF_2	NF_3?	15.9 ± 0.2	EI		1144
			SF_5^+			
SF_5^+	SF_6	F	15.29	PI	(203)	2027
SF_5^+	SF_6	F	15.75 ± 0.05	RPD		3040, 3041
SF_5^+	SF_6	F	15.85 ± 0.15	RPD		196
SF_5^+	SF_5NF_2		12.0 ± 0.2	EI		1144
SF_5^+	SF_5Cl	Cl	13.2 ± 0.2	EI		2777

4.3. The Positive Ion Table—Continued

Ion	Reactant	Other products	Ionization or appearance potential (eV)	Method	Heat of formation (kJ mol^{-1})	Ref.
			SF_6^+			
(SF_6^+)	SF_6		15.35±0.02	PE		2852, 2853
(SF_6^+)	SF_6		15.30	PE		3040, 3041
(SF_6^+)	SF_6		15.69 (V)	PE		3119
(SF_6^+*)	SF_6		16.16	S		3100

This ionization potential is based on maxima of a three–member Rydberg series with very broad bands ~5000–20000 cm^{-1} wide. It is not observed in photoelectron spectra, see refs. 3040, 3041, 3119.

Ion	Reactant	Other products	Ionization or appearance potential (eV)	Method	Heat of formation (kJ mol^{-1})	Ref.
(SF_6^+*)	SF_6		16.71±0.02	PE		2852, 2853
(SF_6^+*)	SF_6		16.70	PE		3040, 3041
(SF_6^+*)	SF_6		16.96 (V)	PE		3119
(SF_6^+*)	SF_6		18.11±0.02	PE		2852, 2853
(SF_6^+*)	SF_6		18.10	PE		3040, 3041
(SF_6^+*)	SF_6		18.40 (V)	PE		3119
(SF_6^+*)	SF_6		18.50	PE		3040, 3041
(SF_6^+*)	SF_6		18.71 (V)	PE		3119
(SF_6^+*)	SF_6		19.50±0.02	PE		2852, 2853
(SF_6^+*)	SF_6		19.30	PE		3040, 3041
(SF_6^+*)	SF_6		19.68 (V)	PE		3119
(SF_6^+*)	SF_6		22.46	S		2653, 3119
(SF_6^+*)	SF_6		22.5 (V)	PE		3119
(SF_6^+*)	SF_6		26.83±0.04	S		2653
(SF_6^+*)	SF_6		26.8 (V)	PE		3119

No significant amount of SF_6^+ is found in the mass spectrum, see for example ref. 2027. The ion is strongly Jahn–Teller distorted. For tentative assignments of energy levels see ref. 3119. At the ionization threshold SF_5^+ is produced; compare PE results with SF_5^+ data above.

Ion	Reactant	Other products	Ionization or appearance potential (eV)	Method	Heat of formation (kJ mol^{-1})	Ref.
			AlS^+			
AlS^+	AlS		9.5±0.5	EI		2449
			Al_2S^+			
Al_2S^+	Al_2S		9.0±0.5	EI		2449
			$Al_2S_2^+$			
$Al_2S_2^+$	Al_2S_2		9.5±0.5	EI		2449

4.3. The Positive Ion Table—Continued

Ion	Reactant	Other products	Ionization or appearance potential (eV)	Method	Heat of formation (kJ mol⁻¹)	Ref.
PS⁺						
PS^+	PSF_3		19.2 ± 0.5	EI		2506
BH₂S₂⁺						
$BH_2S_2^+$	$(HBS_2)_3$	HBS_2+BS_2	11.8 ± 0.3	EI		2483, 3426
B₂HS₂⁺						
$B_2HS_2^+$	$(HBS_2)_3?$		16.6 ± 0.3	EI		2483, 3426
BH₃S₃⁺						
$BH_3S_3^+$	H_3BS_3		9.9 ± 0.3	EI		2483, 3426
B₂HS₃⁺						
$B_2HS_3^+$	$(HBS_2)_3$		13.1 ± 0.3	EI		2483, 3426
B₂HS₄⁺						
$B_2HS_4^+$	$(HBS_2)_3?$		11.4 ± 0.3	EI		2483, 3426
B₂H₂S₄⁺						
$B_2H_2S_4^+$	$(HBS_2)_3$	HBS_2	10.5 ± 0.3	EI		2483, 3426
B₂H₂S₅⁺						
$B_2H_2S_5^+$	$H_2B_2S_5$		8.9 ± 0.3	EI		2483, 3426
B₃HS₅⁺						
$B_3HS_5^+$	$(HBS_2)_3$		11.2 ± 0.3	EI		2483, 3426
B₃H₂S₅⁺						
$B_3H_2S_5^+$	$(HBS_2)_3$		11.5 ± 0.3	EI		2483, 3426
B₃H₂S₆⁺						
$B_3H_2S_6^+$	$(HBS_2)_3$	H	12.3 ± 0.3	EI		2483, 3426

4.3. The Positive Ion Table—Continued

Ion	Reactant	Other products	Ionization or appearance potential (eV)	Method	Heat of formation (kJ mol^{-1})	Ref.
			B$_3$H$_3$S$_6^+$			
B$_3$H$_3$S$_6^+$	(HBS$_2$)$_3$		9.3±0.3	EI	2483,	3426
			CHS$^+$			
CHS$^+$	CH$_3$SH		15.8±0.5	EI		2791
CHS$^+$	(CH$_2$)$_2$S (Ethylene sulfide)	CH$_3$	12.3±0.2	EI		51
CHS$^+$	C$_2$H$_5$SH		17.7±0.3	EI		3286
CHS$^+$	(CH$_3$)$_2$S		14.16±0.08	EI		3202
See also – EI: 84						
CHS$^+$	C$_3$H$_6$S (Propylene sulfide)		14.1±0.2	EI		188
CHS$^+$	(CH$_2$)$_3$S (Trimethylene sulfide)		13.9±0.2	EI		52
CHS$^+$	C$_2$H$_5$SCH$_3$		15.9±0.4	EI		176
CHS$^+$	C$_4$H$_4$S (Thiophene)		13.0±0.2	EI		2166
CHS$^+$	CH$_3$SCH$_2$CH=CH$_2$		13.8±0.3	EI		186
CHS$^+$	(CH$_2$)$_4$S (Tetramethylene sulfide)		13.8±0.2	EI		52
CHS$^+$	n–C$_3$H$_7$SCH$_3$		15.2±0.4	EI		176
CHS$^+$	iso–C$_3$H$_7$SCH$_3$		16.4±0.4	EI		186
CHS$^+$	(C$_2$H$_5$)$_2$S		15.3±0.5	EI		84
CHS$^+$	CH$_3$SCH=CHC≡CH		15.5±0.3	EI		2949
CHS$^+$	CH$_3$SC≡CCH=CH$_2$		14.9±0.3	EI		2949
CHS$^+$	C$_6$H$_5$SD (Mercapto–d_1–benzene)		12.7±0.2	EI		1039
CHS$^+$	C$_2$H$_5$SCH=CHC≡CH		13.8±0.3	EI		2949
CHS$^+$	C$_2$H$_5$SC≡CCH=CH$_2$		14.2±0.3	EI		2949
CHS$^+$	C$_6$H$_5$SCH$_3$ (Methylthiobenzene)		14.5	EI		307
CHS$^+$	C$_6$H$_5$SC$_2$H$_5$ (Ethylthiobenzene)		12.5	EI		307
CHS$^+$	CH$_3$SSCH$_3$		13.43±0.09	EI		3202
See also – EI: 176, 307, 3286						
CHS$^+$	C$_2$H$_5$SSC$_2$H$_5$		16.4±0.3	EI		3286
See also – EI: 186						
CHS$^+$	C$_6$H$_5$SSC$_6$H$_5$ (Diphenyl disulfide)		21.0±0.3	EI		3286
CHS$^+$	CH$_3$SSSCH$_3$		14.5±0.3	EI		84
CHS$^+$	CH$_3$NCS		12.9±0.2	EI		315
CHS$^+$	C$_2$H$_5$NCS		15.2±0.5	EI		315
CHS$^+$	(CH$_3$)$_2$SO		14.8±0.1	EI		3294
CHS$^+$	C$_4$H$_8$OS (1,4–Oxathiane)		14.1	EI		2969

4.3. The Positive Ion Table—Continued

Ion	Reactant	Other products	Ionization or appearance potential (eV)	Method	Heat of formation (kJ mol^{-1})	Ref.
CHS^+	CH_3SCF_3		14.88 ± 0.06	EI		3202
CHS^+	CH_3SSCF_3		14.83 ± 0.08	EI		3202

CDS^+

Ion	Reactant	Other products	Ionization or appearance potential (eV)	Method	Heat of formation (kJ mol^{-1})	Ref.
CDS^+	C_6H_5SD (Mercapto-d_1-benzene)		12.7 ± 0.2	EI		1039

CH_2S^+

Ion	Reactant	Other products	Ionization or appearance potential (eV)	Method	Heat of formation (kJ mol^{-1})	Ref.
CH_2S^+	CH_2S		9.44 ± 0.05	EI		2499
CH_2S^+	$(CH_2)_2S$ (Ethylene sulfide)		12.7 ± 0.2	EI		51
CH_2S^+	C_2H_5SH		11.2 ± 0.3	EI		3286
CH_2S^+	$(CH_3)_2S$		10.97 ± 0.13	EI		3202

See also – EI: 84

Ion	Reactant	Other products	Ionization or appearance potential (eV)	Method	Heat of formation (kJ mol^{-1})	Ref.
CH_2S^+	C_3H_6S (Propylene sulfide)		12.4 ± 0.3	EI		188
CH_2S^+	$(CH_2)_3S$ (Trimethylene sulfide)		10.4 ± 0.1	EI		2499

See also – EI: 52

Ion	Reactant	Other products	Ionization or appearance potential (eV)	Method	Heat of formation (kJ mol^{-1})	Ref.
CH_2S^+	$C_2H_5SCH_3$		13.6 ± 0.3	EI		176
CH_2S^+	$CH_3SCH_2CH=CH_2$		11.4 ± 0.3	EI		186
CH_2S^+	$(CH_2)_4S$ (Tetramethylene sulfide)		13.0 ± 0.2	EI		52
CH_2S^+	$n-C_3H_7SCH_3$		14.1 ± 0.3	EI		176
CH_2S^+	$iso-C_3H_7SCH_3$		15.5 ± 0.4	EI		186
CH_2S^+	$(C_2H_5)_2S$		12.5 ± 0.3	EI		84
CH_2S^+	CH_3SSCH_3		10.61 ± 0.11	EI		3202

See also – EI: 176, 3286

Ion	Reactant	Other products	Ionization or appearance potential (eV)	Method	Heat of formation (kJ mol^{-1})	Ref.
CH_2S^+	$C_4H_8S_2$ (1,4-Dithiane)		13 ± 0.3	EI		2969
CH_2S^+	$C_2H_5SSC_2H_5$		16.6 ± 0.3	EI		3286
CH_2S^+	CH_3SSSCH_3		13.4 ± 0.3	EI		84
CH_2S^+	$(CH_3)_2SO$		11.5 ± 0.1	EI		3294
CH_2S^+	C_4H_8OS (1,4-Oxathiane)		12.2 ± 0.3	EI		2969
CH_2S^+	CH_3SCF_3		~14	EI		3202
CH_2S^+	CH_3SSCF_3		13.43 ± 0.24	EI		3202

4.3. The Positive Ion Table—Continued

Ion	Reactant	Other products	Ionization or appearance potential (eV)	Method	Heat of formation (kJ mol^{-1})	Ref.
			CH$_3$S$^+$			
CH$_3$S$^+$	CH$_3$S		8.06±0.1	EI		120
CH$_3$S$^+$	CH$_3$SH	H	11.6±0.1	RPD		3017
See also – EI: 2685, 2791						
CH$_3$S$^+$	C$_2$H$_5$SH	CH$_3$	11.41±0.1	EI		2504
See also – EI: 2685, 3286						
CH$_3$S$^+$	(CH$_3$)$_2$S	CH$_3$	11.08±0.1	EI		2504
See also – EI: 84, 120, 307, 2685, 3202						
CH$_3$S$^+$	CH$_3$SCD$_3$	CD$_3$	11.0	EI		307
CH$_3$S$^+$	C$_3$H$_6$S (Propylene sulfide)		13.5±0.2	EI		188
CH$_3$S$^+$	(CH$_2$)$_3$S (Trimethylene sulfide)		12.3±0.15	EI		52
CH$_3$S$^+$	C$_2$H$_5$SCH$_3$		14.7±0.2	EI		176
CH$_3$S$^+$	CH$_3$SCH$_2$CH=CH$_2$		11.9±0.2	EI		186
CH$_3$S$^+$	(CH$_2$)$_4$S (Tetramethylene sulfide)		14.0±0.2	EI		52
CH$_3$S$^+$	n–C$_3$H$_7$SCH$_3$		14.0±0.2	EI		176
See also – EI: 2587						
CH$_3$S$^+$	iso–C$_3$H$_7$SCH$_3$		15.3±0.2	EI		186
CH$_3$S$^+$	(C$_2$H$_5$)$_2$S		12.31±0.1	EI		2504
See also – EI: 84, 307, 3286						
CH$_3$S$^+$	(n–C$_3$H$_7$)$_2$S		12.65	EI		307
CH$_3$S$^+$	CH$_3$SSCH$_3$		11.12±0.1	EI		120
See also – EI: 176, 307, 2504, 2685, 3202, 3286						
CH$_3$S$^+$	C$_2$H$_5$SSC$_2$H$_5$		15.5±0.3	EI		3286
CH$_3$S$^+$	CH$_3$SSSCH$_3$		12.9±0.2	EI		84
CH$_3$S$^+$	(CH$_3$)$_2$SO		11.4	EI		2685
See also – EI: 3294						
CH$_3$S$^+$	C$_4$H$_8$OS (1,4–Oxathiane)		12.6	EI		2969
CH$_3$S$^+$	CH$_3$SCF$_3$		11.78±0.03	EI		3202
CH$_3$S$^+$	CH$_3$SSCF$_3$		12.84±0.11	EI		3202
CH$_3$S$^+$	CH$_3$SCH$_2$CH$_2$CH(NH$_2$)COOH		13.0±0.2	EI		2587

4.3. The Positive Ion Table—Continued

Ion	Reactant	Other products	Ionization or appearance potential (eV)	Method	Heat of formation (kJ mol^{-1})	Ref.
			CH$_2$DS$^+$			
CH$_2$DS$^+$	CH$_3$SCD$_3$	CHD$_2$	11.05	EI		307
			CHD$_2$S$^+$			
CHD$_2$S$^+$	CD$_3$SH	D	12.01±0.1	EI		2504
CHD$_2$S$^+$	CH$_3$SCD$_3$	CH$_2$D	11.55	EI		307
			CD$_3$S$^+$			
CD$_3$S$^+$	CD$_3$SH	H	11.76±0.1	EI		2504
CD$_3$S$^+$	CH$_3$SCD$_3$	CH$_3$	11.15	EI		307
CD$_3$S$^+$	C$_2$H$_5$SCD$_3$		12.69±0.1	EI		2504

CH$_3$SH$^+$ $\Delta H^\circ_{f0} = 899$ kJ mol^{-1} (215 kcal mol^{-1})

Ion	Reactant	Other products	Ionization or appearance potential (eV)	Method	Heat of formation (kJ mol^{-1})	Ref.
CH$_4$S$^+$	CH$_3$SH		9.443±0.002	S	899	3317
(Average of three Rydberg series limits)						
CH$_4$S$^+$	CH$_3$SH		9.440±0.005	PI	899	182
See also – EI: 2791						
CH$_4$S$^+$	C$_2$H$_5$SCH$_3$		10.43±0.1	EI		2504
See also – EI: 176						
CH$_4$S$^+$	CH$_3$SCH$_2$CH=CH$_2$		11.5±0.2	EI		186
CH$_4$S$^+$	n–C$_3$H$_7$SCH$_3$		10.45±0.1	EI		2504
See also – EI: 176						
CH$_4$S$^+$	iso–C$_3$H$_7$SCH$_3$		12.0±0.2	EI		186
CH$_4$S$^+$	CH$_3$SSCH$_3$		9.72±0.09	EI		3202
See also – EI: 176, 307, 3286						

Ion	Reactant	Other products	Ionization or appearance potential (eV)	Method	Heat of formation (kJ mol^{-1})	Ref.
			CHD$_3$S$^+$			
CHD$_3$S$^+$	CD$_3$SH		9.54±0.1	EI		2504
CHD$_3$S$^+$	C$_2$H$_5$SCD$_3$		10.48±0.1	EI		2504

4.3. The Positive Ion Table—Continued

Ion	Reactant	Other products	Ionization or appearance potential (eV)	Method	Heat of formation (kJ mol^{-1})	Ref.
			CH$_5$S$^+$			
CH$_5$S$^+$	C$_2$H$_5$SCH$_3$		10.7±0.3	EI		2504
See also – EI:	176					
CH$_5$S$^+$	n–C$_3$H$_7$SCH$_3$		10.23±0.1	EI		2504
See also – EI:	176					
CH$_5$S$^+$	iso–C$_3$H$_7$SCH$_3$		12.1±0.2	EI		186
CH$_5$S$^+$	CH$_3$SSCH$_3$		11.44±0.15	EI		3202
See also – EI:	176, 3286					
			C$_2$HS$^+$			
C$_2$HS$^+$	(CH$_3$)$_2$SO		10.9±0.3	EI		3294
			C$_2$H$_2$S$^+$			
C$_2$H$_2$S$^+$	(CH$_2$)$_2$S (Ethylene sulfide)		15.0±0.2	EI		51
C$_2$H$_2$S$^+$	C$_3$H$_6$S (Propylene sulfide)		15.6±0.4	EI		188
C$_2$H$_2$S$^+$	C$_4$H$_4$S (Thiophene)		10.8±0.2	EI		2166
C$_2$H$_2$S$^+$	(CH$_2$)$_4$S (Tetramethylene sulfide)		17.0±0.3	EI		52
C$_2$H$_2$S$^+$	CH$_3$SCH=CHC≡CH		12.3±0.3	EI		2949
C$_2$H$_2$S$^+$	C$_2$H$_5$SCH=CHC≡CH		12.1±0.3	EI		2949
C$_2$H$_2$S$^+$	C$_2$H$_5$SSC$_2$H$_5$		18.6±0.5	EI		186
C$_2$H$_2$S$^+$	(CH$_3$)$_2$SO		13±0.3	EI		3294
			C$_2$H$_3$S$^+$			
C$_2$H$_3$S$^+$	(CH$_2$)$_2$S (Ethylene sulfide)	H	11.4±0.2	EI		51
C$_2$H$_3$S$^+$	C$_2$H$_5$SH		18.3±0.3	EI		3286
C$_2$H$_3$S$^+$	C$_3$H$_6$S (Propylene sulfide)		12.3±0.3	EI		188
C$_2$H$_3$S$^+$	C$_2$H$_5$SCH$_3$		13.4±0.4	EI		176
C$_2$H$_3$S$^+$	(CH$_2$)$_4$S (Tetramethylene sulfide)		15.7±0.4	EI		52
C$_2$H$_3$S$^+$	iso–C$_3$H$_7$SCH$_3$		16.1±0.3	EI		186
C$_2$H$_3$S$^+$	(C$_2$H$_5$)$_2$S		14.6±0.4	EI		84
See also – EI:	3286					
C$_2$H$_3$S$^+$	CH$_3$SCH=CHC≡CH		10.8±0.3	EI		2949
C$_2$H$_3$S$^+$	C$_2$H$_5$SCH=CHC≡CH		13.8±0.3	EI		2949

4.3. The Positive Ion Table—Continued

Ion	Reactant	Other products	Ionization or appearance potential (eV)	Method	Heat of formation (kJ mol^{-1})	Ref.
$C_2H_3S^+$	$C_2H_5SSC_2H_5$		16.2±0.3	EI		186
See also – EI: 3286						
$C_2H_3S^+$	$C_6H_5SSC_6H_5$ (Diphenyl disulfide)		20.5±0.3	EI		2949
$C_2H_3S^+$	$(CH_3)_2SO$		11±0.3	EI		3294
$C_2H_3S^+$	C_4H_8OS (1,4–Oxathiane)		12.6	EI		2969

$C_2H_4S^+$

Ion	Reactant	Other products	Ionization or appearance potential (eV)	Method	Heat of formation (kJ mol^{-1})	Ref.
$C_2H_4S^+$	$(CH_2)_2S$ (Ethylene sulfide)		8.87±0.15	EI		51
See also – EI: 218						
$C_2H_4S^+$	C_2H_5SH		14.0±0.3	EI		3286
$C_2H_4S^+$	$(CH_2)_4S$ (Tetramethylene sulfide)		11.7±0.3	EI		52
$C_2H_4S^+$	$(C_2H_5)_2S$		11.2±0.2	EI		84
See also – EI: 3286						
$C_2H_4S^+$	$C_4H_8S_2$ (1,4–Dithiane)		10.9±0.2	EI		2969
$C_2H_4S^+$	$C_2H_5SSC_2H_5$		11.6	EI		307
See also – EI: 186						
$C_2H_4S^+$	$(CH_3)_2SO$		11.1±0.5	EI		3294
$C_2H_4S^+$	C_4H_8OS (1,4–Oxathiane)		10.4±0.3	EI		2969

$C_2H_5S^+$

Ion	Reactant	Other products	Ionization or appearance potential (eV)	Method	Heat of formation (kJ mol^{-1})	Ref.
$C_2H_5S^+$	C_2H_5SH	H	11.5	EI		2685
See also – EI: 3286						
$C_2H_5S^+$	$(CH_3)_2S$	H	11.2±0.1	RPD		3017
$C_2H_5S^+$	$(CH_3)_2S$	H	11.50±0.1	EI		2504
See also – EI: 84, 2685, 3202, 3294						
$C_2H_5S^+$	iso–C_3H_7SH	CH_3	10.74±0.1	EI		2504
$C_2H_5S^+$	$C_2H_5SCH_3$	CH_3	10.74	EI		2504
See also – EI: 176						

4.3. The Positive Ion Table—Continued

Ion	Reactant	Other products	Ionization or appearance potential (eV)	Method	Heat of formation (kJ mol⁻¹)	Ref.
$C_2H_5S^+$	$C_2H_5SCD_3$	CD_3	10.75±0.1	EI		2504
$C_2H_5S^+$	$CH_3SCH_2CH=CH_2$		12.0±0.2	EI		186
$C_2H_5S^+$	$n-C_3H_7SCH_3$	C_2H_5	10.97±0.1	EI		2504

See also – EI: 176, 2587

Ion	Reactant	Other products	Ionization or appearance potential (eV)	Method	Heat of formation (kJ mol⁻¹)	Ref.
$C_2H_5S^+$	$iso-C_3H_7SCH_3$		13.5±0.3	EI		186
$C_2H_5S^+$	$(C_2H_5)_2S$	C_2H_5	11.05	EI		307

See also – EI: 84

Ion	Reactant	Other products	Ionization or appearance potential (eV)	Method	Heat of formation (kJ mol⁻¹)	Ref.
$C_2H_5S^+$	$C_2H_5SC_2D_5$	C_2D_5	10.72±0.1	EI		2504
$C_2H_5S^+$	$(n-C_3H_7)_2S$		12.2	EI		307
$C_2H_5S^+$	CH_3SSCH_3		10.0	EI		2685

See also – EI: 176, 307, 3202, 3286

Ion	Reactant	Other products	Ionization or appearance potential (eV)	Method	Heat of formation (kJ mol⁻¹)	Ref.
$C_2H_5S^+$	$C_4H_8S_2$ (1,4–Dithiane)		10.6±0.2	EI		2969
$C_2H_5S^+$	$C_2H_5SSC_2H_5$		10.47	EI		2504

See also – EI: 186, 307, 3286

Ion	Reactant	Other products	Ionization or appearance potential (eV)	Method	Heat of formation (kJ mol⁻¹)	Ref.
$C_2H_5S^+$	$(CH_3)_2SO$		10.5	EI		2685

See also – EI: 3294

Ion	Reactant	Other products	Ionization or appearance potential (eV)	Method	Heat of formation (kJ mol⁻¹)	Ref.
$C_2H_5S^+$	C_4H_8OS (1,4–Oxathiane)		10.3±0.2	EI		2969
$C_2H_5S^+$	$CH_3SCH_2CH_2CH(NH_2)COOH$		12.43±0.10	EI		2587

$C_2H_2D_3S^+$

Ion	Reactant	Other products	Ionization or appearance potential (eV)	Method	Heat of formation (kJ mol⁻¹)	Ref.
$C_2H_2D_3S^+$	$C_2H_5SCD_3$	CH_3	10.84±0.1	EI		2504

$C_2HD_4S^+$

Ion	Reactant	Other products	Ionization or appearance potential (eV)	Method	Heat of formation (kJ mol⁻¹)	Ref.
$C_2HD_4S^+$	C_2D_5SH	D	11.85±0.1	EI		2504

$C_2H_5SH^+$ $\Delta H_{f0}^{\circ} = 867$ kJ mol⁻¹ (207 kcal mol⁻¹)
$(CH_3)_2S^+$ $\Delta H_{f0}^{\circ} = 817$ kJ mol⁻¹ (195 kcal mol⁻¹)

Ion	Reactant	Other products	Ionization or appearance potential (eV)	Method	Heat of formation (kJ mol⁻¹)	Ref.
$C_2H_6S^+$	C_2H_5SH		9.285±0.005	PI	867	182
$C_2H_6S^+$	C_2H_5SH		9.31±0.1	EI		2504

See also – EI: 384, 3286

4.3. The Positive Ion Table—Continued

Ion	Reactant	Other products	Ionization or appearance potential (eV)	Method	Heat of formation (kJ mol^{-1})	Ref.
$C_2H_6S^+$	$(CH_3)_2S$		8.685±0.005	PI	817	182
$C_2H_6S^+$	$(CH_3)_2S$		8.68±0.03	PE		3034
$C_2H_6S^+$	$(CH_3)_2S$		8.71±0.1	EI		2504

See also – S: 3317
PEN: 2430
EI: 84, 307, 2505, 3202, 3338

Ion	Reactant	Other products	Ionization or appearance potential (eV)	Method	Heat of formation (kJ mol^{-1})	Ref.
$C_2H_6S^+$	$(C_2H_5)_2S$		10.4±0.2	EI		84

See also – EI: 307, 3286

Ion	Reactant	Other products	Ionization or appearance potential (eV)	Method	Heat of formation (kJ mol^{-1})	Ref.
$C_2H_6S^+$	$C_2H_5SSC_2H_5$		11.4±0.3	EI		3286
$C_2H_6S^+$	$(CH_3)_2SO$	O	11.6±0.1	EI		3294

$C_2H_3D_3S^+$

Ion	Reactant	Other products	Ionization or appearance potential (eV)	Method	Heat of formation (kJ mol^{-1})	Ref.
$C_2H_3D_3S^+$	CH_3SCD_3		8.72±0.1	EI		2504

See also – EI: 307

C_3HS^+

Ion	Reactant	Other products	Ionization or appearance potential (eV)	Method	Heat of formation (kJ mol^{-1})	Ref.
C_3HS^+	$CH_3SC{\equiv}CCH{=}CH_2$		17.4±0.3	EI		2949
C_3HS^+	C_6H_5SH (Mercaptobenzene)		19.8±0.3	EI		3286
C_3HS^+	$C_2H_5SCH{=}CHC{\equiv}CH$		13.5±0.3	EI		2949
C_3HS^+	$C_2H_5SC{\equiv}CCH{=}CH_2$		13.6±0.3	EI		2949

$C_3H_2S^+$

Ion	Reactant	Other products	Ionization or appearance potential (eV)	Method	Heat of formation (kJ mol^{-1})	Ref.
$C_3H_2S^+$	C_6H_5SH (Mercaptobenzene)		17.0±0.3	EI		3286

$C_3H_3S^+$

Ion	Reactant	Other products	Ionization or appearance potential (eV)	Method	Heat of formation (kJ mol^{-1})	Ref.
$C_3H_3S^+$	C_6H_5SH (Mercaptobenzene)		15.5±0.3	EI		3286
$C_3H_3S^+$	$C_6H_5SSC_6H_5$ (Diphenyl disulfide)		18.6±0.3	EI		3286

$C_3H_5S^+$

Ion	Reactant	Other products	Ionization or appearance potential (eV)	Method	Heat of formation (kJ mol^{-1})	Ref.
$C_3H_5S^+$	C_3H_6S (Propylene sulfide)	H	11.2±0.3	EI		188
$C_3H_5S^+$	$CH_3SCH_2CH{=}CH_2$	CH_3	11.0±0.2	EI		186
$C_3H_5S^+$	$C_4H_8S_2$ (1,4–Dithiane)		10.7±0.1	EI		2969

4.3. The Positive Ion Table—Continued

Ion	Reactant	Other products	Ionization or appearance potential (eV)	Method	Heat of formation (kJ mol⁻¹)	Ref.
	C₃H₆S⁺ (Trimethylene sulfide)		$\Delta H^\circ_{f\,298} = 896$ kJ mol⁻¹ (214 kcal mol⁻¹)			
$C_3H_6S^+$	C_3H_6S (Propylene sulfide)		8.6±0.2	EI		188
$C_3H_6S^+$	$(CH_2)_3S$ (Trimethylene sulfide)		8.65	S	896	2677
(Average of two Rydberg series limits)						
See also – EI: 52, 218, 2499						
$C_3H_6S^+$	$C_4H_8S_2$ (1,4–Dithiane)		11.4±0.2	EI		2969
$C_3H_6S^+$	C_4H_8OS (1,4–Oxathiane)		11±0.1	EI		2969
	C₃H₇S⁺					
$C_3H_7S^+$	$n-C_3H_7SCH_3$	CH_3	11.7±0.2	EI		176
$C_3H_7S^+$	$iso-C_3H_7SCH_3$	CH_3	11.7±0.2	EI		186
$C_3H_7S^+$	$(C_2H_5)_2S$	CH_3	10.65	EI		307
See also – EI: 84, 3286						
$C_3H_7S^+$	$(n-C_3H_7)_2S$		11.55	EI		307
	n–C₃H₇SH⁺		$\Delta H^\circ_{f\,298} = 819$ kJ mol⁻¹ (196 kcal mol⁻¹)			
	C₂H₅SCH₃⁺		$\Delta H^\circ_{f\,298} = 765$ kJ mol⁻¹ (183 kcal mol⁻¹)			
$C_3H_8S^+$	$n-C_3H_7SH$		9.195±0.005	PI	819	182
$C_3H_8S^+$	$C_2H_5SCH_3$		8.55±0.01	PI	765	182
$C_3H_8S^+$	$C_2H_5SCH_3$		8.54±0.1	EI		2504
See also – EI: 176, 2587						
$C_3H_8S^+$	$(n-C_3H_7)_2S$		10.4	EI		307
	C₄HS⁺					
C_4HS^+	$CH_3SCH=CHC\equiv CH$		20.7±0.3	EI		2949
C_4HS^+	$CH_3SC\equiv CCH=CH_2$		19.0±0.3	EI		2949
C_4HS^+	$C_2H_5SC\equiv CCH=CH_2$		21.2±0.3	EI		2949
	C₄H₂S⁺					
$C_4H_2S^+$	$CH_3SCH=CHC\equiv CH$		16.8±0.3	EI		2949
$C_4H_2S^+$	$CH_3SC\equiv CCH=CH_2$		16.1±0.3	EI		2949
$C_4H_2S^+$	C_6H_5SH (Mercaptobenzene)		17.6±0.3	EI		2949
$C_4H_2S^+$	$C_2H_5SCH=CHC\equiv CH$		17.6±0.3	EI		2949
$C_4H_2S^+$	$C_2H_5SC\equiv CCH=CH_2$		17.4±0.3	EI		2949

4.3. The Positive Ion Table—Continued

Ion	Reactant	Other products	Ionization or appearance potential (eV)	Method	Heat of formation (kJ mol^{-1})	Ref.
			$C_4H_3S^+$			
$C_4H_3S^+$	$CH_3SCH=CHC\equiv CH$	CH_3	11.8±0.3	EI		2949
$C_4H_3S^+$	$CH_3SC\equiv CCH=CH_2$	CH_3	11.6±0.3	EI		2949
$C_4H_3S^+$	$C_2H_5SCH=CHC\equiv CH$		13.4±0.3	EI		2949
$C_4H_3S^+$	$C_2H_5SC\equiv CCH=CH_2$		13.2±0.3	EI		2949
$C_4H_3S^+$	$C_6H_5SSC_6H_5$ (Diphenyl disulfide)		19.0±0.3	EI		3286
	$C_4H_4S^+$ (Thiophene)	$\Delta H^\circ_{f298} \sim 972$ kJ mol^{-1} (232 kcal mol^{-1})				
$C_4H_4S^+$	C_4H_4S (Thiophene)		8.95±0.02	S	979	3351
$C_4H_4S^+$	C_4H_4S (Thiophene)		8.860±0.005	PI	971	182
$C_4H_4S^+$	C_4H_4S (Thiophene)		8.872±0.01	PE	972	3381, 3422
$C_4H_4S^+$	C_4H_4S (Thiophene)		8.87±0.05	PE	972	2796
$C_4H_4S^+$	C_4H_4S (Thiophene)		8.80±0.05	PE	965	3246

See also – EI: 2166, 2865, 3233, 3240
　　　　　CTS: 2031, 3369

Ion	Reactant	Other products	Ionization or appearance potential (eV)	Method	Heat of formation	Ref.
$C_4H_4S^+$	C_6H_5SH (Mercaptobenzene)		13.2±0.3	EI		3286
$C_4H_4S^+$	C_6H_5SD (Mercapto-d_1-benzene)		11.8±0.2	EI		1039
$C_4H_4S^+$	$C_2H_5SCH=CHC\equiv CH$		10.8±0.3	EI		2949
$C_4H_4S^+$	$C_2H_5SC\equiv CCH=CH_2$		10.6±0.3	EI		2949
$C_4H_4S^+$	$C_4H_3SC_2H_5$ (2–Ethylthiophene)		11.5±0.2	EI		2166
$C_4H_4S^+$	$C_4H_3SC_3H_7$ (2–Propylthiophene)		11.1±0.2	EI		2166
$C_4H_4S^+$	$C_4H_3SC_4H_9$ (2–Butylthiophene)		11.0±0.2	EI		2166
$C_4H_4S^+$	$C_6H_5SSC_6H_5$ (Diphenyl disulfide)		14.1±0.3	EI		3286
			$C_4H_3DS^+$			
$C_4H_3DS^+$	C_6H_5SD (Mercapto-d_1-benzene)		11.8	EI		1039
			$C_4H_6S^+$			
$C_4H_6S^+$	C_4H_8OS (1,4–Oxathiane)	H_2O	9.9±0.1	EI		2969

4.3. The Positive Ion Table—Continued

Ion	Reactant	Other products	Ionization or appearance potential (eV)	Method	Heat of formation (kJ mol^{-1})	Ref.
			C$_4$H$_7$S$^+$			
C$_4$H$_7$S$^+$	(CH$_2$)$_4$S (Tetramethylene sulfide)	H	12.4±0.3	EI		52
			C$_4$H$_8$S$^+$			
C$_4$H$_8$S$^+$	CH$_3$SCH$_2$CH=CH$_2$		8.70±0.2	EI		186
C$_4$H$_8$S$^+$	(CH$_2$)$_4$S (Tetramethylene sulfide)		8.57±0.15	EI		52
See also – EI: 218						
	n-C$_4$H$_9$SH$^+$	$\Delta H_{f298}^\circ = 794$ kJ mol^{-1} (190 kcal mol^{-1})				
	(C$_2$H$_5$)$_2$S$^+$	$\Delta H_{f298}^\circ = 730$ kJ mol^{-1} (174 kcal mol^{-1})				
C$_4$H$_{10}$S$^+$	n-C$_4$H$_9$SH		9.14±0.02	PI	794	182
C$_4$H$_{10}$S$^+$	n-C$_3$H$_7$SCH$_3$		8.8±0.15	EI		176
See also – EI: 2587						
C$_4$H$_{10}$S$^+$	iso-C$_3$H$_7$SCH$_3$		8.7±0.2	EI		186
C$_4$H$_{10}$S$^+$	(C$_2$H$_5$)$_2$S		8.430±0.005	PI	730	182
See also – EI: 84, 307, 2505, 3286						
			C$_4$H$_5$D$_5$S$^+$			
C$_4$H$_5$D$_5$S$^+$	C$_2$H$_5$SC$_2$D$_5$		8.55±0.1	EI		2504
			C$_5$H$_3$S$^+$			
C$_5$H$_3$S$^+$	C$_6$H$_5$SH (Mercaptobenzene)		15.1±0.3	EI		3286
			C$_5$H$_4$S$^+$			
C$_5$H$_4$S$^+$	C$_6$H$_5$SSC$_6$H$_5$ (Diphenyl disulfide)		15.5±0.3	EI		3286
			C$_5$H$_5$S$^+$			
C$_5$H$_5$S$^+$	CH$_3$SCH=CHC≡CH	H	11.0±0.3	EI		2949
C$_5$H$_5$S$^+$	CH$_3$SC≡CCH=CH$_2$	H	10.8±0.3	EI		2949
C$_5$H$_5$S$^+$	C$_2$H$_5$SCH=CHC≡CH	CH$_3$	10.0±0.3	EI		2949
C$_5$H$_5$S$^+$	C$_2$H$_5$SC≡CCH=CH$_2$	CH$_3$	10.8±0.3	EI		2949
C$_5$H$_5$S$^+$	C$_4$H$_3$SC$_2$H$_5$ (2-Ethylthiophene)	CH$_3$	11.4±0.2	EI		2166

4.3. The Positive Ion Table—Continued

Ion	Reactant	Other products	Ionization or appearance potential (eV)	Method	Heat of formation (kJ mol^{-1})	Ref.
$C_5H_5S^+$	$C_4H_3SC_3H_7$ (2-Propylthiophene)	C_2H_5	11.4±0.2	EI		2166
$C_5H_5S^+$	$C_4H_3SC_4H_9$ (2-Butylthiophene)		11.3±0.2	EI		2166
$C_5H_5S^+$	$C_6H_5SSC_6H_5$ (Diphenyl disulfide)		17.2±0.3	EI		3286
$C_5H_5S^+$	$C_4H_3S(CH_2)_3COOCH_3$ (4-(2-Thienyl)butanoic acid methyl ester)		13.25±0.2	EI		2497

$C_5H_6S^+$ (2-Methylthiophene) $\Delta H^\circ_{f298} \sim 869$ kJ mol^{-1} (208 kcal mol^{-1})
$C_5H_6S^+$ (3-Methylthiophene) $\Delta H^\circ_{f298} \sim 893$ kJ mol^{-1} (214 kcal mol^{-1})

Ion	Reactant	Other products	Ionization or appearance potential (eV)	Method	Heat of formation (kJ mol^{-1})	Ref.
$C_5H_6S^+$	$C_4H_3SCH_3$ (2-Methylthiophene)		8.14	PE	869	3246

See also – EI: 3240
CTS: 3369

Ion	Reactant	Other products	Ionization or appearance potential (eV)	Method	Heat of formation (kJ mol^{-1})	Ref.
$C_5H_6S^+$	$C_4H_3SCH_3$ (3-Methylthiophene)		8.40	PE	893	3246

See also – CTS: 3369

$C_5H_9S^+$

Ion	Reactant	Other products	Ionization or appearance potential (eV)	Method	Heat of formation (kJ mol^{-1})	Ref.
$C_5H_9S^+$	$(n-C_3H_7)_2S$		10.9	EI		307

$C_5H_{11}S^+$

Ion	Reactant	Other products	Ionization or appearance potential (eV)	Method	Heat of formation (kJ mol^{-1})	Ref.
$C_5H_{11}S^+$	$(n-C_3H_7)_2S$	CH_3	11.55	EI		307

$C_5H_{12}S^+$

Ion	Reactant	Other products	Ionization or appearance potential (eV)	Method	Heat of formation (kJ mol^{-1})	Ref.
$C_5H_{12}S^+$	$n-C_3H_7SC_2H_5$		8.50±0.05	RPD		2505

$C_6H_4S^+$

Ion	Reactant	Other products	Ionization or appearance potential (eV)	Method	Heat of formation (kJ mol^{-1})	Ref.
$C_6H_4S^+$	C_6H_5SH (Mercaptobenzene)		17.2±0.3	EI		3286

$C_6H_5S^+$

Ion	Reactant	Other products	Ionization or appearance potential (eV)	Method	Heat of formation (kJ mol^{-1})	Ref.
$C_6H_5S^+$	C_6H_5S (Phenylthio radical)		8.63±0.1	EI		120
$C_6H_5S^+$	C_6H_5SH (Mercaptobenzene)	H	14.7±0.3	EI		3286
$C_6H_5S^+$	C_6H_5SD (Mercapto-d_1-benzene)	D	12.2	EI		1039

4.3. The Positive Ion Table—Continued

Ion	Reactant	Other products	Ionization or appearance potential (eV)	Method	Heat of formation (kJ mol^{-1})	Ref.
$C_6H_5S^+$	$C_6H_5SCH_3$ (Methylthiobenzene)		12.1±0.1	EI		120
See also – EI:	307					
$C_6H_5S^+$	$C_6H_5SC_2H_5$ (Ethylthiobenzene)		12.2	EI		307
$C_6H_5S^+$	$C_6H_5SSC_6H_5$ (Diphenyl disulfide)		14.4±0.3	EI		3286

$C_6H_6S^+$ (Mercaptobenzene) $\Delta H^\circ_{f298} = 914$ kJ mol^{-1} (219 kcal mol^{-1})

Ion	Reactant	Other products	Ionization or appearance potential (eV)	Method	Heat of formation (kJ mol^{-1})	Ref.
$C_6H_6S^+$	C_6H_5SH (Mercaptobenzene)		8.32±0.01	PI	914	190
See also – PI:	182					
EI:	2865, 3286					
$C_6H_6S^+$	$C_6H_5SC_2H_5$ (Ethylthiobenzene)		10.77	EI		2706
See also – EI:	307					
$C_6H_6S^+$	$C_6H_5SC_3H_7$ (Propylthiobenzene)		10.46	EI		2706
$C_6H_6S^+$	$C_6H_5SC_4H_9$ (Butylthiobenzene)		10.33	EI		2706
$C_6H_6S^+$	$C_6H_5SSC_6H_5$ (Diphenyl disulfide)		11.6±0.3	EI		3286
$C_6H_6S^+$	$C_4H_3S(CH_2)_3COOCH_3$ (4–(2–Thienyl)butanoic acid methyl ester)		10.70±0.2	EI		2497

$C_6H_5DS^+$

Ion	Reactant	Other products	Ionization or appearance potential (eV)	Method	Heat of formation (kJ mol^{-1})	Ref.
$C_6H_5DS^+$	C_6H_5SD (Mercapto–d_1–benzene)		8.5±0.1	EI		1039

$C_6H_7S^+$

Ion	Reactant	Other products	Ionization or appearance potential (eV)	Method	Heat of formation (kJ mol^{-1})	Ref.
$C_6H_7S^+$	$C_2H_5SCH=CHC{\equiv}CH$	H	10.3±0.3	EI		2949

$C_6H_8S^+$

Ion	Reactant	Other products	Ionization or appearance potential (eV)	Method	Heat of formation (kJ mol^{-1})	Ref.
$C_6H_8S^+$	$C_4H_3SC_2H_5$ (2–Ethylthiophene)		8.8±0.2	EI		2166

4.3. The Positive Ion Table—Continued

Ion	Reactant	Other products	Ionization or appearance potential (eV)	Method	Heat of formation (kJ mol^{-1})	Ref.
	$(n-C_3H_7)_2S^+$		$\Delta H^\circ_{f298} = 675$ kJ mol^{-1} (161 kcal mol^{-1})			
$C_6H_{14}S^+$	$(n-C_3H_7)_2S$		8.30±0.02	PI	675	182
$C_6H_{14}S^+$	$(n-C_3H_7)_2S$		8.45±0.05	RPD		2505
See also – EI: 307						
$C_6H_{14}S^+$	$(iso-C_3H_7)_2S$		8.38±0.05	RPD		2505
	$C_7H_7S^+$					
$C_7H_7S^+$	$C_6H_5SC_2H_5$ (Ethylthiobenzene)		11.7	EI		307
	$C_7H_8S^+$					
$C_7H_8S^+$	$C_6H_5SCH_3$ (Methylthiobenzene)		8.9	EI		307
$C_7H_8S^+$	$C_6H_5SCH_3$ (Methylthiobenzene)		7.9±0.15	CTS		3373
	$C_7H_{10}S^+$					
$C_7H_{10}S^+$	$C_4H_3SC_3H_7$ (2–Propylthiophene)		8.6±0.2	EI		2166
	$C_7H_{16}S^+$					
$C_7H_{16}S^+$	$(CH_3)_2CHCH_2SCH_2CH_2CH_3$		8.40±0.05	RPD		2505
	$C_8H_6S^+$					
$C_8H_6S^+$	C_8H_6S (2,3–Benzothiophene)		8.17±0.05	PE		2796
	$C_8H_{10}S^+$					
$C_8H_{10}S^+$	$C_6H_5CH_2SCH_3$ (α–Methylthiotoluene)		8.64	EI		3338
$C_8H_{10}S^+$	$C_6H_5SC_2H_5$ (Ethylthiobenzene)		8.8	EI		307
	$C_8H_{12}S^+$					
$C_8H_{12}S^+$	$C_4H_3SC_4H_9$ (2–Butylthiophene)		8.5±0.2	EI		2166

4.3. The Positive Ion Table—Continued

Ion	Reactant	Other products	Ionization or appearance potential (eV)	Method	Heat of forma- tion (kJ mol⁻¹)	Ref.

$$C_8H_{18}S^+$$

Ion	Reactant	Other products	Ionization or appearance potential (eV)	Method	Heat of formation	Ref.
$C_8H_{18}S^+$	$(n-C_4H_9)_2S$		8.40±0.05	RPD		2505
$C_8H_{18}S^+$	$(iso-C_4H_9)_2S$		8.36±0.05	RPD		2505

$$C_9H_{10}S^+$$

Ion	Reactant	Other products	Ionization or appearance potential (eV)	Method	Heat of formation	Ref.
$C_9H_{10}S^+$	$C_9H_{10}S$ (Thiochroman)		8.02	EI		3338
$C_9H_{10}S^+$	$C_9H_{10}S$ (Isothiochroman)		8.70	EI		3338

$$C_{11}H_7S^+$$

Ion	Reactant	Other products	Ionization or appearance potential (eV)	Method	Heat of formation	Ref.
$C_{11}H_7S^+$	$C_6H_5SSC_6H_5$ (Diphenyl disulfide)		13.8±0.3	EI		3286

$$C_{12}H_8S^+$$

Ion	Reactant	Other products	Ionization or appearance potential (eV)	Method	Heat of formation	Ref.
$C_{12}H_8S^+$	$C_{12}H_8S$ (Dibenzothiophene)		8.14	CTS		2911
$C_{12}H_8S^+$	$C_6H_5SSC_6H_5$ (Diphenyl disulfide)		14.8±0.3	EI		3286

$$C_{12}H_9S^+$$

Ion	Reactant	Other products	Ionization or appearance potential (eV)	Method	Heat of formation	Ref.
$C_{12}H_9S^+$	$C_6H_5SSC_6H_5$ (Diphenyl disulfide)		12.6±0.3	EI		3286

$$C_{12}H_{10}S^+$$

Ion	Reactant	Other products	Ionization or appearance potential (eV)	Method	Heat of formation	Ref.
$C_{12}H_{10}S^+$	$(C_6H_5)_2S?$ (Diphenyl sulfide?)		9.9±0.3	EI		3286

(Probably formed by thermal decomposition of diphenyl disulfide)

$$CH_3S_2^+$$

Ion	Reactant	Other products	Ionization or appearance potential (eV)	Method	Heat of formation	Ref.
$CH_3S_2^+$	CH_3SSCH_3	CH_3	11.45	EI		307

See also – EI: 176, 3202, 3286

Ion	Reactant	Other products	Ionization or appearance potential (eV)	Method	Heat of formation	Ref.
$CH_3S_2^+$	$C_2H_5SSC_2H_5$		13.8±0.3	EI		3286

See also – EI: 186

Ion	Reactant	Other products	Ionization or appearance potential (eV)	Method	Heat of formation	Ref.
$CH_3S_2^+$	CH_3SSSCH_3		12.3±0.2	EI		84
$CH_3S_2^+$	CH_3SSCF_3		11.29±0.20	EI		3202

4.3. The Positive Ion Table—Continued

Ion	Reactant	Other products	Ionization or appearance potential (eV)	Method	Heat of formation (kJ mol^{-1})	Ref.
		$CH_4S_2^+$				
$CH_4S_2^+$	CH_3SSSCH_3		10.8±0.2	EI		84
		$C_2H_4S_2^+$				
$C_2H_4S_2^+$	$C_4H_8S_2$ (1,4–Dithiane)		10.95±0.1	EI		2969
		$C_2H_5S_2^+$				
$C_2H_5S_2^+$	$C_2H_5SSC_2H_5$		11.5	EI		307
See also – EI: 3286						
		$C_2H_6S_2^+$				
$C_2H_6S_2^+$	CH_3SSCH_3		8.46±0.03?	PI		182
$C_2H_6S_2^+$	CH_3SSCH_3		8.71±0.03	PE		3034
See also – EI: 176, 307, 411, 2685, 3202, 3286						
$C_2H_6S_2^+$	$C_2H_5SSC_2H_5$		10.8±0.3	EI		186
See also – EI: 307, 3286						
		$C_3H_5S_2^+$				
$C_3H_5S_2^+$	$C_4H_8S_2$ (1,4–Dithiane)		10.2±0.1	EI		2969
		$C_4H_8S_2^+$				
$C_4H_8S_2^+$	$C_4H_8S_2$ (1,4–Dithiane)		8.75±0.1	EI		2969
$C_4H_8S_2^+$	$C_4H_8S_2$? (1,4–Dithiane?)		8.5	EI		218
		$C_4H_{10}S_2^+$				
$C_4H_{10}S_2^+$	$C_2H_5SSC_2H_5$		8.27±0.03?	PI		182
$C_4H_{10}S_2^+$	$C_2H_5SSC_2H_5$		8.30±0.15	EI		186
See also – EI: 307, 3286						

4.3. The Positive Ion Table—Continued

Ion	Reactant	Other products	Ionization or appearance potential (eV)	Method	Heat of formation (kJ mol^{-1})	Ref.
			$C_6H_4S_2^+$			
$C_6H_4S_2^+$	$C_6H_4S_2$ (Thiophthene)		9.15	EI		3284
			$C_8H_6S_2^+$			
$C_8H_6S_2^+$	$(C_4H_3S)_2$ (2,2′–Bithiophene)		7.83	CTS		3369
$C_8H_6S_2^+$	$(C_4H_3S)_2$ (2,3′–Bithiophene)		7.91	CTS		3369
$C_8H_6S_2^+$	$(C_4H_3S)_2$ (3,3′–Bithiophene)		7.99	CTS		3369
			$C_9H_6S_2^+$			
$C_9H_6S_2^+$	$C_9H_6S_2$ (4H–Cyclopenta[2,1–b:3,4–b']dithiophene)		7.42	CTS		3369
$C_9H_6S_2^+$	$C_9H_6S_2$ (7H–Cyclopenta[1,2–b:3,4–b']dithiophene)		7.51	CTS		3369
$C_9H_6S_2^+$	$C_9H_6S_2$ (7H–Cyclopenta[1,2–b:4,3–b']dithiophene)		7.59	CTS		3369
			$C_9H_8S_2^+$			
$C_9H_8S_2^+$	$(C_4H_3S)_2CH_2$ (2,2′–Dithienylmethane)		9.05	CTS		3369
$C_9H_8S_2^+$	$(C_4H_3S)_2CH_2$ (2,3′–Dithienylmethane)		8.97	CTS		3369
$C_9H_8S_2^+$	$(C_4H_3S)_2CH_2$ (3,3′–Dithienylmethane)		8.84	CTS		3369
			$C_{12}H_8S_2^+$			
$C_{12}H_8S_2^+$	$C_{12}H_8S_2$ (Thianthrene)		7.80±0.03	RPD		2538
$C_{12}H_8S_2^+$	$C_{12}H_8S_2$ (Thianthrene)		7.9	CTS		3300
			$C_{12}H_{10}S_2^+$			
$C_{12}H_{10}S_2^+$	$C_6H_5SSC_6H_5$ (Diphenyl disulfide)		9.4±0.3	EI		3286
			$CH_3S_3^+$			
$CH_3S_3^+$	CH_3SSSCH_3	CH_3	11.4±0.2	EI		84

4.3. The Positive Ion Table—Continued

Ion	Reactant	Other products	Ionization or appearance potential (eV)	Method	Heat of formation (kJ mol^{-1})	Ref.
		$C_2H_6S_3^+$				
$C_2H_6S_3^+$	CH_3SSSCH_3		8.73±0.03	PE		3034
$C_2H_6S_3^+$	CH_3SSSCH_3		8.80±0.15	EI		84
		$C_7H_8S_3^+$				
$C_7H_8S_3^+$	$C_7H_8S_3$ (2–Dimethyl–6a–thiathiophthene)		7.47	EI		3284
		$C_{12}H_{10}S_3^+$				
$C_{12}H_{10}S_3^+$	$C_{12}H_{10}S_3$ (2–Methyl–5–phenyl–6a–thiathiophthene)		7.43	EI		3284
		$C_{17}H_{12}S_3^+$				
$C_{17}H_{12}S_3^+$	$C_{17}H_{12}S_3$ (2,5–Diphenyl–6a–thiathiophthene)		7.39	EI		3284
		$C_{12}H_{10}S_4^+$				
$C_{12}H_{10}S_4^+$	$C_{12}H_{10}S_4$ (2–Methylthio–5–phenyl–6a–thiathiophthene)		7.24	EI		3284
		CNS^+				
CNS^+	CH_3NCS		14.9±0.5	EI		315
CNS^+	C_2H_5NCS		14.6±0.4	EI		315
		C_2NS^+				
C_2NS^+	CH_3NCS		14.1±0.3	EI		315
C_2NS^+	C_2H_5NCS		16.3±0.2	EI		315
		HSO^+				
HSO^+	$(CH_3)_2SO$		11.1±0.3	EI		3294
		H_2SO^+				
H_2SO^+	$(CH_3)_2SO$		10.9±0.3	EI		3294
		H_3SO^+				
H_3SO^+	$(CH_3)_2SO$		10.9±0.5	EI		3294

4.3. The Positive Ion Table—Continued

Ion	Reactant	Other products	Ionization or appearance potential (eV)	Method	Heat of formation (kJ mol^{-1})	Ref.
$COS^+(X^2\Pi_{3/2})$		$\Delta H^\circ_{f0} =$ 937 kJ mol^{-1} (224 kcal mol^{-1})				
$COS^+(X^2\Pi_{1/2})$		$\Delta H^\circ_{f0} =$ 941 kJ mol^{-1} (225 kcal mol^{-1})				
$COS^+(A^2\Pi)$		$\Delta H^\circ_{f0} =$ 1313 kJ mol^{-1} (314 kcal mol^{-1})				
$COS^+(B^2\Sigma^+)$		$\Delta H^\circ_{f0} =$ 1406 kJ mol^{-1} (336 kcal mol^{-1})				
$COS^+(C^2\Sigma^+)$		$\Delta H^\circ_{f0} =$ 1590 kJ mol^{-1} (380 kcal mol^{-1})				
$COS^+(X^2\Pi_{3/2})$	COS		11.184±0.01	S	937	2680
$COS^+(X^2\Pi_{1/2})$	COS		11.230±0.01	S	941	2680

See also − S: 149

$COS^+(B^2\Sigma^+)$	COS		16.041	S	1406	149
(Average of four Rydberg series limits)						
$COS^+(C^2\Sigma^+)$	COS		17.938	S	1589	149
$COS^+(X^2\Pi_{3/2})$	COS		11.189±0.005	PE	937	2901
$COS^+(X^2\Pi_{1/2})$	COS		11.233±0.005	PE	942	2901
$COS^+(A^2\Pi)$	COS		15.080±0.005	PE	1313	2901
$COS^+(B^2\Sigma^+)$	COS		16.042±0.005	PE	1406	2901
$COS^+(C^2\Sigma^+)$	COS		17.960±0.005	PE	1591	2901

A band reported in ref. 92 at 19.9 eV is due to a light source impurity, see ref. 2901.

$COS^+(X^2\Pi_{3/2})$	COS		11.175±0.01	PI		2680
$COS^+(X^2\Pi_{3/2})$	COS		11.18±0.01	PI		2624, 2627
$COS^+(X^2\Pi_{1/2})$	COS		11.215±0.01	PI		2680
$COS^+(X^2\Pi_{1/2})$	COS		11.22	PI		2624, 2627

See also − S: 409, 410
       PI: 182, 190
       PE: 92, 2839, 2856, 2875
       PEN: 2430, 2467

CFS^+

CFS^+	CH_3SCF_3		15.32±0.13	EI		3202

CF_2S^+

CF_2S^+	CF_3SSCF_3		10.93±0.17	EI		3202

CF_3S^+

CF_3S^+	CF_3SSCF_3		14.43±0.08	EI		3202

$C_2F_5S^+$

$C_2F_5S^+$	$(CF_3)_2S$	F	~15	EI		3202

4.3. The Positive Ion Table—Continued

Ion	Reactant	Other products	Ionization or appearance potential (eV)	Method	Heat of formation (kJ mol^{-1})	Ref.
			$C_2F_6S^+$			
$C_2F_6S^+$	$(CF_3)_2S$		11.11±0.03	PE		3034
$C_2F_6S^+$	$(CF_3)_2S$		11.28±0.04	EI		3202
			CFS_2^+			
CFS_2^+	CF_3SSCF_3		15.60±0.10	EI		3202
			$CF_2S_2^+$			
$CF_2S_2^+$	CF_3SSCF_3	CF_4	11.28±0.06	EI		3202
			$CF_3S_2^+$			
$CF_3S_2^+$	CF_3SSCF_3	CF_3	13.31±0.05	EI		3202
$CF_3S_2^+$	CH_3SSCF_3	CH_3	14.77±0.12	EI		3202
			$C_2F_5S_2^+$			
$C_2F_5S_2^+$	CF_3SSCF_3	F	14.64±0.07	EI		3202
			$C_2F_6S_2^+$			
$C_2F_6S_2^+$	CF_3SSCF_3		10.60±0.03	PE		3034
$C_2F_6S_2^+$	CF_3SSCF_3		10.68±0.19	EI		3202
			$C_2F_6S_3^+$			
$C_2F_6S_3^+$	CF_3SSSCF_3		10.16±0.03	PE		3034
			$C_2F_6S_4^+$			
$C_2F_6S_4^+$	$CF_3SSSSCF_3$		9.75±0.03	PE		3034
			NSF^+			
NSF^+	NSF		11.36±0.03	RPD		2443
NSF^+	NSF_3		16.82±0.05	RPD		2443
			NSF_2^+			
NSF_2^+	NSF_3	F	15.47±0.04	RPD		2443
NSF_2^+	CF_3NSF_2	CF_3	13.90±0.08	RPD		2443

4.3. The Positive Ion Table—Continued

Ion	Reactant	Other products	Ionization or appearance potential (eV)	Method	Heat of formation (kJ mol^{-1})	Ref.
			NSF$_3^+$			
NSF$_3^+$	NSF$_3$		12.46±0.03	RPD		2443
			SOF$^+$			
SOF$^+$	SO$_2$F$_2$		18.6±0.1	EI		418
			SO$_2$F$^+$			
SO$_2$F$^+$	SO$_2$F$_2$	F	15.1±0.2	EI		418
SO$_2$F$^+$	FSO$_2$NF$_2$		13.1±0.1	EI		1144
SO$_2$F$^+$	FSO$_2$ONF$_2$		13.3±0.2	EI		1144
SO$_2$F$^+$	(FSO$_2$)$_2$NF		13.5±0.1	EI		1144
			SO$_2$F$_2^+$			
SO$_2$F$_2^+$	SO$_2$F$_2$		13.3±0.1	EI		418
			PSF$_2^+$			
PSF$_2^+$	PSF$_3$	F	16.0±0.2	EI		2506
			PSF$_3^+$			
PSF$_3^+$	PSF$_3$		11.1±0.3	EI		2506
			C$_3$H$_9$BS$_3^+$			
C$_3$H$_9$BS$_3^+$	(CH$_3$S)$_3$B		9.24	EI		3227

HNCS$^+$ $\Delta H^\circ_{f298} \sim 1097$ kJ mol^{-1} (262 kcal mol^{-1})

Ion	Reactant	Other products	Ionization or appearance potential (eV)	Method	Heat of formation (kJ mol^{-1})	Ref.
CHNS$^+$	HNCS		10.05±0.1	PE	1097	3067
CHNS$^+$	C$_2$H$_5$NCS	C$_2$H$_4$	11.38±0.15	EI	193,	315
			CH$_2$NS$^+$			
CH$_2$NS$^+$	C$_2$H$_5$NCS		12.0±0.3	EI		315
			C$_2$HNS$^+$			
C$_2$HNS$^+$	C$_2$H$_5$NCS		14.0±0.2	EI		315

4.3. The Positive Ion Table—Continued

Ion	Reactant	Other products	Ionization or appearance potential (eV)	Method	Heat of formation (kJ mol^{-1})	Ref.
		C$_2$HNS^{+2}				
C$_2$HNS^{+2}	CH$_3$NCS		28.0±0.5	EI		315
		C$_2$H$_2$NS$^+$				
C$_2$H$_2$NS$^+$	CH$_3$SCN	H	12.6±0.1	RPD		3017
C$_2$H$_2$NS$^+$	CH$_3$NCS	H	11.9±0.2	EI		315
C$_2$H$_2$NS$^+$	C$_2$H$_5$NCS	CH$_3$	12.5±0.2	EI		315
CH$_3$SCN$^+$	$\Delta H^\circ_{f298} = 1131$ kJ mol^{-1} (270 kcal mol^{-1})					
CH$_3$NCS$^+$	$\Delta H^\circ_{f298} = 1023$ kJ mol^{-1} (245 kcal mol^{-1})					
C$_2$H$_3$NS$^+$	CH$_3$SCN		10.065±0.01	PI	1131	182
C$_2$H$_3$NS$^+$	CH$_3$NCS		9.25±0.03	PI	1023	182
C$_2$H$_3$NS$^+$	CH$_3$NCS		9.13±0.15	EI		315
		C$_3$H$_5$NS$^+$				
C$_3$H$_5$NS$^+$	C$_2$H$_5$SCN		9.89±0.01	PI		182
C$_3$H$_5$NS$^+$	C$_2$H$_5$NCS		9.14±0.03	PI		182
C$_3$H$_5$NS$^+$	C$_2$H$_5$NCS		9.10±0.15	EI		193, 315
		C$_4$H$_{10}$NS$^+$				
C$_4$H$_{10}$NS$^+$	CH$_3$SCH$_2$CH$_2$CH(NH$_2$)COOH		9.68±0.15	EI		2587

See also — EI: 88

Ion	Reactant	Other products	Ionization or appearance potential (eV)	Method	Heat of formation (kJ mol^{-1})	Ref.
		C$_7$H$_5$NS$^+$				
C$_7$H$_5$NS$^+$	C$_6$H$_5$NCS (Isothiocyanic acid phenyl ester)		8.520±0.005	PI		182
C$_7$H$_5$NS$^+$	C$_7$H$_5$NS (Benzothiazole)		8.72±0.05	PE		2796
C$_7$H$_5$NS$^+$	C$_7$H$_5$NS (Benzothiazole)		8.65	CTS		1211
		C$_8$H$_7$NS$^+$				
C$_8$H$_7$NS$^+$	C$_6$H$_5$CH$_2$SCN (Thiocyanic acid benzyl ester)		9.06±0.05	EI		2025

4.3. The Positive Ion Table—Continued

Ion	Reactant	Other products	Ionization or appearance potential (eV)	Method	Heat of formation (kJ mol^{-1})	Ref.
		$C_{12}H_9NS^+$				
$C_{12}H_9NS^+$	$C_{12}H_9NS$ (Phenothiazine)		7.625	S		2661
$C_{12}H_9NS^+$	$C_{12}H_9NS$ (Phenothiazine)		7.7	PI		2661
$C_{12}H_9NS^+$	$C_{12}H_9NS$ (Phenothiazine)		6.96±0.19	CTS		2987
See also – CTS: 2562, 2909						
		$C_{18}H_{19}NS^+$				
$C_{18}H_{19}NS^+$	$C_{18}H_{19}NS$ (Chlorprothixene)		7.68±0.03	CTS		2987
		$CH_4N_2S^+$				
$CH_4N_2S^+$	$(NH_2)_2CS$		8.50±0.05	EI		1390, 2515, 2867
		$C_2H_6N_2S^+$				
$C_2H_6N_2S^+$	$CH_3NHCSNH_2$		8.29±0.05	EI		1390, 2515, 2867
		$C_3H_6N_2S^+$				
$C_3H_6N_2S^+$	$CH_2=CHNHCSNH_2$		8.29±0.05	EI		1390
		$C_3H_8N_2S^+$				
$C_3H_8N_2S^+$	$(CH_3)_2NCSNH_2$		8.34±0.05	EI		1390, 2515, 2867
$C_3H_8N_2S^+$	$(CH_3NH)_2CS$		8.17±0.05	EI		1390, 2515, 2867
		$C_4H_{10}N_2S^+$				
$C_4H_{10}N_2S^+$	$(CH_3)_2NCSNHCH_3$		7.93±0.05	EI		1390, 2515, 2867

4.3. The Positive Ion Table—Continued

Ion	Reactant	Other products	Ionization or appearance potential (eV)	Method	Heat of formation (kJ mol^{-1})	Ref.
			C$_5$H$_{12}$N$_2$S$^+$			
C$_5$H$_{12}$N$_2$S$^+$	C$_5$H$_{12}$N$_2$S (Diethyl–2–thiourea)		7.98±0.05	EI		1390
C$_5$H$_{12}$N$_2$S$^+$	((CH$_3$)$_2$N)$_2$CS		7.95±0.05	EI	1390,	2515, 2867
			C$_{17}$H$_{20}$N$_2$S$^+$			
C$_{17}$H$_{20}$N$_2$S$^+$	C$_{17}$H$_{20}$N$_2$S (Promazine)		7.23±0.12	CTS		2987
C$_{17}$H$_{20}$N$_2$S$^+$	C$_{17}$H$_{20}$N$_2$S (Promethazine)		7.25±0.10	CTS		2987
			C$_{18}$H$_{20}$N$_2$S$^+$			
C$_{18}$H$_{20}$N$_2$S$^+$	C$_{18}$H$_{20}$N$_2$S (Methdilazine)		7.25±0.10	CTS		2987
			C$_{18}$H$_{22}$N$_2$S$^+$			
C$_{18}$H$_{22}$N$_2$S$^+$	C$_{18}$H$_{22}$N$_2$S (Trimeprazine)		7.28±0.03	CTS		2987
			C$_{21}$H$_{26}$N$_2$S$_2^+$			
C$_{21}$H$_{26}$N$_2$S$_2^+$	C$_{21}$H$_{26}$N$_2$S$_2$ (Thioridazine)		7.20±0.05	CTS		2987
			CH$_2$OS$^+$			
CH$_2$OS$^+$	(CH$_3$)$_2$SO	CH$_4$	11.8±0.1	EI		3294
			CH$_3$OS$^+$			
CH$_3$OS$^+$	(CH$_3$)$_2$SO	CH$_3$	11.9±0.1	EI		3294

4.3. The Positive Ion Table—Continued

Ion	Reactant	Other products	Ionization or appearance potential (eV)	Method	Heat of formation (kJ mol^{-1})	Ref.
CH_3COSH^+ $\Delta H^\circ_{f298} = 783$ kJ mol^{-1} (187 kcal mol^{-1})						
$C_2H_4OS^+$	CH_3COSH		10.00 ± 0.02	PI	783	182
$C_2H_4OS^+$	$(CH_3)_2SO$	H_2	10.9 ± 0.5	EI		3294
$C_2H_4OS^+$	C_4H_8OS (1,4−Oxathiane)	C_2H_4	10.2 ± 0.2	EI		2969
$C_2H_5OS^+$						
$C_2H_5OS^+$	$(CH_3)_2SO$	H	12.6 ± 0.1	EI		3294
$C_2H_6OS^+$						
$C_2H_6OS^+$	$(CH_3)_2SO$		9.9 ± 0.1	EI		3294
$C_4H_8OS^+$						
$C_4H_8OS^+$	C_4H_8OS (1,4−Oxathiane)		8.8 ± 0.05	EI		2969
$C_5H_6OS^+$						
$C_5H_6OS^+$	$C_4H_3SOCH_3$ (2−Methoxythiophene)		8.30	EI		3240
$C_8H_9OS^+$						
$C_8H_9OS^+$	$C_4H_3S(CH_2)_3COOCH_3$ (4−(2−Thienyl)butanoic acid methyl ester)	CH_3O	11.05 ± 0.3	EI		2496

See also − EI: 2497

Ion	Reactant	Other products	Ionization or appearance potential (eV)	Method	Heat of formation (kJ mol^{-1})	Ref.
$C_{12}H_8OS^+$						
$C_{12}H_8OS^+$	$C_{12}H_8OS$ (Phenoxathiin)		7.98 ± 0.03	RPD		2538
$C_{12}H_8OS^+$	$C_{12}H_8OS$ (Phenoxathiin)		7.6	CTS		3300
$C_{12}H_8OS^+$	$C_{12}H_8OS$ (Phenoxathiin)		7.85	CTS		3401
$C_{13}H_{10}OS^+$						
$C_{13}H_{10}OS^+$	$C_{13}H_{10}OS$ (2−Methylphenoxathiin)		7.78	CTS		3401

4.3. The Positive Ion Table—Continued

Ion	Reactant	Other products	Ionization or appearance potential (eV)	Method	Heat of formation (kJ mol⁻¹)	Ref.

$C_9H_{12}O_2S^+$

$C_9H_{12}O_2S^+$	$C_4H_3S(CH_2)_3COOCH_3$ (4–(2–Thienyl)butanoic acid methyl ester)		8.43 ± 0.3	EI		2496

See also − EI: 2497

$C_4H_{10}O_3S^+$

$C_4H_{10}O_3S^+$	$(C_2H_5O)_2SO$		9.68?	PI		182

$C_{19}H_{18}O_3S^+$

$C_{19}H_{18}O_3S^+$	$C_6H_4(CH_3)SO_3CH_2CH_2C_{10}H_7$ (4–Toluenesulfonic acid 4–azulylethyl ester)		7.3	EI		3333

$C_4H_3FS^+$

$C_4H_3FS^{+}*$ (F lone pair IP)	C_4H_3SF (2–Fluorothiophene)		13.8 (V)	D		3246

$C_2H_3F_2S^+$

$C_2H_3F_2S^+$	CH_3SCF_3	F	15.7 ± 0.24	EI		3202

CHF_3S^+

CHF_3S^+	CF_3SH		11.35 ± 0.1	EI		3034

$C_2H_3F_3S^+$

$C_2H_3F_3S^+$	CH_3SCF_3		9.88 ± 0.03	PE		3034
$C_2H_3F_3S^+$	CH_3SCF_3		9.75 ± 0.11	EI		3202

$C_2H_3F_2S_2^+$

$C_2H_3F_2S_2^+$	CH_3SSCF_3	F	~15	EI		3202

$C_2H_3F_3S_2^+$

$C_2H_3F_3S_2^+$	CH_3SSCF_3		9.60 ± 0.03	PE		3034
$C_2H_3F_3S_2^+$	CH_3SSCF_3		9.58 ± 0.14	EI		3202

SiH_2FS^+

SiH_2FS^+	SiF_2H_2S	F	12.3 ± 0.3	EI		2896

4.3. The Positive Ion Table—Continued

Ion	Reactant	Other products	Ionization or appearance potential (eV)	Method	Heat of formation (kJ mol^{-1})	Ref.
		SiHF$_2$S$^+$				
SiHF$_2$S$^+$	SiF$_2$H$_2$S	H	11.9±0.3	EI		2896
		SiH$_2$F$_2$S$^+$				
SiH$_2$F$_2$S$^+$	SiF$_2$H$_2$S		10.7±0.3	EI		2896
		Si$_2$H$_2$F$_3$S$^+$				
Si$_2$H$_2$F$_3$S$^+$	Si$_2$F$_4$H$_2$S	F	11.3±0.3	EI		2896
		Si$_2$H$_2$F$_4$S$^+$				
Si$_2$H$_2$F$_4$S$^+$	Si$_2$F$_4$H$_2$S		10.6±0.3	EI		2896
		SiH$_2$F$_2$S$_2^+$				
SiH$_2$F$_2$S$_2^+$	SiF$_2$H$_2$S$_2$		10.7±0.3	EI		2896
		C$_4$H$_3$NO$_2$S$^+$				
C$_4$H$_3$NO$_2$S$^+$	C$_4$H$_3$SNO$_2$ (2−Nitrothiophene)		9.77	EI		3240
		C$_5$H$_{11}$NO$_2$S$^+$				
C$_5$H$_{11}$NO$_2$S$^+$	CH$_3$SCH$_2$CH$_2$CH(NH$_2$)COOH		8.63±0.10	EI		2587
		CH$_3$BF$_2$S$^+$				
CH$_3$BF$_2$S$^+$	CH$_3$SBF$_2$		10.59	EI		3227
		C$_2$H$_6$BFS$_2^+$				
C$_2$H$_6$BFS$_2^+$	(CH$_3$S)$_2$BF		9.79	EI		3227
		C$_{21}$H$_{24}$N$_3$F$_3$S$^+$				
C$_{21}$H$_{24}$N$_3$F$_3$S$^+$	C$_{21}$H$_{24}$N$_3$F$_3$S (Stelazine)		7.43±0.37	CTS		2987

4.3. The Positive Ion Table—Continued

Ion	Reactant	Other products	Ionization or appearance potential (eV)	Method	Heat of formation (kJ mol^{-1})	Ref.
			C$_{12}$H$_7$OFS$^+$			
C$_{12}$H$_7$OFS$^+$	C$_{12}$H$_7$OSF (2–Fluorophenoxathiin)		8.00	CTS		3401

Cl$^+$ $\Delta H^\circ_{f0} = 1371.1$ kJ mol^{-1} (327.7 kcal mol^{-1})

Ion	Reactant	Other products	Ionization or appearance potential (eV)	Method	Heat of formation (kJ mol^{-1})	Ref.
Cl$^+$	Cl		12.967	S	1371.1	2113, 3114

See also – S: 2656, 3089, 3118
      EI: 196, 440, 3178

Ion	Reactant	Other products	Ionization or appearance potential (eV)	Method	Heat of formation (kJ mol^{-1})	Ref.
Cl$^+$	Cl$_2$	Cl$^-$	11.86±0.04	RPD	(1373)	288, 292
Cl$^+$	HCl	H	17.34±0.01	PI		2637

Estimating a 0 K threshold of 17.37 eV one obtains ΔH°_{f0}(Cl$^+$) ≈ 1368 kJ mol^{-1}.

See also – EI: 440

Ion	Reactant	Other products	Ionization or appearance potential (eV)	Method	Heat of formation (kJ mol^{-1})	Ref.
Cl$^+$	CCl$_4$	CCl$_3$	16.10±0.2	RPD		196
Cl$^+$	ClF	F$^-$	12.04	PI	(1358)	3037
Cl$^+$	ClF	F	15.50±0.04	PI	(1364)	3037
(Threshold value approximately corrected to 0 K)						
Cl$^+$	MgCl$_2$		19±1.0	EI		178
Cl$^+$	PCl$_3$		19.8±0.4	EI		1101

See also – EI: 192, 2506

Ion	Reactant	Other products	Ionization or appearance potential (eV)	Method	Heat of formation (kJ mol^{-1})	Ref.
Cl$^+$	CH$_3$Cl	CH$_3$	16.6±0.05	RPD		2154
Cl$^+$	C$_2$H$_5$Cl		23.4±0.3	EI		356
Cl$^+$	CH$_3$C≡CCl		18.4±0.5	EI		13
Cl$^+$	CHCl$_3$		22.0±0.3	EI		43
Cl$^+$	ClCN	CN$^-$	13.6±0.1	EI		73
Cl$^+$	ClCN	CN	17.32±0.02	PI		2621

See also – EI: 73

Ion	Reactant	Other products	Ionization or appearance potential (eV)	Method	Heat of formation (kJ mol^{-1})	Ref.
Cl$^+$	CF$_3$Cl		21±1	EI		24
Cl$^+$	ClO$_3$F		23.0±0.2	EI		53
Cl$^+$	POCl$_3$		19±1	EI		1101
Cl$^+$	SF$_5$Cl		20.8±0.3	EI		2777
Cl$^+$	CHF$_2$Cl		20.5±0.3	EI		43
Cl$^+$	CHFCl$_2$		23.0±0.3	EI		43
Cl$^+$	ZnCl$_2$		19.6±0.3	EI		2506
Cl$^+$	GeCl$_2$		15.5±0.5	EI		2568
Cl$^+$	GeCl$_4$		21±1	EI		2568
Cl$^+$	HgCl$_2$		17.7±0.3	EI		2506
Cl$^+$	TlCl	Tl	16.9±0.1	RPD		2159

4.3. The Positive Ion Table—Continued

Ion	Reactant	Other products	Ionization or appearance potential (eV)	Method	Heat of formation (kJ mol^{-1})	Ref.	
	$Cl_2^+(^2\Pi_{3/2g})$	$\Delta H_{f0}^\circ = 1109$ kJ mol^{-1} (265 kcal mol^{-1})					
	$Cl_2^+(^2\Pi_{1/2g})$	$\Delta H_{f0}^\circ = 1117$ kJ mol^{-1} (267 kcal mol^{-1})					
	$Cl_2^+(^2\Pi_u)$	$\Delta H_{f0}^\circ = 1347$ kJ mol^{-1} (322 kcal mol^{-1})					
	$Cl_2^+(^2\Sigma_g^+)$	$\Delta H_{f0}^\circ = 1517$ kJ mol^{-1} (363 kcal mol^{-1})					
$Cl_2^+(^2\Pi_{3/2g})$	Cl_2		11.48±0.01	PI	1108	182,	416
$Cl_2^+(^2\Pi_{3/2g})$	Cl_2		11.48±0.01	PI	1108		3413
(Threshold value corrected for hot bands)							
$Cl_2^+(^2\Pi_{3/2g})$	Cl_2		11.51±0.01	PE	1111		3409
$Cl_2^+(^2\Pi_{3/2g})$	Cl_2		11.49	PE	1109		3275
$Cl_2^+(^2\Pi_{1/2g})$	Cl_2		11.59±0.01	PE	1118		3409
$Cl_2^+(^2\Pi_{1/2g})$	Cl_2		11.57	PE	1116		3275
$Cl_2^+(^2\Pi_u)$	Cl_2		13.96±0.02	PE	1347		3409
$Cl_2^+(^2\Pi_u)$	Cl_2		14.0	PE			3275
$Cl_2^+(^2\Sigma_g^+)$	Cl_2		15.72±0.02	PE	1517		3409
$Cl_2^+(^2\Sigma_g^+)$	Cl_2		15.8	PE			3275

See also — PE: 2815, 3277
      EI: 75, 292, 440

Cl_2^{+2}

Ion	Reactant	Other products	Ionization or appearance potential (eV)	Method	Heat of formation (kJ mol^{-1})	Ref.
Cl_2^{+2}	Cl_2		32.6	EI		75

Ion	Reactant	Other products	Ionization or appearance potential (eV)	Method	Heat of formation (kJ mol^{-1})	Ref.	
	$HCl^+(^2\Pi_{3/2})$	$\Delta H_{f0}^\circ = 1137$ kJ mol^{-1} (272 kcal mol^{-1})					
	$HCl^+(^2\Pi_{1/2})$	$\Delta H_{f0}^\circ = 1145$ kJ mol^{-1} (274 kcal mol^{-1})					
	$HCl^+(^2\Sigma^+)$	$\Delta H_{f0}^\circ = 1475$ kJ mol^{-1} (353 kcal mol^{-1})					
$HCl^+(^2\Pi_{3/2})$	HCl		12.74±0.01	PI	1137	182,	416
$HCl^+(^2\Pi_{3/2})$	HCl		12.742±0.010	PI	1137		1253
$HCl^+(^2\Pi_{3/2})$	HCl		12.748±0.005	PE	1138		2892
$HCl^+(^2\Pi_{3/2})$	HCl		12.74±0.01	PE	1137		2819
$HCl^+(^2\Pi_{1/2})$	HCl		12.828±0.005	PE	1146		2892
$HCl^+(^2\Pi_{1/2})$	HCl		12.82±0.01	PE	1145		2819
$HCl^+(^2\Sigma^+)$	HCl		16.254±0.005	PE	1476		2892
$HCl^+(^2\Sigma^+)$	HCl		16.23	PE	1474		2819

See also — PI: 2637
       PE: 2815
       EI: 39, 2991

DCl^+

Ion	Reactant	Other products	Ionization or appearance potential (eV)	Method	Heat of formation (kJ mol^{-1})	Ref.
$DCl^+(^2\Pi_{3/2})$	DCl		12.756±0.005	PE		2892
$DCl^+(^2\Pi_{1/2})$	DCl		12.835±0.005	PE		2892
$DCl^+(^2\Sigma^+)$	DCl		16.271±0.005	PE		2892

4.3. The Positive Ion Table—Continued

Ion	Reactant	Other products	Ionization or appearance potential (eV)	Method	Heat of formation (kJ mol^{-1})	Ref.
		HCl^{+2}				
HCl^{+2}	HCl		35.5±0.5	NRE		212
		LiCl$^+$				
LiCl$^+$	LiCl		10.1	EI		2179
		Li$_2$Cl$^+$				
Li$_2$Cl$^+$	Li$_2$Cl$_2$	Cl	10.6	EI		2179
		BeCl$^+$				
BeCl$^+$	BeCl		9.5±0.5	EI		2715
BeCl$^+$	BeCl$_2$	Cl	14.3±0.5	EI		2715
		BeCl$_2^+$				
BeCl$_2^+$	BeCl$_2$		12.4±0.4	EI		2715
		Be$_2$Cl$_3^+$				
Be$_2$Cl$_3^+$	Be$_2$Cl$_4$	Cl	13.8±0.6	EI		2715
		Be$_2$Cl$_4^+$				
Be$_2$Cl$_4^+$	Be$_2$Cl$_4$		12.8±0.4	EI		2715
	BCl$^+$	$\Delta H_{f0}^\circ \sim 1130$ kJ mol^{-1} (270 kcal mol^{-1})				
BCl$^+$	BCl$_3$	2Cl?	18.37±0.02	PI	1130	2628

See also – EI: 206, 440

Ion	Reactant	Other products	Ionization or appearance potential (eV)	Method	Heat of formation (kJ mol^{-1})	Ref.
BCl$^+$	B$_2$Cl$_4$	BCl$_3$	13.71±0.04	PI	(1236)	2628

This process apparently includes excess energy.

Ion	Reactant	Other products	Ionization or appearance potential (eV)	Method	Heat of formation (kJ mol^{-1})	Ref.
	BCl$_2^+$	$\Delta H_{f0}^\circ = 664$ kJ mol^{-1} (159 kcal mol^{-1})				
BCl$_2^+$	BCl$_3$	Cl	12.30±0.02	PI	664	2628

See also – EI: 206, 440

Ion	Reactant	Other products	Ionization or appearance potential (eV)	Method	Heat of formation (kJ mol^{-1})	Ref.
BCl$_2^+$	B$_2$Cl$_4$	BCl$_2$	11.32±0.02	PI		2628

4.3. The Positive Ion Table—Continued

Ion	Reactant	Other products	Ionization or appearance potential (eV)	Method	Heat of formation (kJ mol^{-1})	Ref.
		BCl_2^{+2}				
BCl_2^{+2}	BCl_3	Cl	33.77 ± 0.07	EI		440
		$B_2Cl_2^+$ $\Delta H_{f0}^{\circ} \sim 934$ kJ mol^{-1} (223 kcal mol^{-1})				
$B_2Cl_2^+$	B_2Cl_4	$2Cl?$	17.24 ± 0.03	PI	934	2628
		$BCl_3^+(^2A_2')$ $\Delta H_{f0}^{\circ} \sim 718$ kJ mol^{-1} (172 kcal mol^{-1})				
$BCl_3^+(^2A_2')$	BCl_3		11.60 ± 0.02	PI	716	2628
$BCl_3^+(^2A_2')$	BCl_3		11.64 ± 0.02	PE	720	3375
$BCl_3^+(^2E'')$	BCl_3		12.19 ± 0.04	PE		3375
$BCl_3^+(^2E')$	BCl_3		12.66 ± 0.01 (V)	PE		3375
$BCl_3^+(^2A_2'')$	BCl_3		14.22 ± 0.02	PE		3375
$BCl_3^+(^2E')$	BCl_3		15.32 ± 0.02	PE		3375
$BCl_3^+(^2E')$	BCl_3		15.72 ± 0.04 (V)	PE		3375
$BCl_3^+(^2A_1')$	BCl_3		17.74 ± 0.05	PE		3375

There is disagreement on some of the assignments, see refs. 2834, 3119.
Additional fine structure is resolved in ref. 3119.

See also – PE: 2834, 3119
          EI: 206, 440, 2512, 2513, 3227

Ion	Reactant	Other products	Ionization or appearance potential (eV)	Method	Heat of formation (kJ mol^{-1})	Ref.
		$B_2Cl_3^+$ $\Delta H_{f0}^{\circ} = 502$ kJ mol^{-1} (120 kcal mol^{-1})				
$B_2Cl_3^+$	B_2Cl_4	Cl	11.52 ± 0.02	PI	502	2628
		$B_2Cl_4^+$ $\Delta H_{f0}^{\circ} = 506$ kJ mol^{-1} (121 kcal mol^{-1})				
$B_2Cl_4^+$	B_2Cl_4		10.32 ± 0.02	PI	506	2628

See also – PE: 3417

Ion	Reactant	Other products	Ionization or appearance potential (eV)	Method	Heat of formation (kJ mol^{-1})	Ref.
		CCl^+				
CCl^+	CCl		12.9 ± 0.10	EI		129
CCl^+	CCl_2	Cl	16.3 ± 0.2	EI		319
CCl^+	$CCl{\equiv}CCl$	CCl	15.4 ± 0.1	EDD		3177
CCl^+	CCl_4	$Cl_2+Cl?$	19.35 ± 0.05	EI		129
CCl^+	CCl_4	$Cl_2+Cl?$	19.4 ± 0.1	EI		319
CCl^+	$CH{\equiv}CCl$	CH	17.2 ± 0.1	EDD		3177
CCl^+	$CHCl_3$		16.3 ± 0.2	EI		43
CCl^+	$ClCN$	N	17.2 ± 0.2	EI		73
CCl^+	$CHFCl_2$		18.3 ± 0.2	EI		43
CCl^+	$CCl{\equiv}CBr$	CBr	16.4 ± 0.1	EDD		3177
CCl^+	$CCl{\equiv}CI$	CI	17.5 ± 0.1	EDD		3177

4.3. The Positive Ion Table—Continued

Ion	Reactant	Other products	Ionization or appearance potential (eV)	Method	Heat of formation (kJ mol⁻¹)	Ref.

CCl_2^+

Ion	Reactant	Other products	Ionization or appearance potential (eV)	Method	Heat of formation (kJ mol⁻¹)	Ref.
CCl_2^+	CCl_2		9.76	EI		2578

See also – EI: 129, 319

CCl_2^+	CCl_4		15.4	EI		2578

See also – EI: 129, 319

CCl_2^+	C_2Cl_4		14.7	EI		2578
CCl_2^+	$CHCl_3$		12.2	EI		2578

$C_2Cl_2^+$

Ion	Reactant	Other products	Ionization or appearance potential (eV)	Method	Heat of formation	Ref.
$C_2Cl_2^+(^2\Pi_{3/2u})$	$CCl{\equiv}CCl$		10.09 (V)	PE		3121
$C_2Cl_2^+(^2\Pi_{3/2u})$	$CCl{\equiv}CCl$		10.3±0.1	EDD		3177
$C_2Cl_2^+(^2\Pi_{3/2g})$	$CCl{\equiv}CCl$		13.44 (V)	PE		3121
$C_2Cl_2^+(^2\Pi_{3/2u})$	$CCl{\equiv}CCl$		14.45 (V)	PE		3121
$C_2Cl_2^+(^2\Sigma_g^+)$	$CCl{\equiv}CCl$		16.76 (V)	PE		3121
$C_2Cl_2^+(^2\Sigma_u^+)$	$CCl{\equiv}CCl$		17.81 (V)	PE		3121

$C_2Cl_2^{+2}$

Ion	Reactant	Other products	Ionization or appearance potential (eV)	Method	Heat of formation	Ref.
$C_2Cl_2^{+2}$	$CCl{\equiv}CCl$		27.6±0.1	EDD		3177

CCl_3^+

Ion	Reactant	Other products	Ionization or appearance potential (eV)	Method	Heat of formation	Ref.
CCl_3^+	CCl_3		8.78±0.05	EI		441
CCl_3^+	CCl_4	Cl	11.65±0.10	RPD		196
CCl_3^+	CCl_4	Cl	11.67±0.1	EI		441
CCl_3^+	CCl_4	Cl	11.90±0.07	EI		129
CCl_3^+	$CHCl_3$	H	11.70±0.09?	RPD		1139
CCl_3^+	C_2HCl_5	$CHCl_2$	11.54±0.1	EI		72
CCl_3^+	$CFCl_3$	F	12.77±0.15	RPD		185
CCl_3^+	CCl_3COCHN_2		11.0	EI		2174
CCl_3^+	CCl_3Br	Br	10.90±0.1	EI		441

CCl_4^+ $\Delta H_{f0}^\circ = 1006$ kJ mol⁻¹ (240 kcal mol⁻¹)

Ion	Reactant	Other products	Ionization or appearance potential (eV)	Method	Heat of formation	Ref.
CCl_4^+	CCl_4		11.47±0.01	PI	1006	182, 416
CCl_4^+	CCl_4		11.47±0.08	PE	1006	3362
CCl_4^+	CCl_4		11.47	PE	1006	2843

Orbital assignments are discussed in R. N. Dixon, J. N. Murrell and B. Narayan, Mol. Phys. **20**, 611 (1971).

See also – PE: 3117, 3119
        PEN: 2430

4.3. The Positive Ion Table—Continued

Ion	Reactant	Other products	Ionization or appearance potential (eV)	Method	Heat of formation (kJ mol^{-1})	Ref.

$C_2Cl_4^+$ $\quad \Delta H_{f0}^\circ = 889$ kJ mol^{-1} (212 kcal mol^{-1})

Ion	Reactant	Other products	IP/AP (eV)	Method	ΔH_f	Ref.
$C_2Cl_4^+$	C_2Cl_4		9.32±0.01	PI	888	182
$C_2Cl_4^+$	C_2Cl_4		9.32	PI	888	168
$C_2Cl_4^+$	C_2Cl_4		9.34	PE	890	2885

NCl_3^+

Ion	Reactant	Other products	IP/AP (eV)	Method	ΔH_f	Ref.
NCl_3^+	NCl_3		10.7 (V)	PE		3119

ClO^+

Ion	Reactant	Other products	IP/AP (eV)	Method	ΔH_f	Ref.
ClO^+	ClO		11.1±0.1	EI	2451,	2488
ClO^+	ClO_2	O	13.5±0.1	EI		2488
ClO^+	Cl_2O	Cl	12.5±0.1	EI		2451
ClO^+	ClO_3F		18.0±0.5	EI		53

ClO_2^+

Ion	Reactant	Other products	IP/AP (eV)	Method	ΔH_f	Ref.
ClO_2^+	ClO_2		10.7±0.1	EI	2451,	2488
ClO_2^+	ClO_3F		15.7±0.5	EI		53

ClO_3^+

Ion	Reactant	Other products	IP/AP (eV)	Method	ΔH_f	Ref.
ClO_3^+	Cl_2O_7		13.00±0.05	EI		2451
ClO_3^+	ClO_3F	F	14.3±0.2	EI		53

Cl_2O^+ $\quad \Delta H_{f0}^\circ = 1138$ kJ mol^{-1} (272 kcal mol^{-1})

Ion	Reactant	Other products	IP/AP (eV)	Method	ΔH_f	Ref.
Cl_2O^+	Cl_2O		10.94	PE	1138	3404
Cl_2O^+	Cl_2O		11.16±0.1	EI		2451

ClF^+ $\quad \Delta H_{f0}^\circ = 1166$ kJ mol^{-1} (279 kcal mol^{-1})

Ion	Reactant	Other products	IP/AP (eV)	Method	ΔH_f	Ref.
ClF^+	ClF		12.65±0.01	PI	1166	3037
(Threshold value corrected for hot bands)						
ClF^+	ClF		12.7±0.3	EI		357

ClF_2^+

Ion	Reactant	Other products	IP/AP (eV)	Method	ΔH_f	Ref.
ClF_2^+	ClF_3	F	12.8±0.3	EI		357

ClF_3^+

Ion	Reactant	Other products	IP/AP (eV)	Method	ΔH_f	Ref.
ClF_3^+	ClF_3		13.0±0.2	EI		357

4.3. The Positive Ion Table—Continued

Ion	Reactant	Other products	Ionization or appearance potential (eV)	Method	Heat of formation (kJ mol^{-1})	Ref.
NaCl$^+$		**$\Delta H_{f0}^{\circ} \sim 681$ kJ mol^{-1} (163 kcal mol^{-1})**				
NaCl$^+$	NaCl		8.92±0.06	PI	681	2610
MgCl$^+$						
MgCl$^+$	MgCl		7.49±0.10	EI		2991
MgCl$^+$	MgCl		7.5±0.5	EI		2990
MgCl$^+$	MgCl$_2$	Cl	12.38±0.10	EI		2991
MgCl$^+$	MgCl$_2$	Cl	12.4±0.2	EI		2990
MgCl$^+$	MgCl$_2$	Cl	11.5±0.5	EI		178
MgCl$_2^+$						
MgCl$_2^+$	MgCl$_2$		11.58±0.10	EI		2991
MgCl$_2^+$	MgCl$_2$		11.5±0.2	EI		2990
MgCl$_2^+$	MgCl$_2$		11.1±0.2	EI		178
Mg$_2$Cl$_3^+$						
Mg$_2$Cl$_3^+$	Mg$_2$Cl$_4$	Cl	11.2±0.3	EI		178
AlCl$^+$						
AlCl$^+$	AlCl		9.4±0.4	EI		2715
AlCl$^+$	AlCl		9.5±0.3	EI		2990
AlCl$^+$	AlCl$_3$		19.7±0.5	EI		2897
AlCl$_2^+$						
AlCl$_2^+$	AlCl$_3$	Cl	13.4±0.5	EI		2897
AlCl$_3^+$						
AlCl$_3^+$	AlCl$_3$		12.8±0.5	EI		2897
Al$_2$Cl$_5^+$						
Al$_2$Cl$_5^+$	Al$_2$Cl$_6$	Cl	13.1±0.5	EI		2897
SiCl$_3^+$						
SiCl$_3^+$	SiCl$_4$	Cl	12.48±0.02	EI		2182
SiCl$_3^+$	Si$_2$Cl$_6$	SiCl$_3$	11.55±0.1	EI		2183
SiCl$_3^+$	SiHCl$_3$	H	11.91±0.03	EI		2182
SiCl$_3^+$	CH$_3$SiCl$_3$	CH$_3$	11.90±0.08	EI		2182
SiCl$_3^+$	C$_2$H$_5$SiCl$_3$	C$_2$H$_5$	12.10±0.03	EI		2182
SiCl$_3^+$	iso-C$_3$H$_7$SiCl$_3$		13.1±0.2	EI		2182
SiCl$_3^+$	tert-C$_4$H$_9$SiCl$_3$		13.0±0.1	EI		2182

4.3. The Positive Ion Table—Continued

Ion	Reactant	Other products	Ionization or appearance potential (eV)	Method	Heat of forma- tion (kJ mol⁻¹)	Ref.

$$\text{SiCl}_4^+ \qquad \Delta H_{f0}^\circ = 483 \text{ kJ mol}^{-1} \text{ (115 kcal mol}^{-1}\text{)}$$

Ion	Reactant	Other products	Ionization or appearance potential (eV)	Method	Heat of formation (kJ mol⁻¹)	Ref.
SiCl_4^+	SiCl_4		11.79 ± 0.01	PE	483	3362

See also − PE: 3117

PCl^+

PCl^+	PCl_3		16.83 ± 0.3	EI		192
PCl^+	PCl_3		16.8 ± 0.3	EI		2506
PCl^+	PCl_3		16.5 ± 0.5	EI		1101
PCl^+	P_2Cl_4		15.7 ± 0.3	EI		192
PCl^+	POCl_3		17 ± 1	EI		1101
PCl^+	POCl_3		20.2 ± 0.4	EI		2506

P_2Cl^+

P_2Cl^+	P_2Cl_4	$\text{Cl}_2 + \text{Cl?}$	16.1 ± 0.4	EI		192

PCl_2^+

PCl_2^+	PCl_3	Cl^-	11.8 ± 0.5	EI		1101
PCl_2^+	PCl_3	Cl	12.32 ± 0.2	EI		192
PCl_2^+	PCl_3	Cl	12.3 ± 0.2	EI		2506
PCl_2^+	P_2Cl_4	PCl_2	11.68 ± 0.2	EI		192
PCl_2^+	POCl_3		13.3 ± 0.5	EI		1101

P_2Cl_2^+

P_2Cl_2^+	P_2Cl_4		13.9 ± 0.3	EI		192

PCl_3^+

PCl_3^+	PCl_3		9.91	PI		1091
PCl_3^+	PCl_3		10.18 ± 0.10	PE		3168

See also − PE: 3119
       EI: 192, 1101, 2453, 2506, 2597

PCl_3^+	POCl_3	O	12.3 ± 0.5	EI		1101

P_2Cl_3^+

P_2Cl_3^+	P_2Cl_4	Cl	11.7 ± 0.3	EI		192

4.3. The Positive Ion Table—Continued

Ion	Reactant	Other products	Ionization or appearance potential (eV)	Method	Heat of formation (kJ mol^{-1})	Ref.
			P$_2$Cl$_4^+$			
P$_2$Cl$_4^+$	P$_2$Cl$_4$		9.36±0.2	EI		192
			B$_5$H$_8$Cl$^+$			
B$_5$H$_8$Cl$^+$	1–B$_5$H$_8$Cl		9.48±0.10	RPD		3228
B$_5$H$_8$Cl$^+$	2–B$_5$H$_8$Cl		9.90±0.05	RPD		3228
			CHCl$^+$			
CHCl$^+$	CHCl$_3$		17.5±0.2	EI		43
CHCl$^+$	CHFCl$_2$		19.0±0.2	EI		43
			CH$_2$Cl$^+$			
CH$_2$Cl$^+$	CH$_2$Cl		9.32	EI		141
CH$_2$Cl$^+$	CH$_2$Cl		9.70±0.09	EI		131
CH$_2$Cl$^+$	CH$_3$Cl	H	12.98±0.07	RPD		1139

See also — EI: 72, 160, 356, 3017

Ion	Reactant	Other products	Ionization or appearance potential (eV)	Method	Heat of formation (kJ mol^{-1})	Ref.
CH$_2$Cl$^+$	C$_2$H$_5$Cl	CH$_3$	13.20±0.2	EI		72
CH$_2$Cl$^+$	C$_2$H$_5$Cl	CH$_3$	13.6±0.2	EI		356
CH$_2$Cl$^+$	CH$_2$Cl$_2$	Cl	12.12±0.1	EI		72
CH$_2$Cl$^+$	CH$_2$Cl$_2$	Cl	12.89±0.03	EI		131
CH$_2$Cl$^+$	CH$_2$ClCH$_2$Cl	CH$_2$Cl	12.52±0.1	EI		72
CH$_2$Cl$^+$	CH$_3$COCH$_2$Cl		13.8±0.07	EI		2174
CH$_2$Cl$^+$	C$_3$H$_5$OCl (1–Chloro–2,3–epoxypropane)		12.5±0.1	EI		153
CH$_2$Cl$^+$	CH$_2$ClCOCHN$_2$		12.2±0.1	EI		2174
CH$_2$Cl$^+$	CH$_2$ClBr	Br	11.56±0.1	EI		72

CH$_3$Cl$^+$(^{2}E$_{3/2}$) $\Delta H_{f0}^\circ \sim 1014$ kJ mol^{-1} (242 kcal mol^{-1})

Ion	Reactant	Other products	Ionization or appearance potential (eV)	Method	Heat of formation (kJ mol^{-1})		Ref.
CH$_3$Cl$^+$(^{2}E$_{3/2}$)	CH$_3$Cl		11.220±0.01	S	1010		2064
CH$_3$Cl$^+$(^{2}E$_{3/2}$)	CH$_3$Cl		11.265±0.003	PI	1014		1253
CH$_3$Cl$^+$(^{2}E$_{3/2}$)	CH$_3$Cl		11.28±0.01	PI	1015	182,	416
CH$_3$Cl$^+$(^{2}E$_{3/2}$)	CH$_3$Cl		11.28	PI	1015	1399,	2637
CH$_3$Cl$^+$(^{2}E$_{3/2}$)	CH$_3$Cl		11.26	PE	1014		2843
CH$_3$Cl$^+$(^{2}E$_{3/2}$)	CH$_3$Cl		11.29	PE	1016		3057
CH$_3$Cl$^+$(^{2}E$_{1/2}$)	CH$_3$Cl		11.305±0.01	S			2064
CH$_3$Cl$^+$(^{2}E$_{1/2}$)	CH$_3$Cl		11.37	PE			3057

Additional higher ionization potentials are given in ref. 3057.

See also — PI: 190
          PE: 3119
          PEN: 2430, 2466
          EI: 72, 289, 364, 2146, 2154, 2776, 3201

4.3. The Positive Ion Table—Continued

Ion	Reactant	Other products	Ionization or appearance potential (eV)	Method	Heat of formation (kJ mol^{-1})	Ref.
			CH$_2$DCl$^+$			
CH$_2$DCl$^+$	CH$_2$DCl		11.29	PI		2637
			CHD$_2$Cl$^+$			
CHD$_2$Cl$^+$	CHD$_2$Cl		11.30	PI		2637
			CD$_3$Cl$^+$			
CD$_3$Cl$^+$	CD$_3$Cl		11.30	PI		2637
			C$_2$HCl$^+$			
C$_2$HCl$^+$($^2\Pi$)	CH≡CCl		10.63	PE		3071
C$_2$HCl$^+$($^2\Pi$)	CH≡CCl		10.7±0.1	EDD		3177
See also – S: 2775						
C$_2$HCl$^+$	cis–CHCl=CHCl	HCl	13.27±0.05	EI		114
C$_2$HCl$^+$	trans–CHCl=CHCl	HCl	13.39±0.05	EI		114
			C$_2$HCl^{+2}			
C$_2$HCl^{+2}	CH≡CCl		29.0±0.1	EDD		3177
			C$_2$H$_2$Cl$^+$			
C$_2$H$_2$Cl$^+$	cis–CHCl=CHCl	Cl	12.29±0.05	EI		114
C$_2$H$_2$Cl$^+$	trans–CHCl=CHCl	Cl	12.61±0.05	EI		114

C$_2$H$_3$Cl$^+$ $\Delta H_{f0}^\circ = 1008$ kJ mol^{-1} (241 kcal mol^{-1})

Ion	Reactant	Other products	Ionization or appearance potential (eV)	Method	Heat of formation (kJ mol^{-1})	Ref.
C$_2$H$_3$Cl$^+$	C$_2$H$_3$Cl		10.00±0.01	S	1008	261
(Average of two Rydberg series limits)						
C$_2$H$_3$Cl$^+$	C$_2$H$_3$Cl		10.00±0.01	S	1008	2670
(Average of four Rydberg series limits)						
C$_2$H$_3$Cl$^+$	C$_2$H$_3$Cl		9.995±0.01	PI		182
C$_2$H$_3$Cl$^+$	C$_2$H$_3$Cl		9.995	PI		168
C$_2$H$_3$Cl$^+$	C$_2$H$_3$Cl		10.00±0.02	PI		268
C$_2$H$_3$Cl$^+$	C$_2$H$_3$Cl		10.00±0.01	PI		2670
C$_2$H$_3$Cl$^+$	C$_2$H$_3$Cl		10.00	PE		2885

See also – EI: 268, 2793

4.3. The Positive Ion Table—Continued

Ion	Reactant	Other products	Ionization or appearance potential (eV)	Method	Heat of formation (kJ mol^{-1})	Ref.
		C$_2$D$_3$Cl$^+$				
C$_2$D$_3$Cl$^+$	C$_2$D$_3$Cl		10.02±0.02	PI		268
C$_2$D$_3$Cl$^+$	C$_2$D$_3$Cl		10.10±0.03	EI		268
	C$_2$H$_5$Cl$^+$	**$\Delta H^{\circ}_{f0} = 961$ kJ mol^{-1} (230 kcal mol^{-1})**				
C$_2$H$_5$Cl$^+$	C$_2$H$_5$Cl		10.97±0.02	PI	961	416
C$_2$H$_5$Cl$^+$	C$_2$H$_5$Cl		10.98±0.01	PI	962	190
C$_2$H$_5$Cl$^+$	C$_2$H$_5$Cl		10.98±0.02	PI	962	182

See also – EI: 72, 160, 356, 2146, 3201

		C$_3$H$_3$Cl$^+$				
C$_3$H$_3$Cl$^+$	CH$_3$C≡CCl		9.9±0.1	EI		13
	n–C$_3$H$_7$Cl$^+$	$\Delta H^{\circ}_{f298} \sim 914$ kJ mol^{-1} (218 kcal mol^{-1})				
	iso–C$_3$H$_7$Cl$^+$	$\Delta H^{\circ}_{f298} \sim 894$ kJ mol^{-1} (214 kcal mol^{-1})				
C$_3$H$_7$Cl$^+$	n–C$_3$H$_7$Cl		10.82±0.03	PI	914	182
C$_3$H$_7$Cl$^+$	n–C$_3$H$_7$Cl		10.78±0.04	EI		2146, 3201

See also – EI: 72

C$_3$H$_7$Cl$^+$	iso–C$_3$H$_7$Cl		10.78±0.02	PI	894	182
C$_3$H$_7$Cl$^+$	iso–C$_3$H$_7$Cl		10.77±0.03	EI		2146, 3201

See also – EI: 160, 2776

		C$_4$H$_5$Cl$^+$				
C$_4$H$_5$Cl$^+$	CH$_2$=CClCH=CH$_2$		8.828±0.01	S		3352
(Average of two Rydberg series limits)						

	n–C$_4$H$_9$Cl$^+$	$\Delta H^{\circ}_{f298} \sim 882$ kJ mol^{-1} (211 kcal mol^{-1})				
	sec–C$_4$H$_9$Cl$^+$	$\Delta H^{\circ}_{f298} \sim 866$ kJ mol^{-1} (207 kcal mol^{-1})				
	iso–C$_4$H$_9$Cl$^+$	$\Delta H^{\circ}_{f298} \sim 869$ kJ mol^{-1} (208 kcal mol^{-1})				
	$tert$–C$_4$H$_9$Cl$^+$	$\Delta H^{\circ}_{f298} \sim 840$ kJ mol^{-1} (201 kcal mol^{-1})				
C$_4$H$_9$Cl$^+$	n–C$_4$H$_9$Cl		10.67±0.03	PI	882	182
C$_4$H$_9$Cl$^+$	n–C$_4$H$_9$Cl		10.50±0.07	EI		2146

See also – EI: 3201

C$_4$H$_9$Cl$^+$	sec–C$_4$H$_9$Cl		10.65±0.03	PI	866	182
C$_4$H$_9$Cl$^+$	sec–C$_4$H$_9$Cl		10.52±0.1	EI		2146

See also – EI: 3201

4.3. The Positive Ion Table—Continued

Ion	Reactant	Other products	Ionization or appearance potential (eV)	Method	Heat of formation (kJ mol^{-1})	Ref.
$C_4H_9Cl^+$	iso–C_4H_9Cl		10.66±0.03	PI	869	182
$C_4H_9Cl^+$	iso–C_4H_9Cl		10.48±0.1	EI		2146

See also – EI: 3201

$C_4H_9Cl^+$	tert–C_4H_9Cl		10.61±0.03	PI	840	182
$C_4H_9Cl^+$	tert–C_4H_9Cl		10.3±0.1	EI		2146, 3201

$C_5H_4Cl^+$

$C_5H_4Cl^+$	C_5H_4Cl (Chlorocyclopentadienyl radical)		8.78	EI		126

$C_5H_5Cl^+$

$C_5H_5Cl^+$	$C_6H_4ClNH_2$ (4–Chloroaniline)	HCN	12.24	EI		3238

$C_6H_4Cl^+$

$C_6H_4Cl^+$	$C_6H_4Cl_2$ (1,3–Dichlorobenzene)	Cl	13.29±0.1	EI		2972
$C_6H_4Cl^+$	$C_6H_4Cl_2$ (1,4–Dichlorobenzene)	Cl	13.24±0.1	EI		2972
$C_6H_4Cl^+$	C_6H_4ClBr (1–Bromo–4–chlorobenzene)	Br	12.70	EI		3238

$C_6H_5Cl^+$ (Chlorobenzene) $\Delta H^\circ_{f298} \sim 926$ kJ mol^{-1} (221 kcal mol^{-1})

$C_6H_5Cl^+$	C_6H_5Cl (Chlorobenzene)		9.05	S	925	2666
$C_6H_5Cl^+$	C_6H_5Cl (Chlorobenzene)		9.035	PI	924	2682
$C_6H_5Cl^+$	C_6H_5Cl (Chlorobenzene)		9.08±0.01	PI	928	3212
$C_6H_5Cl^+$	C_6H_5Cl (Chlorobenzene)		9.07±0.02	PI	927	182, 416
$C_6H_5Cl^+$	C_6H_5Cl (Chlorobenzene)		9.07	PI	927	168

See also – S: 3153
       PE: 2015, 2806, 2826, 3212, 3247, 3331
       EI: 301, 1066, 2972, 3223, 3230, 3238

$C_7H_6Cl^+$

$C_7H_6Cl^+$	$C_6H_4ClCH_2$ (4–Chlorobenzyl radical)		7.95±0.1	EI		69
$C_7H_6Cl^+$	$C_6H_5CH_2CH_2C_6H_4Cl$ (1–(3–Chlorophenyl)–2–phenylethane)		~13	EI		3288

4.3. The Positive Ion Table—Continued

Ion	Reactant	Other products	Ionization or appearance potential (eV)	Method	Heat of formation (kJ mol^{-1})	Ref.
	$C_7H_7Cl^+$ (α–Chlorotoluene)	$\Delta H^\circ_{f298} \sim 897$ kJ mol^{-1} (214 kcal mol^{-1})				
$C_7H_7Cl^+$	$C_6H_5CH_2Cl$ (α–Chlorotoluene)		9.10±0.05	PI	897	2612
See also – EI:	2025, 3230					
$C_7H_7Cl^+$	$C_6H_4ClCH_3$ (2–Chlorotoluene)		8.83±0.02	PI		182
See also – EI:	3230					
$C_7H_7Cl^+$	$C_6H_4ClCH_3$ (3–Chlorotoluene)		8.83±0.02	PI		182
See also – EI:	2025, 2972, 3230					
$C_7H_7Cl^+$	$C_6H_4ClCH_3$ (4–Chlorotoluene)		8.69±0.02	PI		416
$C_7H_7Cl^+$	$C_6H_4ClCH_3$ (4–Chlorotoluene)		8.70±0.02	PI		182
See also – PE:	2806					
EI:	1066, 2972, 3223, 3230					
	$C_7H_9Cl^+$					
$C_7H_9Cl^+$	C_7H_9Cl (endo–5–Chlorobicyclo[2.2.1]hept–2–ene)		9.1±0.15	EI		2155
$C_7H_9Cl^+$	C_7H_9Cl (exo–5–Chlorobicyclo[2.2.1]hept–2–ene)		9.15±0.15	EI		2155
$C_7H_9Cl^+$	C_7H_9Cl (3–Chloronortricyclene)		9.51±0.15	EI		2155
	$C_8H_8Cl^+$					
$C_8H_8Cl^+$	$C_6H_4ClCH(CH_3)CH_2COOCH_3$ (3–(4–Chlorophenyl)butanoic acid methyl ester)		11.13±0.2	EI		2497
$C_8H_8Cl^+$	$C_6H_4ClCH_2CH_2Br$ (1–Bromo–2–(3–chlorophenyl)ethane)	Br	10.3	EI		2973
$C_8H_8Cl^+$	$C_6H_4ClCH_2CH_2Br$ (1–Bromo–2–(4–chlorophenyl)ethane)	Br	10.1	EI		2973
	$C_9H_9Cl^+$					
$C_9H_9Cl^+$	$C_6H_4ClCH(CH_3)CH_2COOCH_3$ (3–(4–Chlorophenyl)butanoic acid methyl ester)		9.90±0.2	EI		2497

4.3. The Positive Ion Table—Continued

Ion	Reactant	Other products	Ionization or appearance potential (eV)	Method	Heat of formation (kJ mol^{-1})	Ref.
			C$_{14}$H$_{13}$Cl$^+$			
C$_{14}$H$_{13}$Cl$^+$	C$_6$H$_5$CH$_2$CH$_2$C$_6$H$_4$Cl (1–(3–Chlorophenyl)–2–phenylethane)		8.7±0.1	EI		3288
			CHCl$_2^+$			
CHCl$_2^+$	CHCl$_2$		9.30	EI		141
CHCl$_2^+$	CHCl$_2$		9.54±0.10	EI		131
CHCl$_2^+$	CH$_2$Cl$_2$	H	12.12±0.05?	RPD		1139
CHCl$_2^+$	CH$_2$Cl$_2$	H	13.00±0.10	EI		131
CHCl$_2^+$	CHCl$_3$	Cl	11.64±0.20	EI		43
CHCl$_2^+$	CHCl$_3$	Cl	11.70±0.1	EI		72
CHCl$_2^+$	CHCl$_3$	Cl	12.43±0.02	EI		131
CHCl$_2^+$	CHCl$_2$CHCl$_2$	CHCl$_2$	11.55±0.1	EI		72
CHCl$_2^+$	CHCl$_2$Br	Br	11.02±0.1	EI		72
	CH$_2$Cl$_2^+$ $\Delta H_{f0}^\circ = 1009$ kJ mol^{-1} (241 kcal mol^{-1})					
CH$_2$Cl$_2^+$	CH$_2$Cl$_2$		11.35±0.02	PI	1010	182, 416
CH$_2$Cl$_2^+$	CH$_2$Cl$_2$		11.33	PE	1008	2843
CH$_2$Cl$_2^+$	CH$_2$Cl$_2$		11.36	PEN		2430

Orbital assignments are discussed in R. N. Dixon, J. N. Murrell and B. Narayan, Mol. Phys. **20**, 611 (1971).

See also – PE: 3119
            EI: 72

	CH$_2$=CCl$_2^+$	$\Delta H_{f0}^\circ = 960$ kJ mol^{-1} (229 kcal mol^{-1})				
	cis–CHCl=CHCl$^+$	$\Delta H_{f0}^\circ = 942$ kJ mol^{-1} (225 kcal mol^{-1})				
	trans–CHCl=CHCl$^+$	$\Delta H_{f0}^\circ \sim 943$ kJ mol^{-1} (225 kcal mol^{-1})				
C$_2$H$_2$Cl$_2^+$	CH$_2$=CCl$_2$		9.86	S	960	2675
(Average of four Rydberg series limits)						
C$_2$H$_2$Cl$_2^+$	CH$_2$=CCl$_2$		9.79	PI		168
C$_2$H$_2$Cl$_2^+$	CH$_2$=CCl$_2$		9.83	PE		2885
C$_2$H$_2$Cl$_2^+$	CH$_2$=CCl$_2$		9.74	PE		3001

See also – S: 269

C$_2$H$_2$Cl$_2^+$	cis–CHCl=CHCl		9.66	S	942	2673
C$_2$H$_2$Cl$_2^+$	cis–CHCl=CHCl		9.65±0.01	PI		182
C$_2$H$_2$Cl$_2^+$	cis–CHCl=CHCl		9.65	PI		168
C$_2$H$_2$Cl$_2^+$	cis–CHCl=CHCl		9.66±0.02	PI		114, 268, 1058, 1190
C$_2$H$_2$Cl$_2^+$	cis–CHCl=CHCl		9.65	PE		2885
C$_2$H$_2$Cl$_2^+$	cis–CHCl=CHCl		9.68	PE		3001

See also – S: 261, 3433
            EI: 114, 268, 1058, 1190, 3213

4.3. The Positive Ion Table—Continued

Ion	Reactant	Other products	Ionization or appearance potential (eV)	Method	Heat of formation (kJ mol^{-1})	Ref.
C$_2$H$_2$Cl$_2^+$	trans–CHCl=CHCl		~9.6	S		2672
C$_2$H$_2$Cl$_2^+$	trans–CHCl=CHCl		9.63	PI	941	168
C$_2$H$_2$Cl$_2^+$	trans–CHCl=CHCl		9.64±0.02	PI	942	114, 268, 1058, 1190
C$_2$H$_2$Cl$_2^+$	trans–CHCl=CHCl		9.66±0.03	PI	944	182
C$_2$H$_2$Cl$_2^+$	trans–CHCl=CHCl		9.64	PE	942	2885
C$_2$H$_2$Cl$_2^+$	trans–CHCl=CHCl		9.69	PE	947	3001

See also – S: 261, 3433
         EI: 114, 268, 1058, 1190, 3213

CH$_2$ClCH$_2$Cl$^+$ $\Delta H_{f0}^{\circ} \sim 954$ kJ mol^{-1} (228 kcal mol^{-1})

Ion	Reactant	Other products	Ionization or appearance potential (eV)	Method	Heat of formation (kJ mol^{-1})	Ref.
C$_2$H$_4$Cl$_2^+$	CH$_2$ClCH$_2$Cl		11.12±0.05	PI	954	182

See also – EI: 72

C$_3$H$_4$Cl$_2^+$

Ion	Reactant	Other products	Ionization or appearance potential (eV)	Method	Heat of formation (kJ mol^{-1})	Ref.
C$_3$H$_4$Cl$_2^+$	CH$_2$=CClCH$_2$Cl		9.82±0.03?	PI		182

C$_3$H$_6$Cl$_2^+$

Ion	Reactant	Other products	Ionization or appearance potential (eV)	Method	Heat of formation (kJ mol^{-1})	Ref.
C$_3$H$_6$Cl$_2^+$	CH$_3$CHClCH$_2$Cl		10.87±0.05	PI		182
C$_3$H$_6$Cl$_2^+$	CH$_3$CHClCH$_2$Cl		10.73	PE		3374
C$_3$H$_6$Cl$_2^+$	(CH$_2$Cl)$_2$CH$_2$		10.85±0.05	PI		182
C$_3$H$_6$Cl$_2^+$	(CH$_2$Cl)$_2$CH$_2$		10.93	PE		3374

C$_4$H$_8$Cl$_2^+$

Ion	Reactant	Other products	Ionization or appearance potential (eV)	Method	Heat of formation (kJ mol^{-1})	Ref.
C$_4$H$_8$Cl$_2^+$	CH$_2$ClCH$_2$CH$_2$CH$_2$Cl		11.03	PE		3374

C$_6$H$_4$Cl$_2^+$ (1,2–Dichlorobenzene) $\Delta H_{f298}^{\circ} = 905$ kJ mol^{-1} (216 kcal mol^{-1})
C$_6$H$_4$Cl$_2^+$ (1,3–Dichlorobenzene) $\Delta H_{f298}^{\circ} = 906$ kJ mol^{-1} (217 kcal mol^{-1})
C$_6$H$_4$Cl$_2^+$ (1,4–Dichlorobenzene) $\Delta H_{f298}^{\circ} = 886$ kJ mol^{-1} (212 kcal mol^{-1})

Ion	Reactant	Other products	Ionization or appearance potential (eV)	Method	Heat of formation (kJ mol^{-1})	Ref.
C$_6$H$_4$Cl$_2^+$	C$_6$H$_4$Cl$_2$ (1,2–Dichlorobenzene)		9.06	PI	904	168
C$_6$H$_4$Cl$_2^+$	C$_6$H$_4$Cl$_2$ (1,2–Dichlorobenzene)		9.07±0.01	PI	905	182

See also – S: 3153
         EI: 1066

Ion	Reactant	Other products	Ionization or appearance potential (eV)	Method	Heat of formation (kJ mol^{-1})	Ref.
C$_6$H$_4$Cl$_2^+$	C$_6$H$_4$Cl$_2$ (1,3–Dichlorobenzene)		9.12±0.01	PI	906	182

See also – EI: 2972

4.3. The Positive Ion Table—Continued

Ion	Reactant	Other products	Ionization or appearance potential (eV)	Method	Heat of formation (kJ mol^{-1})	Ref.
$C_6H_4Cl_2^+$	$C_6H_4Cl_2$ (1,4–Dichlorobenzene)		8.94±0.01	PI	886	182
$C_6H_4Cl_2^+$	$C_6H_4Cl_2$ (1,4–Dichlorobenzene)		8.95	PI	887	168

See also – PE: 2806, 2826, 3247
      EI: 2972

$CHCl_3^+$ $\Delta H_{f0}^\circ = 1003$ kJ mol^{-1} (240 kcal mol^{-1})

Ion	Reactant	Other products	Ionization or appearance potential (eV)	Method	Heat of formation (kJ mol^{-1})	Ref.
$CHCl_3^+$	$CHCl_3$		11.42±0.03	PI	1004	182, 416
$CHCl_3^+$	$CHCl_3$		11.40	PE	1002	2843
$CHCl_3^+$	$CHCl_3$		11.50	PEN		2430

Orbital assignments are discussed in R. N. Dixon, J. N. Murrell and B. Narayan, Mol. Phys. **20**, 611 (1971).

See also – PE: 3119
      EI: 43

$C_2HCl_3^+$ $\Delta H_{f0}^\circ = 909$ kJ mol^{-1} (217 kcal mol^{-1})

Ion	Reactant	Other products	Ionization or appearance potential (eV)	Method	Heat of formation (kJ mol^{-1})	Ref.
$C_2HCl_3^+$	C_2HCl_3		9.45±0.01	PI	907	182
$C_2HCl_3^+$	C_2HCl_3		9.47±0.01	PI	909	416
$C_2HCl_3^+$	C_2HCl_3		9.45	PI	907	168
$C_2HCl_3^+$	C_2HCl_3		9.48	PE	910	2885

See also – S: 261, 2674
    PEN: 2430

$C_6H_3Cl_3^+$

Ion	Reactant	Other products	Ionization or appearance potential (eV)	Method	Heat of formation (kJ mol^{-1})	Ref.
$C_6H_3Cl_3^+$	$C_6H_3Cl_3$ (1,3,5–Trichlorobenzene)		9.5±0.15	CTS		3373

$C_2H_2Cl_4^+$

Ion	Reactant	Other products	Ionization or appearance potential (eV)	Method	Heat of formation (kJ mol^{-1})	Ref.
$C_2H_2Cl_4^+$	$CHCl_2CHCl_2$		11.10±0.05	EI		72

4.3. The Positive Ion Table—Continued

Ion	Reactant	Other products	Ionization or appearance potential (eV)	Method	Heat of formation (kJ mol^{-1})	Ref.
	ClCN$^+$($^2\Pi_{3/2}$)	$\Delta H^\circ_{f0} = 1328$ kJ mol^{-1} (317 kcal mol^{-1})				
	ClCN$^+$($^2\Pi_{1/2}$)	$\Delta H^\circ_{f0} = 1331$ kJ mol^{-1} (318 kcal mol^{-1})				
	ClCN$^+$($^2\Sigma^+$)	$\Delta H^\circ_{f0} = 1469$ kJ mol^{-1} (351 kcal mol^{-1})				
	ClCN$^+$($^2\Pi_{3/2}$)	$\Delta H^\circ_{f0} = 1597$ kJ mol^{-1} (382 kcal mol^{-1})				
CNCl$^+$($^2\Pi_{3/2}$)	ClCN		12.34±0.01	PI	1328	2621
CNCl$^+$($^2\Pi_{3/2}$)	ClCN		12.34	PE	1328	3045
CNCl$^+$($^2\Pi_{3/2}$)	ClCN		12.37±0.02	PE		3091
CNCl$^+$($^2\Pi_{1/2}$)	ClCN		12.37	PE	1331	3045
CNCl$^+$($^2\Sigma^+$)	ClCN		13.80	PE	1469	3045
CNCl$^+$($^2\Sigma^+$)	ClCN		13.80±0.02	PE	1469	3091
CNCl$^+$($^2\Pi_{3/2}$)	ClCN		15.13	PE	1597	3045
CNCl$^+$($^2\Pi_{3/2}$)	ClCN		15.37±0.02 (V)	PE		3091
CNCl$^+$($^2\Sigma^+$)	ClCN		19.03 (V)	PE		3045
CNCl$^+$($^2\Sigma^+$)	ClCN		19.0±0.1 (V)	PE		3091

See also − EI: 73

C$_2$NCl$_3^+$

Ion	Reactant	Other products	Ionization or appearance potential (eV)	Method	Heat of formation (kJ mol^{-1})	Ref.
C$_2$NCl$_3^+$	CCl$_3$CN		11.96 (V)	PE		3045

ClOH$^+$

Ion	Reactant	Other products	Ionization or appearance potential (eV)	Method	Heat of formation (kJ mol^{-1})	Ref.
ClOH$^+$	ClOH		11.7±0.2	EI		2488

COCl$_2^+$

Ion	Reactant	Other products	Ionization or appearance potential (eV)	Method	Heat of formation (kJ mol^{-1})	Ref.
COCl$_2^+$	CCl$_2$O		11.7	PEN		2430

4.3. The Positive Ion Table—Continued

Ion	Reactant	Other products	Ionization or appearance potential (eV)	Method	Heat of formation (kJ mol^{-1})	Ref.
			BF$_2$Cl$^+$			
BF$_2$Cl$^+$	BF$_2$Cl		12.43±0.1	EI		2513
BF$_2$Cl$^+$	BF$_2$Cl		13.06±0.11	EI	2512,	3227
			BFCl$_2^+$			
BFCl$_2^+$	BFCl$_2$		12.18±0.10	EI	2512,	2513, 3227
			CFCl$^+$			
CFCl$^+$	CFCl$_3$		17.41±0.15	RPD		185
CFCl$^+$	CHF$_2$Cl		15.9±0.3	EI		43
			CF$_2$Cl$^+$			
CF$_2$Cl$^+$	CF$_3$Cl	F	15.0±0.4	EI		24
CF$_2$Cl$^+$	CF$_3$Cl	F	16.15	EI		2976
CF$_2$Cl$^+$	CF$_2$Cl$_2$	Cl	12.55	EI		2976
CF$_2$Cl$^+$	CF$_2$ClCF$_2$Cl		12.33	EI		2976
CF$_2$Cl$^+$	(CF$_2$Cl)$_2$CO		12.5	EI		2976
			CF$_3$Cl$^+$			
CF$_3$Cl$^+$	CF$_3$Cl		12.39	PI		2643
CF$_3$Cl$^+$	CF$_3$Cl		12.91±0.03	PI		182

A discrepancy exists in ref. 2643 between the threshold wavelength and its equivalent energy; the wavelength has been assumed correct. The disagreement between the two PI values is not understood.

See also – EI: 24, 439

Ion	Reactant	Other products	Ionization or appearance potential (eV)	Method	Heat of formation (kJ mol^{-1})	Ref.
		C$_2$F$_3$Cl$^+$	$\Delta H_{f0}^\circ = 397$ kJ mol^{-1} (95 kcal mol^{-1})			
C$_2$F$_3$Cl$^+$	C$_2$F$_3$Cl		9.84	PE	397	2885

See also – EI: 214

Ion	Reactant	Other products	Ionization or appearance potential (eV)	Method	Heat of formation (kJ mol^{-1})	Ref.
			C$_3$F$_5$Cl$^+$			
C$_3$F$_5$Cl$^+$	CF$_2$ClCF=CF$_2$		10.79	EI		1290
			C$_6$F$_5$Cl$^+$			
C$_6$F$_5$Cl$^+$	C$_6$F$_5$Cl (Chloropentafluorobenzene)		10.4±0.1	EI		301

4.3. The Positive Ion Table—Continued

Ion	Reactant	Other products	Ionization or appearance potential (eV)	Method	Heat of formation (kJ mol⁻¹)	Ref.

$CFCl_2^+$

Ion	Reactant	Other products	Ionization or appearance potential (eV)	Method	Heat of formation (kJ mol⁻¹)	Ref.
$CFCl_2^+$	$CFCl_3$	Cl	11.97±0.07	RPD		185

$CF_2Cl_2^+$ $\Delta H_{f0}^\circ \sim 715$ kJ mol⁻¹ (171 kcal mol⁻¹)

Ion	Reactant	Other products	Ionization or appearance potential (eV)	Method	Heat of formation (kJ mol⁻¹)	Ref.
$CF_2Cl_2^+$	CF_2Cl_2		12.31±0.05	PI	715	182

$CF_2=CCl_2^+$ $\Delta H_{f298}^\circ = 616$ kJ mol⁻¹ (147 kcal mol⁻¹)

Ion	Reactant	Other products	Ionization or appearance potential (eV)	Method	Heat of formation (kJ mol⁻¹)	Ref.
$C_2F_2Cl_2^+$	$CF_2=CCl_2$		9.69±0.01	S	616	3360
(Average of two Rydberg series limits)						
$C_2F_2Cl_2^+$	$CF_2=CCl_2$		9.65	PE		2885

See also – EI: 214

$C_4F_6Cl_2^+$

Ion	Reactant	Other products	Ionization or appearance potential (eV)	Method	Heat of formation (kJ mol⁻¹)	Ref.
$C_4F_6Cl_2^+$	$CF_3CCl=CClCF_3$		10.36±0.01?	PI		182

$CFCl_3^+$ $\Delta H_{f0}^\circ = 863$ kJ mol⁻¹ (206 kcal mol⁻¹)

Ion	Reactant	Other products	Ionization or appearance potential (eV)	Method	Heat of formation (kJ mol⁻¹)	Ref.
$CFCl_3^+$	$CFCl_3$		11.77±0.02	PI	863	182

$CFCl_2CF_2Cl^+$ $\Delta H_{f298}^\circ = 397$ kJ mol⁻¹ (95 kcal mol⁻¹)

Ion	Reactant	Other products	Ionization or appearance potential (eV)	Method	Heat of formation (kJ mol⁻¹)	Ref.
$C_2F_3Cl_3^+$	CF_3CCl_3		11.78±0.03	PI		182
$C_2F_3Cl_3^+$	$CFCl_2CF_2Cl$		11.99±0.02	PI	397	182

ClO_3F^+

Ion	Reactant	Other products	Ionization or appearance potential (eV)	Method	Heat of formation (kJ mol⁻¹)	Ref.
ClO_3F^+	ClO_3F		13.6±0.2	EI		53

$P_3N_3Cl_5^+$

Ion	Reactant	Other products	Ionization or appearance potential (eV)	Method	Heat of formation (kJ mol⁻¹)	Ref.
$P_3N_3Cl_5^+$	$(NPCl_2)_3$ (Cyclo−tris(dichlorophosphonitrile))	Cl	11.06±0.1	EI		2782
$P_3N_3Cl_5^+$	$(NPCl_2)_2(NPClBr)$ (Cyclo−bis(dichlorophosphonitrile)bromochlorophosphonitrile)	Br	10.49±0.1	EI		2782

$P_3N_3Cl_6^+$

Ion	Reactant	Other products	Ionization or appearance potential (eV)	Method	Heat of formation (kJ mol⁻¹)	Ref.
$P_3N_3Cl_6^+$	$(NPCl_2)_3$ (Cyclo−tris(dichlorophosphonitrile))		10.26±0.05	EI		2425, 2952
$P_3N_3Cl_6^+$	$(NPCl_2)_3$ (Cyclo−tris(dichlorophosphonitrile))		10.27±0.1	EI		2782

4.3. The Positive Ion Table—Continued

Ion	Reactant	Other products	Ionization or appearance potential (eV)	Method	Heat of formation (kJ mol⁻¹)	Ref.
		$P_4N_4Cl_8^+$				
$P_4N_4Cl_8^+$	$(NPCl_2)_4$ (Cyclo–tetrakis(dichlorophosphonitrile))		9.80±0.05	EI		2425, 2952
		$P_5N_5Cl_{10}^+$				
$P_5N_5Cl_{10}^+$	$(NPCl_2)_5$ (Cyclo–pentakis(dichlorophosphonitrile))		9.83±0.05	EI		2425, 2952
		$P_6N_6Cl_{12}^+$				
$P_6N_6Cl_{12}^+$	$(NPCl_2)_6$ (Cyclo–hexakis(dichlorophosphonitrile))		9.81±0.05	EI		2425, 2952
		$P_7N_7Cl_{14}^+$				
$P_7N_7Cl_{14}^+$	$(NPCl_2)_7$ (Cyclo–heptakis(dichlorophosphonitrile))		9.80±0.05	EI		2425, 2952
		$POCl^+$				
$POCl^+$	$POCl_3$		15.6±0.3	EI		2506
		$POCl_2^+$				
$POCl_2^+$	$POCl_3$		12.8±0.3	EI		2506
$POCl_2^+$	$POCl_3$		13.3±0.2	EI		1101
		$POCl_3^+$				
$POCl_3^+$	$POCl_3$		11.4±0.3	EI		2506
$POCl_3^+$	$POCl_3$		13.1±0.2	EI		1101
		SF_4Cl^+				
SF_4Cl^+	SF_5Cl	F	15.9±0.1	EI		2777
		$C_4H_{10}BCl^+$				
$C_4H_{10}BCl^+$	$(C_2H_5)_2BCl$		10.28±0.1	EI		2513, 3227

4.3. The Positive Ion Table—Continued

Ion	Reactant	Other products	Ionization or appearance potential (eV)	Method	Heat of formation (kJ mol^{-1})	Ref.
			$C_6H_5BCl^+$			
$C_6H_5BCl^+$	$C_6H_5BCl_2$ (Phenylboron dichloride)		12.88±0.14	EI		2722
			$C_2H_5BCl_2^+$			
$C_2H_5BCl_2^+$	$C_2H_5BCl_2$		10.80±0.3	EI		2513, 3227
			$C_2H_2NCl^+$			
$C_2H_2NCl^+$	CH_2ClCN		12.05 (V)	PE		3045
			$C_5H_4NCl^+$			
$C_5H_4NCl^+$	C_5H_4NCl (2−Chloropyridine)		9.91±0.05	EI		217
$C_5H_4NCl^+$	C_5H_4NCl (4−Chloropyridine)		10.15±0.05	EI		217
			$C_6H_6NCl^+$			
$C_6H_6NCl^+$	$C_6H_4ClNH_2$ (2−Chloroaniline)		7.9	CTS		2978
$C_6H_6NCl^+$	$C_6H_4ClNH_2$ (3−Chloroaniline)		8.09±0.1	EI		2972
$C_6H_6NCl^+$	$C_6H_4ClNH_2$ (4−Chloroaniline)		8.18 (V)	PE		2806
$C_6H_6NCl^+$	$C_6H_4ClNH_2$ (4−Chloroaniline)		7.77±0.1	CTS		2485
$C_6H_6NCl^+$	$C_6H_4ClNH_2$ (4−Chloroaniline)		8.00	CTS		2909

See also − PE: 2826
        EI: 2972, 3231

Ion	Reactant	Other products	Ionization or appearance potential (eV)	Method	Heat of formation (kJ mol^{-1})	Ref.
$C_6H_6NCl^+$	$C_6H_4ClNHCOCH_3$ (N−(3−Chlorophenyl)acetic acid amide)		10.90±0.2	EI		3406
$C_6H_6NCl^+$	$C_6H_4ClNHCOCH_3$ (N−(4−Chlorophenyl)acetic acid amide)		10.60±0.2	EI		3406

4.3. The Positive Ion Table—Continued

Ion	Reactant	Other products	Ionization or appearance potential (eV)	Method	Heat of formation (kJ mol⁻¹)	Ref.

$$C_7H_4NCl^+$$

| $C_7H_4NCl^+$ | C_6H_4ClCN (2–Chlorobenzoic acid nitrile) | | 10.26 ± 0.06 | RPD | | 3223 |
| $C_7H_4NCl^+$ | C_6H_4ClCN (3–Chlorobenzoic acid nitrile) | | 9.93 ± 0.04 | RPD | | 3223 |

See also – EI: 2972

| $C_7H_4NCl^+$ | C_6H_4ClCN (4–Chlorobenzoic acid nitrile) | | 9.94 ± 0.05 | RPD | | 3223 |

See also – EI: 2972

$$C_8H_6NCl^+$$

| $C_8H_6NCl^+$ | $C_6H_4ClCH_2CN$ (3–Chlorophenylacetic acid nitrile) | | 9.48 ± 0.05 | RPD | | 3223 |
| $C_8H_6NCl^+$ | $C_6H_4ClCH_2CN$ (4–Chlorophenylacetic acid nitrile) | | 9.43 ± 0.05 | RPD | | 3223 |

$$C_8H_{10}NCl^+$$

| $C_8H_{10}NCl^+$ | $C_6H_4ClN(CH_3)_2$ (4–Chloro–N,N–dimethylaniline) | | 7.38 | CTS | | 1281 |

$$C_{18}H_{26}N_3Cl^+$$

| $C_{18}H_{26}N_3Cl^+$ | $C_{18}H_{26}N_3Cl$ (Chloroquine) | | 7.84 | CTS | | 2562 |

4.3. The Positive Ion Table—Continued

Ion	Reactant	Other products	Ionization or appearance potential (eV)	Method	Heat of formation (kJ mol^{-1})	Ref.
			C$_2$HNCl$_2^+$			
C$_2$HNCl$_2^+$	CHCl$_2$CN		12.21 (V)	PE		3045
			C$_7$H$_3$NCl$_2^+$			
C$_7$H$_3$NCl$_2^+$	C$_6$H$_3$Cl$_2$CN (2,6–Dichlorobenzoic acid nitrile)		10.09±0.05	RPD		3223
			C$_8$H$_5$NCl$_2^+$			
C$_8$H$_5$NCl$_2^+$	C$_6$H$_3$Cl$_2$CH$_2$CN (2,6–Dichlorophenylacetic acid nitrile)		9.73±0.06	RPD		3223
			C$_2$H$_2$OCl$^+$			
C$_2$H$_2$OCl$^+$	CH$_3$COCH$_2$Cl	CH$_3$	11.97±0.11	EI		2174
C$_2$H$_2$OCl$^+$	CH$_2$ClCOCHN$_2$		11.66±0.04	EI		2174
	CH$_3$COCl$^+$	$\Delta H_{f0}^\circ \sim$ 829 kJ mol^{-1} (198 kcal mol^{-1})				
C$_2$H$_3$OCl$^+$	CH$_2$ClCHO		10.48±0.03	PE		3289
C$_2$H$_3$OCl$^+$	CH$_3$COCl		11.02±0.05	PI	829	182
C$_2$H$_3$OCl$^+$	CH$_3$COCl		11.05	PEN		2430
C$_2$H$_3$OCl$^+$	CH$_3$COCl		11.08±0.06	EI		2026
			C$_2$H$_5$OCl$^+$			
C$_2$H$_5$OCl$^+$	CH$_2$ClCH$_2$OH		10.90	PE		3374

See also – D: 2908

	C$_3$H$_5$OCl$^+$ (1–Chloro–2,3–epoxypropane)	ΔH_{f298}° = 919 kJ mol^{-1} (220 kcal mol^{-1})				
C$_3$H$_5$OCl$^+$	CH$_3$COCH$_2$Cl		9.98±0.01	PE		3289
C$_3$H$_5$OCl$^+$	CH$_3$COCH$_2$Cl		9.98±0.13	EI		2174
C$_3$H$_5$OCl$^+$	CH$_3$COCH$_2$Cl		10.00±0.01	EI		2026
C$_3$H$_5$OCl$^+$	C$_3$H$_5$OCl (1–Chloro–2,3–epoxypropane)		10.64	PE	919	3374

4.3. The Positive Ion Table—Continued

Ion	Reactant	Other products	Ionization or appearance potential (eV)	Method	Heat of formation (kJ mol^{-1})	Ref.
			$C_4H_3OCl^+$			
$C_4H_3OCl^+$	C_4H_3OCl (2–Chlorofuran)		8.75	PE		3331
			$C_4H_6OCl^+$			
$C_4H_6OCl^+$	$CH_2ClCH_2CH_2COOCH_3$	CH_3O	11.28±0.3	EI		2496
			$C_6H_4OCl^+$			
$C_6H_4OCl^+$	$C_6H_5CH_2OC_6H_4Cl$ (Benzyl 3–chlorophenyl ether)		12.6	EI		2737
$C_6H_4OCl^+$	$C_6H_5CH_2OC_6H_4Cl$ (Benzyl 4–chlorophenyl ether)		12.7	EI		2737
			$C_6H_5OCl^+$			
$C_6H_5OCl^+$	C_6H_4ClOH (2–Chlorophenol)		9.28	EI		1066
$C_6H_5OCl^+$	C_6H_4ClOH (4–Chlorophenol)		8.69 (V)	PE		2806
$C_6H_5OCl^+$	C_6H_4ClOH (4–Chlorophenol)		9.07	EI		1066

See also – PE: 2826

Ion	Reactant	Other products	Ionization or appearance potential (eV)	Method	Heat of formation (kJ mol^{-1})	Ref.
$C_6H_5OCl^+$	$C_6H_4ClOC_2H_5$ (1–Chloro–3–ethoxybenzene)	C_2H_4	10.70±0.15	EI		2945
$C_6H_5OCl^+$	$C_6H_4ClOC_2H_5$ (1–Chloro–4–ethoxybenzene)	C_2H_4	10.77±0.15	EI		2945
			$C_7H_4OCl^+$			
$C_7H_4OCl^+$	$C_6H_4ClCOCH_3$ (3–Chloroacetophenone)	CH_3	10.36±0.1	EI		2967
$C_7H_4OCl^+$	$C_6H_4ClCOCH_3$ (4–Chloroacetophenone)	CH_3	10.04	EI		3334
$C_7H_4OCl^+$	$C_6H_4ClCOCH_3$ (4–Chloroacetophenone)	CH_3	10.36±0.1	EI		2967
$C_7H_4OCl^+$	$C_6H_4ClCOCH_3$ (4–Chloroacetophenone)	CH_3	10.69	EI		3238
$C_7H_4OCl^+$	$C_6H_4ClCOOCH_3$ (4–Chlorobenzoic acid methyl ester)	CH_3O	11.02	EI		3238

4.3. The Positive Ion Table—Continued

Ion	Reactant	Other products	Ionization or appearance potential (eV)	Method	Heat of formation (kJ mol^{-1})	Ref.
		$C_7H_5OCl^+$				
$C_7H_5OCl^+$	C_6H_4ClCHO (4–Chlorobenzenecarbonal)		9.59 (V)	PE		2806
$C_7H_5OCl^+$	C_6H_4ClCHO (4–Chlorobenzenecarbonal)		9.61±0.01	EI		2026
$C_7H_5OCl^+$	C_6H_5COCl (Benzoic acid chloride)		9.70±0.01	EI		2026
See also – EI: 308						
		$C_7H_7OCl^+$				
$C_7H_7OCl^+$	$C_6H_4ClOCH_3$ (1–Chloro–2–methoxybenzene)		8.3±0.15	CTS		3373
$C_7H_7OCl^+$	$C_6H_4ClOCH_3$ (1–Chloro–4–methoxybenzene)		8.25±0.15	CTS		3373
		$C_8H_7OCl^+$				
$C_8H_7OCl^+$	$C_6H_5COCH_2Cl$ (α–Chloroacetophenone)		9.44±0.05	EI		2025
$C_8H_7OCl^+$	$C_6H_5COCH_2Cl$ (α–Chloroacetophenone)		9.65±0.01	EI		2026
$C_8H_7OCl^+$	$C_6H_4ClCOCH_3$ (3–Chloroacetophenone)		9.51±0.1	EI		2967
$C_8H_7OCl^+$	$C_6H_4ClCOCH_3$ (4–Chloroacetophenone)		9.47±0.05	EI		2026
$C_8H_7OCl^+$	$C_6H_4ClCOCH_3$ (4–Chloroacetophenone)		9.58±0.1	EI		2967
$C_8H_7OCl^+$	$C_6H_4ClCOCH_3$ (4–Chloroacetophenone)		9.64±0.15	EI		3334
$C_8H_7OCl^+$	$C_6H_4ClCOCH_3$ (4–Chloroacetophenone)		9.63	EI		3238
$C_8H_7OCl^+$	$C_6H_4(CH_3)COCl$ (4–Methylbenzoic acid chloride)		9.37±0.01	EI		2026
		$C_8H_9OCl^+$				
$C_8H_9OCl^+$	$C_6H_4ClOC_2H_5$ (1–Chloro–3–ethoxybenzene)		8.59±0.15	EI		2945
$C_8H_9OCl^+$	$C_6H_4ClOC_2H_5$ (1–Chloro–4–ethoxybenzene)		8.46±0.15	EI		2945
		$C_{10}H_{10}OCl^+$				
$C_{10}H_{10}OCl^+$	$C_6H_4ClCH(CH_3)CH_2COOCH_3$ (3–(4–Chlorophenyl)butanoic acid methyl ester)	CH_3O	11.09±0.2	EI		2497

4.3. The Positive Ion Table—Continued

Ion	Reactant	Other products	Ionization or appearance potential (eV)	Method	Heat of formation (kJ mol^{-1})	Ref.

$C_{10}H_{11}OCl^+$

Ion	Reactant	Other products	Ionization or appearance potential (eV)	Method	Heat of formation	Ref.
$C_{10}H_{11}OCl^+$	$C_6H_4ClCOC_3H_7$ (1-(4-Chlorophenyl)-1-butanone)		9.03±0.2	EI		2534

$C_{10}H_{13}OCl^+$

| $C_{10}H_{13}OCl^+$ | $C_6H_4ClOC_4H_9$
 (1-Chloro-4-tert-butoxybenzene) | | 8.72 (V) | PE | | 2806 |

See also – PE: 2826

$C_{13}H_9OCl^+$

| $C_{13}H_9OCl^+$ | $C_6H_5COC_6H_4Cl$
 (4-Chlorobenzophenone) | | 9.68±0.01 | EI | | 2026 |

$C_{13}H_{11}OCl^+$

| $C_{13}H_{11}OCl^+$ | $C_6H_5CH_2OC_6H_4Cl$
 (Benzyl 3-chlorophenyl ether) | | 8.5 | EI | | 2737 |
| $C_{13}H_{11}OCl^+$ | $C_6H_5CH_2OC_6H_4Cl$
 (Benzyl 4-chlorophenyl ether) | | 8.3 | EI | | 2737 |

$C_2H_4O_2Cl^+$

| $C_2H_4O_2Cl^+$ | $CH_2ClCOOC_2H_5$ | C_2H_3 | 10.97 | EI | | 1059 |

$C_3H_5O_2Cl^+$

| $C_3H_5O_2Cl^+$ | $CH_2ClCOOCH_3$ | | 10.53±0.05 | EI | | 2025 |

$C_5H_9O_2Cl^+$

| $C_5H_9O_2Cl^+$ | $CH_2ClCH_2CH_2COOCH_3$ | | 10.29±0.3 | EI | | 2496 |

$C_8H_7O_2Cl^+$

| $C_8H_7O_2Cl^+$ | $C_6H_4(OCH_3)COCl$
 (4-Methoxybenzoic acid chloride) | | 8.87±0.05 | EI | | 2026 |
| $C_8H_7O_2Cl^+$ | $C_6H_4ClCOOCH_3$
 (4-Chlorobenzoic acid methyl ester) | | 9.57 | EI | | 3238 |

$C_{11}H_{13}O_2Cl^+$

| $C_{11}H_{13}O_2Cl^+$ | $C_6H_4ClCH(CH_3)CH_2COOCH_3$
 (3-(4-Chlorophenyl)butanoic acid methyl ester) | | 8.42±0.2 | EI | | 2497 |

4.3. The Positive Ion Table—Continued

Ion	Reactant	Other products	Ionization or appearance potential (eV)	Method	Heat of formation (kJ mol⁻¹)	Ref.
			$C_3H_4OCl_2^+$			
$C_3H_4OCl_2^+$	$(CH_2Cl)_2CO$		10.03±0.02	PE		3289
			$C_7H_4OCl_2^+$			
$C_7H_4OCl_2^+$	$C_6H_4ClCOCl$ (4–Chlorobenzoic acid chloride)		9.58±0.03	EI		2026
			$C_2H_3OCl_3^+$			
$C_2H_3OCl_3^+$	CCl_3CH_2OH		13.6	D		2908
			$C_{10}H_6O_4Cl_4^+$			
$C_{10}H_6O_4Cl_4^+$	$C_6Cl_4(COOCH_3)_2$ (2,3,5,6–Tetrachloro–1,4–benzenedicarboxylic acid dimethyl ester)		9.57	EI		2718
			$CHFCl^+$			
$CHFCl^+$	CHF_2Cl	F	15.11±0.15	EI		43
$CHFCl^+$	$CHFCl_2$	Cl	12.69±0.15	EI		43
			$C_2H_2FCl^+$			
$C_2H_2FCl^+$	cis–CHF=CHCl		9.87±0.01	PI		182
$C_2H_2FCl^+$	cis–CHF=CHCl		9.86±0.02	PI		268

See also – EI: 268, 3213

Ion	Reactant	Other products	Ionization or appearance potential (eV)	Method	Heat of formation	Ref.
$C_2H_2FCl^+$	trans–CHF=CHCl		9.87±0.01	PI		182
$C_2H_2FCl^+$	trans–CHF=CHCl		9.87±0.02	PI		268

See also – EI: 268, 3213

Ion	Reactant	Other products	Ionization or appearance potential (eV)	Method	Heat of formation	Ref.
			$C_6H_4FCl^+$			
$C_6H_4FCl^+$	C_6H_4FCl (1–Chloro–2–fluorobenzene)		9.155±0.01	PI		182

See also – EI: 1185

Ion	Reactant	Other products	Ionization or appearance potential (eV)	Method	Heat of formation	Ref.
$C_6H_4FCl^+$	C_6H_4FCl (1–Chloro–3–fluorobenzene)		9.21±0.01	PI		182

See also – EI: 1185, 2972

4.3. The Positive Ion Table—Continued

Ion	Reactant	Other products	Ionization or appearance potential (eV)	Method	Heat of formation (kJ mol^{-1})	Ref.
$C_6H_4FCl^+$	C_6H_4FCl (1–Chloro–4–fluorobenzene)		9.26 (V)	PE		2806

See also – PE: 2826
        EI: 1185, 2972

CHF_2Cl^+ $\Delta H^\circ_{f0} \sim 726$ kJ mol^{-1} (174 kcal mol^{-1})

Ion	Reactant	Other products	Ionization or appearance potential (eV)	Method	Heat of formation (kJ mol^{-1})	Ref.
CHF_2Cl^+	CHF_2Cl		12.45±0.05	PI	726	182

See also – EI: 43

$CF_2=CHCl^+$ $\Delta H^\circ_{f0} = 639$ kJ mol^{-1} (153 kcal mol^{-1})

Ion	Reactant	Other products	Ionization or appearance potential (eV)	Method	Heat of formation (kJ mol^{-1})	Ref.
$C_2HF_2Cl^+$	$CF_2=CHCl$		9.84	PE	639	2885
$C_2HF_2Cl^+$	cis–$CHF=CFCl$		9.86±0.02	PI		268

See also – EI: 268

Ion	Reactant	Other products	Ionization or appearance potential (eV)	Method	Heat of formation (kJ mol^{-1})	Ref.
$C_2HF_2Cl^+$	$trans$–$CHF=CFCl$		9.83±0.02	PI		268

See also – EI: 268

$C_2H_3F_2Cl^+$

Ion	Reactant	Other products	Ionization or appearance potential (eV)	Method	Heat of formation (kJ mol^{-1})	Ref.
$C_2H_3F_2Cl^+$	CH_3CF_2Cl		11.98±0.01	PI		182

$C_7H_4F_3Cl^+$

Ion	Reactant	Other products	Ionization or appearance potential (eV)	Method	Heat of formation (kJ mol^{-1})	Ref.
$C_7H_4F_3Cl^+$	$C_6H_4ClCF_3$ (3–Chloro–α,α,α–trifluorotoluene)		9.76±0.1	EI		2972
$C_7H_4F_3Cl^+$	$C_6H_4ClCF_3$ (4–Chloro–α,α,α–trifluorotoluene)		9.80 (V)	PE		2806
$C_7H_4F_3Cl^+$	$C_6H_4ClCF_3$ (4–Chloro–α,α,α–trifluorotoluene)		9.82±0.1	EI		2972

See also – PE: 2826

$C_4H_2F_7Cl^+$

Ion	Reactant	Other products	Ionization or appearance potential (eV)	Method	Heat of formation (kJ mol^{-1})	Ref.
$C_4H_2F_7Cl^+$	n–$C_3F_7CH_2Cl$		11.84±0.02	PI		182

$CHFCl_2^+$

Ion	Reactant	Other products	Ionization or appearance potential (eV)	Method	Heat of formation (kJ mol^{-1})	Ref.
$CHFCl_2^+$	$CHFCl_2$		12.39±0.20	EI		43

$C_3OF_4Cl_2^+$

Ion	Reactant	Other products	Ionization or appearance potential (eV)	Method	Heat of formation (kJ mol^{-1})	Ref.
$C_3OF_4Cl_2^+$	$(CF_2Cl)_2CO$		10.71±0.01	PE		3289

4.3. The Positive Ion Table—Continued

Ion	Reactant	Other products	Ionization or appearance potential (eV)	Method	Heat of formation (kJ mol^{-1})	Ref.
		$C_3OF_3Cl_3^+$				
$C_3OF_3Cl_3^+$	CF_3COCCl_3		10.80 ± 0.01	PE		3289
		$C_2H_6SiCl^+$				
$C_2H_6SiCl^+$	$(CH_3)_3SiCl$	CH_3	11.00 ± 0.16	RPD		1421
$C_2H_6SiCl^+$	$(CH_3)_3SiCl$	CH_3	10.6 ± 0.1	EI		2689
		$C_3H_9SiCl^+$				
$C_3H_9SiCl^+$	$(CH_3)_3SiCl$		10.58 ± 0.04	RPD		1421
$C_3H_9SiCl^+$	$(CH_3)_3SiCl$		9.9 ± 0.1	EI		2689
		$CH_3SiCl_3^+$				
$CH_3SiCl_3^+$	CH_3SiCl_3		11.36 ± 0.03	EI		2182
		$C_2H_3SiCl_3^+$				
$C_2H_3SiCl_3^+$	$C_2H_3SiCl_3$		$10.79\pm0.02?$	PI		182

Because of ambiguity in the nomenclature of ref. 182 the structural formula may be either
$CH_2=CHSiCl_3$ or $CCl_2=CClSiH_3$.

Ion	Reactant	Other products	Ionization or appearance potential (eV)	Method	Heat of formation (kJ mol^{-1})	Ref.
		$C_2H_5SiCl_3^+$				
$C_2H_5SiCl_3^+$	$C_2H_5SiCl_3$		10.74 ± 0.04	EI		2182
		$C_3H_7SiCl_3^+$				
$C_3H_7SiCl_3^+$	$iso-C_3H_7SiCl_3$		10.28 ± 0.1	EI		2182
		$C_{12}H_{10}PCl^+$				
$C_{12}H_{10}PCl^+$	$(C_6H_5)_2PCl$ (Chlorodiphenylphosphine)		8.75 ± 0.05	EI		2481
		$C_6H_5PCl_2^+$				
$C_6H_5PCl_2^+$	$C_6H_5PCl_2$ (Dichloro(phenyl)phosphine)		9.45 ± 0.05	EI		2481
		CH_3SCl^+				
CH_3SCl^+	CH_3SCl		9.2 ± 0.1	EI		3034

4.3. The Positive Ion Table—Continued

Ion	Reactant	Other products	Ionization or appearance potential (eV)	Method	Heat of formation (kJ mol^{-1})	Ref.
			$C_4H_3SCl^+$			
$C_4H_3SCl^+$	C_4H_3SCl (2–Chlorothiophene)		8.68±0.01	PI	182,	416
$C_4H_3SCl^+$	C_4H_3SCl (2–Chlorothiophene)		8.70±0.05	PE		3246
See also – EI: 3240						
			$C_4H_2SCl_2^+$			
$C_4H_2SCl_2^+$	$C_4H_2SCl_2$ (2,5–Dichlorothiophene)		8.60±0.05	PE		3246
			CF_3SCl^+			
CF_3SCl^+	CF_3SCl		10.7±0.1	EI		3034
			$C_4H_{12}BN_2Cl^+$			
$C_4H_{12}BN_2Cl^+$	$((CH_3)_2N)_2BCl$		8.15	EI		3227
See also – EI: 2513, 2863						
			$C_2H_6BNCl_2^+$			
$C_2H_6BNCl_2^+$	$(CH_3)_2NBCl_2$		9.57	EI		3227
See also – EI: 2513, 2863						
			$C_2H_6BO_2Cl^+$			
$C_2H_6BO_2Cl^+$	$(CH_3O)_2BCl$		10.83	EI		3227
			$C_4H_{10}BO_2Cl^+$			
$C_4H_{10}BO_2Cl^+$	$(C_2H_5O)_2BCl$		10.52	EI		3227
			$C_6H_{14}BO_2Cl^+$			
$C_6H_{14}BO_2Cl^+$	$(n-C_3H_7O)_2BCl$		10.45	EI		3227

4.3. The Positive Ion Table—Continued

Ion	Reactant	Other products	Ionization or appearance potential (eV)	Method	Heat of formation (kJ mol^{-1})	Ref.
		CH$_3$BOCl$_2^+$				
CH$_3$BOCl$_2^+$	CH$_3$OBCl$_2$		11.55	EI		3227
		C$_2$H$_5$BOCl$_2^+$				
C$_2$H$_5$BOCl$_2^+$	C$_2$H$_5$OBCl$_2$		11.27	EI		3227
		C$_3$H$_7$BOCl$_2^+$				
C$_3$H$_7$BOCl$_2^+$	n-C$_3$H$_7$OBCl$_2$		11.22	EI		3227
		C$_6$H$_4$NOCl$^+$				
C$_6$H$_4$NOCl$^+$	C$_6$H$_4$ClNO (1−Chloro−4−nitrosobenzene)		9.02 (V)	PE		2806

See also − PE: 2826

Ion	Reactant	Other products	Ionization or appearance potential (eV)	Method	Heat of formation	Ref.
		C$_8$H$_8$NOCl$^+$				
C$_8$H$_8$NOCl$^+$	C$_6$H$_4$ClNHCOCH$_3$ (N−(3−Chlorophenyl)acetic acid amide)		8.65±0.2	EI		3406
C$_8$H$_8$NOCl$^+$	C$_6$H$_4$ClNHCOCH$_3$ (N−(4−Chlorophenyl)acetic acid amide)		8.31±0.2	EI		3406
		C$_3$H$_3$N$_2$OCl$^+$				
C$_3$H$_3$N$_2$OCl$^+$	CH$_2$ClCOCHN$_2$		9.92±0.1	EI		2174
		C$_6$H$_4$NO$_2$Cl$^+$				
C$_6$H$_4$NO$_2$Cl$^+$	C$_6$H$_4$ClNO$_2$ (1−Chloro−4−nitrobenzene)		9.99 (V)	PE		2806
		C$_7$H$_4$NO$_3$Cl$^+$				
C$_7$H$_4$NO$_3$Cl$^+$	C$_6$H$_4$(NO$_2$)COCl (4−Nitrobenzoic acid chloride)		10.66±0.01	EI		2026
		C$_3$HN$_2$OCl$_3^+$				
C$_3$HN$_2$OCl$_3^+$	CCl$_3$COCHN$_2$		9.95±0.06	EI		2174

4.3. The Positive Ion Table—Continued

Ion	Reactant	Other products	Ionization or appearance potential (eV)	Method	Heat of formation (kJ mol^{-1})	Ref.
		$C_2H_6BS_2Cl^+$				
$C_2H_6BS_2Cl^+$	$(CH_3S)_2BCl$		9.64	EI		3227
		$CH_3BSCl_2^+$				
$CH_3BSCl_2^+$	CH_3SBCl_2		10.45	EI		3227
		$C_{17}H_{19}N_2SCl^+$				
$C_{17}H_{19}N_2SCl^+$	$C_{17}H_{19}N_2SCl$ (Chlorpromazine)		7.38±0.13	CTS		2987
		$C_{20}H_{24}N_3SCl^+$				
$C_{20}H_{24}N_3SCl^+$	$C_{20}H_{24}N_3SCl$ (Prochlorperazine)		7.25	CTS		2987
		$C_{12}H_7OSCl^+$				
$C_{12}H_7OSCl^+$	$C_{12}H_7OSCl$ (2–Chlorophenoxathiin)		7.96	CTS		3401
		$C_2H_6BNFCl^+$				
$C_2H_6BNFCl^+$	$(CH_3)_2NBFCl$		9.65	EI		3227
		CH_3BOFCl^+				
CH_3BOFCl^+	CH_3OBFCl		11.96	EI		3227
		$C_{21}H_{24}N_3OSCl^+$				
$C_{21}H_{24}N_3OSCl^+$	$C_{21}H_{24}N_3OSCl$ (Pipamazine)		7.15±0.10	CTS		2987
		$C_{23}H_{28}N_3O_2SCl^+$				
$C_{23}H_{28}N_3O_2SCl^+$	$C_{23}H_{28}N_3O_2SCl$ (Thiopropazate)		7.31±0.14	CTS		2987

4.3. The Positive Ion Table—Continued

Ion	Reactant	Other products	Ionization or appearance potential (eV)	Method	Heat of formation (kJ mol^{-1})	Ref.
$Ar^+(^2P_{3/2})$		$\Delta H^\circ_{f0} = 1520.5$ kJ mol^{-1} (363.4 kcal mol^{-1})				
$Ar^+(^2P_{1/2})$		$\Delta H^\circ_{f0} = 1537.7$ kJ mol^{-1} (367.5 kcal mol^{-1})				
$Ar^+(^2P_{3/2})$	Ar		15.759	S	1520.5	2113, 3055, 3112
$Ar^+(^2P_{3/2})$	Ar		15.757±0.005	PE		2810, 2875
$Ar^+(^2P_{3/2})$	Ar		15.79	PE		248
$Ar^+(^2P_{3/2})$	Ar		15.78±0.03	EDD		2557
$Ar^+(^2P_{3/2})$	Ar		15.74±0.05	EDD		2634
$Ar^+(^2P_{1/2})$	Ar		15.937	S	1537.7	2113, 3055, 3112
$Ar^+(^2P_{1/2})$	Ar		15.93	PE		248

See also – PI: 163, 230, 1118, 2034, 2200
           EI: 3, 35, 52, 224, 2032, 2895, 2991

Ion	Reactant	Other products	Ionization or appearance potential (eV)	Method	Heat of formation (kJ mol^{-1})	Ref.
Ar^{+2}		$\Delta H^\circ_{f0} = 4186.4$ kJ mol^{-1} (1000.6 kcal mol^{-1})				
Ar^{+2}	Ar		43.388	S	4186.4	2113
Ar^{+2}	Ar		41.4	EM		2993
Ar^{+2}	Ar		43.4±0.3	RPD		198
Ar^{+2}	Ar		43.3	NRE		211
Ar^{+2}	Ar		43	EI		1240
Ar^{+2}	Ar^+		27.629	S		2113, 3136
Ar^{+2}	Ar^+		27±2	SEQ		2551

Ion	Reactant	Other products	Ionization or appearance potential (eV)	Method	Heat of formation (kJ mol^{-1})	Ref.
Ar^{+3}		$\Delta H^\circ_{f0} = 8117$ kJ mol^{-1} (1940 kcal mol^{-1})				
Ar^{+3}	Ar		84.13	S	8117	2113
Ar^{+3}	Ar		84.8±0.5	RPD		198
Ar^{+3}	Ar		83.7±0.5	NRE		25
Ar^{+3}	Ar		84.0	NRE		211
Ar^{+3}	Ar		84.3	EI		1040
Ar^{+3}	Ar		85	EI		1240
Ar^{+3}	Ar^{+2}		40.74	S		2113
Ar^{+3}	Ar^{+2}		38±2	SEQ		2551

4.3. The Positive Ion Table—Continued

Ion	Reactant	Other products	Ionization or appearance potential (eV)	Method	Heat of formation (kJ mol^{-1})	Ref.
Ar^{+4}		ΔH°_{f0} = 13888 kJ mol^{-1} (3319 kcal mol^{-1})				
Ar^{+4}	Ar		143.94	S	13888	2113
Ar^{+4}	Ar		150.0±5	RPD		198
Ar^{+4}	Ar		147	EI		1240
Ar^{+4}	Ar^{+3}		59.81	S		2113
Ar^{+4}	Ar^{+3}		55±2	SEQ		2551
Ar^{+5}		ΔH°_{f0} = 21127 kJ mol^{-1} (5049 kcal mol^{-1})				
Ar^{+5}	Ar		218.96	S	21127	2113
Ar^{+5}	Ar		285	EI		1240
Ar^{+5}	Ar^{+4}		75.02	S		2113
Ar^{+5}	Ar^{+4}		73±2	SEQ		2551
Ar^{+6}		ΔH°_{f0} = 29908 kJ mol^{-1} (7148 kcal mol^{-1})				
Ar^{+6}	Ar		309.97	S	29908	2113
Ar^{+6}	Ar		430	EI		1240
Ar^{+6}	Ar^{+5}		91.007	S		2113
Ar^{+6}	Ar^{+5}		89±2	SEQ		2551
Ar$_2^{+}$		ΔH°_{f0} ≤ 1419 kJ mol^{-1} (339 kcal mol^{-1})				
Ar$_2^{+}$	Ar+Ar*		14.710±0.009	PI	≤1419	2763
K^{+}		ΔH°_{f0} = 508.9 kJ mol^{-1} (121.6 kcal mol^{-1})				
K^{+}	K		4.341	S	508.9	2113

See also — EI: 2487, 3189

Ion	Reactant	Other products	Ionization or appearance potential (eV)	Method	Heat of formation	Ref.
K^{+}	KF	F	9.5±0.3	EI		2436
K^{+}	KCl	Cl	9.1±0.3	EI		2406
K^{+}	KCl	Cl	10.6	EI		2860
K^{+}	KOH	OH	7.80±0.15	EI		3189
K^{+}	KI	I	8.6±0.3	EI		2001

K$_2^{+}$

Ion	Reactant	Other products	Ionization or appearance potential (eV)	Method	Heat of formation	Ref.
K$_2^{+}$	K$_2$		~4.1	S		3179
K$_2^{+}$	K$_2$		4.0±0.1	PI		2633

See also — PI: 2615

4.3. The Positive Ion Table—Continued

Ion	Reactant	Other products	Ionization or appearance potential (eV)	Method	Heat of formation (kJ mol^{-1})	Ref.
		K_3^+				
K_3^+	K_3		3.4±0.1	PI		2633
		K_4^+				
K_4^+	K_4		3.6±0.1	PI		2633
		K_2O^+				
K_2O^+	K_2O		~5	EI		2445
K_2O^+	$K_2(OH)_2$	H_2O?	7.6±0.2	EI		3189
K_2O^+	K_2SO_4		13.0±0.5	EI		2487
		$K_2O_2^+$				
$K_2O_2^+$	K_2O_2		~5	EI		2445
		NaK^+				
NaK^+	NaK		4.5±0.1	PI		2633
		Na_2K^+				
Na_2K^+	Na_2K		3.6±0.1	PI		2633
		NaK_2^+				
NaK_2^+	NaK_2		3.4±0.1	PI		2633
		$Na_2K_2^+$				
$Na_2K_2^+$	Na_2K_2		4.1±0.1	PI		2633
		KCl^+				
KCl^+	KCl		8.0±0.3	EI		2406
KCl^+	KCl		10.1	EI		2860
		K_2Cl^+				
K_2Cl^+	K_2Cl_2	Cl	10.4	EI		2860

4.3. The Positive Ion Table—Continued

Ion	Reactant	Other products	Ionization or appearance potential (eV)	Method	Heat of forma-tion (kJ mol^{-1})	Ref.
			KOH$^+$			
KOH$^+$	KOH		7.50±0.15	EI		3189
			K$_2$OH$^+$			
K$_2$OH$^+$	K$_2$(OH)$_2$	OH	7.80±0.15	EI		3189
			K$_2$SO$_4^+$			
K$_2$SO$_4^+$	K$_2$SO$_4$		7.4±0.5	EI		2487
	Ca$^+$	$\Delta H^\circ_{f0} = 767.6$ kJ mol^{-1} (183.5 kcal mol^{-1})				
Ca$^+$	Ca		6.113	S	767.6	2113, 3125
Ca$^+$	Ca		6.21±0.09	SI		3021

See also − S: 3270
         EI: 1297, 2141, 2178, 2532, 2592, 2594, 2595, 2620, 2990, 3203, 3204

Ion	Reactant	Other products	Ionization or appearance potential (eV)	Method	Heat of formation	Ref.
Ca$^+$	CaCl$_2$	2Cl	15.94±0.10	EI		2991
	Ca^{+2}	$\Delta H^\circ_{f0} = 1913.0$ kJ mol^{-1} (457.2 kcal mol^{-1})				
Ca^{+2}	Ca		17.984	S	1913.0	2113

See also − EI: 2178

Ion	Reactant	Other products	IP (eV)	Method	ΔHf	Ref.
Ca^{+2}	Ca$^+$		11.871	S		2113
			CaO$^+$			
CaO$^+$	CaO		6.5	EI		1244, 2123
			CaF$^+$			
CaF$^+$	CaF		6.0±0.5	EI		1297, 2141, 2532, 2620

See also − EI: 2165, 2592, 2594, 2595, 3203

Ion	Reactant	Other products	IP (eV)	Method	ΔHf	Ref.
CaF$^+$	CaF$_2$	F	12.5±0.8	EI		1297, 2141

See also − EI: 2165

4.3. The Positive Ion Table—Continued

Ion	Reactant	Other products	Ionization or appearance potential (eV)	Method	Heat of formation (kJ mol^{-1})	Ref.
			CaCl$^+$			
CaCl$^+$	CaCl		6.01±0.10	EI		2991
CaCl$^+$	CaCl		5.5±0.5	EI		3204
CaCl$^+$	CaCl		5.6±0.5	EI		2990
CaCl$^+$	CaCl$_2$	Cl	10.96±0.10	EI		2991
CaCl$^+$	CaCl$_2$	Cl	11.0±0.2	EI		2990
			CaCl$_2^+$			
CaCl$_2^+$	CaCl$_2$		10.33±0.10	EI		2991
CaCl$_2^+$	CaCl$_2$		10.5±0.3	EI		2990

See also – EI: 3204

			CaOH$^+$			
CaOH$^+$	CaOH		5.90±0.1	D		3242, 3419

			Sc$^+$ $\quad$ $\Delta H^\circ_{f0} = 1007$ kJ mol^{-1} (241 kcal mol^{-1})			
Sc$^+$	Sc		6.54	S	1007	2113

See also – EI: 2594, 2600, 2721

Sc$^+$	ScF$_3$		28.0±0.7	EI		2009, 2600

			ScF$^+$			
ScF$^+$	ScF		6.3	EI		2594
ScF$^+$	ScF		6.5±0.3	EI		2600
ScF$^+$	ScF$_3$		16.0±0.7	EI		2009, 2600

			ScF$_2^+$			
ScF$_2^+$	ScF$_2$		7.0±0.3	EI		2594, 2600
ScF$_2^+$	ScF$_3$	F	13.5±0.7	EI		2009, 2594, 2600

			Ti$^+$ $\quad$ $\Delta H^\circ_{f0} = 1125$ kJ mol^{-1} (269 kcal mol^{-1})			
Ti$^+$	Ti		6.82	S	1125	2113
Ti$^+$	Ti		6.6±0.2	EI		2527
Ti$^+$	TiO?		11.5±0.2	EI		2527
Ti$^+$	TiO$_2$?		20±0.2	EI		2527
Ti$^+$	TiF$_3$		21.3	EI		2592
Ti$^+$	TiCl$_4$	4Cl	25.0±0.3	EI		2506

4.3. The Positive Ion Table—Continued

Ion	Reactant	Other products	Ionization or appearance potential (eV)	Method	Heat of formation (kJ mol⁻¹)	Ref.
		Ti^{+2}	$\Delta H_{f0}^\circ = 2435$ kJ mol⁻¹ (582 kcal mol⁻¹)			
Ti^{+2}	Ti		20.40	S	2435	2113
Ti^{+2}	Ti		21.5±1	EI		2527
Ti^{+2}	Ti^+		13.58	S		2113
Ti^{+2}	$TiCl_4$		39.1±1.3	EI		2506
		Ti^{+3}	$\Delta H_{f0}^\circ = 5088$ kJ mol⁻¹ (1216 kcal mol⁻¹)			
Ti^{+3}	Ti		47.89	S	5088	2113
Ti^{+3}	Ti		50±1	EI		2527
Ti^{+3}	Ti^{+2}		27.491	S		2113
		TiC_2^+				
TiC_2^+	TiC_2		8.7±0.5	EI		3208
		TiC_4^+				
TiC_4^+	TiC_4		9.0±1.0	EI		3208
		TiN^+				
TiN^+	TiN		6±2	EI		3207
		TiO^+				
TiO^+	TiO		5.5±0.5	EI		2527
TiO^+	TiO_2?		8±0.5	EI		2527
		TiO_2^+				
TiO_2^+	TiO_2		9±0.2	EI		2527
		TiF^+				
TiF^+	TiF_3		15.5	EI		2592
		TiF_2^+				
TiF_2^+	TiF_2		12.2±0.5	EI		2592
TiF_2^+	TiF_3	F	14.0	EI		2592
		TiF_3^+				
TiF_3^+	TiF_3		10.5±0.5	EI		2592

4.3. The Positive Ion Table—Continued

Ion	Reactant	Other products	Ionization or appearance potential (eV)	Method	Heat of formation (kJ mol^{-1})	Ref.
		TiCl$^+$				
TiCl$^+$	TiCl$_4$		20.6±0.3	EI		2506
TiCl$^+$	(C$_5$H$_5$)$_2$TiCl$_2$ (Bis(cyclopentadienyl)titanium dichloride)		21.6±0.4	EI		2479
		TiCl^{+2}				
TiCl^{+2}	TiCl$_4$		35.6±0.9	EI		2506
		TiCl$_2^+$				
TiCl$_2^+$	TiCl$_4$		16.7±0.3	EI		2506
TiCl$_2^+$	(C$_5$H$_5$)$_2$TiCl$_2$ (Bis(cyclopentadienyl)titanium dichloride)		19.7±0.4	EI		2479
		TiCl$_2^{+2}$				
TiCl$_2^{+2}$	TiCl$_4$		32.1±0.8	EI		2506
		TiCl$_3^+$				
TiCl$_3^+$	TiCl$_4$	Cl?	13.3±0.3	EI		2506
		TiCl$_3^{+2}$				
TiCl$_3^{+2}$	TiCl$_4$	Cl	30.0±0.5	EI		2506
		TiCl$_4^+$				
TiCl$_4^+$	TiCl$_4$		11.78±0.04 (V)	PE		3079
TiCl$_4^+$	TiCl$_4$		11.7 (V)	PE		3117
TiCl$_4^+$	TiCl$_4$		11.65±0.15	EI		2506
		C$_{10}$H$_{10}$Ti$^+$				
C$_{10}$H$_{10}$Ti$^+$	(C$_5$H$_5$)$_3$Ti (Tris(cyclopentadienyl)titanium)		8.27±0.2	EI		2640
		C$_{15}$H$_{15}$Ti$^+$				
C$_{15}$H$_{15}$Ti$^+$	(C$_5$H$_5$)$_3$Ti (Tris(cyclopentadienyl)titanium)		6.47±0.1	EI		2640

4.3. The Positive Ion Table—Continued

Ion	Reactant	Other products	Ionization or appearance potential (eV)	Method	Heat of formation (kJ mol^{-1})	Ref.
			C$_{10}$H$_{14}$O$_4$Ti$^+$			
C$_{10}$H$_{14}$O$_4$Ti$^+$	(CH$_3$COCHCOCH$_3$)$_3$Ti (Tris(2,4–pentanedionato)titanium)		11.8±0.1	EI		2460
			C$_{15}$H$_{21}$O$_6$Ti$^+$			
C$_{15}$H$_{21}$O$_6$Ti$^+$	(CH$_3$COCHCOCH$_3$)$_3$Ti (Tris(2,4–pentanedionato)titanium)		7.1±0.1	EI		2460
			C$_3$H$_3$ClTi$^+$			
C$_3$H$_3$ClTi$^+$	(C$_5$H$_5$)$_2$TiCl$_2$ (Bis(cyclopentadienyl)titanium dichloride)		~19.5	EI		2479
			C$_5$H$_5$ClTi$^+$			
C$_5$H$_5$ClTi$^+$	(C$_5$H$_5$)$_2$TiCl$_2$ (Bis(cyclopentadienyl)titanium dichloride)		16.3±0.3	EI		2479
			C$_{10}$H$_{10}$ClTi$^+$			
C$_{10}$H$_{10}$ClTi$^+$	(C$_5$H$_5$)$_2$TiCl$_2$ (Bis(cyclopentadienyl)titanium dichloride)	Cl	10.8±0.3	EI		2479
			C$_5$H$_5$Cl$_2$Ti$^+$			
C$_5$H$_5$Cl$_2$Ti$^+$	(C$_5$H$_5$)$_2$TiCl$_2$ (Bis(cyclopentadienyl)titanium dichloride)		11.8±0.2	EI		2479
			C$_{10}$H$_{10}$Cl$_2$Ti$^+$			
C$_{10}$H$_{10}$Cl$_2$Ti$^+$	(C$_5$H$_5$)$_2$TiCl$_2$ (Bis(cyclopentadienyl)titanium dichloride)		8.98±0.16	EI		2479

V$^+$ $\Delta H_{f0}^\circ = 1161$ kJ mol^{-1} (278 kcal mol^{-1})

Ion	Reactant	Other products	Ionization or appearance potential (eV)	Method	Heat of formation (kJ mol^{-1})	Ref.
V$^+$	V		6.74	S	1161	2113
See also – EI: 2530						
V$^+$	(C$_5$H$_5$)$_2$V (Bis(cyclopentadienyl)vanadium)		14.5±0.5	EI		2683
V$^+$	(C$_6$H$_6$)$_2$V (Bis(benzene)vanadium)		13.6±0.3	EI		2530
V$^+$	C$_5$H$_5$VC$_7$H$_7$ (Cycloheptatrienyl(cyclopentadienyl)vanadium)		13.8±0.3	EI		2530

4.3. The Positive Ion Table—Continued

Ion	Reactant	Other products	Ionization or appearance potential (eV)	Method	Heat of formation (kJ mol^{-1})	Ref.
V$^+$	V(CO)$_6$	6CO	15.5±0.2	EI		2403
V$^+$	C$_5$H$_5$V(CO)$_4$		19.4±0.4	EI		1381
	(Cyclopentadienylvanadium tetracarbonyl)					

VC$^+$

VC$^+$	V(CO)$_6$		23.8±0.8	EI		2403

VC$_2^+$

VC$_2^+$	VC$_2$		8.6±0.5	EI		2997

VCl$_4^+$

VCl$_4^+$	VCl$_4$		9.41±0.04 (V)	PE		3079

C$_3$H$_3$V$^+$

C$_3$H$_3$V$^+$	C$_5$H$_5$V(CO)$_4$		18.9±0.3	EI		1381
	(Cyclopentadienylvanadium tetracarbonyl)					

C$_5$H$_5$V$^+$

C$_5$H$_5$V$^+$	C$_5$H$_5$V		7.8±0.5	EI		2530
	(Cyclopentadienylvanadium)					
C$_5$H$_5$V$^+$	(C$_5$H$_5$)$_2$V		12.65±0.1	EI		2683
	(Bis(cyclopentadienyl)vanadium)					
C$_5$H$_5$V$^+$	C$_5$H$_5$VC$_7$H$_7$		9.2±0.8	EI		2530
	(Cycloheptatrienyl(cyclopentadienyl)vanadium)					
C$_5$H$_5$V$^+$	C$_5$H$_5$V(CO)$_4$	4CO	14.2±0.2	EI		1381
	(Cyclopentadienylvanadium tetracarbonyl)					

C$_6$H$_6$V$^+$

C$_6$H$_6$V$^+$	C$_6$H$_6$V		6.3±0.5	EI		2530
	(Benzenevanadium)					
C$_6$H$_6$V$^+$	(C$_6$H$_6$)$_2$V		10.5±0.2	EI		2530
	(Bis(benzene)vanadium)					
C$_6$H$_6$V$^+$	C$_5$H$_5$VC$_7$H$_7$		11.0±0.2	EI		2530
	(Cycloheptatrienyl(cyclopentadienyl)vanadium)					

C$_{10}$H$_{10}$V$^+$

C$_{10}$H$_{10}$V$^+$	(C$_5$H$_5$)$_2$V		7.33±0.1	EI		2683
	(Bis(cyclopentadienyl)vanadium)					

4.3. The Positive Ion Table—Continued

Ion	Reactant	Other products	Ionization or appearance potential (eV)	Method	Heat of formation (kJ mol^{-1})	Ref.
		$C_{12}H_{12}V^+$				
$C_{12}H_{12}V^+$	$(C_6H_6)_2V$ (Bis(benzene)vanadium)		6.26±0.1	EI		2530
$C_{12}H_{12}V^+$	$C_5H_5VC_7H_7$ (Cycloheptatrienyl(cyclopentadienyl)vanadium)		7.24±0.1	EI		2530
		VCO^+				
VCO^+	$V(CO)_6$	5CO	13.8±0.2	EI		2403
		$VC_2O_2^+$				
$VC_2O_2^+$	$V(CO)_6$	4CO	12.3±0.2	EI		2403
		$VC_3O_3^+$				
$VC_3O_3^+$	$V(CO)_6$	3CO	10.98±0.15	EI		2403
		$VC_4O_4^+$				
$VC_4O_4^+$	$V(CO)_6$	2CO	9.70±0.2	EI		2403
		$VC_5O_5^+$				
$VC_5O_5^+$	$V(CO)_6$	CO	8.24±0.15	EI		2403
		$VC_6O_6^+$				
$VC_6O_6^+$	$V(CO)_6$		7.52 (V)	PE		2849
$VC_6O_6^+$	$V(CO)_6$		7.53±0.15	EI		2403
		$C_6H_5OV^+$				
$C_6H_5OV^+$	$C_5H_5V(CO)_4$ (Cyclopentadienylvanadium tetracarbonyl)	3CO	10.7±0.3	EI		1381
		$C_7H_5O_2V^+$				
$C_7H_5O_2V^+$	$C_5H_5V(CO)_4$ (Cyclopentadienylvanadium tetracarbonyl)	2CO	9.7±0.3	EI		1381

4.3. The Positive Ion Table—Continued

Ion	Reactant	Other products	Ionization or appearance potential (eV)	Method	Heat of formation (kJ mol^{-1})	Ref.
$C_9H_5O_4V^+$						
$C_9H_5O_4V^+$	$C_5H_5V(CO)_4$ (Cyclopentadienylvanadium tetracarbonyl)		8.2±0.3	EI		1381
$C_{10}H_{14}O_4V^+$						
$C_{10}H_{14}O_4V^+$	$(CH_3COCHCOCH_3)_3V$ (Tris(2,4–pentanedionato)vanadium)		11.8±0.1	EI		2460
$C_{15}H_{21}O_6V^+$						
$C_{15}H_{21}O_6V^+$	$(CH_3COCHCOCH_3)_3V$ (Tris(2,4–pentanedionato)vanadium)		7.72±0.10	EI		2580
$C_{15}H_{21}O_6V^+$	$(CH_3COCHCOCH_3)_3V$ (Tris(2,4–pentanedionato)vanadium)		7.9±0.1	EI		2460

Cr^+ $\Delta H_{f0}^\circ = 1047.3$ kJ mol^{-1} (250.3 kcal mol^{-1})

Ion	Reactant	Other products	Ionization or appearance potential (eV)	Method	Heat of formation (kJ mol^{-1})	Ref.
Cr^+	Cr		6.766	S	1047.3	2113

See also – EI: 1249, 2530

Ion	Reactant	Other products	Ionization or appearance potential (eV)	Method	Heat of formation (kJ mol^{-1})	Ref.
Cr^+	CrF_3		20.1±0.3	EI		2591
Cr^+	$(C_5H_5)_2Cr$ (Bis(cyclopentadienyl)chromium)		14.6±0.3	EI		2683
Cr^+	$C_5H_5CrC_6H_6$ (Benzenecyclopentadienylchromium)		13.9±0.3	EI		2530
Cr^+	$(C_6H_6)_2Cr$ (Bis(benzene)chromium)		9.9±0.3	EI		2981
Cr^+	$(C_6H_6)_2Cr$ (Bis(benzene)chromium)		10.8±0.3	EI		2530, 3281
Cr^+	$C_5H_5CrC_7H_7$ (Cycloheptatrienyl(cyclopentadienyl)chromium)		12.2±0.3	EI		2530
Cr^+	$C_6H_6Cr(C_6H_5)_2$ (Benzenebiphenylchromium)		11.6±0.2	EI		3281
Cr^+	$(C_6H_3(CH_3)_3)_2Cr$ (Bis(1,3,5–trimethylbenzene)chromium)		12.1±0.2	EI		3281
Cr^+	$((C_6H_5)_2)_2Cr$ (Bis(biphenyl)chromium)		12.3±0.2	EI		3281
Cr^+	$(C_6(CH_3)_6)_2Cr$ (Bis(hexamethylbenzene)chromium)		14.0±0.4	EI		3281
Cr^+	$Cr(CO)_6$	6CO	14.7±0.1	EI		2023
Cr^+	$Cr(CO)_6$	6CO	15.1±0.2	EI		2403
Cr^+	$Cr(CO)_6$	6CO	17.07	EI		2500
Cr^+	$Cr(CO)_6$	6CO	17.7±0.3	EI		1107
Cr^+	CrO_2F_2		22.13±0.07	EI		2788

See also – EI: 30

4.3. The Positive Ion Table—Continued

Ion	Reactant	Other products	Ionization or appearance potential (eV)	Method	Heat of formation (kJ mol^{-1})	Ref.
Cr$^+$	CrO$_2$Cl$_2$		19.15±0.07	EI		2788
See also – EI: 30						
Cr$^+$	C$_6$H$_6$Cr(CO)$_3$ (Benzenechromium tricarbonyl)		12.9±0.2	EI		2608
Cr$^+$	CH$_3$OC(CH$_3$)Cr(CO)$_5$		13.6±0.4	EI		2641

Cr^{+2} $\Delta H^{\circ}_{f0} = 2640$ kJ mol^{-1} (631 kcal mol^{-1})

Ion	Reactant	Other products	Ionization or appearance potential (eV)	Method	Heat of formation (kJ mol^{-1})	Ref.
Cr^{+2}	Cr		23.27	S	2640	2113
Cr^{+2}	Cr$^+$		16.50	S		2113
Cr^{+2}	CrO$_2$F$_2$		45.8	EI		2788
Cr^{+2}	CrO$_2$Cl$_2$		44.2	EI		2788

CrC$^+$

Ion	Reactant	Other products	Ionization or appearance potential (eV)	Method	Heat of formation (kJ mol^{-1})	Ref.
CrC$^+$	Cr(CO)$_6$		23.15±0.3	EI		2403

CrO$^+$

Ion	Reactant	Other products	Ionization or appearance potential (eV)	Method	Heat of formation (kJ mol^{-1})	Ref.
CrO$^+$	CrO		8.4±0.5	EI		2130
CrO$^+$	CrO$_2$	O	13.5	EI		2130
CrO$^+$	Cr(CO)$_6$		23.45±0.3	EI		2403
CrO$^+$	CrO$_2$F$_2$		23.84±0.05	EI		2788
See also – EI: 30						
CrO$^+$	CrO$_2$Cl$_2$		18.30±0.10	EI		2788
See also – EI: 30						

CrO^{+2}

Ion	Reactant	Other products	Ionization or appearance potential (eV)	Method	Heat of formation (kJ mol^{-1})	Ref.
CrO^{+2}	CrO$_2$F$_2$		40.7	EI		2788
CrO^{+2}	CrO$_2$Cl$_2$		39.6	EI		2788

CrO$_2^+$

Ion	Reactant	Other products	Ionization or appearance potential (eV)	Method	Heat of formation (kJ mol^{-1})	Ref.
CrO$_2^+$	CrO$_2$		10.3±0.5	EI		2130
CrO$_2^+$	CrO$_2$F$_2$		19.89±0.08	EI		2788
See also – EI: 30						
CrO$_2^+$	CrO$_2$Cl$_2$		14.18±0.08	EI		2788
See also – EI: 30						

4.3. The Positive Ion Table—Continued

Ion	Reactant	Other products	Ionization or appearance potential (eV)	Method	Heat of formation (kJ mol^{-1})	Ref.
			CrO_2^{+2}			
CrO_2^{+2}	CrO_2F_2		37.6	EI		2788
			CrO_3^+			
CrO_3^+	CrO_3		11.6±0.5	EI		2130
			CrF^+			
CrF^+	CrF		8.4±0.3	EI		1249
CrF^+	CrF_3		14.0±0.3	EI		2591
CrF^+	CrO_2F_2		17.94±0.04	EI		2788
See also – EI: 30						
			CrF^{+2}			
CrF^{+2}	CrO_2F_2		42.6	EI		2788
			CrF_2^+			
CrF_2^+	CrF_2		10.1±0.3	EI		1249
CrF_2^+	CrF_3	F	13.5±0.3	EI		2591
CrF_2^+	CrO_2F_2	O_2	14.24±0.04	EI		2788
See also – EI: 30						
			CrF_3^+			
CrF_3^+	CrF_3		12.2±0.3	EI		2591
			$CrCl^+$			
$CrCl^+$	CrO_2Cl_2		16.52±0.07	EI		2788
See also – EI: 30						
			$CrCl^{+2}$			
$CrCl^{+2}$	CrO_2Cl_2		41.2	EI		2788

4.3. The Positive Ion Table—Continued

Ion	Reactant	Other products	Ionization or appearance potential (eV)	Method	Heat of formation (kJ mol^{-1})	Ref.
			CrCl$_2^+$			
CrCl$_2^+$	CrO$_2$Cl$_2$	O$_2$	14.00±0.10	EI		2788
See also – EI: 30						
			CrCl$_2^{+2}$			
CrCl$_2^{+2}$	CrO$_2$Cl$_2$		36.0	EI		2788
			CH$_3$Cr$^+$			
CH$_3$Cr$^+$	CH$_3$OC(CH$_3$)Cr(CO)$_5$		15.6±0.3	EI		2641
			C$_5$H$_5$Cr$^+$			
C$_5$H$_5$Cr$^+$	C$_5$H$_5$Cr (Cyclopentadienylchromium)		6.4±0.5	EI		2530
C$_5$H$_5$Cr$^+$	C$_5$H$_5$Cr (Cyclopentadienylchromium)		7.0±0.5	EI		2530
C$_5$H$_5$Cr$^+$	(C$_5$H$_5$)$_2$Cr (Bis(cyclopentadienyl)chromium)		12.81±0.1	EI		2683
C$_5$H$_5$Cr$^+$	C$_5$H$_5$CrC$_6$H$_6$ (Benzenecyclopentadienylchromium)		9.3±0.2	EI		2530
C$_5$H$_5$Cr$^+$	C$_5$H$_5$CrC$_7$H$_7$ (Cycloheptatrienyl(cyclopentadienyl)chromium)		10.4±0.6	EI		2530
			C$_6$H$_6$Cr$^+$			
C$_6$H$_6$Cr$^+$	C$_6$H$_6$Cr (Benzenechromium)		6.4±0.5	EI		2530
C$_6$H$_6$Cr$^+$	(C$_6$H$_6$)$_2$Cr (Bis(benzene)chromium)		8.8±0.2	EI		2530, 3281
C$_6$H$_6$Cr$^+$	(C$_6$H$_6$)$_2$Cr (Bis(benzene)chromium)		8.9	EI		2981
C$_6$H$_6$Cr$^+$	(C$_6$H$_6$)$_2$Cr (Bis(benzene)chromium)		9.2±0.2	EI		2545
C$_6$H$_6$Cr$^+$	C$_6$H$_6$Cr(CO)$_3$ (Benzenechromium tricarbonyl)	3CO	10.5±0.2	EI		2608
C$_6$H$_6$Cr$^+$	C$_6$H$_6$Cr(CO)$_3$ (Benzenechromium tricarbonyl)	3CO	10.8±0.2	EI		2545
			C$_9$H$_{12}$Cr$^+$			
C$_9$H$_{12}$Cr$^+$	(C$_6$H$_3$(CH$_3$)$_3$)$_2$Cr (Bis(1,3,5−trimethylbenzene)chromium)		9.3±0.1	EI		3281

4.3. The Positive Ion Table—Continued

Ion	Reactant	Other products	Ionization or appearance potential (eV)	Method	Heat of formation (kJ mol^{-1})	Ref.
		$C_{10}H_{10}Cr^+$				
$C_{10}H_{10}Cr^+$	$(C_5H_5)_2Cr$ (Bis(cyclopentadienyl)chromium)		6.26±0.1	EI		2683
$C_{10}H_{10}Cr^+$	$C_5H_5CrC_7H_7$ (Cycloheptatrienyl(cyclopentadienyl)chromium)	C_2H_2	12.2	EI		2530
		$C_{11}H_{11}Cr^+$				
$C_{11}H_{11}Cr^+$	$C_5H_5CrC_6H_6$ (Benzenecyclopentadienylchromium)		6.13±0.1	EI		2530
		$C_{12}H_{10}Cr^+$				
$C_{12}H_{10}Cr^+$	$C_6H_6Cr(C_6H_5)_2$ (Benzenebiphenylchromium)		8.9±0.1	EI		3281
$C_{12}H_{10}Cr^+$	$((C_6H_5)_2)_2Cr$ (Bis(biphenyl)chromium)		9.5±0.1	EI		3281
		$C_{12}H_{12}Cr^+$				
$C_{12}H_{12}Cr^+$	$(C_6H_6)_2Cr$ (Bis(benzene)chromium)		4.9	EI		2981
$C_{12}H_{12}Cr^+$	$(C_6H_6)_2Cr$ (Bis(benzene)chromium)		5.70±0.1	EI		2545
$C_{12}H_{12}Cr^+$	$(C_6H_6)_2Cr$ (Bis(benzene)chromium)		5.91±0.1	EI	2530,	3281
$C_{12}H_{12}Cr^+$	$C_5H_5CrC_7H_7$ (Cycloheptatrienyl(cyclopentadienyl)chromium)		5.96±0.1	EI		2530
		$C_{12}H_{12}Cr^{+2}$				
$C_{12}H_{12}Cr^{+2}$	$(C_6H_6)_2Cr$ (Bis(benzene)chromium)		15.8	EI		2981
		$C_{12}H_{18}Cr^+$				
$C_{12}H_{18}Cr^+$	$(C_6(CH_3)_6)_2Cr$ (Bis(hexamethylbenzene)chromium)		10.3±0.2	EI		3281
		$C_{18}H_{16}Cr^+$				
$C_{18}H_{16}Cr^+$	$C_6H_6Cr(C_6H_5)_2$ (Benzenebiphenylchromium)		5.94±0.05	EI		3281

4.3. The Positive Ion Table—Continued

Ion	Reactant	Other products	Ionization or appearance potential (eV)	Method	Heat of formation (kJ mol⁻¹)	Ref.
		$C_{18}H_{24}Cr^+$				
$C_{18}H_{24}Cr^+$	$(C_6H_3(CH_3)_3)_2Cr$ (Bis(1,3,5–trimethylbenzene)chromium)		5.47±0.05	EI		3281
		$C_{24}H_{20}Cr^+$				
$C_{24}H_{20}Cr^+$	$((C_6H_5)_2)_2Cr$ (Bis(biphenyl)chromium)		5.87±0.1	EI		3281
		$C_{24}H_{36}Cr^+$				
$C_{24}H_{36}Cr^+$	$(C_6(CH_3)_6)_2Cr$ (Bis(hexamethylbenzene)chromium)		5.19±0.05	EI		3281
		$CrCO^+$				
$CrCO^+$	$Cr(CO)_6$	5CO	13.3±0.2	EI		2023
$CrCO^+$	$Cr(CO)_6$	5CO	13.63±0.2	EI		2403
$CrCO^+$	$Cr(CO)_6$	5CO	14.12	EI		2500
$CrCO^+$	$Cr(CO)_6$	5CO	14.9±0.2	EI		1107
$CrCO^+$	$CH_3OC(CH_3)Cr(CO)_5$		13.9±0.2	EI		2641
		$CrCO^{+2}$				
$CrCO^{+2}$	$Cr(CO)_6$		30.8±1.0	EI		2403
		$CrC_2O_2^+$				
$CrC_2O_2^+$	$Cr(CO)_6$	4CO	11.56±0.2	EI		2023
$CrC_2O_2^+$	$Cr(CO)_6$	4CO	11.94±0.1	EI		2403
$CrC_2O_2^+$	$Cr(CO)_6$	4CO	12.56	EI		2500
$CrC_2O_2^+$	$Cr(CO)_6$	4CO	13.1±0.2	EI		1107
		$CrC_3O_3^+$				
$CrC_3O_3^+$	$Cr(CO)_6$	3CO	~10.42	EI		2500
$CrC_3O_3^+$	$Cr(CO)_6$	3CO	10.62±0.15	EI		2023
$CrC_3O_3^+$	$Cr(CO)_6$	3CO	11.1±0.2	EI		2403
		$CrC_4O_4^+$				
$CrC_4O_4^+$	$Cr(CO)_6$	2CO	~9.52	EI		2500
$CrC_4O_4^+$	$Cr(CO)_6$	2CO	9.64±0.1	EI		2403
$CrC_4O_4^+$	$Cr(CO)_6$	2CO	9.97±0.04	EI		2023

4.3. The Positive Ion Table—Continued

Ion	Reactant	Other products	Ionization or appearance potential (eV)	Method	Heat of formation (kJ mol^{-1})	Ref.
			$CrC_5O_5^+$			
$CrC_5O_5^+$	$Cr(CO)_6$	CO	8.95 ± 0.1	EI		2403
$CrC_5O_5^+$	$Cr(CO)_6$	CO	9.17 ± 0.04	EI		2023
$CrC_5O_5^+$	$Cr(CO)_6$	CO	~9.32	EI		2500

See also – PI: 2886

	$Cr(CO)_6^+$	$\Delta H^\circ_{f298} = -220$ kJ mol^{-1} (-53 kcal mol^{-1})				
$CrC_6O_6^+$	$Cr(CO)_6$		8.142 ± 0.017	PI	-220	2886
(Threshold value approximately corrected for hot bands)						
$CrC_6O_6^+$	$Cr(CO)_6$		8.40 (V)	PE		3029

See also – PI: 1167
       EI: 1107, 2023, 2403, 2500

			$CrOF^+$			
$CrOF^+$	CrO_2F_2		19.09 ± 0.03	EI		2788

See also – EI: 30

			$CrOF^{+2}$			
$CrOF^{+2}$	CrO_2F_2		38.4	EI		2788

			CrO_2F^+			
CrO_2F^+	CrO_2F_2	F	15.72 ± 0.10	EI		2788

See also – EI: 30

			CrO_2F^{+2}			
CrO_2F^{+2}	CrO_2F_2	F	36.4	EI		2788

			$CrOF_2^+$			
$CrOF_2^+$	CrO_2F_2	O	16.29 ± 0.08	EI		2788

See also – EI: 30

4.3. The Positive Ion Table—Continued

Ion	Reactant	Other products	Ionization or appearance potential (eV)	Method	Heat of formation (kJ mol^{-1})	Ref.
			$CrO_2F_2^+$			
$CrO_2F_2^+$	CrO_2F_2		12.91±0.03	EI		2788
See also – EI: 30						
			$CrOCl^+$			
$CrOCl^+$	CrO_2Cl_2		14.80±0.05	EI		2788
See also – EI: 30						
			$CrOCl^{+2}$			
$CrOCl^{+2}$	CrO_2Cl_2		35.7	EI		2788
			CrO_2Cl^+			
CrO_2Cl^+	CrO_2Cl_2	Cl	13.69±0.06	EI		2788
See also – EI: 30						
			CrO_2Cl^{+2}			
CrO_2Cl^{+2}	CrO_2Cl_2	Cl	32.7	EI		2788
			$CrOCl_2^+$			
$CrOCl_2^+$	CrO_2Cl_2	O	15.52±0.11	EI		2788
See also – EI: 30						
			$CrO_2Cl_2^+$			
$CrO_2Cl_2^+$	CrO_2Cl_2		11.99±0.04	EI		2788
See also – EI: 30						
			$CrO_2Cl_2^{+2}$			
$CrO_2Cl_2^{+2}$	CrO_2Cl_2		30.9	EI		2788
			$C_2H_3OCr^+$			
$C_2H_3OCr^+$	$CH_3OC(CH_3)Cr(CO)_5$		13.9±0.2	EI		2641

4.3. The Positive Ion Table—Continued

Ion	Reactant	Other products	Ionization or appearance potential (eV)	Method	Heat of formation (kJ mol^{-1})	Ref.
			$C_3H_6OCr^+$			
$C_3H_6OCr^+$	$CH_3OC(CH_3)Cr(CO)_5$	5CO	12.2±0.2	EI		2641
			$C_7H_6OCr^+$			
$C_7H_6OCr^+$	$C_6H_6Cr(CO)_3$ (Benzenechromium tricarbonyl)	2CO	8.5±0.1	EI		2608
			$C_4H_6O_2Cr^+$			
$C_4H_6O_2Cr^+$	$CH_3OC(CH_3)Cr(CO)_5$	4CO	10.7±0.1	EI		2641
			$C_5H_6O_3Cr^+$			
$C_5H_6O_3Cr^+$	$CH_3OC(CH_3)Cr(CO)_5$	3CO	9.96±0.1	EI		2641
			$C_8H_5O_3Cr^+$			
$C_8H_5O_3Cr^+$	$C_5H_5Cr(CO)_3$ (Cyclopentadienylchromium tricarbonyl)		7.30	EI		3005
			$C_9H_6O_3Cr^+$			
$C_9H_6O_3Cr^+$	$C_6H_6Cr(CO)_3$ (Benzenechromium tricarbonyl)		7.30±0.1	EI		2608
$C_9H_6O_3Cr^+$	$C_6H_6Cr(CO)_3$ (Benzenechromium tricarbonyl)		7.41±0.06	CTS		3403

See also – EI: 2545, 3005

Ion	Reactant	Other products	Ionization or appearance potential (eV)	Method	Heat of formation (kJ mol^{-1})	Ref.
			$C_{10}H_8O_3Cr^+$			
$C_{10}H_8O_3Cr^+$	$C_6H_5CH_3Cr(CO)_3$ (Toluenechromium tricarbonyl)		7.19	EI		3005, 3280
$C_{10}H_8O_3Cr^+$	$C_6H_5CH_3Cr(CO)_3$ (Toluenechromium tricarbonyl)		7.39±0.05	CTS		3403
$C_{10}H_8O_3Cr^+$	$C_7H_8Cr(CO)_3$ (Cycloheptatrienechromium tricarbonyl)		7.10	EI		3005, 3280
			$C_{11}H_{10}O_3Cr^+$			
$C_{11}H_{10}O_3Cr^+$	$C_6H_4(CH_3)_2Cr(CO)_3$ (1,3–Dimethylbenzenechromium tricarbonyl)		7.35±0.04	CTS		3403
$C_{11}H_{10}O_3Cr^+$	$C_6H_4(CH_3)_2Cr(CO)_3$ (1,4–Dimethylbenzenechromium tricarbonyl)		7.35±0.04	CTS		3403

4.3. The Positive Ion Table—Continued

Ion	Reactant	Other products	Ionization or appearance potential (eV)	Method	Heat of formation (kJ mol^{-1})	Ref.
		$C_{12}H_{12}O_3Cr^+$				
$C_{12}H_{12}O_3Cr^+$	$C_6H_3(CH_3)_3Cr(CO)_3$ (1,3,5–Trimethylbenzenechromium tricarbonyl)		7.05	EI		3005
$C_{12}H_{12}O_3Cr^+$	$C_6H_3(CH_3)_3Cr(CO)_3$ (1,3,5–Trimethylbenzenechromium tricarbonyl)		7.33±0.02	CTS		3403
		$C_{13}H_{14}O_3Cr^+$				
$C_{13}H_{14}O_3Cr^+$	$C_6H_5C_4H_9Cr(CO)_3$ (*tert*–Butylbenzenechromium tricarbonyl)		7.08	EI		3005
		$C_{15}H_{10}O_3Cr^+$				
$C_{15}H_{10}O_3Cr^+$	$(C_6H_5)_2Cr(CO)_3$ (Biphenylchromium tricarbonyl)		7.27	EI		3005
$C_{15}H_{10}O_3Cr^+$	$(C_6H_5)_2Cr(CO)_3$ (Biphenylchromium tricarbonyl)		7.35±0.06	CTS		3403
		$C_{15}H_{18}O_3Cr^+$				
$C_{15}H_{18}O_3Cr^+$	$C_6(CH_3)_6Cr(CO)_3$ (Hexamethylbenzenechromium tricarbonyl)		6.88	EI		3005
$C_{15}H_{18}O_3Cr^+$	$C_6(CH_3)_6Cr(CO)_3$ (Hexamethylbenzenechromium tricarbonyl)		7.24±0.05	CTS		3403
		$C_{21}H_{30}O_3Cr^+$				
$C_{21}H_{30}O_3Cr^+$	$C_6(C_2H_5)_6Cr(CO)_3$ (Hexaethylbenzenechromium tricarbonyl)		7.27±0.05	CTS		3403
		$C_6H_6O_4Cr^+$				
$C_6H_6O_4Cr^+$	$CH_3OC(CH_3)Cr(CO)_5$	2CO	8.86±0.1	EI		2641
		$C_{10}H_8O_4Cr^+$				
$C_{10}H_8O_4Cr^+$	$C_6H_5OCH_3Cr(CO)_3$ (Methoxybenzenechromium tricarbonyl)		7.11	EI		3005
$C_{10}H_8O_4Cr^+$	$C_6H_5OCH_3Cr(CO)_3$ (Methoxybenzenechromium tricarbonyl)		7.38±0.05	CTS		3403
		$C_{10}H_{14}O_4Cr^+$				
$C_{10}H_{14}O_4Cr^+$	$(CH_3COCHCOCH_3)_3Cr$ (Tris(2,4–pentanedionato)chromium)		11.3±0.1	EI	2460,	2519, 2959

4.3. The Positive Ion Table—Continued

Ion	Reactant	Other products	Ionization or appearance potential (eV)	Method	Heat of formation (kJ mol⁻¹)	Ref.

$C_{11}H_8O_4Cr^+$

Ion	Reactant	Other products	Ionization or appearance potential (eV)	Method	Heat of formation (kJ mol⁻¹)	Ref.
$C_{11}H_8O_4Cr^+$	$C_6H_5COCH_3Cr(CO)_3$ (Acetophenonechromium tricarbonyl)		7.44	EI		3005

$C_{11}H_{10}O_4Cr^+$

Ion	Reactant	Other products	Ionization or appearance potential (eV)	Method	Heat of formation (kJ mol⁻¹)	Ref.
$C_{11}H_{10}O_4Cr^+$	$C_7H_7OCH_3Cr(CO)_3$ (endo–7–Methoxycycloheptatrienechromium tricarbonyl)		7.03	EI		3280
$C_{11}H_{10}O_4Cr^+$	$C_7H_7OCH_3Cr(CO)_3$ (exo–7–Methoxycycloheptatrienechromium tricarbonyl)		7.16	EI		3280

$C_{12}H_{18}O_4Cr^+$

Ion	Reactant	Other products	Ionization or appearance potential (eV)	Method	Heat of formation (kJ mol⁻¹)	Ref.
$C_{12}H_{18}O_4Cr^+$	$(CH_3COC(CH_3)COCH_3)_3Cr$ (Tris(3–methyl–2,4–pentanedionato)chromium)		10.7±0.10	EI		2519

$C_7H_6O_5Cr^+$

Ion	Reactant	Other products	Ionization or appearance potential (eV)	Method	Heat of formation (kJ mol⁻¹)	Ref.
$C_7H_6O_5Cr^+$	$CH_3OC(CH_3)Cr(CO)_5$	CO	7.92±0.05	EI		2641

$C_{11}H_8O_5Cr^+$

Ion	Reactant	Other products	Ionization or appearance potential (eV)	Method	Heat of formation (kJ mol⁻¹)	Ref.
$C_{11}H_8O_5Cr^+$	$C_6H_5COOCH_3Cr(CO)_3$ (Benzoic acid methyl ester chromium tricarbonyl)		7.41	EI		3005

$C_{13}H_{12}O_5Cr^+$

Ion	Reactant	Other products	Ionization or appearance potential (eV)	Method	Heat of formation (kJ mol⁻¹)	Ref.
$C_{13}H_{12}O_5Cr^+$	$C_7H_7CH_2COOCH_3Cr(CO)_3$ (endo–7–Cycloheptatrienylacetic acid methyl ester chromium tricarbonyl)		7.12	EI		3280
$C_{13}H_{12}O_5Cr^+$	$C_7H_7CH_2COOCH_3Cr(CO)_3$ (exo–7–Cycloheptatrienylacetic acid methyl ester chromium tricarbonyl)		7.21	EI		3280

$C_{13}H_{14}O_5Cr^+$

Ion	Reactant	Other products	Ionization or appearance potential (eV)	Method	Heat of formation (kJ mol⁻¹)	Ref.
$C_{13}H_{14}O_5Cr^+$	$C_6H_2(OCH_3)_2(CH_3)_2Cr(CO)_3$ (1,4–Dimethoxy–2,5–dimethylbenzenechromium tricarbonyl)		7.27±0.02	CTS		3403

$C_{21}H_{28}O_5Cr^+$

Ion	Reactant	Other products	Ionization or appearance potential (eV)	Method	Heat of formation (kJ mol⁻¹)	Ref.
$C_{21}H_{28}O_5Cr^+$	$C_{18}H_{28}O_2Cr(CO)_3$ (15,17–Dimethyl–2,13–dioxabicyclo[12.2.2]octadeca–14,16,17–trienechromium tricarbonyl)		7.29±0.02	CTS		3403

$C_8H_6O_6Cr^+$

Ion	Reactant	Other products	Ionization or appearance potential (eV)	Method	Heat of formation (kJ mol⁻¹)	Ref.
$C_8H_6O_6Cr^+$	$CH_3OC(CH_3)Cr(CO)_5$		7.46±0.05	EI		2641

4.3. The Positive Ion Table

Ion	Reactant	Other products	Ionization or appearance potential (eV)	Method	Heat of formation (kJ mol⁻¹)	Ref.
		$C_{13}H_8O_6Cr^+$				
$C_{13}H_8O_6Cr^+$	$C_6H_5C(OCH_3)Cr(CO)_5$ (α–Methoxybenzylidencchromium pentacarbonyl)		7.26	EI		2641
		$C_{15}H_{21}O_6Cr^+$				
$C_{15}H_{21}O_6Cr^+$	$(CH_3COCHCOCH_3)_3Cr$ (Tris(2,4–pentanedionato)chromium)		7.87±0.12	EI		2580
$C_{15}H_{21}O_6Cr^+$	$(CH_3COCHCOCH_3)_3Cr$ (Tris(2,4–pentanedionato)chromium)		8.10±0.05	EI	2460,	2519, 2959
		$C_{18}H_{27}O_6Cr^+$				
$C_{18}H_{27}O_6Cr^+$	$(CH_3COC(CH_3)COCH_3)_3Cr$ (Tris(3–methyl–2,4–pentanedionato)chromium)		7.81±0.05	EI		2519
		$C_9H_7NO_3Cr^+$				
$C_9H_7NO_3Cr^+$	$C_6H_5NH_2Cr(CO)_3$ (Anilinechromium tricarbonyl)		7.05	EI		3005
		$C_{11}H_{11}NO_3Cr^+$				
$C_{11}H_{11}NO_3Cr^+$	$C_6H_5N(CH_3)_2Cr(CO)_3$ (N,N–Dimethylanilinechromium tricarbonyl)		6.92	EI		3005
$C_{11}H_{11}NO_3Cr^+$	$C_6H_5N(CH_3)_2Cr(CO)_3$ (N,N–Dimethylanilinechromium tricarbonyl)		7.28±0.02	CTS		3403
		$C_{13}H_{16}N_2O_3Cr^+$				
$C_{13}H_{16}N_2O_3Cr^+$	$C_6H_4(N(CH_3)_2)_2Cr(CO)_3$ (1,4–Bis(dimethylamino)benzenechromium tricarbonyl)		6.46	EI		3005
$C_{13}H_{16}N_2O_3Cr^+$	$C_6H_4(N(CH_3)_2)_2Cr(CO)_3$ (1,4–Bis(dimethylamino)benzenechromium tricarbonyl)		7.14±0.05	CTS		3403
		$C_7H_5NO_5Cr^+$				
$C_7H_5NO_5Cr^+$	$NH_2C(CH_3)Cr(CO)_5$		7.35	EI		2641
		$C_8H_7NO_5Cr^+$				
$C_8H_7NO_5Cr^+$	$CH_3NHC(CH_3)Cr(CO)_5$		7.30	EI		2641
		$C_9H_9NO_5Cr^+$				
$C_9H_9NO_5Cr^+$	$C_2H_5NHC(CH_3)Cr(CO)_5$		7.11	EI		2641
$C_9H_9NO_5Cr^+$	$(CH_3)_2NC(CH_3)Cr(CO)_5$		7.15	EI		2641

4.3. The Positive Ion Table—Continued

Ion	Reactant	Other products	Ionization or appearance potential (eV)	Method	Heat of formation (kJ mol^{-1})	Ref.
		$C_{10}H_{11}NO_5Cr^+$				
$C_{10}H_{11}NO_5Cr^+$	iso–$C_3H_7NHC(CH_3)Cr(CO)_5$		7.08	EI		2641
		$C_{11}H_{13}NO_5Cr^+$				
$C_{11}H_{13}NO_5Cr^+$	$tert$–$C_4H_9NHC(CH_3)Cr(CO)_5$		6.98	EI		2641
$C_{11}H_{13}NO_5Cr^+$	$(C_2H_5)_2NC(CH_3)Cr(CO)_5$		7.01	EI		2641
		$C_{12}H_{11}NO_5Cr^+$				
$C_{12}H_{11}NO_5Cr^+$	$C_6H_{11}NCCr(CO)_5$ (Isocyanocyclohexanechromium pentacarbonyl)		7.62	EI		2544
		$C_{13}H_9NO_5Cr^+$				
$C_{13}H_9NO_5Cr^+$	$C_6H_5NHC(CH_3)Cr(CO)_5$ (1–Anilinoethylidenechromium pentacarbonyl)		7.02	EI		2641
		$C_{13}H_{15}NO_5Cr^+$				
$C_{13}H_{15}NO_5Cr^+$	$C_6H_{11}NHC(CH_3)Cr(CO)_5$ (1–Cyclohexylaminoethylidenechromium pentacarbonyl)		7.04	EI		2641
		$C_{14}H_{11}NO_5Cr^+$				
$C_{14}H_{11}NO_5Cr^+$	$C_6H_5CH_2NHC(CH_3)Cr(CO)_5$ (1–Benzylaminoethylidenechromium pentacarbonyl)		7.09	EI		2641
$C_{14}H_{11}NO_5Cr^+$	$C_6H_4(CH_3)NHC(CH_3)Cr(CO)_5$ (1–(4–Methylanilino)ethylidenechromium pentacarbonyl)		6.81	EI		2641
		$C_{14}H_{11}NO_6Cr^+$				
$C_{14}H_{11}NO_6Cr^+$	$C_6H_4(OCH_3)NHC(CH_3)Cr(CO)_5$ (1–(4–Methoxyanilino)ethylidenechromium pentacarbonyl)		6.90	EI		2641
		$C_{10}H_{12}N_2O_8Cr^+$				
$C_{10}H_{12}N_2O_8Cr^+$	$(CH_3COC(NO_2)COCH_3)_3Cr$ (Tris(3–nitro–2,4–pentanedionato)chromium)		11.6±0.10	EI		2519
		$C_{15}H_{18}N_3O_{12}Cr^+$				
$C_{15}H_{18}N_3O_{12}Cr^+$	$(CH_3COC(NO_2)COCH_3)_3Cr$ (Tris(3–nitro–2,4–pentanedionato)chromium)		8.63±0.05	EI		2519

4.3. The Positive Ion Table—Continued

Ion	Reactant	Other products	Ionization or appearance potential (eV)	Method	Heat of formation (kJ mol^{-1})	Ref.

$C_9H_5O_3FCr^+$

$C_9H_5O_3FCr^+$	$C_6H_5FCr(CO)_3$ (Fluorobenzenechromium tricarbonyl)		7.47	EI		3005

$C_{10}H_8O_4F_6Cr^+$

$C_{10}H_8O_4F_6Cr^+$	$(CF_3COCHCOCH_3)_3Cr$ (Tris(1,1,1–trifluoro–2,4–pentanedionato)chromium)		11.9±0.1	EI		2519, 2959

$C_{15}H_{12}O_6F_9Cr^+$

$C_{15}H_{12}O_6F_9Cr^+$	$(CF_3COCHCOCH_3)_3Cr$ (Tris(1,1,1–trifluoro–2,4–pentanedionato)chromium)		9.09±0.05	EI		2519, 2959

$C_{10}H_2O_4F_{12}Cr^+$

$C_{10}H_2O_4F_{12}Cr^+$	$(CF_3COCHCOCF_3)_3Cr$ (Tris(1,1,1,5,5,5–hexafluoro–2,4–pentanedionato)chromium)		14.3±0.1	EI		2519, 2959

$C_{15}H_3O_6F_{18}Cr^+$

$C_{15}H_3O_6F_{18}Cr^+$	$(CF_3COCHCOCF_3)_3Cr$ (Tris(1,1,1,5,5,5–hexafluoro–2,4–pentanedionato)chromium)		9.19±0.02	PE		3168
$C_{15}H_3O_6F_{18}Cr^+$	$(CF_3COCHCOCF_3)_3Cr$ (Tris(1,1,1,5,5,5–hexafluoro–2,4–pentanedionato)chromium)		9.97±0.08	EI		2580
$C_{15}H_3O_6F_{18}Cr^+$	$(CF_3COCHCOCF_3)_3Cr$ (Tris(1,1,1,5,5,5–hexafluoro–2,4–pentanedionato)chromium)		10.13±0.05	EI		2519, 2959

See also – PE: 3169

$C_{12}H_{14}O_3SiCr^+$

$C_{12}H_{14}O_3SiCr^+$	$C_6H_5Si(CH_3)_3Cr(CO)_3$ (Trimethylsilylbenzenechromium tricarbonyl)		7.15	EI		3005

$C_{11}H_{15}O_5PCr^+$

$C_{11}H_{15}O_5PCr^+$	$(C_2H_5)_3PCr(CO)_5$		7.63±0.05	EI		2481

$C_{17}H_{27}O_5PCr^+$

$C_{17}H_{27}O_5PCr^+$	$(n-C_4H_9)_3PCr(CO)_5$		7.37±0.05	EI		2481

$C_8H_9O_8PCr^+$

$C_8H_9O_8PCr^+$	$(CH_3O)_3PCr(CO)_5$		7.80	EI		2544

4.3. The Positive Ion Table—Continued

Ion	Reactant	Other products	Ionization or appearance potential (eV)	Method	Heat of forma- tion (kJ mol^{-1})	Ref.
		C$_{11}$H$_{15}$O$_8$PCr$^+$				
C$_{11}$H$_{15}$O$_8$PCr$^+$	(C$_2$H$_5$O)$_3$PCr(CO)$_5$		7.62	EI		2544
		C$_{17}$H$_{27}$O$_8$PCr$^+$				
C$_{17}$H$_{27}$O$_8$PCr$^+$	(n–C$_4$H$_9$O)$_3$PCr(CO)$_5$		7.63±0.05	EI		2481
		C$_{10}$H$_{18}$O$_{10}$P$_2$Cr$^+$				
C$_{10}$H$_{18}$O$_{10}$P$_2$Cr$^+$	((CH$_3$O)$_3$P)$_2$Cr(CO)$_4$		7.30	EI		2716
		C$_{16}$H$_{30}$O$_{10}$P$_2$Cr$^+$				
C$_{16}$H$_{30}$O$_{10}$P$_2$Cr$^+$	((C$_2$H$_5$O)$_3$P)$_2$Cr(CO)$_4$		7.23	EI		2716
		C$_{13}$H$_8$O$_5$SCr$^+$				
C$_{13}$H$_8$O$_5$SCr$^+$	C$_6$H$_5$SC(CH$_3$)Cr(CO)$_5$ (1–Phenylthioethylidenechromium pentacarbonyl)		7.83	EI		2641
		C$_9$H$_5$O$_3$ClCr$^+$				
C$_9$H$_5$O$_3$ClCr$^+$	C$_6$H$_5$ClCr(CO)$_3$ (Chlorobenzenechromium tricarbonyl)		7.37	EI		3005
		C$_{10}$H$_{12}$O$_4$Cl$_2$Cr$^+$				
C$_{10}$H$_{12}$O$_4$Cl$_2$Cr$^+$	(CH$_3$COCClCOCH$_3$)$_3$Cr (Tris(3–chloro–2,4–pentanedionato)chromium)		11.1±0.10	EI		2519
		C$_{15}$H$_{18}$O$_6$Cl$_3$Cr$^+$				
C$_{15}$H$_{18}$O$_6$Cl$_3$Cr$^+$	(CH$_3$COCClCOCH$_3$)$_3$Cr (Tris(3–chloro–2,4–pentanedionato)chromium)		8.16±0.05	EI		2519
		C$_{14}$H$_8$NO$_5$F$_3$Cr$^+$				
C$_{14}$H$_8$NO$_5$F$_3$Cr$^+$	C$_6$H$_4$(CF$_3$)NHC(CH$_3$)Cr(CO)$_5$ (1–(4–Trifluoromethylanilino)ethylidenechromium pentacarbonyl)		7.85	EI		2641
		C$_{13}$H$_8$NO$_5$ClCr$^+$				
C$_{13}$H$_8$NO$_5$ClCr$^+$	C$_6$H$_4$ClNHC(CH$_3$)Cr(CO)$_5$ (1–(4–Chloroanilino)ethylidenechromium pentacarbonyl)		7.40	EI		2641

4.3. The Positive Ion Table—Continued

Ion	Reactant	Other products	Ionization or appearance potential (eV)	Method	Heat of formation (kJ mol^{-1})	Ref.
		Mn^+	$\Delta H_{f0}^{\circ} = 996.7$ kJ mol^{-1} (238.2 kcal mol^{-1})			
Mn^+	Mn		7.435	S	996.7	2113
See also − EI:	2161, 2523, 2683, 2795, 3023					
Mn^+	MnF_2		13.0±0.5	EI		2859
Mn^+	MnF_3		19.5±0.5	EI		2591
Mn^+	$(C_5H_5)_2Mn$ (Bis(cyclopentadienyl)manganese)		13.6±0.3	EI		2683
Mn^+	$C_5H_5MnC_6H_6$ (Benzenecyclopentadienylmanganese)		14.1±0.3	EI		2530
Mn^+	$C_5H_5MnC_6H_6$ (Benzenecyclopentadienylmanganese)		17.9	EI		2417
Mn^+	$Mn_2(CO)_{10}$		20.8±0.4	EI		2739
Mn^+	$Mn_2(CO)_{10}$		22.13	EI		2563
Mn^+	$C_5H_5Mn(CO)_3$ (Cyclopentadienylmanganese tricarbonyl)		14.4±0.5	EI		2531
See also − EI:	1381					
Mn^+	$Mn(CO)_5Cl$		~14.8	EI		2501
Mn^+	$Mn(CO)_5Br$		16.5	EI		2501
Mn^+	$Mn(CO)_5I$		16.2	EI		2501
Mn^+	$ReMn(CO)_{10}$		25.67	EI		2563
		Mn_2^+				
Mn_2^+	$Mn_2(CO)_{10}$	10CO	18.73	EI		2563
Mn_2^+	$Mn_2(CO)_{10}$	10CO	18.8±0.3	EI		2739
		MnF^+				
MnF^+	MnF		8.1±0.5	EI		3023
MnF^+	MnF		8.7±0.3	EI		2161
MnF^+	MnF		9.5±0.5	EI		2523
MnF^+	MnF_2	F	13.5±0.5	EI		2859
MnF^+	MnF_2	F	14.5±0.5	EI		3023
MnF^+	MnF_2	F	14.5	EI		2161
MnF^+	MnF_3		14.2±0.3	EI		2591
		MnF_2^+				
MnF_2^+	MnF_2		11.5±0.3	EI		2161
MnF_2^+	MnF_2		11.7±0.5	EI		3023
MnF_2^+	MnF_2		12.5±0.5	EI		2859
MnF_2^+	MnF_2		13.4±0.5	EI		2523
MnF_2^+	MnF_3	F	13.7±0.3	EI		2591
		MnF_3^+				
MnF_3^+	MnF_3		12±0.8	EI		2591

4.3. The Positive Ion Table—Continued

Ion	Reactant	Other products	Ionization or appearance potential (eV)	Method	Heat of formation (kJ mol^{-1})	Ref.
			MnCl$^+$			
MnCl$^+$	Mn(CO)$_5$Cl	5CO	13.8	EI		2501
			C$_5$H$_5$Mn$^+$			
C$_5$H$_5$Mn$^+$	(C$_5$H$_5$)$_2$Mn (Bis(cyclopentadienyl)manganese)		11.09±0.1	EI		2683
C$_5$H$_5$Mn$^+$	C$_5$H$_5$MnC$_6$H$_6$ (Benzenecyclopentadienylmanganese)		9.4±0.2	EI		2530
C$_5$H$_5$Mn$^+$	C$_5$H$_5$MnC$_6$H$_6$ (Benzenecyclopentadienylmanganese)		12.3	EI		2417
C$_5$H$_5$Mn$^+$	C$_5$H$_5$Mn(CO)$_3$ (Cyclopentadienylmanganese tricarbonyl)	3CO	11.0±0.3	EI		2531

See also – EI: 1381

Ion	Reactant	Other products	Ionization or appearance potential (eV)	Method	Heat of formation	Ref.
			C$_6$H$_7$Mn$^+$			
C$_6$H$_7$Mn$^+$	C$_5$H$_5$MnC$_6$H$_6$ (Benzenecyclopentadienylmanganese)		12.1	EI		2417
			C$_9$H$_{11}$Mn$^+$			
C$_9$H$_{11}$Mn$^+$	C$_5$H$_5$Mn(CO)C$_4$H$_6$ (1,3–Butadienecyclopentadienylmanganese carbonyl)	CO	7.41±0.1	EI		2531
			C$_{10}$H$_{10}$Mn$^+$			
C$_{10}$H$_{10}$Mn$^+$	(C$_5$H$_5$)$_2$Mn (Bis(cyclopentadienyl)manganese)		7.32±0.1	EI		2683
			C$_{10}$H$_{13}$Mn$^+$			
C$_{10}$H$_{13}$Mn$^+$	C$_5$H$_5$Mn(CO)$_2$C$_5$H$_8$ (Cyclopentadienyl(cyclopentene)manganese dicarbonyl)	2CO	8.11±0.1	EI		2531
			C$_{11}$H$_{11}$Mn$^+$			
C$_{11}$H$_{11}$Mn$^+$	C$_5$H$_5$MnC$_6$H$_6$ (Benzenecyclopentadienylmanganese)		6.92±0.1	EI		2530
C$_{11}$H$_{11}$Mn$^+$	C$_5$H$_5$MnC$_6$H$_6$ (Benzenecyclopentadienylmanganese)		7.1	EI		2417
			C$_{12}$H$_{13}$Mn$^+$			
C$_{12}$H$_{13}$Mn$^+$	C$_5$H$_5$Mn(CO)$_2$C$_7$H$_8$ (Bicyclo[2.2.1]hepta–2,5–dienecyclopentadienylmanganese dicarbonyl)	2CO	7.93±0.1	EI		2531

4.3. The Positive Ion Table—Continued

Ion	Reactant	Other products	Ionization or appearance potential (eV)	Method	Heat of formation (kJ mol^{-1})	Ref.

$C_{12}H_{15}Mn^+$

$C_{12}H_{15}Mn^+$	$C_5H_5Mn(CO)_2C_7H_{10}$ (Bicyclo[2.2.1]hept–2–enecyclopentadienylmanganese dicarbonyl)	2CO	7.80±0.1	EI		2531

$C_{12}H_{17}Mn^+$

$C_{12}H_{17}Mn^+$	$C_5H_5Mn(CO)_2C_7H_{12}$ (Cycloheptenecyclopentadienylmanganese dicarbonyl)	2CO	7.98±0.1	EI		2531

$C_{13}H_{19}Mn^+$

$C_{13}H_{19}Mn^+$	$C_5H_5Mn(CO)_2C_8H_{14}$ (Cyclooctenecyclopentadienylmanganese dicarbonyl)	2CO	7.88±0.1	EI		2531

$MnCO^+$

$MnCO^+$	$Mn_2(CO)_{10}$		17.5±0.2	EI		2739
$MnCO^+$	$Mn_2(CO)_{10}$		18.21	EI		2563
$MnCO^+$	$Mn(CO)_5Cl$		~12.1	EI		2501
$MnCO^+$	$Mn(CO)_5Br$		15.13	EI		2501
$MnCO^+$	$Mn(CO)_5I$		14.4	EI		2501
$MnCO^+$	$ReMn(CO)_{10}$		19.05	EI		2563

$MnC_2O_2^+$

$MnC_2O_2^+$	$Mn_2(CO)_{10}$		14.80	EI		2563

$MnC_3O_3^+$

$MnC_3O_3^+$	$Mn(CO)_5Cl$		~9.7	EI		2501
$MnC_3O_3^+$	$Mn(CO)_5Br$		9.9	EI		2501
$MnC_3O_3^+$	$Mn(CO)_5I$		12.13	EI		2501

$MnC_5O_5^+$

$MnC_5O_5^+$	$Mn(CO)_5$		8.2	EI		3251
$MnC_5O_5^+$	$Mn(CO)_5$		8.44±0.10	EI	2404,	2870
$MnC_5O_5^+$	$Mn_2(CO)_{10}$		9.26±0.1	EI	2404,	2870
$MnC_5O_5^+$	$Mn_2(CO)_{10}$		9.40	EI	2563,	3234, 3251
$MnC_5O_5^+$	$Mn(CO)_5Cl$	Cl	11.4	EI		3251
$MnC_5O_5^+$	$ReMn(CO)_{10}$		10.50	EI		3234

$Mn_2C_2O_2^+$

$Mn_2C_2O_2^+$	$Mn_2(CO)_{10}$	8CO	16.43	EI		2563

4.3. The Positive Ion Table—Continued

Ion	Reactant	Other products	Ionization or appearance potential (eV)	Method	Heat of formation (kJ mol^{-1})	Ref.
		Mn$_2$C$_3$O$_3^+$				
Mn$_2$C$_3$O$_3^+$	Mn$_2$(CO)$_{10}$	7CO	15.34	EI		2563
		Mn$_2$C$_4$O$_4^+$				
Mn$_2$C$_4$O$_4^+$	Mn$_2$(CO)$_{10}$	6CO	13.98	EI		2563
Mn$_2$C$_4$O$_4^+$	Mn$_2$(CO)$_{10}$	6CO	14.0±0.2	EI		2739
		Mn$_2$C$_5$O$_5^+$				
Mn$_2$C$_5$O$_5^+$	Mn$_2$(CO)$_{10}$	5CO	11.91	EI		2563
Mn$_2$C$_5$O$_5^+$	Mn$_2$(CO)$_{10}$	5CO	12.6±0.2	EI		2739
		Mn$_2$C$_{10}$O$_{10}^+$				
Mn$_2$C$_{10}$O$_{10}^+$	Mn$_2$(CO)$_{10}$		8.02 (V)	PE		2879
Mn$_2$C$_{10}$O$_{10}^+$	Mn$_2$(CO)$_{10}$		8.42±0.1	EI		2870
Mn$_2$C$_{10}$O$_{10}^+$	Mn$_2$(CO)$_{10}$		8.46±0.03	EI		3234
Mn$_2$C$_{10}$O$_{10}^+$	Mn$_2$(CO)$_{10}$		8.55±0.10	EI		2739

See also – EI: 2563

		C$_7$H$_{12}$NMn$^+$				
C$_7$H$_{12}$NMn$^+$	C$_5$H$_5$Mn(CO)$_2$(CH$_3$)$_2$NH (Cyclopentadienyl(dimethylamine)manganese dicarbonyl)	2CO	7.42±0.1	EI		2531
		C$_{12}$H$_{16}$NMn$^+$				
C$_{12}$H$_{16}$NMn$^+$	C$_5$H$_5$Mn(CO)$_2$C$_6$H$_{11}$NC (Cyclopentadienyl(isocyanocyclohexane)manganese dicarbonyl)	2CO	7.82±0.1	EI		2531
		C$_6$H$_5$OMn$^+$				
C$_6$H$_5$OMn$^+$	C$_5$H$_5$Mn(CO)$_3$ (Cyclopentadienylmanganese tricarbonyl)	2CO	9.46±0.2	EI		2531

See also – EI: 1381

		C$_{10}$H$_{11}$OMn$^+$				
C$_{10}$H$_{11}$OMn$^+$	C$_5$H$_5$Mn(CO)C$_4$H$_6$ (1,3–Butadienecyclopentadienylmanganese carbonyl)		6.60±0.1	EI		2531
		C$_5$H$_7$O$_2$Mn$^+$				
C$_5$H$_7$O$_2$Mn$^+$	(CH$_3$COCHCOCH$_3$)$_2$Mn (Bis(2,4–pentanedionato)manganese)		13.7±0.1	EI		2731

4.3. The Positive Ion Table—Continued

Ion	Reactant	Other products	Ionization or appearance potential (eV)	Method	Heat of formation (kJ mol^{-1})	Ref.
			C$_7$H$_5$O$_2$Mn$^+$			
C$_7$H$_5$O$_2$Mn$^+$	C$_5$H$_5$Mn(CO)$_2$C$_5$H$_8$ (Cyclopentadienyl(cyclopentene)manganese dicarbonyl)		8.11±0.1	EI		2531
C$_7$H$_5$O$_2$Mn$^+$	C$_5$H$_5$Mn(CO)$_3$ (Cyclopentadienylmanganese tricarbonyl)	CO	8.77±0.1	EI		2531
			C$_{12}$H$_{13}$O$_2$Mn$^+$			
C$_{12}$H$_{13}$O$_2$Mn$^+$	C$_5$H$_5$Mn(CO)$_2$C$_5$H$_8$ (Cyclopentadienyl(cyclopentene)manganese dicarbonyl)		7.29±0.1	EI		2531
			C$_{14}$H$_{13}$O$_2$Mn$^+$			
C$_{14}$H$_{13}$O$_2$Mn$^+$	C$_5$H$_5$Mn(CO)$_2$C$_7$H$_8$ (Bicyclo[2.2.1]hepta−2,5−dienecyclopentadienylmanganese dicarbonyl)		7.27±0.1	EI		2531
			C$_{14}$H$_{15}$O$_2$Mn$^+$			
C$_{14}$H$_{15}$O$_2$Mn$^+$	C$_5$H$_5$Mn(CO)$_2$C$_7$H$_{10}$ (Bicyclo[2.2.1]hept−2−enecyclopentadienylmanganese dicarbonyl)		7.19±0.1	EI		2531
			C$_{14}$H$_{17}$O$_2$Mn$^+$			
C$_{14}$H$_{17}$O$_2$Mn$^+$	C$_5$H$_5$Mn(CO)$_2$C$_7$H$_{12}$ (Cycloheptenecyclopentadienylmanganese dicarbonyl)		7.12±0.1	EI		2531
			C$_{15}$H$_{19}$O$_2$Mn$^+$			
C$_{15}$H$_{19}$O$_2$Mn$^+$	C$_5$H$_5$Mn(CO)$_2$C$_8$H$_{14}$ (Cyclooctenecyclopentadienylmanganese dicarbonyl)		7.00±0.1	EI		2531
			C$_8$H$_5$O$_3$Mn$^+$			
C$_8$H$_5$O$_3$Mn$^+$	C$_5$H$_5$Mn(CO)$_3$ (Cyclopentadienylmanganese tricarbonyl)		8.12±0.1	EI		2531
C$_8$H$_5$O$_3$Mn$^+$	C$_5$H$_5$Mn(CO)$_3$ (Cyclopentadienylmanganese tricarbonyl)		8.3±0.4	EI		1381
			C$_9$H$_7$O$_3$Mn$^+$			
C$_9$H$_7$O$_3$Mn$^+$	C$_5$H$_5$Mn(CO)$_2$C$_4$H$_2$O$_3$ (Cyclopentadienyl(maleic acid anhydride)manganese dicarbonyl)	2CO	8.73±0.1	EI		2531
			C$_9$H$_{11}$O$_4$Mn$^+$			
C$_9$H$_{11}$O$_4$Mn$^+$	(CH$_3$COCHCOCH$_3$)$_2$Mn (Bis(2,4−pentanedionato)manganese)	CH$_3$	11.7±0.1	EI		2731

4.3. The Positive Ion Table—Continued

Ion	Reactant	Other products	Ionization or appearance potential (eV)	Method	Heat of formation (kJ mol^{-1})	Ref.
$C_{10}H_{14}O_4Mn^+$						
$C_{10}H_{14}O_4Mn^+$	$(CH_3COCHCOCH_3)_2Mn$ (Bis(2,4–pentanedionato)manganese)		8.34±0.05	EI		2731
$C_{10}H_{14}O_4Mn^+$	$(CH_3COCHCOCH_3)_3Mn$ (Tris(2,4–pentanedionato)manganese)		⩾8.7±0.1	EI		2460
$C_5HO_5Mn^+$						
$C_5HO_5Mn^+$	$HMn(CO)_5$		9.00 (V)	PE		2879
$C_6H_3O_5Mn^+$						
$C_6H_3O_5Mn^+$	$CH_3Mn(CO)_5$		8.46 (V)	PE		2879
$C_{11}H_5O_5Mn^+$						
$C_{11}H_5O_5Mn^+$	$C_6H_5Mn(CO)_5$ (Phenylmanganese pentacarbonyl)		8.22±0.05	EI		3250
$C_{11}H_7O_5Mn^+$						
$C_{11}H_7O_5Mn^+$	$C_5H_5Mn(CO)_2C_4H_2O_3$ (Cyclopentadienyl(maleic acid anhydride)manganese dicarbonyl)		8.04±0.1	EI		2531
$C_{15}H_{21}O_6Mn^+$						
$C_{15}H_{21}O_6Mn^+$	$(CH_3COCHCOCH_3)_3Mn$ (Tris(2,4–pentanedionato)manganese)		7.85±0.05	EI		2460
$C_{15}H_{21}O_6Mn^+$	$(CH_3COCHCOCH_3)_3Mn$ (Tris(2,4–pentanedionato)manganese)		7.95±0.10	EI		2580
$C_6O_5F_3Mn^+$						
$C_6O_5F_3Mn^+$	$CF_3Mn(CO)_5$		9.20 (V)	PE		2879
$C_7O_6F_3Mn^+$						
$C_7O_6F_3Mn^+$	$CF_3COMn(CO)_5$		9.0 (V)	PE		2879
$C_{14}H_{26}PMn^+$						
$C_{14}H_{26}PMn^+$	$C_5H_5Mn(CO)_2(C_3H_7)_3P$ (Cyclopentadienyl(triisopropylphosphine)manganese dicarbonyl)	2CO	7.80±0.1	EI		2531
$COClMn^+$						
$COClMn^+$	$Mn(CO)_5Cl$	4CO	11.3	EI		2501

4.3. The Positive Ion Table—Continued

Ion	Reactant	Other products	Ionization or appearance potential (eV)	Method	Heat of formation (kJ mol^{-1})	Ref.
$C_2O_2ClMn^+$						
$C_2O_2ClMn^+$	$Mn(CO)_5Cl$	$3CO$	10.6	EI		2501
$C_5O_5ClMn^+$						
$C_5O_5ClMn^+$	$Mn(CO)_5Cl$		8.83 (V)	PE		2879
$C_5O_5ClMn^+$	$Mn(CO)_5Cl$		9.12±0.08	EI		2501
$C_9H_{12}NO_2Mn^+$						
$C_9H_{12}NO_2Mn^+$	$C_5H_5Mn(CO)_2(CH_3)_2NH$ (Cyclopentadienyl(dimethylamine)manganese dicarbonyl)		6.55±0.1	EI		2531
$C_{14}H_{16}NO_2Mn^+$						
$C_{14}H_{16}NO_2Mn^+$	$C_5H_5Mn(CO)_2C_6H_{11}NC$ (Cyclopentadienyl(isocyanocyclohexane)manganese dicarbonyl)		7.01±0.1	EI		2531
$C_7H_8O_2PMn^+$						
$C_7H_8O_2PMn^+$	$C_5H_5Mn(CO)_2PH_3$ (Cyclopentadienyl(phosphine)manganese dicarbonyl)		7.28±0.05	EI		2597
$C_{16}H_{26}O_2PMn^+$						
$C_{16}H_{26}O_2PMn^+$	$C_5H_5Mn(CO)_2(C_3H_7)_3P$ (Cyclopentadienyl(triisopropylphosphine)manganese dicarbonyl)		6.55±0.1	EI		2531
$C_{16}H_{26}O_2PMn^+$	$C_5H_5Mn(CO)_2(C_3H_7)_3P$ (Cyclopentadienyl(triisopropylphosphine)manganese dicarbonyl)		6.90±0.05	EI		2597
$C_{25}H_{20}O_2PMn^+$						
$C_{25}H_{20}O_2PMn^+$	$C_5H_5Mn(CO)_2(C_6H_5)_3P$ (Cyclopentadienyl(triphenylphosphine)manganese dicarbonyl)		6.93±0.05	EI		2597
$C_{16}H_{26}O_5PMn^+$						
$C_{16}H_{26}O_5PMn^+$	$C_5H_5Mn(CO)_2(C_3H_7O)_3P$ (Cyclopentadienyl(triisopropoxyphosphine)manganese dicarbonyl)		7.17±0.05	EI		2597
$C_{25}H_{20}O_5PMn^+$						
$C_{25}H_{20}O_5PMn^+$	$C_5H_5Mn(CO)_2(C_6H_5O)_3P$ (Cyclopentadienyl(triphenoxyphosphine)manganese dicarbonyl)		7.40±0.05	EI		2597

4.3. The Positive Ion Table—Continued

Ion	Reactant	Other products	Ionization or appearance potential (eV)	Method	Heat of formation (kJ mol⁻¹)	Ref.

$C_7H_{11}OSMn^+$

| $C_7H_{11}OSMn^+$ | $C_5H_5Mn(CO)_2(CH_3)_2SO$ (Cyclopentadienyl(dimethylsulfoxide)manganese dicarbonyl) | 2CO | 7.68±0.1 | EI | | 2531 |

$C_9H_{11}O_3SMn^+$

| $C_9H_{11}O_3SMn^+$ | $C_5H_5Mn(CO)_2(CH_3)_2SO$ (Cyclopentadienyl(dimethylsulfoxide)manganese dicarbonyl) | | 7.12±0.1 | EI | | 2531 |

$C_7H_5O_2PCl_3Mn^+$

| $C_7H_5O_2PCl_3Mn^+$ | $C_5H_5Mn(CO)_2PCl_3$ (Cyclopentadienyl(trichlorophosphine)manganese dicarbonyl) | | 8.12±0.05 | EI | | 2597 |

Fe^+ $\quad \Delta H_{f0}^\circ = 1173.3$ kJ mol⁻¹ (280.4 kcal mol⁻¹)

Ion	Reactant	Other products	Ionization or appearance potential (eV)	Method	Heat of formation (kJ mol⁻¹)	Ref.
Fe^+	Fe		7.870	S	1173.3	2113
Fe^+	FeF_2		16.5±0.3	EI		1280
Fe^+	FeF_3		20.5±0.3	EI		2591
Fe^+	$FeCl_2$		16.5±0.5	EI		397
Fe^+	$(C_5H_5)_2Fe$ (Bis(cyclopentadienyl)iron)		14.4±0.5	EI		2683
Fe^+	$Fe(CO)_5$	5CO	14.23±0.1	PI		2956
(Threshold value approximately corrected for hot bands)						
Fe^+	$Fe(CO)_5$	5CO	14.7±0.1	EI		2023
Fe^+	$Fe(CO)_5$	5CO	15.31±0.1	EI		2403
Fe^+	$Fe(CO)_5$	5CO	15.99	EI		2500
Fe^+	$Fe(CO)_5$	5CO	16.1±0.2	EI		112
Fe^+	$FeBr_2$		16.6±0.5	EI		174

Fe_2^+

| Fe_2^+ | Fe_2 | | 5.90±0.2 | EI | | 2917 |

FeC^+

| FeC^+ | $Fe(CO)_5$ | | 23.6±0.3 | EI | | 2403 |

FeO^+

| FeO^+ | FeO | | 8±1 | EI | | 3366 |

FeF^+

| FeF^+ | FeF_2 | F | 12.6±0.3 | EI | | 1280 |
| FeF^+ | FeF_3 | | 16.0±0.3 | EI | | 2591 |

4.3. The Positive Ion Table—Continued

Ion	Reactant	Other products	Ionization or appearance potential (eV)	Method	Heat of formation (kJ mol⁻¹)	Ref.
FeF₂⁺						
FeF₂⁺	FeF₂		11.3±0.3	EI		1280
FeF₂⁺	FeF₃	F	13.7±0.3	EI		2591
FeF₃⁺						
FeF₃⁺	FeF₃		12.5±0.3	EI		2591
Fe₂F₅⁺						
Fe₂F₅⁺	Fe₂F₆	F	12.1±0.5	EI		2591
FeCl⁺						
FeCl⁺	FeCl₂	Cl	12.8±0.5	EI		397
FeCl₂⁺						
FeCl₂⁺	FeCl₂		11.5±0.5	EI		397
Fe₂Cl₃⁺						
Fe₂Cl₃⁺	Fe₂Cl₄	Cl	12.0±1.0	EI		397
Fe₂Cl₄⁺						
Fe₂Cl₄⁺	Fe₂Cl₄		10.5±1.0	EI		397
C₃H₃Fe⁺						
C₃H₃Fe⁺	(C₅H₅)₂Fe (Bis(cyclopentadienyl)iron)		18.9±0.1	EI		2683
C₅H₅Fe⁺						
C₅H₅Fe⁺	(C₅H₅)₂Fe (Bis(cyclopentadienyl)iron)		12.8±1	EI		2545
C₅H₅Fe⁺	(C₅H₅)₂Fe (Bis(cyclopentadienyl)iron)		13.78±0.1	EI		2683
C₅H₅Fe⁺	C₅H₅FeC₄H₄N (Cyclopentadienyl(pyrrolyl)iron)		12.6±0.2	EI		3252
C₈H₈Fe⁺						
C₈H₈Fe⁺	(C₅H₅)₂Fe (Bis(cyclopentadienyl)iron)		13.27±0.1	EI		2683

4.3. The Positive Ion Table—Continued

Ion	Reactant	Other products	Ionization or appearance potential (eV)	Method	Heat of formation (kJ mol^{-1})	Ref.
			$C_{10}H_{10}Fe^+$			
$C_{10}H_{10}Fe^+$	$(C_5H_5)_2Fe$ (Bis(cyclopentadienyl)iron)		6.99	EI		2453
$C_{10}H_{10}Fe^+$	$(C_5H_5)_2Fe$ (Bis(cyclopentadienyl)iron)		7.15±0.1	EI		2683
$C_{10}H_{10}Fe^+$	$(C_5H_5)_2Fe$ (Bis(cyclopentadienyl)iron)		6.97	CTS		3403
(Average of two values)						
			$C_{11}H_{12}Fe^+$			
$C_{11}H_{12}Fe^+$	$C_5H_5FeC_5H_4CH_3$ (Cyclopentadienyl(methylcyclopentadienyl)iron)		6.76	CTS		3403
			$C_{12}H_{14}Fe^+$			
$C_{12}H_{14}Fe^+$	$(C_5H_4CH_3)_2Fe$ (Bis(methylcyclopentadienyl)iron)		6.65	CTS		3403
			$C_{16}H_{22}Fe^+$			
$C_{16}H_{22}Fe^+$	$(C_5H_4C_3H_7)_2Fe$ (Bis(propylcyclopentadienyl)iron)		6.85	CTS		3403
			$FeCO^+$			
$FeCO^+$	$Fe(CO)_5$	4CO	11.53±0.1	PI		2956
$FeCO^+$	$Fe(CO)_5$	4CO	12.9±0.1	EI		2023
$FeCO^+$	$Fe(CO)_5$	4CO	13.39±0.07	EI		2403
$FeCO^+$	$Fe(CO)_5$	4CO	13.76	EI		2500
$FeCO^+$	$Fe(CO)_5$	4CO	14.0±0.2	EI		112
			$FeCO^{+2}$			
$FeCO^{+2}$	$Fe(CO)_5$		30.2±2	EI		112
			$FeC_2O_2^+$			
$FeC_2O_2^+$	$Fe(CO)_5$	3CO	10.68±0.1	PI		2956
$FeC_2O_2^+$	$Fe(CO)_5$	3CO	10.92±0.04	EI		2023
$FeC_2O_2^+$	$Fe(CO)_5$	3CO	11.12	EI		2500
$FeC_2O_2^+$	$Fe(CO)_5$	3CO	11.27±0.05	EI		2403
$FeC_2O_2^+$	$Fe(CO)_5$	3CO	11.8±0.2	EI		112

4.3. The Positive Ion Table—Continued

Ion	Reactant	Other products	Ionization or appearance potential (eV)	Method	Heat of formation (kJ mol^{-1})	Ref.
			$FeC_3O_3^+$			
$FeC_3O_3^+$	$Fe(CO)_5$	2CO	9.87±0.1	PI		2956
$FeC_3O_3^+$	$Fe(CO)_5$	2CO	9.89±0.05	EI		2023
$FeC_3O_3^+$	$Fe(CO)_5$	2CO	10.01±0.04	EI		2403
$FeC_3O_3^+$	$Fe(CO)_5$	2CO	10.04	EI		2500
$FeC_3O_3^+$	$Fe(CO)_5$	2CO	10.3±0.3	EI		112
			$FeC_4O_4^+$			
$FeC_4O_4^+$	$Fe(CO)_4$		8.48	EI		2546
$FeC_4O_4^+$	$Fe(CO)_5$	CO	8.77±0.1	PI		2956
$FeC_4O_4^+$	$Fe(CO)_5$	CO	9.10	EI		2546

See also – PI: 2886
        EI: 112, 2023, 2403, 2500

Ion	Reactant	Other products	Ionization or appearance potential (eV)	Method	Heat of formation (kJ mol^{-1})	Ref.
		$Fe(CO)_5^+$	$\Delta H_{f298}^\circ = 36$ kJ mol^{-1} (9 kcal mol^{-1})			
$FeC_5O_5^+$	$Fe(CO)_5$		7.95±0.03	PI		1167
$FeC_5O_5^+$	$Fe(CO)_5$		7.96±0.02	PI		2886
$FeC_5O_5^+$	$Fe(CO)_5$		7.98±0.01	PI	36	2956
$FeC_5O_5^+$	$Fe(CO)_5$		8.00±0.08	PE		2886

See also – EI: 112, 2023, 2403, 2500

Ion	Reactant	Other products	Ionization or appearance potential (eV)	Method	Heat of formation (kJ mol^{-1})	Ref.
			$Fe_2C_9O_9^+$			
$Fe_2C_9O_9^+$	$Fe_2(CO)_9$		7.91±0.01	EI		3250
			$C_9H_9NFe^+$			
$C_9H_9NFe^+$	$C_5H_5FeC_4H_4N$ (Cyclopentadienyl(pyrrolyl)iron)		7.17±0.1	EI		3252
			$C_5H_7O_2Fe^+$			
$C_5H_7O_2Fe^+$	$(CH_3COCHCOCH_3)_2Fe$ (Bis(2,4–pentanedionato)iron)		13.9±0.1	EI		2731, 2959
			$C_7H_4O_3Fe^+$			
$C_7H_4O_3Fe^+$	$C_4H_4Fe(CO)_3$ (Cyclobutadieneiron tricarbonyl)		8.04	PE		2843, 3056

4.3. The Positive Ion Table—Continued

Ion	Reactant	Other products	Ionization or appearance potential (eV)	Method	Heat of formation (kJ mol^{-1})	Ref.
		$C_7H_6O_3Fe^+$				
$C_7H_6O_3Fe^+$	cis–CH_2=$CHCH$=$CH_2Fe(CO)_3$		8.04	PE		2843, 3056
$C_7H_6O_3Fe^+$	$(CH_2)_3CFe(CO)_3$		8.32±0.02	PE		2845
		$C_8H_8O_3Fe^+$				
$C_8H_8O_3Fe^+$	CH_2=$CHCH$=$CHCH_3Fe(CO)_3$		7.84	PE		2843
		$C_9H_8O_3Fe^+$				
$C_9H_8O_3Fe^+$	$C_6H_8Fe(CO)_3$ (1,3–Cyclohexadieneiron tricarbonyl)		8.0±0.2	EI		2963
		$C_8H_4O_4Fe^+$				
$C_8H_4O_4Fe^+$	$C_4H_3CHOFe(CO)_3$ (Cyclobutadienecarboxaldehydeiron tricarbonyl)		8.32	PE		2843
		$C_9H_6O_4Fe^+$				
$C_9H_6O_4Fe^+$	$C_4H_3COCH_3Fe(CO)_3$ (Cyclobutadienyl methyl ketone iron tricarbonyl)		8.27	PE		2843
		$C_9H_{11}O_4Fe^+$				
$C_9H_{11}O_4Fe^+$	$(CH_3COCHCOCH_3)_2Fe$ (Bis(2,4–pentanedionato)iron)		11.7±0.1	EI		2731, 2959
		$C_{10}H_{14}O_4Fe^+$				
$C_{10}H_{14}O_4Fe^+$	$(CH_3COCHCOCH_3)_2Fe$ (Bis(2,4–pentanedionato)iron)		7.50±0.04	EI		2992
$C_{10}H_{14}O_4Fe^+$	$(CH_3COCHCOCH_3)_2Fe$ (Bis(2,4–pentanedionato)iron)		8.10±0.05	EI	2460	2731, 2959
$C_{10}H_{14}O_4Fe^+$	$(CH_3COCHCOCH_3)_3Fe$ (Tris(2,4–pentanedionato)iron)		≥9.4±0.1	EI	2460	2959
		$C_{15}H_{21}O_6Fe^+$				
$C_{15}H_{21}O_6Fe^+$	$(CH_3COCHCOCH_3)_3Fe$ (Tris(2,4–pentanedionato)iron)		8.45±0.05	EI	2460	2959
$C_{15}H_{21}O_6Fe^+$	$(CH_3COCHCOCH_3)_3Fe$ (Tris(2,4–pentanedionato)iron)		8.64±0.11	EI		2580

4.3. The Positive Ion Table—Continued

Ion	Reactant	Other products	Ionization or appearance potential (eV)	Method	Heat of formation (kJ mol⁻¹)	Ref.

$$CN_2O_3Fe^+$$

$CN_2O_3Fe^+$?	$Fe(CO)_2(NO)_2$		9.46±0.09	PI		2886

Because mass analysis was not employed, identification of this ion is uncertain.

$$C_2N_2O_4Fe^+$$

$C_2N_2O_4Fe^+$	$Fe(CO)_2(NO)_2$		8.25±0.12	PI		2886
$C_2N_2O_4Fe^+$	$Fe(CO)_2(NO)_2$		8.45±0.1	EI		2453

$$C_7H_5NO_3Fe^+$$

$C_7H_5NO_3Fe^+$	$C_4H_3NH_2Fe(CO)_3$ (Aminocyclobutadieneiron tricarbonyl)		7.77	PE		2843

$$C_5H_4O_2F_3Fe^+$$

$C_5H_4O_2F_3Fe^+$	$(CF_3COCHCOCH_3)_2Fe$ (Bis(1,1,1–trifluoro–2,4–pentanedionato)iron)		14.5±0.1	EI		2959

$$C_9H_8O_4F_3Fe^+$$

$C_9H_8O_4F_3Fe^+$	$(CF_3COCHCOCH_3)_2Fe$ (Bis(1,1,1–trifluoro–2,4–pentanedionato)iron)		12.6±0.1	EI		2959

$$C_{10}H_8O_4F_6Fe^+$$

$C_{10}H_8O_4F_6Fe^+$	$(CF_3COCHCOCH_3)_2Fe$ (Bis(1,1,1–trifluoro–2,4–pentanedionato)iron)		8.49±0.03	EI		2992
$C_{10}H_8O_4F_6Fe^+$	$(CF_3COCHCOCH_3)_2Fe$ (Bis(1,1,1–trifluoro–2,4–pentanedionato)iron)		8.75±0.1	EI		2959
$C_{10}H_8O_4F_6Fe^+$	$CH_3COCHCOCH_3FeCF_3COCHCOCF_3$ (1,1,1,5,5,5–Hexafluoro–2,4–pentanedionato(2,4–pentanedionato)iron)		8.70±0.04	EI		2992
$C_{10}H_8O_4F_6Fe^+$	$(CF_3COCHCOCH_3)_3Fe$ (Tris(1,1,1–trifluoro–2,4–pentanedionato)iron)		9.2±0.1?	EI		2959

$$C_9H_2O_4F_9Fe^+$$

$C_9H_2O_4F_9Fe^+$	$(CF_3COCHCOCF_3)_2Fe$ (Bis(1,1,1,5,5,5–hexafluoro–2,4–pentanedionato)iron)		13.2±0.2	EI		2959

$$C_{15}H_{12}O_6F_9Fe^+$$

$C_{15}H_{12}O_6F_9Fe^+$	$(CF_3COCHCOCH_3)_3Fe$ (Tris(1,1,1–trifluoro–2,4–pentanedionato)iron)		9.10±0.05	EI		2959
$C_{15}H_{12}O_6F_9Fe^+$	$(CF_3COCHCOCH_3)_3Fe$ (Tris(1,1,1–trifluoro–2,4–pentanedionato)iron)		9.38±0.11	EI		2580

4.3. The Positive Ion Table—Continued

Ion	Reactant	Other products	Ionization or appearance potential (eV)	Method	Heat of formation (kJ mol^{-1})	Ref.
			$C_{10}H_2O_4F_{12}Fe^+$			
$C_{10}H_2O_4F_{12}Fe^+$	$(CF_3COCHCOCF_3)_2Fe$ (Bis(1,1,1,5,5,5-hexafluoro-2,4-pentanedionato)iron)		9.48±0.07	EI		2992
$C_{10}H_2O_4F_{12}Fe^+$	$(CF_3COCHCOCF_3)_2Fe$ (Bis(1,1,1,5,5,5-hexafluoro-2,4-pentanedionato)iron)		9.7±0.1	EI		2959
$C_{10}H_2O_4F_{12}Fe^+$	$(CF_3COCHCOCF_3)_3Fe$ (Tris(1,1,1,5,5,5-hexafluoro-2,4-pentanedionato)iron)		10.2±0.1?	EI		2959
			$C_{14}H_3O_6F_{15}Fe^+$			
$C_{14}H_3O_6F_{15}Fe^+$	$(CF_3COCHCOCF_3)_3Fe$ (Tris(1,1,1,5,5,5-hexafluoro-2,4-pentanedionato)iron)		11.1±0.1	EI		2959
			$C_{15}H_3O_6F_{18}Fe^+$			
$C_{15}H_3O_6F_{18}Fe^+$	$(CF_3COCHCOCF_3)_3Fe$ (Tris(1,1,1,5,5,5-hexafluoro-2,4-pentanedionato)iron)		8.28 (V)	PE		3169

This value may be spurious, see ref. 3239.

Ion	Reactant	Other products	Ionization or appearance potential (eV)	Method	Heat of formation (kJ mol^{-1})	Ref.
$C_{15}H_3O_6F_{18}Fe^+$	$(CF_3COCHCOCF_3)_3Fe$ (Tris(1,1,1,5,5,5-hexafluoro-2,4-pentanedionato)iron)		10.13±0.03 (V)	PE		3239
$C_{15}H_3O_6F_{18}Fe^+$	$(CF_3COCHCOCF_3)_3Fe$ (Tris(1,1,1,5,5,5-hexafluoro-2,4-pentanedionato)iron)		10.14 (V)	PE		3169
$C_{15}H_3O_6F_{18}Fe^+$	$(CF_3COCHCOCF_3)_3Fe$ (Tris(1,1,1,5,5,5-hexafluoro-2,4-pentanedionato)iron)		10.2±0.1	EI		2959
$C_{15}H_3O_6F_{18}Fe^+$	$(CF_3COCHCOCF_3)_3Fe$ (Tris(1,1,1,5,5,5-hexafluoro-2,4-pentanedionato)iron)		10.34±0.10	EI		2580
			$C_{16}H_{27}O_4PFe^+$			
$C_{16}H_{27}O_4PFe^+$	$(n-C_4H_9)_3PFe(CO)_4$		7.29±0.05	EI		2481, 2716
			$C_7H_9O_7PFe^+$			
$C_7H_9O_7PFe^+$	$(CH_3O)_3PFe(CO)_4$		7.65±0.05	EI		2481, 2716
			$C_{10}H_{15}O_7PFe^+$			
$C_{10}H_{15}O_7PFe^+$	$(C_2H_5O)_3PFe(CO)_4$		7.43±0.05	EI		2481, 2716
			$C_9H_{18}O_9P_2Fe^+$			
$C_9H_{18}O_9P_2Fe^+$	$((CH_3O)_3P)_2Fe(CO)_3$		7.33	EI		2716
			$C_{10}H_{12}O_6P_2Fe_2^+$			
$C_{10}H_{12}O_6P_2Fe_2^+$	$((CH_3)_2P)_2Fe_2(CO)_6$		7.73±0.01	EI		3250

4.3. The Positive Ion Table—Continued

Ion	Reactant	Other products	Ionization or appearance potential (eV)	Method	Heat of formation (kJ mol^{-1})	Ref.
		$C_{14}H_{20}O_6P_2Fe_2^+$				
$C_{14}H_{20}O_6P_2Fe_2^+$	$((C_2H_5)_2P)_2Fe_2(CO)_6$		7.67±0.02	EI		3250
		$C_{30}H_{20}O_6P_2Fe_2^+$				
$C_{30}H_{20}O_6P_2Fe_2^+$	$((C_6H_5)_2P)_2Fe_2(CO)_6$ (Bis(diphenylphosphino)diiron hexacarbonyl)		7.70±0.03	EI		3250
		$C_8H_6O_6S_2Fe_2^+$				
$C_8H_6O_6S_2Fe_2^+$	$(CH_3S)_2Fe_2(CO)_6$		8.07±0.01	EI		3250
		$C_{12}H_{14}O_6S_2Fe_2^+$				
$C_{12}H_{14}O_6S_2Fe_2^+$	$(iso-C_3H_7S)_2Fe_2(CO)_6$		8.05±0.01	EI		3250
		$C_{18}H_{10}O_6S_2Fe_2^+$				
$C_{18}H_{10}O_6S_2Fe_2^+$	$(C_6H_5S)_2Fe_2(CO)_6$ (Bis(phenylthio)diiron hexacarbonyl)		7.90±0.01	EI		3250
		$C_4O_4PCl_3Fe^+$				
$C_4O_4PCl_3Fe^+$	$PCl_3Fe(CO)_4$		8.05±0.05	EI		2481, 2716
		$C_{12}H_{27}N_2O_2PFe^+$				
$C_{12}H_{27}N_2O_2PFe^+$	$(n-C_4H_9)_3PFe(CO)(NO)_2$	CO	7.70	EI		2480
		$C_3H_9N_2O_5PFe^+$				
$C_3H_9N_2O_5PFe^+$	$(CH_3O)_3PFe(CO)(NO)_2$	CO	8.20	EI		2480
$C_3H_9N_2O_5PFe^+$	$((CH_3O)_3P)_2Fe(NO)_2$		9.75	EI		2480
		$C_6H_{15}N_2O_5PFe^+$				
$C_6H_{15}N_2O_5PFe^+$	$(C_2H_5O)_3PFe(CO)(NO)_2$	CO	8.15	EI		2480
$C_6H_{15}N_2O_5PFe^+$	$((C_2H_5O)_3P)_2Fe(NO)_2$		9.45	EI		2480
		$C_{12}H_{27}N_2O_5PFe^+$				
$C_{12}H_{27}N_2O_5PFe^+$	$(n-C_4H_9O)_3PFe(CO)(NO)_2$	CO	7.85	EI		2480

4.3. The Positive Ion Table—Continued

Ion	Reactant	Other products	Ionization or appearance potential (eV)	Method	Heat of formation (kJ mol^{-1})	Ref.
			C$_4$H$_9$N$_2$O$_6$PFe$^+$			
C$_4$H$_9$N$_2$O$_6$PFe$^+$	(CH$_3$O)$_3$PFe(CO)(NO)$_2$		7.66±0.05	EI		2481
			C$_7$H$_{15}$N$_2$O$_6$PFe$^+$			
C$_7$H$_{15}$N$_2$O$_6$PFe$^+$	(C$_2$H$_5$O)$_3$PFe(CO)(NO)$_2$		7.50±0.1	EI		2453, 2716
			C$_{13}$H$_{27}$N$_2$O$_6$PFe$^+$			
C$_{13}$H$_{27}$N$_2$O$_6$PFe$^+$	(n–C$_4$H$_9$O)$_3$PFe(CO)(NO)$_2$		7.52±0.05	EI		2481
C$_{13}$H$_{27}$N$_2$O$_6$PFe$^+$	(n–C$_4$H$_9$O)$_3$PFe(CO)(NO)$_2$		7.57	EI		2716
			C$_6$H$_{18}$N$_2$O$_8$P$_2$Fe$^+$			
C$_6$H$_{18}$N$_2$O$_8$P$_2$Fe$^+$	((CH$_3$O)$_3$P)$_2$Fe(NO)$_2$		7.25	EI		2480
			C$_{12}$H$_{30}$N$_2$O$_8$P$_2$Fe$^+$			
C$_{12}$H$_{30}$N$_2$O$_8$P$_2$Fe$^+$	((C$_2$H$_5$O)$_3$P)$_2$Fe(NO)$_2$		7.02	EI		2480, 2716
			C$_{24}$H$_{15}$O$_6$PSFe$_2^+$			
C$_{24}$H$_{15}$O$_6$PSFe$_2^+$	(C$_6$H$_5$)$_2$P(C$_6$H$_5$S)Fe$_2$(CO)$_6$ (Diphenylphosphino(phenylthio)diiron hexacarbonyl)		7.81±0.05	EI		3250
			Co$^+$ $\Delta H_{f0}^\circ = 1181$ kJ mol^{-1} (282 kcal mol^{-1})			
Co$^+$	Co		7.86	S	1181	2113
See also – EI: 2444						
Co$^+$	(C$_5$H$_5$)$_2$Co (Bis(cyclopentadienyl)cobalt)		14.66±0.2	EI		2683
Co$^+$	Co$_2$(CO)$_8$		16.9±0.4	EI		2739
Co$^+$	C$_5$H$_5$Co(CO)$_2$ (Cyclopentadienylcobalt dicarbonyl)		16.8±0.3	EI		1381
Co$^+$	HCo(CO)$_4$		17.8±0.3	EI		2583
Co$^+$	HCo(PF$_3$)$_4$		17.8±0.3	EI		2583
Co$^+$	SiF$_3$Co(CO)$_4$		17.8±0.2	EI		2582
Co$^+$	SiCl$_3$Co(CO)$_4$		19.2±0.3	EI		3255
Co$^+$	CH$_3$SiF$_2$Co(CO)$_4$		19.5±0.3	EI		2581
Co$^+$	HCo(CO)$_3$PF$_3$		17.8±0.3	EI		2583
Co$^+$	HCo(CO)$_2$(PF$_3$)$_2$		18.2±0.3	EI		2583
Co$^+$	HCo(CO)(PF$_3$)$_3$		17.8±0.3	EI		2583

4.3. The Positive Ion Table—Continued

Ion	Reactant	Other products	Ionization or appearance potential (eV)	Method	Heat of formation (kJ mol^{-1})	Ref.
			Co_2^+			
Co_2^+	$Co_2(CO)_8$	8CO	17.8±0.4	EI		2739
			CoH^+			
CoH^+	$HCo(CO)_4$	4CO	15.2±0.3	EI		2583
CoH^+	$HCo(PF_3)_4$		16.0±0.3	EI		2583
CoH^+	$HCo(CO)_3PF_3$		14.6±0.3	EI		2583
CoH^+	$HCo(CO)_2(PF_3)_2$		14.9±0.3	EI		2583
CoH^+	$HCo(CO)(PF_3)_3$		15.6±0.3	EI		2583
			CoO^+			
CoO^+	CoO		9.0	EI		2444
			CoF^+			
CoF^+	$SiF_3Co(CO)_4$		21.0±0.3	EI		2582
			CH_3Co^+			
CH_3Co^+	$CH_3SiF_2Co(CO)_4$		16.6±0.3	EI		2581
			$C_3H_3Co^+$			
$C_3H_3Co^+$	$(C_5H_5)_2Co$ (Bis(cyclopentadienyl)cobalt)		17.62±0.1	EI		2683
$C_3H_3Co^+$	$C_5H_5Co(CO)_2$ (Cyclopentadienylcobalt dicarbonyl)		16.8±0.3	EI		1381
			$C_5H_5Co^+$			
$C_5H_5Co^+$	C_5H_5Co (Cyclopentadienylcobalt)		10.1	EI		2546
$C_5H_5Co^+$	$(C_5H_5)_2Co$ (Bis(cyclopentadienyl)cobalt)		12.3±1	EI		2545
$C_5H_5Co^+$	$(C_5H_5)_2Co$ (Bis(cyclopentadienyl)cobalt)		14.00±0.1	EI		2683
$C_5H_5Co^+$	$C_5H_5Co(CO)_2$ (Cyclopentadienylcobalt dicarbonyl)	2CO	10.8±0.2	EI	2545,	2546
$C_5H_5Co^+$	$C_5H_5Co(CO)_2$ (Cyclopentadienylcobalt dicarbonyl)	2CO	11.7±0.2	EI		1381

4.3. The Positive Ion Table—Continued

Ion	Reactant	Other products	Ionization or appearance potential (eV)	Method	Heat of formation (kJ mol^{-1})	Ref.
			$C_{10}H_{10}Co^+$			
$C_{10}H_{10}Co^+$	$(C_5H_5)_2Co$ (Bis(cyclopentadienyl)cobalt)		5.95 ± 0.1	EI		2545
$C_{10}H_{10}Co^+$	$(C_5H_5)_2Co$ (Bis(cyclopentadienyl)cobalt)		6.21 ± 0.1	EI		2683
			$LiCoO^+$			
$LiCoO^+$	$LiCoO_2$	O	16.0 ± 0.5	EI		2565
			$LiCoO_2^+$			
$LiCoO_2^+$	$LiCoO_2$		11.0 ± 0.5	EI		2565
			$CoCO^+$			
$CoCO^+$	$Co_2(CO)_8$		14.4 ± 0.5	EI		2739
$CoCO^+$	$C_5H_5Co(CO)_2$ (Cyclopentadienylcobalt dicarbonyl)		16.5 ± 0.4	EI		1381
$CoCO^+$	$HCo(CO)_4$		15.0 ± 0.3	EI		2583
$CoCO^+$	$SiF_3Co(CO)_4$		15.9 ± 0.2	EI		2582
$CoCO^+$	$CH_3SiF_2Co(CO)_4$		16.4 ± 0.4	EI		2581
$CoCO^+$	$HCo(CO)_3PF_3$		15.5 ± 0.2	EI		2583
$CoCO^+$	$HCo(CO)_2(PF_3)_2$		15.3 ± 0.3	EI		2583
$CoCO^+$	$HCo(CO)(PF_3)_3$		15.9 ± 0.2	EI		2583
			$CoC_2O_2^+$			
$CoC_2O_2^+$	$Co_2(CO)_8$		12.7 ± 0.4	EI		2739
$CoC_2O_2^+$	$HCo(CO)_4$		12.9 ± 0.3	EI		2583
$CoC_2O_2^+$	$SiF_3Co(CO)_4$		14.0 ± 0.2	EI		2582
$CoC_2O_2^+$	$SiCl_3Co(CO)_4$		14.3 ± 0.3	EI		3255
$CoC_2O_2^+$	$CH_3SiF_2Co(CO)_4$		14.7 ± 0.3	EI		2581
$CoC_2O_2^+$	$HCo(CO)_3PF_3$		14.0 ± 0.2	EI		2583
$CoC_2O_2^+$	$HCo(CO)_2(PF_3)_2$		13.6 ± 0.2	EI		2583
			$CoC_3O_3^+$			
$CoC_3O_3^+$	$Co_2(CO)_8$		10.9 ± 0.3	EI		2739
$CoC_3O_3^+$	$HCo(CO)_4$		12.1 ± 0.3	EI		2583
$CoC_3O_3^+$	$HCo(CO)_3PF_3$		12.1 ± 0.2	EI		2583
			$CoC_4O_4^+$			
$CoC_4O_4^+$	$Co(CO)_4$		8.30 ± 0.1	EI		2405, 2870
$CoC_4O_4^+$	$Co_2(CO)_8$		8.80 ± 0.1	EI		2405, 2870

4.3. The Positive Ion Table—Continued

Ion	Reactant	Other products	Ionization or appearance potential (eV)	Method	Heat of formation (kJ mol^{-1})	Ref.
		Co_2CO^+				
Co_2CO^+	$Co_2(CO)_8$	7CO	16.7±0.3	EI		2739
		$Co_2C_2O_2^+$				
$Co_2C_2O_2^+$	$Co_2(CO)_8$	6CO	14.7±0.5	EI		2739
		$Co_2C_4O_4^+$				
$Co_2C_4O_4^+$	$Co_2(CO)_8$	4CO	12.2±0.3	EI		2739
		$Co_2C_5O_5^+$				
$Co_2C_5O_5^+$	$Co_2(CO)_8$	3CO	10.1±0.4	EI		2739
		$Co_2C_6O_6^+$				
$Co_2C_6O_6^+$	$Co_2(CO)_8$	2CO	9.4±0.3	EI		2739
		$Co_2C_7O_7^+$				
$Co_2C_7O_7^+$	$Co_2(CO)_8$	CO	8.6±0.3	EI		2739
		$Co_2C_8O_8^+$				
$Co_2C_8O_8^+$	$Co_2(CO)_8$		8.12±0.22	EI		2739
$Co_2C_8O_8^+$	$Co_2(CO)_8$		8.26±0.1	EI		2870
		$CoSiF_3^+$				
$CoSiF_3^+$	$SiF_3Co(CO)_4$	4CO	15.1±0.1	EI		2582
		$CoPF_3^+$				
$CoPF_3^+$	$HCo(PF_3)_4$		15.8±0.3	EI		2583
$CoPF_3^+$	$HCo(CO)_3PF_3$		15.7±0.2	EI		2583
$CoPF_3^+$	$HCo(CO)_2(PF_3)_2$		15.9±0.2	EI		2583
$CoPF_3^+$	$HCo(CO)(PF_3)_3$		15.7±0.5	EI		2583
		$CoP_2F_6^+$				
$CoP_2F_6^+$	$HCo(PF_3)_4$		14.3±0.3	EI		2583
$CoP_2F_6^+$	$HCo(CO)(PF_3)_3$		14.5±0.2	EI		2583

4.3. The Positive Ion Table—Continued

Ion	Reactant	Other products	Ionization or appearance potential (eV)	Method	Heat of formation (kJ mol^{-1})	Ref.
			CoSiCl$_3^+$			
CoSiCl$_3^+$	SiCl$_3$Co(CO)$_4$	4CO	13.3±0.3	EI		3255
			CHOCo$^+$			
CHOCo$^+$	HCo(CO)$_4$	3CO	12.8±0.3	EI		2583
CHOCo$^+$	HCo(CO)$_3$PF$_3$		12.1±0.2	EI		2583
CHOCo$^+$	HCo(CO)$_2$(PF$_3$)$_2$		12.3±0.2	EI		2583
CHOCo$^+$	HCo(CO)(PF$_3$)$_3$		12.0±0.2	EI		2583
			C$_6$H$_5$OCo$^+$			
C$_6$H$_5$OCo$^+$	C$_5$H$_5$Co(CO)$_2$ (Cyclopentadienylcobalt dicarbonyl)	CO	10.1±0.2	EI		1381
			C$_2$HO$_2$Co$^+$			
C$_2$HO$_2$Co$^+$	HCo(CO)$_4$	2CO	11.2±0.2	EI		2583
C$_2$HO$_2$Co$^+$	HCo(CO)$_3$PF$_3$		12.4±0.3	EI		2583
C$_2$HO$_2$Co$^+$	HCo(CO)$_2$(PF$_3$)$_2$		11.6±0.2	EI		2583
			C$_5$H$_7$O$_2$Co$^+$			
C$_5$H$_7$O$_2$Co$^+$	(CH$_3$COCHCOCH$_3$)$_2$Co (Bis(2,4–pentanedionato)cobalt)		13.9±0.2	EI		2731
			C$_7$H$_5$O$_2$Co$^+$			
C$_7$H$_5$O$_2$Co$^+$	C$_5$H$_5$Co(CO)$_2$ (Cyclopentadienylcobalt dicarbonyl)		7.78±0.1	EI		2545
C$_7$H$_5$O$_2$Co$^+$	C$_5$H$_5$Co(CO)$_2$ (Cyclopentadienylcobalt dicarbonyl)		8.3±0.2	EI		1381
			C$_{11}$H$_{19}$O$_2$Co$^+$			
C$_{11}$H$_{19}$O$_2$Co$^+$	((CH$_3$)$_3$CCOCHCOC(CH$_3$)$_3$)$_3$Co (Tris(2,2,6,6–tetramethyl–3,5–heptanedionato)cobalt)		27.0±0.5	EI		2524
			C$_3$HO$_3$Co$^+$			
C$_3$HO$_3$Co$^+$	HCo(CO)$_4$	CO	9.4±0.2	EI		2583
C$_3$HO$_3$Co$^+$	HCo(CO)$_3$PF$_3$		9.9±0.2	EI		2583
			C$_4$HO$_4$Co$^+$			
C$_4$HO$_4$Co$^+$	HCo(CO)$_4$		8.7±0.1	EI		2583

4.3. The Positive Ion Table—Continued

Ion	Reactant	Other products	Ionization or appearance potential (eV)	Method	Heat of formation (kJ mol^{-1})	Ref.
			$C_9H_{11}O_4Co^+$			
$C_9H_{11}O_4Co^+$	$(CH_3COCHCOCH_3)_2Co$ (Bis(2,4–pentanedionato)cobalt)		11.5±0.1	EI		2731
			$C_{10}H_{14}O_4Co^+$			
$C_{10}H_{14}O_4Co^+$	$(CH_3COCHCOCH_3)_2Co$ (Bis(2,4–pentanedionato)cobalt)		8.54±0.05	EI		2731
$C_{10}H_{14}O_4Co^+$	$(CH_3COCHCOCH_3)_3Co$ (Tris(2,4–pentanedionato)cobalt)		10.7±0.3	EI		2460
			$C_{18}H_{29}O_4Co^+$			
$C_{18}H_{29}O_4Co^+$	$((CH_3)_3CCOCHCOC(CH_3)_3)_3Co$ (Tris(2,2,6,6–tetramethyl–3,5–heptanedionato)cobalt)		17.4±0.5	EI		2524
			$C_{22}H_{38}O_4Co^+$			
$C_{22}H_{38}O_4Co^+$	$((CH_3)_3CCOCHCOC(CH_3)_3)_3Co$ (Tris(2,2,6,6–tetramethyl–3,5–heptanedionato)cobalt)		13.2±0.5	EI		2524
			$C_{15}H_{21}O_6Co^+$			
$C_{15}H_{21}O_6Co^+$	$(CH_3COCHCOCH_3)_3Co$ (Tris(2,4–pentanedionato)cobalt)		7.80±0.05	EI		2460
			$C_{33}H_{57}O_6Co^+$			
$C_{33}H_{57}O_6Co^+$	$((CH_3)_3CCOCHCOC(CH_3)_3)_3Co$ (Tris(2,2,6,6–tetramethyl–3,5–heptanedionato)cobalt)		8.4±0.5	EI		2524
			$C_2NO_3Co^+$			
$C_2NO_3Co^+$	$Co(CO)_2NO$		9.30	EI		2546
$C_2NO_3Co^+$	$Co(CO)_3NO$	CO	9.65	EI		2546

See also – PI: 2886

			$C_3NO_4Co^+$			
$C_3NO_4Co^+$	$Co(CO)_3NO$		8.11±0.03	PI		2886
$C_3NO_4Co^+$	$Co(CO)_3NO$		8.75±0.1	EI		2453

4.3. The Positive Ion Table—Continued

Ion	Reactant	Other products	Ionization or appearance potential (eV)	Method	Heat of formation (kJ mol^{-1})	Ref.
			CoHPF$_3^+$			
CoHPF$_3^+$	HCo(PF$_3$)$_4$		14.4±0.3	EI		2583
CoHPF$_3^+$	HCo(CO)$_3$PF$_3$	3CO	13.8±0.2	EI		2583
CoHPF$_3^+$	HCo(CO)$_2$(PF$_3$)$_2$		13.6±0.3	EI		2583
CoHPF$_3^+$	HCo(CO)(PF$_3$)$_3$		13.6±0.2	EI		2583
			CoHP$_2$F$_6^+$			
CoHP$_2$F$_6^+$	HCo(PF$_3$)$_4$		12.7±0.2	EI		2583
CoHP$_2$F$_6^+$	HCo(CO)(PF$_3$)$_3$		12.6±0.2	EI		2583
			CoHP$_3$F$_9^+$			
CoHP$_3$F$_9^+$	HCo(PF$_3$)$_4$		10.4±0.2	EI		2583
CoHP$_3$F$_9^+$	HCo(CO)(PF$_3$)$_3$	CO	11.1±0.2	EI		2583
			CoHP$_4$F$_{12}^+$			
CoHP$_4$F$_{12}^+$	HCo(PF$_3$)$_4$		9.2±0.2	EI		2583
			C$_{15}$H$_3$O$_6$F$_{18}$Co$^+$			
C$_{15}$H$_3$O$_6$F$_{18}$Co$^+$	(CF$_3$COCHCOCF$_3$)$_3$Co (Tris(1,1,1,5,5,5–hexafluoro–2,4–pentanedionato)cobalt)		9.56 (V)	PE		3169
C$_{15}$H$_3$O$_6$F$_{18}$Co$^+$	(CF$_3$COCHCOCF$_3$)$_3$Co (Tris(1,1,1,5,5,5–hexafluoro–2,4–pentanedionato)cobalt)		10.12±0.15	EI		2580
			CH$_3$F$_2$SiCo$^+$			
CH$_3$F$_2$SiCo$^+$	CH$_3$SiF$_2$Co(CO)$_4$	4CO	14.5±0.4	EI		2581
			C$_4$O$_4$F$_2$SiCo$^+$			
C$_4$O$_4$F$_2$SiCo$^+$	CH$_3$SiF$_2$Co(CO)$_4$		13.8±0.2	EI		2581
			COF$_3$SiCo$^+$			
COF$_3$SiCo$^+$	SiF$_3$Co(CO)$_4$	3CO	13.8±0.2	EI		2582
			C$_3$O$_3$F$_3$SiCo$^+$			
C$_3$O$_3$F$_3$SiCo$^+$	SiF$_3$Co(CO)$_4$	CO	10.4±0.1	EI		2582

4.3. The Positive Ion Table—Continued

Ion	Reactant	Other products	Ionization or appearance potential (eV)	Method	Heat of formation (kJ mol^{-1})	Ref.
			$C_4O_4F_3SiCo^+$			
$C_4O_4F_3SiCo^+$	$SiF_3Co(CO)_4$		9.7±0.1	EI		2582
			COF_3PCo^+			
COF_3PCo^+	$HCo(CO)_3PF_3$		14.3±0.3	EI		2583
COF_3PCo^+	$HCo(CO)_2(PF_3)_2$		14.0±0.3	EI		2583
COF_3PCo^+	$HCo(CO)(PF_3)_3$		14.1±0.3	EI		2583
			$COSiCl_3Co^+$			
$COSiCl_3Co^+$	$SiCl_3Co(CO)_4$	3CO	11.8±0.2	EI		3255
			$C_2O_2SiCl_3Co^+$			
$C_2O_2SiCl_3Co^+$	$SiCl_3Co(CO)_4$	2CO	10.6±0.2	EI		3255
			$C_3O_3SiCl_3Co^+$			
$C_3O_3SiCl_3Co^+$	$SiCl_3Co(CO)_4$	CO	9.4±0.1	EI		3255
			$C_4O_4SiCl_3Co^+$			
$C_4O_4SiCl_3Co^+$	$SiCl_3Co(CO)_4$		9.0±0.1	EI		3255
			$C_2H_3OF_2SiCo^+$			
$C_2H_3OF_2SiCo^+$	$CH_3SiF_2Co(CO)_4$	3CO	12.9±0.2	EI		2581
			$C_3H_3O_2F_2SiCo^+$			
$C_3H_3O_2F_2SiCo^+$	$CH_3SiF_2Co(CO)_4$	2CO	10.7±0.3	EI		2581
			$C_4H_3O_3F_2SiCo^+$			
$C_4H_3O_3F_2SiCo^+$	$CH_3SiF_2Co(CO)_4$	CO	10.0±0.1	EI		2581
			$C_5H_3O_4F_2SiCo^+$			
$C_5H_3O_4F_2SiCo^+$	$CH_3SiF_2Co(CO)_4$		9.0±0.1	EI		2581

4.3. The Positive Ion Table—Continued

Ion	Reactant	Other products	Ionization or appearance potential (eV)	Method	Heat of formation (kJ mol^{-1})	Ref.
		$C_8H_{15}NO_3PCo^+$				
$C_8H_{15}NO_3PCo^+$	$(C_2H_5)_3PCo(CO)_2NO$		7.62±0.05	EI		2481
		$C_{14}H_{27}NO_3PCo^+$				
$C_{14}H_{27}NO_3PCo^+$	$(n-C_4H_9)_3PCo(CO)_2NO$		7.51±0.05	EI		2481
		$C_5H_9NO_6PCo^+$				
$C_5H_9NO_6PCo^+$	$(CH_3O)_3PCo(CO)_2NO$		7.92±0.05	EI		2481
		$C_8H_{15}NO_6PCo^+$				
$C_8H_{15}NO_6PCo^+$	$(C_2H_5O)_3PCo(CO)_2NO$		7.82±0.05	EI		2481
		$C_{11}H_{21}NO_6PCo^+$				
$C_{11}H_{21}NO_6PCo^+$	$(iso-C_3H_7O)_3PCo(CO)_2NO$		7.64±0.05	EI		2481
		$C_7H_{18}NO_8P_2Co^+$				
$C_7H_{18}NO_8P_2Co^+$	$((CH_3O)_3P)_2Co(CO)NO$		7.26	EI		2716
		$C_{13}H_{30}NO_8P_2Co^+$				
$C_{13}H_{30}NO_8P_2Co^+$	$((C_2H_5O)_3P)_2Co(CO)NO$		7.13	EI		2716
		$C_{19}H_{42}NO_8P_2Co^+$				
$C_{19}H_{42}NO_8P_2Co^+$	$((iso-C_3H_7O)_3P)_2Co(CO)NO$		7.22	EI		2716
		$CHOF_3PCo^+$				
$CHOF_3PCo^+$	$HCo(CO)_3PF_3$	2CO	12.5±0.3	EI		2583
$CHOF_3PCo^+$	$HCo(CO)_2(PF_3)_2$		12.4±0.2	EI		2583
$CHOF_3PCo^+$	$HCo(CO)(PF_3)_3$		12.3±0.3	EI		2583
		$C_2HO_2F_3PCo^+$				
$C_2HO_2F_3PCo^+$	$HCo(CO)_3PF_3$	CO	10.3±0.3	EI		2583
$C_2HO_2F_3PCo^+$	$HCo(CO)_2(PF_3)_2$		10.1±0.2	EI		2583

4.3. The Positive Ion Table—Continued

Ion	Reactant	Other products	Ionization or appearance potential (eV)	Method	Heat of formation (kJ mol^{-1})	Ref.
			C$_3$HO$_3$F$_3$PCo$^+$			
C$_3$HO$_3$F$_3$PCo$^+$	HCo(CO)$_3$PF$_3$		9.8±0.2	EI		2583
			CHOF$_6$P$_2$Co$^+$			
CHOF$_6$P$_2$Co$^+$	HCo(CO)$_2$(PF$_3$)$_2$	CO	10.3±0.3	EI		2583
CHOF$_6$P$_2$Co$^+$	HCo(CO)(PF$_3$)$_3$		10.4±0.3	EI		2583
			C$_2$HO$_2$F$_6$P$_2$Co$^+$			
C$_2$HO$_2$F$_6$P$_2$Co$^+$	HCo(CO)$_2$(PF$_3$)$_2$		9.6±0.2	EI		2583
			CHOF$_9$P$_3$Co$^+$			
CHOF$_9$P$_3$Co$^+$	HCo(CO)(PF$_3$)$_3$		10.2±0.1	EI		2583
			C$_2$NO$_3$PCl$_3$Co$^+$			
C$_2$NO$_3$PCl$_3$Co$^+$	PCl$_3$Co(CO)$_2$NO		8.40±0.1	EI		2453
	Ni$^+$	$\Delta H_{f0}^\circ = 1164.3$ kJ mol^{-1} (278.3 kcal mol^{-1})				
Ni$^+$	Ni		7.635	S	1164.3	2113
See also – EI:	2125, 2188					
Ni$^+$	NiF$_2$		16.7±0.3	EI		2162
Ni$^+$	NiCl$_2$		15.7±0.5	EI		2125
Ni$^+$	(C$_5$H$_5$)$_2$Ni (Bis(cyclopentadienyl)nickel)		13.65±0.2	EI		2683
Ni$^+$	(C$_5$H$_5$)$_2$Ni (Bis(cyclopentadienyl)nickel)		13.6	EI	2732,	2940
Ni$^+$	Ni(CO)$_4$	4CO	13.75±0.1	PI		2956
(Threshold value approximately corrected for hot bands)						
Ni$^+$	Ni(CO)$_4$	4CO	14.45±0.15	EI		2579
Ni$^+$	Ni(CO)$_4$	4CO	15.1±0.3	EI		2403
Ni$^+$	Ni(CO)$_4$	4CO	15.51	EI		2500
Ni$^+$	Ni(CO)$_4$	4CO	16.0±0.3	EI		112
Ni$^+$	Ni(PF$_3$)$_4$		17.3	EI		2507
			NiC$^+$			
NiC$^+$	Ni(CO)$_4$		22.1±0.3	EI		2579
NiC$^+$	Ni(CO)$_4$		24.2±0.2	EI		2403

4.3. The Positive Ion Table—Continued

Ion	Reactant	Other products	Ionization or appearance potential (eV)	Method	Heat of formation (kJ mol^{-1})	Ref.
		NiC$_2^+$				
NiC$_2^+$	Ni(CO)$_4$		30.1±1	EI		2579
		NiO$^+$				
NiO$^+$	NiO		9.5±0.3	EI		2188
NiO$^+$	Ni(CO)$_4$		26.4±1	EI		2579
		NiF$^+$				
NiF$^+$	NiF$_2$	F	13.0±0.3	EI		2162
		NiF$_2^+$				
NiF$_2^+$	NiF$_2$		11.5±0.3	EI		2162
		NiCl$^+$				
NiCl$^+$	NiCl		11.4±0.5	EI		2125
NiCl$^+$	NiCl$_2$	Cl	12.7±0.5	EI		2125
		NiCl$_2^+$				
NiCl$_2^+$	NiCl$_2$		11.2±0.5	EI		2125
		C$_3$H$_3$Ni$^+$				
C$_3$H$_3$Ni$^+$	(C$_5$H$_5$)$_2$Ni (Bis(cyclopentadienyl)nickel)		17.16±0.2	EI		2683
		C$_5$H$_5$Ni$^+$				
C$_5$H$_5$Ni$^+$	C$_5$H$_5$Ni (Cyclopentadienylnickel)		7.8	EI		2732, 2940
C$_5$H$_5$Ni$^+$	(C$_5$H$_5$)$_2$Ni (Bis(cyclopentadienyl)nickel)		12.59±0.1	EI		2683
C$_5$H$_5$Ni$^+$	(C$_5$H$_5$)$_2$Ni (Bis(cyclopentadienyl)nickel)		11.9±1	EI		2545
C$_5$H$_5$Ni$^+$	(C$_5$H$_5$)$_2$Ni (Bis(cyclopentadienyl)nickel)		12.4	EI		2732, 2940
		C$_8$H$_8$Ni$^+$				
C$_8$H$_8$Ni$^+$	(C$_5$H$_5$)$_2$Ni (Bis(cyclopentadienyl)nickel)		12.19±0.1	EI		2683

4.3. The Positive Ion Table—Continued

Ion	Reactant	Other products	Ionization or appearance potential (eV)	Method	Heat of formation (kJ mol⁻¹)	Ref.
			$C_{10}H_{10}Ni^+$			
$C_{10}H_{10}Ni^+$	$(C_5H_5)_2Ni$ (Bis(cyclopentadienyl)nickel)		6.75	EI		2453
$C_{10}H_{10}Ni^+$	$(C_5H_5)_2Ni$ (Bis(cyclopentadienyl)nickel)		7.16±0.1	EI		2683
$C_{10}H_{10}Ni^+$	$(C_5H_5)_2Ni$ (Bis(cyclopentadienyl)nickel)		6.8	EI	2732,	2940
			$LiNiO^+$			
$LiNiO^+$	$LiNiO_2$	O	14.0±0.5	EI		2565
			$LiNiO_2^+$			
$LiNiO_2^+$	$LiNiO_2$		10.8±0.5	EI		2565
			$NiCO^+$			
$NiCO^+$	$Ni(CO)_4$	3CO	11.65±0.1	PI		2956
$NiCO^+$	$Ni(CO)_4$	3CO	12.17±0.15	EI		2579
$NiCO^+$	$Ni(CO)_4$	3CO	12.84	EI		2500
$NiCO^+$	$Ni(CO)_4$	3CO	12.96±0.10	EI		2403
$NiCO^+$	$Ni(CO)_4$	3CO	13.5±0.2	EI		112
			$NiCO^{+2}$			
$NiCO^{+2}$	$Ni(CO)_4$		30.2±0.5	EI		2579
			NiC_2O^+			
NiC_2O^+	$Ni(CO)_4$		20.6±0.3	EI		2579
			NiC_2O^{+2}			
NiC_2O^{+2}	$Ni(CO)_4$		38.5±0.5	EI		2579
			$NiC_2O_2^+$			
$NiC_2O_2^+$	$Ni(CO)_4$	2CO	10.10±0.1	PI		2956
$NiC_2O_2^+$	$Ni(CO)_4$	2CO	10.21±0.15	EI		2579
$NiC_2O_2^+$	$Ni(CO)_4$	2CO	10.48±0.05	EI		2403
$NiC_2O_2^+$	$Ni(CO)_4$	2CO	10.63	EI		2500
$NiC_2O_2^+$	$Ni(CO)_4$	2CO	10.7±0.2	EI		112

4.3. The Positive Ion Table—Continued

Ion	Reactant	Other products	Ionization or appearance potential (eV)	Method	Heat of formation (kJ mol⁻¹)	Ref.
			$NiC_2O_2^{+2}$			
$NiC_2O_2^{+2}$	$Ni(CO)_4$		27.2±0.5	EI		2579
$NiC_2O_2^{+2}$	$Ni(CO)_4$		28.3±1	EI		112
			$NiC_3O_2^{+}$			
$NiC_3O_2^{+}$	$Ni(CO)_4$		20.1±0.5	EI		2579
			$NiC_3O_3^{+}$			
$NiC_3O_3^{+}$	$Ni(CO)_4$	CO	8.77±0.02	PI		2956
$NiC_3O_3^{+}$	$Ni(CO)_4$	CO	8.89±0.15	EI		2579
$NiC_3O_3^{+}$	$Ni(CO)_4$	CO	9.22±0.10	EI		2403
$NiC_3O_3^{+}$	$Ni(CO)_4$	CO	9.34	EI		2500
$NiC_3O_3^{+}$	$Ni(CO)_4$	CO	9.36±0.15	EI		112

See also – PI:　2886

Ion	Reactant	Other products	Ionization or appearance potential (eV)	Method	Heat of formation (kJ mol⁻¹)	Ref.
			$NiC_3O_3^{+2}$			
$NiC_3O_3^{+2}$	$Ni(CO)_4$		25.2±0.5	EI		2579

$Ni(CO)_4^{+}$ 　 $\Delta H_{f0}^{\circ} = 197 \text{ kJ mol}^{-1} \text{ (47 kcal mol}^{-1})$

Ion	Reactant	Other products	Ionization or appearance potential (eV)	Method	Heat of formation (kJ mol⁻¹)	Ref.
$NiC_4O_4^{+}$	$Ni(CO)_4$		8.28±0.01	PI		2886
$NiC_4O_4^{+}$	$Ni(CO)_4$		8.28±0.03	PI		1167
$NiC_4O_4^{+}$	$Ni(CO)_4$		8.32±0.01	PI	197	2956
$NiC_4O_4^{+}$	$Ni(CO)_4$		8.24±0.14	PE		2886
$NiC_4O_4^{+}$	$Ni(CO)_4$		8.35±0.15	EI		2579
$NiC_4O_4^{+}$	$Ni(CO)_4$		8.57±0.10	EI		2403
$NiC_4O_4^{+}$	$Ni(CO)_4$		8.64±0.15	EI		112
$NiC_4O_4^{+}$	$Ni(CO)_4$		8.75±0.07	EI		2500

Ion	Reactant	Other products	Ionization or appearance potential (eV)	Method	Heat of formation (kJ mol⁻¹)	Ref.
			$NiC_4O_4^{+2}$			
$NiC_4O_4^{+2}$	$Ni(CO)_4$		25.1±0.5	EI		2579

4.3. The Positive Ion Table—Continued

Ion	Reactant	Other products	Ionization or appearance potential (eV)	Method	Heat of formation (kJ mol^{-1})	Ref.
			NiPF$_3^+$			
NiPF$_3^+$	Ni(PF$_3$)$_4$		14.0	EI		2507
			NiP$_2$F$_6^+$			
NiP$_2$F$_6^+$	Ni(PF$_3$)$_4$		11.4	EI		2507
			NiP$_3$F$_9^+$			
NiP$_3$F$_9^+$	Ni(PF$_3$)$_4$		9.7	EI		2507
			NiP$_4$F$_{12}^+$			
NiP$_4$F$_{12}^+$	Ni(PF$_3$)$_4$		9.6 (V)	PE		3070
NiP$_4$F$_{12}^+$	Ni(PF$_3$)$_4$		9.69 (V)	PE		3088
NiP$_4$F$_{12}^+$	Ni(PF$_3$)$_4$		8.7	EI		2507
			C$_5$H$_7$O$_2$Ni$^+$			
C$_5$H$_7$O$_2$Ni$^+$	(CH$_3$COCHCOCH$_3$)$_2$Ni (Bis(2,4−pentanedionato)nickel)		13.5±0.2	EI		2731
			C$_9$H$_{11}$O$_4$Ni$^+$			
C$_9$H$_{11}$O$_4$Ni$^+$	(CH$_3$COCHCOCH$_3$)$_2$Ni (Bis(2,4−pentanedionato)nickel)		11.6±0.1	EI		2731
			C$_{10}$H$_{14}$O$_4$Ni$^+$			
C$_{10}$H$_{14}$O$_4$Ni$^+$	(CH$_3$COCHCOCH$_3$)$_2$Ni (Bis(2,4−pentanedionato)nickel)		8.23±0.05	EI		2731
			C$_5$H$_5$NONi$^+$			
C$_5$H$_5$NONi$^+$	C$_5$H$_5$NiNO (Cyclopentadienylnickel nitrosyl)		8.50±0.1	EI		2453
		Cu$^+$ $\Delta H_{f0}^\circ = 1082.6$ kJ mol^{-1} (258.7 kcal mol^{-1})				
Cu$^+$	Cu		7.726	S	1082.6	2113
See also − EI: 2990						
Cu$^+$	CuF$_2$		16.5±0.3	EI		1458

4.3. The Positive Ion Table—Continued

Ion	Reactant	Other products	Ionization or appearance potential (eV)	Method	Heat of formation (kJ mol^{-1})	Ref.
			CuF$^+$			
CuF$^+$	CuF		8.6±0.3	EI		1458
CuF$^+$	CuF$_2$	F	12.4±0.3	EI		1458
			CuF$_2^+$			
CuF$_2^+$	CuF$_2$		11.3±0.3	EI		1458
			CuCl$^+$			
CuCl$^+$	CuCl		10.7±0.3	EI		2990
			C$_5$H$_7$O$_2$Cu$^+$			
C$_5$H$_7$O$_2$Cu$^+$	(CH$_3$COCHCOCH$_3$)$_2$Cu (Bis(2,4–pentanedionato)copper)		13.1±0.2	EI		2731, 2959, 3405
			C$_6$H$_9$O$_2$Cu$^+$			
C$_6$H$_9$O$_2$Cu$^+$	(CH$_3$COC(CH$_3$)COCH$_3$)$_2$Cu (Bis(3–methyl–2,4–pentanedionato)copper)		10.8±0.1	EI		3405
			C$_{10}$H$_9$O$_2$Cu$^+$			
C$_{10}$H$_9$O$_2$Cu$^+$	(C$_6$H$_5$COCHCOCH$_3$)$_2$Cu (Bis(1–phenyl–1,3–butanedionato)copper)		10.7±0.1	EI		3405
			C$_{15}$H$_{11}$O$_2$Cu$^+$			
C$_{15}$H$_{11}$O$_2$Cu$^+$	(C$_6$H$_5$COCHCOC$_6$H$_5$)$_2$Cu (Bis(1,3–diphenyl–1,3–propanedionato)copper)		10.3±0.1	EI		3405
			C$_9$H$_{11}$O$_4$Cu$^+$			
C$_9$H$_{11}$O$_4$Cu$^+$	(CH$_3$COCHCOCH$_3$)$_2$Cu (Bis(2,4–pentanedionato)copper)		10.9±0.1	EI		2731, 2959
			C$_{10}$H$_{14}$O$_4$Cu$^+$			
C$_{10}$H$_{14}$O$_4$Cu$^+$	(CH$_3$COCHCOCH$_3$)$_2$Cu (Bis(2,4–pentanedionato)copper)		7.75±0.05	EI		2992
C$_{10}$H$_{14}$O$_4$Cu$^+$	(CH$_3$COCHCOCH$_3$)$_2$Cu (Bis(2,4–pentanedionato)copper)		8.31±0.05	EI		2731, 2959, 3405

4.3. The Positive Ion Table—Continued

Ion	Reactant	Other products	Ionization or appearance potential (eV)	Method	Heat of formation (kJ mol^{-1})	Ref.
		$C_{12}H_{18}O_4Cu^+$				
$C_{12}H_{18}O_4Cu^+$	$(CH_3COCHCOC_2H_5)_2Cu$ (Bis(2,4–hexanedionato)copper)		7.68±0.03	EI		2992
$C_{12}H_{18}O_4Cu^+$	$(CH_3COC(CH_3)COCH_3)_2Cu$ (Bis(3–methyl–2,4–pentanedionato)copper)		7.97±0.05	EI		3405
		$C_{14}H_{22}O_4Cu^+$				
$C_{14}H_{22}O_4Cu^+$	$(CH_3COCHCOC_3H_7)_2Cu$ (Bis(5–methyl–2,4–hexanedionato)copper)		7.61±0.06	EI		2992
		$C_{16}H_{26}O_4Cu^+$				
$C_{16}H_{26}O_4Cu^+$	$(CH_3COCHCOC_4H_9)_2Cu$ (Bis(5,5–dimethyl–2,4–hexanedionato)copper)		7.59±0.05	EI		2992
		$C_{20}H_{18}O_4Cu^+$				
$C_{20}H_{18}O_4Cu^+$	$(C_6H_5COCHCOCH_3)_2Cu$ (Bis(1–phenyl–1,3–butanedionato)copper)		8.37±0.05	EI		3405
		$C_{22}H_{22}O_4Cu^+$				
$C_{22}H_{22}O_4Cu^+$	$(CH_3COC(C_6H_5)COCH_3)_2Cu$ (Bis(3–phenyl–2,4–pentanedionato)copper)		8.05±0.05	EI		3405
		$C_{30}H_{22}O_4Cu^+$				
$C_{30}H_{22}O_4Cu^+$	$(C_6H_5COCHCOC_6H_5)_2Cu$ (Bis(1,3–diphenyl–1,3–propanedionato)copper)		8.28±0.05	EI		3405
		$C_5H_4O_2F_3Cu^+$				
$C_5H_4O_2F_3Cu^+$	$(CF_3COCHCOCH_3)_2Cu$ (Bis(1,1,1–trifluoro–2,4–pentanedionato)copper)		13.1±0.1	EI		2959, 3405
		$C_9H_8O_4F_3Cu^+$				
$C_9H_8O_4F_3Cu^+$	$(CF_3COCHCOCH_3)_2Cu$ (Bis(1,1,1–trifluoro–2,4–pentanedionato)copper)		11.5±0.1	EI		2959

4.3. The Positive Ion Table—Continued

Ion	Reactant	Other products	Ionization or appearance potential (eV)	Method	Heat of formation (kJ mol^{-1})	Ref.
$C_{10}H_8O_4F_6Cu^+$						
$C_{10}H_8O_4F_6Cu^+$	$(CF_3COCHCOCH_3)_2Cu$ (Bis(1,1,1−trifluoro−2,4−pentanedionato)copper)		8.61±0.05	EI		2992
$C_{10}H_8O_4F_6Cu^+$	$(CF_3COCHCOCH_3)_2Cu$ (Bis(1,1,1−trifluoro−2,4−pentanedionato)copper)		9.05±0.1	EI		2959, 3405
$C_{10}H_8O_4F_6Cu^+$	$CH_3COCHCOCH_3CuCF_3COCHCOCF_3$ (1,1,1,5,5,5−Hexafluoro−2,4−pentanedionato(2,4−pentanedionato)copper)		8.65±0.01	EI		2992
$C_{10}H_8O_4F_6Cu^+$	$CH_3COCHCOCH_3CuCF_3COCHCOCF_3$ (1,1,1,5,5,5−Hexafluoro−2,4−pentanedionato(2,4−pentanedionato)copper)		9.03±0.05	EI		3405
$C_{20}H_{12}O_4F_6Cu^+$						
$C_{20}H_{12}O_4F_6Cu^+$	$(C_6H_5COCHCOCF_3)_2Cu$ (Bis(1−phenyl−4,4,4−trifluoro−1,3−butanedionato)copper)		9.06±0.05	EI		3405
$C_{28}H_{16}O_4F_6Cu^+$						
$C_{28}H_{16}O_4F_6Cu^+$	$(C_{10}H_7COCHCOCF_3)_2Cu$ (Bis(1−(2−naphthyl)−4,4,4−trifluoro−1,3−butanedionato)copper)		8.39±0.05	EI		3405
$C_{16}H_8O_6F_6Cu^+$						
$C_{16}H_8O_6F_6Cu^+$	$(C_4H_3OCOCHCOCF_3)_2Cu$ (Bis(1−(2−furyl)−4,4,4−trifluoro−1,3−butanedionato)copper)		8.89±0.05	EI		3405
$C_9H_2O_4F_9Cu^+$						
$C_9H_2O_4F_9Cu^+$	$(CF_3COCHCOCF_3)_2Cu$ (Bis(1,1,1,5,5,5−hexafluoro−2,4−pentanedionato)copper)		11.6±0.1	EI		2959
$C_{10}H_2O_4F_{12}Cu^+$						
$C_{10}H_2O_4F_{12}Cu^+$	$(CF_3COCHCOCF_3)_2Cu$ (Bis(1,1,1,5,5,5−hexafluoro−2,4−pentanedionato)copper)		9.68±0.01	EI		2992
$C_{10}H_2O_4F_{12}Cu^+$	$(CF_3COCHCOCF_3)_2Cu$ (Bis(1,1,1,5,5,5−hexafluoro−2,4−pentanedionato)copper)		9.86±0.05	EI		2959, 3405
$C_{16}H_8O_4F_6S_2Cu^+$						
$C_{16}H_8O_4F_6S_2Cu^+$	$(C_4H_3SCOCHCOCF_3)_2Cu$ (Bis(1−(2−thienyl)−4,4,4−trifluoro−1,3−butanedionato)copper)		8.90±0.05	EI		3405

4.3. The Positive Ion Table—Continued

Ion	Reactant	Other products	Ionization or appearance potential (eV)	Method	Heat of forma- tion (kJ mol^{-1})	Ref.
Zn$^+$	$\Delta H_{f0}^\circ = 1036.6$ kJ mol^{-1} (247.7 kcal mol^{-1})					
Zn$^+$	Zn		9.394	S	1036.6	2113, 2659
See also – EI: 3128						
Zn$^+$	ZnCl$_2$		14.6±0.3	EI		2506
Zn$^+$	(CH$_3$)$_2$Zn	2CH$_3$	13.13±0.02	PI		2983
Zn$^+$	(CH$_3$)$_2$Zn	2CH$_3$	13.4±0.3	EI		2556
Zn$^+$	ZnBr$_2$		14.7±0.3	EI		2506
ZnH$^+$						
ZnH$^+$	(CH$_3$)$_2$Zn		13.9±0.4	EI		2556
ZnCl$^+$						
ZnCl$^+$	ZnCl$_2$	Cl	13.7±0.2	EI		2506
ZnCl$_2^+$						
ZnCl$_2^+$	ZnCl$_2$		11.75±0.23	EI		2506
CH$_3$Zn$^+$	$\Delta H_{f298}^\circ = 897$ kJ mol^{-1} (214 kcal mol^{-1})					
CH$_3$Zn$^+$	(CH$_3$)$_2$Zn	CH$_3$	10.22±0.02	PI	897	2983
CH$_3$Zn$^+$	(CH$_3$)$_2$Zn	CH$_3$	11.2±0.2	EI		2556
(CH$_3$)$_2$Zn$^+$	$\Delta H_{f298}^\circ = 921$ kJ mol^{-1} (220 kcal mol^{-1})					
C$_2$H$_6$Zn$^+$	(CH$_3$)$_2$Zn		9.00±0.02	PI	921	2983
(Threshold value approximately corrected for hot bands)						
C$_2$H$_6$Zn$^+$	(CH$_3$)$_2$Zn		8.86±0.15	EI		2556
See also – PE: 2984						
C$_5$H$_7$O$_2$Zn$^+$						
C$_5$H$_7$O$_2$Zn$^+$	(CH$_3$COCHCOCH$_3$)$_2$Zn (Bis(2,4–pentanedionato)zinc)		14.1±0.2	EI		2731, 2959
C$_9$H$_{11}$O$_4$Zn$^+$						
C$_9$H$_{11}$O$_4$Zn$^+$	(CH$_3$COCHCOCH$_3$)$_2$Zn (Bis(2,4–pentanedionato)zinc)		10.9±0.1	EI		2731, 2959

4.3. The Positive Ion Table—Continued

Ion	Reactant	Other products	Ionization or appearance potential (eV)	Method	Heat of formation (kJ mol^{-1})	Ref.
		C$_{10}$H$_{14}$O$_4$Zn$^+$				
C$_{10}$H$_{14}$O$_4$Zn$^+$	(CH$_3$COCHCOCH$_3$)$_2$Zn (Bis(2,4−pentanedionato)zinc)		8.62±0.05	EI		2731, 2959
		C$_5$H$_4$O$_2$F$_3$Zn$^+$				
C$_5$H$_4$O$_2$F$_3$Zn$^+$	(CF$_3$COCHCOCH$_3$)$_2$Zn (Bis(1,1,1−trifluoro−2,4−pentanedionato)zinc)		14.6±0.1	EI		2959
		C$_9$H$_8$O$_4$F$_3$Zn$^+$				
C$_9$H$_8$O$_4$F$_3$Zn$^+$	(CF$_3$COCHCOCH$_3$)$_2$Zn (Bis(1,1,1−trifluoro−2,4−pentanedionato)zinc)		11.3±0.1	EI		2959
		C$_5$HO$_2$F$_6$Zn$^+$				
C$_5$HO$_2$F$_6$Zn$^+$	(CF$_3$COCHCOCF$_3$)$_2$Zn (Bis(1,1,1,5,5,5−hexafluoro−2,4−pentanedionato)zinc)		15.3±0.2	EI		2959
		C$_9$H$_5$O$_4$F$_6$Zn$^+$				
C$_9$H$_5$O$_4$F$_6$Zn$^+$	(CF$_3$COCHCOCH$_3$)$_2$Zn (Bis(1,1,1−trifluoro−2,4−pentanedionato)zinc)		11.7±0.1	EI		2959
		C$_{10}$H$_8$O$_4$F$_6$Zn$^+$				
C$_{10}$H$_8$O$_4$F$_6$Zn$^+$	(CF$_3$COCHCOCH$_3$)$_2$Zn (Bis(1,1,1−trifluoro−2,4−pentanedionato)zinc)		9.40±0.1	EI		2959
		C$_9$H$_2$O$_4$F$_9$Zn$^+$				
C$_9$H$_2$O$_4$F$_9$Zn$^+$	(CF$_3$COCHCOCF$_3$)$_2$Zn (Bis(1,1,1,5,5,5−hexafluoro−2,4−pentanedionato)zinc)		11.35±0.1	EI		2959
		C$_{10}$H$_2$O$_4$F$_{12}$Zn$^+$				
C$_{10}$H$_2$O$_4$F$_{12}$Zn$^+$	(CF$_3$COCHCOCF$_3$)$_2$Zn (Bis(1,1,1,5,5,5−hexafluoro−2,4−pentanedionato)zinc)		10.07±0.05	EI		2959
		Ga$^+$ $\Delta H_{f0}^\circ \sim 855$ kJ mol^{-1} (204 kcal mol^{-1})				
Ga$^+$	Ga		5.999	S	~855	2113, 3167

See also − S: 3270
     EI: 2518, 2523, 2620, 3014, 3244

4.3. The Positive Ion Table—Continued

Ion	Reactant	Other products	Ionization or appearance potential (eV)	Method	Heat of formation (kJ mol^{-1})	Ref.
Ga$^+$	Ga$_2$S		8.7±0.5	EI		2569
Ga$^+$	Ga$_2$Se		8.4±0.5	EI		2569
Ga$^+$	Ga$_2$Te		8.2±0.5	EI		2569

Ga$_2^+$

Ion	Reactant	Other products	Ionization or appearance potential (eV)	Method	Heat of formation (kJ mol^{-1})	Ref.
Ga$_2^+$	Ga$_2$O	O	16±0.5	EI		3014
Ga$_2^+$	Ga$_2$S	S	10.2±0.5	EI		2569
Ga$_2^+$	Ga$_2$Se	Se	10.8±0.5	EI		2569
Ga$_2^+$	Ga$_2$Te	Te	9.8±0.5	EI		2569

GaO$^+$

Ion	Reactant	Other products	Ionization or appearance potential (eV)	Method	Heat of formation (kJ mol^{-1})	Ref.
GaO$^+$	GaO		9.4±0.5	EI		2518
GaO$^+$	Ga$_2$O	Ga	13.1±0.5	EI		3244

Ga$_2$O$^+$

Ion	Reactant	Other products	Ionization or appearance potential (eV)	Method	Heat of formation (kJ mol^{-1})	Ref.
Ga$_2$O$^+$	Ga$_2$O		7.4±0.5	EI		2569
Ga$_2$O$^+$	Ga$_2$O		8.0±0.5	EI		3014
Ga$_2$O$^+$	Ga$_2$O		8.4±0.6	EI		2518
Ga$_2$O$^+$	Ga$_2$O		12.1±0.5	EI		3244

GaF$^+$

Ion	Reactant	Other products	Ionization or appearance potential (eV)	Method	Heat of formation (kJ mol^{-1})	Ref.
GaF$^+$	GaF		9.6±0.5	EI		2523
GaF$^+$	GaF		10.6±0.4	EI		2620

GaF$_2^+$

Ion	Reactant	Other products	Ionization or appearance potential (eV)	Method	Heat of formation (kJ mol^{-1})	Ref.
GaF$_2^+$	GaF$_2$		13.3±0.5	EI		2523

GaS$^+$

Ion	Reactant	Other products	Ionization or appearance potential (eV)	Method	Heat of formation (kJ mol^{-1})	Ref.
GaS$^+$	Ga$_2$S	Ga	11.6±0.3	EI		2569

Ga$_2$S$^+$

Ion	Reactant	Other products	Ionization or appearance potential (eV)	Method	Heat of formation (kJ mol^{-1})	Ref.
Ga$_2$S$^+$	Ga$_2$S		7.5±0.3	EI		2569

LiGaO$^+$

Ion	Reactant	Other products	Ionization or appearance potential (eV)	Method	Heat of formation (kJ mol^{-1})	Ref.
LiGaO$^+$	LiGaO		7.0±0.3	EI		2565

4.3. The Positive Ion Table—Continued

Ion	Reactant	Other products	Ionization or appearance potential (eV)	Method	Heat of formation (kJ mol^{-1})	Ref.
			GaOF$^+$			
GaOF$^+$	GaOF		9.5±0.5	EI		2523

Ge$^+$ $\Delta H^\circ_{f0} = 1135.9$ kJ mol^{-1} (271.5 kcal mol^{-1})

Ion	Reactant	Other products	Ionization or appearance potential (eV)	Method	Heat of formation (kJ mol^{-1})	Ref.
Ge$^+$	Ge		7.899	S	1135.9	2113
See also – EI:	2502, 2714					
Ge$^+$	GeH$_4$	2H$_2$	10.7±0.2	EI		2116
Ge$^+$	Ge$_2$H$_6$		13.3±0.3	EI		2133
Ge$^+$	Ge$_3$H$_8$		16.3±0.3	EI		2133
Ge$^+$	GeO	O	14.0±1	EI		1255
Ge$^+$	GeF$_2$		18.8±0.3	EI		2566
Ge$^+$	GeCl$_2$		16.8±0.5	EI		2568
Ge$^+$	GeCl$_4$		21±1	EI		2568
Ge$^+$	(CH$_3$)$_4$Ge		18.1	EI		2980
Ge$^+$	(CH$_3$)$_4$Ge		19.2±0.5	EI		83
Ge$^+$	(C$_2$H$_5$)$_4$Ge		17.4	EI		2980
Ge$^+$	GeBr$_2$		15.5±0.5	EI		2568
Ge$^+$	GeBr$_4$		20±1	EI		2568
Ge$^+$	GeTe	Te	12.6±0.5	EI		1023
			Ge$_2^+$			
Ge$_2^+$	Ge$_2$		7.9±0.3	EI		2957
Ge$_2^+$	Ge$_2$		8.0±0.5	EI		2502
Ge$_2^+$	Ge$_2$H$_6$	3H$_2$	13.1±0.3	EI		2133
Ge$_2^+$	Ge$_3$H$_8$		15.8±0.3	EI		2133
			Ge$_3^+$			
Ge$_3^+$	Ge$_3$		8.0±0.5	EI		2502
Ge$_3^+$	Ge$_3$		8.4±0.3	EI		2957
Ge$_3^+$	Ge$_3$H$_8$		14.6±0.3	EI		2133
			Ge$_4^+$			
Ge$_4^+$	Ge$_4$		8.0±0.5	EI		2502
Ge$_4^+$	Ge$_4$		8.4±0.3	EI		2957
			GeH$^+$			
GeH$^+$	GeH$_4$	H$_2$+H	11.3±0.3	EI		2116
GeH$^+$	(CH$_3$)$_4$Ge		17.8	EI		2980
GeH$^+$	(C$_2$H$_5$)$_4$Ge		21.6	EI		2980
			GeH$_2^+$			
GeH$_2^+$	GeH$_4$	H$_2$	11.8±0.2	EI		2116

4.3. The Positive Ion Table—Continued

Ion	Reactant	Other products	Ionization or appearance potential (eV)	Method	Heat of formation (kJ mol^{-1})	Ref.
			GeH_3^+			
GeH_3^+	GeH_4	H	10.80±0.07	EI	2002,	2116
GeH_3^+	Ge_2H_6		10.26±0.10	EI		2002
GeH_3^+	$(CH_3)_4Ge$		17.1	EI		2980
GeH_3^+	$(C_2H_5)_4Ge$		16.3	EI		2980
GeH_3^+	H_3GeSiH_3		11.32±0.14	EI		2002
	GeH_4^+ $\Delta H_{f0}^\circ = 1171$ kJ mol^{-1} (280 kcal mol^{-1})					
GeH_4^+	GeH_4		11.31	PE	1171	3116
			Ge_2H^+			
Ge_2H^+	Ge_2H_6	$2H_2+H$	13.0±0.3	EI		2133
			$Ge_2H_2^+$			
$Ge_2H_2^+$	Ge_2H_6	$2H_2$	12.9±0.3	EI		2133
			$Ge_2H_3^+$			
$Ge_2H_3^+$	Ge_2H_6	H_2+H	12.8±0.3	EI		2133
			$Ge_2H_4^+$			
$Ge_2H_4^+$	Ge_2H_6	H_2	12.7±0.3	EI		2133
			$Ge_2H_5^+$			
$Ge_2H_5^+$	Ge_2H_6	H	12.6±0.3	EI		2133
			$Ge_2H_6^+$			
$Ge_2H_6^+$	Ge_2H_6		12.5±0.3	EI		2133
			Ge_3H^+			
Ge_3H^+	Ge_3H_8		11.8±0.3	EI		2133
			$Ge_3H_2^+$			
$Ge_3H_2^+$	Ge_3H_8	$3H_2$	10.7±0.3	EI		2133

4.3. The Positive Ion Table—Continued

Ion	Reactant	Other products	Ionization or appearance potential (eV)	Method	Heat of formation (kJ mol^{-1})	Ref.
		Ge$_3$H$_3^+$				
Ge$_3$H$_3^+$	Ge$_3$H$_8$	2H$_2$+H	10.6±0.3	EI		2133
		Ge$_3$H$_4^+$				
Ge$_3$H$_4^+$	Ge$_3$H$_8$	2H$_2$	10.4±0.3	EI		2133
		Ge$_3$H$_5^+$				
Ge$_3$H$_5^+$	Ge$_3$H$_8$	H$_2$+H	10.1±0.3	EI		2133
		Ge$_3$H$_6^+$				
Ge$_3$H$_6^+$	Ge$_3$H$_8$	H$_2$	10.0±0.3	EI		2133
		Ge$_3$H$_7^+$				
Ge$_3$H$_7^+$	Ge$_3$H$_8$	H	9.9±0.3	EI		2133
		Ge$_3$H$_8^+$				
Ge$_3$H$_8^+$	Ge$_3$H$_8$		9.6±0.3	EI		2133
		GeC$^+$				
GeC$^+$	GeC		10.3±0.3	EI		2957
		GeC$_2^+$				
GeC$_2^+$	GeC$_2$		10.1±0.3	EI		2957
		Ge$_2$C$^+$				
Ge$_2$C$^+$	Ge$_2$C		9.3±0.3	EI		2957
		Ge$_3$C$^+$				
Ge$_3$C$^+$	Ge$_3$C		8.6±0.3	EI		2957
		GeO$^+$				
GeO$^+$	GeO		~11.39	S		2714, 3414
GeO$^+$	GeO		11.10±0.10	EI		3414
GeO$^+$	GeO		11.5±0.5	EI		2714
GeO$^+$	GeO		10.1±0.8	EI		1255

4.3. The Positive Ion Table—Continued

Ion	Reactant	Other products	Ionization or appearance potential (eV)	Method	Heat of formation (kJ mol^{-1})	Ref.
			Ge$_2$O$^+$			
Ge$_2$O$^+$	Ge$_2$O$_2$	O	14.3±1.0	EI		1255
			Ge$_2$O$_2^+$			
Ge$_2$O$_2^+$	Ge$_2$O$_2$		8.7±1.0	EI		1255
			Ge$_3$O$_3^+$			
Ge$_3$O$_3^+$	Ge$_3$O$_3$		8.6±1.0	EI		1255
			GeF$^+$			
GeF$^+$	GeF		~7.28	S		2149
GeF$^+$	GeF		<9.1±0.2	EI		2566
GeF$^+$	GeF$_2$	F	14.0±0.3	EI		2566
			GeF$_2^+$			
GeF$_2^+$	GeF$_2$		11.8±0.1	EI		2566
			GeF$_4^+$			
GeF$_4^+$	GeF$_4$		15.69±0.02	PE		3362

See also – PE: 3059

Ion	Reactant	Other products	Ionization or appearance potential (eV)	Method	Heat of formation	Ref.
			Ge$_2$F$_4^+$			
Ge$_2$F$_4^+$	Ge$_2$F$_4$		10.6±0.3	EI		2566
			Ge$_3$F$_5^+$			
Ge$_3$F$_5^+$	Ge$_3$F$_6$	F	15.6±0.5	EI		2566
			GeSi$^+$			
GeSi$^+$	GeSi		8.2±0.3	EI		2957
			Ge$_2$Si$^+$			
Ge$_2$Si$^+$	Ge$_2$Si		8.4±0.3	EI		2957

4.3. The Positive Ion Table—Continued

Ion	Reactant	Other products	Ionization or appearance potential (eV)	Method	Heat of formation (kJ mol^{-1})	Ref.
			Ge$_3$Si$^+$			
Ge$_3$Si$^+$	Ge$_3$Si		8.6±0.3	EI		2957
			GeCl$^+$			
GeCl$^+$	GeCl$_2$	Cl	11.5±0.5	EI		2568
GeCl$^+$	GeCl$_4$		18±1	EI		2568
			GeCl$_2^+$			
GeCl$_2^+$	GeCl$_2$		10.4±0.3	EI		2568
GeCl$_2^+$	GeCl$_4$		17±1	EI		2568
			GeCl$_3^+$			
GeCl$_3^+$	GeCl$_4$	Cl	12.3±0.3	EI		2568

GeCl$_4^+$	$\Delta H_{f0}^\circ = 652$ kJ mol^{-1} (156 kcal mol^{-1})					
GeCl$_4^+$	GeCl$_4$		11.88±0.02	PE	652	3362
GeCl$_4^+$	GeCl$_4$		11.6±0.3	EI		2568

See also – PE: 3117

			CH$_3$Ge$^+$			
CH$_3$Ge$^+$	(CH$_3$)$_4$Ge		13.8	EI		2980
CH$_3$Ge$^+$	(CH$_3$)$_4$Ge		16.8±0.4	EI		83
			CH$_4$Ge$^+$			
CH$_4$Ge$^+$	(CH$_3$)$_4$Ge		13.2	EI		2980
			CH$_5$Ge$^+$			
CH$_5$Ge$^+$	(CH$_3$)$_4$Ge		13.4	EI		2980
			C$_2$H$_5$Ge$^+$			
C$_2$H$_5$Ge$^+$	(C$_2$H$_5$)$_4$Ge		16.7	EI		2980

4.3. The Positive Ion Table—Continued

Ion	Reactant	Other products	Ionization or appearance potential (eV)	Method	Heat of formation (kJ mol^{-1})	Ref.
			$C_2H_6Ge^+$			
$C_2H_6Ge^+$	$(CH_3)_4Ge$		14.1±0.2	EI		83
$C_2H_6Ge^+$	$(CH_3)_4Ge$		14.2	EI		2980
			$C_2H_7Ge^+$			
$C_2H_7Ge^+$	$(CH_3)_4Ge$		14.2	EI		2980
$C_2H_7Ge^+$	$(C_2H_5)_4Ge$		13.2	EI		2980
			$C_3H_9Ge^+$			
$C_3H_9Ge^+$	$(CH_3)_4Ge$	CH_3	10.05±0.14	EI		2720
$C_3H_9Ge^+$	$(CH_3)_4Ge$	CH_3	10.2±0.1	EI		83
$C_3H_9Ge^+$	$(CH_3)_4Ge$	CH_3	11.4	EI		2980
$C_3H_9Ge^+$	$(CH_3)_3GeGe(CH_3)_3$		11.3±0.2	EI		2700
$C_3H_9Ge^+$	$(CH_3)_3GeGe(C_2H_5)_3$		13.2±0.2	EI		2700
$C_3H_9Ge^+$	$(CH_3)_3SiGe(CH_3)_3$		9.99±0.14	EI		2720
$C_3H_9Ge^+$	$(CH_3)_3GeSn(CH_3)_3$		10.01±0.18	EI		2720
			$C_4H_{11}Ge^+$			
$C_4H_{11}Ge^+$	$(C_2H_5)_4Ge$		11.4	EI		2980
			$C_4H_{12}Ge^+$			
$C_4H_{12}Ge^+$	$(CH_3)_4Ge$		9.2±0.2	EI		83
$C_4H_{12}Ge^+$	$(CH_3)_4Ge$		9.29±0.14	EI		2720
$C_4H_{12}Ge^+$	$(CH_3)_4Ge$		11.2	EI		2980
			$C_6H_{15}Ge^+$			
$C_6H_{15}Ge^+$	$(C_2H_5)_4Ge$		9.6	EI		2980
$C_6H_{15}Ge^+$	$(CH_3)_3GeGe(C_2H_5)_3$		9.1±0.2	EI		2700
$C_6H_{15}Ge^+$	$(C_2H_5)_3GeGe(C_2H_5)_3$		10.9±0.2	EI		2700
			$C_8H_{20}Ge^+$			
$C_8H_{20}Ge^+$	$(C_2H_5)_4Ge$		9.8	EI		2980
			$C_3H_{11}Ge_2^+$			
$C_3H_{11}Ge_2^+$	$(CH_3)_3GeGe(C_2H_5)_3$		12.7±0.2	EI		2700

4.3. The Positive Ion Table—Continued

Ion	Reactant	Other products	Ionization or appearance potential (eV)	Method	Heat of formation (kJ mol^{-1})	Ref.
			$C_4H_{13}Ge_2^+$			
$C_4H_{13}Ge_2^+$	$(CH_3)_3GeGe(C_2H_5)_3$		12.8±0.2	EI		2700
$C_4H_{13}Ge_2^+$	$(C_2H_5)_3GeGe(C_2H_5)_3$		14.9±0.2	EI		2700
			$C_5H_{15}Ge_2^+$			
$C_5H_{15}Ge_2^+$	$(CH_3)_3GeGe(CH_3)_3$	CH_3	10.8±0.2	EI		2700
$C_5H_{15}Ge_2^+$	$(CH_3)_3GeGe(C_2H_5)_3$		11.8±0.2	EI		2700
			$C_6H_{17}Ge_2^+$			
$C_6H_{17}Ge_2^+$	$(CH_3)_3GeGe(C_2H_5)_3$		12.1±0.2	EI		2700
$C_6H_{17}Ge_2^+$	$(C_2H_5)_3GeGe(C_2H_5)_3$		12.9±0.2	EI		2700
			$C_6H_{18}Ge_2^+$			
$C_6H_{18}Ge_2^+$	$(CH_3)_3GeGe(CH_3)_3$		7.76±0.01	EI		2900
$C_6H_{18}Ge_2^+$	$(CH_3)_3GeGe(CH_3)_3$		8.5±0.2	EI		2700
			$C_7H_{19}Ge_2^+$			
$C_7H_{19}Ge_2^+$	$(CH_3)_3GeGe(C_2H_5)_3$		10.7±0.2	EI		2700
			$C_8H_{21}Ge_2^+$			
$C_8H_{21}Ge_2^+$	$(CH_3)_3GeGe(C_2H_5)_3$	CH_3	11.2±0.2	EI		2700
$C_8H_{21}Ge_2^+$	$(C_2H_5)_3GeGe(C_2H_5)_3$		11.7±0.2	EI		2700
			$C_9H_{24}Ge_2^+$			
$C_9H_{24}Ge_2^+$	$(CH_3)_3GeGe(C_2H_5)_3$		7.6±0.2	EI		2700
			$C_{10}H_{25}Ge_2^+$			
$C_{10}H_{25}Ge_2^+$	$(C_2H_5)_3GeGe(C_2H_5)_3$		9.6±0.2	EI		2700
			$C_{12}H_{30}Ge_2^+$			
$C_{12}H_{30}Ge_2^+$	$(C_2H_5)_3GeGe(C_2H_5)_3$		7.4±0.2	EI		2700
$C_{12}H_{30}Ge_2^+$	$(C_2H_5)_3GeGe(C_2H_5)_3$		7.48±0.01	EI		2900

4.3. The Positive Ion Table—Continued

Ion	Reactant	Other products	Ionization or appearance potential (eV)	Method	Heat of formation (kJ mol^{-1})	Ref.
			GeOF$_3^+$			
GeOF$_3^+$	GeOF$_4$	F	15.6±0.5	EI		2566
			GeOF$_4^+$			
GeOF$_4^+$	GeOF$_4$		13.6±0.5	EI		2566
			GeSiH$_6^+$			
GeSiH$_6^+$	H$_3$GeSiH$_3$		10.20±0.03	EI		2002
			GeSiC$^+$			
GeSiC$^+$	GeSiC		9.6±0.3	EI		2957
			Ge$_2$SiC$^+$			
Ge$_2$SiC$^+$	Ge$_2$SiC		8.9±0.3	EI		2957
			C$_6$H$_{18}$SiGe$^+$			
C$_6$H$_{18}$SiGe$^+$	(CH$_3$)$_3$SiGe(CH$_3$)$_3$		8.31±0.10	EI		2720
		As$^+$ $\Delta H_{f0}^\circ = 1245.9$ kJ mol^{-1} (297.8 kcal mol^{-1})				
As$^+$	As		9.7883±0.0002	S	1245.9	3430
See also – S: 2113						
As$^+$	As$_2$	As	13.9±0.4	EI		3194
As$^+$	AsH$_3$	H$_2$+H	14.8±0.2	EI		2116
As$^+$	AsH$_3$	H$_2$+H	15.0±0.3	EI		2584
As$^+$	As$_2$H$_4$		14.3±0.3	EI		2133
			As$_2^+$			
As$_2^+$	As$_2$		10.4±0.4	EI		3194
As$_2^+$	As$_2$H$_4$	2H$_2$	13.0±0.3	EI		2133
			As$_3^+$			
As$_3^+$	As$_4$	As	13.5±0.4	EI		3194

4.3. The Positive Ion Table—Continued

Ion	Reactant	Other products	Ionization or appearance potential (eV)	Method	Heat of formation (kJ mol⁻¹)	Ref.

$$As_4^+$$

Ion	Reactant	Other products	Ionization or appearance potential (eV)	Method	Heat of formation (kJ mol⁻¹)	Ref.
As_4^+	As_4		9.07±0.07	EI		1047
As_4^+	As_4		10.1±0.4	EI		3194

$$AsH^+$$

AsH^+	AsH_3	H_2	12.4±0.2	EI		2116
AsH^+	AsH_3	H_2	12.5±0.2	EI		2584

$$AsH_2^+$$

AsH_2^+	AsH_3	H	13.4±0.2	EI		2584
AsH_2^+	AsH_3	H	14.5±0.2	EI		2116
AsH_2^+	As_2H_4		12.9±0.4	EI		2584
AsH_2^+	H_2AsSiH_3		12.4±0.2	EI		2584

$AsH_3^+(^2A_1)$ $\Delta H_{f0}^\circ = 1043$ kJ mol⁻¹ (249 kcal mol⁻¹)

$AsH_3^+(^2A_1)$	AsH_3		10.03	PI	1042	1091
$AsH_3^+(^2A_1)$	AsH_3		10.06±0.03	PE	1045	3085
$AsH_3^+(^2E)$	AsH_3		11.9±0.1	PE		3085

See also – EI: 1007, 2116, 2584

AsH_3^+	H_2AsSiH_3		13.1±0.6	EI		2584

$$As_2H^+$$

As_2H^+	As_2H_4	H_2+H	12.7±0.3	EI		2133

$$As_2H_2^+$$

$As_2H_2^+$	As_2H_4	H_2	12.6±0.3	EI		2133

$$As_2H_3^+$$

$As_2H_3^+$	As_2H_4	H	12.5±0.3	EI		2133

$$As_2H_4^+$$

$As_2H_4^+$	As_2H_4		12.2±0.3	EI		2133

4.3. The Positive Ion Table—Continued

Ion	Reactant	Other products	Ionization or appearance potential (eV)	Method	Heat of formation (kJ mol^{-1})	Ref.
			AsF$_3^+$			
AsF$_3^+$	AsF$_3$		13.00 (V)	PE		3119
			AsSi$^+$			
AsSi$^+$	H$_2$AsSiH$_3$	2H$_2$+H	16.8±0.3	EI		2584
		AsCl$_3^+$	$\Delta H^\circ_{f0} \sim$ 774 kJ mol^{-1} (185 kcal mol^{-1})			
AsCl$_3^+$	AsCl$_3$		10.72±0.07	PE	774	3168
AsCl$_3^+$	AsCl$_3$		11.7±0.1	EI		1007

See also – PE: 3119

Ion	Reactant	Other products	Ionization or appearance potential (eV)	Method	Heat of formation (kJ mol^{-1})	Ref.
			CH$_5$As$^+$			
CH$_5$As$^+$	CH$_3$AsH$_2$		9.7±0.1	EI		1007
			C$_2$H$_7$As$^+$			
C$_2$H$_7$As$^+$	(CH$_3$)$_2$AsH		9.0±0.1	EI		1007
			C$_3$H$_9$As$^+$			
C$_3$H$_9$As$^+$	(CH$_3$)$_3$As		8.3±0.1	EI		1007
	C$_{18}$H$_{15}$As$^+$ (Triphenylarsine)		$\Delta H^\circ_{f298} \sim$ 1117 kJ mol^{-1} (267 kcal mol^{-1})			
C$_{18}$H$_{15}$As$^+$	(C$_6$H$_5$)$_3$As (Triphenylarsine)		7.34±0.07	PI	1117	1140
			C$_3$F$_9$As$^+$			
C$_3$F$_9$As$^+$	(CF$_3$)$_3$As		11.0±0.1	EI		1007
			AsSiH$^+$			
AsSiH$^+$	H$_2$AsSiH$_3$	2H$_2$	13.4±0.1	EI		2584
			AsSiH$_2^+$			
AsSiH$_2^+$	H$_2$AsSiH$_3$	H$_2$+H	13.7±0.2	EI		2584

4.3. The Positive Ion Table—Continued

Ion	Reactant	Other products	Ionization or appearance potential (eV)	Method	Heat of formation (kJ mol^{-1})	Ref.
		AsSiH$_3^+$				
AsSiH$_3^+$	H$_2$AsSiH$_3$	H$_2$	9.7±0.1	EI		2584
		AsSiH$_4^+$				
AsSiH$_4^+$	H$_2$AsSiH$_3$	H	11.0±0.3	EI		2584
		AsSiH$_5^+$				
AsSiH$_5^+$	H$_2$AsSiH$_3$		10.1±0.1	EI		2584
		C$_3$H$_6$F$_3$As$^+$				
C$_3$H$_6$F$_3$As$^+$	(CH$_3$)$_2$AsCF$_3$		9.2±0.1	EI		1007
		C$_2$HF$_6$As$^+$				
C$_2$HF$_6$As$^+$	(CF$_3$)$_2$AsH		10.9±0.1	EI		1007
		C$_3$H$_3$F$_6$As$^+$				
C$_3$H$_3$F$_6$As$^+$	(CF$_3$)$_2$AsCH$_3$		10.5±0.1	EI		1007
		C$_2$H$_6$ClAs$^+$				
C$_2$H$_6$ClAs$^+$	(CH$_3$)$_2$AsCl		9.9±0.1	EI		1007
		CH$_3$Cl$_2$As$^+$				
CH$_3$Cl$_2$As$^+$	CH$_3$AsCl$_2$		10.4±0.1	EI		1007
		C$_2$F$_6$ClAs$^+$				
C$_2$F$_6$ClAs$^+$	(CF$_3$)$_2$AsCl		11.0±0.1	EI		1007
		C$_7$H$_8$O$_2$MnAs$^+$				
C$_7$H$_8$O$_2$MnAs$^+$	C$_5$H$_5$Mn(CO)$_2$AsH$_3$ (Arsinecyclopentadienylmanganese dicarbonyl)		7.16±0.1	EI		2703

4.3. The Positive Ion Table—Continued

Ion	Reactant	Other products	Ionization or appearance potential (eV)	Method	Heat of forma- tion (kJ mol^{-1})	Ref.

Se$^+$ $\Delta H_{f0}^\circ = 1167.3$ kJ mol^{-1} (279.0 kcal mol^{-1})

Ion	Reactant	Other products	Ionization or appearance potential (eV)	Method	Heat of formation (kJ mol^{-1})	Ref.
Se$^+$	Se		9.752	S	1167.3	2113

See also — EI: 2469

Ion	Reactant	Other products	Ionization or appearance potential (eV)	Method	Heat of formation	Ref.
Se$^+$	Se$_2$	Se	13.16–13.36	PI		2609

(Threshold value approximately corrected to 0 K)

Ion	Reactant	Other products	Ionization or appearance potential (eV)	Method	Heat of formation	Ref.
Se$^+$	Se$_2$	Se	12.6±0.5	EI		2569
Se$^+$	Se$_2$	Se	12.0±0.5	EI		2747
Se$^+$	SeO$_2$		16.8±0.5	EI		3193
Se$^+$	SeO$_3$		18.3±0.5	EI		3193
Se$^+$	Ga$_2$Se		15.6±0.5	EI		2569
Se$^+$	InSe?		14.2±0.5	EI		2469
Se$^+$	SnSe	Sn	12.7±0.5	EI		2063
Se$^+$	SbSe?		13.2±0.3	EI		3249

Se$_2^+$ $\Delta H_{f0}^\circ = 1004$ kJ mol^{-1} (240 kcal mol^{-1})

Ion	Reactant	Other products	Ionization or appearance potential (eV)	Method	Heat of formation	Ref.
Se$_2^+$	Se$_2$		~8.84	S		3218
Se$_2^+$	Se$_2$		8.88±0.03	PI	1004	2609

(Threshold value approximately corrected to 0 K)

See also — EI: 2426, 2469, 2569, 3249

Ion	Reactant	Other products	Ionization or appearance potential (eV)	Method	Heat of formation	Ref.
Se$_2^+$	C$_2$H$_5$SeSeC$_2$H$_5$		11.0±0.3	EI		3285

Se$_5^+$

Ion	Reactant	Other products	Ionization or appearance potential (eV)	Method	Heat of formation	Ref.
Se$_5^+$	Se$_5$		8.63±0.2	EI		2426
Se$_5^+$	Se$_5$		9.2±0.2	EI		2446

Both references report also appearance potentials for the fragment ions Se$_3^+$ and Se$_4^+$. Their parentage is not clear.

Se$_6^+$

Ion	Reactant	Other products	Ionization or appearance potential (eV)	Method	Heat of formation	Ref.
Se$_6^+$	Se$_6$		8.88±0.2	EI		2426
Se$_6^+$	Se$_6$		9.08±0.05	EI		2446

Se$_7^+$

Ion	Reactant	Other products	Ionization or appearance potential (eV)	Method	Heat of formation	Ref.
Se$_7^+$	Se$_7$		8.38±0.2	EI		2426
Se$_7^+$	Se$_7$		8.87±0.05	EI		2446

Se$_8^+$

Ion	Reactant	Other products	Ionization or appearance potential (eV)	Method	Heat of formation	Ref.
Se$_8^+$	Se$_8$		8.63±0.2	EI		2426
Se$_8^+$	Se$_8$		8.97±0.05	EI		2446

4.3. The Positive Ion Table—Continued

Ion	Reactant	Other products	Ionization or appearance potential (eV)	Method	Heat of formation (kJ mol^{-1})	Ref.
	$H_2Se^+(^2B_1)$		$\Delta H_{f0}^\circ = 987$ kJ mol^{-1} (236 kcal mol^{-1})			
$H_2Se^+(^2B_1)$	H_2Se		9.882 ± 0.001	S	987	3317
$H_2Se^+(^2B_1)$	H_2Se		9.93	PE		3170
$H_2Se^+(^2A_1)$	H_2Se		12.17	PE		3170
$H_2Se^+(^2B_2)$	H_2Se		13.61	PE		3170
H_2Se^+	$(C_2H_5)_2Se$		12.8 ± 0.3	EI		3285
		HSe_2^+				
HSe_2^+	$C_2H_5SeSeC_2H_5$		14.5 ± 0.3	EI		3285
		$H_2Se_2^+$				
$H_2Se_2^+$	$C_2H_5SeSeC_2H_5$		12.2 ± 0.3	EI		3285
		BSe^+				
BSe^+	BSe		10.3 ± 1	EI		3206
		SeO^+				
SeO^+	SeO_2	O	13.0 ± 0.5	EI		3193
SeO^+	SeO_3		14.8 ± 0.5	EI		3193
		SeO_2^+				
SeO_2^+	SeO_2		11.5 ± 0.5	EI		3193
SeO_2^+	SeO_3	O	13.8 ± 0.5	EI		3193
		SeO_3^+				
SeO_3^+	SeO_3		11.6 ± 0.5	EI		3193
		$AlSe^+$				
$AlSe^+$	$AlSe$		8.3 ± 0.5	EI		2449
		Al_2Se^+				
Al_2Se^+	Al_2Se		6.0 ± 0.5	EI		2449
		$Al_2Se_2^+$				
$Al_2Se_2^+$	Al_2Se_2		9.0 ± 0.5	EI		2449

4.3. The Positive Ion Table—Continued

Ion	Reactant	Other products	Ionization or appearance potential (eV)	Method	Heat of formation (kJ mol^{-1})	Ref.
		S$_7$Se$^+$				
S$_7$Se$^+$	S$_7$Se		9.35±0.2	EI		2426
		ScSe$^+$				
ScSe$^+$	ScSe		5.62	EI		2721
		GaSe$^+$				
GaSe$^+$	Ga$_2$Se	Ga	11.2±0.3	EI		2569
		Ga$_2$Se$^+$				
Ga$_2$Se$^+$	Ga$_2$Se		7.4±0.3	EI		2569
		C$_2$H$_4$Se$^+$				
C$_2$H$_4$Se$^+$	C$_2$H$_5$SeSeC$_2$H$_5$		14.2±0.3	EI		3285
		C$_2$H$_5$Se$^+$				
C$_2$H$_5$Se$^+$	(C$_2$H$_5$)$_2$Se		11.9±0.3	EI		3285
C$_2$H$_5$Se$^+$	C$_2$H$_5$SeSeC$_2$H$_5$		13.4±0.3	EI		3285
C$_2$H$_5$Se$^+$	CH$_3$SeCH$_2$CH$_2$CH(NH$_2$)COOH		12.05±0.14	EI		2587
		C$_2$H$_6$Se$^+$				
C$_2$H$_6$Se$^+$	(C$_2$H$_5$)$_2$Se		10.6±0.3	EI		3285
C$_2$H$_6$Se$^+$	C$_2$H$_5$SeSeC$_2$H$_5$		10.0±0.3	EI		3285
		C$_3$H$_7$Se$^+$				
C$_3$H$_7$Se$^+$	(C$_2$H$_5$)$_2$Se	CH$_3$	10.5±0.3	EI		3285
		C$_4$H$_4$Se$^+$				
C$_4$H$_4$Se$^+$	C$_4$H$_4$Se (Selenophene)		9.01±0.05	EI		3233
		C$_4$H$_{10}$Se$^+$				
C$_4$H$_{10}$Se$^+$	(C$_2$H$_5$)$_2$Se		8.3±0.3	EI		3285

4.3. The Positive Ion Table—Continued

Ion	Reactant	Other products	Ionization or appearance potential (eV)	Method	Heat of formation (kJ mol^{-1})	Ref.
		$C_2H_5Se_2^+$				
$C_2H_5Se_2^+$	$C_2H_5SeSeC_2H_5$		12.8±0.3	EI		3285
		$C_2H_6Se_2^+$				
$C_2H_6Se_2^+$	$C_2H_5SeSeC_2H_5$		10.2±0.3	EI		3285
		$C_4H_{10}Se_2^+$				
$C_4H_{10}Se_2^+$	$C_2H_5SeSeC_2H_5$		7.4±0.3	EI		3285
		$C_4H_{10}NSe^+$				
$C_4H_{10}NSe^+$	$CH_3SeCH_2CH_2CH(NH_2)COOH$		9.59±0.20	EI		2587
		$C_5H_{11}NO_2Se^+$				
$C_5H_{11}NO_2Se^+$	$CH_3SeCH_2CH_2CH(NH_2)COOH$		8.29±0.15	EI		2587

Br^+ $\Delta H_{f0}^\circ = 1257.8$ kJ mol^{-1} (300.6 kcal mol^{-1})

Ion	Reactant	Other products	Ionization or appearance potential (eV)	Method	Heat of formation (kJ mol^{-1})	Ref.
Br^+	Br		11.814	S	1257.8	2113
See also – EI: 439						
Br^+	Br_2	Br^-	10.31	PI	(1247)	416
Br^+	Br_2	Br^-	10.48±0.02	PI		213
(Position of peak maximum)						
Br^+	Br_2	Br^-	10.38±0.05	RPD		292
See also – EI: 357						
Br^+	CBr_4		18.1±0.2	EI		1246
Br^+	$MgBr_2$		16±1	EI		178
Br^+	PBr_3		17.1±0.5	EI		2506
Br^+	$ZnBr_2$		17.7±0.3	EI		2506
Br^+	$GeBr_2$		17.5±0.5	EI		2568
Br^+	$GeBr_4$		22±1	EI		2568
Br^+	CH_3Br	CH_3	14.7±0.05	RPD		2154
Br^+	C_2H_5Br		18.6±0.3	EI		356
Br^+	$CH_3C{\equiv}CBr$		16.0±0.5	EI		13
Br^+	$BrCN$	CN^-	11.9±0.2	EI		73
Br^+	$BrCN$	CN	15.52±0.02	PI	(1258)	2621
See also – EI: 73						
Br^+	CF_3Br		16.7±0.1	EI		439
See also – EI: 24						
Br^+	$HgBr_2$		16.7±0.2	EI		2506

4.3. The Positive Ion Table—Continued

Ion	Reactant	Other products	Ionization or appearance potential (eV)	Method	Heat of formation (kJ mol⁻¹)	Ref.
$Br_2^+(^2\Pi_{3/2g})$			$\Delta H_{f0}^\circ = 1060$ kJ mol⁻¹ (253 kcal mol⁻¹)			
$Br_2^+(^2\Pi_{1/2g})$			$\Delta H_{f0}^\circ = 1094$ kJ mol⁻¹ (261 kcal mol⁻¹)			
$Br_2^+(^2\Pi_{3/2u})$			$\Delta H_{f0}^\circ = 1243$ kJ mol⁻¹ (297 kcal mol⁻¹)			
$Br_2^+(^2\Sigma_g^+)$			$\Delta H_{f0}^\circ = 1424$ kJ mol⁻¹ (340 kcal mol⁻¹)			
$Br_2^+(^2\Pi_{3/2g})$	Br_2		10.559±0.01	S		2671, 3097
$Br_2^+(^2\Pi_{3/2g})$	Br_2		10.52±0.01	PI	1061	3036, 3413
(Threshold value corrected for hot bands)						
$Br_2^+(^2\Pi_{3/2g})$	Br_2		10.51±0.01	PE	1060	3409
$Br_2^+(^2\Pi_{3/2g})$	Br_2		10.51	PE	1060	3275
$Br_2^+(^2\Pi_{3/2g})$	Br_2		10.7±0.1	EDD		3279

The authors of refs. 3036, 3413 conclude that the spectroscopic measurement refers to the first vibrationally excited state of the ion.

Ion	Reactant	Other products	Ionization or appearance potential (eV)	Method	Heat of formation (kJ mol⁻¹)	Ref.
$Br_2^+(^2\Pi_{1/2g})$	Br_2		10.948±0.01	S		2671, 3097
$Br_2^+(^2\Pi_{1/2g})$	Br_2		10.86±0.01	PE	1094	3409
$Br_2^+(^2\Pi_{1/2g})$	Br_2		10.86	PE	1094	3275
$Br_2^+(^2\Pi_{3/2u})$	Br_2		12.41±0.02	PE	1243	3409
$Br_2^+(^2\Pi_{3/2u})$	Br_2		12.5	PE		3275
$Br_2^+(^2\Pi_{1/2u})$	Br_2		~12.75	PE		3409
$Br_2^+(^2\Pi_{1/2u})$	Br_2		~12.8	PE		3275
$Br_2^+(^2\Sigma_g^+)$	Br_2		14.28±0.02	PE	1424	3409
$Br_2^+(^2\Sigma_g^+)$	Br_2		14.3	PE		3275

See also − PI: 182, 213, 416
PE: 2815, 3277
EI: 75, 292, 357

Ion	Reactant	Other products	Ionization or appearance potential (eV)	Method	Heat of formation (kJ mol⁻¹)	Ref.
Br_2^+	CBr_4		13.3±0.1	EDD		3279
Br_2^+	$MgBr_2$	Mg	17±1	EI		178
Br_2^+	$GeBr_4$		19.5±0.5	EI		2568

Br_2^{+2}

Ion	Reactant	Other products	Ionization or appearance potential (eV)	Method	Heat of formation (kJ mol⁻¹)	Ref.
Br_2^{+2}	Br_2		30.0	EI		75

Ion	Reactant	Other products	Ionization or appearance potential (eV)	Method	Heat of formation (kJ mol⁻¹)	Ref.
$HBr^+(^2\Pi_{3/2})$			$\Delta H_{f0}^\circ = 1097$ kJ mol⁻¹ (262 kcal mol⁻¹)			
$HBr^+(^2\Pi_{1/2})$			$\Delta H_{f0}^\circ = 1131$ kJ mol⁻¹ (270 kcal mol⁻¹)			
$HBr^+(^2\Sigma^+)$			$\Delta H_{f0}^\circ = 1448$ kJ mol⁻¹ (346 kcal mol⁻¹)			
$HBr^+(^2\Pi_{3/2})$	HBr		11.677±0.004	D	1098	3278
$HBr^+(^2\Pi_{3/2})$	HBr		11.62±0.01	PI	1093	182, 416
$HBr^+(^2\Pi_{3/2})$	HBr		11.67±0.01	PE	1097	2819
$HBr^+(^2\Pi_{3/2})$	HBr		11.71±0.01	PE	1101	2815
$HBr^+(^2\Pi_{1/2})$	HBr		12.00±0.01	PE	1129	2819
$HBr^+(^2\Pi_{1/2})$	HBr		12.03±0.01	PE	1132	2815

4.3. The Positive Ion Table—Continued

Ion	Reactant	Other products	Ionization or appearance potential (eV)	Method	Heat of formation (kJ mol^{-1})	Ref.
HBr$^+$($^2\Sigma^+$)	HBr		15.29	PE	1447	2819
HBr$^+$($^2\Sigma^+$)	HBr		15.31±0.01	PE	1449	2815

DBr^{+2}

DBr^{+2}	DBr		33.2±0.3	FD		212

LiBr$^+$

LiBr$^+$	LiBr		9.4	EI		2179

Li$_2$Br$^+$

Li$_2$Br$^+$	Li$_2$Br$_2$	Br	9.9	EI		2179

BBr$^+$

BBr$^+$	BBr?		10.7±0.2	EI		206
(Probably formed by thermal decomposition of BBr$_3$)						
BBr$^+$	BBr$_3$		15.0±0.2	EI		206

BBr$_2^+$

BBr$_2^+$	BBr$_3$	Br	10.7±0.2	EI		206

BBr$_3^+$(^{2}A$_2'$) $\Delta H_{f0}^\circ = 831$ kJ mol^{-1} (199 kcal mol^{-1})

BBr$_3^+$(^{2}A$_2'$)	BBr$_3$		10.51±0.02	PE	831	3375
BBr$_3^+$(^{2}E')	BBr$_3$		11.13±0.03	PE		3375
BBr$_3^+$(^{2}E'')	BBr$_3$		11.71±0.04 (V)	PE		3375
BBr$_3^+$(^{2}A$_2''$)	BBr$_3$		12.89±0.04	PE		3375
BBr$_3^+$(^{2}E')	BBr$_3$		13.67±0.03	PE		3375
BBr$_3^+$(^{2}E')	BBr$_3$		14.46±0.03 (V)	PE		3375
BBr$_3^+$(^{2}A$_1'$)	BBr$_3$		16.63±0.05	PE		3375
BBr$_3^+$*	BBr$_3$		17.14±0.04 (V)	PE		3375

This last band may be due to an impurity. However, it is also observed in ref. 3119 where it is assigned to the ^{2}A$_1'$ state. There is disagreement on some of the assignments, see ref. 2834. Additional fine structure is resolved in ref. 3119.

See also – PE: 2834, 3119
          EI: 206, 2512, 2513, 3227

CBr$^+$

CBr$^+$	CBr		10.43±0.02	EI		129
CBr$^+$	CBr≡CBr	CBr	16.1±0.1	EDD		3177

4.3. The Positive Ion Table—Continued

Ion	Reactant	Other products	Ionization or appearance potential (eV)	Method	Heat of formation (kJ mol^{-1})	Ref.
CBr^+	CBr_4		16.35 ± 0.13	EI		129
CBr^+	CBr_4		17.5 ± 0.2	EI		1246
CBr^+	$CH\equiv CBr$	CH	18.1 ± 0.1	EDD		3177
CBr^+	BrCN	N	17.4 ± 0.2	EI		73
CBr^+	$CCl\equiv CBr$	CCl	16.0 ± 0.1	EDD		3177
CBr^+	$CBr\equiv CI$	CI	16.9 ± 0.1	EDD		3177

CBr_2^+

Ion	Reactant	Other products	Ionization or appearance potential (eV)	Method	Heat of formation (kJ mol^{-1})	Ref.
CBr_2^+	CBr_2		10.11 ± 0.09	EI		129
CBr_2^+	CBr_4		12.30 ± 0.08	EI		129
CBr_2^+	CBr_4		14.6 ± 0.3	EI		1246

$C_2Br_2^+$

Ion	Reactant	Other products	Ionization or appearance potential (eV)	Method	Heat of formation (kJ mol^{-1})	Ref.
$C_2Br_2^+(^2\Pi_{3/2u})$	$CBr\equiv CBr$		9.67	PE		3121
$C_2Br_2^+(^2\Pi_{3/2u})$	$CBr\equiv CBr$		9.7 ± 0.1	EDD		3177
$C_2Br_2^+(^2\Pi_{1/2u})$	$CBr\equiv CBr$		9.87	PE		3121
$C_2Br_2^+(^2\Pi_{3/2g})$	$CBr\equiv CBr$		12.11	PE		3121
$C_2Br_2^+(^2\Pi_{1/2g})$	$CBr\equiv CBr$		12.40	PE		3121
$C_2Br_2^+(^2\Pi_{3/2u})$	$CBr\equiv CBr$		13.31 (V)	PE		3121
$C_2Br_2^+(^2\Pi_{1/2u})$	$CBr\equiv CBr$		13.45 (V)	PE		3121
$C_2Br_2^+(^2\Sigma_g^+)$	$CBr\equiv CBr$		15.64 (V)	PE		3121
$C_2Br_2^+(^2\Sigma_u^+)$	$CBr\equiv CBr$		16.90 (V)	PE		3121

$C_2Br_2^{+2}$

Ion	Reactant	Other products	Ionization or appearance potential (eV)	Method	Heat of formation (kJ mol^{-1})	Ref.
$C_2Br_2^{+2}$	$CBr\equiv CBr$		25.8 ± 0.1	EDD		3177

CBr_3^+

Ion	Reactant	Other products	Ionization or appearance potential (eV)	Method	Heat of formation (kJ mol^{-1})	Ref.
CBr_3^+	CBr_4	Br	9.95 ± 0.05	EI		129
CBr_3^+	CBr_4	Br	11.3 ± 0.2	EI		1246

CBr_4^+

Ion	Reactant	Other products	Ionization or appearance potential (eV)	Method	Heat of formation (kJ mol^{-1})	Ref.
CBr_4^+	CBr_4		10.39 (V)	PE		3119
CBr_4^+	CBr_4		10.40 (V)	PE		3117

Orbital assignments are discussed in R. N. Dixon, J. N. Murrell and B. Narayan, Mol. Phys. **20**, 611 (1971).

See also – EI: 1246

CBr_4^{+2}

Ion	Reactant	Other products	Ionization or appearance potential (eV)	Method	Heat of formation (kJ mol^{-1})	Ref.
CBr_4^{+2}	CBr_4		$\leq28.6\pm0.7$ (V)	AUG		3424

4.3. The Positive Ion Table—Continued

Ion	Reactant	Other products	Ionization or appearance potential (eV)	Method	Heat of formation (kJ mol⁻¹)	Ref.

BrF⁺

Ion	Reactant	Other products	Ionization or appearance potential (eV)	Method	Heat of formation (kJ mol⁻¹)	Ref.
BrF^+	BrF		11.8±0.2	EI		357
BrF^+	BrF_5		~20	EI		357

BrF₂⁺

Ion	Reactant	Other products	Ionization or appearance potential (eV)	Method	Heat of formation (kJ mol⁻¹)	Ref.
BrF_2^+	BrF_3	F	13.5±0.3	EI		357
BrF_2^+	BrF_5		16.1±0.2	EI		357

BrF₃⁺

Ion	Reactant	Other products	Ionization or appearance potential (eV)	Method	Heat of formation (kJ mol⁻¹)	Ref.
BrF_3^+	BrF_3		12.9±0.3	EI		357
BrF_3^+	BrF_5		15.5±0.2	EI		357

BrF₄⁺

Ion	Reactant	Other products	Ionization or appearance potential (eV)	Method	Heat of formation (kJ mol⁻¹)	Ref.
BrF_4^+	BrF_5	F⁻?	14.0±0.3	EI		357

MgBr⁺

Ion	Reactant	Other products	Ionization or appearance potential (eV)	Method	Heat of formation (kJ mol⁻¹)	Ref.
$MgBr^+$	$MgBr_2$	Br	12.0±0.4	EI		178

MgBr₂⁺

Ion	Reactant	Other products	Ionization or appearance potential (eV)	Method	Heat of formation (kJ mol⁻¹)	Ref.
$MgBr_2^+$	$MgBr_2$		10.65±0.15	EI		178

Mg₂Br₃⁺

Ion	Reactant	Other products	Ionization or appearance potential (eV)	Method	Heat of formation (kJ mol⁻¹)	Ref.
$Mg_2Br_3^+$	Mg_2Br_4	Br	10.8±0.3	EI		178

AlBr⁺

Ion	Reactant	Other products	Ionization or appearance potential (eV)	Method	Heat of formation (kJ mol⁻¹)	Ref.
$AlBr^+$	$AlBr_3$		17.7±0.5	EI		2897

AlBr₂⁺

Ion	Reactant	Other products	Ionization or appearance potential (eV)	Method	Heat of formation (kJ mol⁻¹)	Ref.
$AlBr_2^+$	$AlBr_3$	Br	13.3±0.5	EI		2897

AlBr₃⁺

Ion	Reactant	Other products	Ionization or appearance potential (eV)	Method	Heat of formation (kJ mol⁻¹)	Ref.
$AlBr_3^+$	$AlBr_3$		12.2±0.5	EI		2897

4.3. The Positive Ion Table—Continued

Ion	Reactant	Other products	Ionization or appearance potential (eV)	Method	Heat of formation (kJ mol⁻¹)	Ref.
			$Al_2Br_5^+$			
$Al_2Br_5^+$	Al_2Br_6	Br	12.3±0.5	EI		2897
			$SiBr_4^+$			
$SiBr_4^+$	$SiBr_4$		10.8	PE		3117
			PBr^+			
PBr^+	PBr_3		15.6±0.3	EI		2506
			PBr_2^+			
PBr_2^+	PBr_3	Br	11.4±0.2	EI		2506
			PBr_3^+			
PBr_3^+	PBr_3		9.85	EI		2597
PBr_3^+	PBr_3		10.0±0.2	EI		2506
			$BrCl^+$			
$BrCl^+$	$BrCl$		11.1±0.2	EI		357
			$TiBr_4^+$			
$TiBr_4^+$	$TiBr_4$		10.56 (V)	PE		3117
			$MnBr^+$			
$MnBr^+$	$Mn(CO)_5Br$	5CO	12.4	EI		2501
			$FeBr^+$			
$FeBr^+$	$FeBr_2$	Br	12.9±0.5	EI		174
			$FeBr_2^+$			
$FeBr_2^+$	$FeBr_2$		10.7±0.5	EI		174
			$Fe_2Br_3^+$			
$Fe_2Br_3^+$	Fe_2Br_4	Br	13.6±0.5	EI		174

4.3. The Positive Ion Table—Continued

Ion	Reactant	Other products	Ionization or appearance potential (eV)	Method	Heat of formation (kJ mol^{-1})	Ref.
		Fe$_2$Br$_4^+$				
Fe$_2$Br$_4^+$	Fe$_2$Br$_4$		12.6±0.5	EI		174
		ZnBr$^+$				
ZnBr$^+$	ZnBr$_2$	Br	13.4±0.2	EI		2506
		ZnBr$_2^+$				
ZnBr$_2^+$	ZnBr$_2$		10.39±0.33	EI		2506
		GeBr$^+$				
GeBr$^+$	GeBr$_2$	Br	11.0±0.3	EI		2568
GeBr$^+$	GeBr$_4$		15±1	EI		2568
		GeBr$_2^+$				
GeBr$_2^+$	GeBr$_2$		9.5±0.3	EI		2568
GeBr$_2^+$	GeBr$_4$		15±1	EI		2568
		GeBr$_3^+$				
GeBr$_3^+$	GeBr$_4$	Br	11.0±0.3	EI		2568
		GeBr$_4^+$				
GeBr$_4^+$	GeBr$_4$		10.75	PE		3117
GeBr$_4^+$	GeBr$_4$		10.8±0.3	EI		2568
		B$_5$H$_8$Br$^+$				
B$_5$H$_8$Br$^+$	1–B$_5$H$_8$Br		9.58±0.05	RPD		3228

See also – EI: 1102

Ion	Reactant	Other products	Ionization or appearance potential (eV)	Method	Heat of formation (kJ mol^{-1})	Ref.
B$_5$H$_8$Br$^+$	2–B$_5$H$_8$Br		10.48±0.04	RPD		3228
		CH$_2$Br$^+$				
CH$_2$Br$^+$	CH$_2$Br		8.34±0.11	EI		131
CH$_2$Br$^+$	CH$_2$Br		9.30	EI		141
CH$_2$Br$^+$	CH$_3$Br	H	12.12±0.09?	RPD		1139

See also – EI: 160, 356, 3017

4.3. The Positive Ion Table—Continued

Ion	Reactant	Other products	Ionization or appearance potential (eV)	Method	Heat of formation (kJ mol^{-1})	Ref.
CH_2Br^+	C_2H_5Br	CH_3	14.1±0.1	EI		356
CH_2Br^+	CH_2Br_2	Br	10.93±0.04	EI		131

	$CH_3Br^+(^2E_{3/2})$	$\Delta H_{f0}^\circ =$ 997 kJ mol^{-1} (238 kcal mol^{-1})				
	$CH_3Br^+(^2E_{1/2})$	$\Delta H_{f0}^\circ =$ 1028 kJ mol^{-1} (246 kcal mol^{-1})				
$CH_3Br^+(^2E_{3/2})$	CH_3Br		10.541±0.003	S	997	2064
$CH_3Br^+(^2E_{3/2})$	CH_3Br		10.528±0.005	PI		1253
$CH_3Br^+(^2E_{3/2})$	CH_3Br		10.53±0.01	PI	182,	416
$CH_3Br^+(^2E_{3/2})$	CH_3Br		10.53	PI		2637
$CH_3Br^+(^2E_{3/2})$	CH_3Br		10.53	PE		3057
$CH_3Br^+(^2E_{3/2})$	CH_3Br		10.53±0.015	PE		3074
$CH_3Br^+(^2E_{3/2})$	CH_3Br		10.54	PE		3119
$CH_3Br^+(^2E_{3/2})$	CH_3Br		10.53±0.02	RPD		289
$CH_3Br^+(^2E_{1/2})$	CH_3Br		10.856±0.003	S	1028	2064
$CH_3Br^+(^2E_{1/2})$	CH_3Br		10.85	PE		3057
$CH_3Br^+(^2E_{1/2})$	CH_3Br		10.85±0.015	PE		3074
$CH_3Br^+(^2E_{1/2})$	CH_3Br		10.86	PE		3119

Additional higher ionization potentials are given in refs. 3057, 3119.

See also – PI: 190
       PEN: 2430, 2466
       EI: 2154, 2776

CH_3Br^{+2}

Ion	Reactant	Other products	Ionization or appearance potential (eV)	Method	Heat of formation (kJ mol^{-1})	Ref.
CH_3Br^{+2}	CH_3Br		≤29.8±0.7 (V)	AUG		3424

C_2HBr^+

Ion	Reactant	Other products	Ionization or appearance potential (eV)	Method	Heat of formation (kJ mol^{-1})	Ref.
$C_2HBr^+(^2\Pi_{3/2})$	$CH{\equiv}CBr$		10.24	PE		3071
$C_2HBr^+(^2\Pi_{3/2})$	$CH{\equiv}CBr$		10.3±0.1	EDD		3177
$C_2HBr^+(^2\Pi_{1/2})$	$CH{\equiv}CBr$		10.38	PE		3071
$C_2HBr^+(^2\Pi_{3/2})$	$CH{\equiv}CBr$		12.93? (V)	PE		3071
$C_2HBr^+(^2\Pi_{1/2})$	$CH{\equiv}CBr$		13.06? (V)	PE		3071
$C_2HBr^+(^2\Sigma^+)$	$CH{\equiv}CBr$		15.99 (V)	PE		3071
$C_2HBr^+(^2\Sigma^+)$	$CH{\equiv}CBr$		17.6 (V)	PE		3071

See also – S: 3052

C_2HBr^{+2}

Ion	Reactant	Other products	Ionization or appearance potential (eV)	Method	Heat of formation (kJ mol^{-1})	Ref.
C_2HBr^{+2}	$CH{\equiv}CBr$		28.4±0.1	EDD		3177

$C_2H_2Br^+$

Ion	Reactant	Other products	Ionization or appearance potential (eV)	Method	Heat of formation (kJ mol^{-1})	Ref.
$C_2H_2Br^+$	cis–CHBr=CHBr	Br	11.44±0.05	EI		114
$C_2H_2Br^+$	trans–CHBr=CHBr	Br	11.65±0.05	EI		114

4.3. The Positive Ion Table—Continued

Ion	Reactant	Other products	Ionization or appearance potential (eV)	Method	Heat of formation (kJ mol^{-1})	Ref.
	$CH_2=CHBr^+$	$\Delta H^\circ_{f0} = 1039$ kJ mol^{-1} (248 kcal mol^{-1})				
$C_2H_3Br^+$	$CH_2=CHBr$		9.80±0.01	PI	1039	182
$C_2H_3Br^+$	$CH_2=CHBr$		9.80	PI	1039	168
$C_2H_3Br^+$	$CH_2=CHBr$		9.82±0.02	PI	1041	268

See also – EI: 268

Ion	Reactant	Other products	Ionization or appearance potential (eV)	Method	Heat of formation (kJ mol^{-1})	Ref.
	$C_2H_5Br^+(^2E_{3/2})$	$\Delta H^\circ_{f0} = 951$ kJ mol^{-1} (227 kcal mol^{-1})				
	$C_2H_5Br^+(^2E_{1/2})$	$\Delta H^\circ_{f0} = 981$ kJ mol^{-1} (234 kcal mol^{-1})				
$C_2H_5Br^+(^2E_{3/2})$	C_2H_5Br		10.29±0.02	S	950	2065
$C_2H_5Br^+(^2E_{3/2})$	C_2H_5Br		10.29±0.01	PI	950	182, 416
$C_2H_5Br^+(^2E_{3/2})$	C_2H_5Br		10.30±0.015	PE	951	3074
$C_2H_5Br^+(^2E_{3/2})$	C_2H_5Br		10.24±0.03	EDD		3174
$C_2H_5Br^+(^2E_{1/2})$	C_2H_5Br		10.61±0.02	S	981	2065
$C_2H_5Br^+(^2E_{1/2})$	C_2H_5Br		10.61±0.015	PE	981	3074

See also – PI: 190, 297
EI: 160, 356

Ion	Reactant	Other products	Ionization or appearance potential (eV)	Method	Heat of formation (kJ mol^{-1})	Ref.
	$C_3H_3Br^+$					
$C_3H_3Br^+$	$CH_3C\equiv CBr$		10.1±0.1	EI		13

Ion	Reactant	Other products	Ionization or appearance potential (eV)	Method	Heat of formation (kJ mol^{-1})	Ref.
	$C_3H_5Br^+$					
$C_3H_5Br^+$	$CH_3CH=CHBr$		9.30±0.05?	PI		182
$C_3H_5Br^+$	C_3H_5Br (Bromocyclopropane)		9.53	PE		3074

Ion	Reactant	Other products	Ionization or appearance potential (eV)	Method	Heat of formation (kJ mol^{-1})	Ref.
	$n-C_3H_7Br^+(^2E_{3/2})$	$\Delta H^\circ_{f298} = 894$ kJ mol^{-1} (214 kcal mol^{-1})				
	$n-C_3H_7Br^+(^2E_{1/2})$	$\Delta H^\circ_{f298} = 924$ kJ mol^{-1} (221 kcal mol^{-1})				
	$iso-C_3H_7Br^+(^2E_{3/2})$	$\Delta H^\circ_{f298} \sim 877$ kJ mol^{-1} (210 kcal mol^{-1})				
	$iso-C_3H_7Br^+(^2E_{1/2})$	$\Delta H^\circ_{f298} = 907$ kJ mol^{-1} (217 kcal mol^{-1})				
$C_3H_7Br^+(^2E_{3/2})$	$n-C_3H_7Br$		10.18±0.01	PI	894	182
$C_3H_7Br^+(^2E_{3/2})$	$n-C_3H_7Br$		10.18±0.015	PE	894	3074
$C_3H_7Br^+(^2E_{1/2})$	$n-C_3H_7Br$		10.49±0.015	PE	924	3074

See also – EI: 160, 2776

Ion	Reactant	Other products	Ionization or appearance potential (eV)	Method	Heat of formation (kJ mol^{-1})	Ref.
$C_3H_7Br^+(^2E_{3/2})$	$iso-C_3H_7Br$		10.075±0.01	PI	875	182
$C_3H_7Br^+(^2E_{3/2})$	$iso-C_3H_7Br$		10.12±0.015	PE	879	3074
$C_3H_7Br^+(^2E_{1/2})$	$iso-C_3H_7Br$		10.41±0.015	PE	907	3074

4.3. The Positive Ion Table—Continued

Ion	Reactant	Other products	Ionization or appearance potential (eV)	Method	Heat of formation (kJ mol^{-1})	Ref.	
	$n-C_4H_9Br^+(^2E_{3/2})$	$\Delta H^\circ_{f298} = 870$ kJ mol^{-1} (208 kcal mol^{-1})					
	$n-C_4H_9Br^+(^2E_{1/2})$	$\Delta H^\circ_{f298} = 900$ kJ mol^{-1} (215 kcal mol^{-1})					
	$sec-C_4H_9Br^+(^2E_{3/2})$	$\Delta H^\circ_{f298} = 843$ kJ mol^{-1} (201 kcal mol^{-1})					
	$tert-C_4H_9Br^+(^2E_{3/2})$	$\Delta H^\circ_{f298} \sim 823$ kJ mol^{-1} (197 kcal mol^{-1})					
	$tert-C_4H_9Br^+(^2E_{1/2})$	$\Delta H^\circ_{f298} = 854$ kJ mol^{-1} (204 kcal mol^{-1})					
$C_4H_9Br^+(^2E_{3/2})$	$n-C_4H_9Br$		10.125±0.01	PI	870	182,	416
$C_4H_9Br^+(^2E_{3/2})$	$n-C_4H_9Br$		10.13±0.015	PE	870	3074	
$C_4H_9Br^+(^2E_{1/2})$	$n-C_4H_9Br$		10.44±0.015	PE	900	3074	
$C_4H_9Br^+(^2E_{3/2})$	$sec-C_4H_9Br$		9.98±0.01	PI	843	182	
$C_4H_9Br^+(^2E_{3/2})$	$iso-C_4H_9Br$		10.09±0.02	PI		182	
$C_4H_9Br^+(^2E_{3/2})$	$iso-C_4H_9Br$		10.10±0.015	PE		3074	
$C_4H_9Br^+(^2E_{1/2})$	$iso-C_4H_9Br$		10.41±0.015	PE		3074	
$C_4H_9Br^+(^2E_{3/2})$	$tert-C_4H_9Br$		9.89±0.03	PI	820	182	
$C_4H_9Br^+(^2E_{3/2})$	$tert-C_4H_9Br$		9.95±0.015	PE	826	3074	
$C_4H_9Br^+(^2E_{1/2})$	$tert-C_4H_9Br$		10.24±0.015	PE	854	3074	

$C_5H_4Br^+$

Ion	Reactant	Other products	Ionization or appearance potential (eV)	Method	Heat of formation (kJ mol^{-1})	Ref.
$C_5H_4Br^+$	C_5H_4Br (Bromocyclopentadienyl radical)		8.85	EI		126

	$n-C_5H_{11}Br^+(^2E_{3/2})$	$\Delta H^\circ_{f298} = 845$ kJ mol^{-1} (202 kcal mol^{-1})				
$C_5H_{11}Br^+(^2E_{3/2})$	$n-C_5H_{11}Br$		10.10±0.02	PI	845	182
$C_5H_{11}Br^+(^2E_{3/2})$	$neo-C_5H_{11}Br$		10.04±0.015	PE		3074
$C_5H_{11}Br^+(^2E_{1/2})$	$neo-C_5H_{11}Br$		10.34±0.015	PE		3074

$C_6H_4Br^+$

Ion	Reactant	Other products	Ionization or appearance potential (eV)	Method	Heat of formation (kJ mol^{-1})	Ref.
$C_6H_4Br^+$	C_6H_4BrI (1-Bromo-4-iodobenzene)	I	12.04	EI		3238

4.3. The Positive Ion Table—Continued

Ion	Reactant	Other products	Ionization or appearance potential (eV)	Method	Heat of formation (kJ mol^{-1})	Ref.
	$C_6H_5Br^+$ (Bromobenzene)		$\Delta H^\circ_{f298} \sim 972$ kJ mol^{-1} (232 kcal mol^{-1})			
$C_6H_5Br^+$	C_6H_5Br (Bromobenzene)		8.950	PI	969	2682
$C_6H_5Br^+$	C_6H_5Br (Bromobenzene)		8.98±0.02	PI	971	182, 416
$C_6H_5Br^+$	C_6H_5Br (Bromobenzene)		9.03±0.01	PI	976	3212
$C_6H_5Br^+$	C_6H_5Br (Bromobenzene)		8.99	PE	972	3212
$C_6H_5Br^+$	C_6H_5Br (Bromobenzene)		8.98±0.03	EDD		3174

See also – S: 3212
PE: 2806, 3247, 3331
EI: 301, 1066, 3223, 3230, 3238

Ion	Reactant	Other products	Ionization or appearance potential (eV)	Method	Heat of formation (kJ mol^{-1})	Ref.
	$C_6H_{11}Br^+$					
$C_6H_{11}Br^+(^2E_{3/2})$	$C_6H_{11}Br$ (Bromocyclohexane)		9.875	PE		3074
$C_6H_{11}Br^+(^2E_{1/2})$	$C_6H_{11}Br$ (Bromocyclohexane)		10.165	PE		3074

Ion	Reactant	Other products	Ionization or appearance potential (eV)	Method	Heat of formation (kJ mol^{-1})	Ref.
	$C_7H_6Br^+$					
$C_7H_6Br^+$	$C_6H_4BrCH_3$ (4–Bromotoluene)	H	12.48	EI		3238
$C_7H_6Br^+$	$C_6H_4BrC_2H_5$ (1–Bromo–4–ethylbenzene)	CH_3	10.75	EI		3238
$C_7H_6Br^+$	$C_6H_5CH_2CH_2C_6H_4Br$ (1–(4–Bromophenyl)–2–phenylethane)		10.4±0.2	EI		3288

4.3. The Positive Ion Table—Continued

Ion	Reactant	Other products	Ionization or appearance potential (eV)	Method	Heat of formation (kJ mol^{-1})	Ref.
			C$_7$H$_7$Br$^+$			
C$_7$H$_7$Br$^+$	C$_6$H$_5$CH$_2$Br (α–Bromotoluene)		9.10±0.05	EI		2025
C$_7$H$_7$Br$^+$	C$_6$H$_5$CH$_2$Br (α–Bromotoluene)		9.1±0.1	EI		2973
C$_7$H$_7$Br$^+$	C$_6$H$_5$CH$_2$Br (α–Bromotoluene)		8.9	EI		3230
C$_7$H$_7$Br$^+$	C$_6$H$_4$BrCH$_3$ (2–Bromotoluene)		8.78±0.01	PI		416

See also – PI: 182
       EI: 3230

Ion	Reactant	Other products	Ionization or appearance potential (eV)	Method	Heat of formation (kJ mol^{-1})	Ref.
C$_7$H$_7$Br$^+$	C$_6$H$_4$BrCH$_3$ (3–Bromotoluene)		8.81±0.02	PI		182

See also – EI: 3230

Ion	Reactant	Other products	Ionization or appearance potential (eV)	Method	Heat of formation (kJ mol^{-1})	Ref.
C$_7$H$_7$Br$^+$	C$_6$H$_4$BrCH$_3$ (4–Bromotoluene)		8.67±0.02	PI		182, 416
C$_7$H$_7$Br$^+$	C$_6$H$_4$BrCH$_3$ (4–Bromotoluene)		8.71	PE		2806

See also – EI: 1066, 3230

Ion	Reactant	Other products	Ionization or appearance potential (eV)	Method	Heat of formation (kJ mol^{-1})	Ref.
			C$_8$H$_9$Br$^+$			
C$_8$H$_9$Br$^+$	C$_6$H$_5$CH$_2$CH$_2$Br (1–Bromo–2–phenylethane)		9.0±0.1	EI		2973
			C$_9$H$_{11}$Br$^+$			
C$_9$H$_{11}$Br$^+$	C$_6$H$_5$(CH$_2$)$_3$Br (1–Bromo–3–phenylpropane)		8.95±0.1	EI		2973
			C$_{14}$H$_{13}$Br$^+$			
C$_{14}$H$_{13}$Br$^+$	C$_6$H$_5$CH$_2$CH$_2$C$_6$H$_4$Br (1–(4–Bromophenyl)–2–phenylethane)		8.8±0.1	EI		3288
			CHBr$_2^+$			
CHBr$_2^+$	CHBr$_2$		8.13±0.16	EI		131
CHBr$_2^+$	CHBr$_3$	Br	10.80±0.01	EI		131

4.3. The Positive Ion Table—Continued

Ion	Reactant	Other products	Ionization or appearance potential (eV)	Method	Heat of formation (kJ mol^{-1})	Ref.

$CH_2Br_2^+$ $\Delta H_{f298}^\circ = 1008$ kJ mol^{-1} (241 kcal mol^{-1})

Ion	Reactant	Other products	Ionization or appearance potential (eV)	Method	Heat of formation (kJ mol^{-1})	Ref.
$CH_2Br_2^+$	CH_2Br_2		10.49±0.02	PI	1008	182
$CH_2Br_2^+$	CH_2Br_2		10.61 (V)	PE		3119

Orbital assignments are discussed in R. N. Dixon, J. N. Murrell and B. Narayan, Mol. Phys. **20**, 611 (1971).

$CH_2Br_2^{+2}$

Ion	Reactant	Other products	Ionization or appearance potential (eV)	Method	Heat of formation (kJ mol^{-1})	Ref.
$CH_2Br_2^{+2}$	CH_2Br_2		≤29.8±0.7 (V)	AUG		3424

$C_2H_2Br_2^+$

Ion	Reactant	Other products	Ionization or appearance potential (eV)	Method	Heat of formation (kJ mol^{-1})	Ref.
$C_2H_2Br_2^+$	cis–CHBr=CHBr		9.45	PI		168
$C_2H_2Br_2^+$	cis–CHBr=CHBr		9.45±0.02	PI	114, 1058	268, 1190

See also – PI: 182
EI: 114, 268, 1058, 1190, 3213

Ion	Reactant	Other products	Ionization or appearance potential (eV)	Method	Heat of formation (kJ mol^{-1})	Ref.
$C_2H_2Br_2^+$	trans–CHBr=CHBr		9.47	PI		168
$C_2H_2Br_2^+$	trans–CHBr=CHBr		9.46±0.02	PI	114, 1058	268, 1190

See also – PI: 182
EI: 114, 268, 1058, 1190, 3213

$C_2H_4Br_2^+$

Ion	Reactant	Other products	Ionization or appearance potential (eV)	Method	Heat of formation (kJ mol^{-1})	Ref.
$C_2H_4Br_2^+$	CH_3CHBr_2		10.19±0.03	PI		182

$C_3H_6Br_2^+$

Ion	Reactant	Other products	Ionization or appearance potential (eV)	Method	Heat of formation (kJ mol^{-1})	Ref.
$C_3H_6Br_2^+$	$(CH_2Br)_2CH_2$		10.07±0.02	PI		182

$C_6H_4Br_2^+$

Ion	Reactant	Other products	Ionization or appearance potential (eV)	Method	Heat of formation (kJ mol^{-1})	Ref.
$C_6H_4Br_2^+$	$C_6H_4Br_2$ (1,4–Dibromobenzene)		8.97 (V)	PE		2806
$C_6H_4Br_2^+$	$C_6H_4Br_2$ (1,4–Dibromobenzene)		8.9±0.15	CTS		3373

$CHBr_3^+$ $\Delta H_{f0}^\circ \sim 1055$ kJ mol^{-1} (252 kcal mol^{-1})

Ion	Reactant	Other products	Ionization or appearance potential (eV)	Method	Heat of formation (kJ mol^{-1})	Ref.
$CHBr_3^+$	$CHBr_3$		10.51±0.02	PI	1057	182
$CHBr_3^+$	$CHBr_3$		10.47 (V)	PE	1053	3119

Orbital assignments are discussed in R. N. Dixon, J. N. Murrell and B. Narayan, Mol. Phys. **20**, 611 (1971).

4.3. The Positive Ion Table—Continued

Ion	Reactant	Other products	Ionization or appearance potential (eV)	Method	Heat of formation (kJ mol^{-1})	Ref.
			CHBr$_3^{+2}$			
CHBr$_3^{+2}$	CHBr$_3$		≤28.9±0.8 (V)	AUG		3424
			C$_2$HBr$_3^+$			
C$_2$HBr$_3^+$	CHBr=CBr$_2$		9.27±0.01	PI		182
C$_2$HBr$_3^+$	CHBr=CBr$_2$		9.27	PI		168
	BrCN$^+$($^2\Pi_{3/2}$)	$\Delta H_{f0}^\circ \sim 1337$ kJ mol^{-1} (319 kcal mol^{-1})				
	BrCN$^+$($^2\Pi_{1/2}$)	$\Delta H_{f0}^\circ \sim 1355$ kJ mol^{-1} (324 kcal mol^{-1})				
	BrCN$^+$($^2\Sigma^+$)	$\Delta H_{f0}^\circ \sim 1501$ kJ mol^{-1} (359 kcal mol^{-1})				
	BrCN$^+$($^2\Pi_{3/2}$)	$\Delta H_{f0}^\circ = 1562$ kJ mol^{-1} (373 kcal mol^{-1})				
	BrCN$^+$($^2\Pi_{1/2}$)	$\Delta H_{f0}^\circ = 1576$ kJ mol^{-1} (377 kcal mol^{-1})				
CNBr$^+$($^2\Pi_{3/2}$)	BrCN		11.84±0.01	PI	1335	2621
CNBr$^+$($^2\Pi_{3/2}$)	BrCN		11.85±0.02	PE	1336	3091
CNBr$^+$($^2\Pi_{3/2}$)	BrCN		11.88	PE	1339	3045
CNBr$^+$($^2\Pi_{1/2}$)	BrCN		12.03±0.02	PE	1353	3091
CNBr$^+$($^2\Pi_{1/2}$)	BrCN		12.07	PE	1357	3045
CNBr$^+$($^2\Sigma^+$)	BrCN		13.54±0.02	PE	1499	3091
CNBr$^+$($^2\Sigma^+$)	BrCN		13.58	PE	1503	3045
CNBr$^+$($^2\Pi_{3/2}$)	BrCN		14.19	PE	1562	3045
CNBr$^+$($^2\Pi_{3/2}$)	BrCN		14.40±0.02 (V)	PE		3091
CNBr$^+$($^2\Pi_{1/2}$)	BrCN		14.34	PE	1576	3045
CNBr$^+$($^2\Pi_{1/2}$)	BrCN		14.49±0.02 (V)	PE		3091
CNBr$^+$($^2\Sigma^+$)	BrCN		18.07±0.02 (V)	PE		3091
CNBr$^+$($^2\Sigma^+$)	BrCN		18.07 (V)	PE		3045

See also – EI: 73

Ion	Reactant	Other products	Ionization or appearance potential (eV)	Method	Heat of formation (kJ mol^{-1})	Ref.
			BF$_2$Br$^+$			
BF$_2$Br$^+$	BF$_2$Br		11.95±0.21	EI		2512, 3227
			BFBr$_2^+$			
BFBr$_2^+$	BFBr$_2$		11.11±0.10	EI		2512, 3227
			CF$_2$Br$^+$			
CF$_2$Br$^+$	CF$_3$Br	F	15.0±0.1	EI		439
CF$_2$Br$^+$	CF$_3$Br	F	15.0±0.7	EI		24

4.3. The Positive Ion Table—Continued

Ion	Reactant	Other products	Ionization or appearance potential (eV)	Method	Heat of forma- tion (kJ mol^{-1})	Ref.
			CF$_3$Br$^+$			
CF$_3$Br$^+$	CF$_3$Br		11.82±0.02	EI		439
CF$_3$Br$^+$	CF$_3$Br		11.89±0.10	EI		1131
CF$_3$Br$^+$	CF$_3$Br		12.3±0.3	EI		24
			C$_6$F$_5$Br$^+$			
C$_6$F$_5$Br$^+$	C$_6$F$_5$Br (Bromopentafluorobenzene)		9.6±0.1	EI		301
			CF$_2$Br$_2^+$			
CF$_2$Br$_2^+$	CF$_2$Br$_2$		11.07±0.03	PI		182
			CFBr$_3^+$			
CFBr$_3^+$	CFBr$_3$		10.67±0.01	PI		182
			P$_3$N$_3$Br$_5^+$			
P$_3$N$_3$Br$_5^+$	(NPBr$_2$)$_3$ (Cyclo−tris(dibromophosphonitrile))	Br	10.29±0.1	EI		2782
			P$_3$N$_3$Br$_6^+$			
P$_3$N$_3$Br$_6^+$	(NPBr$_2$)$_3$ (Cyclo−tris(dibromophosphonitrile))		9.56±0.1	EI		2782
			P$_4$N$_4$Br$_8^+$			
P$_4$N$_4$Br$_8^+$	(NPBr$_2$)$_4$ (Cyclo−tetrakis(dibromophosphonitrile))		9.21±0.1	EI		2782
			BCl$_2$Br$^+$			
BCl$_2$Br$^+$	BCl$_2$Br		11.13±0.18	EI		2512, 2513, 3227
			BClBr$_2^+$			
BClBr$_2^+$	BClBr$_2$		10.79±0.06	EI		2512, 2513, 3227

4.3. The Positive Ion Table—Continued

Ion	Reactant	Other products	Ionization or appearance potential (eV)	Method	Heat of formation (kJ mol^{-1})	Ref.
			C_2ClBr^+			
$C_2ClBr^+(^2\Pi_{3/2})$	$CCl\equiv CBr$		9.98 (V)	PE		3121
$C_2ClBr^+(^2\Pi_{3/2})$	$CCl\equiv CBr$		10.0±0.1	EDD		3177
$C_2ClBr^+(^2\Pi_{3/2})$	$CCl\equiv CBr$		12.54 (V)	PE		3121
$C_2ClBr^+(^2\Pi_{1/2})$	$CCl\equiv CBr$		12.73 (V)	PE		3121
$C_2ClBr^+(^2\Pi_{3/2})$	$CCl\equiv CBr$		14.08 (V)	PE		3121
$C_2ClBr^+(^2\Sigma^+)$	$CCl\equiv CBr$		16.07 (V)	PE		3121
$C_2ClBr^+(^2\Sigma^+)$	$CCl\equiv CBr$		17.47 (V)	PE		3121
			C_2ClBr^{+2}			
C_2ClBr^{+2}	$CCl\equiv CBr$		26.8±0.1	EDD		3177
			$GeClBr_2^+$			
$GeClBr_2^+$	$GeClBr_3$	Br	13.5±1	EI		2568
			$C_5H_4NBr^+$			
$C_5H_4NBr^+$	C_5H_4NBr (2−Bromopyridine)		9.65±0.05	EI		217
$C_5H_4NBr^+$	C_5H_4NBr (4−Bromopyridine)		9.94±0.05	EI		217
			$C_6H_6NBr^+$			
$C_6H_6NBr^+$	$C_6H_4BrNH_2$ (4−Bromoaniline)		7.74±0.1	CTS		2485
$C_6H_6NBr^+$	$C_6H_4BrNHCOCH_3$ (N−(3−Bromophenyl)acetic acid amide)		10.62±0.2	EI		3406
$C_6H_6NBr^+$	$C_6H_4BrNHCOCH_3$ (N−(4−Bromophenyl)acetic acid amide)		10.58±0.2	EI		3406
			$C_7H_4NBr^+$			
$C_7H_4NBr^+$	C_6H_4BrCN (3−Bromobenzoic acid nitrile)		9.87±0.05	RPD		3223
$C_7H_4NBr^+$	C_6H_4BrCN (4−Bromobenzoic acid nitrile)		9.54 (V)	PE		2806
$C_7H_4NBr^+$	C_6H_4BrCN (4−Bromobenzoic acid nitrile)		9.90±0.05	RPD		3223
			$C_8H_9NBr^+$			
$C_8H_9NBr^+$	$C_6H_4BrN(CH_3)_2$ (4−Bromo−N,N−dimethylaniline)	H	11.15	EI		3238

4.3. The Positive Ion Table—Continued

Ion	Reactant	Other products	Ionization or appearance potential (eV)	Method	Heat of formation (kJ mol^{-1})	Ref.
			C$_8$H$_{10}$NBr$^+$			
C$_8$H$_{10}$NBr$^+$	C$_6$H$_4$BrN(CH$_3$)$_2$ (4−Bromo−N,N−dimethylaniline)		7.33	CTS		1281
C$_8$H$_{10}$NBr$^+$	C$_6$H$_4$(NH$_2$)CH$_2$CH$_2$Br (1−(4−Aminophenyl)−2−bromoethane)		7.8	EI		2973
			C$_{10}$H$_{14}$NBr$^+$			
C$_{10}$H$_{14}$NBr$^+$	C$_6$H$_4$BrN(C$_2$H$_5$)$_2$ (4−Bromo−N,N−diethylaniline)		6.96	CTS		1281
			C$_2$H$_3$OBr$^+$			
C$_2$H$_3$OBr$^+$	CH$_3$COBr		10.55±0.05?	PI		182
			C$_2$H$_5$OBr$^+$			
C$_2$H$_5$OBr$^+$	CH$_2$BrCH$_2$OH		10.63	PE		3374
			C$_3$H$_5$OBr$^+$			
C$_3$H$_5$OBr$^+$	C$_3$H$_5$OBr (1−Bromo−2,3−epoxypropane)		10.46	PE		3374
			C$_4$H$_6$OBr$^+$			
C$_4$H$_6$OBr$^+$	CH$_2$BrCH$_2$CH$_2$COOCH$_3$		11.15±0.3	EI		2496
			C$_6$H$_4$OBr$^+$			
C$_6$H$_4$OBr$^+$	C$_6$H$_4$BrOCH$_3$ (1−Bromo−4−methoxybenzene)	CH$_3$	11.78±0.1	EI		2970
C$_6$H$_4$OBr$^+$	C$_6$H$_4$BrOCH$_3$ (1−Bromo−4−methoxybenzene)	CH$_3$	11.80	EI		3238
			C$_6$H$_5$OBr$^+$			
C$_6$H$_5$OBr$^+$	C$_6$H$_4$BrOH (4−Bromophenol)		8.52 (V)	PE		2806

See also − EI: 1066

4.3. The Positive Ion Table—Continued

Ion	Reactant	Other products	Ionization or appearance potential (eV)	Method	Heat of formation (kJ mol^{-1})	Ref.
			C$_7$H$_4$OBr$^+$			
C$_7$H$_4$OBr$^+$	C$_6$H$_4$BrCOCH$_3$ (3−Bromoacetophenone)	CH$_3$	10.51±0.1	EI		2967
C$_7$H$_4$OBr$^+$	C$_6$H$_4$BrCOCH$_3$ (4−Bromoacetophenone)	CH$_3$	10.35±0.1	EI		2967
C$_7$H$_4$OBr$^+$	C$_6$H$_4$BrCOCH$_3$ (4−Bromoacetophenone)	CH$_3$	10.58	EI		3238
C$_7$H$_4$OBr$^+$	C$_6$H$_4$BrCOOCH$_3$ (4−Bromobenzoic acid methyl ester)		11.11	EI		3238
			C$_7$H$_7$OBr$^+$			
C$_7$H$_7$OBr$^+$	C$_6$H$_4$BrOCH$_3$ (1−Bromo−2−methoxybenzene)		8.3±0.15	CTS		3373
C$_7$H$_7$OBr$^+$	C$_6$H$_4$BrOCH$_3$ (1−Bromo−4−methoxybenzene)		8.49 (V)	PE		2806
C$_7$H$_7$OBr$^+$	C$_6$H$_4$BrOCH$_3$ (1−Bromo−4−methoxybenzene)		8.2±0.15	CTS		3373
			C$_8$H$_7$OBr$^+$			
C$_8$H$_7$OBr$^+$	C$_6$H$_4$BrCOCH$_3$ (3−Bromoacetophenone)		9.60±0.1	EI		2967
C$_8$H$_7$OBr$^+$	C$_6$H$_4$BrCOCH$_3$ (4−Bromoacetophenone)		9.47±0.1	EI		2967
C$_8$H$_7$OBr$^+$	C$_6$H$_4$BrCOCH$_3$ (4−Bromoacetophenone)		9.55	EI		3238
			C$_9$H$_{11}$OBr$^+$			
C$_9$H$_{11}$OBr$^+$	C$_6$H$_4$(OCH$_3$)CH$_2$CH$_2$Br (1−Bromo−2−(3−methoxyphenyl)ethane)		8.5	EI		2973
C$_9$H$_{11}$OBr$^+$	C$_6$H$_4$(OCH$_3$)CH$_2$CH$_2$Br (1−Bromo−2−(4−methoxyphenyl)ethane)		8.2	EI		2973
			C$_{10}$H$_{11}$OBr$^+$			
C$_{10}$H$_{11}$OBr$^+$	C$_6$H$_4$BrCOC$_3$H$_7$ (1−(4−Bromophenyl)−1−butanone)		8.75±0.2	EI		2534
			C$_3$H$_5$O$_2$Br$^+$			
C$_3$H$_5$O$_2$Br$^+$	CH$_2$BrCOOCH$_3$		10.37±0.05	EI		2025

4.3. The Positive Ion Table—Continued

Ion	Reactant	Other products	Ionization or appearance potential (eV)	Method	Heat of formation (kJ mol^{-1})	Ref.
		$C_5H_9O_2Br^+$				
$C_5H_9O_2Br^+$	$CH_2BrCH_2CH_2COOCH_3$		9.85 ± 0.3	EI		2496
		$C_8H_7O_2Br^+$				
$C_8H_7O_2Br^+$	$C_6H_4BrCOOCH_3$ (4–Bromobenzoic acid methyl ester)		9.47	EI		3238
		$C_6H_4FBr^+$				
$C_6H_4FBr^+$	C_6H_4FBr (1–Bromo–4–fluorobenzene)		8.99 ± 0.03	PI		182
		$C_7H_4F_3Br^+$				
$C_7H_4F_3Br^+$	$C_6H_4BrCF_3$ (4–Bromo–α,α,α–trifluorotoluene)		9.55 (V)	PE		2806
		$C_2H_3FBr_2^+$				
$C_2H_3FBr_2^+$	$CH_2BrCHFBr$		10.75 ± 0.02	PI		182
		$C_2H_2F_2Br_2^+$				
$C_2H_2F_2Br_2^+$	CH_2BrCF_2Br		10.83 ± 0.01	PI		182
		$C_2H_6SiBr^+$				
$C_2H_6SiBr^+$	$(CH_3)_3SiBr$	CH_3	10.97 ± 0.02	RPD		1421
$C_2H_6SiBr^+$	$(CH_3)_3SiBr$	CH_3	10.7 ± 0.1	EI		2689
		$C_3H_9SiBr^+$				
$C_3H_9SiBr^+$	$(CH_3)_3SiBr$		10.24 ± 0.02	RPD		1421
$C_3H_9SiBr^+$	$(CH_3)_3SiBr$		9.8 ± 0.1	EI		2689
		$C_{12}H_{10}PBr^+$				
$C_{12}H_{10}PBr^+$	$(C_6H_5)_2PBr$ (Bromodiphenylphosphine)		8.72 ± 0.05	EI		2481

4.3. The Positive Ion Table—Continued

Ion	Reactant	Other products	Ionization or appearance potential (eV)	Method	Heat of formation (kJ mol^{-1})	Ref.
			C$_4$H$_3$SBr$^+$			
C$_4$H$_3$SBr$^+$	C$_4$H$_3$SBr (2–Bromothiophene)		8.63±0.01	PI		182
C$_4$H$_3$SBr$^+$	C$_4$H$_3$SBr (2–Bromothiophene)		8.50	PE		3246
See also – EI: 3240						
C$_4$H$_3$SBr$^+$	C$_4$H$_3$SBr (3–Bromothiophene)		8.90±0.05	PE		3246
			C$_7$H$_7$S$_3$Br$^+$			
C$_7$H$_7$S$_3$Br$^+$	C$_7$H$_7$S$_3$Br (3–Bromo–2,5–dimethyl–6a–thiathiophthene)		7.49	EI		3284
		CH$_2$ClBr$^+$	$\Delta H^\circ_{f298} = 989$ kJ mol^{-1} (236 kcal mol^{-1})			
CH$_2$ClBr$^+$	CH$_2$ClBr		10.77±0.01	PI	989	182
CH$_2$ClBr$^+$	CH$_2$ClBr		10.75±0.05	EI		72
			C$_2$H$_4$ClBr$^+$			
C$_2$H$_4$ClBr$^+$	CH$_2$ClCH$_2$Br		10.63±0.03	PI		182
			C$_6$H$_4$ClBr$^+$			
C$_6$H$_4$ClBr$^+$	C$_6$H$_4$ClBr (1–Bromo–4–chlorobenzene)		9.04 (V)	PE		2806
			C$_8$H$_8$ClBr$^+$			
C$_8$H$_8$ClBr$^+$	C$_6$H$_4$ClCH$_2$CH$_2$Br (1–Bromo–2–(3–chlorophenyl)ethane)		9.1	EI		2973
C$_8$H$_8$ClBr$^+$	C$_6$H$_4$ClCH$_2$CH$_2$Br (1–Bromo–2–(4–chlorophenyl)ethane)		8.8	EI		2973
			CHCl$_2$Br$^+$			
CHCl$_2$Br$^+$	CHCl$_2$Br		10.88±0.05	EI		72
		CHClBr$_2^+$	$\Delta H^\circ_{f298} = 1001$ kJ mol^{-1} (239 kcal mol^{-1})			
CHClBr$_2^+$	CHClBr$_2$		10.59±0.01	PI	1001	182

4.3. The Positive Ion Table—Continued

Ion	Reactant	Other products	Ionization or appearance potential (eV)	Method	Heat of formation (kJ mol^{-1})	Ref.
		BFClBr$^+$				
BFClBr$^+$	BFClBr		11.46±0.02	EI		2512, 3227
		P$_3$N$_3$Cl$_4$Br$^+$				
P$_3$N$_3$Cl$_4$Br$^+$	NPCl$_2$(NPClBr)$_2$ (Cyclo—bis(bromochlorophosphonitrile)dichlorophosphonitrile)	Br	10.54±0.1	EI		2782
		P$_3$N$_3$Cl$_5$Br$^+$				
P$_3$N$_3$Cl$_5$Br$^+$	(NPCl$_2$)$_2$NPClBr (Cyclo—bis(dichlorophosphonitrile)bromochlorophosphonitrile)		9.83±0.1	EI		2782
		P$_3$N$_3$Cl$_3$Br$_2^+$				
P$_3$N$_3$Cl$_3$Br$_2^+$	(NPClBr)$_3$ (Cyclo—tris(bromochlorophosphonitrile))	Br	10.32±0.1	EI		2782
		P$_3$N$_3$Cl$_4$Br$_2^+$				
P$_3$N$_3$Cl$_4$Br$_2^+$	NPCl$_2$(NPClBr)$_2$ (Cyclo—bis(bromochlorophosphonitrile)dichlorophosphonitrile)		9.80±0.1	EI		2782
		P$_3$N$_3$Cl$_2$Br$_3^+$				
P$_3$N$_3$Cl$_2$Br$_3^+$	NPBr$_2$(NPClBr)$_2$ (Cyclo—bis(bromochlorophosphonitrile)dibromophosphonitrile)	Br	10.22±0.1	EI		2782
		P$_3$N$_3$Cl$_3$Br$_3^+$				
P$_3$N$_3$Cl$_3$Br$_3^+$	(NPClBr)$_3$ (Cyclo—tris(bromochlorophosphonitrile))		9.72±0.1	EI		2782
		P$_3$N$_3$ClBr$_4^+$				
P$_3$N$_3$ClBr$_4^+$	(NPBr$_2$)$_2$NPClBr (Cyclo—bis(dibromophosphonitrile)bromochlorophosphonitrile)	Br	10.01±0.1	EI		2782
		P$_3$N$_3$Cl$_2$Br$_4^+$				
P$_3$N$_3$Cl$_2$Br$_4^+$	NPBr$_2$(NPClBr)$_2$ (Cyclo—bis(bromochlorophosphonitrile)dibromophosphonitrile)		9.60±0.1	EI		2782
		P$_3$N$_3$ClBr$_5^+$				
P$_3$N$_3$ClBr$_5^+$	(NPBr$_2$)$_2$NPClBr (Cyclo—bis(dibromophosphonitrile)bromochlorophosphonitrile)		9.47±0.1	EI		2782

4.3. The Positive Ion Table—Continued

Ion	Reactant	Other products	Ionization or appearance potential (eV)	Method	Heat of formation (kJ mol^{-1})	Ref.
COBrMn$^+$						
COBrMn$^+$	Mn(CO)$_5$Br	4CO	11.72	EI		2501
C$_2$O$_2$BrMn$^+$						
C$_2$O$_2$BrMn$^+$	Mn(CO)$_5$Br	3CO	10.47	EI		2501
C$_3$O$_3$BrMn$^+$						
C$_3$O$_3$BrMn$^+$	Mn(CO)$_5$Br	2CO	9.55	EI		2501
C$_5$O$_5$BrMn$^+$						
C$_5$O$_5$BrMn$^+$	Mn(CO)$_5$Br		8.76 (V)	PE		2879
C$_5$O$_5$BrMn$^+$	Mn(CO)$_5$Br		8.97±0.03	EI		2501
C$_4$H$_{12}$BN$_2$Br$^+$						
C$_4$H$_{12}$BN$_2$Br$^+$	((CH$_3$)$_2$N)$_2$BBr		8.05	EI		3227
C$_2$H$_6$BNBr$_2^+$						
C$_2$H$_6$BNBr$_2^+$	(CH$_3$)$_2$NBBr$_2$		9.50	EI		3227
C$_2$H$_6$BO$_2$Br$^+$						
C$_2$H$_6$BO$_2$Br$^+$	(CH$_3$O)$_2$BBr		10.62	EI		3227
CH$_3$BOBr$_2^+$						
CH$_3$BOBr$_2^+$	CH$_3$OBBr$_2$		10.68	EI		3227
C$_8$H$_8$NOBr$^+$						
C$_8$H$_8$NOBr$^+$	C$_6$H$_4$BrNHCOCH$_3$ (N–(3–Bromophenyl)acetic acid amide)		8.56±0.2	EI		3406
C$_8$H$_8$NOBr$^+$	C$_6$H$_4$BrNHCOCH$_3$ (N–(4–Bromophenyl)acetic acid amide)		8.42±0.2	EI		3406
C$_8$H$_8$NO$_2$Br$^+$						
C$_8$H$_8$NO$_2$Br$^+$	C$_6$H$_4$(NO$_2$)CH$_2$CH$_2$Br (1–Bromo–2–(4–nitrophenyl)ethane)		9.6	EI		2973

4.3. The Positive Ion Table—Continued

Ion	Reactant	Other products	Ionization or appearance potential (eV)	Method	Heat of formation (kJ mol^{-1})	Ref.
		$C_2H_6BS_2Br^+$				
$C_2H_6BS_2Br^+$	$(CH_3S)_2BBr$		9.55	EI		3227
		$CH_3BSBr_2^+$				
$CH_3BSBr_2^+$	CH_3SBBr_2		10.25	EI		3227
		$C_9H_5O_3BrCr^+$				
$C_9H_5O_3BrCr^+$	$C_6H_5BrCr(CO)_3$ (Bromobenzenechromium tricarbonyl)		7.40	EI		3005
		$C_{10}H_{12}O_4Br_2Cr^+$				
$C_{10}H_{12}O_4Br_2Cr^+$	$(CH_3COCBrCOCH_3)_3Cr$ (Tris(3−bromo−2,4−pentanedionato)chromium)		11.0±0.10	EI		2519
		$C_{15}H_{18}O_6Br_3Cr^+$				
$C_{15}H_{18}O_6Br_3Cr^+$	$(CH_3COCBrCOCH_3)_3Cr$ (Tris(3−bromo−2,4−pentanedionato)chromium)		8.05±0.05	EI		2519
		$C_2H_6BNFBr^+$				
$C_2H_6BNFBr^+$	$(CH_3)_2NBFBr$		9.68	EI		3227
		CH_3BOFBr^+				
CH_3BOFBr^+	CH_3OBFBr		11.68	EI		3227
		$CH_3BOClBr^+$				
$CH_3BOClBr^+$	$CH_3OBClBr$		11.07	EI		3227
		$C_7H_5O_2PBr_3Mn^+$				
$C_7H_5O_2PBr_3Mn^+$	$C_5H_5Mn(CO)_2PBr_3$ (Cyclopentadienyl(tribromophosphine)manganese dicarbonyl)		8.01±0.05	EI		2597
		$C_{12}H_{27}N_2O_2PBrFe^+$				
$C_{12}H_{27}N_2O_2PBrFe^+$	$(n-C_4H_9)_3PFe(NO)_2Br$		7.85	EI		2799

4.3. The Positive Ion Table—Continued

Ion	Reactant	Other products	Ionization or appearance potential (eV)	Method	Heat of formation (kJ mol⁻¹)	Ref.

$C_{18}H_{15}N_2O_2PBrFe^+$

Ion	Reactant	Other products	IP/AP (eV)	Method	ΔH_f	Ref.
$C_{18}H_{15}N_2O_2PBrFe^+$	$(C_6H_5)_3PFe(NO)_2Br$ (Triphenylphosphineiron dinitrosyl bromide)		7.7	EI		2799

$C_9H_{21}N_2O_5PBrFe^+$

Ion	Reactant	Other products	IP/AP (eV)	Method	ΔH_f	Ref.
$C_9H_{21}N_2O_5PBrFe^+$	$(iso-C_3H_7O)_3PFe(NO)_2Br$		8.05	EI		2799

$C_{12}H_{27}N_2O_2PBrCo^+$

Ion	Reactant	Other products	IP/AP (eV)	Method	ΔH_f	Ref.
$C_{12}H_{27}N_2O_2PBrCo^+$	$(n-C_4H_9)_3PCo(NO)_2Br$		7.9	EI		2799

$C_{18}H_{15}N_2O_2PBrCo^+$

Ion	Reactant	Other products	IP/AP (eV)	Method	ΔH_f	Ref.
$C_{18}H_{15}N_2O_2PBrCo^+$	$(C_6H_5)_3PCo(NO)_2Br$ (Triphenylphosphinecobalt dinitrosyl bromide)		7.7	EI		2799

$C_9H_{21}N_2O_5PBrCo^+$

Ion	Reactant	Other products	IP/AP (eV)	Method	ΔH_f	Ref.
$C_9H_{21}N_2O_5PBrCo^+$	$(iso-C_3H_7O)_3PCo(NO)_2Br$		7.9	EI		2799

Ion	Reactant	Other products	IP/AP (eV)	Method	ΔH_f	Ref.
$Kr^+(^2P_{3/2})$		$\Delta H_{f0}^\circ = 1350.7$ kJ mol⁻¹ (322.8 kcal mol⁻¹)				
$Kr^+(^2P_{1/2})$		$\Delta H_{f0}^\circ = 1415.0$ kJ mol⁻¹ (338.2 kcal mol⁻¹)				
$Kr^+(^2P_{3/2})$	Kr		13.999	S	1350.7	2113
$Kr^+(^2P_{3/2})$	Kr		13.999±0.002	PI		1253
$Kr^+(^2P_{3/2})$	Kr		14.01±0.01	PI		1118
$Kr^+(^2P_{3/2})$	Kr		14.05	PE		248
$Kr^+(^2P_{3/2})$	Kr		14.00±0.05	EDD		2634
$Kr^+(^2P_{1/2})$	Kr		14.665	S	1415.0	2113
$Kr^+(^2P_{1/2})$	Kr		14.69	PE		248

See also – EI: 35, 224, 1012, 1047, 2032, 2776, 2941, 3435

Ion	Reactant	Other products	IP/AP (eV)	Method	ΔH_f	Ref.
Kr^+	KrF_2	F_2?	13.21±0.25	EI		2577

Kr^{+2} $\Delta H_{f0}^\circ = 3701.0$ kJ mol⁻¹ (884.6 kcal mol⁻¹)

Ion	Reactant	Other products	IP/AP (eV)	Method	ΔH_f	Ref.
Kr^{+2}	Kr		38.358	S	3701.0	2113
Kr^{+2}	Kr		38.45±0.1	RPD		198
Kr^{+2}	Kr		38.1±0.1	NRE		211
Kr^{+2}	Kr		38.0±0.5	NRE		201
Kr^{+2}	Kr		38	EI		1240
Kr^{+2}	Kr		38.5	EI		201
Kr^{+2}	Kr		38.5	EI		211

4.3. The Positive Ion Table—Continued

Ion	Reactant	Other products	Ionization or appearance potential (eV)	Method	Heat of formation (kJ mol^{-1})	Ref.
Kr^{+2}	Kr$^+$		24.359	S		2113
Kr^{+2}	Kr$^+$		23±2	SEQ		2551

See also – EI: 218

Kr^{+3} $\Delta H_{f0}^\circ = 7266$ kJ mol^{-1} (1737 kcal mol^{-1})

Ion	Reactant	Other products	Ionization or appearance potential (eV)	Method	Heat of formation (kJ mol^{-1})	Ref.
Kr^{+3}	Kr		75.31	S	7266	2113
Kr^{+3}	Kr		75.6±0.5	RPD		198
Kr^{+3}	Kr		73.3±0.2	NRE		211
Kr^{+3}	Kr		76	EI		1240
Kr^{+3}	Kr		77	EI		211
Kr^{+3}	Kr^{+2}		36.95	S		2113
Kr^{+3}	Kr^{+2}		34±2	SEQ		2551

Kr^{+4} $\Delta H_{f0}^\circ \sim 12331$ kJ mol^{-1} (2947 kcal mol^{-1})

Ion	Reactant	Other products	Ionization or appearance potential (eV)	Method	Heat of formation (kJ mol^{-1})	Ref.
Kr^{+4}	Kr		127.8	S	~12331	2113
Kr^{+4}	Kr		118±1	NRE		211
Kr^{+4}	Kr		117.5±2	NRE		201
Kr^{+4}	Kr		130.0	EI		201
Kr^{+4}	Kr		130	EI		211
Kr^{+4}	Kr		134	EI		1240
Kr^{+4}	Kr		147±2	EI		198
Kr^{+4}	Kr^{+3}		52.5	S		2113
Kr^{+4}	Kr^{+3}		48±2	SEQ		2551

Kr^{+5} $\Delta H_{f0}^\circ \sim 18574$ kJ mol^{-1} (4439 kcal mol^{-1})

Ion	Reactant	Other products	Ionization or appearance potential (eV)	Method	Heat of formation (kJ mol^{-1})	Ref.
Kr^{+5}	Kr		192.5	S	~18574	2113
Kr^{+5}	Kr		181±2	NRE		211
Kr^{+5}	Kr		195	EI		211
Kr^{+5}	Kr		204	EI		1240
Kr^{+5}	Kr		218±10	EI		198
Kr^{+5}	Kr^{+4}		64.7	S		2113
Kr^{+5}	Kr^{+4}		62±2	SEQ		2551

Kr^{+6} $\Delta H_{f0}^\circ \sim 26148$ kJ mol^{-1} (6250 kcal mol^{-1})

Ion	Reactant	Other products	Ionization or appearance potential (eV)	Method	Heat of formation (kJ mol^{-1})	Ref.
Kr^{+6}	Kr		271.0	S	~26148	2113
Kr^{+6}	Kr		275±10	NRE		211
Kr^{+6}	Kr		302	EI		1240
Kr^{+6}	Kr		310	EI		211
Kr^{+6}	Kr		350±10	EI		198
Kr^{+6}	Kr^{+5}		78.5	S		2113
Kr^{+6}	Kr^{+5}		77±2	SEQ		2551

4.3. The Positive Ion Table—Continued

Ion	Reactant	Other products	Ionization or appearance potential (eV)	Method	Heat of formation (kJ mol^{-1})	Ref.
			Kr^{+7} $\Delta H_{f0}^{\circ} \sim 36858$ kJ mol^{-1} (8809 kcal mol^{-1})			
Kr^{+7}	Kr		382.0	S	~36858	2113
Kr^{+7}	Kr		470	EI		1240
Kr^{+7}	Kr^{+6}		111.0	S		2113
Kr^{+7}	Kr^{+6}		107±2	SEQ		2551
			Kr^{+8}			
Kr^{+8}	Kr		508	S	(~49015)	2113
Kr^{+8}	Kr		670	EI		1240
Kr^{+8}	Kr^{+7}		126	S		2113
			Kr$_2^+$			
Kr$_2^+$	Kr+Kr*		12.87	PI	(≤1242)	2650
Kr$_2^+$	Kr+Kr*		13.004±0.007	PI	(≤1255)	2763

The results of these two similar studies cannot be reconciled. For earlier work see references in ref. 2763.

Ion	Reactant	Other products	Ionization or appearance potential (eV)	Method	Heat of formation (kJ mol^{-1})	Ref.
			KrF$^+$			
KrF$^+$	KrF$_2$	F	13.71±0.20	EI		2577
			Rb$^+$			
Rb$^+$	Rb		4.177	S		2113

See also − EI: 2487

Ion	Reactant	Other products	Ionization or appearance potential (eV)	Method	Heat of formation (kJ mol^{-1})	Ref.
Rb$^+$	RbCl	Cl	8.9±0.5	EI		2406
Rb$^+$	RbCl	Cl	9.2±0.5	EI		2711

The possible existence of an ion−pair process in RbCl is discussed in J. Berkowitz and W. A. Chupka, J. Chem. Phys. **29**, 653 (1958).

Ion	Reactant	Other products	Ionization or appearance potential (eV)	Method	Heat of formation (kJ mol^{-1})	Ref.
Rb$^+$	RbI	I	8.1±0.3	EI		2001
			Rb^{+2}			
Rb^{+2}	Rb		31.46	S		2113
Rb^{+2}	Rb$^+$		27.28	S		2113, 3271
			Rb$_2$O$^+$			
Rb$_2$O$^+$	Rb$_2$SO$_4$		11.9±0.5	EI		2487

4.3. The Positive Ion Table—Continued

Ion	Reactant	Other products	Ionization or appearance potential (eV)	Method	Heat of formation (kJ mol^{-1})	Ref.
			Rb$_2$SO$_4^+$			
Rb$_2$SO$_4^+$	Rb$_2$SO$_4$		8.6±0.5	EI		2487
		Sr$^+$ $\Delta H_{f298}^\circ = 714$ kJ mol^{-1} (171 kcal mol^{-1})				
Sr$^+$	Sr		5.695	S	714	2113
See also − S:	3270					
EI:	2178, 2432, 2990					
Sr$^+$	SrCl$_2$		15.36±0.10	EI		2991
		Sr^{+2} $\Delta H_{f298}^\circ = 1778$ kJ mol^{-1} (425 kcal mol^{-1})				
Sr^{+2}	Sr		16.725	S	1778	2113
Sr^{+2}	Sr*?		11.5±0.2	EI		2178
Sr^{+2}	Sr$^+$		11.030	S		2113
			SrO$^+$			
SrO$^+$	SrO		6.1	EI	1244,	2123
See also − EI:	2178					
			Sr$_2$O$^+$			
Sr$_2$O$^+$	Sr$_2$O		4.8	EI	1244,	2123
			SrF$^+$			
SrF$^+$	SrF		4.9±0.3	EI		1105
SrF$^+$	SrF		5.0±0.3	EI		2432
SrF$^+$	SrF		5.2±0.3	EI		1104
SrF$^+$	SrF$_2$	F	13.0±1.0	EI		2432
See also − EI:	1105					
			SrCl$^+$			
SrCl$^+$	SrCl		5.59±0.10	EI		2991
SrCl$^+$	SrCl		5.3±0.5	EI		2990
SrCl$^+$	SrCl$_2$	Cl	10.48±0.10	EI		2991
SrCl$^+$	SrCl$_2$	Cl	10.7±0.3	EI		2990

4.3. The Positive Ion Table—Continued

Ion	Reactant	Other products	Ionization or appearance potential (eV)	Method	Heat of formation (kJ mol^{-1})	Ref.
			SrCl$_2^+$			
SrCl$_2^+$	SrCl$_2$		9.70±0.10	EI		2991
SrCl$_2^+$	SrCl$_2$		10.5±1.0	EI		2990
			SrOH$^+$			
SrOH$^+$	SrOH		5.55±0.1	D		3242, 3419
		Y$^+$ $\Delta H_{f0}^\circ = 1036$ kJ mol^{-1} (248 kcal mol^{-1})				
Y$^+$	Y		6.38	S	1036	2113
See also — EI: 2151, 2167, 2600						
Y$^+$	YO	O	12	EI		2167
Y$^+$	YF$_3$		28.0±0.7	EI		2009, 2600
Y$^+$	YCl$_3$		22.1±0.5	EI		2132
See also — EI: 2594						
		Y^{+2} $\Delta H_{f0}^\circ = 2217$ kJ mol^{-1} (530 kcal mol^{-1})				
Y^{+2}	Y		18.62	S	2217	2113
Y^{+2}	Y$^+$		12.24	S		2113
		Y^{+3} $\Delta H_{f0}^\circ = 4197$ kJ mol^{-1} (1003 kcal mol^{-1})				
Y^{+3}	Y		39.14	S	4197	2113, 3165
Y^{+3}	Y		39.03	BH		3273
Y^{+3}	Y^{+2}		20.5235±0.0002	S		3165
See also — S: 2113						
			YC$^+$			
YC$^+$	YC$_2$	C	13.4±0.5	EI		2151
YC$^+$	YC$_2$	C	14±1	EI		2996
			YC$_2^+$			
YC$_2^+$	YC$_2$		6.7±0.3	EI		2996
YC$_2^+$	YC$_2$		6.8±0.3	EI		2151
			YC$_4^+$			
YC$_4^+$	YC$_4$		7.0±0.3	EI		2996

4.3. The Positive Ion Table—Continued

Ion	Reactant	Other products	Ionization or appearance potential (eV)	Method	Heat of formation (kJ mol^{-1})	Ref.
		YO$^+$				
YO$^+$	YO		5.5	EI		2167
YO$^+$	YO		6.1±1	EI		3206
		YF$^+$				
YF$^+$	YF		6.3±0.3	EI		2600
YF$^+$	YF		7.5	EI		2594
YF$^+$	YF$_3$		21.5±0.7	EI		2009, 2600
		YF$_2^+$				
YF$_2^+$	YF$_2$		7.0±0.3	EI		2594, 2600
YF$_2^+$	YF$_3$	F	13.5±0.7	EI		2009, 2600
YF$_2^+$	YF$_3$	F	14.0	EI		2594
		YS$^+$				
YS$^+$	YS		6.0±1	EI		3206
		YCl$^+$				
YCl$^+$	YCl$_3$		17.3±0.5	EI		2132
		YCl$_2^+$				
YCl$_2^+$	YCl$_3$	Cl	14.5±0.5	EI		2132
		YCl$_3^+$				
YCl$_3^+$	YCl$_3$		12.8±0.5	EI		2132
		Y$_2$Cl$_5^+$				
Y$_2$Cl$_5^+$	Y$_2$Cl$_6$	Cl	13.7±0.5	EI		2132
		YSe$^+$				
YSe$^+$	YSe		6.1±1	EI		3206
		Zr$^+$ $\Delta H_{f0}^\circ = 1267$ kJ mol^{-1} (303 kcal mol^{-1})				
Zr$^+$	Zr		6.84	S	1267	2113
Zr$^+$	ZrF$_4$		29.5±0.5	EI		2532

4.3. The Positive Ion Table—Continued

Ion	Reactant	Other products	Ionization or appearance potential (eV)	Method	Heat of formation (kJ mol^{-1})	Ref.
			ZrN$^+$			
ZrN$^+$	ZrN		7.9±0.4	EI		2490
			ZrO$^+$			
ZrO$^+$	ZrO		~5.5	EI		2783
ZrO$^+$	ZrO		6.6±0.3	EI		2490
			ZrO$_2^+$			
ZrO$_2^+$	ZrO$_2$		8±0.5	EI		2783
			ZrF$^+$			
ZrF$^+$	ZrF$_4$		29.4±0.5	EI		2532
			ZrF$_2^+$			
ZrF$_2^+$	ZrF$_2$		12.0±0.5	EI		2532
ZrF$_2^+$	ZrF$_4$		23.4±0.5	EI		2532
			ZrF$_3^+$			
ZrF$_3^+$	ZrF$_3$		7.5±0.3	EI		2532
ZrF$_3^+$	ZrF$_4$	F	14.5±0.5	EI		2532
			ZrCl$^+$			
ZrCl$^+$	(C$_5$H$_5$)$_2$ZrCl$_2$ (Bis(cyclopentadienyl)zirconium dichloride)		24.3±0.5	EI		2479
			ZrCl$_2^+$			
ZrCl$_2^+$	(C$_5$H$_5$)$_2$ZrCl$_2$ (Bis(cyclopentadienyl)zirconium dichloride)		20.9±0.5	EI		2479
			C$_{33}$H$_{57}$O$_6$Zr$^+$			
C$_{33}$H$_{57}$O$_6$Zr$^+$	((CH$_3$)$_3$CCOCHCOC(CH$_3$)$_3$)$_4$Zr (Tetrakis(2,2,6,6–tetramethyl–3,5–heptanedionato)zirconium)		10.5±0.5	EI		2524

4.3. The Positive Ion Table—Continued

Ion	Reactant	Other products	Ionization or appearance potential (eV)	Method	Heat of formation (kJ mol^{-1})	Ref.
		$C_3H_3ClZr^+$				
$C_3H_3ClZr^+$	$(C_5H_5)_2ZrCl_2$ (Bis(cyclopentadienyl)zirconium dichloride)		~19.9	EI		2479
		$C_5H_5ClZr^+$				
$C_5H_5ClZr^+$	$(C_5H_5)_2ZrCl_2$ (Bis(cyclopentadienyl)zirconium dichloride)		19.8±0.4	EI		2479
		$C_{10}H_{10}ClZr^+$				
$C_{10}H_{10}ClZr^+$	$(C_5H_5)_2ZrCl_2$ (Bis(cyclopentadienyl)zirconium dichloride)	Cl	12.3±0.2	EI		2479
		$C_5H_5Cl_2Zr^+$				
$C_5H_5Cl_2Zr^+$	$(C_5H_5)_2ZrCl_2$ (Bis(cyclopentadienyl)zirconium dichloride)		12.5±0.2	EI		2479
		$C_{10}H_{10}Cl_2Zr^+$				
$C_{10}H_{10}Cl_2Zr^+$	$(C_5H_5)_2ZrCl_2$ (Bis(cyclopentadienyl)zirconium dichloride)		9.37±0.25	EI		2479
		Nb^+ $\Delta H^\circ_{f0} = 1387$ kJ mol^{-1} (331 kcal mol^{-1})				
Nb^+	Nb		6.88	S	1387	2113
Nb^+	$NbCl_5$		28.0±0.7	EI		2506
		$NbCl^+$				
$NbCl^+$	$NbCl_5$		22.8±0.5	EI		2506
		$NbCl_2^+$				
$NbCl_2^+$	$NbCl_5$		18.8±0.3	EI		2506
$NbCl_2^+$	$NbCl_5$		20±1	EI		2861
		$NbCl_3^+$				
$NbCl_3^+$	$NbCl_5$		14.9±0.8	EI		2861

4.3. The Positive Ion Table—Continued

Ion	Reactant	Other products	Ionization or appearance potential (eV)	Method	Heat of formation (kJ mol^{-1})	Ref.
			NbCl$_4^+$			
NbCl$_4^+$	NbCl$_5$	Cl	11.34±0.22	EI		2506
NbCl$_4^+$	NbCl$_5$	Cl	11.6±0.8	EI		2861
		Mo$^+$ $\Delta H_{f0}^\circ = 1341.5$ kJ mol^{-1} (320.6 kcal mol^{-1})				
Mo$^+$	Mo		7.099	S	1341.5	2113
Mo$^+$	Mo(CO)$_6$	6CO	18.3±0.3	EI		2023
Mo$^+$	Mo(CO)$_6$	6CO	18.6±0.2	EI		2403
Mo$^+$	Mo(CO)$_6$	6CO	19.63	EI		2500
Mo$^+$	Mo(CO)$_6$	6CO	20.7±0.5	EI		1107
			MoC$^+$			
MoC$^+$	Mo(CO)$_6$		24.3±1	EI		2403
MoC$^+$	Mo(CO)$_6$		27.2±0.4	EI		1107
			MoO$^+$			
MoO$^+$	MoO		8.0±0.6	EI		2126
MoO$^+$	Mo(CO)$_6$		24.3±1	EI		2403
			MoO$_2^+$			
MoO$_2^+$	MoO$_2$		9.2	EI	1244,	2123
MoO$_2^+$	MoO$_2$		9.4±0.6	EI	2126,	2129
MoO$_2^+$	MoO$_2$		9.7±0.5	EI		3257
			MoO$_3^+$			
MoO$_3^+$	MoO$_3$		11.8±0.5	EI		3257
MoO$_3^+$	MoO$_3$		11.8	EI	1244,	2123
MoO$_3^+$	MoO$_3$		12.0±0.6	EI	2126,	2129
MoO$_3^+$	MoO$_3$		12.2±0.5	EI		3256
			Mo$_2$O$_5^+$			
Mo$_2$O$_5^+$	Mo$_2$O$_5$		10	EI		2129
Mo$_2$O$_5^+$	Mo$_2$O$_6$	O	14.5	EI		2129
			Mo$_2$O$_6^+$			
Mo$_2$O$_6^+$	Mo$_2$O$_6$		12.1±0.6	EI		2129

4.3. The Positive Ion Table—Continued

Ion	Reactant	Other products	Ionization or appearance potential (eV)	Method	Heat of formation (kJ mol^{-1})	Ref.
			$Mo_3O_8^+$			
$Mo_3O_8^+$	Mo_3O_8		12.2	EI		2129
$Mo_3O_8^+$	Mo_3O_9	O	14.5	EI		2129
			$Mo_3O_9^+$			
$Mo_3O_9^+$	Mo_3O_9		12.0±1.0	EI		2129
			$Li_2MoO_4^+$			
$Li_2MoO_4^+$	Li_2MoO_4		9.7±0.5	EI		3257
			$MoCO^+$			
$MoCO^+$	$Mo(CO)_6$	5CO	15.61	EI		2500
$MoCO^+$	$Mo(CO)_6$	5CO	15.7±0.2	EI		2403
$MoCO^+$	$Mo(CO)_6$	5CO	15.8±0.06	EI		2023
$MoCO^+$	$Mo(CO)_6$	5CO	18.1±0.3	EI		1107
			$MoCO^{+2}$			
$MoCO^{+2}$	$Mo(CO)_6$		34.5±0.5	EI		1107
$MoCO^{+2}$	$Mo(CO)_6$		35.7±1	EI		2403
			$MoC_2O_2^+$			
$MoC_2O_2^+$	$Mo(CO)_6$	4CO	13.90±0.3	EI		2023
$MoC_2O_2^+$	$Mo(CO)_6$	4CO	14.5±0.1	EI		2403
$MoC_2O_2^+$	$Mo(CO)_6$	4CO	14.76	EI		2500
$MoC_2O_2^+$	$Mo(CO)_6$	4CO	15.6±0.3	EI		1107
			$MoC_2O_2^{+2}$			
$MoC_2O_2^{+2}$	$Mo(CO)_6$		30.8±0.5	EI		1107
$MoC_2O_2^{+2}$	$Mo(CO)_6$		31.6±1	EI		2403
			$MoC_3O_3^+$			
$MoC_3O_3^+$	$Mo(CO)_6$	3CO	12.36±0.12	EI		2023
$MoC_3O_3^+$	$Mo(CO)_6$	3CO	12.82±0.1	EI		2403
$MoC_3O_3^+$	$Mo(CO)_6$	3CO	13.18	EI		2500
$MoC_3O_3^+$	$Mo(CO)_6$	3CO	13.7±0.3	EI		1107

4.3. The Positive Ion Table—Continued

Ion	Reactant	Other products	Ionization or appearance potential (eV)	Method	Heat of formation (kJ mol^{-1})	Ref.
			MoC$_3$O$_3^{+2}$			
MoC$_3$O$_3^{+2}$	Mo(CO)$_6$		29.1±1.2	EI		1107
			MoC$_4$O$_4^+$			
MoC$_4$O$_4^+$	Mo(CO)$_6$	2CO	10.63±0.15	EI		2403
MoC$_4$O$_4^+$	Mo(CO)$_6$	2CO	~10.72	EI		2500
MoC$_4$O$_4^+$	Mo(CO)$_6$	2CO	11.28±0.14	EI		2023
MoC$_4$O$_4^+$	Mo(CO)$_6$	2CO	11.9±0.2	EI		1107
			MoC$_5$O$_5^+$			
MoC$_5$O$_5^+$	Mo(CO)$_6$	CO	~9.14	EI		2500
MoC$_5$O$_5^+$	Mo(CO)$_6$	CO	9.43±0.1	EI		2403
MoC$_5$O$_5^+$	Mo(CO)$_6$	CO	9.64±0.05	EI		2023
MoC$_5$O$_5^+$	Mo(CO)$_6$	CO	9.80±0.15	EI		1107

See also – PI:　2886

Ion	Reactant	Other products	Ionization or appearance potential (eV)	Method	Heat of formation (kJ mol^{-1})	Ref.
			Mo(CO)$_6^+$　　$\Delta H_{f0}^\circ = -121$ kJ mol^{-1} (-29 kcal mol^{-1})			
MoC$_6$O$_6^+$	Mo(CO)$_6$		8.227±0.011	PI	−121	2886
(Threshold value approximately corrected for hot bands)						
MoC$_6$O$_6^+$	Mo(CO)$_6$		8.12±0.03	PI		1167
MoC$_6$O$_6^+$	Mo(CO)$_6$		8.23±0.12	EI		1107
MoC$_6$O$_6^+$	Mo(CO)$_6$		8.30±0.03	EI		2023
MoC$_6$O$_6^+$	Mo(CO)$_6$		8.43±0.05	EI		2500
MoC$_6$O$_6^+$	Mo(CO)$_6$		8.46±0.08	EI		2403
			MoOF$^+$			
MoOF$^+$	MoO$_2$F$_2$		23.0±0.5	EI		2859
			MoO$_2$F$^+$			
MoO$_2$F$^+$	MoO$_2$F$_2$	F	15.0±0.5	EI		2859
			MoO$_2$F$_2^+$			
MoO$_2$F$_2^+$	MoO$_2$F$_2$		13.0±0.3	EI		2859
			SrMoO$_2^+$			
SrMoO$_2^+$	SrMoO$_4$		11.0	EI		1244

4.3. The Positive Ion Table—Continued

Ion	Reactant	Other products	Ionization or appearance potential (eV)	Method	Heat of formation (kJ mol⁻¹)	Ref.
		$SrMoO_3^+$				
$SrMoO_3^+$	$SrMoO_3$		6.2	EI		1244
		$SrMoO_4^+$				
$SrMoO_4^+$	$SrMoO_4$		9.2	EI		1244
		$C_7H_5NO_3Mo^+$				
$C_7H_5NO_3Mo^+$	$C_5H_5Mo(CO)_2NO$ (Cyclopentadienylmolybdenum dicarbonyl nitrosyl)		8.1±0.2	EI		2963
		$C_{10}H_9NO_5Mo^+$				
$C_{10}H_9NO_5Mo^+$	$n-C_4H_9NCMo(CO)_5$		7.65±0.05	EI		2481
		$C_{12}H_5NO_5Mo^+$				
$C_{12}H_5NO_5Mo^+$	$C_6H_5NCMo(CO)_5$ (Isocyanobenzenemolybdenum pentacarbonyl)		7.88	EI		2544
		$C_{12}H_{11}NO_5Mo^+$				
$C_{12}H_{11}NO_5Mo^+$	$C_6H_{11}NCMo(CO)_5$ (Isocyanocyclohexanemolybdenum pentacarbonyl)		7.72	EI		2544
		$C_{13}H_7NO_5Mo^+$				
$C_{13}H_7NO_5Mo^+$	$C_6H_4(CH_3)NCMo(CO)_5$ (Isocyanotoluenemolybdenum pentacarbonyl)		7.73±0.05	EI		2481
		$C_{11}H_{15}O_5PMo^+$				
$C_{11}H_{15}O_5PMo^+$	$(C_2H_5)_3PMo(CO)_5$		7.73±0.05	EI		2481
		$C_{17}H_{27}O_5PMo^+$				
$C_{17}H_{27}O_5PMo^+$	$(n-C_4H_9)_3PMo(CO)_5$		7.51±0.05	EI		2481
		$C_8H_9O_8PMo^+$				
$C_8H_9O_8PMo^+$	$(CH_3O)_3PMo(CO)_5$		7.89	EI		2544

4.3. The Positive Ion Table—Continued

Ion	Reactant	Other products	Ionization or appearance potential (eV)	Method	Heat of formation (kJ mol^{-1})	Ref.
		C$_{11}$H$_{15}$O$_8$PMo$^+$				
C$_{11}$H$_{15}$O$_8$PMo$^+$	(C$_2$H$_5$O)$_3$PMo(CO)$_5$		7.72	EI		2544
		C$_{17}$H$_{27}$O$_8$PMo$^+$				
C$_{17}$H$_{27}$O$_8$PMo$^+$	(n-C$_4$H$_9$O)$_3$PMo(CO)$_5$		7.71±0.05	EI		2481
		C$_{10}$H$_{18}$O$_{10}$P$_2$Mo$^+$				
C$_{10}$H$_{18}$O$_{10}$P$_2$Mo$^+$	((CH$_3$O)$_3$P)$_2$Mo(CO)$_4$		7.54	EI		2716
		C$_{16}$H$_{30}$O$_{10}$P$_2$Mo$^+$				
C$_{16}$H$_{30}$O$_{10}$P$_2$Mo$^+$	((C$_2$H$_5$O)$_3$P)$_2$Mo(CO)$_4$		7.23	EI		2716
		C$_5$O$_5$PCl$_3$Mo$^+$				
C$_5$O$_5$PCl$_3$Mo$^+$	PCl$_3$Mo(CO)$_5$		8.25	EI		2544
		C$_{11}$H$_5$O$_5$PCl$_2$Mo$^+$				
C$_{11}$H$_5$O$_5$PCl$_2$Mo$^+$	C$_6$H$_5$PCl$_2$Mo(CO)$_5$ (Dichloro(phenyl)phosphinemolybdenum pentacarbonyl)		8.03±0.05	EI		2481
	Tc$^+$ $\Delta H^\circ_{f\,298} = 1380$ kJ mol^{-1} (330 kcal mol^{-1})					
Tc$^+$	Tc		7.28	S	1380	2113
		C$_{10}$H$_{10}$Tc$^+$				
C$_{10}$H$_{10}$Tc$^+$	(C$_5$H$_5$)$_2$TcH (Bis(cyclopentadienyl)technetium hydride)	H	7.86±0.1	EI		2683
		C$_{10}$H$_{11}$Tc$^+$				
C$_{10}$H$_{11}$Tc$^+$	(C$_5$H$_5$)$_2$TcH (Bis(cyclopentadienyl)technetium hydride)		7.13±0.05	EI		2683
		TcC$_5$O$_5^+$				
TcC$_5$O$_5^+$	Tc$_2$(CO)$_{10}$		10.13	EI		3234

4.3. The Positive Ion Table—Continued

Ion	Reactant	Other products	Ionization or appearance potential (eV)	Method	Heat of formation (kJ mol^{-1})	Ref.
			Tc$_2$C$_{10}$O$_{10}^+$			
Tc$_2$C$_{10}$O$_{10}^+$	Tc$_2$(CO)$_{10}$		8.30±0.03	EI		3234
	Ru$^+$	**$\Delta H_{f0}^\circ = 1352$ kJ mol^{-1} (323 kcal mol^{-1})**				
Ru$^+$	Ru		7.37	S	1352	2113
See also – EI: 2537						
Ru$^+$	RuO$_2$		13.0	EI		2537
Ru$^+$	RuO$_4$		22.3±0.3	EI		1284
Ru$^+$	(C$_5$H$_5$)$_2$Ru (Bis(cyclopentadienyl)ruthenium)		16.1±0.5	EI		2683
			RuO$^+$			
RuO$^+$	RuO		8.7	EI		2537
RuO$^+$	RuO$_2$	O	12.8	EI		2537
RuO$^+$	RuO$_4$		18.1±0.3	EI		1284
			RuO$_2^+$			
RuO$_2^+$	RuO$_2$		10.6	EI		2537
RuO$_2^+$	RuO$_4$		14.2±0.2	EI		1284
			RuO$_3^+$			
RuO$_3^+$	RuO$_3$		11.2	EI		2537
RuO$_3^+$	RuO$_4$	O	15.7±0.3	EI		1284
			RuO$_4^+$			
RuO$_4^+$	RuO$_4$		12.33±0.23	EI		1284
RuO$_4^+$	RuO$_4$		12.8	EI		2537
			C$_5$H$_5$Ru$^+$			
C$_5$H$_5$Ru$^+$	(C$_5$H$_5$)$_2$Ru (Bis(cyclopentadienyl)ruthenium)		14.3±0.2	EI		2683
			C$_8$H$_8$Ru$^+$			
C$_8$H$_8$Ru$^+$	(C$_5$H$_5$)$_2$Ru (Bis(cyclopentadienyl)ruthenium)		13.7±0.2	EI		2683

4.3. The Positive Ion Table—Continued

Ion	Reactant	Other products	Ionization or appearance potential (eV)	Method	Heat of formation (kJ mol^{-1})	Ref.
		$C_{10}H_{10}Ru^+$				
$C_{10}H_{10}Ru^+$	$(C_5H_5)_2Ru$ (Bis(cyclopentadienyl)ruthenium)		7.82±0.1	EI		2683
		Rh^+ $\Delta H^{\circ}_{f0} = 1275$ kJ mol^{-1} (305 kcal mol^{-1})				
Rh^+	Rh		7.46	S	1275	2113
See also – EI: 1020						
		RhO^+				
RhO^+	RhO		9.3	EI		1020
		RhO_2^+				
RhO_2^+	RhO_2		10.0	EI		1020
		$C_{16}H_{24}Cl_2Rh_2^+$				
$C_{16}H_{24}Cl_2Rh_2^+$	$(C_8H_{12}RhCl)_2$ (Bis(1,5–cyclooctadienerhodium chloride))		7.1±0.1	EI		2698
		$C_{16}H_{24}Br_2Rh_2^+$				
$C_{16}H_{24}Br_2Rh_2^+$	$(C_8H_{12}RhBr)_2$ (Bis(1,5–cyclooctadienerhodium bromide))		7.2±0.1	EI		2698
		$C_{15}H_3O_6F_{18}Rh^+$				
$C_{15}H_3O_6F_{18}Rh^+$	$(CF_3COCHCOCF_3)_3Rh$ (Tris(1,1,1,5,5,5–hexafluoro–2,4–pentanedionato)rhodium)		10.15±0.12	EI		2580
		Pd^+ $\Delta H^{\circ}_{f0} = 1182$ kJ mol^{-1} (283 kcal mol^{-1})				
Pd^+	Pd		8.34	S	1182	2113
See also – EI: 1020						
		Pd_2^+				
Pd_2^+	Pd_2		7.7±0.3	EI		2441

4.3. The Positive Ion Table—Continued

Ion	Reactant	Other products	Ionization or appearance potential (eV)	Method	Heat of formation (kJ mol⁻¹)	Ref.
PdO⁺						
PdO⁺	PdO		9.1	EI		1020
PdSi⁺						
PdSi⁺	PdSi		8.4±0.5	EI		2943
Ag⁺ $\quad \Delta H^\circ_{f0} = 1015.1$ kJ mol⁻¹ (242.6 kcal mol⁻¹)						
Ag⁺	Ag		7.576	S	1015.1	2113
Ag⁺	Ag		7.53	RPD		2994
See also — EI: 2990						
Ag⁺	AgF	F	11.5±0.3	EI		2596
Ag⁺	AgCl	Cl	11.0±0.4	EI		3205
Ag₂⁺						
Ag₂⁺	Ag₃Cl₂		15.7±1.3	EI		3205
AgF⁺						
AgF⁺	AgF		11.4±0.3	EI		2596
AgCl⁺						
AgCl⁺	AgCl		10.3±0.4	EI		3205
AgCl⁺	AgCl		10.5±0.3	EI		2990
Ag₂Cl⁺						
Ag₂Cl⁺	Ag₃Cl₂		11.6±0.6	EI		3205
Ag₃Cl₂⁺						
Ag₃Cl₂⁺	Ag₃Cl₂		10.7±0.2	EI		3205
Ag₃Cl₃⁺						
Ag₃Cl₃⁺	Ag₃Cl₃		10.3±0.4	EI		3205

4.3. The Positive Ion Table—Continued

Ion	Reactant	Other products	Ionization or appearance potential (eV)	Method	Heat of formation (kJ mol^{-1})	Ref.
Cd$^+$		$\Delta H^\circ_{f0} = 979.8$ kJ mol^{-1} (234.2 kcal mol^{-1})				
Cd$^+$	Cd		8.993	S	979.8	2113
See also – EI: 1047, 2056						
Cd$^+$	(CH$_3$)$_2$Cd	2CH$_3$	12.10±0.02	PI		2983
		CdCl$^+$				
CdCl$^+$	CdCl$_2$	Cl	11.8±0.2	EI		2056
		CdCl$_2^+$				
CdCl$_2^+$	CdCl$_2$		11.2±0.2	EI		2056
CH$_3$Cd$^+$		$\Delta H^\circ_{f298} = 894$ kJ mol^{-1} (214 kcal mol^{-1})				
CH$_3$Cd$^+$	(CH$_3$)$_2$Cd	CH$_3$	9.69±0.02	PI	894	2983
(Threshold value approximately corrected for hot bands)						
(CH$_3$)$_2$Cd$^+$		$\Delta H^\circ_{f298} = 927$ kJ mol^{-1} (222 kcal mol^{-1})				
C$_2$H$_6$Cd$^+$	(CH$_3$)$_2$Cd		8.56±0.02	PI	927	2983
(Threshold value approximately corrected for hot bands)						
In$^+$		$\Delta H^\circ_{f0} = 802.0$ kJ mol^{-1} (191.7 kcal mol^{-1})				
In$^+$	In		5.786	S	802.0	2113, 3167
See also – S: 3270						
EI: 1065, 2138, 2469, 2518, 2565, 2620, 3014, 3244						
In$^+$	In$_2$S		8.7±1.0	EI		2469
In$^+$	In$_2$Se		9.8±0.5	EI		2469
In$^+$	In$_2$Te?		9.3±1.0	EI		2469
		In$_2^+$				
In$_2^+$	In$_2$		5.8±0.3	EI		2138
In$_2^+$	In$_2$O	O	15±0.5	EI		3014
In$_2^+$	In$_2$S	S	10.8±0.5	EI		2469
In$_2^+$	In$_2$Se	Se	11.6±0.5	EI		2469

4.3. The Positive Ion Table—Continued

Ion	Reactant	Other products	Ionization or appearance potential (eV)	Method	Heat of formation (kJ mol⁻¹)	Ref.
			In₂O⁺			
In_2O^+	In_2O		7.8±0.5	EI		1065
In_2O^+	In_2O		8.0±0.5	EI		2518
In_2O^+	In_2O		9±0.5	EI		3014
In_2O^+	In_2O		10.3±0.5	EI		3244
			InF⁺			
InF^+	InF		9.6±0.5	EI		2620
			InS⁺			
InS^+	InS		7.0±0.5	EI		2469
InS^+	In_2S	In	11.7±1.0	EI		2469
			In₂S⁺			
In_2S^+	In_2S		7.6±0.5	EI		2469
			In₂S₂⁺			
$In_2S_2^+$	In_2S_2		6.4±0.5	EI		2469
			InSe⁺			
$InSe^+$	$InSe$		7.1±0.5	EI		2469
$InSe^+$	In_2Se	In	11.9±0.5	EI		2469
			In₂Se⁺			
In_2Se^+	In_2Se		7.5±0.5	EI		2469
			LiInO⁺			
$LiInO^+$	$LiInO$		6.6±0.5	EI		2565

Sn⁺ $\Delta H_{f0}^\circ = 1010.6 \text{ kJ mol}^{-1} (241.5 \text{ kcal mol}^{-1})$

Sn^+	Sn		7.344	S	1010.6	2113

See also – EI: 2595

Sn^+	SnH_4	$2H_2$	9.0±0.3	EI		2116
Sn^+	SnH_4	$2H_2$	11.4±0.2	EI		2137
Sn^+	Sn_2H_6		10.8±0.3	EI		2133
Sn^+	SnO	O	13.0±1	EI		1243

4.3. The Positive Ion Table—Continued

Ion	Reactant	Other products	Ionization or appearance potential (eV)	Method	Heat of formation (kJ mol⁻¹)	Ref.
Sn^+	SnF	F	15.5±0.3	EI		2436
Sn^+	SnF_2		18.5±0.5	EI	2436,	2595
Sn^+	SnS	S	12.5±0.5	EI		2139
Sn^+	$SnCl_2$		12.8±1.0	EI		2871
Sn^+	$SnCl_4$		22.2±1.0	EI		2871
Sn^+	SnSe	Se	12.8±0.5	EI		2063
Sn^+	$SnBr_2$		12.8±1.0	EI		3185
Sn^+	$SnBr_4$		18.5±1.0	EI		3185
Sn^+	$(CH_3)_4Sn$		16.7	EI		2980
Sn^+	$(CH_3)_4Sn$		17.5±0.5	EI		2725
Sn^+	$(CH_3)_4Sn$		18.1±0.3	EI		82
Sn^+	$(C_2H_5)_4Sn$		17.1	EI		2980
Sn^+	$(C_6H_5)_4Sn$ (Tetraphenyltin)		9.4±0.2	EI		2725

$$Sn_2^+$$

Ion	Reactant	Other products	Ionization or appearance potential (eV)	Method	Heat of formation (kJ mol⁻¹)	Ref.
Sn_2^+	Sn_2H_6	$3H_2$	10.7±0.3	EI		2133
Sn_2^+	Sn_2S_2		16.5±1.0	EI		2139

$$SnH^+$$

Ion	Reactant	Other products	Ionization or appearance potential (eV)	Method	Heat of formation (kJ mol⁻¹)	Ref.
SnH^+	SnH_4	H_2+H	10.7±0.3	EI		2116
SnH^+	SnH_4	H_2+H	13.3±0.2	EI		2137
SnH^+	$(CH_3)_4Sn$		16.8	EI		2980
SnH^+	$(C_2H_5)_4Sn$		16.7	EI		2980

$$SnH_2^+$$

Ion	Reactant	Other products	Ionization or appearance potential (eV)	Method	Heat of formation (kJ mol⁻¹)	Ref.
SnH_2^+	SnH_4	H_2	9.5±0.3	EI		2116
SnH_2^+	SnH_4	H_2	12.1±0.2	EI		2137

$$SnH_3^+$$

Ion	Reactant	Other products	Ionization or appearance potential (eV)	Method	Heat of formation (kJ mol⁻¹)	Ref.
SnH_3^+	SnH_4	H	11.31±0.01	RPD		2553
SnH_3^+	SnH_4	H	9.4±0.3	EI		2116
SnH_3^+	SnH_4	H	11.9±0.2	EI		2137
SnH_3^+	$(CH_3)_4Sn$		14.0	EI		2980

$$Sn_2H^+$$

Ion	Reactant	Other products	Ionization or appearance potential (eV)	Method	Heat of formation (kJ mol⁻¹)	Ref.
Sn_2H^+	Sn_2H_6	$2H_2+H$	10.6±0.3	EI		2133

$$Sn_2H_2^+$$

Ion	Reactant	Other products	Ionization or appearance potential (eV)	Method	Heat of formation (kJ mol⁻¹)	Ref.
$Sn_2H_2^+$	Sn_2H_6	$2H_2$	10.5±0.3	EI		2133

4.3. The Positive Ion Table—Continued

Ion	Reactant	Other products	Ionization or appearance potential (eV)	Method	Heat of formation (kJ mol^{-1})	Ref.
			$Sn_2H_3^+$			
$Sn_2H_3^+$	Sn_2H_6	H_2+H	10.4±0.3	EI		2133
			$Sn_2H_4^+$			
$Sn_2H_4^+$	Sn_2H_6	H_2	10.3±0.3	EI		2133
			$Sn_2H_5^+$			
$Sn_2H_5^+$	Sn_2H_6	H	10.0±0.3	EI		2133
			$Sn_2H_6^+$			
$Sn_2H_6^+$	Sn_2H_6		9.0±0.3	EI		2133
			SnO^+			
SnO^+	SnO		10.5±0.5	EI		1243, 1244
			Sn_2O^+			
Sn_2O^+	Sn_2O_2	O	13.8±0.5	EI		1243
Sn_2O^+	Sn_2O_2	O	14.0	EI		1244
			$Sn_2O_2^+$			
$Sn_2O_2^+$	Sn_2O_2		9.8±0.5	EI		1243, 1244
			$Sn_3O_3^+$			
$Sn_3O_3^+$	Sn_3O_3		9.8±0.5	EI		1243, 1244
			$Sn_4O_4^+$			
$Sn_4O_4^+$	Sn_4O_4		9.2±0.5	EI		1243, 1244
			SnF^+			
SnF^+	SnF		~7.04	S		2149
SnF^+	SnF		7.8±0.3	EI		2595
SnF^+	SnF		8.5±0.3	EI		2595
SnF^+	SnF		9.0±0.5	EI		2436
SnF^+	SnF_2	F	12.0±0.3	EI		2595
SnF^+	SnF_2	F	12.5	EI		2436

4.3. The Positive Ion Table—Continued

Ion	Reactant	Other products	Ionization or appearance potential (eV)	Method	Heat of formation (kJ mol^{-1})	Ref.
		SnF$_2^+$				
SnF$_2^+$	SnF$_2$		10.5±0.3	EI		2436
SnF$_2^+$	SnF$_2$		11.5±0.2	EI		2595
		Sn$_2$F$_3^+$				
Sn$_2$F$_3^+$	Sn$_2$F$_4$	F	11.0±0.5	EI		2595
		Sn$_2$F$_4^+$				
Sn$_2$F$_4^+$	Sn$_2$F$_4$		10.5±0.5	EI		2595
		Sn$_3$F$_5^+$				
Sn$_3$F$_5^+$	Sn$_3$F$_6$	F	10.0±0.5	EI		2595
		Sn$_3$F$_6^+$				
Sn$_3$F$_6^+$	Sn$_3$F$_6$		10.0±0.5	EI		2595
		SnS$^+$				
SnS$^+$	SnS		9.7±0.5	EI		2139
		Sn$_2$S$^+$				
Sn$_2$S$^+$	Sn$_2$S$_2$	S	12.4±1.0	EI		2139
		Sn$_2$S$_2^+$				
Sn$_2$S$_2^+$	Sn$_2$S$_2$		9.4±0.5	EI		2139
		SnCl$^+$				
SnCl$^+$	SnCl$_2$	Cl	11.3±0.4	EI		2871
SnCl$^+$	SnCl$_4$		12.5±1.0	EI		2871
		SnCl$_2^+$				
SnCl$_2^+$	SnCl$_2$		10.1±0.4	EI		2871
SnCl$_2^+$	SnCl$_4$		13.6±1.0	EI		2871

4.3. The Positive Ion Table—Continued

Ion	Reactant	Other products	Ionization or appearance potential (eV)	Method	Heat of formation (kJ mol^{-1})	Ref.
			SnCl$_3^+$			
SnCl$_3^+$	SnCl$_4$	Cl	12.2±0.4	EI		2871
	SnCl$_4^+$	$\Delta H_{f0}^\circ = 677$ kJ mol^{-1} (162 kcal mol^{-1})				
SnCl$_4^+$	SnCl$_4$		11.88±0.05	PE	677	3362
SnCl$_4^+$	SnCl$_4$		11.5±0.4	EI		2871
See also – PE: 3117						
			SnSe$^+$			
SnSe$^+$	SnSe		9.7±0.5	EI		2063
			Sn$_2$Se$_2^+$			
Sn$_2$Se$_2^+$	Sn$_2$Se$_2$		9.8±0.5	EI		2063
			SnBr$^+$			
SnBr$^+$	SnBr$_2$	Br	11.0±0.4	EI		3185
SnBr$^+$	SnBr$_4$		13.7±1.0	EI		3185
			SnBr$_2^+$			
SnBr$_2^+$	SnBr$_2$		10.0±0.4	EI		3185
SnBr$_2^+$	SnBr$_4$		12.0±1.0	EI		3185
			SnBr$_3^+$			
SnBr$_3^+$	SnBr$_4$	Br	11.3±0.4	EI		3185
			SnBr$_4^+$			
SnBr$_4^+$	SnBr$_4$		11.0 (V)	PE		3117
SnBr$_4^+$	SnBr$_4$		10.6±0.4	EI		3185
			CH$_3$Sn$^+$			
CH$_3$Sn$^+$	CH$_3$Sn		6.85±0.1	EI		2719
CH$_3$Sn$^+$	(CH$_3$)$_4$Sn		14.2	EI		2980
CH$_3$Sn$^+$	(CH$_3$)$_4$Sn		15.7±0.4	EI		82

ROSENSTOCK ET AL.

4.3. The Positive Ion Table—Continued

Ion	Reactant	Other products	Ionization or appearance potential (eV)	Method	Heat of formation (kJ mol^{-1})	Ref.
			C$_2$H$_5$Sn$^+$			
C$_2$H$_5$Sn$^+$	(C$_2$H$_5$)$_4$Sn		14.4	EI		2980
			C$_2$H$_6$Sn$^+$			
C$_2$H$_6$Sn$^+$	(CH$_3$)$_2$Sn		7.95±0.05	EI		2719
C$_2$H$_6$Sn$^+$	(CH$_3$)$_4$Sn		13.0±0.2	EI		2725
C$_2$H$_6$Sn$^+$	(CH$_3$)$_4$Sn		13.1±0.2	EI		82
C$_2$H$_6$Sn$^+$	(CH$_3$)$_4$Sn		13.6	EI		2980
			C$_2$H$_7$Sn$^+$			
C$_2$H$_7$Sn$^+$	(CH$_3$)$_4$Sn		13.9	EI		2980
C$_2$H$_7$Sn$^+$	(C$_2$H$_5$)$_4$Sn		13.0	EI		2980
			C$_3$H$_9$Sn$^+$			
C$_3$H$_9$Sn$^+$	(CH$_3$)$_3$Sn		7.10±0.05	EI		2719
C$_3$H$_9$Sn$^+$	(CH$_3$)$_4$Sn	CH$_3$	9.65	EM		2553
C$_3$H$_9$Sn$^+$	(CH$_3$)$_4$Sn	CH$_3$	9.72±0.06	RPD	1424,	2553
C$_3$H$_9$Sn$^+$	(CH$_3$)$_4$Sn	CH$_3$	9.58±0.19	EI		2720
See also – EI:	82, 2725, 2980					
C$_3$H$_9$Sn$^+$	(CH$_3$)$_3$SnCH=CH$_2$		10.44±0.11	RPD		2553
C$_3$H$_9$Sn$^+$	(CH$_3$)$_3$SnC$_2$H$_5$		9.50	EM		2553
C$_3$H$_9$Sn$^+$	(CH$_3$)$_3$SnC$_2$H$_5$		9.49±0.07	RPD		2553
See also – EI:	1424					
C$_3$H$_9$Sn$^+$	(CH$_3$)$_3$SnC$_3$H$_5$		8.68±0.02	RPD		2553
C$_3$H$_9$Sn$^+$	n–C$_3$H$_7$Sn(CH$_3$)$_3$		9.50±0.06	RPD	1424,	2553
C$_3$H$_9$Sn$^+$	iso–C$_3$H$_7$Sn(CH$_3$)$_3$		9.17±0.14	RPD		2553
C$_3$H$_9$Sn$^+$	n–C$_4$H$_9$Sn(CH$_3$)$_3$		9.80±0.04	RPD		2553
C$_3$H$_9$Sn$^+$	sec–C$_4$H$_9$Sn(CH$_3$)$_3$		9.20±0.05	RPD		2553
C$_3$H$_9$Sn$^+$	iso–C$_4$H$_9$Sn(CH$_3$)$_3$		9.79±0.12	RPD		2553
C$_3$H$_9$Sn$^+$	tert–C$_4$H$_9$Sn(CH$_3$)$_3$		9.50±0.10	RPD		2553
C$_3$H$_9$Sn$^+$	tert–C$_4$H$_9$Sn(CH$_3$)$_3$		9.32±0.16	EI		2720
C$_3$H$_9$Sn$^+$	(CH$_3$)$_3$SnSn(CH$_3$)$_3$		9.65	EM		2553
C$_3$H$_9$Sn$^+$	(CH$_3$)$_3$SnSn(CH$_3$)$_3$		9.82±0.15	RPD		2553
See also – EI:	1424					
C$_3$H$_9$Sn$^+$	(CH$_3$)$_3$SiSn(CH$_3$)$_3$		9.80±0.24	EI		2720
C$_3$H$_9$Sn$^+$	(CH$_3$)$_3$GeSn(CH$_3$)$_3$		9.85±0.22	EI		2720
			C$_4$H$_9$Sn$^+$			
C$_4$H$_9$Sn$^+$	(CH$_3$)$_3$SnCH=CH$_2$	CH$_3$	9.56±0.08	RPD		2553

4.3. The Positive Ion Table—Continued

Ion	Reactant	Other products	Ionization or appearance potential (eV)	Method	Heat of formation (kJ mol^{-1})	Ref.
			$C_4H_{11}Sn^+$			
$C_4H_{11}Sn^+$	$(CH_3)_3SnC_2H_5$	CH_3	9.88 ± 0.02	RPD		2553
$C_4H_{11}Sn^+$	$(C_2H_5)_4Sn$		12.1	EI		2980
			$C_4H_{12}Sn^+$			
$C_4H_{12}Sn^+$	$(CH_3)_4Sn$		8.25 ± 0.15	EI		82
$C_4H_{12}Sn^+$	$(CH_3)_4Sn$		8.76 ± 0.02	EI		2553
$C_4H_{12}Sn^+$	$(CH_3)_4Sn$		8.76 ± 0.12	EI		2720

See also – EI: 218, 2980

Ion	Reactant	Other products	Ionization or appearance potential (eV)	Method	Heat of formation (kJ mol^{-1})	Ref.
			$C_5H_{11}Sn^+$			
$C_5H_{11}Sn^+$	$(CH_3)_3SnC_3H_5$	CH_3	9.43 ± 0.20	RPD		2553
			$C_5H_{13}Sn^+$			
$C_5H_{13}Sn^+$	$n-C_3H_7Sn(CH_3)_3$	CH_3	9.59 ± 0.07	RPD		2553
$C_5H_{13}Sn^+$	$iso-C_3H_7Sn(CH_3)_3$	CH_3	10.03 ± 0.04	RPD		2553
			$C_6H_5Sn^+$			
$C_6H_5Sn^+$	$(C_6H_5)_4Sn$ (Tetraphenyltin)		16.1 ± 0.5	EI		2725
			$C_6H_{15}Sn^+$			
$C_6H_{15}Sn^+$	$n-C_4H_9Sn(CH_3)_3$	CH_3	9.67 ± 0.09	RPD		2553
$C_6H_{15}Sn^+$	$sec-C_4H_9Sn(CH_3)_3$	CH_3	9.76 ± 0.19	RPD		2553
$C_6H_{15}Sn^+$	$iso-C_4H_9Sn(CH_3)_3$	CH_3	9.62 ± 0.02	RPD		2553
$C_6H_{15}Sn^+$	$tert-C_4H_9Sn(CH_3)_3$	CH_3	10.95 ± 0.19	RPD		2553
$C_6H_{15}Sn^+$	$(C_2H_5)_4Sn$		8.70 ± 0.09	EI		3248
$C_6H_{15}Sn^+$	$(C_2H_5)_4Sn$		10.9	EI		2980
$C_6H_{15}Sn^+$	$(n-C_3H_7)_4Sn$		10.6 ± 0.2	EI		2725
			$C_6H_{16}Sn^+$			
$C_6H_{16}Sn^+$	$n-C_3H_7Sn(CH_3)_3$		8.54 ± 0.01	EI		2553
$C_6H_{16}Sn^+$	$iso-C_3H_7Sn(CH_3)_3$		8.28 ± 0.01	EI		2553
			$C_7H_{18}Sn^+$			
$C_7H_{18}Sn^+$	$sec-C_4H_9Sn(CH_3)_3$		8.27 ± 0.01	EI		2553
$C_7H_{18}Sn^+$	$iso-C_4H_9Sn(CH_3)_3$		8.34 ± 0.02	EI		2553
$C_7H_{18}Sn^+$	$tert-C_4H_9Sn(CH_3)_3$		8.34 ± 0.11	EI		2720

4.3. The Positive Ion Table—Continued

Ion	Reactant	Other products	Ionization or appearance potential (eV)	Method	Heat of formation (kJ mol⁻¹)	Ref.
			$C_8H_{20}Sn^+$			
$C_8H_{20}Sn^+$	$(C_2H_5)_4Sn$		10.3	EI		2980
			$C_9H_{21}Sn^+$			
$C_9H_{21}Sn^+$	$(n-C_3H_7)_4Sn$		8.8±0.2	EI		2725
			$C_{12}H_{10}Sn^+$			
$C_{12}H_{10}Sn^+$	$(C_6H_5)_4Sn$ (Tetraphenyltin)		9.1±0.2	EI		2725
			$C_{18}H_{15}Sn^+$			
$C_{18}H_{15}Sn^+$	$(C_6H_5)_3SnCH_3$ (Methyltriphenyltin)	CH_3	8.7±0.2	EI		3248
$C_{18}H_{15}Sn^+$	$(C_6H_5)_3SnC_2H_5$ (Ethyltriphenyltin)		8.6±0.2	EI		3248
$C_{18}H_{15}Sn^+$	$(C_6H_5)_4Sn$ (Tetraphenyltin)		9.6±0.2	EI		3248
$C_{18}H_{15}Sn^+$	$(C_6H_5)_4Sn$ (Tetraphenyltin)		10.1±0.2	EI		2725
$C_{18}H_{15}Sn^+$	$(C_6H_5)_3SnSn(CH_3)_3$ (1,1,1–Trimethyl–2,2,2–triphenylditin)		8.9±0.2	EI		3248
$C_{18}H_{15}Sn^+$	$(C_6H_5)_3SnSn(C_6H_5)_3$ (Hexaphenylditin)		8.7±0.2	EI		3248
$C_{18}H_{15}Sn^+$	$(C_6H_5)_3SnSC_6H_5$ (Phenylthiotriphenyltin)		9.0±0.2	EI		3248
$C_{18}H_{15}Sn^+$	$(C_6H_5)_3SnGe(CH_3)_3$ (Trimethylgermanyltriphenyltin)		9.1±0.2	EI		3248
$C_{18}H_{15}Sn^+$	$(C_6H_5)_3SnI$ (Triphenyltin iodide)	I	8.6±0.2	EI		3248
			$C_5H_{15}Sn_2^+$			
$C_5H_{15}Sn_2^+$	$(CH_3)_3SnSn(CH_3)_3$	CH_3	8.17±0.03	RPD		2553
			$C_6H_{18}Sn_2^+$			
$C_6H_{18}Sn_2^+$	$(CH_3)_3SnSn(CH_3)_3$		7.42±0.02	EI		2900
$C_6H_{18}Sn_2^+$	$(CH_3)_3SnSn(CH_3)_3$		8.08±0.02	EI		2553
			$C_{12}H_{30}Sn_2^+$			
$C_{12}H_{30}Sn_2^+$	$(C_2H_5)_3SnSn(C_2H_5)_3$		6.60±0.02	EI		2900

4.3. The Positive Ion Table—Continued

Ion	Reactant	Other products	Ionization or appearance potential (eV)	Method	Heat of formation (kJ mol^{-1})	Ref.
			NaSnF$^+$			
NaSnF$^+$	NaSnF$_3$		13.0±1.0	EI		2436
			NaSnF$_2^+$			
NaSnF$_2^+$	NaSnF$_3$	F	9.0±0.3	EI		2436
			NaSnF$_3^+$			
NaSnF$_3^+$	NaSnF$_3$		8.8±0.3	EI		2436
			Na$_2$SnF$_3^+$			
Na$_2$SnF$_3^+$	Na$_2$SnF$_4$	F	9.5±0.3	EI		2436
			Na$_2$SnF$_4^+$			
Na$_2$SnF$_4^+$	Na$_2$SnF$_4$		9.0±0.4	EI		2436
			C$_6$H$_{18}$SiSn$^+$			
C$_6$H$_{18}$SiSn$^+$	(CH$_3$)$_3$SiSn(CH$_3$)$_3$		8.18±0.14	EI		2720
			C$_6$H$_{18}$GeSn$^+$			
C$_6$H$_{18}$GeSn$^+$	(CH$_3$)$_3$GeSn(CH$_3$)$_3$		8.20±0.10	EI		2720

Sb$^+$ $\quad \Delta H_{f0}^\circ = 1095.8$ kJ mol^{-1} (261.9 kcal mol^{-1})

Ion	Reactant	Other products	Ionization or appearance potential (eV)	Method	Heat of formation (kJ mol^{-1})	Ref.
Sb$^+$	Sb		8.641	S	1095.8	2113

See also – EI: 2138

Ion	Reactant	Other products	Ionization or appearance potential (eV)	Method	Heat of formation (kJ mol^{-1})	Ref.
Sb$^+$	Sb$_2$	Sb	11.5±0.3	EI		2138
Sb$^+$	Sb$_2$?		9.3±0.3	EI		2508
Sb$^+$	Sb$_4$		15±0.5	EI		2138
Sb$^+$	SbH$_3$	H$_2$+H	12.1±0.2	EI		2116
Sb$^+$	Sb$_2$H$_4$		11.5±0.3	EI		2133
Sb$^+$	SbF$_3$		20	EI		2696
Sb$^+$	SbS?		16.4±0.3	EI		3249
Sb$^+$	SbSe?		12.8±0.3	EI		3249
Sb$^+$	(CH$_3$)$_3$Sb		14.8±0.4	EI		2556
Sb$^+$	(C$_2$H$_5$)$_3$Sb		19.2±0.3	EI		3285
Sb$^+$	(C$_2$H$_5$)$_2$SbSb(C$_2$H$_5$)$_2$		18.5±0.3	EI		3285

4.3. The Positive Ion Table—Continued

Ion	Reactant	Other products	Ionization or appearance potential (eV)	Method	Heat of formation (kJ mol^{-1})	Ref.
			Sb_2^+			
Sb_2^+	Sb_2		8.4±0.3	EI		2138
Sb_2^+	Sb_2		8.7±0.3	EI		2508
Sb_2^+	Sb_4		11.4±0.3	EI		2138
Sb_2^+	Sb_2H_4	$2H_2$	11.2±0.3	EI		2133
Sb_2^+	Sb_2S_2?		11.8±0.3	EI		3249
Sb_2^+	Sb_2Se_2?		12.8±0.3	EI		3249
Sb_2^+	$(C_2H_5)_2SbSb(C_2H_5)_2$		18.2±0.3	EI		3285
			Sb_3^+			
Sb_3^+	Sb_4	Sb	10.8±0.3	EI		2138
Sb_3^+	Sb_4?		10.4±0.3	EI		2508
Sb_3^+	Sb_3S_2?		13.3±0.3	EI		3249
Sb_3^+	Sb_3Se_2?		13.2±0.3	EI		3249
			Sb_4^+			
Sb_4^+	Sb_4		7.7±0.3	EI		2138
Sb_4^+	Sb_4		8.3±0.3	EI		2508
			SbH^+			
SbH^+	SbH_3	H_2	9.9±0.2	EI		2116
SbH^+	$(CH_3)_3Sb$		14.5±0.5	EI		2556
SbH^+	$(C_2H_5)_3Sb$		18.3±0.3	EI		3285
SbH^+	$(C_2H_5)_2SbSb(C_2H_5)_2$		16.2±0.3	EI		3285
			SbH_2^+			
SbH_2^+	SbH_3	H	11.8±0.3	EI		2116
SbH_2^+	$(CH_3)_3Sb$		13.7±0.2	EI		2556
SbH_2^+	$(C_2H_5)_3Sb$		17.2±0.3	EI		3285
SbH_2^+	$(C_2H_5)_2SbSb(C_2H_5)_2$		15.2±0.3	EI		3285
			SbH_3^+　　$\Delta H_{f0}^\circ = 1078$ kJ mol^{-1} (258 kcal mol^{-1})			
SbH_3^+	SbH_3		9.58	PI	1078	1091
SbH_3^+	SbH_3		9.9±0.3	EI		2116
			Sb_2H^+			
Sb_2H^+	Sb_2H_4	H_2+H	10.9±0.3	EI		2133
Sb_2H^+	$(C_2H_5)_2SbSb(C_2H_5)_2$		18.9±0.3	EI		3285

4.3. The Positive Ion Table—Continued

Ion	Reactant	Other products	Ionization or appearance potential (eV)	Method	Heat of formation (kJ mol^{-1})	Ref.
			$Sb_2H_2^+$			
$Sb_2H_2^+$	Sb_2H_4	H_2	10.7±0.3	EI		2133
$Sb_2H_2^+$	$(C_2H_5)_2SbSb(C_2H_5)_2$		15.4±0.3	EI		3285
			$Sb_2H_3^+$			
$Sb_2H_3^+$	Sb_2H_4	H	10.5±0.3	EI		2133
$Sb_2H_3^+$	$(C_2H_5)_2SbSb(C_2H_5)_2$		17.5±0.3	EI		3285
			$Sb_2H_4^+$			
$Sb_2H_4^+$	Sb_2H_4		10.2±0.3	EI		2133
			SbF^+			
SbF^+	SbF_3		21	EI		2696
			SbF_2^+			
SbF_2^+	SbF_3	F	15	EI		2696
			SbF_3^+			
SbF_3^+	SbF_3		13	EI		2696
			SbS^+			
SbS^+	SbS		8.4±0.3	EI		3249
			SbS_2^+			
SbS_2^+	Sb_2S_2?		10.6±0.3	EI		3249
			Sb_2S^+			
Sb_2S^+	Sb_2S_2?		10.7±0.3	EI		3249
			$Sb_2S_2^+$			
$Sb_2S_2^+$	Sb_2S_2		8.8±0.3	EI		3249

4.3. The Positive Ion Table—Continued

Ion	Reactant	Other products	Ionization or appearance potential (eV)	Method	Heat of formation (kJ mol^{-1})	Ref.
		$Sb_2S_3^+$				
$Sb_2S_3^+$	Sb_2S_3		8.8±0.3	EI		3249
		$Sb_2S_4^+$				
$Sb_2S_4^+$	Sb_2S_4		8.6±0.3	EI		3249
		Sb_3S^+				
Sb_3S^+	Sb_3S_2?		11.3±0.3	EI		3249
		$Sb_3S_2^+$				
$Sb_3S_2^+$	Sb_3S_2		9.6±0.3	EI		3249
		$Sb_3S_3^+$				
$Sb_3S_3^+$	Sb_3S_3		9.5±0.3	EI		3249
		$Sb_3S_4^+$				
$Sb_3S_4^+$	Sb_3S_4		9.3±0.3	EI		3249
		$Sb_4S_3^+$				
$Sb_4S_3^+$	Sb_4S_3		8.5±0.3	EI		3249
		$Sb_4S_4^+$				
$Sb_4S_4^+$	Sb_4S_4		8.5±0.3	EI		3249
		$Sb_4S_5^+$				
$Sb_4S_5^+$	Sb_4S_5		8.5±0.3	EI		3249
		$SbSe^+$				
$SbSe^+$	$SbSe$		8.0±0.3	EI		3249
		$SbSe_2^+$				
$SbSe_2^+$	Sb_2Se_2?		11.0±0.3	EI		3249

4.3. The Positive Ion Table—Continued

Ion	Reactant	Other products	Ionization or appearance potential (eV)	Method	Heat of formation (kJ mol^{-1})	Ref.
		Sb_2Se^+				
Sb_2Se^+	Sb_2Se_2?		10.8±0.3	EI		3249
		$Sb_2Se_2^+$				
$Sb_2Se_2^+$	Sb_2Se_2		8.6±0.3	EI		3249
		$Sb_2Se_3^+$				
$Sb_2Se_3^+$	Sb_2Se_3		8.6±0.3	EI		3249
		$Sb_2Se_4^+$				
$Sb_2Se_4^+$	Sb_2Se_4		8.6±0.3	EI		3249
		Sb_3Se^+				
Sb_3Se^+	Sb_3Se		9.6±0.3	EI		3249
		$Sb_3Se_2^+$				
$Sb_3Se_2^+$	Sb_3Se_2		9.8±0.3	EI		3249
		$Sb_3Se_3^+$				
$Sb_3Se_3^+$	Sb_3Se_3		9.2±0.3	EI		3249
		$Sb_3Se_4^+$				
$Sb_3Se_4^+$	Sb_3Se_4		9.2±0.3	EI		3249
		$Sb_4Se_3^+$				
$Sb_4Se_3^+$	Sb_4Se_3		8.4±0.3	EI		3249
		$Sb_4Se_4^+$				
$Sb_4Se_4^+$	Sb_4Se_4		8.4±0.3	EI		3249
		$InSb^+$				
$InSb^+$	$InSb$		6.6±0.4	EI		2138

4.3. The Positive Ion Table—Continued

Ion	Reactant	Other products	Ionization or appearance potential (eV)	Method	Heat of formation (kJ mol⁻¹)	Ref.

$InSb_2^+$

| $InSb_2^+$ | $InSb_2$ | | 6.6 ± 0.4 | EI | | 2138 |

$CHSb^+$

| $CHSb^+$ | $(CH_3)_3Sb$ | | 13.7 ± 0.4 | EI | | 2556 |

CH_2Sb^+

| CH_2Sb^+ | $(CH_3)_3Sb$ | | 15.1 ± 0.2 | EI | | 2556 |

CH_3Sb^+

| CH_3Sb^+ | $(CH_3)_3Sb$ | | 14.3 ± 0.2 | EI | | 2556 |
| CH_3Sb^+ | $(C_2H_5)_3Sb$ | | 16.5 ± 0.3 | EI | | 3285 |

CH_4Sb^+

| CH_4Sb^+ | $(CH_3)_3Sb$ | | 13.4 ± 0.3 | EI | | 2556 |
| CH_4Sb^+ | $(C_2H_5)_3Sb$ | | 17.4 ± 0.3 | EI | | 3285 |

$C_2H_2Sb^+$

| $C_2H_2Sb^+$ | $(C_2H_5)_3Sb$ | | 13.0 ± 0.3 | EI | | 3285 |

$C_2H_4Sb^+$

| $C_2H_4Sb^+$ | $(CH_3)_3Sb$ | | 12.6 ± 0.3 | EI | | 2556 |
| $C_2H_4Sb^+$ | $(C_2H_5)_3Sb$ | | 14.8 ± 0.3 | EI | | 3285 |

$C_2H_5Sb^+$

| $C_2H_5Sb^+$ | $(C_2H_5)_3Sb$ | | 12.4 ± 0.3 | EI | | 3285 |
| $C_2H_5Sb^+$ | $(C_2H_5)_2SbSb(C_2H_5)_2$ | | 10.7 ± 0.3 | EI | | 3285 |

$C_2H_6Sb^+$

| $C_2H_6Sb^+$ | $(CH_3)_3Sb$ | CH_3 | 10.5 ± 0.2 | EI | | 2556 |
| $C_2H_6Sb^+$ | $(C_2H_5)_3Sb$ | | 12.6 ± 0.3 | EI | | 3285 |

$C_2H_7Sb^+$

| $C_2H_7Sb^+$ | $(C_2H_5)_3Sb$ | | 12.6 ± 0.3 | EI | | 3285 |

4.3. The Positive Ion Table—Continued

Ion	Reactant	Other products	Ionization or appearance potential (eV)	Method	Heat of formation (kJ mol⁻¹)	Ref.
$C_3H_9Sb^+$						
$C_3H_9Sb^+$	$(CH_3)_3Sb$		8.04±0.16	EI		2556
$C_4H_{10}Sb^+$						
$C_4H_{10}Sb^+$	$(C_2H_5)_3Sb$		10.7±0.3	EI		3285
$C_4H_{11}Sb^+$						
$C_4H_{11}Sb^+$	$(C_2H_5)_3Sb$		10.0±0.3	EI		3285
$C_4H_{11}Sb^+$	$(C_2H_5)_2SbSb(C_2H_5)_2$		9.0±0.3	EI		3285
$C_6H_{15}Sb^+$						
$C_6H_{15}Sb^+$	$(C_2H_5)_3Sb$		9.2±0.3	EI		3285
$C_{18}H_{15}Sb^+$ (Triphenylantimony)		$\Delta H^\circ_{f298} \sim 1140$ kJ mol⁻¹ (272 kcal mol⁻¹)				
$C_{18}H_{15}Sb^+$	$(C_6H_5)_3Sb$ (Triphenylantimony)		7.3±0.1	PI	~1140	1140
$C_2H_5Sb_2^+$						
$C_2H_5Sb_2^+$	$(C_2H_5)_2SbSb(C_2H_5)_2$		14.4±0.3	EI		3285
$C_2H_6Sb_2^+$						
$C_2H_6Sb_2^+$	$(C_2H_5)_2SbSb(C_2H_5)_2$		13.5±0.3	EI		3285
$C_4H_{11}Sb_2^+$						
$C_4H_{11}Sb_2^+$	$(C_2H_5)_2SbSb(C_2H_5)_2$		11.5±0.3	EI		3285
$C_6H_{15}Sb_2^+$						
$C_6H_{15}Sb_2^+$	$(C_2H_5)_2SbSb(C_2H_5)_2$		9.6±0.3	EI		3285
$C_8H_{20}Sb_2^+$						
$C_8H_{20}Sb_2^+$	$(C_2H_5)_2SbSb(C_2H_5)_2$		8.3±0.3	EI		3285

4.3. The Positive Ion Table—Continued

Ion	Reactant	Other products	Ionization or appearance potential (eV)	Method	Heat of formation (kJ mol^{-1})	Ref.
Te$^+$			**$\Delta H_{f0}^{\circ} \sim 1066$ kJ mol^{-1} (255 kcal mol^{-1})**			
Te$^+$	Te		9.009	S	~1066	2113
See also – EI:	2063, 2469, 2787					
Te$^+$	Te$_2$	Te	11.71±0.01	PI	(~1104)	2609
(Threshold value approximately corrected to 0 K)						
Te$^+$	Te$_2$	Te	10.8±0.5	EI		2747
Te$^+$	Te$_2$	Te	11.2±0.3	EI		2569
Te$^+$	Te$_2$	Te	11.6	EI		3016
Te$^+$	Te$_2$?		11.0±1.0	EI		2469
Te$^+$	GeTe	Ge	13.3±0.5	EI		1023
Te$^+$	SnTe	Sn	12.1±1.0	EI		2063
Te$^+$	(C$_2$H$_5$)$_2$Te		12.8±0.3	EI		3285
Te$^+$	C$_2$H$_5$TeTeC$_2$H$_5$		14.4±0.3	EI		3285
Te$_2^+$			**$\Delta H_{f0}^{\circ} = 970$ kJ mol^{-1} (232 kcal mol^{-1})**			
Te$_2^+$	Te$_2$		8.29±0.03	PI	970	2609
(Threshold value approximately corrected to 0 K)						
See also – EI:	1023, 2457, 2469, 2569, 3016					
Te$_2^+$	C$_2$H$_5$TeTeC$_2$H$_5$		13.0±0.3	EI		3285
TeH$^+$						
TeH$^+$	(C$_2$H$_5$)$_2$Te		15.2±0.3	EI		3285
TeH$^+$	C$_2$H$_5$TeTeC$_2$H$_5$		15.0±0.3	EI		3285
H$_2$Te$^+$			**$\Delta H_{f298}^{\circ} = 981$ kJ mol^{-1} (235 kcal mol^{-1})**			
TeH$_2^+$	H$_2$Te		9.138±0.005	S	981	3317
TeH$_2^+$	(C$_2$H$_5$)$_2$Te		13.2±0.3	EI		3285
TeH$_2^+$	C$_2$H$_5$TeTeC$_2$H$_5$		12.8±0.3	EI		3285
Te$_2$H$^+$						
Te$_2$H$^+$	C$_2$H$_5$TeTeC$_2$H$_5$		12.8±0.3	EI		3285
Te$_2$H$_2^+$						
Te$_2$H$_2^+$	C$_2$H$_5$TeTeC$_2$H$_5$		12.0±0.3	EI		3285
TeO$^+$						
TeO$^+$	TeO		9.4±0.5	EI		2787

4.3. The Positive Ion Table—Continued

Ion	Reactant	Other products	Ionization or appearance potential (eV)	Method	Heat of formation (kJ mol^{-1})	Ref.
			TeO$_2^+$			
TeO$_2^+$	TeO$_2$		11.3±0.5	EI		2787
			AlTe$^+$			
AlTe$^+$	AlTe		9.0±0.5	EI		2449
			Al$_2$Te$^+$			
Al$_2$Te$^+$	Al$_2$Te		10.0±0.5	EI		2449
			AlTe$_2^+$			
AlTe$_2^+$	AlTe$_2$		6.5±0.5	EI		2449
			Al$_2$Te$_2^+$			
Al$_2$Te$_2^+$	Al$_2$Te$_2$		10.0±0.5	EI		2449
			SiTe$^+$			
SiTe$^+$	SiTe		9.2±0.5	EI		2457
			GaTe$^+$			
GaTe$^+$	GaTe		8.4±0.3	EI		2569
			Ga$_2$Te$^+$			
Ga$_2$Te$^+$	Ga$_2$Te		7.6±0.3	EI		2569
			GaTe$_2^+$			
GaTe$_2^+$	GaTe$_2$		8.3±0.5	EI		2569
			Ga$_2$Te$_2^+$			
Ga$_2$Te$_2^+$	Ga$_2$Te$_2$		8.1±0.5	EI		2569
			GeTe$^+$			
GeTe$^+$	GeTe		10.1±0.5	EI		1023

4.3. The Positive Ion Table—Continued

Ion	Reactant	Other products	Ionization or appearance potential (eV)	Method	Heat of formation (kJ mol^{-1})	Ref.
		GeTe$_2^+$				
GeTe$_2^+$	GeTe$_2$		10.8±0.5	EI		1023
		YTe$^+$				
YTe$^+$	YTe		6.0±1	EI		3206
		InTe$^+$				
InTe$^+$	InTe		7.6±0.5	EI		2469
		In$_2$Te$^+$				
In$_2$Te$^+$	In$_2$Te		7.1±0.5	EI		2469
		InTe$_2^+$				
InTe$_2^+$	InTe$_2$		8.9±0.5	EI		2469
		In$_2$Te$_2^+$				
In$_2$Te$_2^+$	In$_2$Te$_2$		7.2±0.5	EI		2469
		SnTe$^+$				
SnTe$^+$	SnTe		9.1±0.5	EI		2063
		CH$_3$Te$^+$				
CH$_3$Te$^+$	(C$_2$H$_5$)$_2$Te		14.5±0.3	EI		3285
		C$_2$H$_5$Te$^+$				
C$_2$H$_5$Te$^+$	(C$_2$H$_5$)$_2$Te		13.4±0.3	EI		3285
C$_2$H$_5$Te$^+$	C$_2$H$_5$TeTeC$_2$H$_5$		13.0±0.3	EI		3285
		C$_2$H$_6$Te$^+$				
C$_2$H$_6$Te$^+$	(C$_2$H$_5$)$_2$Te		10.9±0.3	EI		3285
C$_2$H$_6$Te$^+$	C$_2$H$_5$TeTeC$_2$H$_5$		10.3±0.3	EI		3285

4.3. The Positive Ion Table—Continued

Ion	Reactant	Other products	Ionization or appearance potential (eV)	Method	Heat of formation (kJ mol^{-1})	Ref.
		C$_4$H$_{10}$Te$^+$				
C$_4$H$_{10}$Te$^+$	(C$_2$H$_5$)$_2$Te		9.2±0.3	EI		3285
		C$_2$H$_5$Te$_2^+$				
C$_2$H$_5$Te$_2^+$	C$_2$H$_5$TeTeC$_2$H$_5$		10.9±0.3	EI		3285
		C$_2$H$_6$Te$_2^+$				
C$_2$H$_6$Te$_2^+$	C$_2$H$_5$TeTeC$_2$H$_5$		10.4±0.3	EI		3285
		C$_4$H$_{10}$Te$_2^+$				
C$_4$H$_{10}$Te$_2^+$	C$_2$H$_5$TeTeC$_2$H$_5$		9.0±0.3	EI		3285
		GaOTe$^+$				
GaOTe$^+$	GaOTe		7.7±0.5	EI		2569

I$^+$ $\Delta H_{f0}^\circ = 1115.6$ kJ mol^{-1} (266.6 kcal mol^{-1})

Ion	Reactant	Other products	Ionization or appearance potential (eV)	Method	Heat of formation (kJ mol^{-1})	Ref.
I$^+$	I		10.451	S	1115.6	2113, 3003

See also – S: 2656, 3089
        EI: 439, 2554

Ion	Reactant	Other products	Ionization or appearance potential (eV)	Method	Heat of formation (kJ mol^{-1})	Ref.
I$^+$	I$_2$	I$^-$	8.922±0.013	PI	(1114)	3102

This appearance potential is corrected for an assumed 0←3 hot band transition, amounting to 0.079 eV.
The actual onset is 8.843 eV. From the appearance potential the authors deduce the electron affinity
of I atoms.

Ion	Reactant	Other products	Ionization or appearance potential (eV)	Method	Heat of formation (kJ mol^{-1})	Ref.
I$^+$	I$_2$	I$^-$	8.83±0.02	PI		416
I$^+$	I$_2$	I$^-$	8.95±0.02	PI		213

(Position of peak maximum)

See also – EI: 288, 292, 357

Ion	Reactant	Other products	Ionization or appearance potential (eV)	Method	Heat of formation (kJ mol^{-1})	Ref.
I$^+$	LiI	Li	14.4±0.3	EI		2001
I$^+$	NaI	Na	14.4±0.3	EI		2001
I$^+$	KI	K	14.6±0.3	EI		2001
I$^+$	RbI	Rb	14.4±0.3	EI		2001
I$^+$	CH$_3$I	CH$_3$	12.9±0.05	RPD		2154
I$^+$	C$_2$H$_5$I		14.8±0.2	EI		356
I$^+$	n–C$_4$H$_9$I		12.45±0.05	RPD		2776

4.3. The Positive Ion Table—Continued

Ion	Reactant	Other products	Ionization or appearance potential (eV)	Method	Heat of formation (kJ mol^{-1})	Ref.
I$^+$	ICN	CN$^-$	9.8±0.1	EI		73
I$^+$	ICN	CN	13.62±0.02	PI	(1108)	2621

From this result the authors deduce $\Delta H^\circ_{f0}(CN) \leqslant 425$ kJ mol^{-1}.

See also − EI: 73

I$^+$	CF$_3$I		13.4±0.1	SD		1111

This fragmentation occurs with little kinetic energy, see refs. 24, 1111.

See also − EI: 24, 439

I$^+$	CsI	Cs	14.1±0.3	EI		2001
I$^+$	HgI$_2$		15.5±0.4	EI		2506
I$^+$	TlI	Tl	13.4±0.1	RPD		2159

$$I_2^+(^2\Pi_{3/2g}) \qquad \Delta H^\circ_{f0} = 971 \text{ kJ mol}^{-1} \text{ (232 kcal mol}^{-1})$$

I$_2^+(^2\Pi_{3/2g})$	I$_2$		9.3995±0.0012	S	972	2906
I$_2^+(^2\Pi_{3/2g})$	I$_2$		9.331	PI		3102
(Threshold value corrected for hot bands)						
I$_2^+(^2\Pi_{3/2g})$	I$_2$		~9.37	PI	~970	3413
I$_2^+(^2\Pi_{3/2g})$	I$_2$		9.22±0.01	PE		3409
I$_2^+(^2\Pi_{3/2g})$	I$_2$		9.34 (V)	PE		3275

The disagreement among these values is partly in the varying corrections for hot bands. The adiabatic ionization potential is most likely either 9.37 or 9.3995 eV.

I$_2^+(^2\Pi_{1/2g})$	I$_2$		10.029±0.006	S		2906
I$_2^+(^2\Pi_{1/2g})$	I$_2$		9.87±0.01	PE		3409
I$_2^+(^2\Pi_{1/2g})$	I$_2$		9.97 (V)	PE		3275
I$_2^+(^2\Pi_{3/2u})$	I$_2$		10.74±0.02	PE		3409
I$_2^+(^2\Pi_{3/2u})$	I$_2$		10.8	PE		3275
I$_2^+(^2\Pi_{1/2u})$	I$_2$		11.54±0.02	PE		3409
I$_2^+(^2\Pi_{1/2u})$	I$_2$		11.6	PE		3275
I$_2^+(^2\Sigma_g^+)$	I$_2$		12.66±0.02	PE		3409
I$_2^+(^2\Sigma_g^+)$	I$_2$		12.7	PE		3275

See also − PI: 182, 213, 416
         PE: 2815, 3277
         EI: 288, 292, 2554, 3279

I$_2^+$	CI$_4$		11.3±0.1	EDD		3279
I$_2^+$	CHI$_3$		12.0±0.1	EDD		3279

4.3. The Positive Ion Table—Continued

Ion	Reactant	Other products	Ionization or appearance potential (eV)	Method	Heat of formation (kJ mol⁻¹)	Ref.
HI⁺($^2\Pi_{3/2}$)			$\Delta H^\circ_{f0} = 1031$ kJ mol^{-1} (247 kcal mol^{-1})			
HI⁺($^2\Pi_{1/2}$)			$\Delta H^\circ_{f0} = 1095$ kJ mol^{-1} (262 kcal mol^{-1})			
HI⁺($^2\Pi_{3/2}$)	HI		~10.39	S	~1031	3150
HI⁺($^2\Pi_{3/2}$)	HI		10.38±0.02	PI	1030	182, 416
HI⁺($^2\Pi_{3/2}$)	HI		10.38±0.01	PE	1030	2819
HI⁺($^2\Pi_{3/2}$)	HI		10.42±0.01	PE	1034	2815
HI⁺($^2\Pi_{1/2}$)	HI		11.050±0.005	S	1095	3150
HI⁺($^2\Pi_{1/2}$)	HI		11.05±0.01	PE		2819
HI⁺($^2\Pi_{1/2}$)	HI		11.08±0.01	PE		2815
HI⁺($^2\Sigma^+$)	HI		13.85	PE		2819
HI⁺($^2\Sigma^+$)	HI		14.03±0.01	PE		2815

See also – PI: 213
        EI: 463, 2001

Ion	Reactant	Other products	Ionization or appearance potential (eV)	Method	Heat of formation (kJ mol⁻¹)	Ref.
HI⁺	C$_2$H$_5$I		11.7±0.1	EI		356

HI⁺²

Ion	Reactant	Other products	Ionization or appearance potential (eV)	Method	Heat of formation (kJ mol⁻¹)	Ref.
HI⁺²	HI		30.0±0.5	FD		212

LiI⁺

Ion	Reactant	Other products	Ionization or appearance potential (eV)	Method	Heat of formation (kJ mol⁻¹)	Ref.
LiI⁺	LiI		8.6±0.3	EI		2001

Li$_2$I⁺

Ion	Reactant	Other products	Ionization or appearance potential (eV)	Method	Heat of formation (kJ mol⁻¹)	Ref.
Li$_2$I⁺	Li$_2$I$_2$	I	9.2±0.3	EI		2001

Li$_3$I$_2^+$

Ion	Reactant	Other products	Ionization or appearance potential (eV)	Method	Heat of formation (kJ mol⁻¹)	Ref.
Li$_3$I$_2^+$	Li$_3$I$_3$	I	9.2±0.3	EI		2001

BI⁺

Ion	Reactant	Other products	Ionization or appearance potential (eV)	Method	Heat of formation (kJ mol⁻¹)	Ref.
BI⁺	BI$_3$		14.4±0.2	EI		206

BI$_2^+$

Ion	Reactant	Other products	Ionization or appearance potential (eV)	Method	Heat of formation (kJ mol⁻¹)	Ref.
BI$_2^+$	BI$_3$	I	9.7±0.2	EI		206

4.3. The Positive Ion Table—Continued

Ion	Reactant	Other products	Ionization or appearance potential (eV)	Method	Heat of formation (kJ mol⁻¹)	Ref.

$BI_3^+(^2A_2')$　　　　$\Delta H_{f0}^\circ = 968$ kJ mol⁻¹ (231 kcal mol⁻¹)

Ion	Reactant	Other products	Ionization or appearance potential (eV)	Method	Heat of formation (kJ mol⁻¹)	Ref.
$BI_3^+(^2A_2')$	BI_3		9.25±0.03	PE	968	3375
$BI_3^+(^2E')$	BI_3		9.85±0.02	PE		3375
$BI_3^+(^2E')$	BI_3		10.03±0.02	PE		3375
$BI_3^+(^2E'')$	BI_3		10.28±0.02	PE		3375
$BI_3^+(^2E'')$	BI_3		10.44±0.02	PE		3375
$BI_3^+(^2A_2'')$	BI_3		11.56±0.02	PE		3375
$BI_3^+(^2E')$	BI_3		12.28±0.03	PE		3375
$BI_3^+(^2E')$	BI_3		12.83±0.03	PE		3375
$BI_3^+(^2A_1')$	BI_3		15.10±0.02	PE		3375

See also − PE:　3119
　　　　　EI:　206, 2512, 3227

CI^+

Ion	Reactant	Other products	Ionization or appearance potential (eV)	Method	Heat of formation (kJ mol⁻¹)	Ref.
CI^+	$CI\equiv CI$	CI	16.8±0.1	EDD		3177
CI^+	$CH\equiv CI$	CH	18.1±0.1	EDD		3177
CI^+	ICN	N	17.6±0.3	EI		73
CI^+	$CCl\equiv CI$	CCl	16.4±0.1	EDD		3177
CI^+	$CBr\equiv CI$	CBr	16.9±0.1	EDD		3177

$C_2I_2^+$

Ion	Reactant	Other products	Ionization or appearance potential (eV)	Method	Heat of formation (kJ mol⁻¹)	Ref.
$C_2I_2^+(^2\Pi_{3/2u})$	$CI\equiv CI$		9.03	PE		3121
$C_2I_2^+(^2\Pi_{3/2u})$	$CI\equiv CI$		9.2±0.1	EDD		3177
$C_2I_2^+(^2\Pi_{1/2u})$	$CI\equiv CI$		9.47	PE		3121
$C_2I_2^+(^2\Pi_{3/2g})$	$CI\equiv CI$		10.63	PE		3121
$C_2I_2^+(^2\Pi_{1/2g})$	$CI\equiv CI$		11.23	PE		3121
$C_2I_2^+(^2\Pi_{3/2u})$	$CI\equiv CI$		12.17 (V)	PE		3121
$C_2I_2^+(^2\Pi_{1/2u})$	$CI\equiv CI$		12.38 (V)	PE		3121
$C_2I_2^+(^2\Sigma_g^+)$	$CI\equiv CI$		14.22 (V)	PE		3121
$C_2I_2^+(^2\Sigma_u^+)$	$CI\equiv CI$		15.48 (V)	PE		3121

$C_2I_2^{+2}$

Ion	Reactant	Other products	Ionization or appearance potential (eV)	Method	Heat of formation (kJ mol⁻¹)	Ref.
$C_2I_2^{+2}$	$CI\equiv CI$		23.7±0.1	EDD		3177

IF^+

Ion	Reactant	Other products	Ionization or appearance potential (eV)	Method	Heat of formation (kJ mol⁻¹)	Ref.
IF^+	IF		10.5±0.3	EI		357
IF^+	IF_5		~24	EI		357

IF_2^+

Ion	Reactant	Other products	Ionization or appearance potential (eV)	Method	Heat of formation (kJ mol⁻¹)	Ref.
IF_2^+	IF_5		15.1±0.3	EI		357

4.3. The Positive Ion Table—Continued

Ion	Reactant	Other products	Ionization or appearance potential (eV)	Method	Heat of formation (kJ mol^{-1})	Ref.
		IF$_3^+$				
IF$_3^+$	IF$_5$		11.5±0.3	EI		357
		IF$_4^+$				
IF$_4^+$	IF$_5$	F$^-$?	13.6±0.3	EI		357
		IF$_5^+$				
IF$_5^+$	IF$_5$		13.5±0.2	EI		357
		NaI$^+$				
NaI$^+$	NaI		7.64	PI		2610
(Threshold value corrected for hot bands)						
NaI$^+$	NaI		8.0±0.3	EI		2610
NaI$^+$	NaI		8.7±0.3	EI		2001
		Na$_2$I$^+$				
Na$_2$I$^+$	Na$_2$I$_2$	I	~7.80±0.05	PI		2610
(Threshold value corrected for hot bands)						
Na$_2$I$^+$	Na$_2$I$_2$	I	9.1±0.3	EI		2001
		Na$_2$I$_2^+$				
Na$_2$I$_2^+$	Na$_2$I$_2$		7.64	PI		2610
		Na$_3$I$_2^+$				
Na$_3$I$_2^+$	Na$_3$I$_3$	I	8.6±0.3	EI		2001
		MgI$^+$				
MgI$^+$	MgI$_2$	I	~10.12±0.12	PI		2610
MgI$^+$	MgI$_2$	I	11.5±0.5	EI		178

MgI$_2^+$ $\quad \Delta H_{f298}^{\circ} \sim 752$ kJ mol^{-1} (180 kcal mol^{-1})

Ion	Reactant	Other products	Ionization or appearance potential (eV)	Method	Heat of formation (kJ mol^{-1})	Ref.
MgI$_2^+$	MgI$_2$		9.57	PI	~752	2610
MgI$_2^+$	MgI$_2$		10.0±0.4	EI		178

4.3. The Positive Ion Table—Continued

Ion	Reactant	Other products	Ionization or appearance potential (eV)	Method	Heat of formation (kJ mol^{-1})	Ref.
Mg$_2$I$_3^+$						
Mg$_2$I$_3^+$	Mg$_2$I$_4$	I	10.0±0.4	EI		178
ICl$^+$ $\quad$ $\Delta H_{f0}^\circ = 991$ kJ mol^{-1} (237 kcal mol^{-1})						
ICl$^+$	ICl		10.07±0.01	PI	991	3413
(Threshold value corrected for hot bands)						
See also – PE: 3277, 3409						
$\quad\quad\quad\quad$ EI: 292, 357						
KI$^+$						
KI$^+$	KI		8.2±0.3	EI		2001
K$_2$I$^+$						
K$_2$I$^+$	K$_2$I$_2$	I	8.2±0.3	EI		2001
MnI$^+$						
MnI$^+$	Mn(CO)$_5$I	5CO	14.0	EI		2501
IBr$^+$ $\quad$ $\Delta H_{f0}^\circ = 994$ kJ mol^{-1} (238 kcal mol^{-1})						
IBr$^+$	IBr		9.79±0.01	PI	994	3413
(Threshold value corrected for hot bands)						
See also – PE: 3277, 3409						
$\quad\quad\quad\quad$ EI: 292, 357						
RbI$^+$						
RbI$^+$	RbI		8.0±0.3	EI		2001
Rb$_2$I$^+$						
Rb$_2$I$^+$	Rb$_2$I$_2$	I	8.2±0.3	EI		2001

4.3. The Positive Ion Table—Continued

Ion	Reactant	Other products	Ionization or appearance potential (eV)	Method	Heat of formation (kJ mol^{-1})	Ref.
$B_5H_8I^+$						
$B_5H_8I^+$	$1-B_5H_8I$		8.49 ± 0.05	RPD		3228
See also – EI:	103, 1102					
$B_5H_8I^+$	$2-B_5H_8I$		8.82 ± 0.02	RPD		3228
CH_2I^+						
CH_2I^+	CH_3I	H	12.08 ± 0.09	RPD		1139
See also – EI:	160, 356, 3017					
CH_2I^+	C_2H_5I	CH_3	13.7 ± 0.3	EI		356
$CH_3I^+(^2E_{3/2})$ $\Delta H^\circ_{f0} =$ 943 kJ mol^{-1} (225 kcal mol^{-1})						
$CH_3I^+(^2E_{1/2})$ $\Delta H^\circ_{f0} =$ 1003 kJ mol^{-1} (240 kcal mol^{-1})						
$CH_3I^+(^2E_{3/2})$	CH_3I		9.538 ± 0.003	S	943	2064
$CH_3I^+(^2E_{3/2})$	CH_3I		9.534 ± 0.005	RPI	942	2866
$CH_3I^+(^2E_{3/2})$	CH_3I		9.54	PE		3119
$CH_3I^+(^2E_{3/2})$	CH_3I		9.50	PE		3057
$CH_3I^+(^2E_{1/2})$	CH_3I		10.165 ± 0.003	S	1003	2064
$CH_3I^+(^2E_{1/2})$	CH_3I		10.151 ± 0.015	RPI	1002	2866
$CH_3I^+(^2E_{1/2})$	CH_3I		10.16	PE		3119
$CH_3I^+(^2E_{1/2})$	CH_3I		10.13	PE		3057

Additional higher ionization potentials are given in refs. 3057, 3119.

See also – PI: 158, 182, 190, 213, 297, 416, 1253
PEN: 2430, 2466
EI: 289, 2154, 2776, 3201

Ion	Reactant	Other products	Ionization or appearance potential (eV)	Method	Heat of formation (kJ mol^{-1})	Ref.
C_2HI^+						
$C_2HI^+(^2\Pi_{3/2})$	$CH{\equiv}CI$		9.73	PE		3071
$C_2HI^+(^2\Pi_{3/2})$	$CH{\equiv}CI$		9.9 ± 0.1	EDD		3177
$C_2HI^+(^2\Pi_{1/2})$	$CH{\equiv}CI$		10.14	PE		3071
$C_2HI^+(^2\Pi_{3/2})$	$CH{\equiv}CI$		11.96 (V)	PE		3071
$C_2HI^+(^2\Pi_{1/2})$	$CH{\equiv}CI$		12.19 (V)	PE		3071
$C_2HI^+(^2\Sigma^+)$	$CH{\equiv}CI$		14.86 (V)	PE		3071
$C_2HI^+(^2\Sigma^+)$	$CH{\equiv}CI$		17.4 (V)	PE		3071
C_2HI^{+2}						
C_2HI^{+2}	$CH{\equiv}CI$		26.9 ± 0.1	EDD		3177

4.3. The Positive Ion Table—Continued

Ion	Reactant	Other products	Ionization or appearance potential (eV)	Method	Heat of formation (kJ mol^{-1})	Ref.
	$C_2H_5I^+(^2E_{3/2})$	$\Delta H_{f0}^\circ = 910$ kJ mol^{-1} (217 kcal mol^{-1})				
	$C_2H_5I^+(^2E_{1/2})$	$\Delta H_{f0}^\circ = 966$ kJ mol^{-1} (231 kcal mol^{-1})				
$C_2H_5I^+(^2E_{3/2})$	C_2H_5I		9.346 ± 0.005	S	910	2065
$C_2H_5I^+(^2E_{3/2})$	C_2H_5I		9.33 ± 0.01	PI	182,	416
$C_2H_5I^+(^2E_{3/2})$	C_2H_5I		9.37	PE		3374
$C_2H_5I^+(^2E_{3/2})$	C_2H_5I		9.35 ± 0.02	RPD		224
$C_2H_5I^+(^2E_{1/2})$	C_2H_5I		9.928 ± 0.005	S	966	2065
$C_2H_5I^+(^2E_{1/2})$	C_2H_5I		9.93	PE		3374

Additional higher ionization potentials are given in ref. 3374.

See also – PI: 190
          EI: 160, 356

Ion	Reactant	Other products	Ionization or appearance potential (eV)	Method	Heat of formation (kJ mol^{-1})	Ref.
	$n-C_3H_7I^+(^2E_{3/2})$	$\Delta H_{f298}^\circ = 863$ kJ mol^{-1} (206 kcal mol^{-1})				
	$n-C_3H_7I^+(^2E_{1/2})$	$\Delta H_{f298}^\circ = 919$ kJ mol^{-1} (220 kcal mol^{-1})				
	$iso-C_3H_7I^+(^2E_{3/2})$	$\Delta H_{f298}^\circ = 844$ kJ mol^{-1} (202 kcal mol^{-1})				
	$iso-C_3H_7I^+(^2E_{1/2})$	$\Delta H_{f298}^\circ \sim 896$ kJ mol^{-1} (214 kcal mol^{-1})				
$C_3H_7I^+(^2E_{3/2})$	$n-C_3H_7I$		9.26 ± 0.01	PI	863	182
$C_3H_7I^+(^2E_{3/2})$	$n-C_3H_7I$		9.26 ± 0.005	PE	863	3289
$C_3H_7I^+(^2E_{3/2})$	$n-C_3H_7I$		9.27	PE	864	3374
$C_3H_7I^+(^2E_{1/2})$	$n-C_3H_7I$		9.84 ± 0.005	PE	919	3289
$C_3H_7I^+(^2E_{1/2})$	$n-C_3H_7I$		9.84	PE	919	3374

Additional higher ionization potentials are given in ref. 3374.

See also – EI: 160

Ion	Reactant	Other products	Ionization or appearance potential (eV)	Method	Heat of formation (kJ mol^{-1})	Ref.
$C_3H_7I^+(^2E_{3/2})$	$iso-C_3H_7I$		9.17 ± 0.02	PI	843	182
$C_3H_7I^+(^2E_{3/2})$	$iso-C_3H_7I$		9.18 ± 0.005	PE	844	3289
$C_3H_7I^+(^2E_{3/2})$	$iso-C_3H_7I$		9.19	PE	845	3374
$C_3H_7I^+(^2E_{1/2})$	$iso-C_3H_7I$		9.69 ± 0.005	PE	893	3289
$C_3H_7I^+(^2E_{1/2})$	$iso-C_3H_7I$		9.74	PE	898	3374

Additional higher ionization potentials are given in ref. 3374.

4.3. The Positive Ion Table—Continued

Ion	Reactant	Other products	Ionization or appearance potential (eV)	Method	Heat of formation (kJ mol^{-1})	Ref.
	$tert-C_4H_9I^+(^2E_{3/2})$	$\Delta H^\circ_{f298} = 797$ kJ mol^{-1} (190 kcal mol^{-1})				
	$tert-C_4H_9I^+(^2E_{1/2})$	$\Delta H^\circ_{f298} = 845$ kJ mol^{-1} (202 kcal mol^{-1})				
$C_4H_9I^+(^2E_{3/2})$	$n-C_4H_9I$		9.21±0.01	PI		182
$C_4H_9I^+(^2E_{3/2})$	$n-C_4H_9I$		9.23±0.005	PE		3289
$C_4H_9I^+(^2E_{1/2})$	$n-C_4H_9I$		9.78±0.005	PE		3289
$C_4H_9I^+(^2E_{3/2})$	$sec-C_4H_9I$		9.09±0.02	PI		182
$C_4H_9I^+(^2E_{3/2})$	$iso-C_4H_9I$		9.18±0.02	PI		182
$C_4H_9I^+(^2E_{3/2})$	$iso-C_4H_9I$		9.18±0.005	PE		3289
$C_4H_9I^+(^2E_{1/2})$	$iso-C_4H_9I$		9.73±0.005	PE		3289
$C_4H_9I^+(^2E_{3/2})$	$tert-C_4H_9I$		9.02±0.03	PI	797	182
$C_4H_9I^+(^2E_{3/2})$	$tert-C_4H_9I$		9.02±0.01	PE	797	3289
$C_4H_9I^+(^2E_{1/2})$	$tert-C_4H_9I$		9.52±0.01	PE	845	3289
	$C_5H_{11}I^+$					
$C_5H_{11}I^+(^2E_{3/2})$	$n-C_5H_{11}I$		9.19±0.01	PI		182, 416
$C_5H_{11}I^+(^2E_{3/2})$	$n-C_5H_{11}I$		9.18±0.01	PE		3289
$C_5H_{11}I^+(^2E_{3/2})$	$n-C_5H_{11}I$		9.21	PE		3374
$C_5H_{11}I^+(^2E_{1/2})$	$n-C_5H_{11}I$		9.72±0.01	PE		3289
$C_5H_{11}I^+(^2E_{1/2})$	$n-C_5H_{11}I$		9.77	PE		3374

Additional higher ionization potentials are given in ref. 3374.

Ion	Reactant	Other products	Ionization or appearance potential (eV)	Method	Heat of formation (kJ mol^{-1})	Ref.
$C_5H_{11}I^+(^2E_{3/2})$	$iso-C_5H_{11}I$		9.17±0.01	PE		3289
$C_5H_{11}I^+(^2E_{1/2})$	$iso-C_5H_{11}I$		9.72±0.01	PE		3289
$C_5H_{11}I^+(^2E_{3/2})$	$(CH_3)_2CIC_2H_5$		8.93±0.01	PE		3289
$C_5H_{11}I^+(^2E_{1/2})$	$(CH_3)_2CIC_2H_5$		9.50±0.01	PE		3289
	$C_6H_5I^+$ (Iodobenzene)	$\Delta H^\circ_{f298} \sim 1003$ kJ mol^{-1} (240 kcal mol^{-1})				
$C_6H_5I^+$	C_6H_5I (Iodobenzene)		8.685	PI	1001	2682
$C_6H_5I^+$	C_6H_5I (Iodobenzene)		8.73±0.01	PI	1005	3212
$C_6H_5I^+$	C_6H_5I (Iodobenzene)		8.73±0.03	PI	1005	182, 416

The information on higher ionization potentials is quite confused, see refs. 2682, 2806, 3212, 3247, 3331.

See also – PE: 2806, 3212, 3247, 3331

EI: 3230, 3238

1.3. The Positive Ion Table—Continued

Ion	Reactant	Other products	Ionization or appearance potential (eV)	Method	Heat of formation (kJ mol^{-1})	Ref.
		$C_7H_6I^+$				
$C_7H_6I^+$	$C_6H_5CH_2CH_2C_6H_4I$ (1-(4-Iodophenyl)-2-phenylethane)		10.3±0.2	EI		3288
	$C_7H_7I^+$ (2-Iodotoluene)	$\Delta H^\circ_{f298} = 964$ kJ mol^{-1} (230 kcal mol^{-1})				
	$C_7H_7I^+$ (3-Iodotoluene)	$\Delta H^\circ_{f298} = 964$ kJ mol^{-1} (230 kcal mol^{-1})				
	$C_7H_7I^+$ (4-Iodotoluene)	$\Delta H^\circ_{f298} = 942$ kJ mol^{-1} (225 kcal mol^{-1})				
$C_7H_7I^+$	$C_6H_5CH_2I$ (α-Iodotoluene)		8.8	EI		3230
$C_7H_7I^+$	$C_6H_4ICH_3$ (2-Iodotoluene)		8.62±0.01	PI	964	182
$C_7H_7I^+$	$C_6H_4ICH_3$ (3-Iodotoluene)		8.61±0.03	PI	964	182
$C_7H_7I^+$	$C_6H_4ICH_3$ (4-Iodotoluene)		8.50±0.01	PI	942	182
		$C_{14}H_{13}I^+$				
$C_{14}H_{13}I^+$	$C_6H_5CH_2CH_2C_6H_4I$ (1-(4-Iodophenyl)-2-phenylethane)		8.7±0.1	EI		3288
		$CH_2I_2^+$				
$CH_2I_2^+$	CH_2I_2		9.46 (V)	PE		3119
	$ICN^+(^2\Pi_{3/2})$	$\Delta H^\circ_{f0} = 1277$ kJ mol^{-1} (305 kcal mol^{-1})				
	$ICN^+(^2\Pi_{1/2})$	$\Delta H^\circ_{f0} = 1330$ kJ mol^{-1} (318 kcal mol^{-1})				
	$ICN^+(^2\Sigma^+)$	$\Delta H^\circ_{f0} = 1496$ kJ mol^{-1} (358 kcal mol^{-1})				
	$ICN^+(^2\Pi_{3/2})$	$\Delta H^\circ_{f0} = 1514$ kJ mol^{-1} (362 kcal mol^{-1})				
$CNI^+(^2\Pi_{3/2})$	ICN		10.870±0.001	S	1275	2887
$CNI^+(^2\Pi_{3/2})$	ICN		10.87±0.02	PI	1275	2621
$CNI^+(^2\Pi_{3/2})$	ICN		10.91±0.02	PE	1279	3091
$CNI^+(^2\Pi_{3/2})$	ICN		10.91	PE	1279	3045
$CNI^+(^2\Pi_{1/2})$	ICN		11.44±0.02	PE	1330	3091
$CNI^+(^2\Pi_{1/2})$	ICN		11.45	PE	1331	3045
$CNI^+(^2\Sigma^+)$	ICN		13.15±0.02	PE	1495	3091
$CNI^+(^2\Sigma^+)$	ICN		13.17	PE	1497	3045
$CNI^+(^2\Pi_{3/2})$	ICN		13.35	PE	1514	3045
$CNI^+(^2\Pi_{3/2})$	ICN		13.41±0.02 (V)	PE		3091
$CNI^+(^2\Pi_{1/2})$	ICN		13.56±0.02 (V)	PE		3091

4.3. The Positive Ion Table—Continued

Ion	Reactant	Other products	Ionization or appearance potential (eV)	Method	Heat of formation (kJ mol^{-1})	Ref.
$CNI^+(^2\Sigma^+)$	ICN		16.69 ± 0.02 (V)	PE		3091
$CNI^+(^2\Sigma^+)$	ICN		16.71 (V)	PE		3045

See also $-$ S: 2660
EI: 73

BF_2I^+

BF_2I^+	BF_2I		10.42 ± 0.08	EI		2512, 3227

BFI_2^+

BFI_2^+	BFI_2		9.69 ± 0.06	EI		2512, 3227

CF_2I^+

CF_2I^+	CF_3I	F	14.58 ± 0.06	EI		439
CF_2I^+	CF_3I	F	15.3 ± 0.3	EI		24

CF_3I^+ $\Delta H_{f0}^\circ = 404$ kJ mol^{-1} (97 kcal mol^{-1})

CF_3I^+	CF_3I		10.23	PI	404	2643
CF_3I^+	CF_3I		10.5 ± 0.1	SD		1111
CF_3I^+	CF_3I		10.0 ± 0.3	EI		24
CF_3I^+	CF_3I		10.64 ± 0.02	EI		439

$C_6F_5I^+$

$C_6F_5I^+$	C_6F_5I (Iodopentafluorobenzene)		9.5 ± 0.1	EI		301

$C_3F_7I^+$

$C_3F_7I^+$	$n-C_3F_7I$		10.36 ± 0.01	PI		182

$LiNaI^+$

$LiNaI^+$	$LiNaI_2$	I	9.0 ± 0.3	EI		2001

BCl_2I^+

BCl_2I^+	BCl_2I		9.93 ± 0.10	EI		2512, 3227

$BClI_2^+$

$BClI_2^+$	$BClI_2$		9.49 ± 0.14	EI		2512, 3227

4.3. The Positive Ion Table—Continued

Ion	Reactant	Other products	Ionization or appearance potential (eV)	Method	Heat of formation (kJ mol^{-1})	Ref.
C$_2$ClI$^+$						
C$_2$ClI$^+$($^2\Pi_{3/2}$)	CCl≡CI		9.44	PE		3121
C$_2$ClI$^+$($^2\Pi_{3/2}$)	CCl≡CI		9.7±0.1	EDD		3177
C$_2$ClI$^+$($^2\Pi_{1/2}$)	CCl≡CI		9.75	PE		3121
C$_2$ClI$^+$($^2\Pi_{3/2}$)	CCl≡CI		11.48 (V)	PE		3121
C$_2$ClI$^+$($^2\Pi_{1/2}$)	CCl≡CI		11.84 (V)	PE		3121
C$_2$ClI$^+$($^2\Pi_{3/2}$)	CCl≡CI		13.85 (V)	PE		3121
C$_2$ClI$^+$($^2\Sigma^+$)	CCl≡CI		14.88 (V)	PE		3121
C$_2$ClI$^+$($^2\Sigma^+$)	CCl≡CI		17.21 (V)	PE		3121
C$_2$ClI^{+2}						
C$_2$ClI^{+2}	CCl≡CI		25.5±0.1	EDD		3177
LiKI$^+$						
LiKI$^+$	LiKI$_2$	I	9.0±0.3	EI		2001
BBr$_2$I$^+$						
BBr$_2$I$^+$	BBr$_2$I		9.74±0.11	EI		2512, 3227
BBrI$_2^+$						
BBrI$_2^+$	BBrI$_2$		9.40±0.11	EI		2512, 3227
C$_2$BrI$^+$						
C$_2$BrI$^+$($^2\Pi_{3/2}$)	CBr≡CI		9.34	PE		3121
C$_2$BrI$^+$($^2\Pi_{3/2}$)	CBr≡CI		9.4±0.1	EDD		3177
C$_2$BrI$^+$($^2\Pi_{1/2}$)	CBr≡CI		9.68	PE		3121
C$_2$BrI$^+$($^2\Pi_{3/2}$)	CBr≡CI		11.24	PE		3121
C$_2$BrI$^+$($^2\Pi_{1/2}$)	CBr≡CI		11.67	PE		3121
C$_2$BrI$^+$($^2\Pi_{3/2}$)	CBr≡CI		13.03 (V)	PE		3121
C$_2$BrI$^+$($^2\Sigma^+$)	CBr≡CI		14.71 (V)	PE		3121
C$_2$BrI$^+$($^2\Sigma^+$)	CBr≡CI		16.35 (V)	PE		3121
C$_2$BrI^{+2}						
C$_2$BrI^{+2}	CBr≡CI		24.7±0.1	EDD		3177

4.3. The Positive Ion Table—Continued

Ion	Reactant	Other products	Ionization or appearance potential (eV)	Method	Heat of formation (kJ mol⁻¹)	Ref.

LiRbI⁺

| LiRbI⁺ | LiRbI$_2$ | I | 8.4±0.3 | EI | | 2001 |

C₆H₆NI⁺

| C$_6$H$_6$NI$^+$ | C$_6$H$_4$INH$_2$ (4–Iodoaniline) | | 7.66±0.1 | CTS | | 2485 |

C₈H₁₀NI⁺

| C$_8$H$_{10}$NI$^+$ | C$_6$H$_4$IN(CH$_3$)$_2$ (N,N–Dimethyl–4–iodoaniline) | | 7.29 | CTS | | 1281 |

C₂H₅OI⁺

| C$_2$H$_5$OI$^+$ | CH$_2$ICH$_2$OH | | 9.62 | PE | | 3374 |

C₇H₄OI⁺

| C$_7$H$_4$OI$^+$ | C$_6$H$_4$ICOOCH$_3$ (4–Iodobenzoic acid methyl ester) | | 11.08 | EI | | 3238 |

C₈H₇O₂I⁺

| C$_8$H$_7$O$_2$I$^+$ | C$_6$H$_4$ICOOCH$_3$ (4–Iodobenzoic acid methyl ester) | | 9.10 | EI | | 3238 |

C₂H₂F₃I⁺

| C$_2$H$_2$F$_3$I$^+$ | CF$_3$CH$_2$I | | 10.00±0.01 | PI | | 182 |

C₄H₂F₇I⁺

| C$_4$H$_2$F$_7$I$^+$ | n–C$_3$F$_7$CH$_2$I | | 9.96±0.02 | PI | | 182 |

C₂H₆SiI⁺

| C$_2$H$_6$SiI$^+$ | (CH$_3$)$_3$SiI | CH$_3$ | 10.3±0.1 | EI | | 2689 |

4.3. The Positive Ion Table—Continued

Ion	Reactant	Other products	Ionization or appearance potential (eV)	Method	Heat of formation (kJ mol^{-1})	Ref.
			C$_3$H$_9$SiI$^+$			
C$_3$H$_9$SiI$^+$	(CH$_3$)$_3$SiI		8.9±0.1	EI		2689
			C$_4$H$_3$SI$^+$			
C$_4$H$_3$SI$^+$	C$_4$H$_3$SI (2−Iodothiophene)		8.55±0.05	PE		3246
			C$_6$H$_4$ClI$^+$			
C$_6$H$_4$ClI$^+$	C$_6$H$_4$ClI (1−Chloro−2−iodobenzene)		8.35±0.1?	PI		416
			BFClI$^+$			
BFClI$^+$	BFClI		10.18±0.09	EI		2512, 3227
			COIMn$^+$			
COIMn$^+$	Mn(CO)$_5$I	4CO	12.37	EI		2501
			C$_2$O$_2$IMn$^+$			
C$_2$O$_2$IMn$^+$	Mn(CO)$_5$I	3CO	10.46	EI		2501
			C$_3$O$_3$IMn$^+$			
C$_3$O$_3$IMn$^+$	Mn(CO)$_5$I	2CO	9.49	EI		2501
			C$_5$O$_5$IMn$^+$			
C$_5$O$_5$IMn$^+$	Mn(CO)$_5$I		8.35 (V)	PE		2879
C$_5$O$_5$IMn$^+$	Mn(CO)$_5$I		8.55±0.02	EI		2501
			BFBrI$^+$			
BFBrI$^+$	BFBrI		10.11±0.08	EI		2512, 3227
			BClBrI$^+$			
BClBrI$^+$	BClBrI		9.81±0.02	EI		2512, 3227

4.3. The Positive Ion Table—Continued

Ion	Reactant	Other products	Ionization or appearance potential (eV)	Method	Heat of formation (kJ mol^{-1})	Ref.
		$C_4H_{12}BN_2I^+$				
$C_4H_{12}BN_2I^+$	$((CH_3)_2N)_2BI$		7.97	EI		3227
		$C_2H_6BNI_2^+$				
$C_2H_6BNI_2^+$	$(CH_3)_2NBI_2$		8.99	EI		3227
		$C_2H_6BO_2I^+$				
$C_2H_6BO_2I^+$	$(CH_3O)_2BI$		9.63	EI		3227
		$CH_3BOI_2^+$				
$CH_3BOI_2^+$	CH_3OBI_2		9.21	EI		3227
		$C_2H_6BS_2I^+$				
$C_2H_6BS_2I^+$	$(CH_3S)_2BI$		8.90	EI		3227
		$CH_3BSI_2^+$				
$CH_3BSI_2^+$	CH_3SBI_2		9.26	EI		3227
		$C_9H_5O_3ICr^+$				
$C_9H_5O_3ICr^+$	$C_6H_5ICr(CO)_3$ (Iodobenzenechromium tricarbonyl)		7.36	EI		3005
		$C_{10}H_{12}O_4I_2Cr^+$				
$C_{10}H_{12}O_4I_2Cr^+$	$(CH_3COCICOCH_3)_3Cr$ (Tris(3−iodo−2,4−pentanedionato)chromium)		10.8±0.10	EI		2519
		$C_{15}H_{18}O_6I_3Cr^+$				
$C_{15}H_{18}O_6I_3Cr^+$	$(CH_3COCICOCH_3)_3Cr$ (Tris(3−iodo−2,4−pentanedionato)chromium)		8.03±0.05	EI		2519
		$C_2H_6BNFI^+$				
$C_2H_6BNFI^+$	$(CH_3)_2NBFI$		9.61	EI		3227

4.3. The Positive Ion Table—Continued

Ion	Reactant	Other products	Ionization or appearance potential (eV)	Method	Heat of formation (kJ mol^{-1})	Ref.

CH_3BOFI^+

Ion	Reactant	Other products	Ionization or appearance potential (eV)	Method	Heat of formation	Ref.
CH_3BOFI^+	CH_3OBFI		9.96	EI		3227

$C_{18}H_{15}N_2O_2PIFe^+$

| $C_{18}H_{15}N_2O_2PIFe^+$ | $(C_6H_5)_3PFe(NO)_2I$ (Triphenylphosphineiron dinitrosyl iodide) | | 7.5 | EI | | 2799 |

$C_{12}H_{27}N_2O_2PICo^+$

| $C_{12}H_{27}N_2O_2PICo^+$ | $(n-C_4H_9)_3PCo(NO)_2I$ | | 7.7 | EI | | 2799 |

$C_{18}H_{15}N_2O_2PICo^+$

| $C_{18}H_{15}N_2O_2PICo^+$ | $(C_6H_5)_3PCo(NO)_2I$ (Triphenylphosphinecobalt dinitrosyl iodide) | | 7.6 | EI | | 2799 |

$C_9H_{21}N_2O_5PICo^+$

| $C_9H_{21}N_2O_5PICo^+$ | $(iso-C_3H_7O)_3PCo(NO)_2I$ | | 7.95 | EI | | 2799 |

$Xe^+(^2P_{3/2})$　　　$\Delta H^\circ_{f0} = 1170.4$ kJ mol^{-1} (279.7 kcal mol^{-1})
$Xe^+(^2P_{1/2})$　　　$\Delta H^\circ_{f0} = 1296.4$ kJ mol^{-1} (309.8 kcal mol^{-1})

Ion	Reactant	Other products	Ionization or appearance potential (eV)	Method	Heat of formation	Ref.
$Xe^+(^2P_{3/2})$	Xe		12.130	S	1170.4	2113
$Xe^+(^2P_{3/2})$	Xe		12.129±0.002	PI	1032,	1253
$Xe^+(^2P_{3/2})$	Xe		12.12±0.01	PI		1118
$Xe^+(^2P_{3/2})$	Xe		12.17	PE		248
$Xe^+(^2P_{3/2})$	Xe		12.15±0.03	EDD		2557
$Xe^+(^2P_{3/2})$	Xe		12.09±0.03	EDD		3174
$Xe^+(^2P_{1/2})$	Xe		13.436	S	1296.4	2113
$Xe^+(^2P_{1/2})$	Xe		13.49	PE		248

See also − EI:　35, 52, 2032, 2053, 3201

| Xe^+ | XeF_2 | $F+F^-$ | 11.481 | PI | | 3276 |

(Threshold value corrected to 0 K)

The measurement was made upon the F^- ion. From this result the authors deduce $\Delta H^\circ_{f0}(XeF_2) = -117.2$ kJ mol^{-1} (-28.0 kcal mol^{-1}).

| Xe^+ | XeF_2 | $F+F^-$? | 12.0±0.1 | EI | | 2053 |

See also − PI:　3217

| Xe^+ | XeF_2 | $2F$ | ~15.3 | PI | | 3276 |
| Xe^+ | XeF_4 | F_2+F+F^-? | 12.4±0.1 | EI | | 2053 |

4.3. The Positive Ion Table—Continued

Ion	Reactant	Other products	Ionization or appearance potential (eV)	Method	Heat of formation (kJ mol^{-1})	Ref.
Xe^{+2}	$\Delta H_{f0}^\circ = 3217$ kJ mol^{-1} (769 kcal mol^{-1})					
Xe^{+2}	Xe		33.34	S	3217	2113
Xe^{+2}	Xe		33.3±0.1	RPD		3137
Xe^{+2}	Xe		33.4±0.2	EDD		2634
Xe^{+2}	Xe		33.5±0.2	NRE		25, 211
Xe^{+2}	Xe		33.3±0.5	NRE		201
Xe^{+2}	Xe		33	EI		1240
Xe^{+2}	Xe		33.3	EI		211
Xe^{+2}	Xe		33.5	EI		201
Xe^{+2}	Xe^+		21.21	S		2113
Xe^{+2}	Xe^+		20±2	SEQ		2551
Xe^{+3}	$\Delta H_{f0}^\circ \sim 6310$ kJ mol^{-1} (1508 kcal mol^{-1})					
Xe^{+3}	Xe		65.4	S	~6310	2113
Xe^{+3}	Xe		66.2±0.3	RPD		3137
Xe^{+3}	Xe		64.8±0.5	NRE		25, 211
Xe^{+3}	Xe		64.3±1	NRE		201
Xe^{+3}	Xe		65	EI		1240
Xe^{+3}	Xe		65.4	EI		211
Xe^{+3}	Xe		66.5	EI		201
Xe^{+3}	Xe^{+2}		32.1	S		2113
Xe^{+3}	Xe^{+2}		29±2	SEQ		2551
Xe^{+4}						
Xe^{+4}	Xe		110±1	RPD		3137
Xe^{+4}	Xe		107±1	NRE		25, 211
Xe^{+4}	Xe		103±2	NRE		201
Xe^{+4}	Xe		107	EI		201
Xe^{+4}	Xe		110	EI		1240
Xe^{+4}	Xe		111	EI		211
Xe^{+4}	Xe^{+3}		40±2	SEQ		2551
Xe^{+5}						
Xe^{+5}	Xe		170±2	RPD		3137
Xe^{+5}	Xe		160±1	NRE		25, 211
Xe^{+5}	Xe		157±3	NRE		201
Xe^{+5}	Xe		170	EI		201
Xe^{+5}	Xe		171	EI		211
Xe^{+5}	Xe		172	EI		1240
Xe^{+5}	Xe^{+4}		51±2	SEQ		2551

4.3. The Positive Ion Table—Continued

Ion	Reactant	Other products	Ionization or appearance potential (eV)	Method	Heat of formation (kJ mol^{-1})	Ref.
			Xe^{+6}			
Xe^{+6}	Xe		218±1	NRE		25, 211
Xe^{+6}	Xe		246	EI		211
Xe^{+6}	Xe		248	EI		1240
Xe^{+6}	Xe		255±3	EI		3137
Xe^{+6}	Xe^{+5}		64±2	SEQ		2551
			Xe^{+7}			
Xe^{+7}	Xe		362	EI		1240
Xe^{+7}	Xe		390±10	EI		3137
Xe^{+7}	Xe^{+6}		92±2	SEQ		2551
			Xe^{+8}			
Xe^{+8}	Xe		535	EI		1240
Xe^{+8}	Xe^{+7}		105±2	SEQ		2551
			Xe^{+9}			
Xe^{+9}	Xe		820	EI		1240
Xe^{+9}	Xe^{+8}		183±2	SEQ		2551
			Xe^{+10}			
Xe^{+10}	Xe^{+9}		203±2	SEQ		2551

Xe$_2^+$ $\Delta H_{f0}^{\circ} = 1075 \text{ kJ mol}^{-1}$ (257 kcal mol^{-1})

Ion	Reactant	Other products	Ionization or appearance potential (eV)	Method	Heat of formation (kJ mol^{-1})	Ref.
Xe$_2^+$	Xe$_2$?		11.14±0.02	PI	1075	2650
Xe$_2^+$		Xe+Xe*	11.162	PI	(≤1077)	2650
Xe$_2^+$		Xe+Xe*	11.162±0.005	PI	(≤1077)	2763

XeF$^+$ $\Delta H_{f0}^{\circ} \sim 1050 \text{ kJ mol}^{-1}$ (251 kcal mol^{-1})

Ion	Reactant	Other products	Ionization or appearance potential (eV)	Method	Heat of formation (kJ mol^{-1})	Ref.
XeF$^+$	XeF$_2$	F	12.90	PI	1051	3276
(Threshold value corrected to 0 K)						
XeF$^+$	XeF$_2$	F	12.86	PI	1047	2639

See also — EI: 2053

Ion	Reactant	Other products	Ionization or appearance potential (eV)	Method	Heat of formation (kJ mol^{-1})	Ref.
XeF$^+$	XeF$_4$	3F	15.81	PI	1054	3276
(Threshold value corrected to 0 K)						

See also — EI: 2053

4.3. The Positive Ion Table—Continued

Ion	Reactant	Other products	Ionization or appearance potential (eV)	Method	Heat of formation (kJ mol^{-1})	Ref.
\multicolumn{7}{c}{$XeF_2^+(^2\Pi_{3/2u})$ $\Delta H_{f0}^\circ = 1074$ kJ mol^{-1} (257 kcal mol^{-1})}						
$XeF_2^+(^2\Pi_{3/2u})$	XeF_2		12.35±0.01	PI	1074	3217, 3276
(Threshold value corrected to 0 K)						
$XeF_2^+(^2\Pi_{3/2u})$	XeF_2		12.28	PI		2639
$XeF_2^+(^2\Pi_{3/2u})$	XeF_2		12.35±0.01	PE		2912
$XeF_2^+(^2\Pi_{3/2u})$	XeF_2		12.33±0.02	PE		3062
$XeF_2^+(^2\Pi_{1/2u})$	XeF_2		12.89±0.01	PE		2912
$XeF_2^+(^2\Pi_{1/2u})$	XeF_2		12.83±0.02	PE		3062
$XeF_2^+(^2\Sigma_g^+)$	XeF_2		~13.5	PE		2912
$XeF_2^+(^2\Sigma_g^+)$	XeF_2		13.58±0.05	PE		3062
$XeF_2^+(^2\Pi_g)$	XeF_2		14.00±0.05	PE		2912
$XeF_2^+(^2\Pi_g)$	XeF_2		14.06±0.05	PE		3062
$XeF_2^+(^2\Pi_u)$	XeF_2		15.25±0.05	PE		2912
$XeF_2^+(^2\Pi_u)$	XeF_2		15.40±0.05	PE		3062
$XeF_2^+(^2\Sigma_u^+)$	XeF_2		16.80±0.05	PE		2912
$XeF_2^+(^2\Sigma_u^+)$	XeF_2		17.10±0.1	PE		3062
$XeF_2^+(^2\Sigma_g)$	XeF_2		20.05±0.1	PE		3062
$XeF_2^+(^2\Sigma_g)$	XeF_2		~22.5 (V)	PE		2912

See also – S: 2181
        EI: 2053

Ion	Reactant	Other products	Ionization or appearance potential (eV)	Method	Heat of formation (kJ mol^{-1})	Ref.
XeF_2^+	XeF_4	2F	≤15.22	PI		3276
(Threshold value corrected to 0 K)						

From this result the authors deduce $\Delta H_{f0}^\circ(XeF_4) = -241.0$ kJ mol^{-1} (-57.6 kcal mol^{-1}).

See also – EI: 2053

Ion	Reactant	Other products	Ionization or appearance potential (eV)	Method	Heat of formation (kJ mol^{-1})	Ref.
\multicolumn{7}{c}{XeF_3^+ $\Delta H_{f0}^\circ = 946$ kJ mol^{-1} (226 kcal mol^{-1})}						
XeF_3^+	XeF_4	F	13.10	PI	946	3276
(Threshold value corrected to 0 K)						
XeF_3^+	XeF_4	F	13.1±0.1	EI		2053
XeF_3^+	XeF_6	3F	15.9–16.1	PI		3276

4.3. The Positive Ion Table—Continued

Ion	Reactant	Other products	Ionization or appearance potential (eV)	Method	Heat of formation (kJ mol^{-1})	Ref.

$$XeF_4^+ \qquad \Delta H_{f0}^\circ = 980 \text{ kJ mol}^{-1} \text{ (234 kcal mol}^{-1})$$

Ion	Reactant	Other products	Ionization or appearance potential (eV)	Method	Heat of formation (kJ mol^{-1})	Ref.
XeF_4^+	XeF_4		12.65 ± 0.1	PI	980	3276
XeF_4^+	XeF_4		$\leqslant12.72$	PE		3364
XeF_4^+	XeF_4		12.9 ± 0.1	EI		2053
XeF_4^+	XeF_6	2F	15.67 ± 0.05	PI		3276

(Threshold value corrected to 0 K)

$$XeF_5^+$$

Ion	Reactant	Other products	Ionization or appearance potential (eV)	Method	Heat of formation (kJ mol^{-1})	Ref.
XeF_5^+	XeF_6	F	12.56	PI		3276

(Threshold value corrected to 0 K)

$$XeF_6^+$$

Ion	Reactant	Other products	Ionization or appearance potential (eV)	Method	Heat of formation (kJ mol^{-1})	Ref.
XeF_6^+	XeF_6		12.19 ± 0.02	PI		3276
(Threshold value corrected to 0 K)						
XeF_6^+	XeF_6		$\geqslant11.96$	PE		3364

$$Cs^+ \qquad \Delta H_{f0}^\circ = 453.9 \text{ kJ mol}^{-1} \text{ (108.5 kcal mol}^{-1})$$

Ion	Reactant	Other products	Ionization or appearance potential (eV)	Method	Heat of formation (kJ mol^{-1})	Ref.
Cs^+	Cs		3.894	S	453.9	2113

See also – S: 3160
        EI: 2487, 3221

| Cs^+ | CsF | F | 8.75 ± 0.1 | PI | (413) | 2611 |

(Threshold value corrected for hot bands)

See also – EI: 2406

| Cs^+ | CsCl | Cl | 8.47 ± 0.07 | PI | (459) | 2611 |

(Threshold value corrected for hot bands)

See also – EI: 2056, 2406

Cs^+	CsBr	Br	8.06 ± 0.05	PI		2611
(Threshold value corrected for hot bands)						
Cs^+	CsI	I	7.46 ± 0.05	PI		2611

(Threshold value corrected for hot bands)

See also – EI: 1239, 2001

| Cs^+ | CsOH | OH | 7.46 ± 0.14 | RPD | | 2985 |
| Cs^+ | CsOH | OH | 7.60 ± 0.15 | EI | | 3189 |

$$Cs_2^+$$

Ion	Reactant	Other products	Ionization or appearance potential (eV)	Method	Heat of formation (kJ mol^{-1})	Ref.
Cs_2^+	Cs_2		3.80 ± 0.1	PI		2633
Cs_2^+	Cs_2CO_3		11.4	EI		3243

4.3. The Positive Ion Table—Continued

Ion	Reactant	Other products	Ionization or appearance potential (eV)	Method	Heat of formation (kJ mol^{-1})	Ref.
		Cs_3^+				
Cs_3^+	Cs_3		3.2±0.1	PI		2633
		Cs_4^+				
Cs_4^+	Cs_4		3.2±0.1	PI		2633
		Cs_2O^+				
Cs_2O^+	Cs_2O		4.45±0.06	EI		2985
Cs_2O^+	Cs_2O		4.5	EI		3243
Cs_2O^+	$Cs_2(OH)_2$		7.52±0.14	EI		2985
Cs_2O^+	Cs_2SO_4		10.6±0.5	EI		2487
		Cs_2F^+				
Cs_2F^+	Cs_2F_2	F	9.64±0.05	PI		2611
(Threshold value corrected for hot bands)						
		$CsCl^+$				
$CsCl^+$	$CsCl$		8.3±0.3	EI		2406
		Cs_2Cl^+				
Cs_2Cl^+	Cs_2Cl_2	Cl	8.48±0.05	PI		2611
(Threshold value corrected for hot bands)						
		$CsBr^+$				
$CsBr^+$	$CsBr$		7.72±0.05	PI		2611
(Threshold value corrected for hot bands)						
		Cs_2Br^+				
Cs_2Br^+	Cs_2Br_2	Br	8.14±0.06	PI		2611
(Threshold value corrected for hot bands)						
		CsI^+				
CsI^+	CsI		7.25±0.05	PI		2611
(Threshold value corrected for hot bands)						

See also – EI: 1239, 2001

4.3. The Positive Ion Table—Continued

Ion	Reactant	Other products	Ionization or appearance potential (eV)	Method	Heat of formation (kJ mol⁻¹)	Ref.
			Cs$_2$I$^+$			
Cs$_2$I$^+$	Cs$_2$I$_2$	I	8.1±0.3	EI		2001
Cs$_2$I$^+$	Cs$_2$I$_2$	I	10.7±1	EI		1239
			CsOH$^+$			
CsOH$^+$	CsOH		7.21±0.14	EI		2985
CsOH$^+$	CsOH		7.40±0.15	EI		3189
			Cs$_2$OH$^+$			
Cs$_2$OH$^+$	Cs$_2$(OH)$_2$	OH	7.10±0.12	RPD		2985
			Cs$_2$CO$_3^+$			
Cs$_2$CO$_3^+$	Cs$_2$CO$_3$		5.5	EI		3243
			Cs$_2$SO$_4^+$			
Cs$_2$SO$_4^+$	Cs$_2$SO$_4$		8.9±0.5	EI		2487
			CsCdCl$^+$			
CsCdCl$^+$	CsCdCl$_3$		10.5±0.2	EI		2056
			LiCsI$^+$			
LiCsI$^+$	LiCsI$_2$	I	8.6±0.3	EI		2001

Ba$^+$ $\Delta H_{f0}^{\circ} = 684$ kJ mol^{-1} (163 kcal mol^{-1})

Ion	Reactant	Other products	Ionization or appearance potential (eV)	Method	Heat of formation (kJ mol⁻¹)	Ref.
Ba$^+$	Ba		5.212	S	684	2113

See also − S: 3270
         EI: 2178, 2432, 2527, 2589, 2990, 3204

Ion	Reactant	Other products	Ionization or appearance potential (eV)	Method	Heat of formation (kJ mol⁻¹)	Ref.
Ba$^+$	BaO	O	13.2±0.2	EI		2178
Ba$^+$	BaO	O	13.7±0.5	EI		2527
Ba$^+$	BaCl$_2$		15.43±0.10	EI		2991
Ba$^+$	BaI$_2$		12.8±1	EI		1239

Ba^{+2} $\Delta H_{f0}^{\circ} = 1649$ kJ mol^{-1} (394 kcal mol^{-1})

Ion	Reactant	Other products	Ionization or appearance potential (eV)	Method	Heat of formation (kJ mol⁻¹)	Ref.
Ba^{+2}	Ba		15.216	S	1649	2113
Ba^{+2}	Ba*?		11.5±0.3	EI		2178
Ba^{+2}	Ba$^+$		10.004	S		2113
Ba^{+2}	BaO?		20.6±0.5	EI		2178

4.3. The Positive Ion Table—Continued

Ion	Reactant	Other products	Ionization or appearance potential (eV)	Method	Heat of formation (kJ mol^{-1})	Ref.
			Ba^{+3}			
Ba^{+3}	Ba		57.5	EI		2178
			BaO$^+$			
BaO$^+$	BaO		6.5±0.3	EI		2178
BaO$^+$	BaO		6.8±0.5	EI		3256
See also − EI: 2052, 2527						
BaO$^+$	BaMoO$_4$		12.6±0.5	EI		3256
BaO$^+$	BaWO$_4'$		12.6±0.5	EI		3256
			BaF$^+$			
BaF$^+$	BaF		4.8±0.3	EI		2432
BaF$^+$	BaF		4.8±0.3	EI		2589
BaF$^+$	BaF		4.9±0.3	EI		1104
BaF$^+$	BaF$_2$	F	13.5±1.0	EI		2432
			BaCl$^+$			
BaCl$^+$	BaCl		5.01±0.10	EI		2991
BaCl$^+$	BaCl		5.0±0.5	EI		2990
BaCl$^+$	BaCl		5.5±0.5	EI		3204
BaCl$^+$	BaCl$_2$	Cl	10.03±0.10	EI		2991
BaCl$^+$	BaCl$_2$	Cl	10.3±0.3	EI		2990
			BaCl$_2^+$			
BaCl$_2^+$	BaCl$_2$		9.18±0.10	EI		2991
BaCl$_2^+$	BaCl$_2$		10.0±1.0	EI		2990
BaCl$_2^+$	BaCl$_2$		11.0±1	EI		3204
			BaI$^+$			
BaI$^+$	BaI$_2$	I	9.6±1	EI		1239
			BaI$_2^+$			
BaI$_2^+$	BaI$_2$		8.1±1	EI		1239
			BaOH$^+$			
BaOH$^+$	BaOH		4.5±1	EI		2052
BaOH$^+$	BaOH		5.25±0.1	D		3242, 3419

4.3. The Positive Ion Table—Continued

Ion	Reactant	Other products	Ionization or appearance potential (eV)	Method	Heat of formation (kJ mol^{-1})	Ref.
		BaMoO$_3^+$				
BaMoO$_3^+$	BaMoO$_3$		8.5±0.5	EI		3256
BaMoO$_3^+$	BaMoO$_4$	O	14.2±0.5	EI		3256
		BaMoO$_4^+$				
BaMoO$_4^+$	BaMoO$_4$		9.6±0.5	EI		3256
	La$^+$	$\Delta H_{f0}^\circ = 969.4$ kJ mol^{-1} (231.7 kcal mol^{-1})				
La$^+$	La		5.577	S	969.4	2113
La$^+$	La		5.55±0.05	SI		2495
See also – EI: 2600, 2779, 3195						
La$^+$	LaF$_3$		26.5±0.7	EI	2009,	2600
La$^+$	LaS?		~12	EI		3198
La$^+$	LaCl$_3$		23±1	EI		2435
La$^+$	(C$_5$H$_5$)$_3$La (Tris(cyclopentadienyl)lanthanum)		24.4±0.4	EI		3298
	La^{+2}	$\Delta H_{f0}^\circ \sim 2037$ kJ mol^{-1} (487 kcal mol^{-1})				
La^{+2}	La		16.64±0.08	S	2037	2113
La^{+2}	La$^+$		11.06±0.08	S		2113, 3265
	La^{+3}	$\Delta H_{f0}^\circ \sim 3887$ kJ mol^{-1} (929 kcal mol^{-1})				
La^{+3}	La		35.81±0.08	S	3887	2113
La^{+3}	La^{+2}		19.175	S		2113, 3262
See also – S: 3263						
		LaO$^+$				
LaO$^+$	LaO		4.8±0.5	EI		3195
LaO$^+$	LaO		~5	EI		3198
		LaF$^+$				
LaF$^+$	LaF		6.3±0.3	EI		2600
LaF$^+$	LaF$_3$		18.5±0.7	EI	2009,	2600
		LaF$_2^+$				
LaF$_2^+$	LaF$_2$		6.8±0.3	EI		2600
LaF$_2^+$	LaF$_3$	F	12.0±0.7	EI	2009,	2600

4.3. The Positive Ion Table—Continued

Ion	Reactant	Other products	Ionization or appearance potential (eV)	Method	Heat of formation (kJ mol⁻¹)	Ref.
		LaS⁺				
LaS⁺	LaS		~5	EI		3198
		LaCl⁺				
LaCl⁺	LaCl₃		17.5±1	EI		2435
		LaCl₂⁺				
LaCl₂⁺	LaCl₃	Cl	13.6±0.5	EI		2435
		LaCl₃⁺				
LaCl₃⁺	LaCl₃		13.8±1	EI		2435
		La₂Cl₅⁺				
La₂Cl₅⁺	La₂Cl₆	Cl	15±1	EI		2435
		C₅H₅La⁺				
C₅H₅La⁺	(C₅H₅)₃La (Tris(cyclopentadienyl)lanthanum)		17.3±0.3	EI		3298
		C₁₀H₁₀La⁺				
C₁₀H₁₀La⁺	(C₅H₅)₃La (Tris(cyclopentadienyl)lanthanum)		10.2±0.3	EI		3298
		C₁₅H₁₅La⁺				
C₁₅H₁₅La⁺	(C₅H₅)₃La (Tris(cyclopentadienyl)lanthanum)		7.9±0.3	EI		3298
		C₂₂H₃₈O₄La⁺				
C₂₂H₃₈O₄La⁺	((CH₃)₃CCOCHCOC(CH₃)₃)₃La (Tris(2,2,6,6–tetramethyl–3,5–heptanedionato)lanthanum)		14.9±0.5	EI		2524
		C₂₉H₄₈O₆La⁺				
C₂₉H₄₈O₆La⁺	((CH₃)₃CCOCHCOC(CH₃)₃)₃La (Tris(2,2,6,6–tetramethyl–3,5–heptanedionato)lanthanum)		11.1±0.5	EI		2524

4.3. The Positive Ion Table—Continued

Ion	Reactant	Other products	Ionization or appearance potential (eV)	Method	Heat of formation (kJ mol^{-1})	Ref.
			$C_{33}H_{57}O_6La^+$			
$C_{33}H_{57}O_6La^+$	$((CH_3)_3CCOCHCOC(CH_3)_3)_3La$ (Tris(2,2,6,6–tetramethyl–3,5–heptanedionato)lanthanum)		8.0 ± 0.5	EI		2524
		Ce^+ $\Delta H^\circ_{f0} = 951$ kJ mol^{-1} (227 kcal mol^{-1})				
Ce^+	Ce		5.47 ± 0.02	S	951	2113, 3051
Ce^+	Ce		5.54 ± 0.06	SI		2495
Ce^+	Ce		5.60 ± 0.05	SI		2882

See also – S: 2667
        EI: 2517, 2600, 2690, 2734, 2779

Ion	Reactant	Other products	Ionization or appearance potential (eV)	Method	Heat of formation (kJ mol^{-1})	Ref.
Ce^+	CeF_3		28.0 ± 0.7	EI		2600
		Ce^{+2} $\Delta H^\circ_{f0} \sim 1998$ kJ mol^{-1} (478 kcal mol^{-1})				
Ce^{+2}	Ce		16.32 ± 0.10	S	1998	2113
Ce^{+2}	Ce^+		10.85 ± 0.08	S		2113, 3265
		Ce^{+3} $\Delta H^\circ_{f0} \sim 3947$ kJ mol^{-1} (943 kcal mol^{-1})				
Ce^{+3}	Ce		36.52 ± 0.11	S	3947	2113
Ce^{+3}	Ce		36.4	BH		3259
Ce^{+3}	Ce		36.56	BH		3273
Ce^{+3}	Ce^{+2}		20.20 ± 0.01	S		2113

See also – S: 3264

Ion	Reactant	Other products	Ionization or appearance potential (eV)	Method	Heat of formation (kJ mol^{-1})	Ref.
			Ce_2^+			
Ce_2^+	Ce_2		5.2 ± 0.5	EI		2690
			CeC^+			
CeC^+	CeC		7.5 ± 1.0	EI		2491
CeC^+	CeC_2	C	~14	EI		2491
			CeC_2^+			
CeC_2^+	CeC_2		5.6 ± 0.5	EI		2517
			CeC_4^+			
CeC_4^+	CeC_4		6.2 ± 0.5	EI		2517

4.3. The Positive Ion Table—Continued

Ion	Reactant	Other products	Ionization or appearance potential (eV)	Method	Heat of formation (kJ mol^{-1})	Ref.
			CeO$^+$			
CeO$^+$	CeO		5.2±0.5	EI		2734
CeO$^+$	CeO$_2$	O	11.6±0.5	EI		2734
			CeO$_2^+$			
CeO$_2^+$	CeO$_2$		9.5±0.5	EI		2734
			CeF$^+$			
CeF$^+$	CeF		6.5±0.3	EI		2600
CeF$^+$	CeF$_3$		20.5±0.7	EI		2600
			CeF$_2^+$			
CeF$_2^+$	CeF$_2$		6.0±0.3	EI		2600
CeF$_2^+$	CeF$_3$	F	13.0±0.7	EI		2600

Pr$^+$ $\Delta H_{f0}^\circ = 880 \text{ kJ mol}^{-1} (210 \text{ kcal mol}^{-1})$

Ion	Reactant	Other products	Ionization or appearance potential (eV)	Method	Heat of formation (kJ mol^{-1})	Ref.	
Pr$^+$	Pr		5.42±0.02	S	880	2113,	2667
Pr$^+$	Pr		5.40±0.05	SI			2495
Pr$^+$	Pr		5.48±0.01	SI			2881
Pr$^+$	Pr		5.61±0.12	SI		2478,	2982
Pr$^+$	Pr		5.6±0.2	SI			3024

See also − EI: 2600, 2779

Ion	Reactant	Other products	Ionization or appearance potential (eV)	Method	Heat of formation	Ref.
Pr$^+$	PrO$_2$?		12.9±0.5	EI		2734
Pr$^+$	PrF$_3$		27.5±0.7	EI		2600
Pr$^+$	(C$_5$H$_5$)$_3$Pr		24.1±0.4	EI		3298
	(Tris(cyclopentadienyl)praseodymium)					

Pr^{+2} $\Delta H_{f0}^\circ \sim 1898 \text{ kJ mol}^{-1} (454 \text{ kcal mol}^{-1})$

Ion	Reactant	Other products	Ionization or appearance potential (eV)	Method	Heat of formation	Ref.	
Pr^{+2}	Pr		15.97±0.10	S	1898		2113
Pr^{+2}	Pr$^+$		10.55±0.08	S		2113,	3265

Pr^{+3} $\Delta H_{f0}^\circ \sim 3984 \text{ kJ mol}^{-1} (952 \text{ kcal mol}^{-1})$

Ion	Reactant	Other products	Ionization or appearance potential (eV)	Method	Heat of formation	Ref.	
Pr^{+3}	Pr		37.59±0.12	S	3984		2113
Pr^{+3}	Pr		37.55	BH			3259
Pr^{+3}	Pr		37.62	BH			3273
Pr^{+3}	Pr^{+2}		21.62±0.02	S		2113,	3282

See also − S: 3260, 3267

4.3. The Positive Ion Table—Continued

Ion	Reactant	Other products	Ionization or appearance potential (eV)	Method	Heat of formation (kJ mol^{-1})	Ref.

Pr^{+4} $\Delta H_{f0}^{\circ} \sim 7742$ kJ mol^{-1} (1850 kcal mol^{-1})

Ion	Reactant	Other products	Ionization or appearance potential (eV)	Method	Heat of formation (kJ mol^{-1})	Ref.
Pr^{+4}	Pr		76.54±0.14	S	7742	2113
Pr^{+4}	Pr^{+3}		38.95±0.02	S		2113, 3268

PrO$^+$

Ion	Reactant	Other products	Ionization or appearance potential (eV)	Method	Heat of formation (kJ mol^{-1})	Ref.
PrO$^+$	PrO		4.9±0.5	EI		2734
PrO$^+$	PrO$_2$	O	10.9±0.5	EI		2734

PrO$_2^+$

Ion	Reactant	Other products	Ionization or appearance potential (eV)	Method	Heat of formation (kJ mol^{-1})	Ref.
PrO$_2^+$	PrO$_2$		9.6±0.5	EI		2734

PrF$^+$

Ion	Reactant	Other products	Ionization or appearance potential (eV)	Method	Heat of formation (kJ mol^{-1})	Ref.
PrF$^+$	PrF$_3$		19.0±0.7	EI		2600

PrF$_2^+$

Ion	Reactant	Other products	Ionization or appearance potential (eV)	Method	Heat of formation (kJ mol^{-1})	Ref.
PrF$_2^+$	PrF$_3$	F	12.5±0.7	EI		2600

C$_5$H$_5$Pr$^+$

Ion	Reactant	Other products	Ionization or appearance potential (eV)	Method	Heat of formation (kJ mol^{-1})	Ref.
C$_5$H$_5$Pr$^+$	(C$_5$H$_5$)$_3$Pr (Tris(cyclopentadienyl)praseodymium)		17.0±0.4	EI		2640
C$_5$H$_5$Pr$^+$	(C$_5$H$_5$)$_3$Pr (Tris(cyclopentadienyl)praseodymium)		17±0.4	EI		3298

C$_{10}$H$_{10}$Pr$^+$

Ion	Reactant	Other products	Ionization or appearance potential (eV)	Method	Heat of formation (kJ mol^{-1})	Ref.
C$_{10}$H$_{10}$Pr$^+$	(C$_5$H$_5$)$_3$Pr (Tris(cyclopentadienyl)praseodymium)		8.48±0.2	EI		2640
C$_{10}$H$_{10}$Pr$^+$	(C$_5$H$_5$)$_3$Pr (Tris(cyclopentadienyl)praseodymium)		10±0.2	EI		3298

C$_{15}$H$_{15}$Pr$^+$

Ion	Reactant	Other products	Ionization or appearance potential (eV)	Method	Heat of formation (kJ mol^{-1})	Ref.
C$_{15}$H$_{15}$Pr$^+$	(C$_5$H$_5$)$_3$Pr (Tris(cyclopentadienyl)praseodymium)		7.68±0.1	EI		2640
C$_{15}$H$_{15}$Pr$^+$	(C$_5$H$_5$)$_3$Pr (Tris(cyclopentadienyl)praseodymium)		8.2±0.2	EI		3298

4.3. The Positive Ion Table—Continued

Ion	Reactant	Other products	Ionization or appearance potential (eV)	Method	Heat of formation (kJ mol^{-1})	Ref.

Nd$^+$ $\Delta H_{f0}^\circ = 858$ kJ mol^{-1} (205 kcal mol^{-1})

Nd$^+$	Nd		5.49±0.02	S	858	2113, 2667
Nd$^+$	Nd		5.49±0.05	SI		2495
Nd$^+$	Nd		5.51±0.02	SI		2881
Nd$^+$	Nd		5.5±0.1	SI		3024

See also – S: 2177
EI: 2459, 2589, 2600, 2779, 3210

Nd$^+$	NdO$_2$?		11.8±0.5	EI		2734
Nd$^+$	NdF$_3$		25.5±0.3	EI		2589, 2600
Nd$^+$	(C$_5$H$_5$)$_3$Nd (Tris(cyclopentadienyl)neodymium)		23.2±0.4	EI		3269
Nd$^+$	(C$_5$H$_5$)$_3$Nd (Tris(cyclopentadienyl)neodymium)		23.6±0.4	EI		3298

Nd^{+2} $\Delta H_{f0}^\circ \sim 1894$ kJ mol^{-1} (453 kcal mol^{-1})

Nd^{+2}	Nd		16.22±0.10	S	1894	2113
Nd^{+2}	Nd$^+$		10.73±0.08	S		2113, 3265

Nd^{+3}

Nd^{+3}	Nd		38.27	BH		3273
Nd^{+3}	Nd		38.3	BH		3261
Nd^{+3}	Nd		38.4	BH		3259

NdC$_2^+$

NdC$_2^+$	NdC$_2$		6.5±0.3	EI		3210
NdC$_2^+$	NdC$_2$		6.6±0.3	EI		2459

NdC$_4^+$

NdC$_4^+$	NdC$_4$		7.6±0.7	EI		2459

NdO$^+$

NdO$^+$	NdO		5.0±0.5	EI		2734

NdO$_2^+$

NdO$_2^+$	NdO$_2$		9.5±0.5	EI		2734

4.3. The Positive Ion Table—Continued

Ion	Reactant	Other products	Ionization or appearance potential (eV)	Method	Heat of formation (kJ mol^{-1})	Ref.
NdF$^+$						
NdF$^+$	NdF		5.0±0.3	EI	2589,	2600
NdF$^+$	NdF$_2$	F	~12.5±0.5	EI		2600
NdF$^+$	NdF$_3$		19.8±0.3	EI	2589,	2600
NdF$_2^+$						
NdF$_2^+$	NdF$_2$		5.6±0.3	EI	2589,	2600
NdF$_2^+$	NdF$_3$	F	12.9±0.3	EI	2589,	2600
C$_5$H$_5$Nd$^+$						
C$_5$H$_5$Nd$^+$	(C$_5$H$_5$)$_3$Nd (Tris(cyclopentadienyl)neodymium)		16.8±0.2	EI		3298
C$_5$H$_5$Nd$^+$	(C$_5$H$_5$)$_3$Nd (Tris(cyclopentadienyl)neodymium)		17.8±0.3	EI		3269
C$_{10}$H$_{10}$Nd$^+$						
C$_{10}$H$_{10}$Nd$^+$	(C$_5$H$_5$)$_3$Nd (Tris(cyclopentadienyl)neodymium)		9.8±0.2	EI		3298
C$_{10}$H$_{10}$Nd$^+$	(C$_5$H$_5$)$_3$Nd (Tris(cyclopentadienyl)neodymium)		10.0±0.2	EI		3269
C$_{15}$H$_{15}$Nd$^+$						
C$_{15}$H$_{15}$Nd$^+$	(C$_5$H$_5$)$_3$Nd (Tris(cyclopentadienyl)neodymium)		8.0±0.2	EI		3298
C$_{15}$H$_{15}$Nd$^+$	(C$_5$H$_5$)$_3$Nd (Tris(cyclopentadienyl)neodymium)		8.3±0.1	EI		3269
Pm$^+$						
Pm$^+$	Pm		5.55±0.02	S	2113,	2667
Pm^{+2}						
Pm^{+2}	Pm		16.45±0.10	S		2113
Pm^{+2}	Pm$^+$		10.90±0.08	S	2113,	3265

4.3. The Positive Ion Table—Continued

Ion	Reactant	Other products	Ionization or appearance potential (eV)	Method	Heat of formation (kJ mol⁻¹)	Ref.

$$\text{Sm}^+ \qquad \Delta H^\circ_{f0} = 749 \text{ kJ mol}^{-1} \text{ (179 kcal mol}^{-1}\text{)}$$

Ion	Reactant	Other products	IP/AP (eV)	Method	ΔH_f	Ref.
Sm^+	Sm		5.63±0.02	S	749	2113, 2667
Sm^+	Sm		5.56±0.10	EI		2588
Sm^+	Sm		5.61±0.05	SI		2495
Sm^+	Sm		5.70±0.02	SI		1165
Sm^+	Sm		6.15±0.05	SI		3024

See also – EI: 2593, 2600

Ion	Reactant	Other products	IP/AP (eV)	Method	ΔH_f	Ref.
Sm^+	SmF_3		26.0±0.7	EI		2600
Sm^+	$(\text{C}_5\text{H}_5)_3\text{Sm}$ (Tris(cyclopentadienyl)samarium)		22.2±0.4	EI		3269

$$\text{Sm}^{+2} \qquad \Delta H^\circ_{f0} \sim 1817 \text{ kJ mol}^{-1} \text{ (434 kcal mol}^{-1}\text{)}$$

Ion	Reactant	Other products	IP/AP (eV)	Method	ΔH_f	Ref.
Sm^{+2}	Sm		16.70±0.10	S	1817	2113
Sm^{+2}	Sm^+		11.07±0.08	S	2113,	3265

$$\text{Sm}^{+3}$$

Ion	Reactant	Other products	IP/AP (eV)	Method	ΔH_f	Ref.
Sm^{+3}	Sm		40.38	BH		3273
Sm^{+3}	Sm		40.4	BH		3259
Sm^{+3}	Sm		40.4	BH		3261

$$\text{SmF}^+$$

Ion	Reactant	Other products	IP/AP (eV)	Method	ΔH_f	Ref.
SmF^+	SmF		5.7±0.3	EI		2593, 2600
SmF^+	SmF_3		19.0±0.7	EI		2600

$$\text{SmF}_2^+$$

Ion	Reactant	Other products	IP/AP (eV)	Method	ΔH_f	Ref.
SmF_2^+	SmF_3	F	13.0±0.7	EI		2600

$$\text{C}_5\text{H}_5\text{Sm}^+$$

Ion	Reactant	Other products	IP/AP (eV)	Method	ΔH_f	Ref.
$\text{C}_5\text{H}_5\text{Sm}^+$	$(\text{C}_5\text{H}_5)_3\text{Sm}$ (Tris(cyclopentadienyl)samarium)		14.4±0.3	EI		3269

$$\text{C}_{10}\text{H}_{10}\text{Sm}^+$$

Ion	Reactant	Other products	IP/AP (eV)	Method	ΔH_f	Ref.
$\text{C}_{10}\text{H}_{10}\text{Sm}^+$	$(\text{C}_5\text{H}_5)_3\text{Sm}$ (Tris(cyclopentadienyl)samarium)		10.0±0.2	EI		3269

4.3. The Positive Ion Table—Continued

Ion	Reactant	Other products	Ionization or appearance potential (eV)	Method	Heat of formation (kJ mol^{-1})	Ref.
			$C_{15}H_{15}Sm^+$			
$C_{15}H_{15}Sm^+$	$(C_5H_5)_3Sm$ (Tris(cyclopentadienyl)samarium)		8.0 ± 0.1	EI		3269

	Eu^+	**$\Delta H_{f0}^\circ = 724$ kJ mol^{-1} (173 kcal mol^{-1})**				
Eu^+	Eu		5.67 ± 0.01	S	724	2113
Eu^+	Eu		5.68 ± 0.02	S	725	2667
Eu^+	Eu		5.664 ± 0.008	PI	723	3258
Eu^+	Eu		5.61 ± 0.10	EI		2588
Eu^+	Eu		5.64 ± 0.05	SI		2495
Eu^+	Eu		5.68 ± 0.03	SI		1165

See also – EI: 2593, 2600, 2710, 3190

Ion	Reactant	Other products	Ionization or appearance potential (eV)	Method	Heat of formation	Ref.
Eu^+	EuF_3		27.0 ± 0.7	EI		2600
Eu^+	$EuCl_2$		15.0 ± 0.5	EI		2435

	Eu^{+2}	**$\Delta H_{f0}^\circ \sim 1809$ kJ mol^{-1} (432 kcal mol^{-1})**				
Eu^{+2}	Eu		16.92 ± 0.09	S	1809	2113
Eu^{+2}	Eu^+		11.25 ± 0.08	S		2113, 3265
Eu^{+2}	$EuCl_2$		26.0 ± 0.5	EI		2435

			Eu^{+3}			
Eu^{+3}	Eu		41.8	BH		3259
Eu^{+3}	Eu		41.8	BH		3261
Eu^{+3}	Eu		42.05	BH		3273

			EuF^+			
EuF^+	EuF		5.9 ± 0.3	EI		2593, 2600
EuF^+	EuF_3		19.5 ± 0.7	EI		2600

			EuF_2^+			
EuF_2^+	EuF_3	F	13.5 ± 0.7	EI		2600

			$EuCl^+$			
$EuCl^+$	$EuCl_2$	Cl	10.3 ± 0.5	EI		2435

			$EuCl_2^+$			
$EuCl_2^+$	$EuCl_2$		10.5 ± 0.5	EI		2435

4.3. The Positive Ion Table—Continued

Ion	Reactant	Other products	Ionization or appearance potential (eV)	Method	Heat of formation (kJ mol^{-1})	Ref.
		EuCl$_3^+$				
EuCl$_3^+$	EuCl$_3$		13±1	EI		2435
		EuBr$^+$				
EuBr$^+$	EuBr$_2$	Br	10.4	EI		2903, 3188
		C$_{11}$H$_{19}$O$_2$Eu$^+$				
C$_{11}$H$_{19}$O$_2$Eu$^+$	((CH$_3$)$_3$CCOCHCOC(CH$_3$)$_3$)$_3$Eu (Tris(2,2,6,6–tetramethyl–3,5–heptanedionato)europium)		24.0±0.5	EI		2524
		C$_{22}$H$_{38}$O$_4$Eu$^+$				
C$_{22}$H$_{38}$O$_4$Eu$^+$	((CH$_3$)$_3$CCOCHCOC(CH$_3$)$_3$)$_3$Eu (Tris(2,2,6,6–tetramethyl–3,5–heptanedionato)europium)		16.5±0.5	EI		2524
		C$_{29}$H$_{48}$O$_6$Eu$^+$				
C$_{29}$H$_{48}$O$_6$Eu$^+$	((CH$_3$)$_3$CCOCHCOC(CH$_3$)$_3$)$_3$Eu (Tris(2,2,6,6–tetramethyl–3,5–heptanedionato)europium)		12.7±0.5	EI		2524
		C$_{33}$H$_{57}$O$_6$Eu$^+$				
C$_{33}$H$_{57}$O$_6$Eu$^+$	((CH$_3$)$_3$CCOCHCOC(CH$_3$)$_3$)$_3$Eu (Tris(2,2,6,6–tetramethyl–3,5–heptanedionato)europium)		9.8±0.5	EI		2524
		Gd$^+$ $\quad \Delta H_{f0}^\circ = 991$ kJ mol^{-1} (237 kcal mol^{-1})				
Gd$^+$	Gd		6.14±0.02	S	991	2113, 3051
Gd$^+$	Gd		5.98±0.10	EI		2588
Gd$^+$	Gd		6.16±0.05	SI		2495
Gd$^+$	Gd		6.73±0.09	SI		2781
Gd$^+$	Gd		6.73±0.09	SI	2477,	2982

See also – EI: 2593, 2600

Ion	Reactant	Other products	Ionization or appearance potential (eV)	Method	Heat of formation (kJ mol^{-1})	Ref.
Gd$^+$	GdF$_3$		26.5±0.7	EI		2600
		Gd^{+2}				
Gd^{+2}	Gd		18.2±0.4	S		2113
Gd^{+2}	Gd$^+$		12.1±0.4	S	2113,	3265

4.3. The Positive Ion Table—Continued

Ion	Reactant	Other products	Ionization or appearance potential (eV)	Method	Heat of formation (kJ mol^{-1})	Ref.
			Gd^{+3}			
Gd^{+3}	Gd		38.8	BH		3259
Gd^{+3}	Gd		39.03	BH		3273
Gd^{+3}	Gd		39.1	BH		3261
			GdF^{+}			
GdF^{+}	GdF		6.2±0.3	EI		2593, 2600
GdF^{+}	GdF$_3$		20.0±0.7	EI		2600
			GdF$_2^{+}$			
GdF$_2^{+}$	GdF$_2$		6.5±0.3	EI		2593, 2600
GdF$_2^{+}$	GdF$_3$	F	12.2±0.5	EI		2593
GdF$_2^{+}$	GdF$_3$	F	14.5±0.7	EI		2600

Tb^{+} $\Delta H_{f0}^{\circ} = 955$ kJ mol^{-1} (228 kcal mol^{-1})

Ion	Reactant	Other products	Ionization or appearance potential (eV)	Method	Heat of formation (kJ mol^{-1})	Ref.
Tb^{+}	Tb		5.85±0.02	S	955	2113, 2667
Tb^{+}	Tb		5.89±0.04	SI		2495
Tb^{+}	Tb		5.98±0.02	SI		2882

See also — EI: 2600

Ion	Reactant	Other products	Ionization or appearance potential (eV)	Method	Heat of formation (kJ mol^{-1})	Ref.
Tb^{+}	TbF$_3$		27.0±0.7	EI		2600

Tb^{+2} $\Delta H_{f0}^{\circ} \sim 2067$ kJ mol^{-1} (494 kcal mol^{-1})

Ion	Reactant	Other products	Ionization or appearance potential (eV)	Method	Heat of formation (kJ mol^{-1})	Ref.
Tb^{+2}	Tb		17.37±0.10	S	2067	2113
Tb^{+2}	Tb^{+}		11.52±0.08	S		2113, 3265
			Tb^{+3}			
Tb^{+3}	Tb		39.1	BH		3261
Tb^{+3}	Tb		39.3	BH		3259
Tb^{+3}	Tb		39.41	BH		3273
			TbF^{+}			
TbF^{+}	TbF$_3$		19.5±0.7	EI		2600
			TbF$_2^{+}$			
TbF$_2^{+}$	TbF$_3$	F	13.0±0.7	EI		2600

4.3. The Positive Ion Table—Continued

Ion	Reactant	Other products	Ionization or appearance potential (eV)	Method	Heat of formation (kJ mol⁻¹)	Ref.
Dy^+		$\Delta H_{f0}^\circ = 865$ kJ mol⁻¹ (207 kcal mol⁻¹)				
Dy^+	Dy		5.93±0.02	S	865	2113, 2667
Dy^+	Dy		5.78±0.10	EI		2588
Dy^+	Dy		5.72±0.10	SI	2478,	2982
Dy^+	Dy		5.80±0.02	SI		1165
Dy^+	Dy		5.82±0.03	SI		2495
See also – EI:	2590, 2600, 2755					
Dy^+	DyF_3		25.0±0.3	EI		2402
Dy^+	DyF_3		27.4±0.7	EI		2600
Dy^{+2}		$\Delta H_{f0}^\circ \sim 1991$ kJ mol⁻¹ (476 kcal mol⁻¹)				
Dy^{+2}	Dy		17.60±0.10	S	1991	2113
Dy^{+2}	Dy^+		11.67±0.08	S	2113,	3265
Dy^{+3}						
Dy^{+3}	Dy		40.4	BH		3259
Dy^{+3}	Dy		40.5	BH		3261
Dy^{+3}	Dy		40.66	BH		3273
DyC_2^+						
DyC_2^+	DyC_2		8.3±0.5	EI		2755
DyC_4^+						
DyC_4^+	DyC_4		8.8±0.5	EI		2755
DyF^+						
DyF^+	DyF		6.0±0.3	EI		2590, 2600
DyF^+	DyF_3		19.5±0.3	EI		2402
DyF^+	DyF_3		20.0±0.7	EI		2600
DyF_2^+						
DyF_2^+	DyF_2		6.7±0.3	EI		2600
DyF_2^+	DyF_3	F	13.5±0.7	EI		2600

4.3. The Positive Ion Table—Continued

Ion	Reactant	Other products	Ionization or appearance potential (eV)	Method	Heat of formation (kJ mol^{-1})	Ref.
			Ho$^+$ $\quad$ $\Delta H^\circ_{f0} = 883$ kJ mol^{-1} (211 kcal mol^{-1})			
Ho$^+$	Ho		6.02±0.02	S	883	2113, 2667
Ho$^+$	Ho		5.85±0.10	EI		2588
Ho$^+$	Ho		5.89±0.03	SI		2495
Ho$^+$	Ho		6.08±0.09	SI	2477,	2982
Ho$^+$	Ho		6.19±0.02	SI		1165
See also — EI: $\quad$ 2590, 2593, 2600, 2755						
Ho$^+$	HoF$_3$		27.5±0.5	EI		2402
Ho$^+$	HoF$_3$		28.0±0.7	EI		2600
			Ho^{+2} $\quad$ $\Delta H^\circ_{f0} \sim 2022$ kJ mol^{-1} (483 kcal mol^{-1})			
Ho^{+2}	Ho		17.82±0.10	S	2022	2113
Ho^{+2}	Ho$^+$		11.80±0.08	S	2113,	3265
			Ho^{+3}			
Ho^{+3}	Ho		40.6±0.7	S	2113,	3261
Ho^{+3}	Ho		40.77	BH		3273
Ho^{+3}	Ho		40.8	BH		3259
Ho^{+3}	Ho^{+2}		22.8±0.6	S		3261
			HoC$_2^+$			
HoC$_2^+$	HoC$_2$		7.3±0.5	EI		2755
			HoC$_4^+$			
HoC$_4^+$	HoC$_4$		7.7+0.5	EI		2755
			HoF$^+$			
HoF$^+$	HoF		6.0±0.5	EI		2755
HoF$^+$	HoF		6.1±0.3	EI	2590,	2593, 2600
HoF$^+$	HoF$_3$		21.0±0.5	EI	2402,	2600
			HoF$_2^+$			
HoF$_2^+$	HoF$_2$		6.9±0.3	EI	2590,	2593, 2600
HoF$_2^+$	HoF$_3$	F	12.5±0.5	EI		2590
HoF$_2^+$	HoF$_3$	F	13.3±0.5	EI		2593
HoF$_2^+$	HoF$_3$	F	14.5±0.5	EI	2402,	2600

4.3. The Positive Ion Table—Continued

Ion	Reactant	Other products	Ionization or appearance potential (eV)	Method	Heat of formation (kJ mol^{-1})	Ref.
			$C_{10}H_{10}Ho^+$			
$C_{10}H_{10}Ho^+$	$(C_5H_5)_3Ho$ (Tris(cyclopentadienyl)holmium)		8.92 ± 0.3	EI		2640
			$C_{15}H_{15}Ho^+$			
$C_{15}H_{15}Ho^+$	$(C_5H_5)_3Ho$ (Tris(cyclopentadienyl)holmium)		7.46 ± 0.1	EI		2640
		Er^+ $\Delta H^\circ_{f0} = 907$ kJ mol^{-1} (217 kcal mol^{-1})				
Er^+	Er		6.10 ± 0.02	S	907	2113, 2667
Er^+	Er		6.11 ± 0.10	EI		2588
Er^+	Er		5.95 ± 0.03	SI		2495
Er^+	Er		6.08 ± 0.03	SI		2882
Er^+	Er		6.23 ± 0.09	SI		2781
Er^+	Er		6.36 ± 0.10	SI		2982

See also – S: 1286
      EI: 2590, 2600, 2754, 3384

Ion	Reactant	Other products	Ionization or appearance potential (eV)	Method	Heat of formation (kJ mol^{-1})	Ref.
Er^+	ErF_3		26.5 ± 0.3	EI		2402
Er^+	ErF_3		27.5 ± 0.7	EI		2600
		Er^{+2} $\Delta H^\circ_{f0} \sim 2058$ kJ mol^{-1} (492 kcal mol^{-1})				
Er^{+2}	Er		18.03 ± 0.10	S	2058	2113
Er^{+2}	Er^+		11.93 ± 0.08	S	2113,	3265
			Er^{+3}			
Er^{+3}	Er		40.4	BH		3261
Er^{+3}	Er		40.5	BH		3259
Er^{+3}	Er		40.83	BH		3273
			ErC_2^+			
ErC_2^+	ErC_2		7.0 ± 0.5	EI		2754
			ErF^+			
ErF^+	ErF		6.3 ± 0.3	EI		2590, 2600
ErF^+	ErF_3		20.0 ± 0.7	EI		2600
ErF^+	ErF_3		20.5 ± 0.3	EI		2402

4.3. The Positive Ion Table—Continued

Ion	Reactant	Other products	Ionization or appearance potential (eV)	Method	Heat of formation (kJ mol^{-1})	Ref.	
			ErF$_2^+$				
ErF$_2^+$	ErF$_2$		7.0±0.3	EI	2590,	2600	
ErF$_2^+$	ErF$_3$	F	12.5±0.5	EI		2590	
ErF$_2^+$	ErF$_3$	F	13.8±0.3	EI		2402	
ErF$_2^+$	ErF$_3$	F	14.0±0.7	EI		2600	
			Tm$^+$ $\Delta H_{f0}^\circ = 831$ kJ mol^{-1} (199 kcal mol^{-1})				
Tm$^+$	Tm		6.18±0.02	S	830	2113,	2667
Tm$^+$	Tm		6.22±0.02	S	834		1448
Tm$^+$	Tm		6.180±0.008	PI	830		3258
Tm$^+$	Tm		5.87±0.10	EI			2588
Tm$^+$	Tm		6.03±0.04	SI			2495
Tm$^+$	Tm		6.15±0.02	SI			1165

See also — S: 1286, 2186
 EI: 2600

Ion	Reactant	Other products	Ionization or appearance potential (eV)	Method	Heat of formation (kJ mol^{-1})	Ref.	
Tm$^+$	TmF$_3$		26.5±0.5	EI		2598	
Tm$^+$	TmF$_3$		27.0±0.7	EI		2600	
			Tm^{+2} $\Delta H_{f0}^\circ \sim 1992$ kJ mol^{-1} (476 kcal mol^{-1})				
Tm^{+2}	Tm		18.23±0.10	S	1992		2113
Tm^{+2}	Tm$^+$		12.05±0.08	S		2113,	3265
			Tm^{+3} $\Delta H_{f0}^\circ \sim 4280$ kJ mol^{-1} (1023 kcal mol^{-1})				
Tm^{+3}	Tm		41.94±0.16	S	4280		2113
Tm^{+3}	Tm		41.7	BH			3261
Tm^{+3}	Tm		41.85	BH			3259
Tm^{+3}	Tm		42.04	BH			3273
Tm^{+3}	Tm^{+2}		23.71±0.06	S		2113,	3283
			TmF$^+$				
TmF$^+$	TmF$_3$		20.0±0.5	EI		2598	
TmF$^+$	TmF$_3$		20.8±0.7	EI		2600	
			TmF$_2^+$				
TmF$_2^+$	TmF$_3$	F	13.0±0.5	EI		2598	
TmF$_2^+$	TmF$_3$	F	13.5±0.7	EI		2600	
			C$_{15}$H$_{15}$Tm$^+$				
C$_{15}$H$_{15}$Tm$^+$	(C$_5$H$_5$)$_3$Tm (Tris(cyclopentadienyl)thulium)		7.43±0.1	EI		2640	

4.3. The Positive Ion Table—Continued

Ion	Reactant	Other products	Ionization or appearance potential (eV)	Method	Heat of formation (kJ mol^{-1})	Ref.

Yb^+ $\Delta H^\circ_{f0} = 756.1$ kJ mol^{-1} (180.7 kcal mol^{-1})

Ion	Reactant	Other products	Ionization or appearance potential (eV)	Method	Heat of formation (kJ mol^{-1})	Ref.
Yb^+	Yb		6.254	S	756.1	2113, 2953
Yb^+	Yb		6.25±0.01	S		2645
Yb^+	Yb		6.25±0.02	S		2667
Yb^+	Yb		5.90±0.10	EI		2588
Yb^+	Yb		6.04±0.04	SI		2495
Yb^+	Yb		6.69±0.08	SI		2781

See also – EI: 2600

Ion	Reactant	Other products	Ionization or appearance potential (eV)	Method	Heat of formation (kJ mol^{-1})	Ref.
Yb^+	YbF_3		26.0±0.5	EI		2598
Yb^+	YbF_3		27.0±0.7	EI		2600
Yb^+	$(C_5H_5)_2Yb$ (Bis(cyclopentadienyl)ytterbium)		15.6±0.3	EI		3269
Yb^+	$(C_5H_5)_3Yb$ (Tris(cyclopentadienyl)ytterbium)		17.8±0.4	EI		3269
Yb^+	$(C_5H_5)_3Yb$ (Tris(cyclopentadienyl)ytterbium)		19.0±0.5	EI		2640

(Misprint corrected, see ref. 3269)

Yb^{+2} $\Delta H^\circ_{f0} \sim 1927$ kJ mol^{-1} (460 kcal mol^{-1})

Ion	Reactant	Other products	Ionization or appearance potential (eV)	Method	Heat of formation (kJ mol^{-1})	Ref.
Yb^{+2}	Yb		18.42±0.08	S	1930	2113
Yb^{+2}	Yb		18.35±0.10	PI	1923	2958
Yb^{+2}	Yb^+		12.17±0.08	S	2113,	3265

Yb^{+3}

Ion	Reactant	Other products	Ionization or appearance potential (eV)	Method	Heat of formation (kJ mol^{-1})	Ref.
Yb^{+3}	Yb		43.6	S		2113
Yb^{+3}	Yb		43.39	BH		3273
Yb^{+3}	Yb		43.5	BH		3259
Yb^{+3}	Yb^{+2}		25.2	S		2113
Yb^{+3}	Yb^{+2}		25.0±0.7	S		3261

See also – S: 3266

YbF^+

Ion	Reactant	Other products	Ionization or appearance potential (eV)	Method	Heat of formation (kJ mol^{-1})	Ref.
YbF^+	YbF_3		20.0±0.5	EI		2598
YbF^+	YbF_3		20.5±0.7	EI		2600

YbF_2^+

Ion	Reactant	Other products	Ionization or appearance potential (eV)	Method	Heat of formation (kJ mol^{-1})	Ref.
YbF_2^+	YbF_3	F	13.5±0.5	EI		2598
YbF_2^+	YbF_3	F	14.0±0.7	EI		2600

4.3. The Positive Ion Table—Continued

Ion	Reactant	Other products	Ionization or appearance potential (eV)	Method	Heat of formation (kJ mol^{-1})	Ref.
			C$_5$H$_5$Yb$^+$			
C$_5$H$_5$Yb$^+$	(C$_5$H$_5$)$_2$Yb (Bis(cyclopentadienyl)ytterbium)		10.4±0.2	EI		3269
C$_5$H$_5$Yb$^+$	(C$_5$H$_5$)$_3$Yb (Tris(cyclopentadienyl)ytterbium)		12.0±0.4	EI		2640
C$_5$H$_5$Yb$^+$	(C$_5$H$_5$)$_3$Yb (Tris(cyclopentadienyl)ytterbium)		12.2±0.2	EI		3269
			C$_{10}$H$_{10}$Yb$^+$			
C$_{10}$H$_{10}$Yb$^+$	(C$_5$H$_5$)$_2$Yb (Bis(cyclopentadienyl)ytterbium)		7.62±0.09	EI		3269
C$_{10}$H$_{10}$Yb$^+$	(C$_5$H$_5$)$_3$Yb (Tris(cyclopentadienyl)ytterbium)		8.94±0.2	EI		2640
C$_{10}$H$_{10}$Yb$^+$	(C$_5$H$_5$)$_3$Yb (Tris(cyclopentadienyl)ytterbium)		10.1±0.2	EI		3269
			C$_{15}$H$_{15}$Yb$^+$			
C$_{15}$H$_{15}$Yb$^+$	(C$_5$H$_5$)$_3$Yb (Tris(cyclopentadienyl)ytterbium)		7.30±0.1	EI		2640
C$_{15}$H$_{15}$Yb$^+$	(C$_5$H$_5$)$_3$Yb (Tris(cyclopentadienyl)ytterbium)		7.72±0.09	EI		3269
			C$_{20}$H$_{24}$N$_2$Yb$_2^+$			
C$_{20}$H$_{24}$N$_2$Yb$_2^+$	((C$_5$H$_5$)$_2$YbNH$_2$)$_2$ (Bis(bis(cyclopentadienyl)ytterbium amide))		7.87±0.1	EI		2640
			C$_{10}$H$_{10}$ClYb$^+$			
C$_{10}$H$_{10}$ClYb$^+$	(C$_5$H$_5$)$_2$YbCl (Bis(cyclopentadienyl)ytterbium chloride)		8.65±0.1	EI		2640
Lu$^+$		**ΔH_{f0}° = 951.3 kJ mol^{-1} (227.4 kcal mol^{-1})**				
Lu$^+$	Lu		5.426	S	951.3	2113
Lu$^+$	Lu		5.32±0.05	SI		2495
Lu$^+$	Lu		5.41±0.02	SI		1165

See also – EI: 2600

Ion	Reactant	Other products	Ionization or appearance potential (eV)	Method	Heat of formation (kJ mol^{-1})	Ref.
Lu$^+$	LuF$_3$		27.0±0.5	EI		2598
Lu$^+$	LuF$_3$		28.0±0.7	EI		2600
Lu$^+$	LuCl$_3$		20.5±0.5	EI		2435

4.3. The Positive Ion Table—Continued

Ion	Reactant	Other products	Ionization or appearance potential (eV)	Method	Heat of formation (kJ mol^{-1})	Ref.
			Lu^{+2}			
Lu^{+2}	Lu		19.3±0.4	S		2113
Lu^{+2}	Lu$^+$		13.9±0.4	S		2113, 3265
			Lu^{+3}			
Lu^{+3}	Lu		40.4	BH		3259
Lu^{+3}	Lu		40.51	BH		3273
			LuF$^+$			
LuF$^+$	LuF$_3$		20.5±0.5	EI		2598
LuF$^+$	LuF$_3$		21.0±0.7	EI		2600
			LuF$_2^+$			
LuF$_2^+$	LuF$_3$	F	14.0±0.5	EI		2598
LuF$_2^+$	LuF$_3$	F	14.3±0.7	EI		2600
			LuCl$^+$			
LuCl$^+$	LuCl$_3$		16.0±0.5	EI		2435
			LuCl$_2^+$			
LuCl$_2^+$	LuCl$_3$	Cl	12.5±0.3	EI		2435
			LuCl$_3^+$			
LuCl$_3^+$	LuCl$_3$		11.5±0.5	EI		2435
			Lu$_2$Cl$_5^+$			
Lu$_2$Cl$_5^+$	Lu$_2$Cl$_6$	Cl	12.5±0.5	EI		2435
			C$_5$H$_5$Lu$^+$			
C$_5$H$_5$Lu$^+$	(C$_5$H$_5$)$_3$Lu (Tris(cyclopentadienyl)lutetium)		18.3±0.2	EI		2640
			C$_{10}$H$_{10}$Lu$^+$			
C$_{10}$H$_{10}$Lu$^+$	(C$_5$H$_5$)$_3$Lu (Tris(cyclopentadienyl)lutetium)		8.94±0.2	EI		2640

4.3. The Positive Ion Table—Continued

Ion	Reactant	Other products	Ionization or appearance potential (eV)	Method	Heat of formation (kJ mol⁻¹)	Ref.
			$C_{15}H_{15}Lu^+$			
$C_{15}H_{15}Lu^+$	$(C_5H_5)_3Lu$ (Tris(cyclopentadienyl)lutetium)		7.36±0.1	EI		2640
			Ta^+ $\Delta H_{f0}^\circ = 1543$ kJ mol⁻¹ (369 kcal mol⁻¹)			
Ta^+	Ta		7.89	S	1543	2113
Ta^+	TaF_5		39.5±1.0	EI		3023
Ta^+	$TaCl_5$		29.5±0.4	EI		2506
			TaO^+			
TaO^+	TaO		6±0.5	EI		2050
			TaO_2^+			
TaO_2^+	TaO_2		9±0.5	EI		2050
			TaF^+			
TaF^+	TaF_5		36.5±1.0	EI		3023
			TaF_2^+			
TaF_2^+	TaF_5		27.3±0.8	EI		3023

See also – EI: 2859

Ion	Reactant	Other products	Ionization or appearance potential (eV)	Method	Heat of formation (kJ mol⁻¹)	Ref.
			TaF_3^+			
TaF_3^+	TaF_5		22.5±0.5	EI		3023

See also – EI: 2859

Ion	Reactant	Other products	Ionization or appearance potential (eV)	Method	Heat of formation (kJ mol⁻¹)	Ref.
			TaF_4^+			
TaF_4^+	TaF_5	F	14.8±0.3	EI		3023
TaF_4^+	TaF_5	F	15.0±0.3	EI		2859
			$TaCl^+$			
$TaCl^+$	$TaCl_5$		25.0±0.5	EI		2506
			$TaCl_2^+$			
$TaCl_2^+$	$TaCl_5$		19.6±0.6	EI		2506

4.3. The Positive Ion Table—Continued

Ion	Reactant	Other products	Ionization or appearance potential (eV)	Method	Heat of formation (kJ mol^{-1})	Ref.
		TaCl$_3^+$				
TaCl$_3^+$	TaCl$_5$		15.0±0.4	EI		2506
		TaCl$_4^+$				
TaCl$_4^+$	TaCl$_5$	Cl	12.0±0.28	EI		2506
		TaOF$^+$				
TaOF$^+$	TaOF$_3$		20.5±0.5	EI		3023
TaOF$^+$	TaOF$_3$		21±1	EI		2859
		TaOF$_2^+$				
TaOF$_2^+$	TaOF$_3$	F	14.8±0.3	EI		3023
TaOF$_2^+$	TaOF$_3$	F	15.0±0.5	EI		2859
		TaOF$_3^+$				
TaOF$_3^+$	TaOF$_3$		12.5±0.5	EI		2859
TaOF$_3^+$	TaOF$_3$		12.7±0.3	EI		3023
W$^+$	ΔH_{i0}° = 1618 kJ mol^{-1} (387 kcal mol^{-1})					
W$^+$	W		7.98	S	1618	2113
W$^+$	W(CO)$_6$	6CO	20.6±0.2	EI		2023
W$^+$	W(CO)$_6$	6CO	21.7±0.3	EI		2403
W$^+$	W(CO)$_6$	6CO	22.25	EI		2500
W$^+$	W(CO)$_6$	6CO	22.9±0.6	EI		1107
W$^+$	WO$_2$F$_2$?		37.4±1.0	EI		2859
		WC$^+$				
WC$^+$	W(CO)$_6$		28.8±0.5	EI		1107
		WO$^+$				
WO$^+$	WO		9.1±1	EI		2126

4.3. The Positive Ion Table—Continued

Ion	Reactant	Other products	Ionization or appearance potential (eV)	Method	Heat of formation (kJ mol^{-1})	Ref.
			WO$_2^+$			
WO$_2^+$	WO$_2$		9.8	EI		1244, 2123
WO$_2^+$	WO$_2$		9.9±0.6	EI		2126
WO$_2^+$	WO$_2$		9.9±1.0	EI		3257
WO$_2^+$	WO$_3$?		14.2	EI		2131
WO$_2^+$	WO$_2$F$_2$		21.0±0.5	EI		2859
			WO$_3^+$			
WO$_3^+$	WO$_3$		11.7±0.6	EI		2126
WO$_3^+$	WO$_3$		11.9±0.5	EI		3257
WO$_3^+$	WO$_3$		11.9	EI		1244, 2123
WO$_3^+$	WO$_3$		12.0±0.5	EI		3256
WO$_3^+$	WO$_3$		12.1	EI		2131
			W$_2$O$_5^+$			
W$_2$O$_5^+$	W$_2$O$_6$?		15.8	EI		2131
			W$_2$O$_6^+$			
W$_2$O$_6^+$	W$_2$O$_6$		13.4	EI		2131
			W$_3$O$_8^+$			
W$_3$O$_8^+$	W$_3$O$_9$	O	15.5	EI		2131
			W$_3$O$_9^+$			
W$_3$O$_9^+$	W$_3$O$_9$		13.3	EI		2131
			WF$^+$			
WF$^+$	WO$_2$F$_2$?		32.1±0.8	EI		2859
			WF$_2^+$			
WF$_2^+$	WO$_2$F$_2$?		27.6±0.5	EI		2859
			WF$_3^+$			
WF$_3^+$	WOF$_4$		22.8±1.0	EI		2859

4.3. The Positive Ion Table—Continued

Ion	Reactant	Other products	Ionization or appearance potential (eV)	Method	Heat of formation (kJ mol^{-1})	Ref.
		WCl$_3^+$				
WCl$_3^+$	WCl$_4$?		11.8	EI		2552
		WCl$_4^+$				
WCl$_4^+$	WCl$_4$		8.0	EI		2552
		WCl$_5^+$				
WCl$_5^+$	WCl$_5$		9.1	EI		2552
		W$_2$Cl$_5^+$				
W$_2$Cl$_5^+$	W$_2$Cl$_6$?		11.9	EI		2552
		W$_2$Cl$_6^+$				
W$_2$Cl$_6^+$	W$_2$Cl$_6$		9.5	EI		2552
		W$_3$Cl$_9^+$				
W$_3$Cl$_9^+$	W$_3$Cl$_9$		9.3	EI		2552
		C$_{10}$H$_{12}$W$^+$				
C$_{10}$H$_{12}$W$^+$	(C$_5$H$_5$)$_2$WH$_2$ (Bis(cyclopentadienyl)tungsten dihydride)		6.49±0.1	EI		2683
		Li$_2$WO$_4^+$				
Li$_2$WO$_4^+$	Li$_2$WO$_4$		9.2±0.5	EI		3257
		COW$^+$				
COW$^+$	W(CO)$_6$	5CO	18.5±0.16	EI		2023
COW$^+$	W(CO)$_6$	5CO	18.51	EI		2500
COW$^+$	W(CO)$_6$	5CO	18.7±0.3	EI		2403
COW$^+$	W(CO)$_6$	5CO	20.2±0.3	EI		1107
		COW^{+2}				
COW^{+2}	W(CO)$_6$		31.7±1	EI		2403

4.3. The Positive Ion Table—Continued

Ion	Reactant	Other products	Ionization or appearance potential (eV)	Method	Heat of formation (kJ mol^{-1})	Ref.
			C_2OW^+			
C_2OW^+	$W(CO)_6$		25.9 ± 0.6	EI		1107
			$C_2O_2W^+$			
$C_2O_2W^+$	$W(CO)_6$	4CO	15.8 ± 0.3	EI		2403
$C_2O_2W^+$	$W(CO)_6$	4CO	16.07 ± 0.04	EI		2023
$C_2O_2W^+$	$W(CO)_6$	4CO	16.08	EI		2500
$C_2O_2W^+$	$W(CO)_6$	4CO	17.6 ± 0.2	EI		1107
			$C_2O_2W^{+2}$			
$C_2O_2W^{+2}$	$W(CO)_6$		35.0 ± 1	EI		2403
			$C_3O_3W^+$			
$C_3O_3W^+$	$W(CO)_6$	3CO	13.60 ± 0.02	EI		2023
$C_3O_3W^+$	$W(CO)_6$	3CO	13.70 ± 0.15	EI		2403
$C_3O_3W^+$	$W(CO)_6$	3CO	13.87	EI		2500
$C_3O_3W^+$	$W(CO)_6$	3CO	14.9 ± 0.2	EI		1107
			$C_4O_4W^+$			
$C_4O_4W^+$	$W(CO)_6$	2CO	11.82 ± 0.02	EI		2023
$C_4O_4W^+$	$W(CO)_6$	2CO	11.93 ± 0.15	EI		2403
$C_4O_4W^+$	$W(CO)_6$	2CO	12.05	EI		2500
$C_4O_4W^+$	$W(CO)_6$	2CO	12.7 ± 0.2	EI		1107
			$C_5O_5W^+$			
$C_5O_5W^+$	$W(CO)_6$	CO	~9.21	EI		2500
$C_5O_5W^+$	$W(CO)_6$	CO	9.80 ± 0.17	EI		1107
$C_5O_5W^+$	$W(CO)_6$	CO	9.86 ± 0.1	EI		2403
$C_5O_5W^+$	$W(CO)_6$	CO	9.97 ± 0.04	EI		2023

See also – PI: 2886

$W(CO)_6^+$ $\quad\Delta H^\circ_{1298} = -76 \text{ kJ mol}^{-1} (-18 \text{ kcal mol}^{-1})$

Ion	Reactant	Other products	Ionization or appearance potential (eV)	Method	Heat of formation (kJ mol^{-1})	Ref.
$C_6O_6W^+$	$W(CO)_6$		8.242 ± 0.006	PI	-76	2886
(Threshold value approximately corrected for hot bands)						
$C_6O_6W^+$	$W(CO)_6$		8.18 ± 0.03	PI		1167
$C_6O_6W^+$	$W(CO)_6$		8.46 ± 0.02	EI		2023
$C_6O_6W^+$	$W(CO)_6$		8.47 ± 0.1	EI		2403
$C_6O_6W^+$	$W(CO)_6$		8.48 ± 0.05	EI		2500
$C_6O_6W^+$	$W(CO)_6$		8.56 ± 0.13	EI		1107

4.3. The Positive Ion Table—Continued

Ion	Reactant	Other products	Ionization or appearance potential (eV)	Method	Heat of formation (kJ mol^{-1})	Ref.
			WOF$^+$			
WOF$^+$	WO$_2$F$_2$?		23.6±0.5	EI		2859
			WO$_2$F$^+$			
WO$_2$F$^+$	WO$_2$F$_2$	F	16.3±0.5	EI		2859
			WOF$_2^+$			
WOF$_2^+$	WO$_2$F$_2$?		18.4±0.5	EI		2859
			WO$_2$F$_2^+$			
WO$_2$F$_2^+$	WO$_2$F$_2$		13.0±0.3	EI		2859
			WOF$_3^+$			
WOF$_3^+$	WOF$_4$	F	14.6±0.5	EI		2859
			CaWO$_3^+$			
CaWO$_3^+$	CaWO$_3$		6.7	EI		1244
			CaWO$_4^+$			
CaWO$_4^+$	CaWO$_4$		9.8	EI		1244
			SrWO$_3^+$			
SrWO$_3^+$	SrWO$_3$		6.4	EI		1244
			SrWO$_4^+$			
SrWO$_4^+$	SrWO$_4$		9.4	EI		1244
			SnWO$_4^+$			
SnWO$_4^+$	SnWO$_4$		10.8	EI		1244
			Sn$_2$WO$_5^+$			
Sn$_2$WO$_5^+$	Sn$_2$WO$_5$		8.4	EI		1244

4.3. The Positive Ion Table—Continued

Ion	Reactant	Other products	Ionization or appearance potential (eV)	Method	Heat of formation (kJ mol^{-1})	Ref.
			WO$_2$I$^+$			
WO$_2$I$^+$	WO$_2$I$_2$	I	14.4	EI		2554
			WO$_2$I$_2^+$			
WO$_2$I$_2^+$	WO$_2$I$_2$		13.4	EI		2554
			BaWO$_3^+$			
BaWO$_3^+$	BaWO$_3$		8.5±0.5	EI		3256
BaWO$_3^+$	BaWO$_4$	O	14.0±0.5	EI		3256
			BaWO$_4^+$			
BaWO$_4^+$	BaWO$_4$		9.8±0.5	EI		3256
			C$_{10}$H$_5$NO$_5$W$^+$			
C$_{10}$H$_5$NO$_5$W$^+$	C$_5$H$_5$NW(CO)$_5$ (Pyridinetungsten pentacarbonyl)		7.6±0.2	EI		2481
			C$_{10}$H$_9$NO$_5$W$^+$			
C$_{10}$H$_9$NO$_5$W$^+$	n-C$_4$H$_9$NCW(CO)$_5$		7.60±0.05	EI		2481
			C$_{12}$H$_5$NO$_5$W$^+$			
C$_{12}$H$_5$NO$_5$W$^+$	C$_6$H$_5$NCW(CO)$_5$ (Isocyanobenzenetungsten pentacarbonyl)		8.03	EI		2544
			C$_{12}$H$_{11}$NO$_5$W$^+$			
C$_{12}$H$_{11}$NO$_5$W$^+$	C$_6$H$_{11}$NCW(CO)$_5$ (Isocyanocyclohexanetungsten pentacarbonyl)		7.75	EI		2544
			C$_{11}$H$_{15}$O$_5$PW$^+$			
C$_{11}$H$_{15}$O$_5$PW$^+$	(C$_2$H$_5$)$_3$PW(CO)$_5$		7.82±0.05	EI		2481
			C$_{17}$H$_{27}$O$_5$PW$^+$			
C$_{17}$H$_{27}$O$_5$PW$^+$	(n-C$_4$H$_9$)$_3$PW(CO)$_5$		7.63±0.05	EI		2481

4.3. The Positive Ion Table—Continued

Ion	Reactant	Other products	Ionization or appearance potential (eV)	Method	Heat of formation (kJ mol^{-1})	Ref.
		C$_8$H$_9$O$_8$PW$^+$				
C$_8$H$_9$O$_8$PW$^+$	(CH$_3$O)$_3$PW(CO)$_5$		7.96	EI		2544
		C$_{11}$H$_{15}$O$_8$PW$^+$				
C$_{11}$H$_{15}$O$_8$PW$^+$	(C$_2$H$_5$O)$_3$PW(CO)$_5$		7.8	EI		2544
		C$_{17}$H$_{27}$O$_8$PW$^+$				
C$_{17}$H$_{27}$O$_8$PW$^+$	(n–C$_4$H$_9$O)$_3$PW(CO)$_5$		7.85±0.05	EI		2481
		C$_5$O$_5$PCl$_3$W$^+$				
C$_5$O$_5$PCl$_3$W$^+$	PCl$_3$W(CO)$_5$		8.50±0.05	EI		2481
		C$_{11}$H$_5$O$_5$PCl$_2$W$^+$				
C$_{11}$H$_5$O$_5$PCl$_2$W$^+$	C$_6$H$_5$PCl$_2$W(CO)$_5$ (Dichloro(phenyl)phosphinetungsten pentacarbonyl)		8.20±0.05	EI		2481
	Re$^+$ $\quad$ $\Delta H^\circ_{f0} = 1529$ kJ mol^{-1} (366 kcal mol^{-1})					
Re$^+$	Re		7.88	S	1529	2113
Re$^+$	Re$_2$(CO)$_{10}$		37.55	EI		2563
Re$^+$	Re(CO)$_5$Cl		23.1	EI		2501
Re$^+$	ReMn(CO)$_{10}$		30.93	EI		2563
Re$^+$	Re(CO)$_5$Br		>23	EI		2501
Re$^+$	Re(CO)$_5$I		>23	EI		2501
		Re$_2^+$				
Re$_2^+$	Re$_2$(CO)$_{10}$	10CO	28.96	EI		2563
		ReO$_2^+$				
ReO$_2^+$	Re$_2$O$_7$?		~20	EI		2461
		ReO$_3^+$				
ReO$_3^+$	Re$_2$O$_7$?		15.9±0.5	EI		2461
		Re$_2$O$_5^+$				
Re$_2$O$_5^+$	Re$_2$O$_7$		~19	EI		2461

4.3. The Positive Ion Table—Continued

Ion	Reactant	Other products	Ionization or appearance potential (eV)	Method	Heat of formation (kJ mol^{-1})	Ref.
		Re$_2$O$_6^+$				
Re$_2$O$_6^+$	Re$_2$O$_7$	O	17.2±0.5	EI		2461
		Re$_2$O$_7^+$				
Re$_2$O$_7^+$	Re$_2$O$_7$		13.0±0.5	EI		2461
		ReCl$^+$				
ReCl$^+$	Re(CO)$_5$Cl	5CO	19.27	EI		2501
		ReCl$_3^+$				
ReCl$_3^+$	Re$_3$Cl$_9$		16±0.5	EI		2140
		ReCl$_4^+$				
ReCl$_4^+$	Re$_3$Cl$_9$		16±0.5	EI		2140
		Re$_2$Cl$_5^+$				
Re$_2$Cl$_5^+$	Re$_3$Cl$_9$		13.5±0.5	EI		2140
		Re$_2$Cl$_6^+$				
Re$_2$Cl$_6^+$	Re$_3$Cl$_9$		13.5±0.5	EI		2140
		Re$_3$Cl$_8^+$				
Re$_3$Cl$_8^+$	Re$_3$Cl$_9$	Cl	13±0.5	EI		2140
		Re$_3$Cl$_9^+$				
Re$_3$Cl$_9^+$	Re$_3$Cl$_9$		10.5±0.5	EI		2140
		ReMn$^+$				
ReMn$^+$	ReMn(CO)$_{10}$	10CO	25.98	EI		2563
		ReBr$^+$				
ReBr$^+$	Re(CO)$_5$Br	5CO	19.51	EI		2501

4.3. The Positive Ion Table—Continued

Ion	Reactant	Other products	Ionization or appearance potential (eV)	Method	Heat of formation (kJ mol⁻¹)	Ref.

$Re_3Br_8^+$

Ion	Reactant	Other products	IP/AP	Method		Ref.
$Re_3Br_8^+$	Re_3Br_9	Br	12.5±1	EI		3159
$Re_3Br_8^+$	Re_3Br_9	Br	12.5	EI		2140

$Re_3Br_9^+$

Ion	Reactant	Other products	IP/AP	Method		Ref.
$Re_3Br_9^+$	Re_3Br_9		10	EI		2140
$Re_3Br_9^+$	Re_3Br_9		11.0±1	EI		3159

ReI^+

Ion	Reactant	Other products	IP/AP	Method		Ref.
ReI^+	$Re(CO)_5I$	5CO	19.20	EI		2501

$C_{10}H_{10}Re^+$

Ion	Reactant	Other products	IP/AP	Method		Ref.
$C_{10}H_{10}Re^+$	$(C_5H_5)_2ReH$ (Bis(cyclopentadienyl)rhenium hydride)	H	7.85±0.1	EI		2683

$C_{10}H_{11}Re^+$

Ion	Reactant	Other products	IP/AP	Method		Ref.
$C_{10}H_{11}Re^+$	$(C_5H_5)_2ReH$ (Bis(cyclopentadienyl)rhenium hydride)		6.76±0.05	EI		2683

$C_3O_3Re^+$

Ion	Reactant	Other products	IP/AP	Method		Ref.
$C_3O_3Re^+$	$Re(CO)_5Br$		15.51	EI		2501
$C_3O_3Re^+$	$Re(CO)_5I$		14.71	EI		2501

$C_4O_4Re^+$

Ion	Reactant	Other products	IP/AP	Method		Ref.
$C_4O_4Re^+$	$Re_2(CO)_{10}$		13.30	EI		2563

$C_5O_5Re^+$

Ion	Reactant	Other products	IP/AP	Method		Ref.
$C_5O_5Re^+$	$Re_2(CO)_{10}$		10.34	EI		3234

See also – EI: 2563

Ion	Reactant	Other products	IP/AP	Method		Ref.
$C_5O_5Re^+$	$ReMn(CO)_{10}$		10.80	EI		2563, 3234

$C_6O_6Re^+$

Ion	Reactant	Other products	IP/AP	Method		Ref.
$C_6O_6Re^+$	$ReMn(CO)_{10}$		9.36	EI		2563

4.3. The Positive Ion Table—Continued

Ion	Reactant	Other products	Ionization or appearance potential (eV)	Method	Heat of formation (kJ mol⁻¹)	Ref.

$$CORe_2^+$$

Ion	Reactant	Other products	Ionization or appearance potential (eV)	Method	Heat of formation (kJ mol⁻¹)	Ref.
$CORe_2^+$	$Re_2(CO)_{10}$	9CO	26.26	EI		2563

$$C_2O_2Re_2^+$$

| $C_2O_2Re_2^+$ | $Re_2(CO)_{10}$ | 8CO | 23.55 | EI | | 2563 |

$$C_3O_3Re_2^+$$

| $C_3O_3Re_2^+$ | $Re_2(CO)_{10}$ | 7CO | 21.46 | EI | | 2563 |

$$C_4O_4Re_2^+$$

| $C_4O_4Re_2^+$ | $Re_2(CO)_{10}$ | 6CO | 19.31 | EI | | 2563 |

$$C_5O_5Re_2^+$$

| $C_5O_5Re_2^+$ | $Re_2(CO)_{10}$ | 5CO | 16.71 | EI | | 2563 |

$$C_6O_6Re_2^+$$

| $C_6O_6Re_2^+$ | $Re_2(CO)_{10}$ | 4CO | 15.01 | EI | | 2563 |

$$C_7O_7Re_2^+$$

| $C_7O_7Re_2^+$ | $Re_2(CO)_{10}$ | 3CO | 13.55 | EI | | 2563 |

$$C_8O_8Re_2^+$$

| $C_8O_8Re_2^+$ | $Re_2(CO)_{10}$ | 2CO | 10.89 | EI | | 2563 |

$$C_9O_9Re_2^+$$

| $C_9O_9Re_2^+$ | $Re_2(CO)_{10}$ | CO | 9.57 | EI | | 2563 |

$$C_{10}O_{10}Re_2^+$$

| $C_{10}O_{10}Re_2^+$ | $Re_2(CO)_{10}$ | | 8.36±0.03 | EI | | 3234 |

See also – EI: 2563

Ion	Reactant	Other products	Ionization or appearance potential (eV)	Method	Heat of formation (kJ mol⁻¹)	Ref.
		NaReO₄⁺				
NaReO₄⁺	NaReO₄		~9.5	EI		2573
		Na₂ReO₄⁺				
Na₂ReO₄⁺	(NaReO₄)₂		~10.5	EI		2573
		KReO₄⁺				
KReO₄⁺	KReO₄		~9.5	EI		2573
		K₂ReO₄⁺				
K₂ReO₄⁺	(KReO₄)₂		~10.5	EI		2573
		COClRe⁺				
COClRe⁺	Re(CO)₅Cl	4CO	16.84	EI		2501
		C₂O₂ClRe⁺				
C₂O₂ClRe⁺	Re(CO)₅Cl	3CO	14.85	EI		2501
		C₃O₃ClRe⁺				
C₃O₃ClRe⁺	Re(CO)₅Cl	2CO	11.92	EI		2501
		C₄O₄ClRe⁺				
C₄O₄ClRe⁺	Re(CO)₅Cl	CO	10.45	EI		2501
		C₅O₅ClRe⁺				
C₅O₅ClRe⁺	Re(CO)₅Cl		9.18±0.03	EI		2501
		COMnRe⁺				
COMnRe⁺	ReMn(CO)₁₀	9CO	23.00	EI		2563
		C₂O₂MnRe⁺				
C₂O₂MnRe⁺	ReMn(CO)₁₀	8CO	19.75	EI		2563

4.3. The Positive Ion Table—Continued

Ion	Reactant	Other products	Ionization or appearance potential (eV)	Method	Heat of formation (kJ mol^{-1})	Ref.
		$C_3O_3MnRe^+$				
$C_3O_3MnRe^+$	$ReMn(CO)_{10}$	7CO	16.94	EI		2563
		$C_4O_4MnRe^+$				
$C_4O_4MnRe^+$	$ReMn(CO)_{10}$	6CO	14.65	EI		2563
		$C_5O_5MnRe^+$				
$C_5O_5MnRe^+$	$ReMn(CO)_{10}$	5CO	12.12	EI		2563
		$C_{10}O_{10}MnRe^+$				
$C_{10}O_{10}MnRe^+$	$ReMn(CO)_{10}$		8.14±0.01	EI		3234

See also — EI: 2563

Ion	Reactant	Other products	Ionization or appearance potential (eV)	Method	Heat of formation (kJ mol^{-1})	Ref.
		$COBrRe^+$				
$COBrRe^+$	$Re(CO)_5Br$	4CO	16.94	EI		2501
		$C_2O_2BrRe^+$				
$C_2O_2BrRe^+$	$Re(CO)_5Br$	3CO	15.02	EI		2501
		$C_3O_3BrRe^+$				
$C_3O_3BrRe^+$	$Re(CO)_5Br$	2CO	11.97	EI		2501
		$C_4O_4BrRe^+$				
$C_4O_4BrRe^+$	$Re(CO)_5Br$	CO	10.50	EI		2501
		$C_5O_5BrRe^+$				
$C_5O_5BrRe^+$	$Re(CO)_5Br$		9.07±0.02	EI		2501
		$COIRe^+$				
$COIRe^+$	$Re(CO)_5I$	4CO	16.69	EI		2501

4.3. The Positive Ion Table—Continued

Ion	Reactant	Other products	Ionization or appearance potential (eV)	Method	Heat of formation (kJ mol^{-1})	Ref.
			$C_2O_2IRe^+$			
$C_2O_2IRe^+$	$Re(CO)_5I$	3CO	14.65	EI		2501
			$C_3O_3IRe^+$			
$C_3O_3IRe^+$	$Re(CO)_5I$	2CO	12.04	EI		2501
			$C_4O_4IRe^+$			
$C_4O_4IRe^+$	$Re(CO)_5I$	CO	10.29	EI		2501
			$C_5O_5IRe^+$			
$C_5O_5IRe^+$	$Re(CO)_5I$		8.64±0.03	EI		2501
			Os^+			
Os^+	Os		8.7	S	~1630*	2113
Os^+	OsO_4		26.8±0.5	EI		1284
*ΔH°_{f298}						
			OsO^+			
OsO^+	OsO_4		21.2±0.2	EI		1284
			OsO_2^+			
OsO_2^+	OsO_4		17.1±0.2	EI		1284
			OsO_3^+			
OsO_3^+	OsO_3		12.3±1	EI		2127
OsO_3^+	OsO_4	O	17.00±0.10	EI		1284
			OsO_4^+			
OsO_4^+	OsO_4		12.97±0.12	EI		1284
OsO_4^+	OsO_4		12.6±1	EI		2127
			$C_{10}H_{10}Os^+$			
$C_{10}H_{10}Os^+$	$(C_5H_5)_2Os$ (Bis(cyclopentadienyl)osmium)		7.59±0.1	EI		2683

4.3. The Positive Ion Table—Continued

Ion	Reactant	Other products	Ionization or appearance potential (eV)	Method	Heat of formation (kJ mol^{-1})	Ref.
			Ir$^+$			
Ir$^+$	Ir		9.1±0.1	S	~1542	2113
See also − EI: 1124, 2526						
Ir$^+$	IrO$_2$		13.1	EI		1124
			IrC$^+$			
IrC$^+$	IrC		9.5±1	EI		2526
			IrO$^+$			
IrO$^+$	IrO		10.1	EI		1124
IrO$^+$	IrO$_2$	O	15.1	EI		1124
			IrO$_2^+$			
IrO$_2^+$	IrO$_2$		10.9	EI		1124
			IrO$_3^+$			
IrO$_3^+$	IrO$_3$		11.9	EI		1124
			Pt$^+$			
Pt$^+$	Pt		9.0	S	~1433	2113
See also − EI: 2526, 2536						
			PtB$^+$			
PtB$^+$	PtB		10.0±1	EI		2526
			PtO$^+$			
PtO$^+$	PtO		10.1±0.3	EI		2536
PtO$^+$	PtO$_2$	O	14.8±0.3	EI		2536
			PtO$_2^+$			
PtO$_2^+$	PtO$_2$		11.2±0.3	EI		2536

4.3. The Positive Ion Table—Continued

Ion	Reactant	Other products	Ionization or appearance potential (eV)	Method	Heat of formation (kJ mol^{-1})	Ref.
		PtSi$^+$				
PtSi$^+$	PtSi		7.9±0.5	EI		2943
		PtF$_{12}$P$_4^+$				
PtF$_{12}$P$_4^+$	Pt(PF$_3$)$_4$		9.83 (V)	PE		3088
PtF$_{12}$P$_4^+$	Pt(PF$_3$)$_4$		9.8 (V)	PE		3070
	Au$^+$ $\Delta H_{f0}^\circ = 1256.0$ kJ mol^{-1} (300.2 kcal mol^{-1})					
Au$^+$	Au		9.225	S	1256.0	2113
		Au$_2^+$				
Au$_2^+$	Au$_2$		9.2±0.4	EI		2707
Au$_2^+$	Au$_2$		9.7±0.3	EI		2779
		AuSi$^+$				
AuSi$^+$	AuSi		7.7±0.4	EI		2707
AuSi$^+$	AuSi		8.9±0.5	EI		2943
		AuLa$^+$				
AuLa$^+$	AuLa		5.9±0.5	EI		2779
		AuCe$^+$				
AuCe$^+$	AuCe		6.0±0.3	EI		2779
		AuPr$^+$				
AuPr$^+$	AuPr		5.4±0.8	EI		2779
		AuNd$^+$				
AuNd$^+$	AuNd		5.8±0.8	EI		2779

4.3. The Positive Ion Table—Continued

Ion	Reactant	Other products	Ionization or appearance potential (eV)	Method	Heat of formation (kJ mol^{-1})	Ref.
$Hg^+(^2S_{1/2})$		$\Delta H_{f0}^\circ = 1071.5$ kJ mol^{-1} (256.1 kcal mol^{-1})				
$Hg^+(^2D_{5/2})$		$\Delta H_{f0}^\circ = 1496.3$ kJ mol^{-1} (357.6 kcal mol^{-1})				
$Hg^+(^2D_{3/2})$		$\Delta H_{f0}^\circ = 1676.2$ kJ mol^{-1} (400.6 kcal mol^{-1})				
$Hg^+(^2S_{1/2})$	Hg		10.437	S	1071.5	2113
$Hg^+(^2S_{1/2})$	Hg		10.443±0.009	PE		2814, 2853, 3109
$Hg^+(^2D_{5/2})$	Hg		14.8396±0.0004	S	1496.3	3377
$Hg^+(^2D_{3/2})$	Hg		16.7043±0.0004	S	1676.2	3377
$Hg^+(^2D_{3/2})$	Hg		16.715±0.004	PE		2814, 2853, 3109

The $^2D_{5/2}$ level was also observed in the PE study, but used for calibration. A fourth ionization
potential at 20.725 eV is given in refs. 2814, 2853, 3109 but is spurious, see P. Mitchell and M. Wilson,
Chem. Phys. Letters **3**, 389 (1969) and V. Fuchs and H. Hotop, Chem. Phys. Letters **4**, 71 (1969).

See also – PEN: 2468

Ion	Reactant	Other products	Ionization or appearance potential (eV)	Method	Heat of formation	Ref.
Hg^+	$(CH_3)_2Hg$	$2CH_3$	13.05±0.02	PI		2983

Hg^{+2}		$\Delta H_{f0}^\circ = 2881.2$ kJ mol^{-1} (688.6 kcal mol^{-1})				
Hg^{+2}	Hg		29.193	S	2881.2	2113
Hg^{+2}	Hg		29.0±0.2	NRE		211
Hg^{+2}	Hg		29.8	EI		211
Hg^{+2}	Hg^+		18.756	S		2113

Hg^{+3}						
Hg^{+3}	Hg		63.4	S	~6182	2113
Hg^{+3}	Hg		63.5±0.5	NRE		211
Hg^{+3}	Hg		68.5	EI		211
Hg^{+3}	Hg^{+2}		34.2	S		2113

Hg^{+4}						
Hg^{+4}	Hg		113±1	NRE		211
Hg^{+4}	Hg		122.5	EI		211
Hg^{+4}	Hg^{+3}		46.0±0.2	SEQ		2730

4.3. The Positive Ion Table—Continued

Ion	Reactant	Other products	Ionization or appearance potential (eV)	Method	Heat of forma- tion (kJ mol^{-1})	Ref.
Hg^{+5}						
Hg^{+5}	Hg		158±2	NRE		211
Hg^{+5}	Hg		175	EI		211
Hg^{+5}	Hg^{+4}		61.2±0.2	SEQ		2730
Hg^{+6}						
Hg^{+6}	Hg		225±10	NRE		211
Hg^{+6}	Hg		300	EI		211
Hg^{+6}	Hg^{+5}		76.6±0.5	SEQ		2730
Hg^{+7}						
Hg^{+7}	Hg^{+6}		93±2	SEQ		2730
Hg^{+8}						
Hg^{+8}	Hg^{+7}		122±2	SEQ		2730
Hg^{+9}						
Hg^{+9}	Hg^{+8}		141±2	SEQ		2730
Hg$_2^+$						
Hg$_2^+$	Hg$_2$		9.52±0.15	EI		3297
Hg$_3^+$						
Hg$_3^+$	Hg$_3$		9.32±0.20	EI		3297
HgCl$^+$						
HgCl$^+$	HgCl$_2$	Cl	12.06±0.26	EI		2506
HgCl^{+2}						
HgCl^{+2}	HgCl$_2$	Cl	32.0±0.5	EI		2506

4.3. The Positive Ion Table—Continued

Ion	Reactant	Other products	Ionization or appearance potential (eV)	Method	Heat of formation (kJ mol^{-1})	Ref.
$HgCl_2^+(^2\Pi_{3/2g})$			$\Delta H_{f298}^\circ = 951$ kJ mol^{-1} (227 kcal mol^{-1})			
$HgCl_2^+(^2\Pi_{1/2g})$			$\Delta H_{f298}^\circ = 963$ kJ mol^{-1} (230 kcal mol^{-1})			
$HgCl_2^+(^2\Pi_{3/2g})$	$HgCl_2$		11.37	PE	951	2984
$HgCl_2^+(^2\Pi_{1/2g})$	$HgCl_2$		11.50	PE	963	2984
$HgCl_2^+(^2\Pi_u)$	$HgCl_2$		12.13 (V)	PE		2984
$HgCl_2^+(^2\Sigma_u)$	$HgCl_2$		12.74 (V)	PE		2984
$HgCl_2^+(^2\Sigma_g)$	$HgCl_2$		13.74 (V)	PE		2984

Additional higher ionization potentials are given in ref. 2984.

See also − EI: 2506

$HgCl_2^{+2}$

Ion	Reactant	Other products	Ionization or appearance potential (eV)	Method	Heat of formation (kJ mol^{-1})	Ref.
$HgCl_2^{+2}$	$HgCl_2$		28.3±0.5	EI		2506

$HgBr^+$

Ion	Reactant	Other products	Ionization or appearance potential (eV)	Method	Heat of formation (kJ mol^{-1})	Ref.
$HgBr^+$	$HgBr_2$	Br	12.09±0.17	EI		2506

$HgBr^{+2}$

Ion	Reactant	Other products	Ionization or appearance potential (eV)	Method	Heat of formation (kJ mol^{-1})	Ref.
$HgBr^{+2}$	$HgBr_2$	Br	31.1±0.7	EI		2506

Ion	Reactant	Other products	Ionization or appearance potential (eV)	Method	Heat of formation (kJ mol^{-1})	Ref.
$HgBr_2^+(^2\Pi_{3/2g})$			$\Delta H_{f298}^\circ = 939$ kJ mol^{-1} (224 kcal mol^{-1})			
$HgBr_2^+(^2\Pi_{1/2g})$			$\Delta H_{f298}^\circ = 972$ kJ mol^{-1} (232 kcal mol^{-1})			
$HgBr_2^+(^2\Pi_{3/2g})$	$HgBr_2$		10.62	PE	939	2984
$HgBr_2^+(^2\Pi_{1/2g})$	$HgBr_2$		10.96	PE	972	2984
$HgBr_2^+(^2\Pi_{3/2u})$	$HgBr_2$		11.20 (V)	PE		2984
$HgBr_2^+(^2\Pi_{1/2u})$	$HgBr_2$		11.54 (V)	PE		2984
$HgBr_2^+(^2\Sigma_u)$	$HgBr_2$		12.09 (V)	PE		2984
$HgBr_2^+(^2\Sigma_g)$	$HgBr_2$		13.39 (V)	PE		2984

Additional higher ionization potentials are given in ref. 2984.

See also − EI: 2506

$HgBr_2^{+2}$

Ion	Reactant	Other products	Ionization or appearance potential (eV)	Method	Heat of formation (kJ mol^{-1})	Ref.
$HgBr_2^{+2}$	$HgBr_2$		25.7±0.3	EI		2506

HgI^+

Ion	Reactant	Other products	Ionization or appearance potential (eV)	Method	Heat of formation (kJ mol^{-1})	Ref.
HgI^+	HgI_2	I	11.3±0.4	EI		2506

4.3. The Positive Ion Table—Continued

Ion	Reactant	Other products	Ionization or appearance potential (eV)	Method	Heat of formation (kJ mol^{-1})	Ref.
	$HgI_2^+(^2\Pi_{3/2g})$	$\Delta H_{f0}^\circ = 906$ kJ mol^{-1} (216 kcal mol^{-1})				
	$HgI_2^+(^2\Pi_{1/2g})$	$\Delta H_{f0}^\circ = 969$ kJ mol^{-1} (232 kcal mol^{-1})				
$HgI_2^+(^2\Pi_{3/2g})$	HgI_2		9.50	PE	906	2984
$HgI_2^+(^2\Pi_{1/2g})$	HgI_2		10.16	PE	969	2984
$HgI_2^+(^2\Pi_{3/2u})$	HgI_2		10.00 (V)	PE		2984
$HgI_2^+(^2\Pi_{1/2u})$	HgI_2		10.40 (V)	PE		2984
$HgI_2^+(^2\Sigma_u)$	HgI_2		11.29 (V)	PE		2984
$HgI_2^+(^2\Sigma_g)$	HgI_2		12.85 (V)	PE		2984

Additional higher ionization potentials are given in ref. 2984.

See also – EI: 2506

		HgI_2^{+2}				
HgI_2^{+2}	HgI_2		21.4±0.5	EI		2506

	CH_3Hg^+	$\Delta H_{f298}^\circ = 927$ kJ mol^{-1} (221 kcal mol^{-1})				
CH_3Hg^+	$(CH_3)_2Hg$	CH_3	10.10±0.02	PI	927	2983

See also – EI: 83, 306

CH_3Hg^+	CH_3HgCl	Cl	12.35±0.2	EI		306

		$C_2H_5Hg^+$				
$C_2H_5Hg^+$	$(C_2H_5)_2Hg$	C_2H_5	9.65±0.1	EI		306

	$(CH_3)_2Hg^+$	$\Delta H_{f298}^\circ = 972$ kJ mol^{-1} (232 kcal mol^{-1})				
$C_2H_6Hg^+$	$(CH_3)_2Hg$		9.10±0.05	PI	972	2983

(Threshold value approximately corrected for hot bands)

See also – PE: 2984

EI: 83, 306

		$C_3H_7Hg^+$				
$C_3H_7Hg^+$	$(iso-C_3H_7)_2Hg$	C_3H_7	9.1±0.1	EI		306

4.3. The Positive Ion Table—Continued

Ion	Reactant	Other products	Ionization or appearance potential (eV)	Method	Heat of formation (kJ mol⁻¹)	Ref.

$C_4H_{10}Hg^+$

| $C_4H_{10}Hg^+$ | $(C_2H_5)_2Hg$ | | 8.5±0.1 | EI | | 306 |

See also − PE: 2984

$C_6H_{14}Hg^+$

| $C_6H_{14}Hg^+$ | $(iso-C_3H_7)_2Hg$ | | 7.6±0.1 | EI | | 306 |

CH_3HgCl^+

$CH_3HgCl^+(^2E, ^2A_1)$	CH_3HgCl		10.88 (V)	PE		2984
$CH_3HgCl^+(^2A_1)$	CH_3HgCl		12.70 (V)	PE		2984
$CH_3HgCl^+(^2E)$	CH_3HgCl		14.1 (V)	PE		2984

Additional higher ionization potentials are given in ref. 2984.

See also − EI: 306

CH_3HgBr^+

$CH_3HgBr^+(^2E_{3/2})$	CH_3HgBr		10.16 (V)	PE		2984
$CH_3HgBr^+(^2E_{1/2})$	CH_3HgBr		10.43 (V)	PE		2984
$CH_3HgBr^+(^2A_1)$	CH_3HgBr		10.66 (V)	PE		2984
$CH_3HgBr^+(^2A_1)$	CH_3HgBr		12.52 (V)	PE		2984
$CH_3HgBr^+(^2E)$	CH_3HgBr		13.9 (V)	PE		2984

Additional higher ionization potentials are given in ref. 2984.

CH_3HgI^+

$CH_3HgI^+(^2E_{3/2})$	CH_3HgI		9.25 (V)	PE		2984
$CH_3HgI^+(^2E_{1/2})$	CH_3HgI		9.68 (V)	PE		2984
$CH_3HgI^+(^2A_1)$	CH_3HgI		10.29 (V)	PE		2984
$CH_3HgI^+(^2A_1)$	CH_3HgI		12.21 (V)	PE		2984
$CH_3HgI^+(^2E)$	CH_3HgI		13.6 (V)	PE		2984

Additional higher ionization potentials are given in ref. 2984.

4.3. The Positive Ion Table—Continued

Ion	Reactant	Other products	Ionization or appearance potential (eV)	Method	Heat of formation (kJ mol^{-1})	Ref.

Tl$^+$ $\Delta H^\circ_{f0} = 772.2$ kJ mol^{-1} (184.6 kcal mol^{-1})

Ion	Reactant	Other products	IP/AP	Method	ΔH_f	Ref.
Tl$^+$	Tl		6.108	S	772.2	2113

See also – S: 3270
         EI: 2620, 3244

Ion	Reactant	Other products	IP/AP	Method	ΔH_f	Ref.
Tl$^+$	Tl$_2$O		12±1	EI		2955
Tl$^+$	TlF	F$^-$	7.59	PI		2604

This is the energy of the maximum ion–pair formation cross section, and is about 0.3 eV above the thermochemical threshold. The measurement was made upon the F$^-$ ion.

Ion	Reactant	Other products	IP/AP	Method	ΔH_f	Ref.
Tl$^+$	TlF	F	10.68	PI		2604
(Threshold value corrected for hot bands)						
Tl$^+$	TlCl	Cl$^-$	7.01	PI		2604

This is the energy of the maximum ion–pair formation cross section, and is about 0.7 eV above the thermochemical threshold. The measurement was made upon the Cl$^-$ ion.

Ion	Reactant	Other products	IP/AP	Method	ΔH_f	Ref.
Tl$^+$	TlCl	Cl	9.925±0.02	PI		2604
(Threshold value corrected for hot bands)						

See also – EI: 2159

Ion	Reactant	Other products	IP/AP	Method	ΔH_f	Ref.
Tl$^+$	TlBr	Br$^-$	6.3	PI		2604

This is the energy of the maximum ion–pair formation cross section, and is about 0.1 eV above the thermochemical threshold. The measurement was made upon the Br$^-$ ion.

See also – EI: 2159

Ion	Reactant	Other products	IP/AP	Method	ΔH_f	Ref.
Tl$^+$	TlBr	Br	9.53±0.02	PI		2604
(Threshold value corrected for hot bands)						
Tl$^+$	TlI	I$^-$	5.799±0.005	PI		2610
(Threshold value corrected for hot bands)						

See also – EI: 2159

Ion	Reactant	Other products	IP/AP	Method	ΔH_f	Ref.
Tl$^+$	TlI	I	8.875	PI		2610
(Threshold value corrected for hot bands)						

Tl$_2^+$

Ion	Reactant	Other products	IP/AP	Method	ΔH_f	Ref.
Tl$_2^+$	Tl$_2$O	O	9.7±0.5	EI		3244

See also – EI: 2955

Ion	Reactant	Other products	IP/AP	Method	ΔH_f	Ref.
Tl$_2^+$	Tl$_2$F$_2$		~18	EI		2920

4.3. The Positive Ion Table—Continued

Ion	Reactant	Other products	Ionization or appearance potential (eV)	Method	Heat of formation (kJ mol^{-1})	Ref.
			Tl$_2$O$^+$			
Tl$_2$O$^+$	Tl$_2$O		7.5±0.3	EI		2955
Tl$_2$O$^+$	Tl$_2$O		8.8±0.5	EI		3244
			TlF$^+$			
TlF$^+$	TlF		11.2±0.5	EI		2620
			Tl$_2$F$^+$			
Tl$_2$F$^+$	Tl$_2$F$_2$	F	9.97±0.02	PI		2604
(Threshold value corrected for hot bands)						
			Tl$_2$F$_2^+$			
Tl$_2$F$_2^+$	Tl$_2$F$_2$		9.71±0.02	PI		2604
(Threshold value corrected for hot bands)						
		TlCl$^+$ $\Delta H_{f298}^\circ = 868$ kJ mol^{-1} (207 kcal mol^{-1})				
TlCl$^+$	TlCl		9.70±0.03	PI	868	2604
(Threshold value corrected for hot bands)						
		TlBr$^+$ $\Delta H_{f298}^\circ = 844$ kJ mol^{-1} (202 kcal mol^{-1})				
TlBr$^+$	TlBr		9.14±0.02	PI	844	2604
(Threshold value corrected for hot bands)						
		TlI$^+$ $\Delta H_{f298}^\circ = 824$ kJ mol^{-1} (197 kcal mol^{-1})				
TlI$^+$	TlI		8.469	PI	824	2610
(Threshold value corrected for hot bands)						
		Pb$^+$ $\Delta H_{f0}^\circ = 911.2$ kJ mol^{-1} (217.8 kcal mol^{-1})				
Pb$^+$	Pb		7.416	S	911.2	2113, 3113
See also – EI: 1245, 2139, 2595, 2747						
Pb$^+$	PbH$_4$		11.2	EI		2116
Pb$^+$	PbO?		11±1	EI		2747
See also – EI: 1245						
Pb$^+$	PbS	S	11.6±0.5	EI		2139
See also – EI: 2747						

4.3. The Positive Ion Table—Continued

Ion	Reactant	Other products	Ionization or appearance potential (eV)	Method	Heat of formation (kJ mol^{-1})	Ref.
Pb$^+$	PbCl$_2$		12.0±0.2	EI		2056
See also – EI:	2434, 2462					
Pb$^+$	PbSe	Se	11±1	EI		2747
Pb$^+$	(CH$_3$)$_4$Pb		14.4	EI		2980
See also – EI:	82					
Pb$^+$	(C$_2$H$_5$)$_4$Pb		11.6	EI		2980
		PbH$^+$				
PbH$^+$	PbH$_4$		11.1	EI		2116
PbH$^+$	(CH$_3$)$_4$Pb		15.3	EI		2980
PbH$^+$	(C$_2$H$_5$)$_4$Pb		14.4	EI		2980
		PbH$_2^+$				
PbH$_2^+$	PbH$_4$		10.1	EI		2116
		PbH$_3^+$				
PbH$_3^+$	PbH$_4$	H	9.6	EI		2116
		PbO$^+$				
PbO$^+$	PbO		9.0±0.5	EI		1245
PbO$^+$	PbO		9.1±0.5	EI		2747
		Pb$_2$O$^+$				
Pb$_2$O$^+$	Pb$_2$O$_2$?		11.9±1.0	EI		1245
		Pb$_2$O$_2^+$				
Pb$_2$O$_2^+$	Pb$_2$O$_2$		8.8±0.5	EI		1245
Pb$_2$O$_2^+$	Pb$_2$O$_2$		9.1±0.5	EI		2747
		Pb$_3$O$_2^+$				
Pb$_3$O$_2^+$	Pb$_3$O$_3$?		14.6±1.0	EI		1245

4.3. The Positive Ion Table—Continued

Ion	Reactant	Other products	Ionization or appearance potential (eV)	Method	Heat of formation (kJ mol^{-1})	Ref.
		$Pb_3O_3^+$				
$Pb_3O_3^+$	Pb_3O_3		9.7 ± 1.0	EI		1245
		$Pb_4O_4^+$				
$Pb_4O_4^+$	Pb_4O_4		8.5 ± 1.0	EI		1245
		PbF^+				
PbF^+	PbF		7.5 ± 0.3	EI		2595
PbF^+	PbF_2?		12.1 ± 0.3	EI		2595
		PbF_2^+				
PbF_2^+	PbF_2		11.6 ± 0.3	EI		2595
		PbF_4^+				
PbF_4^+	PbF_4		10.4 ± 0.3	EI		2595
		PbS^+				
PbS^+	PbS		8.4 ± 0.5	EI		2747
PbS^+	PbS		8.6 ± 0.5	EI		2139
		$Pb_2S_2^+$				
$Pb_2S_2^+$	Pb_2S_2		9.2 ± 0.5	EI		2139
		$PbCl^+$				
$PbCl^+$	$PbCl_2$	Cl^-	~ 7.1	EI		2434, 2462
$PbCl^+$	$PbCl_2$	Cl	10.7 ± 0.1	EI		2434, 2462
$PbCl^+$	$PbCl_2$	Cl	11.7 ± 0.2	EI		2056
		$PbCl_2^+$				
$PbCl_2^+$	$PbCl_2$		10.3 ± 0.1	EI		2434, 2462
$PbCl_2^+$	$PbCl_2$		11.2 ± 0.2	EI		2056
		$PbSe^+$				
$PbSe^+$	$PbSe$		8.4 ± 0.5	EI		2747

4.3. The Positive Ion Table—Continued

Ion	Reactant	Other products	Ionization or appearance potential (eV)	Method	Heat of formation (kJ mol^{-1})	Ref.
			PbBr$^+$			
PbBr$^+$	PbBr$_2$	Br	10.5±0.2	EI		2434
			PbBr$_2^+$			
PbBr$_2^+$	PbBr$_2$		10.2±0.2	EI		2434
			PbTe$^+$			
PbTe$^+$	PbTe		8.3±0.5	EI		2747
			CH$_3$Pb$^+$			
CH$_3$Pb$^+$	(CH$_3$)$_4$Pb		12.4±0.2	EI		82
CH$_3$Pb$^+$	(CH$_3$)$_4$Pb		13.1	EI		2980
			C$_2$H$_5$Pb$^+$			
C$_2$H$_5$Pb$^+$	(C$_2$H$_5$)$_4$Pb		12.2	EI		2980
			C$_2$H$_6$Pb$^+$			
C$_2$H$_6$Pb$^+$	(CH$_3$)$_4$Pb		11.6±0.2	EI		82
C$_2$H$_6$Pb$^+$	(CH$_3$)$_4$Pb		12.7	EI		2980
			C$_3$H$_9$Pb$^+$			
C$_3$H$_9$Pb$^+$	(CH$_3$)$_4$Pb	CH$_3$	8.9±0.1	EI		82
C$_3$H$_9$Pb$^+$	(CH$_3$)$_4$Pb	CH$_3$	10.1	EI		2980
			C$_4$H$_{10}$Pb$^+$			
C$_4$H$_{10}$Pb$^+$	(C$_2$H$_5$)$_4$Pb		12.2	EI		2980
			C$_4$H$_{12}$Pb$^+$			
C$_4$H$_{12}$Pb$^+$	(CH$_3$)$_4$Pb		8.0±0.4	EI		82
C$_4$H$_{12}$Pb$^+$	(CH$_3$)$_4$Pb		9.3	EI		2980
			C$_6$H$_{15}$Pb$^+$			
C$_6$H$_{15}$Pb$^+$	(C$_2$H$_5$)$_4$Pb	C$_2$H$_5$	10.8	EI		2980

4.3. The Positive Ion Table—Continued

Ion	Reactant	Other products	Ionization or appearance potential (eV)	Method	Heat of formation (kJ mol^{-1})	Ref.
		$C_8H_{20}Pb^+$				
$C_8H_{20}Pb^+$	$(C_2H_5)_4Pb$		11.1	EI		2980
		$NaPbCl_2^+$				
$NaPbCl_2^+$	$NaPbCl_3$	Cl	9.8±0.2	EI		2462
		$KPbCl_2^+$				
$KPbCl_2^+$	$KPbCl_3$	Cl	9.6±0.2	EI		2462
		$KPbCl_3^+$				
$KPbCl_3^+$	$KPbCl_3$		9.3±0.2	EI		2462
		$PbClBr^+$				
$PbClBr^+$	$PbClBr$		10.4±0.2	EI		2434
		$RbPbCl^+$				
$RbPbCl^+$	$RbPbCl_3$		~13.3±0.4	EI		2462
		$RbPbCl_2^+$				
$RbPbCl_2^+$	$RbPbCl_3$	Cl^-	~5.7	EI		2462
$RbPbCl_2^+$	$RbPbCl_3$	Cl	9.3±0.2	EI		2462
$RbPbCl_2^+$	$RbPbCl_3$	Cl	11.6±0.2	EI		2056
		$RbPbCl_3^+$				
$RbPbCl_3^+$	$RbPbCl_3$		9.2±0.2	EI		2462
		$CsPbCl^+$				
$CsPbCl^+$	$CsPbCl_3$		11.5±0.2	EI		2056
$CsPbCl^+$	$CsPbCl_3$		13.3±0.2	EI		2462
		$CsPbCl_2^+$				
$CsPbCl_2^+$	$CsPbCl_3$	Cl	9.4±0.2	EI		2462

4.3. The Positive Ion Table—Continued

Ion	Reactant	Other products	Ionization or appearance potential (eV)	Method	Heat of formation (kJ mol^{-1})	Ref.

Bi^+ $\quad \Delta H_{f0}^{\circ} = 910.7 \text{ kJ mol}^{-1} (217.7 \text{ kcal mol}^{-1})$

Ion	Reactant	Other products	Ionization or appearance potential (eV)	Method	Heat of formation (kJ mol^{-1})	Ref.
Bi^+	Bi		7.289	S	910.7	2113

See also – EI: 2049, 2747

Ion	Reactant	Other products	Ionization or appearance potential (eV)	Method	Heat of formation (kJ mol^{-1})	Ref.
Bi^+	Bi_2	Bi	9.7±0.1	RPD		2509
Bi^+	BiH_3		13.4	EI		2116
Bi^+	BiO?		10±1	EI		2747

Bi_2^+

Ion	Reactant	Other products	Ionization or appearance potential (eV)	Method	Heat of formation (kJ mol^{-1})	Ref.
Bi_2^+	Bi_2		7.4±0.1	RPD		2509
Bi_2^+	Bi_2		7.1±0.3	EI		2509
Bi_2^+	Bi_2		7.2±0.5	EI		2747

See also – EI: 2049, 2508

Bi_3^+

Ion	Reactant	Other products	Ionization or appearance potential (eV)	Method	Heat of formation (kJ mol^{-1})	Ref.
Bi_3^+	Bi_4?		7.7±0.5	EI		2509

Bi_4^+

Ion	Reactant	Other products	Ionization or appearance potential (eV)	Method	Heat of formation (kJ mol^{-1})	Ref.
Bi_4^+	Bi_4		7.6±0.3	EI		2508, 2509

BiH^+

Ion	Reactant	Other products	Ionization or appearance potential (eV)	Method	Heat of formation (kJ mol^{-1})	Ref.
BiH^+	BiH_3		12.2	EI		2116

BiH_2^+

Ion	Reactant	Other products	Ionization or appearance potential (eV)	Method	Heat of formation (kJ mol^{-1})	Ref.
BiH_2^+	BiH_3	H	12.4	EI		2116

BiH_3^+

Ion	Reactant	Other products	Ionization or appearance potential (eV)	Method	Heat of formation (kJ mol^{-1})	Ref.
BiH_3^+	BiH_3		10.1	EI		2116

BiO^+

Ion	Reactant	Other products	Ionization or appearance potential (eV)	Method	Heat of formation (kJ mol^{-1})	Ref.
BiO^+	BiO		9.0±0.5	EI		2747

4.3. The Positive Ion Table—Continued

Ion	Reactant	Other products	Ionization or appearance potential (eV)	Method	Heat of formation (kJ mol^{-1})	Ref.
		BiS$^+$				
BiS$^+$	BiS		8.7±0.5	EI		2747
See also – EI: 2049						
		Bi$_2$S$^+$				
Bi$_2$S$^+$	Bi$_2$S$_2$	S	14	EI		2049
		BiSb$^+$				
BiSb$^+$	BiSb		8.0±0.3	EI		2508
		BiSb$_2^+$				
BiSb$_2^+$	Bi$_2$Sb$_2$?		10.5±0.3	EI		2508
		BiSb$_3^+$				
BiSb$_3^+$	BiSb$_3$		7.8±0.3	EI		2508
		Bi$_2$Sb$^+$				
Bi$_2$Sb$^+$	Bi$_2$Sb$_2$?		10.6±0.3	EI		2508
		Bi$_2$Sb$_2^+$				
Bi$_2$Sb$_2^+$	Bi$_2$Sb$_2$		7.7±0.3	EI		2508
		Bi$_3$Sb$^+$				
Bi$_3$Sb$^+$	Bi$_3$Sb		7.2±0.3	EI		2508
		BiTe$^+$				
BiTe$^+$	BiTe		8.4±0.5	EI		2747
	C$_{18}$H$_{15}$Bi$^+$ (Triphenylbismuth)	$\Delta H^\circ_{f298} \sim 1284$ kJ mol^{-1} (307 kcal mol^{-1})				
C$_{18}$H$_{15}$Bi$^+$	(C$_6$H$_5$)$_3$Bi (Triphenylbismuth)		7.3±0.1	PI	~1284	1140

4.3. The Positive Ion Table—Continued

Ion	Reactant	Other products	Ionization or appearance potential (eV)	Method	Heat of formation (kJ mol^{-1})	Ref.
			Po$^+$			
Po$^+$	Po		8.42	S		2113
Po$^+$	Po		8.2±0.4	S		3378
			H$_2$Po$^+$			
H$_2$Po$^+$	H$_2$Po		8.6	D		3317
			At$^+$			
At$^+$	At		9.2±0.4	S		3378
			At$_2^+$			
At$_2^+$	At$_2$		8.3	D		104
			Fr$^+$			
Fr$^+$	Fr		3.98±0.10	S		3378
			Ac$^+$			
Ac$^+$	Ac		6.89±0.6	S		2113, 3378
			Th$^+$			
Th$^+$	Th		6.95±0.06	SI		2882
Th$^+$	Th		7.5±0.3	SI		2756
Th$^+$	ThF$_4$		39±1.0	EI		3203
			ThO$^+$			
ThO$^+$	ThO		8.1±0.1	RPD		2994
			ThO$_2^+$			
ThO$_2^+$	ThO$_2$		10.9	RPD		2994
			ThF$^+$			
ThF$^+$	ThF$_4$		30±1.0	EI		3203

4.3. The Positive Ion Table—Continued

Ion	Reactant	Other products	Ionization or appearance potential (eV)	Method	Heat of formation (kJ mol⁻¹)	Ref.

ThF_2^+

| ThF_2^+ | ThF_2? | | 13.1±0.6 | EI | | 3203 |
| ThF_2^+ | ThF_4 | | 23.2±0.6 | EI | | 3203 |

ThF_3^+

| ThF_3^+ | ThF_3 | | 7.8±0.5 | EI | | 3203 |
| ThF_3^+ | ThF_4 | F | 14.5±0.5 | EI | | 3203 |

$C_{15}H_{15}Th^+$

| $C_{15}H_{15}Th^+$ | $(C_5H_5)_4Th$ (Tetrakis(cyclopentadienyl)thorium) | | 7.73±0.1 | EI | | 2640 |

$C_{20}H_{20}Th^+$

| $C_{20}H_{20}Th^+$ | $(C_5H_5)_4Th$ (Tetrakis(cyclopentadienyl)thorium) | | 7.41±0.1 | EI | | 2640 |

$C_{10}H_{10}FTh^+$

| $C_{10}H_{10}FTh^+$ | $(C_5H_5)_3ThF$ (Tris(cyclopentadienyl)thorium fluoride) | | 9.18±0.2 | EI | | 2640 |

$C_{15}H_{15}FTh^+$

| $C_{15}H_{15}FTh^+$ | $(C_5H_5)_3ThF$ (Tris(cyclopentadienyl)thorium fluoride) | | 8.06±0.1 | EI | | 2640 |

U^+

U^+	U		6.11±0.05	RPD		16
U^+	U		6.08±0.08	SI		317
U^+	U		6.19±0.06	SI	2494,	2756

See also − EI: 2054, 2126, 3196

| U^+ | UO | O | 13.4 | EI | | 2054 |

UC^+

| UC^+ | UC | | 6.1±0.5 | EI | | 2491 |

4.3. The Positive Ion Table—Continued

Ion	Reactant	Other products	Ionization or appearance potential (eV)	Method	Heat of formation (kJ mol^{-1})	Ref.
		UC$_2^+$				
UC$_2^+$	UC$_2$		5.7±0.5	EI		2491
		UO$^+$				
UO$^+$	UO		5.72±0.06	RPD		16
UO$^+$	UO		4.7±0.6	EI		2126
See also – EI: 2054						
		UO$_2^+$				
UO$_2^+$	UO$_2$		5.5±0.1	RPD		16
UO$_2^+$	UO$_2$		4.3±0.6	EI		2126
See also – EI: 2054						
		UO$_3^+$				
UO$_3^+$	UO$_3$		10.4±0.6	EI		2126
		UF$_6^+$				
UF$_6^+$	UF$_6$		14.14 (V)	PE		3119
		UP$^+$				
UP$^+$	UP		7.3±0.4	EI		2440
		US$^+$				
US$^+$	US		6.3	EI		3196
		C$_{10}$H$_{10}$U$^+$				
C$_{10}$H$_{10}$U$^+$	(C$_5$H$_5$)$_4$U (Tetrakis(cyclopentadienyl)uranium)		15.1±0.5	EI		2640
		C$_{15}$H$_{15}$U$^+$				
C$_{15}$H$_{15}$U$^+$	(C$_5$H$_5$)$_4$U (Tetrakis(cyclopentadienyl)uranium)		7.29±0.1	EI		2640

4.3. The Positive Ion Table—Continued

Ion	Reactant	Other products	Ionization or appearance potential (eV)	Method	Heat of formation (kJ mol^{-1})	Ref.
		$C_{20}H_{20}U^+$				
$C_{20}H_{20}U^+$	$(C_5H_5)_4U$ (Tetrakis(cyclopentadienyl)uranium)		6.50±0.1	EI		2640
		HUO_3^+				
HUO_3^+	HUO_3		<3.89	D		3419
		$C_{10}H_{10}FU^+$				
$C_{10}H_{10}FU^+$	$(C_5H_5)_3UF$ (Tris(cyclopentadienyl)uranium fluoride)		10.18±0.2	EI		2640
		$C_{15}H_{15}FU^+$				
$C_{15}H_{15}FU^+$	$(C_5H_5)_3UF$ (Tris(cyclopentadienyl)uranium fluoride)		7.53±0.2	EI		2640
		Np^+				
Np^+	Np		6.16±0.06	SI		2756
		Pu^+				
Pu^+	Pu		5.8	S		2113
Pu^+	Pu		5.71±0.06	SI		2756

See also – EI: 2503, 2944
      SI: 1149

Ion	Reactant	Other products	Ionization or appearance potential (eV)	Method	Heat of formation (kJ mol^{-1})	Ref.
Pu^+	PuF_3		25.0±1.0	EI		2503
		PuF^+				
PuF^+	PuF		5.9±0.5	EI		2503
PuF^+	PuF_3		17.8±0.5	EI		2503
		PuF_2^+				
PuF_2^+	PuF_2		6.4±0.5	EI		2503
PuF_2^+	PuF_3	F	12.6±0.5	EI		2503
		Am^+				
Am^+	Am		6.0	S		341, 2113
		Cm^+				
Cm^+	Cm		6.18±0.09	SI		3272

4.4. Bibliography

[1] Berkowitz, J., Bafus, D. A., and Brown, T. L., The mass spectrum of ethyllithium vapor, J. Phys. Chem. **65**, 1380 (1961).

[2] Berkowitz, J., Chupka, W. A., and Kistiakowsky, G. B., Mass spectrometric study of the kinetics of nitrogen afterglow, J. Chem. Phys. **25**, 457 (1956).

[3] Asundi, R. K., and Kurepa, M. V., The calibration of the electron energy scale for ionization potential measurements, J. Sci. Instr. **40**, 183 (1963).

[6] Branscomb, L. M., Calculated calibration points for negative ion appearance potentials, J. Chem. Phys. **29**, 452 (1958).

[8] Carette, J.–D., and Kerwin, L., Une étude du phosphore rouge par spectrométrie de masse, Can. J. Phys. **39**, 1300 (1961).

[11] Chupka, W. A., Effect of unimolecular decay kinetics on the interpretation of appearance potentials, J. Chem. Phys. **30**, 191 (1959).

[13] Coats, F. H., and Anderson, R. C., Thermodynamic data from electron–impact measurements on acetylene and substituted acetylenes, J. Am. Chem. Soc. **79**, 1340 (1957).

[14] Collin, J., Ionization potentials of methylamines and ethylamines by electron impact, Can. J. Chem. **37**, 1053 (1959).

[16] Mann, J. B., Ionization of U, UO, and UO_2 by electron impact, J. Chem. Phys. **40**, 1632 (1964).

[17] Collin, J., and Lossing, F. P., Mass spectra of propyne and propyne–d_3, and the appearance potentials of $C_3H_4^+$, $C_3H_3^+$ and equivalent deuterated ions, J. Am. Chem. Soc. **80**, 1568 (1958).

[20] Curran, R. K., Negative ion formation in ozone, J. Chem. Phys. **35**, 1849 (1961).

[24] Dibeler, V. H., Reese, R. M., and Mohler, F. L., Ionization and dissociation of the trifluoromethyl halides by electron impact, J. Res. NBS **57**, 113 (1956).

[25] Dorman, F. H., Morrison, J. D., and Nicholson, A. J. C., Probability of multiple ionization by electron impact, J. Chem. Phys. **31**, 1335 (1959).

[28] Sjögren, H., and Lindholm, E., Higher ionisation potentials in alcohols and water measured by electron impact, Phys. Letters **4**, 85 (1963).

[29] Fisher, I. P., and Lossing, F. P., Ionization potential of benzyne, J. Am. Chem. Soc. **85**, 1018 (1963).

[30] Flesch, G. D., and Svec, H. J., The mass spectra of chromyl chloride, chromyl chlorofluoride and chromyl fluoride, J. Am. Chem. Soc. **81**, 1787 (1959).

[31] Foner, S. N., and Hudson, R. L., Ionization potential of the free HO_2 radical and the $H–O_2$ bond dissociation energy, J. Chem. Phys. **23**, 1364 (1955).

[33] Foner, S. N., and Hudson, R. L., Diimide–identification and study by mass spectrometry, J. Chem. Phys. **28**, 719 (1958).

[34] Foner, S. N., and Hudson, R. L., Mass spectrometric detection of triazene and tetrazene and studies of the free radicals NH_2 and N_2H_3, J. Chem. Phys. **29**, 442 (1958).

[35] Foner, S. N., and Nall, B. H., Structure in the ionization near threshold of rare gases by electron impact, Phys. Rev. **122**, 512 (1961).

[36] Foner, S. N., and Hudson, R. L., Mass spectrometry of the HO_2 free radical, J. Chem. Phys. **36**, 2681 (1962).

[37] Foner, S. N., and Hudson, R. L., Ionization and dissociation of hydrogen peroxide by electron impact, J. Chem. Phys. **36**, 2676 (1962).

[39] Fox, R. E., Threshold ionization of HCl by electron impact, J. Chem. Phys. **32**, 385 (1960).

[42] Malone, T. J., and McGee, H. A., Jr., Mass spectrometric investigations of the synthesis, stability, and energetics of the low–temperature oxygen fluorides. I. Dioxygen difluoride, J. Phys. Chem. **69**, 4338 (1965).

43] Hobrock, D. L., and Kiser, R. W., Electron impact studies of some trihalomethanes: trichloromethane, dichlorofluoromethane, chlorodifluoromethane, and trifluoromethane, J. Phys. Chem. **68**, 575 (1964).

[46] Friedman, L., Long, F. A., and Wolfsberg, M., Study of the mass spectra of the lower aliphatic alcohols, J. Chem. Phys. **27**, 613 (1957).

[49] Frost, D. C., and McDowell, C. A., The dissociation energy of the nitrogen molecule, Proc. Roy. Soc. (London) **A236**, 278 (1956).

[50] Gallegos, E. J., and Kiser, R. W., Electron impact spectroscopy of ethylene oxide and propylene oxide, J. Am. Chem. Soc. **83**, 773 (1961).

[51] Gallegos, E., and Kiser, R. W., Electron impact spectroscopy of ethylene sulfide and ethylenimine, J. Phys. Chem. **65**, 1177 (1961).

[52] Gallegos, E. J., and Kiser, R. W., Electron impact spectroscopy of the four– and five–membered, saturated heterocyclic compounds containing nitrogen, oxygen and sulfur, J. Phys. Chem. **66**, 136 (1962).

[53] Dibeler, V. H., Reese, R. M., and Mann, D. E., Ionization and dissociation of perchlorylfluoride by electron impact, J. Chem. Phys. **27**, 176 (1957).

[54] Dibeler, V. H., and Reese, R. M., Mass spectrometric study of photoionization. I. Apparatus and initial observations on acetylene, acetylene–d_2, benzene, and benzene–d_6, J. Res. NBS **68A**, 409 (1964).

[58] Collin, J., and Lossing, F. P., Ionization and dissociation of NO_2 and N_2O by electron impact, J. Chem. Phys. **28**, 900 (1958).

[59] Farmer, J. B., and Lossing, F. P., Free radicals by mass spectrometry. VII. The ionization potentials of ethyl, isopropyl, and propargyl radicals and the appearance potentials of the radical ions in some derivatives, Can. J. Chem. **33**, 861 (1955).

[61] Frost, D. C., Mak, D., and McDowell, C. A., The photoionization of nitrogen dioxide, Can. J. Chem. **40**, 1064 (1962).

[62] Collin, J., and Lossing, F. P., Ionization potentials of some olefins, di–olefins and branched paraffins, J. Am. Chem. Soc. **81**, 2064 (1959).

[67] Foner, S. N., and Hudson, R. L., Mass spectrometry of inorganic free radicals, Advan. Chem. Ser. **36**, 34 (1962).

[68] Harrison, A. G., Honnen, L. R., Dauben, H. J., Jr., and Lossing, F. P., Free radicals by mass spectrometry. XX. Ionization potentials of cyclopentadienyl and cycloheptatrienyl radicals, J. Am. Chem. Soc. **82**, 5593 (1960).

[69] Harrison, A. G., Kebarle, P., and Lossing, F. P., Free radicals by mass spectrometry. XXI. The ionization potentials of some *meta* and *para* substituted benzyl radicals, J. Am. Chem. Soc. **83**, 777 (1961).

[70] Harrison, A. G., and Lossing, F. P., Free radicals by mass spectrometry. XVII. Ionization potential and heat of formation of vinyl radical, J. Am. Chem. Soc. **82**, 519 (1960).

[71] Harrison, A. G., and Lossing, F. P., Free radicals by mass spectrometry. XVIII. The ionization potentials of conjugated hydrocarbon radicals and the resonance energies of radicals and carbonium ions, J. Am. Chem. Soc. **82**, 1052 (1960).

[72] Harrison, A. G., and Shannon, T. W., An electron impact study of chloromethyl and dichloromethyl derivatives, Can. J. Chem. **40**, 1730 (1962).

[73] Herron, J. T., and Dibeler, V. H., Electron impact study of the cyanogen halides, J. Am. Chem. Soc. **82**, 1555 (1960).

[74] Herron, J. T., and Dibeler, V. H., Mass spectrum and appearance potentials of tetrafluorohydrazine, J. Chem. Phys. **33**, 1595 (1960).

[75] Herron, J. T., and Dibeler, V. H., Ionization potential of fluorine, J. Chem. Phys. **32**, 1884 (1960).

[76] Herron, J. T., and Dibeler, V. H., Mass spectrometric study of NF_2, NF_3, N_2F_2, and N_2F_4, J. Res. NBS **65A**, 405 (1961).

[77] Herron, J. T., and Schiff, H. I., Mass spectrometry of ozone, J. Chem. Phys. **24**, 1266 (1956).

[78] Herron, J. T., Rate of the reaction NO+N, and some heterogeneous reactions observed in the ion source of a mass spectrometer, J. Res. NBS **65A**, 411 (1961).

[79] Herron, J. T., and Schiff, H. I., A mass spectrometric study of normal oxygen and oxygen subjected to electrical discharge, Can. J. Chem. **36**, 1159 (1958).

[82] Hobrock, B. G., and Kiser, R. W., Electron impact spectroscopy of tetramethylsilicon, –tin and –lead, J. Phys. Chem. **65**, 2186 (1961).

[83] Hobrock, B. G., and Kiser, R. W., Electron impact spectroscopy of tetramethylgermanium, trimethylsilane, and dimethylmercury, J. Phys. Chem. **66**, 155 (1962).

[84] Hobrock, B. G., and Kiser, R. W., Electron impact investigations of sulfur compounds. III. 2–Thiapropane, 3–thiapentane, and 2,3,4–trithiapentane, J. Phys. Chem. **67**, 1283 (1963).

[86] Hurzeler, H., Inghram, M. G., and Morrison, J. D., Photon impact studies of molecules using a mass spectrometer, J. Chem. Phys. **28**, 76 (1958).

[87] Beck, D., Neutral fragments from hydrocarbons under electron impact, Discussions Faraday Soc. **36**, 56 (1963).

[88] Junk, G., and Svec, H., The mass spectra of the α–amino acids, J. Am. Chem. Soc. **85**, 839 (1963).

[90] Kandel, R. J., Appearance potential studies. II. Nitromethane, J. Chem. Phys. **23**, 84 (1955).

[92] Al–Joboury, M. I., May, D. P., and Turner, D. W., Molecular photoelectron spectroscopy. Part IV. The ionisation potentials and configurations of carbon dioxide, carbon oxysulphide, carbon disulphide, and nitrous oxide, J. Chem. Soc., 6350 (1965).

[95] Murad, E., and Inghram, M. G., Photoionization of aliphatic ketones, J. Chem. Phys. **40**, 3263 (1964).

[97] Lambdin, W. J., Tuffly, B. L., and Yarborough, V. A., Appearance potentials as obtained with an analytical mass spectrometer, Appl. Spectry. **13**, 71 (1959).

[99] Dibeler, V. H., and Reese, R. M., Multiple ionization of sodium vapor by electron impact, J. Chem. Phys. **31**, 282 (1959).

[100] Loughran, E. D., and Mader, C., Appearance potential study of tetrafluorohydrazine, J. Chem. Phys. **32**, 1578 (1960).

[101] Lowrey, A., III, and Watanabe, K., Absorption and ionization coefficients of ethylene oxide, J. Chem. Phys. **28**, 208 (1958).

[102] Margrave, J. L., A mass spectrometric appearance potential study of diborane, J. Phys. Chem. **61**, 38 (1957).

[103] Margrave, J. L., Ionization potentials of B_5H_9, B_5H_8I, $B_{10}H_{14}$, and $B_{10}H_{13}C_2H_5$ from electron impact studies, J. Chem. Phys. **32**, 1889 (1960).

[104] Kiser, R. W., Estimation of the ionization potential and dissociation energy of molecular astatine, J. Chem. Phys. **33**, 1265 (1960).

[108] McDowell, C. A., Lossing, F. P., Henderson, I. H. S., and Farmer, J. B., Free radicals by mass spectrometry. X. The ionization potentials of methyl substituted allyl radicals, Can. J. Chem. **34**, 345 (1956).

[112] Winters, R. E., and Kiser, R. W., A mass spectrometric investigation of nickel tetracarbonyl and iron pentacarbonyl, Inorg. Chem. **3**, 699 (1964).

[114] Momigny, J., Comportement des dihalogénoéthylènes cis et trans sous l'impact électronique, Bull. Soc. Chim. Belges **70**, 241 (1961).

[115] Wada, Y., and Kiser, R. W., A mass spectrometric study of trimethyl borate, J. Phys. Chem. **68**, 1588 (1964).

[116] Brion, C. E., Ionization of oxygen by "monoenergetic" electrons, J. Chem. Phys. **40**, 2995 (1964).

[117] Nakayama, T., Kitamura, M. Y., and Watanabe, K., Ionization potential and absorption coefficients of nitrogen dioxide, J. Chem. Phys. **30**, 1180 (1959).

[119] Melton, C. E., and Hamill, W. H., Appearance potentials of positive and negative ions by mass spectrometry, J. Chem. Phys. **41**, 546 (1964).

[120] Palmer, T. F., and Lossing, F. P., Free radicals by mass spectrometry. XXVIII. The HS, CH_3S, and phenyl–S radicals: ionization potentials and heats of formation, J. Am. Chem. Soc. **84**, 4661 (1962).

[122] Meyer, F., and Harrison, A. G., An electron impact study of methyl phenyl ethers, Can. J. Chem. **42**, 2008 (1064).

[123] Pottie, R. F., Harrison, A. G., and Lossing, F. P., Free radicals by mass spectrometry. XXIV. Ionization potentials of cycloalkyl free radicals and cycloalkanes, J. Am. Chem. Soc. **83**, 3204 (1961).

[124] Pottie, R. F., and Lossing, F. P., Free radicals by mass spectrometry. XXIII. Mass spectra of benzyl and $\alpha-d_2-$benzyl free radicals, J. Am. Chem. Soc. **83**, 2634 (1961).

[125] Pottie, R. F., and Lossing, F. P., Free radicals by mass spectrometry. XXV. Ionization potentials of cyanoalkyl radicals, J. Am. Chem. Soc. **83**, 4737 (1961).

[126] Pottie, R. F., and Lossing, F. P., Free radicals by mass spectrometry. XXIX. Ionization potentials of substituted cyclopentadienyl radicals, J. Am. Chem. Soc. **85**, 269 (1963).

[127] Reed, R. I., Studies in electron impact methods. Part 1.–Formaldehyde, deuteroformaldehyde, and some related compounds, Trans. Faraday Soc. **52**, 1195 (1956).

[128] Reed, R. I., and Brand, J. C. D., Electron impact studies. Part 4.–Glyoxal, methylglyoxal and diacetyl, Trans. Faraday Soc. **54**, 478 (1958).

[129] Reed, R. I., and Snedden, W., Studies in electron impact methods. Part 2.–The latent heat of sublimation of carbon, Trans. Faraday Soc. **54**, 301 (1958).

[130] Reed, R. I., and Thornley, M. B., Studies in electron impact methods. Part 5.–Acetaldehyde, acrolein, benzaldehyde, and propionaldehyde, Trans. Faraday Soc. **54**, 949 (1958).

[131] Reed, R. I., and Snedden, W., Studies in electron impact methods. Part 6.–The formation of the methine and carbon ions, Trans. Faraday Soc. **55**, 876 (1956).

[132] Reed, R. I., and Snedden, W., The ionisation potential of NH, J. Chem. Soc., 4132 (1959).

[133] Samson, J. A. R., Marmo, F. F., and Watanabe, K., Absorption and photoionization coefficients of propylene and butene–1 in the vacuum ultraviolet, J. Chem. Phys. **36**, 783 (1962).

[135] Schissler, D. O., and Stevenson, D. P., The benzyl–hydrogen bond dissociation energy from electron impact measurements, J. Chem. Phys. **22**, 151 (1954).

[138] Sun, H., and Weissler, G. L., Absorption cross sections of methane and ammonia in the vacuum ultraviolet, J. Chem. Phys. **23**, 1160 (1955).

[139] Kiser, R. W., and Hisatsune, I. C., Electron impact spectroscopy of nitrogen dioxide, J. Phys. Chem. **65**, 1444 (1961).

[141] Lossing, F. P., Kebarle, P., and DeSousa, J. B., Ionization potentials of alkyl and halogenated alkyl free radicals, Advan. Mass Spectrom. **1**, 431 (1959).

[145] Lossing, F. P., and DeSousa, J. B., Free radicals by mass spectrometry. XIV. Ionization potentials of propyl and butyl free radicals, J. Am. Chem. Soc. **81**, 281 (1959).

[148] Tanaka, Y., Jursa, A. S., and LeBlanc, F. J., Higher ionization potentials of linear triatomic molecules. I. CO_2, J. Chem. Phys. **32**, 1199 (1960).

[149] Tanaka, Y., Jursa, A. S., and LeBlanc, F. J., Higher ionization potentials of linear triatomic molecules. II. CS_2, COS, and N_2O, J. Chem. Phys. **32**, 1205 (1960).

[151] Taubert, R., and Lossing, F. P., Free radicals by mass spectrometry. XXVII. Ionization potentials of four pentyl radicals, J. Am. Chem. Soc. **84**, 1523 (1962).

[153] Wada, Y., and Kiser, R. W., Electron impact spectroscopy of some substituted oxiranes, J. Phys. Chem. **66**, 1652 (1962).

[154] Dibeler, V. H., Reese, R. M., and Franklin, J. L., Mass spectrometric study of cyanogen and cyanoacetylenes, J. Am. Chem. Soc. **83**, 1813 (1961).

[155] Walker, W. C., and Weissler, G. L., Photoionization efficiencies and cross sections in NH_3, J. Chem. Phys. **23**, 1540 (1955).

[156] Walker, W. C., and Weissler, G. L., Preliminary data on photoionization efficiencies and cross sections in C_2H_4 and C_2H_2, J. Chem. Phys. **23**, 1547 (1955).

[157] Walker, W. C., and Weissler, G. L., Photoionization efficiencies and cross sections in N_2O and NO, J. Chem. Phys. **23**, 1962 (1955).

[158] Watanabe, K., Photoionization and total absorption cross section of gases. I. Ionization potentials of several molecules. Cross sections of NH_3 and NO, J. Chem. Phys. **22**, 1564 (1954).

[159] Watanabe, K., and Mottl, J. R., Ionization potentials of ammonia and some amines, J. Chem. Phys. **26**, 1773 (1957).

[160] Tsuda, S., and Hamill, W. H., Structure in ionization-efficiency curves near threshold from alkanes and alkyl halides, J. Chem. Phys. **41**, 2713 (1964).

[161] Watanabe, K., and Nakayama, T., Absorption and photoionization coefficients of furan vapor, J. Chem. Phys. **29**, 48 (1958).

[162] Watanabe, K., and Namioka, T., Ionization potential of propyne, J. Chem. Phys. **24**, 915 (1956).

[163] Weissler, G. L., Samson, J. A. R., Ogawa, M., and Cook, G. R., Photoionization analysis by mass spectroscopy, J. Opt. Soc. Am. **49**, 338 (1959).

[164] Collin, J., Etude des états excités des ions par spectrométrie de masse, Bull. Classe Sci. Acad. Roy. Belg. **45**, 734 (1959).

[165] Wiberg, K. B., Bartley, W. J., and Lossing, F. P., On the strain energy in cyclopropene and the heat of formation of the $C_3H_3^+$ ion, J. Am. Chem. Soc. **84**, 3980 (1962).

[166] Collin, J. E., Ionization and dissociation of molecules by monoenergetic electrons. V. Acetylene and ethylene, Bull. Soc. Chim. Belges **71**, 15 (1962).

[168] Bralsford, R., Harris, P. V., and Price, W. C., The effect of fluorine on the electronic spectra and ionization potentials of molecules, Proc. Roy. Soc. (London) **A258**, 459 (1960).

[169] Collin, J., L'ionisation et la dissociation des molécules par des électrons monoenergetiques. II.—Etats excités de l'ion moléculaire de CO_2 et CS_2, J. Chim. Phys. **57**, 424 (1960).

[171] Hirota, K., Nagoshi, K., and Hatada, M., Studies on mass spectra and appearance potentials of acetic acid and deuteroacetic acid CD_3COOH, Bull. Chem. Soc. Japan **34**, 226 (1961).

[173] Saalfeld, F. E., and Svec, H. J., Mass spectra of volatile hydrides. III. Silylphosphine, Inorg. Chem. **3**, 1442 (1964).

[174] Porter, R. F., and Schoonmaker, R. C., A mass spectrometric study of the vaporization of ferrous bromide, J. Phys. Chem. **63**, 626 (1959).

[176] Hobrock, B. G., and Kiser, R. W., Electron impact spectroscopy of sulfur compounds. I. 2–Thiabutane, 2–thiapentane, and 2,3–dithiabutane, J. Phys. Chem. **66**, 1648 (1962).

[178] Berkowitz, J., and Marquart, J. R., Mass–spectrometric study of the magnesium halides, J. Chem. Phys. **37**, 1853 (1962).

[182] Watanabe, K., Nakayama, T., and Mottl, J., Ionization potentials of some molecules, J. Quant. Spectry. Radiative Transfer **2**, 369 (1962).

[185] Curran, R. K., Positive and negative ion formation in CCl_3F, J. Chem. Phys. **34**, 2007 (1961).

[186] Hobrock, B. G., and Kiser, R. W., Electron impact investigations of sulfur compounds. II. 3–Methyl–2–thiabutane, 4–thia–1–pentene, and 3,4–dithiahexane, J. Phys. Chem. **67**, 648 (1963).

[188] Hobrock, B. G., and Kiser, R. W., Electron impact spectroscopy of propylene sulfide, J. Phys. Chem. **66**, 1551 (1962).

[189] Kaufman, J. J., Koski, W. S., Kuhns, L. J., and Law, R. W., Appearance and ionization potentials of selected fragments from decaborane, $B^{11}_{10}H_{14}$, J. Am. Chem. Soc. **84**, 4198 (1962).

[190] Matsunaga, F. M., Photoionization yield of several molecules in the Schumann region, Contribution No. 27, Hawaii Institute of Geophysics, Honolulu (1961).

[191] Kiser, R. W., and Hobrock, B. G., The ionization potential of hydrogen disulfide (H_2S_2), J. Phys. Chem. **66**, 1214 (1962).

[192] Sandoval, A. A., Moser, H. C., and Kiser, R. W., Ionization and dissociation processes in phosphorus trichloride and diphosphorus tetrachloride, J. Phys. Chem. **67**, 124 (1963).

[193] Shenkel, R. C., Hobrock, B. G., and Kiser, R. W., The ionization potential of (iso)thiocyanic acid, J. Phys. Chem. **66**, 2074 (1962).

[194] Omura, I., Study on unimolecular decomposition of excited olefin ions, Bull. Chem. Soc. Japan **35**, 1845 (1962).

[195] Omura, I., Mass spectra at low ionizing voltage and bond dissociation energies of molecular ions from hydrocarbons, Bull. Chem. Soc. Japan **34**, 1227 (1961).

[196] Fox, R. E., and Curran, R. K., Ionization processes in CCl_4 and SF_6 by electron beams, J. Chem. Phys. **34**, 1595 (1961).

[198] Fox, R. E., Multiple ionization in argon and krypton by electron impact, J. Chem. Phys. **33**, 200 (1960).

[199] Fox, R. E., Negative ion formation in hydrogen chloride by electron impact, J. Chem. Phys. **26**, 1281 (1957).

[200] Curran, R. K., Negative ion formation in various gases at pressures up to .5 mm of Hg, Scientific Paper 62–908–113–P7, Westinghouse Research Laboratories, Pittsburgh (1962).

[201] Kiser, R. W., Studies of the shapes of ionization–efficiency curves of multiply charged monatomic ions. I. Instrumentation and relative electronic–transition probabilities for krypton and xenon ions, J. Chem. Phys. **36**, 2964 (1962).

[202] Kiser, R. W., and Hobrock, B. G., The ionization potentials of cyclopropyl radical and cyclopropyl cyanide, J. Phys. Chem. **66**, 957 (1962).

[204] Brand, J. C. D., and Reed, R. I., The electronic spectrum of formaldehyde. Part II. Mechanisms of dissociation of formaldehyde and the formaldehyde molecular ion, J. Chem. Soc., 2386 (1957).

[205] Kaufman, J. J., Koski, W. S., Kuhns, L. J., and Wright, S. S., Appearance and ionization potentials of selected fragments from isotopically labeled pentaboranes, J. Am. Chem. Soc. **85**, 1369 (1963).

[206] Koski, W. S., Kaufman, J. J., and Pachucki, C. F., A mass spectroscopic appearance potential study of some boron trihalides, J. Am. Chem. Soc. **81**, 1326 (1959).

[209] Koski, W. S., Kaufman, J. J., Pachucki, C. F., and Shipko, F. J., A mass spectrometric appearance potential study of isotopically labeled diboranes, J. Am. Chem. Soc. **80**, 3202 (1958).

[210] Van Raalte, D., and Harrison, A. G., Ionization and dissociation of formate esters by electron impact, Can. J. Chem. **41**, 2054 (1963).

[211] Dorman, F. H., and Morrison, J. D., Ionization potentials of multiply charged krypton, xenon, and mercury, J. Chem. Phys. **34**, 1407 (1961).

[212] Dorman, F. H., and Morrison, J. D., Double and triple ionization in molecules induced by electron impact, J. Chem. Phys. **35**, 575 (1961).

[213] Morrison, J. D., Hurzeler, H., Inghram, M. G., and Stanton, H. E., Threshold law for the probability of excitation of molecules by photon impact. A study of the photoionization efficiencies of Br_2, I_2, HI, and CH_3I, J. Chem. Phys. **33**, 821 (1960).

[214] Margrave, J. L., Ionization potentials for C_2F_4, C_2F_3Cl, and $C_2F_2Cl_2$ and the appearance potential of CF_2^+ from C_2F_4, J. Chem. Phys. **31**, 1432 (1959).

[217] Basila, M. R., and Clancy, D. J., The ionization potentials of monosubstituted pyridines by electron impact, J. Phys. Chem. **67**, 1551 (1963).

[218] Kiser, R. W., and Gallegos, E. J., A technique for the rapid determination of ionization and appearance potentials, J. Phys. Chem. **66**, 947 (1962).

[219] Lifshitz, C., and Bauer, S. H., Mass spectra of valence tautomers, J. Phys. Chem. **67**, 1629 (1963).

[224] Melton, C. E., and Hamill, W. H., Appearance potentials by the retarding potential–difference method for secondary ions produced by excited–neutral, excited ion–neutral, and ion–neutral reactions, J. Chem. Phys. **41**, 1469 (1964).

[227] Watanabe, K., Marmo, F., and Inn, E. C. Y., Formation of the D layer, Phys. Rev. **90**, 155 (1953).

[228] Watanabe, K., Marmo, F. F., and Inn, E. C. Y., Photoionization cross section of nitric oxide, Phys. Rev. **91**, 1155 (1953).

[230] Wainfan, N., Walker, W. C., and Weissler, G. L., Photoionization efficiencies and cross sections in O_2, N_2, CO_2, A, H_2O, H_2, and CH_4, Phys. Rev. **99**, 542 (1955).

[248] Al–Joboury, M. I., and Turner, D. W., Molecular photo–electron spectroscopy. Part I. The hydrogen and nitrogen molecules, J. Chem. Soc., 5141 (1963).

[261] Walsh, A. D., The absorption spectra of the chloro ethylenes in the vacuum ultra–violet, Trans. Faraday Soc. **41**, 35 (1945).

[268] Momigny, J., Ionization potentials and the structures of the photo–ionization yield curves of ethylene and its halogeno derivatives, Nature **199**, 1179 (1963).

[269] Teegan, J. P., and Walsh, A. D., The absorption spectrum of 1:1–dichloroethylene in the vacuum ultra–violet, Trans. Faraday Soc. **47**, 1 (1951).

[282] Morrison, J. D., and Nicholson, A. J. C., Studies of ionization efficiency. Part II. The ionization potentials of some organic molecules, J. Chem. Phys. **20**, 1021 (1952).

[286] Frost, D. C., and McDowell, C. A., The determination of ionization and dissociation potentials of molecules by radiation with electrons, Report No. AFCRL–TR–60–423, University of British Columbia, Vancouver (1960).

[287] Frost, D. C., and McDowell, C. A., The ionization and dissociation of oxygen by electron impact, J. Am. Chem. Soc. **80**, 6183 (1958).

[288] Frost, D. C., and McDowell, C. A., Recent electron impact studies on simple molecules (O_2, Cl_2, I_2), Advan. Mass Spectrom. **1**, 413 (1959).

[289] Frost, D. C., and McDowell, C. A., Studies of the ionization of molecules by electron impact. II. Excited states of the molecular ions of methane and the methyl halides, Proc. Roy. Soc. (London) **A241**, 194 (1957).

[292] Frost, D. C., and McDowell, C. A., The ionization and dissociation of some halogen molecules by electron impact, Can. J. Chem. **38**, 407 (1960).

[297] Inn, E. C. Y., The photoionization of molecules in the vacuum ultraviolet, Phys. Rev. **91**, 1194 (1953).

[298] Majer, J. R., Patrick, C. R., and Robb, J. C., Appearance potentials of the acetyl radical–ion, Trans. Faraday Soc. **57**, 14 (1961).

[299] Majer, J. R., and Patrick, C. R., Ionization potentials of perfluorocycloalkanes, Nature **193**, 161 (1962).

[301] Majer, J. R., and Patrick, C. R., Electron impact on some halogenated aromatic compounds, Trans. Faraday Soc. **58**, 17 (1962).

[303] Gowenlock, B. G., Jones, P. P., and Majer, J. R., Bond dissociation energies in some molecules containing alkyl substituted CH_3, NH_2, and OH, Trans. Faraday Soc. **57**, 23 (1961).

[304] Gowenlock, B. G., Majer, J. R., and Snelling, D. R., Bond dissociation energies in some azo compounds, Trans. Faraday Soc. **58**, 670 (1962).

[305] Brion, C. E., and Dunning, W. J., Electron impact studies of simple carboxylic esters, Trans. Faraday Soc. **59**, 647 (1963).

[306] Gowenlock, B. G., Haynes, R. M., and Majer, J. R., Electron impact measurement on some organomercury compounds, Trans. Faraday Soc. **58**, 1905 (1962).

[307] Gowenlock, B. G., Kay, J., and Majer, J. R., Electron impact studies of some sulphides and disulphides, Trans. Faraday Soc. **59**, 2463 (1963).

[308] Majer, J. R., and Patrick, C. R., Appearance potentials of the benzoyl radical–ion, Trans. Faraday Soc. **59**, 1274 (1963).

[314] Paulett, G. S., and Ettinger, R., Mass spectra and appearance potentials of diazirine and diazomethane, J. Chem. Phys. **39**, 825 (1963).

[315] Hobrock, B. G., Shenkel, R. C., and Kiser, R. W., Singly– and doubly–charged ions from methyl and ethyl isothiocyanates by electron impact, J. Phys. Chem. **67**, 1684 (1963).

[317] Bakulina, I. N., and Ionov, N. I., Determination of the ionization potential of uranium atoms by the surface ionization method, Zh. Eksperim. i Teor. Fiz. **36**, 1001 (1959).

[318] Berkowitz, J., Chupka, W. A., Blue, G. D., and Margrave, J. L., Mass spectrometric study of the sublimation of lithium oxide, J. Phys. Chem. **63**, 644 (1959).

[319] Blanchard, L. P., and Le Goff, P., Mass spectrometric study of the species CS, SO, and CCl_2 produced in primary heterogeneous reactions, Can. J. Chem. **35**, 89 (1957).

[328] Cloutier, G. G., and Schiff, H. I., Electron impact study of nitric oxide using a modified retarding potential difference method, J. Chem. Phys. **31**, 793 (1959).

[331] Ditchburn, R. W., Absorption cross–sections in the vacuum ultra–violet. III. Methane, Proc. Roy. Soc. (London) **A229**, 44 (1955).

[333] Drowart, J., De Maria, G., and Inghram, M. G., Thermodynamic study of SiC utilizing a mass spectrometer, J. Chem. Phys. **29**, 1015 (1958).

[339] Van Raalte, D., and Harrison, A. G., Energetics and mechanism of hydronium ion formation by electron impact, Can. J. Chem. **41**, 3118 (1963).

[340] Franklin, J. L., Dibeler, V. H., Reese, R. M., and Krauss, M., Ionization and dissociation of hydrazoic acid and methyl azide by electron impact, J. Am. Chem. Soc. **80**, 298 (1958).

[341] Fred, M., and Tomkins, F. S., Preliminary term analysis of Am I and Am II spectra, J. Opt. Soc. Am. **47**, 1076 (1957).

[344] Hammond, V. J., Price, W. C., Teegan, J. P., and Walsh, A. D., The absorption spectra of some substituted benzenes and naphthalenes in the vacuum ultra-violet, Discussions Faraday Soc. **9**, 53 (1950).

[349] Herzberg, G., and Shoosmith, J., Absorption spectrum of free CH_3 and CD_3 radicals, Can. J. Phys. **34**, 523 (1956).

[355] Iczkowski, R. P., and Margrave, J. L., Absorption spectrum of fluorine in the vacuum ultraviolet, J. Chem. Phys. **30**, 403 (1959).

[356] Irsa, A. P., Electron impact studies on C_2H_5Cl, C_2H_5Br, and C_2H_5I, J. Chem. Phys. **26**, 18 (1957).

[357] Irsa, A. P., and Friedman, L., Mass spectra of halogen fluorides, J. Inorg. Nucl. Chem. **6**, 77 (1958).

[362] Kreuzer, H., Appearance-Potentiale von BF_3^+ und BF_2^+ aus BF_3 bei Elektronenstoss, Z. Naturforsch. **12a**, 519 (1957).

[364] Law, R. W., and Margrave, J. L., Mass spectrometer appearance potentials for positive ion fragments from BF_3, $B(CH_3)_3$, $B(C_2H_5)_3$, $B(OCH_3)_3$, and $HB(OCH_3)_2$, J. Chem. Phys. **25**, 1086 (1956).

[381] Omura, I., Baba, H., and Higasi, K., The ionization potentials of some conjugated molecules, J. Phys. Soc. Japan **10**, 317 (1955).

[383] Omura, I., Higasi, K., and Baba, H., Ionization potentials of some organic molecules. I. Apparatus, Bull. Chem. Soc. Japan **29**, 501 (1956).

[384] Omura, I., Higasi, K., and Baba, H., Ionization potentials of some organic molecules. II. Aliphatic compounds, Bull. Chem. Soc. Japan **29**, 504 (1956).

[387] Price, W. C., The far ultraviolet absorption spectra and ionization potentials of H_2S, CS_2, and SO_2, Bull. Am. Phys. Soc. **10**, 9 (1935).

[397] Schoonmaker, R. C., and Porter, R. F., Mass spectrometric study of ferrous chloride vapor, J. Chem. Phys. **29**, 116 (1958).

[401] Reese, R. M., and Dibeler, V. H., Ionization and dissociation of nitrogen trifluoride by electron impact, J. Chem. Phys. **24**, 1175 (1956).

[409] Tanaka, Y., Jursa, A. S., and LeBlanc, F. J., Higher ionization potentials of linear triatomic molecules, Spectrochim. Acta **10**, 233 (1957).

[410] Tanaka, Y., Jursa, A. S., and LeBlanc, F. J., Higher ionization potentials of linear triatomic molecules CO_2, CS_2, COS, N_2O, J. Chem. Phys. **28**, 350 (1958).

[411] Varsel, C. J., Morrell, F. A., Resnik, F. E., and Powell, W. A., Qualitative and quantitative analysis of organic compounds. Use of low-voltage mass spectrometry, Anal. Chem. **32**, 182 (1960).

[413] Wacks, M. E., and Dibeler, V. H., Electron impact studies of aromatic hydrocarbons. I. Benzene, naphthalene, anthracene, and phenanthrene, J. Chem. Phys. **31**, 1557 (1959).

[414] Waldron, J. D., The ionization and dissociation of methyl radicals on electron impact, Metropolitan Vickers Gaz. **27**, 66 (1956).

[416] Watanabe, K., Ionization potentials of some molecules, J. Chem. Phys. **26**, 542 (1957).

[418] Reese, R. M., Dibeler, V. H., and Franklin, J. L., Electron impact studies of sulfur dioxide and sulfuryl fluoride, J. Chem. Phys. **29**, 880 (1958).

[419] Lifshitz, C., and Long, F. A., Appearance potentials and mass spectra of fluorinated ethylenes. I. Decomposition mechanisms and their energetics, J. Phys. Chem. **67**, 2463 (1963).

[422] Wilkinson, P. G., Absorption spectra and ionization potentials of benzene and benzene-d_6. J. Chem. Phys. **24**, 917 (1956).

[423] Wilkinson, P. G., Absorption spectra of benzene and benzene-d_6 in the vacuum ultra-violet, Spectrochim. Acta **8**, 283 (1956).

[424] Dibeler, V. H., Franklin, J. L., and Reese, R. M., Electron impact studies of hydrazine and the methyl-substituted hydrazines, J. Am. Chem. Soc. **81**, 68 (1959).

[427] Samson, J. A. R., and Cook, G. R., Photoionization of water vapor with mass analysis of ions, Bull. Am. Phys. Soc. **4**, 459 (1959).

[431] Higasi, K., Nozoe, T., and Omura, I., Ionization potentials of some organic molecules. IV. Troponoid compounds, Bull. Chem. Soc. Japan **30**, 408 (1957).

[439] Marriott, J., and Craggs, J. D., Ionization and dissociation by electron impact. I. Trifluoromethyl halides, J. Electron. **1**, 405 (1956).

[440] Marriott, J., and Craggs, J. D., Ionization and dissociation by electron impact. II. Boron trifluoride and boron trichloride, J. Electron. Control **3**, 194 (1957).

[441] Farmer, J. B., Henderson, I. H. S., Lossing, F. P., and Marsden, D. G. H., Free radicals by mass spectrometry. IX. Ionization potentials of CF_3 and CCl_3 radicals and bond dissociation energies in some derivatives, J. Chem. Phys. **24**, 348 (1956).

[462] Collin, J., and Lossing, F. P., Ionization and dissociation of allene, propyne, 1-butyne, and 1,2- and 1,3-butadienes by electron impact; the $C_3H_3^+$ ion, J. Am. Chem. Soc. **79**, 5848 (1957).

[463] Frost, D. C., and McDowell, C. A., Excited states of the molecular ions of hydrogen fluoride, hydrogen iodide, water, hydrogen sulphide, and ammonia, Can. J. Chem. **36**, 39 (1958).

[464] Berkowitz, J., and Wexler, S., On the ionization potential of the CH_2 radical, J. Chem. Phys. **37**, 1476 (1962).

[1004] Collin, J. E., Ionization potential of NO_2, Nature **196**, 373 (1962).

[1007] Cullen, W. R., and Frost, D. C., Ionization potentials of some perfluoroalkyl arsines, Can. J. Chem. **40**, 390 (1962).

[1011] Palmer, T. F., and Lossing, F. P., Free radicals by mass spectrometry. XXX. Ionization potentials of anilino and 2-, 3-, and 4-pyridylmethyl radicals, J. Am. Chem. Soc. **85**, 1733 (1963).

[1012] Kaneko, Y., Ionization efficiency curves for A^+, Kr^+, N_2^+, and CO^+ by electron impact, J. Phys. Soc. Japan **16**, 1587 (1960).

[1013] Collin, J. E., L'ionisation et la dissociation du nitrate d'éthyle sous l'impact électronique, et le potentiel d'ionisation du dioxyde d'azote, Bull. Soc. Roy. Sci. Liège **32**, 133 (1963).

[1019] Dibeler, V. H., and Reese, R. M., Mass spectrometric study of the photoionization of acetylene and acetylene-d_2, J. Chem. Phys. **40**, 2034 (1964).

[1020] Norman, J. H., Staley, H. G., and Bell, W. E., Mass spectrometric study of gaseous oxides of rhodium and palladium, J. Phys. Chem. **68**, 662 (1964).

[1022] Nakayama, T., and Watanabe, K., Absorption and photoionization coefficients of acetylene, propyne, and 1-butyne, J. Chem. Phys. **40**, 558 (1964).

[1023] Colin, R., and Drowart, J., Thermodynamic study of germanium monotelluride using a mass spectrometer, J. Phys. Chem. **68**, 428 (1964).

[1024] Fehlner, T. P., and Koski, W. S., The fragmentation of some boron hydrides by electron impact, J. Am. Chem. Soc. **86**, 581 (1964).

[1029] Asundi, R. K., Craggs, J. D., and Kurepa, M. V., Electron attachment and ionization in oxygen, carbon monoxide and carbon dioxide, Proc. Phys. Soc. (London) **82**, 967 (1963).

[1032] Nicholson, A. J. C., Photo–ionization efficiency curves. Measurement of ionization potentials and interpretation of fine structure, J. Chem. Phys. **39**, 954 (1963).

[1033] Wada, Y., and Kiser, R. W., Mass spectrometric study of phosphine and diphosphine, Inorg. Chem. **3**, 174 (1964).

[1035] Berkowitz, J., and Chupka, W. A., Vaporization processes involving sulfur, J. Chem. Phys. **40**, 287 (1964).

[1036] Fischler, J., and Halmann, M., Electron–impact studies of phosphorus compounds, J. Chem. Soc., 31 (1964).

[1039] Earnshaw, D. G., Cook, G. L., and Dinneen, G. U., A study of selected ions in the mass spectra of benzenethiol and deuterated benzenethiol, J. Phys. Chem. **68**, 296 (1964).

[1040] Newton, A. S., Triple ionization in small molecules, J. Chem. Phys. **40**, 607 (1964).

[1047] Westmore, J. B., Mann, K. H., and Tickner, A. W., Mass spectrometric study of the nonstoichiometric vaporization of cadmium arsenide, J. Phys. Chem. **68**, 606 (1964).

[1050] Frost, D. C., McDowell, C. A., and Vroom, D. A., Photoelectron spectroscopy with a spherical analyzer. The vibrational energy levels of H_2^+, Phys. Rev. Letters **15**, 612 (1965).

[1051] Momigny, J., Measurement of molecular excitation potentials by means of an ion source, J. Chem. Phys. **25**, 787 (1956).

[1058] Momigny, J., Ionization potentials of *cis* and *trans* dichloro–and dibromo–ethylenes, Nature **191**, 1089 (1961).

[1059] Godbole, E. W., and Kebarle, P., Ionization and dissociation of deuterated ethyl and isopropyl acetates and ethyl formate under electron impact, Trans. Faraday Soc. **58**, 1897 (1962).

[1062] Bibby, M. M., and Carter, G., Ionization and dissociation in some fluorocarbon gases, Trans. Faraday Soc. **59**, 2455 (1963).

[1064] Kuroda, H., Ionization potentials of polycyclic aromatic hydrocarbons, Nature **201**, 1214 (1964).

[1065] Burns, R. P., DeMaria, G., Drowart, J., and Inghram, M. G., Mass spectrometric investigation of the vaporization of In_2O_3, J. Chem. Phys. **38**, 1035 (1963).

[1066] Crable, G. F., and Kearns, G. L., Effect of substituent groups on the ionization potentials of benzenes, J. Phys. Chem. **66**, 436 (1962).

[1067] Chelobov, F. N., Dubov, S. S., Tikhomirov, M. I., and Dobrovitskii, M. I., Ionization and dissociation of hexafluoropropylene by electrons of different energies, Dokl. Akad. Nauk SSSR **151**, 631 (1963) [Engl. transl.: Dokl. Phys. Chem. **151**, 670 (1963)].

[1068] Elder, F. A., Giese, C., Steiner, B., and Inghram, M., Photo–ionization of alkyl free radicals, J. Chem. Phys. **36**, 3292 (1962).

[1069] Wacks, M. E., Electron–impact studies of aromatic hydrocarbons. II. Naphthacene, naphthaphene, chrysene, triphenylene, and pyrene, J. Chem. Phys. **41**, 1661 (1964).

[1070] Walsh, A. D., and Warsop, P. A., The ultra–violet absorption spectrum of ammonia, Trans. Faraday Soc. **57**, 345 (1961).

[1072] Collin, J. E., Ionisation et dissociation du méthane, de la méthylamine et de l'alcool méthylique par des électrons monoénergétiques, Colloq. Spectros. Intern., 9th, Lyons, 1961 (1962) **3**, p. 596.

[1075] Steele, W. C., and Stone, F. G. A., An electron impact study of 1,1,1–trifluoroethane, 1,1,1–trifluoropropane and 3,3,3–trifluoropropene, J. Am. Chem. Soc. **84**, 3450 (1962).

[1076] Steele, W. C., Nichols, L. D., and Stone, F. G. A., An electron impact investigation of some organoboron difluorides, J. Am. Chem. Soc. **84**, 1154 (1962).

[1077] Kaufman, J. J., The effect of substitution on the ionization potentials of free radicals and molecules. IV. δ_K values for alcohols, ethers, thiols, and sulfides, J. Phys. Chem. **66**, 2269 (1962).

[1078] Herzberg, G., The ionization potential of CH_2, Can J. Phys. **39**, 1511 (1961).

[1079] Fisher, I. P., Palmer, T. F., and Lossing, F. P., The vertical ionization potentials of phenyl and phenoxy radicals, J. Am. Chem. Soc. **86**, 2741 (1964).

[1091] Price, W. C., and Passmore, T. R., Discussions Faraday Soc. **35**, 232 (1963).

[1094] McGowan, J. W., Clarke, E. M., Hanson, H. P., and Stebbings, R. F., Electron– and photon–impact ionization of O_2, Phys. Rev. Letters **13**, 620 (1964).

[1097] Tanaka, Y., and Jursa, A. S., Higher ionization potentials of NO_2, J. Chem. Phys. **36**, 2493 (1962).

[1099] Murad, E., and Inghram, M. G., Thermodynamic properties of the acetyl radical and bond dissociation energies in aliphatic carbonyl compounds, J. Chem. Phys. **41**, 404 (1964).

[1100] Munson, M. S. B., and Franklin, J. L., Energetics of some gaseous oxygenated organic ions, J. Phys. Chem. **68**, 3191 (1964).

[1101] Halmann, M., and Klein, Y., Electron–impact spectroscopy of phosphorus compounds. Part III. Positive– and negative–ion formation from phosphorus trichloride and phosphoryl chloride, J. Chem. Soc., 4324 (1964).

[1102] Hall, L. H., Subbanna, V. V., and Koski, W. S., Ionization and appearance potentials of selected ions from decaborane–16, B_5H_8I and B_5H_8Br, J. Am. Chem. Soc. **86**, 3969 (1964).

[1103] Watanabe, K., and Jursa, A. S., Absorption and photoionization cross sections of H_2O and H_2S, J. Chem. Phys. **41**, 1650 (1964).

[1104] Ehlert, T. C., Blue, G. D., Green, J. W., and Margrave, J. L., Mass spectrometric studies at high temperatures. IV. Dissociation energies of the alkaline earth monofluorides, J. Chem. Phys. **41**, 2250 (1964).

[1105] Green, J. W., Blue, G. D., Ehlert, T. C., and Margrave, J. L., Mass spectrometric studies at high temperatures. III. The sublimation pressures of magnesium, strontium, and barium fluorides, J. Chem. Phys. **41**, 2245 (1964).

[1106] Theard, L. P., and Hildenbrand, D. L., Heat of formation of $Be_2O(g)$ by mass spectrometry, J. Chem. Phys. **41**, 3416 (1964).

[1107] Winters, R. E., and Kiser, R. W., Mass spectrometric studies of chromium, molybdenum, and tungsten hexacarbonyls, Inorg. Chem. **4**, 157 (1965).

[1108] Al–Joboury, M. I., May, D. P., and Turner, D. W., Molecular photoelectron spectroscopy. Part III. The ionization potentials of oxygen, carbon monoxide, nitric oxide, and acetylene, J. Chem. Soc., 616 (1965).

[1109] Bauer, S. H., Herzberg, G., and Johns, J. W. C., The absorption spectrum of BH and BD in the vacuum ultraviolet, J. Mol. Spectry. **13**, 256 (1964).

[1110] Kiser, R. W., and Hobrock, D. L., The ionization potential of carbon tetrafluoride, J. Am. Chem. Soc. **87**, 922 (1965).

[1111] Dorman, F. H., Appearance potentials of the fragment ions from CF_3I, J. Chem. Phys. **41**, 2857 (1964).

[1112] White, D., Seshadri, K. S., Dever, D. F., Mann, D. E., and Linevski, M. J., Infrared spectra and the structures and thermodynamics of gaseous LiO, Li_2O, and Li_2O_2, J. Chem. Phys. **39**, 2463 (1963).

[1114] Price, W. C., and Wood, R. W., The far ultraviolet absorption spectra and ionization potentials of C_6H_6 and C_6D_6, J. Chem. Phys. **3**, 439 (1935).

[1115] El–Sayed, M. F. A., Kasha, M., and Tanaka, Y., Ionization potentials of benzene, hexadeuterobenzene, and pyridine from their observed Rydberg series in the region 600–2000 Å, J. Chem. Phys. **34**, 334 (1961).

[1116] Verhaegen, G., Stafford, F. E., and Drowart, J., Mass spectrometric study of the systems boron–carbon and boron–carbon–silicon, J. Chem. Phys. **40**, 1622 (1964).

[1118] Dibeler, V. H., Reese, R. M., and Krauss, M., Mass spectrometric study of the photoionization of small molecules, Advan. Mass Spectrom. **3**, 471 (1966).

[1119] Fehlner, T. P., and Koski, W. S., Fragmentation of tetraborane by electron impact, J. Am. Chem. Soc. **85**, 1905 (1963).

[1120] Steiner, B., Giese, C. F., and Inghram, M. G., Photoionization of alkanes. Dissociation of excited molecular ions, J. Chem. Phys. **34**, 189 (1961).

[1121] Briglia, D. D., and Rapp, D., Ionization of the hydrogen molecule by electron impact near threshold, Phys. Rev. Letters **14**, 245 (1965).

[1122] Meyer, F., Haynes, P., McLean, S., and Harrison, A. G., An electron impact study of some C_8H_{10} isomers, Can. J. Chem. **43**, 211 (1965).

[1124] Norman, J. H., Staley, H. G., and Bell, W. E., Mass-spectrometric study of gaseous oxides of iridium, J. Chem. Phys. **42**, 1123 (1965).

[1125] Cotter, J. L., Electron–impact fragmentation patterns of 3,5–diphenyl–1,2,4–oxadiazole and 2,5–diphenyl–1,3,4–oxadiazole, J. Chem. Soc., 5491 (1964).

[1126] Cotter, J. L., Electron impact fragmentation patterns of acetanilide and benzamide, J. Chem. Soc., 5477 (1964).

[1127] Cotter, J. L., Electron impact studies of some aromatic fluorocarbons, J. Chem. Soc., 1520 (1965).

[1128] Dibeler, V. H., Krauss, M., Reese, R. M., and Harllee, F. N., Mass–spectrometric study of photoionization. III. Methane and methane–d_4, J. Chem. Phys. **42**, 3791 (1965).

[1129] Beck, D., and Osberghaus, O., Das Massenspektrum der ungeladenen Bruchstücke bei Elektronenbeschuss von Molekülen, Z. Physik **160**, 406 (1960).

[1130] Al–Joboury, M. I., and Turner, D. W., Molecular photoelectron spectroscopy. Part II. A summary of ionization potentials, J. Chem. Soc., 4434 (1964).

[1131] Jacobson, A., Steigman, J., Strakna, R. A., and Friedland, S. S., Appearance potential study of $CBrF_3$, J. Chem. Phys. **24**, 637 (1956).

[1132] Dibeler, V. H., Reese, R. M., and Mohler, F. L., Ionization and dissociation of hexafluorobenzene by electron impact, J. Chem. Phys. **26**, 304 (1957).

[1133] Jackson, D. S., and Schiff, H. I., Mass spectrometric investigation of active nitrogen, J. Chem. Phys. **23**, 2333 (1955).

[1136] Dibeler, V. H., and Reese, R. M., Selected positive and negative ions in the mass spectra of the monohalomethanes, J. Res. NBS **54**, 127 (1955).

[1139] Martin, R. H., Lampe, F. W., and Taft, R. W., An electron–impact study of ionization and dissociation in methoxy–and halogen–substituted methanes, J. Am. Chem. Soc. **88**, 1353 (1966).

[1140] Vilesov, F. I., and Zaitsev, V. M., Photoionization of phenyl derivatives of group 5 elements, Dokl. Akad. Nauk SSSR **154**, 886 (1964) [Engl. transl.: Dokl. Phys. Chem. **154**, 117 (1964)].

[1141] Akopyan, M. E., and Vilesov, F. I., Decay of the molecular ions formed in photoionization of hydrazine and some of its alkyl derivatives, Kinetika i Kataliz **4**, 39 (1963) [Engl. transl.: Kinetics Catalysis (USSR) **4**, 32 (1963)].

[1142] Vilesov, F. I., Photoionisation of organic vapours in the vacuum ultra–violet, Zh. Fiz. Khim. **35**, 2010 (1961) [Engl. transl.: Russ. J. Phys. Chem. **35**, 986 (1961)].

[1143] Dibeler, V. H., Reese, R. M., and Krauss, M., Mass–spectrometric study of photoionization. II. H_2, HD, and D_2, J. Chem. Phys. **42**, 2045 (1965).

[1144] Paulett, G. S., and Lustig, M., Sulfur–nitrogen and oxygen–nitrogen bond dissociation energies of some N–fluorinated amines, J. Am. Chem. Soc. **87**, 1020 (1965).

[1145] Natalis, P., Étude du comportement d'isomères géométriques sous l'impact électronique. V. Comparaison des données obtenues, par différentes méthodes, pour les 1,2–diméthylcyclohexanes cis et trans, Bull. Soc. Chim. Belges **73**, 961 (1964).

[1146] Natalis, P., and Laune, J., Étude du comportement d'isomères géométriques sous l'impact électronique. IV. Cyclopropane, méthylcyclopropane et diméthylcyclopropanes gem, cis et trans, Bull. Soc. Chim. Belges **73**, 944 (1964).

[1147] Akopyan, M. E., Vilesov, F. I., and Terenin, A. N., A mass–spectroscopic study of the spectral dependence of the efficiency of photoionization of benzene derivatives, Dokl. Akad. Nauk SSSR **140**, 1037 (1961) [Engl. transl.: Soviet Phys. – Dokl. **6**, 890 (1962)].

[1148] Dressler, K., and Miescher, E., Absorption spectrum of the NO molecule. V. Survey of excited states and their interactions, Astrophys. J. **141**, 1266 (1965).

[1149] Dawton, R. H. V. M., and Wilkinson, K. L., Plutonium: evaporation tests, ionisation potential and electron emission, Report No. A.E.R.E. GP/R 1906, Atomic Energy Research Establishment, Harwell (1956).

[1155] Drowart, J., Burns, R. P., DeMaria, G., and Inghram, M. G., Mass spectrometric study of carbon vapor, J. Chem. Phys. **31**, 1131 (1959).

[1159] Kurbatov, B. L., Vilesov, F. I., and Terenin, A. N., Kinetic energy distribution of electrons during the photoionization of methyl derivatives of benzene, Dokl. Akad. Nauk SSSR **140**, 797 (1961) [Engl. transl.: Soviet Phys. – Dokl. **6**, 883 (1962)].

[1160] Akopyan, M. E., and Vilesov, F. I., Excited states of positive ions and dissociative photoionization of aromatic amines, Dokl. Akad. Nauk SSSR **158**, 1386 (1964) [Engl. transl.: Dokl. Phys. Chem. **158**, 965 (1964)].

[1165] Alekseev, N. I., and Kaminskii, D. L., Ionization of some rare earth elements on the surfaces of tungsten, rhenium, and iridium, Zh. Tekhn. Fiz. **34**, 1521 (1964) [Engl. transl.: Soviet Phys. – Tech. Phys. **9**, 1177 (1965)].

[1166] Vilesov, F. I., and Terenin, A. N., The photoionization of the vapors of certain organic compounds, Dokl. Akad. Nauk SSSR **115**, 744 (1957) [Engl. transl.: Proc. Acad. Sci. USSR, Phys. Chem. Sect. **115**, 539 (1957)].

[1167] Vilesov, F. I., and Kurbatov, B. L., Photoionization of esters and metal carbonyls in the gaseous phase, Dokl. Akad. Nauk SSSR **140**, 1364 (1961) [Engl. transl.: Proc. Acad. Sci. USSR, Phys. Chem. Sect. **140**, 792 (1961)].

[1168] Cotter, J. L., Electron–impact fragmentation of pentafluorobenzamide, J. Chem. Soc., 5742 (1965).

[1169] Merer, A. J., The vacuum ultraviolet absorption spectrum of diazomethane, Can. J. Phys. **42**, 1242 (1964).

[1172] Momigny, J., Mesure de potentiels d'excitation moléculaires au moyen d'une source d'ions, Bull. Soc. Chim. Belges **66**, 33 (1957).

[1179] Tanaka, Y., and Ogawa, M., Rydberg absorption series of CO_2 converging to the $^2\Pi_u$ state of CO_2^+, Can. J. Phys. **40**, 879 (1962).

[1182] Natalis, P., Note sur le comportement des isomères cis et trans de la décaline soumis a l'impact électronique, Bull. Soc. Roy. Sci. Liège **31**, 803 (1962).

[1183] Natalis, P., Behaviour of *cis* and *trans* isomers of decaline under electron impact, Nature **197**, 284 (1963).

[1184] Natalis, P., Étude du comportement d'isomères géométriques sous l'impact électronique. II. Les cis et trans hexahydroindanes, Bull. Soc. Chim. Belges **72**, 374 (1963).

[1185] Momigny, J., and Wirtz–Cordier, A. M., Les effets de l'impact électronique sur des dérivés dihalogénés du benzène, Ann. Soc. Sci. Bruxelles **76**, 164 (1962).

[1189] Hudson, R. D., Measurements of the molecular absorption cross section and the photoionization of sodium vapor between 1600 and 3700 Å, J. Chem. Phys. **43**, 1790 (1965).

[1190] Momigny, J., Les potentiels d'ionisation des dihalogénoéthylènes cis et trans, Bull. Classe Sci. Acad. Roy. Belg. **46**, 686 (1960).

[1197] Momigny, J., Brakier, L., and D'Or, L., Comparaison entre les effets de l'impact électronique sur le benzène et sur les isomères du benzène en chaîne ouverte, Bull. Classe Sci. Acad. Roy. Belg. **48**, 1002 (1962).

[1211] Collin, J., and Nagels, M., Ionization potential and molecular complex of benzothiazol with iodine, Nature **190**, 82 (1961).

[1217] Huber, K. P., Die Rydberg—Serien im Absorptions–spektrum des NO—Moleküls, Helv. Phys. Acta **34**, 929 (1961).

[1235] Cook, G. R., and Ching, B. K., Photoionization and absorption cross sections and fluorescence of CF_4, J. Chem. Phys. **43**, 1794 (1965).

[1237] Natalis, P., and Franklin, J. L., Ionization and dissociation of diphenyl and condensed ring aromatics by electron impact. II. Diphenylcarbonyls and ethers, J. Phys. Chem. **69**, 2943 (1965).

[1238] Natalis, P., and Franklin, J. L., Ionization and dissociation of diphenyl and condensed ring aromatics by electron impact. I. Biphenyl, diphenylacetylene, and phenanthrene, J. Phys. Chem. **69**, 2935 (1965).

[1239] Winchell, P., Mass spectrometric investigation of barium iodide and caesium iodide vaporizations, Nature **206**, 1252 (1965).

[1240] Stuber, F. A., Multiple ionization in neon, argon, krypton, and xenon, J. Chem. Phys. **42**, 2639 (1965).

[1241] Collin, J., Ionization and dissociation of molecules by monoenergetic electrons. III. On the existence of a bent excited state of NO_2^+, J. Chem. Phys. **30**, 1621 (1959).

[1243] Colin, R., Drowart, J., and Verhaegen, G., Mass–spectrometric study of the vaporization of tin oxides. Dissociation energy of SnO, Trans. Faraday Soc. **61**, 1364 (1965).

[1244] Verhaegen, G., Colin, R., Exsteen, G., and Drowart, J., Mass spectrometric determination of the stability of gaseous molybdites, tungstites, molybdates and tungstates of magnesium, calcium, strontium and tin, Trans. Faraday Soc. **61**, 1372 (1965).

[1245] Drowart, J., Colin, R., and Exsteen, G., Mass–spectrometric study of the vaporization of lead monoxide. Dissociation energy of PbO, Trans. Faraday Soc. **61**, 1376 (1965).

[1246] Thorburn, R., Ionization and dissociation by electron impact in carbon tetrabromide, Brit. J. Appl. Phys. **16**, 1397 (1965).

[1249] Kent, R. A., and Margrave, J. L., Mass spectrometric studies at high temperatures. VII. The sublimation pressure of chromium(II) fluoride and the dissociation energy of chromium(I) fluoride, J. Am. Chem. Soc. **87**, 3582 (1965).

[1251] McGowan, J. W., and Fineman, M. A., Rotational structure at threshold in the H_2^+ electron–impact ionization spectrum, Phys. Rev. Letters **15**, 179 (1965).

[1252] Pottie, R. F., Ionization potential and heat of formation of the difluoromethylene radical, J. Chem. Phys. **42**, 2607 (1965).

[1253] Nicholson, A. J. C., Photoionization–efficiency curves. II. False and genuine structure, J. Chem. Phys. **43**, 1171 (1965).

[1254] Hatada, M., and Hirota, K., The ionization and dissociation of aliphatic ketones under electron impact. II. Methyl–sec–butylketone and methyl–tert–butylketone, Z. Physik. Chem. (Frankfurt) **44**, 328 (1965).

[1255] Drowart, J., Degrève, F., Verhaegen, G., and Colin, R., Thermochemical study of the germanium oxides using a mass spectrometer. Dissociation energy of the molecule GeO, Trans. Faraday Soc. **61**, 1072 (1965).

[1256] Hatada, M., and Hirota, K., The ionization and dissociation of methyl butyl ketones under electron impact. I. Methyl n–butyl ketone and methyl isobutyl ketone, Bull. Chem. Soc. Japan **38**, 599 (1965).

[1264] Olmsted, J., III, Street, K., Jr., and Newton, A. S., Excess–kinetic–energy ions in organic mass spectra, J. Chem. Phys. **40**, 2114 (1964).

[1280] Kent, R. A., and Margrave, J. L., Mass spectrometric studies at high temperatures. VIII. The sublimation pressure of iron(II) fluoride, J. Am. Chem. Soc. **87**, 4754 (1965).

[1281] Farrell, P. G., and Newton, J., Ionization potentials of aromatic amines, J. Phys. Chem. **69**, 3506 (1965).

[1284] Dillard, J. G., and Kiser, R. W., Ionization and dissociation of ruthenium and osmium tetroxides, J. Phys. Chem. **69**, 3893 (1965).

[1286] Murakawa, K., The ionization potential of Tm I, J. Phys. Soc. Japan **20**, 1733 (1965).

[1288] Lifshitz, C., and Long, F. A., Appearance potentials and mass spectra of fluorinated ethylenes. II. Heats of formation of fluorinated species and their positive ions, J. Phys. Chem. **69**, 3731 (1965).

[1290] Lifshitz, C., and Long, F. A., Appearance potentials and mass spectra of C_3F_6, C_3F_5Cl, and $c-C_3F_6$, J. Phys. Chem. **69**, 3741 (1965).

[1297] Hildenbrand, D. L., and Murad, E., Dissociation energy of boron monofluoride from mass–spectrometric studies, J. Chem. Phys. **43**, 1400 (1965).

[1372] Martignoni, P., Morgan, R. L., and Cason, C., Rapid determination of electron impact ionization and appearance potentials, Rev. Sci. Instr. **36**, 1783 (1965).

[1378] Dorman, F. H., First differential ionization efficiency curves for fragment ions by electron impact, J. Chem. Phys. **44**, 35 (1966).

[1381] Winters, R. E., and Kiser, R. W., Ions produced in the mass spectrometer from cyclopentadienylmetal carbonyl compounds of cobalt, manganese and vanadium, J. Organometal. Chem. **4**, 190 (1965).

[1382] Cook, G. R., Metzger, P. H., and Ogawa, M., Photoionization and absorption coefficients of CO in the 600 to 1000 Å region, Can. J. Phys. **43**, 1706 (1965).

[1390] Baldwin, M., Maccoll, A., Kirkien–Konasiewicz, A., and Saville, B., Ionisation potentials of the N–methylated thioureas, Chem. Ind., 286 (1966).

[1399] Dibeler, V. H., and Walker, J. A., Ion–pair process in CH_3Cl by photoionization, J. Chem. Phys. **43**, 1842 (1965).

[1400] Botter, R., Dibeler, V. H., Walker, J. A., and Rosenstock, H. M., Experimental and theoretical studies of photoionization–efficiency curves for C_2H_2 and C_2D_2, J. Chem. Phys. **44**, 1271 (1966).

[1404] Dorman, F. H., Fragment ions from CH_3CHO and $(CH_3)_2CO$ by electron impact, J. Chem. Phys. **42**, 65 (1965).

[1406] Momigny, J., Urbain, J., and Wankenne, H., Les effets de l'impact électronique sur la pyridine et les diazines isomères, Bull. Soc. Roy. Sci. Liège **34**, 337 (1965).

[1407] Fehlner, T. P., The identification of the P_2H_2 molecule in the pyrolysis of diphosphine, J. Am. Chem. Soc. **88**, 1819 (1966).

[1408] Ehrhardt, H., and Tekaat, T., Auftrittspotentialmessungen an ionisierten Molekülbruchstücken mit kinetischer Anfangsenergie, Z. Naturforsch. **19a**, 1382 (1964).

[1413] Harrison, A. G., and Jones, E. G., Rearrangement reactions following electron impact on ethyl and isopropyl esters, Can. J. Chem. **43**, 960 (1965).

[1414] Waddington, T. C., Lattice energies of phosphonium bromide and iodide and the proton affinity of phosphine, Trans. Faraday Soc. **61**, 2652 (1965).

[1415] Doolittle, P. H., and Schoen, R. I., Electron retarding potential studies of photoionization, Report No. D1-82-0463, Boeing Scientific Research Laboratories, Seattle (1965).

[1418] Dewar, M. J. S., and Rona, P., Doubly charged ions in the mass spectra of some organoboron derivatives, J. Am. Chem. Soc. **87**, 5510 (1965).

[1419] Lifshitz, C., and Long, F. A., Some observations concerning the positive ion decomposition of C_2F_6 and C_3F_8 in the mass spectrometer, J. Phys. Chem. **69**, 3746 (1965).

[1420] Clark, L. B., Ionization potential of azulene, J. Chem. Phys. **43**, 2566 (1965).

[1421] Hess, G. G., Lampe, F. W., and Sommer, L. H., An electron impact study of ionization and dissociation of trimethylsilanes, J. Am. Chem. Soc. **87**, 5327 (1965).

[1424] Yergey, A. L., and Lampe, F. W., An electron impact study of the bond dissociation energies of some trimethyltin compounds, J. Am. Chem. Soc. **87**, 4204 (1965).

[1441] Radwan, T. N., and Turner, D. W., Molecular photoelectron spectroscopy. Part V. Ozone, J. Chem. Soc. (A), 85 (1966).

[1448] Camus, P., and Blaise, J., Nouvelle détermination du potentiel d'ionisation du spectre d'arc du thulium, Compt. Rend. **261**, 4359 (1965).

[1451] Dorman, F. H., Second differential ionization-efficiency curves for fragment ions by electron impact, J. Chem. Phys. **43**, 3507 (1965).

[1455] Fisher, I. P., and Heath, G. A., Dissociation energy of the N–H bond in hydrazine, Nature **208**, 1199 (1965).

[1458] Kent, R. A., McDonald, J. D., and Margrave, J. L., Mass spectrometric studies at high temperatures. IX. The sublimation pressure of copper(II) fluoride, J. Phys. Chem. **70**, 874 (1966).

[1459] Kitagawa, T., Harada, Y., Inokuchi, H., and Kodera, K., Absorption spectrum of vapor phase azulene in vacuum ultraviolet region, J. Mol. Spectry. **19**, 1 (1966).

[2001] Platel, G., Mesures des potentiels d'apparition des ions obtenus par impact électronique dans la phase vapeur des iodures alcalins et des mélanges LiI–MI, J. Chim. Phys. **62**, 1176 (1965).

[2002] Saalfeld, F. E., and Svec, H. J., Mass spectra of volatile hydrides. IV. Silylgermane, J. Phys. Chem. **70**, 1753 (1966).

[2009] Kent, R. A., Zmbov, K. F., Kana'an, A. S., Besenbruch, G., McDonald, J. D., and Margrave, J. L., Mass spectrometric studies at high temperatures. X. The sublimation pressures of scandium(III), yttrium(III) and lanthanum(III), trifluorides, J. Inorg. Nucl. Chem. **28**, 1419 (1966).

[2013] Brehm, B., Massenspektrometrische Untersuchung der Photoionisation von Molekülen, Z. Naturforsch. **21a**, 196 (1966).

[2014] Dorman, F. H., Morrison, J. D., and Nicholson, A. J. C., Threshold law for the probability of excitation by electron impact, J. Chem. Phys. **32**, 378 (1960).

[2015] May, D. P., and Turner, D. W., The higher ionization potentials of chlorobenzene, Chem. Commun., 199 (1966).

[2016] Cuthbert, J., Farren, J., Prahallada Rao, B. S., and Preece, E. R., Appearance potentials and transition probabilities for electron impact ionization of CO and CO^+, Proc. Phys. Soc. (London) **88**, 91 (1966).

[2018] Tsuda, S., and Hamill, W. H., Ionization efficiency measurements by the retarding potential difference method, Advan. Mass Spectrom. **3**, 249 (1966).

[2021] Appell, J., Durup, J., and Heitz, F., Sur le seuil d'apparition des ions fragments produits avec excès d'énergie cinetique, Advan. Mass Spectrom. **3**, 457 (1966).

[2022] Botter, R., Hagemann, R., Nief, G., and Roth, E., Etude thermodynamique de l'hemioxyde de soufre par spectrometrie de masse, Advan. Mass Spectrom. **3**, 951 (1966).

[2023] Foffani, A., Pignataro, S., Cantone, B., and Grasso, F., Mass spectra of metal hexacarbonyls, Z. Physik. Chem. (Frankfurt) **45**, 79 (1965).

[2025] Pignataro, S., Foffani, A., Innorta, G., and Distefano, G., Molecular structural effects on the ionization potentials for metasubstituted aromatic compounds and for compounds of the type $X–CH_2–R$, Z. Physik. Chem. (Frankfurt) **49**, 291 (1966).

[2026] Foffani, A., Pignataro, S., Cantone, B., and Grasso, F., Ionization potentials and substituent effects for aromatic carbonyl compounds, Z. Physik. Chem. (Frankfurt) **42**, 221 (1964).

[2027] Dibeler, V. H., and Walker, J. A., Photoionization efficiency curve for SF_6 in the wavelength region 1050 to 600 Å, J. Chem. Phys. **44**, 4405 (1966).

[2028] Natalis, P., Effect of geometrical isomerism on the behaviour of *cis* and *trans* hexahydroindanes under electron impact, Nature **200**, 881 (1963).

[2029] Lorquet–Julien, J. C., Comportement du 1 fluoro– et du 1 chloropentane normal et des dérivés halogénés correspondants du cyclopentane sous l'impact électronique, Bull. Soc. Roy. Sci. Liège **30**, 170 (1961).

[2030] Fehlner, T. P., and Koski, W. S., Direct detection of the borane molecule and the boryl radical by mass spectrometry, J. Am. Chem. Soc. **86**, 2733 (1964).

[2031] Collin, J. E., Relations between charge–transfer spectra and ionization potentials of some electron–donor organic molecules, Z. Elektrochem. **64**, 936 (1960).

[2032] Asundi, R. K., and Kurepa, M. V., Ionization cross sections in He, Ne, A, Kr and Xe by electron impact, J. Electron. Control **15**, 41 (1963).

[2033] Comes, F. J., and Lessmann, W., Messung von Anregungszuständen des Stickstoffmoleküls mit Hilfe der Photoionisation, Z. Naturforsch. **16a**, 1038 (1961).

[2034] Comes, F. J., and Lessmann, W., Neue Anregungszustände des Argons oberhalb der Ionisationsgrenze $^2P_{3/2}$, Z. Naturforsch. **16a**, 1396 (1961).

[2036] Finan, P. A., Reed, R. I., Snedden, W., and Wilson, J. M., Electron impact and molecular dissociation. Part X. Some studies of glycosides, J. Chem. Soc., 5945 (1963).

[2037] Finch, A. C. M., Charge–transfer spectra and the ionization energy of azulene, J. Chem. Soc., 2272 (1964).

[2040] Hildenbrand, D. L., Theard, L. P., and Saul, A. M., Transpiration and mass spectrometric studies of equilibria involving BOF(g) and $(BOF)_3$(g), J. Chem. Phys. **39**, 1973 (1963).

[2045] Wada, Y., and Kiser, R. W., A mass spectrometric study of some alkyl–substituted phosphines, J. Phys. Chem. **68**, 2290 (1964).

[2047] Dibeler, V. H., Reese, R. M., and Franklin, J. L., Ionization and dissociation of oxygen difluoride by electron impact, J. Chem. Phys. **27**, 1296 (1957).

[2048] Samson, J. A. R., and Cairns, R. B., Ionization potential of O_2, J. Opt. Soc. Am. **56**, 769 (1966).

[2049] Cubicciotti, D., Mass spectrometric study of the vapors over bismuth–sulfur melts, J. Phys. Chem. **67**, 1385 (1963).

[2050] Inghram, M. G., Chupka, W. A., and Berkowitz, J., Thermodynamics of the Ta–O system: the dissociation energies of TaO and TaO_2, J. Chem. Phys. **27**, 569 (1957).

[2052] Stafford, F. E., and Berkowitz, J., Mass–spectrometric study of the reaction of water vapor with solid barium oxide, J. Chem. Phys. **40**, 2963 (1964).

[2053] Svec, H. J., and Flesch, G. D., Thermochemical properties of xenon difluoride and xenon tetrafluoride from mass spectra, Science **142**, 954 (1963).

[2054] Gingerich, K. A., and Lee, P. K., Mass spectrometric study of the vaporization of uranium monophosphide, J. Chem. Phys. **40**, 3520 (1964).

[2055] Connor, J. A., Finney, G., Leigh, G. J., Haszeldine, R. N., Robinson, P. J., Sedgwick, R. D., and Simmons, R. F., Bond dissociation energies in organosilicon compounds, Chem. Commun., 178 (1966).

[2056] Bloom, H., and Hastie, J. W., Vapour phase composition above molten salt mixtures: a mass spectrometric study, Australian J. Chem. **19**, 1003 (1966).

[2059] Zelikoff, M., and Watanabe, K., Absorption coefficients of ethylene in the vacuum ultraviolet, J. Opt. Soc. Am. **43**, 756 (1953).

[2060] Sjögren, H., Ionization potentials, Arkiv Fysik **22**, 437 (1962).

[2063] Colin, R., and Drowart, J., Thermodynamic study of tin selenide and tin telluride using a mass spectrometer, Trans. Faraday Soc. **60**, 673 (1964).

[2064] Price, W. C., The far ultraviolet absorption spectra and ionization potentials of the alkyl halides. Part I, J. Chem. Phys. **4**, 539 (1936).

[2065] Price, W. C., The far ultraviolet absorption spectra and ionization potentials of the alkyl halides. Part II, J. Chem. Phys. **4**, 547 (1936).

[2066] Lindeman, L. P., and Guffy, J. C., Determination of the O–O bond energy in hydrogen peroxide by electron impact, J. Chem. Phys. **29**, 247 (1958).

[2095] Beutler, H., and Jünger, H.-O., Über das Absorptionsspektrum des Wasserstoffs. III. Die Autoionisierung im Term $3p\pi$ $^1\Pi_u$ des H_2 und ihre Auswahlgesetze. Bestimmung der Ionisierungsenergie des H_2, Z. Physik **100**, 80 (1936).

[2098] Krupenie, P. H., The band spectrum of carbon monoxide, Natl. Stand. Ref. Data Ser., Natl. Bur. Stand. NSRDS–NBS 5 (1966).

[2099] Namioka, T., Ogawa, M., and Tanaka, Y., The $b^4\Sigma_g^- \leftarrow X^3\Sigma_g^-$ Rydberg series of the oxygen molecule, Proc. Intern. Symp. Mol. Struct. Spectry., Tokyo, 1962, p. B 208–1.

[2100] Lofthus, A., The molecular spectrum of nitrogen, Spectroscopic Report No. 2, University of Oslo, Blindern, Norway (1960).

[2101] Lampe, F. W., and Field, F. H., The decomposition of neopentane under electron impact, J. Am. Chem. Soc. **81**, 3238 (1959).

[2102] Field, F. H., Franklin, J. L., and Lampe, F. W., Reactions of gaseous ions. II. Acetylene, J. Am. Chem. Soc. **79**, 2665 (1957).

[2103] Momigny, J., Détermination et discussion des potentiels d'apparition d'ions fragmentaires dans le benzène et ses dérivés monohalogénés, Bull. Soc. Roy. Sci. Liège **28**, 251 (1959).

[2105] Kybett, B. D., Carroll, S., Natalis, P., Bonnell, D. W., Margrave, J. L., and Franklin, J. L., Thermodynamic properties of cubane, J. Am. Chem. Soc. **88**, 626 (1966).

[2108] Meyerson, S., and Rylander, P. N., Organic ions in the gas phase. IV. $C_7H_7^+$ and $C_5H_5^+$ ions from alkylbenzenes and cycloheptatriene, J. Chem. Phys. **27**, 901 (1957).

[2109] Meyerson, S., McCollum, J. D., and Rylander, P. N., Organic ions in the gas phase. VIII. Bicycloheptadiene, J. Am. Chem. Soc. **83**, 1401 (1961).

[2112] Van Brunt, R. J., and Wacks, M. E., Electron–impact studies of aromatic hydrocarbons. III. Azulene and naphthalene, J. Chem. Phys. **41**, 3195 (1964).

[2113] Moore, C. E., Ionization potentials and ionization limits derived from the analyses of optical spectra, Natl. Stand. Ref. Data Ser., Natl. Bur. Stand. NSRDS–NBS 34 (1970).

[2116] Saalfeld, F. E., and Svec, H. J., The mass spectra of volatile hydrides. I. The monoelemental hydrides of the group IVB and VB elements, Inorg. Chem. **2**, 46 (1963).

[2123] Drowart, J., Exsteen, G., and Verhaegen, G., Mass spectrometric determination of the dissociation energy of the molecules MgO, CaO, SrO and Sr_2O, Trans. Faraday Soc. **60**, 1920 (1964).

[2125] McKinley, J. D., Mass–spectrometric investigation of the high–temperature reaction between nickel and chlorine, J. Chem. Phys. **40**, 120 (1964).

[2126] DeMaria, G., Burns, R. P., Drowart, J., and Inghram, M. G., Mass spectrometric study of gaseous molybdenum, tungsten, and uranium oxides, J. Chem. Phys. **32**, 1373 (1960).

[2127] Grimley, R. T., Burns, R. P., and Inghram, M. G., Mass–spectrometric study of the osmium–oxygen system, J. Chem. Phys. **33**, 308 (1960).

[2128] Drowart, J., DeMaria, G., Burns, R. P., and Inghram, M. G., Thermodynamic study of Al_2O_3 using a mass spectrometer, J. Chem. Phys. **32**, 1366 (1960).

[2129] Burns, R. P., DeMaria, G., Drowart, J., and Grimley, R. T., Mass spectrometric investigation of the sublimation of molybdenum dioxide, J. Chem. Phys. **32**, 1363 (1960).

[2130] Grimley, R. T., Burns, R. P., and Inghram, M. G., Thermodynamics of the vaporization of Cr_2O_3: dissociation energies of CrO, CrO_2, and CrO_3, J. Chem. Phys. **34**, 664 (1961).

[2131] Norman, J. H., and Staley, H. G., Thermodynamics of the dimerization and trimerization of gaseous tungsten trioxide and molybdenum trioxide, J. Chem. Phys. **43**, 3804 (1965).

[2132] McKinley, J. D., Mass spectrum of yttrium chloride vapor, J. Chem. Phys. **42**, 2245 (1965).

[2133] Saalfeld, F. E., and Svec, H. J., The mass spectra of volatile hydrides. II. Some higher hydrides of the group IVB and VB elements, Inorg. Chem. **2**, 50 (1963).

[2136] Franklin, J. L., and Munson, M. S. B., Ion–molecule reactions in methane–oxygen and acetylene–oxygen systems, Symp. Combust., 10th, Univ. Cambridge, Cambridge, Engl., 1964 (1965) p. 561.

[2137] Saalfeld, F. E., and Svec, H. J., The mass spectrum of stannane, J. Inorg. Nucl. Chem. **18**, 98 (1961).

[2138] DeMaria, G., Drowart, J., and Inghram, M. G., Thermodynamic study of InSb with a mass spectrometer, J. Chem. Phys. **31**, 1076 (1959).

[2139] Colin, R., and Drowart, J., Thermodynamic study of tin sulfide and lead sulfide using a mass spectrometer, J. Chem. Phys. **37**, 1120 (1962).

[2140] Büchler, A., Blackburn, P. E., and Stauffer, J. L., The vaporization of rhenium trichloride and rhenium tribromide, J. Phys. Chem. **70**, 685 (1966).

[2141] Hildenbrand, D. L., and Murad, E., Mass–spectrometric determination of the dissociation energy of beryllium monofluoride, J. Chem. Phys. **44**, 1524 (1966).

[2142] Hildenbrand, D. L., and Theard, L. P., Effusion studies, mass spectra, and thermodynamics of beryllium fluoride vapor, J. Chem. Phys. **42**, 3230 (1965).

[2143] Malone, T. J., and McGee, H. A., Jr., Ionization potentials of the dioxygen fluoride free radical and the dioxygen difluoride molecule, J. Phys. Chem. **70**, 316 (1966).

[2144] Meyer, F., and Harrison, A. G., A mechanism for tropylium ion formation by electron impact, J. Am. Chem. Soc. **86**, 4757 (1964).

[2145] Berkowitz, J., Heat of formation of the CN radical, J. Chem. Phys. **36**, 2533 (1962).

[2146] Baldwin, M., Maccoll, A., and Miller, S. I., Appearance potentials of the lower chloroalkanes, J. Am. Chem. Soc. **86**, 4498 (1964).

[2147] Greene, F. T., and Gilles, P. W., High molecular weight boron sulfides. II. Identification, relative intensities, appearance potentials, and origins of the ions, J. Am. Chem. Soc. **86**, 3964 (1964).

[2148] Ehlert, T. C., and Margrave, J. L., The heat of atomization of aluminum difluoride, J. Am. Chem. Soc. **86**, 3901 (1964).

[2149] Johns, J. W. C., and Barrow, R. F., The band spectrum of silicon monofluoride, SiF, Proc. Phys. Soc. (London) **71**, 476 (1958).

[2150] Barrow, R. F., and Rowlinson, H. C., The absorption spectra of the gaseous monoxides of silicon, germanium and tin in the Schumann region, Proc. Roy. Soc. (London) **A224**, 374 (1954).

[2151] De Maria, G., Guido, M., Malaspina, L., and Pesce, B., Mass–spectrometric study of the yttrium–carbon system, J. Chem. Phys. **43**, 4449 (1965).

[2152] De Maria, G., Gingerich, K. A., Malaspina, L., and Piacente, V., Dissociation energy of the gaseous AlP molecule, J. Chem. Phys. **44**, 2531 (1966).

[2153] King, A. B., Mass–spectral fragmentation of phenyl–*n*– alkanes, J. Chem. Phys. **42**, 3526 (1965).

[2154] Tsuda, S., Melton, C. E., and Hamill, W. H., Ionization– efficiency curves for molecular and fragment ions from methane and the methyl halides, J. Chem. Phys. **41**, 689 (1964).

[2155] Steele, W. C., Jennings, B. H., Botyos, G. L., and Dudek, G. O., An electron impact study of norbornenyl and nortricyclyl chlorides, J. Org. Chem. **30**, 2886 (1965).

[2156] Cotter, J. L., The mass spectra of some fluorine– containing 1,3,4–oxadiazoles, J. Chem. Soc., 6842 (1965).

[2157] Bibby, M. M., Toubelis, B. J., and Carter, G., Ionization and dissociation in CF_4, Electron. Letters **1**, 50 (1965).

[2158] Melton, C. E., and Hamill, W. H., Ionization potentials of alkyl free radicals and of benzene, toluene, and triethylamine by the RPD technique, J. Chem. Phys. **41**, 3464 (1964).

[2159] Khvostenko, V. I., and Sultanov, A. Sh., Ionisation of thallium(I) chloride, bromide, and iodide molecules by electron bombardment, Zh. Fiz. Khim. **39**, 475 (1965) [Engl. transl.: Russ. J. Phys. Chem. **39**, 252 (1965)].

[2160] Steele, W. C., Appearance potentials of the difluoromethylene positive ion, J. Phys. Chem. **68**, 2359 (1964).

[2161] Kent, R. A., Ehlert, T. C., and Margrave, J. L., Mass spectrometric studies at high temperatures. V. The sublimation pressure of manganese(II) fluoride and the dissociation energy of manganese(I) fluoride, J. Am. Chem. Soc. **86**, 5090 (1964).

[2162] Ehlert, T. C., Kent, R. A., and Margrave, J. L., Mass spectrometric studies at high temperatures. VI. The sublimation pressure of nickel(II) fluoride, J. Am. Chem. Soc. **86**, 5093 (1964).

[2163] Meyer, F., and Harrison, A. G., Ionization potentials of methyl–substituted benzenes and cyclopentadienes, Can. J. Chem. **42**, 2256 (1964).

[2164] Fisher, I. P., Homer, J. B., and Lossing, F. P., Free radicals by mass spectrometry. XXXIII. Ionization potentials of CF_2, CF_3CF_2, CF_3CH_2, n–C_3F_7, and i–C_3F_7 radicals, J. Am. Chem. Soc. **87**, 957 (1965).

[2165] Blue, G. D., Green, J. W., Bautista, R. G., and Margrave, J. L., The sublimation pressure of calcium(II) fluoride and the dissociation energy of calcium(I) fluoride, J. Phys. Chem. **67**, 877 (1963).

[2166] Khvostenko, V. I., Ionisation of thiophen and some of its derivatives by electron impact, Zh. Fiz. Khim. **36**, 384 (1962) [Engl. transl.: Russ. J. Phys. Chem. **36**, 197 (1962)].

[2167] Ackermann, R. J., Rauh, E. G., and Thorn, R. J., Thermodynamic properties of gaseous yttrium monoxide. Correlation of bonding in group III transition–metal monoxides, J. Chem. Phys. **40**, 883 (1964).

[2169] Hernandez, G. J., Vacuum–ultraviolet absorption spectra of the cyclic ethers: trimethylene oxide, tetrahydrofuran, and tetrahydropyran, J. Chem. Phys. **38**, 2233 (1963).

[2170] Hernandez, G. J., Vacuum ultraviolet absorption spectrum of dimethyl ether, J. Chem. Phys. **38**, 1644 (1963).

[2171] Gillis, R. G., and Occolowitz, J. L., Isocyanides. III. Electron impact study of aliphatic isocyanides, J. Org. Chem. **28**, 2924 (1963).

[2172] Hagemann, R., Détermination de la chaleur de formation de S_2O par spectrométrie de masse, Compt. Rend. **255**, 1102 (1962).

[2173] Akopyan, M. E., Vilesov, F. I., and Terenin, A. N., Mass– spectrometric investigation of photoionization of molecules and dissociation of excited molecular ions, Izv. Akad. Nauk SSSR, Ser. Fiz. **27**, 1083 (1963) [Engl. transl.: Bull. Acad. Sci. USSR, Phys. Ser. **27**, 1054 (1963)].

[2174] Foffani, A., Pignataro, S., Cantone, B., and Grasso, F., Mass spectra of diazocompounds. I. Diazocarbonyl compounds, Nuovo Cimento **29**, 918 (1963).

[2175] Sholette, W. P., and Porter, R. F., Mass spectrometric study of high temperature reactions in the boron– hydrogen–oxygen system, J. Phys. Chem. **67**, 177 (1963).

[2176] Verhaegen, G., Stafford, F. E., Ackerman, M., and Drowart, J., Mass spectrometric investigation of gaseous species in the system boron–carbon, Nature **193**, 1280 (1962).

[2177] Hassan, G. E. M. A., The atomic spectrum of neodymium. I. Preliminary analysis and ionization potential of Nd I, Physica **29**, 1119 (1963).

[2178] Mesnard, G., Uzan, R., and Cabaud, B., Potentiels d'apparition des ions obtenus dans un spectromètre de masse a partir d'oxydes alcalino–terreux, Cahiers Phys. **17**, 333 (1963).

[2179] Berkowitz, J., Tasman, H. A., and Chupka, W. A., Double–oven experiments with lithium halide vapors, J. Chem. Phys. **36**, 2170 (1962).

[2180] Fineman, M. A., and Petrocelli, A. W., Electron impact study of CO using a Lozier apparatus, J. Chem. Phys. **36**, 25 (1962).

[2181] Wilson, E. G., Jortner, J., and Rice, S. A., A far– ultraviolet spectroscopic study of xenon difluoride, J. Am. Chem. Soc. **85**, 813 (1963).

[2182] Steele, W. C., Nichols, L. D., and Stone, F. G. A., The determination of silicon–carbon and silicon–hydrogen bond dissociation energies by electron impact, J. Am. Chem. Soc. **84**, 4441 (1962).

[2183] Steele, W. C., and Stone, F. G. A., Silicon–silicon bond dissociation energies in disilane and hexachlorodisilane, J. Am. Chem. Soc. **84**, 3599 (1962).

[2185] Dolejšek, Z., Hanuš, V., and Prinzbach, H., Das massenspektrometrische Verhalten von Quadricyclen, Angew. Chem. **74**, 902 (1962).

[2186] Blaise, J., and Vetter, R., Détermination du potentiel d'ionisation du spectre d'arc du thulium, Compt. Rend. **256**, 630 (1963).

[2187] Kuppermann, A., and Raff, L. M., Determination of electronic energy levels of molecules by low–energy electron impact spectroscopy, J. Chem. Phys. **37**, 2497 (1962).

[2188] Grimley, R. T., Burns, R. P., and Inghram, M. G., Thermodynamics of the vaporization of nickel oxide, J. Chem. Phys. **35**, 551 (1961).

[2189] Thrush, B. A., and Zwolenik, J. J., Spectrum of the tropyl radical, Discussions Faraday Soc. **35**, 196 (1963).

[2191] Fineman, M. A., and Petrocelli, A. W., Molecular studies with a Lozier electron impact apparatus, Planetary Space Sci. **3**, 187 (1961).

[2192] James, L. H., and Carter, G., A mass spectrometric study of ionization and dissociation by electron impact of perfluoro–methyl cyclohexane, J. Electron. Control **13**, 213 (1962).

[2194] Cantone, B., Grasso, F., Foffani, A., and Pignataro, S., Calculation of ionization potentials for aromatic compounds, Z. Physik. Chem. (Frankfurt) **42**, 236 (1964).

[2195] Ryabchikov, L. N., and Tikhinskii, G. F., Mass spectrometer study of the sublimation of beryllium chloride, Fiz. Metal. i Metalloved. **10**, 635 (1960) [Engl. transl.: Phys. Metals Metallog. USSR **10**(4), 146 (1960)].

[2196] Kanomata, I., Mass–spectrometric study on ionization and dissociation of diethyl ether by electron impact, Bull. Chem. Soc. Japan **34**, 1596 (1961).

[2199] Eriksson, K. B. S., and Isberg, H. B. S., The spectrum of atomic aluminium, Al I, Arkiv Fysik **23**, 527 (1963).

[2200] Schönheit, E., Massenspektrometrische Untersuchung der Photoionisation von Argon, Z. Naturforsch. **16a**, 1094 (1961).

[2401] Band, S. J., Davidson, I. M. T., Lambert, C. A., and Stephenson, I. L., Bond dissociation energies and heats of formation of trimethylsilyl compounds, Chem. Commun., 723 (1967).

[2402] Besenbruch, G., Charlu, T. V., Zmbov, K. F., and Margrave, J. L., Mass spectrometric studies at high temperatures. XVII. Sublimation and vapor pressures of Dy(III), Ho(III) and Er(III) fluorides, J. Less–Common Metals **12**, 375 (1967).

[2403] Bidinosti, D. R., and McIntyre, N. S., Electron–impact study of some binary metal carbonyls, Can. J. Chem. **45**, 641 (1967).

[2404] Bidinosti, D. R., and McIntyre, N. S., The metal–metal bond dissociation energy in manganese carbonyl, Chem. Commun., 555 (1966).

[2405] Bidinosti, D. R., and McIntyre, N. S., The metal–metal bond dissociation energy in cobalt octacarbonyl, Chem. Commun., 1 (1967).

[2406] Bloom, H., Hastie, J. W., and Morrison, J. D., Ionization and dissociation of the alkali halides by electron impact, J. Phys. Chem. **72**, 3041 (1968).

[2407] Bock, H., Seidl, H., and Fochler, M., d–Orbitaleffekte in silicium–substituierten π–Elektronensystemen. X. Vertikale Ionisierungsenergien von Alkyl– und Silyl–benzolen, Chem. Ber. **101**, 2815 (1968).

[2408] Bock, H., and Seidl, H., d–Orbital effects in silicon–substituted π–electron systems. Part XII. Some spectroscopic properties of alkyl and silyl acetylenes and polyacetylenes, J. Chem. Soc. (B), 1158 (1968).

[2409] Burns, R. P., Jason, A. J., and Inghram, M. G., Evaporation coefficient of boron, J. Chem. Phys. **46**, 394 (1967).

[2410] Bock, H., and Seidl, H., d–Orbitaleffekte in silizium–substituierten π–Elektronensystemen. VI. Spektroskopische Untersuchungen an Alkyl– und Silyläthylenen, J. Organometal. Chem. **13**, 87 (1968).

[2411] Brittain, E. F. H., Wells, C. H. J., and Paisley, H. M., Mass spectra of isomers. Part I. Cyclobutanes and cyclohexenes of molecular formula $C_{10}H_{16}$, J. Chem. Soc. (B), 304 (1968).

[2412] Cotter, J. L., and Dine–Hart, R. A., Ionization and dissociation of some aromatic imides under electron impact, Chem. Commun., 809 (1966).

[2413] Connor, J. A., Haszeldine, R. N., Leigh, G. J., and Sedgwick, R. D., Organosilicon chemistry. Part II. The thermal decomposition of hexamethyldisilane, the Me_3Si–$SiMe_3$ bond dissociation energy, and the ionisation potential of the trimethylsilyl radical, J. Chem. Soc. (A), 768 (1967).

[2414] Collin, J. E., and Delwiche, J., Ionization of methane and its electronic energy levels, Can. J. Chem. **45**, 1875 (1967).

[2416] Arnett, E. M., Sanda, J. C., Bollinger, J. M., and Barber, M., Crowded benzenes. VI. The strain energy in o–di–t–butylbenzenes, J. Am. Chem. Soc. **89**, 5389 (1967).

[2417] Denning, R. G., and Wentworth, R. A. D., A mass spectroscopic study of benzene(cyclopentadienyl)manganese(I), J. Am. Chem. Soc. **88**, 4619 (1966).

[2418] Davidson, I. M. T., and Stephenson, I. L., Appearance potentials of trialkylsilyl ions, J. Organometal. Chem. **7**, 24 (1967).

[2420] Hertel, I., and Ottinger, C., Appearance potentials of metastable molecular ions, Z. Naturforsch. **22a**, 40 (1967).

[2421] D'Or, L., Collin, J. E., and Longrée, J., Ionisation et dissociation de l'éthane sous l'impact électronique. Spectres de masse et phénomènes d'échange dans C_2H_6, C_2H_5D, CH_3CD_3 et C_2D_6, Bull. Classe Sci. Acad. Roy. Belg. **52**, 518 (1966).

[2422] Collin, J. E., and Condé, G., L'ionisation et la dissociation des polyéthers cycliques soumis a l'impact électronique, Bull. Classe Sci. Acad. Roy. Belg. **52**, 978 (1966).

[2423] Collin, J. E., and Cahay, R., L'étude des complexes moléculaires de transfert de charge en phase gaseuze par spectrométrie de masse. Complexe 1:2 de la pyridine avec l'iode, Bull. Classe Sci. Acad. Roy. Belg. **52**, 606 (1966).

[2424] Chambers, D. B., Glockling, F., and Weston, M., Mass spectra of organo–stannanes and –plumbanes, J. Chem. Soc. (A), 1759 (1967).

[2425] Brion, C. E., Oldfield, D. J., and Paddock, N. L., Hückel–type interactions in phosphonitrilic derivatives, Chem. Commun., 226 (1966).

[2426] Berkowitz, J., and Chupka, W. A., Equilibrium composition of selenium vapor; the thermodynamics of the vaporization of HgSe, CdSe, and SrSe, J. Chem. Phys. **45**, 4289 (1966).

[2427] Carette, J.–D., Ionisation par impact électronique de CO_2 et N_2O, Can. J. Phys. **45**, 2931 (1967).

[2428] Collin, J. E., and Franskin, M. J., Ionisation, dissociation et réarrangements intramoléculaires dans les amines aliphatiques soumises a l'impact électronique. Cas de la diéthylamine et de la diéthylamine–Nd, Bull. Soc. Roy. Sci. Liège **35**, 285 (1966).

[2429] Collin, J. E., and Franskin, M. J., Ionisation, dissociation et réarrangements intramoléculaires dans les amines aliphatiques par impact électronique. Cas de la méthylamine et de la méthylamine–Nd_2, Bull. Soc. Roy. Sci. Liège **35**, 267 (1966).

[2430] Čermák, V., Penning ionization electron spectroscopy. I. Determination of ionization potentials of polyatomic molecules, Collection Czech. Chem. Commun. **33**, 2739 (1968).

[2431] Hierl, P. M., and Franklin, J. L., Appearance potentials and kinetic energies of ions from N_2, CO, and NO, J. Chem. Phys. **47**, 3154 (1967).

[2432] Hildenbrand, D. L., Mass–spectrometric studies of bonding in the group IIA fluorides, J. Chem. Phys. **48**, 3657 (1968).

[2433] Hirota, K., and Hatada, M., Mass–spectra of dialkyl ketones and their theoretical interpretation, Kinetika i Kataliz **8**, 748 (1967) [Engl. transl.: Kinetics Catalysis (USSR) **8**, 634 (1967)].

[2434] Hastie, J. W., Bloom, H., and Morrison, J. D., Electron–impact studies of $PbCl_2$, $PbBr_2$, and $PbClBr$, J. Chem. Phys. **47**, 1580 (1967).

[2435] Hastie, J. W., Ficalora, P., and Margrave, J. L., Mass spectrometric studies at high temperatures. XXV. Vapor composition over $LaCl_3$, $EuCl_3$ and $LuCl_3$ and stabilities of the trichloride dimers, J. Less–Common Metals **14**, 83 (1968).

[2436] Hastie, J. W., Zmbov, K. F., and Margrave, J. L., Mass spectrometric studies at high temperatures. XXIII. Vapor equilibria over molten $NaSnF_3$ and $KSnF_3$, J. Inorg. Nucl. Chem. **30**, 729 (1968).

[2438] Gallegos, E. J., Mass spectrometry of some polyphenyls, J. Phys. Chem. **71**, 1647 (1967).

[2439] Gingerich, K. A., The AlCN molecule and its possible isomers, Naturwiss. **54**, 646 (1967).

[2440] Gingerich, K. A., On the existence of gaseous uranium monophosphide, Naturwiss. **53**, 525 (1966).

[2441] Gingerich, K. A., On the equilibrium between monatomic and diatomic palladium and the appearance potential of Pd_2, Naturwiss. **54**, 43 (1967).

[2442] Gingerich, K. A., Dissociation energy of gaseous uranium mononitride, J. Chem. Phys. **47**, 2192 (1967).

[2443] Glemser, O., Müller, A., Böhler, D., and Krebs, B., Die N–S–Bindung: Bindungslängen, Kraftkonstanten, Bindungsgrade und Bindungsenergien, Z. Anorg. Allgem. Chem. **357**, 184 (1968).

[2444] Grimley, R. T., Burns, R. P., and Inghram, M. G., Mass–spectrometric study of the vaporization of cobalt oxide, J. Chem. Phys. **45**, 4158 (1966).

[2445] Gusarov, A. V., and Gorokhov, L. N., Vapor composition over potassium oxide, Teplofiz. Vysokikh Temperatur **4**, 590 (1966) [Engl. transl.: High Temp. (USSR) **4**, 558 (1966)].

[2446] Fujisaki, H., Westmore, J. B., and Tickner, A. W., Mass spectrometric study of subliming selenium, Can. J. Chem. **44**, 3063 (1966).

[2447] Fuchs, R., and Taubert, R., Die kinetische Energie ionisierter Molekülfragmente. IV. Ionen hoher Anfangsenergie in den Massenspektren einfacher Paraffine, Z. Naturforsch. **19a**, 1181 (1964).

[2449] Ficalora, P. J., Hastie, J. W., and Margrave, J. L., Mass spectrometric studies at high temperatures. XXVII. The reactions of aluminum vapor with $S_2(g)$, $Se_2(g)$, and $Te_2(g)$, J. Phys. Chem. **72**, 1660 (1968).

[2450] Fiquet–Fayard, F., Importance des prédissociations dans la fragmentation de l'ion acétylène, J. Chim. Phys. **64**, 320 (1967).

[2451] Fisher, I. P., Intermediates in the pyrolysis and mass spectrometry of chlorine monoxide and chlorine heptoxide, Trans. Faraday Soc. **64**, 1852 (1968).

[2452] Fisher, I. P., and Henderson, E., Mass spectrometry of free radicals, Trans. Faraday Soc. **63**, 1342 (1967).

[2453] Foffani, A., Pignataro, S., Distefano, G., and Innorta, G., Influence of the ligand donor ability on the ionization potentials and fragmentation patterns of transition–metal nitrosyl complexes, J. Organometal. Chem. **7**, 473 (1967).

[2454] Foner, S. N., and Hudson, R. L., Mass spectrometry of free radicals and vibronically excited molecules produced by pulsed electrical discharges, J. Chem. Phys. **45**, 40 (1966).

[2455] Franklin, J. L., and Mogenis, A., An electron impact study of ions from several dienes, J. Phys. Chem. **71**, 2820 (1967).

[2456] Fraser, R. T. M., and Paul, N. C., The mass spectrometry of nitrate esters and related compounds. Part I, J. Chem. Soc. (B), 659 (1968).

[2457] Exsteen, G., Drowart, J., Auwera–Mahieu, A. V., and Callaerts, R., Thermodynamic study of silicon sesquitelluride using a mass spectrometer, J. Phys. Chem. **71**, 4130 (1967).

[2458] Eland, J. H. D., Shepherd, P. J., and Danby, C. J., Ionization potentials of aromatic molecules determined by analytical interpretation of electron impact data, Z. Naturforsch. **21a**, 1580 (1966).

[2459] Balducci, G., Capalbi, A., De Maria, G., and Guido, M., Atomization energy of the NdC_4 molecule, J. Chem. Phys. **48**, 5275 (1968).

[2460] Bancroft, G. M., Reichert, C., and Westmore, J. B., Mass spectral studies of metal chelates. II. Mass spectra and appearance potentials of acetylacetonates of trivalent metals of the first transition series, Inorg. Chem. **7**, 870 (1968).

[2461] Battles, J. E., Gundersen, G. E., and Edwards, R. K., A mass spectrometric study of the rhenium–oxygen system, J. Phys. Chem. **72**, 3963 (1968).

[2462] Bloom, H., and Hastie, J. W., Mass spectrometry of the vapors over $PbCl_2 + ACl$ mixtures (A = Na, K, Rb, or Cs). II. Electron–impact studies, J. Chem. Phys. **49**, 2230 (1968).

[2463] Bock, H., Alt, H., and Seidl, H., d–Orbital effects in silicon–substituted π–electron systems. XV. The color of silyl ketones, J. Am. Chem. Soc. **91**, 355 (1969).

[2464] Burgess, A. R., Sen Sharma, D. K., and White, M. J. D., The mass spectrometry of some alkyl peroxy radicals and related alkyl hydroperoxides, Advan. Mass Spectrom. **4**, 345 (1968).

[2465] Carlson, K. D., Kohl, F. J., and Uy, O. M., Mass spectrometry of molecules of the nitrogen family, Advan. Chem. Ser. **72**, 245 (1968).

[2466] Čermák, V., Penning ionization electron spectroscopy, Advan. Mass Spectrom. **4**, 697 (1968).

[2467] Čermák, V., Retarding–potential measurement of the kinetic energy of electrons released in Penning ionization, J. Chem. Phys. **44**, 3781 (1966).

[2468] Čermák, V., and Herman, Z., Penning ionization electron spectroscopy: ionization of mercury, Chem. Phys. Letters **2**, 359 (1968).

[2469] Colin, R., and Drowart, J., Mass spectrometric determination of dissociation energies of gaseous indium sulphides, selenides and tellurides, Trans. Faraday Soc. **64**, 2611 (1968).

[2470] Collin, J. E., Franskin, M. J., and Hyatt, D., Étude par spectrométrie de masse des méchanismes de dissociation dans l'éthylamine et ses homologues deutérés, Bull. Soc. Roy. Sci. Liège **36**, 318 (1967).

[2471] Cuthbert, J., Farren, J., and Prahallada Rao, B. S., Sequential mass spectrometry: nitrogen and the determination of appearance potentials, Proc. Phys. Soc. (London) **91**, 63 (1967).

[2472] Cuthbert, J., Farren, J., Prahallada Rao, B. S., and Preece, E. R., Sequential mass spectrometry. III. Ions and fragments from carbon dioxide and disulphide, J. Phys. B **1**, 62 (1968).

[2473] Daly, N. R., and Powell, R. E., Electron ionization of He^+ and the destruction of He^{2+} by charge transfer processes in a space–charge trap, Proc. Phys. Soc. (London) **89**, 281 (1966).

[2474] Daly, N. R., and Powell, R. E., Electron collisions in oxygen, Proc. Phys. Soc. (London) **90**, 629 (1967).

[2475] Davidson, I. M. T., and Stephenson, I. L., Bond dissociation energies of trimethylsilyl compounds, Chem. Commun., 746 (1966).

[2476] Newton, A. S., and Sciamanna, A. F., Metastable state of the doubly charged carbon dioxide ion, J. Chem. Phys. **40**, 718 (1964); Erratum, *ibid.*, **44**, 2830 (1966).

[2477] Dey, S. D., and Karmohapatro, S. B., Determination of ionization potential of Gd and Ho by surface ionization method, Indian J. Phys. **40**, 151 (1966).

[2478] Dey, S. D., and Karmohapatro, S. B., First atomic ionization potentials of dysprosium and praseodymium, J. Phys. Soc. Japan **22**, 682 (1967).

[2479] Dillard, J. G., and Kiser, R. W., The formation of gaseous ions from dicyclopentadienyltitanium dichloride and dicyclopentadienylzirconium dichloride upon electron impact, J. Organometal. Chem. **16**, 265 (1969).

[2480] Distefano, G., and Innorta, G., Iron–phosphorus bond dissociation energies in tetrahedral iron carbonyl nitrosyl complexes, J. Organometal. Chem. **14**, 465 (1968).

[2481] Distefano, G., Innorta, G., Pignataro, S., and Foffani, A., Correlation between the ionization potentials of transition metal complexes and of the corresponding ligands, J. Organometal. Chem. **14**, 165 (1968).

[2482] Drowart, J., Goldfinger, P., Detry, D., Rickert, H., and Keller, H., Mass spectrometric study of the equilibria in sulphur vapour generated by an electrochemical knudsen cell, Advan. Mass Spectrom. **4**, 499 (1968).

[2483] Edwards, J. G., and Gilles, P. W., High molecular weight boron sulfides. IV. Mass spectrometric investigation of the conversion of metathioboric acid to boron sulfide, Advan. Chem. Ser. **72**, 211 (1968).

[2484] Ehrhardt, H., and Kresling, A., Die dissoziative Ionisation von N_2, O_2, H_2O, CO_2 und Äthan, Z. Naturforsch. **22a**, 2036 (1967).

[2485] Farrell, P. G., and Newton, J., Ionization potentials of primary aromatic amines and aza–hydrocarbons, Tetrahedron Letters, 5517 (1966).

[2486] Fehlner, T. P., and Callen, R. B., Mass spectrometry of phosphorus hydrides, Advan. Chem. Ser. **72**, 181 (1968).

[2487] Ficalora, P. J., Uy, O. M., Muenow, D. W., and Margrave, J. L., Mass spectrometric studies at high temperatures: XXIX. Thermal decomposition and sublimation of alkali metal sulfates, J. Am. Ceram. Soc. **51**, 574 (1968).

[2488] Fisher, I. P., Mass spectrometry study of intermediates in thermal decomposition of perchloric acid and chlorine dioxide, Trans. Faraday Soc. **63**, 684 (1967).

[2489] Gallegos, E. J., Mass spectrometry and ionization energies of some condensed–ring aromatic and heterocyclic compounds, J. Phys. Chem. **72**, 3452 (1968).

[2490] Gingerich, K. A., Gaseous metal nitrides. II. The dissociation energy, heat of sublimation, and heat of formation of zirconium mononitride, J. Chem. Phys. **49**, 14 (1968).

[2491] Gingerich, K. A., Mass–spectrometric evidence for the molecules UC and CeC and predicted stability of diatomic carbides of electropositive transition metals, J. Chem. Phys. **50**, 2255 (1969).

[2492] Grützmacher, H.–F., and Lohmann, J., Massenspektrometrie instabiler organischer Moleküle. I. Ionisationspotential und Bildungsenthalpie von Dehydrobenzol, Ann. Chem. **705**, 81 (1967).

[2493] Halmann, M., and Platzner, I., Ion–molecule reactions of phosphine in the mass spectrometer, J. Phys. Chem. **71**, 4522 (1967).

[2494] Hertel, G. R., Surface ionization. II. The first ionization potential of uranium, J. Chem. Phys. **47**, 335 (1967).

[2495] Hertel, G. R., Surface ionization. III. The first ionization potentials of the lanthanides, J. Chem. Phys. **48**, 2053 (1968).

[2496] Howe, I., and Williams, D. H., Studies in mass spectrometry. XXX. Substituent effects in mass spectrometry. Comparison of charge localization and quasi–equilibrium theories, J. Am. Chem. Soc. **90**, 5461 (1968).

[2497] Howe, I., Williams, D. H., Kingston, D. G. I., and Tannenbaum, H. P., Substituent effects in the mass spectra of some γ– and β–substituted methyl butyrates, J. Chem. Soc. (B), 439 (1969).

[2498] Johnstone, R. A. W., and Millard, B. J., Some novel eliminations of neutral fragments from ions in mass spectrometry. Part III. Fragmentation of the 1,3–diphenylpropene ion–radical, J. Chem. Soc. (C), 1955 (1966).

[2499] Jones, A., and Lossing, F. P., The ionization potential and heat of formation of thioformaldehyde, J. Phys. Chem. **71**, 4111 (1967).

[2500] Junk, G. A., and Svec, H. J., Energetics of the ionization and dissociation of $Ni(CO)_4$, $Fe(CO)_5$, $Cr(CO)_6$, $Mo(CO)_6$ and $W(CO)_6$, Z. Naturforsch. **23b**, 1 (1968).

[2501] Junk, G. A., Svec, H. J., and Angelici, R. J., Electron impact studies of manganese and rhenium pentacarbonyl halides, J. Am. Chem. Soc. **90**, 5758 (1968).

[2502] Kant, A., and Strauss, B. H., Atomization energies of the polymers of germanium, Ge_2 to Ge_7, J. Chem. Phys. **45**, 882 (1966).

[2503] Kent, R. A., Mass spectrometric studies of plutonium compounds at high temperatures. II. The enthalpy of sublimation of plutonium(III) fluoride and the dissociation energy of plutonium(I) fluoride, J. Am. Chem. Soc. **90**, 5657 (1968).

[2504] Keyes, B. G., and Harrison, A. G., The fragmentation of aliphatic sulfur compounds by electron impact, J. Am. Chem. Soc. **90**, 5671 (1968).

[2505] Khvostenko, V. I., and Furlei, I. I., Ionisation potentials of sulphides, Zh. Fiz. Khim. **42**, 13 (1968) [Engl. transl.: Russ. J. Phys. Chem. **42**, 5 (1968)].

[2506] Kiser, R. W., Dillard, J. G., and Dugger, D. L., Mass spectrometry of inorganic halides, Advan. Chem. Ser. **72**, 153 (1969).

[2507] Kiser, R. W., Krassoi, M. A., and Clark, R. J., Mass spectrometric study of tetrakis(trifluorophosphine) nickel(0), J. Am. Chem. Soc. **89**, 3653 (1967).

[2508] Kohl, F. J., Prusaczyk, J. E., and Carlson, K. D., New gaseous molecules of the pnictides, J. Am. Chem. Soc. **89**, 5501 (1967).

[2509] Kohl, F. J., Uy, O. M., and Carlson, K. D., Cross sections for electron–impact fragmentation and dissociation energies of the dimer and tetramer of bismuth, J. Chem. Phys. **47**, 2667 (1967).

[2510] Kohout, F. C., and Lampe, F. W., Ionization potential of the nitroxyl molecule, J. Chem. Phys. **45**, 1074 (1966).

[2511] Kuznesof, P. M., Stafford, F. E., and Shriver, D. F., Electron impact ionization potentials of methyl-substituted borazines, J. Phys. Chem. **71**, 1939 (1967).

[2512] Lappert, M. F., Litzow, M. R., Pedley, J. B., Riley, P. N. K., and Tweedale, A., Bonding studies of compounds of boron and the group IV elements. Part II. Ionisation potentials of boron halides and mixed halides by electron impact and by molecular orbital calculations, J. Chem. Soc. (A), 3105 (1968).

[2513] Lappert, M. F., Pedley, J. B., Riley, P. N. K., and Tweedale, A., Ionisation potentials and electronic spectra of halogeno– and amino–boranes, and a study of some redistribution reactions, Chem. Commun., 788 (1966).

[2514] Lifschitz, C., Bergmann, E. D., and Pullman, B., The ionization potentials of biological purines and pyrimidines, Tetrahedron Letters, 4583 (1967).

[2515] Loudon, A. G., Maccoll, A., and Webb, K. S., Charge location and fragmentation in mass spectrometry. II. The mass spectra and ionization potentials of some methylguanidines, Advan. Mass Spectrom. **4**, 223 (1968).

[2516] Malone, T. J., and McGee, H. A., Jr., Mass spectrometric investigations of the synthesis, stability, and energetics of the low–temperature oxygen fluorides. II. Ozone difluoride, J. Phys. Chem. **71**, 3060 (1967).

[2517] Balducci, G., Capalbi, A., De Maria, G., and Guido, M., Thermodynamics of rare–earth–carbon systems. I. The cerium–carbon system, J. Chem. Phys. **50**, 1969 (1969).

[2518] Burns, R. P., Systematics of the evaporation coefficient Al_2O_3, Ga_2O_3, In_2O_3, J. Chem. Phys. **44**, 3307 (1966).

[2519] Bancroft, G. M., Reichert, C., Westmore, J. B., and Gesser, H. D., Mass spectral studies of metal chelates. III. Mass spectra and appearance potentials of substituted acetylacetonates of trivalent chromium. Comparison with other trivalent metals of the first transition series, Inorg. Chem. **8**, 474 (1969).

[2520] Lifshitz, C., and Reuben, B. G., Ion–molecule reactions in aromatic systems. I. Secondary ions and reaction rates in benzene, J. Chem. Phys. **50**, 951 (1969).

[2521] Lifshitz, C., and Shapiro, M., Isotope effects on metastable transitions: C_3H_8 and C_3D_8, J. Chem. Phys. **45**, 4242 (1966).

[2522] Occolowitz, J. L., Charge localization and the electron–impact fragmentation of carbonyl compounds, Australian J. Chem. **20**, 2387 (1967).

[2523] Zmbov, K. F., and Margrave, J. L., Mass spectrometric studies at high temperatures. XXI. The heat of atomization of galium oxyfluoride, J. Inorg. Nucl. Chem. **29**, 2649 (1967).

[2524] McDonald, J. D., and Margrave, J. L., Mass–spectrometric studies of volatile Al, Sc, Co, Y, Zr and rare–earth chelates, J. Less–Common Metals **14**, 236 (1968).

[2525] McDonald, J. D., Williams, C. H., Thompson, J. C., and Margrave, J. L., Appearance potentials, ionization potentials and heats of formation for perfluorosilanes and perfluoroborosilanes, Advan. Chem. Ser. **72**, 261 (1968).

[2526] McIntyre, N. S., Auwera–Mahieu, A. V., and Drowart, J., Mass spectrometric determination of the dissociation energies of gaseous RuC, IrC and PtB, Trans. Faraday Soc. **64**, 3006 (1968).

[2527] Mesnard, G., Uzan, R., and Cabaud, B., Étude au spectromètre de masse des produits d'évaporation du bioxyde de titane et du titanate de baryum, Rev. Phys. Appl. **1**, 123 (1966).

[2528] Momigny, J., and Delwiche, J., Photoionisation et impact électronique dans le disulfure de carbone, J. Chim. Phys. **65**, 1213 (1968).

[2529] Momigny, J., and Lorquet, J. C., On the position of the electronic states of the $C_6H_6^+$ and $C_6H_5F^+$ ions, Chem. Phys. Letters **1**, 505 (1968).

[2530] Müller, J., and Göser, P., Massenspektroskopische Untersuchungen an Übergangsmetall–π–Komplexen mit fünf–, sechs– und siebengliedrigen aromatischen Ringliganden, J. Organometal. Chem. **12**, 163 (1968).

[2531] Müller, J., and Herberhold, M., Massenspektroskopische Untersuchungen an $C_5H_5Mn(CO)_3$ und $C_5H_5Mn(CO)_2L$–Komplexen, J. Organometal. Chem. **13**, 399 (1968).

[2532] Murad, E., and Hildenbrand, D. L., Mass–spectrometric study of the thermodynamic properties of Zr–F species, J. Chem. Phys. **45**, 4751 (1966).

[2533] Natalis, P., and Franklin, J. L., Étude du comportement d'isomères géométriques sous l'impact électronique. VI. Les cis– et trans–1,2–di–terbutyl–éthylènes, Bull. Soc. Chim. Belges **75**, 328 (1966).

[2534] McLafferty, F. W., and Wachs, T., Substituent effects in unimolecular ion decompositions. IX. Specific hydrogen rearrangement in butyrophenones, J. Am. Chem. Soc. **89**, 5043 (1967).

[2535] Niehaus, A., Anregung und Dissoziation von Molekülen beim Elektronenbeschuss. Messung der Bildungswahrscheinlichkeit für neutrale Fragmente als Funktion der Elektronenenergie, Z. Naturforsch. **22a**, 690 (1967).

[2536] Norman, J. H., Staley, H. G., and Bell, W. E., Mass spectrometric–Knudsen cell study of the gaseous oxides of platinum, J. Phys. Chem. **71**, 3686 (1967).

[2537] Norman, J. H., Staley, H. G., and Bell, W. E., Mass spectrometric study of the noble metal oxides. Ruthenium–oxygen system, Advan. Chem. Ser. **72**, 101 (1968).

[2538] Nounou, P., Étude des composés aromatiques par spectrométrie de masse. I. Mesure des potentiels d'ionisation et d'apparition par la méthode du potentiel retardateur et interprétation des courbes d'ionisation différentielle, J. Chim. Phys. **63**, 994 (1966).

[2539] Nounou, P., Étude des composés aromatiques par spectrométrie de masse. III. Interprétation des processus d'ionisation et de dissociation des molécules de phénanthrène et de méthyl–phénanthrène par impact électronique, J. Chim. Phys. **65**, 700 (1968).

[2540] Nounou, P., Application of the quasi–equilibrium theory of mass spectra to large aromatic molecules, Advan. Mass Spectrom. **4**, 551 (1968).

[2541] Occolowitz, J. L., and White, G. L., Energetic considerations in the assignment of some fragment ion structures, Australian J. Chem. **21**, 997 (1968).

[2542] Peers, A. M., and Vigny, P., Réactions molécule–ion dans le propylène, J. Chim. Phys. **65**, 805 (1968).

[2543] Pignataro, S., Cassuto, A., and Lossing, F. P., Free radicals by mass spectrometry. XXXVI. Ionization potentials of conjugated and nonconjugated radicals, J. Am. Chem. Soc. **89**, 3693 (1967).

[2544] Pignataro, S., Foffani, A., Innorta, G., and Distefano, G., Mass spectra and ionization potentials of group VI–B substituted transition–metal carbonyls, Advan. Mass Spectrom. **4**, 323 (1968).

[2545] Pignataro, S., and Lossing, F. P., Mass spectra and ionization potentials of $C_6H_6CrC_6H_6$ and $C_6H_6Cr(CO)_3$, J. Organometal. Chem. **10**, 531 (1967).

[2546] Pignataro, S., and Lossing, F. P., Thermal decomposition of organometallic compounds in the ion source of a mass spectrometer, J. Organometal. Chem. **11**, 571 (1968).

[2547] Pitt, C. G., Habercom, M. S., Bursey, M. M., and Rogerson, P. F., The electronic absorption and ionization potentials of carbosilanes. The question of $1,3(d–d)$ interaction in the excited state, J. Organometal. Chem. **15**, 359 (1968).

[2548] Potapov, V. K., and Shigorin, D. N., Relation between nature of electronic states of the acetone molecule and mechanism of its breakdown on electron bombardment, Zh. Fiz. Khim. **40**, 200 (1966) [Engl. transl.: Russ. J. Phys. Chem. **40**, 101 (1966)].

[2549] Prášil, Z., and Forst, W., Mass spectrometry of azomethane, J. Am. Chem. Soc. **90**, 3344 (1968).

[2550] Pritchard, H., and Harrison, A. G., Heat of formation of CHO^+, J. Chem. Phys. **48**, 2827 (1968).

[2551] Redhead, P. A., Multiple ionization of the rare gases by successive electron impacts (0–250 eV). I. Appearance potentials and metastable ion formation, Can. J. Phys. **45**, 1791 (1967).

[2552] Rinke, K., and Schäfer, H., Detection of the molecules W_2Cl_6 and W_3Cl_9 in the gaseous state, Angew. Chem. Intern. Ed. **6**, 637 (1967).

[2553] Yergey, A. L., and Lampe, F. W., An electron impact study of ionization and dissociation of trimethylstannanes, J. Organometal. Chem. **15**, 339 (1968).

[2554] Schäfer, H., Giegling, D., and Rinke, K., Zum System W/O/J. III. Das thermische Verhalten von WO_2J_2, Z. Anorg. Allgem. Chem. **357**, 25 (1968).

[2555] Yamdagni, R., and Porter, R. F., Mass spectrometric and torsion effusion studies of the evaporation of liquid selenium, J. Electrochem. Soc. **115**, 601 (1968).

[2556] Winters, R. E., and Kiser, R. W., Ionization and fragmentation of dimethylzinc, trimethylaluminum, and trimethylantimony, J. Organometal. Chem. **10**, 7 (1967).

[2557] Winters, R. E., Collins, J. H., and Courchene, W. L., Resolution of fine structure in ionization–efficiency curves, J. Chem. Phys. **45**, 1931 (1966).

[2558] Winters, R. E., and Collins, J. H., Mass spectrometric studies of structural isomers. I. Mono– and bicyclic C_7H_{12} molecules, J. Am. Chem. Soc. **90**, 1235 (1968).

[2559] Wilson, J. H., and McGee, H. A., Jr., Mass–spectrometric studies of the synthesis, energetics, and cryogenic stability of the lower boron hydrides, J. Chem. Phys. **46**, 1444 (1967).

[2560] Wiberg, K. B., and Connor, D. S., Bicyclo[1.1.1]pentane, J. Am. Chem. Soc. **88**, 4437 (1966).

[2561] Steck, S. J., Pressley, G. A., Jr., Lin, S.–S., and Stafford, F. E., Mass–spectrometric investigation of the reaction of hydrogen with graphite at 1900°–2400°K, J. Chem. Phys. **50**, 3196 (1969).

[2562] Slifkin, M. A., and Allison, A. C., Measurement of ionization potentials from contact charge transfer spectra, Nature **215**, 949 (1967).

[2563] Svec, H. J., and Junk, G. A., Energetics of the ionization and dissociation of $Mn_2(CO)_{10}$, $Re_2(CO)_{10}$, and $ReMn(CO)_{10}$, J. Am. Chem. Soc. **89**, 2836 (1967).

[2564] Zapesochnyĭ, I. P., and Shpenik, O. B., Atomic excitation by monoenergetic electron beams, Zh. Eksperim. i Teor. Fiz. **50**, 890 (1966) [Engl. transl.: Soviet Phys. JETP **23**, 592 (1966)].

[2565] Zmbov, K. F., Ficalora, P., and Margrave, J. L., Mass spectrometric studies at high temperatures. XXVIII. Gaseous ternary oxides, LiMO and $LiMO_2$, J. Inorg. Nucl. Chem. **30**, 2059 (1968).

[2566] Zmbov, K. F., Hastie, J. W., Hauge, R., and Margrave, J. L., Formation of polymeric $(GeF_2)_n$ in the vapor phase over GeF_2, Inorg. Chem. **7**, 608 (1968).

[2567] Wachs, T., and McLafferty, F. W., The influence of the charge and radical sites in unimolecular ion decomposition, J. Am. Chem. Soc. **89**, 5044 (1967).

[2568] Uy, O. M., Muenow, D. W., and Margrave, J. L., Mass spectrometric studies at high temperatures. Part 35. Stabilities of gaseous $GeCl_2$ and $GeBr_2$, Trans. Faraday Soc. **65**, 1296 (1969).

[2569] Uy, O. M., Muenow, D. W., Ficalora, P. J., and Margrave, J. L., Mass spectrometric studies at high temperatures. Part 30. Vaporization of Ga_2S_3, Ga_2Se_3 and Ga_2Te_3, and stabilities of the gaseous gallium chalcogenides, Trans. Faraday Soc. **64**, 2998 (1968).

[2570] Uy, O. M., Kohl, F. J., and Carlson, K. D., Dissociation energy of PN and other thermodynamic properties for the vaporization of P_3N_5, J. Phys. Chem. **72**, 1611 (1968).

[2571] Tikhomirov, M. V., and Komarov, V. N., Effect of the surface on the mass spectrum of tetrafluoroethylene and the appearance potential of F^+, Zh. Fiz. Khim. **40**, 1392 (1966) [Engl. transl.: Russ. J. Phys. Chem. **40**, 751 (1966)].

[2572] Smith, D., and Kevan, L., Dissociative charge exchange of rare–gas ions with C_2F_6 and C_3F_8, J. Chem. Phys. **46**, 1586 (1967).

[2573] Skudlarski, K., Drowart, J., Exsteen, G., and Auwera–Mahieu, A. V., Thermochemical study of vaporization of sodium and potassium perrhenate using a mass spectrometer, Trans. Faraday Soc. **63**, 1146 (1967).

[2574] Sjögren, H., Formation of carbon dioxide ions after electron and ion impact, Arkiv Fysik **32**, 529 (1966).

[2575] Sjögren, H., The ionization of methane and methyl halides using electron and ion impact, Arkiv Fysik **31**, 159 (1966).

[2576] Shigorin, D. N., Filyugina, A. D., and Potapov, V. K., Ionization and dissociation of molecules of acetaldehyde, acetone, and acetic acid on electron impact, Teor. i Eksperim. Khim. **2**, 554 (1966) [Engl. transl.: Theoret. Exptl. Chem. **2**, 417 (1966)].

[2577] Sessa, P. A., and McGee, H. A., Jr., Mass spectrum and molecular energetics of krypton difluoride, J. Phys. Chem. **73**, 2078 (1969).

[2578] Shapiro, J. S., and Lössing, F. P., Free radicals by mass spectrometry. XXXVII. The ionization potential and heat of formation of dichlorocarbene, J. Phys. Chem. **72**, 1552 (1968).

[2579] Schildcrout, S. M., Pressley, G. A., Jr., and Stafford, F. E., Pyrolysis and molecular–beam mass spectrum of tetracarbonylnickel(0), J. Am. Chem. Soc. **89**, 1617 (1967).

[2580] Schildcrout, S. M., Pearson, R. G., and Stafford, F. E., Ionization potentials of tris(β–diketonate)metal(III) complexes and Koopmans' theorem, J. Am. Chem. Soc. **90**, 4006 (1968).

[2581] Saalfeld, F. E., McDowell, M. V., Gondal, S. K., and MacDiarmid, A. G., The mass spectrum of methyldifluorosilyltetracarbonylcobalt, Inorg. Chem. **7**, 1465 (1968).

[2582] Saalfeld, F. E., McDowell, M. V., Hagen, A. P., and MacDiarmid, A. G., The mass spectrum of trifluorosilyltetracarbonylcobalt, Inorg. Chem. **7**, 1665 (1968).

[2583] Saalfeld, F. E., McDowell, M. V., Gondal, S. K., and MacDiarmid, A. G., The mass spectra of trifluorophosphinecarbonylcobalt hydrides, J. Am. Chem. Soc. **90**, 3684 (1968).

[2584] Saalfeld, F. E., and McDowell, M. V., The mass spectra of volatile hydrides. V. Silylarsine, Inorg. Chem. **6**, 96 (1967).

[2585] Robbins, E. J., Leckenby, R. E., and Willis, P., The ionization potentials of clustered sodium atoms, Advan. Phys. **16**, 739 (1967).

[2586] Tikhomirov, M. V., and Komarov, V. N., Temperature dependence of mass spectra. II. Mass spectra of tetrafluoroethylene, hexafluoropropene, and trifluoroethylene, Zh. Fiz. Khim. **41**, 1065 (1967) [Engl. transl.: Russ. J. Phys. Chem. **41**, 561 (1967)].

[2587] Svec, H. J., and Junk, G. A., Electron–impact studies of substituted alkanes, J. Am. Chem. Soc. **89**, 790 (1967).

[2588] Zmbov, K. F., and Margrave, J. L., The first ionization potentials of samarium, europium, gadolinium, dysprosium, holmium, erbium, thulium, and ytterbium by the electron–impact method, J. Phys. Chem. **70**, 3014 (1966).

[2589] Zmbov, K. F., and Margrave, J. L., Mass–spectrometric studies at high temperatures. XI. The sublimation pressure of NdF_3 and the stabilities of gaseous NdF_2 and NdF, J. Chem. Phys. **45**, 3167 (1966).

[2590] Zmbov, K. F., and Margrave, J. L., Mass spectrometric studies at high temperatures. XII. Stabilities of dysprosium, holmium, and erbium subfluorides, J. Phys. Chem. **70**, 3379 (1966).

[2591] Zmbov, K. F., and Margrave, J. L., Mass spectrometric studies at high temperatures. XV. Sublimation pressures of chromium, manganese and iron trifluorides and the heat of dissociation of $Fe_2F_6(g)$, J. Inorg. Nucl. Chem. **29**, 673 (1967).

[2592] Zmbov, K. F., and Margrave, J. L., Mass spectrometric studies at high temperatures. XVI. Sublimation pressures for TiF_3 and the stabilities of $TiF_2(g)$ and TiF(g), J. Phys. Chem. **71**, 2893 (1967).

[2593] Zmbov, K. F., and Margrave, J. L., Mass spectrometric studies at high temperatures. XIII. Stabilities of samarium, europium and gadolinium mono– and difluorides, J. Inorg. Nucl. Chem. **29**, 59 (1967).

[2594] Zmbov, K. F., and Margrave, J. L., Mass spectrometry at high temperatures. XVIII. The stabilities of the mono– and difluorides of scandium and yttrium, J. Chem. Phys. **47**, 3122 (1967).

[2595] Zmbov, K. F., Hastie, J. W., and Margrave, J. L., Mass spectrometric studies at high temperatures. Part 24. Thermodynamics of vaporization of SnF_2 and PbF_2 and the dissociation energies of SnF and PbF, Trans. Faraday Soc. **64**, 861 (1968).

[2596] Zmbov, K. F., and Margrave, J. L., Mass spectrometric studies at high temperatures. XIV. The vapor pressure and dissociation energy of silver monofluoride, J. Phys. Chem. **71**, 446 (1967).

[2597] Müller, J., and Fenderl, K., Massenspektren und Ionisierungspotentiale von C_5H_5Mn $(CO)_2PX_3$–Komplexen, J. Organomet. Chem. **19**, 123 (1969).

[2598] Zmbov, K. F., and Margrave, J. L., Mass spectrometric studies at high temperatures. XIX. Sublimation pressures and heats of sublimation of TmF_3, YbF_3 and LuF_3, J. Less–Common Metals **12**, 494 (1967).

[2599] Zmbov, K. F., and Margrave, J. L., Mass spectrometric evidence for the gaseous Si_2N molecule, J. Am. Chem. Soc. **89**, 2492 (1967).

[2600] Zmbov, K. F., and Margrave, J. L., Mass spectrometric studies of scandium, yttrium, lanthanum, and rare–earth fluorides, Advan. Chem. Ser. **72**, 267 (1968).

[2601] Zmbov, K. F., Uy, O. M., and Margrave, J. L., Mass spectrometric study of the high–temperature equilibrium $C_2F_4 \rightleftharpoons 2CF_2$ and the heat of formation of the CF_2 radical, J. Am. Chem. Soc. **90**, 5090 (1968).

[2602] Berkowitz, J., Chupka, W. A., and Walter, T. A., Photoionization of HCN: the electron affinity and heat of formation of CN, J. Chem. Phys. **50**, 1497 (1969).

[2603] Berkowitz, J., and Lifshitz, C., Photoionization of high–temperature vapors. II. Sulfur molecular species, J. Chem. Phys. **48**, 4346 (1968).

[2604] Berkowitz, J., and Walter, T. A., Photoionization of high-temperature vapors. IV. TlF, TlCl, and TlBr, J. Chem. Phys. **49**, 1184 (1968).

[2605] Chupka, W. A., Mass–spectrometric study of the photoionization of methane, J. Chem. Phys. **48**, 2337 (1968).

[2606] Chupka, W. A., and Berkowitz, J., Photoionization of ethane, propane, and n–butane with mass analysis, J. Chem. Phys. **47**, 2921 (1967).

[2607] Botter, R., Dibeler, V. H., Walker, J. A., and Rosenstock, H. M., Mass–spectrometric study of photoionization. IV. Ethylene and 1,2–dideuteroethylene, J. Chem. Phys. **45**, 1298 (1966).

[2608] Müller, J., and Göser, P., Massenspektroskopische Untersuchungen an substituierten Benzol–chromtricarbonyl–Komplexen, Chem. Ber. **102**, 3314 (1969).

[2609] Berkowitz, J., and Chupka, W. A., Photoionization of high–temperature vapors. VI. S_2, Se_2, and Te_2, J. Chem. Phys. **50**, 4245 (1969).

[2610] Berkowitz, J., and Chupka, W. A., Photoionization of high–temperature vapors. I. The iodides of sodium, magnesium, and thallium, J. Chem. Phys. **45**, 1287 (1966).

[2611] Berkowitz, J., Photoionization of high–temperature vapors. V. Cesium halides; chemical shift of autoionization, J. Chem. Phys. **50**, 3503 (1969).

[2612] Akopyan, M. E., and Vilesov, F. I., A mass–spectrometric study of the photo–ionisation of benzene derivatives at wavelengths up to 885 Å, Zh. Fiz. Khim. **40**, 125 (1966) [Engl. transl.: Russ. J. Phys. Chem. **40**, 63 (1966)].

[2613] Allinson, I. I. O., and Sedgwick, R. D., A computer programme for the resolution of fine structure in ionization efficiency curves, Advan. Mass Spectrom. **4**, 99 (1968).

[2614] Elder, F. A., Villarejo, D., and Inghram, M. G., Electron affinity of oxygen, J. Chem. Phys. **43**, 758 (1965).

[2615] Williams, R. A., Photoionization of potassium vapor, J. Chem. Phys. **47**, 4281 (1967).

[2616] Chupka, W. A., and Berkowitz, J., Photoionization of the H_2 molecule near threshold, J. Chem. Phys. **48**, 5726 (1968).

[2617] Chupka, W. A., Berkowitz, J., and Refaey, K. M. A., Photoionization of ethylene with mass analysis, J. Chem. Phys. **50**, 1938 (1969).

[2618] Chupka, W. A., and Lifshitz, C., Photoionization of CH_3; heat of formation of CH_2, J. Chem. Phys. **48**, 1109 (1968).

[2619] Dibeler, V. H., N_2O bond dissociation energy by photon impact, J. Chem. Phys. **47**, 2191 (1967).

[2620] Murad, E., Hildenbrand, D. L., and Main, R. P., Dissociation energies of group IIIA monofluorides–the possibility of potential maxima in their excited Π states, J. Chem. Phys. **45**, 263 (1966).

[2621] Dibeler, V. H., and Liston, S. K., Mass–spectrometric study of photoionization. VIII. Dicyanogen and the cyanogen halides, J. Chem. Phys. **47**, 4548 (1967).

[2622] Dibeler, V. H., and Liston, S. K., Mass–spectrometric study of photoionization. XI. Hydrogen sulfide and sulfur dioxide, J. Chem. Phys. **49**, 482 (1968).

[2623] Dibeler, V. H., and Liston, S. K., Mass–spectrometric study of photoionization. IX. Hydrogen cyanide and acetonitrile, J. Chem. Phys. **48**, 4765 (1968).

[2624] Dibeler, V. H., and Walker, J. A., Mass spectrometric study of the photoionization of small polyatomic molecules, Advan. Mass Spectrom. **4**, 767 (1967).

[2626] Dibeler, V. H., and Liston, S. K., Mass spectrometric study of photoionization. XII. Boron trifluoride and diboron tetrafluoride, Inorg. Chem. **7**, 1742 (1968).

[2627] Dibeler, V. H., and Walker, J. A., Mass–spectrometric study of photoionization. VI. O_2, CO_2, COS, and CS_2, J. Opt. Soc. Am. **57**, 1007 (1967).

[2628] Dibeler, V. H., and Walker, J. A., Mass spectrometric study of photoionization. XIII. Boron trichloride and diboron tetrachloride, Inorg. Chem. **8**, 50 (1969).

[2629] Dibeler, V. H., Walker, J. A., and Liston, S. K., Mass spectrometric study of photoionization. VII. Nitrogen dioxide and nitrous oxide, J. Res. NBS **71A**, 371 (1967).

[2630] Dibeler, V. H., Walker, J. A., and McCulloh, K. E., Dissociation energy of fluorine, J. Chem. Phys. **50**, 4592 (1969).

[2631] Dibeler, V. H., Walker, J. A., and Rosenstock, H. M., Mass spectrometric study of photoionization. V. Water and ammonia, J. Res. NBS **70A**, 459 (1966).

[2632] Elder, F. A., and Parr, A. C., Photoionization of the cycloheptatrienyl radical, J. Chem. Phys. **50**, 1027 (1969).

[2633] Foster, P. J., Leckenby, R. E., and Robbins, E. J., The ionization potentials of clustered alkali metal atoms, J. Phys. B **2**, 478 (1969).

[2634] Gallegos, E. J., and Klaver, R. F., Automatic voltage scanner for a peak switching mass spectrometer, J. Sci. Instr. **44**, 427 (1967).

[2635] Genzel, H., and Osberghaus, O., Ungeladene Bruchstücke aus einfachen aromatischen Molekülen bei Elektronenstoss, Z. Naturforsch. **22a**, 331 (1967).

[2636] Kitagawa, T., Inokuchi, H., and Kodera, K., Photoionization of polycyclic aromatic compounds in vacuum ultraviolet region. Azulene, J. Mol. Spectry. **21**, 267 (1966).

[2637] Krauss, M., Walker, J. A., and Dibeler, V. H., Mass spectrometric study of photoionization. X. Hydrogen chloride and methyl halides, J. Res. NBS **72A**, 281 (1968).

[2638] Lifshitz, C., and Chupka, W. A., Photoionization of the CF_3 free radical, J. Chem. Phys. **47**, 3439 (1967).

[2639] Morrison, J. D., Nicholson, A. J. C., and O'Donnell, T. A., Ionization and dissociation of xenon difluoride induced by photon impact, J. Chem. Phys. **49**, 959 (1968).

[2640] Müller, J., Massenspektroskopische Untersuchungen an Tri– und Tetra–cyclopentadienyl–metall–Komplexen, Chem. Ber. **102**, 152 (1969).

[2641] Müller, J., and Connor, J. A., Massenspektroskopische Untersuchungen an Pentacarbonylchrom–carben–Komplexen, Chem. Ber. **102**, 1148 (1969).

[2642] Kaul, W., and Fuchs, R., Massenspektrometrische Untersuchungen von Argon–Stickstoff–Gemischen und Stickstoff, Z. Naturforsch. **15a**, 326 (1960).

[2643] Noutary, C. J., Mass spectrometric study of some fluorocarbons and trifluoromethyl halides, J. Res. NBS **72A**, 479 (1968).

[2644] Parr, A. C., and Elder, F. A., Photoionization of 1,3–butadiene, 1,2–butadiene, allene, and propyne, J. Chem. Phys. **49**, 2659 (1968).

[2645] Parr, A. C., and Elder, F. A., Photoionization of ytterbium: 1350–2000 Å, J. Chem. Phys. **49**, 2665 (1968).

[2646] Pritchard, H., Thynne, J. C. J., and Harrison, A. G., Ion–molecule reactions in formic–d acid and methyl formate, Can. J. Chem. **46**, 2141 (1968).

[2647] Refaey, K. M. A., and Chupka, W. A., Photoionization of the lower aliphatic alcohols with mass analysis, J. Chem. Phys. **48**, 5205 (1968).

[2648] Reese, R. M., and Rosenstock, H. M., Photoionization mass spectrometry of NO, J. Chem. Phys. **44**, 2007 (1966).

[2649] Higasi, K., Omura, I., and Baba, H., Ionization potentials of carbonyl molecules, Nature **178**, 652 (1956).

[2650] Samson, J. A. R., and Cairns, R. B., Ionization potential of molecular xenon and krypton, J. Opt. Soc. Am. **56**, 1140 (1966).

[2651] Yencha, A. J., and El–Sayed, M. A., Lowest ionization potentials of some nitrogen heterocyclics, J. Chem. Phys. **48**, 3469 (1968).

[2652] Angus, J. A., and Morris, G. C., Ionization potential of the anthracene molecule from Rydberg absorption bands, J. Mol. Spectry. **21**, 310 (1966).

[2653] Codling, K., Structure in the photoionization continuum of SF_6 below 630 Å, J. Chem. Phys. **44**, 4401 (1966).

[2654] Codling, K., Structure in the photo–ionization continuum of N_2 near 500 Å, Astrophys. J. **143**, 552 (1966).

[2655] Comes, F. J., and Wellern, H. O., Die Spektroskopie des Wasserstoffmoleküls in der Nähe seiner Ionisierungsgrenze. Z. Naturforsch. **23a**, 881 (1968).

[2656] Huffman, R. E., Larrabee, J. C., and Tanaka, Y., New absorption series and ionization potentials of atomic fluorine, chlorine, bromine, and iodine, J. Chem. Phys. **47**, 856 (1967).

[2657] Huffman, R. E., Larrabee, J. C., and Tanaka, Y., New absorption spectra of atomic and molecular oxygen in the vacuum ultraviolet. I. Rydberg series from O I ground state and new excited O_2 bands, J. Chem. Phys. **46**, 2213 (1967).

[2658] Humphries, C. M., Walsh, A. D., and Warsop, P. A., Ultra–violet absorption spectrum of tetrachloro ethylene, Trans. Faraday Soc. **63**, 513 (1967).

[2659] Johansson, I., and Contreras, R., New measurements in the arc spectrum of zinc, Arkiv Fysik **37**, 513 (1968).

[2660] King, G. W., and Richardson, A. W., The ultraviolet absorption of cyanogen halides. Part I. Identification and correlation, J. Mol. Spectry. **21**, 339 (1966).

[2661] Kitagawa, T., Absorption spectra and photoionization of polycyclic aromatics in vacuum ultraviolet region, J. Mol. Spectry. **26**, 1 (1968).

[2662] La Paglia, S. R., and Duncan, A. B. F., Vacuum ultraviolet absorption spectrum and dipole moment of nitrogen trifluoride, J. Chem. Phys. **34**, 1003 (1961).

[2663] McDiarmid, R., Rydberg progressions in $cis-$ and $trans-$butene, J. Chem. Phys. **50**, 2328 (1969).

[2664] Metzger, P. H., Cook, G. R., and Ogawa, M., Photoionization and absorption coefficients of NO in the 600 to 950 Å region, Can. J. Phys. **45**, 203 (1967).

[2665] Morrow, B. A., The absorption spectrum of SH and SD in the vacuum ultraviolet, Can. J. Phys. **44**, 2447 (1966).

[2666] Quemerais, A., Morlais, M., and Robin, S., Spectres d'absorption du benzène et du monochlorobenzène dans l'ultraviolet de 1300 a 2300 Å, Compt. Rend. **265**, 649 (1967).

[2667] Reader, J., and Sugar, J., Ionization energies of the neutral rare earths, J. Opt. Soc. Am. **56**, 1189 (1966).

[2668] Roebber, J. L., Larrabee, J. C., and Huffman, R. E., Vacuum–ultraviolet absorption spectrum of carbon suboxide, J. Chem. Phys. **46**, 4594 (1967).

[2669] Smith, W. L., The absorption spectrum of diacetylene in the vacuum ultraviolet, Proc. Roy. Soc. (London) **A300**, 519 (1967).

[2670] Sood, S. P., and Watanabe, K., Absorption and ionization coefficients of vinyl chloride, J. Chem. Phys. **45**, 2913 (1966).

[2671] Venkateswarlu, P., Vacuum ultraviolet spectrum of bromine molecule, Bull. Am. Phys. Soc. **13**, 1666 (1968).

[2672] Walsh, A. D., and Warsop, P. A., Ultra–violet absorption of trans–dichloro ethylene, Trans. Faraday Soc. **63**, 524 (1967).

[2673] Walsh, A. D., and Warsop, P. A., Ultra–violet absorption spectrum of cis–1,2–dichloro ethylene, Trans. Faraday Soc. **64**, 1418 (1968).

[2674] Walsh, A. D., and Warsop, P. A., Ultra–violet absorption spectrum of trichloro ethylene, Trans. Faraday Soc. **64**, 1425 (1968).

[2675] Walsh, A. D., Warsop, P. A., and Whiteside, J. A. B., Ultra–violet absorption spectrum of 1,1–dichloro ethylene, Trans. Faraday Soc. **64**, 1432 (1968).

[2676] Watanabe, K., Matsunaga, F. M., and Sakai, H., Absorption coefficient and photoionization yield of NO in the region 580–1350 Å, Appl. Opt. **6**, 391 (1967).

[2677] Whiteside, J. A. B., and Warsop, P. A., The electronic spectrum of trimethylene sulfide, J. Mol. Spectry. **29**, 1 (1969).

[2678] Yoshino, K., and Tanaka, Y., Rydberg absorption series and ionization energies of the oxygen molecule. I, J. Chem. Phys. **48**, 4859 (1968).

[2679] Eriksson, K. B. S., and Isberg, H. B. S., Comments on the sharp and diffuse series in the spectrum of atomic aluminium, Arkiv Fysik **33**, 593 (1967).

[2680] Matsunaga, F. M., and Watanabe, K., Ionization potential and absorption coefficient of COS, J. Chem. Phys. **46**, 4457 (1967).

[2681] McConkey, J. W., and Kernahan, J. A., The ionization limits of N I, Phys. Letters **27A**, 82 (1968).

[2682] Momigny, J., Goffart, C., and D'Or, L., Photoionization studies by total ionization measurements. I. Benzene and its monohalogeno derivatives, Intern. J. Mass Spectrom. Ion Phys. **1**, 53 (1968).

[2683] Müller, J., and D'Or, L., Massenspektrometrische Untersuchungen an Dicyclopentadienylkomplexen von Übergangsmetallen, J. Organometal. Chem. **10**, 313 (1967).

[2684] Asundi, R. K., Schulz, G. J., and Chantry, P. J., Studies of N_4^+ and N_3^+ ion formation in nitrogen using high-pressure mass spectrometry, J. Chem. Phys. **47**, 1584 (1967).

[2685] Amos, D., Gillis, R. G., Occolowitz, J. L., and Pisani, J. F., The ions $[CH_3S]^+$, $[C_2H_5S]^+$ and $[CH_3O]^+$ formed by electron–impact, Org. Mass Spectrom. **2**, 209 (1969).

[2686] Angus, J. G., Christ, B. J., and Morris, G. C., Absorption spectra in the vacuum ultraviolet and the ionization potentials of naphthalene and naphthalene–d_8 molecules, Australian J. Chem. **21**, 2153 (1968).

[2687] Botter, R., Carbon sub–oxide, Advan. Mass Spectrom. **2**, 540 (1963).

[2688] Berkowitz, J., and Marquart, J. R., Equilibrium composition of sulfur vapor, J. Chem. Phys. **39**, 275 (1963).

[2689] Band, S. J., Davidson, I. M. T., and Lambert, C. A., Bond dissociation energies: electron impact studies on some trimethylsilyl compounds, J. Chem. Soc. (A), 2068 (1968).

[2690] Balducci, G., De Maria, G., and Guido, M., Dissociation energy of Ce_2, J. Chem. Phys. **50**, 5424 (1969).

[2693] Coleman, R. J., Delderfield, J. S., and Reuben, B. G., The gas–phase decomposition of the nitrous oxide ion, Intern. J. Mass Spectrom. Ion Phys. **2**, 25 (1969).

[2694] Collin, J. E., and Conde–Caprace, G., Ionization and dissociation of cyclic ethers by electron impact, Intern. J. Mass Spectrom. Ion Phys. **1**, 213 (1966).

[2695] Collins, J. H., Winters, R. E., and Engerholm, G. G., Fine structure in energy–distribution–difference ionization-efficiency curves, J. Chem. Phys. **49**, 2469 (1968).

[2696] Cubicciotti, D., Some thermodynamic properties of antimony trifluoride, High Temp. Sci. **1**, 268 (1969).

[2697] Curran, R. K., and Fox, R. E., Mass spectrometer investigation of ionization of N_2O by electron impact, J. Chem. Phys. **34**, 1590 (1961).

[2698] Davidson, I. M. T., Jones, M., and Kemmitt, R. D. W., Mass spectra and appearance potentials of some 1,5-cyclooctadiene–rhodium complexes, J. Organometal. Chem. **17**, 169 (1969).

[2699] De Maria, G., Gingerich, K. A., and Piacente, V., Vaporization of aluminum phosphide, J. Chem. Phys. **49**, 4705 (1968).

[2700] de Ridder, J. J., and Dijkstra, G., On the mass spectra of some alkyl digermanium compounds, Org. Mass Spectrom. **1**, 647 (1968).

[2701] Dewar, M. J. S., Shanshal, M., and Worley, S. D., Calculated and observed ionization potentials of nitroalkanes and of nitrous and nitric acids and esters. Extension of the MINDO method to nitrogen–oxygen compounds, J. Am. Chem. Soc. **91**, 3590 (1969).

[2702] Fallon, P. J., Kelly, P., and Lockhart, J. C., Fragmentation of some alkyl borates under electron impact, Intern. J. Mass Spectrom. Ion Phys. **1**, 133 (1968).

[2703] Fischer, E. O., Bathelt, W., Herberhold, M., and Müller, J., Arsinecyclopentadienyldicarbonylmanganese(I), Angew. Chem. Intern. Ed. **7**, 634 (1968).

[2704] Franklin, J. L., Wada, Y., Natalis, P., and Hierl, P. M., Ion–molecule reactions in acetonitrile and propionitrile, J. Phys. Chem. **70**, 2353 (1966).

[2705] Ganguli, P. S., and McGee, H. A., Jr., Molecular energetics of borane carbonyl and the symmetric dissociation energy of diborane, J. Chem. Phys. **50**, 4658 (1969).

[2706] Gillis, R. G., Long, G. J., Moritz, A. G., and Occolowitz, J. L., Energetics of the electron–impact fragmentation of alkoxybenzenes and alkylthiobenzenes, Org. Mass Spectrom. **1**, 527 (1968).

[2707] Gingerich, K. A., Gaseous metal silicides. I. Dissociation energy of the molecule AuSi, J. Chem. Phys. **50**, 5426 (1969).

[2708] Grishin, N. N., Bogolyubov, G. M., and Petrov, A. A., Compounds containing P–P bonds. IV. Mass–spectrometric investigation of phosphines. P–P bond energy in diphosphines, Zh. Obshch. Khim. **38**, 2683 (1968) [Engl. transl.: J. Gen. Chem. USSR **38**, 2595 (1968)].

[2709] Harrison, A. G., Ivko, A., and Van Raalte, D., Energetics of formation of some oxygenated ions and the proton affinities of carbonyl compounds, Can. J. Chem. **44**, 1625 (1966).

[2710] Haschke, J. M., and Eick, H. A., The vaporization thermodynamics of europium monoxide, J. Phys. Chem. **73**, 374 (1969).

[2711] Hastie, J. W., and Swingler, D. L., Use of the quadrupole mass filter for high–temperature studies, High Temp. Sci. **1**, 46 (1969).

[2712] Hedaya, E., Miller, R. D., McNeil, D. W., D'Angelo, P. F., and Schissel, P., Flash vacuum pyrolysis. V. Cyclobutadiene, J. Am. Chem. Soc. **91**, 1875 (1969).

[2714] Hildenbrand, D. L., and Murad, E., Dissociation energy and ionization potential of silicon monoxide, J. Chem. Phys. **51**, 807 (1969).

[2715] Hildenbrand, D. L., and Theard, L. P., Mass–spectrometric measurement of the dissociation energies of BeCl and $BeCl_2$, J. Chem. Phys. **50**, 5350 (1969).

[2716] Innorta, G., Distefano, G., and Pignataro, S., Equivalent orbital calculations of the ionization energies of zerovalent transition–metal complexes, Intern. J. Mass Spectrom. Ion Phys. **1**, 435 (1968).

[2717] Kim, H. H., and Roebber, J. L., Vacuum–ultraviolet absorption spectrum of carbon suboxide, J. Chem. Phys. **44**, 1709 (1966).

[2718] Kuhn, W. F., Levins, R. J., and Lilly, A. C., Jr., Electron affinities and ionization potentials of phthalate compounds, J. Chem. Phys. **49**, 5550 (1968).

[2719] Lampe, F. W., and Niehaus, A., Ionization potentials of free radicals formed by electron impact. Methylstannyl radicals, J. Chem. Phys. **49**, 2949 (1968).

[2720] Lappert, M. F., Simpson, J., and Spalding, T. R., Bonding studies of compounds of the group IV elements: ionisation potentials of the Me_3M radicals, J. Organometal. Chem. **17**, P1 (1969).

[2721] Leary, H. J., Jr., and Wahlbeck, P. G., The vaporization process and rate of effusion of crystalline Sc_2Se_3, High Temp. Sci. **1**, 277 (1969).

[2722] Lockhart, J. C., and Kelly, P., Mass spectrum of phenylboron dichloride and the ionic bond energy $D(PhBCl^+Cl^-)$, Intern. J. Mass Spectrom. Ion Phys. **1**, 209 (1968).

[2723] Li, P. H., and McGee, H. A., Jr., Mass spectrum and ionization potential of condensed cyclobutadiene, Chem. Commun., 592 (1969).

[2724] Matthews, C. S., and Warneck, P., Heats of formation of CHO^+ and $C_3H_3^+$ by photoionization, J. Chem. Phys. **51**, 854 (1969).

[2725] Occolowitz, J. L., Electron impact fragmentation of some organotin compounds, Tetrahedron Letters, 5291 (1966).

[2726] Omura, I., Kaneko, T., Yamada, Y., and Tanaka, K., Mass spectrometric studies of photoionization. IV. Acetylene and propyne, J. Phys. Soc. Japan **27**, 178 (1969).

[2727] Potapov, V. K., Mechanism of ionic–molecular reactions, Dokl. Akad. Nauk SSSR **183**, 386 (1968) [Engl. transl.: Dokl. Phys. Chem. **183**, 843 (1968)].

[2728] Potapov, V. K., Filyugina, A. D., Shigorin, D. N., and Ozerova, G. A., Photoionization of some compounds containing the carbonyl and amino groups, Dokl. Akad. Nauk SSSR **180**, 398 (1968) [Engl. transl.: Dokl. Phys. Chem. **180**, 352 (1968)].

[2729] Praet, M.-T., and Delwiche, J., Photoionization of formaldehyde, Intern. J. Mass Spectrom. Ion Phys. **1**, 321 (1968).

[2730] Redhead, P. A., and Feser, S., Multiple ionization of mercury by successive electron impacts, Can. J. Phys. **46**, 1905 (1968).

[2731] Reichert, C., and Westmore, J. B., Mass spectral studies of metal chelates. IV. Mass spectra, appearance potentials, and coordinate bond energies of bis(acetylacetonate)metal(II) complexes of the first transition series, Inorg. Chem. **8**, 1012 (1969).

[2732] Schissel, P., McAdoo, D. J., Hedaya, E., and McNeil, D. W., Flash vacuum pyrolysis. III. Formation and ionization of cyclopentadienyl, cyclopentadienyl nickel, and dihydrofulvalene (dicyclopentadienyl) derived from nickelocene, J. Chem. Phys. **49**, 5061 (1968).

[2734] Staley, H. G., and Norman, J. H., Thermodynamics of gaseous monoxide–dioxide equilibria for cerium, praseodymium and neodymium, Intern. J. Mass Spectrom. Ion Phys. **2**, 35 (1969).

[2735] Steck, S. J., Pressley, G. A., Jr., and Stafford, F. E., Mass spectrometric investigation of the high–temperature reaction of hydrogen with boron carbide, J. Phys. Chem. **73**, 1000 (1969).

[2736] Steck, S. J., Pressley, G. A., Jr., Stafford, F. E., Dobson, J., and Schaeffer, R., Molecular beam mass spectrum, pyrolysis, and structure of octaborane(18), Inorg. Chem. **8**, 830 (1969).

[2737] Ward, R. S., Cooks, R. G., and Williams, D. H., Substituent effects in mass spectrometry. Mass spectra of substituted phenyl benzyl ethers, J. Am. Chem. Soc. **91**, 2727 (1969).

[2738] Winters, R. E., and Collins, J. H., Mass spectrometric studies of structural isomers. II. Mono– and bicyclic C_6H_{10} molecules, Org. Mass Spectrom. **2**, 299 (1969).

[2739] Winters, R. E., and Kiser, R. W., Ions produced by electron impact with the dimetallic carbonyls of cobalt and manganese, J. Phys. Chem. **69**, 1618 (1965).

[2740] Yeo, A. N. H., and Williams, D. H., Internal hydrogen rearrangement as a function of ion lifetime in the mass spectra of aliphatic ketones, J. Am. Chem. Soc. **91**, 3582 (1969).

[2741] Zandberg, É. Ya., and Rasulev, U. Kh., Surface ionization of aniline molecules, Zh. Tekhn. Fiz. **38**, 1798 (1968) [Engl. transl.: Soviet Phys. – Tech. Phys. **13**, 1450 (1969)].

[2742] Hughes, B. M., and Tiernan, T. O., Ionic reactions in gaseous cyclobutane, J. Chem. Phys. **51**, 4373 (1969).

[2744] Dibeler, V. H., Walker, J. A., and McCulloh, K. E., Photoionization study of the dissociation energy of fluorine and the heat of formation of hydrogen fluoride, J. Chem. Phys. **51**, 4230 (1969).

[2745] Fehsenfeld, F. C., Ferguson, E. E., and Mosesman, M., Measurement of the thermal energy reaction $NO_2^+ + NO \rightarrow NO^+ + NO_2$, Chem. Phys. Letters **4**, 73 (1969).

[2746] Walter, T. A., Lifshitz, C., Chupka, W. A., and Berkowitz, J., Mass–spectrometric study of the photoionization of C_2F_4 and CF_4, J. Chem. Phys. **51**, 3531 (1969).

[2747] Uy, O. M., and Drowart, J., Mass spectrometric determination of the dissociation energies of the molecules BiO, BiS, BiSe and BiTe, Trans. Faraday Soc. **65**, 3221 (1969).

[2748] Chan, R. K., and Liao, S. C., Dipole moments, charge–transfer parameters, and ionization potentials of the methyl-substituted benzene-tetracyanoethylene complexes, Can. J. Chem. **48**, 299 (1970).

[2749] DeJongh, D. C., and Shrader, S. R., A mass spectral study of *exo–* and *endo–*2-norbornyl bromides, J. Am. Chem. Soc. **88**, 3881 (1966).

[2750] Muenow, D. W., Hastie, J. W., Hauge, R., Bautista, R., and Margrave, J. L., Vaporization, thermodynamics and structures of species in the tellurium+oxygen system, Trans. Faraday Soc. **65**, 3210 (1969).

[2751] Franklin, J. L., and Carroll, S. R., Effect of molecular structure on ionic decomposition. II. An electron–impact study of 1,3– and 1,4–cyclohexadiene and 1,3,5–hexatriene, J. Am. Chem. Soc. **91**, 6564 (1969).

[2752] Hedaya, E., Krull, I. S., Miller, R. D., Kent, M. E., D'Angelo, P. F., and Schissel, P., Flash vacuum pyrolysis. VI. Cyclobutadieneiron tricarbonyl, J. Am. Chem. Soc. **91**, 6880 (1969).

[2753] Fehlner, T. P., The preparation and mass spectrometry of triphosphine-5, J. Am. Chem. Soc. **90**, 6062 (1968).

[2754] Balducci, G., De Maria, G., and Guido, M., Thermodynamics of rare–earth–carbon systems. III. The erbium–carbon system, J. Chem. Phys. **51**, 2876 (1969).

[2755] Balducci, G., Capalbi, A., De Maria, G., and Guido, M., Thermodynamics of rare–earth–carbon systems. II. The holmium–carbon and dysprosium–carbon systems, J. Chem. Phys. **51**, 2871 (1969).

[2756] Smith, D. H., and Hertel, G. R., First ionization potentials of Th, Np, and Pu by surface ionization, J. Chem. Phys. **51**, 3105 (1969).

[2757] Potzinger, P., and Lampe, F. W., Ionic reactions in gaseous methylsilane, J. Phys. Chem. **74**, 587 (1970).

[2758] Asundi, R. K., The first ionization potential of oxygen molecule, Current Sci. **37**, 160 (1968).

[2759] Collin, J. E., and Delwiche, J., Ionization of acetylene and its electronic energy levels, Can. J. Chem. **45**, 1883 (1967).

[2760] Comes, F. J., and Elzer, A., Photoionisationsuntersuchungen an Atomstrahlen. III. Der Ionisierungsquerschnitt des atomaren Stickstoffs, Z. Naturforsch. **23a**, 133 (1968).

[2761] Donovan, R. J., Husain, D., and Jackson, P. T., Spectroscopic and kinetic studies of the SO radical and the photolysis of thionyl chloride, Trans. Faraday Soc. **65**, 2930 (1969).

[2762] Herzberg, G., Dissociation energy and ionization potential of molecular hydrogen, Phys. Rev. Letters **23**, 1081 (1969).

[2763] Huffman, R. E., and Katayama, D. H., Photoionization study of diatomic–ion formation in argon, krypton, and xenon, J. Chem. Phys. **45**, 138 (1966).

[2764] Jungen, C., and Miescher, E., Absorption spectrum of the NO molecule. IX. The structure of the f complexes, the ionization potential of NO, and the quadrupole moment of NO^+, Can. J. Phys. **47**, 1769 (1969).

[2765] Kieffer, L. J., and Van Brunt, R. J., Energetic ions from N_2 produced by electron impact, J. Chem. Phys. **46**, 2728 (1967).

[2766] McLafferty, F. W., McAdoo, D. J., and Smith, J. S., Unimolecular gaseous ion reactions of low activation energy. Five–membered–ring formation, J. Am. Chem. Soc. **91**, 5400 (1969).

[2767] McNeal, R. J., and Cook, G. R., Photoionization of O_2 in the metastable $a^1\Delta_g$ state, J. Chem. Phys. **45**, 3469 (1966).

[2768] Melton, C. E., Study by mass spectrometry of the decomposition of ammonia by ionizing radiation in a wide–range radiolysis source, J. Chem. Phys. **45**, 4414 (1966).

[2769] Melton, C. E., and Rudolph, P. S., The mass spectrometer as a research laboratory, Naturwiss. **54**, 297 (1967).

[2770] Melton, C. E., and Joy, H. W., Mass–spectrometric and theoretical evidence for NH_4 and H_3O, J. Chem. Phys. **46**, 4275 (1967).

[2771] Melton, C. E., The mass spectrometer as a radiolytic and a catalytic laboratory, Advan. Chem. Ser. **72**, 48 (1968).

[2772] Melton, C. E., The mass spectrometer as a radiolytic and catalytic laboratory, Intern. J. Mass Spectrom. Ion Phys. **1**, 353 (1968).

[2773] Peatman, W. B., Borne, T. B., and Schlag, E. W., Photoionization resonance spectra. I. Nitric oxide and benzene, Chem. Phys. Letters **3**, 492 (1969).

[2774] Schönheit, E., Massenspektrometrische Untersuchung der Photoionisation von Wasserstoff, Z. Naturforsch. **15a**, 841 (1960).

[2775] Thomson, R., and Warsop, P. A., Electronic spectrum of the haloacetylenes. Part I. Chloroacetylene, Trans. Faraday Soc. **65**, 2806 (1969).

[2776] Williams, J. M., and Hamill, W. H., Ionization potentials of molecules and free radicals and appearance potentials by electron impact in the mass spectrometer, J. Chem. Phys. **49**, 4467 (1968).

[2777] Harland, P., and Thynne, J. C. J., Ionization and dissociation of pentafluorosulfur chloride by electron impact, J. Phys. Chem. **73**, 4031 (1969).

[2778] Haney, M. A., and Franklin, J. L., Mass spectrometric determination of the proton affinities of various molecules, J. Phys. Chem. **73**, 4328 (1969).

[2779] Gingerich, K. A., and Finkbeiner, H. C., Mass spectrometric determination of the dissociation energies of LaAu, CeAu, PrAu, and NdAu and predicted stability of gaseous monoaurides of electropositive metals, J. Chem. Phys. **52**, 2956 (1970).

[2780] Foner, S. N., and Hudson, R. L., Mass spectrometric studies of metastable nitrogen atoms and molecules in active nitrogen, J. Chem. Phys. **37**, 1662 (1962).

[2781] Dresser, M. J., and Hudson, D. E., Surface ionization of some rare earths on tungsten, Phys. Rev. **137**, A673 (1965).

[2782] Coxon, G. E., Palmer, T. F., and Sowerby, D. B., Cyclic inorganic compounds. Part VII. The mass spectra of the trimeric chlorobromophosphonitriles, J. Chem. Soc. (A), 358 (1969).

[2783] Chupka, W. A., Berkowitz, J., and Inghram, M. G., Thermodynamics of the Zr–ZrO$_2$ system: the dissociation energies of ZrO and ZrO$_2$, J. Chem. Phys. **26**, 1207 (1957).

[2785] Dorman, F. H., and Morrison, J. D., Ionization potentials of doubly charged oxygen and nitrogen, J. Chem. Phys. **39**, 1906 (1963).

[2786] Foner, S. N., and Hudson, R. L., Ionization potential of the OH free radical by mass spectrometry, J. Chem. Phys. **25**, 602 (1956).

[2787] Staley, H. G., Mass–spectrometric Knudsen–cell study of the gaseous oxides of tellurium, J. Chem. Phys. **52**, 4311 (1970).

[2788] Flesch, G. D., White, R. M., and Svec, H. J., The positive and negative ion mass spectra of chromyl chloride and chromyl fluoride, Intern. J. Mass Spectrom. Ion Phys. **3**, 339 (1969).

[2789] Goffart, C., Momigny, J., and Natalis, P., Photoionization studies by total ionization measurements and photoelectron spectra. II. Pyridine, Intern. J. Mass Spectrom. Ion Phys. **3**, 371 (1969).

[2790] Lifshitz, C., and Grajower, R., Dissociative electron capture and dissociative ionization in perfluoropropane, Intern. J. Mass Spectrom. Ion Phys. **3**, 211 (1969).

[2791] Ruska, W. E. W., and Franklin, J. L., Ion–molecule reactions in hydrogen sulfide, methanethiol and 2–thiapropane, Intern. J. Mass Spectrom. Ion Phys. **3**, 221 (1969).

[2792] Collin, J. E., and Natalis, P., Ionic states and photon impact–enhanced vibrational excitation in diatomic molecules by photoelectron spectroscopy. Photoelectron spectra of N$_2$, CO and O$_2$, Intern. J. Mass Spectrom. Ion Phys. **2**, 231 (1969).

[2793] Hughes, B. M., Tiernan, T. O., and Futrell, J. H., Ionic reactions in unsaturated compounds. IV. Vinyl chloride, J. Phys. Chem. **73**, 829 (1969).

[2795] Ehlert, T. C., Bonding in C$_1$ and C$_2$ fluorides, J. Phys. Chem. **73**, 949 (1969).

[2796] Eland, J. H. D., Photoelectron spectra of conjugated hydrocarbons and heteromolecules, Intern. J. Mass Spectrom. Ion Phys. **2**, 471 (1969).

[2797] Rowland, C. G., Eland, J. H. D., and Danby, C. J., Kinetic energy distributions of fragment ions in the mass spectrum of isocyanic acid, Intern. J. Mass Spectrom. Ion Phys. **2**, 457 (1969).

[2798] Lossing, F. P., and Semeluk, G. P., Threshold ionization efficiency curves for monoenergetic electron impact on H$_2$, D$_2$, CH$_4$ and CD$_4$, Intern. J. Mass Spectrom. Ion Phys. **2**, 408 (1969).

[2799] Pignataro, S., Distefano, G., Nencini, G., and Foffani, A., Mass spectrometric study of paramagnetic and diamagnetic transition metal complexes, Intern. J. Mass Spectrom. Ion Phys. **3**, 479 (1970).

[2800] Collin, J. E., and Locht, R., Positive and negative ion formation in ketene by electron impact, Intern. J. Mass Spectrom. Ion Phys. **3**, 465 (1970).

[2801] Al–Joboury, M. I., and Turner, D. W., Molecular photoelectron spectroscopy. Part VI. Water, methanol, methane, and ethane, J. Chem. Soc. (B), 373 (1967).

[2802] Bassett, P. J., and Lloyd, D. R., The photoelectron spectrum of boron trifluoride, Chem. Commun., 36 (1970).

[2803] Baker, A. D., Baker, C., Brundle, C. R., and Turner, D. W., The electronic structures of methane, ethane, ethylene and formaldehyde studied by high–resolution molecular photoelectron spectroscopy, Intern. J. Mass Spectrom. Ion Phys. **1**, 285 (1968).

[2804] Baker, C., and Turner, D. W., Photoelectron spectra of acetylene, diacetylene, and their deutero–derivatives, Chem. Commun., 797 (1967).

[2805] Baker, C., and Turner, D. W., High resolution molecular photoelectron spectroscopy. III. Acetylenes and aza–acetylenes, Proc. Roy. Soc. (London) **A308**, 19 (1968).

[2806] Baker, A. D., May, D. P., and Turner, D. W., Molecular photoelectron spectroscopy. Part VII. The vertical ionisation potentials of benzene and some of its monosubstituted and 1,4–disubstituted derivatives, J. Chem. Soc. (B), 22 (1968).

[2807] Baker, C., and Turner, D. W., The photoelectron spectrum and ionisation potentials of carbon suboxide, Chem. Commun., 400 (1968).

[2808] Basch, H., Robin, M. B., Kuebler, N. A., Baker, C., and Turner, D. W., Optical and photoelectron spectra of small rings. III. The saturated three–membered rings, J. Chem. Phys. **51**, 52 (1969).

[2809] Collin, J. E., and Natalis, P., Electronic states of the nitric oxide ion, Chem. Phys. Letters **2**, 194 (1968).

[2810] Collin, J. E., and Natalis, P., Vibrational and electronic ionic states of nitric oxide. An accurate method for measuring ionization potentials by photoelectron spectroscopy, Intern. J. Mass Spectrom. Ion Phys. **1**, 483 (1968).

[2811] Chupka, W. A., and Berkowitz, J., High–resolution photoionization study of the H$_2$ molecule near threshold, J. Chem. Phys. **51**, 4244 (1969).

[2812] Berkowitz, J., Ehrhardt, H., and Tekaat, T., Spektren und Winkelverteilungen der Photoelektronen von Atomen und Molekülen, Z. Physik **200**, 69 (1967).

[2813] Blake, A. J., and Carver, J. H., Determination of partial photoionization cross sections by photoelectron spectroscopy, J. Chem. Phys. **47**, 1038 (1967).

[2814] Frost, D. C., McDowell, C. A., and Vroom, D. A., Ionization potentials of mercury by photoelectron spectrometry, Chem. Phys. Letters **1**, 93 (1967).

[2815] Frost, D. C., McDowell, C. A., and Vroom, D. A., Photoelectron spectra of the halogens and the hydrogen halides, J. Chem. Phys. **46**, 4255 (1967).

[2816] Frost, D. C., McDowell, C. A., and Vroom, D. A., Photoelectron kinetic energy analysis in gases by means of a spherical analyser, Proc. Roy. Soc. (London) **A296**, 566 (1967).

[2817] Frost, D. C., McDowell, C. A., and Vroom, D. A., The photoionization of oxygen, nitrogen and nitric oxide at 584 Å, Bull. Am. Phys. Soc. **12**, 238 (1967).

[2819] Lempka, H. J., Passmore, T. R., and Price, W. C., The photoelectron spectra and ionized states of the halogen acids, Proc. Roy. Soc. (London) **A304**, 53 (1968).

[2820] Lempka, H. J., and Price, W. C., Ionization energies of HF and DF, J. Chem. Phys. **48**, 1875 (1968).

[2821] Natalis, P., and Collin, J. E., The first ionization potential of nitrogen dioxide, Chem. Phys. Letters **2**, 79 (1968).

[2822] Natalis, P., Collin, J. E., and Momigny, J., Energy levels of benzene and fluorobenzene ions by photoelectron spectroscopy, Intern. J. Mass Spectrom. Ion Phys. **1**, 327 (1968).

[2823] Newton, A. S., and Sciamanna, A. F., Metastable peaks in the mass spectra of N$_2$ and NO, J. Chem. Phys. **50**, 4868 (1969).

[2824] Samson, J. A. R., Higher ionization potentials of molecules determined by photoelectron spectroscopy, Bull. Am. Phys. Soc. **14**, 266 (1969).

[2825] Samson, J. A. R., Higher ionization potentials of nitric oxide, Phys. Letters **28A**, 391 (1968).

[2826] Turner, D. W., High resolution molecular photo–electron spectroscopy, Advan. Mass Spectrom. **4**, 755 (1968).

[2827] Turner, D. W., High resolution molecular photoelectron spectroscopy. I. Fine structure in the spectra of hydrogen and oxygen, Proc. Roy. Soc. (London) **A307**, 15 (1968).

[2829] Turner, D. W., and Al–Joboury, M. I., Molecular photoelectron spectroscopy, Bull. Soc. Chim. Belges **73**, 428 (1964).

[2830] Turner, D. W., and May, D. P., Franck–Condon factors in ionization: experimental measurement using molecular photoelectron spectroscopy, J. Chem. Phys. **45**, 471 (1966).

[2831] Villarejo, D., Measurement of threshold electrons in the photoionization of H_2 and D_2, J. Chem. Phys. **48**, 4014 (1968).

[2833] Beynon, J. H., Hopkinson, J. A., and Lester, G. R., Mass spectrometry—the appearance potentials of "meta–stable peaks" in some aromatic nitro compounds—a chemical reaction in the mass spectrometer, Intern. J. Mass Spectrom. Ion Phys. **2**, 291 (1969).

[2834] Boyd, R. J., and Frost, D. C., The ionization potentials of BF_3, BCl_3 and BBr_3, Chem. Phys. Letters **1**, 649 (1968).

[2835] Brundle, C. R., and Turner, D. W., The carbonyl π–ionization potential of formaldehyde, Chem. Commun., 314 (1967); Erratum, ibid., 472 (1967).

[2836] Brundle, C. R., and Turner, D. W., High resolution molecular photoelectron spectroscopy. II. Water and deuterium oxide, Proc. Roy. Soc. (London) **A307**, 27 (1968).

[2837] Brundle, C. R., Turner, D. W., Robin, M. B., and Basch, H., Photoelectron spectroscopy of simple amides and carboxylic acids, Chem. Phys. Letters **3**, 292 (1969).

[2838] Clark, I. D., and Frost, D. C., A study of the energy levels in benzene and some fluorobenzenes by photoelectron spectroscopy, J. Am. Chem. Soc. **89**, 244 (1967).

[2839] Collin, J. E., and Natalis, P., Ionization, preionization and internal energy conversion in CO_2, COS and CS_2 by photoelectron spectroscopy, Intern. J. Mass Spectrom. Ion Phys. **1**, 121 (1968).

[2840] Davis, F. A., Dewar, M. J. S., Jones, R., and Worley, S. D., New heteroaromatic compounds. XXXII. Properties of 10,9–borazaronaphthalene and 9–aza–10–boradecalin, J. Am. Chem. Soc. **91**, 2094 (1969).

[2841] Dewar, M. J. S., and Worley, S. D., Unpublished results reported in: Dewar, M. J. S., Harget, A., and Haselbach, E., Cyclooctatetraene and ions derived from it, J. Am. Chem. Soc. **91**, 7521 (1969).

[2842] Dewar, M. J. S., and Worley, S. D., Ionization potential of cis–1,3–butadiene, J. Chem. Phys. **49**, 2454 (1968).

[2843] Dewar, M. J. S., and Worley, S. D., Photoelectron spectra of molecules. I. Ionization potentials of some organic molecules and their interpretation, J. Chem. Phys. **50**, 654 (1969).

[2844] Dewar, M. J. S., and Worley, S. D., Photoelectron spectra of molecules. II. The ionization potentials of azabenzenes and azanaphthalenes, J. Chem. Phys. **51**, 263 (1969).

[2845] Dewar, M. J. S., and Worley, S. D., Photoelectron spectrum of trimethylenemethane iron tricarbonyl and a theoretical study of trimethylenemethane, J. Chem. Phys. **51**, 1672 (1969).

[2846] Ehrhardt, H., and Linder, F., Die Streuung niederenergetischer Elektronen an Methan. Das zweite Ionisierungspotential, Z. Naturforsch. **22a**, 11 (1967).

[2847] Eland, J. H. D., and Danby, C. J., Inner ionization potentials of aromatic compounds, Z. Naturforsch. **23a**, 355 (1968).

[2848] Eland, J. H. D., and Danby, C. J., Photoelectron spectra and ionic structure of carbon dioxide, carbon disulphide and sulphur dioxide, Intern. J. Mass Spectrom. Ion Phys. **1**, 111 (1968).

[2849] Evans, S., Green, J. C., Orchard, A. F., Saito, T., and Turner, D. W., The photoelectron spectrum of vanadium hexacarbonyl, Chem. Phys. Letters **4**, 361 (1969).

[2850] Frost, D. C., Herring, F. G., McDowell, C. A., Mustafa, M. R., and Sandhu, J. S., The ionization potentials of carbon tetrafluoride as determined by photoelectron spectroscopy, Chem. Phys. Letters **2**, 663 (1968).

[2851] Frost, D. C., Herring, F. G., McDowell, C. A., and Stenhouse, I. A., The ionization potentials of methyl cyanide and methyl acetylene by photoelectron spectroscopy and semi–rigorous LCAO SCF calculations, Chem. Phys. Letters **4**, 533 (1970).

[2852] Frost, D. C., McDowell, C. A., Sandhu, J. S., and Vroom, D. A., Photoelectron spectrum of sulfur hexafluoride at 584 Å, J. Chem. Phys. **46**, 2008 (1967).

[2853] Frost, D. C., McDowell, C. A., Sandhu, J. S., and Vroom, D. A., 584 Å photo–electron spectra of Hg, NH_3, PH_3, and SF_6, Advan. Mass Spectrom. **4**, 781 (1968).

[2854] Frost, D. C., McDowell, C. A., and Vroom, D. A., Ionization potentials of ammonia, Can. J. Chem. **45**, 1343 (1967).

[2855] Spohr, R., and Puttkamer, E. v., Energiemessung von Photoelektronen und Franck–Condon–Faktoren der Schwingungsübergänge einiger Molekülionen, Z. Naturforsch. **22a**, 705 (1967).

[2856] Turner, D. W., and May, D. P., Franck–Condon factors in ionization: experimental measurement using molecular photoelectron spectroscopy. II, J. Chem. Phys. **46**, 1156 (1967).

[2857] Villarejo, D., Stockbauer, R., and Inghram, M. G., Measurement of threshold electrons in the photoionization of small molecules, Bull. Am. Phys. Soc. **13**, 39 (1968).

[2858] Villarejo, D., Stockbauer, R., and Inghram, M. G., Measurements of threshold electrons in the photoionization of methane, J. Chem. Phys. **50**, 4599 (1969).

[2859] Zmbov, K. F., Uy, O. M., and Margrave, J. L., Mass spectrometric studies at high temperatures. XXXI. Stabilities of tungsten and molybdenum oxyfluorides, J. Phys. Chem. **73**, 3008 (1969).

[2860] Grimley, R. T., and Joyce, T. E., A technique for the calibration of high–temperature mass spectrometers, J. Phys. Chem. **73**, 3047 (1969).

[2861] Keneshea, F. J., and Cubicciotti, D., The thermodynamic properties of gaseous niobium chlorides, J. Phys. Chem. **73**, 3054 (1969).

[2862] McAllister, T., and Lossing, F. P., Free radicals by mass spectrometry. XLI. Ionization potential and heat of formation of PH_2 radical, J. Phys. Chem. **73**, 2996 (1969).

[2863] Baldwin, J. C., Lappert, M. F., Pedley, J. B., Riley, P. N. K., and Sedgwick, R. D., Ionisation potentials and π–bonding in the series $Cl_nB(NMe_2)_{3-n}$, Inorg. Nucl. Chem. Letters **1**, 57 (1965).

[2864] Harland, P., and Thynne, J. C. J., Positive and negative ion formation in hexafluoroacetone by electron impact, J. Phys. Chem. **74**, 52 (1970).

[2865] Baba, H., Omura, I., and Higasi, K., Ionization potentials of some organic molecules. III. Aromatic and conjugated compounds, Bull. Chem. Soc. Japan **29**, 521 (1956).

[2866] Baer, T., Peatman, W. B., and Schlag, E. W., Photoionization resonance studies with a steradiancy analyzer. II. The photoionization of CH_3I, Chem. Phys. Letters **4**, 243 (1969).

[2867] Baldwin, M., Kirkien–Konasiewicz, A., Loudon, A. G., Maccoll, A., and Smith, D., Localised or delocalised charges in molecule–ions?, Chem. Commun., 574 (1966).

[2868] Beggs, D. P., and Lampe, F. W., Ionic reactions in gaseous mixtures of monosilane with methane and benzene, J. Phys. Chem. **73**, 4194 (1969).

[2869] Herstad, O., Pressley, G. A., Jr., and Stafford, F. E., Mass spectrometric investigation of the fragmentation pattern and the pyrolysis of borane carbonyl, J. Phys. Chem. **74**, 874 (1970).

[2870] Bidinosti, D. R., and McIntyre, N. S., Mass spectrometric study of the thermal decomposition of dimanganese decacarbonyl and dicobalt octacarbonyl, Can. J. Chem. **48,** 593 (1970).

[2871] Buchanan, A. S., Knowles, D. J., and Swingler, D. L., Electron impact studies of stannous chloride and stannic chloride, J. Phys. Chem. **73,** 4394 (1969).

[2872] Caton, R. B., and Douglas, A. E., Electronic spectrum of the BF molecule, Can. J. Phys. **48,** 432 (1970).

[2873] Čermák, V., and Herman, Z., Ionizing reactions of noble gas atoms in metastable states with polyatomic molecules, Collection Czech. Chem. Commun. **30,** 169 (1965).

[2874] Chambers, D. B., Coates, G. E., and Glockling, F., Electron impact studies on organo-beryllium and -aluminium compounds, Discussions Faraday Soc. **47,** 157 (1969).

[2875] Collin, J. E., and Natalis, P., Détermination des états électroniques et des niveaux de vibration des ions moléculaires par spectroscopie de photoélectrons, Bull. Classe Sci. Acad. Roy. Belg. **55,** 352 (1969).

[2877] Demeo, D. A., and El–Sayed, M. A., Ionization potential and structure of olefins, J. Chem. Phys. **52,** 2622 (1970).

[2878] Donovan, R. J., Husain, D., and Stevenson, C. D., Vacuum ultra–violet spectra of transient molecules and radicals. Part 1. CS and S_2, Trans. Faraday Soc. **66,** 1 (1970).

[2879] Evans, S., Green, J. C., Green, M. L. H., Orchard, A. F., and Turner, D. W., Study of the bonding in pentacarbonylmanganese derivatives by photoelectron spectroscopy, Discussions Faraday Soc. **47,** 112 (1969).

[2880] Watanabe, K., Unpublished results reported in: Hart, R. R., Robin, M. B., and Kuebler, N. A., $3p$ Orbitals, bent bonds, and the electronic spectrum of the P_4 molecule, J. Chem. Phys. **42,** 3631 (1965).

[2881] Ionov, N. I., and Mittsev, M. A., Determination of the first ionization potentials of neodymium and praseodymium by the surface ionization method, Zh. Eksperim. i Teor. Fiz. **38,** 1350 (1960) [Engl. transl.: Soviet Phys. JETP **11,** 972 (1960)].

[2882] Ionov, N. I., and Mittsev, M. A., Atomic first ionization potentials determined by the method of surface ionization, Zh. Eksperim. i Teor. Fiz. **40,** 741 (1961) [Engl. transl.: Soviet Phys. JETP **13,** 518 (1961).

[2883] Kanomata, I., Mass–spectrometric study on ionization and dissociation of formaldehyde, acetaldehyde, acetone and ethyl methyl ketone by electron impact, Bull. Chem. Soc. Japan **34,** 1864 (1961).

[2884] Kant, A., Dissociation energies of diatomic molecules of the transition elements. I. Nickel, J. Chem. Phys. **41,** 1872 (1964).

[2885] Lake, R. F., and Thompson, H., Photoelectron spectra of halogenated ethylenes, Proc. Roy. Soc. (London) **A315,** 323 (1970).

[2886] Lloyd, D. R., and Schlag, E. W., Photoionization studies of metal carbonyls. I. Ionization potentials and the bonding in group VI metal hexacarbonyls and in mononuclear carbonyls and nitrosyl carbonyls of iron, cobalt, and nickel, Inorg. Chem. **8,** 2544 (1969).

[2887] Myer, J. A., and Samson, J. A. R., Vacuum–ultraviolet absorption cross sections of CO, HCl, and ICN between 1050 and 2100 Å, J. Chem. Phys. **52,** 266 (1970).

[2888] O'Hare, P. A. G., Dissociation energies, enthalpies of formation, ionization potentials, and dipole moments of NS and NS^+, J. Chem. Phys. **52,** 2992 (1970).

[2889] Ölme, A., The spectrum of doubly ionized boron, B III, Arkiv Fysik **40,** 35 (1969).

[2890] Samson, J. A. R., The electronic states of $C_6H_6^+$, Chem. Phys. Letters **4,** 257 (1969).

[2891] Takezawa, S., Absorption spectrum of H_2 in the vacuum–uv region. I. Rydberg states and ionization energies, J. Chem. Phys. **52,** 2575 (1970).

[2892] Weiss, M. J., Lawrence, G. M., and Young, R. A., Photoelectron spectroscopy of HCl and DCl using molecular beams, J. Chem. Phys. **52,** 2867 (1970).

[2893] Workman, G. L., and Duncan, A. B. F., Electronic spectrum of carbonyl fluoride, J. Chem. Phys. **52,** 3204 (1970).

[2894] Thomassy, F. A., and Lampe, F. W., A mass spectrometric study of the dimerization of nitrosomethane, J. Phys. Chem. **74,** 1188 (1970).

[2895] Saporoschenko, M., Ions in nitrogen, Phys. Rev. **111,** 1550 (1958).

[2896] Sharp, K. G., and Margrave, J. L., Silicon–fluorine chemistry. VII. The reaction of silicon difluoride with hydrogen sulfide, Inorg. Chem. **8,** 2655 (1969).

[2897] Porter, R. F., and Zeller, E. E., Mass spectra of aluminum(III) halides and the heats of dissociation of $Al_2F_6(g)$ and $LiF \cdot AlF_3(g)$, J. Chem. Phys. **33,** 858 (1960).

[2898] Potzinger, P., and Lampe, F. W., Thermochemistry of simple alkylsilanes, J. Phys. Chem. **74,** 719 (1970).

[2899] Potzinger, P., and Lampe, F. W., An electron impact study of ionization and dissociation of monosilane and disilane, J. Phys. Chem. **73,** 3912 (1969).

[2900] Pitt, C. G., Bursey, M. M., and Rogerson, P. F., Catenates of the group IV elements. Correlation of σ electron energies, J. Am. Chem. Soc. **92,** 519 (1970).

[2901] Brundle, C. R., and Turner, D. W., Studies on the photoionisation of the linear triatomic molecules: N_2O, COS, CS_2 and CO_2 using high–resolution photoelectron spectroscopy, Intern. J. Mass Spectrom. Ion Phys. **2,** 195 (1969).

[2902] Natalis, P., and Collin, J. E., Vibrational and electronic states of N_2O^+ by photoelectron spectroscopy. Dissociation processes in N_2O^+, Intern. J. Mass Spectrom. Ion Phys. **2,** 221 (1969).

[2903] Haschke, J. M., and Eick, H. A., The vaporization thermodynamics of europium dibromide, J. Phys. Chem. **74,** 1806 (1970).

[2904] Preston, F. J., Tsuchiya, M., and Svec, H. J., A mass spectrometer for the detection of neutral species formed by electron impact, Intern. J. Mass Spectrom. Ion Phys. **3,** 323 (1969).

[2905] Lifshitz, C., Shapiro, M., and Sternberg, R., Isotope effects on metastable transitions. IV. Isotopic methanols, Israel J. Chem. **7,** 391 (1969).

[2906] Venkateswarlu, P., Vacuum ultraviolet spectrum of the iodine molecule, Can. J. Phys. **48,** 1055 (1970).

[2907] Lifshitz, C., and Shapiro, M., Isotope effects on metastable transitions. II. $CH_3CD_2CH_3$ and $CD_3CH_2CD_3$, J. Chem. Phys. **46,** 4912 (1967).

[2908] Levitt, L. S., and Levitt, B. W., Evaluation of the basic ionization constants of water and alcohols from their ionization potentials, J. Phys. Chem. **74,** 1812 (1970).

[2909] Kinoshita, M., The absorption spectra of the molecular complexes of aromatic compounds with p–bromanil, Bull. Chem. Soc. Japan **35,** 1609 (1962).

[2910] Briegleb, G., Electron affinity of organic molecules, Angew. Chem. Intern. Ed. **3,** 617 (1964).

[2911] Mukherjee, T. K., Charge–transfer donor abilities of o,o'–bridged biphenyls, J. Phys. Chem. **73,** 3442 (1969).

[2912] Brundle, C. R., Robin, M. B., and Jones, G. R., High-resolution He I and He II photoelectron spectra of xenon difluoride, J. Chem. Phys. **52,** 3383 (1970).

[2913] Chambers, D. B., Coates, G. E., and Glockling, F., Electron impact studies on beryllium dialkyls, J. Chem. Soc. (A), 741 (1970).

[2914] Franklin, J. L., and Carroll, S. R., The effect of molecular structure on ionic decomposition. I. An electron impact study of seven C_8H_8 isomers, J. Am. Chem. Soc. **91,** 5940 (1969).

[2915] Omura, I., Kaneko, T., Yamada, Y., and Tanaka, K., Mass spectrometric studies of photoionization. V. Methanol and methanol–d_1, J. Phys. Soc. Japan **27,** 981 (1969).

[2916] Gingerich, K. A., Gaseous phosphorus compounds. III. Mass spectrometric study of the reaction between diatomic nitrogen and phosphorus vapor and dissociation energy of phosphorus mononitride and diatomic phosphorus, J. Phys. Chem. **73**, 2734 (1969).

[2917] Lin, S.–S., and Kant, A., Dissociation energy of Fe_2, J. Phys. Chem. **73**, 2450 (1969).

[2918] Shapiro, R. H., Turk, J., and Serum, J. W., The effect of substituents on the average internal energy of benzoyl ions generated from N–(substituted–phenyl)benzamides, Org. Mass Spectrom. **3**, 171 (1970).

[2919] Yeo, A. N. H., Cooks, R. G., and Williams, D. H., The question of hydrogen and randomization in phenyl isocyanide, Org. Mass Spectrom. **1**, 910 (1968).

[2920] Cubicciotti, D., The gaseous trimer and tetramer of thallous fluoride, High Temp. Sci. **2**, 65 (1970).

[2940] Hedaya, E., The techniques of flash vacuum pyrolysis. The cyclopentadienyl radical and its dimer, Acct. Chem. Res. **2**, 367 (1969).

[2941] Meisels, G. G., Park, J. Y., and Giessner, B. G., Ionization and dissociation of C_4H_8 isomers, J. Am. Chem. Soc. **92**, 254 (1970).

[2942] Dewar, M. J. S., Haselbach, E., and Worley, S. D., Calculated and observed ionization potentials of unsaturated polycyclic hydrocarbons; calculated heats of formation by several semiempirical s.c.f. m.o. methods, Proc. Roy. Soc. (London) **A315**, 431 (1970).

[2943] Auwera–Mahieu, A. V., Peeters, R., McIntyre, N. S., and Drowart, J., Mass spectrometric determination of dissociation energies of the borides and silicides of some transition metals, Trans. Faraday Soc. **66**, 809 (1970).

[2944] Battles, J. E., Shinn, W. A., Blackburn, P. E., and Edwards, R. K., A mass spectrometric study of the volatilization behavior in the plutonium–carbon system, High Temp. Sci. **2**, 80 (1970).

[2945] McLafferty, F. W., and Schiff, L. J., The structure of $[YC_6H_5O]^{+\cdot}$ ions in the spectra of aryl alkyl ethers, Org. Mass Spectrom. **2**, 757 (1969).

[2946] Barnes, C. S., and Occolowitz, J. L., The mass spectra of some naturally occurring oxygen heterocycles and related compounds, Australian J. Chem. **17**, 975 (1964).

[2947] Birks, J. B., and Slifkin, M. A., π–Electronic excitation and ionization energies of condensed ring aromatic hydrocarbons, Nature **191**, 761 (1961).

[2948] Bogolyubov, G. M., Grishin, N. N., and Petrov, A. A., Organic derivatives of group V and group VI elements. VIII. Mass spectra of phosphines and diphosphines, Zh. Obshch. Khim. **39**, 1808 (1969) [Engl. transl.: J. Gen. Chem. USSR **39**, 1772 (1969)].

[2949] Bogolyubov, G. M., Plotnikov, V. F., Boiko, Yu. A., and Petrov, A. A., Organic derivatives of elements of groups V and VI. IX. Mass spectra of vinylacetylene sulfides, Zh. Obshch. Khim. **39**, 2467 (1969) [Engl. transl.: J. Gen. Chem. USSR **39**, 2407 (1969)].

[2950] Bock, H., and Alt, H., d–Orbital effects in silicon–substituted π–electron systems. XXIV. Charge–transfer studies of silyl– and alkylbenzenes, J. Am. Chem. Soc. **92**, 1569 (1970).

[2951] Bodor, N., Dewar, M. J. S., and Worley, S. D., Photoelectron spectra of molecules. III. Ionization potentials of some cyclic hydrocarbons and their derivatives, and heats of formation and ionization potentials calculated by the MINDO SCF MO method, J. Am. Chem. Soc. **92**, 19 (1970).

[2952] Branton, G. R., Brion, C. E., Frost, D. C., Mitchell, K. A. R., and Paddock, N. L., Phosphonitrilic derivatives. Part XVIII. Ionisation potentials, orbital symmetry, and π–electron interactions, J. Chem. Soc. (A), 151 (1970).

[2953] Camus, P., and Tomkins, F. S., Spectre d'absorption de Yb I, J. Phys. (Paris) **30**, 545 (1969).

[2954] Chelobov, F. N., Dubov, S. S., Tikhomirov, M. V., Dobrovitskii, M. I., Gitel', P. O., and Rozenshtein, S. M., Processes in fluoroacrylonitriles under electron impact, Zh. Fiz. Khim. **43**, 33 (1969) [Engl. transl.: Russ. J. Phys. Chem. **43**, 15 (1969)].

[2955] Cubicciotti, D., Some thermodynamic properties of the thallium oxides, High Temp. Sci. **1**, 11 (1969).

[2956] Distefano, G., Photoionization study of $Fe(CO)_5$ and $Ni(CO)_4$, J. Res. NBS **74A**, 233 (1970).

[2957] Drowart, J., De Maria, G., Boerboom, A. J. H., and Inghram, M. G., Mass spectrometric study of inter–group IVB molecules, J. Chem. Phys. **30**, 308 (1959).

[2958] Parr, A. C., and Inghram, M. G., Threshold behavior for photo double ionization, J. Chem. Phys. **52**, 4916 (1970).

[2959] Reichert, C., Bancroft, G. M., and Westmore, J. B., Mass spectral studies of metal chelates. V. Mass spectra and appearance potentials of some fluorine–substituted acetylacetonates, Can. J. Chem. **48**, 1362 (1970).

[2960] Rose, T., Frey, R., and Brehm, B., The electronic states of the diborane ion determined by photoelectron spectroscopy, Chem. Commun., 1518 (1969); Erratum, ibid., 460 (1970).

[2961] Tsuchiya, S., Ishihara, F., Tashiro, S., and Hikita, T., Mass spectroscopic detection of methyl radical in the reaction of hydrogen atom with ethylene, Bull. Chem. Soc. Japan **42**, 847 (1969).

[2962] Bischof, P., Hashmall, J. A., Heilbronner, E., and Hornung, V., Nitrogen lone pair interaction in 1,4–diaza–bicyclo[2.2.2]octane (DABCO), Tetrahedron Letters, 4025 (1969).

[2963] Winters, R. E., and Kiser, R. W., Ionization potentials and mass spectra of cyclopentadienylmolybdenum dicarbonyl nitrosyl and 1,3–cyclohexadieneiron tricarbonyl, J. Phys. Chem. **69**, 3198 (1965).

[2964] Wünsche, C., Ege, G., Beisiegel, E., and Pasedach, F., Skelettumlagerungen unter Elektronenbeschuss–III. Benzotriazinone und 1,3–Diphenyltriazene, Tetrahedron **25**, 5869 (1969).

[2965] Omura, I., Kaneko, T., and Yamada, Y., Mass spectrometric study of the photoionization of small polyatomic molecules, Org. Mass Spectrom. **2**, 847 (1969).

[2966] Heerma, W., de Ridder, J. J., and Dijkstra, G., The electron–impact–induced fragmentation of n–alkyl cyanides, Org. Mass Spectrom. **2**, 1103 (1969).

[2967] Chin, M. S., and Harrison, A. G., Substituent effects on ion abundances and energetics in substituted acetophenones, Org. Mass Spectrom. **2**, 1073 (1969).

[2968] Turner, D. W., The photoelectron spectra of benzene, hexafluorobenzene and pyridine, Tetrahedron Letters, 3419 (1967).

[2969] Condé–Caprace, G., and Collin, J. E., Ionization and dissociation of cyclic ethers and thioethers by electron–impact. A comparison between 1,4 dioxane, 1,4 dithiane and 1,4 oxathiane, Org. Mass Spectrom. **2**, 1277 (1969).

[2970] Tait, J. M. S., Shannon, T. W., and Harrison, A. G., The structure of substituted C_7 ions from benzyl derivatives at the appearance potential threshold, J. Am. Chem. Soc. **84**, 4 (1962).

[2971] Tsang, C. W., and Harrison, A. G., Four–centred rearrangements in the mass spectra of aliphatic ethers, Org. Mass Spectrom. **3**, 647 (1970).

[2972] Brown, P., Kinetic studies in mass spectrometry–IV. The $[M - Cl]$ reaction in substituted chlorobenzenes and the question of molecular ion isomerization, Org. Mass Spectrom. **3**, 639 (1970).

[2973] Grützmacher, H.-F., Zum Mechanismus massenspektrometrischer Fragmentierungsreaktionen–IV: zur Bildung von Phenonium-Ionen bei der Elektronenstoss-Fragmentierung von β-Phenyläthylbromiden, Org. Mass Spectrom. **3**, 131 (1970).

[2974] Rapp, U., Staab, H. A., and Wünsche, C., Skelettumlagerungen unter Elektronenbeschuss—IV: zur Struktur der $C_{13}H_9$– und $C_{12}H_9N$–Ionen bei Benzylidenaminobenztriazolen, Org. Mass Spectrom. **3**, 45 (1970).

[2975] Baker, A. D., Betteridge, D., Kemp, N. R., and Kirby, R. E., Nitrogen lone pairs and the ionization potentials of azines and azoles, Chem. Commun., 286 (1970).

[2976] Leyland, L. M., Majer, J. R., and Robb, J. C., Heat of formation of the $CF_2Cl\cdot$ radical, Trans. Faraday Soc. **66**, 898 (1970).

[2977] Potzinger, P., and Bünau, G. v., Empirische Berücksichtigung von Überschussenergien bei der Auftrittspotentialbestimmung, Ber. Bunsenges. Physik. Chem. **73**, 466 (1969).

[2978] Briegleb, G., and Czekalla, J., Die Bestimmung von Ionisierungsenergien aus den Spektren von Elektronenübergangskomplexen, Z. Elektrochem. **63**, 6 (1959).

[2979] Bünau, G. v., Schade, G., and Gollnick, K., Massenspektrometrische Untersuchungen von Terpenen. I. Monoterpenaldehyde und –ketone, Z. Anal. Chem. **227**, 173 (1967).

[2980] de Ridder, J. J., and Dijkstra, G., Mass spectra of the tetramethyl and tetraethyl compounds of carbon, silicon, germanium, tin and lead, Rec. Trav. Chim. **86**, 737 (1967).

[2981] Devyatykh, G. G., Larin, N. V., and Gaivoronskii, P. E., Mass spectrum of bis(benzene)chromium and the appearance potentials of ions, Zh. Obshch. Khim. **39**, 1823 (1969) [Engl. transl.: J. Gen. Chem. USSR **39**, 1786 (1969)].

[2982] Dey, S. D., and Karmohapatro, S. B., Measurement of ionization potentials of the rare earth elements by surface ionization method, Proc. Intern. Conf. Spectry., 1st, Bombay, 1967, **1**, p. 38.

[2983] Distefano, G., and Dibeler, V. H., Photoionization study of the dimethyl compounds of zinc, cadmium, and mercury, Intern. J. Mass Spectrom. Ion Phys. **4**, 59 (1970).

[2984] Eland, J. H. D., Photoelectron spectra and chemical bonding of mercury(II) compounds, Intern. J. Mass Spectrom. Ion Phys. **4**, 37 (1970).

[2985] Emel'yanov, A. M., Gusarov, A. V., Gorokhov, L. N., and Sadovnikova, N. A., Electron–impact studies of the vapors of cesium oxide and hydroxide, Teor. i Eksperim. Khim. **3**, 226 (1967) [Engl. transl.: Theoret. Exptl. Chem. **3**, 126 (1967)].

[2986] Fisher, I. P., The qualitative detection of free radicals using a modified commercial mass spectrometer, J. Sci. Instr. **43**, 633 (1966).

[2987] Fulton, A., and Lyons, L. E., The ionization energies of some phenothiazine tranquillizers and molecules of similar structure, Australian J. Chem. **21**, 873 (1968).

[2988] Gingerich, K. A., Mass spectrometric evidence for the gaseous AlOCN molecule, J. Am. Chem. Soc. **91**, 4302 (1969).

[2989] Higasi, K., Omura, I., and Baba, H., Ionization potentials of xylenes and picolines, J. Chem. Phys. **24**, 623 (1956).

[2990] Hildenbrand, D. L., Dissociation energies and chemical bonding in the alkaline–earth chlorides from mass spectrometric studies, J. Chem. Phys. **52**, 5751 (1970).

[2991] Hildenbrand, D. L., Electron impact studies of the IIA metal chlorides, Intern. J. Mass Spectrom. Ion Phys. **4**, 75 (1970).

[2992] Holtzclaw, H. F., Jr., Lintvedt, R. L., Baumgarten, H. E., Parker, R. G., Bursey, M. M., and Rogerson, P. F., Mass spectra of metal chelates. I. Substituent effects on ionization potentials and fragmentation patterns of some 1–methyl–3–alkyl–1,3–dione–copper(II) chelates, J. Am. Chem. Soc. **91**, 3774 (1969).

[2993] Hutchison, D. A., Critical ionization potentials using a parallel–plate energy selector, J. Chem. Phys. **24**, 628 (1956).

[2994] Il'ina, G. G., Rutgaizer, Yu. S., and Semenov, G. A., Ion source for mass spectral study of the energy characteristics of molecules, Pribory i Tekhn. Eksperim., 151 (1967) [Engl. transl.: Instr. Exptl. Tech. (USSR), 158 (1967)].

[2995] Kinson, P., and Trost, B. M., Decomposition of diazoketones under electron impact conditions, Tetrahedron Letters, 1075 (1969).

[2996] Kohl, F. J., and Stearns, C. A., Vaporization thermodynamics of yttrium dicarbide–carbon system and dissociation energy of yttrium dicarbide and tetracarbide, J. Chem. Phys. **52**, 6310 (1970).

[2997] Kohl, F. J., and Stearns, C. A., Dissociation energy of vanadium and chromium dicarbide and vanadium tetracarbide, J. Phys. Chem. **74**, 2714 (1970).

[2998] Lambert, R. M., Christie, M. I., Golesworthy, R. C., and Linnett, J. W., Mass spectrometric study of the reaction of nitrogen atoms with acetaldehyde, Proc. Roy. Soc. (London) **A302**, 167 (1968).

[2999] Lewis, D., and Hamill, W. H., Excited states of neutral molecular fragments from appearance potentials by electron impact in a mass spectrometer, J. Chem. Phys. **52**, 6348 (1970).

[3000] Matsen, F. A., Electron affinities, methyl affinities, and ionization energies of condensed ring aromatic hydrocarbons, J. Chem. Phys. **24**, 602 (1956).

[3001] Jonathan, N., Ross, K., and Tomlinson, V., The photoelectron spectra of dichloroethylenes, Intern. J. Mass Spectrom. Ion Phys. **4**, 51 (1970).

[3002] McAllister, T., Dolešek, Z., Lossing, F. P., Gleiter, R., and Schleyer, P. v. R., Thermal instability of 1– and 2–norbornyl and of 1–bicyclo[2.2.2]octyl radicals in a mass spectrometer, J. Am. Chem. Soc. **89**, 5982 (1967).

[3003] Minnhagen, L., The energy levels of neutral atomic iodine, Arkiv Fysik **21**, 415 (1962).

[3004] Muenow, D. W., and Margrave, J. L., Mass spectrometric determination of the heat of atomization of the molecule SiCN, J. Phys. Chem. **74**, 2577 (1970).

[3005] Müller, J., Ionisierungspotentiale von substituierten Benzolchrom–tricarbonyl–Komplexen, J. Organometal. Chem. **18**, 321 (1969).

[3006] Ogawa, M., Tanaka–Takamine Rydberg series of O_2, Can. J. Phys. **46**, 312 (1968).

[3007] Polyakova, A. A., Khmel'nitskii, R. A., and Petrov, A. A., Mass spectra and the structure of organic compounds. VIII. Mass spectra of some allene hydrocarbons containing t–butyl groups, Zh. Obshch. Khim. **33**, 2518 (1963) [Engl. transl.: J. Gen. Chem. USSR **33**, 2455 (1963)].

[3008] Polyakova, A. A., Zimina, K. I., Petrov, A. A., and Khmel'nitskii, R. A., Mass spectra and structure of some allenes, Zh. Obshch. Khim. **30**, 2977 (1960) [Engl. transl.: J. Gen. Chem. USSR **30**, 2949 (1960)].

[3009] Polyakova, A. A., Zimina, K. I., Petrov, A. A., and Khmel'nitskii, R. A., Mass spectra of vinylalkylacetylenes, Zh. Obshch. Khim. **30**, 912 (1960) [Engl. transl.: J. Gen. Chem. USSR **30**, 927 (1960)].

[3010] Price, W. C., and Walsh, A. D., The absorption spectra of hexatriene and divinyl acetylene in the vacuum ultra–violet, Proc. Roy. Soc. (London) **A185**, 182 (1946).

[3011] Riepe, W., and Zander, M., Massenspektroskopische Untersuchungen an Carbazolen, Z. Naturforsch. **24a**, 2017 (1969).

[3012] Rowland, C. G., Eland, J. H. D., and Danby, C. J., A spin–forbidden predissociation in the mass spectrum of isocyanic acid, Chem. Commun., 1535 (1968).

[3013] Schulz, W., Drost, H., and Klotz, H.-D., Massenspektrometrische Untersuchungen von ionischen Anlagerungsreaktionen des Pyridin, Z. Physik. Chem. (Leipzig) 237, 319 (1968).

[3014] Shchukarev, S. A., Semenov, G. A., and Rat'kovskii, I. A., A thermal investigation of the evaporation of gallium and indium oxides, Zh. Neorgan. Khim. 14, 3 (1969) [Engl. transl.: Russ. J. Inorg. Chem. 14, 1 (1969)].

[3015] Shigorin, D. N., Filyugina, A. D., and Potapov, V. K., Role of intramolecular hydrogen bonding in the ionisation and dissociation of compounds, Zh. Fiz. Khim. 41, 2336 (1967) [Engl. transl.: Russ. J. Phys. Chem. 41, 1255 (1967)].

[3016] Strel'chenko, S. S., Bondar', S. A., Molodyk, A. D., Balanevskaya, A. E., and Berger, L. I., Mass-spectrometric study of the sublimation processes of a group of ternary compounds of the type $A^IB^{III}C_2^{VI}$, Zh. Fiz. Khim. 41, 3118 (1967) [Engl. transl.: Russ. J. Phys. Chem. 41, 1679 (1967)].

[3017] Taft, R. W., Martin, R. H., and Lampe, F. W., Stabilization energies of substituted methyl cations. The effect of strong demand on the resonance order, J. Am. Chem. Soc. 87, 2490 (1965).

[3018] Takezawa, S., Absorption spectrum of H_2 in the vacuum-uv region. II. Rydberg series converging to the first six vibrational levels of the H_2^+ ground state, J. Chem. Phys. 52, 5793 (1970).

[3019] Villarejo, D., Stockbauer, R., and Inghram, M. G., Photoionization of CO_2 and the Franck–Condon principle for polyatomic molecules, J. Chem. Phys. 48, 3342 (1968).

[3020] Walsh, A. D., The absorption spectrum of acetaldehyde in the vacuum ultra–violet, Proc. Roy. Soc. (London) A185, 176 (1946).

[3021] Zinkiewicz, J. M., Determination of the ionization potential of Ca and Mg atoms by the surface ionization method using a mass spectrometer, Proc. Colloq. Spectros. Intern., 14th, Debrecen, Hung., 1967, 3, p. 1553.

[3022] Ward, R. S., and Williams, D. H., A study of water elimination as a function of ion lifetime in the mass spectrum of cyclohexanol, J. Organometal. Chem. 34, 3373 (1969).

[3023] Zmbov, K. F., and Margrave, J. L., Mass spectrometric studies at high temperatures. XXII. The stabilities of tantalum pentafluoride and tantalum oxytrifluoride, J. Phys. Chem. 72, 1099 (1968).

[3024] Weiershausen, W., Oberflächenionisation und Oberflächenoxydation einiger Seltener Erden an W und Re. Ann. Physik 15, 252 (1965).

[3025] Akopyan, M. E., and Vilesov, F. I., Mass-spectrometric investigation of the photo–ionization of benzene and its methyl derivatives, Khim. Vysokikh Energ. 2, 107 (1968) [Engl. transl.: High Energy Chem. (USSR) 2, 89 (1968)].

[3026] Bahr, J. L., Blake, A. J., Carver, J. H., and Kumar, V., Photoelectron spectra and partial photoionization cross-sections for carbon dioxide, J. Quant. Spectry. Radiative Transfer 9, 1359 (1969).

[3027] Branton, G. R., Frost, D. C., Herring, F. G., McDowell, C. A., and Stenhouse, I. A., The ionization potentials of ammonia and ammonia-d_3, measured by photoelectron spectroscopy, and an INDO calculation of these values, Chem. Phys. Letters 3, 581 (1969).

[3028] Branton, G. R., Frost, D. C., Makita, T., McDowell, C. A., and Stenhouse, I. A., Photoelectron spectra of ethylene and ethylene-d_4, J. Chem. Phys. 52, 802 (1970).

[3029] Price, W. C., Unpublished results reported in: Caulton, K. G., and Fenske, R. F., Electronic structure and bonding in $V(CO)_6^-$, $Cr(CO)_6$, and $Mn(CO)_6^+$, Inorg. Chem. 7, 1273 (1968).

[3030] Chupka, W. A., and Russell, M. E., Ion–molecule reactions of NH_3^+ by photoionization, J. Chem. Phys. 48, 1527 (1968).

[3031] Cook, G. R., and Metzger, P. H., Photoionization and absorption cross sections of O_2 and N_2 in the 600– to 1000–Å region, J. Chem. Phys. 41, 321 (1964).

[3032] Cook, G. R., Metzger, P. H., and Ogawa, M., Absorption, photoionization, and fluorescence of CO_2, J. Chem. Phys. 44, 2935 (1966).

[3033] Cook, G. R., Metzger, P. H., and Ogawa, M., Photoionization and absorption coefficients of N_2O, J. Opt. Soc. Am. 58, 129 (1968).

[3034] Cullen, W. R., Frost, D. C., and Vroom, D. A., Ionization potentials of some sulfur compounds, Inorg. Chem. 8, 1803 (1969).

[3035] Delwiche, J., and Natalis, P., Photoelectron spectrometry of hydrogen sulfide, Chem. Phys. Letters 5, 564 (1970).

[3036] Dibeler, V. H., Walker, J. A., and McCulloh, K. E., Threshold for molecular photoionization of bromine, J. Chem. Phys. 53, 4715 (1970).

[3037] Dibeler, V. H., Walker, J. A., and McCulloh, K. E., Photoionization study of chlorine monofluoride and the dissociation energy of fluorine, J. Chem. Phys. 53, 4414 (1970).

[3038] Dibeler, V. H., and Walker, J. A., Mass spectrometric study of photoionization. XIV. Nitrogen trifluoride and trifluoramine oxide, Inorg. Chem. 8, 1728 (1969).

[3039] Distefano, G., Photoionization study of tetramethylsilicon, Inorg. Chem. 9, 1919 (1970).

[3040] Delwiche, J., Ionization of sulphur hexafluoride by photon and electron impact, Bull. Classe Sci. Acad. Roy. Belg. 55, 215 (1969).

[3041] Delwiche, J., Photoelectron spectrometry and R.P.D. measurements on sulphur hexafluoride, in: Dynamic Mass Spectrometry, Vol. 1, ed. D. Price and J. E. Williams (Heyden and Son, London, 1970) p. 71.

[3042] Herzberg, G., and Johns, J. W. C., New spectra of the CH molecule, Astrophys. J. 158, 399 (1969).

[3043] Hillier, I. H., Marriott, J. C., Saunders, V. R., Ware, M. J., Lloyd, D. R., and Lynaugh, N., A theoretical and experimental study of the bonding in $PF_3 \cdot BH_3$, Chem. Commun., 1586 (1970).

[3044] Lloyd, D. R., and Lynaugh, N., The photoelectron spectra of NH_3BH_3 and BH_3CO, Chem. Commun., 1545 (1970).

[3045] Lake, R. F., and Thompson, H., The photoelectron spectra of some molecules containing the $C \equiv N$ group, Proc. Roy. Soc. (London) A317, 187 (1970).

[3046] Morris, G. C., The intensity of absorption of naphthacene vapor from 20 000 to 54 000 cm^{-1}, J. Mol. Spectry. 18, 42 (1965).

[3047] Potapov, V. K., and Sorokin, V. V., Investigation of molecular–ion reactions upon photoionization of methanol and ethanol, Dokl. Akad. Nauk SSSR 192, 590 (1970) [Engl. transl.: Dokl. Phys. Chem. 192, 412 (1970)].

[3048] Nakata, R. S., Watanabe, K., and Matsunaga, F. M., Absorption and photoionization coefficients of CO_2 in the region 580–1670 Å, Sci. Light (Tokyo) 14, 54 (1965).

[3049] Ogawa, M., and Chang, H. C., Absorption spectrum of CS_2 in the region from 600 to 1015 Å, Can. J. Phys. 48, 2455 (1970).

[3050] Rose, T. L., Frey, R., and Brehm, B., Photoelectron spectra of diborane, Bull. Am. Phys. Soc. 15, 430 (1970).

[3051] Reader, J., and Sugar, J., Ionization energies of Ce I and Gd I, J. Opt. Soc. Am. 60, 1421 (1970).

[3052] Thomson, R., and Warsop, P. A., Electronic spectrum of the haloacetylenes. Part 2. Bromoacetylene, Trans. Faraday Soc. 66, 1871 (1970).

[3053] Watanabe, K., and Sood, S. P., Absorption and photoionization coefficients of NH_3 in the 580–1650 Å region, Sci. Light (Tokyo) 14, 36 (1965).

[3054] Weiss, M. J., and Lawrence, G. M., Photoelectron spectroscopy of NH_3 and ND_3 using molecular beams, J. Chem. Phys. **53**, 214 (1970).

[3055] Yoshino, K., Absorption spectrum of the argon atom in the vacuum–ultraviolet region, J. Opt. Soc. Am. **60**, 1220 (1970).

[3056] Worley, S. D., Ionization potentials of cyclobutadiene, Chem. Commun., 980 (1970).

[3057] Ragle, J. L., Stenhouse, I. A., Frost, D. C., and McDowell, C. A., Valence–shell ionization potentials of halomethanes by photoelectron spectroscopy. I. CH_3Cl, CH_3Br, CH_3I. Vibrational frequencies and vibronic interaction in CH_3Br^+ and CH_3Cl^+, J. Chem. Phys. **53**, 178 (1970).

[3058] Baker, C., and Turner, D. W., Photoelectron spectra of allene and keten; Jahn–Teller distortion in the ionisation of allene, Chem. Commun., 480 (1969).

[3059] Bassett, P. J., and Lloyd, D. R., He(I) resonance photoelectron spectra of group IV tetrafluorides, Chem. Phys. Letters **3**, 22 (1969).

[3060] Price, W. C., Unpublished results reported in: Brailsford, D. F., and Ford, B., Calculated ionization potentials of the linear alkanes, Mol. Phys. **18**, 621 (1970).

[3061] Branton, G. R., Frost, D. C., Makita, T., McDowell, C. A., and Stenhouse, I. A., Photoelectron spectra of some polyatomic molecules, Phil. Trans. Roy. Soc. (London) **A268**, 77 (1970).

[3062] Brehm, B., Menzinger, M., and Zorn, C., The photo–electron spectrum of XeF_2, Can. J. Chem. **48**, 3193 (1970).

[3063] Bull, W. E., Pullen, B. P., Grimm, F. A., Moddeman, W. E., Schweitzer, G. K., and Carlson, T. A., High-resolution photoelectron spectroscopy of carbon and silicon tetrafluorides, Inorg. Chem. **9**, 2474 (1970).

[3064] Price, W. C., Private communication reported in: Dixon, R. N., and Hull, S. E., The photo–ionization of π electrons from O_2, Chem. Phys. Letters **3**, 367 (1969).

[3065] Duncan, A. B. F., The absorption spectrum of acetone vapor in the far ultraviolet, J. Chem. Phys. **3**, 131 (1935).

[3066] Edlén, B., Ölme, A., Herzberg, G., and Johns, J. W. C., Ionization potential of boron, and the isotopic and fine structure of $2s2p^2$ ^2D, J. Opt. Soc. Am. **60**, 889 (1970).

[3067] Eland, J. H. D., The photoelectron spectra of isocyanic acid and related compounds, Phil. Trans. Roy. Soc. (London) **A268**, 87 (1970).

[3068] Edqvist, O., Lindholm, E., Selin, L. E., and Åsbrink, L., On the photoelectron spectrum of O_2, Physica Scripta **1**, 25 (1970).

[3069] Edqvist, O., Lindholm, E., Selin, L. E., and Åsbrink, L., The photoelectron spectra of O_2, N_2 and CO, Phys. Letters **31A**, 292 (1970).

[3070] Green, J. C., King, D. I., and Eland, J. H. D., Photoelectron spectra of trifluorophosphine and its complexes $Ni(PF_3)_4$ and $Pt(PF_3)_4$, Chem. Commun., 1121 (1970).

[3071] Haink, H. J., Heilbronner, E., Hornung, V., and Kloster–Jensen, E., Die Photoelektron–Spektren der Monohalogenacetylene, Helv. Chim. Acta **53**, 1073 (1970).

[3072] Hamrin, K., Johansson, G., Gelius, U., Fahlman, A., Nordling, C., and Siegbahn, K., Ionization energies in methane and ethane measured by means of ESCA, Chem. Phys. Letters **1**, 613 (1968).

[3073] Haselbach, E., Hashmall, J. A., Heilbronner, E., and Hornung, V., The interaction between the lone pairs in azomethane, Angew. Chem. Intern. Ed. **8**, 878 (1969).

[3074] Hashmall, J. A., and Heilbronner, E., n–Ionization potentials of alkyl bromides, Angew. Chem. Intern. Ed. **9**, 305 (1970).

[3075] Lossing, F. P., Threshold ionization of acetylene by monoenergetic electron impact, Intern. J. Mass Spectrom. Ion Phys. **5**, 190 (1970).

[3076] Jonathan, N., Morris, A., Smith, D. J., and Ross, K. J., Photoelectron spectra of ground state atomic hydrogen, nitrogen and oxygen, Chem. Phys. Letters **7**, 497 (1970).

[3077] Herzberg, G., The dissociation energy of the hydrogen molecule, J. Mol. Spectry. **33**, 147 (1970).

[3078] Frost, D. C., Herring, F. G., McDowell, C. A., and Stenhouse, I. A., The ionization potentials of borazine by photoelectron spectrometry and INDO theory, Chem. Phys. Letters **5**, 291 (1970).

[3079] Cox, P. A., Evans, S., Hamnett, A., and Orchard, A. F., The helium–(I) photoelectron spectrum of vanadium tetrachloride, Chem. Phys. Letters **7**, 414 (1970).

[3080] Åsbrink, L., Lindholm, E., and Edqvist, O., Jahn–Teller effect in the vibrational structure of the photoelectron spectrum of benzene, Chem. Phys. Letters **5**, 609 (1970).

[3081] Åsbrink, L., Edqvist, O., Lindholm, E., and Selin, L. E., The electronic structure of benzene, Chem. Phys. Letters **5**, 192 (1970).

[3082] Barrow, R. F., and Beale, J. R., Rotational analysis of electronic bands of gaseous MgF, Proc. Phys. Soc. (London) **91**, 483 (1967).

[3083] Bassett, P. J., and Lloyd, D. R., The photoelectron spectra of nitrogen trifluoride and nitrogen oxide trifluoride, and a reassignment of the spectra of tetrafluorides of group IV, Chem. Phys. Letters **6**, 166 (1970).

[3084] Bassett, P. J., Lloyd, D. R., Hillier, I. H., and Saunders, V. R., A theoretical and experimental study of the electronic structure of PF_3O and the ligand properties of PF_3, Chem. Phys. Letters **6**, 253 (1970).

[3085] Branton, G. R., Frost, D. C., McDowell, C. A., and Stenhouse, I. A., The photoelectron spectra of phosphine and arsine, Chem. Phys. Letters **5**, 1 (1970).

[3086] Brundle, C. R., Ionization and dissociation energies of HF and DF and their bearing on $D_0^0(F_2)$, Chem. Phys. Letters **7**, 317 (1970).

[3087] Haselbach, E., and Heilbronner, E., Die Photoelektron–Spektren von CH_3–CH'=CH–CH_3, CH_3–N'=CH–CH_3, CH_3–N'=N–CH_3, ein Beitrag zur Frage nach der Wechselwirkung zwischen den einsamen Elektronenpaaren der *trans*–konfigurierten Azogruppe, Helv. Chim. Acta **53**, 684 (1970).

[3088] Hillier, I. H., Saunders, V. R., Ware, M. J., Bassett, P. J., Lloyd, D. R., and Lynaugh, N., On the bonding and the photoelectron spectra of $Ni(PF_3)_4$ and $Pt(PF_3)_4$, Chem. Commun., 1316 (1970).

[3089] Huffman, R. E., Larrabee, J. C., and Tanaka, Y., Comment on absorption series and ionization potentials of atomic chlorine and iodine, J. Chem. Phys. **48**, 3835 (1968).

[3090] Brundle, C. R., Neumann, D., Price, W. C., Evans, D., Potts, A. W., and Streets, D. G., Electronic structure of NO_2 studied by photoelectron and vacuum–uv spectroscopy and Gaussian orbital calculations, J. Chem. Phys. **53**, 705 (1970).

[3091] Heilbronner, E., Hornung, V., and Muszkat, K. A., Die Photoelektron–Spektren von Chlor–, Brom– und Jodcyan, Helv. Chim. Acta **53**, 347 (1970).

[3092] Brundle, C. R., Robin, M. B., and Basch, H., Electronic energies and electronic structures of the fluoromethanes, J. Chem. Phys. **53**, 2196 (1970).

[3093] Watanabe, K., and Marmo, F. F., Photoionization and total absorption cross section of gases. II. O_2 and N_2 in the region 850–1500 Å, J. Chem. Phys. **25**, 965 (1956).

[3094] Thomas, T. D., X–ray photoelectron spectroscopy of carbon monoxide, J. Chem. Phys. **53**, 1744 (1970).

[3095] Samson, J. A. R., Angular distributions of photoelectrons and partial photoionization cross-sections, Phil. Trans. Roy. Soc. (London) **A268,** 141 (1970).

[3096] Thomas, T. D., X-ray photoelectron spectroscopy of simple hydrocarbons, J. Chem. Phys. **52,** 1373 (1970).

[3097] Venkateswarlu, P., The vacuum ultraviolet spectrum of the bromine molecule, Can. J. Phys. **47,** 2525 (1969).

[3098] Person, J. C., and Nicole, P. P., Isotope effects in the photoionization yields and the absorption cross sections for acetylene, propyne, and propene, J. Chem. Phys. **53,** 1767 (1970).

[3099] Price, W. C., Passmore, T. R., and Roessler, D. M., Discussions Faraday Soc. **35,** 238 (1963).

[3100] Nostrand, E. D., and Duncan, A. B. F., Effect of pressure on intensity of some electronic transitions in SF_6, C_2H_2 and C_2D_2 vapors in the vacuum ultraviolet region, J. Am. Chem. Soc. **76,** 3377 (1954).

[3101] Turner, D. W., and May, D. P., Unpublished results reported in: Brundle, C. R., and Turner, D. W., The carbonyl π-ionization potential of formaldehyde, Chem. Commun., 314 (1967); Erratum, *ibid.*, 472 (1967).

[3102] Myer, J. A., and Samson, J. A. R., Absorption cross section and photoionization yield of I_2 between 1050 and 2200 Å, J. Chem. Phys. **52,** 716 (1970).

[3103] McDonald, J. R., Rabalais, J. W., and McGlynn, S. P., Electronic spectra of the azide ion, hydrazoic acid, and azido molecules, J. Chem. Phys. **52,** 1332 (1970).

[3104] Lossing, F. P., and Semeluk, G. P., Free radicals by mass spectrometry. XLII. Ionization potentials and ionic heats of formation for C_1-C_4 alkyl radicals, Can. J. Chem. **48,** 955 (1970).

[3105] Lloyd, D. R., and Lynaugh, N., Photoelectron studies of boron compounds. I. Diborane, borazine and *B*-trifluoroborazine, Phil. Trans. Roy. Soc. (London) **A268,** 97 (1970).

[3106] Kerwin, L., Marmet, P., and Clarke, E. M., Recent work with the electrostatic electron selector, Advan. Mass Spectrom. **2,** 522 (1963).

[3107] Jonathan, N., Smith, D. J., and Ross, K. J., High-resolution vacuum ultraviolet photoelectron spectroscopy of transient species: $O_2(^1\Delta_g)$, J. Chem. Phys. **53,** 3758 (1970).

[3108] Hunt, H. D., and Simpson, W. T., Spectra of simple amides in the vacuum ultraviolet, J. Am. Chem. Soc. **75,** 4540 (1953).

[3109] Frost, D. C., Ionization potentials and transition probabilities by photoelectron spectrometry, Intern. Conf. Phys. Electron. At. Collisions, 5th, Leningrad, 1967, Abstr. Pap., p. 615.

[3110] Carroll, P. K., and Grennan, T. P., The B-X system of CF, J. Phys. B **3,** 865 (1970).

[3111] Barrow, R. F., du Parcq, R. P., and Ricks, J. M., Rotational analysis of the C $^3\Sigma_u^- $-X $^3\Sigma_g^-$ system of $^{32}S_2$, J. Phys. B **2,** 413 (1969).

[3112] Yoshino, K., Absorption spectrum of the argon atom in the vacuum-ultraviolet region, J. Cpt. Soc. Am. **59,** 1525 (1969).

[3113] Wood, D. R., and Andrew, K. L., Arc spectrum of lead, J. Opt. Soc. Am. **58,** 818 (1968).

[3114] Radziemski, L. J., Jr., and Kaufman, V., Wavelengths, energy levels, and analysis of neutral atomic chlorine (Cl I), J. Opt. Soc. Am. **59,** 424 (1969).

[3115] Vlaskov, V. A., and Ovchinnikov, A. A., The temperature dependence of the photoionization cross-section of polyatomic molecules, Opt. i Spektroskopiya **27,** 748 (1969) [Engl. transl.: Opt. Spectry. (USSR) **27,** 408 (1969)].

[3116] Pullen, B. P., Carlson, T. A., Moddeman, W. E., Schweitzer, G. K., Bull, W. E., and Grimm, F. A., Photoelectron spectra of methane, silane, germane, methyl fluoride, difluoromethane, and trifluoromethane, J. Chem. Phys. **53,** 768 (1970).

[3117] Green, J. C., Green, M. L. H., Joachim, P. J., Orchard, A. F., and Turner, D. W., A study of the bonding in the group IV tetrahalides by photoelectron spectroscopy, Phil. Trans. Roy. Soc. (London) **A268,** 111 (1970).

[3118] Minnhagen, L., $(^3P)nf$ levels of Cl I, J. Opt. Soc. Am. **51,** 298 (1961).

[3119] Potts, A. W., Lempka, H. J., Streets, D. G., and Price, W. C., Photoelectron spectra of the halides of elements in groups III, IV, V and VI, Phil. Trans. Roy. Soc. (London) **A268,** 59 (1970).

[3120] Brundle, C. R., and Robin, M. B., Nonplanarity in hexafluorobutadiene as revealed by photoelectron and optical spectroscopy, J. Am. Chem. Soc. **92,** 5550 (1970).

[3121] Heilbronner, E., Hornung, V., and Kloster-Jensen, E., Die Photoelektron-Spektren der Dihalogen-acetylene, Helv. Chim. Acta **53,** 331 (1970).

[3122] Brundle, C. R., On the first ionization potentials of nitrogen dioxide and nitric oxide, Chem. Phys. Letters **5,** 410 (1970).

[3123] Edqvist, O., Lindholm, E., Selin, L. E., Sjögren, H., and Åsbrink, L., Rydberg series in small molecules. VII. Rydberg series and photoelectron spectroscopy of NO, Arkiv Fysik **40,** 439 (1970).

[3124] Sutcliffe, L. H., and Walsh, A. D., The absorption spectrum of allene in the vacuum ultra-violet, J. Chem. Soc., 899 (1952).

[3125] Risberg, G., The spectrum of atomic calcium, Ca I, and extensions to the analysis of Ca II, Arkiv Fysik **37,** 231 (1968).

[3126] Risberg, G., The spectrum of atomic magnesium, Mg I, Arkiv Fysik **28,** 381 (1965).

[3127] Toresson, Y. G., Spectrum and term system of doubly ionized silicon, Si III, Arkiv Fysik **18,** 389 (1961).

[3128] White, D., Sommer, A., Walsh, P. N., and Goldstein, H. W., The application of the time-of-flight mass spectrometer to the study of inoganic materials at elevated temperatures, Advan. Mass Spectrom. **2,** 110 (1963).

[3129] Schulz, G. J., Excitation and negative ions in H_2O, J. Chem. Phys. **33,** 1661 (1960).

[3130] Morrison, J. D., The study of molecular energy states, Advan. Mass Spectrom. **2,** 479 (1963).

[3131] Momigny, J., and Derouane, E., Fine structure in the first derivative of ionization curves obtained under electron impact, Advan. Mass Spectrom. **4,** 607 (1968).

[3132] Puttkamer, E. v., Koinzidenzmessung von Photoionen und Photoelektronen, Z. Naturforsch. **25a,** 1062 (1970).

[3133] Cloutier, G. G., and Schiff, H. I., A modification of the R.P.D. method for measuring appearance potentials, Advan. Mass Spectrom. **1,** 473 (1959).

[3134] Dibeler, V. H., and Rosenstock, H. M., Mass spectra and metastable transitions of H_2S, HDS, and D_2S, J. Chem. Phys. **39,** 3106 (1963).

[3135] Ferreira, M. A. A., Mass spectrometric study of ionization and dissociation of N_2O by electron impact, Rev. Port. Quím. **10,** 168 (1968).

[3136] Minnhagen, L., The *nf* and *ng* levels of Ar II, Arkiv Fysik **18,** 97 (1960).

[3137] Fox, R. E., Study of multiple ionization in helium and xenon by electron impact, Advan. Mass Spectrom. **1,** 397 (1959).

[3138] Newton, A. S., and Sciamanna, A. F., Metastable peaks in the mass spectra of N_2O and NO_2. II, J. Chem. Phys. **52,** 327 (1970).

[3139] Newton, A. S., and Sciamanna, A. F., Metastable dissociation of the doubly charged carbon monoxide ion, J. Chem. Phys. **53,** 132 (1970).

[3140] Price, W. C., The far ultraviolet absorption spectra and ionization potentials of H_2O and H_2S, J. Chem. Phys. **4,** 147 (1936).

[3141] Price, W. C., The far ultraviolet absorption spectra of formaldehyde and the alkyl derivatives of H_2O and H_2S, J. Chem. Phys. **3**, 256 (1935).

[3142] Price, W. C., and Simpson, D. M., The absorption spectra of nitrogen dioxide, ozone and nitrosyl chloride in the vacuum ultra–violet, Trans. Faraday Soc. **37**, 106 (1941).

[3143] Ogawa, M., and Tanaka, Y., Rydberg absorption series of N_2, Can. J. Phys. **40**, 1593 (1962).

[3144] Cottin, M., Étude des ions produits par impact électronique dans la vapeur d'eau, J. Chim. Phys. **56**, 1024 (1959).

[3145] Price, W. C., The absorption spectra of acetylene, ethylene and ethane in the far ultraviolet, Phys. Rev. **47**, 444 (1935).

[3146] Cook, G. R., Photoionization and absorption cross sections of ozone, Trans. Am. Geophys. Union **49**, 736 (1968).

[3147] Brundle, C. R., Robin, M. B., Basch, H., Pinsky, M., and Bond, A., Experimental and theoretical comparison of the electronic structures of ethylene and diborane, J. Am. Chem. Soc. **92**, 3863 (1970).

[3148] Price, W. C., and Potts, A. W., Private communication reported in: Lorquet, J. C., and Cadet, C., Configuration interaction intensity borrowing in photoelectron spectroscopy, Chem. Phys. Letters **6**, 198 (1970).

[3149] DeCorpo, J. J., Steiger, R. P., Franklin, J. L., and Margrave, J. L., Dissociation energy of F_2, J. Chem. Phys. **53**, 936 (1970).

[3150] Price, W. C., The absorption spectra of the halogen acids in the vacuum ultra–violet, Proc. Roy. Soc. (London) **A167**, 216 (1938).

[3151] Price, W. C., Teegan, J. P., and Walsh, A. D., The absorption spectrum of keten in the far ultra–violet, J. Chem. Soc., 920 (1951).

[3152] Price, W. C., and Walsh, A. D., The absorption spectra of triple bond molecules in the vacuum ultra violet, Trans. Faraday Soc. **41**, 381 (1945).

[3153] Price, W. C., and Walsh, A. D., The absorption spectra of benzene derivatives in the vacuum ultra–violet. I, Proc. Roy. Soc. (London) **A191**, 22 (1947).

[3154] Price, W. C., Bralsford, R., Harris, P. V., and Ridley, R. G., Ultra–violet spectra and ionization potentials of hydrocarbon molecules, Spectrochim. Acta **14**, 45 (1959).

[3155] Bradt, P., Mohler, F. L., and Dibeler, V. H., Mass spectrum of sulfur vapor, J. Res. NBS **57**, 223 (1956).

[3156] Bloch, A., Mass spectra of acetylene under high pressure in the ion source, Advan. Mass Spectrom. **2**, 48 (1963).

[3157] Bhattacharya, A. K., Smithson, L. D., and Hurley, D. W., Modification of a CEC 21–110B mass spectrometer for appearance potential measurements, Rev. Sci. Instr. **41**, 1111 (1970).

[3158] Praet, M.–T., and Delwiche, J., Ionization energies of some cyclic molecules, Chem. Phys. Letters **5**, 546 (1970).

[3159] Nikolaev, E. N., Ovchinnikov, K. V., and Semenov, G. A., Mass–spectrometric investigation of gaseous rhenium bromide, Zh. Obshch. Khim. **40**, 1302 (1970) [Engl. transl.: J. Gen. Chem. USSR **40**, 1294 (1970)].

[3160] Eriksson, K. B. S., and Wenåker, I., New wavelength measurements in Cs I, Physica Scripta **1**, 21 (1970).

[3161] Palenius, H. P., Spectrum and term system of doubly ionized fluorine, F III, Physica Scripta **1**, 113 (1970).

[3162] Dick, K. A., The spark spectra of zinc. II. Zinc III, Can. J. Phys. **46**, 1291 (1968).

[3163] Iglesias, L., The fourth and fifth spectra of vanadium (V IV and V V), J. Res. NBS **72A**, 295 (1968).

[3164] Palenius, H. P., Spectrum and term system of singly ionized fluorine, F II, Arkiv Fysik **39**, 15 (1969).

[3165] Epstein, G. L., and Reader, J., Analysis of the spectrum of doubly ionized yttrium (Y III), J. Opt. Soc. Am. **60**, 1556 (1970).

[3166] Bromander, J., Johansson, B., and Bockasten, K., Spectra of O IV and O V, J. Opt. Soc. Am. **57**, 1158 (1967).

[3167] Johansson, I., and Litzén, U., The term systems of the neutral gallium and indium atoms derived from new measurements in the infrared region, Arkiv Fysik **34**, 573 (1967).

[3168] Lloyd, D. R., The failure of electron impact measurements to detect first ionization potentials in some molecules, Intern. J. Mass Spectrom. Ion Phys. **4**, 500 (1970).

[3169] Lloyd, D. R., Photoelectron spectra of transition–metal hexafluoroacetylacetonates and the supposed breakdown of Koopmans' theorem in these compounds, Chem. Commun., 868 (1970).

[3170] Delwiche, J., Natalis, P., and Collin, J. E., High resolution photoelectron spectrometry of H_2S and H_2Se, Intern. J. Mass Spectrom. Ion Phys. **5**, 443 (1970).

[3171] Hotop, H., and Niehaus, A., Reactions of excited atoms and molecules with atoms and molecules. V. Comparison of Penning electron and photoelectron spectra of H_2, N_2 and CO, Intern. J. Mass Spectrom. Ion Phys. **5**, 415 (1970).

[3172] Cristy, S. S., and Mamantov, G., Cryogenic mass spectrometry of reactive fluorine–containing species. I. The mass spectra of sulfur hexafluoride, chlorine trifluoride, chlorine monofluoride, nitrosyl fluoride and tetrafluorohydrazine, Intern. J. Mass Spectrom. Ion Phys. **5**, 309 (1970).

[3173] Cristy, S. S., and Mamantov, G., Cryogenic mass spectrometry of reactive fluorine–containing species. II. Applications to synthesis via pyrolysis, photolysis and microwave discharge, Intern. J. Mass Spectrom. Ion Phys. **5**, 319 (1970).

[3174] Johnstone, R. A. W., Mellon, F. A., and Ward, S. D., On–line acquisition of ionization efficiency data, Intern. J. Mass Spectrom. Ion Phys. **5**, 241 (1970).

[3175] Newton, A. S., Sciamanna, A. F., and Thomas, G. E., The occurrence of the H_3^+ ion in the mass spectra of organic compounds, Intern. J. Mass Spectrom. Ion Phys. **5**, 465 (1970).

[3176] Friedland, S. S., and Strakna, R. E., Appearance potential studies. I, J. Phys. Chem. **60**, 815 (1956).

[3177] Kloster–Jensen, E., Pascual, C., and Vogt, J., Mass spectrometric studies of mono– and di–haloacetylenes, Helv. Chim. Acta **53**, 2109 (1970).

[3178] Foner, S. N., and Hudson, R. L., Mass spectrometric studies of atom–molecule reactions using high–intensity crossed molecular beams, J. Chem. Phys. **53**, 4377 (1970).

[3179] Robertson, E. W., and Barrow, R. F., Rotational analysis of the $C^1\Pi_u$–$X^1\Sigma_g^+$ system of K_2, and the ionisation potential of K_2, Proc. Chem. Soc., 329 (1961).

[3180] Eastmond, G. B. M., and Pratt, G. L., Pyrolyses in the presence of nitric oxide. Part III. Acetaldehyde oxime and nitrosoethane, J. Chem. Soc. (A), 2337 (1970).

[3181] Ölme, A., The spectrum of singly ionized boron B II, Physica Scripta **1**, 256 (1970).

[3182] Foner, S. N., and Hudson, R. L., Mass spectrometry of very fast reactions: identification of free radicals and unstable molecules formed in atom–molecule reactions, J. Chem. Phys. **49**, 3724 (1968).

[3183] Barrow, R. F., Travis, N., and Wright, C. V., Excited electronic states of lithium and sodium molecules, Nature **187**, 141 (1960).

[3184] Kwon, C. T., and McGee, H. A., Jr., Cryochemical preparation of monomeric aminoborane, Inorg. Chem. **9**, 2458 (1970).

[3185] Knowles, D. J., Nicholson, A. J. C., and Swingler, D. L., Electron impact studies. II. Stannous bromide and stannic bromide, J. Phys. Chem. **74**, 3642 (1970).

[3186] Hildenbrand, D. L., and Murad, E., Dissociation energy of NaO(g) and the heat of atomization of $Na_2O(g)$, J. Chem. Phys. **53**, 3403 (1970).

[3187] Hess, G. G., Lampe, F. W., and Sommer, L. H., Bond dissociation energies and ion energetics in organosilicon compounds by electron impact, J. Am. Chem. Soc. 86, 3174 (1964).

[3188] Haschke, J. M., and Eick, H. A., Preparation and vaporization thermodynamics of europium oxide bromides, J. Am. Chem. Soc. 92, 4550 (1970).

[3189] Gorokhov, L. N., Gusarov, A. V., and Panchenkov, I. G., Determination of dissociation energies of potassium and caesium hydroxides by electron bombardment, Zh. Fiz. Khim. 44, 269 (1970) [Engl. transl.: Russ. J. Phys. Chem. 44, 150 (1970)].

[3190] Gebelt, R. E., and Eick, H. A., Vaporization behavior of europium dicarbide, J. Chem. Phys. 44, 2872 (1966).

[3191] Harrison, A. G., and Tait, J. M. S., Concurrent ion-molecule reactions leading to the same product ion, Can. J. Chem. 40, 1986 (1962).

[3192] Franklin, J. L., Lampe, F. W., and Lumpkin, H. E., The proton affinity of benzene, J. Am. Chem. Soc. 81, 3152 (1959).

[3193] Ficalora, P. J., Thompson, J. C., and Margrave, J. L., Mass spectrometric studies at high temperatures—XXVI. The sublimation of SeO$_2$ and SeO$_3$, J. Inorg. Nucl. Chem. 31, 3771 (1969).

[3194] De Maria, G., Malaspina, L., and Piacente, V., Mass spectrometric study of the GaAs system, J. Chem. Phys. 52, 1019 (1970).

[3195] Chupka, W. A., Inghram, M. G., and Porter, R. F., Dissociation energy of gaseous LaO, J. Chem. Phys. 24, 792 (1956).

[3196] Cater, E. D., Rauh, E. G., and Thorn, R. J., Uranium monosulfide. III. Thermochemistry, partial pressures, and dissociation energies of US and US$_2$, J. Chem. Phys. 44, 3106 (1966).

[3197] Chupka, W. A., Berkowitz, J., and Giese, C. F., Vaporization of beryllium oxide and its reaction with tungsten, J. Chem. Phys. 30, 827 (1959).

[3198] Cater, E. D., Lee, T. E., Johnson, E. W., Rauh, E. G., and Eick, H. A., Vaporization, thermodynamics, and dissociation energy of lanthanum monosulfide, J. Phys. Chem. 69, 2684 (1965).

[3199] Blackburn, P. E., Büchler, A., and Stauffer, J. L., Thermodynamics of vaporization in the aluminum oxide–boron oxide system, J. Phys. Chem. 70, 2469 (1966).

[3200] Bidinosti, D. R., and Coatsworth, L. L., Mass spectrometric study of the reaction of BF$_3$ with B$_2$O$_3$; the identification and heat of formation of B$_2$OF$_4$, Can. J. Chem. 48, 2484 (1970).

[3201] Baldwin, M., Maccoll, A., and Miller, S. I., Ionization and appearance potentials from a study of alkyl chlorides, Advan. Mass Spectrom. 3, 259 (1966).

[3202] Cullen, W. R., Frost, D. C., and Pun, M. T., Mass spectra, appearance potentials, heats of formation, and bond energies of some alkyl and perfluoroalkyl sulfides, Inorg. Chem. 9, 1976 (1970).

[3203] Zmbov, K. F., Heats of formation of gaseous ThF$_3$ and ThF$_2$ from mass spectrometric studies, J. Inorg. Nucl. Chem. 32, 1381 (1970).

[3204] Zmbov, K. F., Mass spectrometric determination of the dissociation energies of calcium and barium chlorides, Chem. Phys. Letters 4, 191 (1969).

[3205] Visnapuu, A., and Jensen, J. W., Composition and properties of vapors over molten silver chloride, J. Less–Common Metals 20, 141 (1970).

[3206] Uy, O. M., and Drowart, J., Mass spectrometric determination of the dissociation energies of the boron monochalcogenides, High Temp. Sci. 2, 293 (1970).

[3207] Stearns, C. A., and Kohl, F. J., The dissociation energy of gaseous titanium mononitride, High Temp. Sci. 2, 146 (1970).

[3208] Stearns, C. A., and Kohl, F. J., The dissociation energies of titanium dicarbide and titanium tetracarbide, High Temp. Sci. 2, 274 (1970).

[3209] Meschi, D. J., Chupka, W. A., and Berkowitz, J., Heterogeneous reactions studied by mass spectrometry. I. Reaction of B$_2$O$_3$(s) with H$_2$O(g), J. Chem. Phys. 33, 530 (1960).

[3210] De Maria, G., Balducci, G., Capalbi, A., and Guido, M., High–temperature mass spectrometric study of the system neodymium–carbon, Proc. Brit. Ceram. Soc., 127 (1967).

[3211] Bafus, D. A., Gallegos, E. J., and Kiser, R. W., An electron impact investigation of some alkyl phosphate esters, J. Phys. Chem. 70, 2614 (1966).

[3212] Sergeev, Yu. L., Akopyan, M. E., Vilesov, F. I., and Kleimenov, V. I., Photoionization processes in phenyl halides, Opt. i Spektroskopiya 29, 119 (1970) [Engl. transl.: Opt. Spectry. (USSR) 29, 63 (1970)].

[3213] D'Or, L., Momigny, J., and Natalis, P., Mass spectra and geometrical isomerism, Advan. Mass Spectrom. 2, 370 (1963).

[3214] Heerma, W., and de Ridder, J. J., The electron–impact-induced fragmentation of some alkyl isocyanides and α–branched alkyl cyanides, Org. Mass Spectrom. 3, 1439 (1970).

[3215] Edqvist, O., Lindholm, E., Selin, L. E., Åsbrink, L., Kuyatt, C. E., Mielczarek, S. R., Simpson, J. A., and Fischer–Hjalmars, I., Rydberg series of small molecules. VIII. Photoelectron spectroscopy and electron spectroscopy of NO$_2$, Physica Scripta 1, 172 (1970).

[3216] Dibeler, V. H., Franklin, J. L., and Reese, R. M., Electron impact studies of hydrazine and the methyl–substituted hydrazines, Advan. Mass Spectrom. 1, 443 (1959).

[3217] Berkowitz, J., Chupka, W. A., Guyon, P. M., Holloway, J., and Spohr, R., Photo–ionization studies of F$_2$, HF, DF, and the xenon fluorides, Advan. Mass Spectrom. 5, 112 (1971).

[3218] Barrow, R. F., Burton, W. G., and Callomon, J. H., Absorption spectrum of gaseous ^{80}Se$_2$ in the region 51500–55000 cm^{-1}, Trans. Faraday Soc. 66, 2685 (1970).

[3219] Douglas, A. E., and Lutz, B. L., Spectroscopic identification of the SiH$^+$ molecule: the A$^1\Pi$–X$^1\Sigma^+$ system, Can. J. Phys. 48, 247 (1970).

[3220] Di Lonardo, G., and Trombetti, A., Spectrum of SF, Trans. Faraday Soc. 66, 2694 (1970).

[3221] Emel'yanov, A. M., Khodeyev, Yu. S., and Gorokhov, L. N., Electron impact ionization of Cs$^+$ ions, Intern. Conf. Phys. Electron. At. Collisions, 5th, Leningrad, 1967, Abstr. Pap., p. 46.

[3222] Briggs, P. R., and Shannon, T. W., The heat of formation of the methoxycarbonyl ion, J. Am. Chem. Soc. 91, 4307 (1969).

[3223] Buchs, A., Etude par spectrométrie de masse de l'ionisation de benzonitriles, de phénylacétonitriles et de N,N–diméthylanilines substitués, Helv. Chim. Acta 53, 2026 (1970).

[3224] King, A. B., and Long, F. A., Mass spectra of some simple esters and their interpretation by quasi–equilibrium theory, J. Chem. Phys. 29, 374 (1958).

[3225] Steck, S. J., Pressley, G. A., Jr., Stafford, F. E., Dobson, J., and Schaeffer, R., Molecular–beam mass spectra of hexaborane(12) and octaborane(12), Inorg. Chem. 9, 2452 (1970).

[3226] Hollins, R. E., and Stafford, F. E., Molecular beam mass spectra and pyrolysis of pentaborane(9), tetraborane carbonyl, and pentaborane(11). Formation and mass spectrum of tetraborane(8), Inorg. Chem. 9, 877 (1970).

[3227] Lappert, M. F., Litzow, M. R., Pedley, J. B., Riley, P. N. K., Spalding, T. R., and Tweedale, A., Bonding studies of compounds of boron and the group IV elements. Part III. First ionisation potentials of some simple boron compounds by electron impact and by a new empirical molecular orbital method, J. Chem. Soc. (A), 2320 (1970).

[3228] Murphy, C. B., Jr., and Enrione, R. E., Ionization potentials of pentaborane(9) derivatives by electron impact and molecular orbital calculations, Intern. J. Mass Spectrom. Ion Phys. 5, 157 (1970).

[3229] Zandberg, É. Ya., and Rasulev, U. Kh., Surface ionization of the $C_4H_{10}N$ radical, Dokl. Akad. Nauk SSSR 178, 327 (1968) [Engl. transl.: Soviet Phys. – Dokl. 13, 35 (1968)].

[3230] Yeo, A. N. H., and Williams, D. H., Rearrangement in the molecular ions of halogenotoluenes prior to fragmentation in the mass spectrometer, Chem. Commun., 886 (1970).

[3231] Yeo, A. N. H., and Williams, D. H., Calculation of partial mass spectra of some organic compounds undergoing competing reactions from the molecular ions, J. Am. Chem. Soc. 92, 3984 (1970).

[3232] Omura, I., Baba, H., Higasi, K., and Kanaoka, Y., Ionization potentials of some organic molecules. V. Heterocyclic compounds containing nitrogen, Bull. Chem. Soc. Japan 30, 633 (1957).

[3233] Linda, P., Marino, G., and Pignataro, S., Ionization potentials and relative rates of electrophilic substitution in five–membered heteroaromatic rings, Ric. Sci. 39, 666 (1969).

[3234] Junk, G. A., and Svec, H. J., The mass spectra, ionization potentials, and bond energies of the group VIIA decacarbonyls, J. Chem. Soc. (A), 2102 (1970).

[3235] Bodor, N., Dewar, M. J. S., Jennings, W. B., and Worley, S. D., Photoelectron spectra of molecules—IV. Ionization potentials and heats of formation of some hydrazines and amines, Tetrahedron 26, 4109 (1970).

[3236] Thynne, J. C. J., and MacNeil, K. A. G., Ionisation and dissociation of carbonyl fluoride and trifluoromethyl hypofluorite by electron impact, Intern. J. Mass Spectrom. Ion Phys. 5, 95 (1970).

[3237] Meyerson, S., Rylander, P. N., Eliel, E. L., and McCollum, J. D., Organic ions in the gas phase. VII. Tropylium ion from benzyl chloride and benzyl alcohol, J. Am. Chem. Soc. 81, 2606 (1959).

[3238] Howe, I., and Williams, D. H., Calculation and qualitative predictions of mass spectra. Mono– and para–disubstituted benzenes, J. Am. Chem. Soc. 91, 7137 (1969).

[3239] Evans, S., Hamnett, A., and Orchard, A. F., The helium–(I) photoelectron spectrum of tris(hexafluoroacetylacetonato)iron(III), Chem. Commun., 1282 (1970).

[3240] Pignataro, S., Linda, P., and Marino, G., The ionization potential of 2–substituted thiophenes, Ric. Sci. 39, 668 (1969).

[3241] Jonathan, N., Smith, D. J., and Ross, K. J., The high resolution photoelectron spectra of transient species: sulphur monoxide, Chem. Phys. Letters 9, 217 (1971).

[3242] Kelly, R., and Padley, P. J., Ionization potentials of alkaline–earth monohydroxides, Chem. Commun., 1606 (1970).

[3243] Gusarov, A. V., Gorokhov, L. N., and Efimova, A. G., Mass–spectrometer study of the evaporation products of cesium carbonate, Teplofiz. Vysokikh Temperatur 5, 783 (1967) [Engl. transl.: High Temp. (USSR) 5, 699 (1967)].

[3244] Shchukarev, S. A., Semenov, G. A., and Rat'kovskii, I. A., Study of the evaporation of gallium, indium, and thallium with the aid of the mass spectrometer, Zh. Prikl. Khim. 35, 1454 (1962) [Engl. transl.: J. Appl. Chem. USSR 35, 1401 (1962)].

[3245] Borisov, Yu. A., Gusarov, A. V., and Gorokhov, L. N., Mass spectrometer investigation of the evaporation of cesium peroxide, Teplofiz. Vysokikh Temperatur 2, 487 (1964) [Engl. transl.: High Temp. (USSR) 2, 440 (1964)].

[3246] Baker, A. D., Betteridge, D., Kemp, N. R., and Kirby, R. E., Application of photoelectron spectrometry to pesticide analysis. Photoelectron spectra of five–membered heterocycles and related molecules, Anal. Chem. 42, 1064 (1970).

[3247] Baker, A. D., and Turner, D. W., Unpublished results reported in: Baker, A. D., Betteridge, D., Kemp, N. R., and Kirby, R. E., Application of photoelectron spectrometry to pesticide analysis. Photoelectron spectra of five–membered heterocycles and related molecules, Anal. Chem. 42, 1064 (1970).

[3248] Chambers, D. B., and Glockling, F., Electron impact determination of heats of formation and bond energies in triphenyltin compounds, Inorg. Chim. Acta 4, 150 (1970).

[3249] Sullivan, C. L., Prusaczyk, J. E., and Carlson, K. D., Molecules in the equilibrium vaporization of antimony sulfide and selenide, J. Chem. Phys. 53, 1289 (1970).

[3250] Junk, G. A., Preston, F. J., Svec, H. J., and Thompson, D. T., Ligand effects on the ionization potentials and bond energies of some $L^1L^2Fe_2(CO)_6$ complexes, J. Chem. Soc. (A), 3171 (1970).

[3251] Svec, H. J., and Junk, G. A., Thermal reactions in the mass spectrometry of organometallic compounds, Inorg. Chem. 7, 1688 (1968).

[3252] Cataliotti, R., Foffani, A., and Pignataro, S., Infrared and mass spectral studies on azaferrocene (π–cyclopentadienyl–π–pyrrolyliron), Inorg. Chem. 9, 2594 (1970).

[3253] Reese, R. M., and Dibeler, V. H., Private communication reported in: Hobrock, B. G., and Kiser, R. W., Electron impact spectroscopy of tetramethylgermanium, trimethylsilane and dimethylmercury, J. Phys. Chem. 66, 155 (1962).

[3254] Seidl, H., Bock, H., Wiberg, N., and Veith, M., The color of trimethylsilyl derivatives of diimine, Angew. Chem. Intern. Ed. 9, 69 (1970).

[3255] Saalfeld, F. E., McDowell, M. V., and MacDiarmid, A. G., Nature of the bonding between silicon and the cobalt tetracarbonyl group in silylcobalt tetracarbonyls. II. Mass spectral evidence, J. Am. Chem. Soc. 92, 2324 (1970).

[3256] Pupp, C., Yamdagni, R., and Porter, R. F., Mass spectrometric study of the evaporation of $BaMoO_4$ and $BaWO_4$, J. Inorg. Nucl. Chem. 31, 2021 (1969).

[3257] Yamdagni, R., Pupp, C., and Porter, R. F., Mass spectrometric study of the evaporation of lithium and sodium molybdates and tungstates, J. Inorg. Nucl. Chem. 32, 3509 (1970).

[3258] Parr, A. C., Photoionization of europium and thulium: threshold to 1350 Å, J. Chem. Phys. 54, 3161 (1971).

[3259] Faktor, M. M., and Hanks, R., Calculation of the third ionisation potentials of the lanthanons, J. Inorg. Nucl. Chem. 31, 1649 (1969).

[3260] Reader, J., and Sugar, J., Nuclear magnetic moment of Pr^{141} from the hyperfine structure of doubly ionized praseodymium, Phys. Rev. 137, B784 (1965).

[3261] Johnson, D. A., Third ionization potentials and sublimation energies of the lanthanides, J. Chem. Soc. (A), 1525 (1969).

[3262] Odabasi, H., Spectrum of doubly ionized lanthanum (La III), J. Opt. Soc. Am. 57, 1459 (1967).

[3263] Sugar, J., and Kaufman, V., Spectrum of doubly ionized lanthanum (La III), J. Opt. Soc. Am. 55, 1283 (1965).

[3264] Sugar, J., Description and analysis of the third spectrum of cerium (Ce III), J. Opt. Soc. Am. 55, 33 (1965).

[3265] Sugar, J., and Reader, J., Ionization energies of the singly ionized rare earths, J. Opt. Soc. Am. 55, 1286 (1965).

[3266] Bryant, B. W., Spectra of doubly and triply ionized ytterbium, Yb III and Yb IV, J. Opt. Soc. Am. 55, 771 (1965).

[3267] Sugar, J., Analysis of the third spectrum of praseodymium, J. Opt. Soc. Am. 53, 831 (1963).

[3268] Sugar, J., New energy levels of triply ionized praseodymium, J. Opt. Soc. Am. 61, 727 (1971).

[3269] Thomas, J. L., and Hayes, R. G., Mass spectra and bond energies of lanthanide tricyclopentadienyl complexes, J. Organometal. Chem. 23, 487 (1970).

[3270] Penkin, N. P., and Shabanova, L. N., Absorption spectra of aluminum, gallium, indium, and thallium atoms, Opt. i Spektroskopiya 18, 749 (1965) [Engl. transl.: Opt. Spectry. (USSR) 18, 425 (1965)].

[3271] Epstein, G. L., and Reader, J., Zeeman effect and revised analysis of Rb II, J. Opt. Soc. Am. 61, 673 (1971).

[3272] Smith, D. H., Mass spectrometric investigation of surface ionization. VII. The first ionization potential of curium, J. Chem. Phys. 54, 1424 (1971).

[3273] Morss, L. R., Thermochemistry of some chlorocomplex compounds of the rare earths. Third ionization potentials and hydration enthalpies of the trivalent ions, J. Phys. Chem. 75, 392 (1971).

[3274] Berkowitz, J., Chupka, W. A., Guyon, P. M., Holloway, J. H., and Spohr, R., Photoionization mass spectrometric study of F_2, HF, and DF, J. Chem. Phys. 54, 5165 (1971).

[3275] Cornford, A. B., Frost, D. C., McDowell, C. A., Ragle, J. L., and Stenhouse, I. A., Photoelectron spectra of the halogens, J. Chem. Phys. 54, 2651 (1971).

[3276] Berkowitz, J., Chupka, W. A., Guyon, P. M., Holloway, J. H., and Spohr, R., Photoionization mass spectrometric study of XeF_2, XeF_4, and XeF_6, J. Phys. Chem. 75, 1461 (1971).

[3277] Evans, S., and Orchard, A. F., The helium–(I) photoelectron spectra of some halogens and diatomic interhalogens, Inorg. Chim. Acta 5, 81 (1971).

[3278] Haugh, M. J., and Bayes, K. D., Predissociation and dissociation energy of HBr^+, J. Phys. Chem. 75, 1472 (1971).

[3279] DeCorpo, J. J., and Franklin, J. L., Electron affinities of the halogen molecules by dissociative electron attachment, J. Chem. Phys. 54, 1885 (1971).

[3280] Müller, J., and Fenderl, K., Massenspektroskopische Untersuchungen an π Komplexen des Chroms, Chem. Ber. 103, 3128 (1970).

[3281] Herberich, G. E., and Müller, J., Massenspektren und Stabilitäten von Diaromatenchrom–Komplexen, J. Organometal. Chem. 16, 111 (1969).

[3282] Sugar, J., The third spectrum of praseodymium (Pr III) in the vacuum ultraviolet, J. Res. NBS 73A, 333 (1969).

[3283] Sugar, J., Spectrum of doubly ionized thulium (Tm III), J. Opt. Soc. Am. 60, 454 (1970).

[3284] Johnstone, R. A. W., and Ward, S. D., Ultraviolet spectra of, and SCF MO calculations on 6a–thiathiophthens, Theoret. Chim. Acta 14, 420 (1969).

[3285] Bogolyubov, G. M., Grishin, N. N., and Petrov, A. A., Organic derivatives of group V and group VI elements. X. Mass spectra of tetraethyldistibine, diethyl diselenide, diethyl ditelluride, and the corresponding monoderivatives. Interpretation of mass–spectral intensities, Zh. Obshch. Khim. 39, 2244 (1969) [Engl. transl.: J. Gen. Chem. USSR 39, 2190 (1969)].

[3286] Gal'perin, Ya. V., Bogolyubov, G. M., Grishin, N. N., and Petrov, A. A., Organic derivatives of elements of groups V and VI. VI. Mass spectra of compounds with S–S bonds, Zh. Obshch. Khim. 39, 1599 (1969) [Engl. transl.: J. Gen. Chem. USSR 39, 1567 (1969)].

[3287] Hoffman, M. K., and Bursey, M. M., Structural characteristics of non–decomposing $C_7H_7^+$ ions from some methyl ethers on electron impact, Chem. Commun., 824 (1971).

[3288] McLafferty, F. W., Wachs, T., Lifshitz, C., Innorta, G., and Irving, P., Substituent effects in unimolecular ion decompositions. XV. Mechanistic interpretations and the quasi–equilibrium theory, J. Am. Chem. Soc. 92, 6867 (1970).

[3289] Cocksey, B. J., Eland, J. H. D., and Danby, C. J., The effect of alkyl substitution on ionisation potential, J. Chem. Soc. (B), 790 (1971).

[3290] Pignataro, S., Mancini, V., Ridyard, J. N. A., and Lempka, H. J., Photoelectron energy spectra of molecules having classically non–conjugated π–systems, Chem. Commun., 142 (1971).

[3291] Haselbach, E., Heilbronner, E., and Schröder, G., The interaction of π–orbitals in barrelene, Helv. Chim. Acta 54, 153 (1971).

[3292] Heilbronner, E., Gleiter, R., Hopf, H., Hornung, V., and de Meijere, A., Photoelectron–spectroscopic evidence for the orbital sequence in fulvene and 3,4–dimethylene–cyclobutene, Helv. Chim. Acta 54, 783 (1971).

[3293] Stockbauer, R., and Inghram, M. G., Experimental relative Franck–Condon factors for the ionization of methane, ethane, and propane, J. Chem. Phys. 54, 2242 (1971).

[3294] Blais, J.–C., Cottin, M., and Gitton, B., Ionisation positive et négative dans le diméthylsulfoxyde en phase gazeuse, J. Chim. Phys. 67, 1475 (1970).

[3295] Martínez de Bertorello, M., Bertorello, H. E., and García–Martínez, N., Descomposicion por impacto electronico del dianhidrido del acido 3,4,3',4'–tetracarboxibifenilo y del 4,4'–dicianobifenilo, Anales Asoc. Quím. Arg. 58, 291 (1970).

[3296] Mamer, O. A., Lossing, F. P., Hedaya, E., and Kent, M. E., Pyrolysis of bicyclo[2.2.1]hepta–2,5–diene–2,3–dicarboxylic anhydride, Can. J. Chem. 48, 3606 (1970).

[3297] Harbour, P. J., The detection and ionization potential of mercury clusters, J. Phys. B 4, 528 (1971).

[3298] Devyatykh, G. G., Krasnova, S. G., Borisov, G. K., Larin, N. V., and Gaivoronskii, P. E., Preparation and investigation by mass spectrometry of cyclopentadienyl π–complexes of lanthanum, praseodymium, and neodymium, Dokl. Akad. Nauk SSSR 193, 1069 (1970) [Engl. transl.: Dokl. Chem. 193, 580 (1970)].

[3299] Sharp, J. H., Charge–transfer complexes of N–isopropylcarbazole, J. Phys. Chem. 70, 584 (1966).

[3300] Kuboyama, A., Molecular complexes and their spectra. XVII. The iodine and the chloranil complexes with thianthrene analogs, J. Am. Chem. Soc. 86, 164 (1964).

[3301] Worley, R. E., Absorption spectrum of N_2 in the extreme ultraviolet, Phys. Rev. 64, 207 (1943).

[3302] Janin, J., and d'Incan, J., Observation d'un nouveau système de bandes de la molécule d'azote ionisée, Compt. Rend. 246, 3436 (1958).

[3303] Gilmore, F. R., Potential energy curves for N_2, NO, O_2 and corresponding ions, J. Quant. Spectry. Radiative Transfer 5, 369 (1965).

[3304] Moddeman, W. E., Carlson, T. A., Krause, M. O., Pullen, B. P., Bull, W. E., and Schweitzer, G. K., Determination of the K–LL Auger spectra of N_2, O_2, CO, NO, H_2O, and CO_2, J. Chem. Phys. 55, 2317 (1971).

[3305] Albritton, D. L., Schmeltekopf, A. L., and Zare, R. N., Evidence in support of the vibrational renumbering of the $O_2^{+2}\Pi_g$ ground state, J. Chem. Phys. 51, 1667 (1969).

[3306] Bhale, G. L., and Rao, P. R., Isotope shifts in the second negative bands of O_2^+, Proc. Indian Acad. Sci. A67, 350 (1968).

[3307] Codling, K., and Madden, R. P., New Rydberg series in molecular oxygen near 500 Å, J. Chem. Phys. **42**, 3935 (1965).

[3308] Cook, G. R., Photoionization of ozone with mass analysis, Proc. Intern. Conf. Mass Spectry., Kyoto, 1969 (1970) p. 761.

[3309] Herzberg, G., Molecular Spectra and Molecular Structure. I. Spectra of Diatomic Molecules, 2nd Ed. (D. Van Nostrand Co., New York, 1950).

[3310] Herzberg, G., Molecular Spectra and Molecular Structure. III. Electronic Spectra and Electronic Structure of Polyatomic Molecules (D. Van Nostrand Co., Princeton, N. J., 1966).

[3311] Johns, J. W. C., On the absorption spectrum of H_2O and D_2O in the vacuum ultraviolet, Can. J. Phys. **41**, 209 (1963).

[3312] Natalis, P., Delwiche, J., and Collin, J. E., The first ionization energy of NO_2. What answer can be brought at present by photoelectron spectroscopy?, Chem. Phys. Letters **9**, 139 (1971).

[3313] Price, W. C., and Simpson, D. M., The absorption spectra of sulphur dioxide and carbon disulphide in the vacuum ultra–violet, Proc. Roy. Soc. (London) **A165**, 272 (1938).

[3317] Price, W. C., Teegan, J. P., and Walsh, A. D., The far ultra–violet absorption spectra of the hydrides and deuterides of sulphur, selenium and tellurium and of the methyl derivatives of hydrogen sulphide, Proc. Roy. Soc. (London) **A201**, 600 (1950).

[3318] Golomb, D., Watanabe, K., and Marmo, F. F., Absorption coefficients of sulfur dioxide in the vacuum ultraviolet, J. Chem. Phys. **36**, 958 (1962).

[3319] Maier, W. B., II, and Holland, R. F., Emission from long–lived states in ion beams. New band systems of NO^+, J. Chem. Phys. **54**, 2693 (1971).

[3320] Cornford, A. B., Frost, D. C., Herring, F. G., and McDowell, C. A., Electronic levels of methyl amines by photoelectron spectroscopy and an i.n.d.o. calculation, Can. J. Chem. **49**, 1135 (1971).

[3321] Turner, D. W., Ionization potentials, Advan. Phys. Org. Chem. **4**, 31 (1966).

[3322] Hammond, V. J., Price, W. C., Teegan, J. P., and Walsh, A. D., The absorption spectra of some substituted benzenes and naphthalenes in the vacuum ultra–violet, Discussions Faraday Soc. **9**, 53 (1950).

[3323] Momigny, J., Goffart, C., and Natalis, P., On the position of electronic states of the pyridine molecular ion, Bull. Soc. Chim. Belges **77**, 533 (1968).

[3324] Kumakura, M., Sugiura, T., and Okamura, S., Characteristics of RPD ion source of time–of–flight mass spectrometer, and measurement of the ionization potential and the appearance potentials of trioxymethylene, Mass Spectry. (Tokyo) **16**, 16 (1968).

[3325] Potapov, V. K., and Sorokin, V. V., Investigation of ionic-molecular reactions proceeding during photoionization of aromatic compounds and alcohols, Dokl. Akad. Nauk SSSR **195**, 616 (1970) [Engl. transl.: Dokl. Chem. **195**, 848 (1970)].

[3326] Bischof, P., Hashmall, J. A., Heilbronner, E., and Hornung, V., Interaction of nonconjugated double bonds in 1,4,5,8–tetrahydronaphthalene, Tetrahedron Letters, 1033 (1970).

[3327] Bischof, P., Hashmall, J. A., Heilbronner, E., and Hornung, V., Photoelektronspektroskopische Bestimmung der Wechselwirkung zwischen nicht–konjugierten Doppelbindungen, Helv. Chim. Acta **52**, 1745 (1969).

[3328] Bischof, P., Gleiter, R., and Heilbronner, E., Photoelectron–spectroscopic evidence concerning "homo–aromaticity", Helv. Chim. Acta **53**, 1425 (1970).

[3329] Bischof, P., Gleiter, R., Heilbronner, E., Hornung, V., and Schröder, G., The conjugative interaction between π–orbitals and $Walsh$–e–orbitals in bullvalene and related systems, Helv. Chim. Acta **53**, 1645 (1970).

[3330] Bischof, P., and Heilbronner, E., Photoelektron–Spektren von Cycloalkenen und Cycloalkadienen, Helv. Chim. Acta **53**, 1677 (1970).

[3331] Baker, A. D., and Turner, D. W., Orderings of π and σ ionization potentials in carbocyclic and heterocyclic aromatic compounds, Phil. Trans. Roy. Soc. (London) **A268**, 131 (1970).

[3332] Parkin, J. E., and Innes, K. K., The vacuum ultraviolet spectra of pyrazine, pyrimidine, and pyridazine vapors. Part I. Spectra between 1550 Å and 2000 Å, J. Mol. Spectry. **15**, 407 (1965).

[3333] Cooks, R. G., Wolfe, N. L., Curtis, J. R., Petty, H. E., and McDonald, R. N., Neighboring–group participation reactions in the mass spectral fragmentations of some azulenes. Comparisons with solvolytic processes, J. Org. Chem. **35**, 4048 (1970).

[3334] Buchs, A., Rossetti, G. P., and Susz, B. P., Étude, en fonction de la constante σ_p de Hammett, du spectre de masse d'acétophénones p–substituées, Helv. Chim. Acta **47**, 1563 (1964).

[3335] Harrison, A. G., Haynes, P., McLean, S., and Meyer, F., The mass spectra of methyl-substituted cyclopentadienes, J. Am. Chem. Soc. **87**, 5099 (1965).

[3336] Bergmann, E. D., Ionization potentials of biological molecules, Ann. N. Y. Acad. Sci. **158**, 140 (1969).

[3337] Natalis, P., Étude du comportement d'isomères géométriques sous l'impact électronique. I. Les cis et trans décalines, Bull. Soc. Chim. Belges **72**, 264 (1963).

[3338] Loudon, A. G., Maccoll, A., and Wong, S. K., Comparison between unimolecular gas phase pyrolysis and electron impact fragmentation. Part I. The mass spectra of tetralin and some related heterocycles, J. Chem. Soc. (B), 1727 (1970).

[3339] Gorfinkel', M. I., Isaev, I. S., Shleider, I. A., and Koptyug, V. A., Structure of $C_6H_7^+$ ions in the mass spectra of cyclohexadienes and their monomethyl derivatives. III. $C_6H_7^+$ ions from 1–methyl–^{13}C-1,3– and -1,4–cyclohexadienes and 5-methyl-1,3-cyclohexadiene, Zh. Obshch. Khim. **39**, 1363 (1969) [Engl. transl.: J. Gen. Chem. USSR **39**, 1333 (1969)].

[3340] Grützmacher, H.-F., and Lohmann, J., Massenspektrometrie instabiler organischer Moleküle. II. Nachweis von 9.10-Dehydro-phenanthren durch Pyrolyse-Massenspektrometrie, Ann. Chem. **726**, 47 (1969).

[3341] Winstein, S., and Lossing, F. P., On the question of homoconjugation in 1,4,7–cyclononatriene, J. Am. Chem. Soc. **86**, 4485 (1964).

[3342] Meier, H., Heiss, J., Suhr, H., and Müller, E., Energetische Untersuchungen zum Mills–Nixon–Effekt. Ionisierungsenergien von Benzolmolekülen mit ankondensierten gesättigten Ringen, Tetrahedron **24**, 2307 (1968).

[3343] Demeo, D. A., and Yencha, A. J., Photoelectron spectra of bicyclic and exocyclic olefins, J. Chem. Phys. **53**, 4536 (1970).

[3344] Majer, J. R., and Patrick, C. R., Electron impact studies on aromatic halogen compounds, Advan. Mass Spectrom. **2**, 555 (1963).

[3345] Haney, M. A., and Franklin, J. L., Correlation of excess energies of electron–impact dissociations with the translational energies of the products, J. Chem. Phys. **48**, 4093 (1968).

[3346] Stokes, S., and Duncan, A. B. F., Electronic transitions in methyl fluoride and in fluoroform, J. Am. Chem. Soc. **80**, 6177 (1958).

[3347] Haney, M. A., and Franklin, J. L., Excess energies in mass spectra of some oxygen–containing organic compounds, Trans. Faraday Soc. **65**, 1794 (1969).

[3348] Jones, E. G., and Harrison, A. G., A study of Penning ionization reactions using a single–source mass spectrometer, Intern. J. Mass Spectrom. Ion Phys. **5**, 137 (1970).

[3349] Dewar, M. J. S., Harget, A. J., Trinajstić, N., and Worley, S. D., Ground states of conjugated molecules–XXI. Benzofurans and benzopyrroles, Tetrahedron **26**, 4505 (1970).

[3350] Lossing, F. P., Free radicals by mass spectrometry. XLIII. Ionization potentials and ionic heats of formation for vinyl, allyl, and benzyl radicals, Can. J. Chem. **49**, 357 (1971).

[3351] Price, W. C., and Walsh, A. D., The absorption spectra of the cyclic dienes in the vacuum ultra–violet, Proc. Roy. Soc. (London) **A179**, 201 (1941).

[3352] Price, W. C., and Walsh, A. D., The absorption spectra of conjugated dienes in the vacuum ultra–violet (1), Proc. Roy. Soc. (London) **A174**, 220 (1940).

[3353] Price, W. C., and Tutte, W. T., The absorption spectra of ethylene, deutero–ethylene and some alkyl–substituted ethylenes in the vacuum ultra–violet, Proc. Roy. Soc. (London) **A174**, 207 (1940).

[3354] Potapov, V. K., and Yuzhakova, O. A., Photoionization and electronic structure of pyrrole and its methyl derivatives, Dokl. Akad. Nauk SSSR **192**, 131 (1970) [Engl. transl.: Dokl. Phys. Chem. **192**, 365 (1970)].

[3355] Klöpffer, W., Über die elektronischen Energieniveaus des N–Isopropylcarbazols und seiner Radikalionen. 1. Teil: Ionisierungsenergie, Elektronenaffinität und Radikalionenspektren, Z. Naturforsch. **24a**, 1923 (1969).

[3356] Mitani, E., Okamoto, J., and Omura, I., Ionization probability curves of n–butane and iso–butane by RPD method, Mass Spectry. (Tokyo) **15**, 1 (1967).

[3357] Majer, J. R., and Patrick, C. R., Appearance potentials and apparent heats of formation of the methylene and difluoromethylene ions, Nature **201**, 1022 (1964).

[3358] Frost, D. C., Herring, F. G., Mitchell, K. A. R., and Stenhouse, I. A., Photoelectron spectra and electronic structures of trifluoramine oxide and trifluorophosphine oxide, J. Am. Chem. Soc. **93**, 1596 (1971).

[3359] Holliday, A. K., Reade, W., Johnstone, R. A. W., and Neville, A. F., Photo–electron spectrum of trivinylboron, Chem. Commun., 51 (1971).

[3360] Scott, J. D., and Russell, B. R., The vacuum ultraviolet absorption spectrum of 1,1-dichloro-2,2-difluoroethylene, Chem. Phys. Letters **9**, 375 (1971).

[3361] Whitlock, R. F., and Duncan, A. B. F., Electronic spectrum of cyclobutanone, J. Chem. Phys. **55**, 218 (1971).

[3362] Bassett, P. J., and Lloyd, D. R., Photoelectron spectra of halides. Part I. Tetrafluorides and tetrachlorides of group IVB, J. Chem. Soc. (A), 641 (1971).

[3363] Cornford, A. B., Frost, D. C., Herring, F. G., and McDowell, C. A., Ionization potentials of the difluoroamino radical by photoelectron spectroscopy and INDO calculations, J. Chem. Phys. **54**, 1872 (1971).

[3364] Brundle, C. R., Jones, G. R., and Basch, H., He I and He II photoelectron spectra and the electronic structures of XeF_2, XeF_4, and XeF_6, J. Chem. Phys. **55**, 1098 (1971).

[3365] Bogan, D. J., and Hand, C. W., Mass spectrum of isocyanic acid, J. Phys. Chem. **75**, 1532 (1971).

[3366] Balducci, G., De Maria, G., Guido, M., and Piacente, V., Dissociation energy of FeO, J. Chem. Phys. **55**, 2596 (1971).

[3367] Verhoeven, J. W., Dirkx, I. P., and de Boer, T. J., Studies of inter– and intra–molecular donor–acceptor interactions–II. Intermolecular charge transfer involving substituted pyridinium ions, Tetrahedron **25**, 3395 (1969).

[3368] Hart, P. J., and Friedli, H. R., Desorption of allyl radicals in the heterogeneously–catalysed oxidation of propene: mass spectrometric study, Chem. Commun., 621 (1970).

[3369] Kraak, A., and Wynberg, H., Charge–transfer interaction of dithienyls and cyclopentadithiophenes with 1,3,5–trinitrobenzene (TNB), Tetrahedron **24**, 3881 (1968).

[3370] Muenow, D. W., Uy, O. M., and Margrave, J. L., Mass spectrometric studies of the vaporization of phosphorus oxides, J. Inorg. Nucl. Chem. **32**, 3459 (1970).

[3371] Gilbert, R., and Sandorfy, C., The vacuum–ultraviolet spectrum of fluorobenzene, Chem. Phys. Letters **9**, 121 (1971).

[3372] Nakato, Y., Ozaki, M., Egawa, A., and Tsubomura, H., Organic amino compounds with very low ionization potentials, Chem. Phys. Letters **9**, 615 (1971).

[3373] Voigt, E. M., and Reid, C., Ionization potentials of substituted benzenes and their charge–transfer spectra with tetracyanoethylene, J. Am. Chem. Soc. **86**, 3930 (1964).

[3374]. Baker, A. D., Betteridge, D., Kemp, N. R., and Kirby, R. E., Application of photoelectron spectrometry to pesticide analysis. II. Photoelectron spectra of hydroxy–, and halo–alkanes and halohydrins, Anal. Chem. **43**, 375 (1971).

[3375] Bassett, P. J., and Lloyd, D. R., Photoelectron spectra of halides. Part II. High–resolution spectra of the boron trihalides, J. Chem. Soc. (A), 1551 (1971).

[3376] Wilkinson, P. G., Absorption spectra of benzene and benzene–d_6 in the vacuum ultraviolet, Can. J. Phys. **34**, 596 (1956).

[3377] Beutler, H., Über Absorptionsspektren aus der Anregung innerer Elektronen. II. Das Quecksilberspektrum zwischen 1190 und 600 Å aus der Anregung der $(5\ d)^{10}$–Schale (Hg I^b), Z. Physik **86**, 710 (1933).

[3378] Finkelnburg, W., and Humbach, W., Ionisierungsenergien von Atomen und Atomionen, Naturwiss. **42**, 35 (1955).

[3379] Lossing, F. P., Maeda, K., and Semeluk, G. P., Ionization of alkyl free radicals using a double–hemispherical electron velocity selector, Proc. Intern. Conf. Mass Spectry., Kyoto, 1969 (1970) p. 791.

[3380] Lossing, F. P., Free radicals by mass spectrometry. XLV. Ionization potentials and heats of formation of C_3H_3, C_3H_5, and C_4H_7 radicals and ions, Can. J. Chem. **50**, 3973 (1972).

[3381] Derrick, P. J., Åsbrink, L., Edqvist, O., and Lindholm, E., Photoelectron–spectroscopical study of the vibrations of furan, thiophene, pyrrole and *cyclo*pentadiene, Spectrochim. Acta **27A**, 2525 (1971).

[3382] Brundle, C. R., and Brown, D. B., The vibrational structure in the photoelectron spectra of ethylene and ethylene–d_4, and its relationship to the vibrational spectrum of Zeise's salt $K[PtCl_3(C_2H_4)]\cdot H_2O$, Spectrochim. Acta **27A**, 2491 (1971).

[3383] Walsh, A. D., The absorption spectra of acrolein, crotonaldehyde and mesityl oxide in the vacuum ultra–violet, Trans. Faraday Soc. **41**, 498 (1945).

[3384] Sivgals, P., and Hudson, D. E., Unpublished results reported in: Dresser, M. J., and Hudson, D. E., Surface ionization of some rare earths on tungsten, Phys. Rev. **137**, A673 (1965).

[3401] Hillebrand, M., Maior, O., Sahini, V. E., and Volanschi, E., Spectral study of some phenoxathiin derivatives and their positive ions, J. Chem. Soc. (B), 755 (1969).

[3402] Ferreira, M. A. A., and Fronteira e Silva, M. E., Ionização e dissociação do di–sulfureto de carbono por impacto electrónico, Rev. Port. Quím. **12**, 70 1970).

[3403] Huttner, G., and Fischer, E. O., Über Aromatenkomplexe von Metallen. XCVIII. Spectroskopische Untersuchungen an Charge–Transfer–Komplexen von Aromaten–Chrom–Tricarbonylen mit 1,3,5–Trinitrobenzol, J. Organometal. Chem. **8**, 299 (1967).

[3404] Cornford, A. B., Frost, D. C., Herring, F. G., and McDowell, C. A., Photoelectron spectra of F_2O and Cl_2O, J. Chem. Phys. **55**, 2820 (1971).

[3405] Reichert, C., and Westmore, J. B., Mass spectral studies of metal chelates. VI. Mass spectra and appearance potentials of β–diketonates of copper(II), Can. J. Chem. **48**, 3213 (1970).

[3406] Gamble, A. A., Gilbert, J. R., and Tillett, J. G., Substituent effects in the fragmentation of acetanilides, J. Chem. Soc. (B), 1231 (1970).

[3407] Chadwick, D., Frost, D. C., and Weiler, L., Photoelectron spectra of cyclopentanone and cyclopentenones, J. Am. Chem. Soc. **93**, 4320 (1971).

[3408] Chadwick, D., Frost, D. C., and Weiler, L., Photoelectron spectra of norbornanones and norbornenones, J. Am. Chem. Soc. **93**, 4962 (1971).

[3409] Potts, A. W., and Price, W. C., Photoelectron spectra of the halogens and mixed halides ICl and IBr, Trans. Faraday Soc. **67**, 1242 (1971).

[3410] Lake, R. F., The photoelectron spectra of complexes of boron trifluoride and amines, Spectrochim. Acta **27A**, 1220 (1971).

[3411] Derrick, P. J., Åsbrink, L., Edqvist, O., Jonsson, B.–Ö., and Lindholm, E., Rydberg series in small molecules. XIII. Photoelectron spectroscopy and electronic structure of cyclopentadiene, Intern. J. Mass Spectrom. Ion Phys. **6**, 203 (1971).

[3412] Eland, J. H. D., Unpublished results reported in: Rowland, C. G., Kinetic energy distributions of $C_{12}H_8$ fragment ions in the mass spectra of anthracene, phenanthrene and diphenylacetylene, Intern. J. Mass Spectrom. Ion Phys. **7**, 79 (1971).

[3413] Dibeler, V. H., Walker, J. A., McCulloh, K. E., and Rosenstock, H. M., Effect of hot bands on the ionization threshold of some diatomic halogen molecules, Intern. J. Mass Spectrom. Ion Phys. **7**, 209 (1971).

[3414] Hildenbrand, D. L., First ionization potentials of the molecules BF, SiO and GeO, Intern. J. Mass Spectrom. Ion Phys. **7**, 255 (1971).

[3415] Chupka, W. A., and Berkowitz, J., Photoionization of methane: ionization potential and proton affinity of CH_4, J. Chem. Phys. **54**, 4256 (1971).

[3416] Emel'yanov, A. M., Peredvigina, V. A., and Gorokhov, L. N., Ionization potentials of Li_2 and Na_2 molecules and the dissociation energy of Li_2^+ and Na_2^+ ions, Teplofiz. Vysokikh Temperatur **9**, 190 (1971) [Engl. transl.: High Temp. (USSR) **9**, 164 (1971)].

[3417] Lloyd, D. R., and Lynaugh, N., Photoelectron spectrum of B_4Cl_4, Chem. Commun., 627 (1971).

[3418] Collin, J. E., Delwiche, J., and Natalis, P., Energy levels of NO^+ ion by He and Ar resonance lines photoelectron spectrometry, Intern. J. Mass Spectrom. Ion Phys. **7**, 19 (1971).

[3419] Kelly, R., and Padley, P. J., Use of a rotating single probe in studies of ionization of additives to premixed flames. Part 4.–The alkaline earths and uranium, Trans. Faraday Soc. **67**, 1384 (1971).

[3420] Lake, R. F., and Thompson, H. W., The photoelectron spectra of methyl isocyanide and trideuteromethyl isocyanide, Spectrochim. Acta **27A**, 783 (1971).

[3421] Derrick, P. J., Åsbrink, L., Edqvist, O., Jonsson, B.–Ö., and Lindholm, E., Rydberg series in small molecules. X. Photoelectron spectroscopy and electronic structure of furan, Intern. J. Mass Spectrom. Ion Phys. **6**, 161 (1971).

[3422] Derrick, P. J., Åsbrink, L., Edqvist, O., Jonsson, B.–Ö., and Lindholm, E., Rydberg series in small molecules. XI. Photoelectron spectroscopy and electronic structure of thiophene, Intern. J. Mass Spectrom. Ion Phys. **6**, 177 (1971).

[3423] Derrick, P. J., Åsbrink, L., Edqvist, O., Jonsson, B.–Ö., and Lindholm, E., Rydberg series in small molecules. XII. Photoelectron spectroscopy and electronic structure of pyrrole, Intern. J. Mass Spectrom. Ion Phys. **6**, 191 (1971).

[3424] Spohr, R., Bergmark, T., Magnusson, N., Werme, L. O., Nordling, C., and Siegbahn, K., Electron spectroscopic investigation of Auger processes in bromine substituted methanes and some hydrocarbons, Physica Scripta **2**, 31 (1970).

[3425] Mentall, J. E., Gentieu, E. P., Krauss, M., and Neumann, D., Photoionization and absorption spectrum of formaldehyde in the vacuum ultraviolet, J. Chem. Phys. **55**, 5471 (1971).

[3426] Edwards, J. G., Wiedemeier, H., and Gilles, P. W., High molecular weight boron sulfides. III. A mass spectrometric study of the vaporization and decomposition of metathioboric acid, J. Am. Chem. Soc. **88**, 2935 (1966).

[3427] Edwards, J. G., Leitnaker, J. M., Wiedemeier, H., and Gilles, P. W., High molecular weight boron sulfides. VII. Lower temperature studies and metastable decompositions, J. Phys. Chem. **75**, 2410 (1971).

[3428] Chen, H., and Gilles, P. W., High molecular weight boron sulfides. V. Vaporization behavior of the boron–sulfur system, J. Am. Chem. Soc. **92**, 2309 (1970).

[3429] Natalis, P., Steiner, B., and Inghram, M. G., Private communication reported in: Doepker, R. D., and Ausloos, P., Gas–phase radiolysis of cyclobutane, J. Chem. Phys. **44**, 1641 (1966).

[3430] Bhatia, K. S., and Jones, W. E., Autoionized series in the arc spectrum of arsenic, Can. J. Phys. **49**, 1773 (1971).

[3431] Fehlner, T. P., The preparation of triphosphine. An intermediate in the pyrolysis of diphosphine, J. Am. Chem. Soc. **88**, 2613 (1966).

[3432] Garcia, J. D., and Mack, J. E., Energy level and line tables for one–electron atomic spectra, J. Opt. Soc. Am. **55**, 654 (1965).

[3433] Mahncke, H. E., and Noyes, W. A., Jr., The ultraviolet absorption spectra of cis– and transdichloroethylenes, J. Chem. Phys. **3**, 536 (1935).

[3434] Price, W. C., and Evans, W. M., The absorption spectrum of formic acid in the vacuum ultra–violet, Proc. Roy. Soc. (London) **A162**, 110 (1937).

[3435] Bernecker, R. R., and Long, F. A., Heats of formation of some organic positive ions and their parent radicals and molecules, J. Phys. Chem. **65**, 1565 (1961).

4.5. Author Index

Chang, H. C.   3049

Chantry, P. J.   2684

Charlu, T. V.   2402

Chelobov, F. N.   1067, 2954

Chen, H.   3428

Chin, M. S.   2967

Ching, B. K.   1235

Christ, B. J.   2686

Christie, M. I.   2998

Chupka, W. A.   2, 11, 318, 1035, 2050, 2179, 2426, 2602, 2605,
2606, 2609, 2610, 2616, 2617, 2618, 2638, 2647, 2746,
2783, 2811, 3030, 3195, 3197, 3209, 3217, 3274, 3276,
3415

Clancy, D. J.   217

Clark, I. D.   2838

Clark, L. B.   1420

Clark, R. J.   2507

Clarke, E. M.   1094, 3106

Clarke, E. W. C.   327

Cloutier, G. G.   328, 3133

Coates, G. E.   2874, 2913

Coats, F. H.   13

Coatsworth, L. L.   3200

Cocksey, B. J.   3289

Codling, K.   2653, 2654, 3307

Coleman, R. J.   2693

Colin, R.   1023, 1243, 1244, 1245, 1255, 2063, 2139, 2469

Collin, J. E.   14, 17, 58, 62, 164, 166, 169, 462, 1004, 1013, 1072,
1211, 1241, 2031, 2414, 2421, 2422, 2423, 2428, 2429,
2470, 2694, 2759, 2792, 2800, 2809, 2810, 2821, 2822,
2839, 2875, 2902, 2969, 3170, 3312, 3418

Collins, J. H.   2557, 2558, 2695, 2738

Comes, F. J.   2033, 2034, 2655, 2760

Condé, G.   2422

Condé–Caprace, G.   2694, 2969

Connor, D. S.   2560

Connor, J. A.   2055, 2413, 2641

Contreras, R.   2659

Cook, G. L.   1039

Cook, G. R.   163, 427, 1235, 1382, 2664, 2767, 3031, 3032, 3033,
3146, 3308

Cooks, R. G.   2737, 2919, 3333

Cornford, A. B.   3275, 3320, 3363, 3404

Cotter, J. L.   1125, 1126, 1127, 1168, 2156, 2412

Cottin, M.   3144, 3294

Courchene, W. L.   2557

Cox, P. A.   3079

Coxon, G. E.   2782

Crable, G. F.   1066

Craggs, J. D.   439, 440, 1029

Cristy, S. S.   3172, 3173

Cubicciotti, D.   2049, 2696, 2861, 2920, 2955

Cullen, W. R.   1007, 3034, 3202

Curran, R. K.   20, 185, 196, 200, 2697

Curtis, J. R.   3333

Cuthbert, J.   2016, 2471, 2472

Czekalla, J.   2978

D

Daly, N. R.   2473, 2474

Danby, C. J.   2458, 2797, 2847, 2848, 3012, 3289

D'Angelo, P. F.   2712, 2752

Dauben, H. J., Jr.   68

Davidson, I. M. T.   2401, 2418, 2475, 2689, 2698

Davis, F. A.   2840

Dawton, R. H. V. M.   1149

de Boer, T. J.   3367

DeCorpo, J. J.   3149, 3279

Degrève, F.   1255

DeJongh, D. C.   2749

Delderfield, J. S.   2693

Delwiche, J.   2414, 2528, 2729, 2759, 3035, 3040, 3041, 3158, 3170,
3312, 3418

De Maria, G.   333, 1065, 1155, 2126, 2128, 2129, 2138, 2151, 2152,
2459, 2517, 2690, 2699, 2754, 2755, 2957, 3194, 3210,
3366

de Meijere, A.   3292

Demeo, D. A.   2877, 3343

Denning, R. G.   2417

de Ridder, J. J.   2700, 2966, 2980, 3214

Derouane, E.   3131

Derrick, P. J.   3381, 3411, 3421, 3422, 3423

DeSousa, J. B.   141, 145

Detry, D.   2482

Dever, D. F.   1112

Devyatykh, G. G.   2981, 3298

Dewar, M. J. S.   1418, 2701, 2840, 2841, 2842, 2843, 2844, 2845,
2942, 2951, 3235, 3349

Dey, S. D.   2477, 2478, 2982

Dibeler, V. H.   24, 53, 54, 73, 74, 75, 76, 99, 154, 340, 401, 413,
418, 424, 1019, 1118, 1128, 1132, 1136, 1143, 1399, 1400,
2027, 2047, 2607, 2619, 2621, 2622, 2623, 2624, 2626,
2627, 2628, 2629, 2630, 2631, 2637, 2744, 2983, 3036,
3037, 3038, 3134, 3155, 3216, 3253, 3413

Dick, K. A.   3162

Dijkstra, G.   2700, 2966, 2980

Dillard, J. G.   1284, 2479, 2506

Di Lonardo, G.   3220

d'Incan, J.   3302

Dine–Hart, R. A.   2412

Dinneen, G. U.   1039

Dirkx, I. P.   3367

Distefano, G.   2025, 2453, 2480, 2481, 2544, 2716, 2799, 2956,
2983, 3039

Ditchburn, R. W.   331

Dixon, R. N.   3064

Dobrovitskii, M. I.   1067, 2954

Dobson, J.   2736, 3225

Doepker, R. D.   3429

Dolejšek, Z.   2185

Dolešek, Z.   3002

Donovan, R. J.   2761, 2878

Doolittle, P. H.   1415

D'Or, L.   1197, 2421, 2682, 2683, 3213

Dorman, F. H.   25, 211, 212, 1111, 1378, 1404, 1451, 2014, 2785

Douglas, A. E.   2872, 3219

Dresser, M. J.   2781, 3384

Dressler, K.   1148

Drost, H.   3013

Drowart, J.   333, 1023, 1065, 1116, 1155, 1243, 1245, 1255, 2063,
2123, 2126, 2128, 2129, 2138, 2139, 2176, 2457, 2469,
2482, 2526, 2573, 2747, 2943, 2957, 3206

Dubov, S. S.   1067, 2954

Dudek, G. O.   2155

Dugger, D. L.   2506

Duncan, A. B. F.   2662, 2893, 3065, 3100, 3346, 3361

Dunning, W. J.   305

du Parcq, R. P.   3111

Durup, J.   2021

E

Earnshaw, D. G.   1039

Edlén, B.   3066

Edqvist, O.   3068, 3069, 3080, 3081, 3123, 3215, 3381, 3411, 3421,
3422, 3423

Edwards, J. G.   2483, 3426, 3427

Edwards, R. K.   2461, 2944

Efimova, A. G.   3243

Egawa, A.   3372

Ege, G.   2964

Ehlert, T. C.   1104, 1105, 2148, 2161, 2162, 2795
Ehrhardt, H.   1408, 2812, 2846
Eick, H. A.   2710, 2903, 3188, 3190, 3198
Eland, J. H. D.   2458, 2796, 2797, 2847, 2848, 2984, 3012, 3067, 3070, 3289, 3412
Elder, F. A.   1068, 2614, 2632, 2644, 2645
Eliel, E. L.   3237
El–Sayed, M. F. A.   1115, 2651, 2877
Elzer, A.   2760
Emel'yanov, A. M.   2985, 3221, 3416
Engerholm, G. G.   2695
Enrione, R. E.   3228
Epstein, G. L.   3165, 3271
Eriksson, K. B. S.   2199, 2679, 3160
Ettinger, R.   314
Evans, S.   2849, 2879, 3079, 3090, 3239, 3277
Evans, W. M.   3434
Exsteen, G.   1244, 1245, 2123, 2457, 2573

F

Fahlman, A.   3072
Faktor, M. M.   3259
Fallon, P. J.   2702
Farmer, J. B.   59, 108, 441
Farrell, P. G.   1281, 2485
Farren, J.   2016, 2471, 2472
Fehlner, T. P.   1024, 1119, 1407, 2030, 2486, 2753, 3431
Fehsenfeld, F. C.   2745
Fenderl, K.   2597, 3280
Fenske, R. F.   3029
Ferguson, E. E.   2745
Ferreira, M. A. A.   3135, 3402
Feser, S.   2730
Ficalora, P. J.   2435, 2449, 2487, 2565, 2569, 3193
Field, F. H.   2101, 2102
Filyugina, A. D.   2576, 2728, 3015
Finan, P. A.   2036
Finch, A. C. M.   2037
Fineman, M. A.   1251, 2180, 2191
Finkbeiner, H. C.   2779
Finkelnburg, W.   3378
Finney, G.   2055
Fiquet–Fayard, F.   2450
Fischer, E. O.   2703, 3403
Fischer–Hjalmars, I.   3215
Fischler, J.   1036
Fisher, I. P.   29, 1079, 1455, 2164, 2451, 2452, 2488, 2986
Flesch, G. D.   30, 2053, 2788
Fochler, M.   2407
Foffani, A.   2023, 2025, 2026, 2174, 2194, 2453, 2481, 2544, 2799, 3252
Foner, S. N.   31, 33, 34, 35, 36, 37, 67, 2454, 2780, 2786, 3182
Ford, B.   3060
Forst, W.   2549
Foster, P. J.   2633
Fox, R. E.   39, 196, 198, 199, 2697, 3137
Franklin, J. L.   154, 340, 418, 424, 1100, 1237, 1238, 2047, 2102, 2105, 2136, 2431, 2455, 2533, 2704, 2751, 2778, 2791, 2914, 3149, 3192, 3216, 3279, 3345, 3347
Franskin, M. J.   2428, 2429, 2470
Fraser, R. T. M.   2456
Fred, M.   341
Frey, R.   2960, 3050
Friedland, S. S.   1131, 3176
Friedli, H. R.   3368
Friedman, L.   46, 357
Fronteira e Silva, M. E.   3402

Frost, D. C.   49, 61, 286, 287, 288, 289, 292, 463, 1007, 1050, 2814, 2815, 2816, 2817, 2834, 2838, 2850, 2851, 2852, 2853, 2854, 2952, 3027, 3028, 3034, 3057, 3061, 3078, 3085, 3109, 3202, 3275, 3320, 3358, 3363, 3404, 3407, 3408
Fuchs, R.   2447, 2642
Fujisaki, H.   2446
Fulton, A.   2987
Furlei, I. I.   2505
Futrell, J. H.   2793

G

Gaivoronskii, P. E.   2981, 3298
Gallegos, E. J.   50, 51, 52, 218, 2438, 2489, 2634, 3211
Gal'perin, Ya. V.   3286
Gamble, A. A.   3406
Ganguli, P. S.   2705
Garcia, J. D.   3432
García–Martínez, N.   3295
Gebelt, R. E.   3190
Gelius, U.   3072
Gentieu, E. P.   3425
Genzel, H.   2635
Gesser, H. D.   2519
Giegling, D.   2554
Giese, C. F.   1068, 1120, 3197
Giessner, B. G.   2941
Gilbert, J. R.   3406
Gilbert, R.   3371
Gilles, P. W.   2147, 2483, 3426, 3427, 3428
Gillis, R. G.   2171, 2685, 2706
Gilmore, F. R.   3303
Gingerich, K. A.   2054, 2152, 2439, 2440, 2441, 2442, 2490, 2491, 2699, 2707, 2779, 2916, 2988
Gitel', P. O.   2954
Gitton, B.   3294
Gleiter, R.   3002, 3292, 3328, 3329
Glemser, O.   2443
Glockling, F.   2424, 2874, 2913, 3248
Godbole, E. W.   1059
Goffart, C.   2682, 2789, 3323
Goldfinger, P.   2482
Goldstein, H. W.   3128
Golesworthy, R. C.   2998
Gollnick, K.   2979
Golomb, D.   3318
Gondal, S. K.   2581, 2583
Gorfinkel', M. I.   3339
Gorokhov, L. N.   2445, 2985, 3189, 3221, 3243, 3245, 3416
Göser, P.   2530, 2608
Gowenlock, B. G.   303, 304, 306, 307
Grajower, R.   2790
Grasso, F.   2023, 2026, 2174, 2194
Green, J. C.   2849, 2879, 3070, 3117
Green, J. W.   1104, 1105, 2165
Green, M. L. H.   2879, 3117
Greene, F. T.   2147
Grennan, T. P.   3110
Grimley, R. T.   2127, 2129, 2130, 2188, 2444, 2860
Grimm, F. A.   3063, 3116
Grishin, N. N.   2708, 3285, 3286
Grützmacher, H.–F.   2492, 2973, 3340
Guffy, J. C.   2066
Guido, M.   2151, 2459, 2517, 2690, 2754, 2755, 3210, 3366
Gundersen, G. E.   2461
Gusarov, A. V.   2445, 2985, 3189, 3243, 3245
Guyon, P. M.   3217, 3274, 3276

H

Habercom, M. S. 2547
Hagemann, R. 2022, 2172
Hagen, A. P. 2582
Haink, H. J. 3071
Hall, L. H. 1102
Halmann, M. 1036, 1101, 2493
Hamill, W. H. 119, 160, 224, 2018, 2154, 2158, 2776, 2999
Hammond, V. J. 344, 3322
Hamnett, A. 3079, 3239
Hamrin, K. 3072
Hand, C. W. 3365
Haney, M. A. 2778, 3345, 3347
Hanks, R. 3259
Hanson, H. P. 1094
Hanuš, V. 2185
Harada, Y. 1459
Harbour, P. J. 3297
Harget, A. J. 2841, 3349
Harland, P. 2777, 2864
Harllee, F. N. 1128
Harris, P. V. 168
Harrison, A. G. 68, 69, 70, 71, 72, 122, 123, 210, 339, 1122, 1413,
        2144, 2163, 2504, 2550, 2646, 2709, 2967, 2970, 2971,
        3191, 3335, 3348
Hart, P. J. 3368
Hart, R. R. 2880
Haschke, J. M. 2710, 2903, 3188
Haselbach, E. 2841, 2942, 3073, 3087, 3291
Hashmall, J. A. 2962, 3073, 3074, 3326, 3327
Hassan, G. E. M. A. 2177
Hastie, J. W. 2056, 2406, 2434, 2435, 2436, 2449, 2462, 2566,
        2595, 2711, 2750
Haszeldine, R. N. 2055, 2413
Hatada, M. 171, 1254, 1256, 2433
Hauge, R. 2566, 2750
Haugh, M. J. 3278
Hayes, R. G. 3269
Haynes, P. 1122, 3335
Haynes, R. M. 306
Heath, G. A. 1455
Hedaya, E. 2712, 2732, 2752, 2940, 3296
Heerma, W. 2966, 3214
Heilbronner, E. 2962, 3071, 3073, 3074, 3087, 3091, 3121, 3291,
        3292, 3326, 3327, 3329, 3330, 3328
Heiss, J. 3342
Heitz, F. 2021
Henderson, E. 2452
Henderson, I. H. S. 108, 441
Herberhold, M. 2531, 2703
Herberich, G. E. 3281
Herman, Z. 2468, 2873
Hernandez, G. J. 2169, 2170
Herring, F. G. 2850, 2851, 3027, 3078, 3320, 3358, 3363, 3404
Herron, J. T. 73, 74, 75, 76, 77, 78, 79
Herstad, O. 2869
Hertel, G. R. 2494, 2495, 2756
Hertel, I. 2420
Herzberg, G. 349, 1078, 1109, 2762, 3042, 3066, 3077, 3309, 3310
Hess, G. G. 1421, 3187
Hierl, P. M. 2431, 2704
Higasi, K. 381, 383, 384, 431, 2649, 2865, 2989, 3232
Hikita, T. 2961
Hildenbrand, D. L. 1106, 1297, 2040, 2141, 2142, 2432, 2532,
        2714, 2715, 2990, 2991, 3186, 3414
Hillebrand, M. 3401
Hillier, I. H. 3043, 3084, 3088
Hirota, K. 171, 1254, 1256, 2433
Hisatsune, I. C. 139
Hobrock, B. G. 82, 83, 84, 176, 186, 188, 191, 193, 202, 315, 3253

Hobrock, D. L. 43, 1110
Hoffman, M. K. 3287
Holland, R. F. 3319
Holliday, A. K. 3359
Hollins, R. E. 3226
Holloway, J. H. 3217, 3274, 3276
Holtzclaw, H. F., Jr. 2992
Homer, J. B. 2164
Honnen, L. R. 68
Hopf, H. 3292
Hopkinson, J. A. 2833
Hornung, V. 2962, 3071, 3073, 3091, 3121, 3292, 3326, 3327, 3329
Hotop, H. 3171
Howe, I. 2496, 2497, 3238
Huber, K. P. 1217
Hudson, D. E. 2781, 3384
Hudson, R. D. 1189
Hudson, R. L. 31, 33, 34, 36, 37, 67, 2454, 2780, 2786, 3182
Huffman, R. E. 2657, 2668, 2763, 3089
Hughes, B. M. 2742, 2793
Hull, S. E. 3064
Humbach, W. 3378
Humphries, C. M. 2658
Hunt, H. D. 3108
Hurley, D. W. 3157
Hurzeler, H. 86, 213
Husain, D. 2761, 2878
Hutchison, D. A. 2993
Huttner, G. 3403
Hyatt, D. 2470

I

Iczkowski, R. P. 355
Iglesias, L. 3163
Il'ina, G. G. 2994
Inghram, M. G. 86, 95, 213, 333, 1065, 1068, 1099, 1120, 1155,
        2050, 2126, 2127, 2128, 2130, 2138, 2188, 2409, 2444,
        2614, 2783, 2857, 2858, 2957, 2958, 3019, 3195, 3293,
        3429
Inn, E. C. Y. 227, 228, 297
Innes, K. K. 3332
Innorta, G. 2025, 2453, 2480, 2481, 2544, 2716, 3288
Inokuchi, H. 1459, 2636
Ionov, N. I. 317, 2881, 2882
Irsa, A. P. 356, 357
Irving, P. 3288
Isaev, I. S. 3339
Isberg, H. B. S. 2199, 2679
Ishihara, F. 2961
Ivko, A. 2709

J

Jackson, D. S. 1133
Jackson, P. T. 2761
Jacobson, A. 1131
James, L. H. 2192
Janin, J. 3302
Jason, A. J. 2409
Jennings, B. H. 2155
Jennings, W. B. 3235
Jensen, J. W. 3205
Joachim, P. J. 3117
Johansson, B. 3166
Johansson, G. 3072
Johansson, I. 2659, 3167
Johns, J. W. C. 1109, 2149, 3042, 3066, 3311
Johnson, D. A. 3261
Johnson, E. W. 3198
Johnstone, R. A. W. 2498, 3174, 3284, 3359

Lindeman, L. P. 2066
Linder, F. 2846
Lindholm, E. 28, 3068, 3069, 3080, 3081, 3123, 3215, 3381, 3411, 3421, 3422, 3423
Linevski, M. J. 1112
Linnett, J. W. 2998
Lintvedt, R. L. 2992
Liston, S. K. 2621, 2622, 2623, 2626, 2629
Litzén, U. 3167
Litzow, M. R. 2512, 3227
Lloyd, D. R. 2802, 2886, 3043, 3044, 3083, 3084, 3088, 3105, 3168, 3169, 3362, 3375, 3417
Locht, R. 2800
Lockhart, J. C. 2702, 2722
Lofthus, A. 2100
Lohmann, J. 2492, 3340
Long, F. A. 46, 419, 1288, 1290, 1419, 3224, 3435
Long, G. J. 2706
Longrée, J. 2421
Lorquet, J. C. 2029, 2529, 3148
Lossing, F. P. 17, 29, 58, 59, 62, 70, 71, 108, 120, 123, 124, 125, 126, 141, 145, 151, 165, 441, 462, 1011, 1079, 2164, 2499, 2543, 2545, 2546, 2578, 2798, 2862, 3002, 3075, 3104, 3296, 3341, 3350, 3379, 3380
Loudon, A. G. 2515, 2867, 3338
Loughran, E. D. 100
Lowrey, A., III 101
Lumpkin, H. E. 3192
Lustig, M. 1144
Lutz, B. L. 3219
Lynaugh, N. 3043, 3044, 3088, 3105, 3417
Lyons, L. E. 2987

M

Maccoll, A. 1390, 2146, 2515, 2867, 3201, 3338
MacDiarmid, A. G. 2581, 2582, 2583, 3255
Mack, J. E. 3432
MacNeil, K. A. G. 3236
Mader, C. 100
Madden, R. P. 3307
Maeda, K. 3379
Magnusson, N. 3424
Mahncke, H. E. 3433
Maier, W. B., II 3319
Maior, O. 3401
Majer, J. R. 298, 299, 301, 303, 304, 306, 307, 308, 2976, 3344, 3357
Makita, T. 3028, 3061
Malaspina, L. 2151, 2152, 3194
Malone, T. J. 42, 2143, 2516
Mamantov, G. 2172, 3173
Mamer, O. A. 3296
Mancini, V. 3290
Mann, D. E. 53, 1112
Mann, J. B. 16
Mann, K. H. 1047
Margrave, J. L. 102, 103, 214, 318, 355, 364, 1104, 1105, 1249, 1280, 1458, 2009, 2105, 2148, 2161, 2162, 2165, 2402, 2435, 2436, 2449, 2487, 2523, 2524, 2525, 2565, 2566, 2568, 2569, 2588, 2589, 2590, 2591, 2592, 2593, 2594, 2595, 2596, 2598, 2599, 2600, 2601, 2750, 2859, 2896, 3004, 3023, 3149, 3193, 3370
Marino, G. 3233, 3240
Marmet, P. 3106
Marmo, F. F. 133, 227, 228, 3093, 3318
Marquart, J. R. 178, 2688
Marriott, J. C. 439, 440, 3043
Marsden, D. G. H. 441
Martignoni, P. 1372
Martin, R. H. 1139, 3017

Martínez de Bertorello, M. 3295
Matsen, F. A. 3000
Matsunaga, F. M. 190, 2676, 2680, 3048
Matthews, C. S. 2724
May, D. P. 92, 1108, 2015, 2806, 2830, 2856, 3101
McAdoo, D. J. 2732, 2766
McAllister, T. 2862, 3002
McCollum, J. D. 2109, 3237
McConkey, J. W. 2681
McCulloh, K. E. 2630, 2744, 3036, 3037, 3413
McDiarmid, R. 2663
McDonald, J. D. 1458, 2009, 2524, 2525
McDonald, R. N. 3333
McDowell, C. A. 49, 61, 108, 286, 287, 288, 289, 292, 327, 463, 1050, 2814, 2815, 2816, 2817, 2850, 2851, 2852, 2853, 2854, 3027, 3028, 3057, 3061, 3078, 3085, 3275, 3320, 3363, 3404
McDowell, M. V. 2581, 2582, 2583, 2584, 3255
McGee, H. A., Jr. 42, 2143, 2516, 2559, 2577, 2705, 2723, 3184
McGlynn, S. P. 3103
McGowan, J. W. 1094, 1251
McIntyre, N. S. 2403, 2404, 2405, 2526, 2870, 2943
McKinley, J. D. 2125, 2132
McLafferty, F. W. 2534, 2567, 2766, 2945, 3288
McLean, S. 1122, 3335
McNeal, R. J. 2767
McNeil, D. W. 2712, 2732
Meier, H. 3342
Meisels, G. G. 2941
Mellon, F. A. 3174
Melton, C. E. 119, 224, 2154, 2158, 2768, 2769, 2770, 2771, 2772
Mentall, J. E. 3425
Menzinger, M. 3062
Merer, A. J. 1169
Meschi, D. J. 3209
Mesnard, G. 2178, 2527
Metzger, P. H. 1382, 2664, 3031, 3032, 3033
Meyer, F. 122, 1122, 2144, 2163, 3335
Meyerson, S. 2108, 2109, 3237
Mielczarek, S. R. 3215
Miescher, E. 1148, 2764
Millard, B. J. 2498
Miller, R. D. 2712, 2752
Miller, S. I. 2146, 3201
Minnhagen, L. 3003, 3118, 3136
Mitani, E. 3356
Mitchell, K. A. R. 2952, 3358
Mittsev, M. A. 2881, 2882
Moddeman, W. E. 3063, 3116, 3304
Mogenis, A. 2455
Mohler, F. L. 24, 1132, 3155
Molodyk, A. D. 3016
Momigny, J. 114, 268, 1051, 1058, 1172, 1185, 1190, 1197, 1406, 2103, 2528, 2529, 2682, 2789, 2822, 3131, 3213, 3323
Moore, C. E. 2113
Morgan, R. L. 1372
Moritz, A. G. 2706
Morlais, M. 2666
Morrell, F. A. 411
Morris, A. 3076
Morris, G. C. 2652, 2686, 3046
Morrison, J. D. 25, 86, 211, 212, 213, 282, 2014, 2406, 2434, 2639, 2785, 3130
Morrow, B. A. 2665
Morss, L. R. 3273
Moser, H. C. 192
Mosesman, M. 2745
Mottl, J. R. 159, 182
Muenow, D. W. 2487, 2568, 2569, 2750, 3004, 3370
Mukherjee, T. K. 2911
Müller, A. 2443

Müller, E. 3342
Müller, J. 2530, 2531, 2597, 2608, 2640, 2641, 2683, 2703, 3005, 3280, 3281
Munson, M. S. B. 1100, 2136
Murad, E. 95, 1099, 1297, 2141, 2532, 2620, 2714, 3186
Murakawa, K. 1286
Murphy, C. B., Jr. 3228
Mustafa, M. R. 2850
Muszkat, K. A. 3091
Myer, J. A. 2887, 3102

N

Nagels, M. 1211
Nagoshi, K. 171
Nakata, R. S. 3048
Nakato, Y. 3372
Nakayama, T. 117, 161, 182, 1022
Nall, B. H. 35
Namioka, T. 162, 2099
Natalis, P. 1145, 1146, 1182, 1183, 1184, 1237, 1238, 2028, 2105, 2533, 2704, 2789, 2792, 2809, 2810, 2821, 2822, 2839, 2875, 2902, 3035, 3170, 3213, 3312, 3323, 3337, 3418, 3429
Nencini, G. 2799
Neumann, D. 3090, 3425
Neville, A. F. 3359
Newton, A. S. 1040, 1264, 2476, 2823, 3138, 3139, 3175
Newton, J. 1281, 2485
Nichols, L. D. 1076, 2182
Nicholson, A. J. C. 25, 282, 1032, 1253, 2014, 2639, 3185
Nicole, P. P. 3098
Nief, G. 2022
Niehaus, A. 2535, 2719, 3171
Nikolaev, E. N. 3159
Nordling, C. 3072, 3424
Norman, J. H. 1020, 1124, 2131, 2536, 2537, 2734
Nostrand, E. D. 3100
Nounou, P. 2538, 2539, 2540
Noutary, C. J. 2643
Noyes, W. A., Jr. 3433
Nozoe, T. 431

O

Occolowitz, J. L. 2171, 2522, 2541, 2685, 2706, 2725, 2946
Odabasi, H. 3262
O'Donnell, T. A. 2639
Ogawa, M. 163, 1179, 1382, 2099, 2664, 3006, 3032, 3033, 3049, 3143
O'Hare, P. A. G. 2888
Okamoto, J. 3356
Okamura, S. 3324
Oldfield, D. J. 2425
Ölme, A. 2889, 3066, 3181
Olmsted, J., III 1264
Omura, I. 194, 195, 381, 383, 384, 431, 2649, 2726, 2865, 2915, 2965, 2989, 3232, 3356
Orchard, A. F. 2849, 2879, 3079, 3117, 3239, 3277
Osberghaus, O. 1129, 2635
Ottinger, C. 2420
Ovchinnikov, A. A. 3115
Ovchinnikov, K. V. 3159
Ozaki, M. 3372
Ozerova, G. A. 2728

P

Pachucki, C. F. 206, 209
Paddock, N. L. 2425, 2952
Padley, P. J. 3242, 3419

Paisley, H. M. 2411
Palenius, H. P. 3161, 3164
Palmer, T. F. 120, 1011, 1079, 2782
Panchenkov, I. G. 3189
Park, J. Y. 2941
Parker, R. G. 2992
Parkin, J. E. 3332
Parr, A. C. 2632, 2644, 2645, 2958, 3258
Pascual, C. 3177
Pasedach, F. 2964
Passmore, T. R. 1091, 2819, 3099
Patrick, C. R. 298, 299, 301, 308, 3344, 3357
Paul, N. C. 2456
Paulett, G. S. 314, 1144
Pearson, R. G. 2580
Peatman, W. B. 2773, 2866
Pedley, J. B. 2512, 2513, 2863, 3227
Peers, A. M. 2542
Peeters, R. 2943
Penkin, N. P. 3270
Peredvigina, V. A. 3416
Person, J. C. 3098
Pesce, B. 2151
Petrocelli, A. W. 2180, 2191
Petrov, A. A. 2708, 2949, 3007, 3008, 3009, 3285, 3286
Petty, H. E. 3333
Piacente, V. 2152, 2699, 3194, 3366
Pignataro, S. 2023, 2025, 2026, 2174, 2194, 2453, 2481, 2543, 2544, 2545, 2546, 2716, 2799, 3233, 3240, 3252, 3290
Pinsky, M. 3147
Pisani, J. F. 2685
Pitt, C. G. 2547, 2900
Platel, G. 2001
Platzner, I. 2493
Plotnikov, V. F. 2949
Polyakova, A. A. 3007, 3008, 3009
Porter, R. F. 174, 397, 2175, 2555, 2897, 3195, 3256, 3257
Potapov, V. K. 2548, 2576, 2727, 2728, 3015, 3047, 3325, 3354
Pottie, R. F. 123, 124, 125, 126, 1252
Potts, A. W. 3090, 3119, 3148, 3409
Potzinger, P. 2757, 2898, 2899, 2977
Powell, R. E. 2473, 2474
Powell, W. A. 411
Praet, M.–T. 2729, 3158
Prahallada Rao, B. S. 2016, 2471, 2472
Prášil, Z. 2549
Preece, E. R. 2016, 2472
Pressley, G. A., Jr. 2561, 2579, 2735, 2736, 2869, 3225
Preston, F. J. 2904, 3250
Price, W. C. 168, 344, 387, 1091, 1114, 2064, 2065, 2819, 2820, 3010, 3029, 3060, 3064, 3090, 3099, 3119, 3140, 3141, 3142, 3145, 3148, 3150, 3151, 3152, 3153, 3313, 3317, 3322, 3351, 3352, 3353, 3409, 3434
Prinzbach, H. 2185
Pritchard, H. 2550, 2646
Prusaczyk, J. E. 2508, 3249
Pullen, B. P. 3063, 3116, 3304
Pullman, B. 2514
Pun, M. T. 3202
Pupp, C. 3256, 3257
Puttkamer, E. v. 2855, 3132

Q

Quemerais, A. 2666

R

Rabalais, J. W. 3103
Radwan, T. N. 1441
Radziemski, L. J., Jr. 3114

Raff, L. M.   2187
Ragle, J. L.   3057, 3275
Rao, P. R.   3306
Rapp, D.   1121
Rapp, U.   2974
Rasulev, U. Kh.   2741, 3229
Rat'kovskii, I. A.   3014, 3244
Rauh, E. G.   2167, 3196, 3198
Reade, W.   3359
Reader, J.   2667, 3051, 3165, 3260, 3265, 3271
Redhead, P. A.   2551, 2730
Reed, R. I.   127, 128, 129, 130, 131, 132, 204, 2036
Reese, R. M.   24, 53, 54, 99, 154, 340, 401, 418, 424, 1019, 1118,
        1132, 1136, 1143, 2047, 2648, 3216, 3253
Refaey, K. M. A.   2617, 2647
Reichert, C.   2460, 2519, 2731, 2959, 3405
Reid, C.   3373
Resnik, F. E.   411
Reuben, B. G.   2520, 2693
Rice, S. A.   2181
Richardson, A. W.   2660
Ricks, J. M.   3111
Ridyard, J. N. A.   3290
Riepe, W.   3011
Riley, P. N. K.   2512, 2513, 2863, 3227
Rinke, K.   2552, 2554
Risberg, G.   3125, 3126
Robb, J. C.   298, 2976
Robbins, E. J.   2585, 2633
Robin, M. B.   2808, 2837, 2880, 2912, 3092, 3120, 3147
Robin, S.   2666
Robinson, P. J.   2055
Roebber, J. L.   2668, 2717
Roessler, D. M.   3099
Rogerson, P. F.   2547, 2900, 2992
Rona, P.   1418
Rose, T. L.   2960, 3050
Rosenstock, H. M.   1400, 2607, 2648, 3134, 3413
Ross, K. J.   3001, 3076, 3107, 3241
Rossetti, G. P.   3334
Roth, E.   2022
Rowland, C. G.   2797, 3012, 3412
Rowlinson, H. C.   2150
Rozenshtein, S. M.   2954
Rudolph, P. S.   2769
Ruska, W. E. W.   2791
Russell, B. R.   3360
Russell, M. E.   3030
Rutgaizer, Yu. S.   2994
Ryabchikov, L. N.   2195
Rylander, P. N.   2108, 2109, 3237

                    S

Saalfeld, F. E.   173, 2002, 2116, 2133, 2137, 2581, 2582, 2583,
        2584, 3255
Sadovnikova, N. A.   2985
Sahini, V. E.   3401
Saito, T.   2849
Sakai, H.   2676
Samson, J. A. R.   133, 163, 427, 2048, 2650, 2824, 2825, 2887,
        2890, 3095, 3102
Sanda, J. C.   2416
Sandhu, J. S.   2850, 2852, 2853
Sandorfy, C.   3371
Sandoval, A. A.   192
Saporoschenko, M.   2895
Saul, A. M.   2040
Saunders, V. R.   3043, 3084, 3088
Saville, B.   1390
Schade, G.   2979

Schaeffer, R.   2736, 3225
Schäfer, H.   2552, 2554
Schiff, H. I.   77, 79, 328, 1133, 3133
Schiff, L. J.   2945
Schildcrout, S. M.   2579, 2580
Schissel, P.   2712, 2732, 2752
Schissler, D. O.   135
Schlag, E. W.   2773, 2866, 2886
Schleyer, P. v. R.   3002
Schmeltekopf, A. L.   3305
Schoen, R. I.   1415
Schönheit, E.   2200, 2774
Schoonmaker, R. C.   174, 397
Schröder, G.   3291, 3329
Schulz, G. J.   2684, 3129
Schulz, W.   3013
Schweitzer, G. K.   3063, 3116, 3304
Sciamanna, A. F.   2476, 2823, 3138, 3139, 3175
Scott, J. D.   3360
Sedgwick, R. D.   2055, 2413, 2613, 2863
Seidl, H.   2407, 2408, 2410, 2463, 3254
Selin, L. E.   3068, 3069, 3081, 3123, 3215
Semeluk, G. P.   2798, 3379
Semenov, G. A.   2994, 3014, 3159, 3244
Sen Sharma, D. K.   2464
Sergeev, Yu. L.   3212
Serum, J. W.   2918
Seshadri, K. S.   1112
Sessa, P. A.   2577
Shabanova, L. N.   3270
Shannon, T. W.   72, 2970, 3222
Shanshal, M.   2701
Shapiro, J. S.   2578
Shapiro, M.   2521, 2905, 2907
Shapiro, R. H.   2918
Sharp, J. H.   3299
Sharp, K. G.   2896
Shchukarev, S. A.   3014, 3244
Shenkel, R. C.   193, 315
Shepherd, P. J.   2458
Shigorin, D. N.   2548, 2576, 2728, 3015
Shinn, W. A.   2944
Shipko, F. J.   209
Shleider, I. A.   3339
Sholette, W. P.   2175
Shoosmith, J.   349
Shpenik, O. B.   2564
Shrader, S. R.   2749
Shriver, D. F.   2511
Siegbahn, K.   3072, 3424
Simmons, R. F.   2055
Simpson, D. M.   3142, 3313
Simpson, J.   2720
Simpson, J. A.   3215
Simpson, W. T.   3108
Sivgals, P.   3384
Sjögren, H.   28, 2060, 2574, 2575, 3123
Skudlarski, K.   2573
Slifkin, M. A.   2562, 2947
Smith, D.   2572, 2867
Smith, D. H.   2756, 3272
Smith, D. J.   3076, 3107, 3241
Smith, J. S.   2766
Smith, W. L.   2669
Smithson, L. D.   3157
Snedden, W.   129, 131, 132, 2036
Snelling, D. R.   304
Sommer, A.   3128
Sommer, L. H.   1421, 3187
Sood, S. P.   2670, 3053
Sorokin, V. V.   3047, 3325

Sowerby, D. B. 2782
Spalding, T. R. 2720, 3227
Spohr, R. 2855, 3217, 3274, 3276, 3424
Staab, H. A. 2974
Stafford, F. E. 1116, 2052, 2176, 2511, 2561, 2579, 2580, 2735, 2736, 2869, 3225, 3226
Staley, H. G. 1020, 1124, 2131, 2536, 2537, 2734, 2787
Stanton, H. E. 213
Stauffer, J. L. 2140, 3199
Stearns, C. A. 2996, 2997, 3207, 3208
Stebbings, R. F. 1094
Steck, S. J. 2561, 2735, 2736, 3225
Steele, W. C. 1075, 1076, 2155, 2160, 2182, 2183
Steiger, R. P. 3149
Steigman, J. 1131
Steiner, B. 1068, 1120, 3429
Stenhouse, I. A. 2851, 3027, 3028, 3057, 3061, 3078, 3085, 3275, 3358
Stephenson, I. L. 2401, 2418, 2475
Sternberg, R. 2905
Stevenson, C. D. 2878
Stevenson, D. P. 135
Stockbauer, R. 2857, 2858, 3019, 3293
Stokes, S. 3346
Stone, F. G. A. 1075, 1076, 2182, 2183
Strakna, R. A. 1131
Strakna, R. E. 3176
Strauss, B. H. 2502
Street, K., Jr. 1264
Streets, D. G. 3090, 3119
Strel'chenko, S. S. 3016
Stuber, F. A. 1240
Subbanna, V. V. 1102
Sugar, J. 2667, 3051, 3260, 3263, 3264, 3265, 3267, 3268, 3282, 3283
Sugiura, T. 3324
Suhr, H. 3342
Sullivan, C. L. 3249
Sultanov, A. Sh. 2159
Sun, H. 138
Susz, B. P. 3334
Sutcliffe, L. H. 3124
Svec, H. J. 30, 88, 173, 2002, 2053, 2116, 2133, 2137, 2500, 2501, 2563, 2587, 2788, 2904, 3234, 3250, 3251
Swingler, D. L. 2711, 3185

T

Taft, R. W. 1139, 3017
Tait, J. M. S. 2970, 3191
Takezawa, S. 2891, 3018
Tanaka, K. 2726, 2915
Tanaka, Y. 148, 149, 409, 410, 1097, 1115, 1179, 2099, 2657, 2678, 3089, 3143
Tannenbaum, H. P. 2497
Tashiro, S. 2961
Tasman, H. A. 2179
Taubert, R. 151, 2447
Teegan, J. P. 269, 344, 3151, 3317, 3322
Tekaat, T. 1408, 2812
Terenin, A. N. 1147, 1159, 1166, 2173
Theard, L. P. 1106, 2040, 2142, 2715
Thomas, G. E. 3175
Thomas, J. L. 3269
Thomas, T. D. 3094, 3096
Thomassy, F. A. 2894
Thompson, D. T. 3250
Thompson, H. W. 2885, 3045, 3420
Thompson, J. C. 2525, 3193
Thomson, R. 2775, 3052
Thorburn, R. 1246

Thorn, R. J. 2167, 3196
Thornley, M. B. 130
Thrush, B. A. 2189
Thynne, J. C. J. 2646, 2777, 2864, 3236
Tickner, A. W. 1047, 2446
Tiernan, T. O. 2742, 2793
Tikhinskii, G. F. 2195
Tikhomirov, M. V. 1067, 2571, 2586, 2954
Tillett, J. G. 3406
Tomkins, F. S. 341, 2953
Tomlinson, V. 3001
Toresson, Y. G. 3127
Toubelis, B. J. 2157
Travis, N. 3183
Trinajstić, N. 3349
Trombetti, A. 3220
Trost, B. M. 2995
Tsang, C. W. 2971
Tsubomura, H. 3372
Tsuchiya, M. 2904
Tsuchiya, S. 2961
Tsuda, S. 160, 2018, 2154
Tuffly, B. L. 97
Turk, J. 2918
Turner, D. W. 92, 248, 1108, 1130, 1441, 2015, 2801, 2803, 2804, 2805, 2806, 2807, 2808, 2826, 2827, 2829, 2830, 2835, 2836, 2837, 2849, 2856, 2879, 2901, 2968, 3058, 3101, 3117, 3247, 3331
Tutte, W. T. 3353
Tweedale, A. 2512, 2513, 3227

U

Urbain, J. 1406
Uy, O. M. 2465, 2487, 2509, 2568, 2569, 2570, 2601, 2747, 2859, 3206, 3370
Uzan, R. 2178, 2527

V

Van Brunt, R. J. 2112, 2765
Van Raalte, D. 210, 339, 2709
Varsel, C. J. 411
Veith, M. 3254
Venkateswarlu, P. 2671, 2906, 3097
Verhaegen, G. 1116, 1243, 1244, 1255, 2123, 2176
Verhoeven, J. W. 3367
Vetter, R. 2186
Vigny, P. 2542
Vilesov, F. I. 1140, 1141, 1142, 1147, 1159, 1160, 1166, 1167, 2173, 2612, 3025, 3212
Villarejo, D. 2614, 2831, 2857, 2858, 3019
Visnapuu, A. 3205
Vlaskov, V. A. 3115
Vogt, J. 3177
Voigt, E. M. 3373
Volanschi, E. 3401
Vroom, D. A. 1050, 2814, 2815, 2816, 2817, 2852, 2853, 2854, 3034

W

Wachs, T. 2534, 2567, 3288
Wacks, M. E. 413, 1069, 2112
Wada, Y. 115, 153, 1033, 2045, 2704
Waddington, T. C. 1414
Wahlbeck, P. G. 2721
Wainfan, N. 230
Waldron, J. D. 414
Walker, J. A. 1399, 1400, 2027, 2607, 2624, 2627, 2628, 2629, 2630, 2631, 2637, 2744, 3036, 3037, 3038, 3413

Walker, W. C.   155, 156, 157, 230
Walsh, A. D.   261, 269, 344, 1070, 2658, 2672, 2673, 2674, 2675, 3010, 3020, 3124, 3151, 3152, 3153, 3317, 3320, 3351, 3352, 3383
Walsh, P. N.   3128
Walter, T. A.   2602, 2604, 2746
Wankenne, H.   1406
Ward, R. S.   2737, 3022
Ward, S. D.   3174, 3284
Ware, M. J.   3043, 3088
Warneck, P.   2724
Warsop, P. A.   1070, 2658, 2672, 2673, 2674, 2675, 2677, 2775, 3052
Watanabe, K.   101, 117, 133, 158, 159, 161, 162, 182, 227, 228, 416, 1022, 1103, 2059, 2670, 2676, 2680, 2880, 3048, 3053, 3093, 3318
Webb, K. S.   2515
Weiershausen, W.   3024
Weiler, L.   3407, 3408
Weiss, M. J.   2892, 3054
Weissler, G. L.   138, 155, 156, 157, 163, 230
Wellern, H. O.   2655
Wells, C. H. J.   2411
Wenåker, I.   3160
Wentworth, R. A. D.   2417
Werme, L. O.   3424
Westmore, J. B.   1047, 2446, 2460, 2519, 2731, 2959, 3405
Weston, M.   2424
Wexler, S.   464
White, D.   1112, 3128
White, G. L.   2541
White, M. J. D.   2464
White, R. M.   2788
Whiteside, J. A. B.   2675, 2677
Whitlock, R. F.   3361
Wiberg, K. B.   165, 2560
Wiberg, N.   3254
Wiedemeier, H.   3426, 3427
Wilkinson, K. L.   1149
Wilkinson, P. G.   422, 423, 3376
Williams, C. H.   2525
Williams, D. H.   2496, 2497, 2737, 2740, 2919, 3022, 3230, 3231, 3238
Williams, J. M.   2776
Williams, R. A.   2615
Willis, P.   2585
Wilson, E. G.   2181
Wilson, J. H.   2559

Wilson, J. M.   2036
Winchell, P.   1239
Winstein, S.   3341
Winters, R. E.   112, 1107, 1381, 2556, 2557, 2558, 2695, 2738, 2739, 2963
Wirtz–Cordier, A. M.   1185
Wolfe, N. L.   3333
Wolfsberg, M.   46
Wong, S. K.   3338
Wood, D. R.   3113
Wood, R. W.   1114
Workman, G. L.   2893
Worley, R. E.   3301
Worley, S. D.   2701, 2840, 2841, 2842, 2843, 2844, 2845, 2942, 2951, 3056, 3235, 3349
Wright, C. V.   3183
Wright, S. S.   205
Wünsche, C.   2964, 2974
Wynberg, H.   3369

Y

Yamada, Y.   2726, 2915, 2965
Yamdagni, R.   2555, 3256, 3257
Yarborough, V. A.   97
Yencha, A. J.   2651, 3343
Yeo, A. N. H.   2740, 2919, 3230, 3231
Yergey, A. L.   1424, 2553
Yoshino, K.   2678, 3055, 3112
Young, R. A.   2892
Yuzhakova, O. A.   3354

Z

Zaitsev, V. M.   1140
Zandberg, É. Ya.   2741, 3229
Zander, M.   3011
Zapesochnyĭ, I. P.   2564
Zare, R. N.   3305
Zelikoff, M.   2059
Zeller, E. E.   2897
Zimina, K. I.   3008, 3009
Zinkiewicz, J. M.   3021
Zmbov, K. F.   2402, 2436, 2523, 2565, 2566, 2588, 2589, 2590, 2591, 2592, 2593, 2594, 2595, 2596, 2598, 2599, 2600, 2601, 2859, 3023, 3203, 3204
Zorn, C.   3062
Zwolenik, J. J.   2189

5. Energetics of Gaseous Negative Ions

The electron affinity of an atom or molecule is the difference in energy between the unexcited negative ion and a reference state of the neutral species, usually the ground state. Where the neutral reference is not the ground state, the state in question is identified. The electron affinities of both atomic and molecular species evaluated before the end of 1973 are included in the Negative Ion Table (section 6). The relatively weak binding of negative ions, however, leads to a status for their study different from that for positive ions. Several of these differences are reflected in the table.

In the first place, the difficulty of experimental work has long encouraged the wide use of approximate methods, whose results differ greatly from those of recent, more sophisticated techniques. The coverage of the Negative Ion Table is therefore not exhaustive. Early values never intended as precise estimates and more recent values by relatively imprecise techniques have not been included, since they would add little useful information.

Moreover, because no reliable experimental values are yet available for many species, theoretical calculations and empirical extrapolations are more crucial to the selection of the best electron affinities than are the corresponding techniques in the evaluation of ionization potentials. The most reliable of these calculated affinities are thus included in the table, both where no experimental data exist and in some other instances, as well. In this way an estimate of the reliability of the calculated values can be made by the user.

The Negative Ion Table contains all values, consistent with the preceding criteria, from previous reviews. The most recent of these are by Compton [1] and Steiner [2]. Another review by Hotop and Lineberger [3] covers more recent literature, but deals only with atomic negative ions.

5.1. Experimental Techniques

Values obtained by nearly two dozen experimental techniques are included in the Negative Ion Table. The accuracy of the results from a given experiment will be a function of three sets of conditions. The first set is associated with experimental imprecision and with systematic deviations from an ideal thought experiment. Some of these factors are inherent in the given technique, while others are associated with the care exercised by the individual experimenter. A second set of conditions is associated with the energy level structure of the negative ion and/or neutral molecule. The primary considerations here have been described in section 3.2. Transitions close in energy to the one to be determined will complicate the analysis of negative ion spectra and, typically, reduce its accuracy. A third set of conditions affecting accuracy is associated with the presence of interfering species that have transitions similar in energy to the one to be measured.

The accuracy of a given experiment may be dominated by any of these three sources of uncertainty. Since only the first source, and only part of that source, is associated generally with a given technique, a general assessment of the relative experimental accuracy of each of the approaches cannot be precise. Nevertheless, the following techniques are described roughly in order of decreasing accuracy for transitions where molecular dynamics do not dominate. Laser photodetachment can have an accuracy less than one millielectron volt. Other techniques using photons can have an accuracy of a few millielectron volts. The remaining experimental techniques considered here attain at best an accuracy of several tenths of an electron volt.

a. Dye Laser Photodetachment (LP)

The precision of crossed negative ion photon beam electron detachment experiments

$$X^- + h\nu \rightarrow X + e^-,$$

has been greatly increased with the incorporation of a tunable dye laser. The formation of either the electrons or the neutralized ions is followed as a function of wavelength [4–6]. Dye lasers have also been applied to experiments utilizing ion-cyclotron-resonance-concentrated negative ions [7]. The primary problem in dye laser photodetachment experiments is the determination of the states between which transitions are being observed. (Compare section c.) The accuracy can exceed one millielectron volt.

b. Spectroscopy (S)

An initial, extremely difficult problem preventing the direct spectroscopic observation of negative ions was the achievement of an ion density sufficient to absorb a measurable fraction of a photon beam. Berry and co-workers pioneered in the solution of this problem with shock wave techniques [8]. Both absorption [9] and emission [10] in shock waves have now been observed. Popp has extended these techniques to other plasmas [11, 12]. Care must be exercised that unexpected species do not confuse the experimenter with misleading thresholds, since the plasmas contain many species. Accuracies of 2 millielectron volts have been achieved.

c. Incoherent Photon Detachment (P)

The classical ion photon crossed beam photodetachment experiment

$$X^- + h\nu \rightarrow X + e^-,$$

as developed first by Branscomb and colleagues [13] has been described in detail [2]. A mass analyzed ion beam traverses spectrally resolved radiation from an arc, now typically a xenon arc. The energy of the detached electrons is measured as a function of photon wavelength. The identification of the transitions is substantially less precise, greater than 2 millielectron volts, than for similar experiments employing a dye laser. The applicability of threshold behavior theory is thus more questionable.

d. Photoelectron Spectroscopy (PE)

Crossed beam photodetachment with a fixed frequency laser has also been performed. The energy of the detached electrons is measured with an electron spectrometer [14–16]. Energy resolution of the detached electrons can be achieved sufficient to resolve transitions between various vibrational levels of the ion and neutral molecule to a precision of several millielectron volts. This technique is thus particularly useful for polyatomic negative ions. Molecules with much smaller affinities can be studied than can those using a threshold technique. The energy determined for the electrons can be influenced by the surface chemistry of the detachment region, however.

e. Photoionization (PI)

Dissociative pair production from a molecule

$$XY + h\nu \rightarrow X^+ + Y^-,$$

occurs at a lower energy than simple dissociative ionization

$$XY + h\nu \rightarrow X^+ + Y + e^-,$$

by an amount equal to the electron affinity where the degree of internal excitation is the same. The difference in the thresholds for these two processes leading to unexcited species is the electron affinity. Although the initial application of the technique [17] was not sufficiently precise (> 0.1 eV) to distinguish errors in prevailing values, more recent applications approaching a precision of 1 millielectron volt have been extremely useful in choosing among conflicting values for electron affinities [18]. Unless the kinetic energy of the products at each threshold is measured, however, the result represents only an upper bound on the electron affinity.

f. Electrostatic Field Detachment (FD)

The ions of atoms with very small electron affinities (several millielectron volts) may be detached in a strong electric field [19]. From the threshold strength of this field and a detailed knowledge of the ionic structure, the electron affinity may be calculated. The reliability, estimated to approach 2 millielectron volts, will thus depend strongly on the theoretical analysis on which it is based.

g. Heavy Particle Collisions (HPC)

Atomic collisions have been used to determine negative ion bond dissociation energies from reaction thresholds. These may be combined with molecular dissociation energies and appropriate atomic electron affinities to determine polyatomic electron affinities [20, 21]. The accuracy is thought to be several tenths of an electron volt.

h. Charge Exchange (CE)

Observation of the cross section for the interaction

$$X + Y^- \rightarrow X^- + Y,$$

as a function of Y^- energy can be used to determine a difference between two electron affinities if the interaction is endothermic at threshold and to bracket the affinity for exothermic reactions [22–24]. Internal and kinetic energies severely complicate the interpretation. The energy resolution, several tenths of an electron volt, is limited partially by the energy spread of the electron beam used to generate the initial negative ion.

i. Dissociative Electron Attachment (ATT)

The threshold energy for the interaction

$$XY + e^- \rightarrow X + Y^-,$$

has been used to determine electron affinities where the relevant dissociation energy is known [25]. Appreciable internal energies can severely complicate the analysis [26–28]. Energy resolution, several tenths of an electron volt, is limited here also by the energy spread of the electron beam.

j. Thermochemistry (T)

The temperature dependence of the concentration of negative ions in equilibrium with neutral species, combined with the laws of thermodynamics, leads to ion heats of formation from those of the neutral species. A number of techniques for generating a quasi-equilibrium population are used, a notable one of which is plasmas [29–31]. The identity of the species being observed is sometimes uncertain. The achievement of equilibrium is also difficult to ascertain. An uncertainty of several tenths of an electron volt is achieved.

k. Surface Ionization (SI)

Surface ionization has been employed for negative ion study by a number of workers, particularly successfully by Scheer and Fine [32]. A number of parameters must be carefully evaluated, especially the work function of the surface under the particular conditions. If mass analysis is not provided, the identity of the ions observed is in doubt. Accuracies are thought to approach several tenths of an electron volt.

l. Exploding Wire (EW)

Ya'akobi has interpreted the power balance in exploding wires and the observation of an anomaly as due to evaporating negative ions. From the temperature for onset of the anomaly, he evaluates the electron affinity associated with the negative ion to several tenths of an electron volt [33]. Evidence for the validity of the supporting assumptions is indirect.

m. Electron Scattering (ES)

The energies of temporary (excited) negative ion states can be explored by changes in the differential elastic electron scattering cross section [34–37]. For molecular ions with small affinities, features of the elastic [38] and total [39] electron scattering cross sections may be extrapolated to the lowest vibrational level of the bound ion. Since the threshold is not observed directly, the analysis is crucial to the success of this technique. The accuracy is of the order of a tenth of an electron volt.

n. Electron Impact with Analytic Deconvolution (DEC)

Dissociative ionization under electron impact

$$XY + e^- \rightarrow X^+ + Y + 2e^-,$$

can be observed as a function of incident electron energy to identify the threshold for the process. When a negative ion as well as a positive ion is formed,

$$XY + e^- \rightarrow X^+ + Y^- + e^-,$$

the electron affinity is obtained from the difference in positive ion thresholds for the two types of process. In addition to internal and kinetic energy changes in the molecular particles, the energy spread of the electron beam obscures the threshold identification. Various analytical techniques have been used to reduce this latter source of imprecision [40–43] to achieve a resolution of several tenths of an electron volt. Because of the necessity for a mathematical model, unexpected transitions cannot be allowed for in such an analysis.

o. Electron Impact with Retarding Potential Difference (RPD)

The threshold behavior for negative ion formation on electron impact can be clarified by varying the electron energy slightly and observing the correlated modification in the generation of negative ions. The technique developed by Fox for positive ions has been applied to negative ions only recently by other workers [44, 45]. The primary goal of this negative ion work is the survey of ions not previously studied by other techniques. The sensitivity of this approach is limited by the presence of noise from the entire electron beam. Moreover, Fox's successors have not always applied his care in the utilization of this technique. Accuracies of several tenths of an electron volt are possible in favorable cases.

p. Electron Impact (EI)

Simple electron impact leading to ion-pair production has generally been superseded by more sophisticated handling of the electron beam energy spread, either by analytic deconvolution (see section n) or by the retarding potential difference method (see section o). Occasionally, however [46], no other data are available. Accuracies of about one half an electron volt can then be achieved where a fortuitous reference ion is available.

q. Born-Haber Cycle Calculations (BH)

The earliest measure of electron affinities came from application of the Born-Haber thermodynamic analysis to ionic crystals. One must know the energy of formation of the salt with respect to the elements, the lattice energy, the ionization energy of the alkali, the sublimation energy of the alkali, and the dissociation energy of the molecular electronegative component [47, 48]. The uncertainty in one or more of these quantities is typically of a size that is a substantial fraction of the electron affinity to be determined. An accuracy greater than one half an electron volt seems unlikely.

r. Dipole Moment Extrapolation (DME)

From uv absorption spectra, Sklar has observed changes in the dipole moment of a given molecular bond as a function of the electron affinity of various electronegative atoms. This is extrapolated to zero moment to obtain the affinity of the radical on which the substitution was observed [49]. Since the method is indirect, it is difficult to evaluate quantitatively. Its accuracy is certainly not better than one half an electron volt.

s. Dissociation Energy (DE and AA)

Mulliken's thesis that the dissociation energy of a molecular negative ion should be about half that of the corresponding neutral species has been applied by Geiger [50] to the determination of electron affinities (DE). He has estimated homoatomic molecular halogen affinities from the corresponding atomic halogen affinities.

Gaines and Page [51], on the other hand, have estimated certain molecular electron affinities by assuming that the dissociation energies of both negative ion and corresponding neutral species are identical (AA). The molecular affinity can then be calculated where appropriate molecular and atomic negative ion heats of formation are known. The basic assumption here clearly conflicts with that of Geiger. The uncertainty in both cases is certainly no better than one half an electron volt.

t. Lyotropic Series Analysis (L)

The sum of the heat of hydration of a negative ion and its electron affinity has been postulated to be a constant. The heat of hydration is determined from a plot of the relative efficiencies of salts in precipitating gelatine from a solution [52]. In this manner, comparisons of electron affinities are made. The validity of the assumptions is difficult to assess. The accuracy is probably less than one half an electron volt.

u. Observation in Mass Spectrum (OBS)

Ions observed in a negative ion mass spectrum have a lifetime of at least a microsecond. When no other evidence is available, the affinity of such species is taken to be positive. Autodetachment can in principle occur, however.

5.2. Theoretical Techniques

Calculations employing ab initio techniques are proving increasingly accurate. Theoretical techniques which first utilized empirical correlation terms, now are being made entirely ab initio. Accuracy better than a millielectron volt has been attained for the variational calculation of H^-, which is a special case. Hartree-Fock and configuration interaction calculations now agree with experiments to better than a tenth of an electron volt upon occasion. The other approaches can be as good as several tenths of an electron volt.

a. Variational Calculations (VAR)

For sufficiently small atoms and ions, it is feasible to vary the parameters of an energy calculation in order to produce an energy minimum, which, following Ritz, is always greater than the true value. Pekeris has employed this technique in a definitive fashion to calculate the electron affinity for H, probably the most precisely known affinity [53]. It approaches a microelectron volt in accuracy. More recently, Nesbet has performed Bethe-Goldstone variational calculations on more complex atomic species [54].

b. Hartree-Fock Calculations (HF)

The Hartree-Fock self-consistent field ab initio calculations previously performed widely for neutral atoms and positive ions have now been used to calculate negative ion energies as well. However, since the electron correlation energy is of the order of the electron affinity, the procedure used to estimate this correction is crucial to the reliability of the technique when applied to negative ions. Clementi and colleagues have employed isoelectronic ionization potential extrapolations [55–57] to estimate the correlation energy. Weiss has employed an ab initio superposition of configuration technique [58, 59]. Cade has extended this method to molecular ions [60, 61].

c. Configuration Interaction Calculations (CI)

Various configuration interaction calculations have been used for molecular and atomic species [62–69]. This theoretical approach has been limited to systems with few electrons. Even for these, the reliability is uncertain, probably several tenths of an electron volt.

d. Many Electron Theory (MET)

Öksüz and Sinanoğlu have developed an ab initio theory incorporating electron correlation [70]. Although the results are initially less accurate than those of some experiments and some empirical extrapolation approaches, the method is promising because of its ab initio character for relatively heavy negative ions. An accuracy of several tenths of an electron volt is achieved.

e. Rydberg-Klein-Rees Calculations (RKR)

Rydberg-Klein-Rees calculations have been used to generate molecular negative ion potential energy curves (dissociation energies) as well as those for neutral mole-cules. The electron affinity can then be derived indirectly where the separated atom affinities are known [71]. The uncertainty is of the order of one half an electron volt.

5.3. Empirical Extrapolation Techniques

The use of regularities in ionization energies or other periodic parameters for atoms in the periodic table to predict atomic electron affinities has become more reliable as the experimental values on which the techniques are based have improved. The approach most widely used recently, the estimation of affinities from the location in a row of the periodic table, is probably reliable to several tenths of an electron volt. The others may be somewhat less reliable.

a. Horizontal Analysis of the Periodic Table (HA)

Probably the most generally applicable and reliable empirical extrapolation technique involves analysis of regular changes in electron affinities of atoms along rows of the periodic table. The analysis depends on the prior knowledge of at least one affinity in each row. Such analyses have been performed by Ginsberg and Miller [72], Edie and Rohrlich [73], Crossley [74], Zollweg [75], and most recently by Hotop, Bennett, and Lineberger [76]. As noted an accuracy of several tenths of an electron volt may be achieved.

b. Extrapolation of Ionization Potentials (IE)

One of the earliest approaches to electron affinity determinations was an empirical extrapolation developed by Glockler [77]. The ionization energies of isoelectronic ions of differing charge are fit to the equation

$$E = a + bZ + cZ^2,$$

where Z is the nuclear charge. The parabola so determined is extrapolated to find the ionization energy of the isoelectronic singly-charged negative ion, i.e. the electron affinity of the corresponding neutral atom. This technique has been extended by Charkin and Dyatkina [78], Geltman [79], Johnson and Rohrlich [80], Edlén [81], and Kaufman [82]. An accuracy of several tenths of an electron volt may be obtained.

c. Vertical Analysis of the Periodic Table (VA)

A logarithmic plot of the difference between the first ionization potential and the electron affinity of an atom against the atomic radius has been observed to form a straight line [83]. Politzer has used this line to estimate the electron affinity for atoms in the lower part of the periodic table from those in the first rows. An accuracy of better than one half an electron volt may be achieved.

d. Isoelectronic Model (IM)

Gilmore [34] has referred to isoelectronic molecules in order to estimate the bond energies of molecular negative ions. These values are then combined with the neutral molecular bond energies and atomic electron affinities to estimate molecular electron affinities to within perhaps one half an electron volt.

5.4. Order of Tabular Listing

In the Negative Ion Table the various species are listed, as are the positive ions, in order of increasing atomic number of the highest-atomic-number constituent. Polyatomic ions are arranged after the atomic ions in order of increasing atomic numbers of other constituents.

Where an electron affinity is thought to be known to better than 0.1 eV, a heat of formation has been calculated from the first listed affinity and presented in the ion heading.

For a given negative ion, the various determinations of the electron affinity are listed in order of estimated decreasing reliability. The abbreviations for the techniques used are listed alphabetically in table 1 and explained in the preceding text.

TABLE 1.   Techniques for negative ion energetics

Abbreviation	Technique
AA	Constituent atom electron affinity (dissociation energy)
ATT	Dissociative electron attachment
BH	Born-Haber cycle calculation
CE	Charge exchange
CI	Configuration interaction calculation
DE	Dissociation energy
DEC	Electron impact with analytic deconvolution
DME	Dipole moment extrapolation
EI	Electron impact
ES	Electron scattering
EW	Exploding wire
FD	Electrostatic field detachment
HA	Horizontal analysis of the periodic table
HF	Hartree-Fock calculation
HPC	Heavy particle collisions
IE	Extrapolation of ionization potentials
IM	Isoelectronic model
L	Lyotropic series analysis
LP	Dye laser photodetachment
MET	Many electron theory
OBS	Observation in mass spectrum
P	Incoherent photon detachment
PE	Photoelectron spectroscopy with fixed frequency laser
PI	Photoionization
RKR	Rydberg-Klein-Rees calculation
RPD	Electron impact with retarding potential difference
S	Spectroscopy
SI	Surface ionization
T	Thermochemistry
VA	Vertical analysis of the periodic table
VAR	Variational calculation

5.5. References for Section 5

[1] Compton, R. N., and Huebner, R. H., Collisions of low-energy electrons with molecules: threshold excitation and negative–ion formation, in: Advances in Radiation Chemistry, Vol. 2, ed. M. Burton and J. Magee (Wiley, New York, 1970) p. 281.

[2] Steiner, B., Photodetachment: cross sections and electron affinities, in: Case Studies in Atomic Collision Physics, Vol. 2, ed. E. W. McDaniel and M. R. C. McDowell (North–Holland Pub. Co., Amsterdam, 1972) p. 483.

[3] Hotop, H., and Lineberger, W. C., Binding energies in atomic negative ions, J. Phys. Chem. Ref. Data **4**, 539 (1975).

[4] Lineberger, W. C., and Woodward, B. W., High resolution photodetachment of S$^-$ near threshold, Phys. Rev. Letters **25**, 424 (1970).

[5] Lineberger, W. C., Photodetachment of negative ions, in: Energy, Structure, and Reactivity, ed. D. W. Smith and W. B. McRae (Wiley, New York, 1973).

[6] Hotop, H., and Lineberger, W. C., Dye–laser photodetachment studies of Au$^-$, Pt$^-$, PtN$^-$, and Ag$^-$, J. Chem. Phys. **58**, 2379 (1973).

[7] Smyth, K. C., McIver, R. T., Jr., Brauman, J. I., and Wallace, R. W., Photodetachment of negative ions using a continuously tunable laser and an ion cyclotron resonance spectrometer, J. Chem. Phys. **54**, 2758 (1971).

[8] Berry, R. S., Small free negative ions, Chem. Rev. **69**, 533 (1969).

[9] Berry, R. S., Reimann, C. W., and Spokes, G. N., Absorption spectra of gaseous halide ions and halogen electron affinities: chlorine, bromine, and iodine, J. Chem. Phys. **37**, 2278 (1962).

[10] Berry, R. S., Mackie, J. C., Taylor, R. L., and Lynch, R., Spin–orbit coupling and electron–affinity determinations from radiative capture of electrons by oxygen atoms, J. Chem. Phys. **43**, 3067 (1965).

[11] Popp, H.–P., Quantitative Ausmessung des Fluor–Affinitätskontinuums, Z. Naturforsch. **22a**, 254 (1967).

[12] Popp, H.–P., Nachweis des Fluor–Affinitätskontinuums in Emission, Z. Naturforsch. **20a**, 642 (1965).

[13] Branscomb, L. M., Burch, D. S., Smith, S. J., and Geltman, S., Photodetachment cross section and the electron affinity of atomic oxygen, Phys. Rev. **111**, 504 (1958).

[14] Brehm, B., Gusinow, M. A., and Hall, J. L., Electron affinity of helium via laser photodetachment of its negative ion, Phys. Rev. Letters **19**, 737 (1967).

[15] Celotta, R. J., Bennett, R. A., Hall, J. L., Siegel, M. W., and Levine, J., Molecular photodetachment spectrometry. II. The electron affinity of O$_2$ and the structure of O$_2^-$, Phys. Rev. A **6**, 631 (1972).

[16] Siegel, M. W., Celotta, R. J., Hall, J. L., Levine, J., and Bennett, R. A., Molecular photodetachment spectrometry. I. The electron affinity of nitric oxide and the molecular constants of NO$^-$, Phys. Rev. A **6**, 607 (1972).

[17] Morrison, J. D., Hurzeler, H., Inghram, M. G., and Stanton, H. E., Threshold law for the probability of excitation of molecules by photon impact. A study of the photoionization efficiencies of Br$_2$, I$_2$, HI, and CH$_3$I, J. Chem. Phys. **33**, 821 (1960).

[18] Berkowitz, J., Chupka, W. A., and Walter, T. A., Photoionization of HCN: the electron affinity and heat of formation of CN, J. Chem. Phys. **50**, 1497 (1969).

[19] Smirnov, B. M., and Chibisov, M. I., The breaking up of atomic particles by an electric field and by electron collisions, Zh. Eksperim. i Teor. Fiz. **49**, 841 (1965) [Engl. transl.: Soviet Phys. JETP **22**, 585 (1966)].

[20] Edwards, A. K., Risley, J. S., and Geballe, R., Two autodetaching states of O$^-$, Phys. Rev. A **3**, 583 (1971).

[21] De Paz, M., Giardini, A. G., and Friedman, L., Tandem–mass–spectrometer study of solvated derivatives of OD$^-$. Total hydration energy of the proton, J. Chem. Phys. **52**, 687 (1970).

[22] Berkowitz, J., Chupka, W. A., and Gutman, D., Electron affinities of O$_2$, O$_3$, NO, NO$_2$, NO$_3$ by endothermic charge transfer, J. Chem. Phys. **55**, 2733 (1971).

[23] Fehsenfeld, F. C., Ferguson, E. E., and Schmeltekopf, A. L., Thermal–energy associative–detachment reactions of negative ions, J. Chem. Phys. **45**, 1844 (1966).

[24] Chupka, W. A., Berkowitz, J., and Gutman, D., Electron affinities of halogen diatomic molecules as determined by endoergic charge transfer, J. Chem. Phys. **55**, 2724 (1971).

[25] Dorman, F. H., Negative fragment ions from resonance capture processes, J. Chem. Phys. **44**, 3856 (1966).

[26] Chantry, P. J., and Schulz, G. J., Kinetic–energy distribution of negative ions formed by dissociative attachment and the measurement of the electron affinity of oxygen, Phys. Rev. **156**, 134 (1967).

[27] Henderson, W. R., Fite, W. L., and Brackmann, R. T., Dissociative attachment of electrons to hot oxygen, Phys. Rev. **183**, 157 (1969).

[28] Schulz, G. J., and Spence, D., Temperature dependence of the onset for negative–ion formation in CO_2, Phys. Rev. Letters **22**, 47 (1969).

[29] Pack, J. L., and Phelps, A. V., Electron attachment and detachment. I. Pure O_2 at low energy, J. Chem. Phys. **44**, 1870 (1966).

[30] Pack, J. L., and Phelps, A. V., Electron attachment and detachment. II. Mixtures of O_2 and CO_2 and of O_2 and H_2O, J. Chem. Phys. **45**, 4316 (1966).

[31] Ferguson, E. E., Fehsenfeld, F. C., and Phelps, A. V., Comment on photodetachment cross sections for CO_3^- and its first hydrate, J. Chem. Phys. **59**, 1565 (1973).

[32] Scheer, M. D., Positive and negative ion sublimation from transition metal surfaces: a review of some recent results, J. Res. NBS **74A**, 37 (1970).

[33] Ya'akobi, B., An experimental estimate of the electron affinity of lithium, Phys. Letters **23**, 655 (1966).

[34] Gilmore, F. R., Potential energy curves for N_2, NO, O_2 and corresponding ions, J. Quant. Spectry. Radiative Transfer **5**, 369 (1965).

[35] Comer, J., and Read, F. H., Electron impact studies of a resonant state of N_2^-, J. Phys. B **4**, 1055 (1971).

[36] Rempt, R. D., Electron–impact excitation of carbon monoxide near threshold in the 1.5– to 3–eV incident energy range, Phys. Rev. Letters **22**, 1034 (1969).

[37] Bardsley, J. N., Negative ions of N_2O and CO_2, J. Chem. Phys. **51**, 3384 (1969).

[38] Boness, M. J. W., and Schulz, G. J., Structure of O_2, Phys. Rev. A **2**, 2182 (1970).

[39] Spence, D., and Schulz, G. J., Vibrational excitation by electron impact in O_2, Phys. Rev. A **2**, 1802 (1970).

[40] Thynne, J. C. J., and MacNeil, K. A. G., Negative ion formation by ethylene and 1,1–difluoroethylene, J. Phys. Chem. **75**, 2584 (1971).

[41] Thynne, J. C. J., and MacNeil, K. A. G., Ionisation of tetrafluoroethylene by electron impact, Intern. J. Mass Spectrom. Ion Phys. **5**, 329 (1970).

[42] Harland, P. W., and Thynne, J. C. J., Dissociative electron capture in perfluoropropylene and perfluoropropane, Intern. J. Mass Spectrom. Ion Phys. **9**, 253 (1972).

[43] Harland, P. W., and Thynne, J. C. J., Autodetachment lifetimes, attachment cross sections, and negative ions formed by sulfur hexafluoride and sulfur tetrafluoride, J. Phys. Chem. **75**, 3517 (1971).

[44] Williams, J. M., and Hamill, W. H., Ionization potentials of molecules and free radicals and appearance potentials by electron impact in the mass spectrometer, J. Chem. Phys. **49**, 4467 (1968).

[45] Lifshitz, C., Peers, A. M., Grajower, R., and Weiss, M., Breakdown curves for polyatomic negative ions, J. Chem. Phys. **53**, 4605 (1970).

[46] Locht, R., and Momigny, J., Mass spectrometric determination of the electronaffinities of radicals, Chem. Phys. Letters **6**, 273 (1970).

[47] Born, M., Problems of Atomic Dynamics (F. Ungar Pub. Co., New York, 1960) p. 168.

[48] West, C. D., Thermochemistry and physical properties of bromides and hydrosulfides, J. Phys. Chem. **39**, 493 (1935).

[49] Sklar, A. L., The near ultraviolet absorption of substituted benzenes, J. Chem. Phys. **7**, 984 (1939).

[50] Geiger, W., Die Emission negativer Ladungsträger bei Einwirkung von Halogenmolekülen auf Alkalimetall-oberflächen, Z. Physik **140**, 608 (1955).

[51] Gaines, A. F., and Page, F. M., Semi–empirical prediction of the electron affinities of gaseous radicals, Trans. Faraday Soc. **62**, 3086 (1966).

[52] Buchner, E. H., The azide ion in the lyotropic series, Rec. Trav. Chim. **69**, 329 (1950).

[53] Pekeris, C. L., 1^1S, 2^1S, and 2^3S states of H^- and of He, Phys. Rev. **126**, 1470 (1962).

[54] Moser, C. M., and Nesbet, R. K., Atomic Bethe–Goldstone calculations of term splittings, ionization potentials, and electron affinities for B, C, N, O, F, and Ne, Phys. Rev. A **4**, 1336 (1971).

[55] Clementi, E., and McLean, A. D., Atomic negative ions, Phys. Rev. **133**, A419 (1964).

[56] Clementi, E., McLean, A. D., Raimondi, D. L., and Yoshimine, M., Atomic negative ions. Second period, Phys. Rev. **133**, A1274 (1964).

[57] Clementi, E., Atomic negative ions: the iron series, Phys. Rev. **135**, A980 (1964).

[58] Weiss, A. W., Theoretical electron affinities for some of the alkali and alkaline–earth elements, Phys. Rev. **166**, 70 (1968).

[59] Weiss, A. W., Symmetry–adapted pair correlations in Ne, F^-, Ne^+, and F, Phys. Rev. A **3**, 126 (1971).

[60] Cade, P. E., The electron affinities of the diatomic hydrides CH, NH, SiH and PH, Proc. Phys. Soc. (London) **91**, 842 (1967).

[61] Cade, P. E., Hartree–Fock wavefunctions, potential curves, and molecular properties for $OH^-(^1\Sigma^+)$ and $SH^-(^1\Sigma^+)$, J. Chem. Phys. **47**, 2390 (1967).

[62] Taylor, H. S., The ground state of H_2^-, Proc. Phys. Soc. (London) **90**, 877 (1967).

[63] Somerville, W. B., The ground state of H_2^-, Proc. Phys. Soc. (London) **89**, 185 (1966).

[64] Taylor, H. S., and Harris, F. E., Potential curve for the $^2\Sigma_u^+$ state of H_2^-, J. Chem. Phys. **39**, 1012 (1963).

[65] Dalgarno, A., and McDowell, M. R. C., Charge transfer and the mobility of H^- ions in atomic hydrogen, Proc. Phys. Soc. (London) **A69**, 615 (1956).

[66] Victor, G. A., and Laughlin, C., Model potential calculation of the electron affinity of lithium, Chem. Phys. Letters **14**, 74 (1972).

[67] Schwarz, W. H. E., Electron affinities of the alkali atoms. A model–potential calculation, Chem. Phys. Letters **10**, 478 (1971).

[68] Grün, N., Genäherte natürliche Orbitale für die Grundzustände von Li und Li^-; Elektronenaffinität von Li^-, Z. Naturforsch. **27a**, 843 (1972).

[69] Norcross, D. W., and Moores, D. L., Photodetachment of Li^- and Na^-, in: Atomic Physics, Vol. 3, ed. S. J. Smith and G. K. Walters (Plenum Press, New York, 1972) p. 261.

[70] Öksüz, İ., and Sinanoğlu, O., Theory of atomic structure including electron correlation. II. All–external pair correlations in the various states and ions of B, C, N, O, F, Ne, and Na, and prediction of electron affinities and atomic excitation energies, Phys. Rev. **181**, 54 (1969).

[71] Sharp, T. E., Potential energy diagram for molecular hydrogen and its ions, Lockheed Report LMSC 5–10–69–9 (1969).

[72] Ginsberg, A. P., and Miller, J. M., An empirical method for estimating the electron affinities of atoms, J. Inorg. Nucl. Chem. **7**, 351 (1958).

[73] Edie, J. W., and Rohrlich, F., Negative atomic ions. II, J. Chem. Phys. **36**, 623 (1962); Erratum, *ibid.*, **37**, 1151 (1962).

[74] Crossley, R. J. S., Glockler's equation for ionization potentials and electron affinities: II, Proc. Phys. Soc. (London) **83**, 375 (1964).

[75] Zollweg, R. J., Electron affinities of the heavy elements, J. Chem. Phys. **50**, 4251 (1969).

[76] Hotop, H., Bennett, R. A., and Lineberger, W. C., Electron affinities of Cu and Ag, J. Chem. Phys. **58**, 2373 (1973).

[77] Glockler, G., Estimated electron affinities of the light elements, Phys. Rev. **46**, 111 (1934).

[78] Charkin, O. P., and Dyatkina, M. E., Calculation of the electron affinity of transitional elements by the Glockler method, Zh. Strukt. Khim. **6**, 422 (1965) [Engl. transl.: J. Struct. Chem. (USSR) **6**, 397 (1965)].

[79] Geltman, S., Determination of electron affinities by extrapolation, J. Chem. Phys. **25**, 782 (1956).

[80] Johnson, H. R., and Rohrlich, F., Negative atomic ions, J. Chem. Phys. **30**, 1608 (1959).

[81] Edlén, B., Isoelectronic extrapolation of electron affinities, J. Chem. Phys. **33**, 98 (1960).

[82] Kaufman, M., Semiempirical electron affinities, Astrophys. J. **137**, 1296 (1963).

[83] Politzer, P., Electron affinities of atoms, Trans. Faraday Soc. **64**, 2241 (1968).

6. Tabulation of Negative Ion Data

6.1. Index

6.2. The Negative Ion Table

Ion	State	Electron affinity (eV)	Method	Ref.
	H$^-$(^{1}S$_0$) ΔH°_{f0} = 143.2 kJ mol^{-1} (34.2 kcal mol^{-1})			
H$^-$	1s^2 ^{1}S$_0$	0.7542	VAR	1
		0.756±0.013	S	2
		0.776±0.020	P	3
	2p^2 ^{3}P ($\rightarrow$2p ^{2}P°)	>0.0095	VAR	4
	H$_2^-$			
H$_2^-$	$^2\Sigma_u^+$	−2.5	RKR–CI	5
		−2.85	VAR	169
		<0	CI	6
		<0	CI	7
		<0	CI	8
		<0	CI	9
		>0	OBS	10
		>0	OBS	11
	He$^-$			
He$^-$	1s 2s 2p ^{4}P ($\rightarrow$1s 2s ^{3}S$_1$)	0.080±0.008	PE	12
		0.060±0.005	FD	13
		≥0.033	VAR	14
	Li$^-$(^{1}S$_0$) ΔH°_{f0} ~ 99 kJ mol^{-1} (24 kcal mol^{-1})			
Li$^-$	2s^2 ^{1}S$_0$	0.62±0.10	HF	15
		0.58±0.05	HF	16
		0.591	CI	17
		0.62	CI	18
		0.593±0.021	CI	19
		0.85+0.20	SI	20
		0.59	HA	21
		0.64	IE	22
		0.82	IE	23
		0.6	EW	24
	Be$^-$			
Be$^-$	2s^2 3s ^{2}S$_{1/2}$	<0	HF	15
		0.38±0.20	HA	25
		>0	OBS	26
	2s 2p^2 ^{4}P ($\rightarrow$2s 2p ^{3}P°)	0.24±0.10	HF	15
	2s^2 2p ^{2}P°$_{1/2}$	−0.65	HA	25
		−0.68	HA	21
		−0.38	HA	27
		−0.19	IE	23

6.2. The Negative Ion Table—Continued

Ion	State	Electron affinity (eV)	Method	Ref.
BeH$^-$		1.00	AA	28

BeH$_3^-$

BeH$_3^-$		3.8	HF	29

B$^-$

Ion	State	Electron affinity (eV)	Method	Ref.
B$^-$	2p^2 ^{3}P$_0$	0.223	HF–VAR	30
		0.187±0.100	HF	31
		0.30±0.05	HF	16
		0.18±0.20	HA	25
		0.16	HA	21
		0.33	HA	27
		0.33	IE	23
		>0	OBS	32

BH$^-$

BH$^-$		0.09	AA	28

BH$_2^-$

BH$_2^-$		1.39	AA	28

C$^-$(^{4}S$^\circ_{3/2}$) $\Delta H^\circ_{f0} = 588.7$ kJ mol^{-1} (140.7 kcal mol^{-1})

Ion	State	Electron affinity (eV)	Method	Ref.
C$^-$	2p^3 ^{4}S$^\circ_{3/2}$	1.270±0.010	PE	33
		1.25±0.03	P	34
		1.29	HF–VAR	30
		1.24±0.10	HF	31
		1.17±0.06	HF	16
		1.17±>0.10	MET	35
		1.29±0.20	HA	25
		1.33	HA	21
		1.38	HA	27
		1.24	IE	23
		1.2	SI	36
	2p^3 ^{2}D$^\circ$	0.050±0.025	PE	37
		0.062	PE	33
		>0	P	34
		<0.5	P	3
		−0.105±0.100	HF	31
		−0.08±0.05	HF	16
		−0.13±>0.10	MET	35

6.2. The Negative Ion Table—Continued

Ion	State	Electron affinity (eV)	Method	Ref.
$C_2^-(^2\Sigma_g^+)$	$\Delta H_{f0}^\circ \geq 488$ kJ mol^{-1} (117 kcal mol^{-1})			
C_2^-	$^2\Sigma_g^+$	$\leq 3.54 \pm 0.05$	P	3
		4.0	SI	36
		3.3 ± 0.2	EI	38
		$> 2.9 \pm 0.5$	DEC	39
C_3^-				
C_3^-		2.5	SI	36
$CH^-(^3\Sigma^-)$	$\Delta H_{f0}^\circ = 521$ kJ mol^{-1} (125 kcal mol^{-1})			
CH^-	$^3\Sigma^-$	0.74 ± 0.05	P	3
		1.61 ± 0.20	HF	40
		2.6 ± 0.3	EI	38
		$\geq 4.1 \pm 0.4$	DEC	39
CH_3^-				
CH_3^-		1.8	DME	41
C_2H^-	$\Delta H_{f0}^\circ \geq 114$ kJ mol^{-1} (27 kcal mol^{-1})			
C_2H^-		$\leq 3.73 \pm 0.05$	P	3
		2.1 ± 0.3	EI	38
		$\geq 2.3 \pm 0.7$	DEC	39
$C_5H_5^-$				
$C_5H_5^-$ (Cyclopentadienyl)		$\leq 2.2 \pm 0.3$	DEC	42
N^-				
N^-	$2p^4\ ^3P_2$	-0.12	HF–VAR	30
		-0.213	HF	31
		0.05	IE	23
		>0	CE	43
N_2^-				
N_2^-	$^2\Pi_g$	-1.6 ± 1.0	ES	44
	$^2\Sigma_g^+$	-11.345	ES	45

6.2. The Negative Ion Table—Continued

Ion	State	Electron affinity (eV)	Method	Ref.
		N_3^-		
N_3^-	$^1\Sigma_g^+$	1.8	HF	46
		3.5±0.2	S	47
		2.3	L	48
		NH^-		
NH^-	$^2\Pi_i$	0.22±0.20	HF	40
		NH_2^- $\quad \Delta H^\circ_{f\,298} \sim$ **100 kJ mol^{-1} (24 kcal mol^{-1})**		
NH_2^-		0.744±0.022	P	49
		0.74	P	50
		0.76±0.04	P	51
		0.6	ATT	52
		1.21	SI	53
		$CN^-(^1\Sigma^+)$ $\quad \Delta H^\circ_{f0} =$ **63 kJ mol^{-1} (15 kcal mol^{-1})**		
CN^-	$^1\Sigma^+$	3.82±0.02	PI	54
		3.6±0.3	ATT	54
		3.17±0.04	SI	55
		2.797±0.022	SI	170
		$O^-(^2P^\circ_{3/2})$ $\quad \Delta H^\circ_{f0} =$ **105.4 kJ mol^{-1} (25.2 kcal mol^{-1})**		
O^-	$2p^5\ ^2P^\circ_{3/2}$	1.465±0.005	P	56
		1.478±0.002	S	57
		1.461±0.024	PI	58
		1.5±0.1	ATT	59
		1.43	HF–VAR	30
		1.46±0.10	HF	31
		1.31±0.15	ATT	60
		1.22±0.14	HF	16
		1.24±>0.10	MET	35
		1.39	HA	21
		1.47	IE	23
		−10.112±0.010	HPC	61
		−12.115±0.010	HPC	61

6.2. The Negative Ion Table—Continued

Ion	State	Electron affinity (eV)	Method	Ref.
$O_2^-(^2\Pi_{3/2g})$		$\Delta H_{f0}^\circ = -42.5$ kJ mol^{-1} (-10.1 kcal mol^{-1})		
O_2^-	$^2\Pi_{3/2g}$	0.440 ± 0.008	PE	62
		0.43 ± 0.02	T	63
		0.45 ± 0.10	CE	64
		0.5 ± 0.1	CE	65
		$\geqslant0.48$	CE	66
		0.42	DE	67
		0.15 ± 0.05	P	68
		$\geqslant0.21$	PI	69
O_3^-				
O_3^-		1.92 ± 0.13	S	70
		1.91 ± 0.43	BH	71
		>1.5	S	130
		$\geqslant1.96$	CE	66
O_4^-				
O_4^-		>0	OBS	72
$OH^-(^1\Sigma^+)$		$\Delta H_{f0}^\circ = -137.4$ kJ mol^{-1} (-32.8 kcal mol^{-1})		
OH^-	$^1\Sigma^+$	1.825 ± 0.002	LP	171
		1.826 ± 0.002	LP	73
		1.83 ± 0.04	P	74
		1.91 ± 0.10	HF	75
$OD^-(^1\Sigma^+)$		$\Delta H_{f0}^\circ = -139.4$ kJ mol^{-1} (-33.3 kcal mol^{-1})		
OD^-	$^1\Sigma^+$	1.823 ± 0.002	LP	73
H_2O^-				
H_2O^-		0.9	S	47
HO_2^-				
HO_2^-		4.6	BH	76
$H_2O_2^-$				
$H_2O_2^-$		>0	OBS	77

6.2. The Negative Ion Table—Continued

Ion	State	Electron affinity (eV)	Method	Ref.
		$H_3O_2^-$		
$H_3O_2^-$		$\leq 2.95 \pm 0.15$	P	78
		3.35 ± 0.15	HPC	79
		$H_2O_3^-$		
$H_2O_3^-$		>0	OBS	77
		BO^-		
BO^-		3.12 ± 0.09	SI	80
		>2.5	T	81
		2.12	AA	28
		BO_2^-		
BO_2^-		3.57 ± 0.13	SI	80
		4.07 ± 0.21	T	178
		CO^-		
CO^-		$<-1.8 \pm 0.1$	ES	82
		CO_2^-		
CO_2^-		>-0.9	ES	83
		CO_3^-		
CO_3^-		>2.6	T	172
		1.8 ± 0.2	P	173
		>0	OBS	77
		CO_4^-		
CO_4^-		1.22 ± 0.07	T	84
		>0	OBS	77

6.2. The Negative Ion Table—Continued

Ion	State	Electron affinity (eV)	Method	Ref.
$NO^-(^3\Sigma_g^-)$		$\Delta H^\circ_{f0} = 87.4$ kJ mol^{-1} (20.9 kcal mol^{-1})		
NO^-	$^3\Sigma_g^-$	$0.024^{+0.010}_{-0.005}$	PE	85
		$\geqslant 0.09$	CE	66
		$\leqslant 0.440$	CE	86
		0.15	IM	44
		0.89 ± 0.11	SI	87
		0.85; 0.86	RPD	88
N_2O^-				
N_2O^-		0.27 ± 0.17	ATT	89
		<1.5	CE	90
		>0.5	ATT	91
		>0	CE	92
		~ 0	ATT	93
		~ 0	CE	94
		~ 0	ATT	83
NO_2^-		$\Delta H^\circ_{f0} = -263$ kJ mol^{-1} (-63 kcal mol^{-1})		
NO_2^-		3.10 ± 0.05	P	95
		2.38 ± 0.06	CE	96
		2.5 ± 0.1	CE	65
		$\geqslant 2.04$	CE	66
		$\leqslant 2.6$	CE	97
		>3.6	CE	98
		3.99 ± 0.16	SI	87
NO_3^-				
NO_3^-		3.9 ± 0.2	CE	97
		$3.7 \pm > 0.2$	CE	99
		$\geqslant 2.48$	CE	66
		4.22 ± 0.22	BH	100
$C_2H_5O^-$				
$C_2H_5O^-$		1.68	RPD	88
$C_3H_7O^-$				
$n-C_3H_7O^-$		1.87	RPD	88
$iso-C_3H_7O^-$		1.73	RPD	88

6.2. The Negative Ion Table—Continued

Ion	State	Electron affinity (eV)	Method	Ref.
		$C_4H_9O^-$		
$n-C_4H_9O^-$		1.90	RPD	88
		$C_4H_6O_2^-$		
$(CH_3CO)_2^-$		1.1	ATT	174
		$CH_2O_4^-$		
$CO_3(H_2O)^-$		>3.1	T	172
		2.1±0.2	P	173
		$C_6H_5NO_2^-$		
$C_6H_5NO_2^-$ (Nitrobenzene)		0.4	ATT	174

$F^-(^1S_0)$ $\Delta H_{f0}^{\circ} = -251.2$ kJ mol^{-1} (-60.0 kcal mol^{-1})

Ion	State	Electron affinity (eV)	Method	Ref.
F^-	$2p^6\ ^1S_0$	3.400±0.002	S	101, 102
		3.398±0.002	S	103
		3.39±0.02	PI	104
		3.47±0.03	PI	105
		3.37	HF–VAR	30
		3.47	HF	106
		3.45±0.10	HF	31
		3.37±0.08	HF	16
		3.23±>0.10	MET	35
		3.50±0.20	HA	25
		3.47	HA	27
		3.50	IE	23

Ion	State	Electron affinity (eV)	Method	Ref.
		F_2^-		
F_2^-		3.08±0.10	CE	107
		HF^-		
HF^-	$^2\Sigma^+$	<0	HF	108
		HF_2^-		
HF_2^-		3	HF	109

6.2. The Negative Ion Table—Continued

Ion	State	Electron affinity (eV)	Method	Ref.
CF^-				
CF^-	$^3\Sigma$	1.06 ± 0.20	HF	110
		$\geq 3.3 \pm 1.1$	DEC	111
C_2F^-				
C_2F^-		$\geq 3.4 \pm 0.8$	DEC	39
CF_2^-				
CF_2^-		≥ 0.2	DEC	111
CF_3^-				
CF_3^-		2.1 ± 0.3	DEC	111
$C_2F_3^-$				
$C_2F_3^-$		2.0 ± 0.2	DEC	111
$C_2F_5^-$				
$C_2F_5^-$		2.1	DEC	112
$C_3F_5^-$				
$C_3F_5^-$		2.7	DEC	112
$C_3F_7^-$				
$C_3F_7^-$		2.3	DEC	112
$C_5F_9^-$				
$C_5F_9^-$		3.1	RPD	113
$C_6F_{11}^-$				
$C_6F_{11}^-$		3.5	RPD	113
$C_7F_{13}^-$				
$C_7F_{13}^-$		3.9	RPD	113

6.2. The Negative Ion Table—Continued

Ion	State	Electron affinity (eV)	Method	Ref.
		NF⁻		
NF⁻	$^2\Pi$	0.6±0.5	HF	114
		OF⁻		
OF⁻	$^1\Sigma$	<1.4	HF	175
	Na⁻($^1S_0$) $\quad \Delta H^\circ_{f0} = 55.6$ kJ mol⁻¹ (13.3 kcal mol⁻¹)			
Na⁻	$3s^2\ ^1S_0$	0.543±0.010	LP	171
		0.542±0.100	LP	115
		0.54±0.10	HF	15
		0.541±0.004	CI	116
		0.54	CI	18
		0.78±0.06	HF	118
		$0.41^{+0.06}_{-0.02}$	CE	119
		0.22	HA	21
		0.47	IE	23
		NaO⁻		
NaO⁻		1.1±0.5	HF	117
		Mg⁻		
Mg⁻	$3s^2\ 4s\ ^2S_{1/2}$	<0	HF	15
		−0.22	HA	25
	$3s\ 3p^2\ ^4P\ (\to 3s\ 3p\ ^3P^\circ)$	0.32±0.10	HF	15
		>0	OBS	26
	$3s^2\ 3p\ ^2P^\circ_{1/2}$	−0.52	HA	25
		−0.69	HA	21
		−0.61	HA	27
		−0.32	IE	23
		MgH⁻		
MgH⁻		1.08	AA	28

6.2. The Negative Ion Table—Continued

Ion	State	Electron affinity (eV)	Method	Ref.
		Al⁻		
Al⁻	$3p^2\ ^3P_0$	0.52	HF	118
		0.20±0.20	HA	25
		0.27	HA	21
		0.28	HA	27
		0.52	IE	23
		>0	OBS	120
		>0	OBS	32
	$3p^2\ ^1D_2$	0.23	HF	118
		AlH⁻		
AlH⁻		0.04	AA	28
		AlH₂⁻		
AlH₂⁻		2.12	AA	28
		AlO⁻		
AlO⁻		3.68±0.13	SI	121
		2.60	AA	28
		AlO₂⁻		
AlO₂⁻		4.11±0.13	SI	121
		Si⁻		
Si⁻	$3p^3\ ^4S^\circ_{3/2}$	1.39	HF	118
		1.36±0.20	HA	25
		1.40	HA	21
		1.36	HA	27
		1.46	IE	23
		>0	OBS	122
	$3p^3\ ^2D^\circ$	0.56±0.04	P	51
		0.58	HF	118
	$3p^3\ ^2P^\circ$	0.08	HF	118

6.2. The Negative Ion Table—Continued

Ion	State	Electron affinity (eV)	Method	Ref.
SiH⁻				
SiH⁻	$^3\Sigma^-$	1.46±0.20	HF	40
SiH₂⁻				
SiH₂⁻		3.47	AA	28
SiH₃⁻				
SiH₃⁻		2.73	AA	28
Si₂H⁻				
Si₂H⁻		4.08	AA	28
Si₂H₃⁻				
Si₂H₃⁻		3.17	AA	28
SiF⁻				
SiF⁻	$^3\Sigma$	1.0±0.2	HF	110
P⁻				
P⁻	$3p^4\ {}^3P_2$	0.78	HF	118
		0.71±0.20	HA	25
		0.62	HA	21
		0.72	HA	27
		0.77	IE	23
		>0	OBS	32
	$3p^4\ {}^1D_2$	0.01	HF	118
PH⁻				
PH⁻	$^2\Pi_i$	0.93±0.20	HF	40

PH₂⁻ $\Delta H^\circ_{f0} = 7$ kJ mol⁻¹ (2 kcal mol⁻¹)

Ion	State	Electron affinity (eV)	Method	Ref.
PH₂⁻		1.25±0.03	P	123
		1.26±0.03	LP	50
		>0	OBS	124
		1.60	AA	28

6.2. The Negative Ion Table—Continued

Ion	State	Electron affinity (eV)	Method	Ref.
		PO⁻		
PO^-	$^3\Sigma^-$	<1.13	HF	125
		PF⁻		
PF^-	$^2\Pi$	≤1.4±0.3	HF	114
	S⁻($^2P^\circ_{3/2}$) $\Delta H^\circ_{f0} = 76$ kJ mol⁻¹ (18 kcal mol⁻¹)			
S^-	$3p^5\ ^2P^\circ_{3/2}$	2.0772±0.0005	LP	126
		2.095±0.015	P	127
		2.07±0.07	P	32
		2.090±0.065	SI	128
		2.12	HF	118
		2.04±0.20	HA	25
		2.03	HA	21
		2.08	HA	27
		2.15	IE	23
		S₂⁻		
S_2^-	$^2\Pi_g$	>2.0	ATT	129
	HS⁻($^1\Sigma^+$) $\Delta H^\circ_{f0} \sim -81$ kJ mol⁻¹ (−19 kcal mol⁻¹)			
HS^-	$^1\Sigma^+$	2.319±0.010	P	131
		2.28±0.15	P	132
		2.25±0.10	HF	75
		2.30±0.04	T	128
		H₂S⁻		
H_2S^-		1.11±0.09	SI	128
	SO⁻($^2\Pi$) $\Delta H^\circ_{f0} = -102.4$ kJ mol⁻¹ (−24.5 kcal mol⁻¹)			
SO^-	$^2\Pi$	1.126±0.013	PE	171
		≤1.09±0.05	P	3
		>1.0; <1.27	CE	176
		SO₂⁻		
SO_2^-		<1.0±0.1	P	3
		>1.0; <1.27	CE	176

6.2. The Negative Ion Table—Continued

Ion	State	Electron affinity (eV)	Method	Ref.
		SF⁻		
SF⁻	$^1\Sigma$	≤2.5±0.5	HF	133
		SF$_3^-$		
SF$_3^-$		2.95±0.05	DEC	134
		SF$_5^-$		
SF$_5^-$		3.66±0.04	SI	135
		SF$_6^-$		
SF$_6^-$		>0.432; <1.470	CE	127
		1.49±0.22	SI	135
		1.1	ATT	174
		CNS⁻		
CNS⁻		2.164±0.022	SI	170
		1.99	SI	53

Cl⁻($^1S_0$) $\Delta H_{f0}^\circ = -228.6$ kJ mol⁻¹ (-54.6 kcal mol⁻¹)

Ion	State	Electron affinity (eV)	Method	Ref.
Cl⁻	$3p^6\ ^1S_0$	3.613±0.003	S	136
		3.616±0.003	S	137
		3.628±0.005	S	138
		3.56	HF	118
		3.70	IE	23
		Cl$_2^-$		
Cl$_2^-$		2.38±0.10	CE	107
		2.45±0.15	CE	65
		2.46±0.14	CE	96
		<1.7	ATT	139
		2.54	DE	140
		CCl$_3^-$		
CCl$_3^-$		>2.10±0.35	ATT	141
		1.6±0.2	ATT	142
		1.44±0.05	SI	143

6.2. The Negative Ion Table—Continued

Ion	State	Electron affinity (eV)	Method	Ref.
		CCl_4^-		
CCl_4^-		2.12 ± 0.10	SI	143
		ClO^-		
ClO^-	$^1\Sigma$	$\leq 2.2\pm0.5$	HF	144
		$SiCl_2^-$		
$SiCl_2^-$		≥ 2.6	ATT	145
		$CHCl_3^-$		
$CHCl_3^-$		1.76 ± 0.05	SI	143
$K^-(^1S_0)$		**$\Delta H_{f0}^\circ = 41.7$ kJ mol^{-1} (10.0 kcal mol^{-1})**		
K^-	$4s^2\ ^1S_0$	0.5012 ± 0.0005	LP	171
		0.47 ± 0.10	HF	15
		0.90 ± 0.05	HF	146
		0.40 ± 0.20	HA	25
		$0.22^{+0.08}_{-0.06}$	CE	119
		0.5	IE	147
		0.51	CI	18
		Ca^-		
Ca^-	$3d\ 4s^2\ ^2D_{3/2}$	-1.62 ± 0.10	HA	148
		-1.88 ± 0.20	HA	25
		1.6	IE	147
		Sc^-		
Sc^-	$3d^2\ 4s^2\ ^3F_2$	-0.14 ± 0.10	HF	146
		-0.80 ± 0.10	HA	148
		-0.70 ± 0.20	HA	25
		-0.4	IE	147
		Ti^-		
Ti^-	$3d^3\ 4s^2\ ^4F_{3/2}$	0.39 ± 0.20	HF	146
		-0.11 ± 0.10	HA	148
		-0.03 ± 0.20	HA	25
		0.15	IE	147

6.2. The Negative Ion Table—Continued

Ion	State	Electron affinity (eV)	Method	Ref.
		V⁻		
V^-	$3d^4\ 4s^2\ {}^5D_0$	0.94 ± 0.25	HF	146
		0.51 ± 0.10	HA	148
		0.56 ± 0.20	HA	25
		0.65	IE	147
		Cr⁻		
Cr^-	$3d^5\ 4s^2\ {}^6S_{5/2}$	0.98 ± 0.35	HF	146
		0.85 ± 0.10	HA	148
		0.85 ± 0.20	HA	25
		0.85	IE	147
		Mn⁻		
Mn^-	$3d^6\ 4s^2\ {}^5D_4$	-1.07 ± 0.20	HF	146
		-1.19 ± 0.10	HA	148
		-1.07 ± 0.20	HA	25
		-1.2	IE	147
		Fe⁻		
Fe^-	$3d^7\ 4s^2\ {}^4F_{9/2}$	0.58 ± 0.20	HF	146
		0.14 ± 0.10	HA	148
		0.40 ± 0.20	HA	25
		0.1	IE	147
		Co⁻		
Co^-	$3d^8\ 4s^2\ {}^3F_4$	0.94 ± 0.15	HF	146
		0.65 ± 0.10	HA	148
		1.03 ± 0.20	HA	25
		0.7	IE	147
		Ni⁻		
Ni^-	$3d^9\ 4s^2\ {}^2D_{5/2}$	1.28 ± 0.20	HF	146
		1.13 ± 0.10	HA	148
		1.62 ± 0.20	HA	25
		1.1	IE	147

6.2. The Negative Ion Table—Continued

Ion	State	Electron affinity (eV)	Method	Ref.
		Cu⁻($^1S_0$)　　$\Delta H_{f0}^\circ = 218.9$ kJ mol⁻¹ (52.3 kcal mol⁻¹)		
Cu⁻	$3d^{10}\ 4s^2\ {}^1S_0$	1.226±0.010	PE	148
		1.80±0.10	HF	146
		0.7	CI	18
		1.80±0.20	HA	25
		1.5±0.5	T	149
		1.4	IE	147
		Zn⁻		
Zn⁻	$5s\ {}^2S_{1/2}$	0.09±0.20	HA	25
	$4p\ {}^2P^\circ_{1/2}$	−0.67±0.20	HA	25
		Ga⁻		
Ga⁻	$4p^2\ {}^3P_0$	0.37±0.20	HA	25
		0.50	VA	150
		>0	OBS	120
		Ge⁻		
Ge⁻	$4p^3\ {}^4S^\circ_{3/2}$	1.44±0.20	HA	25
		1.37	VA	150
		>0	OBS	122
		As⁻		
As⁻	$4p^4\ {}^3P_2$	1.07±0.20	HA	25
		0.74	VA	150
		AsH₂⁻		
AsH₂⁻		1.27±0.03	P	49
		Se⁻($^2P^\circ_{3/2}$)　　$\Delta H_{f0}^\circ = 31.4$ kJ mol⁻¹ (7.5 kcal mol⁻¹)		
Se⁻	$4p^5\ {}^2P^\circ_{3/2}$	2.0206±0.0004	LP	171
		2.0204±0.0004	LP	151
		2.12±0.20	HA	25
		2.11	VA	150
		5.4±2.0	ATT	152

6.2. The Negative Ion Table—Continued

Ion	State	Electron affinity (eV)	Method	Ref.
		Se_2^-		
Se_2^-		>0	OBS	153
		Se_3^-		
Se_3^-		>0	OBS	153
		Se_4^-		
Se_4^-		>0	OBS	153
		SeH^-		
SeH^-	$^1\Sigma^+$	2.21±0.03	P	154
		>0	OBS	124
		2.0	BH	152
		SeF^-		
SeF^-	$^1\Sigma$	≤2.8±0.5	HF	133
	$Br^-(^1S_0)$	$\Delta H^\circ_{f0} = -206.8$ kJ mol^{-1} (-49.4 kcal mol^{-1})		
Br^-	$4p^6\ ^1S_0$	3.366±0.003	S	155
		3.363±0.003	S	136
		3.378±0.005	S	138
		3.53±0.12	PI	156
		Br_2^-		
Br_2^-		2.51±0.10	CE	107
		2.55±0.10	CE	65
		>0	OBS	157
		~2.62	DE	140
		Rb^-		
Rb^-	$5s^2\ ^1S_0$	0.48	CI	18
		0.40±0.20	HA	25
		0.16±0.06	CE	119
		>0.20	PI	158
		0.6	IE	147

6.2. The Negative Ion Table—Continued

Ion	State	Electron affinity (eV)	Method	Ref.
		Sr⁻		
Sr⁻	$4d\ 5s^2\ ^2D_{3/2}$	-1.74 ± 0.10	HA	148
		-1.33 ± 0.20	HA	25
		-0.5	IE	147
		Y⁻		
Y⁻	$4d^2\ 5s^2\ ^3F_2$	-0.76 ± 0.10	HA	148
		-0.15 ± 0.20	HA	25
		0.3	IE	147
		Zr⁻		
Zr⁻	$4d^3\ 5s^2\ ^4F_{3/2}$	0.08 ± 0.10	HA	148
		0.62 ± 0.20	HA	25
		1.0	IE	147
		Nb⁻		
Nb⁻	$4d^4\ 5s^2\ ^5D_0$	0.77 ± 0.10	HA	148
		1.23 ± 0.20	HA	25
		1.3	IE	147
		Mo⁻		
Mo⁻	$4d^5\ 5s^2\ ^6S_{5/2}$	0.86 ± 0.10	HA	148
		1.0 ± 0.2	SI	159, 160
		1.15 ± 0.20	HA	25
		1.3	IE	147
		Tc⁻		
Tc⁻	$4d^6\ 5s^2\ ^5D_4$	0.63 ± 0.10	HA	148
		0.94 ± 0.20	HA	25
		1.0	IE	147
		Ru⁻		
Ru⁻	$4d^7\ 5s^2\ ^4F_{9/2}$	1.04 ± 0.10	HA	148
		1.49 ± 0.20	HA	25
		1.45	IE	147

6.2. The Negative Ion Table—Continued

Ion	State	Electron affinity (eV)	Method	Ref.
		Rh⁻		
Rh⁻	$4d^8\ 5s^2\ {}^3F_4$	1.12 ± 0.10	HA	148
		1.68 ± 0.20	HA	25
		1.35	IE	147
		Pd⁻		
Pd⁻	$4d^9\ 5s^2\ {}^2D_{5/2}$	0.40 ± 0.10	HA	148
		1.01 ± 0.20	HA	25
		1.4	IE	147
	Ag⁻ (${}^1S_0$) $\Delta H_{f0}^\circ = 158.4\ \text{kJ mol}^{-1}\ (37.9\ \text{kcal mol}^{-1})$			
Ag⁻	$4d^{10}\ 5s^2\ {}^1S_0$	$1.303^{+0.007}_{-0.011}$	PE	148
		<1.78	LP	161
		2.0 ± 0.2	SI	149
		2.00 ± 0.20	HA	25
		1.5	IE	147
		Cd⁻		
Cd⁻	$6s\ {}^2S_{1/2}$	-0.27 ± 0.20	HA	25
	$5p\ {}^2P_{1/2}^\circ$	-0.78 ± 0.20	HA	25
		In⁻		
In⁻	$5p^2\ {}^3P_0$	0.20 ± 0.20	HA	25
		0.72	VA	150
		>0	OBS	120
		Sn⁻		
Sn⁻	$5p^3\ {}^4S_{3/2}^\circ$	1.03 ± 0.20	HA	25
		1.47	VA	150
		>0	OBS	122
		Sb⁻		
Sb⁻	$5p^4\ {}^3P_2$	0.94 ± 0.20	HA	25
		0.61	VA	150
		>0	OBS	153
		Sb₂⁻		
Sb₂⁻		>0	OBS	153

6.2. The Negative Ion Table—Continued

Ion	State	Electron affinity (eV)	Method	Ref.
		Sb_3^-		
Sb_3^-		>0	OBS	153
		Te^-		
Te^-	$5p^5\ ^2P^\circ_{3/2}$	1.96 ± 0.20	HA	25
		2.21	VA	150
		>0	OBS	153
		Te_2^-		
Te_2^-		>0	OBS	153
	$I^-(^1S_0)$	$\Delta H^\circ_{f0} = -187.9\ kJ\ mol^{-1}\ (-44.9\ kcal\ mol^{-1})$		
I^-	$5p^6\ ^1S_0$	3.059 ± 0.002	P	162
		3.063 ± 0.003	S	136
		3.078 ± 0.005	S	138
		$\leqslant3.076\pm0.005$	P	163
		3.073 ± 0.014	PI	179
		3.13 ± 0.12	PI	156
		I_2^-		
I_2^-		2.58 ± 0.10	CE	107
		2.55 ± 0.05	CE	65
		>0	OBS	164
		2.42	DE	140
		I_3^-		
I_3^-		>0	OBS	164
		IBr^-		
IBr^-		2.7 ± 0.2	CE	107
		2.55 ± 0.10	CE	65
	$Cs^-(^1S_0)$	$\Delta H^\circ_{f0} = 32.8\ kJ\ mol^{-1}\ (7.8\ kcal\ mol^{-1})$		
Cs^-	$6s^2\ ^1S_0$	0.4705 ± 0.0015	LP	171
		0.40 ± 0.30	HA	25
		$0.13^{+0.07}_{-0.06}$	CE	119
		>0.19	PI	158

6.2. The Negative Ion Table—Continued

Ion	State	Electron affinity (eV)	Method	Ref.
Ba⁻				
Ba⁻	5d 6s² ²D₃/₂	−0.54±0.10	HA	148
		−0.48±0.30	HA	25
La⁻				
La⁻	5d² 6s² ³F₂	0.44±0.10	HA	148
		0.55±0.30	HA	25
Hf⁻				
Hf⁻	5d³ 6s² ⁴F₃/₂	−0.78±0.10	HA	148
		−0.63±0.30	HA	25
Ta⁻				
Ta⁻	5d⁴ 6s² ⁵D₀	0.8±0.3	SI	159
		−0.05±0.10	HA	148
		0.15±0.30	HA	25
W⁻				
W⁻	5d⁵ 6s² ⁶S₅/₂	0.5±0.3	SI	159, 165, 166
		0.98±0.10	HA	148
		1.23±0.30	HA	25
WO₃⁻				
WO₃⁻		3.6	T	167
HWO₄⁻				
HWO₄⁻		4.4	T	167
Re⁻				
Re⁻	5d⁶ 6s² ⁵D₄	0.15±0.10	SI	159, 165
		0.09±0.10	HA	148
		0.38±0.30	HA	25
Os⁻				
Os⁻	5d⁷ 6s² ⁴F₉/₂	1.10±0.10	HA	148
		1.44±0.30	HA	25

6.2. The Negative Ion Table—Continued

Ion	State	Electron affinity (eV)	Method	Ref.

Ir⁻

Ion	State	Electron affinity (eV)	Method	Ref.
Ir⁻	$5d^8\ 6s^2\ ^3F_4$	1.58 ± 0.10	HA	148
		1.97 ± 0.30	HA	25

Pt⁻($^2D_{5/2}$) $\Delta H_{f0}^\circ = 359.1$ kJ mol⁻¹ (85.8 kcal mol⁻¹)

Ion	State	Electron affinity (eV)	Method	Ref.
Pt⁻	$5d^9\ 6s^2\ ^2D_{5/2}$	2.128 ± 0.002	LP	161
		2.12 ± 0.10	HA	148
		2.56 ± 0.30	HA	25

PtN⁻

Ion	State	Electron affinity (eV)	Method	Ref.
PtN⁻		>0	OBS	161

Au⁻($^1S_0$) $\Delta H_{f0}^\circ = 143.2$ kJ mol⁻¹ (34.2 kcal mol⁻¹)

Ion	State	Electron affinity (eV)	Method	Ref.
Au⁻	$5d^{10}\ 6s^2\ ^1S_0$	2.3086 ± 0.0007	LP	161
		2.8 ± 0.1	SI	149
		2.80 ± 0.30	HA	25

Hg⁻

Ion	State	Electron affinity (eV)	Method	Ref.
Hg⁻	$7s\ ^2S_{1/2}$	-0.19 ± 0.30	HA	25
	$6p\ ^2P^\circ_{1/2}$	-0.67 ± 0.30	HA	25

Tl⁻

Ion	State	Electron affinity (eV)	Method	Ref.
Tl⁻	$6p^2\ ^3P_0$	0.5 ± 0.1	ATT	168
		0.32 ± 0.30	HA	25
		1.21	VA	150
		>0	OBS	120

Pb⁻

Ion	State	Electron affinity (eV)	Method	Ref.
Pb⁻	$6p^3\ ^4S^\circ_{3/2}$	1.03 ± 0.30	HA	25
		1.79	VA	150
		>0	OBS	122

Bi⁻

Ion	State	Electron affinity (eV)	Method	Ref.
Bi⁻	$6p^4\ ^3P_2$	0.95 ± 0.30	HA	25
		-0.34	VA	150
		>0	OBS	153

6.2. The Negative Ion Table—Continued

Ion	State	Electron affinity (eV)	Method	Ref.
		Bi_2^-		
Bi_2^-		>0	OBS	153
		Bi_3^-		
Bi_3^-		>0	OBS	153
		Po^-		
Po^-	$6p^5\ {}^2P^o_{3/2}$	1.32 ± 0.30	HA	25
		1.97	VA	150

6.3. Bibliography

[1] Pekeris, C. L., $1\,{}^1S$, $2\,{}^1S$, and $2\,{}^3S$ states of H^- and of He, Phys. Rev. **126**, 1470 (1962).

[2] Berry, R. S., Small free negative ions, Chem. Rev. **69**, 533 (1969).

[3] Feldmann, D., Photoablösung von Elektronen bei einigen stabilen negativen Ionen, Z. Naturforsch. **25a**, 621 (1970).

[4] Drake, G. W. F., Second bound state for the hydrogen negative ion, Phys. Rev. Letters **24**, 126 (1970).

[5] Sharp, T. E., Potential energy diagram for molecular hydrogen and its ions, Lockheed Report LMSC 5–10–69–9 (1969).

[6] Taylor, H. S., The ground state of H_2^-, Proc. Phys. Soc. (London) **90**, 877 (1967).

[7] Somerville, W. B., The ground state of H_2^-, Proc. Phys. Soc. (London) **89**, 185 (1966).

[8] Taylor, H. S., and Harris, F. E., Potential curve for the ${}^2\Sigma_u^+$ state of H_2^-, J. Chem. Phys. **39**, 1012 (1963).

[9] Dalgarno, A., and McDowell, M. R. C., Charge transfer and the mobility of H^- ions in atomic hydrogen, Proc. Phys. Soc. (London) **A69**, 615 (1956).

[10] Carter, E. B., and Davis, R. H., He^-, H_2^-, and other negative ion beams available from a duoplasmatron ion source with gas charge exchange, Rev. Sci. Instr. **34**, 93 (1963).

[11] Khvostenko, V. I., and Dukel'skii, V. M., The negative ion H_2^-, Zh. Eksperim. i Teor. Fiz. **34**, 1026 (1958) [Engl. transl.: Soviet Phys. JETP **7**, 709 (1958)].

[12] Brehm, B., Gusinow, M. A., and Hall, J. L., Electron affinity of helium via laser photodetachment of its negative ion, Phys. Rev. Letters **19**, 737 (1967).

[13] Smirnov, B. M., and Chibisov, M. I., The breaking up of atomic particles by an electric field and by electron collisions, Zh. Eksperim. i Teor. Fiz. **49**, 841 (1965) [Engl. transl.: Soviet Phys. JETP **22**, 585 (1966)].

[14] Holøien, E., and Geltman, S., Variational calculations for quartet states of three–electron atomic systems, Phys. Rev. **153**, 81 (1967).

[15] Weiss, A. W., Theoretical electron affinities for some of the alkali and alkaline–earth elements, Phys. Rev. **166**, 70 (1968).

[16] Clementi, E., and McLean, A. D., Atomic negative ions, Phys. Rev. **133**, A419 (1964).

[17] Victor, G. A., and Laughlin, C., Model potential calculation of the electron affinity of lithium, Chem. Phys. Letters **14**, 74 (1972).

[18] Schwarz, W. H. E., Electron affinities of the alkali atoms. A model–potential calculation, Chem. Phys. Letters **10**, 478 (1971).

[19] Grün, N., Genäherte natürliche Orbitale für die Grundzustände von Li und Li^-; Elektronenaffinität von Li^-, Z. Naturforsch. **27a**, 843 (1972).

[20] Scheer, M. D., and Fine, J., Electron affinity of lithium, J. Chem. Phys. **50**, 4343 (1969).

[21] Crossley, R. J. S., Glockler's equation for ionization potentials and electron affinities: II, Proc. Phys. Soc. (London) **83**, 375 (1964).

[22] Edlén, B., Ionization potentials of atoms and ions containing one to ten electrons, in: Topics in Modern Physics, ed. W. E. Brittin and H. Odabasi (Colo. Assoc. Univ. Press, Boulder, Colo., 1971) p. 133.

[23] Edlén, B., Isoelectronic extrapolation of electron affinities, J. Chem. Phys. **33**, 98 (1960).

[24] Ya'akobi, B., An experimental estimate of the electron affinity of lithium, Phys. Letters **23**, 655 (1966).

[25] Zollweg, R. J., Electron affinities of the heavy elements, J. Chem. Phys. **50**, 4251 (1969).

[26] Bethge, K., Heinicke, E., and Baumann, H., On the existence of negative beryllium and magnesium ions, Phys. Letters **23**, 542 (1966).

[27] Edie, J. W., and Rohrlich, F., Negative atomic ions. II, J. Chem. Phys. **36**, 623 (1962); Erratum, ibid., **37**, 1151 (1962).

[28] Gaines, A. F., and Page, F. M., Semi–empirical prediction of the electron affinities of gaseous radicals, Trans. Faraday Soc. **62**, 3086 (1966).

[29] Joshi, B. D., Study of BeH_3^-, BH_3, CH_3^+, NH_3^{++}, and OH_3^{3+} by one–center–expansion, self–consistent–field method, J. Chem. Phys. **46**, 875 (1967).

[30] Moser, C. M., and Nesbet, R. K., Atomic Bethe–Goldstone calculations of term splittings, ionization potentials, and electron affinities for B, C, N, O, F, and Ne, Phys. Rev. A **4**, 1336 (1971).

[31] Schaefer, H. F., III, Klemm, R. A., and Harris, F. E., First–order wavefunctions, orbital correlation energies, and electron affinities of first–row atoms, J. Chem. Phys. **51**, 4643 (1969).

[32] Branscomb, L. M., and Smith, S. J., Electron affinity of atomic sulfur and empirical affinities of the light elements, J. Chem. Phys. **25**, 598 (1956).

J. Phys. Chem. Ref. Data, Vol. 6, Suppl. 1, 1977

[33] Hall, J. L., and Siegel, M. W., Angular dependence of the laser photodetachment of the negative ions of carbon, oxygen, and hydrogen, J. Chem. Phys. **48**, 943 (1968).

[34] Seman, M. L., and Branscomb, L. M., Structure and photodetachment spectrum of the atomic carbon negative ion, Phys. Rev. **125**, 1602 (1962).

[35] Öksüz, İ., and Sinanoğlu, O., Theory of atomic structure including electron correlation. II. All–external pair correlations in the various states and ions of B, C, N, O, F, Ne, and Na, and prediction of electron affinities and atomic excitation energies, Phys. Rev. **181**, 54 (1969).

[36] Honig, R. E., Mass spectrometric study of the molecular sublimation of graphite, J. Chem. Phys. **22**, 126 (1954).

[37] Hall, J. L., private communication.

[38] Locht, R., and Momigny, J., Mass spectrometric determination of the electronaffinities of radicals, Chem. Phys. Letters **6**, 273 (1970).

[39] Thynne, J. C. J., and MacNeil, K. A. G., Negative ion formation by ethylene and 1,1–difluoroethylene, J. Phys. Chem. **75**, 2584 (1971).

[40] Cade, P. E., The electron affinities of the diatomic hydrides CH, NH, SiH and PH, Proc. Phys. Soc. (London) **91**, 842 (1967).

[41] Sklar, A. L., The near ultraviolet absorption of substituted benzenes, J. Chem. Phys. **7**, 984 (1939).

[42] di Domenico, A., Harland, P. W., and Franklin, J. L., Negative ion formation and negative ion–molecule reactions in cyclopentadiene, J. Chem. Phys. **56**, 5299 (1972).

[43] Fogel', Ya. M., Kozlov, V. F., and Kalmykov, A. A., On the existence of the negative nitrogen ion, Zh. Eksperim. i Teor. Fiz. **36**, 1354 (1959) [Engl. transl.: Soviet Phys. JETP **36**, 963 (1959)].

[44] Gilmore, F. R., Potential energy curves for N_2, NO, O_2 and corresponding ions, J. Quant. Spectry. Radiative Transfer **5**, 369 (1965).

[45] Comer, J., and Read, F. H., Electron impact studies of a resonant state of N_2^-, J. Phys. B **4**, 1055 (1971).

[46] Archibald, T. W., and Sabin, J. R., Theoretical investigation of the electronic structure and properties of N_3^-, N_3, and N_3^+, J. Chem. Phys. **55**, 1821 (1971).

[47] Gray, P., and Waddington, T. C., Thermochemistry and reactivity of the azides. II. Lattice energies of ionic azides, electron affinity and heat of formation of the azide radical and related properties, Proc. Roy. Soc. (London) **A235**, 481 (1956).

[48] Buchner, E. H., The azide ion in the lyotropic series, Rec. Trav. Chim. **69**, 329 (1950).

[49] Smyth, K. C., and Brauman, J. I., Photodetachment of electrons from amide and arsenide ions: the electron affinities of $NH_2 \cdot$ and $AsH_2 \cdot$, J. Chem. Phys. **56**, 4620 (1972).

[50] Smyth, K. C., McIver, R. T., Jr., Brauman, J. I., and Wallace, R. W., Photodetachment of negative ions using a continuously tunable laser and an ion cyclotron resonance spectrometer, J. Chem. Phys. **54**, 2758 (1971).

[51] Feldmann, D., Photoablösung von Elektronen bei Si^- und NH_2^-, Z. Naturforsch. **26a**, 1100 (1971).

[52] Dorman, F. H., Negative fragment ions from resonance capture processes, J. Chem. Phys. **44**, 3856 (1966).

[53] Page, F. M., Experimental determination of the electron affinities of inorganic radicals, Advan. Chem. Ser. **36**, 68 (1962).

[54] Berkowitz, J., Chupka, W. A., and Walter, T. A., Photoionization of HCN: the electron affinity and heat of formation of CN, J. Chem. Phys. **50**, 1497 (1969).

[55] Page, F. M., Comment on the paper by V. H. Dibeler and S. K. Liston, J. Chem. Phys. **49**, 2466 (1968).

[56] Branscomb, L. M., Burch, D. S., Smith, S. J., and Geltman, S., Photodetachment cross section and the electron affinity of atomic oxygen, Phys. Rev. **111**, 504 (1958).

[57] Berry, R. S., Mackie, J. C., Taylor, R. L., and Lynch, R., Spin–orbit coupling and electron–affinity determinations from radiative capture of electrons by oxygen atoms, J. Chem. Phys. **43**, 3067 (1965).

[58] Elder, F. A., Villarejo, D., and Inghram, M. G., Electron affinity of oxygen, J. Chem. Phys. **43**, 758 (1965).

[59] Chantry, P. J., and Schulz, G. J., Kinetic–energy distribution of negative ions formed by dissociative attachment and the measurement of the electron affinity of oxygen, Phys. Rev. **156**, 134 (1967).

[60] Henderson, W. R., Fite, W. L., and Brackmann, R. T., Dissociative attachment of electrons to hot oxygen, Phys. Rev. **183**, 157 (1969).

[61] Edwards, A. K., Risley, J. S., and Geballe, R., Two autodetaching states of O^-, Phys. Rev. A **3**, 583 (1971).

[62] Celotta, R. J., Bennett, R. A., Hall, J. L., Siegel, M. W., and Levine, J., Molecular photodetachment spectrometry. II. The electron affinity of O_2 and the structure of O_2^-, Phys. Rev. A **6**, 631 (1972).

[63] Pack, J. L., and Phelps, A. V., Electron attachment and detachment. I. Pure O_2 at low energy, J. Chem. Phys. **44**, 1870 (1966).

[64] Tiernan, T. O., Hughes, B. M., and Lifshitz, C., Electron affinities from endothermic negative–ion charge–transfer reactions. II. O_2, J. Chem. Phys. **55**, 5692 (1971).

[65] Baede, A. P. M., The adiabatic electron affinities of Cl_2, Br_2, I_2, IBr, NO_2 and O_2, Physica **59**, 541 (1972).

[66] Berkowitz, J., Chupka, W. A., and Gutman, D., Electron affinities of O_2, O_3, NO, NO_2, NO_3 by endothermic charge transfer, J. Chem. Phys. **55**, 2733 (1971).

[67] Zemke, W. T., Das, G., and Wahl, A. C., Theoretical determination of the electron affinity of O_2 molecule from the binding energy of O_2^-, Chem. Phys. Letters **14**, 310 (1972).

[68] Burch, D. S., Smith, S. J., and Branscomb, L. M., Photodetachment of O_2^-, Phys. Rev. **112**, 171 (1958).

[69] Samson, J. A. R., and Cairns, R. B., Ionization potential of O_2, J. Opt. Soc. Am. **56**, 769 (1966).

[70] Wong, S. F., Vorburger, T. V., and Woo, S. B., Photodetachment of O_3^-, Ann. Gaseous Electron. Conf., 23rd, Hartford, 1970, Abstr. Pap., p. 82.

[71] Wood, R. H., and D'Orazio, L. A., Thermodynamics of the higher oxides. III. The lattice energy of potassium ozonide and the electron affinity of ozone, J. Phys. Chem. **69**, 2562 (1965).

[72] Conway, D. C., and Nesbitt, L. E., Stability of O_4^-, J. Chem Phys. **48**, 509 (1968).

[73] Hotop, H., Patterson, T. A., and Lineberger, W. C., Dye–laser photodetachment of OH^- and OD^-, Ann. Gaseous Electron. Conf., 25th, London, Ont., 1972, Abstr. Pap.

[74] Branscomb, L. M., Photodetachment cross section, electron affinity, and structure of the negative hydroxyl ion, Phys. Rev. **148**, 11 (1966).

[75] Cade, P. E., Hartree–Fock wavefunctions, potential curves, and molecular properties for $OH^-(^1\Sigma^+)$ and $SH^-(^1\Sigma^+)$, J. Chem. Phys. **47**, 2390 (1967).

[76] Weiss, J., The electron affinity of the radicals HO_2 and OH, and the oxygen molecule, Trans. Faraday Soc. **31**, 966 (1935).

[77] Moruzzi, J. L., and Phelps, A. V., Survey of negative–ion–molecule reactions in O_2, CO_2, H_2O, CO, and mixtures of these gases at high pressures, J. Chem. Phys. **45**, 4617 (1966).

[78] Golub, S., and Steiner, B., Photodetachment of $[OH(H_2O)]^-$, J. Chem. Phys. **49**, 5191 (1968).

[79] De Paz, M., Giardini, A. G., and Friedman, L., Tandem–mass–spectrometer study of solvated derivatives of OD^-. Total hydration energy of the proton, J. Chem. Phys. **52**, 687 (1970).

[80] Srivastava, R. D., Uy, O. M., and Farber, M., Effusion–mass spectrometric study of the thermodynamic properties of BO^- and BO_2^-, Trans. Faraday Soc. **67**, 2941 (1971).

[81] Jensen, D. E., Electron attachment and compound formation in flames. II. Mass spectrometry of boron–containing flames, J. Chem. Phys. **52**, 3305 (1970).

[82] Rempt, R. D., Electron–impact excitation of carbon monoxide near threshold in the 1.5– to 3–eV incident energy range, Phys. Rev. Letters **22**, 1034 (1969).

[83] Bardsley, J. N., Negative ions of N_2O and CO_2, J. Chem. Phys. **51**, 3384 (1969).

[84] Pack, J. L., and Phelps, A. V., Electron attachment and detachment. II. Mixtures of O_2 and CO_2 and of O_2 and H_2O, J. Chem. Phys. **45**, 4316 (1966).

[85] Siegel, M. W., Celotta, R. J., Hall, J. L., Levine, J., and Bennett, R. A., Molecular photodetachment spectrometry. I. The electron affinity of nitric oxide and the molecular constants of NO^-, Phys. Rev. A **6**, 607 (1972).

[86] Fehsenfeld, F. C., Ferguson, E. E., and Schmeltekopf, A. L., Thermal–energy associative–detachment reactions of negative ions, J. Chem. Phys. **45**, 1844 (1966).

[87] Farragher, A. L., Page, F. M., and Wheeler, R. C., Electron affinities of the oxides of nitrogen, Discussions Faraday Soc. **37**, 203 (1964).

[88] Williams, J. M., and Hamill, W. H., Ionization potentials of molecules and free radicals and appearance potentials by electron impact in the mass spectrometer, J. Chem. Phys. **49**, 4467 (1968).

[89] Wentworth, W. E., Chen, E., and Freeman, R., Thermal electron attachment to nitrous oxide, J. Chem. Phys. **55**, 2075 (1971).

[90] Paulson, J. F., Some negative ion reactions in simple gases, Advan. Chem. Ser. **58**, 28 (1966).

[91] Ferguson, E. E., The N_2O^- potential curves arising from $O^-(^2P)+N_2(X^1\Sigma_g^+)$, Bull. Am. Phys. Soc. **14**, 266 (1969).

[92] Ferguson, E. E., Fehsenfeld, F. C., and Schmeltekopf, A. L., Geometrical considerations for negative ion processes, J. Chem. Phys. **47**, 3085 (1967).

[93] Chantry, P. J., Temperature dependence of dissociative attachment in N_2O, J. Chem. Phys. **51**, 3369 (1969).

[94] Chantry, P. J., Formation of N_2O^- via ion–molecule reactions in N_2O, J. Chem. Phys. **51**, 3380 (1969).

[95] Warneck, P., Photodetachment of NO_2^-, Chem. Phys. Letters **3**, 532 (1969).

[96] Dunkin, D. B., Fehsenfeld, F. C., and Ferguson, E. E., Thermal energy rate constants for the reactions $NO_2^-+Cl_2\rightarrow Cl_2^-+NO_2$, $Cl_2^-+NO_2\rightarrow Cl^-+NO_2Cl$, $SH^-+NO_2\rightarrow NO_2^-+SH$, $SH^-+Cl_2\rightarrow Cl_2^-+SH$, and $S^-+NO_2\rightarrow NO_2^-+S$, Chem. Phys. Letters **15**, 257 (1972).

[97] Ferguson, E. E., Dunkin, D. B., and Fehsenfeld, F. C., Reactions of NO_2^- and NO_3^- with HCl and HBr, J. Chem. Phys. **57**, 1459 (1972).

[98] Curran, R. K., Formation of NO_2^- by charge transfer at very low energies, Phys. Rev. **125**, 910 (1962).

[99] McFarland, M., Dunkin, D. B., Fehsenfeld, F. C., Schmeltekopf, A. L., and Ferguson, E. E., Collisional detachment studies of NO^-, J. Chem. Phys. **56**, 2358 (1972).

[100] Jenkins, H. D. B., and Waddington, T. C., Enthalpy of formation, ΔH_f° $NO_3^-(g)$ and solvation, ΔH_{solv}° $NO_3^-(g)$, of the gaseous nitrate ion. Charge distribution in the NO_3^- ion and lattice energies of $LiNO_3$, $NaNO_3$ and KNO_3, J. Inorg. Nucl. Chem. **34**, 2465 (1972).

[101] Popp, H.–P., Quantitative Ausmessung des Fluor–Affinitätskontinuums, Z. Naturforsch. **22a**, 254 (1967).

[102] Popp, H.–P., Nachweis des Fluor–Affinitätskontinuums in Emission, Z. Naturforsch. **20a**, 642 (1965).

[103] Milstein, R., and Berry, R. S., On the electron affinity of fluorine, J. Chem. Phys. **55**, 4146 (1971).

[104] Chupka, W. A., and Berkowitz, J., Kinetic energy of ions produced by photoionization of HF and F_2, J. Chem. Phys. **54**, 5126 (1971).

[105] Dibeler, V. H., Walker, J. A., and McCulloh, K. E., Photoionization study of the dissociation energy of fluorine and the heat of formation of hydrogen fluoride, J. Chem. Phys. **51**, 4230 (1969).

[106] Weiss, A. W., Symmetry–adapted pair correlations in Ne, F^-, Ne^+, and F, Phys. Rev. A **3**, 126 (1971).

[107] Chupka, W. A., Berkowitz, J., and Gutman, D., Electron affinities of halogen diatomic molecules as determined by endoergic charge transfer, J. Chem. Phys. **55**, 2724 (1971).

[108] Bondybey, V., Pearson, P. K., and Schaefer, H. F., III, Theoretical potential energy curves for OH, HF^+, HF, HF^-, NeH^+, and NeH, J. Chem. Phys. **57**, 1123 (1972).

[109] Noble, P. N., and Kortzeborn, R. N., LCAO–MO–SCF studies of HF_2^- and the related unstable systems HF_2^0 and HeF_2, J. Chem. Phys. **52**, 5375 (1970).

[110] O'Hare, P. A. G., and Wahl, A. C., Molecular orbital investigation of CF and SiF and their positive and negative ions, J. Chem. Phys. **55**, 666 (1971).

[111] Thynne, J. C. J., and MacNeil, K. A. G., Ionisation of tetrafluoroethylene by electron impact, Intern. J. Mass Spectrom. Ion Phys. **5**, 329 (1970).

[112] Harland, P. W., and Thynne, J. C. J., Dissociative electron capture in perfluoropropylene and perfluoropropane, Intern. J. Mass Spectrom. Ion Phys. **9**, 253 (1972).

[113] Lifshitz, C., Peers, A. M., Grajower, R., and Weiss, M., Breakdown curves for polyatomic negative ions, J. Chem. Phys. **53**, 4605 (1970).

[114] O'Hare, P. A. G., and Wahl, A. C., Quantum–chemical study of some pnicogen monofluorides, J. Chem. Phys. **54**, 4563 (1971).

[115] Patterson, T. A., Hotop, H., and Lineberger, W. C., Photodetachment of Na^-, Bull. Am. Phys. Soc. **17**, 1128 (1972).

[116] Norcross, D. W., and Moores, D. L., Photodetachment of Li^- and Na^-, in: Atomic Physics, Vol. 3, ed. S. J. Smith and G. K. Walters (Plenum Press, New York, 1972) p. 261.

[117] O'Hare, P. A. G., and Wahl, A. C., Thermochemical and theoretical investigations of the sodium–oxygen system. II. Properties of NaO and its ions from Hartree–Fock molecular orbital studies, J. Chem. Phys. **56**, 4516 (1972).

[118] Clementi, E., McLean, A. D., Raimondi, D. L., and Yoshimine, M., Atomic negative ions. Second period, Phys. Rev. **133**, A1274 (1964).

[119] Bydin, Yu. F., Resonance charge exchange of negative alkali metal ions, Zh. Eksperim. i Teor. Fiz. **46**, 1612 (1964) [Engl. transl.: Soviet Phys. JETP **19**, 1091 (1964)].

[120] Khvostenko, V. I., and Sultanov, A. Sh., Formation of negative aluminum, gallium, indium and thallium ions on the capture of electrons by molecules of halides of these elements, Zh. Eksperim. i Teor. Fiz. **46**, 1605 (1964) [Engl. transl.: Soviet Phys. JETP **19**, 1086 (1964)].

[121] Srivastava, R. D., Uy, O. M., and Farber, M., Effusion–mass spectrometric study of the thermodynamic properties of AlO^- and AlO_2^-, J. Chem. Soc. Faraday Trans. II **68**, 1388 (1972).

[122] Dukel'skii, V. M., and Sokolov, V. M., Negative ions of silicon, germanium, tin and lead, Zh. Eksperim. i Teor. Fiz. **32**, 394 (1957) [Engl. transl.: Soviet Phys. JETP **5**, 306 (1957)].

[123] Smyth, K. C., and Brauman, J. I., Photodetachment of electrons from phosphide ion; the electron affinity of PH_2, J. Chem. Phys. **56**, 1132 (1972).

[124] Neuert, H., and Clasen, H., Massenspektrometrische Untersuchung von SH_2, SeH_2, PH_3, SiH_4 und GeH_4, Z. Naturforsch. **7a**, 410 (1952).

[125] Boyd, D. B., and Lipscomb, W. N., Molecular SCF calculations on PH_3, PO, PO^-, and P_2, J. Chem. Phys. **46**, 910 (1967).

[126] Lineberger, W. C., and Woodward, B. W., High resolution photodetachment of S⁻ near threshold, Phys. Rev. Letters **25**, 424 (1970).

[127] Steiner, B., Photodetachment of S⁻: behavior of the cross section above threshold, Intern. Conf. Phys. Electron. At. Collisions, 6th, Cambridge, Mass., 1969, Abstr. Pap., p. 535.

[128] Ansdell, D. A., and Page, F. M., Determination of electron affinities. Part 4.—Electron affinities of the sulphydryl radical and the sulphur atom, Trans. Faraday Soc. **58**, 1084 (1962).

[129] Jäger, K., and Henglein, A., Negative Ionen durch Elektronenstoss an einigen organischen Schwefelverbindungen, Z. Naturforsch. **21a**, 1251 (1966).

[130] Herman, K., and Giguère, P. A., Le spectre infrarouge de l'ozonide d'ammonium et la structure de l'ion O₃⁻, Can. J. Chem. **43**, 1746 (1965).

[131] Steiner, B., Photodetachment of electrons from SH⁻, J. Chem. Phys. **49**, 5097 (1968).

[132] Brauman, J. I., and Smyth, K. C., Photodetachment energies of negative ions by ion cyclotron resonance spectroscopy. Electron affinities of neutral radicals, J. Am. Chem. Soc. **91**, 7778 (1969).

[133] O'Hare, P. A. G., and Wahl, A. C., Hartree–Fock wavefunctions and computed properties for the ground (²Π) states of SF and SeF and their positive and negative ions. A comparison of the theoretical results with experiment, J. Chem. Phys. **53**, 2834 (1970).

[134] Harland, P. W., and Thynne, J. C. J., Autodetachment lifetimes, attachment cross sections, and negative ions formed by sulfur hexafluoride and sulfur tetrafluoride, J. Phys. Chem. **75**, 3517 (1971).

[135] Kay, J., and Page, F. M., Determination of electron affinities. Part 7.—Sulphur hexafluoride and disulphur decafluoride, Trans. Faraday Soc. **60**, 1042 (1964).

[136] Berry, R. S., and Reimann, C. W., Absorption spectrum of gaseous F⁻ and electron affinities of the halogen atoms, J. Chem. Phys. **38**, 1540 (1963).

[137] Mück, G., and Popp, H.–P., Quantitative Ausmessung des Chlor–Affinitätskontinuums, Z. Naturforsch. **23a**, 1213 (1968).

[138] Berry, R. S., Reimann, C. W., and Spokes, G. N., Absorption spectra of gaseous halide ions and halogen electron affinities: chlorine, bromine, and iodine, J. Chem. Phys. **37**, 2278 (1962).

[139] Baker, R. F., and Tate, J. T., Ionization and dissociation by electron impact in CCl₂F₂ and in CCl₄ vapor, Phys. Rev. **53**, 683 (1938).

[140] Geiger, W., Die Emission negativer Ladungsträger bei Einwirkung von Halogenmolekülen auf Alkalimetalloberflächen, Z. Physik **140**, 608 (1955).

[141] Curran, R. K., Positive and negative ion formation in CCl₃F, J. Chem. Phys. **34**, 2007 (1961).

[142] Reese, R. M., Dibeler, V. H., and Mohler, F. L., A survey of negative ions in mass spectra of polyatomic molecules, J. Res. NBS **57**, 367 (1956).

[143] Gaines, A. F., Kay, J., and Page, F. M., Determination of electron affinities. Part 8.—Carbon tetrachloride, chloroform and hexachloroethane, Trans. Faraday Soc. **62**, 874 (1966).

[144] O'Hare, P. A. G., and Wahl, A. C., Dissociation energy, ionization potential, electron affinity, dipole and quadrupole moments of chlorine monoxide (ClO, ²Π) from ab initio molecular orbital calculations, J. Chem. Phys. **54**, 3770 (1971).

[145] Vought, R. H., Molecular dissociation by electron bombardment: a study of SiCl₄, Phys. Rev. **71**, 93 (1947).

[146] Clementi, E., Atomic negative ions: the iron series, Phys. Rev. **135**, A980 (1964).

[147] Charkin, O. P., and Dyatkina, M. E., Calculation of the electron affinity of transitional elements by the Glockler method, Zh. Strukt. Khim. **6**, 422 (1965) [Engl. transl.: J. Struct. Chem. (USSR) **6**, 397 (1965)].

[148] Hotop, H., Bennett, R. A., and Lineberger, W. C., Electron affinities of Cu and Ag, J. Chem. Phys. **58**, 2373 (1973).

[149] Bakulina, I. N., and Ionov, N. I., A determination of the electron affinity of atoms of copper, silver, and gold by the surface ionization method, Dokl. Akad. Nauk SSSR **155**, 309 (1964) [Engl. transl.: Soviet Phys. – Dokl. **9**, 217 (1964)].

[150] Politzer, P., Electron affinities of atoms, Trans. Faraday Soc. **64**, 2241 (1968).

[151] Lineberger, W. C., Photodetachment of negative ions, in: Energy, Structure, and Reactivity, ed. D. W. Smith and W. B. McRae (Wiley, New York, 1973).

[152] Pritchard, H. O., The determination of electron affinities, Chem. Rev. **52**, 529 (1953).

[153] Dukel'skii, V. M., and Ionov, N. I., Negative ions of selenium, tellurium, antimony and bismuth, Dokl. Akad. Nauk SSSR **81**, 767 (1951).

[154] Smyth, K. C., and Brauman, J. I., Photodetachment of an electron from selenide ion; the electron affinity and spin-orbit coupling constant for SeH•, J. Chem. Phys. **56**, 5993 (1972).

[155] Frank, H., Neiger, M., and Popp, H.–P., Quantitative Ausmessung des Brom–Affinitätskontinuums und Bestimmung der Detachment– und Attachment–Querschnitte, Z. Naturforsch. **25a**, 1617 (1970).

[156] Morrison, J. D., Hurzeler, H., Inghram, M. G., and Stanton, H. E., Threshold law for the probability of excitation of molecules by photon impact. A study of the photoionization efficiencies of Br₂, I₂, HI, and CH₃I, J. Chem. Phys. **33**, 821 (1960).

[157] Blewett, J. P., Mass spectrograph analysis of bromine, Phys. Rev. **49**, 900 (1936).

[158] Lee, Y., and Mahan, B. H., Photosensitized ionization of alkali–metal vapors, J. Chem. Phys. **42**, 2893 (1965).

[159] Scheer, M. D., Positive and negative ion sublimation from transition metal surfaces: a review of some recent results, J. Res. NBS **74A**, 37 (1970).

[160] Fine, J., and Scheer, M. D., Positive and negative self–surface ionization of molybdenum, J. Chem. Phys. **47**, 4267 (1967).

[161] Hotop, H., and Lineberger, W. C., Dye–laser photodetachment studies of Au⁻, Pt⁻, PtN⁻, and Ag⁻, J. Chem. Phys. **58**, 2379 (1973).

[162] Steiner, B., Seman, M. L., and Branscomb, L. M., Energy dependence for the photodetachment of I⁻ near threshold, in: Atomic Collision Processes, ed. M. R. C. McDowell (North–Holland Pub. Co., Amsterdam, 1964) p. 537.

[163] Steiner, B., Seman, M. L., and Branscomb, L. M., Electron affinity of atomic iodine, J. Chem. Phys. **37**, 1200 (1962).

[164] Hogness, T. R., and Harkness, R. W., The ionization processes of iodine interpreted by the mass-spectrograph, Phys. Rev. **32**, 784 (1928).

[165] Scheer, M. D., and Fine, J., Positive and negative self-surface ionization of tungsten and rhenium, J. Chem. Phys. **46**, 3998 (1967).

[166] Scheer, M. D., and Fine, J., Electron affinity of tungsten determined by its positive and negative self–surface ionization, Phys. Rev. Letters **17**, 283 (1966).

[167] Jensen, D. E., and Miller, W. J., Electron attachment and compound formation in flames. III. Negative ion and compound formation in flames containing tungsten and potassium, J. Chem. Phys. **53**, 3287 (1970).

[168] Khvostenko, V. I., and Sultanov, A. Sh., Ionisation of thallium(I) chloride, bromide, and iodide molecules by electron bombardment, Zh. Fiz. Khim. **39**, 475 (1965) [Engl. transl.: Russ. J. Phys. Chem. **39**, 252 (1965)].

[169] Eyring, H., Hirschfelder, J. O., and Taylor, H. S., The theoretical treatment of chemical reactions produced by ionization processes. Part I. The ortho–para hydrogen conversion by alpha–particles, J. Chem. Phys. **4**, 479 (1936).

[170] Napper, R., and Page, F. M., Determination of electron affinities. Part 5.—Cyanide and thiocyanate radicals, Trans. Faraday Soc. **59**, 1086 (1963).

[171] Lineberger, W. C., Laser negative ion spectroscopy, in: Laser Spectroscopy, ed. R. G. Brewer and A. Mooradian (Plenum Press, New York, 1974).

[172] Ferguson, E. E., Fehsenfeld, F. C., and Phelps, A. V., Comment on photodetachment cross sections for CO_3^- and its first hydrate, J. Chem. Phys. **59**, 1565 (1973).

[173] Burt, J. A., Photodetachment cross sections for CO_3^- and its first hydrate, J. Chem. Phys. **57**, 4649 (1972).

[174] Compton, R. N., Christophorou, L. G., Hurst, G. S., and Reinhardt, P. W., Nondissociative electron capture in complex molecules and negative–ion lifetimes, J. Chem. Phys. **45**, 4634 (1966).

[175] O'Hare, P. A. G., and Wahl, A. C., Oxygen monofluoride (OF, $^2\Pi$): Hartree–Fock wavefunction, binding energy, ionization potential, electron affinity, dipole and quadrupole moments, and spectroscopic constants. A comparison of theoretical and experimental results, J. Chem. Phys. **53**, 2469 (1970).

[176] Kraus, K., Müller–Duysing, W., and Neuert, H., Über Stösse langsamer negativer Ionen mit Ladungs-übertragung, Z. Naturforsch. **16a**, 1385 (1961).

[177] Fehsenfeld, F. C., Ion chemistry of SF_6, J. Chem. Phys. **54**, 438 (1971).

[178] Jensen, D. E., Electron attachment and compound formation in flames. Part 1.—Electron affinity of BO_2 and heats of formation of alkali metal metaborates, Trans. Faraday Soc. **65**, 2123 (1969).

[179] Myer, J. A., and Samson, J. A. R., Absorption cross section and photoionization yield of I_2 between 1050 and 2200 Å, J. Chem. Phys. **52**, 716 (1970).

6.4. Author Index

Golub, S. 78
Grajower, R. 113
Gray, P. 47
Grün, N. 19
Gusinow, M. A. 12
Gutman, D. 66, 107

H

Hall, J. L. 12, 33, 37, 62, 85
Hamill, W. H. 88
Harkness, R. W. 164
Harland, P. W. 42, 112, 134
Harris, F. E. 8, 31
Heinicke, E. 26
Henderson, W. R. 60
Henglein, A. 129
Herman, K. 130
Hirschfelder, J. O. 169
Hogness, T. R. 164
Holøien, E. 14
Honig, R. E. 36
Hotop, H. 73, 115, 148, 161
Hughes, B. M. 64
Hurst, G. S. 174
Hurzeler, H. 156

I

Inghram, M. G. 58, 156
Ionov, N. I. 149, 153

J

Jäger, K. 129
Jenkins, H. D. B. 100
Jensen, D. E. 81, 167, 178
Joshi, B. D. 29

K

Kalmykov, A. A. 43
Kay, J. 135, 143
Khvostenko, V. I. 11, 120, 168
Klemm, R. A. 31
Kortzeborn, R. N. 109
Kozlov, V. F. 43
Kraus, K. 176

L

Laughlin, C. 17
Lee, Y. 158
Levine, J. 62, 85
Lifshitz, C. 64, 113
Lineberger, W. C. 73, 115, 126, 148, 151, 161, 171
Lipscomb, W. N. 125
Locht, R. 38
Lynch, R. 57

M

Mackie, J. C. 57
MacNeil, K. A. G. 39, 111
Mahan, B. H. 158
McCulloh, K. E. 105
McDowell, M. R. C. 9
McIver, R. T., Jr. 50
McLean, A. D. 16, 118
Miller, W. J. 167
Milstein, R. 103

Mohler, F. L. 142
Momigny, J. 38
Moores, D. L. 116
Morrison, J. D. 156
Moruzzi, J. L. 77
Moser, C. M. 30
Mück, G. 137
Myer, J. A. 179

N

Napper, R. 170
Neiger, M. 155
Nesbet, R. K. 30
Nesbitt, L. E. 72
Neuert, H. 124
Noble, P. N. 109
Norcross, D. W. 116

O

O'Hare, P. A. G. 110, 114, 117, 133, 144, 175
Öksüz, İ. 35

P

Pack, J. L. 63, 84
Page, F. M. 28, 53, 55, 87, 128, 135, 143, 170
Patterson, T. A. 73, 115
Paulson, J. F. 90
Pearson, P. K. 108
Peers, A. M. 113
Pekeris, C. L. 1
Phelps, A. V. 63, 77, 84, 172
Politzer, P. 150
Popp, H.-P. 101, 102, 137, 155
Pritchard, H. O. 152

R

Raimondi, D. L. 118
Read, F. H. 45
Reese, R. M. 142
Reimann, C. W. 136, 138
Reinhardt, P. W. 174
Rempt, R. D. 82
Risley, J. S. 61
Rohrlich, F. 27

S

Sabin, J. R. 46
Samson, J. A. R. 69, 179
Schaefer, H. F., III 31, 108
Scheer, M. D. 20, 159, 160, 165, 166
Schmeltekopf, A. L. 86, 92
Schulz, G. J. 59
Schwarz, W. H. E. 18
Seman, M. L. 34, 162, 163
Sharp, T. E. 5
Siegel, M. W. 33, 62, 85
Sinanoğlu, O. 35
Sklar, A. L. 41
Smirnov, B. M. 13
Smith, S. J. 32, 56, 68
Smyth, K. C. 49, 50, 123, 132, 154
Sokolov, V. M. 122
Somerville, W. B. 7
Spokes, G. N. 138
Srivastava, R. D. 80, 121
Stanton, H. E. 156

7. Auxiliary Thermochemistry

7.1. Introduction

In this section are given the auxiliary heats of formation of the species on which the ionic heats of formation are based. In contrast to the previous edition (NSRDS-NBS 26) we list only values based on analysis of experimental thermochemical information and no longer include estimated heats of formation based on empirical correlation schemes. As noted in section 3.3 we have striven wherever possible to calculate ΔH_{f0}° rather than ΔH_{f298}° in the belief that the ionization experiments are more easily relatable to the former quantity, even if only approximately. However, where only ΔH_{f298}° is available it has been used to arrive at room temperature heats of formation. For the convenience of the reader both absolute zero and room temperature heats of formation have been tabulated, and are given both in kJ mol^{-1} and kcal mol^{-1}. Further, no attempt has been made to further modify or refine the evaluations embodied in the source material used here. We are aware of occasional significant differences among the three sources of heats of for-

mation for organic molecules. We have arbitrarily chosen the values given in order of decreasing preference in NBS Technical Note 270-3 [1], the compilation of Stull, Westrum, and Sinke [2] and the compilation of Cox and Pilcher [3]. As for heats of formation of radicals we have emphasized the use of heats of formation obtained by non-mass spectrometric methods in order to minimize circular reasoning.

The compounds are listed in order of increasing complexity and increasing atomic number, as in the ion tables.

References for Section 7.1

[1] Wagman, D. D., Evans, W. H., Parker, V. B., Halow, I., Bailey, S. M., and Schumm, R. H., Selected Values of Chemical Thermodynamic Properties. Tables for the First Thirty-Four Elements in the Standard Order of Arrangement, NBS Tech. Note 270–3 (U.S. Government Printing Office, Washington, D.C., 1968).

[2] Stull, D. R., Westrum, E. F., Jr., and Sinke, G. C., The Chemical Thermodynamics of Organic Compounds (John Wiley and Sons, New York, 1969).

[3] Cox, J. D., and Pilcher, G., Thermochemistry of Organic and Organometallic Compounds (Academic Press, New York, 1970).

7.2. Table of Auxiliary Heats of Formation

Empirical formula	Description	$\Delta H^\circ_{f0}(g)$ (kJ mol^{-1})	$\Delta H^\circ_{f298}(g)$ (kJ mol^{-1})	$\Delta H^\circ_{f0}(g)$ (kcal mol^{-1})	$\Delta H^\circ_{f298}(g)$ (kcal mol^{-1})	Ref.
H	H	216.003	217.965	51.626	52.095	1
D	D	219.760	221.673	52.524	52.981	1
H$^-$	H$^-$	143.22		34.23		6
Li	Li	159.134	160.71	38.034	38.41	11
Be	Be	320.03	324.3	76.49	77.5	4
B	B	557.64	562.7	133.28	134.5	1
BH	BH	446.4	449.61	106.7	107.46	1
B$_2$H$_6$	B$_2$H$_6$	51.42	35.6	12.29	8.5	1
C	C	711.20	716.682	169.98	171.291	1
C$^-$	C$^-$	588.65		140.69		6
C$_2$	C$_2$	829.3	837.6	198.2	200.2	11
C$_3$	C$_3$	812	820	194	196	11
CH	CH	592.5	595.8	141.6	142.4	1
CH$_2$	CH$_2$	392.9	392.0	93.9	93.7	1
CH$_3$	CH$_3$	145.6	142.3	34.8	34.0	12
CH$_4$	CH$_4$	−66.818	−74.81	−15.970	−17.88	1
C$_2$H	C$_2$H	474.0	477	113.3	114	11
C$_2$H$_2$	C$_2$H$_2$	227.292	226.73	54.324	54.19	1
C$_2$H$_4$	C$_2$H$_4$	60.731	52.26	14.515	12.49	1
C$_2$H$_5$	C$_2$H$_5$		107.5		25.7	9
C$_2$H$_6$	C$_2$H$_6$	−69.132	−84.68	−16.523	−20.24	1
C$_3$H$_3$	CH$_2$C≡CH		337.6		80.7	14
C$_3$H$_4$	CH$_2$=C=CH$_2$		192.13		45.92	7
C$_3$H$_4$	CH$_3$C≡CH		185.43		44.32	7
C$_3$H$_4$	Cyclopropene		276.1		66.0	7
C$_3$H$_5$	CH$_2$=CHCH$_2$		169.9		40.6	9
C$_3$H$_6$	CH$_3$CH=CH$_2$		20.42		4.88	7
C$_3$H$_6$	Cyclopropane		53.30		12.74	7
C$_3$H$_7$	n–C$_3$H$_7$		86.6		20.7	9
C$_3$H$_7$	iso–C$_3$H$_7$		73.6		17.6	9
C$_3$H$_8$	C$_3$H$_8$		−103.85		−24.82	7
C$_4$H$_2$	CH≡CC≡CH		472.79		113.00	7
C$_4$H$_6$	CH$_3$CH=C=CH$_2$		162.21		38.77	7
C$_4$H$_6$	CH$_2$=CHCH=CH$_2$		110.16		26.33	7
C$_4$H$_6$	C$_2$H$_5$C≡CH		165.18		39.48	7
C$_4$H$_6$	CH$_3$C≡CCH$_3$		146.31		34.97	7
C$_4$H$_6$	Cyclobutene		129.70		31.00	7
C$_4$H$_7$	CH$_3$CHCH=CH$_2$		127.6		30.5	9
C$_4$H$_8$	1–C$_4$H$_8$		−0.13		−0.03	7
C$_4$H$_8$	cis–2–C$_4$H$_8$		−6.99		−1.67	7
C$_4$H$_8$	trans–2–C$_4$H$_8$		−11.17		−2.67	7
C$_4$H$_8$	iso–C$_4$H$_8$		−16.90		−4.04	7
C$_4$H$_8$	Cyclobutane		26.65		6.37	7
C$_4$H$_9$	n–C$_4$H$_9$		65.7		15.7	9
C$_4$H$_9$	sec–C$_4$H$_9$		52.7		12.6	9
C$_4$H$_9$	iso–C$_4$H$_9$		57.3		13.7	9
C$_4$H$_9$	tert–C$_4$H$_9$		28.5		6.8	9
C$_4$H$_{10}$	n–C$_4$H$_{10}$		−126.15		−30.15	7
C$_4$H$_{10}$	iso–C$_4$H$_{10}$		−134.52		−32.15	7
C$_5$H$_6$	Cyclopentadiene		133.89		32.00	7
C$_5$H$_8$	cis–CH$_2$=CHCH=CHCH$_3$		78.24		18.70	7
C$_5$H$_8$	trans–CH$_2$=CHCH=CHCH$_3$		77.82		18.60	7
C$_5$H$_8$	CH$_2$=C(CH$_3$)CH=CH$_2$		75.73		18.10	7
C$_5$H$_8$	Cyclopentene		32.93		7.87	7
C$_5$H$_8$	Spiropentane		185.23		44.27	7
C$_5$H$_{10}$	CH$_3$CH$_2$CH$_2$CH=CH$_2$		−20.92		−5.00	7
C$_5$H$_{10}$	trans–C$_2$H$_5$CH=CHCH$_3$		−31.76		−7.59	7
C$_5$H$_{10}$	(CH$_3$)$_2$CHCH=CH$_2$		−28.95		−6.92	7
C$_5$H$_{10}$	C$_2$H$_5$C(CH$_3$)=CH$_2$		−36.32		−8.68	7

7.2. Table of Auxiliary Heats of Formation—Continued

Empirical formula	Description	$\Delta H_{f0}^{\circ}(g)$ (kJ mol^{-1})	$\Delta H_{f298}^{\circ}(g)$ (kJ mol^{-1})	$\Delta H_{f0}^{\circ}(g)$ (kcal mol^{-1})	$\Delta H_{f298}^{\circ}(g)$ (kcal mol^{-1})	Ref.
C_5H_{10}	$(CH_3)_2C=CHCH_3$		−42.55		−10.17	7
C_5H_{10}	Cyclopentane		−77.24		−18.46	7
C_5H_{12}	$n-C_5H_{12}$		−146.44		−35.00	7
C_5H_{12}	$iso-C_5H_{12}$		−154.47		−36.92	7
C_5H_{12}	$neo-C_5H_{12}$		−165.98		−39.67	7
C_6H_6	Benzene		82.93		19.82	7
C_6H_8	1,3–Cyclohexadiene		108.4		25.9	7
C_6H_{10}	$CH_2=C(CH_3)C(CH_3)=CH_2$		47.7		11.4	7
C_6H_{10}	1–Methylcyclopentene		−5.4		−1.3	7
C_6H_{10}	Cyclohexene		−5.36		−1.28	7
C_6H_{12}	$1-C_6H_{12}$		−41.67		−9.96	7
C_6H_{12}	$trans-3-C_6H_{12}$		−54.43		−13.01	7
C_6H_{12}	$(CH_3)_3CCH=CH_2$		−43.14		−10.31	7
C_6H_{12}	$(CH_3)_2C=C(CH_3)_2$		−59.20		−14.15	7
C_6H_{12}	Cyclohexane		−123.14		−29.43	7
C_6H_{14}	$n-C_6H_{14}$		−167.19		−39.96	7
C_6H_{14}	$iso-C_6H_{14}$		−174.31		−41.66	7
C_6H_{14}	$(C_2H_5)_2CHCH_3$		−171.63		−41.02	7
C_6H_{14}	$C_2H_5C(CH_3)_3$		−185.56		−44.35	7
C_6H_{14}	$(CH_3)_2CHCH(CH_3)_2$		−177.78		−42.49	7
C_7H_8	Toluene		50.00		11.95	7
C_7H_8	Cycloheptatriene		181.88		43.47	7
C_7H_{12}	Cycloheptene		−9.33		−2.23	7
C_7H_{14}	Methylcyclohexane		−154.77		−36.99	7
C_7H_{16}	$n-C_7H_{16}$		−187.78		−44.88	7
C_8H_6	Ethynylbenzene		327.27		78.22	7
C_8H_8	Ethenylbenzene		147.36		35.22	7
C_8H_8	Cyclooctatetraene		298.02		71.23	7
C_8H_{10}	Ethylbenzene		29.79		7.12	7
C_8H_{10}	1,2–Dimethylbenzene		19.00		4.54	7
C_8H_{10}	1,3–Dimethylbenzene		17.24		4.12	7
C_8H_{10}	1,4–Dimethylbenzene		17.95		4.29	7
C_8H_{16}	$1-C_8H_{16}$		−82.93		−19.82	7
C_8H_{18}	$n-C_8H_{18}$		−208.45		−49.82	7
C_8H_{18}	$(CH_3)_2CHCH_2C(CH_3)_3$		−224.14		−53.57	7
C_9H_8	Indene		163.34		39.04	8
C_9H_{10}	Isopropenylbenzene		112.97		27.00	7
C_9H_{12}	Propylbenzene		7.82		1.87	7
C_9H_{12}	Isopropylbenzene		3.93		0.94	7
C_9H_{12}	1,2,3–Trimethylbenzene		−9.58		−2.29	7
C_9H_{12}	1,2,4–Trimethylbenzene		−13.93		−3.33	7
C_9H_{12}	1,3,5–Trimethylbenzene		−16.07		−3.84	7
C_9H_{20}	$n-C_9H_{20}$		−229.03		−54.74	7
$C_{10}H_8$	Naphthalene		150.96		36.08	7
$C_{10}H_8$	Azulene		279.91		66.90	7
$C_{10}H_{14}$	Butylbenzene		−13.81		−3.30	7
$C_{10}H_{14}$	$sec-$Butylbenzene		−17.45		−4.17	7
$C_{10}H_{14}$	Isobutylbenzene		−21.55		−5.15	7
$C_{10}H_{14}$	$tert-$Butylbenzene		−22.68		−5.42	7
$C_{10}H_{14}$	1,2,4,5–Tetramethylbenzene		−45.27		−10.82	7
$C_{10}H_{18}$	$cis-$Decalin		−168.95		−40.38	7
$C_{10}H_{18}$	$trans-$Decalin		−182.30		−43.57	7
$C_{11}H_{10}$	1–Methylnaphthalene		116.86		27.93	7
$C_{11}H_{10}$	2–Methylnaphthalene		116.11		27.75	7
$C_{11}H_{16}$	Pentamethylbenzene		−74.48		−17.80	7
$C_{12}H_8$	Biphenylene		482.0		115.2	7
$C_{12}H_8$	Acenaphthylene		258.2		61.7	7
$C_{12}H_{10}$	Biphenyl		182.09		43.52	7
$C_{12}H_{10}$	Acenaphthene		156.5		37.4	7

7.2. Table of Auxiliary Heats of Formation—Continued

Empirical formula	Description	$\Delta H^\circ_{f0}(g)$ (kJ mol⁻¹)	$\Delta H^\circ_{f298}(g)$ (kJ mol⁻¹)	$\Delta H^\circ_{f0}(g)$ (kcal mol⁻¹)	$\Delta H^\circ_{f298}(g)$ (kcal mol⁻¹)	Ref.
$C_{12}H_{12}$	1,4–Dimethylnaphthalene		82.51		19.72	7
$C_{12}H_{12}$	2,3–Dimethylnaphthalene		83.55		19.97	7
$C_{12}H_{18}$	Hexamethylbenzene		−105.69		−25.26	7
$C_{14}H_{10}$	Anthracene		230.83		55.17	7
$C_{14}H_{10}$	Phenanthrene		206.94		49.46	7
$C_{16}H_{10}$	Fluoranthene		294.6		70.4	7
N	N	470.842	472.704	112.534	112.979	1
NH	NH	331	331	79	79	1
NH_2	NH_2		172		41	1
NH_3	NH_3	−39.08	−46.11	−9.34	−11.02	1
N_3H	HN_3	300.49	294.1	71.82	70.3	1
CN	CN	431.8	435.1	103.2	104.0	11
C_2N_2	C_2N_2	307.047	308.95	73.386	73.84	1
CHN	HCN	135.52	135.1	32.39	32.3	1
CH_5N	CH_3NH_2		−22.97		−5.49	1
C_2H_2N	CH_2CN		213.8		51.1	9
C_2H_3N	CH_3CN	94.47	87.4	22.58	20.9	1
C_2H_3N	CH_3NC	155.44	149.0	37.15	35.6	1
C_2H_5N	Ethylenimine		126.4		30.2	1
C_2H_7N	$C_2H_5NH_2$		−47.15		−11.27	1
C_2H_7N	$(CH_3)_2NH$		−18.45		−4.41	1
C_3H_3N	$CH_2=CHCN$		184.93		44.20	7
C_3H_5N	C_2H_5CN		50.63		12.10	7
C_3H_9N	$n–C_3H_7NH_2$		−70.17		−16.77	8
C_3H_9N	$iso–C_3H_7NH_2$		−83.76		−20.02	8
C_3H_9N	$(CH_3)_3N$		−23.85		−5.70	7
C_4H_5N	Pyrrole		108.28		25.88	8
C_4H_7N	$n–C_3H_7CN$		34.06		8.14	7
$C_4H_{11}N$	$n–C_4H_9NH_2$		−92.05		−22.00	7
$C_4H_{11}N$	$sec–C_4H_9NH_2$		−104.18		−24.90	7
$C_4H_{11}N$	$tert–C_4H_9NH_2$		−119.87		−28.65	7
$C_4H_{11}N$	$(C_2H_5)_2NH$		−72.38		−17.30	7
C_5H_5N	Pyridine		140.16		33.50	7
C_6H_7N	Aniline		86.86		20.76	7
C_6H_7N	2–Methylpyridine		98.95		23.65	7
C_6H_7N	3–Methylpyridine		106.15		25.37	7
C_6H_7N	4–Methylpyridine		102.22		24.43	7
$C_6H_{15}N$	$(C_2H_5)_3N$		−99.58		−23.80	7
C_7H_5N	Benzoic acid nitrile		218.82		52.30	7
C_8H_7N	Indole		186.6		44.6	8
$C_{12}H_{11}N$	Diphenylamine		201.67		48.20	7
CH_2N_2	Diazomethane		192.5		46.0	7
$C_2H_8N_2$	$(CH_3)_2NNH_2$		84.43		20.18	1
$C_2H_8N_2$	$CH_3NHNHCH_3$		94.98		22.70	1
$C_3H_4N_2$	Pyrazole		181.2		43.3	7
$C_4H_4N_2$	1,2–Diazine		278.36		66.53	7
$C_4H_4N_2$	1,3–Diazine		196.61		46.99	7
$C_4H_4N_2$	1,4–Diazine		196.06		46.86	7
$C_6H_8N_2$	Phenylhydrazine		203.84		48.72	8
O	O	246.785	249.170	58.983	59.553	1
O^-	O^-	105.44		25.20		6
O_3	O_3	145.35	142.7	34.74	34.1	1
OH	OH	38.70	38.95	9.25	9.31	1
OD	OD	36.48	36.86	8.72	8.81	1
H_2O	H_2O	−238.915	−241.818	−57.102	−57.796	1
D_2O	D_2O	−246.249	−249.199	−58.855	−59.560	1
CO	CO	−113.801	−110.525	−27.199	−26.416	1
CO_2	CO_2	−393.141	−393.509	−93.963	−94.051	1
C_3O_2	C_3O_2	−96.82	−93.64	−23.14	−22.38	11

7.2. Table of Auxiliary Heats of Formation—Continued

Empirical formula	Description	$\Delta H_{f0}^{\circ}(g)$ (kJ mol^{-1})	$\Delta H_{f298}^{\circ}(g)$ (kJ mol^{-1})	$\Delta H_{f0}^{\circ}(g)$ (kcal mol^{-1})	$\Delta H_{f298}^{\circ}(g)$ (kcal mol^{-1})	Ref.
NO	NO	89.75	90.25	21.45	21.57	1
N$_2$O	N$_2$O	85.500	82.05	20.435	19.61	1
NO$_2$	NO$_2$	35.98	33.18	8.60	7.93	1
CH$_2$O	HCHO	−113.4	−117	−27.1	−28	1
CH$_4$O	CH$_3$OH	−189.765	−200.66	−45.355	−47.96	1
C$_2$H$_2$O	CH$_2$=CO	−57.99	−61.1	−13.86	−14.6	1
C$_2$H$_3$O	CH$_3$CO		−22.6		−5.4	9
C$_2$H$_4$O	CH$_3$CHO	−155.39	−166.19	−37.14	−39.72	1
C$_2$H$_4$O	1,2-Epoxyethane	−40.120	−52.63	−9.589	−12.58	1
C$_2$H$_6$O	C$_2$H$_5$OH	−217.438	−235.10	−51.969	−56.19	1
C$_2$H$_6$O	(CH$_3$)$_2$O	−166.293	−184.05	−39.745	−43.99	1
C$_3$H$_6$O	C$_2$H$_5$CHO		−192.05		−45.90	7
C$_3$H$_6$O	(CH$_3$)$_2$CO		−217.57		−52.00	7
C$_3$H$_6$O	1,2-Epoxypropane		−92.76		−22.17	7
C$_3$H$_6$O	1,3-Epoxypropane		−80.54		−19.25	7
C$_3$H$_8$O	n−C$_3$H$_7$OH	−233.09	−257.53	−55.71	−61.55	7, 10
C$_3$H$_8$O	iso−C$_3$H$_7$OH	−248.24	−272.59	−59.33	−65.15	7, 10
C$_4$H$_4$O	Furan		−34.69		−8.29	7
C$_4$H$_6$O	CH$_3$CH=CHCHO		−100.54		−24.03	8
C$_4$H$_8$O	n−C$_3$H$_7$CHO		−205.02		−49.00	7
C$_4$H$_8$O	iso−C$_3$H$_7$CHO		−218.61		−52.25	8
C$_4$H$_8$O	C$_2$H$_5$COCH$_3$		−238.36		−56.97	7
C$_4$H$_8$O	1,4-Epoxybutane		−184.22		−44.03	7
C$_4$H$_{10}$O	n−C$_4$H$_9$OH	−246.02	−274.43	−58.80	−65.59	7, 10
C$_4$H$_{10}$O	iso−C$_4$H$_9$OH		−283.84		−67.84	8
C$_4$H$_{10}$O	tert−C$_4$H$_9$OH	−282.13	−325.81	−67.43	−77.87	7, 10
C$_4$H$_{10}$O	(C$_2$H$_5$)$_2$O		−252.21		−60.28	7
C$_5$H$_8$O	Cyclopentanone		−192.59		−46.03	8
C$_5$H$_8$O	Dihydropyran		−125.1		−29.9	8
C$_5$H$_{10}$O	n−C$_4$H$_9$CHO		−227.82		−54.45	7
C$_5$H$_{10}$O	n−C$_3$H$_7$COCH$_3$		−258.65		−61.82	7
C$_5$H$_{10}$O	iso−C$_3$H$_7$COCH$_3$		−262.59		−62.76	8
C$_5$H$_{10}$O	(C$_2$H$_5$)$_2$CO		−258.65		−61.82	7
C$_5$H$_{10}$O	1,5-Epoxypentane		−223.84		−53.50	7
C$_6$H$_6$O	Phenol		−96.36		−23.03	7
C$_6$H$_{10}$O	Cyclohexanone		−230.12		−55.00	7
C$_6$H$_{12}$O	n−C$_4$H$_9$COCH$_3$		−279.78		−66.87	8
C$_6$H$_{12}$O	tert−C$_4$H$_9$COCH$_3$		−289.87		−69.28	8
C$_6$H$_{14}$O	(n−C$_3$H$_7$)$_2$O		−292.88		−70.00	7
C$_6$H$_{14}$O	(iso−C$_3$H$_7$)$_2$O		−318.82		−76.20	7
C$_7$H$_8$O	Methoxybenzene		−72.26		−17.27	8
C$_7$H$_{12}$O	Cycloheptanone		−247.3		−59.1	8
C$_7$H$_{14}$O	(iso−C$_3$H$_7$)$_2$CO		−311.29		−74.40	8
C$_8$H$_8$O	Acetophenone		−86.86		−20.76	7
C$_8$H$_{10}$O	Ethoxybenzene		−110.0		−26.3	8
C$_8$H$_{18}$O	(tert−C$_4$H$_9$)$_2$O		−364.84		−87.20	7
C$_9$H$_{18}$O	(tert−C$_4$H$_9$)$_2$CO		−345.77		−82.64	8
C$_{12}$H$_8$O	Dibenzofuran		83.3		19.9	8
CH$_2$O$_2$	HCOOH		−378.57		−90.48	1
C$_2$H$_4$O$_2$	CH$_3$COOH	−418.283	−432.25	−99.972	−103.31	1
C$_2$H$_4$O$_2$	HCOOCH$_3$		−350.2		−83.7	1
C$_3$H$_4$O$_2$	3-Hydroxypropanoic acid lactone		−282.88		−67.61	7
C$_3$H$_6$O$_2$	C$_2$H$_5$COOH		−453.5		−108.4	8
C$_3$H$_6$O$_2$	HCOOC$_2$H$_5$		−371.29		−88.74	7
C$_3$H$_6$O$_2$	CH$_3$COOCH$_3$		−409.6		−97.9	7
C$_3$H$_8$O$_2$	(CH$_3$O)$_2$CH$_2$		−348.40		−83.27	8
C$_4$H$_6$O$_2$	CH$_3$COCOCH$_3$		−327.19		−78.20	8
C$_4$H$_6$O$_2$	CH$_3$COOCH=CH$_2$		−315.72		−75.46	8

7.2. Table of Auxiliary Heats of Formation—Continued

Empirical formula	Description	$\Delta H_{f0}^{\circ}(g)$ (kJ mol^{-1})	$\Delta H_{f298}^{\circ}(g)$ (kJ mol^{-1})	$\Delta H_{f0}^{\circ}(g)$ (kcal mol^{-1})	$\Delta H_{f298}^{\circ}(g)$ (kcal mol^{-1})	Ref.
$C_4H_8O_2$	$n-C_3H_7COOH$		-470.3		-112.4	8
$C_4H_8O_2$	$CH_3COOC_2H_5$		-442.92		-105.86	7
$C_4H_8O_2$	1,4-Dioxane		-315.06		-75.30	7
$C_4H_{10}O_2$	$(CH_3O)_2CHCH_3$		-390.20		-93.26	8
$C_5H_4O_2$	Furfural		-151.0		-36.1	8
$C_5H_8O_2$	$CH_3COCH_2COCH_3$		-378.53		-90.47	8
$C_5H_{10}O_2$	$CH_3COOCH(CH_3)_2$		-481.66		-115.12	8
$C_5H_{10}O_2$	$C_2H_5COOC_2H_5$		-425.5		-101.7	8
$C_6H_4O_2$	1,4-Benzoquinone		-118.4		-28.3	7
$C_6H_{12}O_2$	$CH_3COO(CH_2)_3CH_3$		-485.76		-116.10	8
$C_4H_6O_3$	$CH_3COOCOCH_3$		-575.72		-137.60	7
CH_3NO	$HCONH_2$		-186.2		-44.5	7
C_3H_7NO	$HCON(CH_3)_2$		-191.6		-45.8	8
$C_6H_{11}NO$	6-Aminohexanoic acid lactam		-246.0		-58.8	8
CH_3NO_2	CH_3NO_2	-60.860	-74.73	-14.546	-17.86	1
$C_2H_5NO_2$	$C_2H_5NO_2$		-98.58		-23.56	1
$C_2H_5NO_2$	C_2H_5ONO		-104.2		-24.9	1
$C_3H_7NO_2$	$n-C_3H_7NO_2$		-124.68		-29.80	7
$C_3H_7NO_2$	$iso-C_3H_7NO_2$		-140.16		-33.50	7
$C_4H_9NO_2$	$n-C_4H_9NO_2$		-143.93		-34.40	7
$C_4H_9NO_2$	$sec-C_4H_9NO_2$		-163.59		-39.10	7
CH_3NO_3	CH_3ONO_2		-124.7		-29.8	1
$C_3H_7NO_3$	$n-C_3H_7ONO_2$		-174.05		-41.60	7
F	F	76.90	78.99	18.38	18.88	1
F^-	F^-	-251.17		-60.03		6
HF	HF	-271.077	-271.1	-64.789	-64.8	1
BF	BF	-125.23	-122.2	-29.93	-29.2	1
BF_3	BF_3	-1134.207	-1137.00	-271.082	-271.75	1
B_2F_4	B_2F_4	-1437.12	-1440.1	-343.48	-344.2	1
CF	CF	251.5	255	60.1	61	11
CF_2	CF_2	-182.4	-182.0	-43.6	-43.5	11
CF_3	CF_3	-467.4	-470.3	-111.7	-112.4	11
CF_4	CF_4	-927.22	-933.20	-221.61	-223.04	11
C_2F_4	C_2F_4	-655.2	-658.6	-156.6	-157.4	11
C_2F_6	C_2F_6	-1335.5	-1343.9	-319.2	-321.2	11
C_6F_6	Hexafluorobenzene		-956.63		-228.64	7
C_3F_8	C_3F_8		-1748.5		-417.9	7
C_4F_{10}	$n-C_4F_{10}$		-2158.9		-516.0	7
NF_2	NF_2	45.69	43.1	10.92	10.3	1
NF_3	NF_3	-118.95	-124.7	-28.43	-29.8	1
N_2F_4	N_2F_4	3.68	-7.1	0.88	-1.7	1
OF_2	OF_2	-19.7	-21.8	-4.7	-5.2	1
CH_3F	CH_3F		-233.89		-55.90	7
C_6H_5F	Fluorobenzene		-116.57		-27.86	7
C_7H_7F	4-Fluorotoluene		-148.03		-35.38	7
CH_2F_2	CH_2F_2	-439.19	-446.9	-104.97	-106.8	1
$C_2H_2F_2$	$CH_2=CF_2$	-321.96	-328.9	-76.95	-78.6	1
$C_6H_4F_2$	1,2-Difluorobenzene		-294.51		-70.39	7
$C_6H_4F_2$	1,4-Difluorobenzene		-307.23		-73.43	7
CHF_3	CHF_3	-681.32	-688.3	-162.84	-164.5	1
C_2HF_3	C_2HF_3		-495.80		-118.50	7
$C_7H_5F_3$	$\alpha,\alpha,\alpha-$Trifluorotoluene		-600.07		-143.42	7
C_6HF_5	Pentafluorobenzene		-806.05		-192.65	8
CNF	FCN	23.4		5.6		16
Na	Na	108.010	107.759	25.815	25.755	11
Na_2	Na_2	140.026	137.53	33.467	32.87	11
Mg	Mg	146.499	147.70	35.014	35.30	4
Al	Al	324.01	326.4	77.44	78.0	1
Si	Si	451.29	455.6	107.86	108.9	1

7.2. Table of Auxiliary Heats of Formation—Continued

Empirical formula	Description	$\Delta H_{f0}^{\circ}(g)$ (kJ mol^{-1})	$\Delta H_{f298}^{\circ}(g)$ (kJ mol^{-1})	$\Delta H_{f0}^{\circ}(g)$ (kcal mol^{-1})	$\Delta H_{f298}^{\circ}(g)$ (kcal mol^{-1})	Ref.
SiH$_4$	SiH$_4$	43.10	34.3	10.30	8.2	1
SiO	SiO	−100.75	−99.6	−24.08	−23.8	1
SiF	SiF	4	7.1	1	1.7	1
SiF$_4$	SiF$_4$	−1609.42	−1614.94	−384.66	−385.98	1
C$_4$H$_{12}$Si	(CH$_3$)$_4$Si		−239.111		−57.149	1
P	P	314	314.64	75	75.20	1
P$_4$	P$_4$	66.23	58.91	15.83	14.08	1
PH$_2$	PH$_2$	128.0	125.9	30.6	30.1	11
PH$_3$	PH$_3$	13.39	5.4	3.20	1.3	1
PF$_3$	PF$_3$	−913.16	−918.8	−218.25	−219.6	1
C$_{18}$H$_{15}$P	Triphenylphosphine		328.4		78.5	8
POF$_3$	POF$_3$	−1202.44	−1211.3	−287.39	−289.5	1
S	S	276.6	278.805	66.1	66.636	1
S$_2$	S$_2$	128.227	128.37	30.647	30.68	1
S$_3$	S$_3$		132.6		31.7	1
S$_4$	S$_4$		136.8		32.7	1
S$_5$	S$_5$		123.8		29.6	1
S$_6$	S$_6$		102.5		24.5	1
S$_7$	S$_7$		113.4		27.1	1
S$_8$	S$_8$	106.06	102.30	25.35	24.45	1
HS	HS	142	142.67	34	34.10	1
H$_2$S	H$_2$S	−17.707	−20.63	−4.232	−4.93	1
CS	CS	269.4		64.4		15
CS$_2$	CS$_2$	116.57	117.36	27.86	28.05	1
SO	SO	6.3	6.259	1.5	1.496	1
SO$_2$	SO$_2$	−294.286	−296.830	−70.336	−70.944	1
SF$_4$	SF$_4$	−767.3	−774.9	−183.4	−185.2	1
SF$_6$	SF$_6$	−1195.4	−1209	−285.7	−289	1
CH$_4$S	CH$_3$SH	−12.071	−22.34	−2.885	−5.34	1
C$_2$H$_6$S	C$_2$H$_5$SH	−29.037	−45.81	−6.940	−10.95	1
C$_2$H$_6$S	(CH$_3$)$_2$S	−21.058	−37.24	−5.033	−8.90	1
C$_3$H$_6$S	Trimethylene sulfide		61.13		14.61	7
C$_3$H$_8$S	n−C$_3$H$_7$SH		−67.86		−16.22	7
C$_3$H$_8$S	C$_2$H$_5$SCH$_3$		−59.62		−14.25	7
C$_4$H$_4$S	Thiophene		115.73		27.66	7
C$_4$H$_{10}$S	n−C$_4$H$_9$SH		−88.07		−21.05	7
C$_4$H$_{10}$S	(C$_2$H$_5$)$_2$S		−83.47		−19.95	7
C$_5$H$_6$S	2−Methylthiophene		83.68		20.00	7
C$_5$H$_6$S	3−Methylthiophene		82.80		19.79	7
C$_6$H$_6$S	Mercaptobenzene		111.55		26.66	7
C$_6$H$_{14}$S	(n−C$_3$H$_7$)$_2$S		−125.35		−29.96	7
COS	COS	−142.218	−142.09	−33.991	−33.96	1
CHNS	HNCS		127.6		30.5	1
C$_2$H$_3$NS	CH$_3$SCN		160.2		38.3	1
C$_2$H$_3$NS	CH$_3$NCS	140.00	131.0	33.46	31.3	1
C$_2$H$_4$OS	CH$_3$COSH		−181.96		−43.49	7
Cl	Cl	120.00	121.679	28.68	29.082	1
Cl$^-$	Cl$^-$	−228.61		−54.64		6
HCl	HCl	−92.132	−92.307	−22.020	−22.062	1
BCl$_3$	BCl$_3$	−402.84	−403.76	−96.28	−96.50	1
B$_2$Cl$_4$	B$_2$Cl$_4$	−489.90	−490.4	−117.09	−117.2	1
CCl$_4$	CCl$_4$	−100.75	−102.9	−24.08	−24.6	1
C$_2$Cl$_4$	C$_2$Cl$_4$	−11.30	−12.1	−2.70	−2.9	1
Cl$_2$O	Cl$_2$O	82.47	80.3	19.71	19.2	1
ClF	ClF	−54.4	−54.48	−13.0	−13.02	1
NaCl	NaCl	−180.00	−181.42	−43.02	−43.36	11
SiCl$_4$	SiCl$_4$	−654.829	−657.01	−156.508	−157.03	1
CH$_3$Cl	CH$_3$Cl	−72.910	−80.83	−17.426	−19.32	1
C$_2$H$_3$Cl	C$_2$H$_3$Cl	43.14	35.6	10.31	8.5	1

7.2. Table of Auxiliary Heats of Formation — Continued

Empirical formula	Description	$\Delta H^\circ_{f0}(g)$ (kJ mol^{-1})	$\Delta H^\circ_{f298}(g)$ (kJ mol^{-1})	$\Delta H^\circ_{f0}(g)$ (kcal mol^{-1})	$\Delta H^\circ_{f298}(g)$ (kcal mol^{-1})	Ref.
C_2H_5Cl	C_2H_5Cl	−97.617	−112.17	−23.331	−26.81	1
C_3H_7Cl	$n−C_3H_7Cl$		−130.12		−31.10	7
C_3H_7Cl	$iso−C_3H_7Cl$		−146.44		−35.00	7
C_4H_9Cl	$n−C_4H_9Cl$		−147.28		−35.20	7
C_4H_9Cl	$sec−C_4H_9Cl$		−161.50		−38.60	7
C_4H_9Cl	$iso−C_4H_9Cl$		−159.41		−38.10	7
C_4H_9Cl	$tert−C_4H_9Cl$		−183.26		−43.80	7
C_6H_5Cl	Chlorobenzene		51.84		12.39	7
C_7H_7Cl	$\alpha−$Chlorotoluene		18.8		4.5	8
CH_2Cl_2	CH_2Cl_2	−85.613	−92.47	−20.462	−22.10	1
$C_2H_2Cl_2$	$CH_2=CCl_2$	8.401	2.43	2.008	0.58	1
$C_2H_2Cl_2$	$cis−CHCl=CHCl$	9.749	3.77	2.330	0.90	1
$C_2H_2Cl_2$	$trans−CHCl=CHCl$	11.556	6.15	2.762	1.47	1
$C_2H_4Cl_2$	CH_2ClCH_2Cl	−118.646	−129.79	−28.357	−31.02	1
$C_6H_4Cl_2$	1,2−Dichlorobenzene		29.96		7.16	7
$C_6H_4Cl_2$	1,3−Dichlorobenzene		26.44		6.32	7
$C_6H_4Cl_2$	1,4−Dichlorobenzene		23.01		5.50	7
$CHCl_3$	$CHCl_3$	−98.265	−103.14	−23.486	−24.65	1
C_2HCl_3	C_2HCl_3	−4.318	−7.78	−1.032	−1.86	1
$CNCl$	$ClCN$	137.252	137.95	32.804	32.97	1
CF_3Cl	CF_3Cl	−689.5	−695	−164.8	−166	1
C_2F_3Cl	C_2F_3Cl	−552.46	−555.2	−132.04	−132.7	1
CF_2Cl_2	CF_2Cl_2	−472.8	−477	−113.0	−114	1
$C_2F_2Cl_2$	$CF_2=CCl_2$		−319.2		−76.3	1
$CFCl_3$	$CFCl_3$	−272.8	−276	−65.2	−66	1
$C_2F_3Cl_3$	$CFCl_2CF_2Cl$		−759.4		−181.5	1
C_2H_3OCl	CH_3COCl	−234.530	−243.51	−56.054	−58.20	1
C_3H_5OCl	1−Chloro−2,3−epoxypropane		−107.9		−25.8	8
CHF_2Cl	CHF_2Cl	−475.3	−481.6	−113.6	−115.1	11
C_2HF_2Cl	$CF_2=CHCl$	−310.66	−315.5	−74.25	−75.4	1
K	K	90.048	89.16	21.522	21.31	11
Ca	Ca	177.74	178.2	42.48	42.6	4
Sc	Sc	376.02	377.8	89.87	90.3	3
Ti	Ti	467.14	469.9	111.65	112.3	3
V	V	510.95	514.21	122.12	122.90	3
Cr	Cr	394.51	396.6	94.29	94.8	2
CrC_6O_6	$Cr(CO)_6$		−1005.8		−240.4	2
Mn	Mn	279.37	280.7	66.77	67.1	2
Fe	Fe	413.96	416.3	98.94	99.5	2
FeC_5O_5	$Fe(CO)_5$		−733.9		−175.4	2
Co	Co	423.082	424.7	101.119	101.5	2
Ni	Ni	427.659	429.7	102.213	102.7	2
NiC_4O_4	$Ni(CO)_4$	−606.165	−602.91	−144.877	−144.10	2
Cu	Cu	337.15	338.32	80.58	80.86	2
Zn	Zn	130.181	130.729	31.114	31.245	1
C_2H_6Zn	$(CH_3)_2Zn$		53.01		12.67	1
Ga	Ga	276	277.0	66	66.2	1
Ge	Ge	373.80	376.6	89.34	90.0	1
GeH_4	GeH_4	79.96	90.8	19.11	21.7	1
GeO	GeO	−46	−46.19	−11	−11.04	1
$GeCl_4$	$GeCl_4$	−493.96	−495.8	−118.06	−118.5	1
As	As	301.42	302.5	72.04	72.3	1
AsH_3	AsH_3	74.06	66.44	17.70	15.88	1
$AsCl_3$	$AsCl_3$	−259.91	−261.5	−62.12	−62.5	1
$C_{18}H_{15}As$	Triphenylarsine		408.4		97.6	8
Se	Se	226.40	227.07	54.11	54.27	1
Se_2	Se_2	147.53	146.0	35.26	34.9	1
H_2Se	H_2Se	33.68	29.7	8.05	7.1	1
Br	Br	117.943	111.884	28.189	26.741	1

7.2. Table of Auxiliary Heats of Formation—Continued

Empirical formula	Description	$\Delta H^{\circ}_{f0}(g)$ (kJ mol^{-1})	$\Delta H^{\circ}_{f298}(g)$ (kJ mol^{-1})	$\Delta H^{\circ}_{f0}(g)$ (kcal mol^{-1})	$\Delta H^{\circ}_{f298}(g)$ (kcal mol^{-1})	Ref.
Br$^-$	Br$^-$	−206.82		−49.43		6
Br$_2$	Br$_2$	45.702	30.907	10.923	7.387	1
HBr	HBr	−28.560	−36.40	−6.826	−8.70	1
BBr$_3$	BBr$_3$	−183.38	−205.64	−43.83	−49.15	1
CH$_3$Br	CH$_3$Br	−19.75	−35.1	−4.72	−8.4	1
C$_2$H$_3$Br	CH$_2$=CHBr	93.14	78.2	22.26	18.7	1
C$_2$H$_5$Br	C$_2$H$_5$Br	−42.627	−64.52	−10.188	−15.42	1
C$_3$H$_7$Br	n−C$_3$H$_7$Br		−87.86		−21.00	7
C$_3$H$_7$Br	iso−C$_3$H$_7$Br		−97.07		−23.20	7
C$_4$H$_9$Br	n−C$_4$H$_9$Br		−107.32		−25.65	7
C$_4$H$_9$Br	sec−C$_4$H$_9$Br		−120.08		−28.70	7
C$_4$H$_9$Br	$tert$−C$_4$H$_9$Br		−133.89		−32.00	7
C$_5$H$_{11}$Br	n−C$_5$H$_{11}$Br		−129.16		−30.87	7
C$_6$H$_5$Br	Bromobenzene		105.02		25.10	7
CH$_2$Br$_2$	CH$_2$Br$_2$		−4.2		−1.0	7
CHBr$_3$	CHBr$_3$	42.84	17	10.24	4	1
CNBr	BrCN	192.76	186.2	46.07	44.5	1
CF$_3$Br	CF$_3$Br	−630.61	−642.7	−150.72	−153.6	1
CH$_2$ClBr	CH$_2$ClBr		−50.2		−12.0	7
CHClBr$_2$	CHClBr$_2$		−20.9		−5.0	7
Sr	Sr		164.4		39.3	4
Y	Y	420.45	421.3	100.49	100.7	3
Zr	Zr	607.47	608.8	145.19	145.5	3
Nb	Nb	722.819	725.9	172.758	173.5	3
Mo	Mo	656.55	658.1	156.92	157.3	2
MoC$_6$O$_6$	Mo(CO)$_6$	−914.6	−912.1	−218.6	−218.0	2
Tc	Tc		678		162	2
Ru	Ru	641.031	642.7	153.210	153.6	2
Rh	Rh	555.59	556.9	132.79	133.1	2
Pd	Pd	377.4	378.2	90.2	90.4	2
Ag	Ag	284.09	284.55	67.90	68.01	2
Cd	Cd	112.05	112.01	26.78	26.77	1
C$_2$H$_6$Cd	(CH$_3$)$_2$Cd		101.55		24.27	1
In	In	243.72	243.30	58.25	58.15	1
Sn	Sn	302.00	302.1	72.18	72.2	1
SnCl$_4$	SnCl$_4$	−469.28	−471.5	−112.16	−112.7	1
Sb	Sb	262.04	262.3	62.63	62.7	1
SbH$_3$	SbH$_3$	153.239	145.105	36.625	34.681	1
C$_{18}$H$_{15}$Sb	Triphenylantimony		435.6		104.1	8
Te	Te	197	196.73	47	47.02	1
Te$_2$	Te$_2$	170.50	168.2	40.75	40.2	1
TeH$_2$	H$_2$Te		99.6		23.8	1
I	I	107.240	106.838	25.631	25.535	1
I$^-$	I$^-$	−187.90		−44.91		6
I$_2$	I$_2$	65.517	62.438	15.659	14.923	1
HI	HI	28.660	26.48	6.850	6.33	1
BI$_3$	BI$_3$	75	71.13	18	17.00	1
MgI$_2$	MgI$_2$		−172		−41	4
ICl	ICl	19.41	17.78	4.64	4.25	1
IBr	IBr	49.79	40.84	11.90	9.76	1
CH$_3$I	CH$_3$I	22.51	13.0	5.38	3.1	1
C$_2$H$_5$I	C$_2$H$_5$I	8.16	−7.70	1.95	−1.84	1
C$_3$H$_7$I	n−C$_3$H$_7$I		−30.54		−7.30	7
C$_3$H$_7$I	iso−C$_3$H$_7$I		−41.84		−10.00	7
C$_4$H$_9$I	$tert$−C$_4$H$_9$I		−73.64		−17.60	7
C$_6$H$_5$I	Iodobenzene		162.55		38.85	7
C$_7$H$_7$I	2−Iodotoluene		132.6		31.7	8
C$_7$H$_7$I	3−Iodotoluene		133.5		31.9	8
C$_7$H$_7$I	4−Iodotoluene		121.8		29.1	8

7.2. Table of Auxiliary Heats of Formation—Continued

Empirical formula	Description	$\Delta H^\circ_{f0}(g)$ (kJ mol^{-1})	$\Delta H^\circ_{f298}(g)$ (kJ mol^{-1})	$\Delta H^\circ_{f0}(g)$ (kcal mol^{-1})	$\Delta H^\circ_{f298}(g)$ (kcal mol^{-1})	Ref.
CNI	ICN	226.10	225.5	54.04	53.9	1
CF$_3$I	CF$_3$I	−583.2	−589.1	−139.4	−140.8	11
XeF$_2$	XeF$_2$	−117.2		−28.0		13
XeF$_4$	XeF$_4$	−241.0		−57.6		13
Cs	Cs	78.16	76.65	18.68	18.32	11
CsF	CsF	−354.0	−356.5	−84.6	−85.2	11
CsCl	CsCl	−238.1	−240.2	−56.9	−57.4	11
Ba	Ba	180.7	180	43.2	43	4
La	La	431.303		103.084		5
Ce	Ce	423.4		101.2		5
Pr	Pr	356.69		85.25		5
Nd	Nd	328.57		78.53		5
Sm	Sm	206.10		49.26		5
Eu	Eu	176.699		42.232		5
Gd	Gd	398.957		95.353		5
Tb	Tb	390.62		93.36		5
Dy	Dy	293.05		70.04		5
Ho	Ho	302.63		72.33		5
Er	Er	318.32		76.08		5
Tm	Tm	233.409		55.786		5
Yb	Yb	152.628		36.479		5
Lu	Lu	427.781		102.242		5
Ta	Ta	781.425	782.0	186.765	186.9	3
W	W	848.10	849.4	202.70	203.0	2
C$_6$O$_6$W	W(CO)$_6$		−871.5		−208.3	2
Re	Re	769.0	769.9	183.8	184.0	2
Os	Os		791		189	2
Ir	Ir	664.34	665.3	158.78	159.0	2
Pt	Pt	564.42	565.3	134.90	135.1	2
Au	Au	365.93	366.1	87.46	87.5	2
Hg	Hg	64.463	61.317	15.407	14.655	2
HgCl$_2$	HgCl$_2$		−146.294		−34.965	11
HgBr$_2$	HgBr$_2$		−85.454		−20.424	11
HgI$_2$	HgI$_2$	−10.88	−17.2	−2.60	−4.1	2
C$_2$H$_6$Hg	(CH$_3$)$_2$Hg		94.39		22.56	2
Tl	Tl	182.845	182.21	43.701	43.55	1
TlF	TlF		−182.4		−43.6	1
TlCl	TlCl		−67.8		−16.2	1
TlBr	TlBr		−37.7		−9.0	1
TlI	TlI		7.1		1.7	1
Pb	Pb	195.64	195.0	46.76	46.6	1
Bi	Bi	207.36	207.1	49.56	49.5	1
C$_{18}$H$_{15}$Bi	Triphenylbismuth		579.9		138.6	8

7.3. Bibliography

[1] Wagman, D. D., Evans, W. H., Parker, V. B., Halow, I., Bailey, S. M., and Schumm, R. H., Selected Values of Chemical Thermodynamic Properties. Tables for the First Thirty–Four Elements in the Standard Order of Arrangement, NBS Tech. Note 270–3 (U.S. Government Printing Office, Washington, D.C., 1968).

[2] Wagman, D. D., Evans, W. H., Parker, V. B., Halow, I., Bailey, S. M., and Schumm, R. H., Selected Values of Chemical Thermodynamic Properties. Tables for Elements 35 Through 53 in the Standard Order of Arrangement, NBS Tech. Note 270–4 (U.S. Government Printing Office, Washington, D.C., 1969).

[3] Wagman, D. D., Evans, W. H., Parker, V. B., Halow, I., Bailey, S. M., Schumm, R. H., and Churney, K. L., Selected Values of Chemical Thermodynamic Properties. Tables for Elements 54 Through 61 in the Standard Order of Arrangement, NBS Tech. Note 270–5 (U.S. Government Printing Office, Washington, D.C., 1971).

[4] Parker, V. B., Wagman, D. D., and Evans, W. H., Selected Values of Chemical Thermodynamic Properties. Tables for the Alkaline Earth Elements (Elements 92 Through 97 in the Standard Order of Arrangement), NBS Tech. Note 270–6 (U.S. Government Printing Office, Washington, D.C., 1971).

[5] Schumm, R. H., Wagman, D. D., Bailey, S., Evans, W. H., and Parker, V. B., Selected Values of Chemical Thermodynamic Properties. Tables for the Lanthanide (Rare Earth) Elements (Elements 62 Through 76 in the Standard Order of Arrangement), NBS Tech. Note 270–7 (U.S. Government Printing Office, Washington, D.C., 1973).

[6] Table 6.2, this work.

[7] Stull, D. R., Westrum, E. F., Jr., and Sinke, G. C., The Chemical Thermodynamics of Organic Compounds (John Wiley and Sons, New York, 1969).

[8] Cox, J. D., and Pilcher, G., Thermochemistry of Organic and Organometallic Compounds (Academic Press, New York, 1970).

[9] Benson, S. W., and O'Neal, H. E., Kinetic Data on Gas Phase Unimolecular Reactions, NSRDS–NBS 21 (U.S. Government Printing Office, Washington, D.C., 1970).

[10] Wilhoit, R. C., and Zwolinski, B. J., Physical and Thermodynamic Properties of Aliphatic Alcohols, J. Phys. Chem. Ref. Data, Vol. 2, Suppl. 1 (1973).

[11] Stull, D. R., and Prophet, H., JANAF Thermochemical Tables, Second Edition, NSRDS–NBS 37 (U.S. Government Printing Office, Washington, D.C., 1971).

[12] Using $\Delta H^{\circ}_{f298}(CH_3)$ from reference [9] and temperature correction from reference [1].

[13] Berkowitz, J., Chupka, W. A., Guyon, P. M., Holloway, J. H., and Spohr, R., Photoionization mass spectrometric study of XeF_2, XeF_4, and XeF_6, J. Phys. Chem. **75**, 1461 (1971).

[14] Tsang, W., Rate and mechanism of thermal decomposition of 4–methyl–1–pentyne in a single pulse shock tube, Intern. J. Chem. Kinetics **2**, 23 (1970).

[15] Dibeler, V. H., and Walker, J. A., Mass–spectrometric study of photoionization. VI. O_2, CO_2, COS, and CS_2, J. Opt. Soc. Am. **57**, 1007 (1967).

[16] Dibeler, V. H., and Liston, S. K., Mass–spectrometric study of photoionization. VIII. Dicyanogen and the cyanogen halides, J. Chem. Phys. **47**, 4548 (1967).

(Continuation of Cumulative Listing of Reprints)

(Continuation of Cumulative Listing of Reprints)

Journal of **Physical and Chemical Reference Data**
Cumulative Listing of Reprints and Supplements

(Cumulative Listing of Reprints continued on next page)

To: **American Chemical Society**
Business Operations
Books and Journals Division
1155 Sixteenth Street N. W.
Washington, D. C. 20036

Please ship the following reprints and supplements:

Reprint No. _____ , _____ copies, $ _____

Reprint No. _____ , _____ copies, $ _____

Reprint No. _____ , _____ copies, $ _____

Reprint No. _____ , _____ copies, $ _____

Reprint No. _____ , _____ copies, $ _____

Reprint No. _____ , _____ copies, $ _____

Supplement No. _____ to Vol. _____ $ _____

Supplement No. _____ to Vol. _____ $ _____

Other: _____

BULK RATES: Subtract 20% from the listed price for orders of 50 or more of any one item.

*ORDERS FOR REPRINTS MUST BE PREPAID

Please enter my subscription for the calendar year 1977 as follows:

Journal of Physical and Chemical Reference Data	Non-Member Domestic°	Foreign	Member* Domestic°	Foreign
	☐ $90	☐ $94	☐ $25	☐ $29
Supplements	☐	☐	☐	☐

(Check appropriate box if you wish to enter a standing order for all **1977** Supplements. Prices to be announced.)

*Member of **ACS** or member of **AIP** member or affiliated society
°Domestic rates apply to U.S. and possessions, Canada and Mexico.

Name _____

Address _____

City _____ State _____ Zip _____

Country _____

I am a member of _____
ACS or AIP Member or Affiliated Society

☐ Check Enclosed   ☐ Please Bill Me   ☐ Purchase Order Enclosed
(Make checks payable to the American Chemical Society)